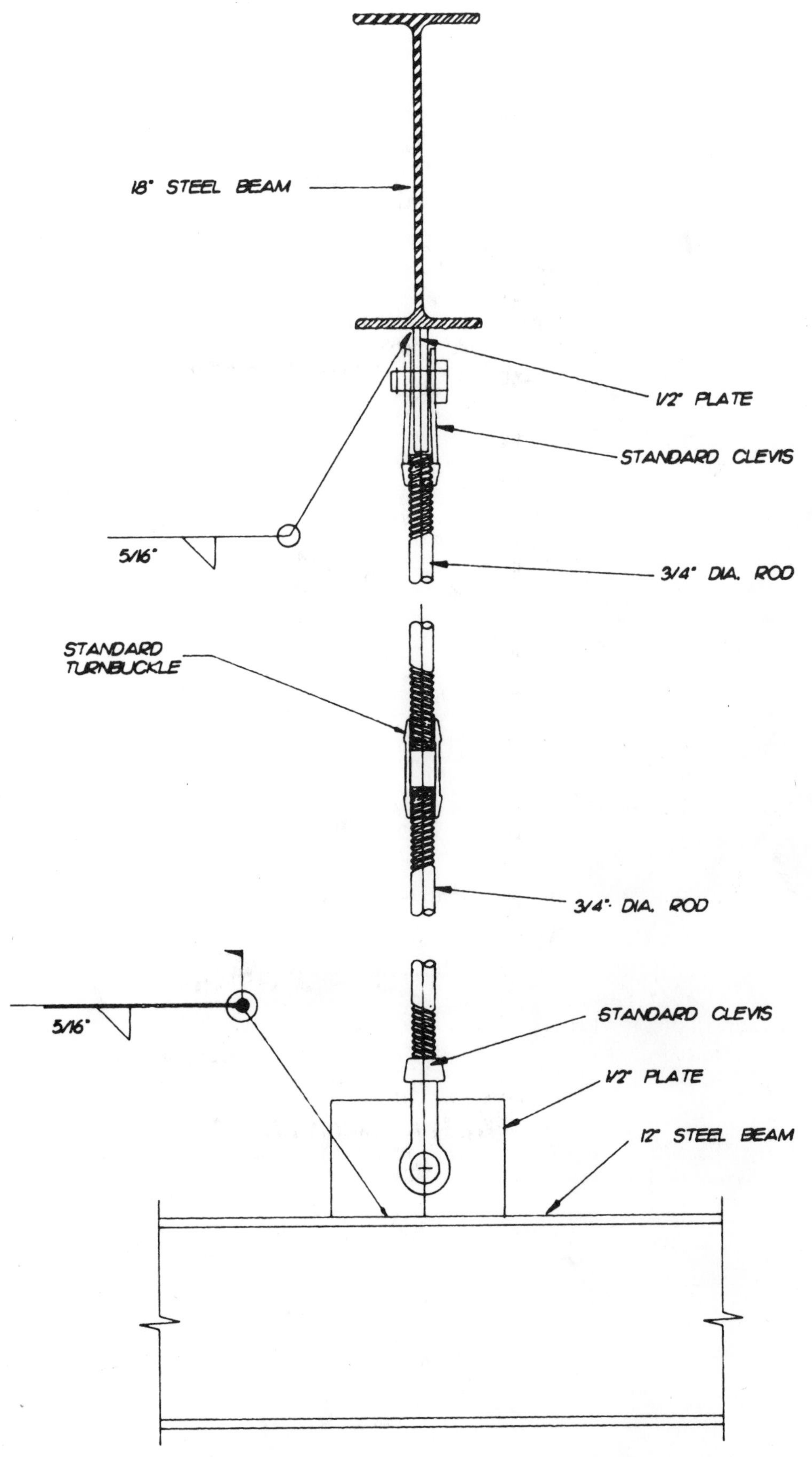
18" STEEL BEAM
1/2" PLATE
STANDARD CLEVIS
5/16"
3/4" DIA. ROD
STANDARD TURNBUCKLE
3/4" DIA. ROD
5/16"
STANDARD CLEVIS
1/2" PLATE
12" STEEL BEAM

STANDARD HANDBOOK OF STRUCTURAL DETAILS FOR BUILDING CONSTRUCTION

STANDARD HANDBOOK OF STRUCTURAL DETAILS FOR BUILDING CONSTRUCTION

Morton Newman
Civil Engineer

Second Edition

McGraw-Hill, Inc.
New York St. Louis San Francisco Auckland Bogotá
Caracas Lisbon London Madrid Mexico Milan
Montreal New Delhi Paris San Juan São Paulo
Singapore Sydney Tokyo Toronto

Library of Congress Cataloging-in-Publication Data

Newman, Morton.
Standard handbook of structural details for building construction / Morton Newman.
p. cm.
Rev. ed. of: Standard structural details for building construction. 1st ed. 1968
Includes index.
ISBN 0-07-046352-2
1. Building—Details—Drawings—Handbooks, manuals, etc. 2. Structural drawing—Handbooks, manuals, etc. I. Newman, Morton. Standard structural details for building construction. II. Title.
TH2031.N4 1992
692'.1—dc20 92-35654
CIP

The first edition of this book was published under the title *Standard Structural Details for Building Construction*.

1 2 3 4 5 6 7 8 9 0 9 8 7 6 5 4 3 2

ISBN 0-07-046352-2

The sponsoring editor for this book was Larry Hager, the editing supervisor was Jim Halston, the designer was Susan Maksuta, and the production supervisor was Donald F. Schmidt.

To the Newmans

CONTENTS

PREFACE

The original purpose of this book was to provide a graphic means of communication between architects, engineers, contractors, and students of the design and construction of buildings. The original work was presented in a single volume; however, it was subsequently determined that many design professionals usually work within the context of only one or two structural materials. This was taken into consideration and the original book was then brought out as four separate books, each related to one of the four basic structural materials, that is, wood, masonry, concrete, and steel. These individual versions also provided a format for the addition of new or revised details for the structural materials.

The original concept of the book has not changed; however, the scope of the detailed information displayed in this second edition has been greatly expanded to meet the current needs of the design and construction professions. The increase in the complexity of building designs and the stringent requirements of the governing criteria have made it necessary for field construction documents to be more accurate and specifically detailed. Also, in view of the increase in the potential for lawsuits concerning construction delays and extra costs, designers realize that it is imperative to produce complete drawings to avoid these situations.

The details shown in the following chapters are not actual structural designs; however, they are suggested representations for the utilization of the materials. This book should not be used as a "cook book" or as a shortcut substitute for structural design. It is most important that a structural engineer review the final drawings to ensure the design's adequacy to sustain the required design loads. Architectural and engineering students can use these drawings as practical examples for translating structural calculations into real applications. No claim is made for the originality of the details in this book; nevertheless, they represent generally accepted field standards. Because of the variety of loading conditions and configurations, it is not possible to depict a detail for every design or field condition that might be encountered. The original work has been expanded to demonstrate numerous variations of configurations to meet probable design requirements. Also, this edition specifies sizes and dimension to add to the feasibility of its utilization.

Ease of locating a particular detail is essential to the use of this book. The two key words that facilitate this procedure are "what" and "where." The placement of the details within the chapters is organized to coordinate with their function in the construction process. Designation charts are presented after the text material of a chapter to assist the reader in this endeavor, and the functional division of each chapter is outlined at the beginning of its text.

The engineering facts given in this book are in accordance with the requirements of the American Institute of Steel Construction, the American Concrete Institute, the International Conference of Building Officials Uniform Building Code, the West Coast Lumbermen's Douglas Fir Use Book, and the Concrete Masonry Association of Southern California.

There are a great many people who gave me valuable encouragement in the compilation of this book. First, I would like to express my thanks to Mr. Loren Bersack for his very able assistance and his knowledge of computer technology. On a personal note, in the course of the work, I experienced a physical setback that caused a four-month delay. I would like to thank the many friends who helped me to get back on line: to name only a few; Mr. Louis Rogers, Ms. Susan Grossinger, Mrs. Brenda Atkins, Dr. Marvin Abrams, Mr. Angelo Caponigro, Architect Earl Rubenstein, Mr. Paul DeJoseph, Attorney Ed Siegel, and my entire family—especially David Newman and Bill Newman.

I would like to take this opportunity to express my thanks to the staff of the West Los Angeles Veteran's Hospital for their care, their concern, and their professional excellence. Thank you Dr. Steven Kaye, Therapist Lisa Grod, Mary Sanchez, Dr. William Janes, John Ewart, Connie Brown, and Betty Baldwin.

Our flag is still there.

Morton Newman

LIST OF TABLES

ABBREVIATIONS

ADJUSTABLE	ADJUST.
ALTERNATE	ALT.
AMERICAN CONCRETE INSTITUTE	ACI
AMERICAN INSTITUTE OF STEEL CONSTRUCTION	AISC
AMERICAN SOCIETY OF TESTING AND MATERIALS	ASTM
ARCHITECT	ARCH.
AREA	A.
BEAM	BM.
BLOCK	BLK.
BLOCKING	BLKG.
BOTTOM	BOTT.
BUILDING	BLDG.
CALCULATIONS	CALCS.
CEILING	CEIL.
CEMENT	CEM.
CENTER LINE	C.L.
CHANNEL STUD	C.S.
CIVIL ENGINEER	C.E.
CLEAR	CLR.
COLUMN	COL.
CONCRETE	CONC.
CONNECTION	CONN.
CONSTRUCTION	CONSTR.
CONTINUOUS	CONT.
CUBIC	CU.
DEFLECTION	DEFL.
DEPRESSION	DEPR.
DETAIL	DET.
DIAGONAL	DIAG.
DIAMETER	DIA.
DIMENSION	DIM.
DISCONTINUOUS	DISC.
DOUBLE	DBL.
DRAWING	DRWG.
EACH	EA.
ELEVATION	EL. OR ELEV.
ENGINEER	ENGR.
EQUAL	EQ.
EQUIPMENT	EQUIP.
EXISTING	EXIST.
EXPAND	EXP.
EXPOSE	EXPO.
EXTERIOR	EXT.

FILLET	FILL.
FINISH	FIN.
FLOOR	FLR.
FOOT	FT.
FOOTING	FTG.
FOUNDATION	FDN.
FRAMING	FRMG.
GAUGE	GA.
GLUED LAMINATED	GL. LAM.
GRADE	GR.
GROUT	GRT.
GYPSUM	GYP.
HANGER	HNGR.
HEIGHT	HT.
HOOK	HK.
HORIZONTAL	HORIZ.
INCH	IN.
INCLUSIVE	INCL.
INSIDE DIAMETER	I.D.
INTERIOR	INT.
JOINT	JNT.
JOIST	JST.
LAG SCREW	L.S.
LAMINATED	LAM.
LATERAL	LAT.
LIGHT WEIGHT	LT. WT.
LONG LEG VERTICAL	LLV
MACHINE	MACH.
MASONRY	MAS.
MAXIMUM	MAX.
MEMBRANE	MEMB.
METAL	MET. OR M'
MINIMUM	MIN.
MOMENT OF INERTIA	I
NAILS	D (PENNY)
NATURAL	NAT.
NUMBER	NO. OR #
ON CENTER	O.C.
OPENING	OPNG.
OPPOSITE	OPP.
OUTSIDE DIAMETER	O.D.
PANELS	PNLS.
PARTITION	PART.
PENETRATION	PEN.
PLASTER	PLAS.
PLATE	PL.
PLYWOOD	PLYWD.
POUNDS PER CUBIC FOOT	P.C.F.
POUNDS PER SQUARE FOOT	P.S.F.

POUNDS PER SQUARE INCH	P.S.I.
PRESSURE	PRESS.
RADIUS	R.
RAFTER	RFTR.
RECTANGULAR	RECT.
REINFORCING	REINF.
REQUIRED	REQD.
RISER	R.
ROOF	RF.
ROOM	RM.
ROUND	Φ
SCHEDULE	SCHED.
SECTION	SECT.
SECTION MODULUS	S.
SEISMIC	SEIS.
SHEATHING	SHTG.
SHEET	SHT.
SHORT LEG VERTICAL	SLV
SPACING	SPCG.
SPECIFICATION	SPEC.
SPIRAL	SP.
STAGGER	STGR.
STANDARD	STD.
STEEL	STL.
STEEL JOIST	S.J.
STIFFENER	STIFF.
STIRRUP	STIRR.
STRUCTURAL	STRUCT.
STRUCTURAL STEEL TUBE	S.S.T.
SQUARE	SQ.
SYMMETRICAL	SYM.
THICK	THK.
THROUGH	THRU.
TREAD	TR.
ULTIMATE	ULT.
ULTIMATE STRESS DESIGN	U.S.D.
UNIFORM BUILDING CODE	U.B.C.
UTILITY	UTIL.
VERTICAL	VERT.
VOLUME	VOL.
WATERPROOF	W.P.
WEIGHT	WT.
WELDED WIRE FABRIC	W.W.F.
WIDE FLANGE	W.F.
WITH	W/
WORKING STRESS DESIGN	W.S.D.

STANDARD HANDBOOK OF STRUCTURAL DETAILS FOR BUILDING CONSTRUCTION

Wood Structural Details

The details presented in this chapter are composed primarily of wood components. There are other details in this book that display wood framing or parts; however, wood is not the primary element of these details, and therefore they are not included in this chapter. Most of the wood drawings provide specific design data, such as the sizes of members, nails, bolts, and sheet metal connecting hardware. This was done to display a feasible example of the use of the detail. For actual use, a detail should be checked for its capacity to resist expected dead and live loads. This prerequisite must be determined by structural calculations according to the requirements of a governing criterion or building code.

Wood is generally used for the construction of relatively light structures, such as residences and small office buildings, primarily because it is an inexpensive material in certain geographic areas. The use of wood as a construction material is also governed by its availability; there are places in the country where wood is quite scarce, and in those areas it is not considered as a feasible material. Structural parts of wood structures are usually comparatively easy to fabricate and erect at the job site; for this reason, dimensional accuracy is not a major factor, as it is in steel or concrete work. It can be said that a wood structure is a product that is manufactured and consumed at the same place. There are, however, certain elements of a wood structure that are not manufactured at the job site, such as glued laminated beams or columns. The degree of accuracy in wood construction at the job site should be sufficient to interconnect the members and parts with a minimum of adjustment; that is to say, the elements must fit together to a reasonable degree of tightness. In this respect wood differs from other construction materials, which cannot be easily altered at the job site to achieve a specific configuration. Quality of workmanship can eliminate much of the reworking of wood members in the field, which can be expensive and time-consuming. Dealing with deflection and shrinkage of wood requires quality workmanship. Wood joists, beams, and posts will deflect or compress when they are required to sustain a heavy compressive loading for a long duration. This is a consequence of the material's low modulus of elasticity and its loss of internal moisture content over a long period of time. Special consideration should be given in the design calculations and drawings of wood members to allow for these possible dimensional changes, particularly in instances where the members are connected to plumbing lines or are supported by other types of structural materials such as steel. In cases where plumbing lines are connected to wood studs or floor joists, some allowance for independent movement should be made so that the lines will not separate as a result of material shrinkage. Posts and flexural members that support a load for a long period of time will also deflect from compression and material "creep."

This chapter presents drawings of the various alternative accepted methods of connecting the component members of a wood structure. Each type of connection and its combinations of different sizes of members are detailed and shown individually. The accessibility of the information in the chapter is facilitated by arranging the material in a logical, relevant sequence so that the reader may readily locate a particular detail. The basic concept of this arrangement places the drawings in an assigned hierarchy which starts with a general condition, then progresses to a discrete number of associated explicit conditions.

Each drawing is identified by a coded designator starting with a material identification character and followed by a four-digit number. Since this chapter is concerned with wood details, the first character of the detail designator is the letter W. The detail numbers which follow are coded to represent a construction function section and the detail number in that particular function section. The last digit of the detail number represents the number of the variation of the original detail. This arrangement allows for some degree of parallelism between the different chapters. For example, the numbers for the wood columns are similar to those for the concrete and masonry columns. However, an exact comparison between chapters will reveal that the drawings are not identical.

The details are organized in a sequence of sections, each of which pertains to a particular function in the construction process. The first digit of the detail number is used to describe the particular construction purpose or function of the detail. This chapter consists of seven function sections, which are defined as follows:

Foundations	W1020 to W1xxx
Columns	W2020 to W2xxx
Walls and wood floors	W3020 to W3xxx
Walls and wood roofs	W4020 to W4xxx
Glued laminated wood	W7020 to W7xxx
Connections	W8020 to W8xxx
Miscellaneous	W9020 to W9xxx

ıe following is an example of the number coding system:

Given No. W3145

ıe number 3 indicates that the detail is in the "Walls and Wood Floors" ction; the number 14 indicates that it is the fourteenth drawing in this ction; and the number 5 indicates that it is the fifth variation of the iginal configuration. A look at Detail W3145 shows a wood exterior wall ıpporting a wood floor. The catalog charts prior to the drawings are esented to assist the reader in locating details. The construction function ction numbers at the top of the charts are arranged in numerical order, outlined above. The detail descriptions and their respective particulars e stated in the left-hand columns, and the designation numbers are stated the right-hand column of the chart.

The species of lumber is not given in the drawing assemblies, since the ailability of different types and grades of lumber varies for each geo- raphic region of the country. Local building codes or the applicable design iteria specify the allowable working stress values for each species of ımber used in that particular region. Building codes also regulate quality ontrol for structural lumber members by requiring that these members e inspected, and certified by a mark, by a responsible lumber grading rganization. The allowable stress grade of wood members is only one of ıe factors that is involved in the use of structural lumber. Protection against ıoisture and decay is an important consideration, depending upon re- ional climate conditions and the presence of insects. Wood members are sually not subject to decay when the internal moisture content is main- ined at or below 20 percent of the volume; however, decay will become significant factor when the moisture content exceeds 25 percent of the olume. This condition usually occurs in members used for mudsills, in dgers in contact with masonry or concrete, or in any situation in which ıe member is exposed to the weather. Structural members that might be xposed to moisture or possible termite intrusion should be protected or e constructed with a chemically pressure-treated lumber. Some regional uilding codes permit the use of a naturally durable species of lumber, uch as redwood, for these conditions. Although joists and rafters are not irectly exposed to moisture, floors and attics should be ventilated to revent moisture accumulation through long-term condensation.

Another source of decay of wood members is the presence of termites. he termite is an insect that destroys wood by devouring the interior ellulose material. It is not unusual to find a structural member that shows ıo damage on its exterior surfaces, yet discover that the structural capacity ıas been considerably reduced through internal decay caused by the pres- nce of termites. It is difficult to totally eliminate the presence of termites n certain areas of the country; however, attacks by large numbers of ermites can be reasonably controlled by creating a barrier between the djacent moist ground and wood mudsills. A termite barrier can be made y installing a galvanized sheet metal flashing between the lumber and he exposed surface. There are also many commercial chemical solutions vailable that can be added to the ground adjacent to the mudsills to revent termite intrusion. Since lumber is susceptible to decay from mois- ure and termites, it should be protected when it is stored at the job site. t is good practice to cover structural lumber with a water-resistant tar- aulin and place it at least 6 in above the surface of the ground.

The details of floors and roofs are shown framed with 2-in-wide mem- bers spaced at 16 in on center (o.c.). Joists or rafters may also be spaced at 12 or 16 in on center, depending on the design requirements. Roof o floor plywood is placed with its surface grain perpendicular to the framin members; therefore, these dimensions are an even module of the 8′-0″ o 10′-0″ length of standard plywood sheets so that the edges will lap on rafter or joist. There are instances, however, in which the joist or rafte spacing may be 19.2 in on center for the purpose of economy. The di mension of 19.2 in represents one-fifth of an 8′-0″ plywood sheet lengt Also, there are situations in which the floor or roof framing members ar exposed for the sake of architectural design; however, in any case th member spacing should be an even module of an 8′-0″ or 10′-0″ plywoo sheet length to allow for a nailing edge where two plywood sheets abu The size and spacing of joists or rafters depend on the dead load, th imposed live load, the allowable shear and bending stress of the lumbe and the allowable vertical deflection. The designer should use some judg ment in determining the effect of the loads for both bending and deflectio The design criteria may specify the floor or roof live loads; however, ther are times in the life of a building when these values may be exceeded Roof members may be overloaded by high or unusual snow loads; floor of offices may be overstressed by high concentrated live loads such a machinery or file cabinets placed at the mid-span of the floor joists. Also it is possible for a floor structure to meet all the design criteria and ye bounce or vibrate from normal use. This condition does not inspire muc confidence in the occupants. Good judgment, field and design experience and good workmanship are important factors in the use of structura lumber.

Wood structural framing requires that the members be cut to size an fit together with even bearing surfaces to prevent movement. The woo members that are specified in the details of this chapter are designate by their nominal or rough sizes; however, they are drawn using their mi cut net dimensions. Table 1-1 lists the standard nominal wood membe dimensions and their respective net dimensions for S4S (surfaced fou sides) seasoned and unseasoned lumber. Certain minimum dimension are required in assembling wood structures. Floor joists and rafters mus have at least a 1½-in length of bearing on wood or steel supports an 3 in on masonry or concrete supports. These are only minimum dimensio requirements, and the designer should check to be sure that the shea and end bearing stresses meet the allowable stress requirements. It is ofte found that the shear and bearing stresses are quite high in short span

Table 1-1. *American Standard Lumber Dimensions S4S*

Nominal, or rough, dimensions, in	Standard surfaced four sides, or actual net dimension, in
1	¾
2	1⅝
3	2⅝
4	3⅝
6	5½
8	7½
10	9½
12	11½
14	13½
16	15½
18	17½

that support high uniform or concentrated loads, and so these spans may not meet design criteria. In all the details in this book, the ends and the intersections of rafters and joists are continuously blocked with lumber pieces that are the same size as the framing joists. In the case where a joist or rafter ends at a stud wall, a continuous rim joist may be used in lieu of continuous blocking. This configuration braces the members against lateral movement in the horizontal plane and allows for a nailed connection between the sheathing and the support wall or beam. The 1991 Uniform Building Code requires that continuous rows of blocking shall be installed at 8′-0″ intervals of the joist or rafter spans when the ratio of depth to width (based on nominal lumber dimensions) exceeds 6 to 1.

Often it is necessary to notch a framing member or to drill holes through it to accommodate plumbing pipes or electrical conduits. Usually the locations of these notches or holes are not determined until long after the structural design has been completed. The UBC recommends that notches in sawn lumber shall not be greater than one-sixth of the depth of the members and shall not be located in the middle one-third of the span. Notches at ends of members shall not be greater than one-quarter of the joist depth. Holes may be bored through floor or roof joists provided that the edge of the hole is not less than 2 in from the top or the bottom of the member. Holes and notches can negate much of the quality of the design process and leave a relatively weakened structure. In reality, a design is not much better than the workmanship and knowledge of the field people at the job site. The engineer must exercise some control over the quality of workmanship to ensure that the building will perform as he or she designed it. The problems created by random location of holes and notches are easier to prevent during construction than to resolve after the damage is done.

Interior and exterior walls of wood buildings are constructed with evenly spaced vertical wood studs. The size and spacing of studs depends on vertical and lateral loading, the stress grade of the lumber, and the unbraced height of the wall. Interior partitions or nonbearing walls are usually constructed with 2 × 4 studs spaced at 16 or 24 in on center; however, in cases where the studs are used in walls of heights exceeding 14′-0″, the studs should be either 3 × 4s or 2 × 6s at 16 in o.c. Wall studs are usually spaced at 16 in o.c. to accommodate the interior and exterior wall covering materials' nailing edges. As a rule, exterior and interior bearing walls supporting two stories of load should be 2 × 4s at 16 in o.c. for a wall height not exceeding 10′-0″. Walls that support three stories of load and exceed 6′-0″ in height from sole plate to the top double plate should be framed with either 3 × 4s or 2 × 6s at 16 in o.c. The details in this chapter show wood stud walls with mudsills and sole plates the same size as the studs. Building codes require that studs be connected at the top of the wall with a double flat 2 × 4 or 2 × 6 plate. The UBC allows some nonbearing partitions to use only a single flat 2 × 4 plate at the top. Stud walls are continuously blocked with 2 × blocking at mid-height to maintain the height-to-least-thickness ratio equal to 50 as required by the UBC. When a nonbearing partition is supported by a wood floor, the sole plate should be nailed to either double joists, if the wall is parallel to the floor joists, or double blocking, if the floor joists are perpendicular to the wall. During construction the vertical plane of stud walls can be maintained square by using 1 × 4 let-in diagonal braces on the exterior surface of the wall. Building codes require that these braces be placed at a minimum of 25′-0″ o.c. at the intersections of walls and within 6′-0″ from the ends of a wall.

All the details shown in this book use plywood as roof or floor sheathing. Also, there are instances in which plywood is used as a covering for seismic-resistant shear walls. Straight or diagonal sheathing lumber may be used in lieu of plywood; however, it has a lesser capacity to resist seismic or wind forces in vertical and horizontal diaphragms.

Douglas fir plywood panels are manufactured by laminating and gluing together an odd number of layers of Douglas fir veneer sheets. The veneer sheets are usually ⅛ in thick and are laminated so that the surface grain of each layer is perpendicular to that of the adjacent layers. Standard plywood panels are 48 × 96 in, with the exposed surface grain on each side running in the direction of the length of the panel. The strength of the wood running in the direction of the surface grain of the panel is much greater than the strength of the cross-grain plies, which only serve as a filler material between the laminations; therefore, plywood should be placed so that the surface grain is perpendicular to the framing members. Two types of Douglas fir plywood are manufactured, interior type and exterior type, their classification depending on the glue used in their fabrication. The exterior-type plywood uses a waterproof adhesive. Structural plywood is designated as Structural I or II C-D INT APA for interiors and Structural I or II C-C EXT APA for exterior use. Vertical and horizontal structural diaphragms are constructed by nailing the plywood panels to the framing members. The capacity of the diaphragm to resist lateral force depends on the thickness of the plywood, the size and spacing of the nails to the panel edges and the diaphragm edges, and the panel-edge blocking.

Building codes define heavy timber as structural wood members of sufficient width and depth to qualify as slow-burning construction. The fire rating for heavy timber varies in different building codes; however, the National Board of Fire Underwriters recommends the following nominal dimension for heavy timber construction: columns shall not be less than 8 in in any dimension; beams and girders shall not be less than 6 in in width or 10 in in depth; floors shall be constructed of tongue-and-grooved planks not less than 3 in thick and covered with 1 in thick flooring laid perpendicular or diagonal to the subfloor planks; roof sheathing shall be not less than 3-in-thick tongue-and-groove planks.

Glued laminated wood members are often used in heavy timber construction, since their nominal dimensions can readily conform to building code requirements. These members are factory fabricated and consist of vertically laminated, nominal 2-in-thick boards in various combinations, each board being glued to the adjacent board. Two types of adhesives are used in the manufacture of glued laminated structural members. When the moisture content of a member exceeds 15 percent or when the member is exposed to the weather, the boards are laminated with an exterior-type phenol-resorcinol glue. A fortified casein glue is used to fabricate members that are used in the interior of a building or that are not exposed to excessive moisture. The allowable working stresses for glued laminated structural members depend on the number of laminations and the structural grade of the laminated boards. Since these members are manufactured to meet the engineer's design requirements, it is recommended that they be shop drawn and detailed before they are fabricated. Quite often glued laminated wood members are left exposed to attain an architectural effect. In such instances, the finished appearance of the wood is important and should be specified as either architectural or premium finished. Also, the member should be delivered to the job site in a protective wrapping, which should not be removed until after the member is in place. Glued laminated members that are not used in an exposed situation may have n industrial-finished appearance and need only a protective wrapping gainst excessive moisture.

Wood framing members and floor sheathing that are nominally 1 and in in width are connected by wire nails. Common wire nails are most ften used in framed construction, although other types of nails are available where stronger withdrawal-type connections may be required. Nails re generally employed to react principally in shear caused by lateral oading of the connection. The lateral resistance value of a nail depends n the nail diameter and the depth of penetration into the member to be oined. When a force is applied parallel to the nail, a withdrawal resistance vill occur which is much less than the lateral resistance; therefore, this ype of connection should be avoided. Nails that are driven into a member t an angle of approximately 30 degrees to the surface grain are called oenails and will have two-thirds of the normal lateral resistance value. ails should penetrate the connected member at least one-half the length f the nail, all connections should have at least two nails, and nails should e at least one-half the nail length apart and not less than one-fourth of he nail length from the edge of the member. Many of the drawings in his volume call for metal clips. These clips are a commercially standard iece of hardware and are commonly used to join wood members when high connection resistance is required. The clips are made of 18-gauge heet metal in various left- and right-hand configurations and are basically sed for light wood framed connections. Table 1-2 shows the various ommon wire nail sizes and lengths. Table 1-3 gives the recommended ght-framed nailing schedule for the different types of connections. Plywood roof and floor sheathing nails are not specified in this table, since hey are determined by the shear to be resisted by the diaphragm.

The use of metal bolts to connect wood members to each other and o other structural materials such as concrete, masonry, and steel is a ommon and economical method of construction. Bolts connecting wood nembers are capable of resisting forces applied both parallel and lateral o the shank of the bolt by bearing and shear. In the drawings in this hapter, the bolts are used to connect metal plates to wood posts and eams, to connect wood members to each other, and to connect wood raming to steel beams. Flat washers should be used in bolted wood onnections when metal side plates are not specifically designated.

The washers are either round or square and are made of malleable ron. The size and thickness of a washer are determined by calculation nd depend on the bolt tension when the bolt is tightened or loaded and n the bearing stress normal to the wood surface. Heavy timber construc-

able 1-2. *Common Wire-nail Sizes*

Size of nail, d	Standard length, in	Wire gauge
6	2	11½
8	2½	10¼
10	3	9
12	3¼	9
16	3½	8
20	4	6
30	4½	5
40	5	4
50	5½	3
60	6	2

Table 1-3. *Recommended Nailing Schedule*

Connection	Nailing	Nail size, d
Joist to sill or girder	Toenail	2–16
Bridging to joist	Toenail	2–8
1 × 6 subfloor to joist	Face nail	2–8
2 subfloor to joist or girder	—	2–16
Plate to joist or blocking	—	16 at 16″ o.c.
Stud to plate	End nail	2–16
Stud to plate	Toenail	3–16 or 4–8
Top plates spike together:		
Laps and intersections	—	16 at 24″ o.c.
Ceiling joists:		
To plate	Toenail	2–16
Laps over partitions	Toenail	3–16
To parallel alternate rafters	Toenail	3–16
Rafter to plate	—	3–16
Continuous 1″ brace to stud	—	2–8
2″ cut in bracing to stud	—	2–16
1″ sheathing to bearing	—	2–8
Corner studs and angles	—	16 at 30″ o.c.

tion often requires that the washers be made of cast iron; however, th washers are not designated in the following wood details. The strength c a bolted wood connection depends on the thickness of the members, th number and size of the bolts, the species of the wood, the angle of th resisted force to the grain of the members, the use of metal side plate or other standard hardware connectors, and the arrangement of the bol spacing in the connection. It is important that the bolt spacing and th edge distance be sufficient to ensure that the wood will not split and t allow enough bearing area. Each bolt should be installed through a pre drilled hole $\frac{1}{16}$ in larger than the bolt shank diameter.

Lag screws are often used in lieu of bolts when it is not possible o convenient to obtain full penetration through a wood member by a bol The capacity of lag screws to resist lateral forces and withdrawal depend on the same factors as that of bolts in wood members; however, lag screw will not resist as much force as an equal-size bolt. A lag screw should b installed in a predrilled hole approximately 70 percent of the shank di ameter of the screw, and it should penetrate the member to be joined a least one-half the screw length or eight shank diameters.

In general, wood is a highly versatile and inexpensive constructio material. Except for glued laminated members, wood structures can b fabricated and assembled on the job site. The contractor should have clea and complete structural drawings of the framing and connections to avoi inefficient use of labor and materials.

Section 1: Foundations. Details W1020 to W1502 show continuou footings supporting wood stud walls. The shapes of the footings are rec tangular, L-shaped, or T-shaped. Details W1540 to W1561 show wood post on spread footings.

Section 2: Columns. Details W2020 to W2121 show wood posts sup porting wood or steel beams.

Section 3: Walls and Wood Floors. Details W3020 to W3703 show series of exterior or interior wood stud walls supporting wood floors.

ection 4: Walls and Wood Roofs. Details W4020 to W4903 show a eries of exterior and interior wood stud walls supporting wood roofs. etails W4920 and W4921 show roof equipment platforms. Details W4940 W4943 show roof rafter slope scarfing. Details W4960 and W4961 show ethods of connecting partition walls to rafters.

ection 7: Glued Laminated Wood. Details W7020 to W7320 show ethods of connecting and supporting glued laminated wood members.

ection 8: Connections. Details W8020 to W8103 show methods of onnecting wood members supported by wood beams.

ection 9: Miscellaneous. This section contains details of wood parts nd connections such as double plate splices, diagonal sheathing diagrams, eismic splices and hold-downs, stair stringer connections, and veneer ipport on a wood stud wall.

CHAPTER 1 - WOOD SECTION 1 - FOUNDATIONS			
CONTINUOUS FOOTING EXTERIOR WALL SLAB ON GRADE	RECTANGULAR SHAPE	2 X 4 STUDS	W1020 - W1023
		2 X 6 STUDS	W1024 - W1027
CONTINUOUS FOOTING EXTERIOR WALL SLAB ON GRADE WITH A CURB	RECTANGULAR SHAPE	2 X 4 STUDS	W1040 - W1043
		2 X 6 STUDS	W1044 - W1047
CONTINUOUS FOOTING EXTERIOR WALL SLAB ON GRADE WITH VENEER	RECTANGULAR SHAPE	2 X 4 STUDS	W1060 - W1063
		2 X 6 STUDS	W1064 - W1067
CONTINUOUS FOOTING EXTERIOR WALL SLAB ON GRADE WITH VENEER & A CURB	RECTANGULAR SHAPE	2 X 4 STUDS	W1080 - W1083
		2 X 6 STUDS	W1084 - W1087
CONTINUOUS FOOTING EXTERIOR WALL SLAB ON GRADE	L SHAPE	2 X 4 STUDS	W1100 - W1103
		2 X 6 STUDS	W1104 - W1107
CONTINUOUS FOOTING EXTERIOR WALL SLAB ON GRADE	T SHAPE	2 X 4 STUDS	W1120 - W1123
		2 X 6 STUDS	W1124 - W1127
CONTINUOUS FOOTING EXTERIOR WALL WOOD FLOOR JOISTS PERPENDICULAR TO FTG	L SHAPE	2 X 4 STUDS	W1140 - W1143
		________	________
CONTINUOUS FOOTING EXTERIOR WALL WOOD FLOOR JOISTS PARALELL TO FTG.	L SHAPE	2 X 4 STUDS	W1144 - W1147
		________	________
CONTINUOUS FOOTING EXTERIOR WALL - WITH VENEER WOOD FLOOR JOISTS PERPENDICULAR TO FTG	L SHAPE	2 X 4 STUDS	W1160 - W1162
		________	________
CONTINUOUS FOOTING EXTERIOR WALL - WITH VENEER WOOD FLOOR JOISTS PARALELL TO FTG.	L SHAPE	2 X 4 STUDS	W1163 - W1165
		________	________
CONTINUOUS FOOTING EXTERIOR WALL WOOD FLOOR JOISTS PERPENDICULAR TO FTG		2 X 4 STUDS	W1180 - W1183
		________	________
CONTINUOUS FOOTING EXTERIOR WALL WOOD FLOOR JOISTS PARALELL TO FTG.	T SHAPE	2 X 4 STUDS	W1184 - W1187
		________	________
CONTINUOUS FOOTING EXTERIOR WALL- CRIPPLE STUDS WOOD FLOOR JOISTS PERPENDICULAR TO FTG	L SHAPE	2 X 4 STUDS	W1200 - W1203
		________	________
CONTINUOUS FOOTING EXTERIOR WALL- CRIPPLE STUDS WOOD FLOOR JOISTS PARALELL TO FTG.	L SHAPE	2 X 4 STUDS	W1204 - W1207
		________	________
CONTINUOUS FOOTING EXTERIOR WALL- CRIPPLE STUDS WOOD FLOOR JOISTS PERPENDICULAR TO FTG	T SHAPE	2 X 4 STUDS	W1220 - W1223
		________	________
CONTINUOUS FOOTING EXTERIOR WALL- CRIPPLE STUDS WOOD FLOOR JOISTS PARALELL TO FTG.	T SHAPE	2 X 4 STUDS	W1224 - W1227
		________	________

CHAPTER 1 - WOOD	SECTION 1 - FOUNDATIONS		
CONTINUOUS FOOTING INTERIOR WALL SLAB ON GRADE	RECTANGULAR SHAPE	2 X 4 STUDS	W1240 - W1243
		2 X 6 STUDS	W1244 - W1247
CONTINUOUS FOOTING INTERIOR WALL SLAB ON GRADE	T SHAPE	2 X 4 STUDS	W1260 - W1263
		2 X 6 STUDS	W1264 - W1267
CONTINUOUS FOOTING INTERIOR WALL WOOD FLOOR JOISTS PERPENDICULAR TO FTG	T SHAPE	2 X 4 STUDS	W1280 - W1283
		———	———
CONTINUOUS FOOTING INTERIOR WALL WOOD FLOOR JOISTS PARALELL BOTH SIDES	T SHAPE	2 X 4 STUDS	W1284 - W1287
		———	———
CONTINUOUS FOOTING INTERIOR WALL WOOD FLOOR JOISTS PERP. & PAR. ALT SIDES	T SHAPE	2 X 4 STUDS	W1300 - W1303
		———	———
CONTINUOUS FOOTING INTERIOR WALL - CRIPPLE STUDS WOOD FLOOR JOISTS PERPENDICULAR TO FTG	T SHAPE	2 X 4 STUDS	W1320 - W1323
		———	———
CONTINUOUS FOOTING INTERIOR WALL - CRIPPLE STUDS WOOD FLOOR JOISTS PARALELL TO FTG.	T SHAPE	2 X 4 STUDS	W1324 - W1327
		———	———
CONTINUOUS FOOTING INTERIOR WALL - CRIPPLE STUDS WOOD FLOOR JOISTS PERP. & PAR. ALT. SIDES	T SHAPE	2 X 4 STUDS	W1340 - W1343
		———	———
CONTINUOUS FOOTING INTERIOR DIVISION WALL SEPARATED STUDS SLAB ON GRADE	RECTANGULAR SHAPE	2 X 4 STUDS	W1360 - W1361
		———	———
CONTINUOUS FOOTING INTERIOR DIVISION WALL SEPARATED STUDS WOOD FLOOR JSTS PERP. TO FTG	T SHAPE	2 X 4 STUDS	W1380 - W1381
		———	———
CONTINUOUS FOOTING INTERIOR DIVISION WALL SEPARATED STUDS WOOD FLOOR JOISTS PERP. & PAR. ALT. SIDES	T SHAPE	2 X 4 STUDS	W1400 - W1401
		———	———
CONTINUOUS FOOTING INTERIOR DIVISION WALL SEPARATED STUDS WOOD FLOOR JOISTS PARALELL TO FTG.	T SHAPE	2 X 4 STUDS	W1420 - W1421
		———	———
CONTINUOUS FOOTING INTERIOR DIVISION WALL STAGGERED STUDS SLAB ON GRADE	RECTANGULAR SHAPE	2 X 4 STUDS	W1440 - W1441
		———	———
CONTINUOUS FOOTING INTERIOR DIVISION WALL STAGGERED STUDS WOOD FLOOR JOISTS PERPENDICULAR TO FTG	T SHAPE	2 X 4 STUDS	W1460 - W1461
		———	———
CONTINUOUS FOOTING INTERIOR DIVISION WALL STAGGERED STUDS WOOD FLOOR JOISTS PERP. & PAR. ALT. SIDES	T SHAPE	2 X 4 STUDS	W1480 - W1481
		———	———
CONTINUOUS FOOTING INTERIOR DIVISION WALL STAGGERED STUDS WOOD FLOOR JOISTS PARALELL TO FTG.	T SHAPE	2 X 4 STUDS	W1500 - W1501
		———	———

CHAPTER 1 - WOOD SECTION 1 - FOUNDATIONS			
PARTITION WALL	SLAB ON GRADE	2 X 4 STUDS	W1520
SPREAD FOOTING	SLAB ON GRADE	WOOD POST	W1540
WOOD FLOOR PIER SUPPORT	WOOD FLOOR	————	W1560

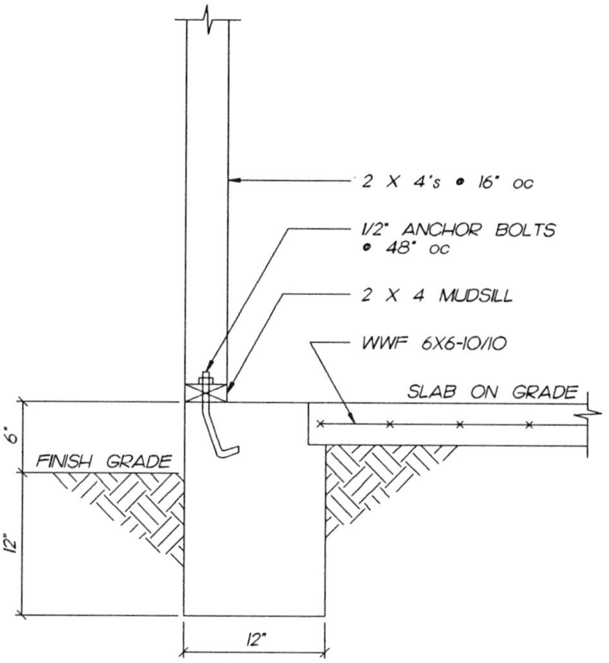

2 X 4's @ 16" oc
1/2" ANCHOR BOLTS @ 48" oc
2 X 4 MUDSILL
WWF 6X6-10/10
SLAB ON GRADE
FINISH GRADE
6"
12"
12"

W1020

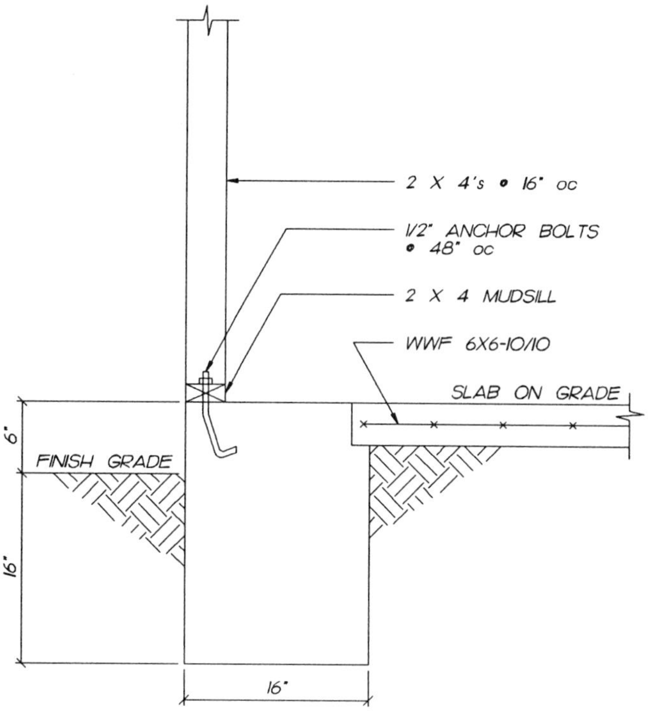

2 X 4's @ 16" oc
1/2" ANCHOR BOLTS @ 48" oc
2 X 4 MUDSILL
WWF 6X6-10/10
SLAB ON GRADE
FINISH GRADE
6"
16"
16"

W1021

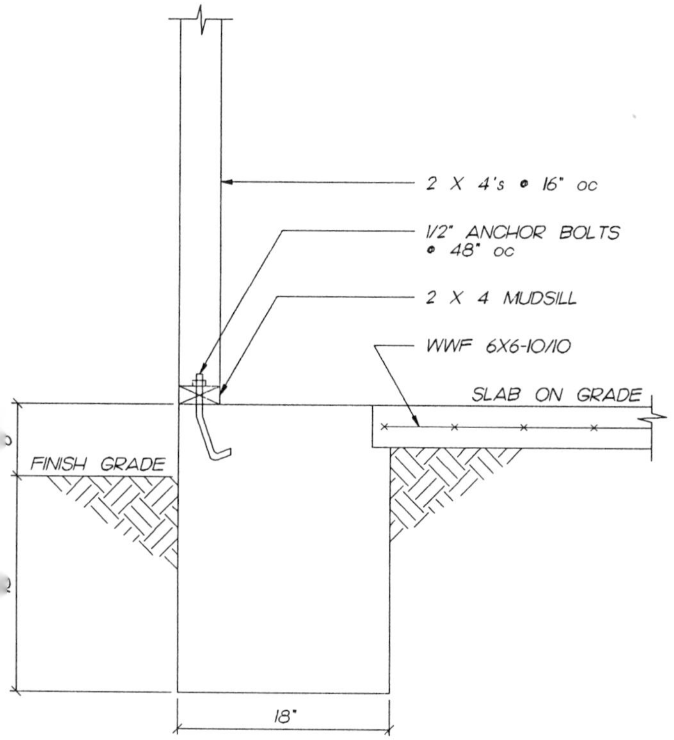

2 X 4's @ 16" oc
1/2" ANCHOR BOLTS @ 48" oc
2 X 4 MUDSILL
WWF 6X6-10/10
SLAB ON GRADE
FINISH GRADE
18"

W1022

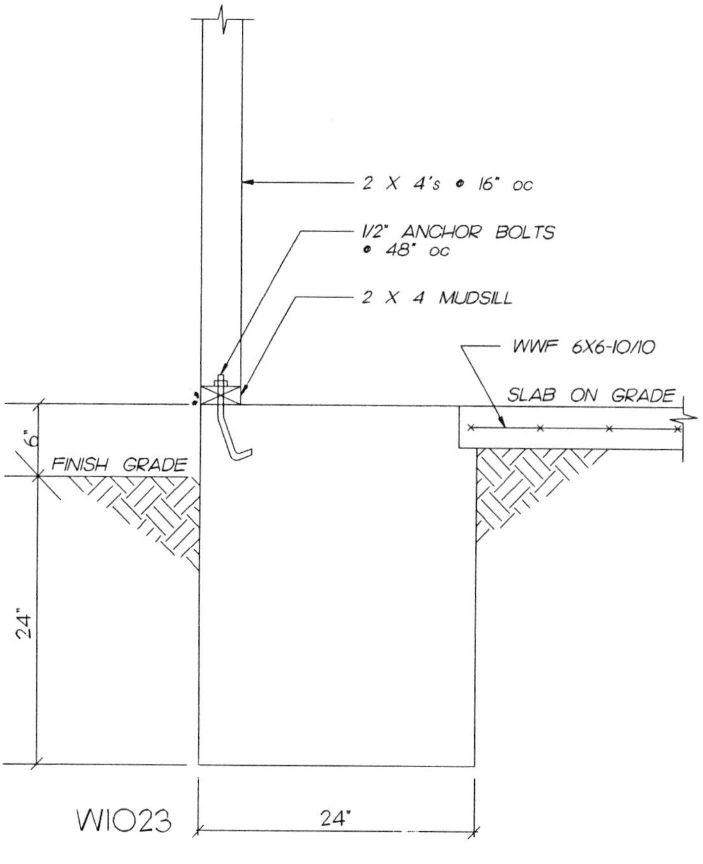

2 X 4's @ 16" oc
1/2" ANCHOR BOLTS @ 48" oc
2 X 4 MUDSILL
WWF 6X6-10/10
SLAB ON GRADE
FINISH GRADE
6"
24"
24"

W1023

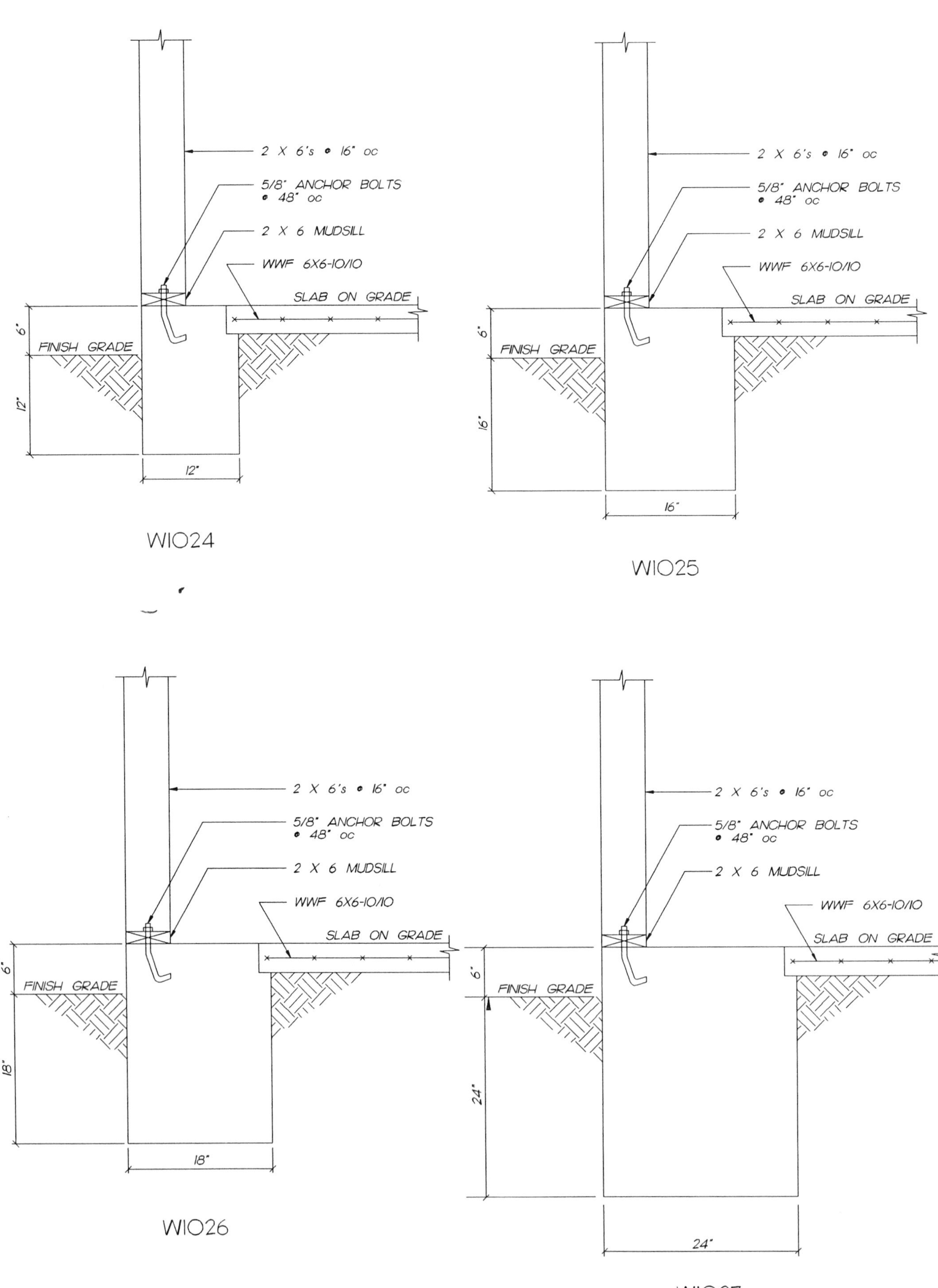
2 X 6's @ 16" oc
5/8" ANCHOR BOLTS
@ 48" oc
2 X 6 MUDSILL
WWF 6X6-10/10
SLAB ON GRADE
FINISH GRADE
6"
12"
12"
WIO24
2 X 6's @ 16" oc
5/8" ANCHOR BOLTS
@ 48" oc
2 X 6 MUDSILL
WWF 6X6-10/10
SLAB ON GRADE
FINISH GRADE
6"
16"
16"
WIO25
2 X 6's @ 16" oc
5/8" ANCHOR BOLTS
@ 48" oc
2 X 6 MUDSILL
WWF 6X6-10/10
SLAB ON GRADE
FINISH GRADE
6"
18"
18"
WIO26
2 X 6's @ 16" oc
5/8" ANCHOR BOLTS
@ 48" oc
2 X 6 MUDSILL
WWF 6X6-10/10
SLAB ON GRADE
FINISH GRADE
6"
24"
24"
WIO27

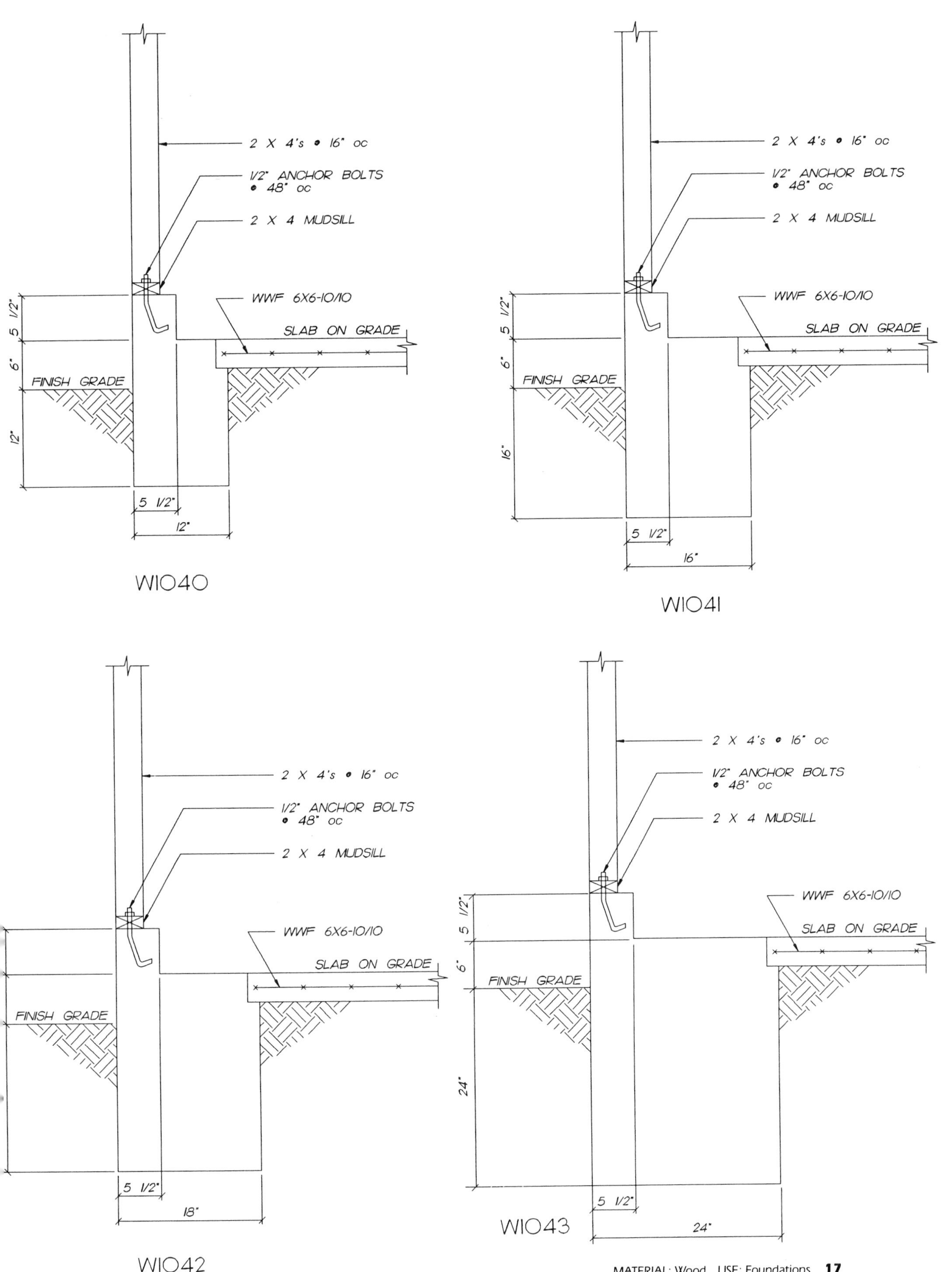
2 X 4's @ 16" oc
1/2" ANCHOR BOLTS @ 48" oc
2 X 4 MUDSILL
WWF 6X6-10/10
SLAB ON GRADE
FINISH GRADE
5 1/2"
6"
12"
5 1/2"
12"
W1040
2 X 4's @ 16" oc
1/2" ANCHOR BOLTS @ 48" oc
2 X 4 MUDSILL
WWF 6X6-10/10
SLAB ON GRADE
FINISH GRADE
5 1/2"
6"
16"
5 1/2"
16"
W1041
2 X 4's @ 16" oc
1/2" ANCHOR BOLTS @ 48" oc
2 X 4 MUDSILL
WWF 6X6-10/10
SLAB ON GRADE
FINISH GRADE
5 1/2"
18"
W1042
2 X 4's @ 16" oc
1/2" ANCHOR BOLTS @ 48" oc
2 X 4 MUDSILL
WWF 6X6-10/10
SLAB ON GRADE
FINISH GRADE
5 1/2"
6"
24"
5 1/2"
24"
W1043

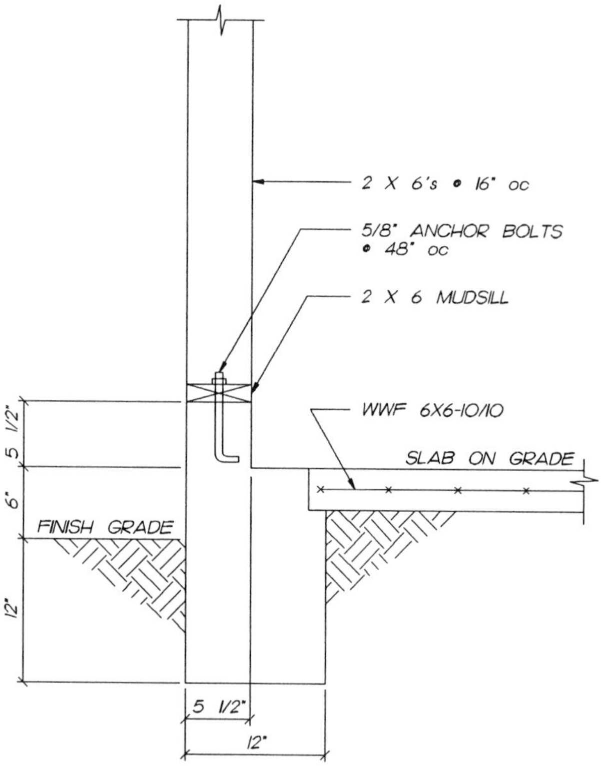

2 X 6's @ 16" oc
5/8" ANCHOR BOLTS
@ 48" oc
2 X 6 MUDSILL
WWF 6X6-10/10
SLAB ON GRADE
FINISH GRADE
5 1/2"
6"
12"
5 1/2"
12"

WI044

2 X 6's @ 16" oc
5/8" ANCHOR BOLTS
@ 48" oc
2 X 6 MUDSILL
WWF 6X6-10/10
SLAB ON GRADE
FINISH GRADE
5 1/2"
6"
16"
5 1/2"
16"

WI045

2 X 6's @ 16" oc
5/8" ANCHOR BOLTS
@ 48" oc
2 X 6 MUDSILL
WWF 6X6-10/10
SLAB ON GRADE
FINISH GRADE
5 1/2"
6"
18"
5 1/2"
18"

WI046

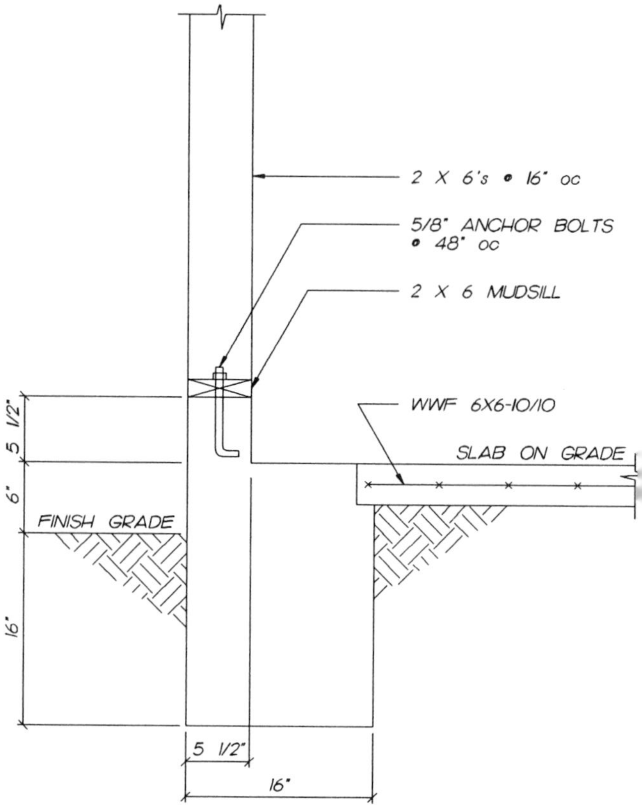

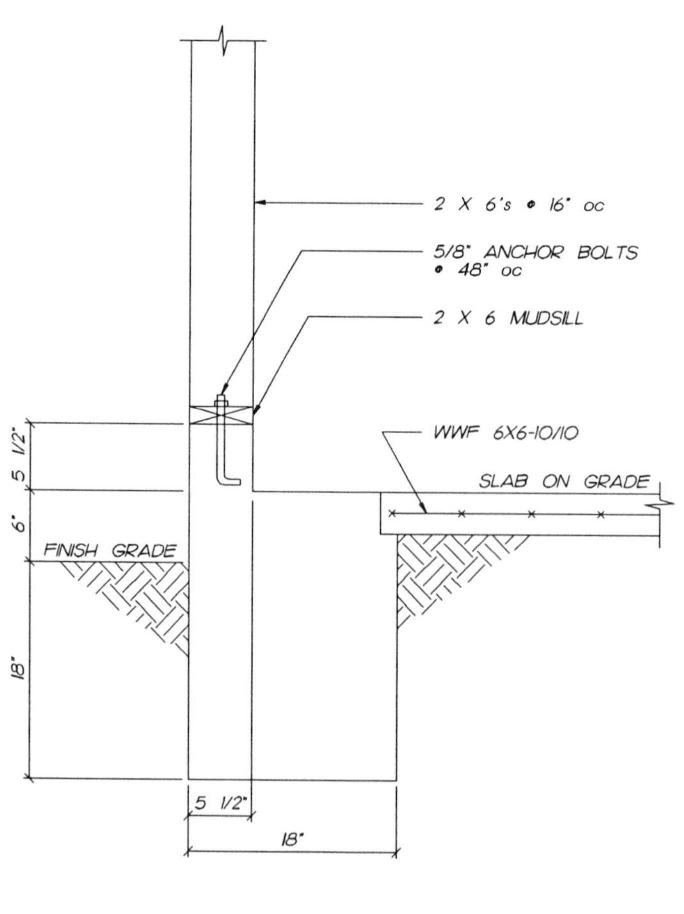

2 X 6's @ 16" oc
5/8" ANCHOR BOLTS @ 48" oc
2 X 6 MUDSILL
WWF 6X6-10/10
SLAB ON GRADE
5 1/2"
6"
FINISH GRADE
24"
5 1/2"
24"

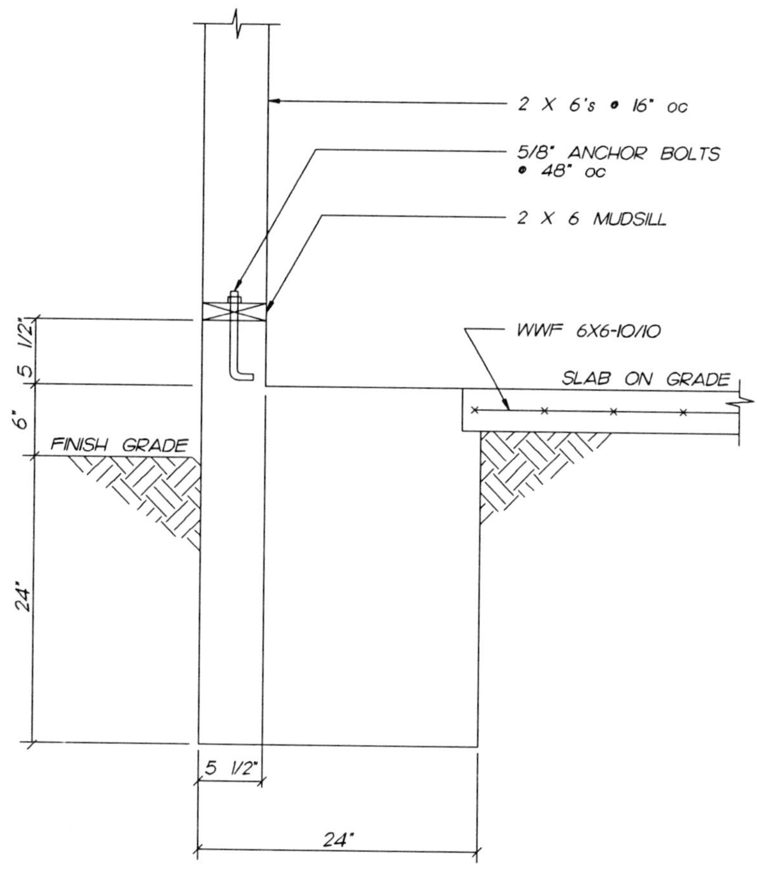

W1047

1"
WATER PROOF PAPER AND WIRE MESH
2 X 4's @ 16" oc
2 X 4 MUDSILL 1/2" DIA. A. BOLTS AT 6'-0" oc
WWF 6X6 - 10/10
SLAB ON GRADE
6"
FIN. GRADE
12"
6"
4 1/2"
12"

W1060

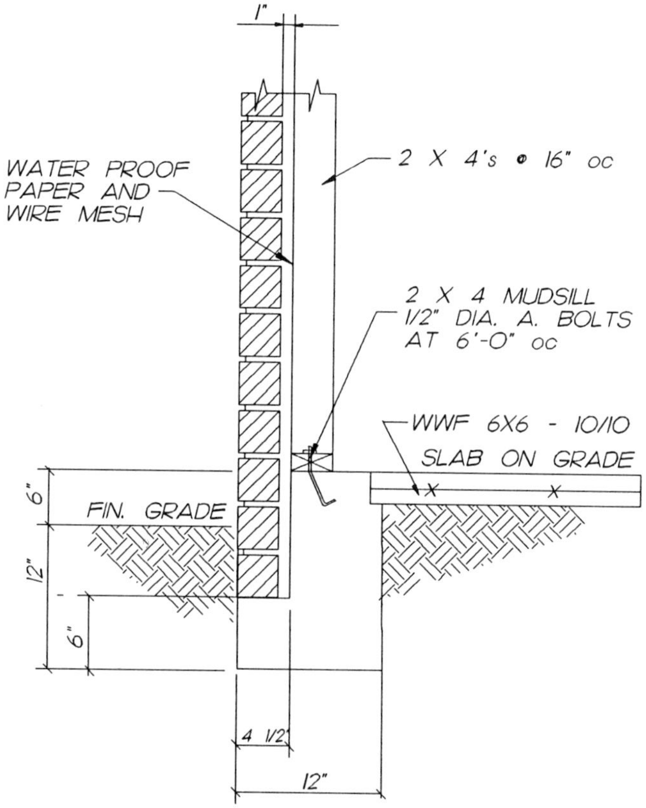

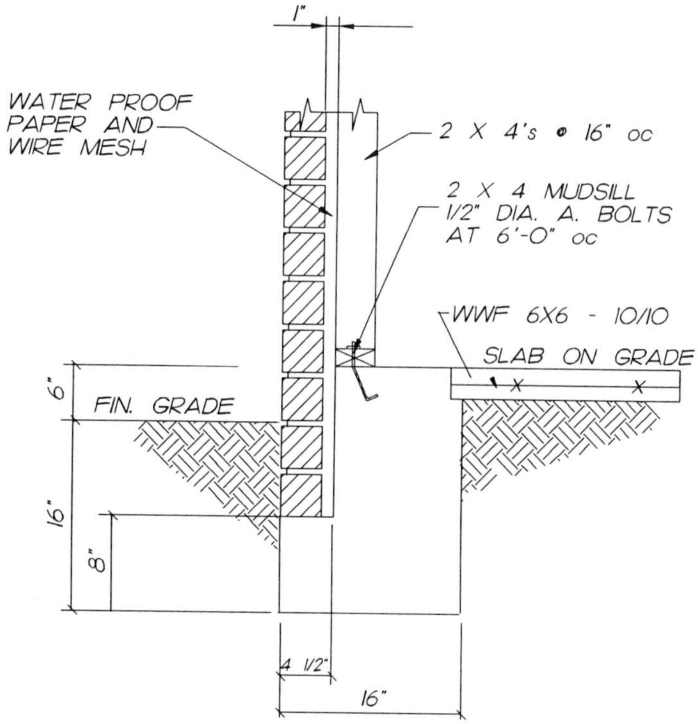

1"
WATER PROOF PAPER AND WIRE MESH
2 X 4's @ 16" oc
2 X 4 MUDSILL 1/2" DIA. A. BOLTS AT 6'-0" oc
WWF 6X6 - 10/10
SLAB ON GRADE
6"
FIN. GRADE
16"
8"
4 1/2"
16"

W1061

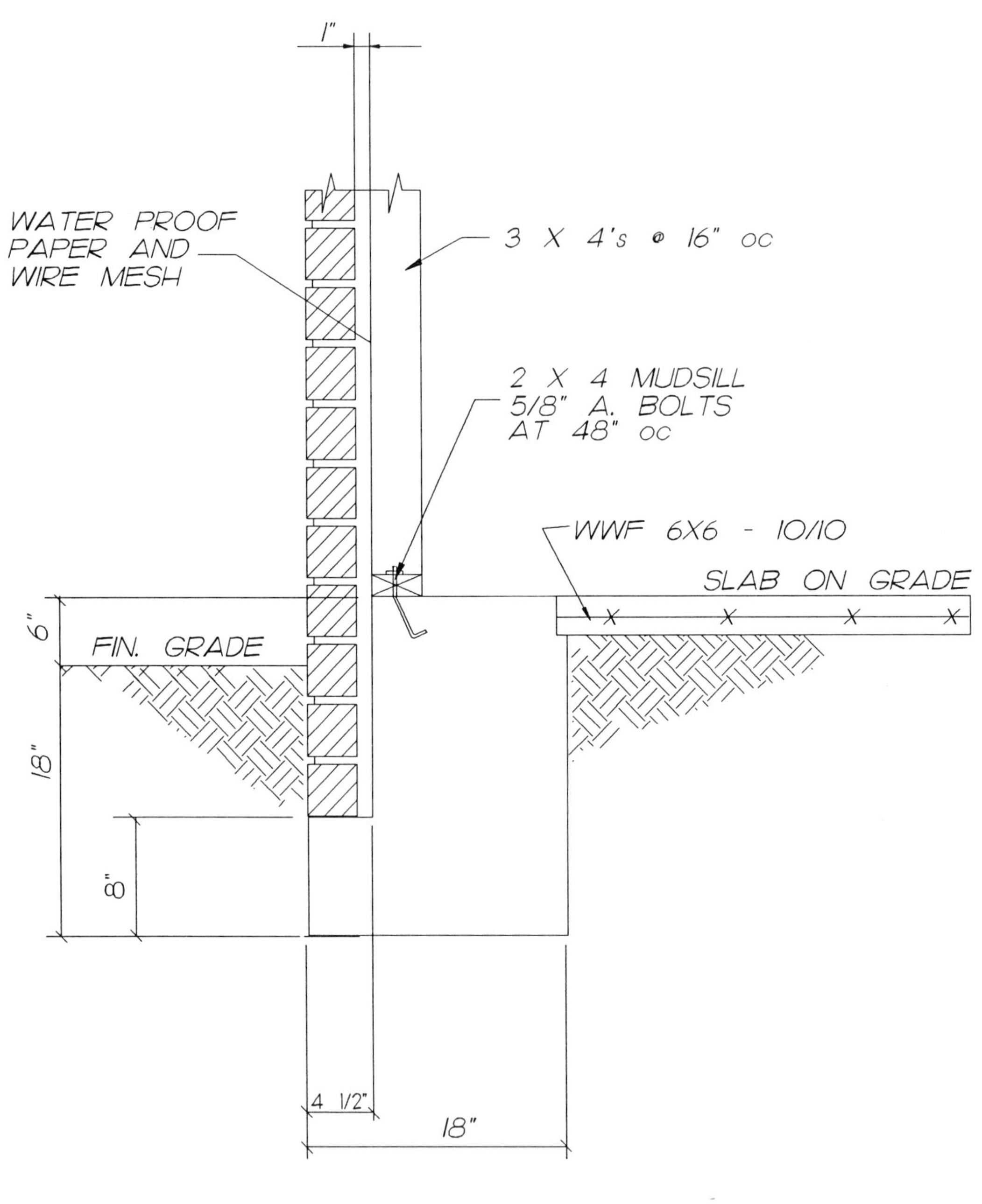

1"
WATER PROOF
PAPER AND
WIRE MESH
3 X 4's @ 16" oc
2 X 4 MUDSILL
5/8" A. BOLTS
AT 48" oc
WWF 6X6 - 10/10
SLAB ON GRADE
6"
FIN. GRADE
18"
8"
4 1/2"
18"

WIO62

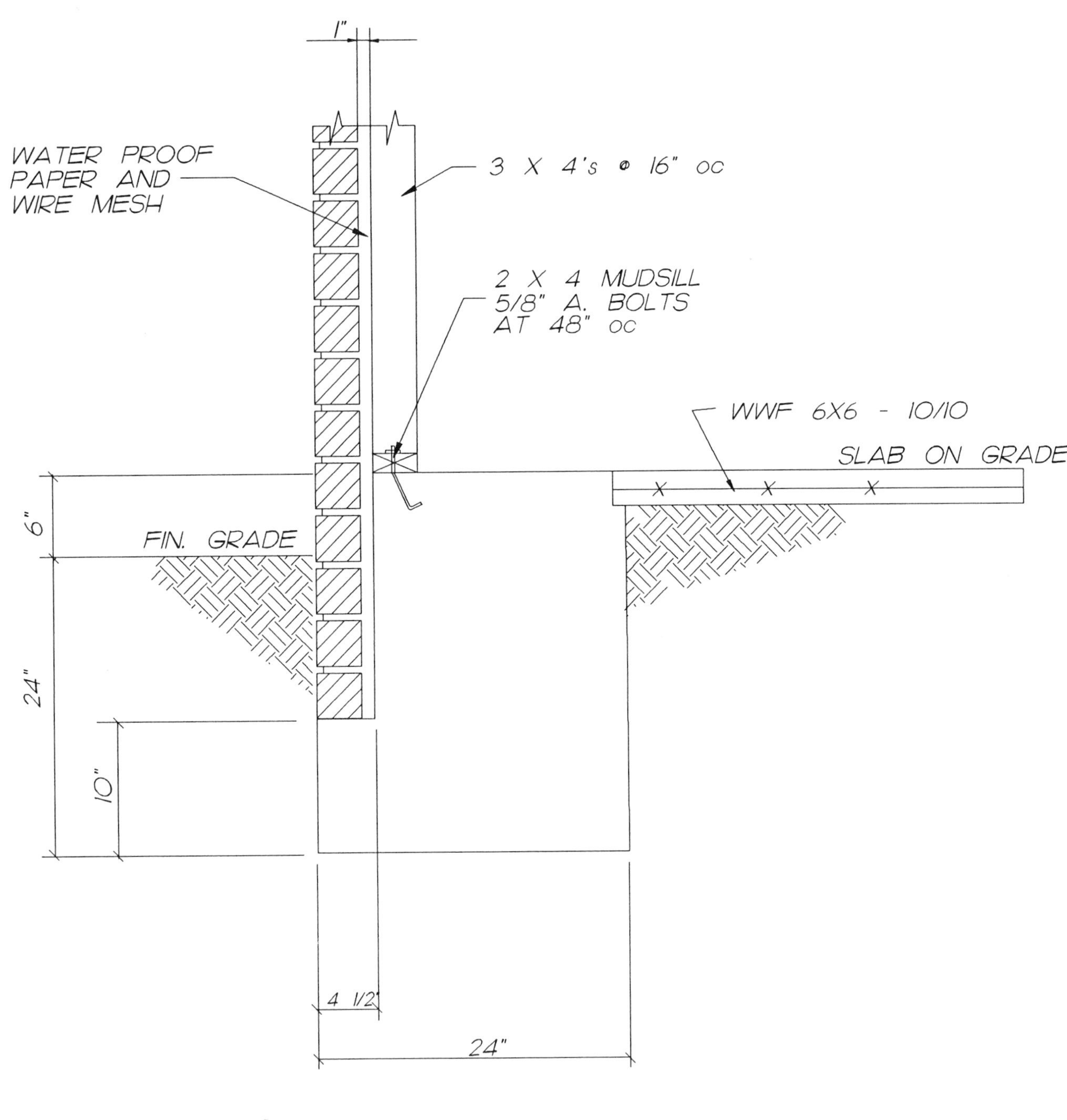
1"
WATER PROOF
PAPER AND
WIRE MESH
3 X 4's @ 16" OC
2 X 4 MUDSILL
5/8" A. BOLTS
AT 48" OC
WWF 6X6 - 10/10
SLAB ON GRADE
6"
FIN. GRADE
24"
10"
4 1/2"
24"

W1063

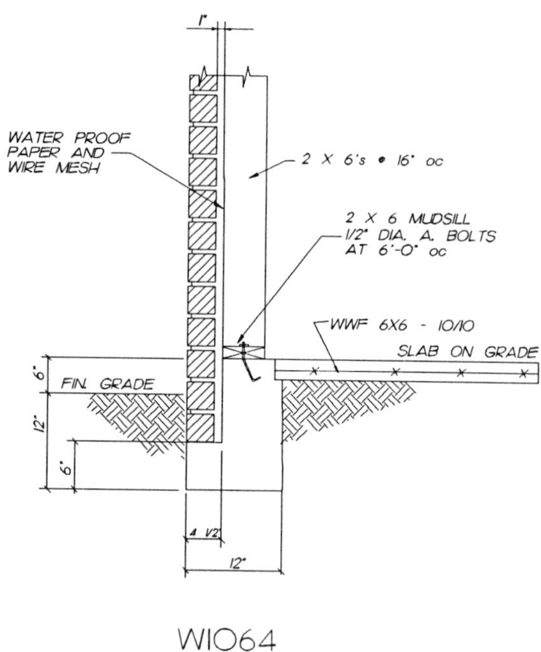
1"
WATER PROOF PAPER AND WIRE MESH
2 X 6's @ 16" oc
2 X 6 MUDSILL
1/2" DIA. A. BOLTS AT 6'-0" oc
WWF 6X6 - 10/10
SLAB ON GRADE
FIN. GRADE
6"
12"
6"
4 1/2"
12"

W1064

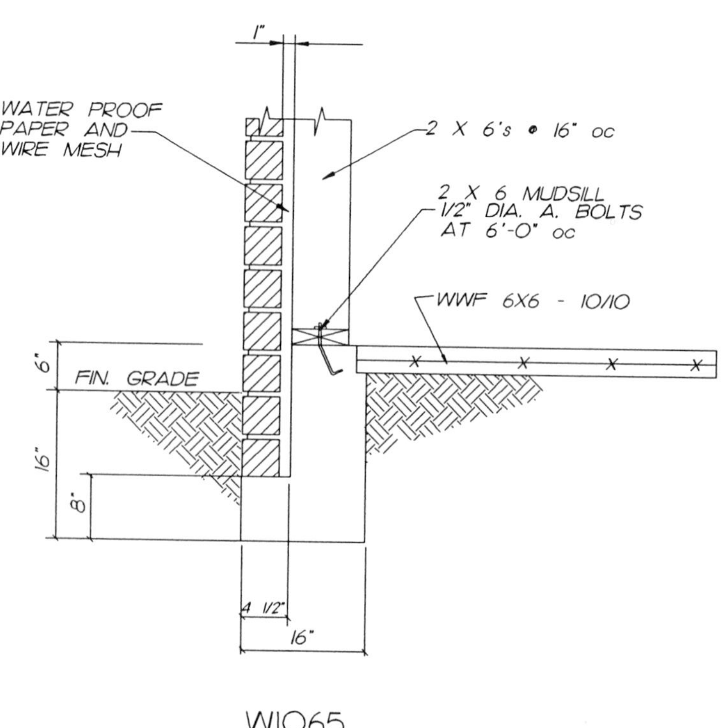
1"
WATER PROOF PAPER AND WIRE MESH
2 X 6's @ 16" oc
2 X 6 MUDSILL
1/2" DIA. A. BOLTS AT 6'-0" oc
WWF 6X6 - 10/10
FIN. GRADE
6"
16"
8"
4 1/2"
16"

W1065

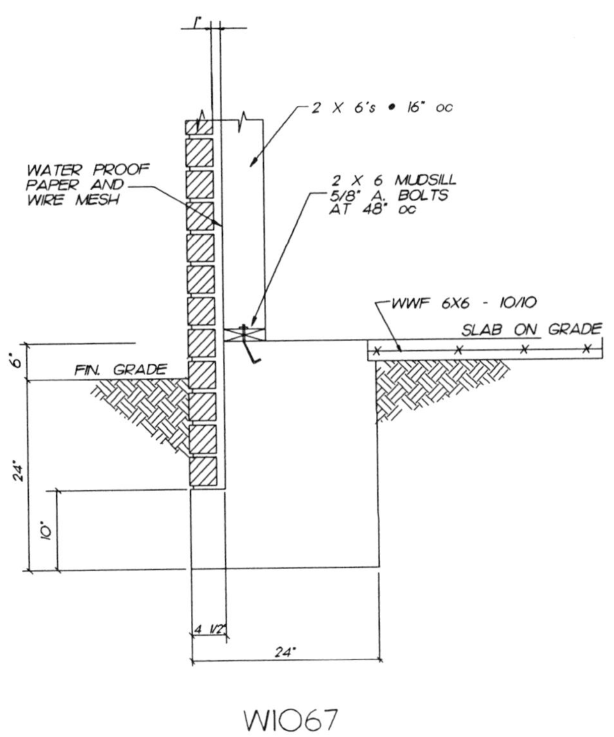

1"
2 X 6's @ 16" oc
WATER PROOF
PAPER AND
WIRE MESH
2 X 6 MUDSILL
5/8" A. BOLTS
AT 48" oc
WWF 6X6 - 10/10
SLAB ON GRADE
6"
FIN. GRADE
24"
10"
4 1/2"
24"

W1067

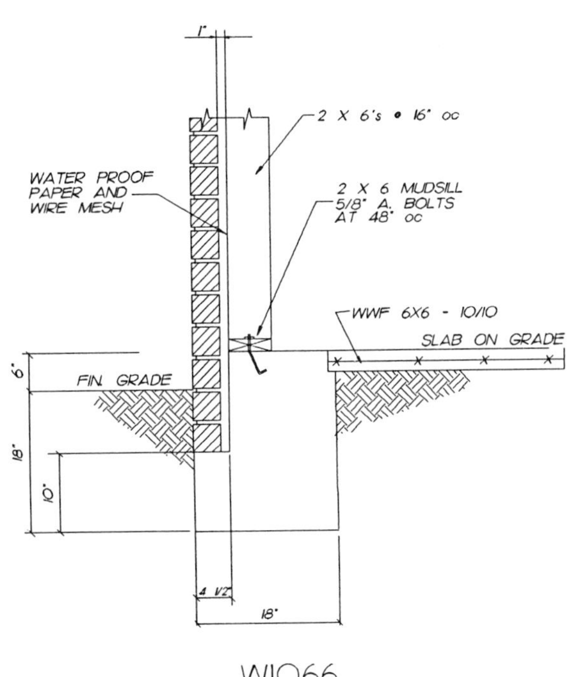

1"
2 X 6's @ 16" oc
WATER PROOF
PAPER AND
WIRE MESH
2 X 6 MUDSILL
5/8" A. BOLTS
AT 48" oc
WWF 6X6 - 10/10
SLAB ON GRADE
6"
FIN. GRADE
18"
10"
4 1/2"
18"

W1066

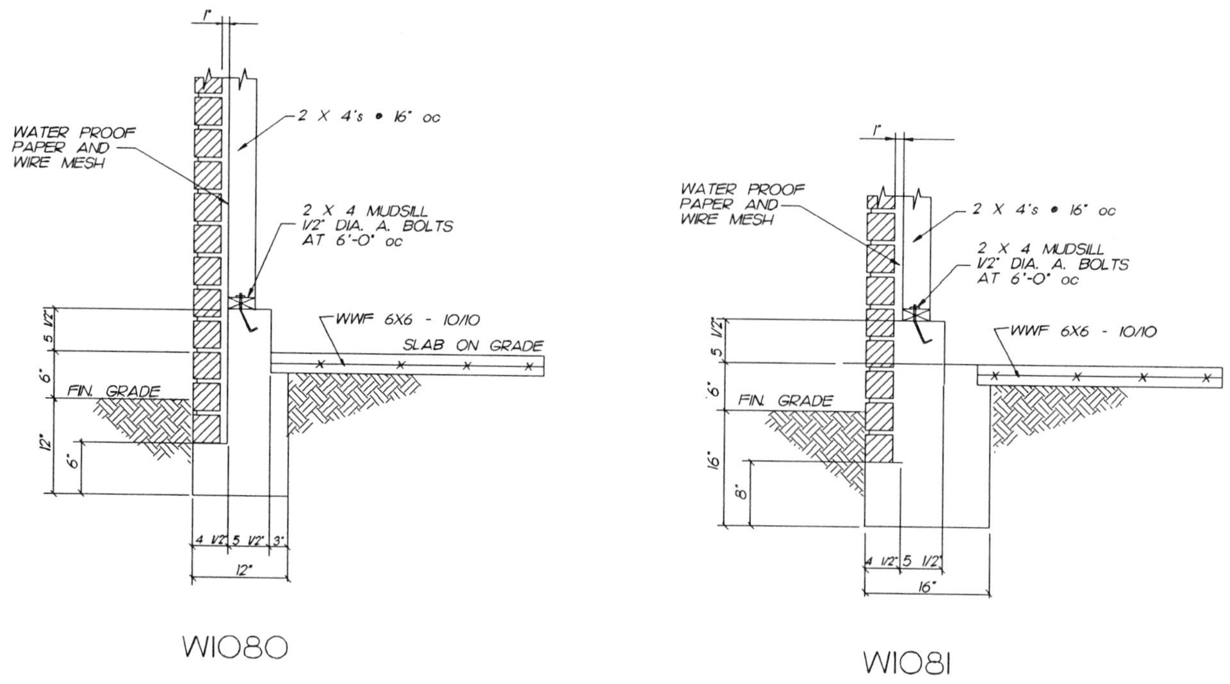
1"
WATER PROOF PAPER AND WIRE MESH
2 X 4's @ 16" oc
2 X 4 MUDSILL 1/2" DIA. A. BOLTS AT 6'-0" oc
WWF 6X6 - 10/10
SLAB ON GRADE
FIN. GRADE
5 1/2"
6"
12"
6"
4 1/2"
5 1/2"
3"
12"
W1080
1"
WATER PROOF PAPER AND WIRE MESH
2 X 4's @ 16" oc
2 X 4 MUDSILL 1/2" DIA. A. BOLTS AT 6'-0" oc
WWF 6X6 - 10/10
FIN. GRADE
5 1/2"
6"
16"
8"
4 1/2"
5 1/2"
16"
W1081

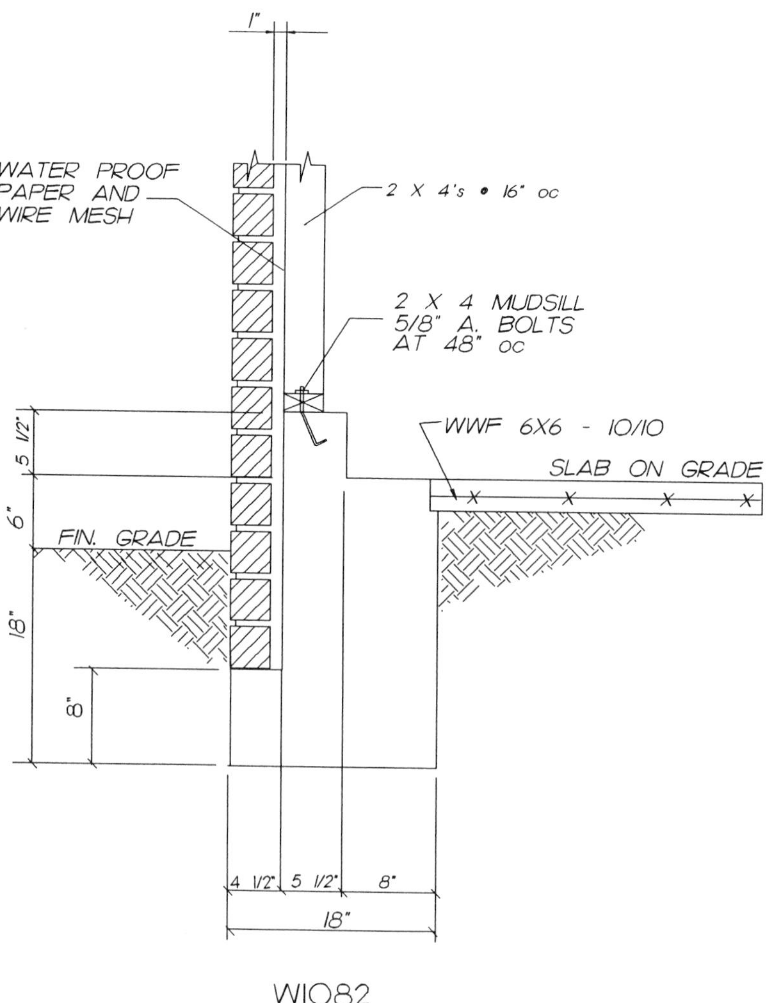
1"
WATER PROOF PAPER AND WIRE MESH
2 X 4's @ 16" oc
2 X 4 MUDSILL 5/8" A. BOLTS AT 48" oc
WWF 6X6 - 10/10
SLAB ON GRADE
FIN. GRADE
5 1/2"
6"
18"
8"
4 1/2"
5 1/2"
8"
18"
W1082

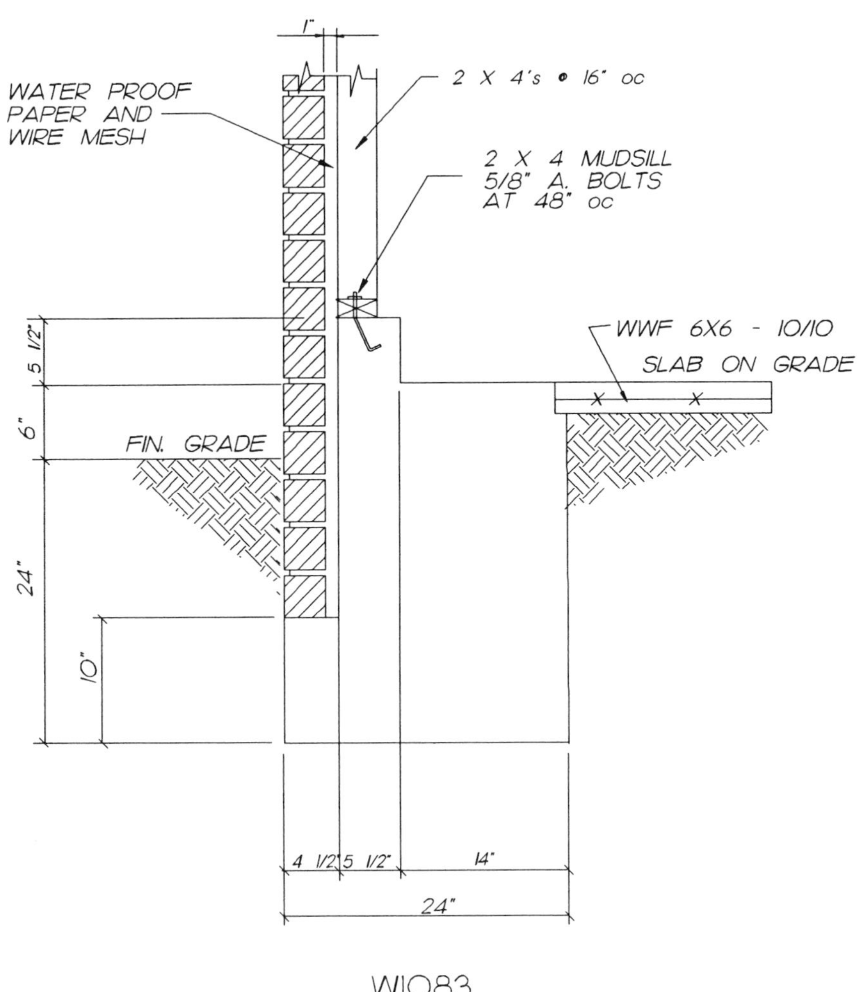

W1083

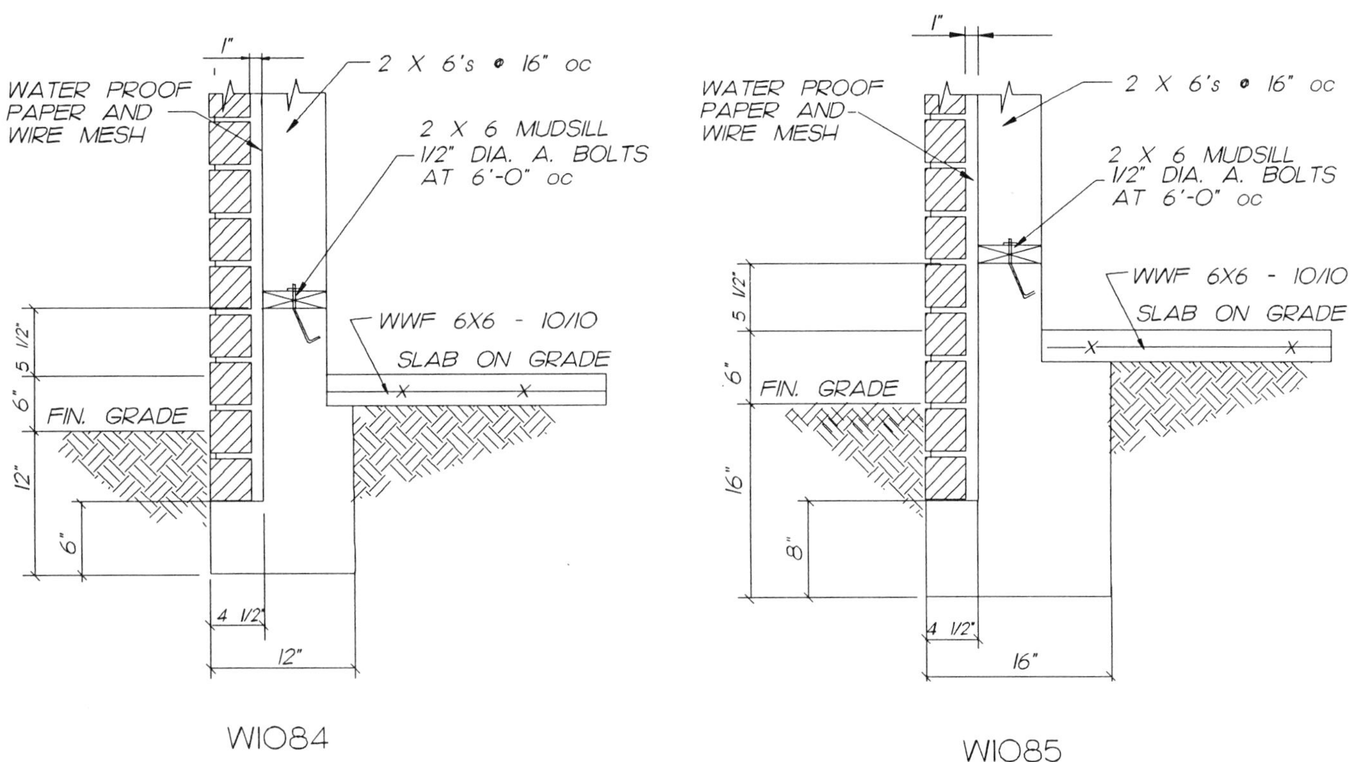

W1084

W1085

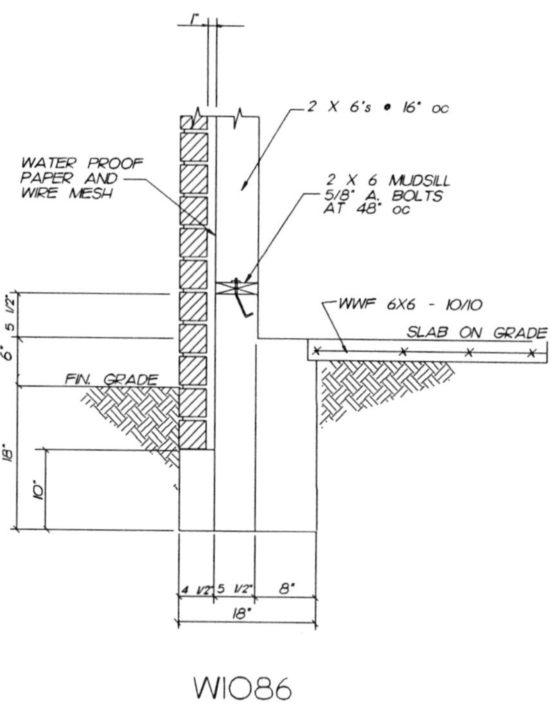

W1086

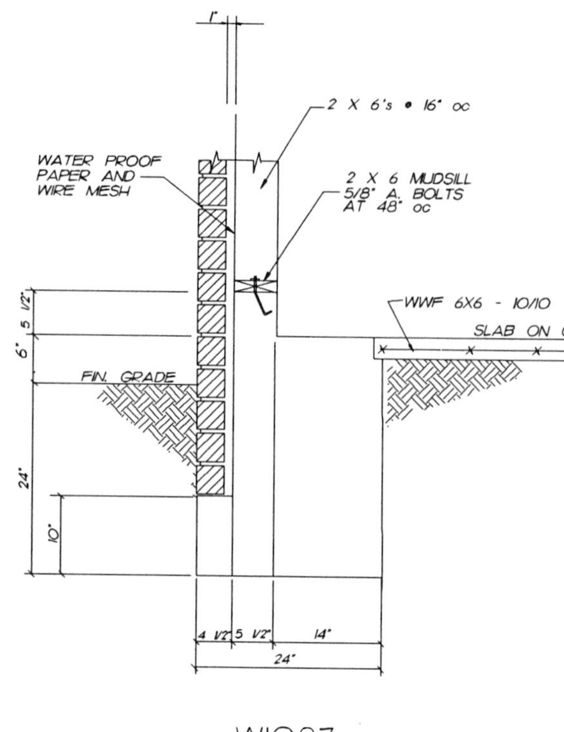

W1087

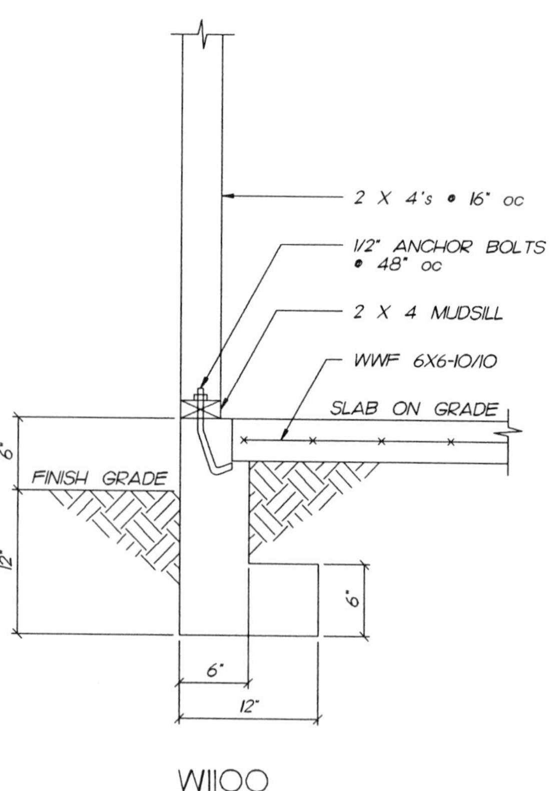

W1100

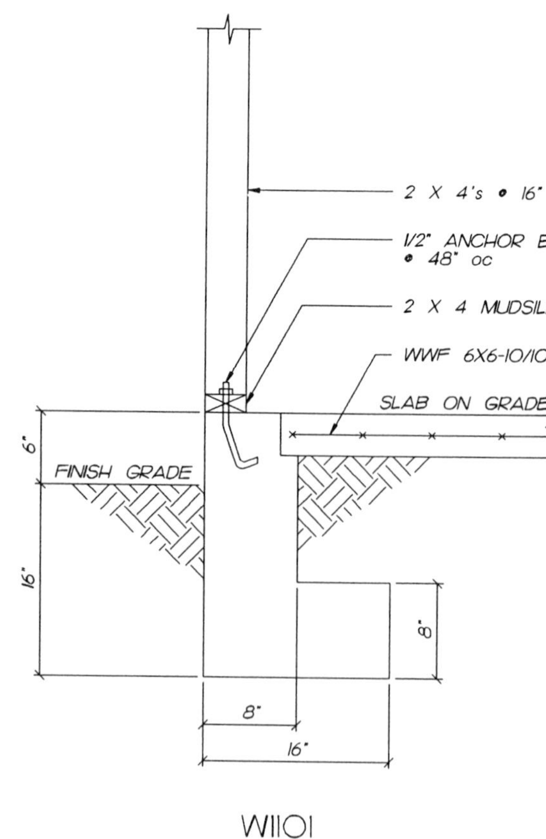

W1101

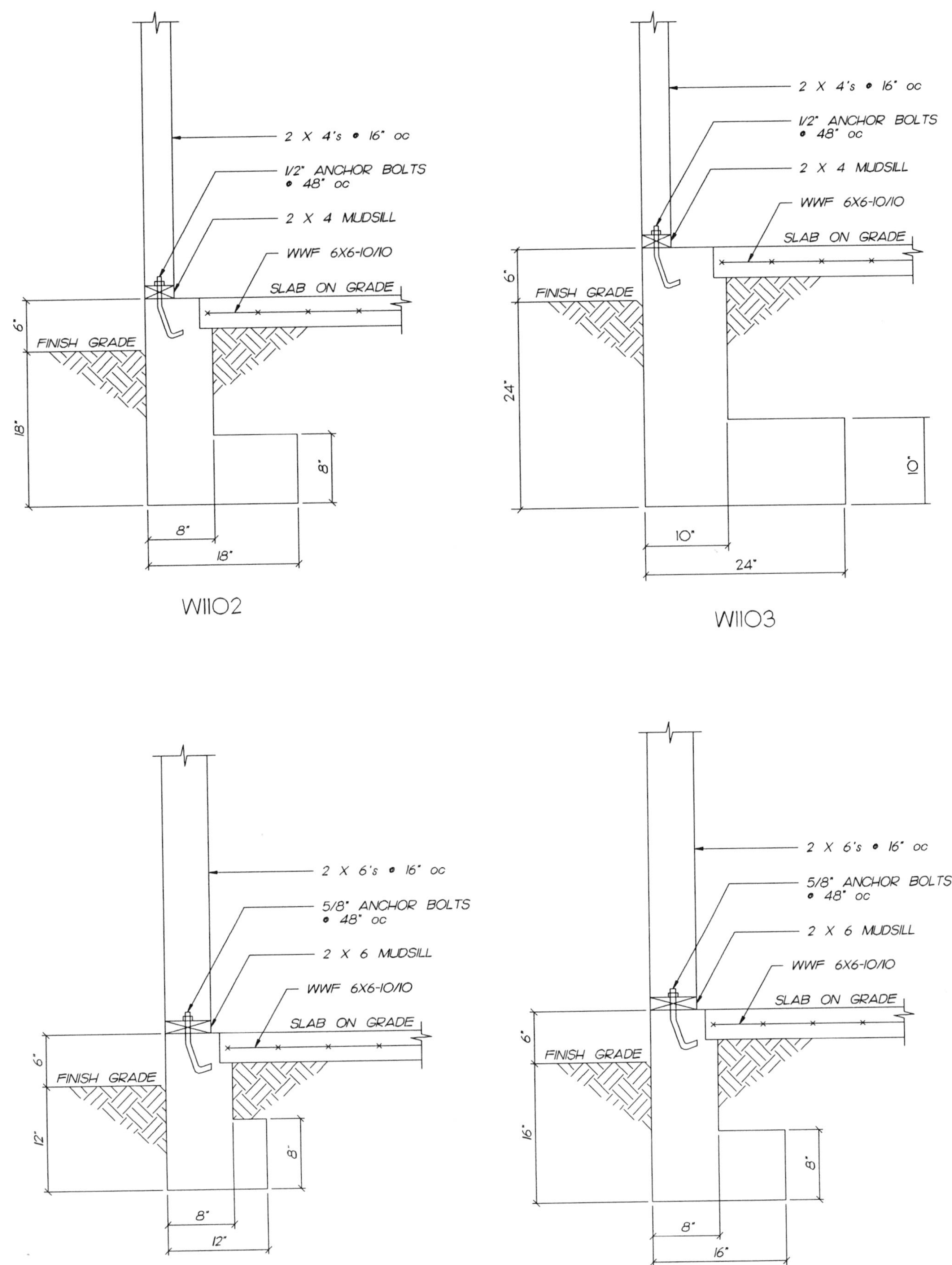
2 X 4's @ 16" oc
1/2" ANCHOR BOLTS @ 48" oc
2 X 4 MUDSILL
WWF 6X6-10/10
SLAB ON GRADE
FINISH GRADE
6"
18"
8"
8"
18"
W1102
2 X 4's @ 16" oc
1/2" ANCHOR BOLTS @ 48" oc
2 X 4 MUDSILL
WWF 6X6-10/10
SLAB ON GRADE
FINISH GRADE
6"
24"
10"
10"
24"
W1103
2 X 6's @ 16" oc
5/8" ANCHOR BOLTS @ 48" oc
2 X 6 MUDSILL
WWF 6X6-10/10
SLAB ON GRADE
FINISH GRADE
6"
12"
8"
8"
12"
W1104
2 X 6's @ 16" oc
5/8" ANCHOR BOLTS @ 48" oc
2 X 6 MUDSILL
WWF 6X6-10/10
SLAB ON GRADE
FINISH GRADE
6"
16"
8"
8"
16"
W1105

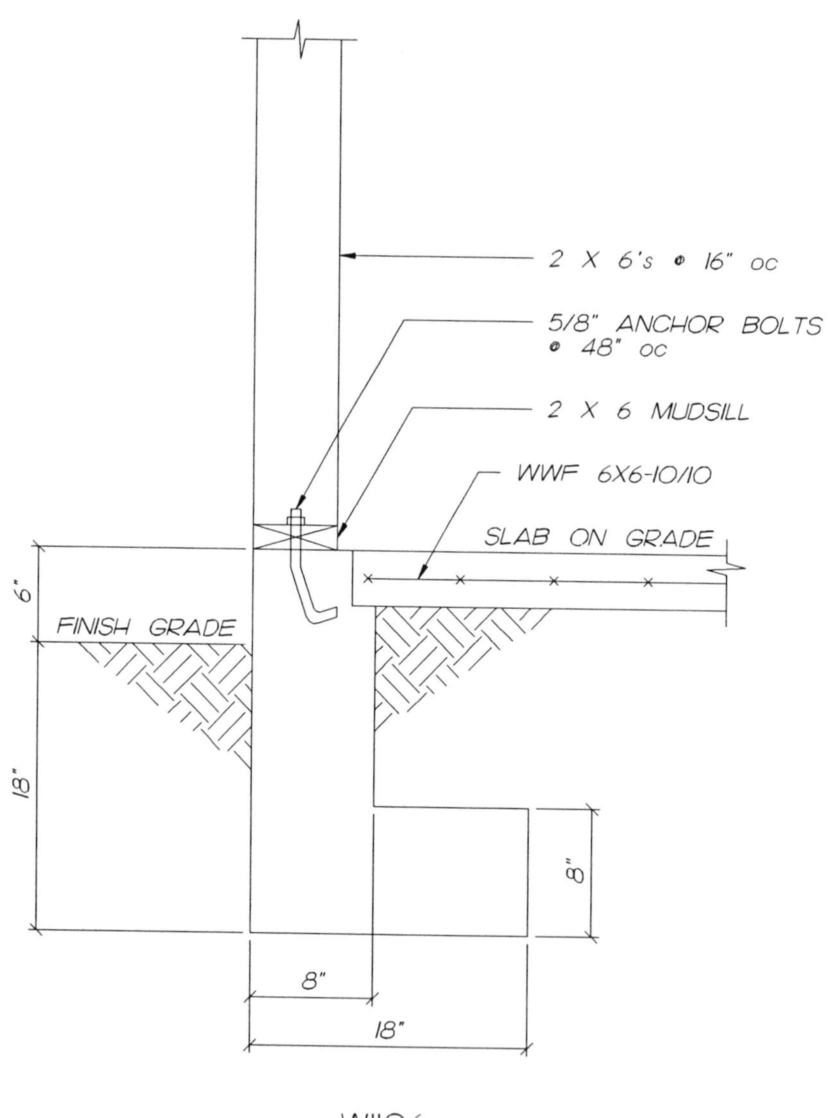
2 X 6's @ 16" OC
5/8" ANCHOR BOLTS
@ 48" OC
2 X 6 MUDSILL
WWF 6X6-10/10
SLAB ON GRADE
FINISH GRADE
6"
18"
8"
8"
18"

W1106

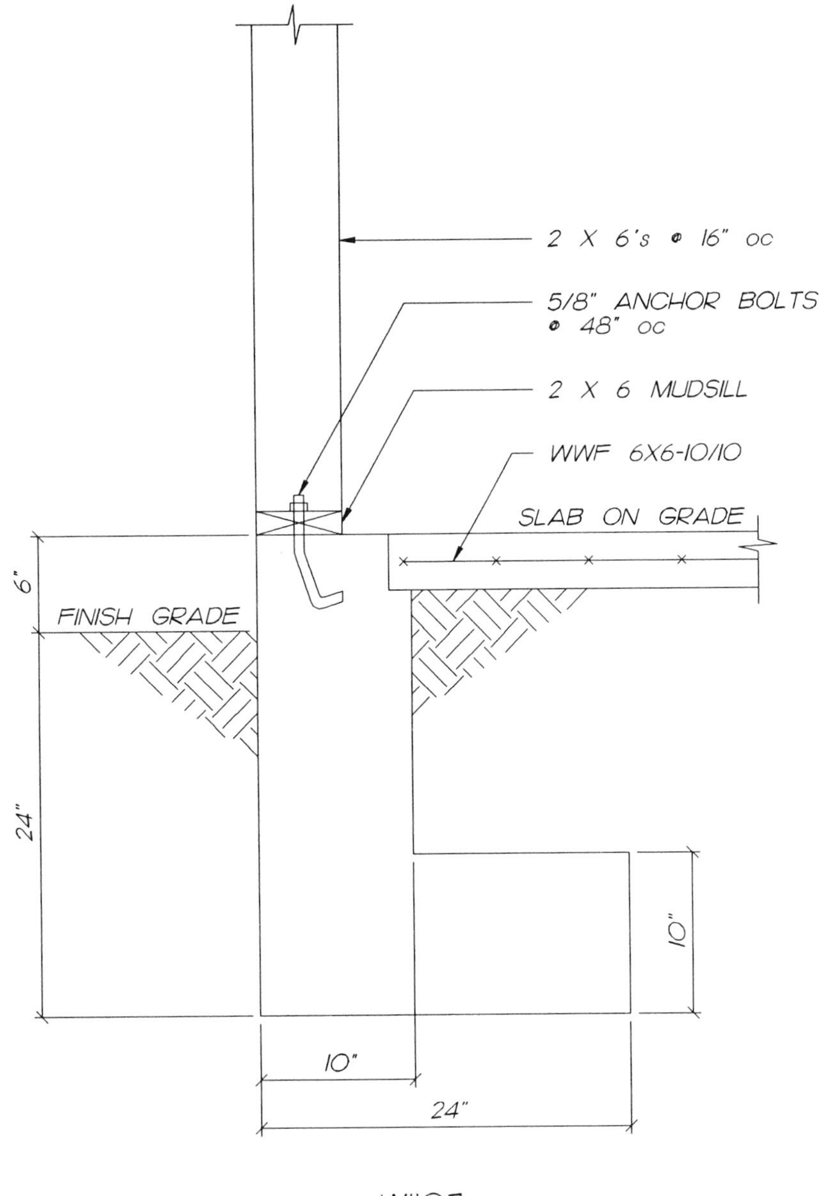
2 X 6's @ 16" oc
5/8" ANCHOR BOLTS
@ 48" oc
2 X 6 MUDSILL
WWF 6X6-10/10
SLAB ON GRADE
FINISH GRADE
6"
24"
10"
10"
24"

W1107

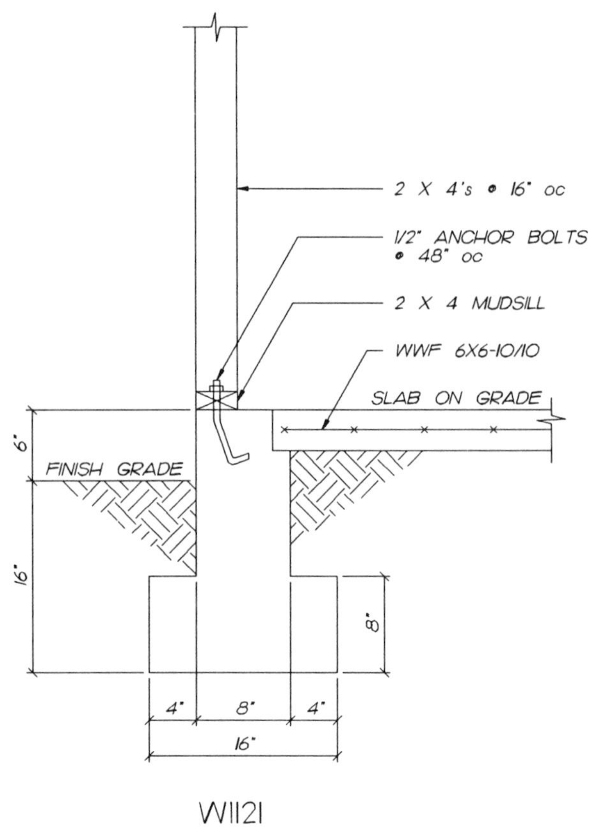
2 X 4's @ 16" oc
1/2" ANCHOR BOLTS @ 48" oc
2 X 4 MUDSILL
WWF 6X6-10/10
SLAB ON GRADE
FINISH GRADE
6"
16"
8"
4"
8"
4"
16"

W1121

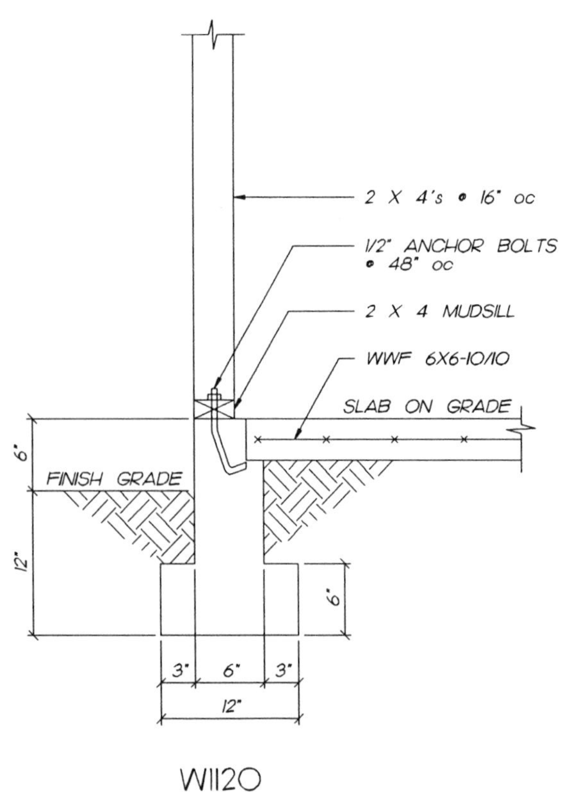
2 X 4's @ 16" oc
1/2" ANCHOR BOLTS @ 48" oc
2 X 4 MUDSILL
WWF 6X6-10/10
SLAB ON GRADE
FINISH GRADE
6"
12"
6"
3"
6"
3"
12"

W1120

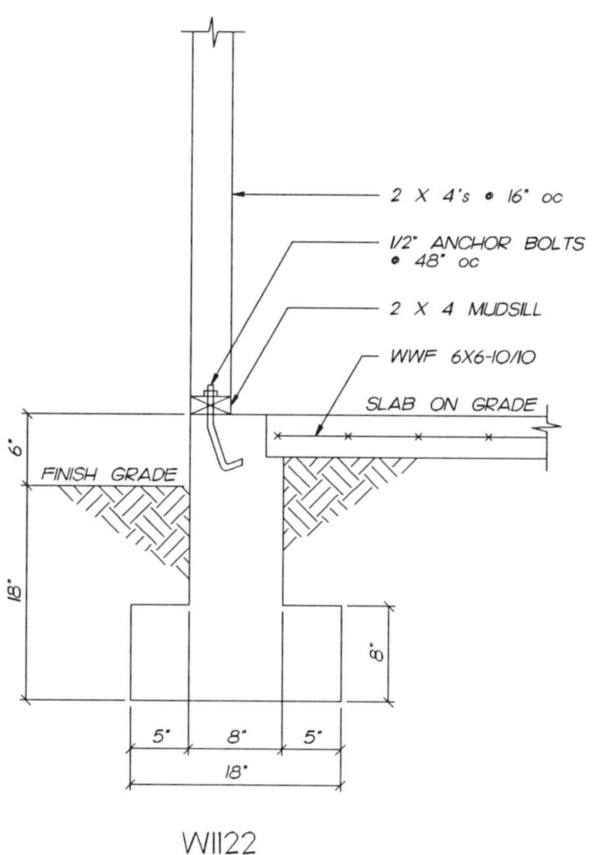
2 X 4's @ 16" oc
1/2" ANCHOR BOLTS @ 48" oc
2 X 4 MUDSILL
WWF 6X6-10/10
SLAB ON GRADE
FINISH GRADE
6"
18"
8"
5"
8"
5"
18"

WII22

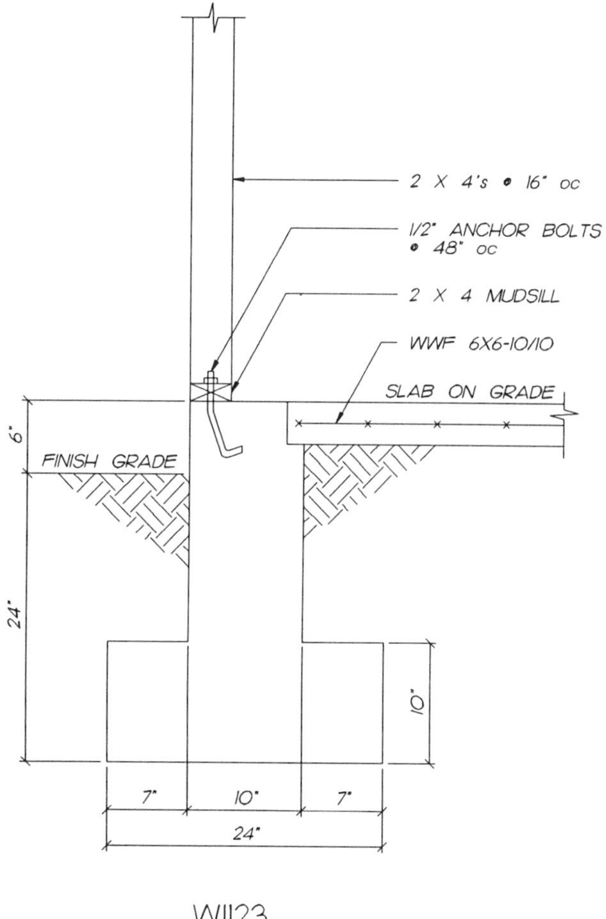
2 X 4's @ 16" oc
1/2" ANCHOR BOLTS @ 48" oc
2 X 4 MUDSILL
WWF 6X6-10/10
SLAB ON GRADE
FINISH GRADE
6"
24"
10"
7"
10"
7"
24"

WII23

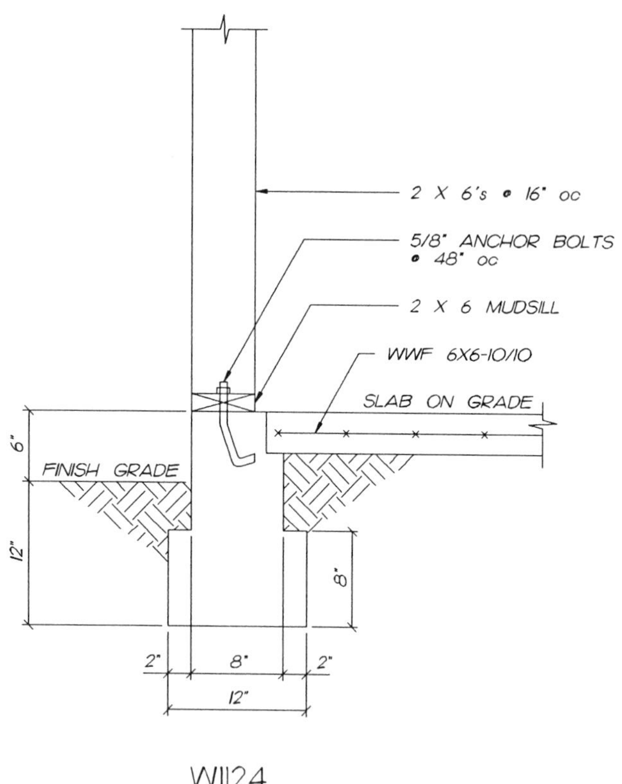
2 X 6's @ 16" oc
5/8" ANCHOR BOLTS
@ 48" oc
2 X 6 MUDSILL
WWF 6X6-10/10
SLAB ON GRADE
FINISH GRADE
6"
12"
8"
2"
8"
2"
12"

W1124

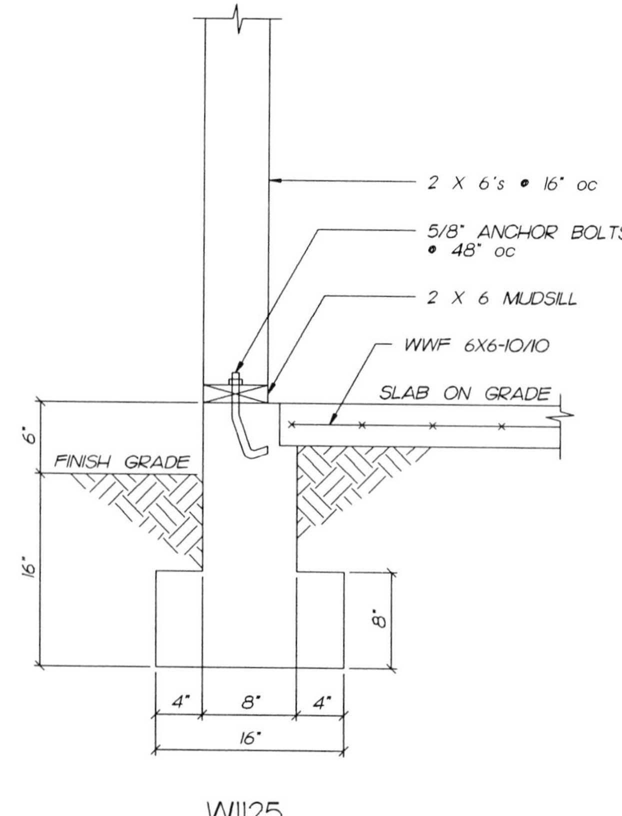
2 X 6's @ 16" oc
5/8" ANCHOR BOLTS
@ 48" oc
2 X 6 MUDSILL
WWF 6X6-10/10
SLAB ON GRADE
FINISH GRADE
6"
16"
8"
4"
8"
4"
16"

W1125

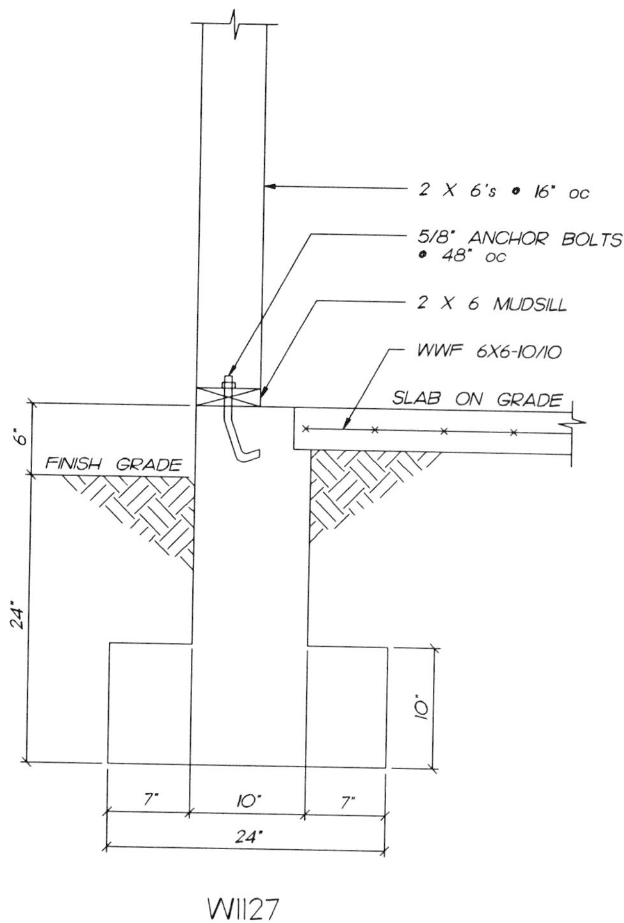
2 X 6's @ 16" oc
5/8" ANCHOR BOLTS @ 48" oc
2 X 6 MUDSILL
WWF 6X6-10/10
SLAB ON GRADE
FINISH GRADE
6"
24"
10"
7"
10"
7"
24"

W1127

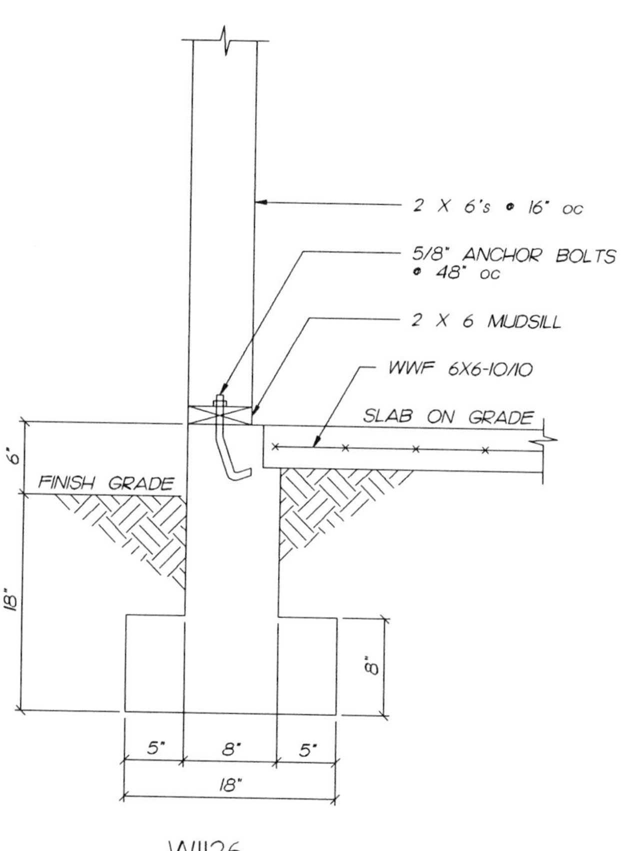
2 X 6's @ 16" oc
5/8" ANCHOR BOLTS @ 48" oc
2 X 6 MUDSILL
WWF 6X6-10/10
SLAB ON GRADE
FINISH GRADE
6"
18"
8"
5"
8"
5"
18"

W1126

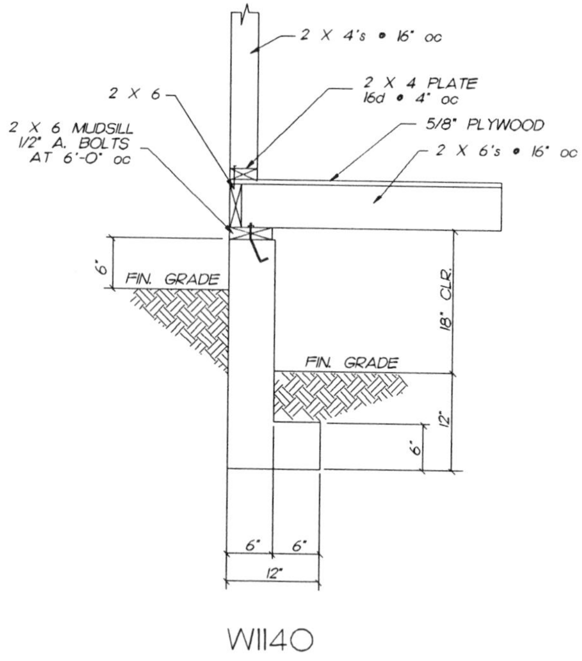

W1140

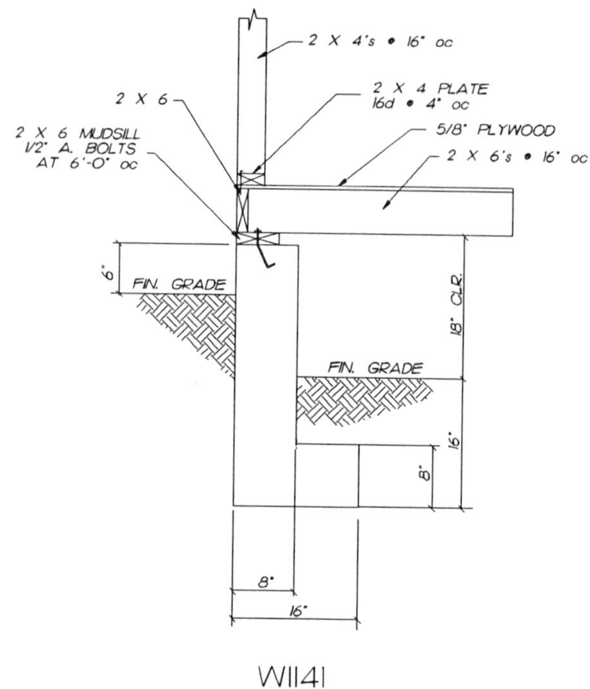

W1141

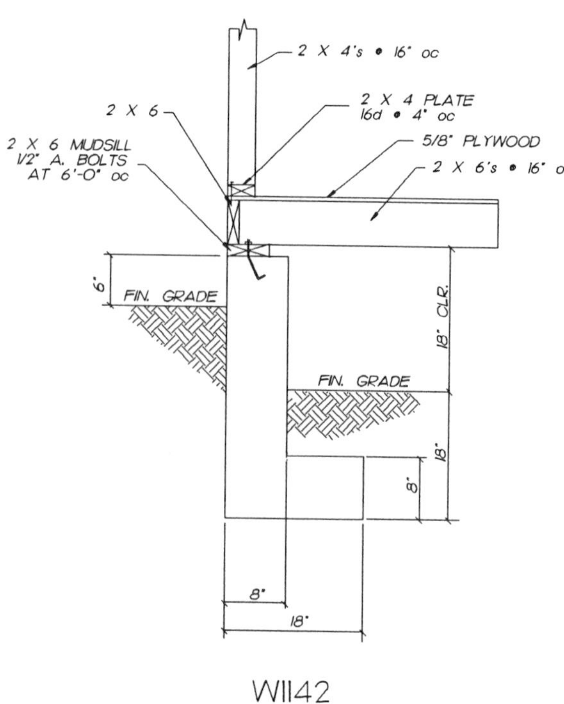

W1142

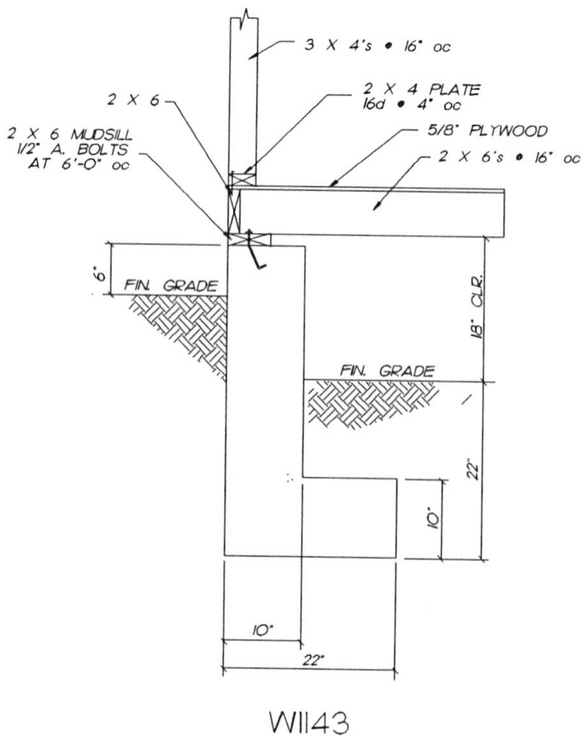

W1143

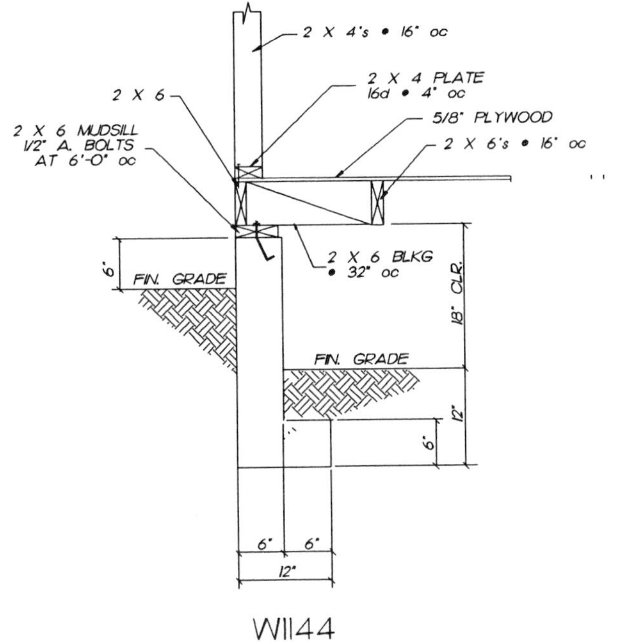

W1144

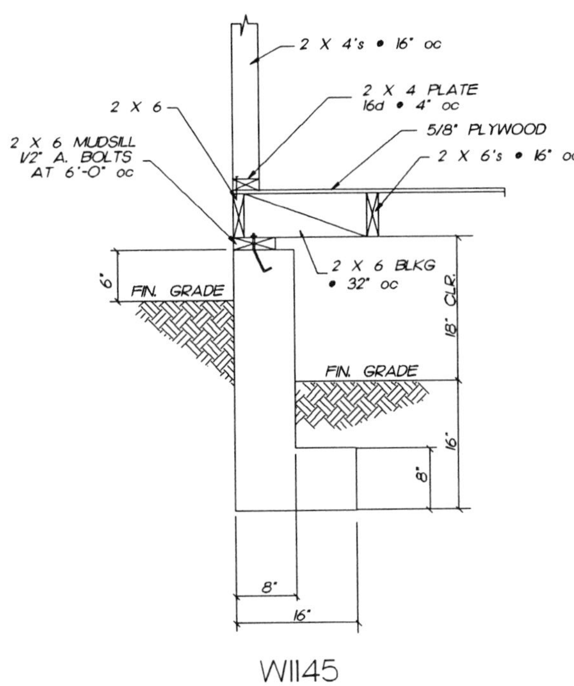

W1145

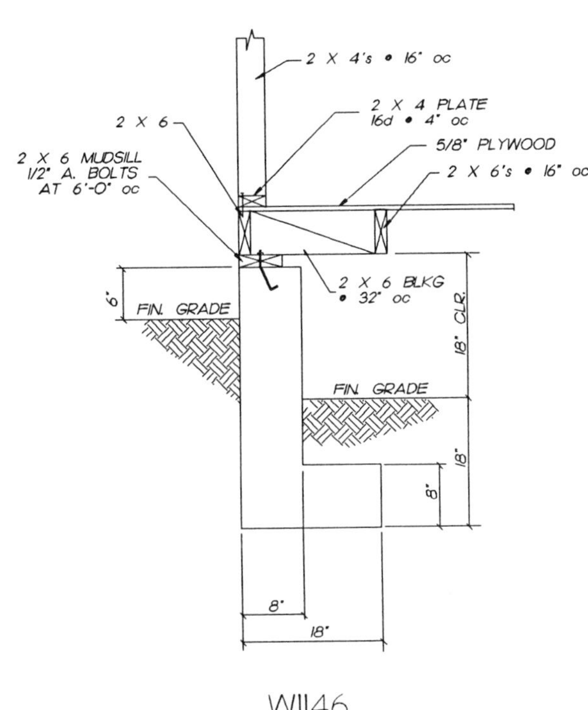

W1146

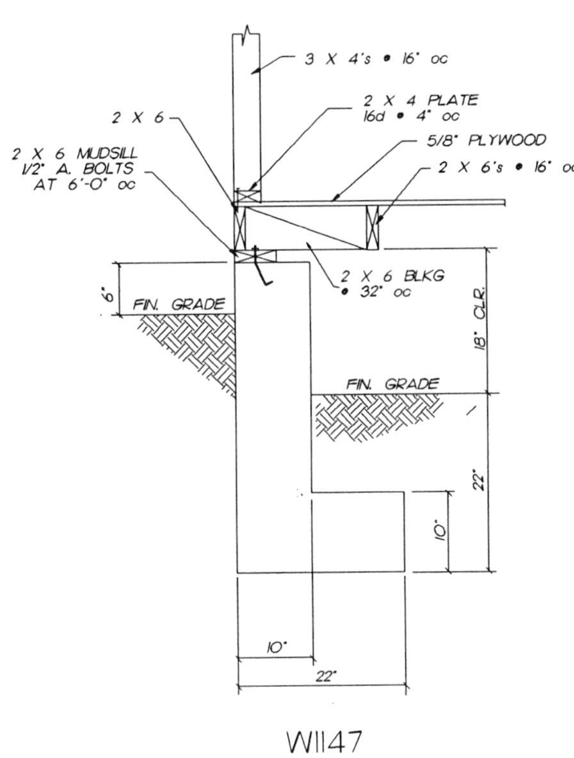

W1147

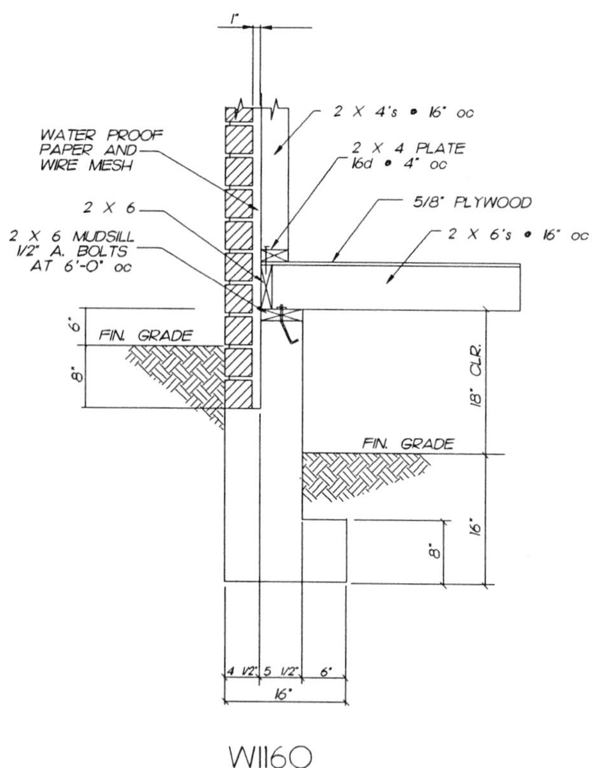

W1160

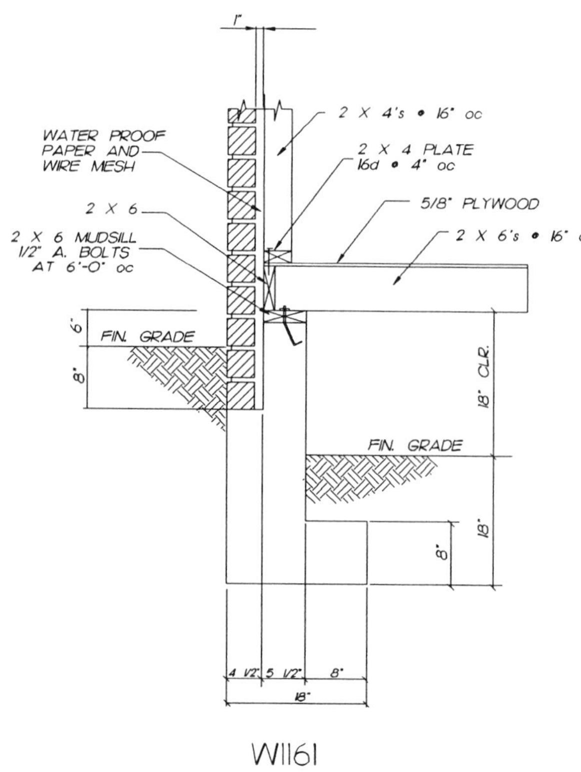

W1161

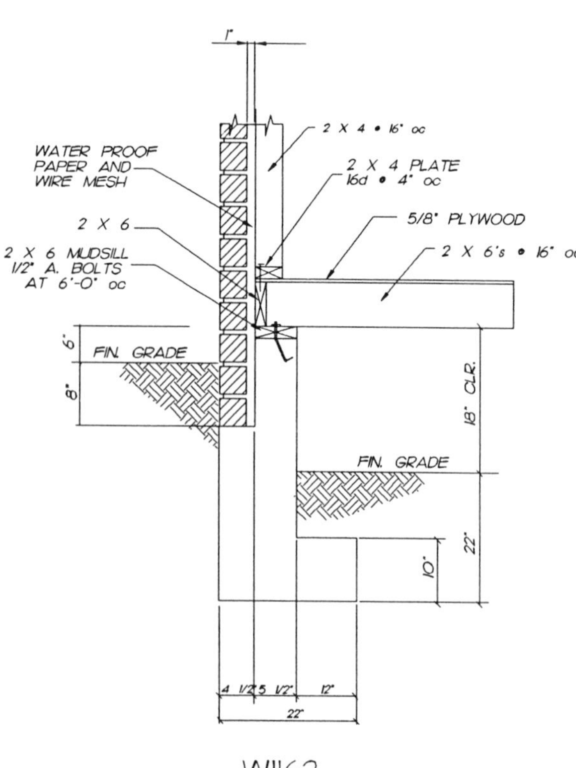

W1162

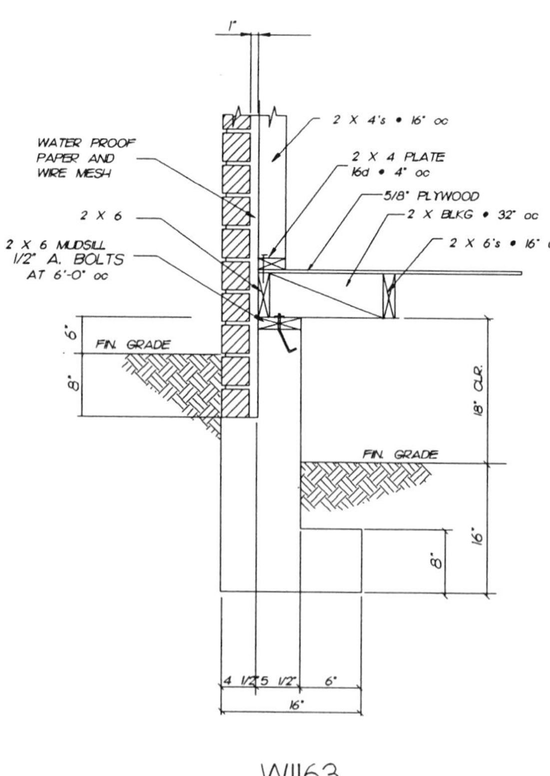

W1163

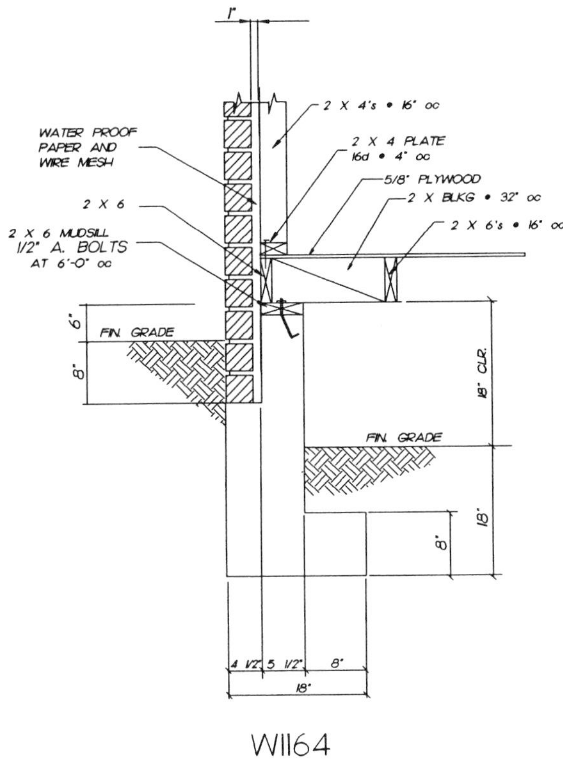

WII64

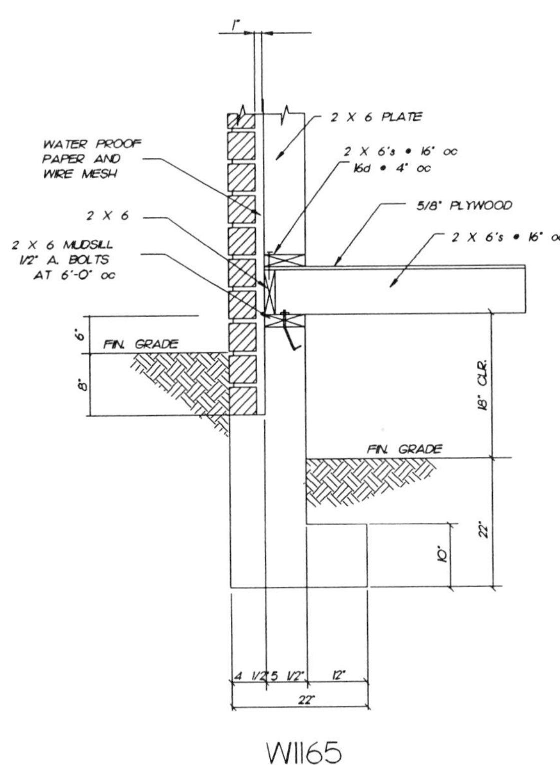

WII65

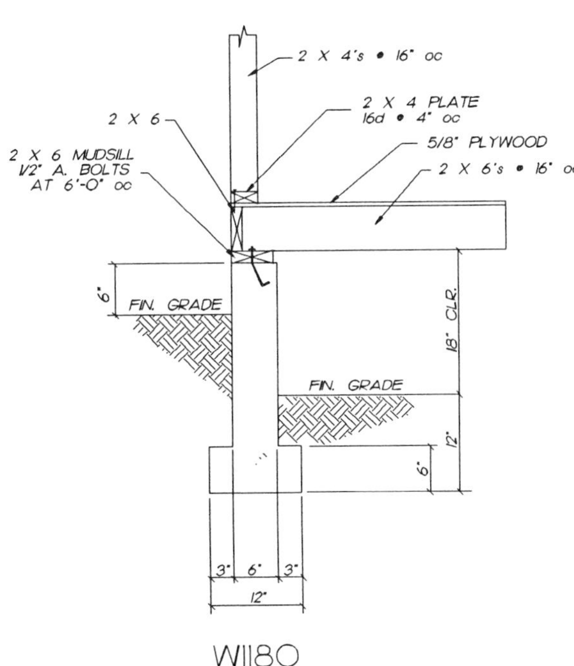

WII80

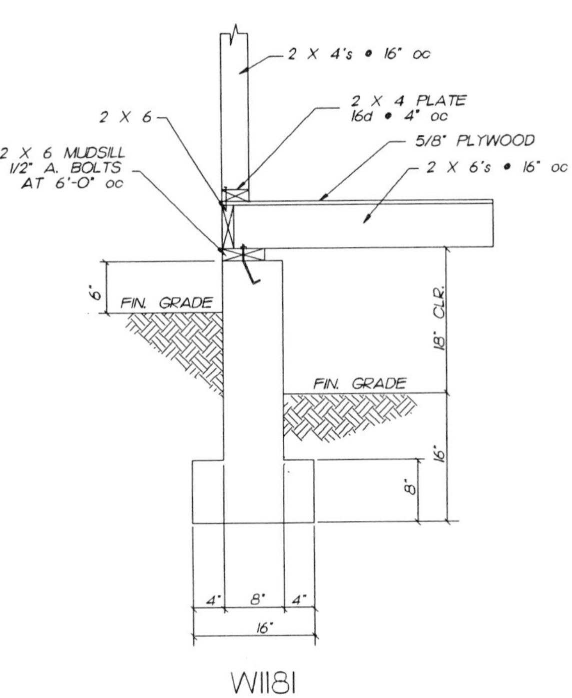

WII81

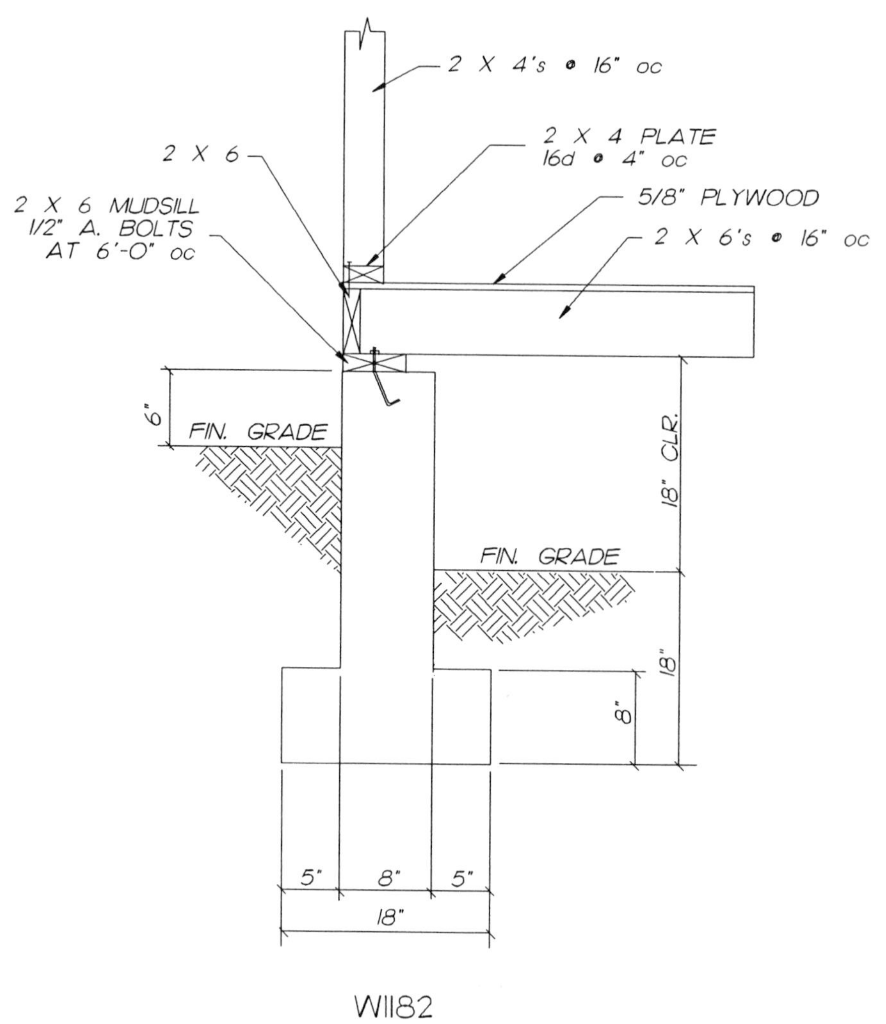

W1182

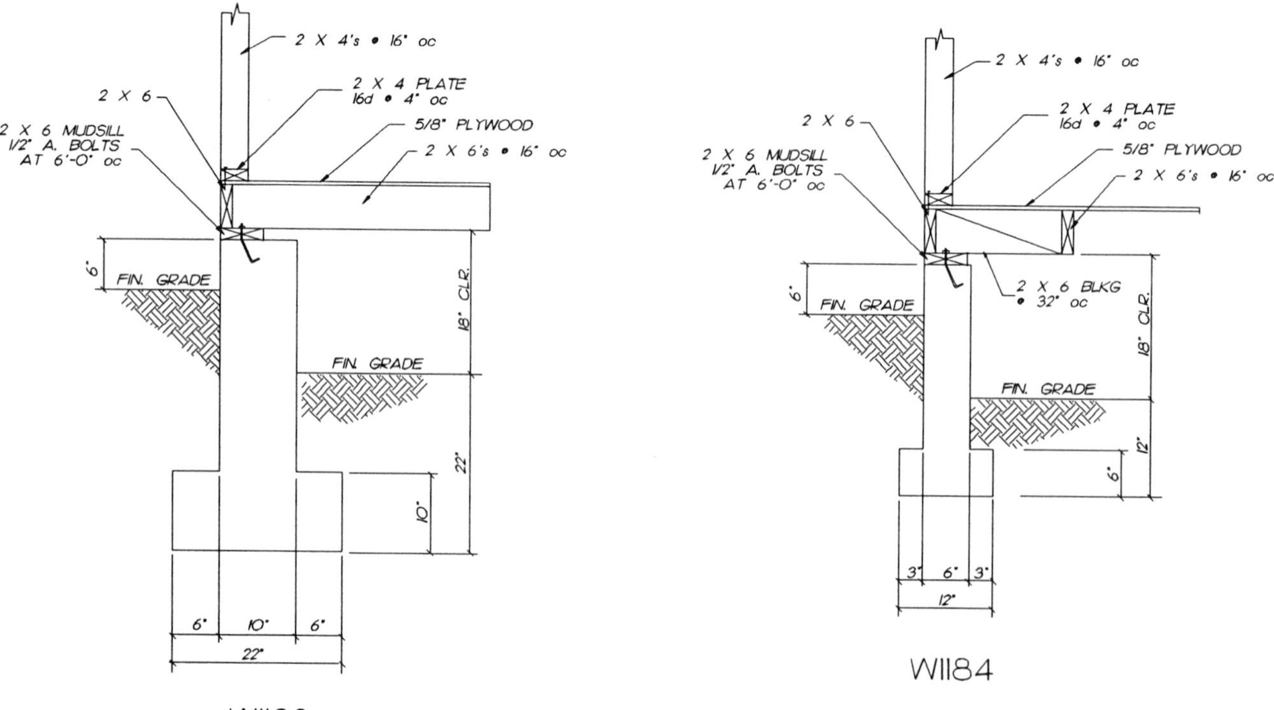

W1183

W1184

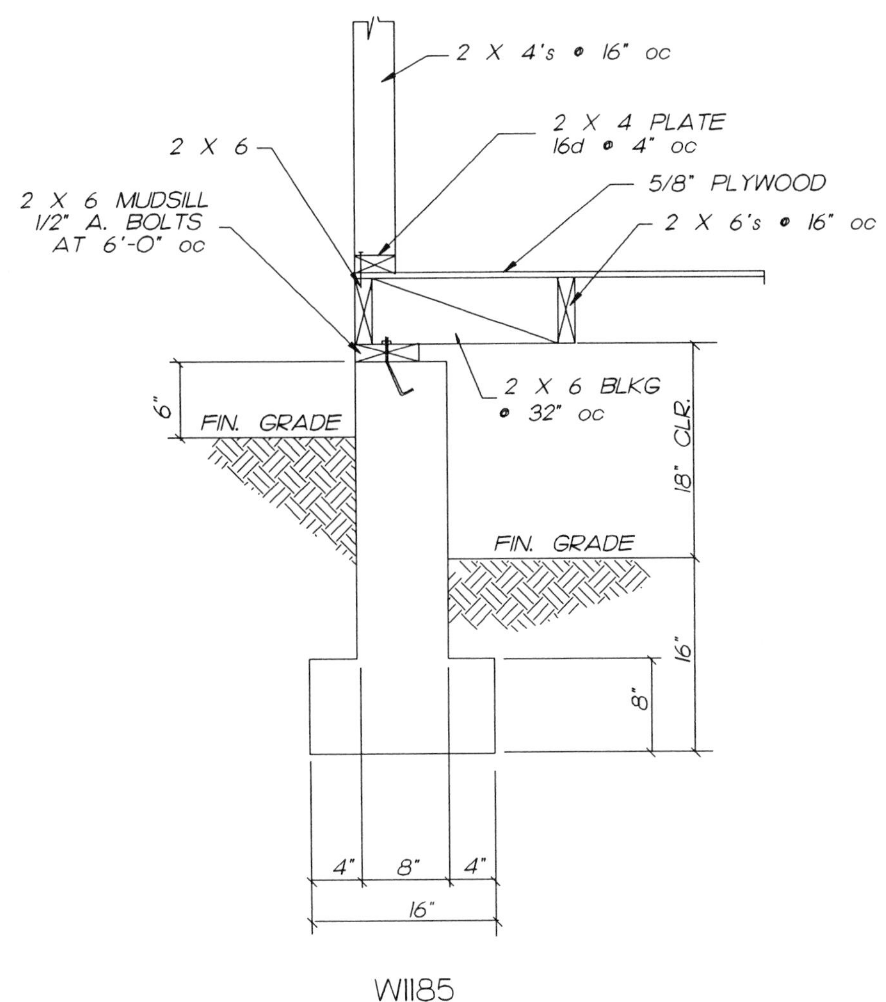

W1185

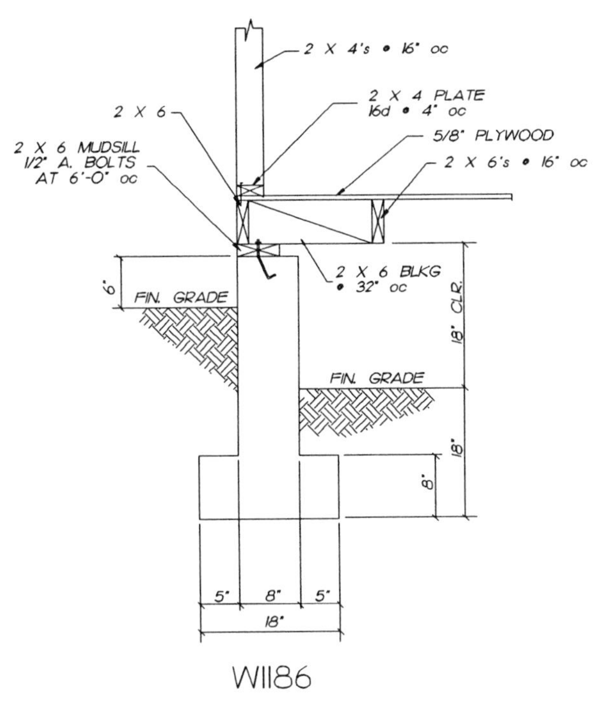

W1186

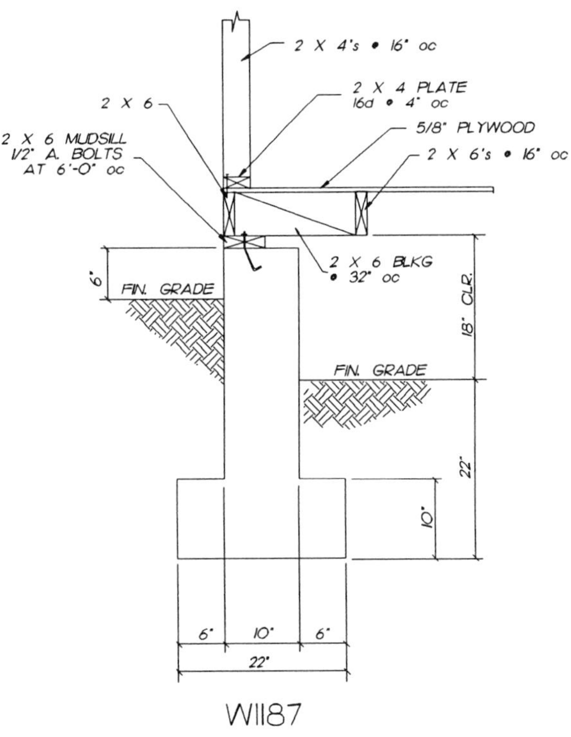

W1187

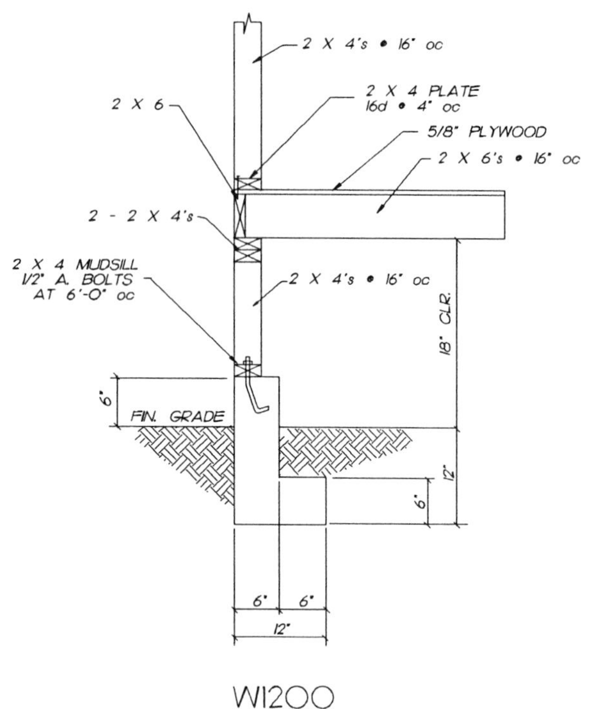

W1200

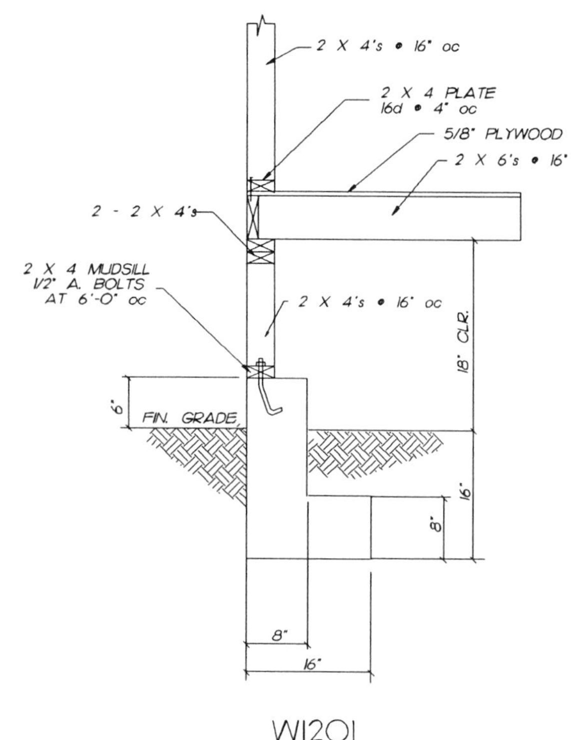

W1201

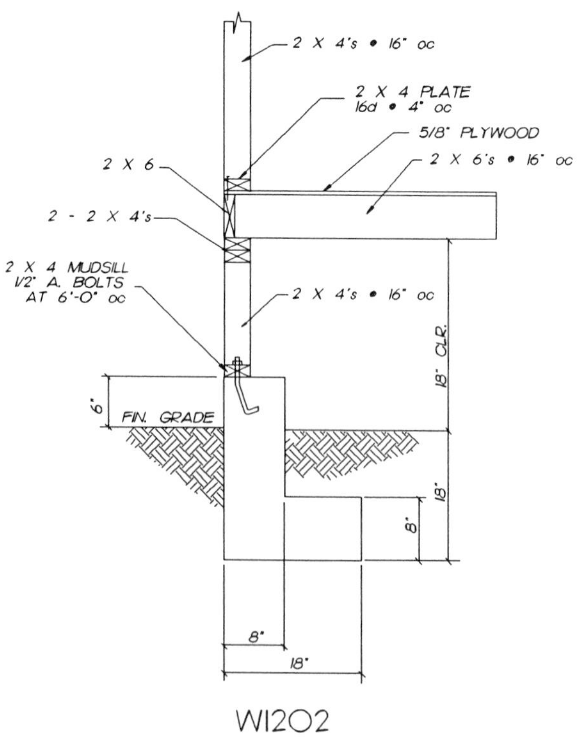

W1202

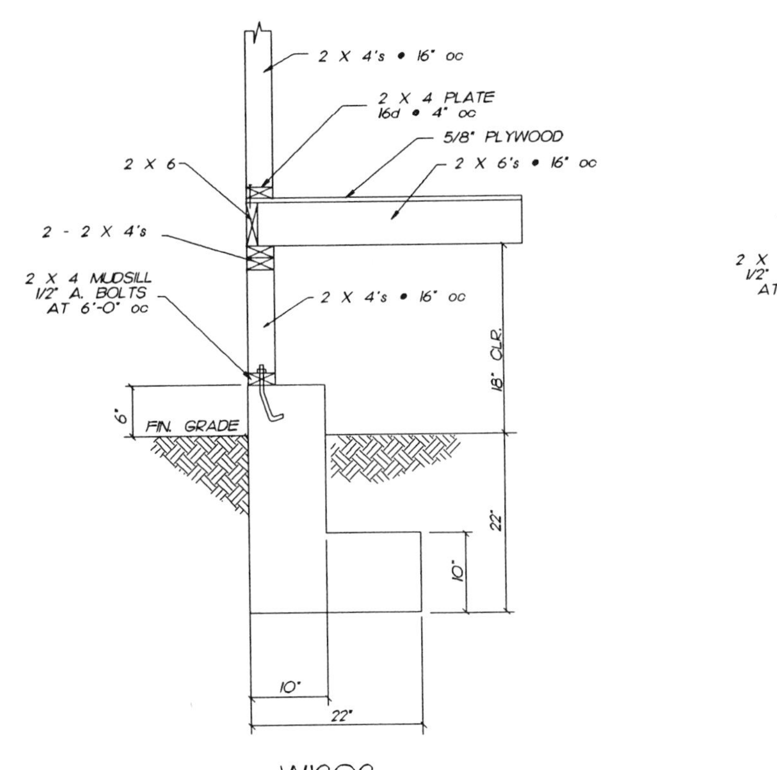

W1203

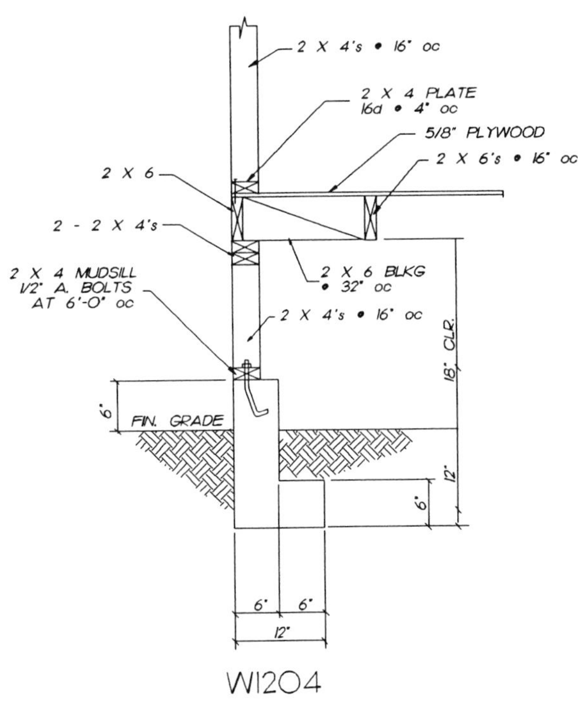

W1204

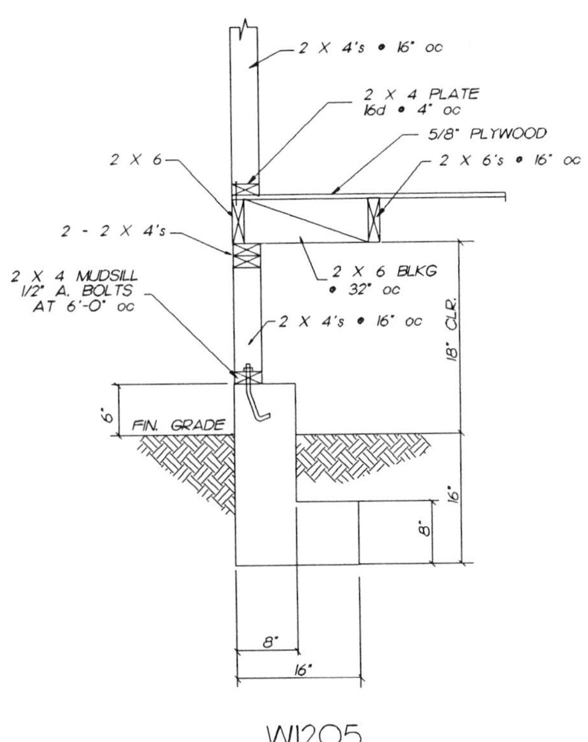

W1205

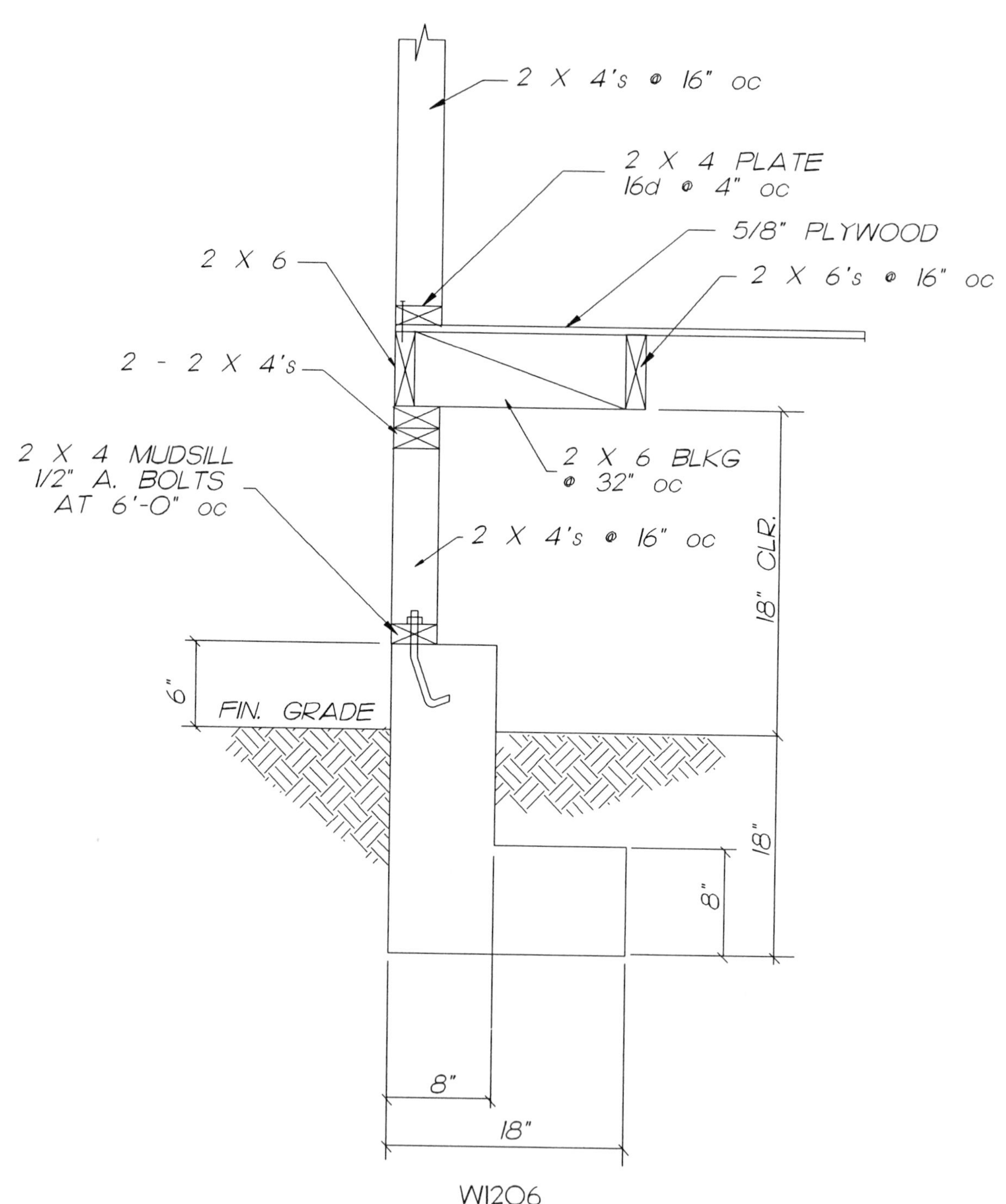
2 X 4's @ 16" oc
2 X 4 PLATE
16d @ 4" oc
5/8" PLYWOOD
2 X 6
2 X 6's @ 16" oc
2 - 2 X 4's
2 X 4 MUDSILL
1/2" A. BOLTS
AT 6'-0" oc
2 X 6 BLKG
@ 32" oc
2 X 4's @ 16" oc
18" CLR.
6"
FIN. GRADE
18"
8"
8"
18"

W1206

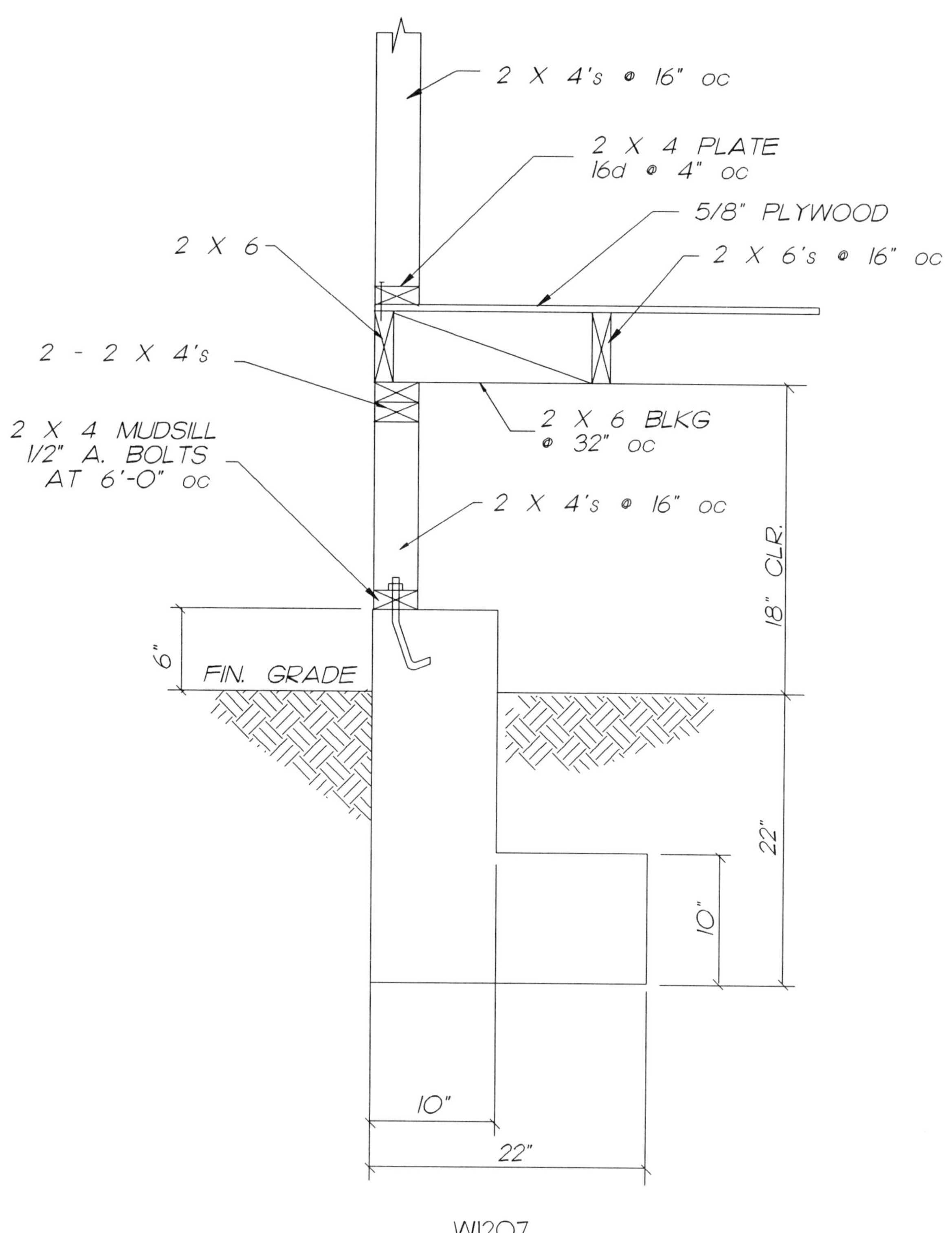
2 X 4's @ 16" oc
2 X 4 PLATE
16d @ 4" oc
5/8" PLYWOOD
2 X 6
2 X 6's @ 16" oc
2 - 2 X 4's
2 X 6 BLKG
@ 32" oc
2 X 4 MUDSILL
1/2" A. BOLTS
AT 6'-0" oc
2 X 4's @ 16" oc
18" CLR.
6"
FIN. GRADE
22"
10"
10"
22"

WI207

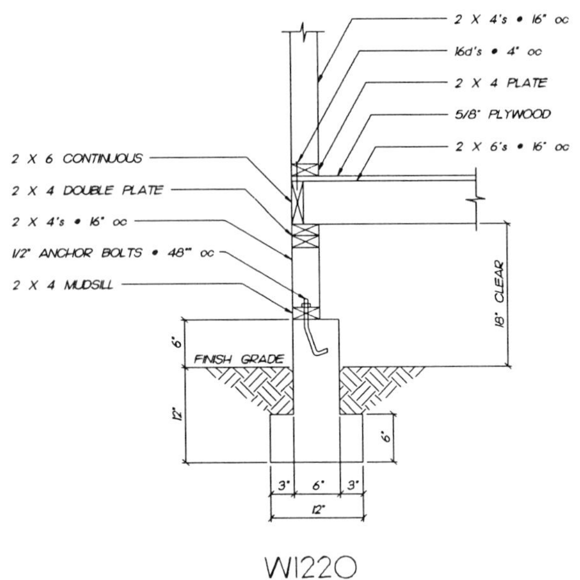

W1220

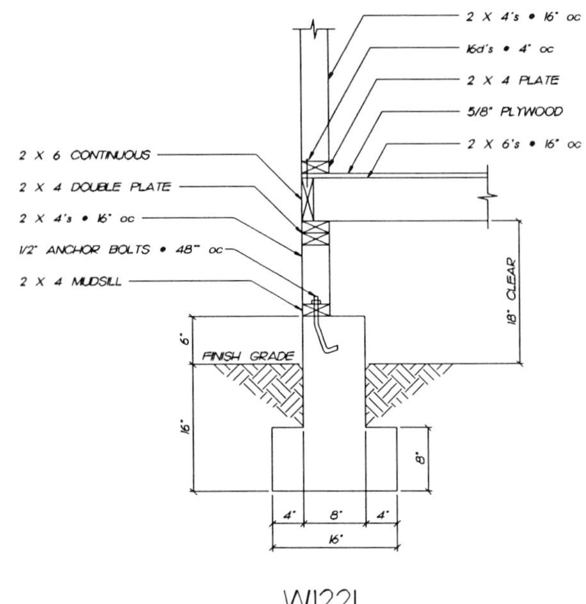

W1221

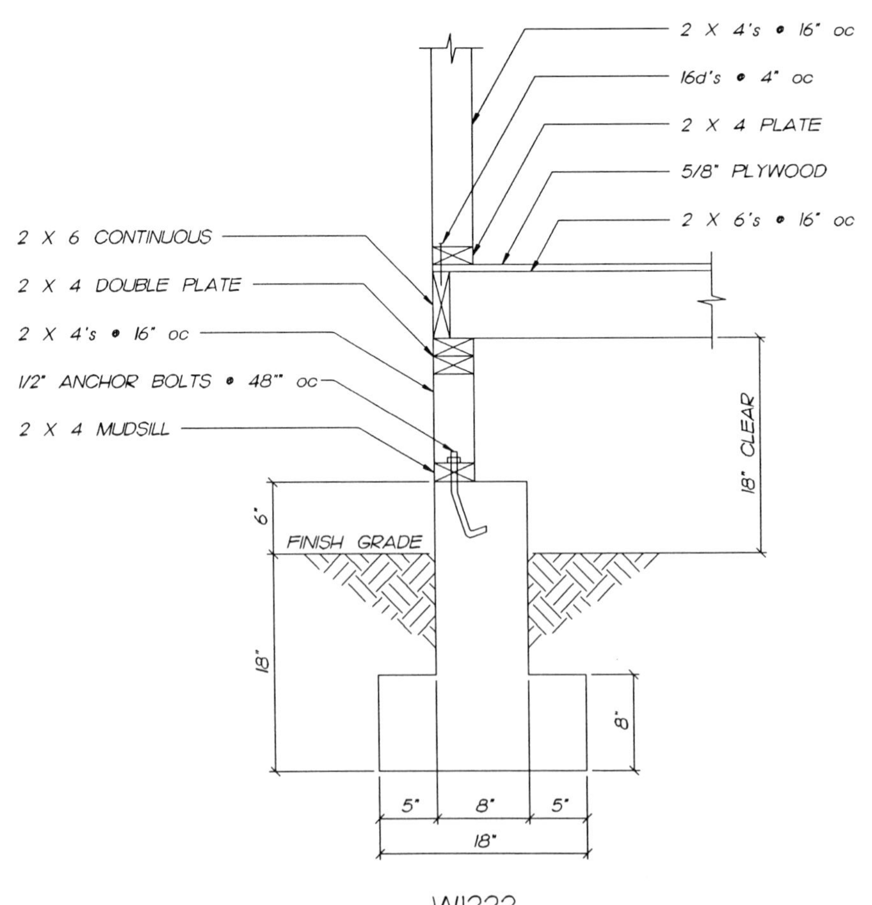

W1222

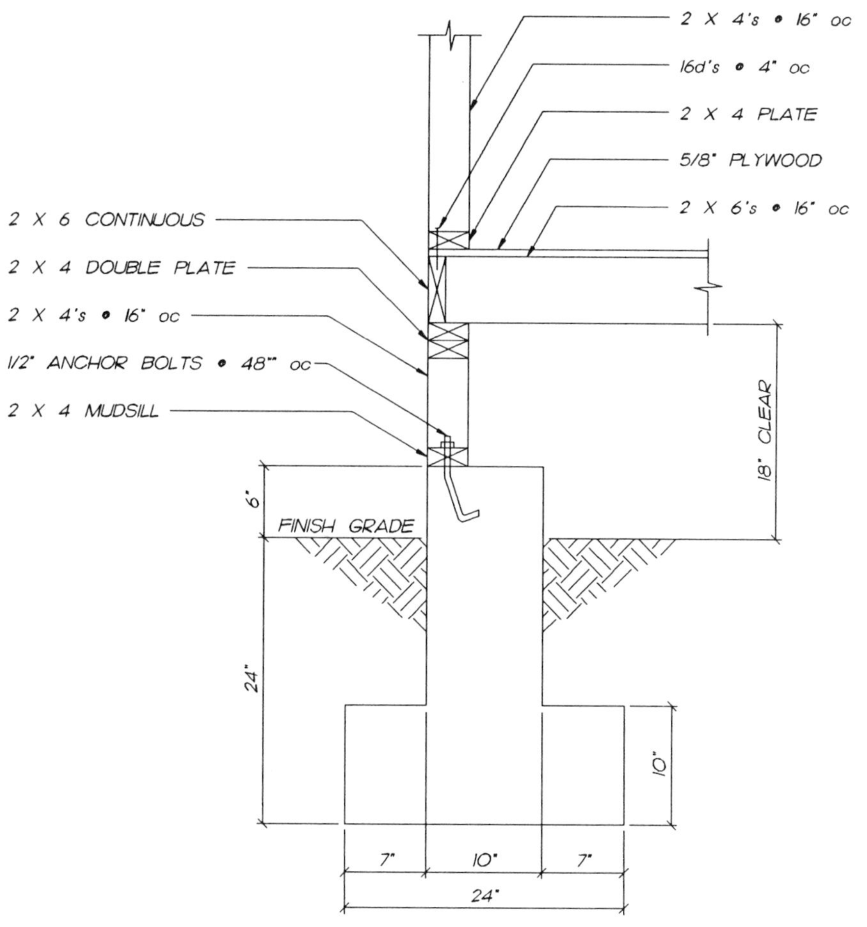

WI223

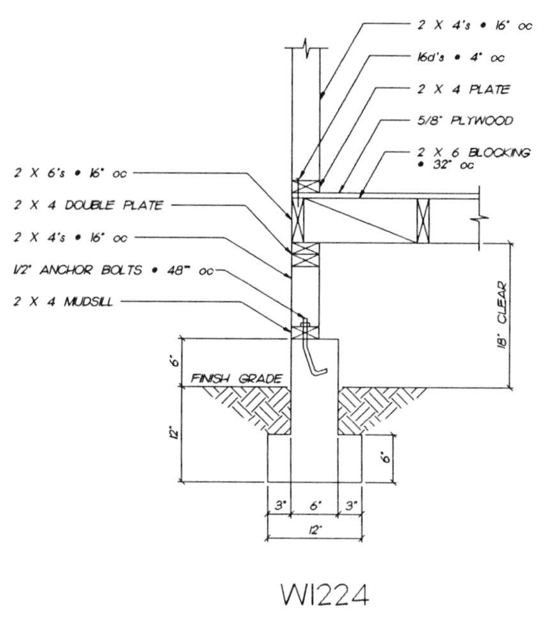

WI224

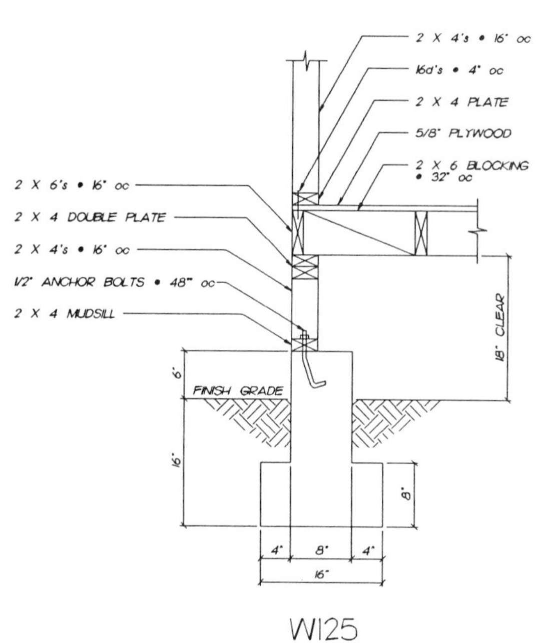

WI25

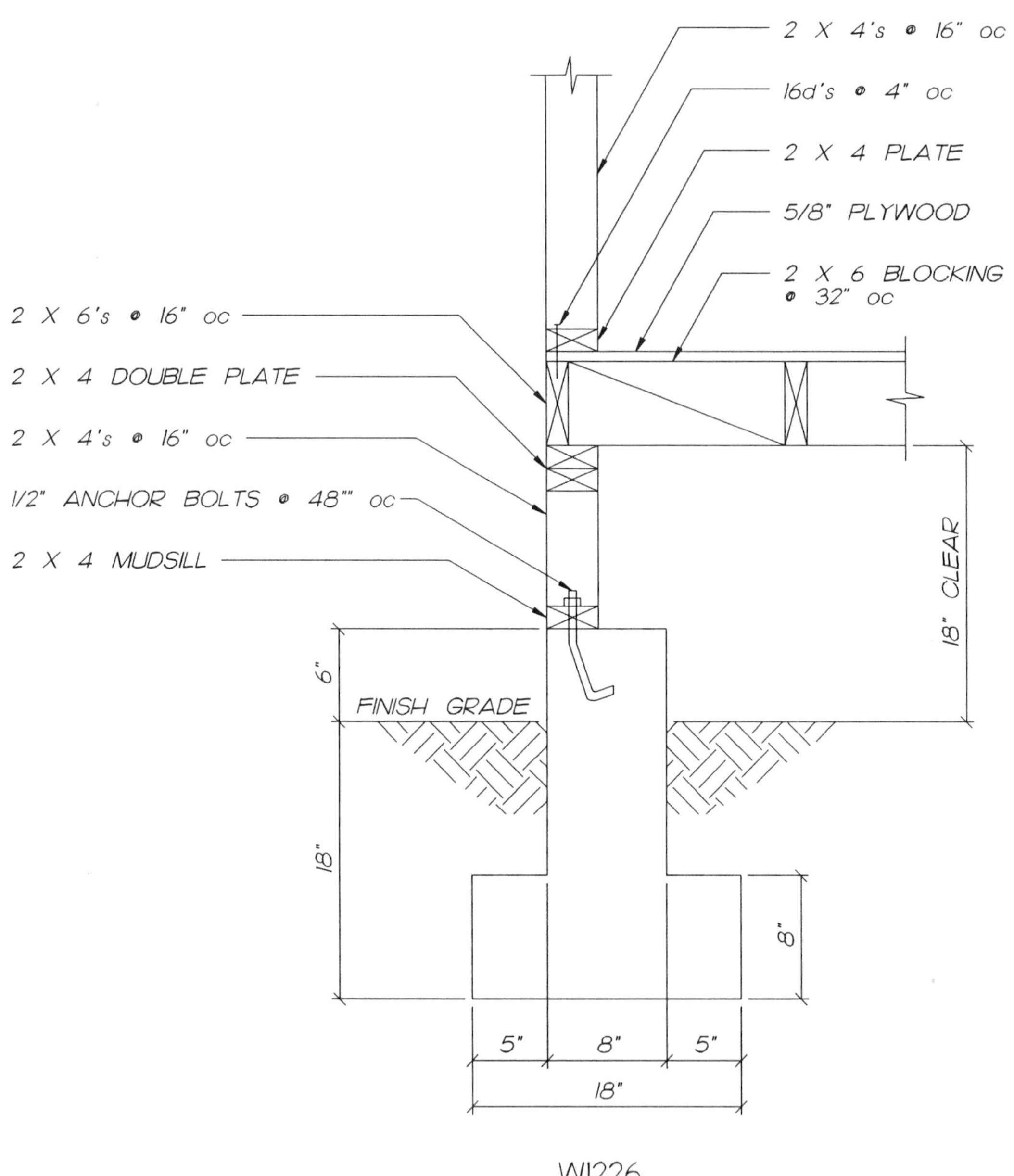
2 X 4's @ 16" oc
16d's @ 4" oc
2 X 4 PLATE
5/8" PLYWOOD
2 X 6 BLOCKING @ 32" oc
2 X 6's @ 16" oc
2 X 4 DOUBLE PLATE
2 X 4's @ 16" oc
1/2" ANCHOR BOLTS @ 48"" oc
2 X 4 MUDSILL
18" CLEAR
6"
FINISH GRADE
18"
8"
5"
8"
5"
18"

WI226

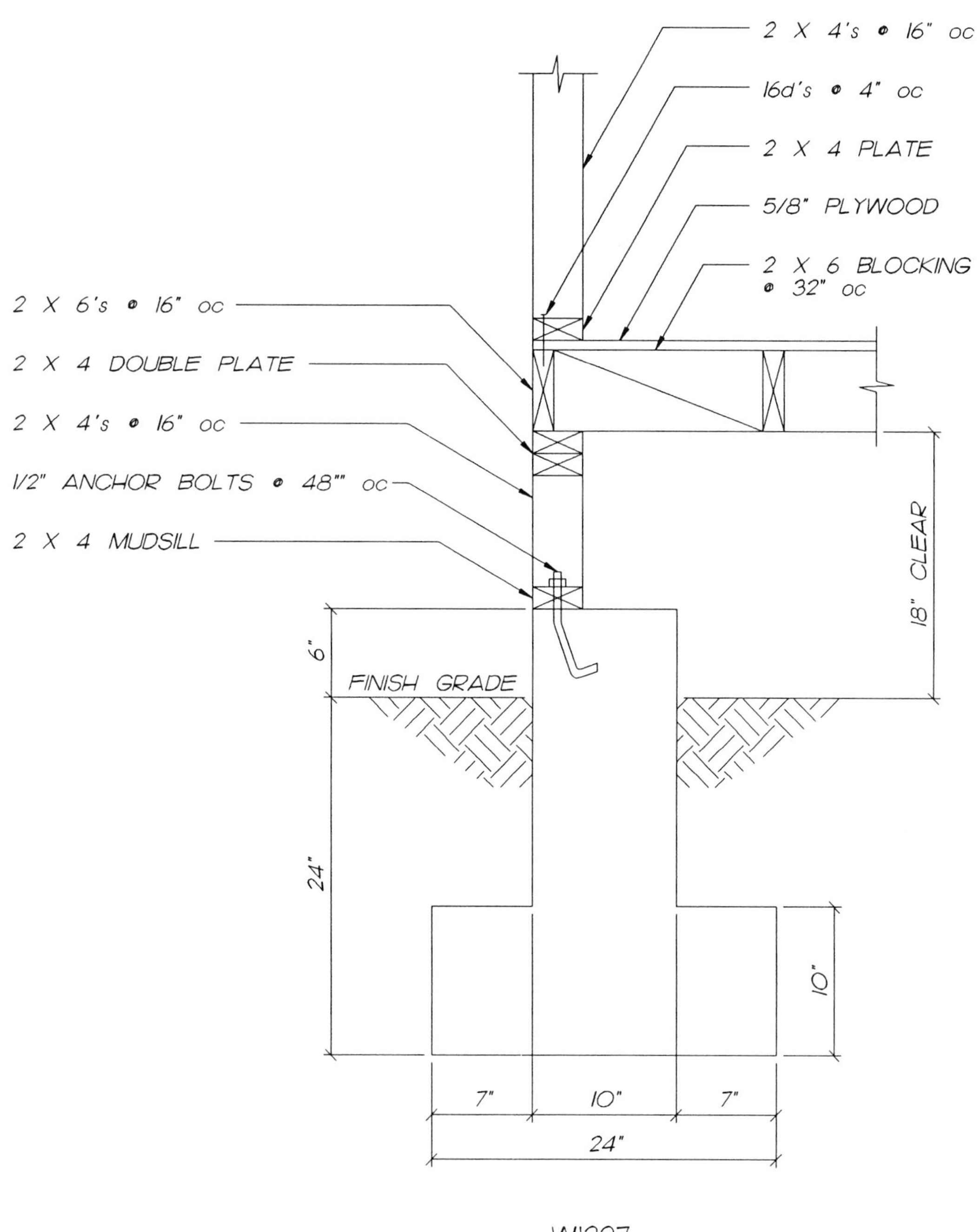
2 X 4's @ 16" oc
16d's @ 4" oc
2 X 4 PLATE
5/8" PLYWOOD
2 X 6 BLOCKING @ 32" oc
2 X 6's @ 16" oc
2 X 4 DOUBLE PLATE
2 X 4's @ 16" oc
1/2" ANCHOR BOLTS @ 48"" oc
2 X 4 MUDSILL
18" CLEAR
6"
FINISH GRADE
24"
10"
7"
10"
7"
24"

W1227

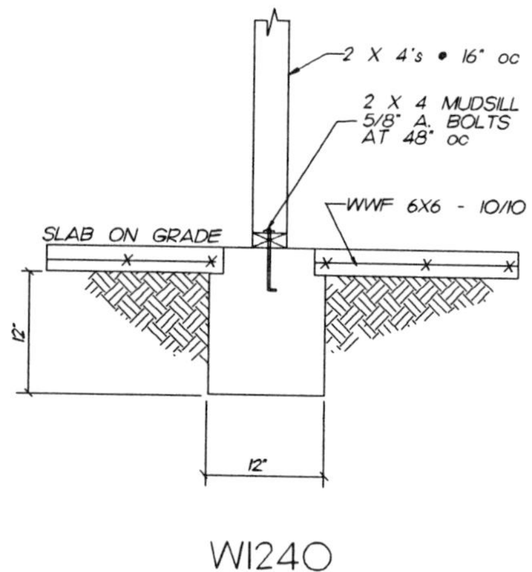

W1240

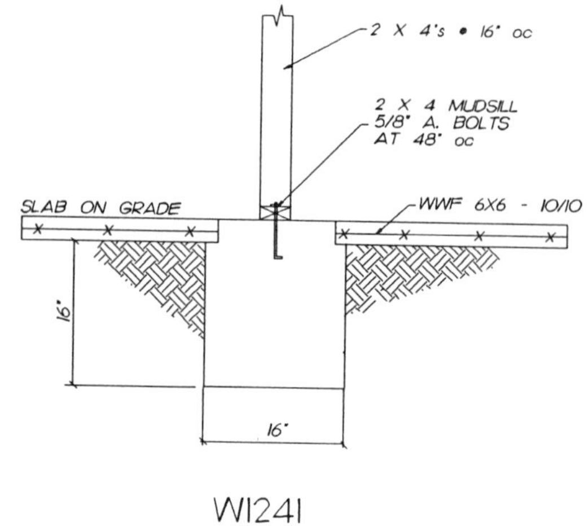

W1241

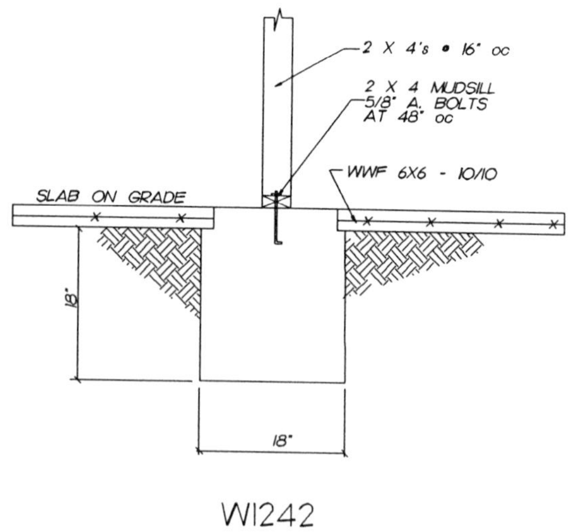

W1242

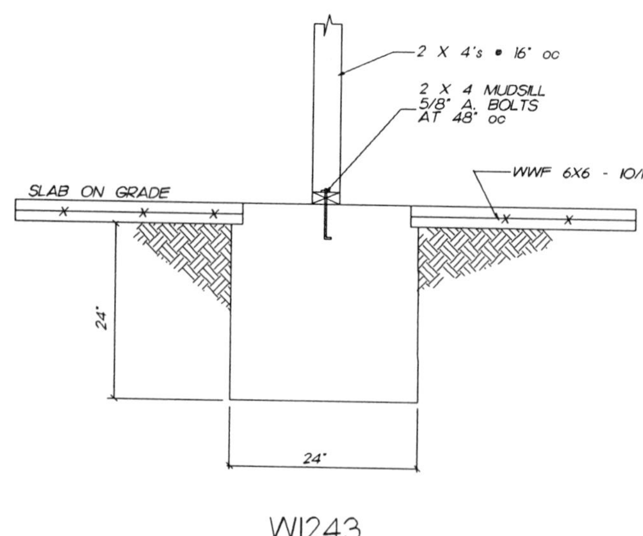

W1243

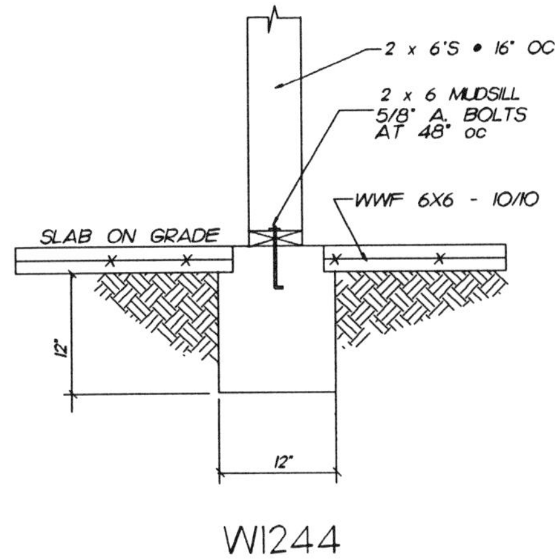

WI244

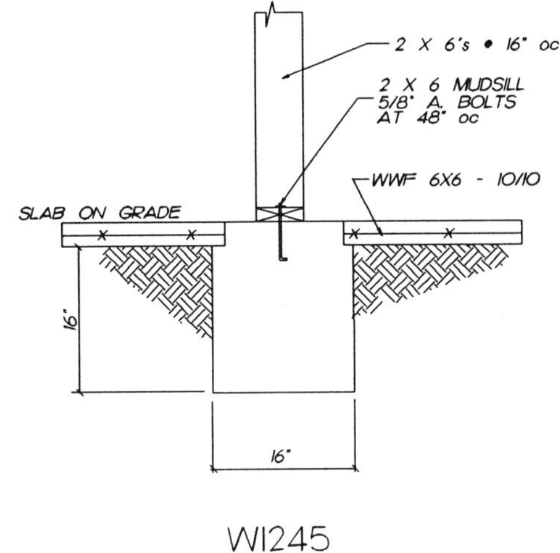

WI245

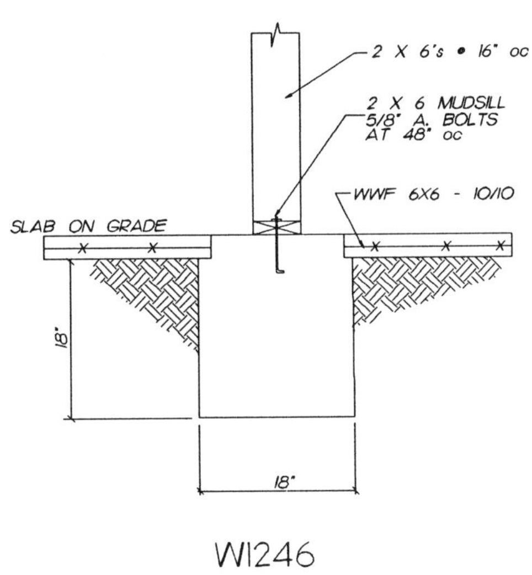

WI246

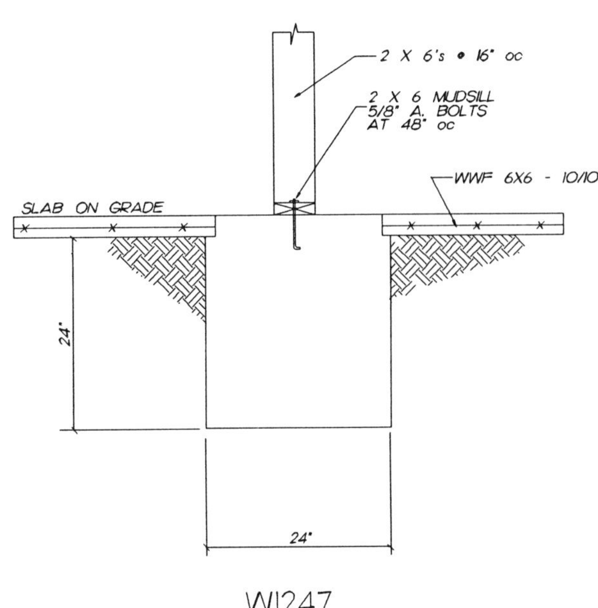

WI247

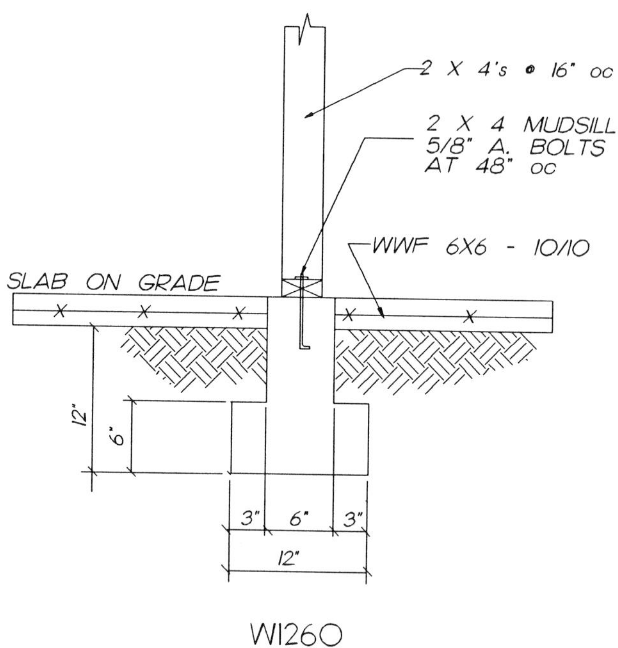
2 X 4's @ 16" oc
2 X 4 MUDSILL
5/8" A. BOLTS
AT 48" oc
WWF 6X6 - 10/10
SLAB ON GRADE
12"
6"
3"
6"
3"
12"

W1260

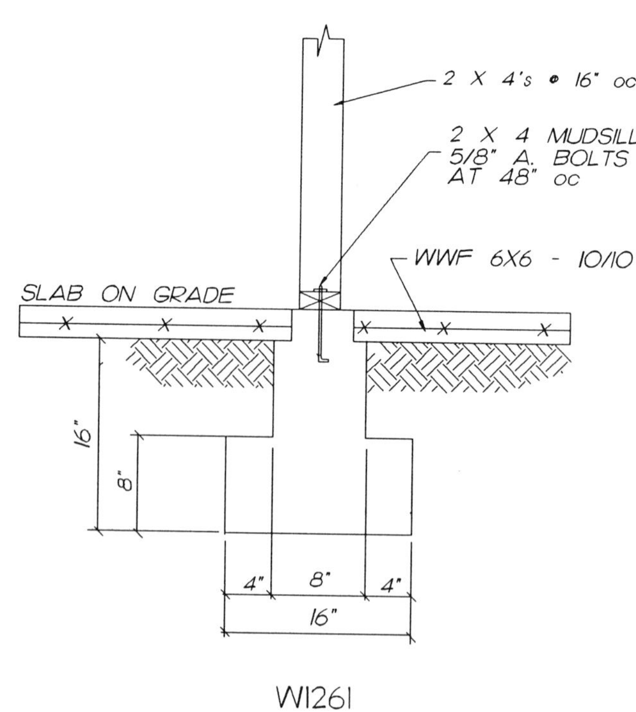
2 X 4's @ 16" oc
2 X 4 MUDSILL
5/8" A. BOLTS
AT 48" oc
WWF 6X6 - 10/10
SLAB ON GRADE
16"
8"
4"
8"
4"
16"

W1261

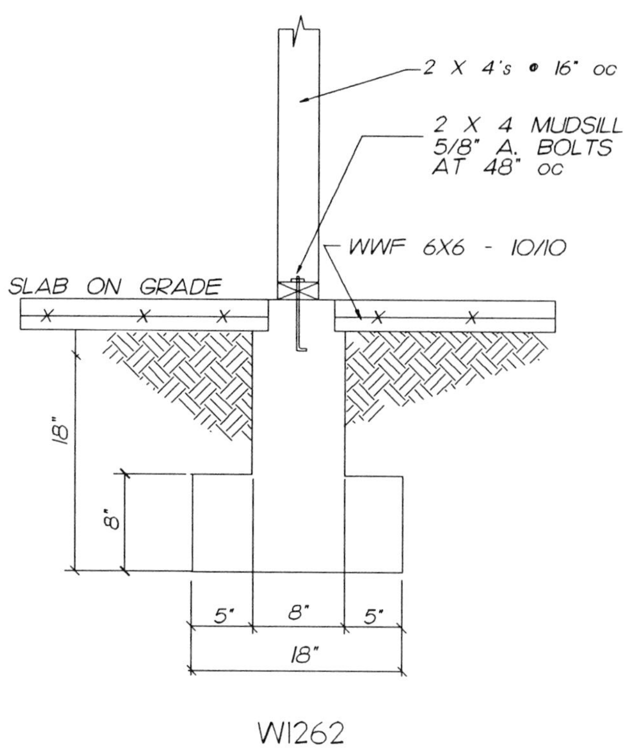
2 X 4's @ 16" oc
2 X 4 MUDSILL
5/8" A. BOLTS
AT 48" oc
WWF 6X6 - 10/10
SLAB ON GRADE
18"
8"
5"
8"
5"
18"

W1262

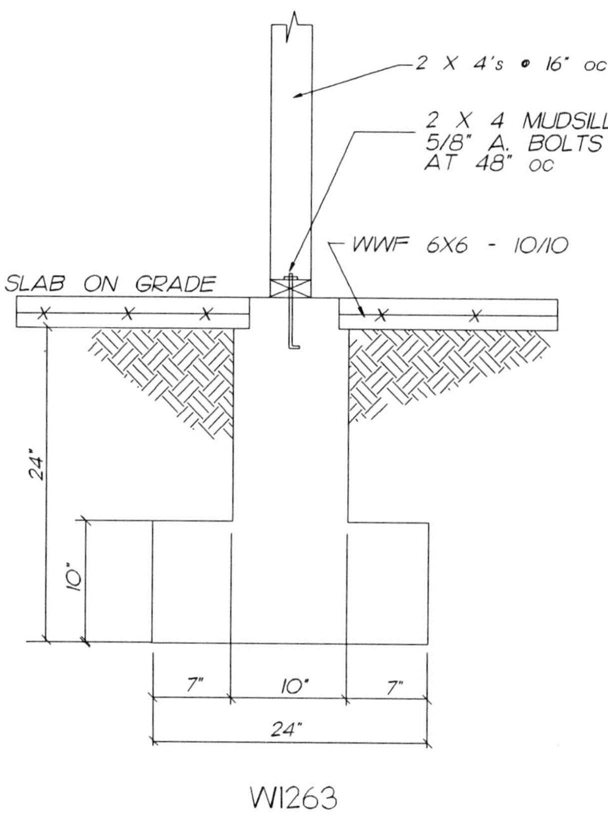
2 X 4's @ 16" oc
2 X 4 MUDSILL
5/8" A. BOLTS
AT 48" oc
WWF 6X6 - 10/10
SLAB ON GRADE
24"
10"
7"
10"
7"
24"

W1263

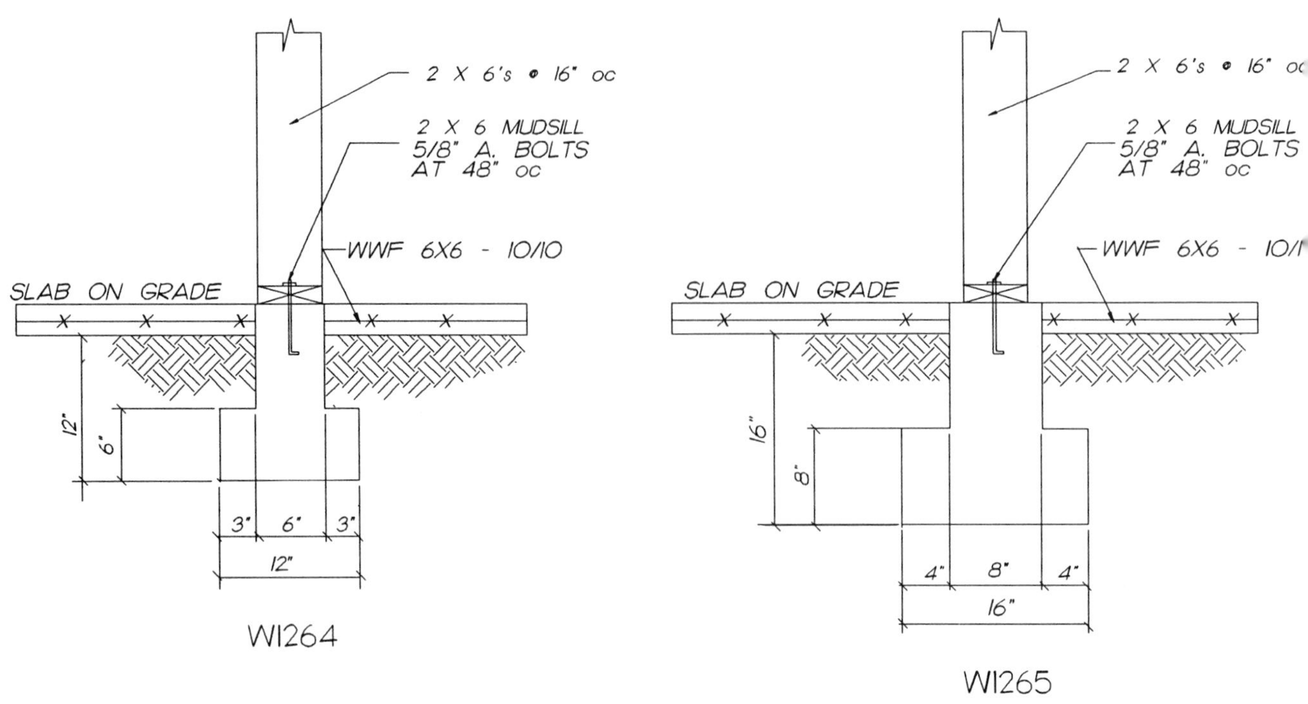

2 X 6's @ 16" oc
2 X 6 MUDSILL
5/8" A. BOLTS
AT 48" oc
WWF 6X6 - 10/10
SLAB ON GRADE
12"
6"
3"
6"
3"
12"
W1264
2 X 6's @ 16" oc
2 X 6 MUDSILL
5/8" A. BOLTS
AT 48" oc
WWF 6X6 - 10/1
SLAB ON GRADE
16"
8"
4"
8"
4"
16"
W1265

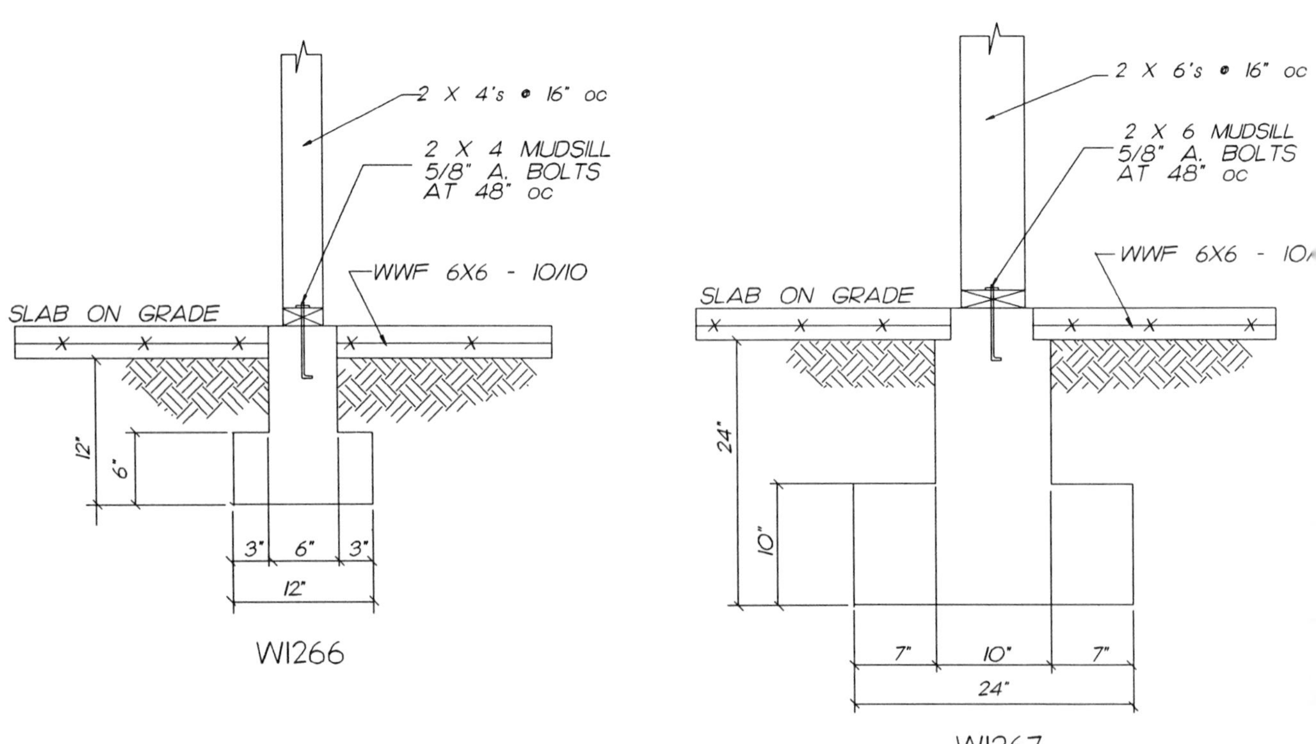

2 X 4's @ 16" oc
2 X 4 MUDSILL
5/8" A. BOLTS
AT 48" oc
WWF 6X6 - 10/10
SLAB ON GRADE
12"
6"
3"
6"
3"
12"
W1266
2 X 6's @ 16" oc
2 X 6 MUDSILL
5/8" A. BOLTS
AT 48" oc
WWF 6X6 - 10/
SLAB ON GRADE
24"
10"
7"
10"
7"
24"
W1267

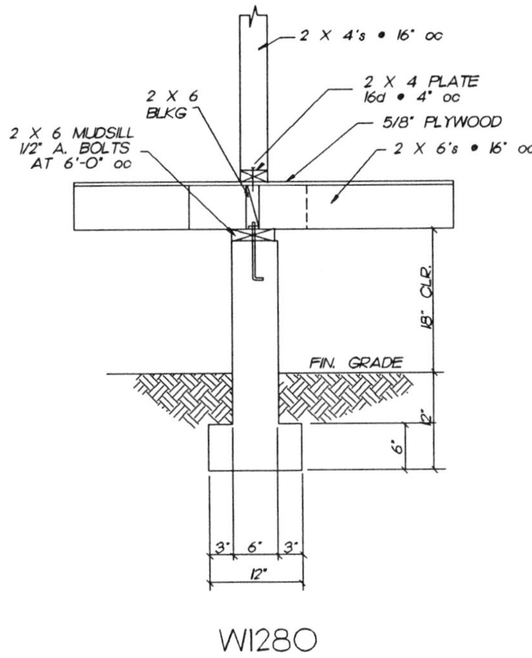

W1280

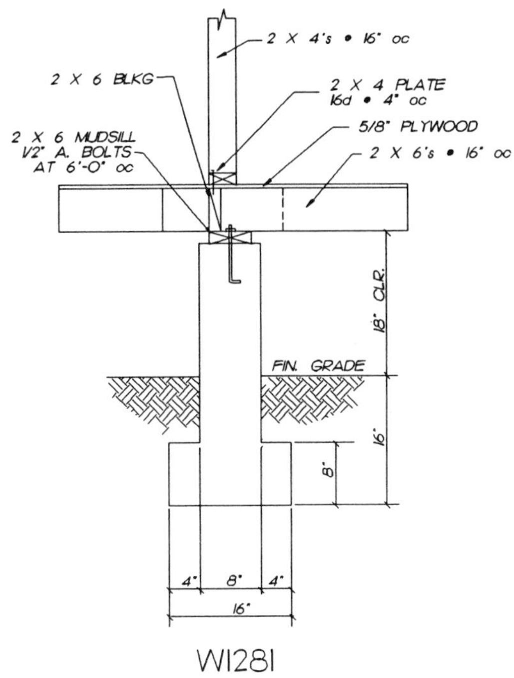

W1281

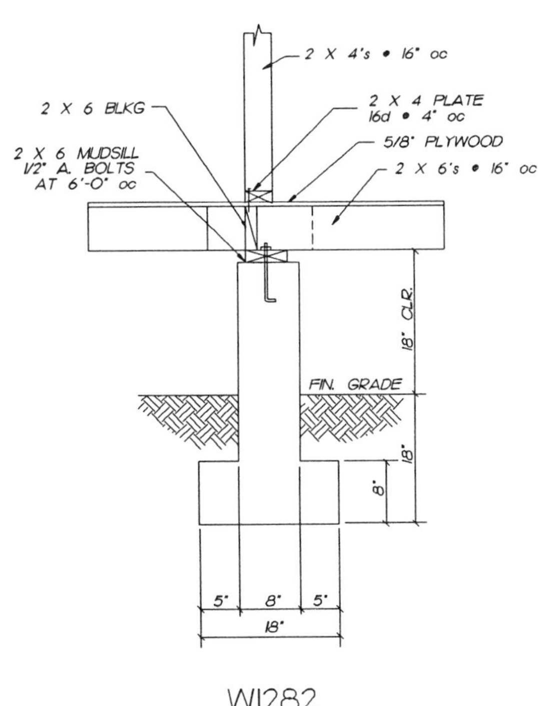

W1282

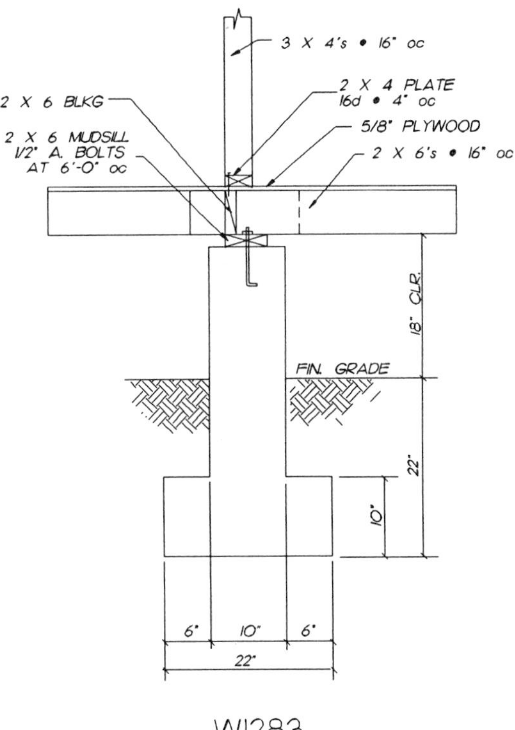

W1283

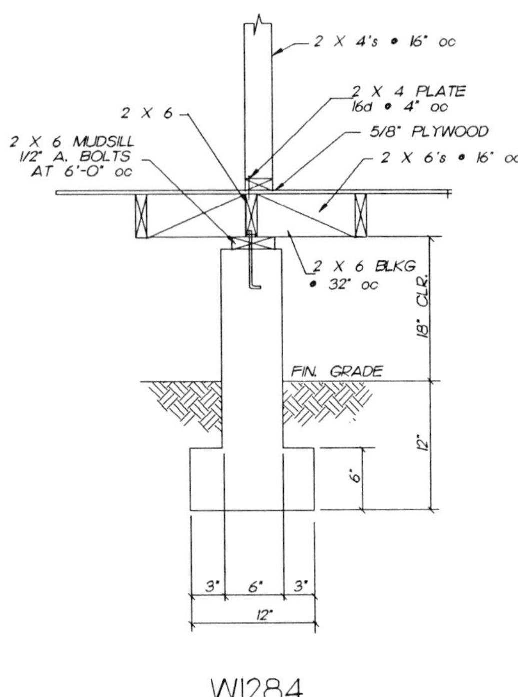

WI284

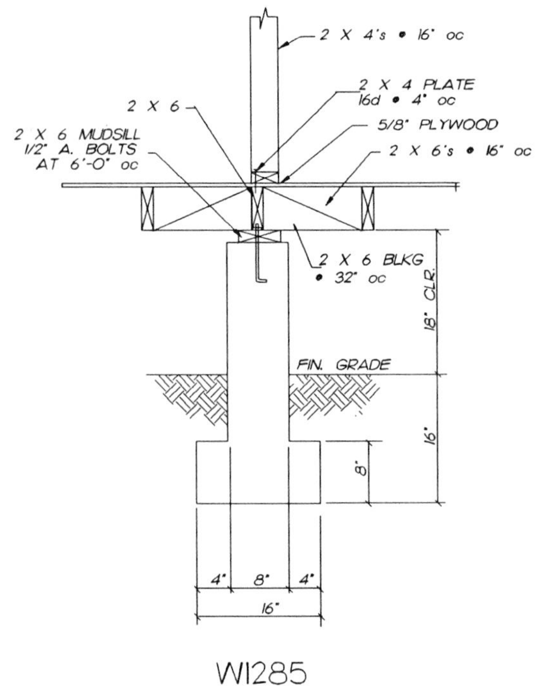

WI285

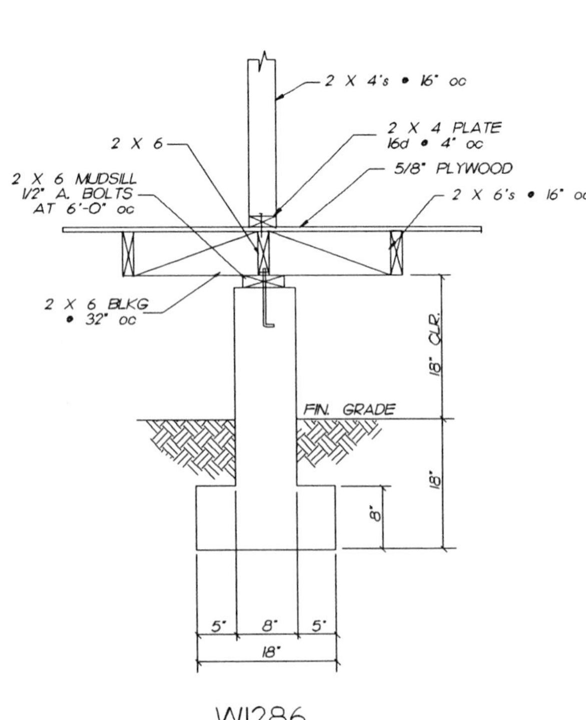

WI286

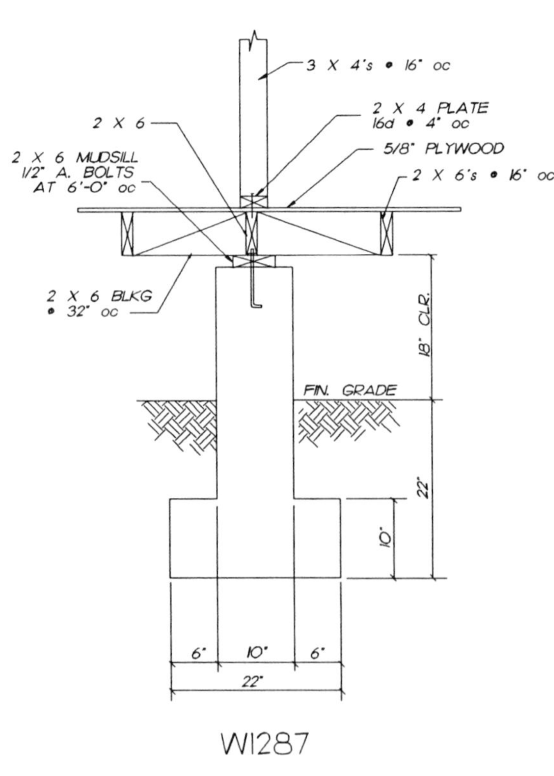

WI287

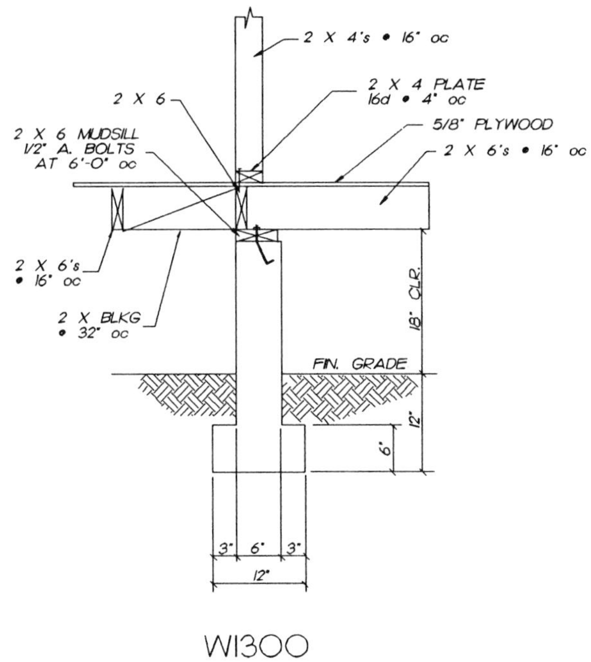

W1300

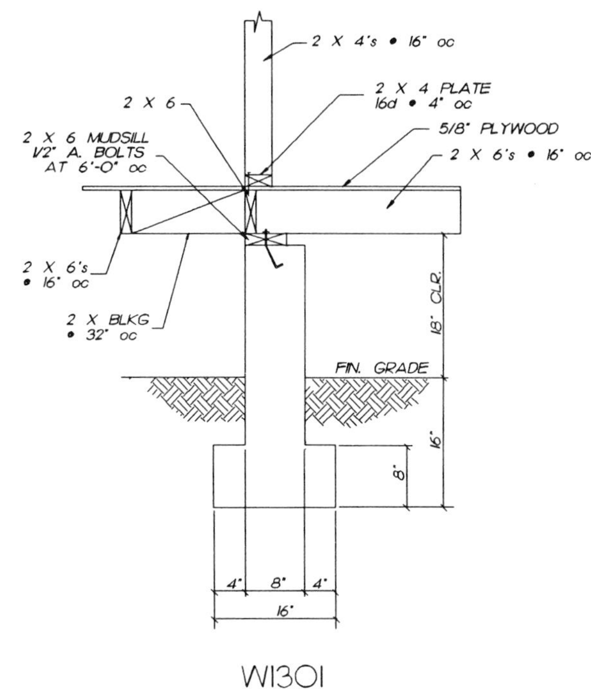

W1301

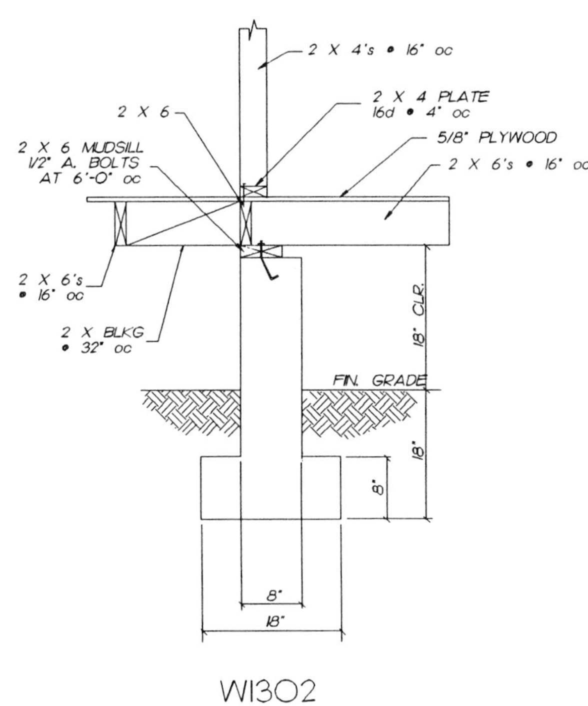

W1302

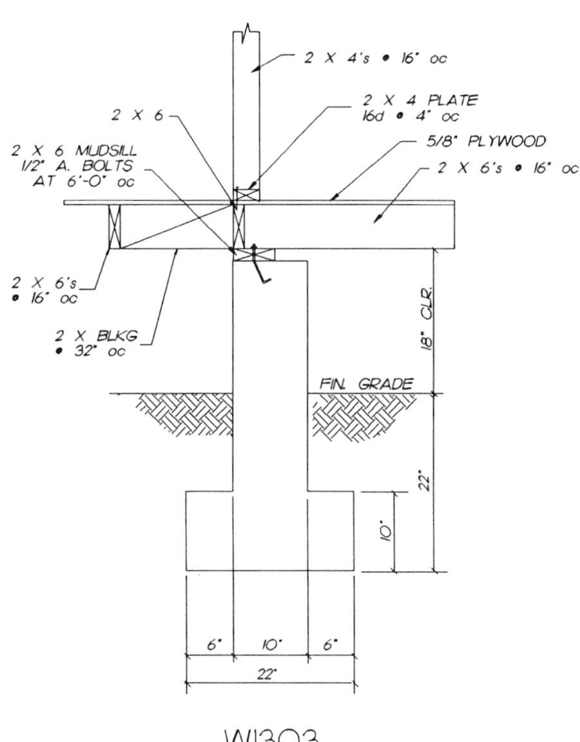

W1303

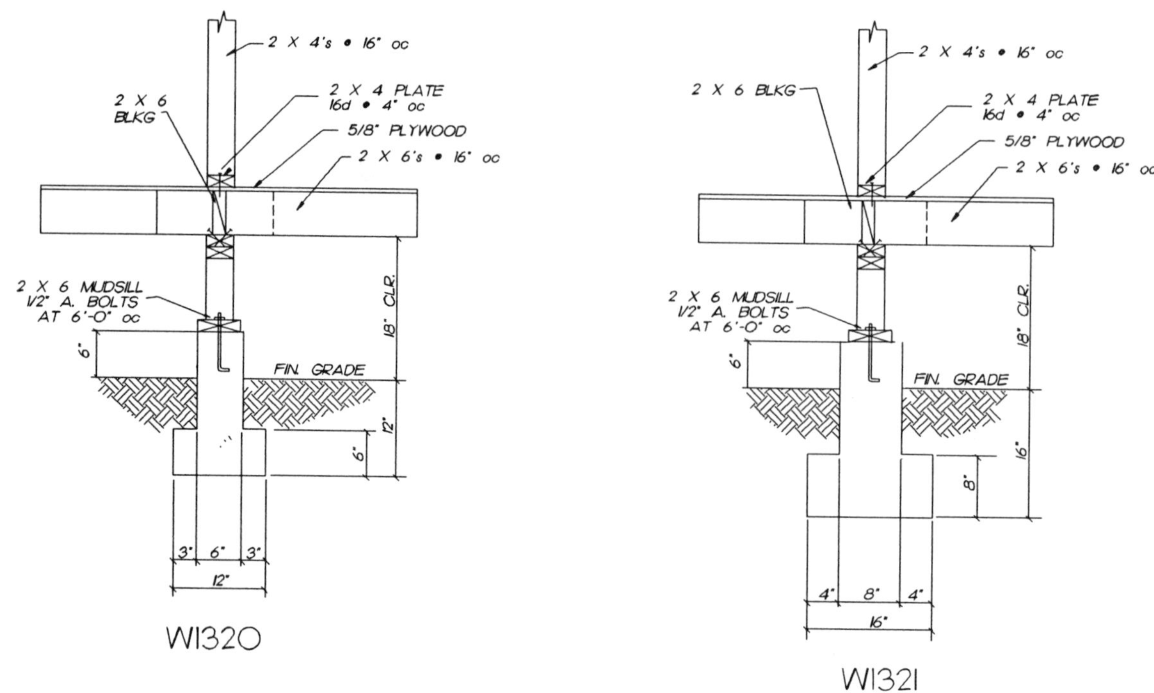

2 X 4's • 16" oc
2 X 6 BLKG
2 X 4 PLATE 16d • 4" oc
5/8" PLYWOOD
2 X 6's • 16" oc
2 X 6 MUDSILL 1/2" A. BOLTS AT 6'-0" oc
18" CLR.
6"
FIN. GRADE
12"
6"
3"
6"
3"
12"
W1320
2 X 4's • 16" oc
2 X 6 BLKG
2 X 4 PLATE 16d • 4" oc
5/8" PLYWOOD
2 X 6's • 16" oc
2 X 6 MUDSILL 1/2" A. BOLTS AT 6'-0" oc
18" CLR.
6"
FIN. GRADE
16"
8"
4"
8"
4"
16"
W1321

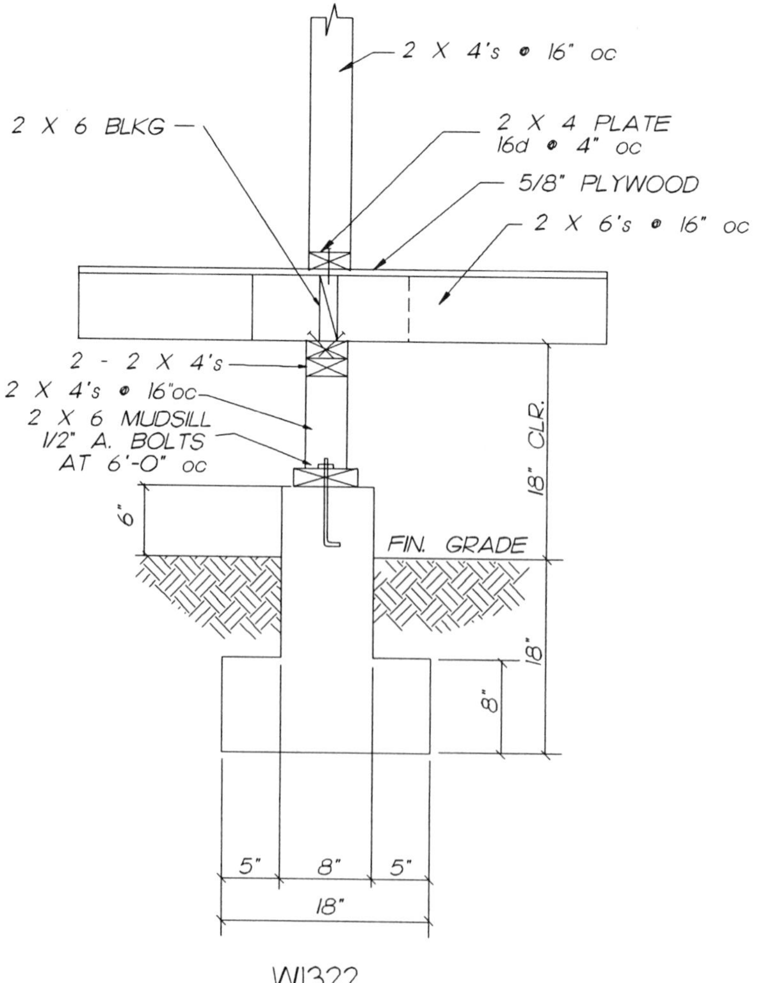

2 X 4's • 16" oc
2 X 6 BLKG
2 X 4 PLATE 16d • 4" oc
5/8" PLYWOOD
2 X 6's • 16" oc
2 - 2 X 4's
2 X 4's • 16"oc
2 X 6 MUDSILL 1/2" A. BOLTS AT 6'-0" oc
18" CLR.
6"
FIN. GRADE
18"
8"
5"
8"
5"
18"
W1322

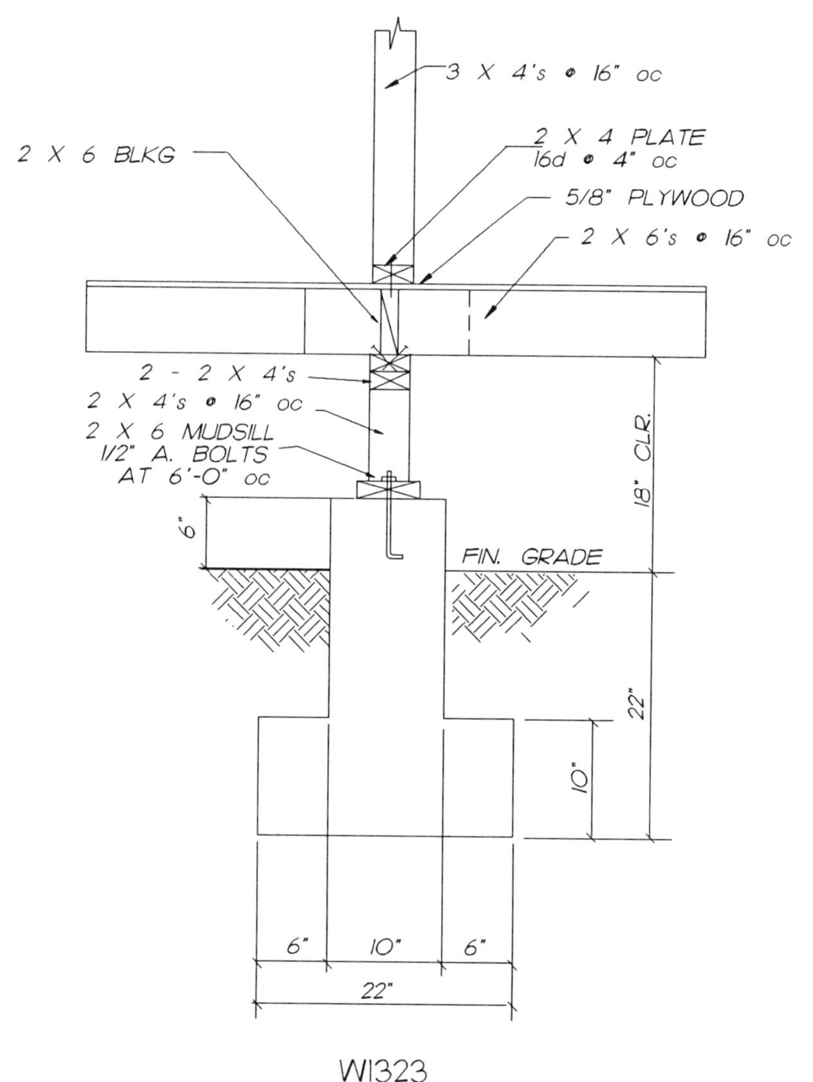

WI323

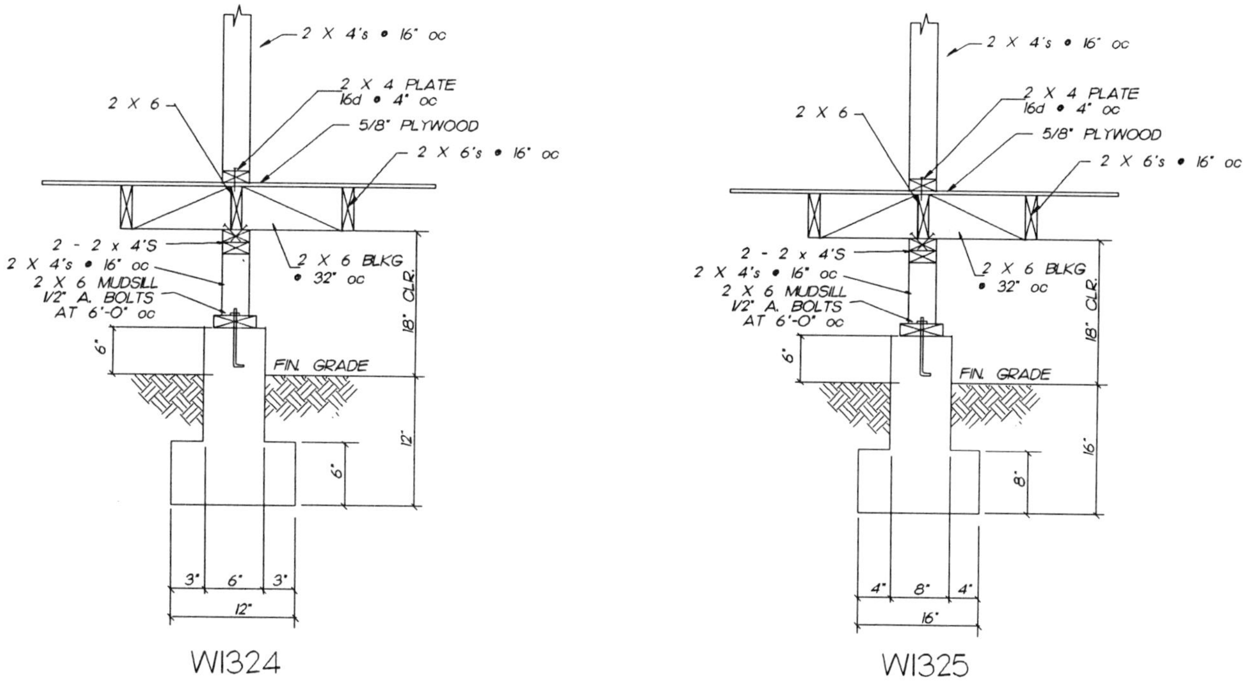

WI324

WI325

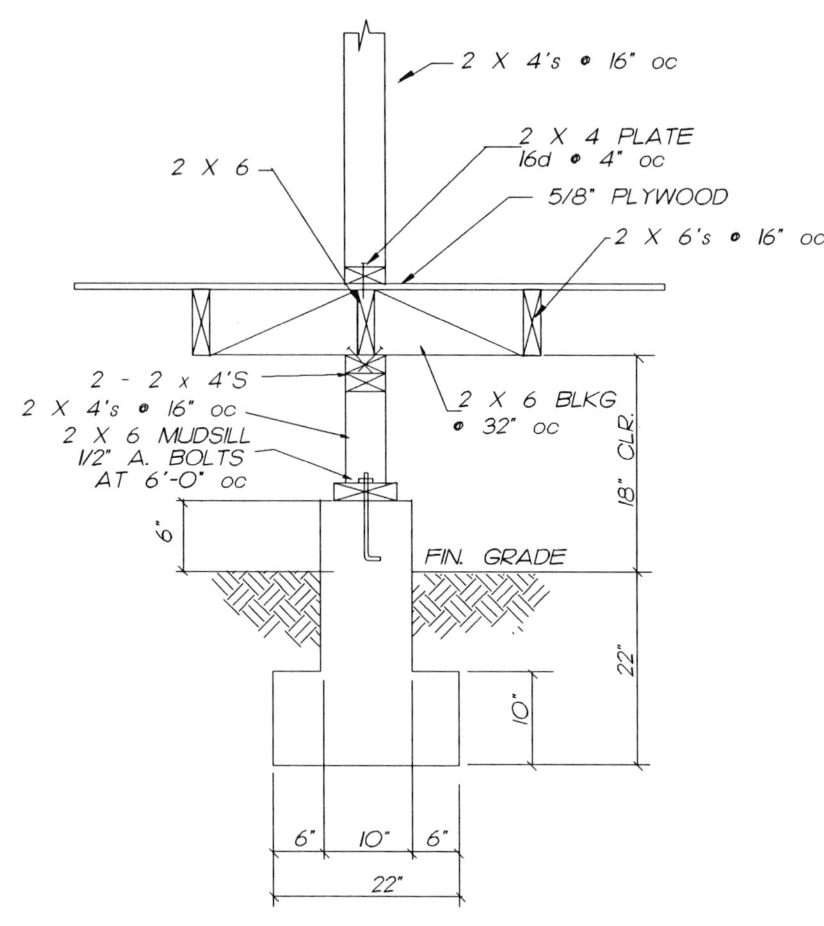
2 X 4's @ 16" oc
2 X 4 PLATE
16d @ 4" oc
2 X 6
5/8" PLYWOOD
2 X 6's @ 16" oc
2 - 2 x 4'S
2 X 4's @ 16" oc
2 X 6 MUDSILL
1/2" A. BOLTS
AT 6'-0" oc
2 X 6 BLKG
@ 32" oc
18" CLR.
6"
FIN. GRADE
22"
10"
6"
10"
6"
22"

W1327

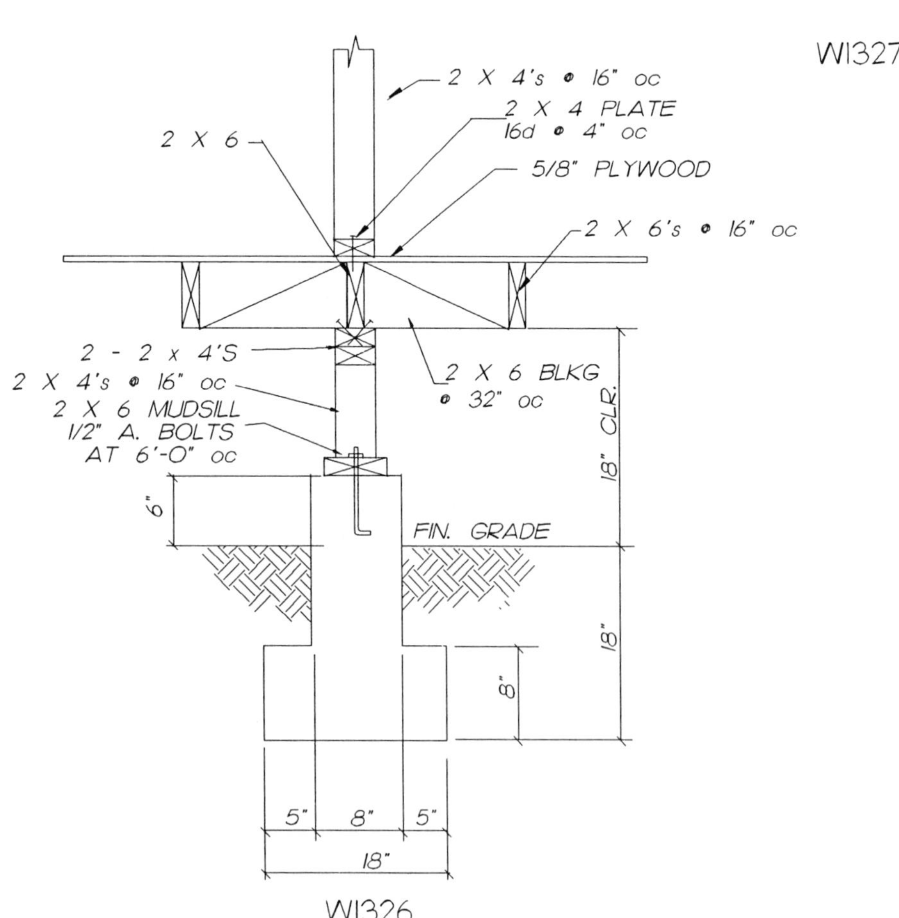
2 X 4's @ 16" oc
2 X 4 PLATE
16d @ 4" oc
2 X 6
5/8" PLYWOOD
2 X 6's @ 16" oc
2 - 2 x 4'S
2 X 4's @ 16" oc
2 X 6 MUDSILL
1/2" A. BOLTS
AT 6'-0" oc
2 X 6 BLKG
@ 32" oc
18" CLR.
6"
FIN. GRADE
18"
8"
5"
8"
5"
18"

W1326

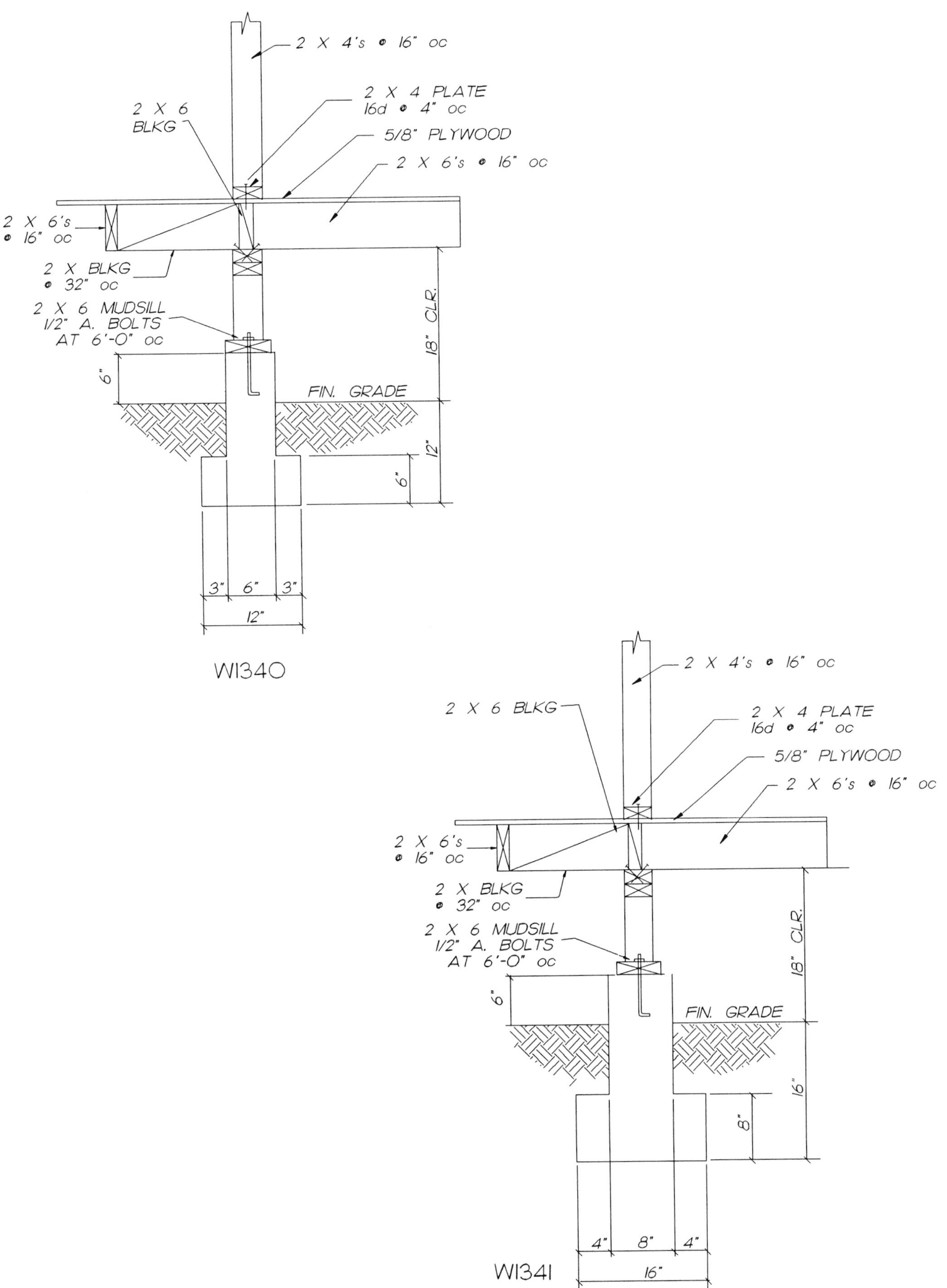
2 X 4's @ 16" oc
2 X 4 PLATE
16d @ 4" oc
2 X 6
BLKG
5/8" PLYWOOD
2 X 6's @ 16" oc
2 X 6's
@ 16" oc
2 X BLKG
@ 32" oc
2 X 6 MUDSILL
1/2" A. BOLTS
AT 6'-0" oc
18" CLR.
6"
FIN. GRADE
12"
6"
3"
6"
3"
12"
W1340
2 X 4's @ 16" oc
2 X 6 BLKG
2 X 4 PLATE
16d @ 4" oc
5/8" PLYWOOD
2 X 6's @ 16" oc
2 X 6's
@ 16" oc
2 X BLKG
@ 32" oc
2 X 6 MUDSILL
1/2" A. BOLTS
AT 6'-0" oc
18" CLR.
6"
FIN. GRADE
16"
8"
4"
8"
4"
16"
W1341

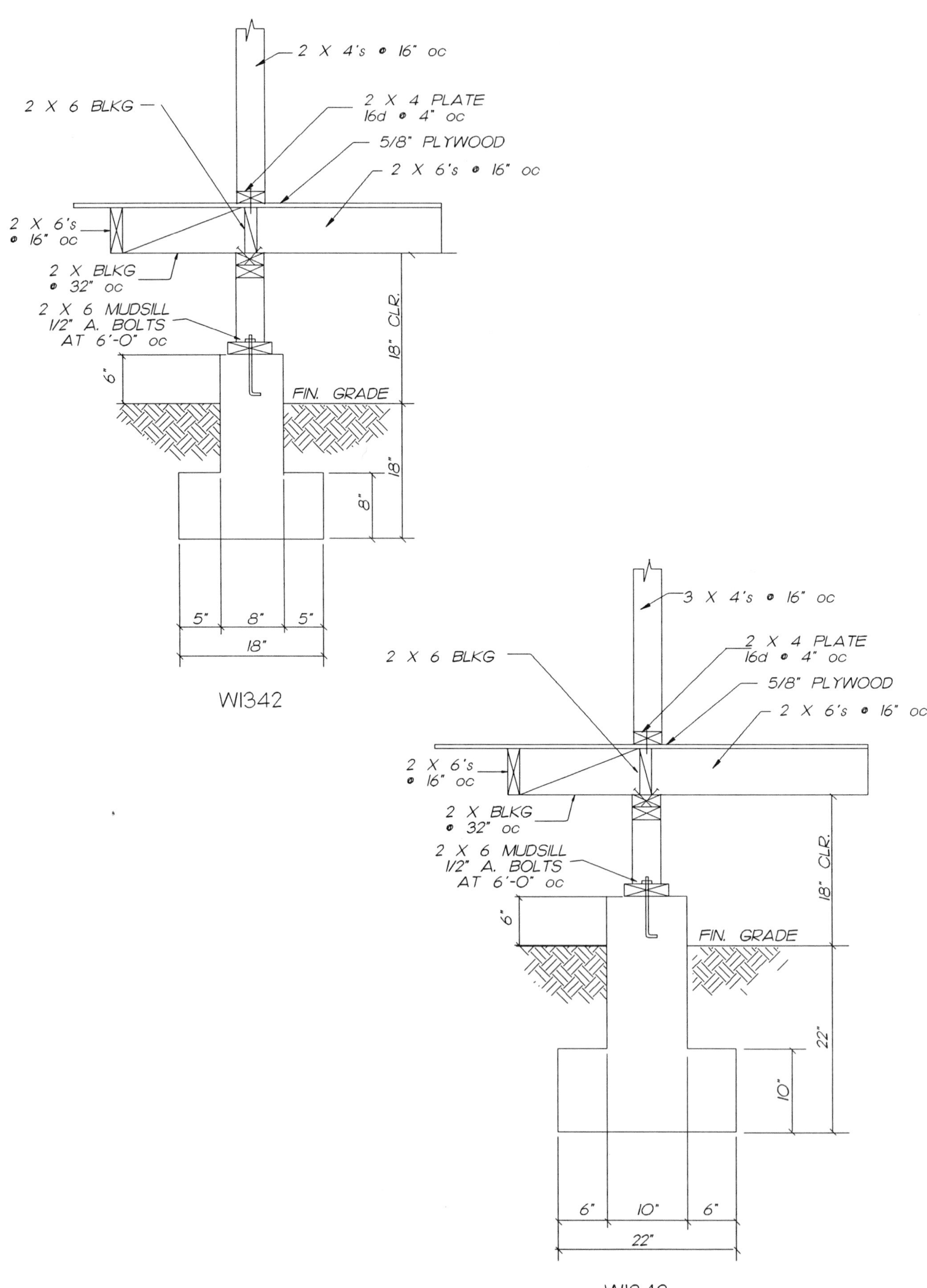
2 X 4's @ 16" oc
2 X 6 BLKG
2 X 4 PLATE
16d @ 4" oc
5/8" PLYWOOD
2 X 6's @ 16" oc
2 X 6's
@ 16" oc
2 X BLKG
@ 32" oc
2 X 6 MUDSILL
1/2" A. BOLTS
AT 6'-0" oc
6"
18" CLR.
FIN. GRADE
18"
8"
5"
8"
5"
18"
WI342
3 X 4's @ 16" oc
2 X 6 BLKG
2 X 4 PLATE
16d @ 4" oc
5/8" PLYWOOD
2 X 6's @ 16" oc
2 X 6's
@ 16" oc
2 X BLKG
@ 32" oc
2 X 6 MUDSILL
1/2" A. BOLTS
AT 6'-0" oc
6"
18" CLR.
FIN. GRADE
22"
10"
6"
10"
6"
22"
WI343

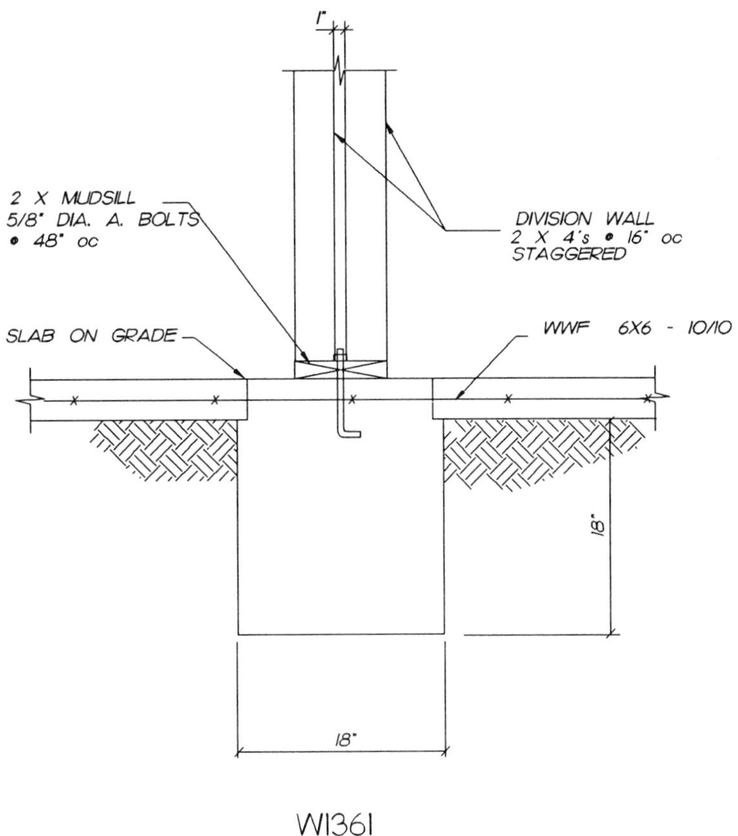
1"
2 X MUDSILL
5/8" DIA. A. BOLTS
@ 48" oc
DIVISION WALL
2 X 4's @ 16" oc
STAGGERED
SLAB ON GRADE
WWF 6X6 - 10/10
18"
18"

W1361

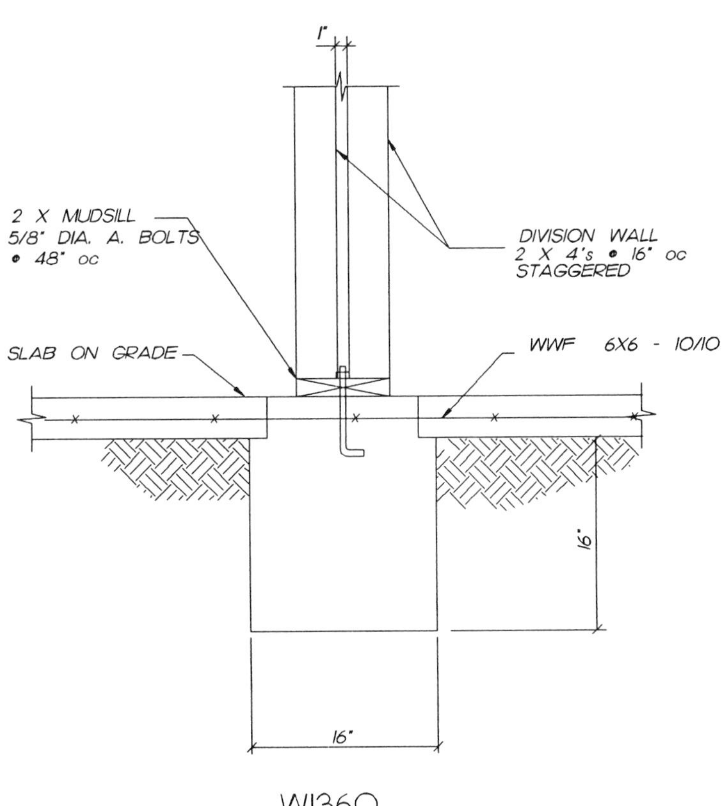
1"
2 X MUDSILL
5/8" DIA. A. BOLTS
@ 48" oc
DIVISION WALL
2 X 4's @ 16" oc
STAGGERED
SLAB ON GRADE
WWF 6X6 - 10/10
16"
16"

W1360

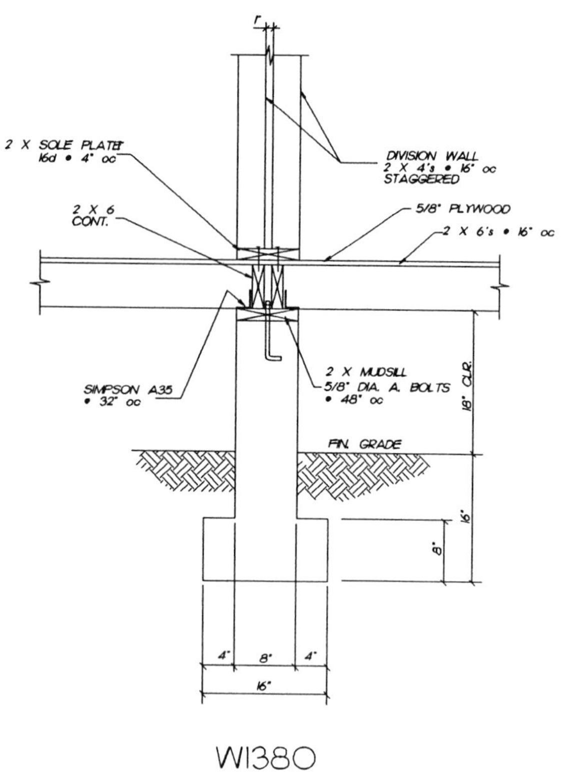

W1380

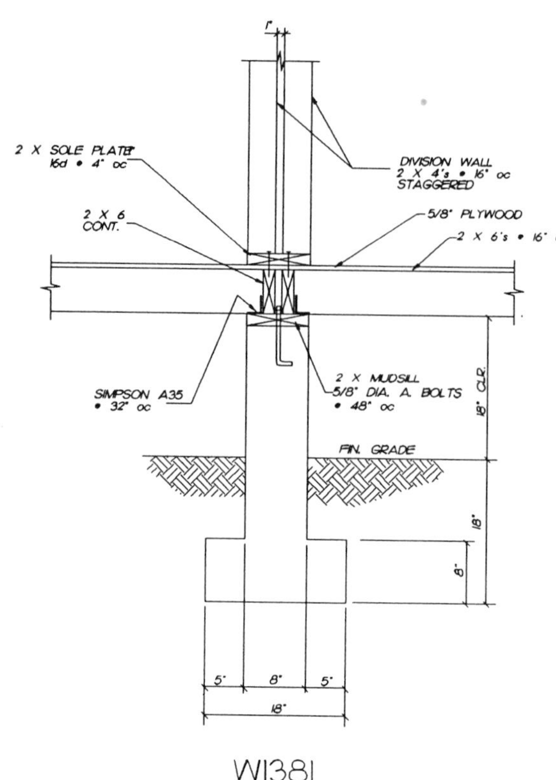

W1381

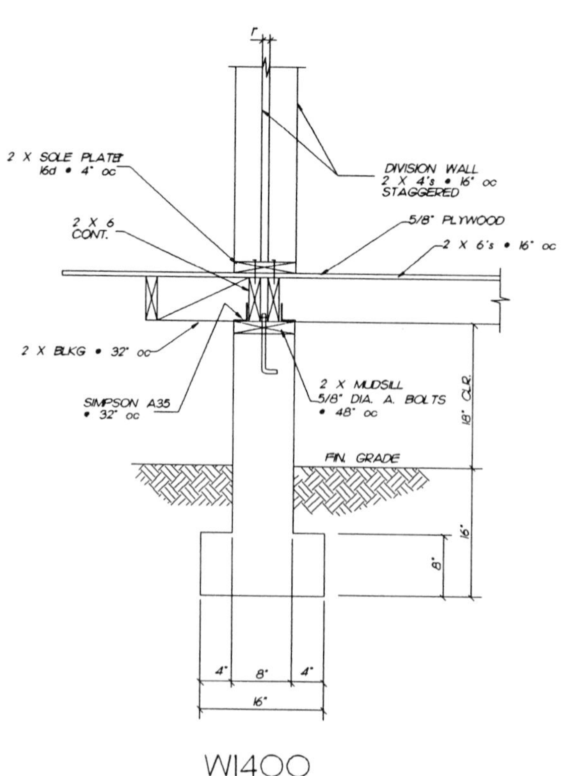

W1400

2 X 6 MUDSILL
5/8" DIA. A. BOLTS
@ 48" oc
DIVISION WALL
2 X 4's @ 16" oc
STAGGERED
SLAB ON GRADE
WWF 6X6 - 10/10
16"
16"

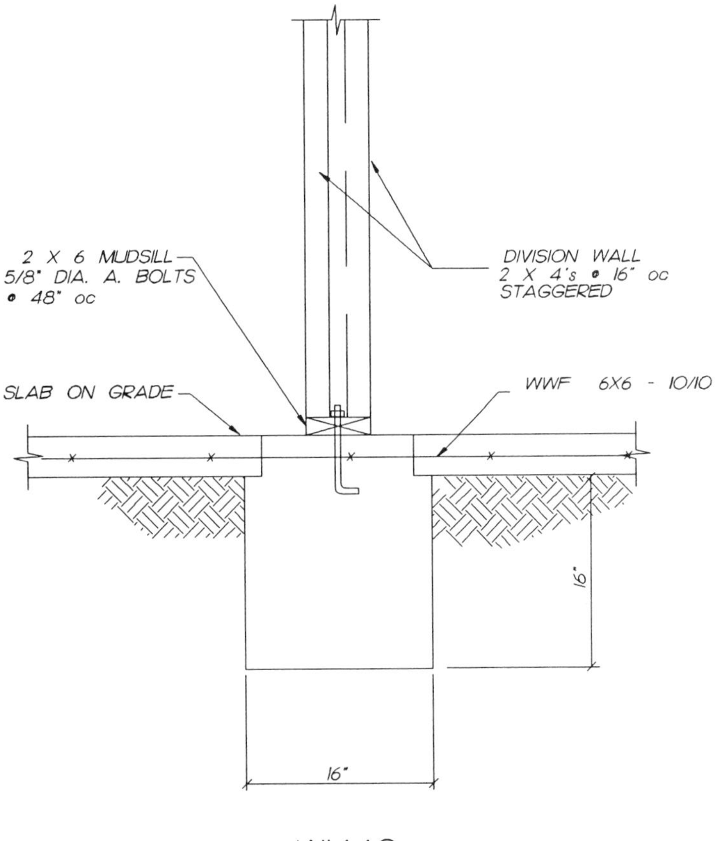

W1440

2 X 6 MUDSILL
5/8" DIA. A. BOLTS
@ 48" oc
DIVISION WALL
2 X 4's @ 16" oc
STAGGERED
SLAB ON GRADE
WWF 6X6 - 10/10
18"
18"

W1441

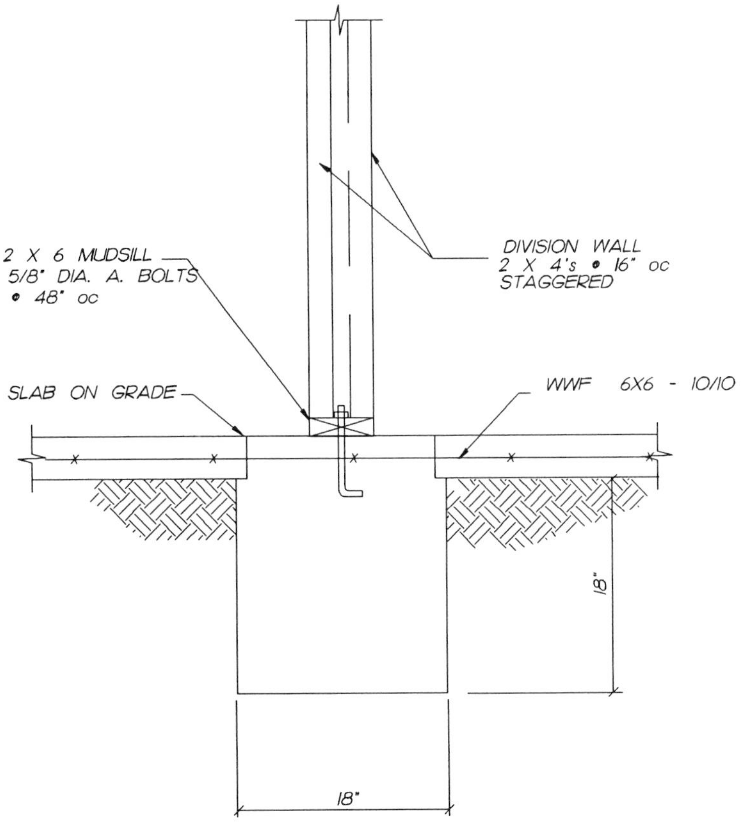

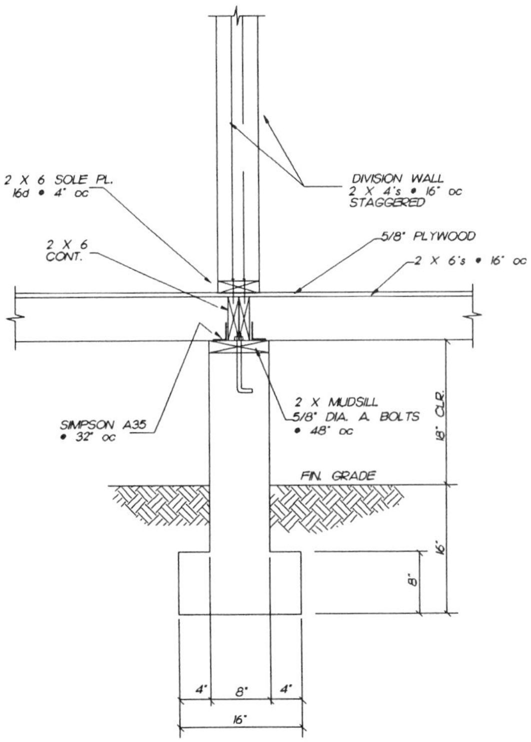

2 X 6 SOLE PL.
16d @ 4" oc
DIVISION WALL
2 X 4's @ 16" oc
STAGGERED
2 X 6
CONT.
5/8" PLYWOOD
2 X 6's @ 16" oc
SIMPSON A35
@ 32" oc
2 X MUDSILL
5/8" DIA. A. BOLTS
@ 48" oc
18" CLR.
FIN. GRADE
16"
8"
4"
8"
4"
16"

W1460

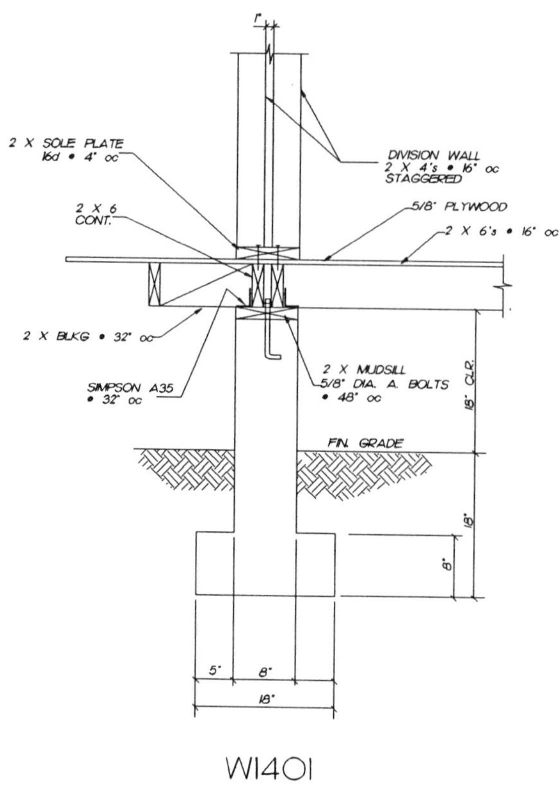

W1401

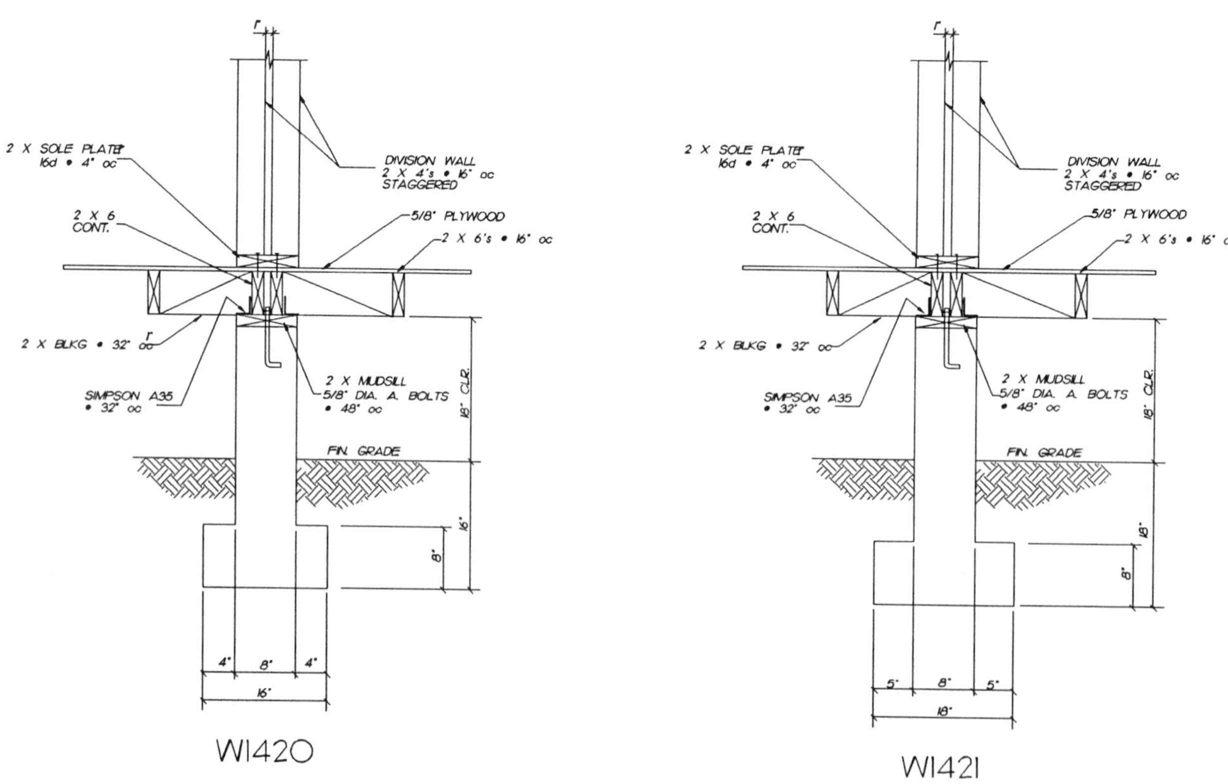

W1420

W1421

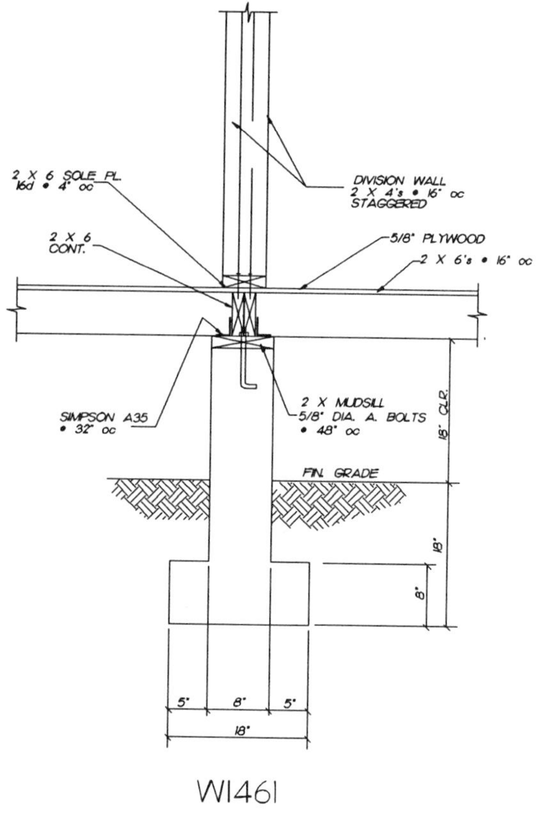

WI461

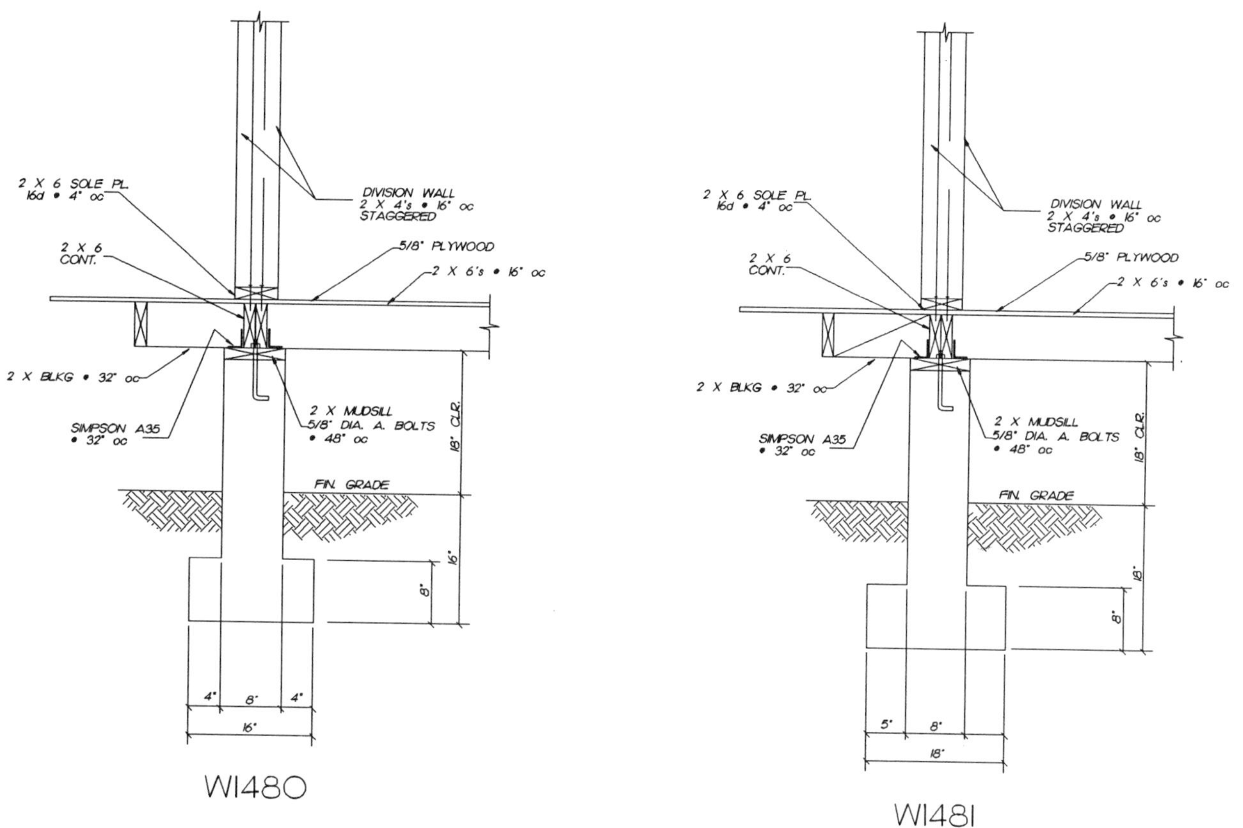

WI480

WI481

2 X 6 SOLE PL.
16d • 4" oc
DIVISION WALL
2 X 4's • 16" oc
STAGGERED
2 X 6
CONT.
5/8" PLYWOOD
2 X 6's • 16" oc
2 X BLKG • 32" oc
2 X MUDSILL
5/8" DIA. A. BOLTS
• 48" oc
SIMPSON A35
• 32" oc
18" CLR.
FIN. GRADE
16"
8"
4"
8"
4"
16"

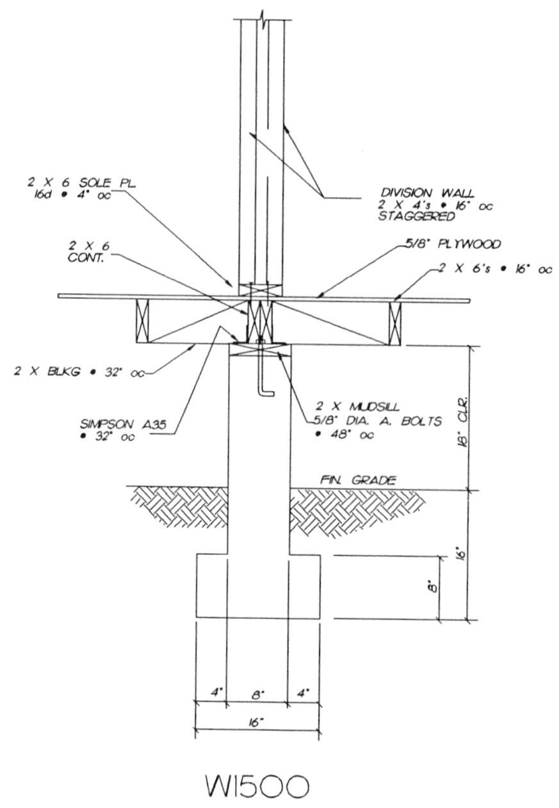

W1500

2 X 6 SOLE PL.
16d • 4" oc
DIVISION WALL
2 X 4's • 16" oc
STAGGERED
2 X 6
CONT.
5/8" PLYWOOD
2 X 6's • 16" oc
2 X BLKG • 32" oc
2 X MUDSILL
5/8" DIA. A. BOLTS
• 48" oc
SIMPSON A35
• 32" oc
18" CLR.
FIN. GRADE
18"
8"
5"
8"
5"
18"

W1501

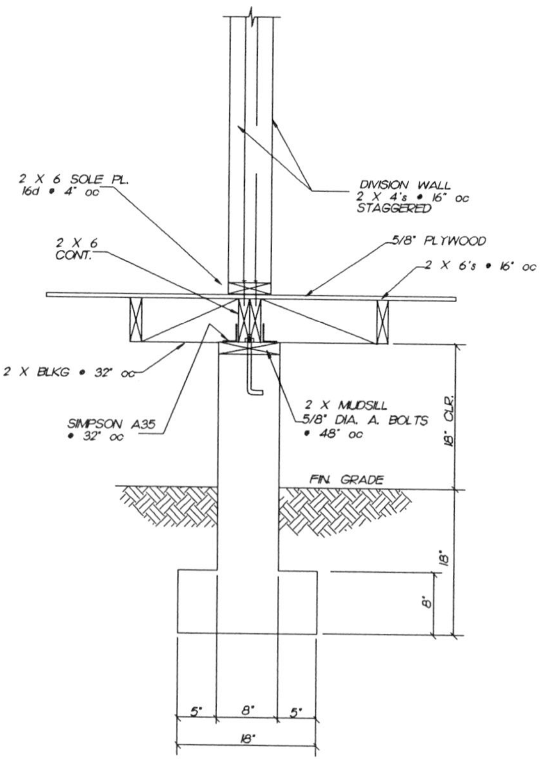

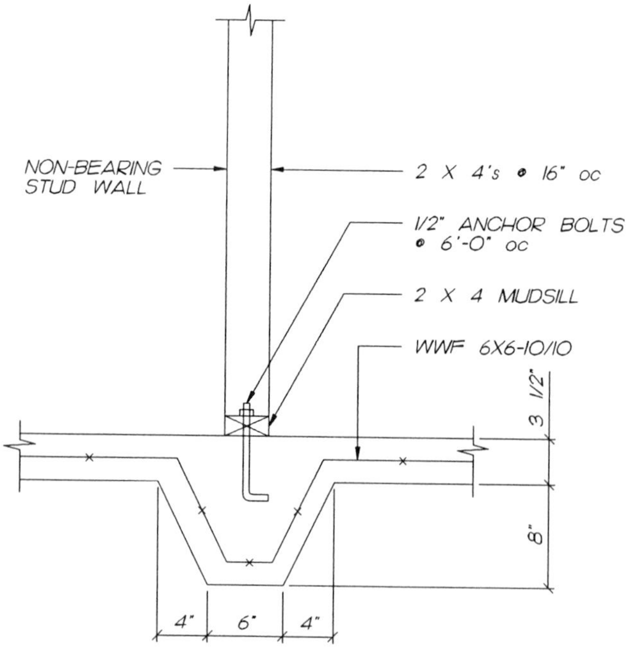

NON-BEARING
STUD WALL
2 X 4's • 16" oc
1/2" ANCHOR BOLTS
• 6'-0" oc
2 X 4 MUDSILL
WWF 6X6-10/10
3 1/2"
8"
4"
6"
4"

W1520

℄ WOOD POST
3 1/2" CONC.
SLAB ON GRADE
W/6x6-10/10
SIMPSON POST BASE
12"
3 - #4's
EA. WAY
3'-0" SQ.
3" CLR.

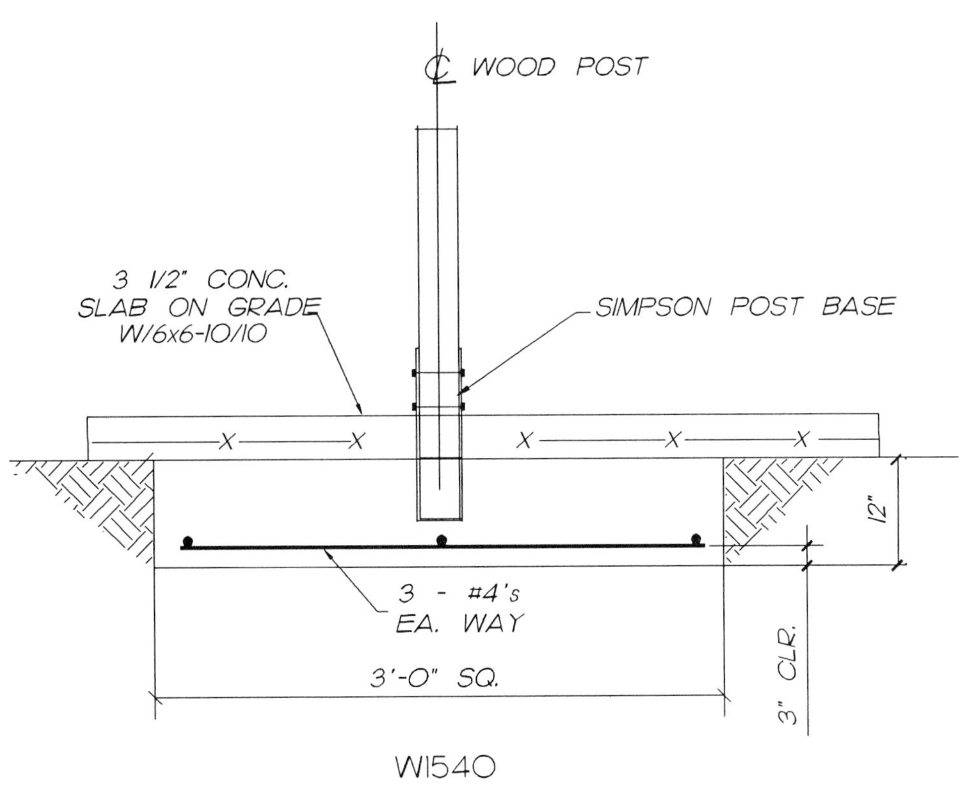

W1540

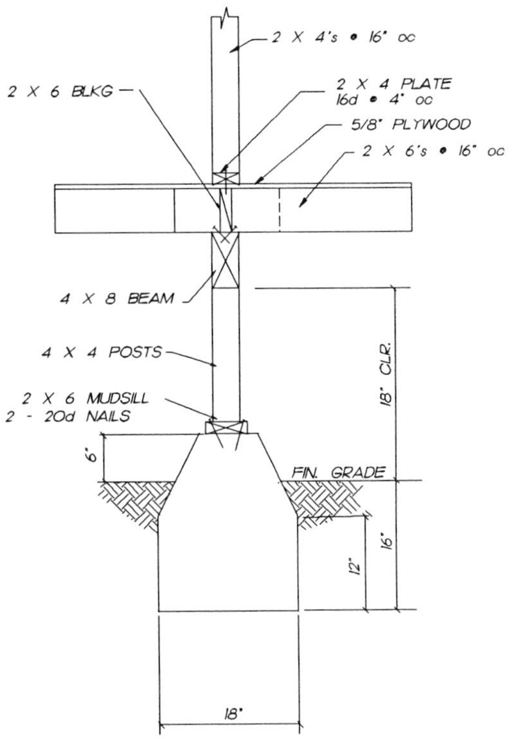

2 X 4's @ 16" oc
2 X 6 BLKG
2 X 4 PLATE
16d @ 4" oc
5/8" PLYWOOD
2 X 6's @ 16" oc
4 X 8 BEAM
4 X 4 POSTS
2 X 6 MUDSILL
2 - 20d NAILS
18" CLR.
6"
FIN. GRADE
16"
12"
18"

W1560

2 X 4's @ 16" oc
2 X 4 PLATE
16d @ 4" oc
5/8" PLYWOOD
2 X 6's @ 16" oc
4 X 8 BEAM
1" X BRACING
4 X 4 POSTS
2 X 6 MUDSILL
2 - 20d NAILS
18" CLR.
6"
FIN. GRADE
16"
12"
18"

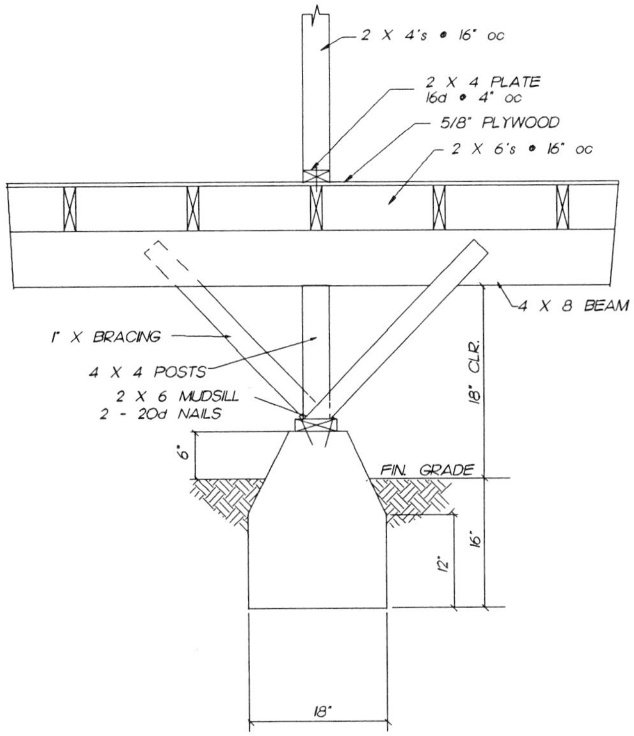

W1561

CHAPTER 1	WOOD	SECTION 2	COLUMNS
KING POST STUD	WOOD BEAM	STUD WALL	W2020
END POST	WOOD BEAM	SIMPSON ECC HARDWARE	W2040
END POST ABOVE & BELOW	STEEL BEAM	METAL PLATE CONNECTION	W2060 - W2061
INTERIOR POST	WOOD BEAM	SIMPSON CC HARDWARE	W2080
INTERIOR POST	STEEL BEAM	STEEL ANGLE CONNECTION	W2100 - W2101
WOOD POSTS	SUPPORTED ON CONC.	METAL PLATE CONNECTION	W2120 - W2121

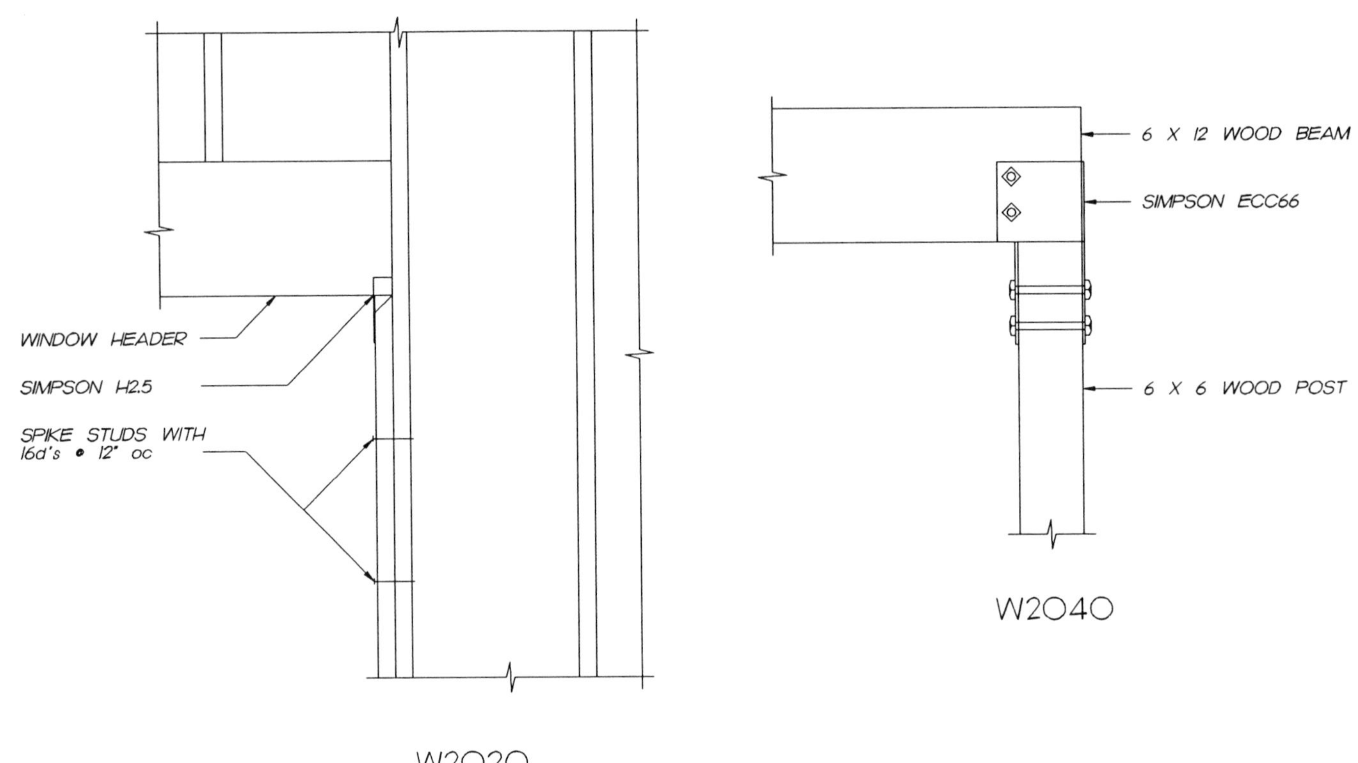
WINDOW HEADER
SIMPSON H2.5
SPIKE STUDS WITH
16d's @ 12" oc
W2020
6 X 12 WOOD BEAM
SIMPSON ECC66
6 X 6 WOOD POST
W2040

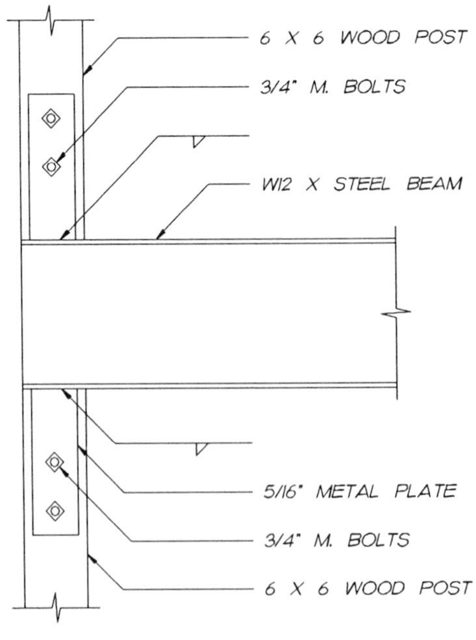
6 X 6 WOOD POST
3/4" M. BOLTS
W12 X STEEL BEAM
5/16" METAL PLATE
3/4" M. BOLTS
6 X 6 WOOD POST
W2060

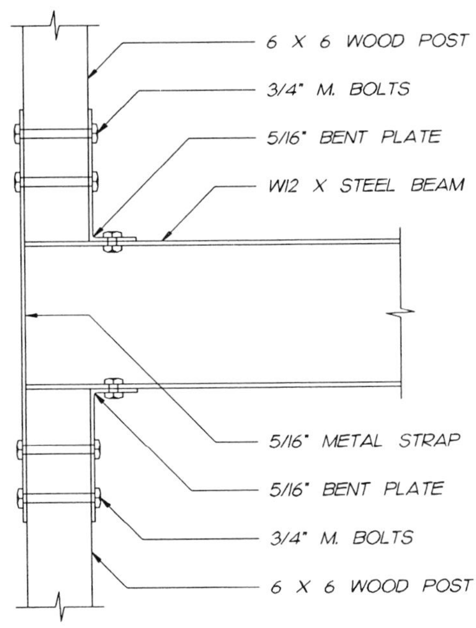

W2061

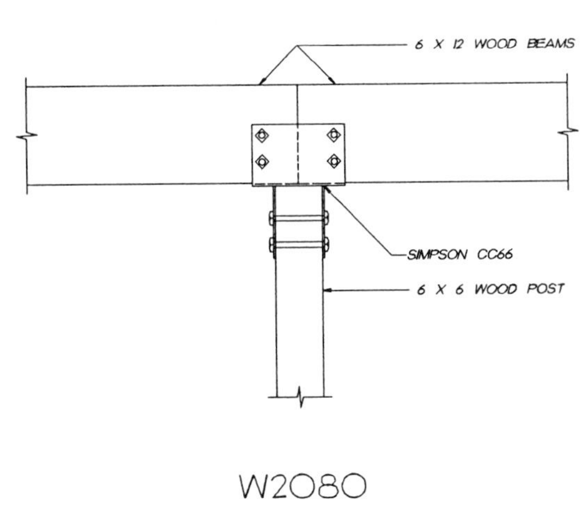

W2080

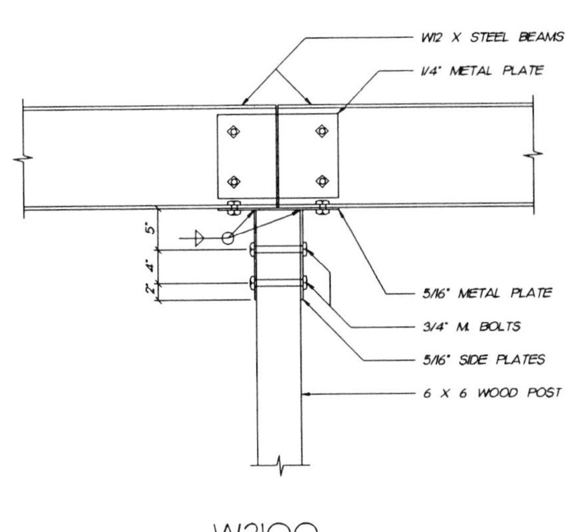

W2100

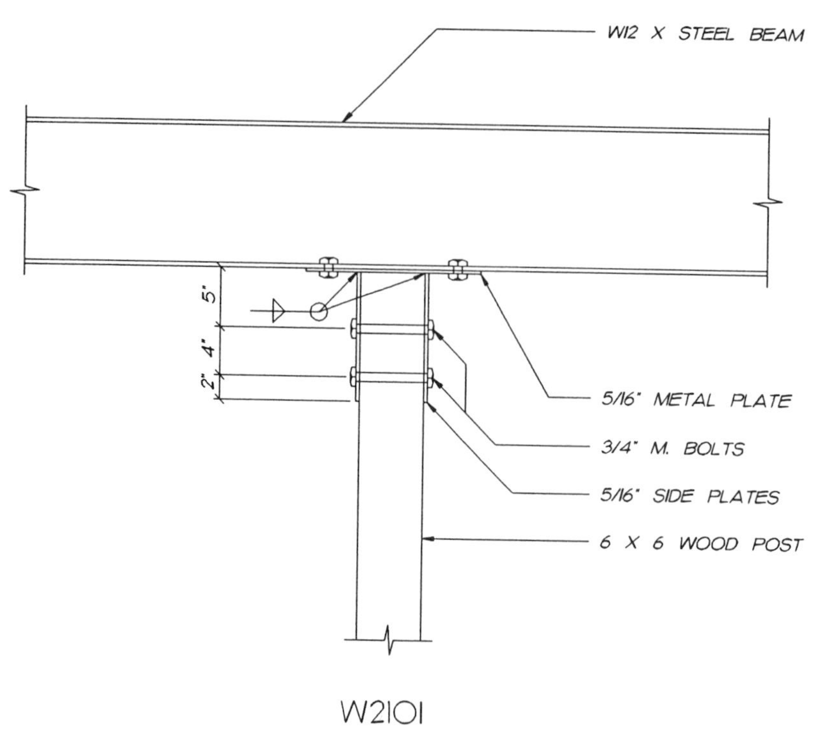

W2101

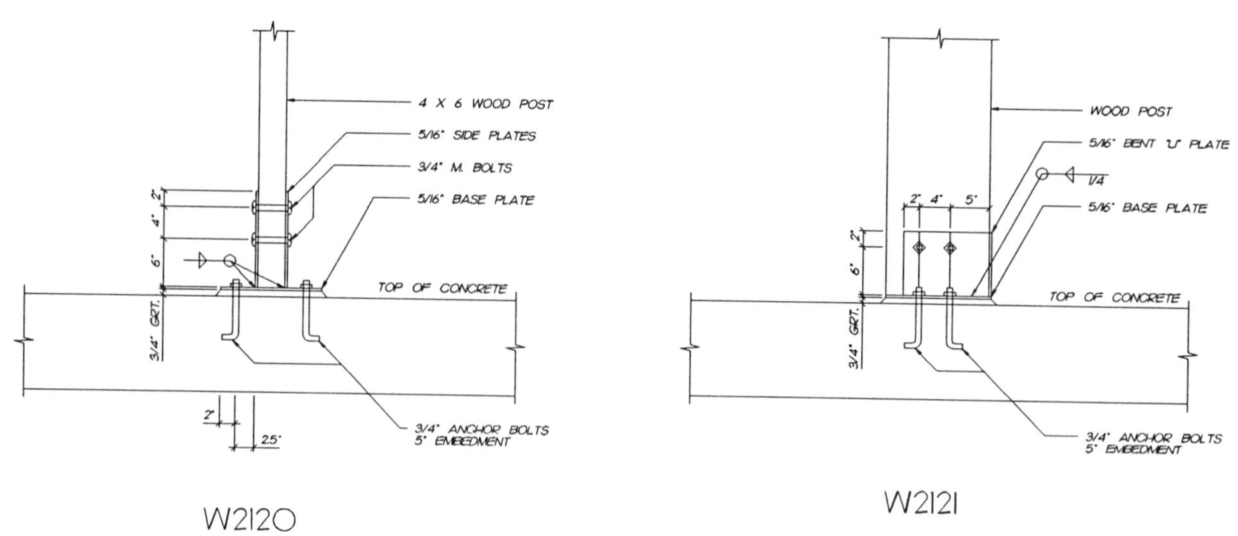

W2120

W2121

CHAPTER 1 WOOD	SECTION 3	WALLS & FLOORS
EXTERIOR WALL JOIST PERP. TO WALL WITH LT. WT. CONCRETE METAL SHEAR RESIST. CLIPS	2 X 4 STUDS ABOVE THE FLOOR 2 X 4 STUDS BELOW THE FLOOR	W3020 - W3024
	2 X 4 STUDS ABOVE FLOOR 2 X 6 STUDS BELOW FLOOR	W3025 - W3029
EXTERIOR WALL JOISTS PARALELL TO WALL WITH LT. WT. CONCRETE METAL SHEAR RESIST. CLIPS	2 X 4 STUDS ABOVE THE FLOOR 2 X 4 STUDS BELOW THE FLOOR	W3040 - W3044
	2 X 4 STUDS ABOVE FLOOR 2 X 6 STUDS BELOW FLOOR	W3045 - W3049
EXTERIOR WALL JOISTS PARALELL TO WALL NO LT. WT. CONCRETE METAL SHEAR RESIST. CLIPS	2 X 4 STUDS ABOVE THE FLOOR 2 X 4 STUDS BELOW THE FLOOR	W3060 - W3064
	2 X 4 STUDS ABOVE FLOOR 2 X 6 STUDS BELOW FLOOR	W3065 - W3069
EXTERIOR WALL JOISTS PARALELL TO WALL NO LT. WT. CONCRETE METAL SHEAR RESIST. CLIPS	2 X 4 STUDS ABOVE THE FLOOR 2 X 4 STUDS BELOW THE FLOOR	W3080 - W3084
	2 X 4 STUDS ABOVE FLOOR 2 X 6 STUDS BELOW FLOOR	W3085 - W3089
EXTERIOR WALL JOISTS PARALELL TO WALL WITH LT. WT. CONCRETE TOE NAIL SHEAR RESISTANCE	2 X 4 STUDS ABOVE THE FLOOR 2 X 4 STUDS BELOW THE FLOOR	W3100 - W3104
	2 X 4 STUDS ABOVE FLOOR 2 X 6 STUDS BELOW FLOOR	W3105 - W3109
EXTERIOR WALL JOISTS PARALELL TO WALL WITH LT. WT. CONCRETE TOE NAIL SHEAR RESISTANCE	2 X 4 STUDS ABOVE THE FLOOR 2 X 4 STUDS BELOW THE FLOOR	W3120 - W3124
	2 X 4 STUDS ABOVE FLOOR 2 X 6 STUDS BELOW FLOOR	W3125 - W3129
EXTERIOR WALL JOISTS PARALELL TO WALL NO LT. WT. CONCRETE TOE NAIL SHEAR RESISTANCE	2 X 4 STUDS ABOVE THE FLOOR 2 X 4 STUDS BELOW THE FLOOR	W3140 - W3144
	2 X 4 STUDS ABOVE FLOOR 2 X 6 STUDS BELOW FLOOR	W3145 - W3149
EXTERIOR WALL JOISTS PARALELL TO WALL NO LT. WT. CONCRETE TOE NAIL SHEAR RESISTANCE	2 X 4 STUDS ABOVE THE FLOOR 2 X 4 STUDS BELOW THE FLOOR	W3160 - W3164
	2 X 4 STUDS ABOVE FLOOR 2 X 6 STUDS BELOW FLOOR	W3165 - W3169
EXTERIOR WALL JOISTS PARALELL TO WALL WITH LT. WT. CONCRETE SHEAR BLOCKING RESISTANCE	2 X 4 STUDS ABOVE THE FLOOR 2 X 4 STUDS BELOW THE FLOOR	W3180 - W3184
	2 X 4 STUDS ABOVE FLOOR 2 X 6 STUDS BELOW FLOOR	W3185 - W3189
EXTERIOR WALL - OVERHANG JOISTS PARALELL TO WALL WITH LT. WT. CONCRETE METAL SHEAR RESIST. CLIPS	2 X 4 STUDS ABOVE THE FLOOR 2 X 4 STUDS BELOW THE FLOOR	W3200 - W3204
	2 X 4 STUDS ABOVE FLOOR 2 X 6 STUDS BELOW FLOOR	W3205 - W3209
EXTERIOR WALL - OVERHANG JOISTS PARALELL TO WALL WITH LT. WT. CONCRETE METAL SHEAR RESIST. CLIPS	2 X 4 STUDS ABOVE THE FLOOR 2 X 4 STUDS BELOW THE FLOOR	W3220 - W3224
	2 X 4 STUDS ABOVE FLOOR 2 X 6 STUDS BELOW FLOOR	W3225 - W3229
EXTERIOR WALL - OVERHANG JOIST PERP. TO WALL WITH LT. WT. CONCRETE TOE NAIL SHEAR RESISTANCE	2 X 4 STUDS ABOVE THE FLOOR 2 X 4 STUDS BELOW THE FLOOR	W3240 - W3244
	2 X 4 STUDS ABOVE FLOOR 2 X 6 STUDS BELOW FLOOR	W3245 - W3249
EXTERIOR WALL - OVERHANG JOISTS PARALELL TO WALL WITH LT. WT. CONCRETE TOE NAIL SHEAR RESISTANCE	2 X 4 STUDS ABOVE THE FLOOR 2 X 4 STUDS BELOW THE FLOOR	W3260 - W3264
	2 X 4 STUDS ABOVE FLOOR 2 X 6 STUDS BELOW FLOOR	W3265 - W3269
EXTERIOR WALL- CANTILEVER JOISTS PERP. TO WALL WITH LT. WT. CONCRETE METAL SHEAR RESIST. CLIPS	2 X 4 STUDS ABOVE THE FLOOR 2 X 4 STUDS BELOW THE FLOOR	W3280 - W3284
	2 X 4 STUDS ABOVE FLOOR 2 X 6 STUDS BELOW FLOOR	W3285 - W3289
EXTERIOR WALL- CANTILEVER JOISTS PERP. TO WALL NO LT. WT. CONCRETE METAL SHEAR RESIST. CLIPS	2 X 4 STUDS ABOVE THE FLOOR 2 X 4 STUDS BELOW THE FLOOR	W3300 - W3304
	2 X 4 STUDS ABOVE FLOOR 2 X 6 STUDS BELOW FLOOR	W3305 - W3309
EXTERIOR WALL- CANTILEVER JOISTS PERP. TO WALL WITH LT. WT. CONCRETE TOE NAIL SHEAR RESISTANCE	2 X 4 STUDS ABOVE THE FLOOR 2 X 4 STUDS BELOW THE FLOOR	W3320 - W3324
	2 X 4 STUDS ABOVE FLOOR 2 X 6 STUDS BELOW FLOOR	W3325 - W3329
EXTERIOR WALL- CANTILEVER JOISTS PERP. TO WALL NO LT. WT. CONCRETE TOE NAIL SHEAR RESISTANCE	2 X 4 STUDS ABOVE THE FLOOR 2 X 4 STUDS BELOW THE FLOOR	W3340 - W3344
	2 X 4 STUDS ABOVE FLOOR 2 X 6 STUDS BELOW FLOOR	W3345 - W3349

CHAPTER 1 WOOD	SECTION 3	WALLS & FLOORS
EXTERIOR WALL - BALCONY JOIST PERP. TO WALL WITH LT. WT. CONCRETE METAL SHEAR RESIST. CLIPS	2 X 4 STUDS ABOVE THE FLOOR 2 X 4 STUDS BELOW THE FLOOR	W3360 - W3364
	2 X 4 STUDS ABOVE FLOOR 2 X 6 STUDS BELOW FLOOR	W3365 - W3369
INTERIOR WALL JOIST PERP. TO WALL WITH LT. WT. CONCRETE METAL SHEAR RESIST. CLIPS	2 X 4 STUDS ABOVE THE FLOOR 2 X 4 STUDS BELOW THE FLOOR	W3380 - W3384
	2 X 4 STUDS ABOVE FLOOR 2 X 6 STUDS BELOW FLOOR	W3385 - W3389
INTERIOR WALL JOISTS PARALELL TO WALL WITH LT. WT. CONCRETE METAL SHEAR RESIST. CLIPS	2 X 4 STUDS ABOVE THE FLOOR 2 X 4 STUDS BELOW THE FLOOR	W3400 - W3404
	2 X 4 STUDS ABOVE FLOOR 2 X 6 STUDS BELOW FLOOR	W3405 - W3409
INTERIOR WALL JSTS PERP. & PARALELL ALT. SIDES NO LT. WT. CONCRETE METAL SHEAR RESIST. CLIPS	2 X 4 STUDS ABOVE THE FLOOR 2 X 4 STUDS BELOW THE FLOOR	W3420 - W3424
	2 X 4 STUDS ABOVE FLOOR 2 X 6 STUDS BELOW FLOOR	W3425 - W3429
INTERIOR WALL JOIST PERP. TO WALL WITH LT. WT. CONCRETE METAL SHEAR RESIST. CLIPS	2 X 4 STUDS ABOVE THE FLOOR 2 X 4 STUDS BELOW THE FLOOR	W3440 - W3444
	2 X 4 STUDS ABOVE FLOOR 2 X 6 STUDS BELOW FLOOR	W3445 - W3449
INTERIOR WALL JOISTS PARALELL TO WALL NO LT. WT. CONCRETE METAL SHEAR RESIST. CLIPS	2 X 4 STUDS ABOVE THE FLOOR 2 X 4 STUDS BELOW THE FLOOR	W3460 - W3464
	2 X 4 STUDS ABOVE FLOOR 2 X 6 STUDS BELOW FLOOR	W3465 - W3469
INTERIOR WALL JSTS PERP. & PARALELL ALT. SIDES NO LT. WT. CONCRETE TOE NAIL SHEAR RESISTANCE	2 X 4 STUDS ABOVE THE FLOOR 2 X 4 STUDS BELOW THE FLOOR	W3480 - W3484
	2 X 4 STUDS ABOVE FLOOR 2 X 6 STUDS BELOW FLOOR	W3485 - W3489
INTERIOR WALL JOISTS PARALELL TO WALL WITH LT. WT. CONCRETE SHEAR BLOCKING RESISTANCE	2 X 4 STUDS ABOVE THE FLOOR 2 X 4 STUDS BELOW THE FLOOR	W3500 - W3504
	2 X 4 STUDS ABOVE FLOOR 2 X 6 STUDS BELOW FLOOR	W3505 - W3509
INTERIOR WALL- UNEQUAL JOISTS JOIST PERP. TO WALL WITH LT. WT. CONCRETE METAL SHEAR RESIST. CLIPS	2 X 4 STUDS ABOVE THE FLOOR 2 X 4 STUDS BELOW THE FLOOR	W3520 - W3522
	2 X 4 STUDS ABOVE FLOOR 2 X 6 STUDS BELOW FLOOR	W3523 - W3525
INTERIOR WALL- UNEQUAL JOISTS JOISTS PARALELL TO WALL WITH LT. WT. CONCRETE METAL SHEAR RESIST. CLIPS	2 X 4 STUDS ABOVE THE FLOOR 2 X 4 STUDS BELOW THE FLOOR	W3540 - W3542
	2 X 4 STUDS ABOVE FLOOR 2 X 6 STUDS BELOW FLOOR	W3543 - W3545
INTERIOR WALL- UNEQUAL JOISTS JSTS PERP. & PARALELL ALT. SIDES WITH LT. WT. CONCRETE METAL SHEAR RESIST. CLIPS	2 X 4 STUDS ABOVE THE FLOOR 2 X 4 STUDS BELOW THE FLOOR	W3560 - W3562
	2 X 4 STUDS ABOVE FLOOR 2 X 6 STUDS BELOW FLOOR	W3563 - W3565
INTERIOR WALL- UNEQUAL JOISTS JSTS PERP. & PARALELL ALT. SIDES WITH LT. WT. CONCRETE METAL SHEAR RESIST. CLIPS	2 X 4 STUDS ABOVE THE FLOOR 2 X 4 STUDS BELOW THE FLOOR	W3580 - W3582
	2 X 4 STUDS ABOVE FLOOR 2 X 6 STUDS BELOW FLOOR	W3583 - 3585
INTERIOR DIVISION WALL JOISTS PERP. TO WALL WITH LT. WT. CONCRETE METAL SHEAR RESIST. CLIPS	2 X 4 STUDS ABOVE THE FLOOR 2 X 4 STUDS BELOW THE FLOOR SEPARATED STUD WALL	W3600 - W3603
INTERIOR DIVISION WALL JOISTS PERP. TO WALL WITH LT. WT. CONCRETE METAL SHEAR RESIST. CLIPS	2 X 4 STUDS ABOVE THE FLOOR 2 X 4 STUDS BELOW THE FLOOR SEPARATED STUD WALL	W3620 - W3623
INTERIOR DIVISION WALL JSTS PERP. & PARALELL ALT. SIDES WITH LT. WT. CONCRETE METAL SHEAR RESIST. CLIPS	2 X 4 STUDS ABOVE THE FLOOR 2 X 4 STUDS BELOW THE FLOOR SEPARATED STUD WALL	W3640 - W3643
INTERIOR DIVISION WALL JOISTS PERP. TO WALL WITH LT. WT. CONCRETE METAL SHEAR RESIST. CLIPS	2 X 4 STUDS ABOVE THE FLOOR 2 X 4 STUDS BELOW THE FLOOR STAGGERED STUD WALL	W3660 - W3663
INTERIOR DIVISION WALL JOISTS PARALELL TO WALL WITH LT. WT. CONCRETE METAL SHEAR RESIST. CLIPS	2 X 4 STUDS ABOVE THE FLOOR 2 X 4 STUDS BELOW THE FLOOR STAGGERED STUD WALL	W3680 - W3683

NTERIOR DIVISION WALL STS PERP. & PARALELL ALT. SIDES WITH LT. WT. CONCRETE METAL SHEAR RESIST. CLIPS	2 X 4 STUDS ABOVE THE FLOOR 2 X 4 STUDS BELOW THE FLOOR STAGGERED STUD WALL	W3700 - W3703

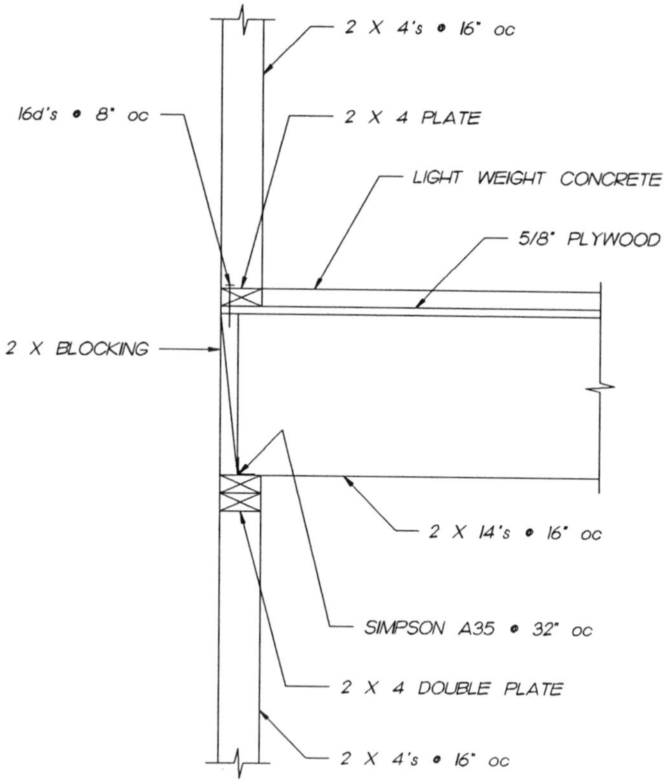

2 X 4's @ 16" oc

16d's @ 8" oc

2 X 4 PLATE

LIGHT WEIGHT CONCRETE

5/8" PLYWOOD

2 X BLOCKING

2 X 14's @ 16" oc

SIMPSON A35 @ 32" oc

2 X 4 DOUBLE PLATE

2 X 4's @ 16" oc

W3020

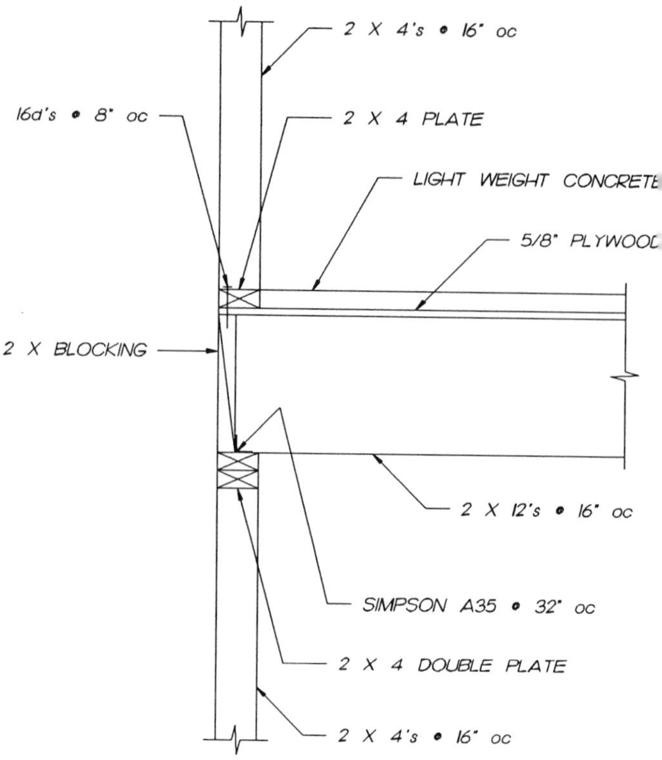

2 X 4's @ 16" oc

16d's @ 8" oc

2 X 4 PLATE

LIGHT WEIGHT CONCRETE

5/8" PLYWOOD

2 X BLOCKING

2 X 12's @ 16" oc

SIMPSON A35 @ 32" oc

2 X 4 DOUBLE PLATE

2 X 4's @ 16" oc

W3021

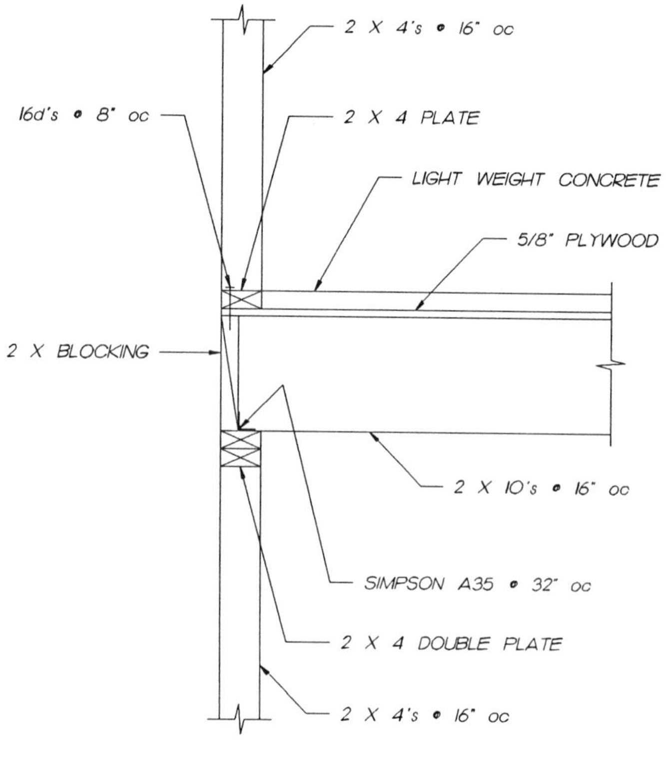

2 X 4's @ 16" oc

16d's @ 8" oc

2 X 4 PLATE

LIGHT WEIGHT CONCRETE

5/8" PLYWOOD

2 X BLOCKING

2 X 10's @ 16" oc

SIMPSON A35 @ 32" oc

2 X 4 DOUBLE PLATE

2 X 4's @ 16" oc

W3122

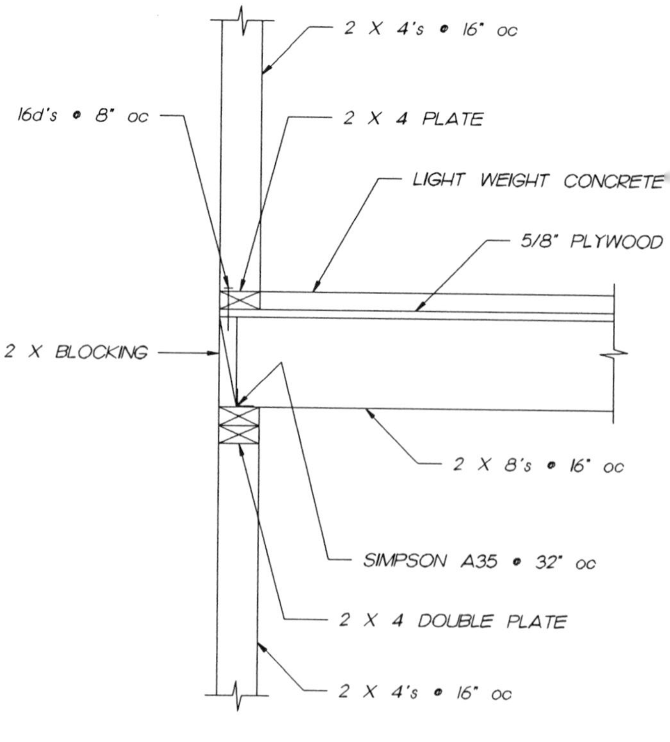

2 X 4's @ 16" oc

16d's @ 8" oc

2 X 4 PLATE

LIGHT WEIGHT CONCRETE

5/8" PLYWOOD

2 X BLOCKING

2 X 8's @ 16" oc

SIMPSON A35 @ 32" oc

2 X 4 DOUBLE PLATE

2 X 4's @ 16" oc

W3023

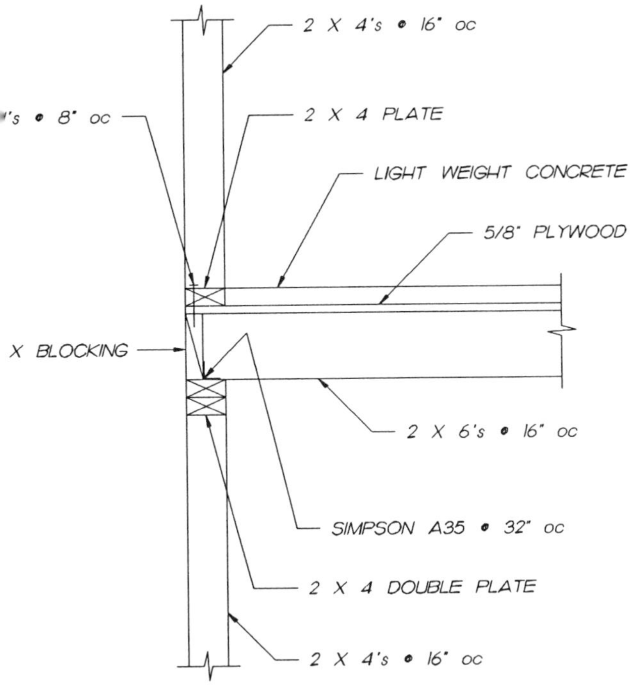

2 X 4's @ 16" oc

's @ 8" oc

2 X 4 PLATE

LIGHT WEIGHT CONCRETE

5/8" PLYWOOD

X BLOCKING

2 X 6's @ 16" oc

SIMPSON A35 @ 32" oc

2 X 4 DOUBLE PLATE

2 X 4's @ 16" oc

W3024

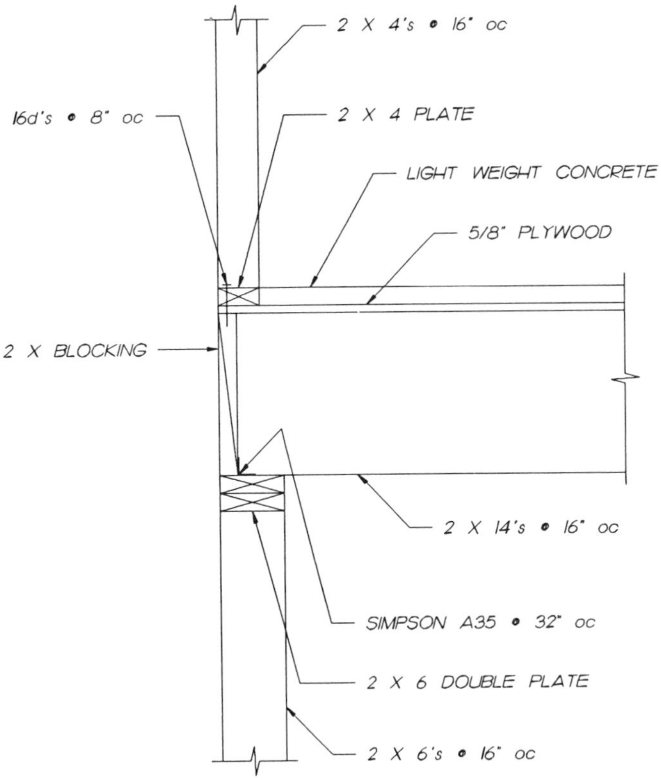

2 X 4's @ 16" oc

16d's @ 8" oc

2 X 4 PLATE

LIGHT WEIGHT CONCRETE

5/8" PLYWOOD

2 X BLOCKING

2 X 14's @ 16" oc

SIMPSON A35 @ 32" oc

2 X 6 DOUBLE PLATE

2 X 6's @ 16" oc

W3025

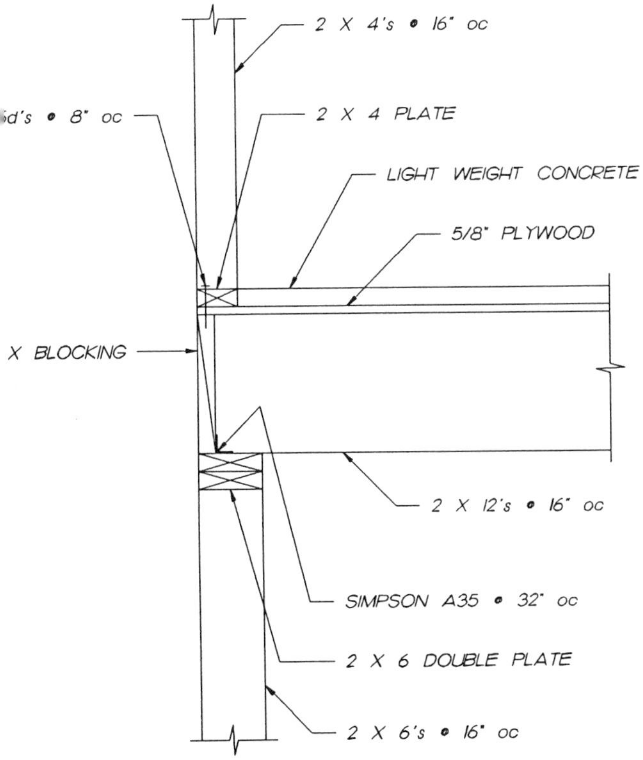

2 X 4's @ 16" oc

d's @ 8" oc

2 X 4 PLATE

LIGHT WEIGHT CONCRETE

5/8" PLYWOOD

X BLOCKING

2 X 12's @ 16" oc

SIMPSON A35 @ 32" oc

2 X 6 DOUBLE PLATE

2 X 6's @ 16" oc

W3026

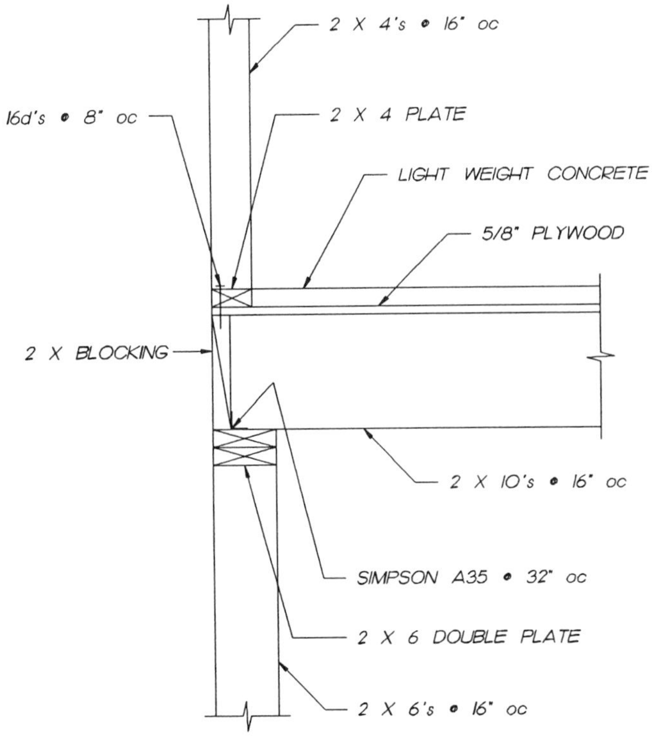

2 X 4's @ 16" oc

16d's @ 8" oc

2 X 4 PLATE

LIGHT WEIGHT CONCRETE

5/8" PLYWOOD

2 X BLOCKING

2 X 10's @ 16" oc

SIMPSON A35 @ 32" oc

2 X 6 DOUBLE PLATE

2 X 6's @ 16" oc

W3027

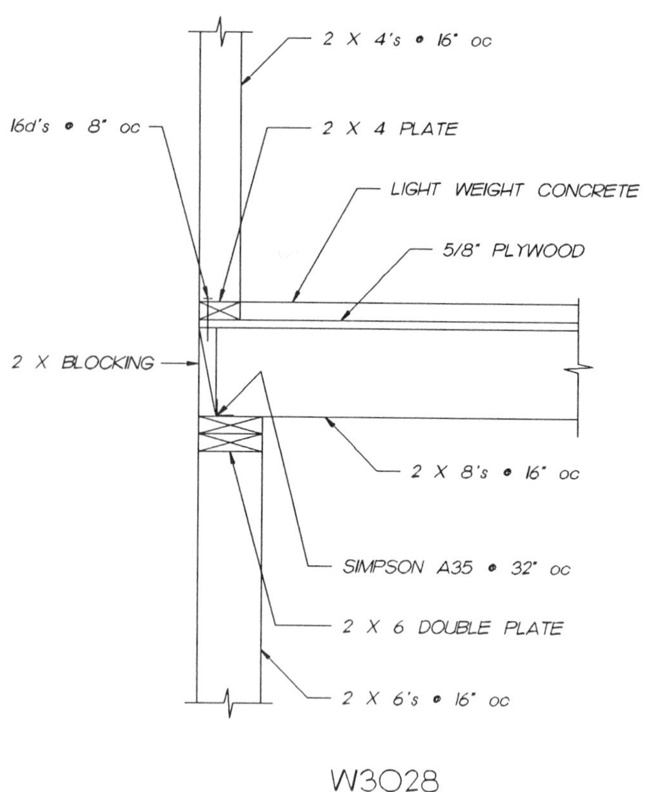

2 X 4's @ 16" oc
16d's @ 8" oc
2 X 4 PLATE
LIGHT WEIGHT CONCRETE
5/8" PLYWOOD
2 X BLOCKING
2 X 8's @ 16" oc
SIMPSON A35 @ 32" oc
2 X 6 DOUBLE PLATE
2 X 6's @ 16" oc

W3028

2 X 4's @ 16" oc
16d's @ 8" oc
2 X 4 PLATE
LIGHT WEIGHT CONCRETE
5/8" PLYWOOD
2 X BLOCKING
2 X 6's @ 16" oc
SIMPSON A35 @ 32" oc
2 X 6 DOUBLE PLATE
2 X 6's @ 16" oc

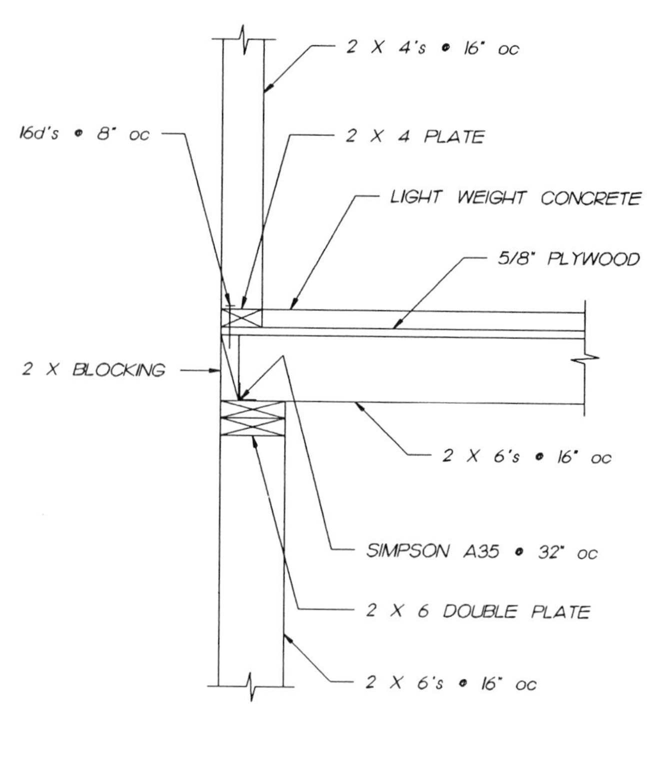

W3029

2 X 4's @ 16" oc
16d's @ 8" oc
2 X 4 PLATE
LIGHT WEIGHT CONCRETE
5/8" PLYWOOD
2 X 14's @ 16" oc
2 X BLOCKING @ 32" oc
SIMPSON A35 @ 32" oc
2 X 4 DOUBLE PLATE
2 X 4's @ 16" oc

W3041

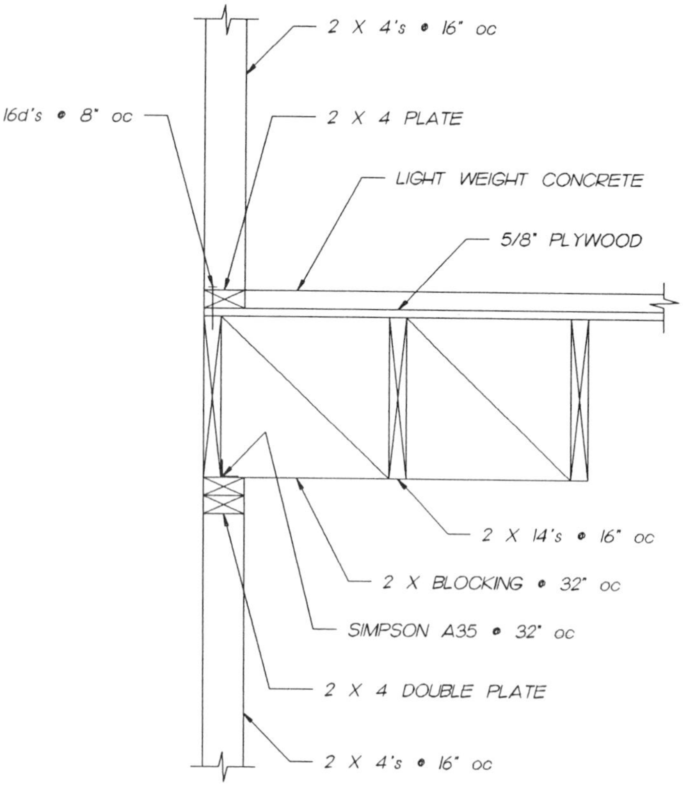

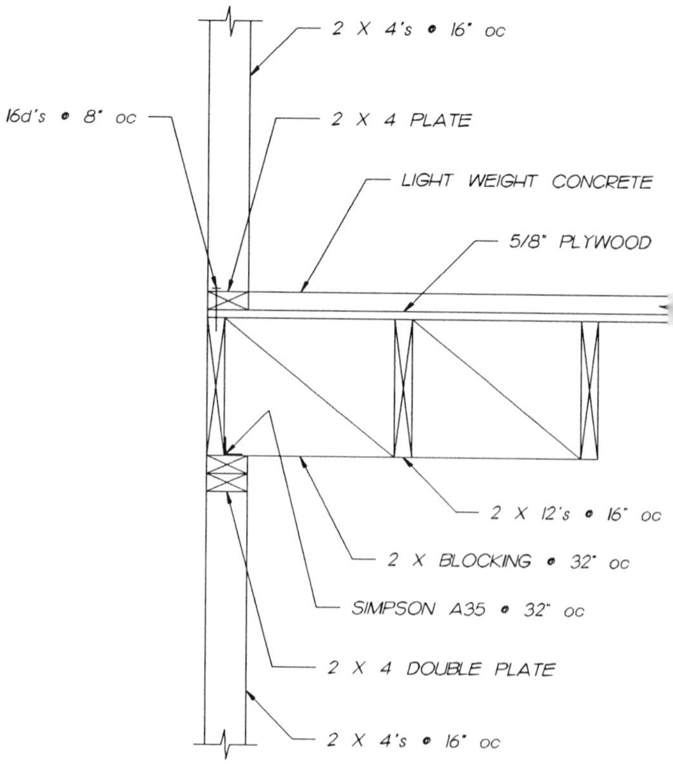

2 X 4's @ 16" oc
16d's @ 8" oc
2 X 4 PLATE
LIGHT WEIGHT CONCRETE
5/8" PLYWOOD
2 X 12's @ 16" oc
2 X BLOCKING @ 32" oc
SIMPSON A35 @ 32" oc
2 X 4 DOUBLE PLATE
2 X 4's @ 16" oc

W3040

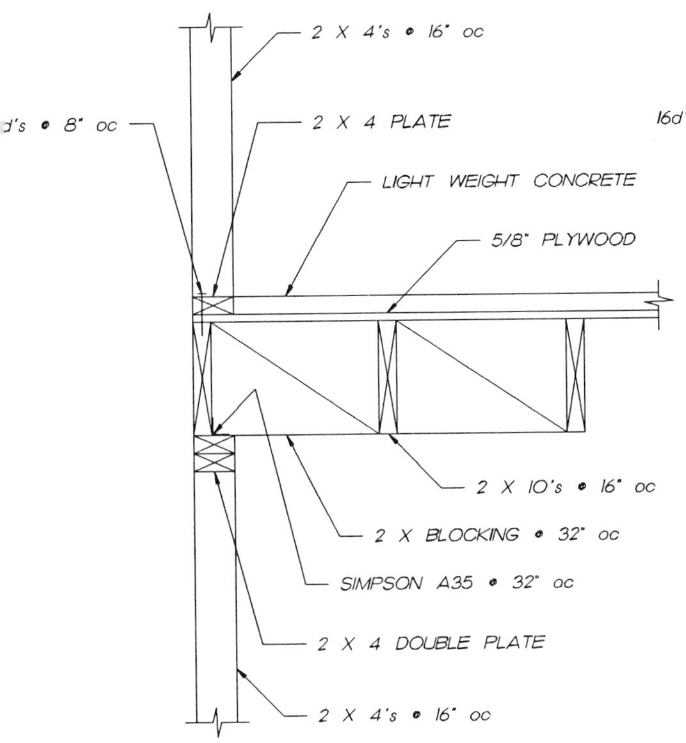

2 X 4's • 16" oc

d's • 8" oc

2 X 4 PLATE

LIGHT WEIGHT CONCRETE

5/8" PLYWOOD

2 X 10's • 16" oc

2 X BLOCKING • 32" oc

SIMPSON A35 • 32" oc

2 X 4 DOUBLE PLATE

2 X 4's • 16" oc

W3042

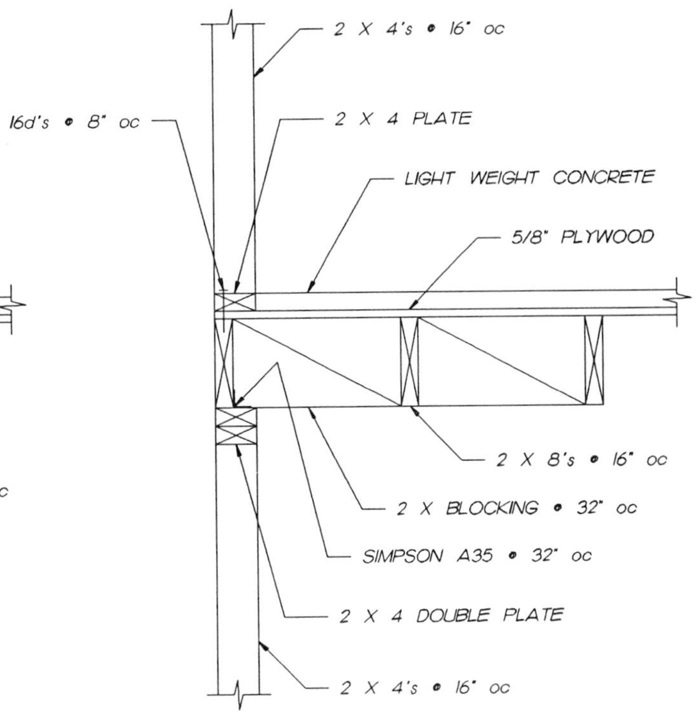

2 X 4's • 16" oc

16d's • 8" oc

2 X 4 PLATE

LIGHT WEIGHT CONCRETE

5/8" PLYWOOD

2 X 8's • 16" oc

2 X BLOCKING • 32" oc

SIMPSON A35 • 32" oc

2 X 4 DOUBLE PLATE

2 X 4's • 16" oc

W3043

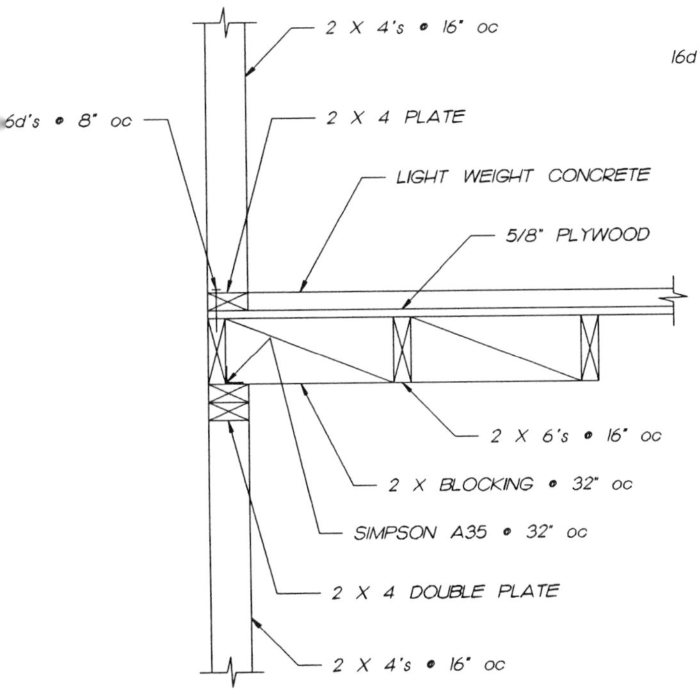

2 X 4's • 16" oc

6d's • 8" oc

2 X 4 PLATE

LIGHT WEIGHT CONCRETE

5/8" PLYWOOD

2 X 6's • 16" oc

2 X BLOCKING • 32" oc

SIMPSON A35 • 32" oc

2 X 4 DOUBLE PLATE

2 X 4's • 16" oc

W3044

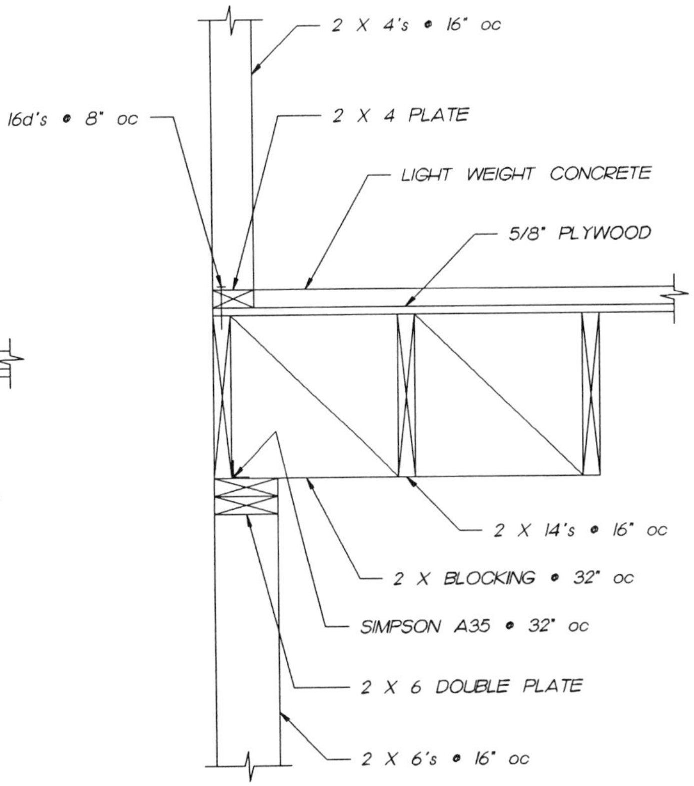

2 X 4's • 16" oc

16d's • 8" oc

2 X 4 PLATE

LIGHT WEIGHT CONCRETE

5/8" PLYWOOD

2 X 14's • 16" oc

2 X BLOCKING • 32" oc

SIMPSON A35 • 32" oc

2 X 6 DOUBLE PLATE

2 X 6's • 16" oc

W3045

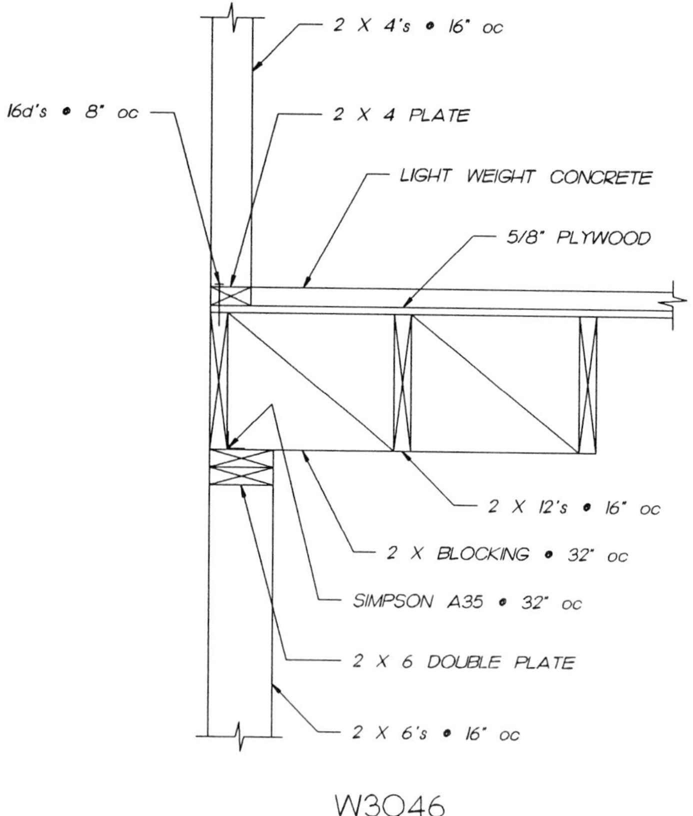
2 X 4's @ 16" oc
16d's @ 8" oc
2 X 4 PLATE
LIGHT WEIGHT CONCRETE
5/8" PLYWOOD
2 X 12's @ 16" oc
2 X BLOCKING @ 32" oc
SIMPSON A35 @ 32" oc
2 X 6 DOUBLE PLATE
2 X 6's @ 16" oc

W3046

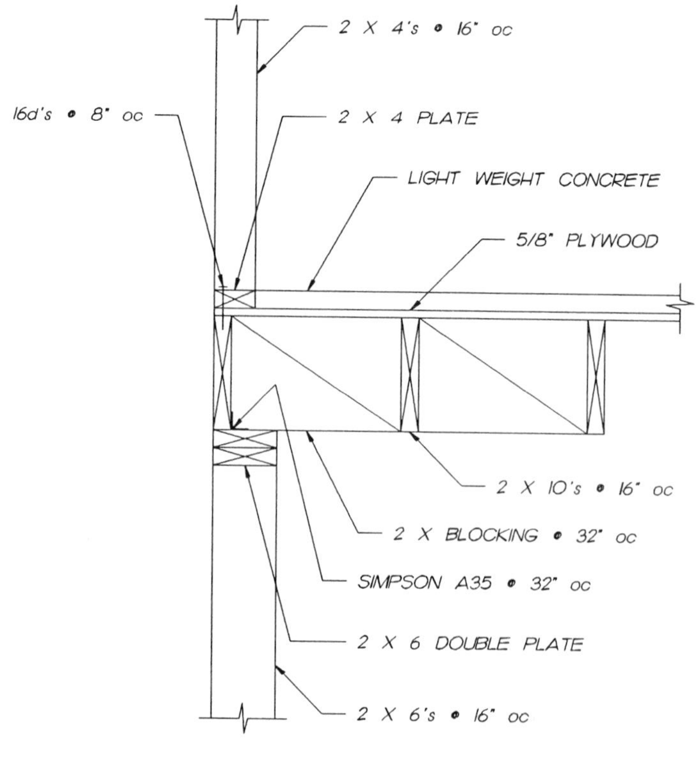
2 X 4's @ 16" oc
16d's @ 8" oc
2 X 4 PLATE
LIGHT WEIGHT CONCRETE
5/8" PLYWOOD
2 X 10's @ 16" oc
2 X BLOCKING @ 32" oc
SIMPSON A35 @ 32" oc
2 X 6 DOUBLE PLATE
2 X 6's @ 16" oc

W3047

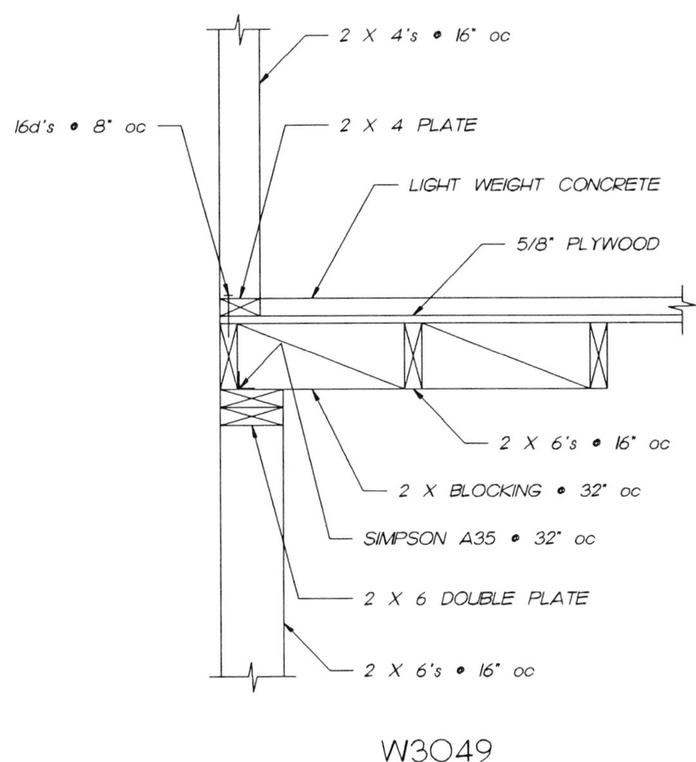
2 X 4's @ 16" oc
16d's @ 8" oc
2 X 4 PLATE
LIGHT WEIGHT CONCRETE
5/8" PLYWOOD
2 X 6's @ 16" oc
2 X BLOCKING @ 32" oc
SIMPSON A35 @ 32" oc
2 X 6 DOUBLE PLATE
2 X 6's @ 16" oc

W3049

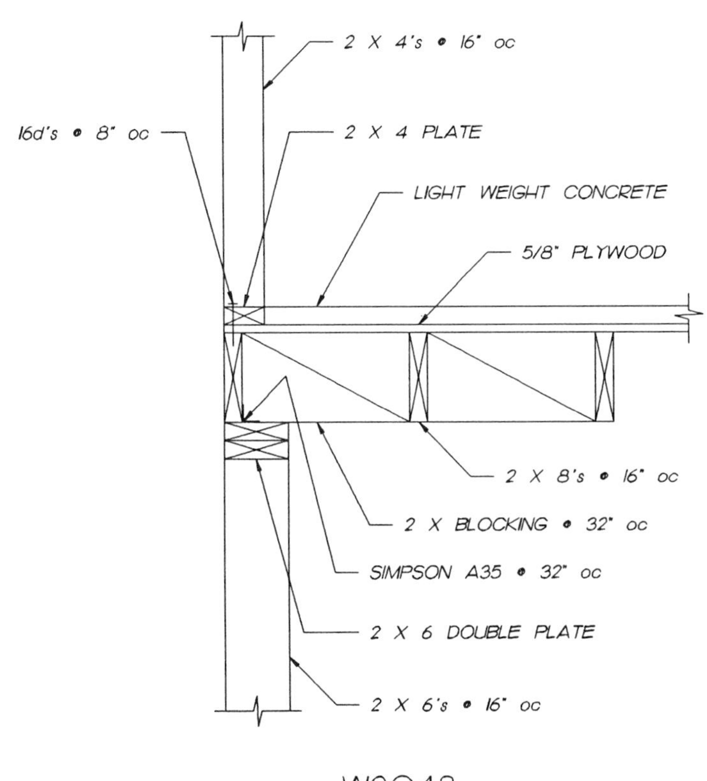
2 X 4's @ 16" oc
16d's @ 8" oc
2 X 4 PLATE
LIGHT WEIGHT CONCRETE
5/8" PLYWOOD
2 X 8's @ 16" oc
2 X BLOCKING @ 32" oc
SIMPSON A35 @ 32" oc
2 X 6 DOUBLE PLATE
2 X 6's @ 16" oc

W3048

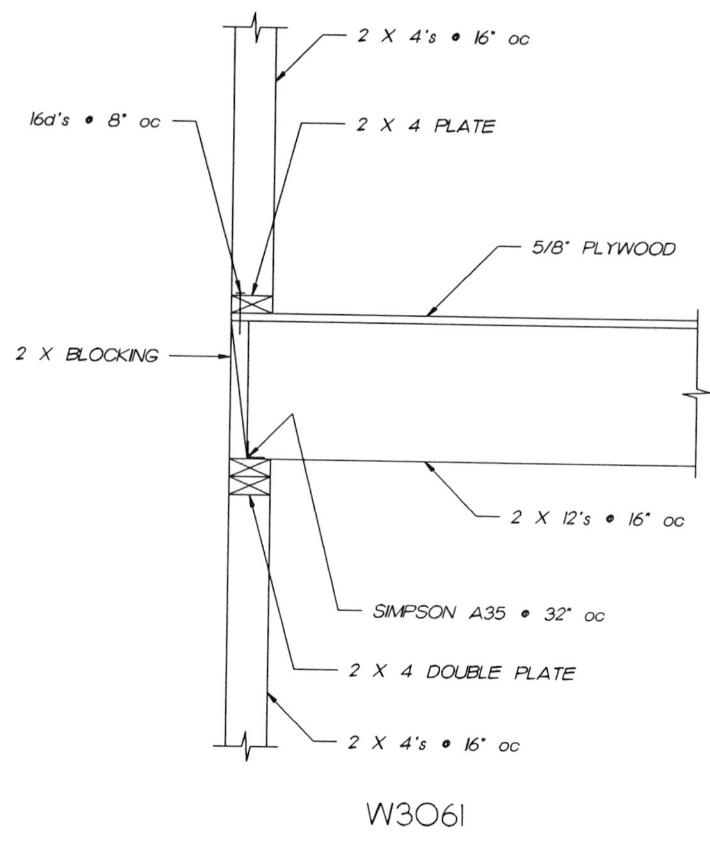
2 X 4's @ 16" oc
16d's @ 8" oc
2 X 4 PLATE
5/8" PLYWOOD
2 X BLOCKING
2 X 12's @ 16" oc
SIMPSON A35 @ 32" oc
2 X 4 DOUBLE PLATE
2 X 4's @ 16" oc

W3061

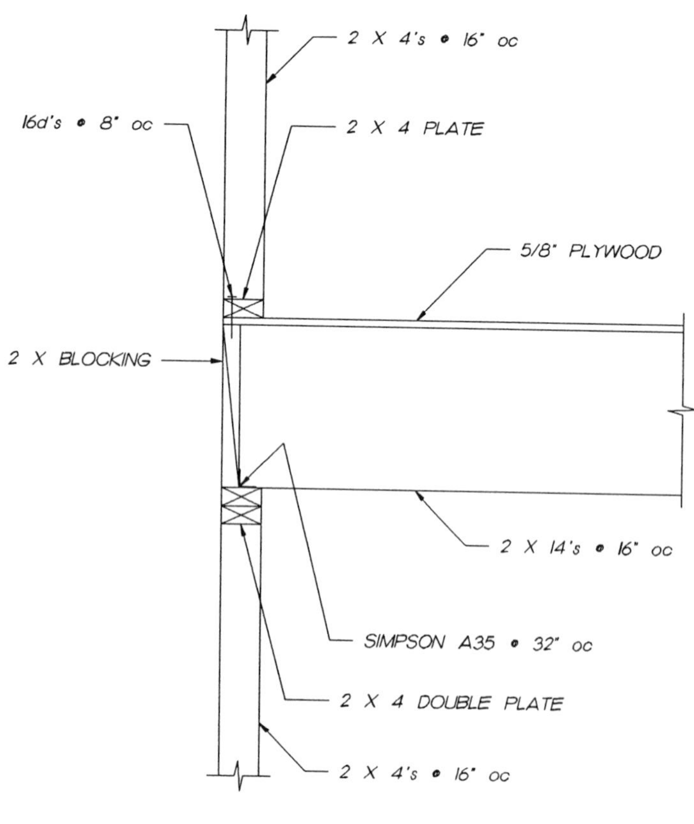
2 X 4's @ 16" oc
16d's @ 8" oc
2 X 4 PLATE
5/8" PLYWOOD
2 X BLOCKING
2 X 14's @ 16" oc
SIMPSON A35 @ 32" oc
2 X 4 DOUBLE PLATE
2 X 4's @ 16" oc

W3060

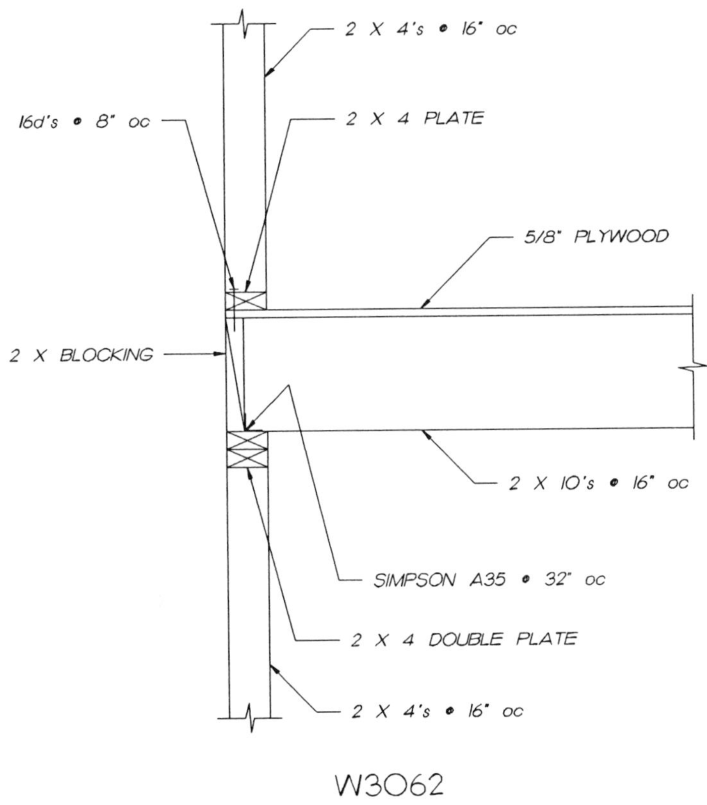
2 X 4's @ 16" oc
16d's @ 8" oc
2 X 4 PLATE
5/8" PLYWOOD
2 X BLOCKING
2 X 10's @ 16" oc
SIMPSON A35 @ 32" oc
2 X 4 DOUBLE PLATE
2 X 4's @ 16" oc

W3062

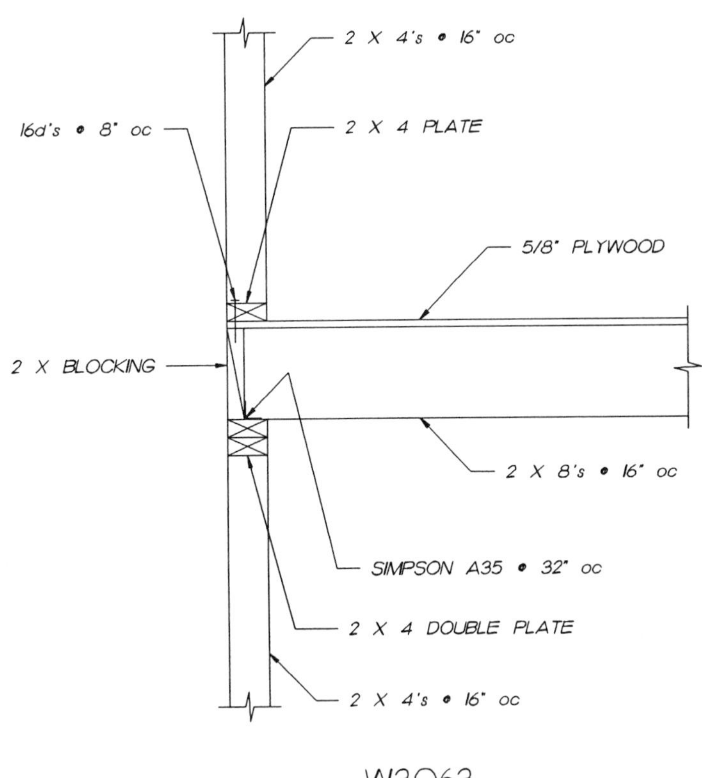
2 X 4's @ 16" oc
16d's @ 8" oc
2 X 4 PLATE
5/8" PLYWOOD
2 X BLOCKING
2 X 8's @ 16" oc
SIMPSON A35 @ 32" oc
2 X 4 DOUBLE PLATE
2 X 4's @ 16" oc

W3063

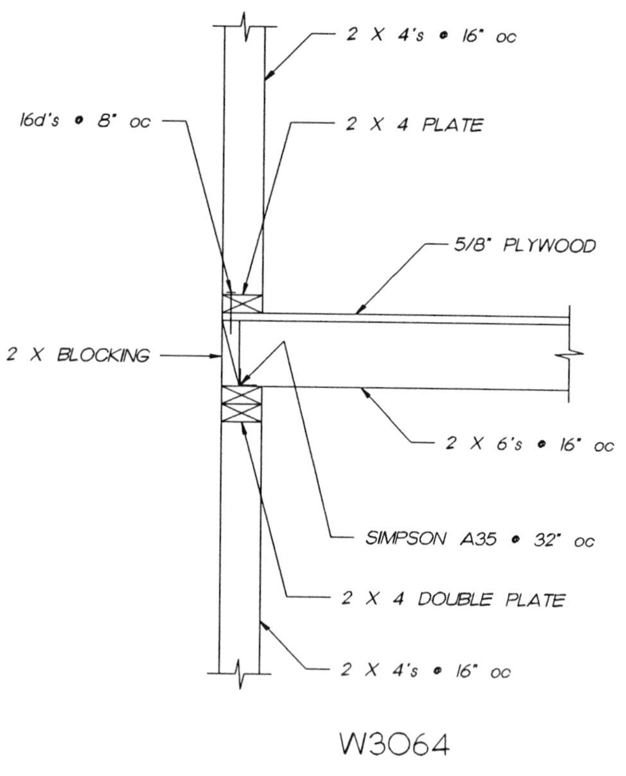

W3064

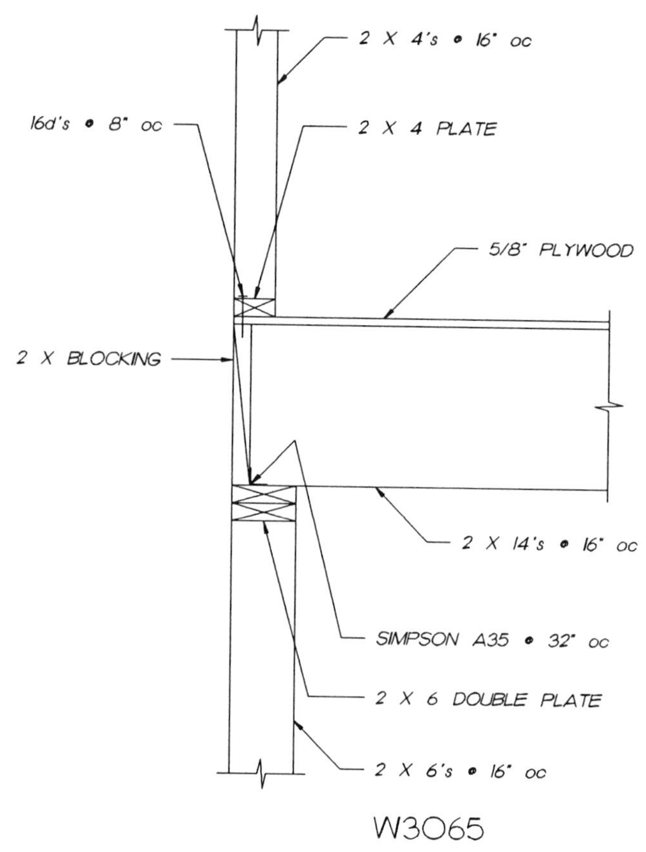

W3065

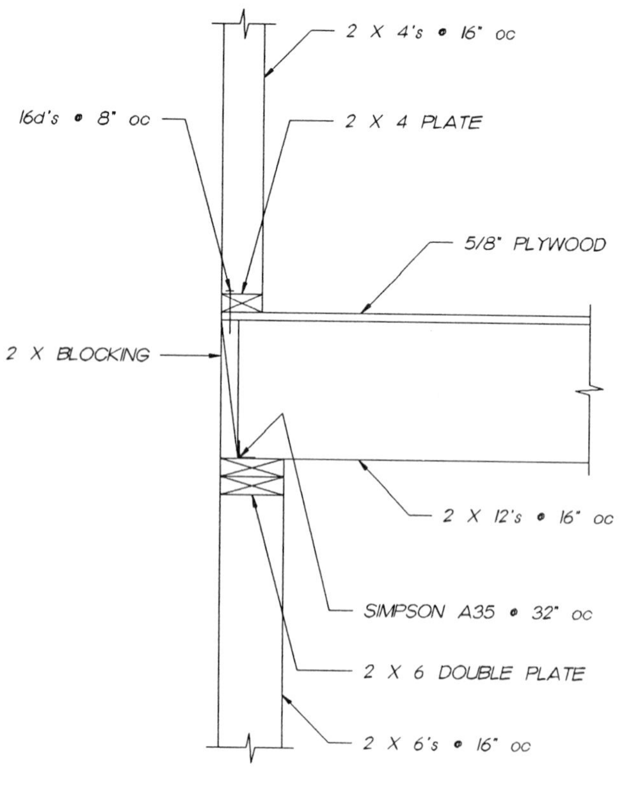

W3066

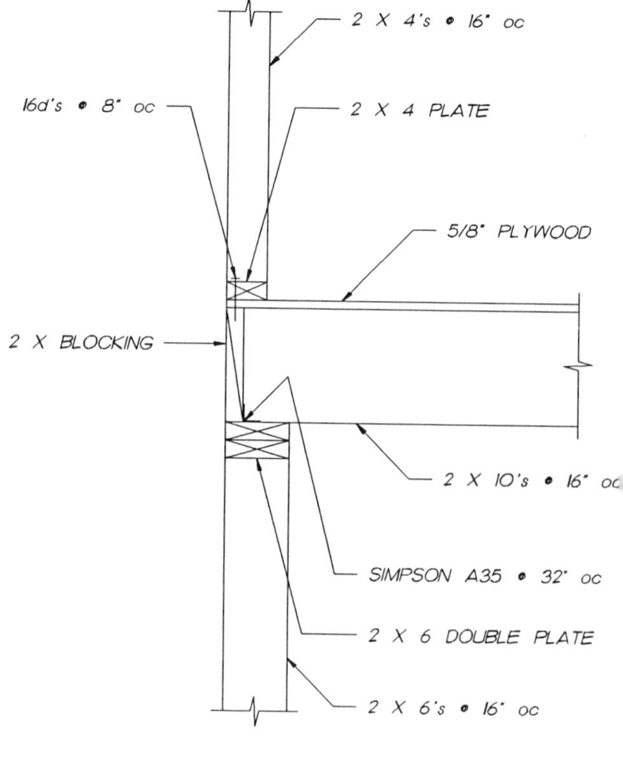

W3067

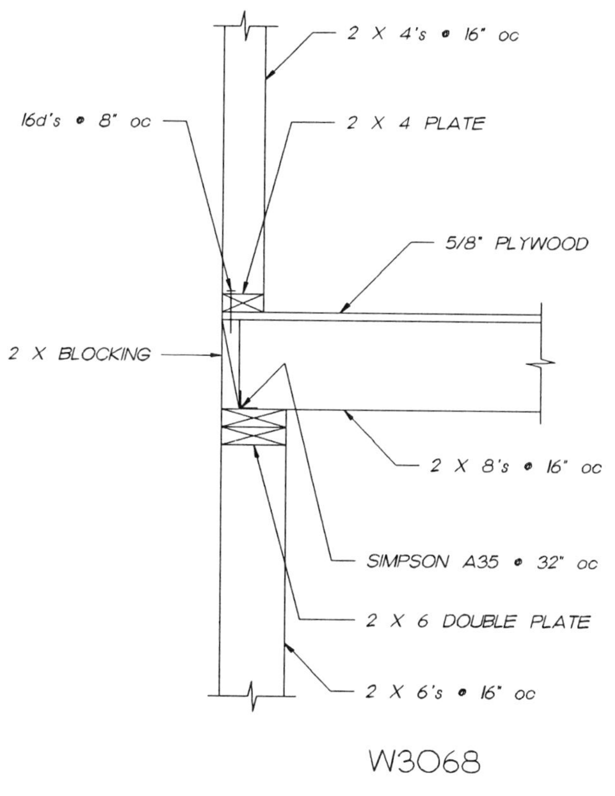

2 X 4's @ 16" oc
16d's @ 8" oc
2 X 4 PLATE
5/8" PLYWOOD
2 X BLOCKING
2 X 8's @ 16" oc
SIMPSON A35 @ 32" oc
2 X 6 DOUBLE PLATE
2 X 6's @ 16" oc

W3068

2 X 4's @ 16" oc
16d's @ 8" oc
2 X 4 PLATE
5/8" PLYWOOD
2 X BLOCKING
2 X 6's @ 16" oc
SIMPSON A35 @ 32" oc
2 X 6 DOUBLE PLATE
2 X 6's @ 16" oc

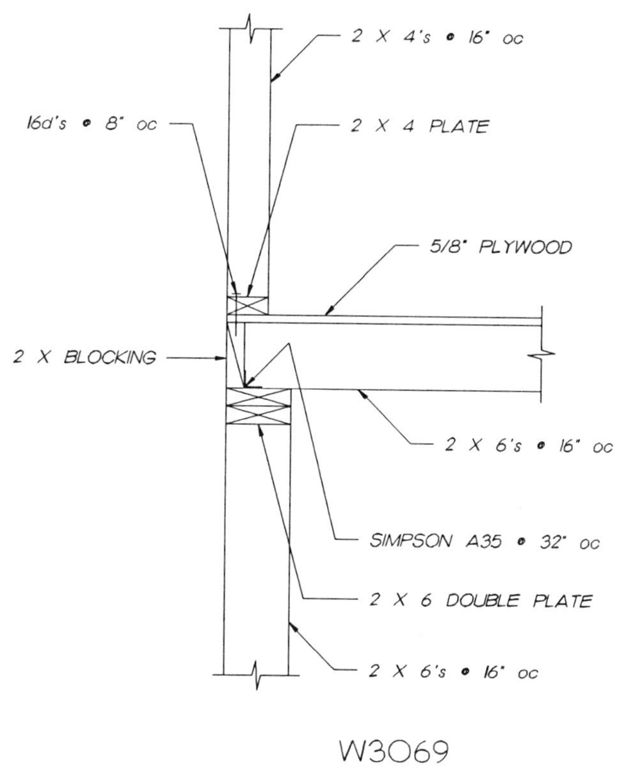

W3069

2 X 4's @ 16" oc
16d's @ 8" oc
2 X 4 PLATE
5/8" PLYWOOD
2 X 14's @ 16" oc
2 X BLOCKING @ 32" oc
SIMPSON A35 @ 32" oc
2 X 4 DOUBLE PLATE
2 X 4's @ 16" oc

W3080

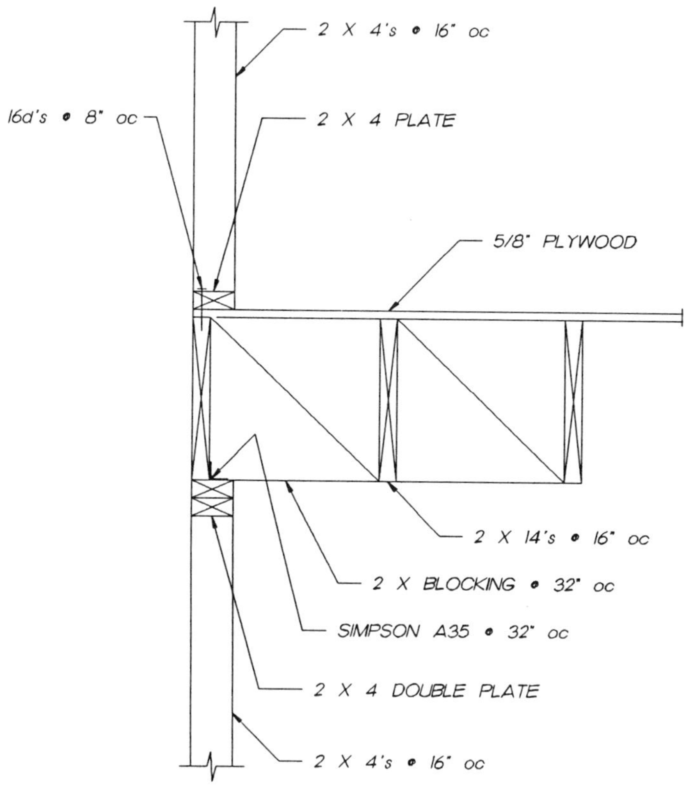

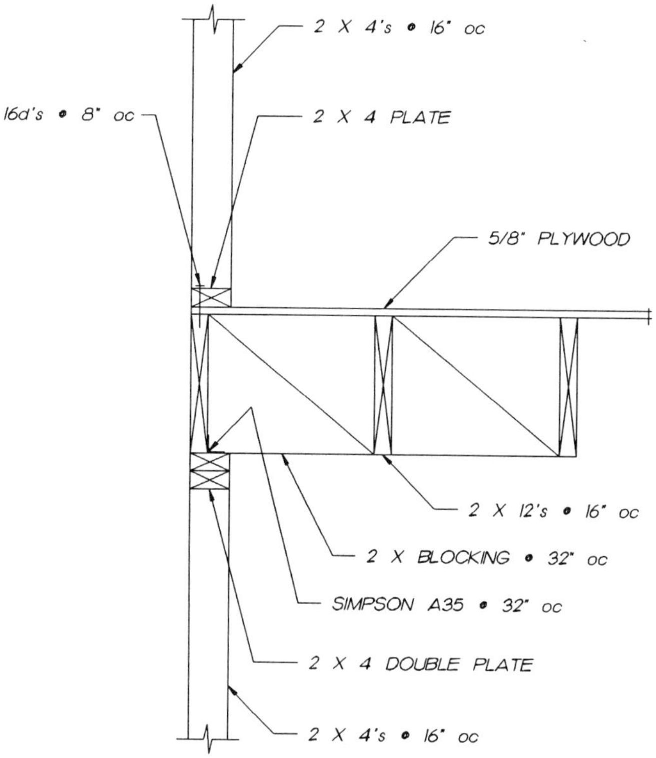

2 X 4's @ 16" oc
16d's @ 8" oc
2 X 4 PLATE
5/8" PLYWOOD
2 X 12's @ 16" oc
2 X BLOCKING @ 32" oc
SIMPSON A35 @ 32" oc
2 X 4 DOUBLE PLATE
2 X 4's @ 16" oc

W3081

2 X 4's @ 16" oc
16d's @ 8" oc
2 X 4 PLATE
5/8" PLYWOOD
2 X 10's @ 16" oc
2 X BLOCKING @ 32" oc
SIMPSON A35 @ 32" oc
2 X 4 DOUBLE PLATE
2 X 4's @ 16" oc

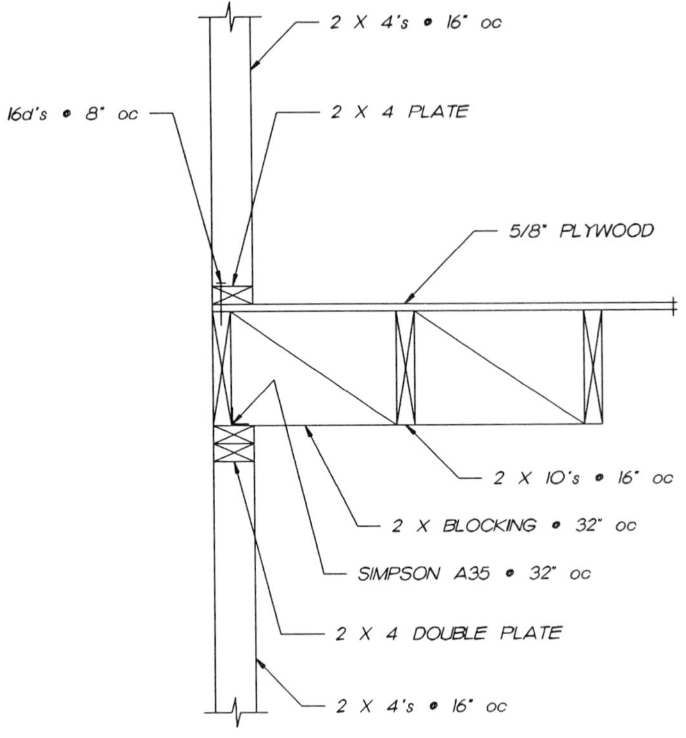

W3082

2 X 4's @ 16" oc
16d's @ 8" oc
2 X 4 PLATE
5/8" PLYWOOD
2 X 8's @ 16" oc
2 X BLOCKING @ 32" oc
SIMPSON A35 @ 32" oc
2 X 4 DOUBLE PLATE
2 X 4's @ 16" oc

W3083

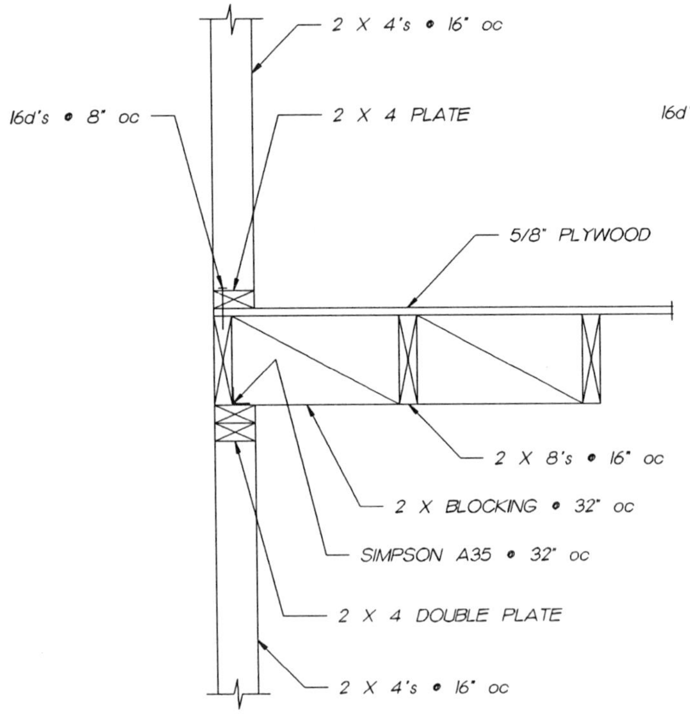

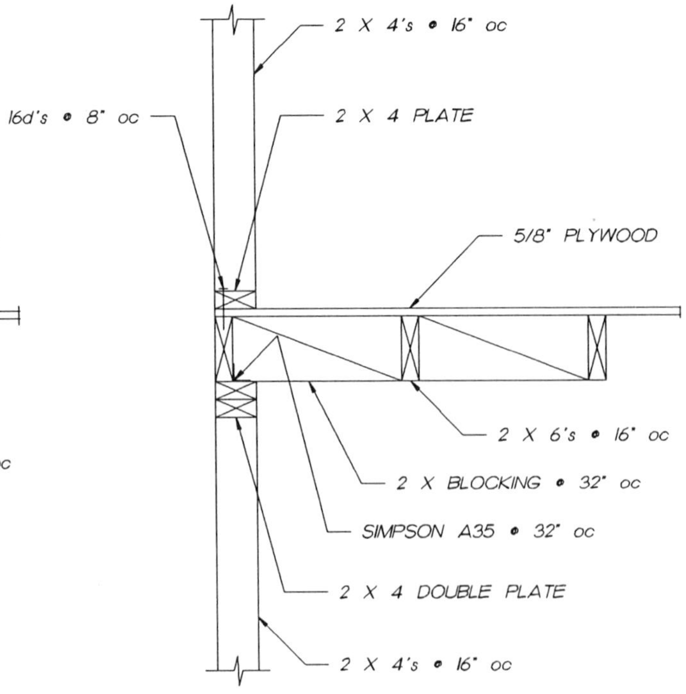

2 X 4's @ 16" oc
16d's @ 8" oc
2 X 4 PLATE
5/8" PLYWOOD
2 X 6's @ 16" oc
2 X BLOCKING @ 32" oc
SIMPSON A35 @ 32" oc
2 X 4 DOUBLE PLATE
2 X 4's @ 16" oc

W3084

2 X 4's @ 16" oc
16d's @ 8" oc
2 X 4 PLATE
5/8" PLYWOOD
2 X 14's @ 16" oc
2 X BLOCKING @ 32" oc
SIMPSON A35 @ 32" oc
2 X 6 DOUBLE PLATE
2 X 6's @ 16" oc

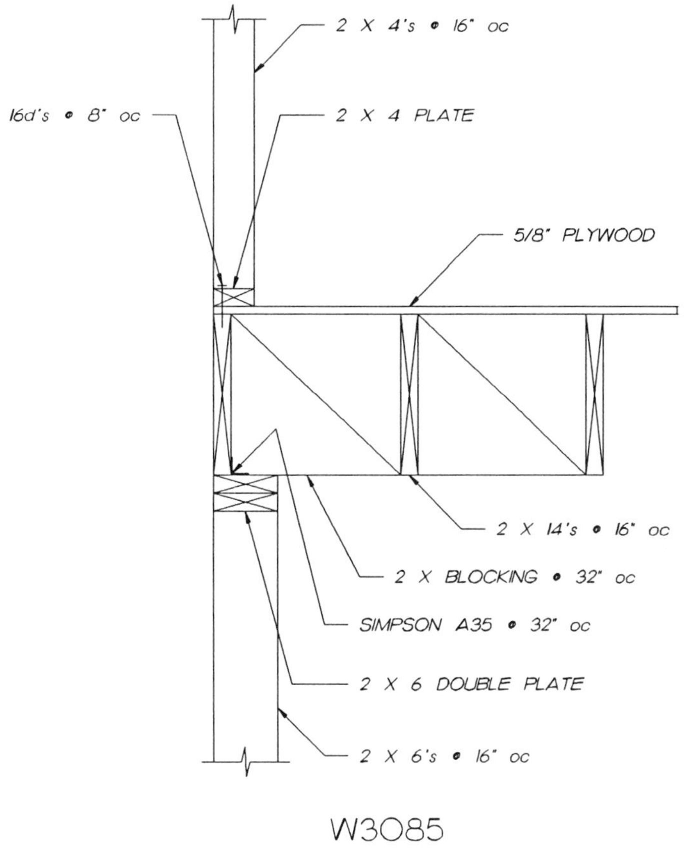

W3085

2 X 4's @ 16" oc
16d's @ 8" oc
2 X 4 PLATE
5/8" PLYWOOD
2 X 12's @ 16" oc
2 X BLOCKING @ 32" oc
SIMPSON A35 @ 32" oc
2 X 6 DOUBLE PLATE
2 X 6's @ 16" oc

W3086

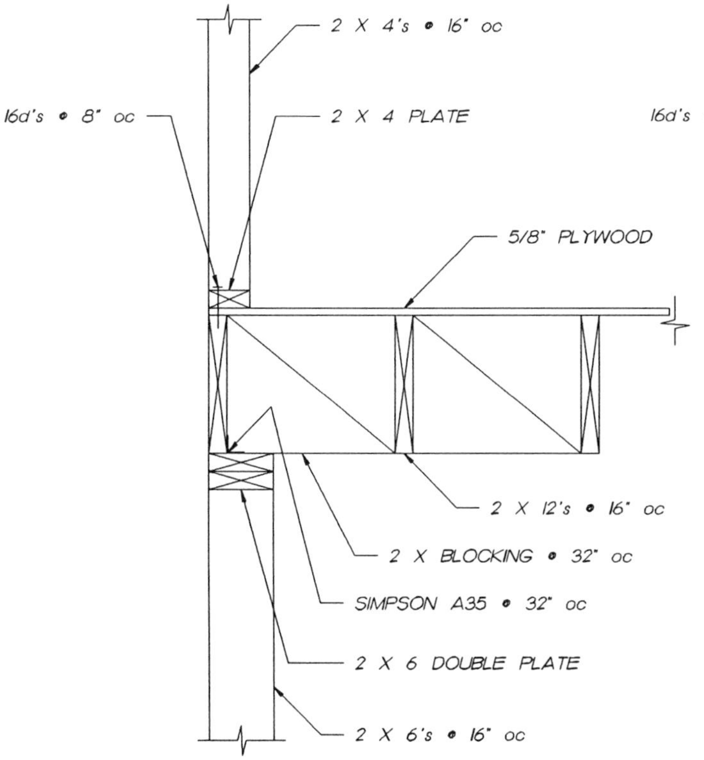

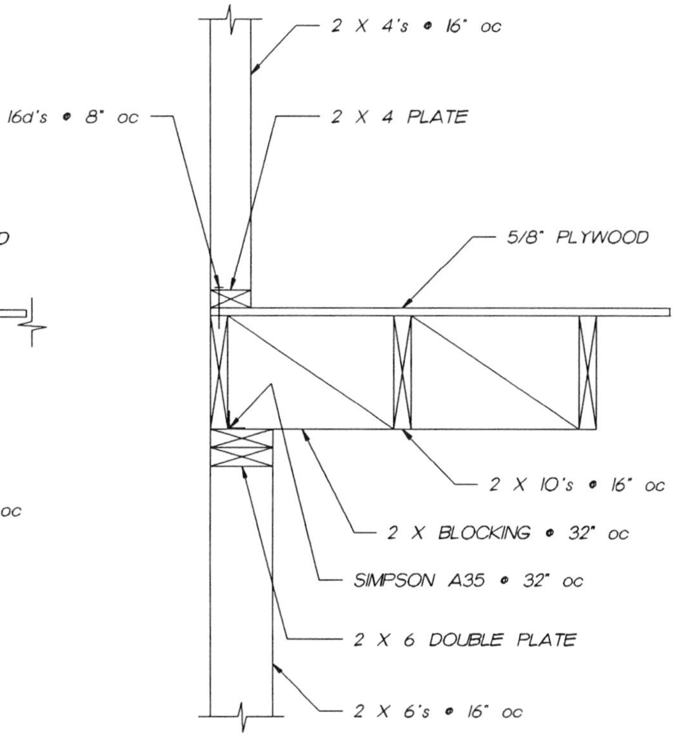

2 X 4's @ 16" oc
16d's @ 8" oc
2 X 4 PLATE
5/8" PLYWOOD
2 X 10's @ 16" oc
2 X BLOCKING @ 32" oc
SIMPSON A35 @ 32" oc
2 X 6 DOUBLE PLATE
2 X 6's @ 16" oc

W3087

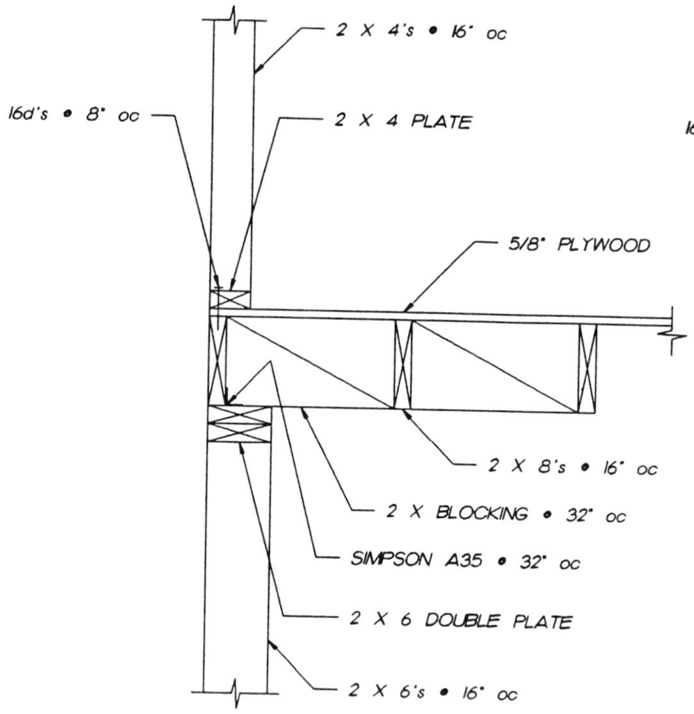

2 X 4's • 16" oc

16d's • 8" oc

2 X 4 PLATE

5/8" PLYWOOD

2 X 8's • 16" oc

2 X BLOCKING • 32" oc

SIMPSON A35 • 32" oc

2 X 6 DOUBLE PLATE

2 X 6's • 16" oc

W3088

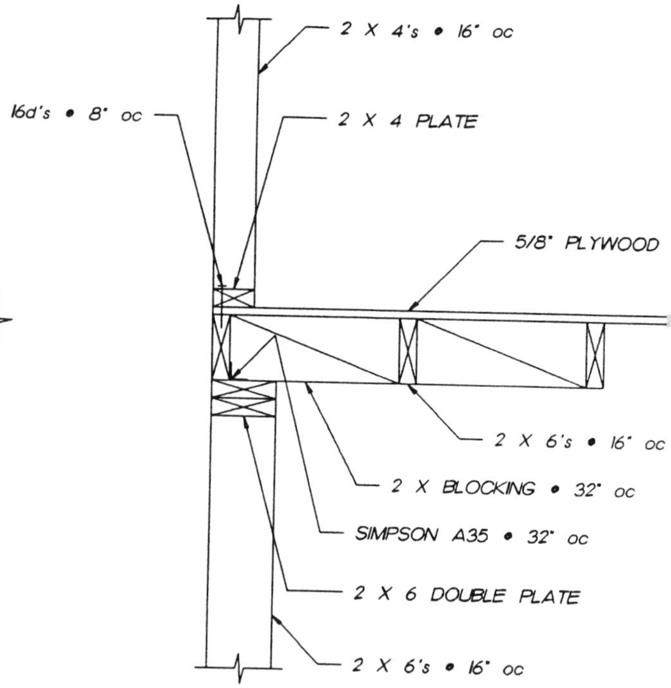

2 X 4's • 16" oc

16d's • 8" oc

2 X 4 PLATE

5/8" PLYWOOD

2 X 6's • 16" oc

2 X BLOCKING • 32" oc

SIMPSON A35 • 32" oc

2 X 6 DOUBLE PLATE

2 X 6's • 16" oc

W3089

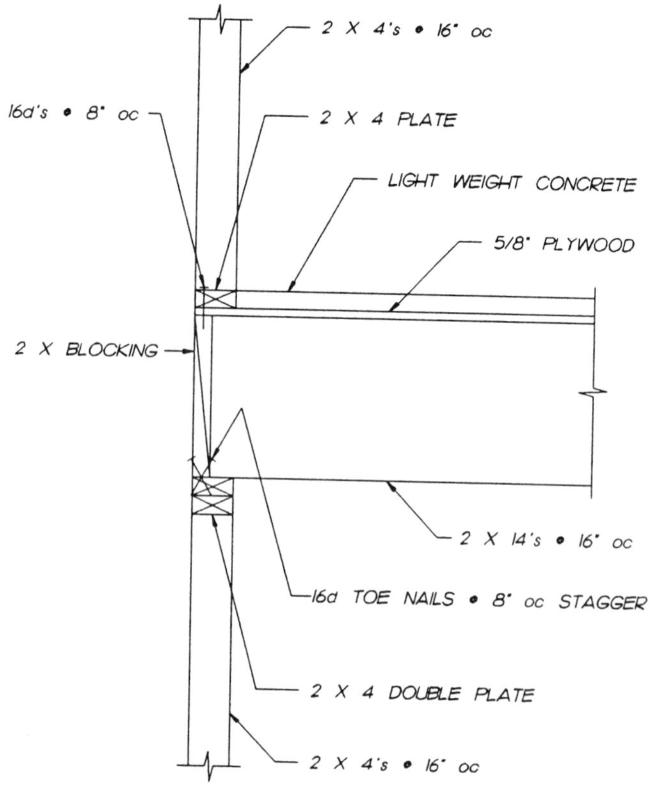

2 X 4's • 16" oc

16d's • 8" oc

2 X 4 PLATE

LIGHT WEIGHT CONCRETE

5/8" PLYWOOD

2 X BLOCKING

2 X 14's • 16" oc

16d TOE NAILS • 8" oc STAGGER

2 X 4 DOUBLE PLATE

2 X 4's • 16" oc

W3100

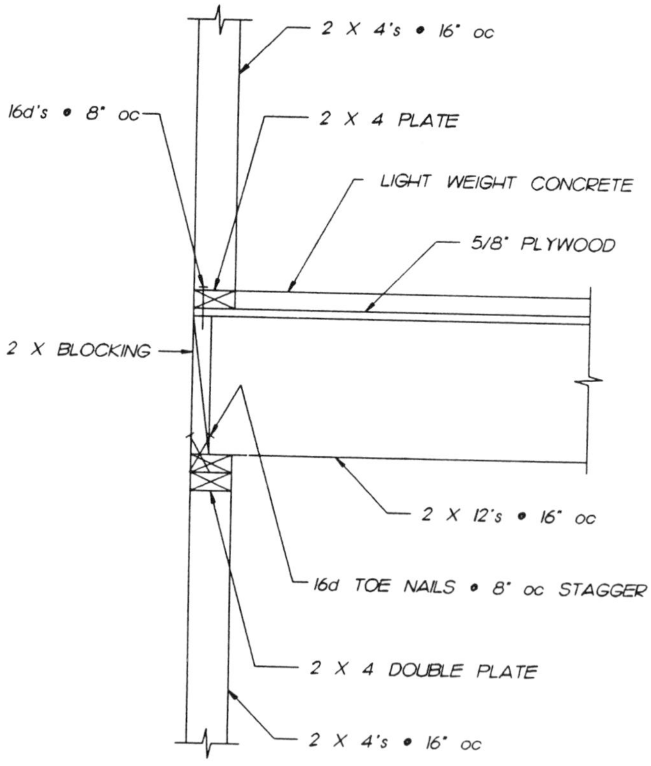

2 X 4's • 16" oc

16d's • 8" oc

2 X 4 PLATE

LIGHT WEIGHT CONCRETE

5/8" PLYWOOD

2 X BLOCKING

2 X 12's • 16" oc

16d TOE NAILS • 8" oc STAGGER

2 X 4 DOUBLE PLATE

2 X 4's • 16" oc

W3102

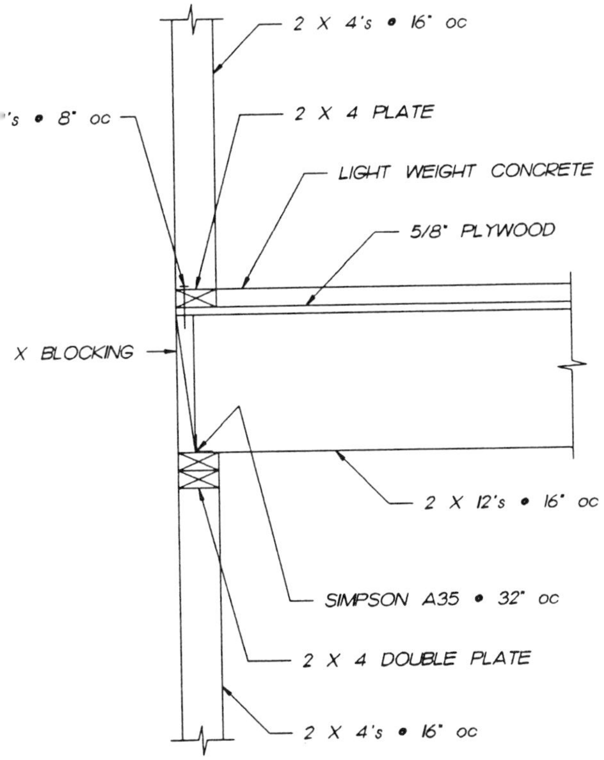

2 X 4's • 16" oc

·'s • 8" oc

2 X 4 PLATE

LIGHT WEIGHT CONCRETE

5/8" PLYWOOD

X BLOCKING

2 X 12's • 16" oc

SIMPSON A35 • 32" oc

2 X 4 DOUBLE PLATE

2 X 4's • 16" oc

W3102

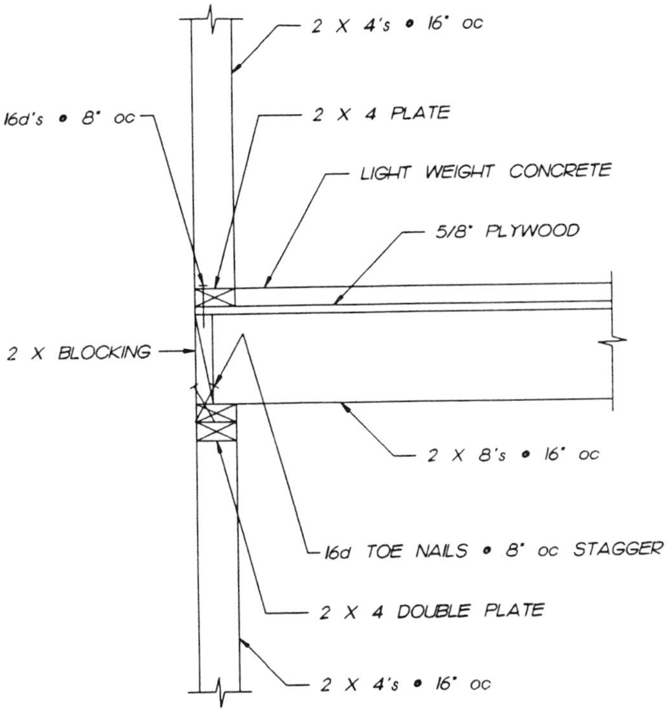

2 X 4's • 16" oc

16d's • 8" oc

2 X 4 PLATE

LIGHT WEIGHT CONCRETE

5/8" PLYWOOD

2 X BLOCKING

2 X 8's • 16" oc

16d TOE NAILS • 8" oc STAGGER

2 X 4 DOUBLE PLATE

2 X 4's • 16" oc

W3103

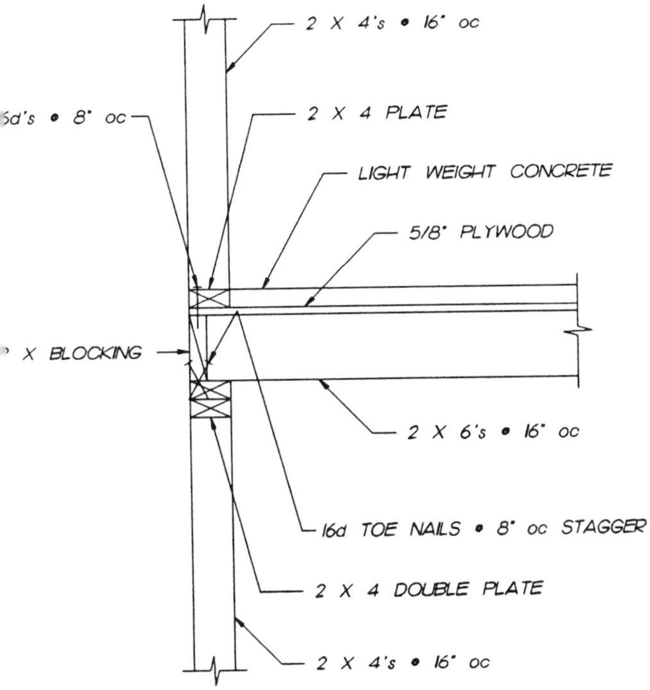

2 X 4's • 16" oc

6d's • 8" oc

2 X 4 PLATE

LIGHT WEIGHT CONCRETE

5/8" PLYWOOD

· X BLOCKING

2 X 6's • 16" oc

16d TOE NAILS • 8" oc STAGGER

2 X 4 DOUBLE PLATE

2 X 4's • 16" oc

W3104

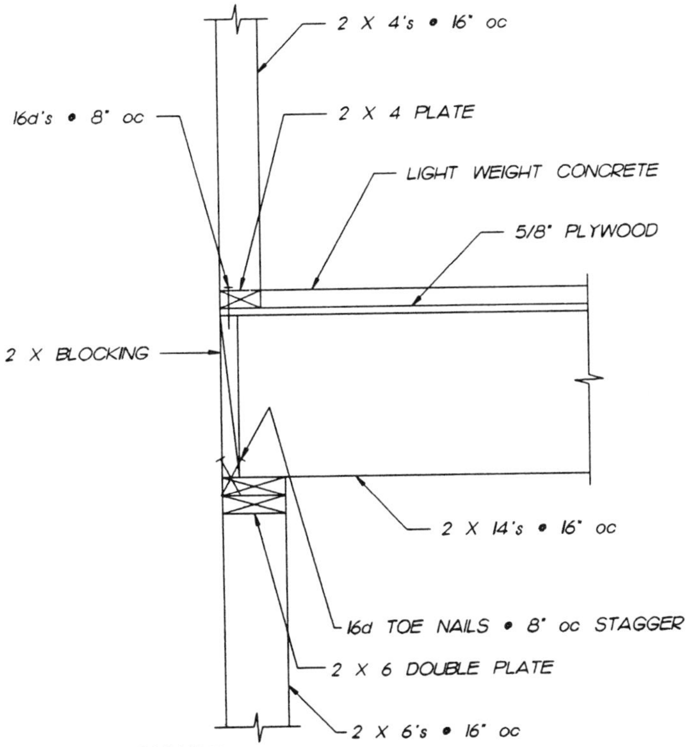

2 X 4's • 16" oc

16d's • 8" oc

2 X 4 PLATE

LIGHT WEIGHT CONCRETE

5/8" PLYWOOD

2 X BLOCKING

2 X 14's • 16" oc

16d TOE NAILS • 8" oc STAGGER

2 X 6 DOUBLE PLATE

2 X 6's • 16" oc

W3105

2 X 4's • 16" oc
16d's • 8" oc
2 X 4 PLATE
LIGHT WEIGHT CONCRETE
5/8" PLYWOOD
2 X BLOCKING
2 X 12's • 16" oc
16d TOE NAILS • 8" oc STAGGER
2 X 6 DOUBLE PLATE
2 X 6's • 16" oc

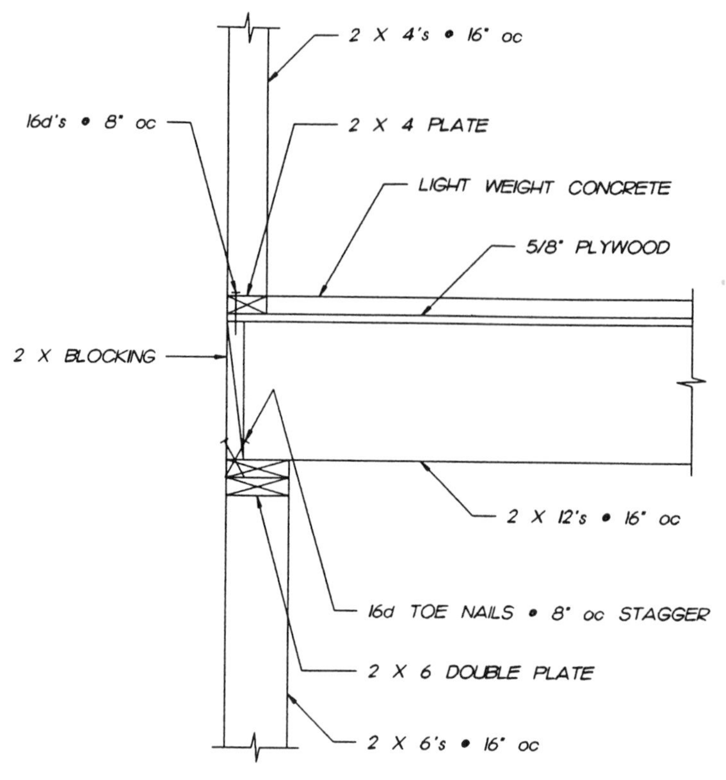

W3106

2 X 4's • 16" oc
16d's • 8" oc
2 X 4 PLATE
LIGHT WEIGHT CONCRETE
5/8" PLYWOOD
2 X BLOCKING
2 X 10's • 16" oc
16d TOE NAILS • 8" oc STAGGER
2 X 6 DOUBLE PLATE
2 X 6's • 16" oc

W3107

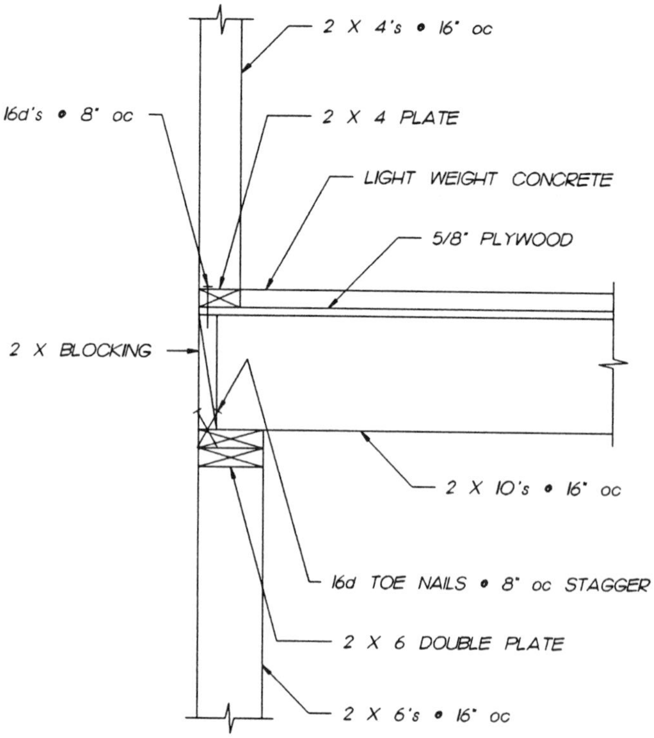

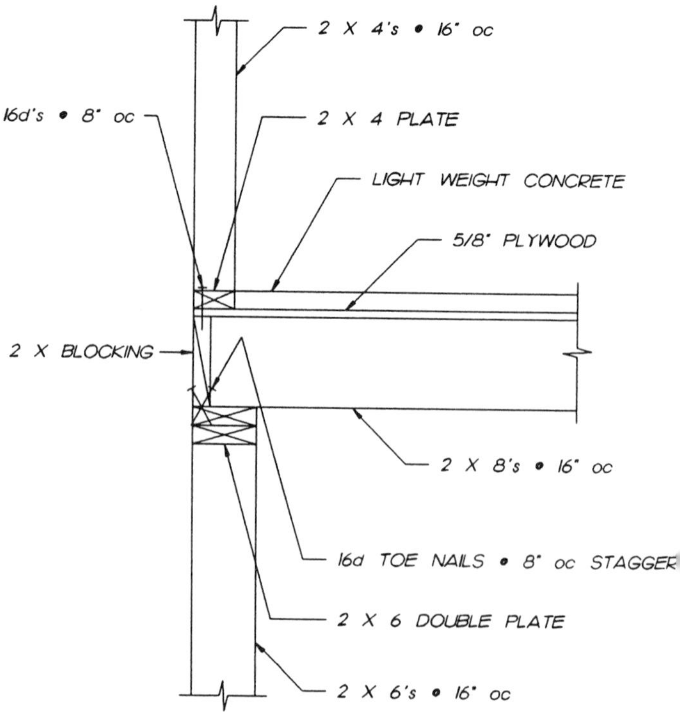

2 X 4's • 16" oc
16d's • 8" oc
2 X 4 PLATE
LIGHT WEIGHT CONCRETE
5/8" PLYWOOD
2 X BLOCKING
2 X 8's • 16" oc
16d TOE NAILS • 8" oc STAGGER
2 X 6 DOUBLE PLATE
2 X 6's • 16" oc

W3108

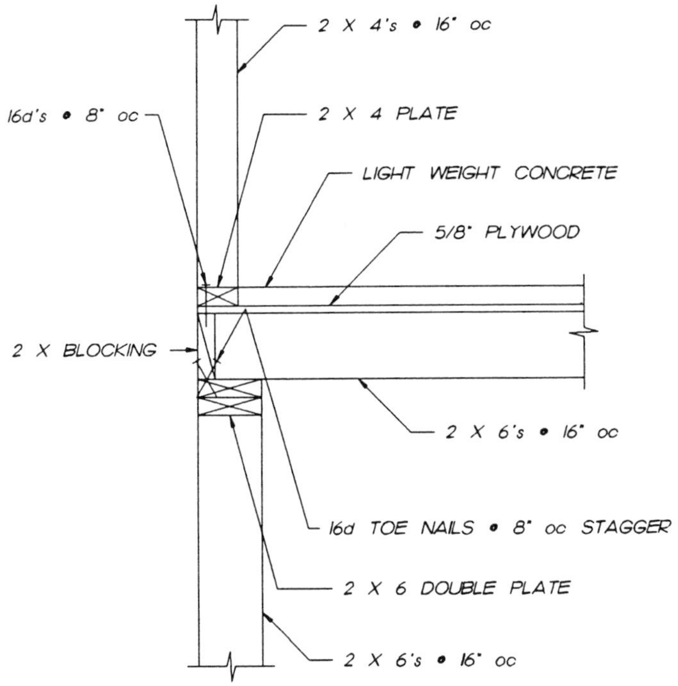

2 X 4's • 16" oc
16d's • 8" oc
2 X 4 PLATE
LIGHT WEIGHT CONCRETE
5/8" PLYWOOD
2 X BLOCKING
2 X 6's • 16" oc
16d TOE NAILS • 8" oc STAGGER
2 X 6 DOUBLE PLATE
2 X 6's • 16" oc

W3109

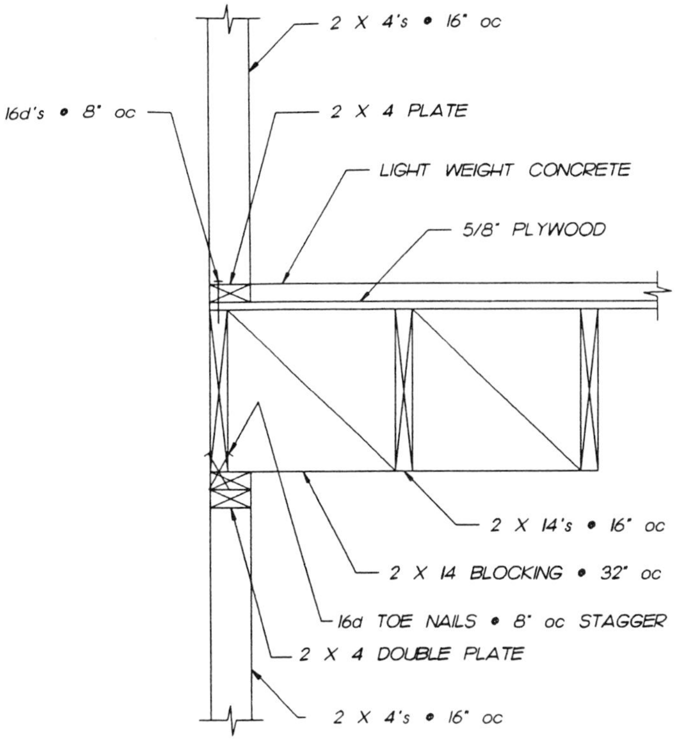

2 X 4's • 16" oc
16d's • 8" oc
2 X 4 PLATE
LIGHT WEIGHT CONCRETE
5/8" PLYWOOD
2 X 14's • 16" oc
2 X 14 BLOCKING • 32" oc
16d TOE NAILS • 8" oc STAGGER
2 X 4 DOUBLE PLATE
2 X 4's • 16" oc

W3120

2 X 4's • 16" oc
16d's • 8" oc
2 X 4 PLATE
LIGHT WEIGHT CONCRETE
5/8" PLYWOOD
2 X 12's • 16" oc
2 X BLOCKING • 32" oc
16d TOE NAILS • 8" oc STAGGER
2 X 4 DOUBLE PLATE
2 X 4's • 16" oc

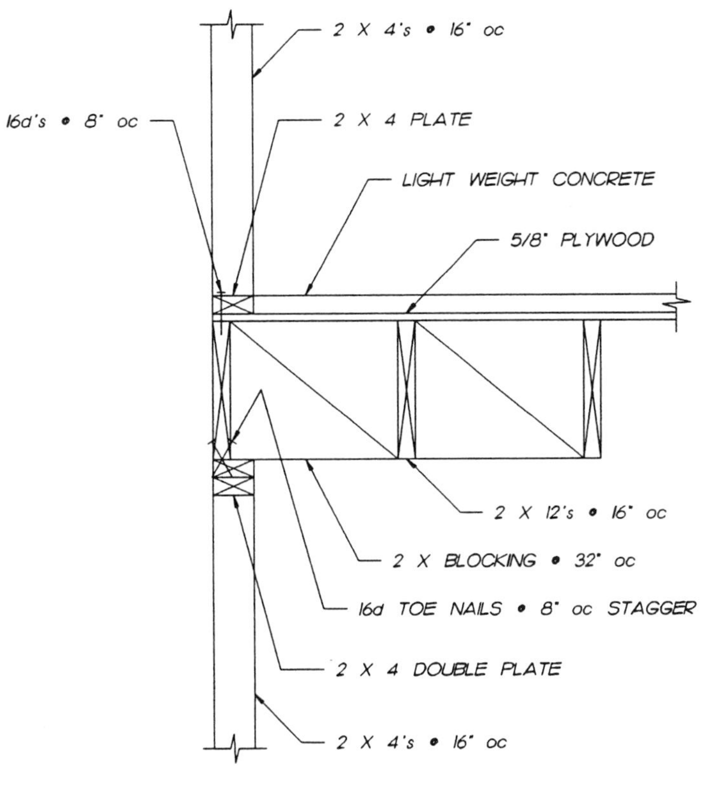

W3121

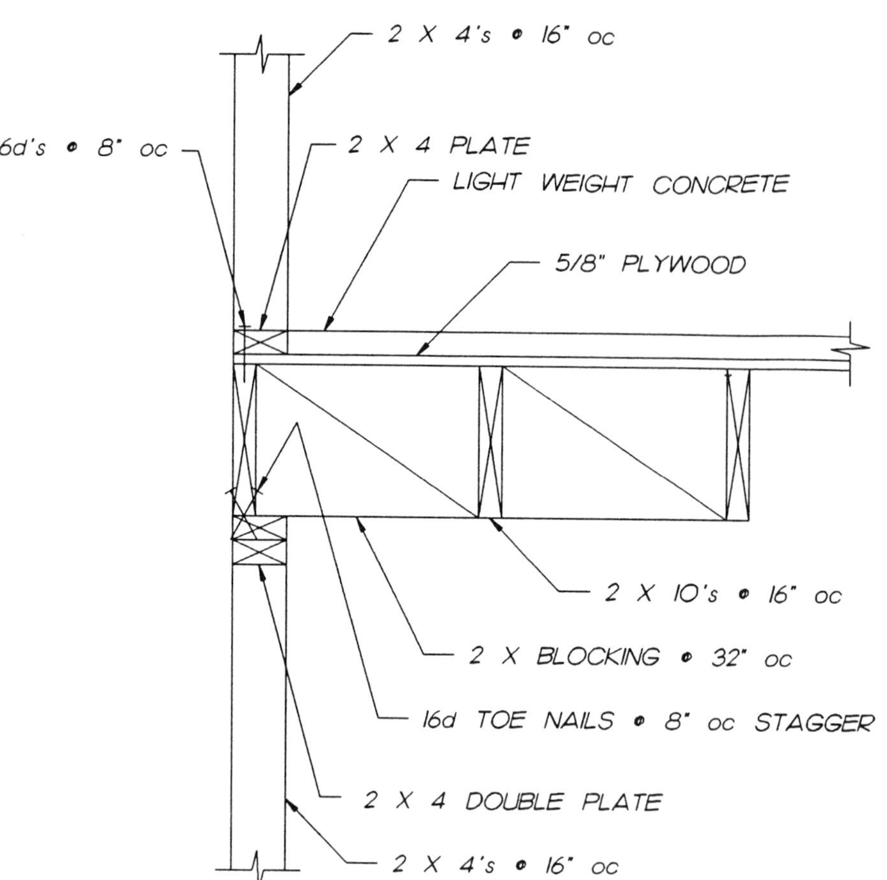
2 X 4's @ 16" oc
16d's @ 8" oc
2 X 4 PLATE
LIGHT WEIGHT CONCRETE
5/8" PLYWOOD
2 X 10's @ 16" oc
2 X BLOCKING @ 32" oc
16d TOE NAILS @ 8" oc STAGGER
2 X 4 DOUBLE PLATE
2 X 4's @ 16" oc

W3122

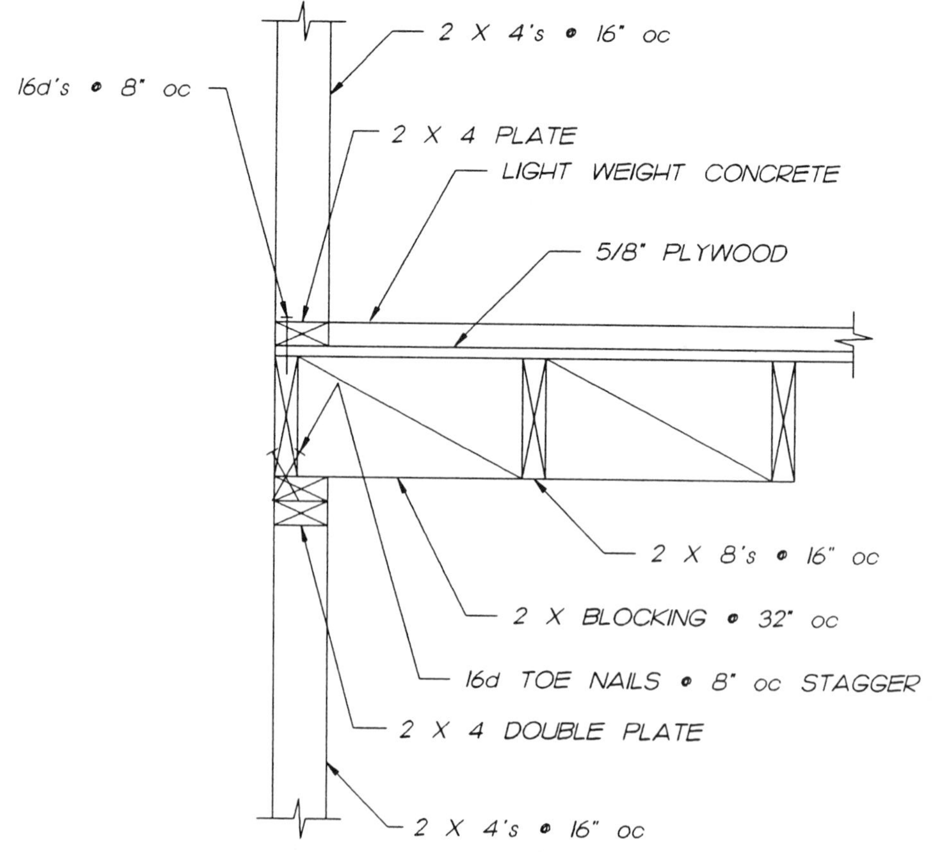
2 X 4's @ 16" oc
16d's @ 8" oc
2 X 4 PLATE
LIGHT WEIGHT CONCRETE
5/8" PLYWOOD
2 X 8's @ 16" oc
2 X BLOCKING @ 32" oc
16d TOE NAILS @ 8" oc STAGGER
2 X 4 DOUBLE PLATE
2 X 4's @ 16" oc

W3123

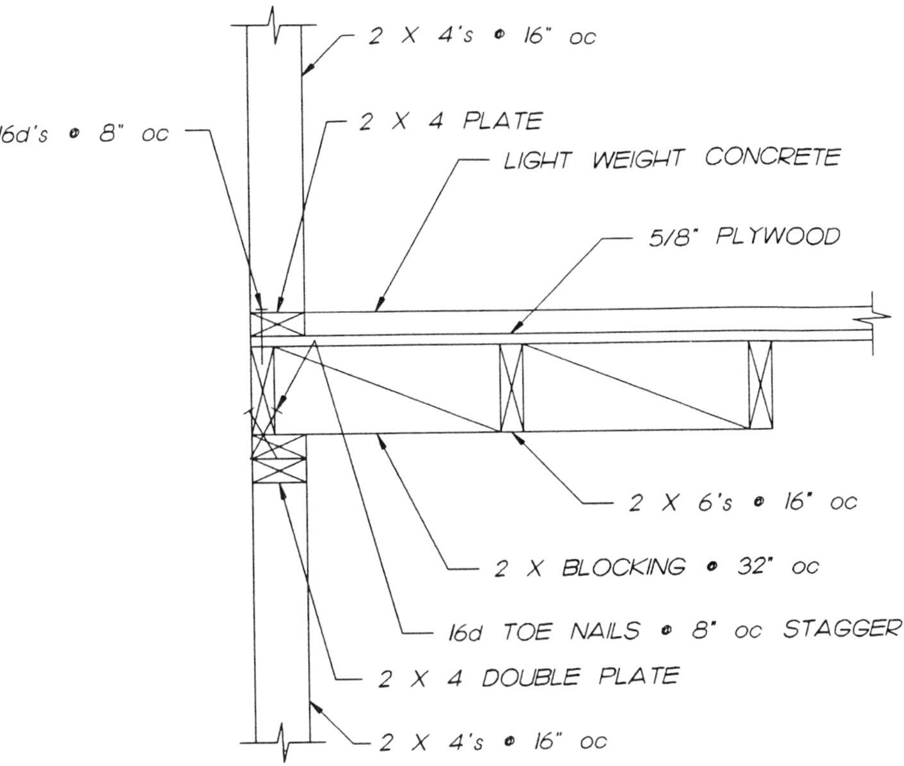
2 X 4's @ 16" oc
16d's @ 8" oc
2 X 4 PLATE
LIGHT WEIGHT CONCRETE
5/8" PLYWOOD
2 X 6's @ 16" oc
2 X BLOCKING @ 32" oc
16d TOE NAILS @ 8" oc STAGGER
2 X 4 DOUBLE PLATE
2 X 4's @ 16" oc

W3124

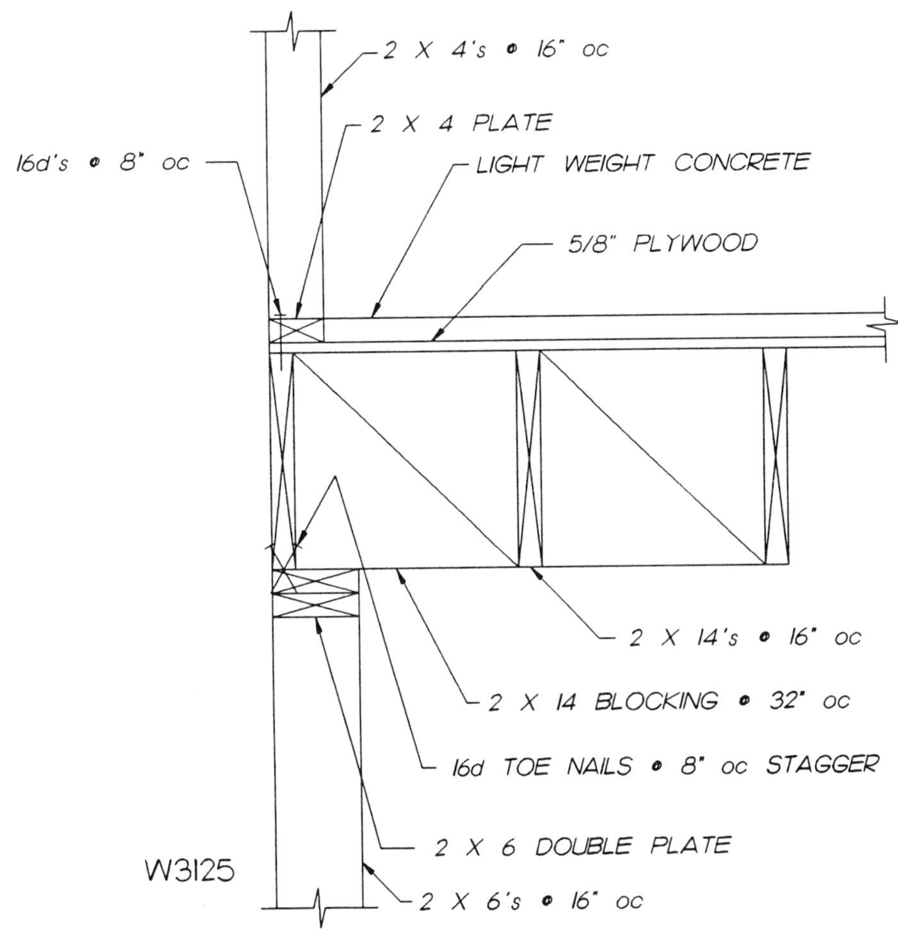
2 X 4's @ 16" oc
2 X 4 PLATE
16d's @ 8" oc
LIGHT WEIGHT CONCRETE
5/8" PLYWOOD
2 X 14's @ 16" oc
2 X 14 BLOCKING @ 32" oc
16d TOE NAILS @ 8" oc STAGGER
2 X 6 DOUBLE PLATE
2 X 6's @ 16" oc

W3125

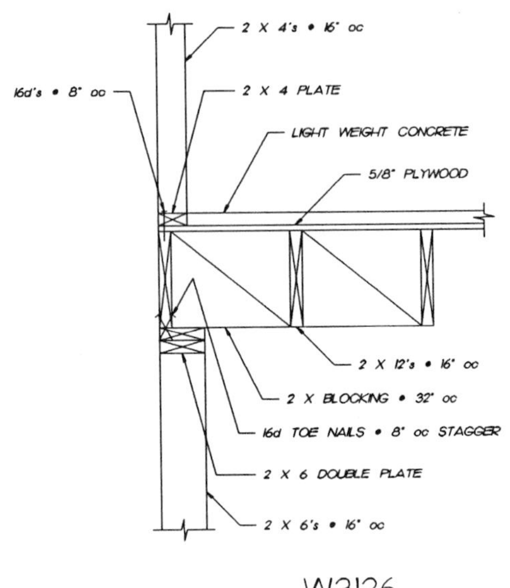

W3126

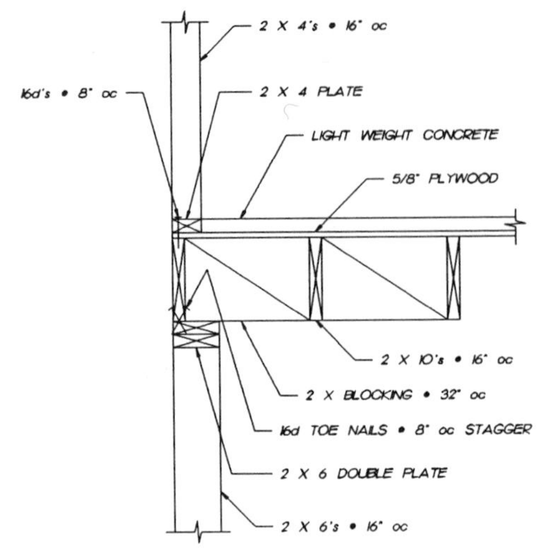

W3127

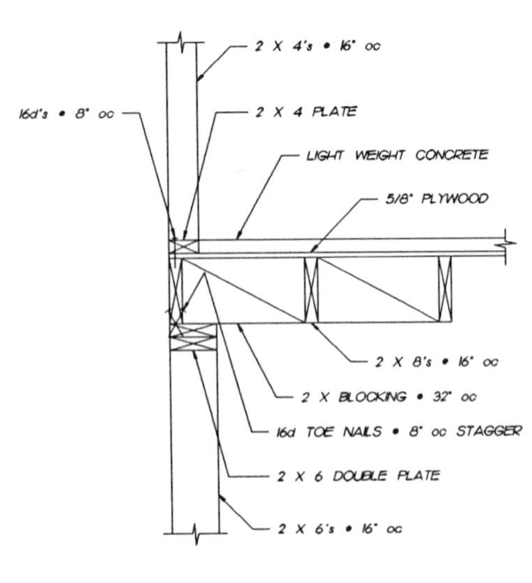

W3128

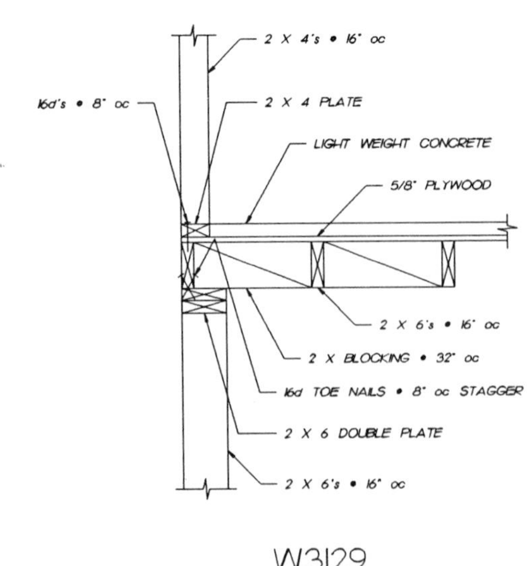

W3129

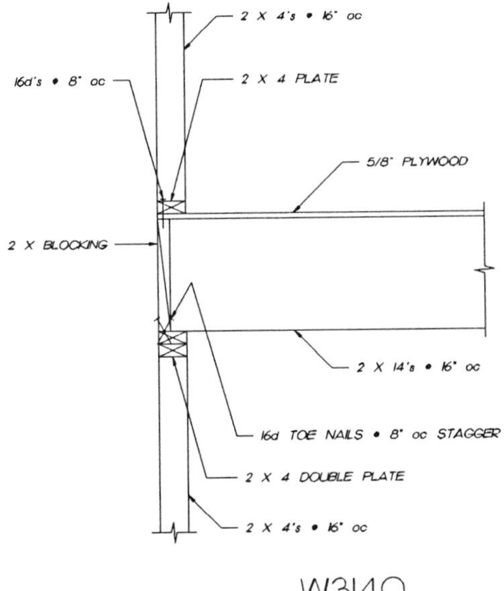

W3140

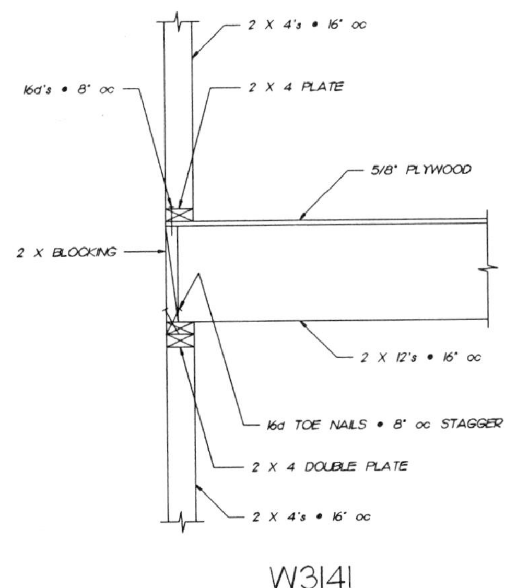

W3141

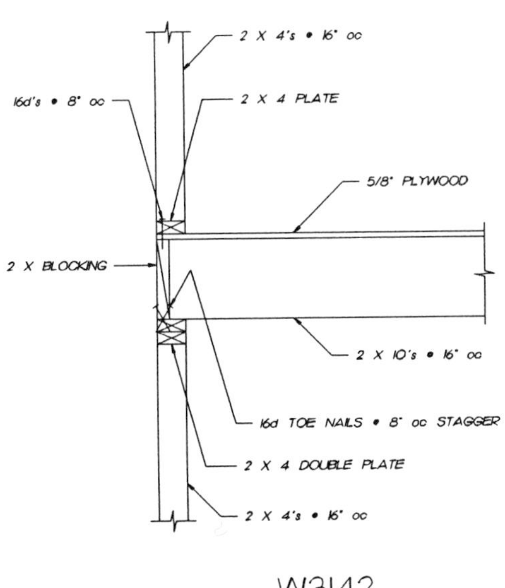

W3142

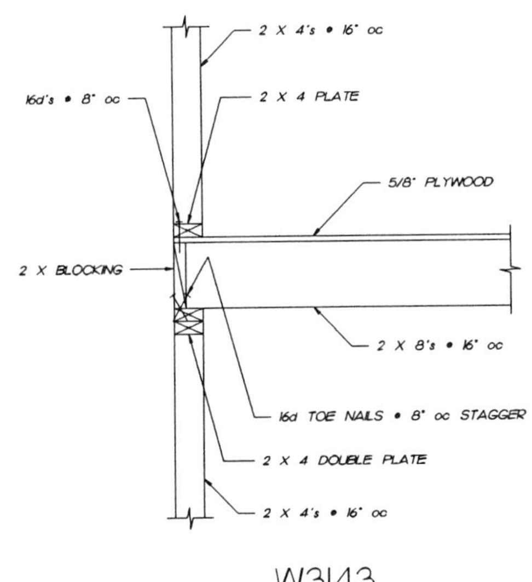

W3143

2 X 4's @ 16" oc
16d's @ 8" oc
2 X 4 PLATE
5/8" PLYWOOD
2 X BLOCKING
2 X 6's @ 16" oc
16d TOE NAILS @ 8" oc STAGGER
2 X 4 DOUBLE PLATE
2 X 4's @ 16" oc

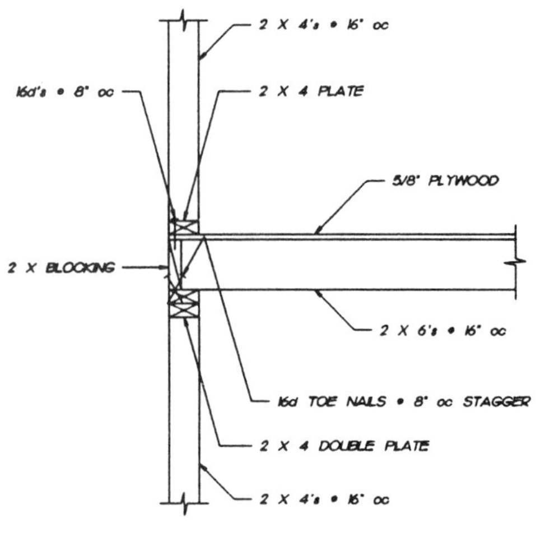

W3144

2 X 4's @ 16" oc
16d's @ 8" oc
2 X 4 PLATE
5/8" PLYWOOD
2 X BLOCKING
2 X 14's @ 16" oc
16d TOE NAILS @ 8" oc STAGGER
2 X 6 DOUBLE PLATE
2 X 6's @ 16" oc

W3145

2 X 4's @ 16" oc
16d's @ 8" oc
2 X 4 PLATE
5/8" PLYWOOD
2 X BLOCKING
2 X 12's @ 16" oc
16d TOE NAILS @ 8" oc STAGGER
2 X 6 DOUBLE PLATE
2 X 6's @ 16" oc

W3146

2 X 4's @ 16" oc
16d's @ 8" oc
2 X 4 PLATE
5/8" PLYWOOD
2 X BLOCKING
2 X 10's @ 16" oc
16d TOE NAILS @ 8" oc STAGGER
2 X 6 DOUBLE PLATE
2 X 6's @ 16" oc

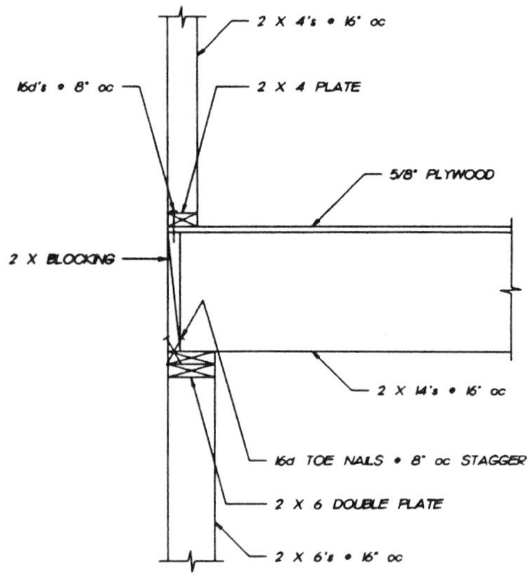

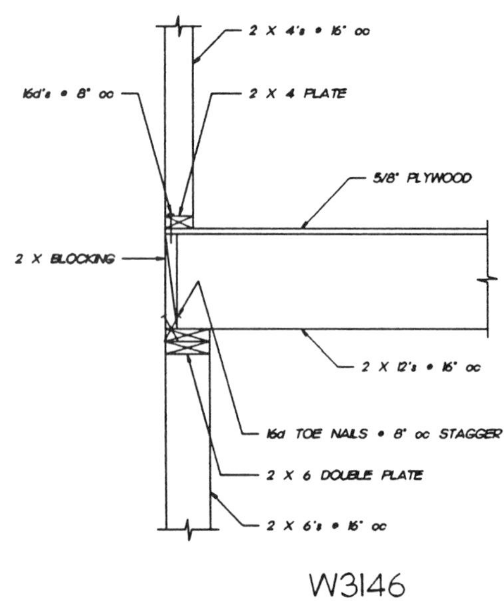

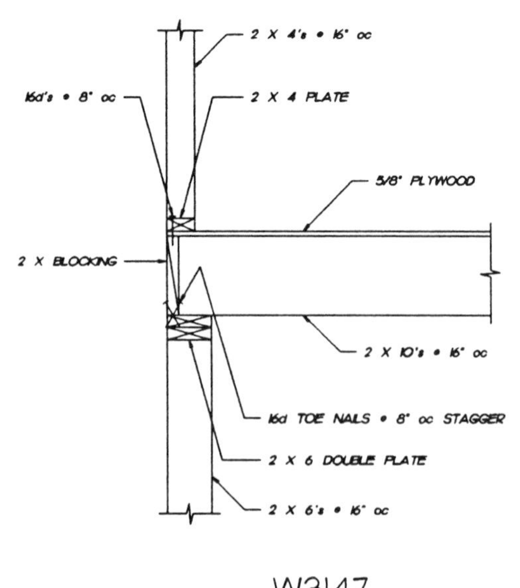

W3147

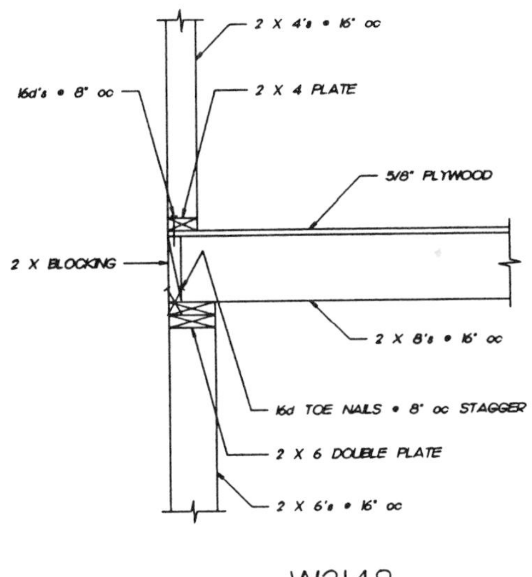

W3148

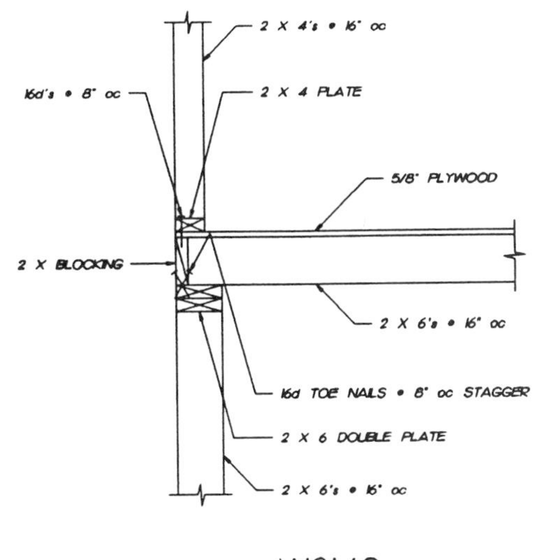

W3149

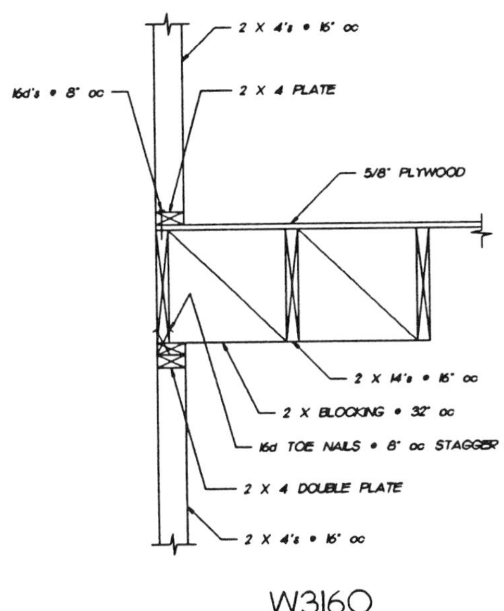

W3160

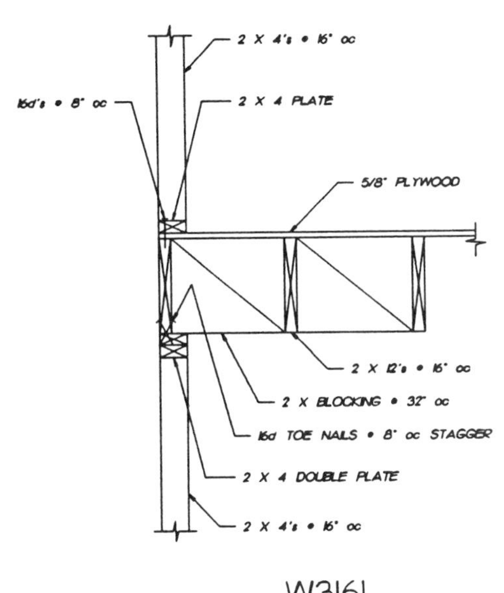

W3161

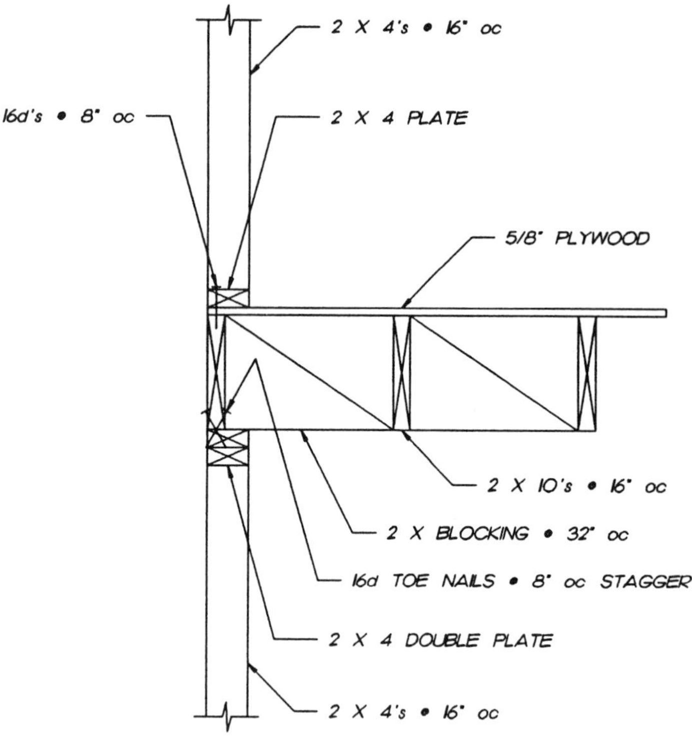
2 X 4's • 16" oc
16d's • 8" oc
2 X 4 PLATE
5/8" PLYWOOD
2 X 10's • 16" oc
2 X BLOCKING • 32" oc
16d TOE NAILS • 8" oc STAGGER
2 X 4 DOUBLE PLATE
2 X 4's • 16" oc

W3162

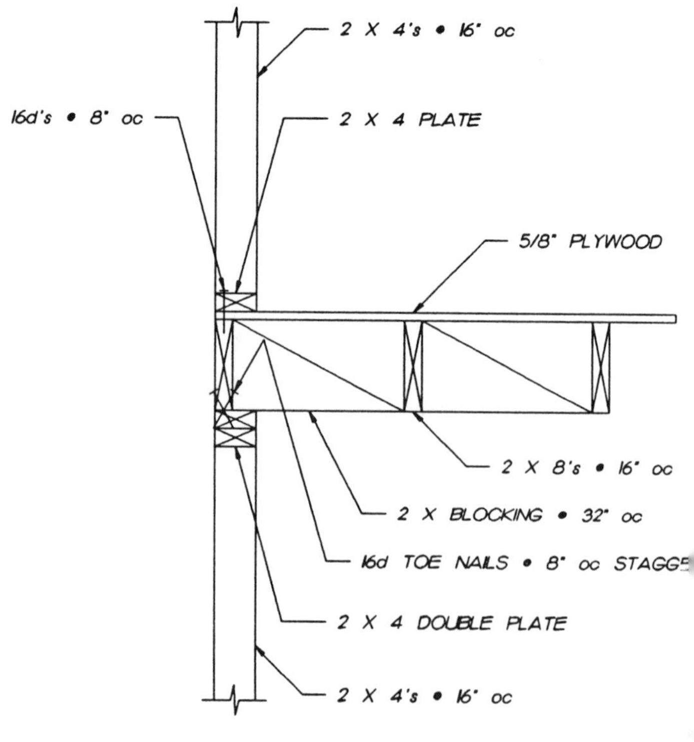
2 X 4's • 16" oc
16d's • 8" oc
2 X 4 PLATE
5/8" PLYWOOD
2 X 8's • 16" oc
2 X BLOCKING • 32" oc
16d TOE NAILS • 8" oc STAGG
2 X 4 DOUBLE PLATE
2 X 4's • 16" oc

W3163

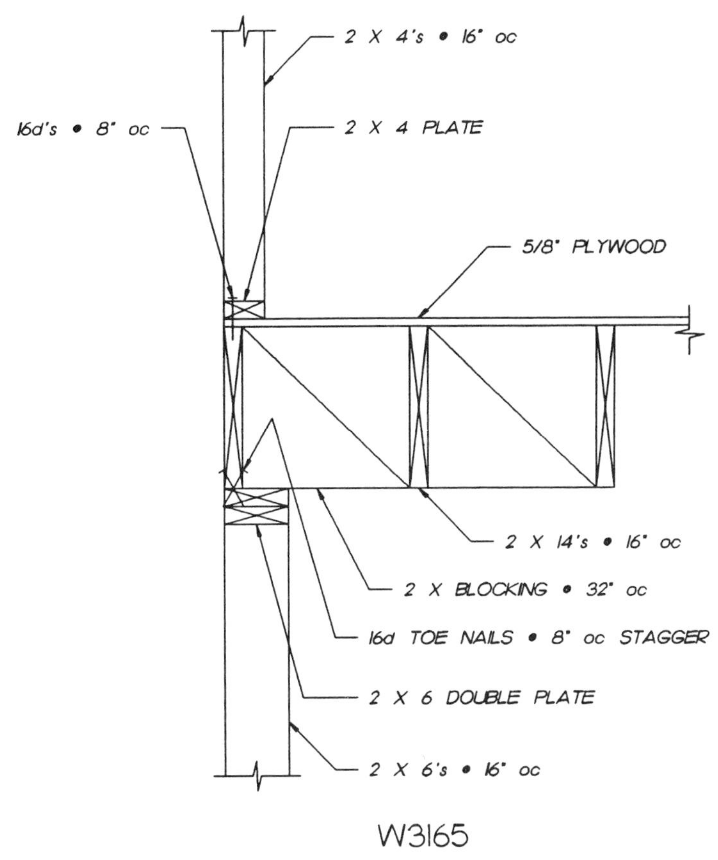
2 X 4's • 16" oc
16d's • 8" oc
2 X 4 PLATE
5/8" PLYWOOD
2 X 14's • 16" oc
2 X BLOCKING • 32" oc
16d TOE NAILS • 8" oc STAGGER
2 X 6 DOUBLE PLATE
2 X 6's • 16" oc

W3165

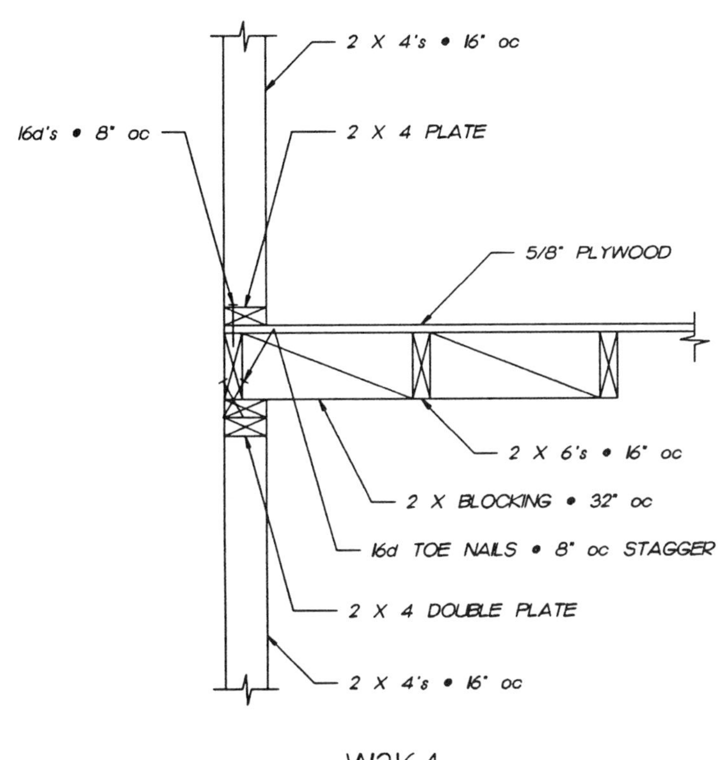
2 X 4's • 16" oc
16d's • 8" oc
2 X 4 PLATE
5/8" PLYWOOD
2 X 6's • 16" oc
2 X BLOCKING • 32" oc
16d TOE NAILS • 8" oc STAGGER
2 X 4 DOUBLE PLATE
2 X 4's • 16" oc

W3164

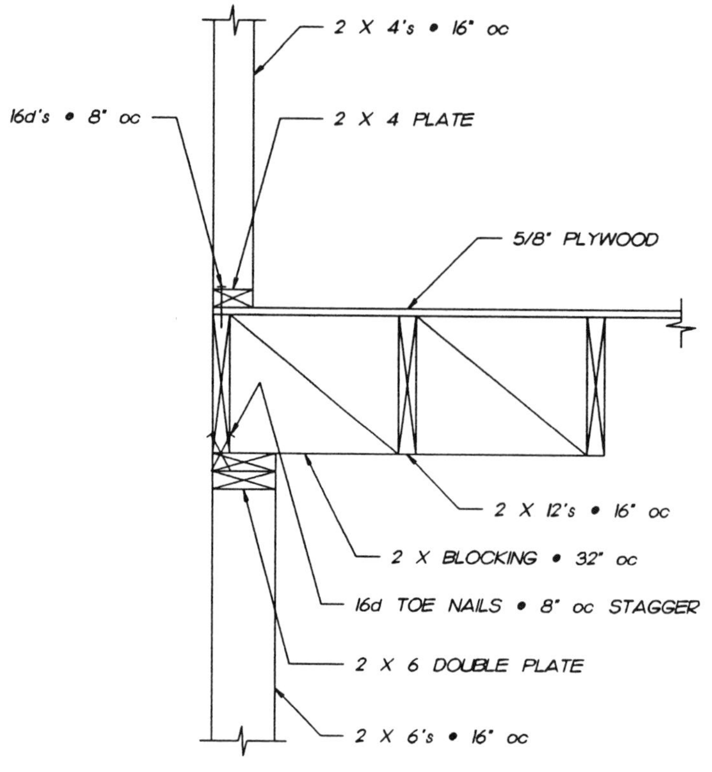

W3166

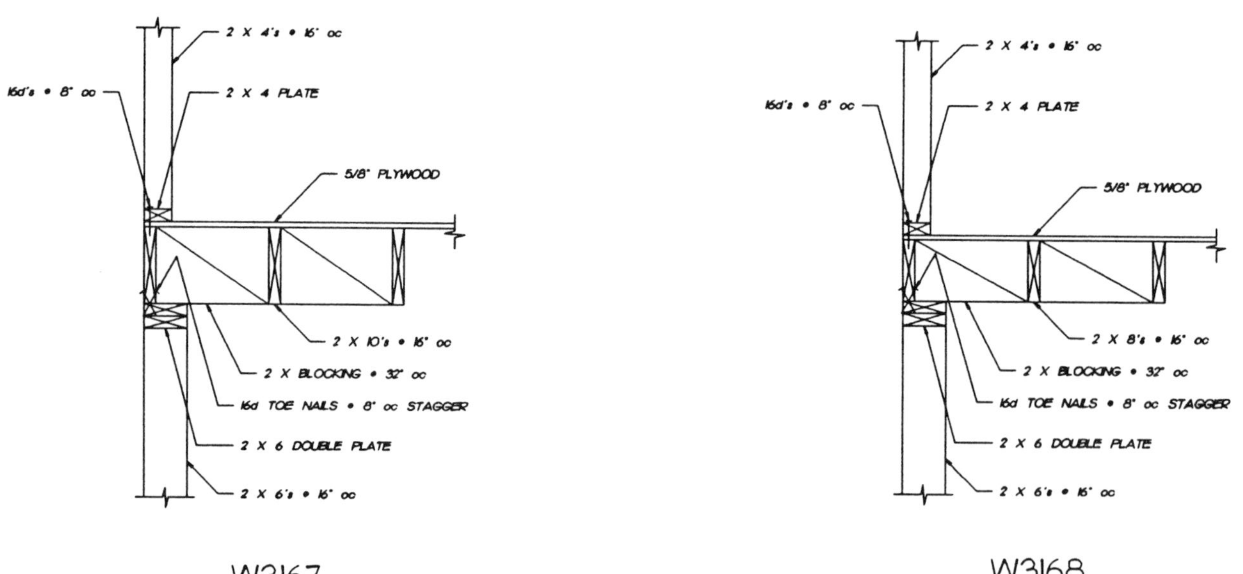

W3167

W3168

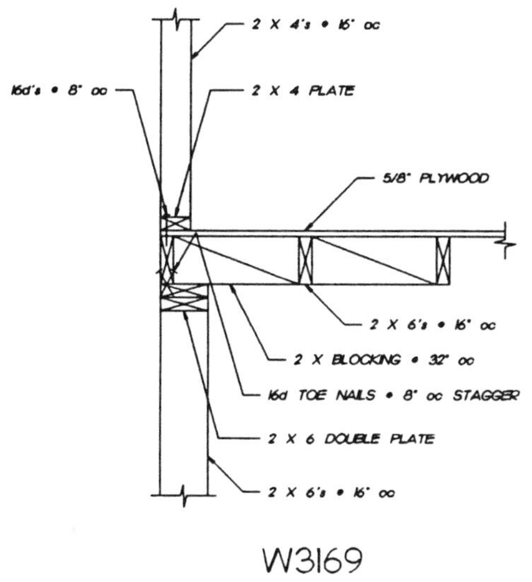

W3169

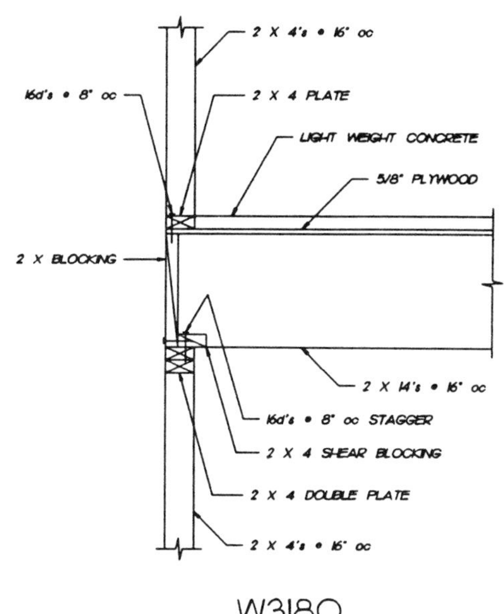

W3180

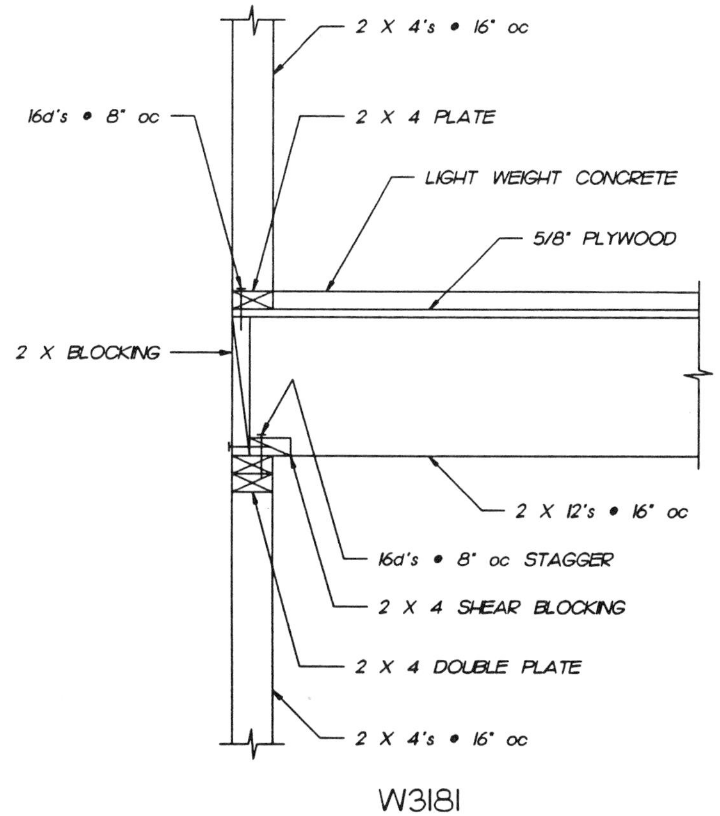

W3181

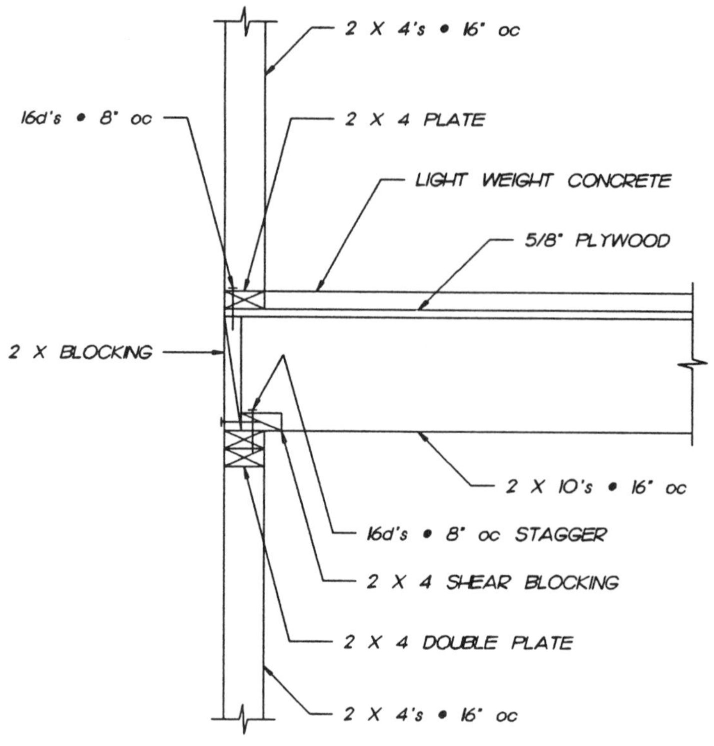

W3182

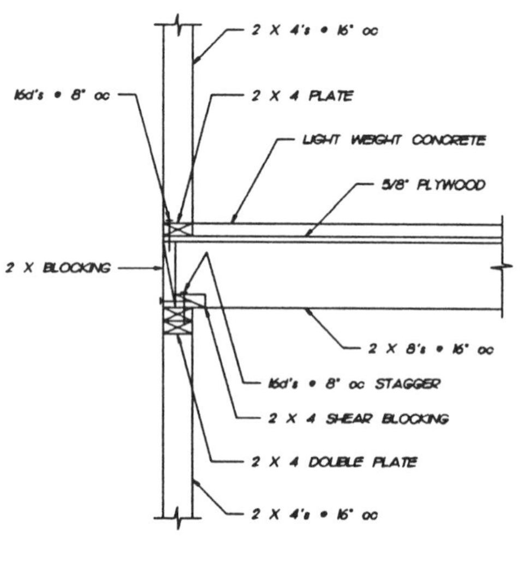

W3183

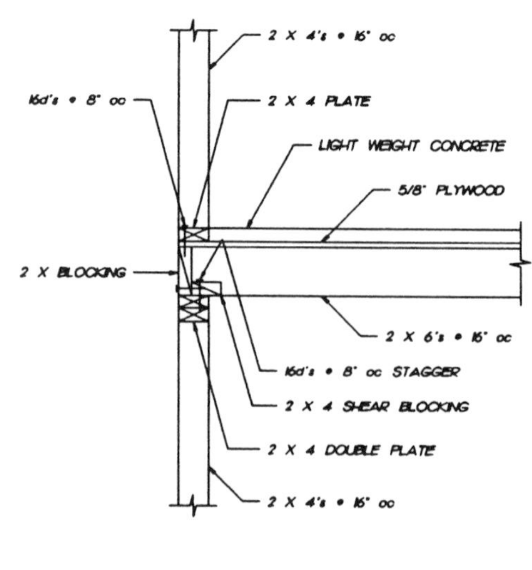

W3184

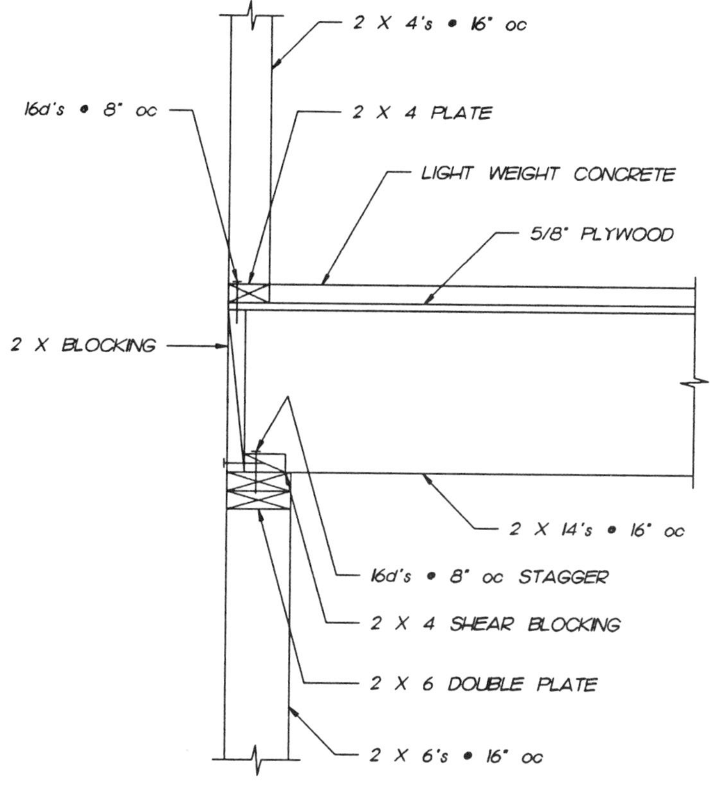

W3185

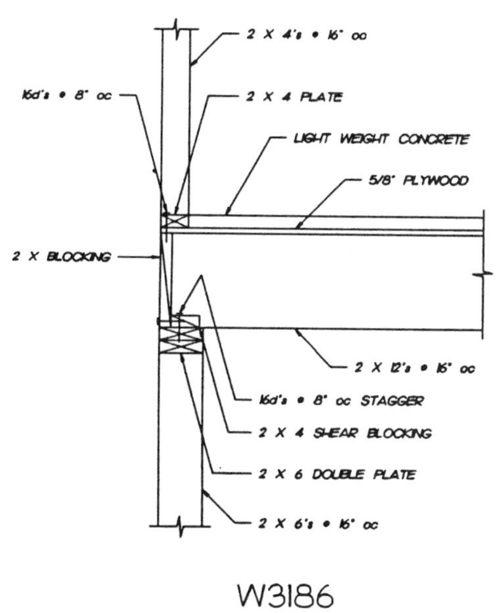

W3186

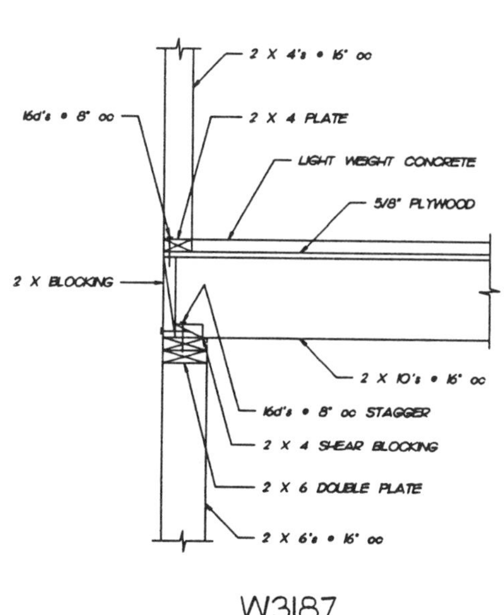

W3187

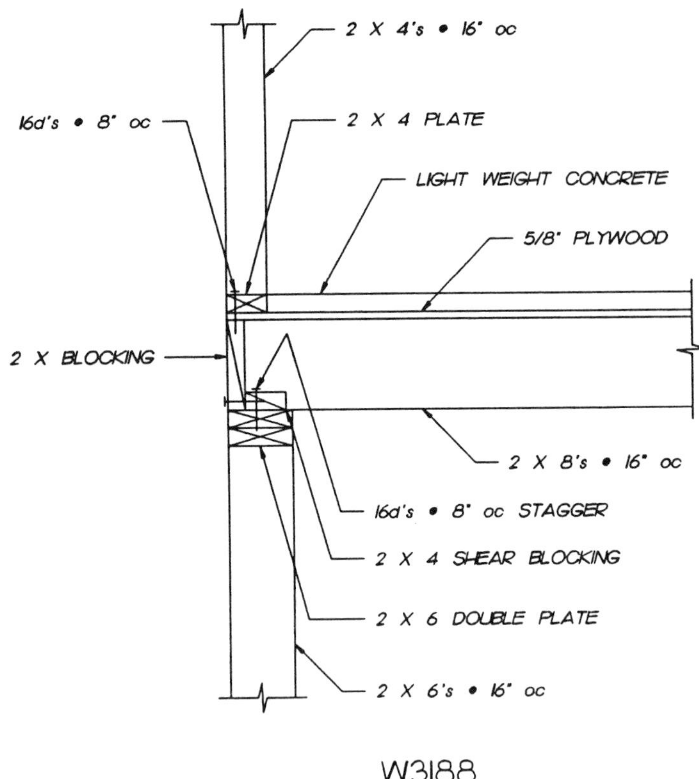
2 X 4's • 16" oc
16d's • 8" oc
2 X 4 PLATE
LIGHT WEIGHT CONCRETE
5/8" PLYWOOD
2 X BLOCKING
2 X 8's • 16" oc
16d's • 8" oc STAGGER
2 X 4 SHEAR BLOCKING
2 X 6 DOUBLE PLATE
2 X 6's • 16" oc

W3188

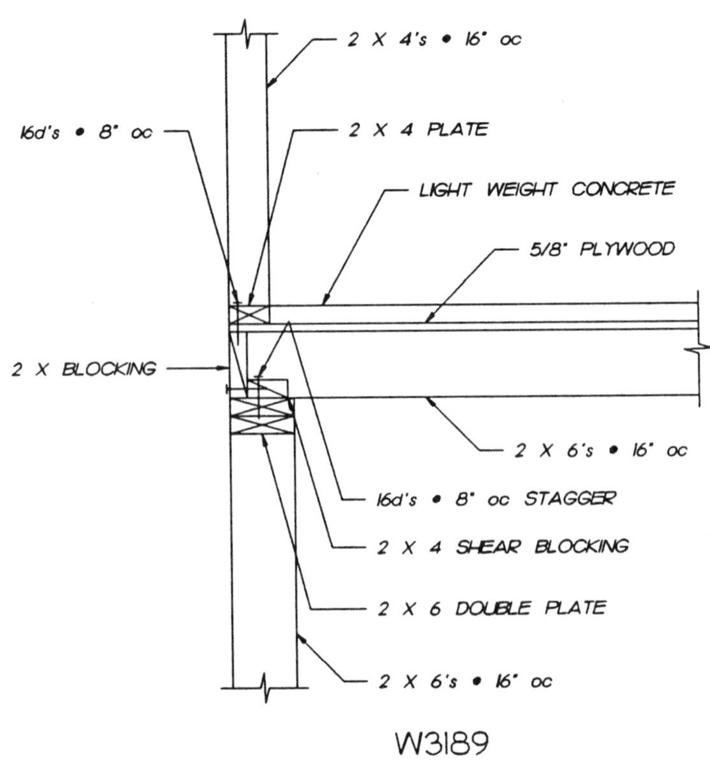
2 X 4's • 16" oc
16d's • 8" oc
2 X 4 PLATE
LIGHT WEIGHT CONCRETE
5/8" PLYWOOD
2 X BLOCKING
2 X 6's • 16" oc
16d's • 8" oc STAGGER
2 X 4 SHEAR BLOCKING
2 X 6 DOUBLE PLATE
2 X 6's • 16" oc

W3189

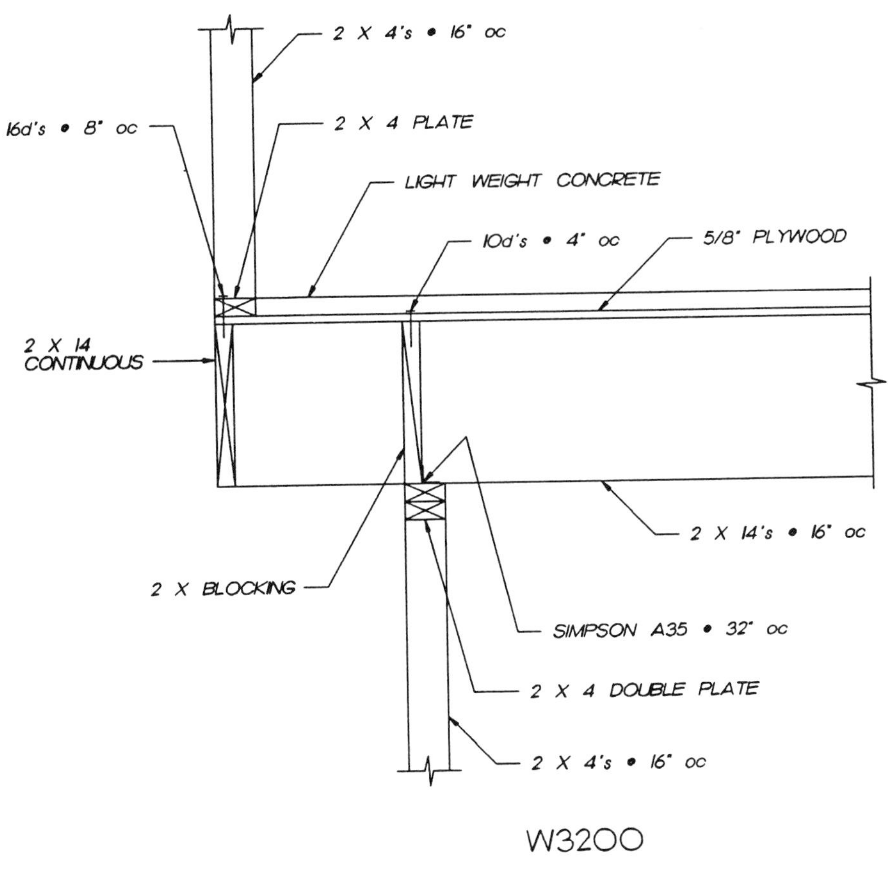
2 X 4's • 16" oc
16d's • 8" oc
2 X 4 PLATE
LIGHT WEIGHT CONCRETE
10d's • 4" oc
5/8" PLYWOOD
2 X 14
CONTINUOUS
2 X 14's • 16" oc
2 X BLOCKING
SIMPSON A35 • 32" oc
2 X 4 DOUBLE PLATE
2 X 4's • 16" oc

W3200

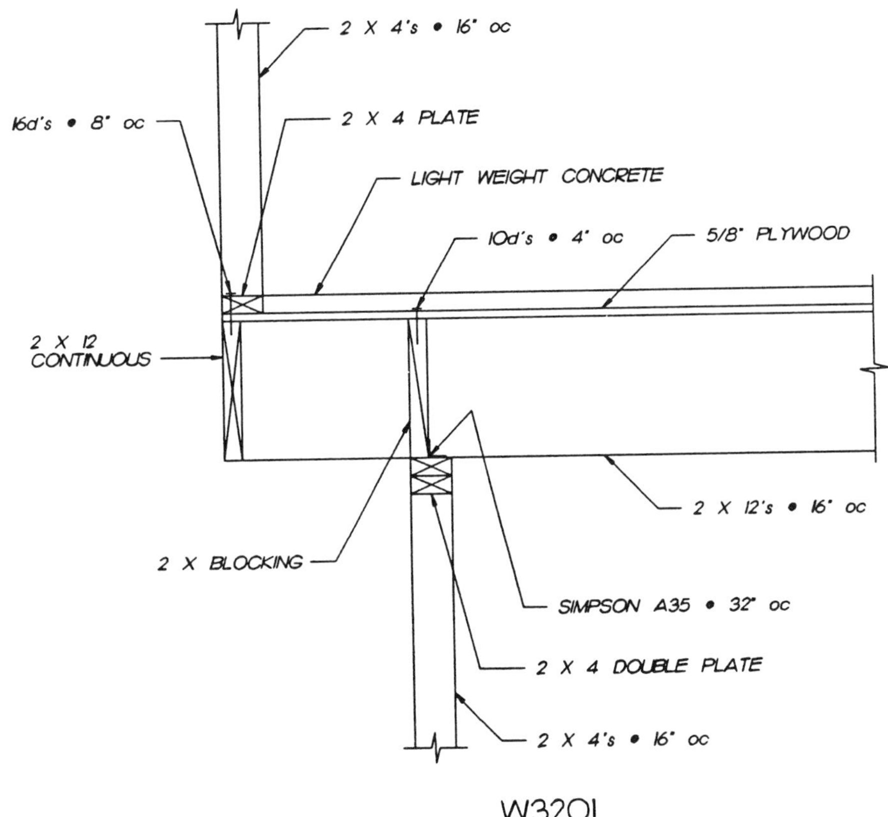
2 X 4's • 16" oc
16d's • 8" oc
2 X 4 PLATE
LIGHT WEIGHT CONCRETE
10d's • 4" oc
5/8" PLYWOOD
2 X 12
CONTINUOUS
2 X 12's • 16" oc
2 X BLOCKING
SIMPSON A35 • 32" oc
2 X 4 DOUBLE PLATE
2 X 4's • 16" oc

W3201

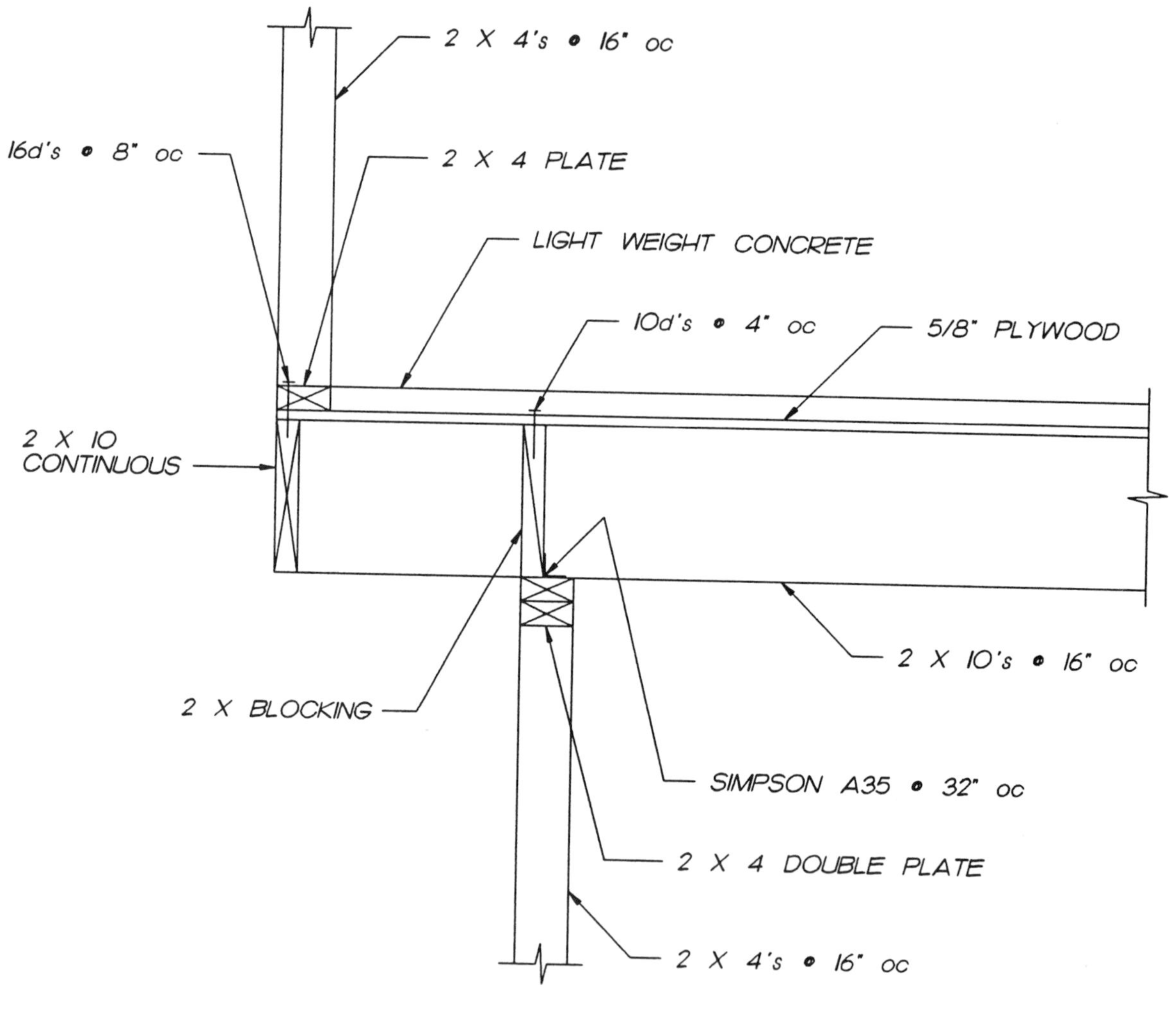
2 X 4's @ 16" oc
16d's @ 8" oc
2 X 4 PLATE
LIGHT WEIGHT CONCRETE
10d's @ 4" oc
5/8" PLYWOOD
2 X 10
CONTINUOUS
2 X 10's @ 16" oc
2 X BLOCKING
SIMPSON A35 @ 32" oc
2 X 4 DOUBLE PLATE
2 X 4's @ 16" oc

W3202

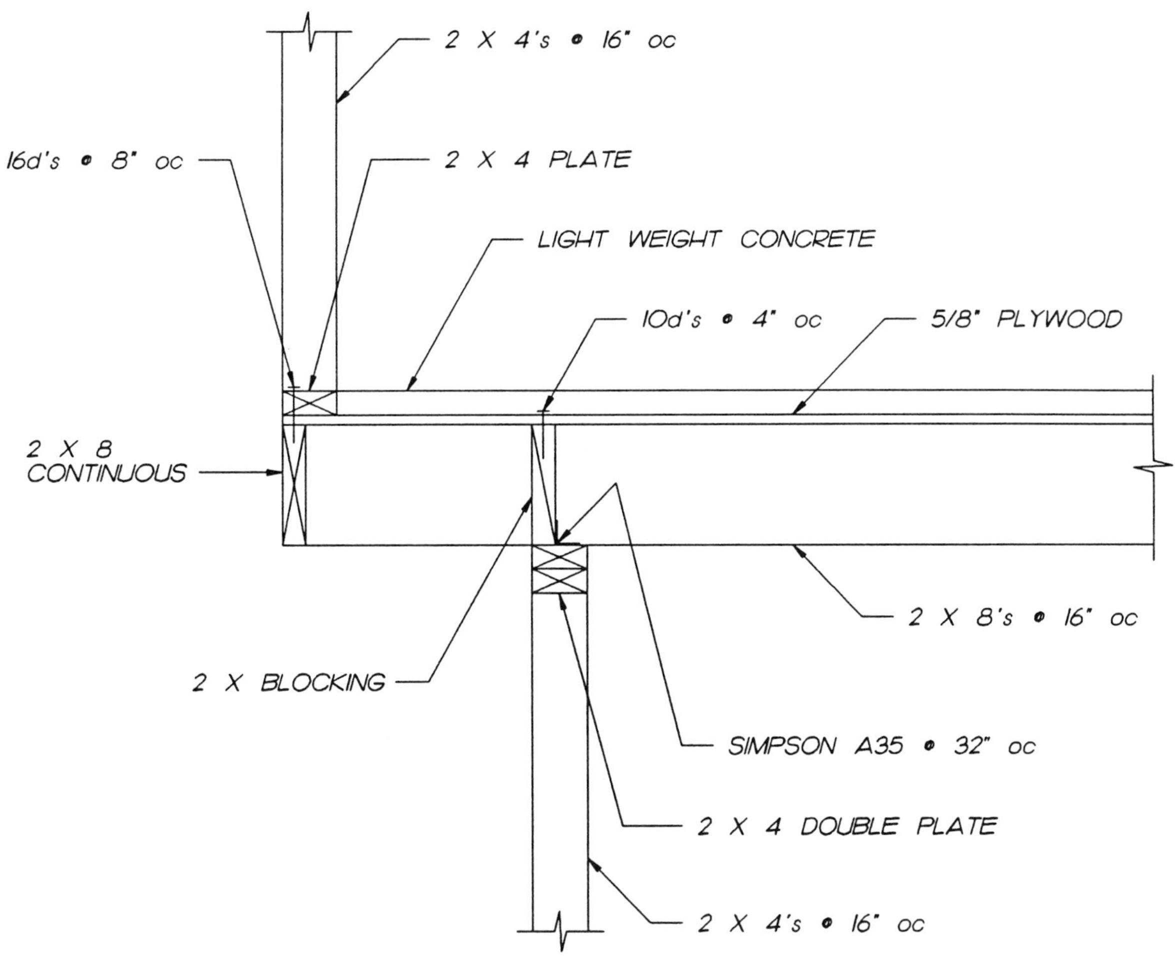
2 X 4's @ 16" oc
16d's @ 8" oc
2 X 4 PLATE
LIGHT WEIGHT CONCRETE
10d's @ 4" oc
5/8" PLYWOOD
2 X 8
CONTINUOUS
2 X 8's @ 16" oc
2 X BLOCKING
SIMPSON A35 @ 32" oc
2 X 4 DOUBLE PLATE
2 X 4's @ 16" oc

W3203

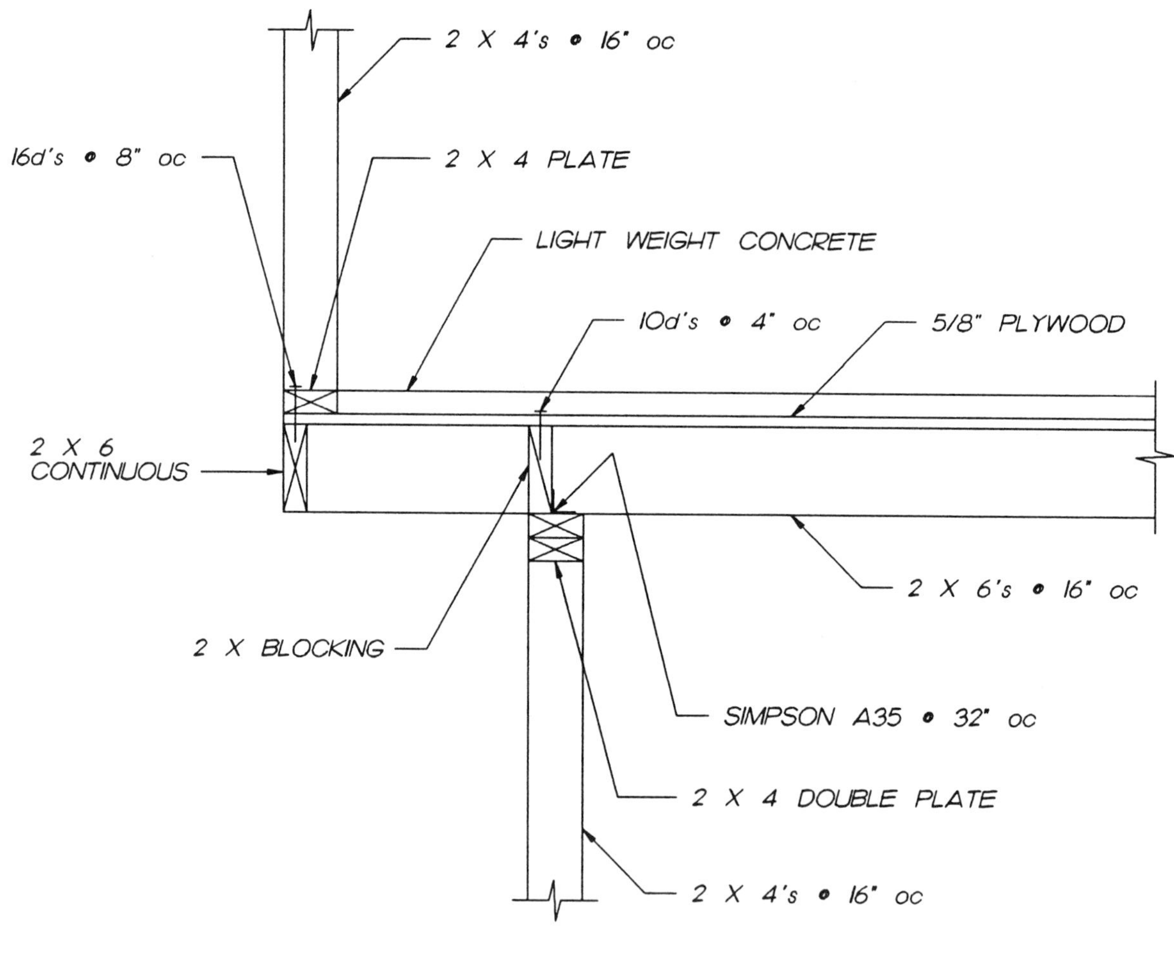
2 X 4's @ 16" oc
16d's @ 8" oc
2 X 4 PLATE
LIGHT WEIGHT CONCRETE
10d's @ 4" oc
5/8" PLYWOOD
2 X 6
CONTINUOUS
2 X 6's @ 16" oc
2 X BLOCKING
SIMPSON A35 @ 32" oc
2 X 4 DOUBLE PLATE
2 X 4's @ 16" oc

W3204

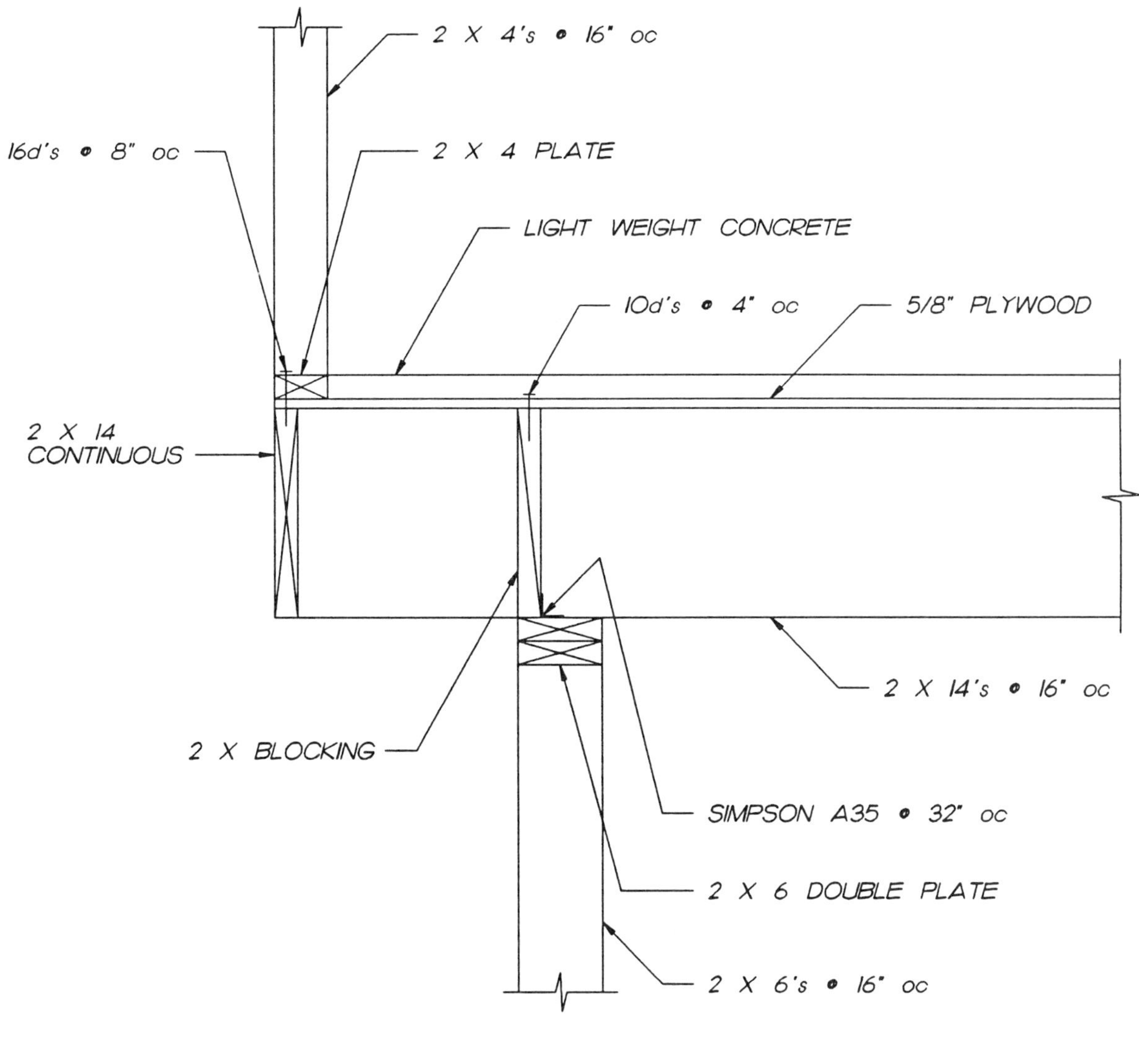
2 X 4's @ 16" oc
16d's @ 8" oc
2 X 4 PLATE
LIGHT WEIGHT CONCRETE
10d's @ 4" oc
5/8" PLYWOOD
2 X 14
CONTINUOUS
2 X 14's @ 16" oc
2 X BLOCKING
SIMPSON A35 @ 32" oc
2 X 6 DOUBLE PLATE
2 X 6's @ 16" oc

W3205

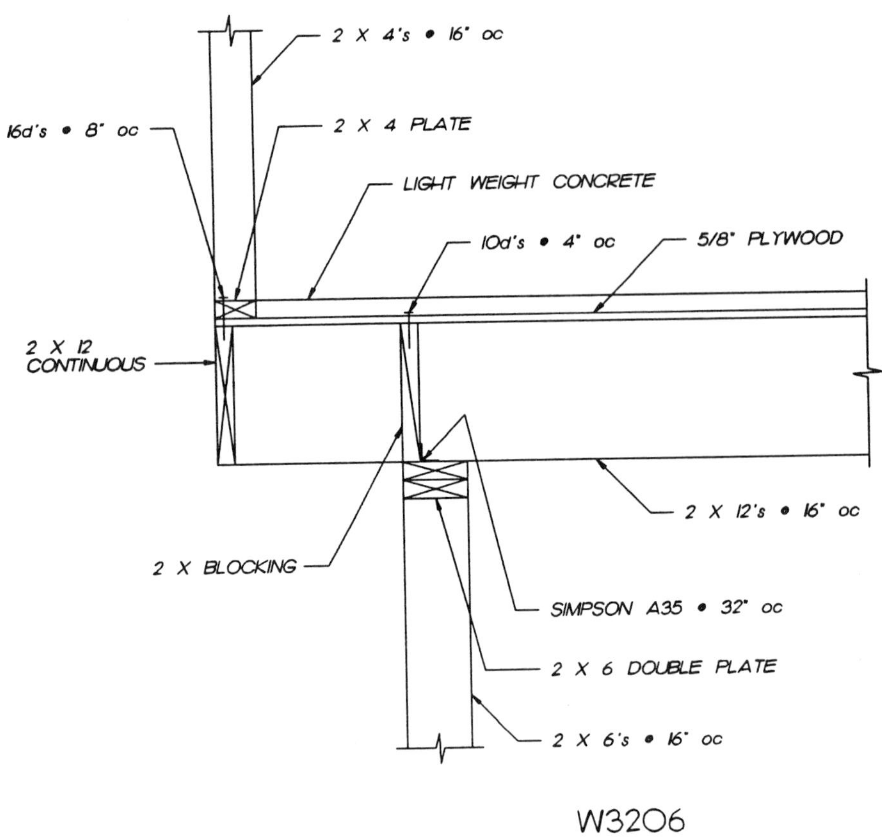
2 X 4's • 16" oc
16d's • 8" oc
2 X 4 PLATE
LIGHT WEIGHT CONCRETE
10d's • 4" oc
5/8" PLYWOOD
2 X 12
CONTINUOUS
2 X 12's • 16" oc
2 X BLOCKING
SIMPSON A35 • 32" oc
2 X 6 DOUBLE PLATE
2 X 6's • 16" oc

W3206

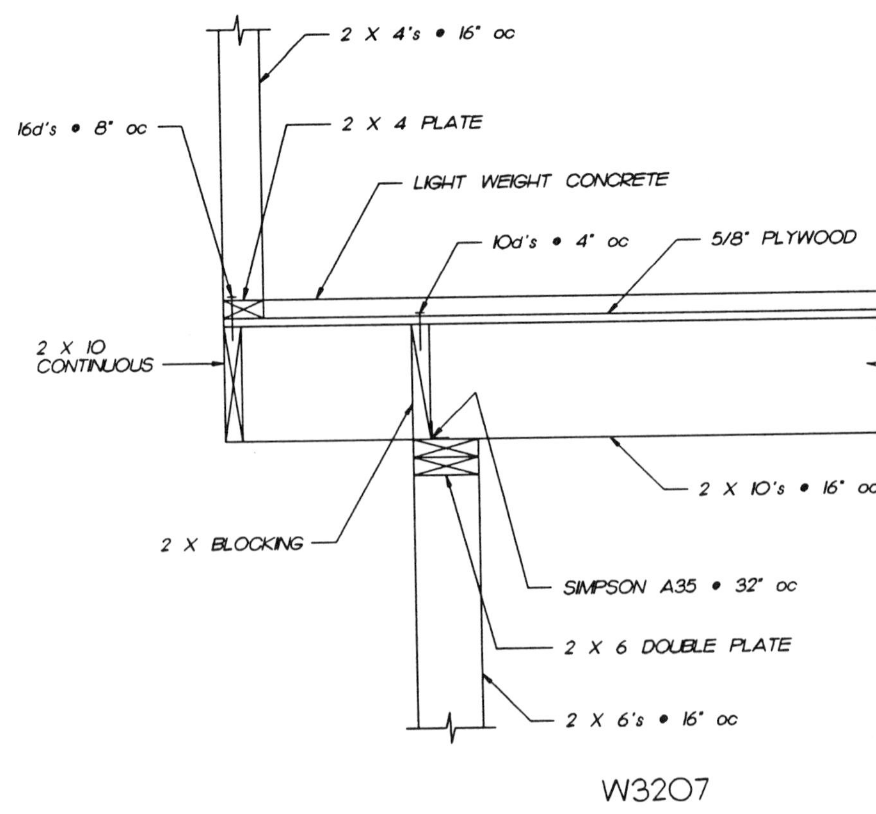
2 X 4's • 16" oc
16d's • 8" oc
2 X 4 PLATE
LIGHT WEIGHT CONCRETE
10d's • 4" oc
5/8" PLYWOOD
2 X 10
CONTINUOUS
2 X 10's • 16" oc
2 X BLOCKING
SIMPSON A35 • 32" oc
2 X 6 DOUBLE PLATE
2 X 6's • 16" oc

W3207

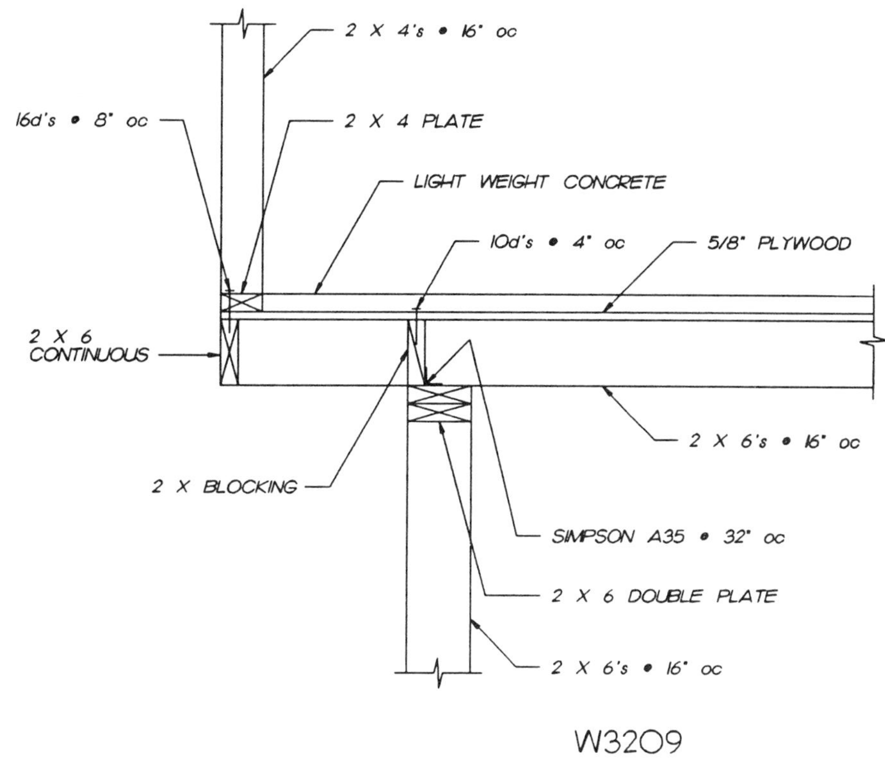
2 X 4's • 16" oc
16d's • 8" oc
2 X 4 PLATE
LIGHT WEIGHT CONCRETE
10d's • 4" oc
5/8" PLYWOOD
2 X 6
CONTINUOUS
2 X 6's • 16" oc
2 X BLOCKING
SIMPSON A35 • 32" oc
2 X 6 DOUBLE PLATE
2 X 6's • 16" oc

W3209

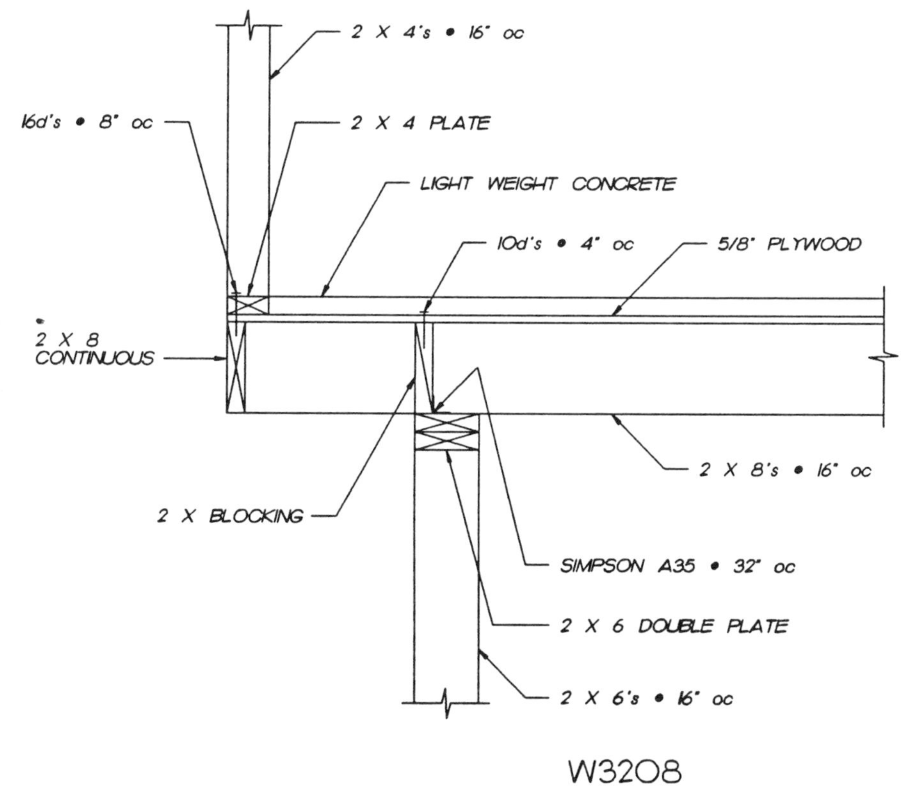
2 X 4's • 16" oc
16d's • 8" oc
2 X 4 PLATE
LIGHT WEIGHT CONCRETE
10d's • 4" oc
5/8" PLYWOOD
2 X 8
CONTINUOUS
2 X 8's • 16" oc
2 X BLOCKING
SIMPSON A35 • 32" oc
2 X 6 DOUBLE PLATE
2 X 6's • 16" oc

W3208

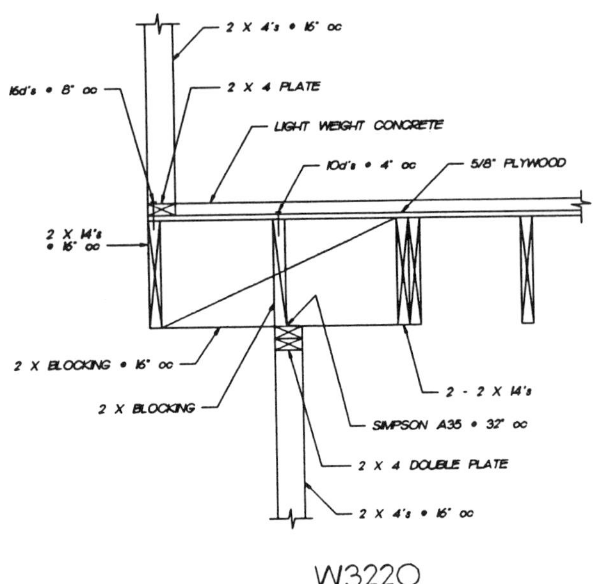

W3220

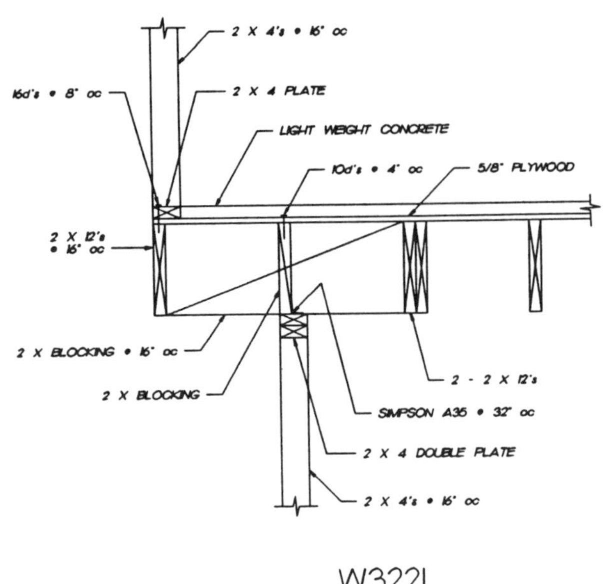

W3221

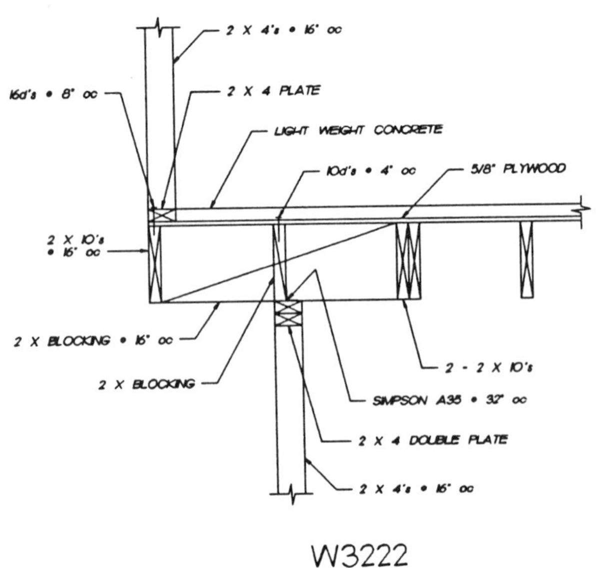

W3222

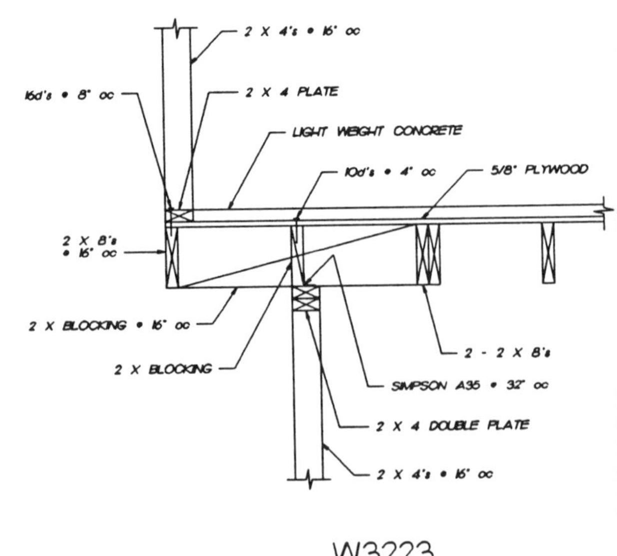

W3223

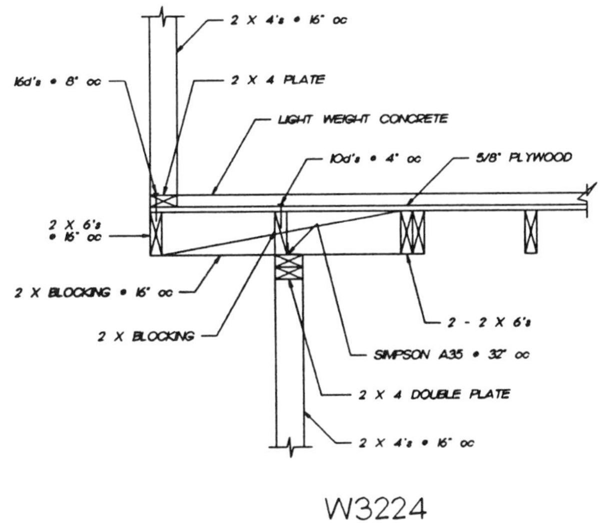

W3224

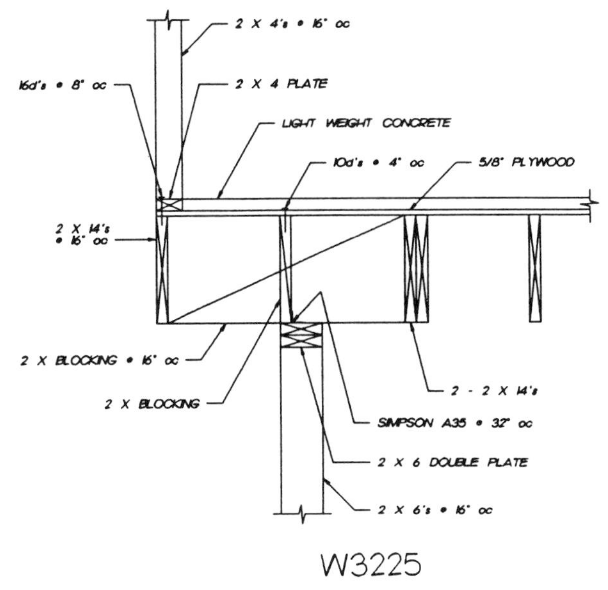

W3225

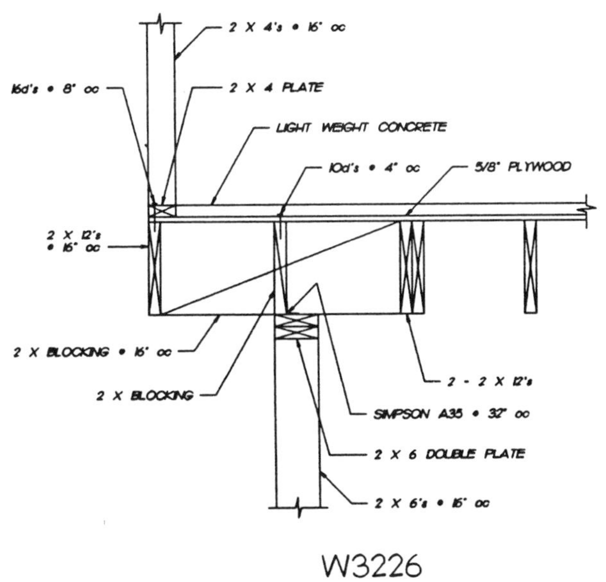

W3226

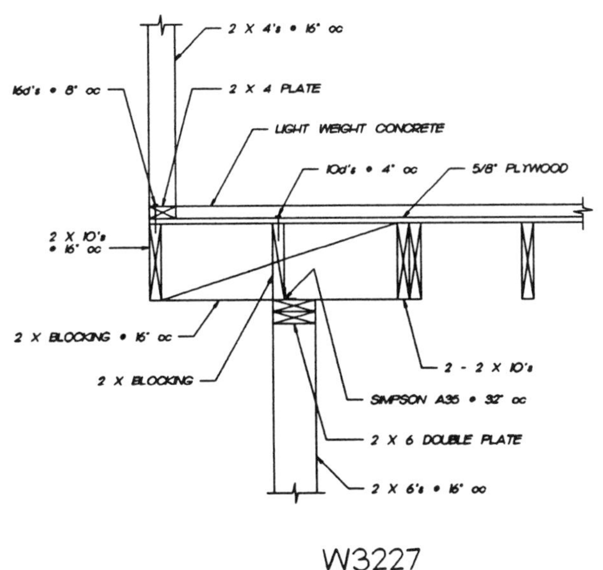

W3227

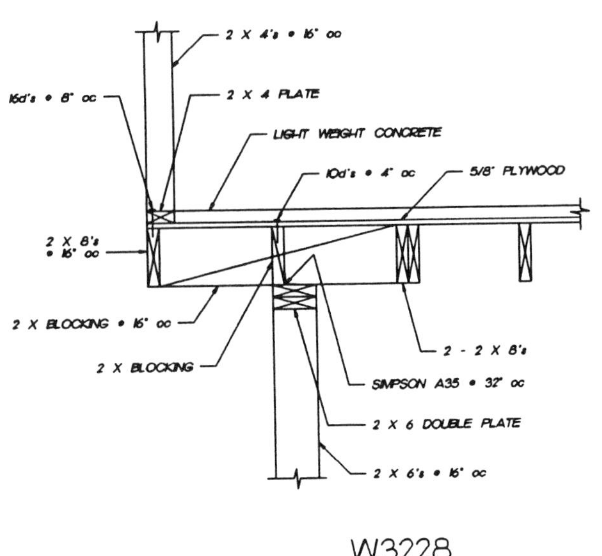

W3228

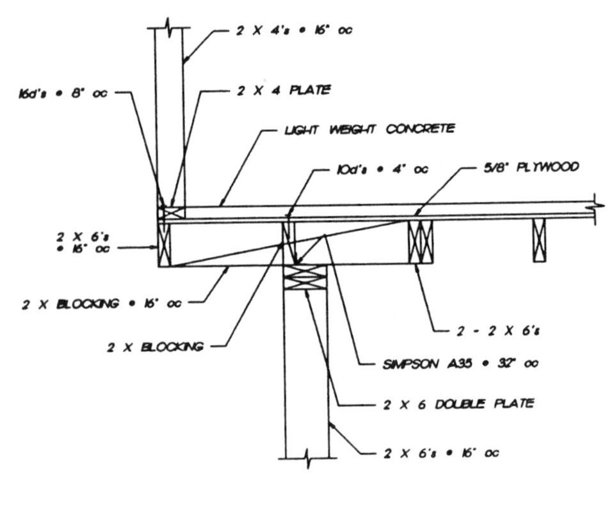

W3229

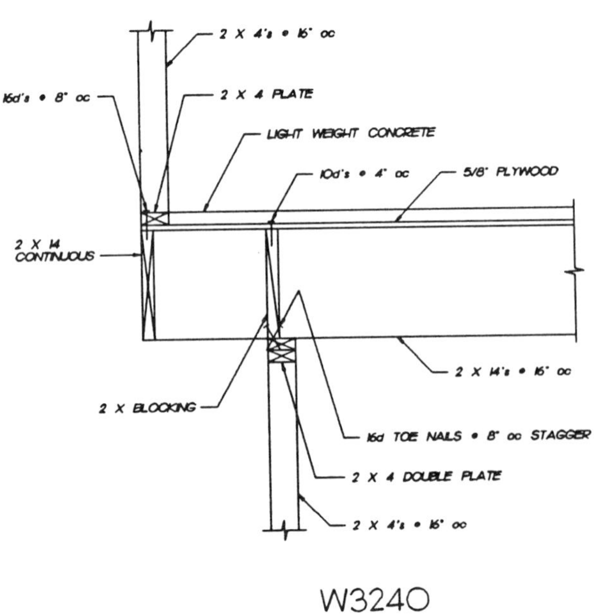

W3240

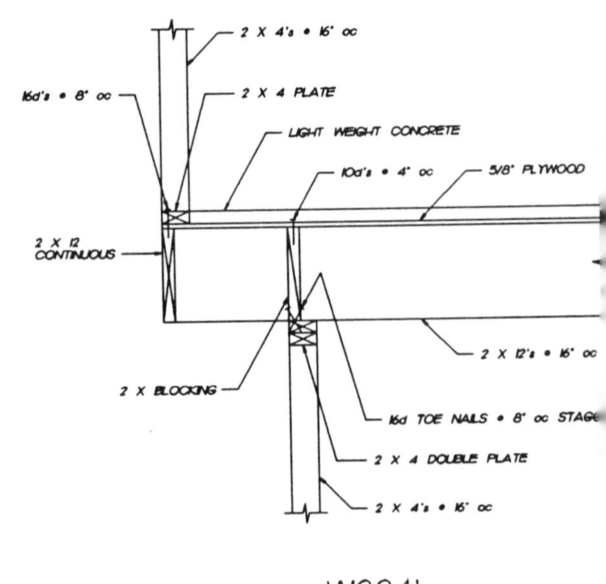

W3241

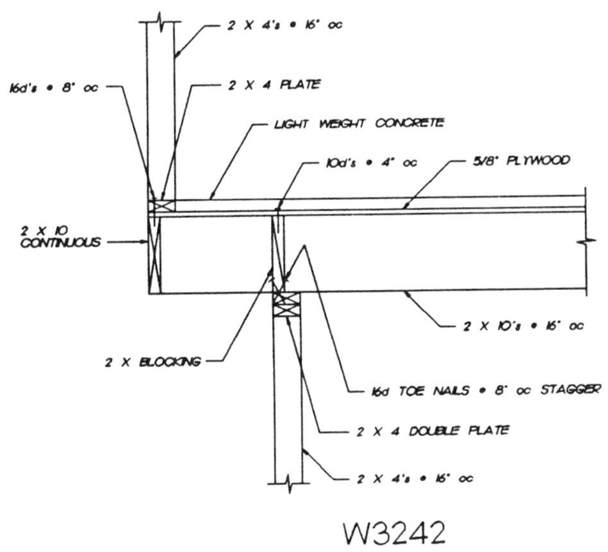

W3242

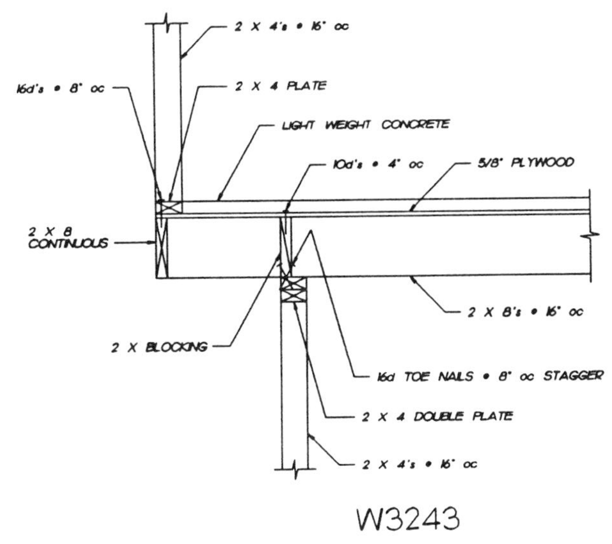

W3243

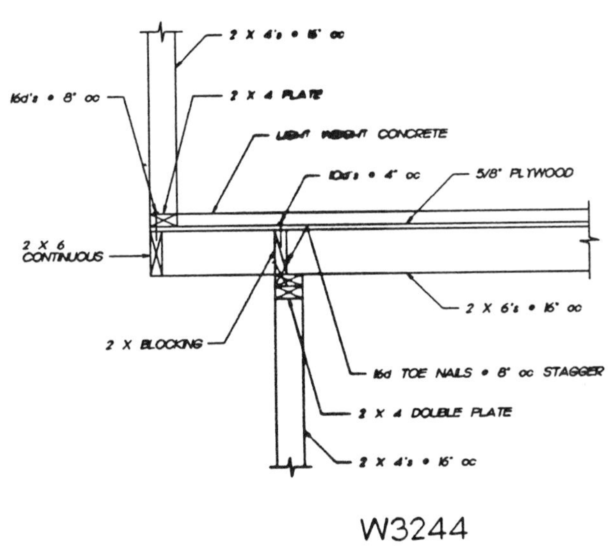

W3244

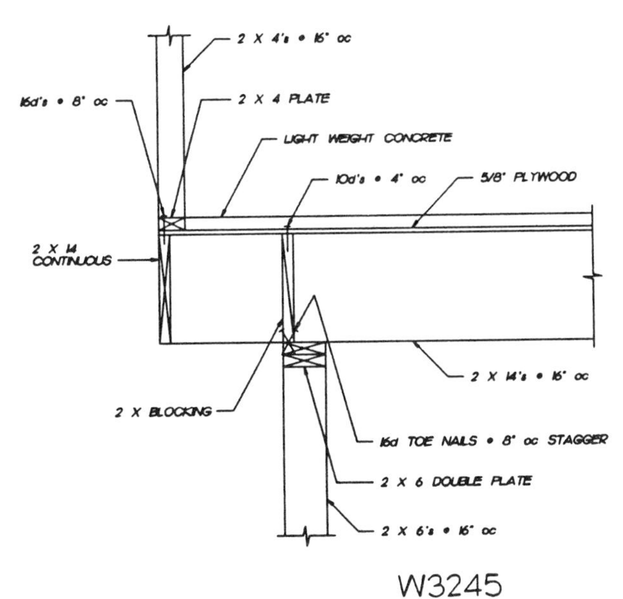

W3245

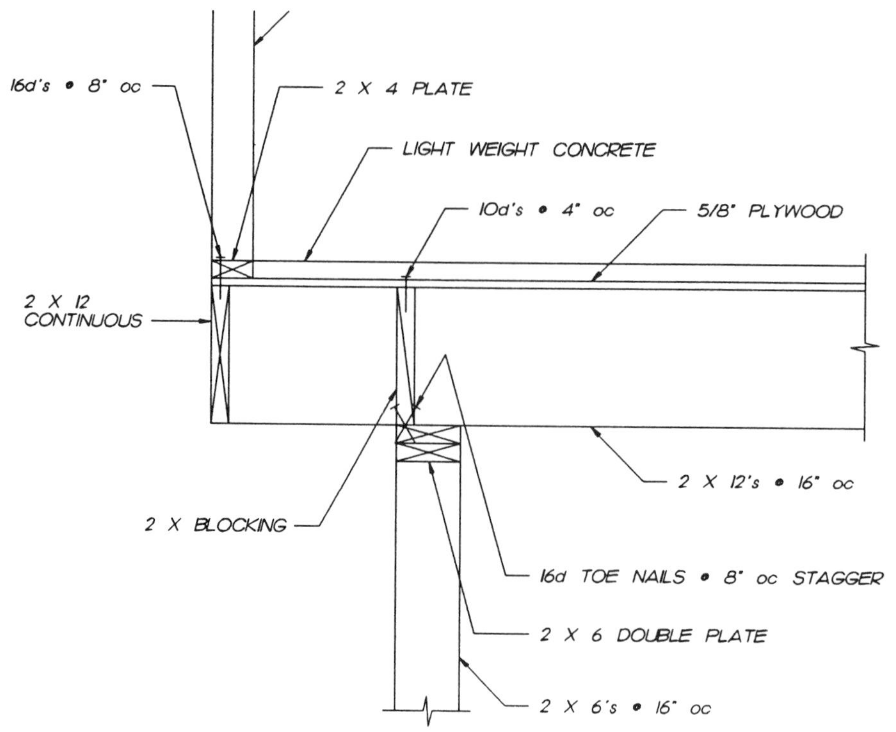

W3246

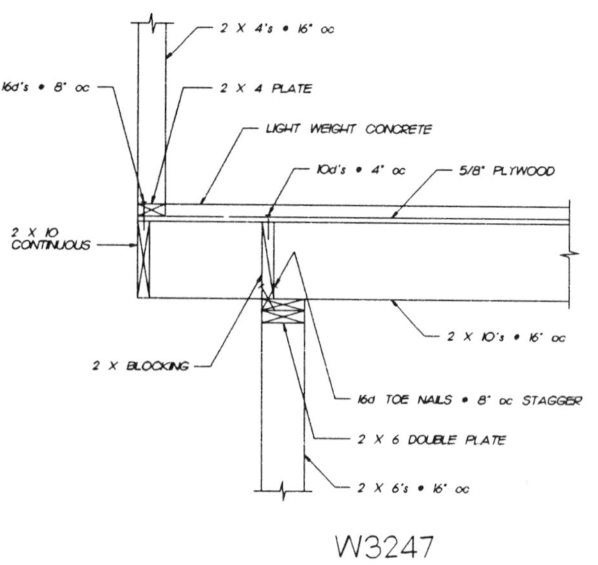

W3247

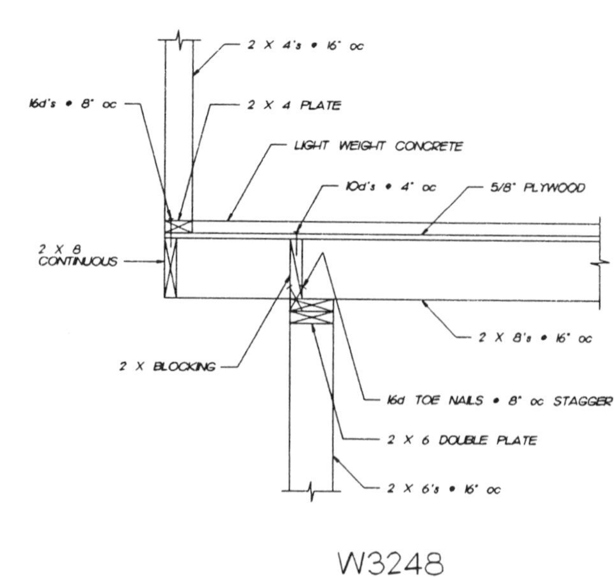

W3248

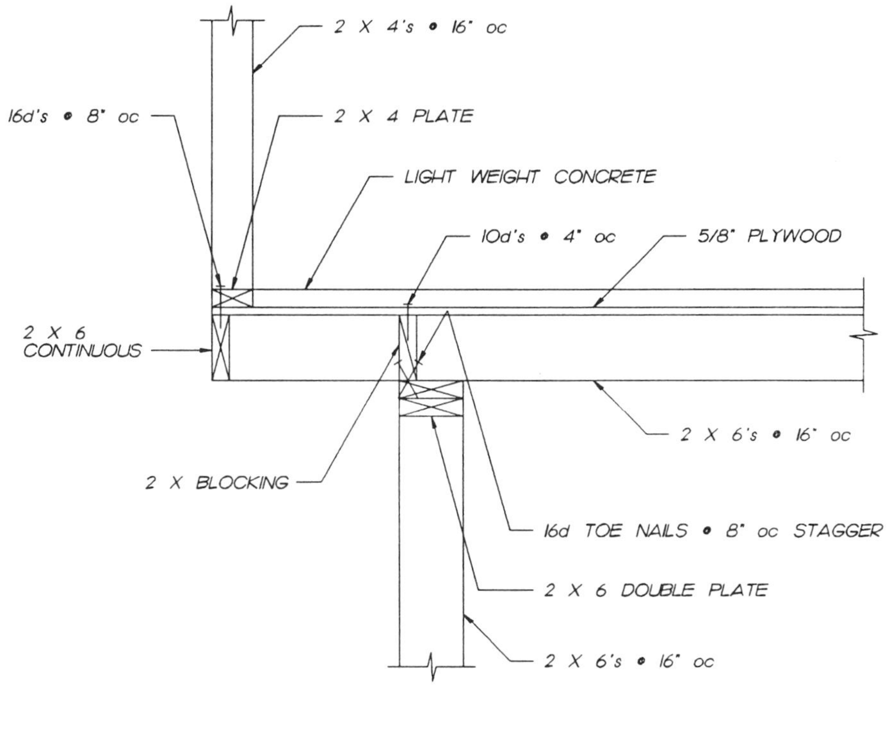

W3249

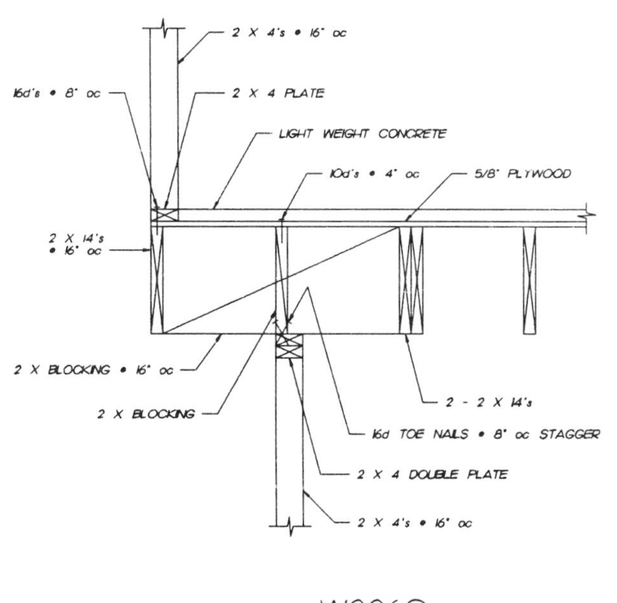

W3260

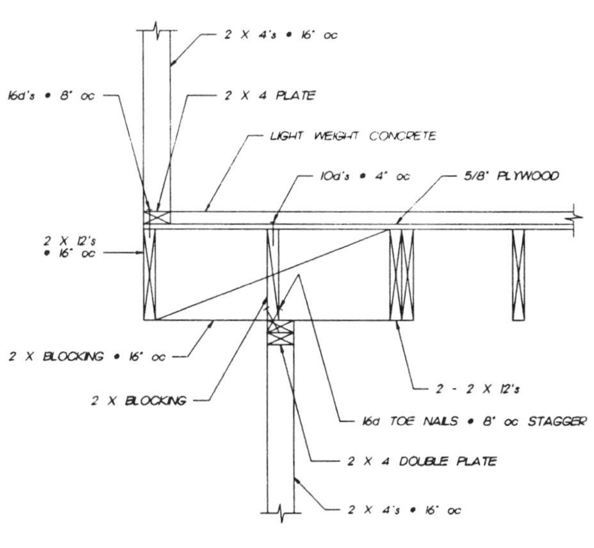

W3261

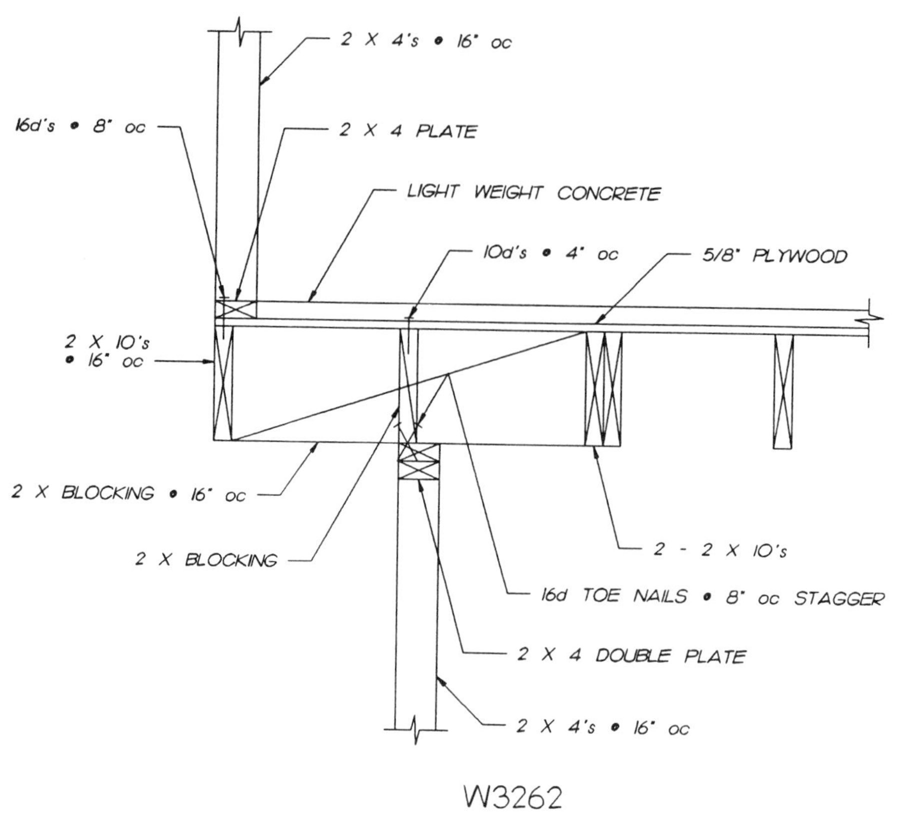

W3262

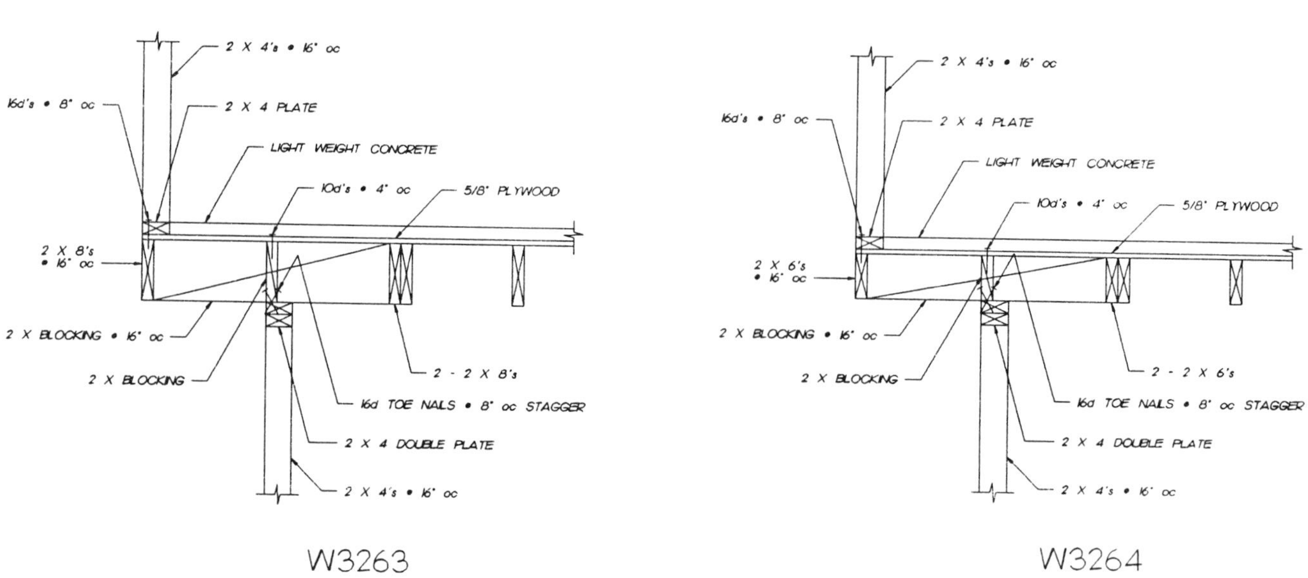

W3263

W3264

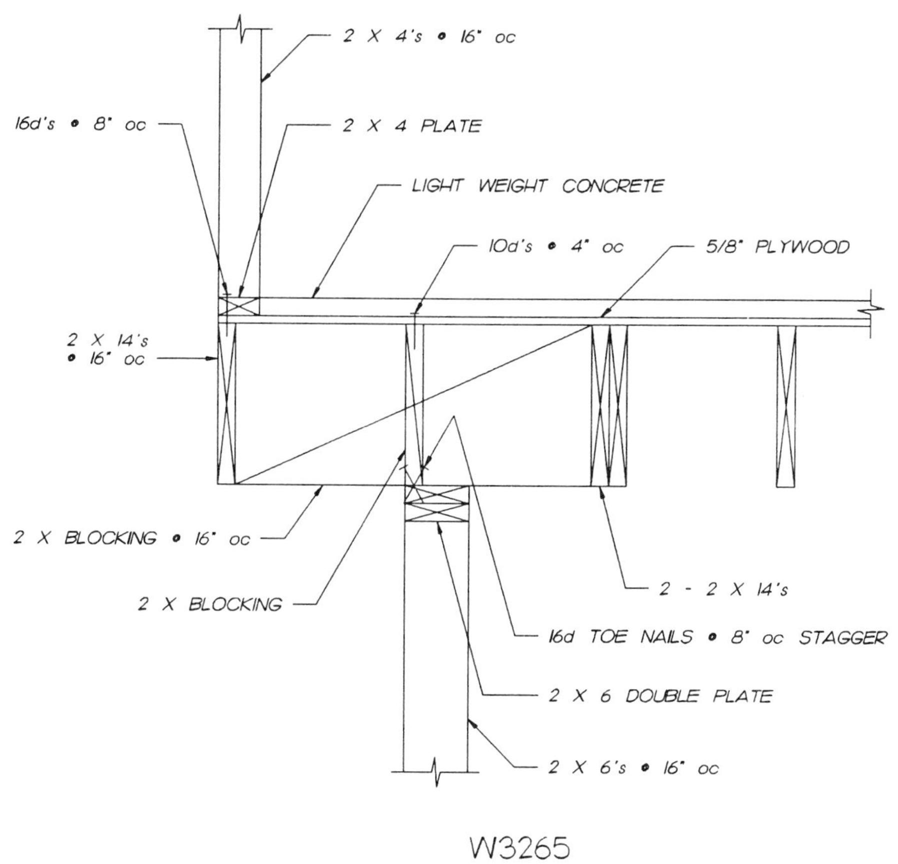

W3265

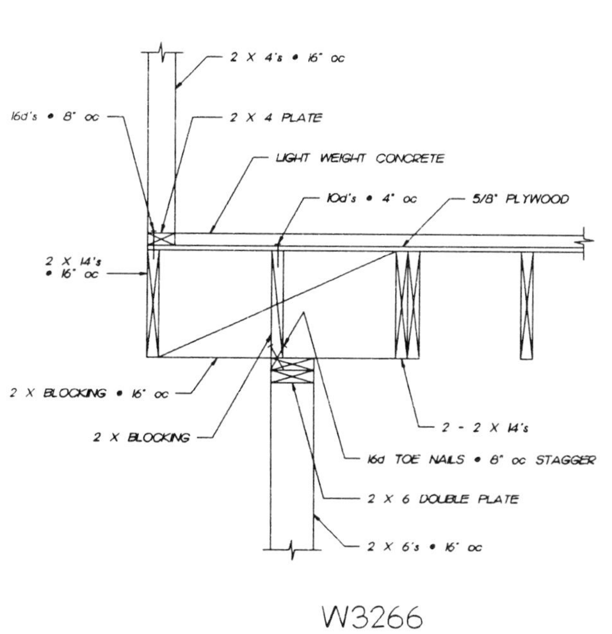

W3266

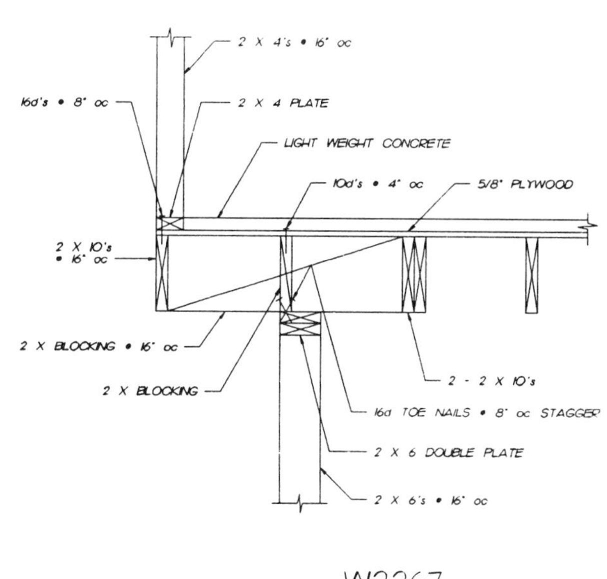

W3267

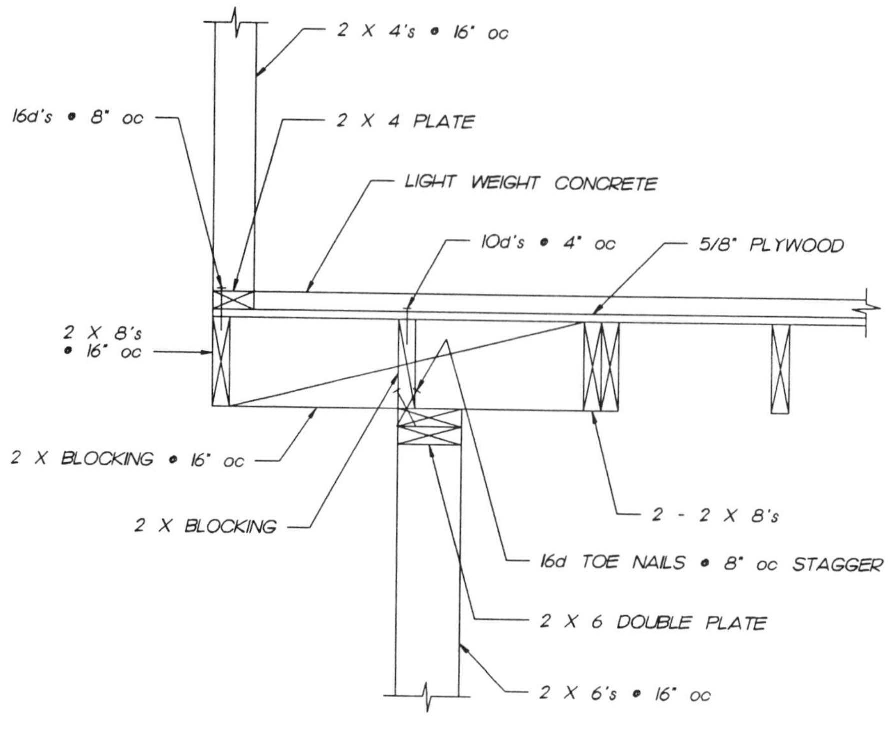

W3268

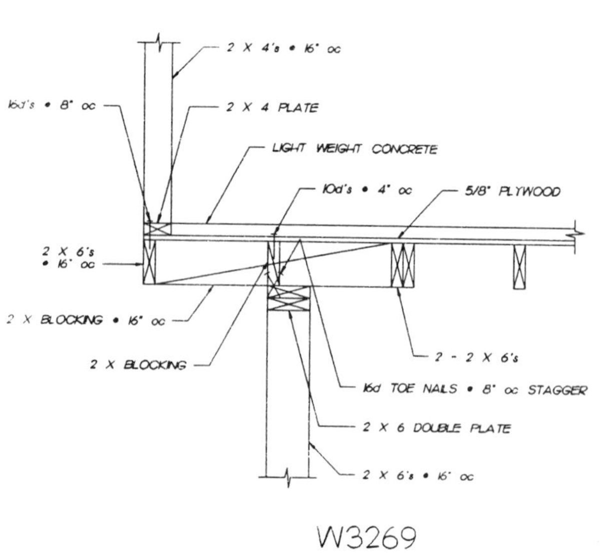

W3269

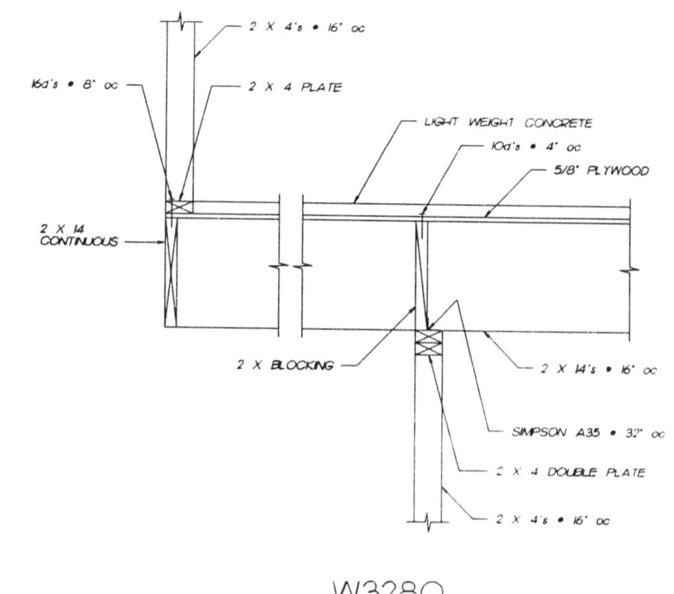

W3280

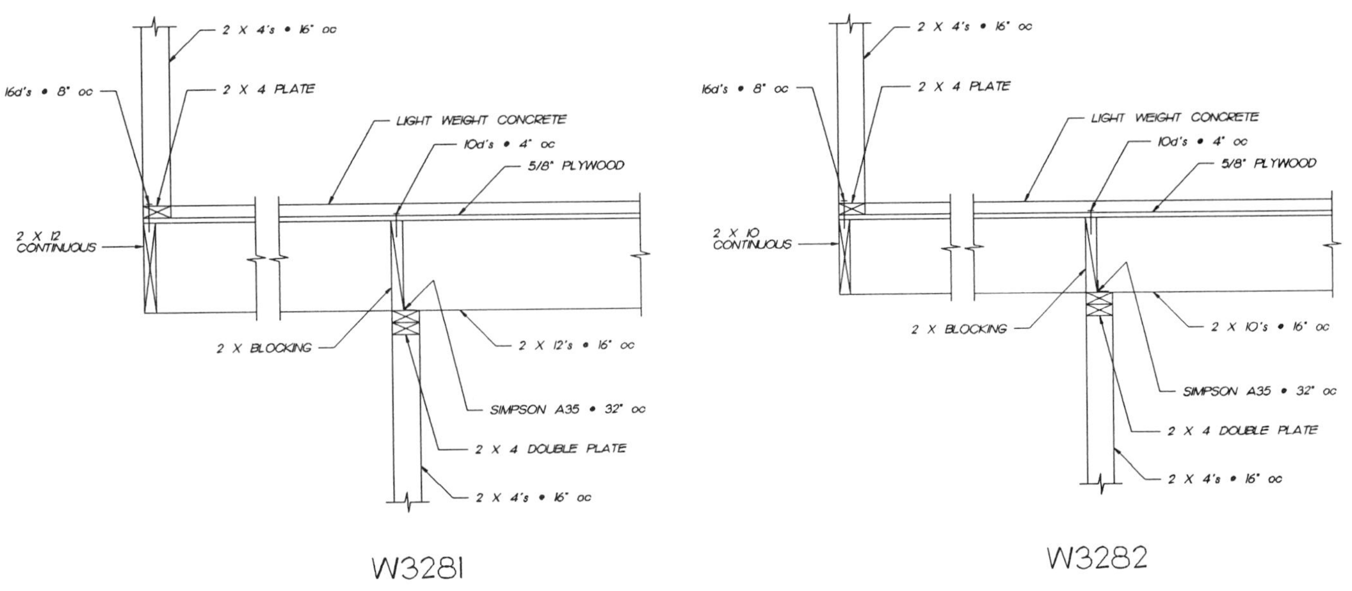
2 X 4's • 16" oc
16d's • 8" oc
2 X 4 PLATE
LIGHT WEIGHT CONCRETE
10d's • 4" oc
5/8" PLYWOOD
2 X 12 CONTINUOUS
2 X BLOCKING
2 X 12's • 16" oc
SIMPSON A35 • 32" oc
2 X 4 DOUBLE PLATE
2 X 4's • 16" oc
W3281
2 X 4's • 16" oc
16d's • 8" oc
2 X 4 PLATE
LIGHT WEIGHT CONCRETE
10d's • 4" oc
5/8" PLYWOOD
2 X 10 CONTINUOUS
2 X BLOCKING
2 X 10's • 16" oc
SIMPSON A35 • 32" oc
2 X 4 DOUBLE PLATE
2 X 4's • 16" oc
W3282

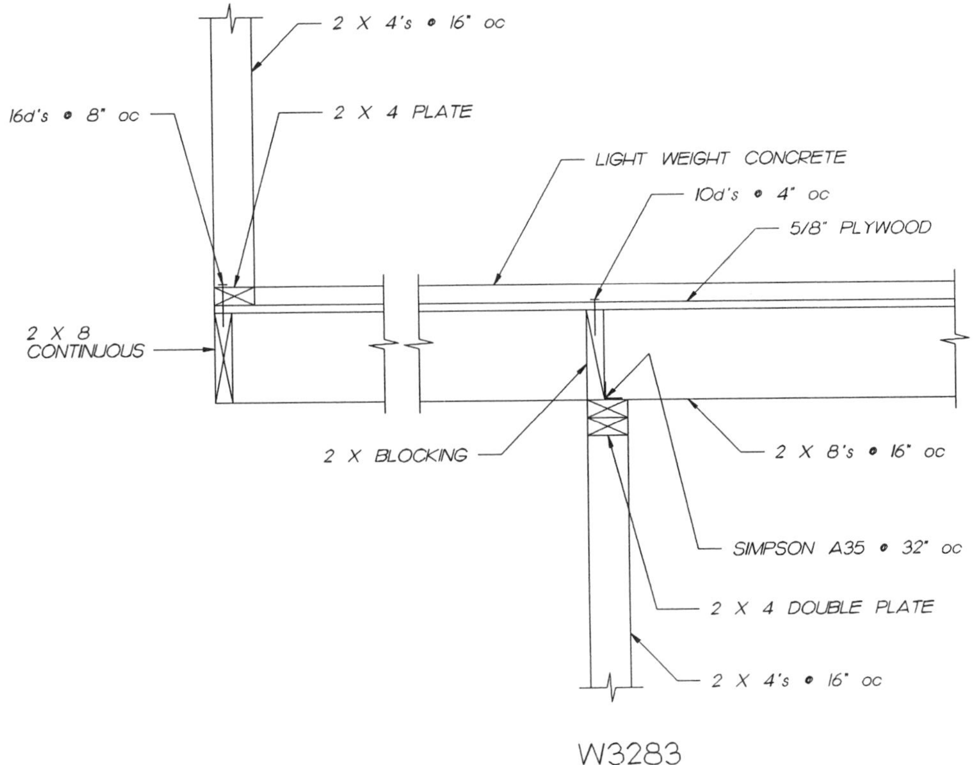
2 X 4's • 16" oc
16d's • 8" oc
2 X 4 PLATE
LIGHT WEIGHT CONCRETE
10d's • 4" oc
5/8" PLYWOOD
2 X 8 CONTINUOUS
2 X BLOCKING
2 X 8's • 16" oc
SIMPSON A35 • 32" oc
2 X 4 DOUBLE PLATE
2 X 4's • 16" oc
W3283

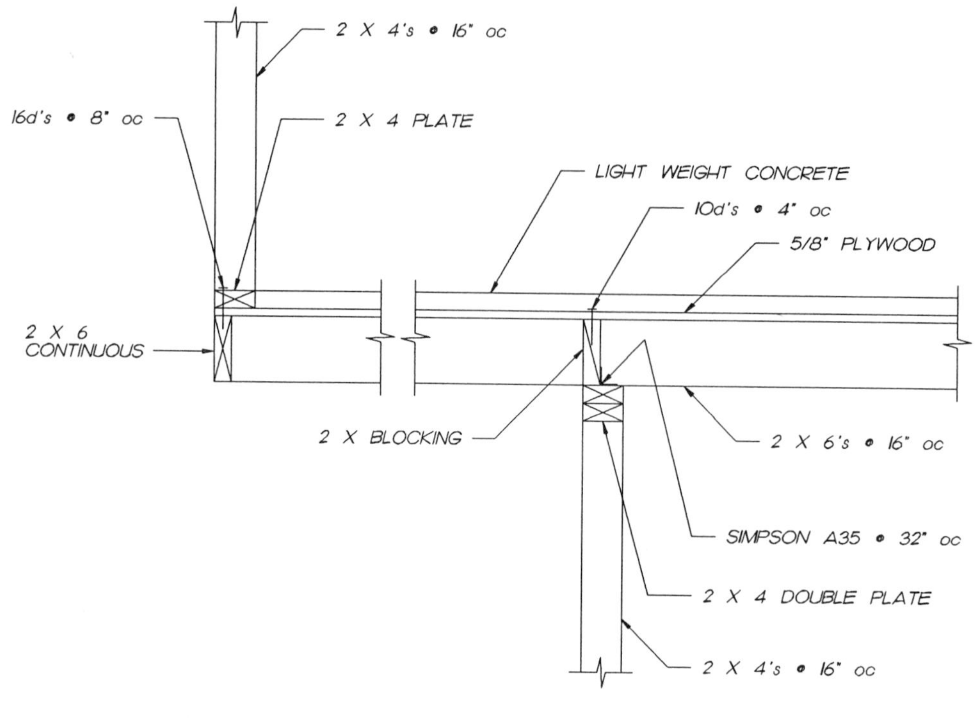

W3284

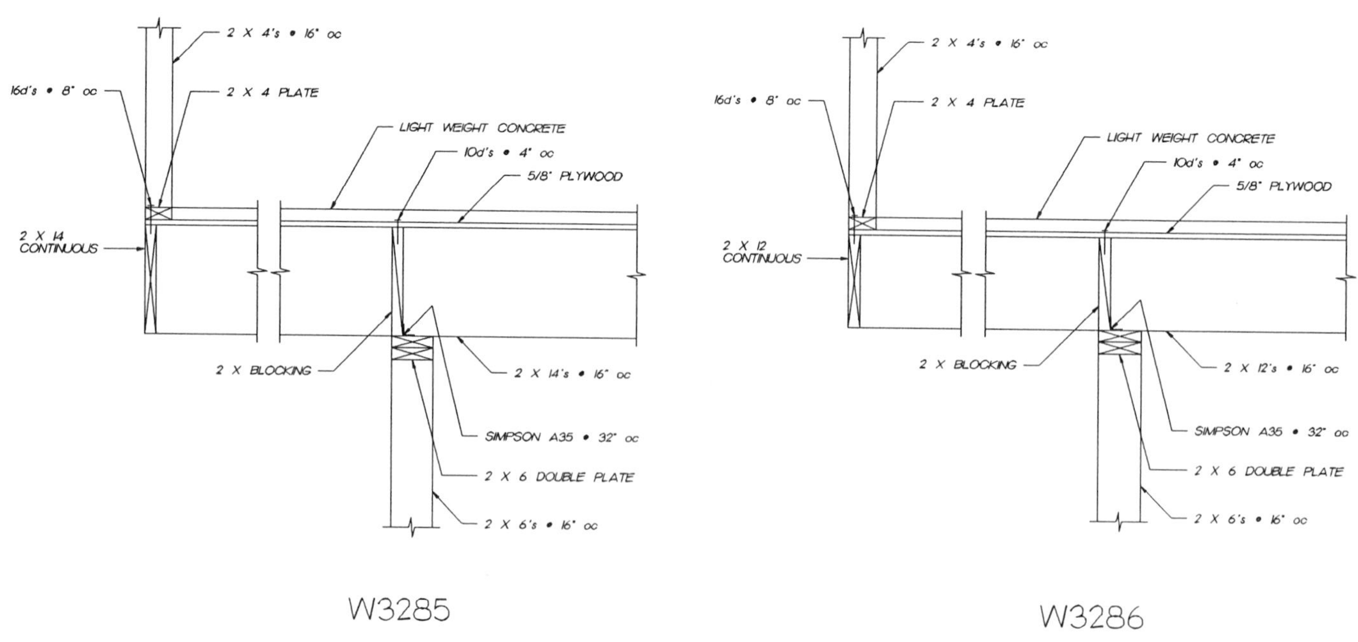

W3285

W3286

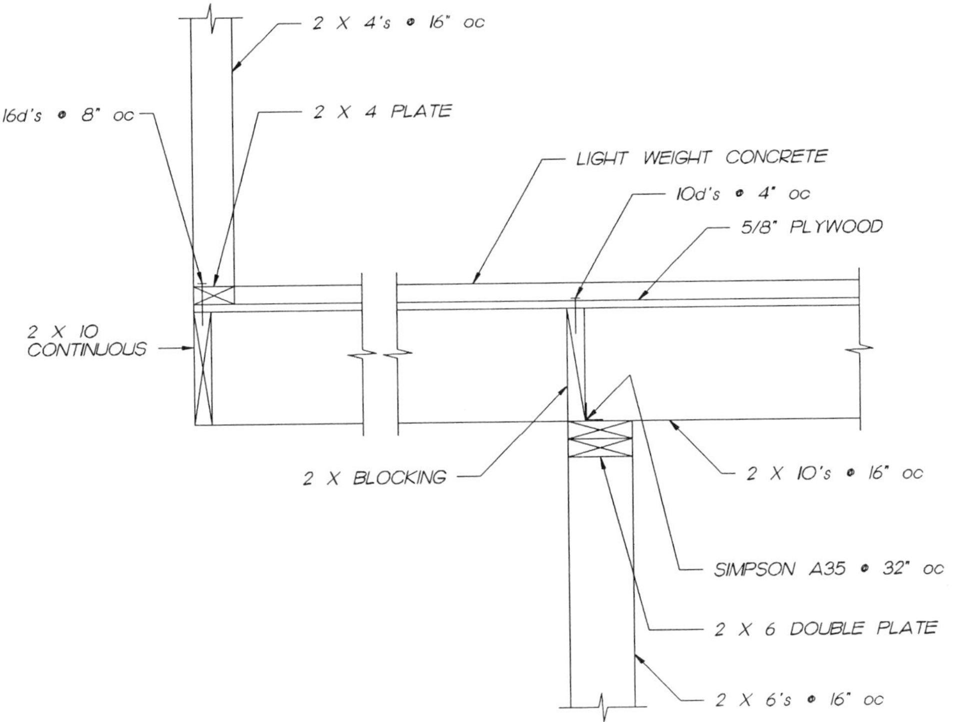

W3287

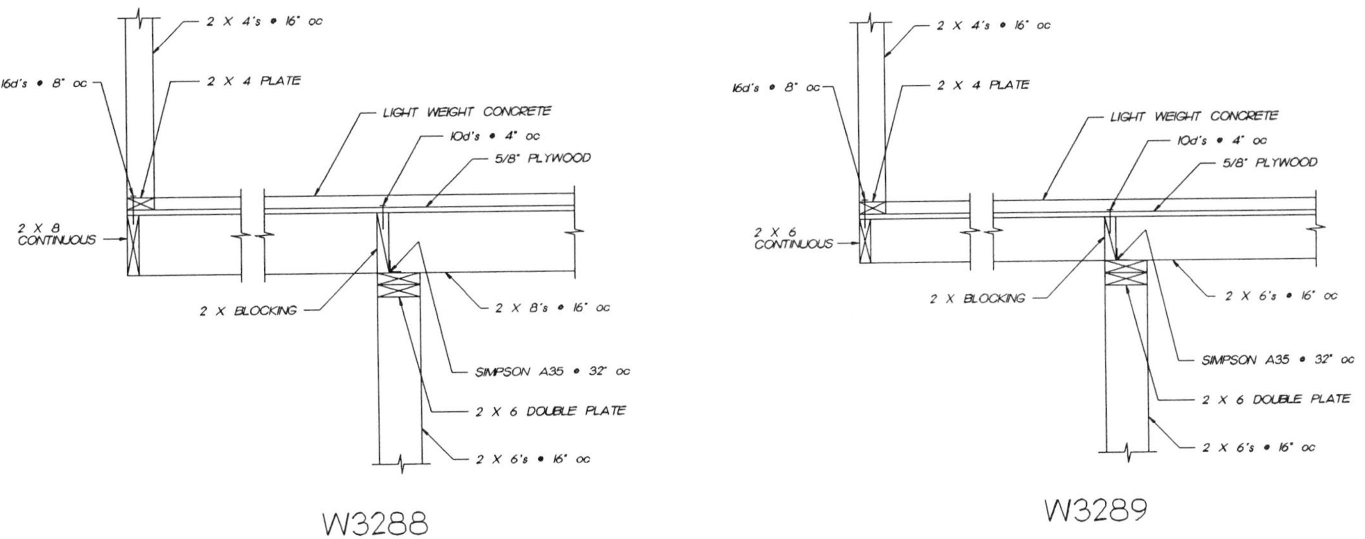

W3288

W3289

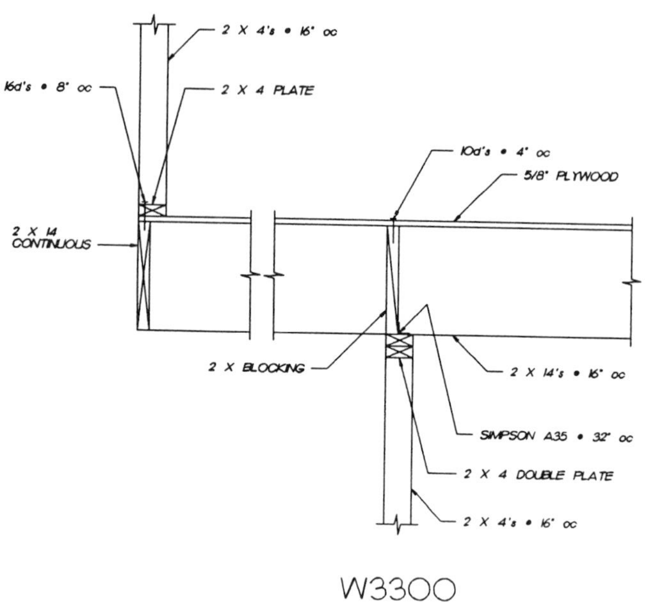

W3300

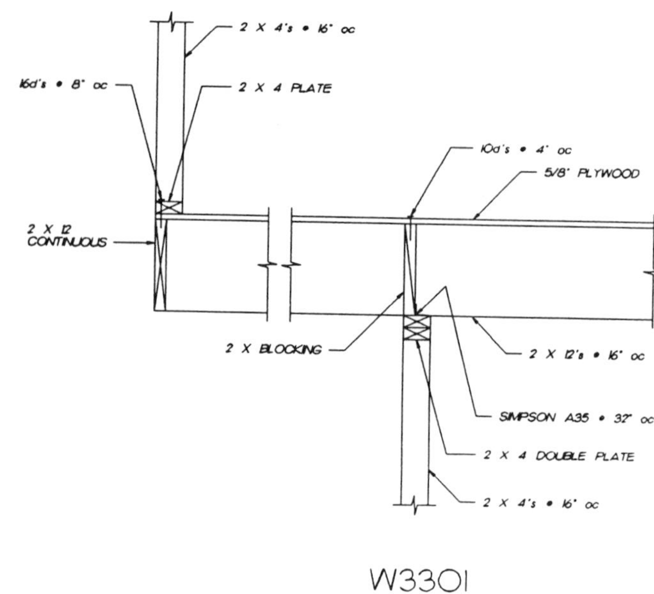

W3301

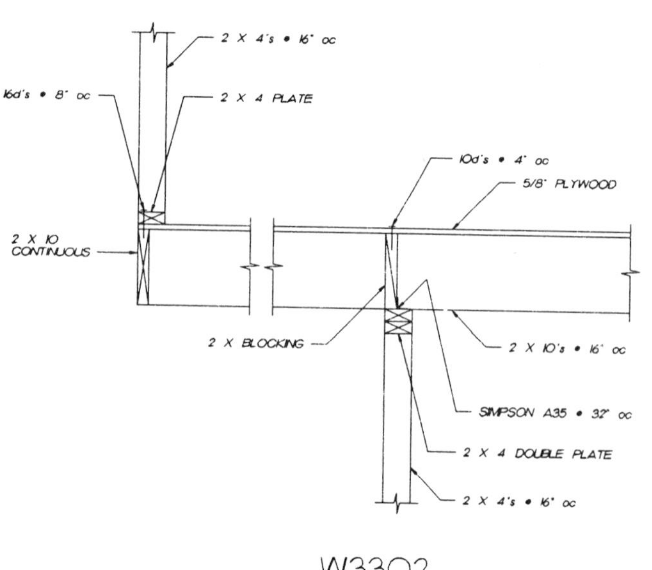

W3302

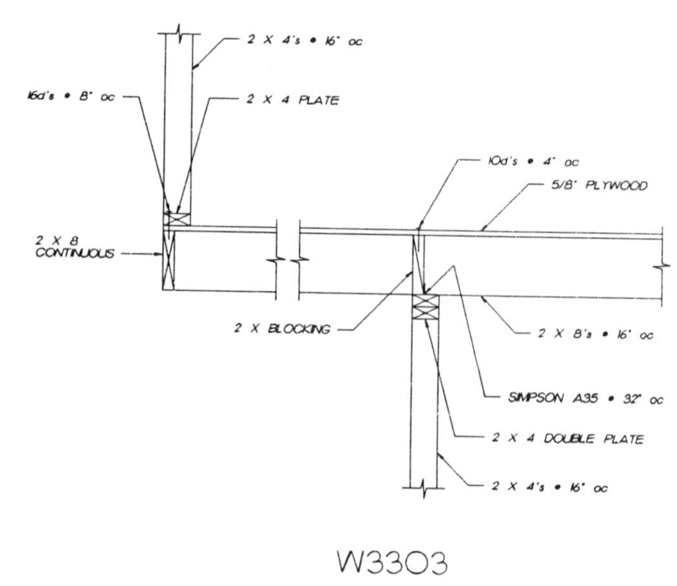

W3303

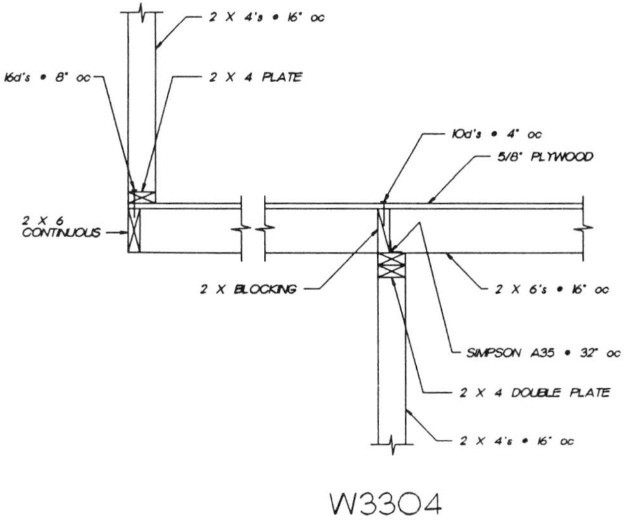

W3304

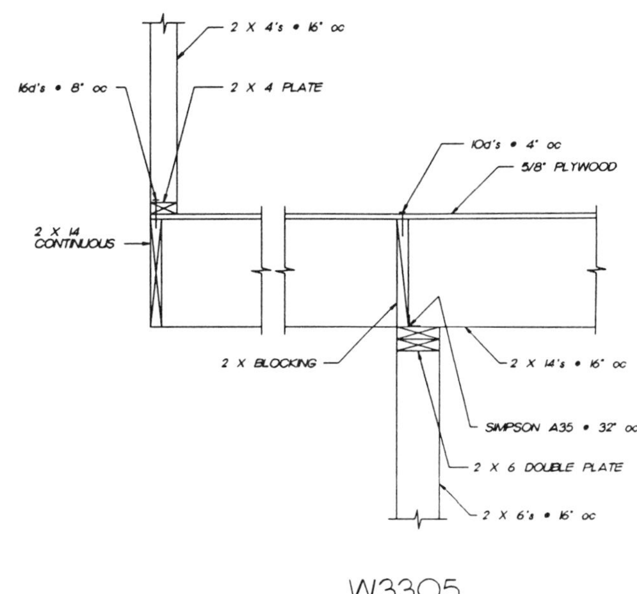

W3305

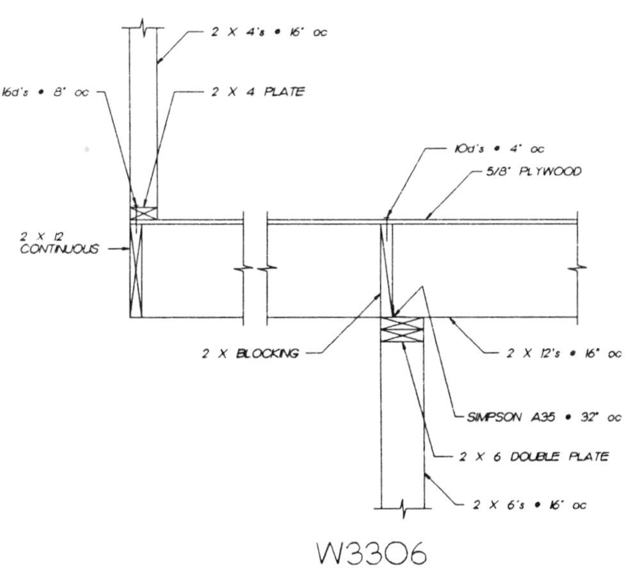

W3306

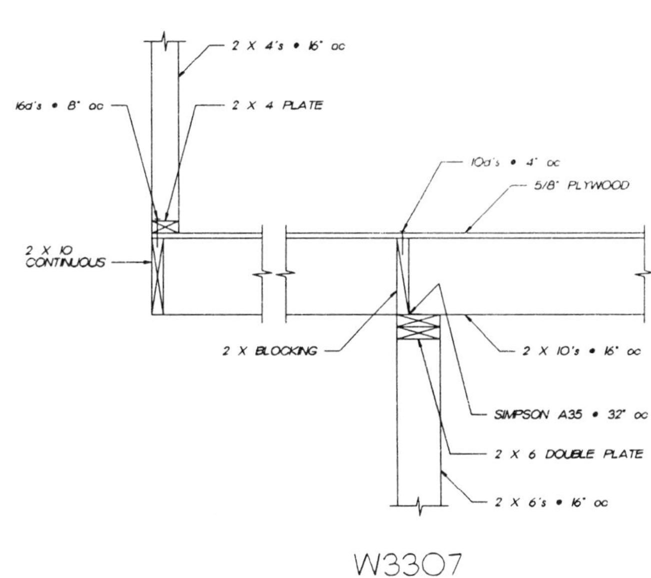

W3307

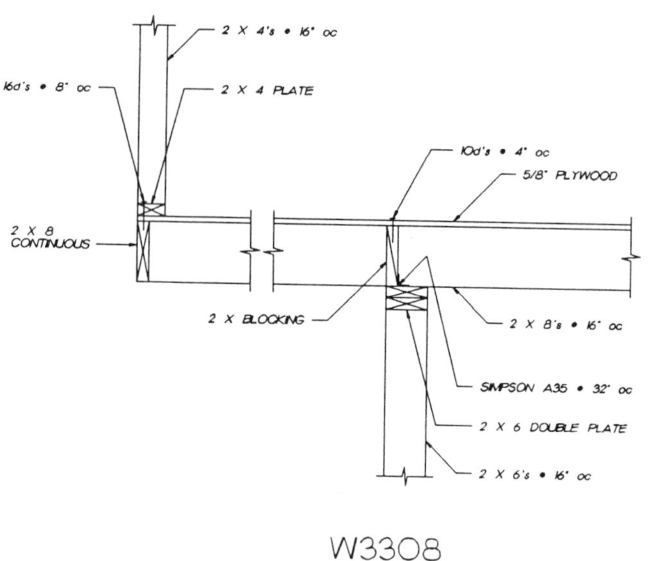

W3308

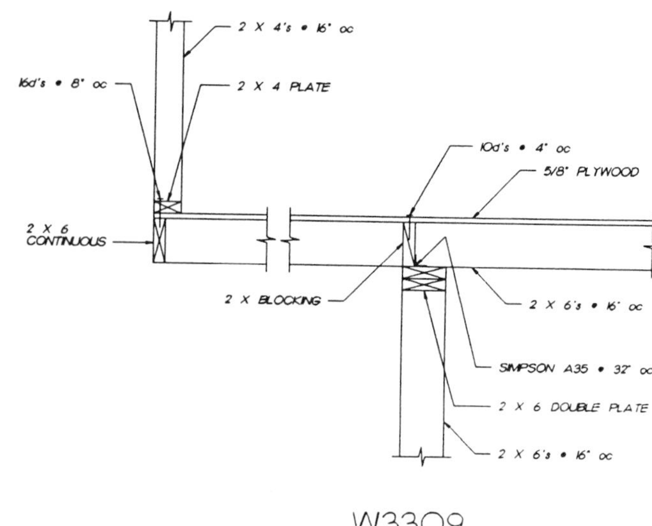

W3309

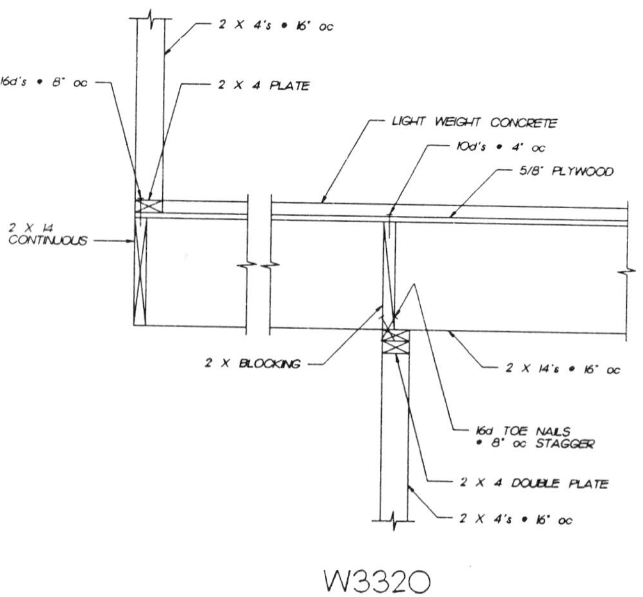

W3320

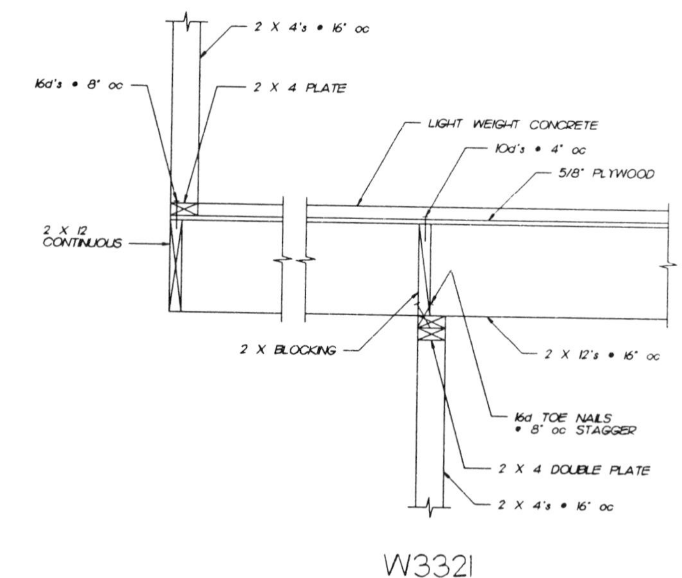

W3321

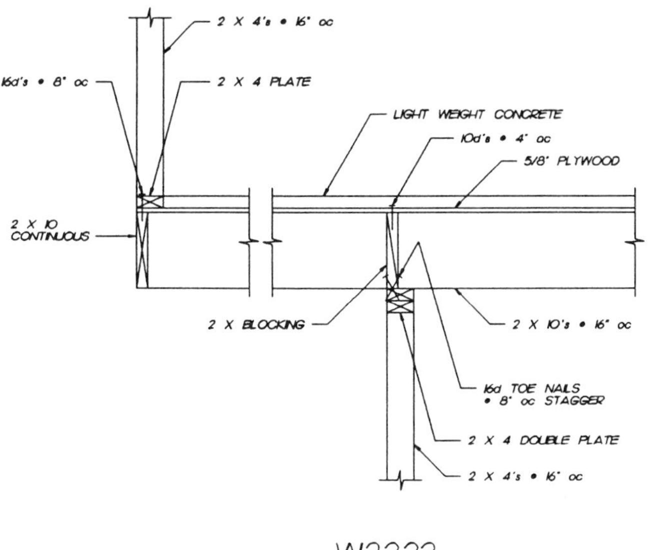

W3322

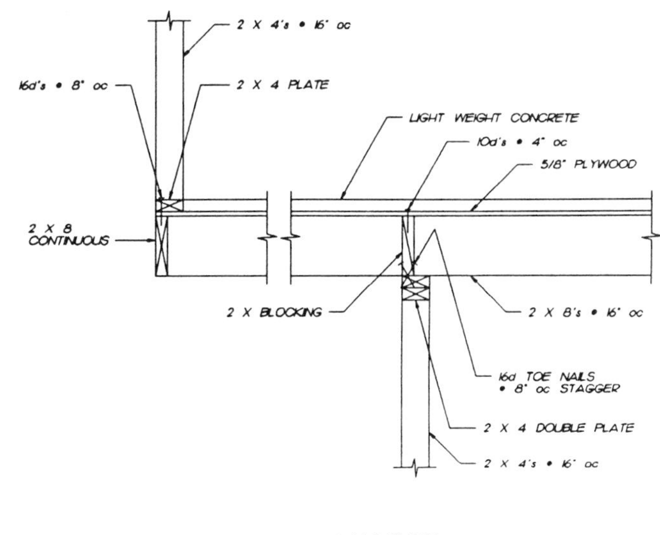

W3323

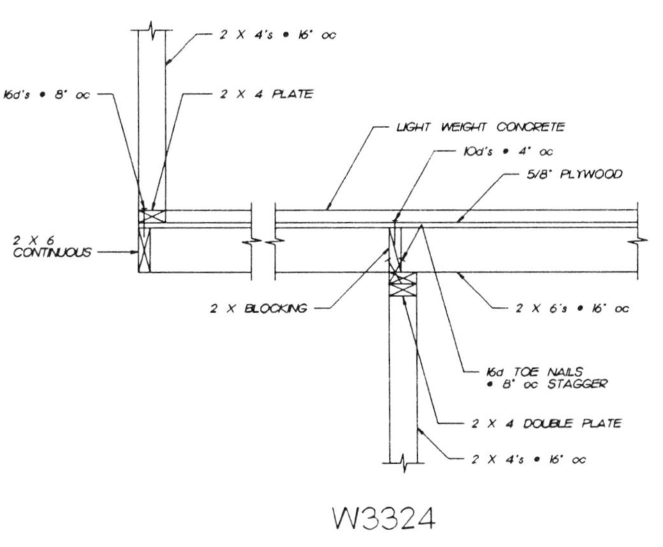

W3324

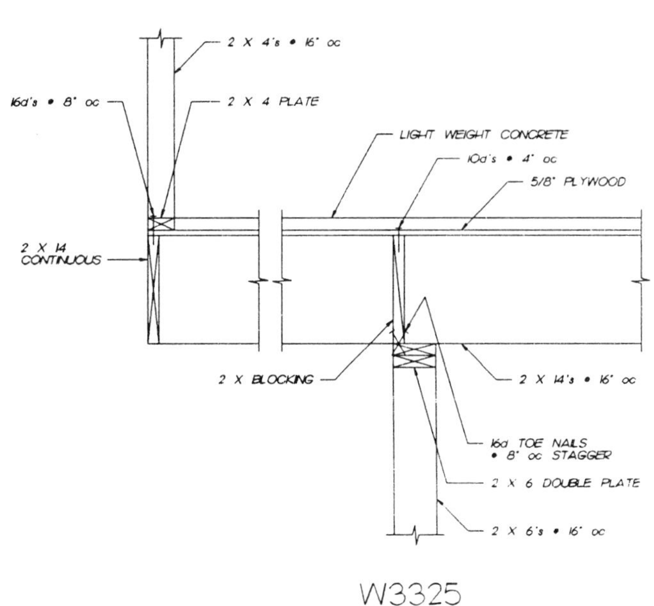

W3325

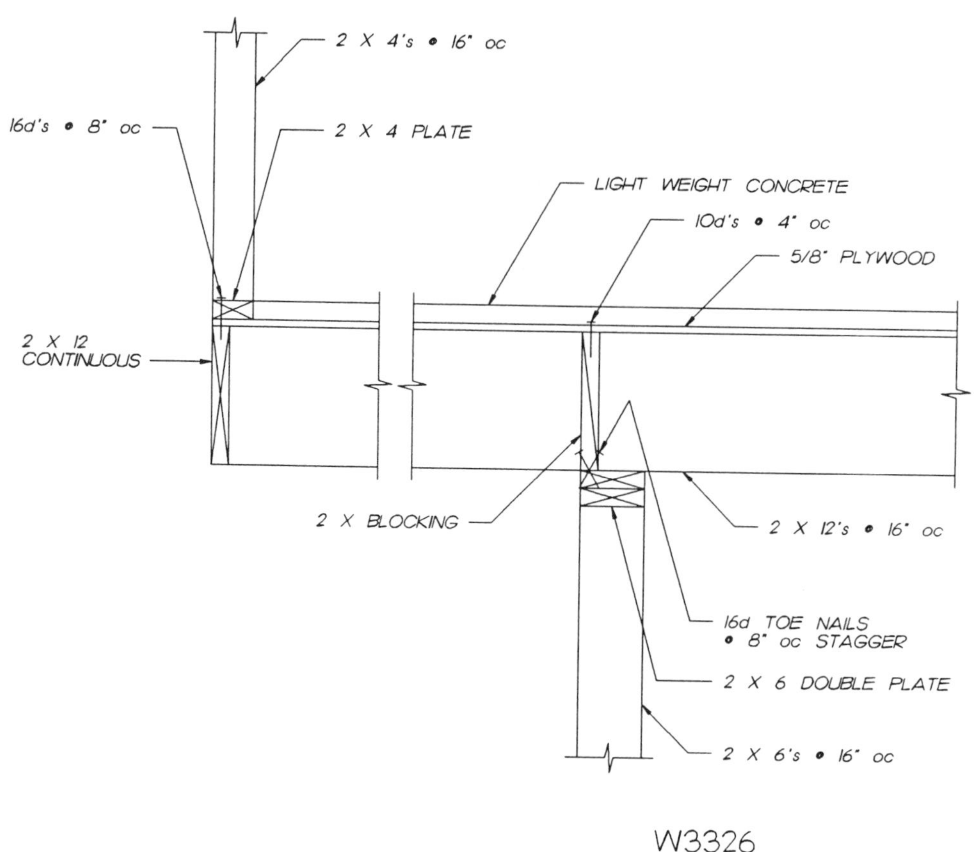

W3326

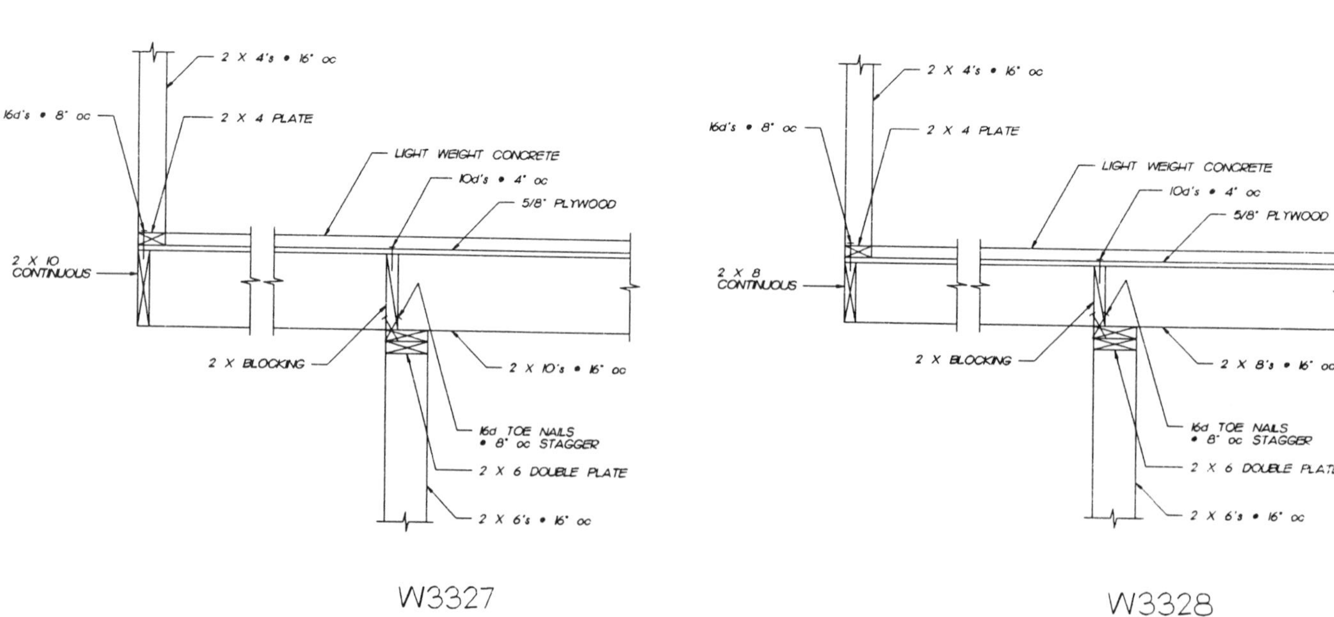

W3327

W3328

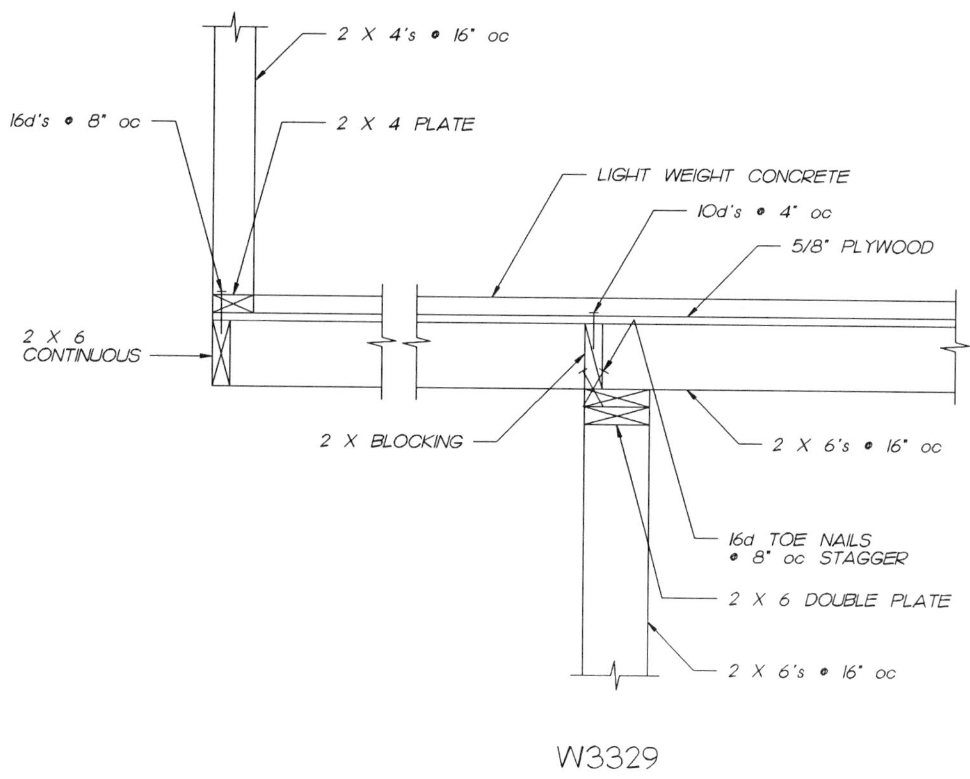

W3329

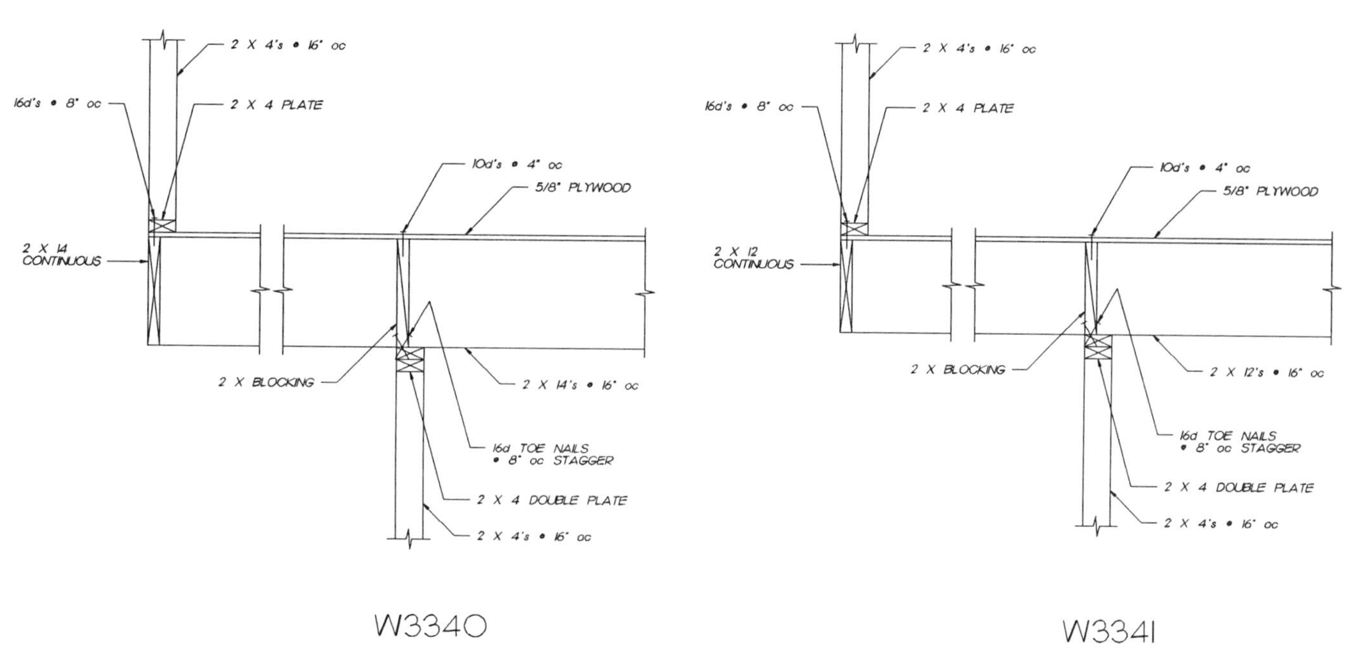

W3340

W3341

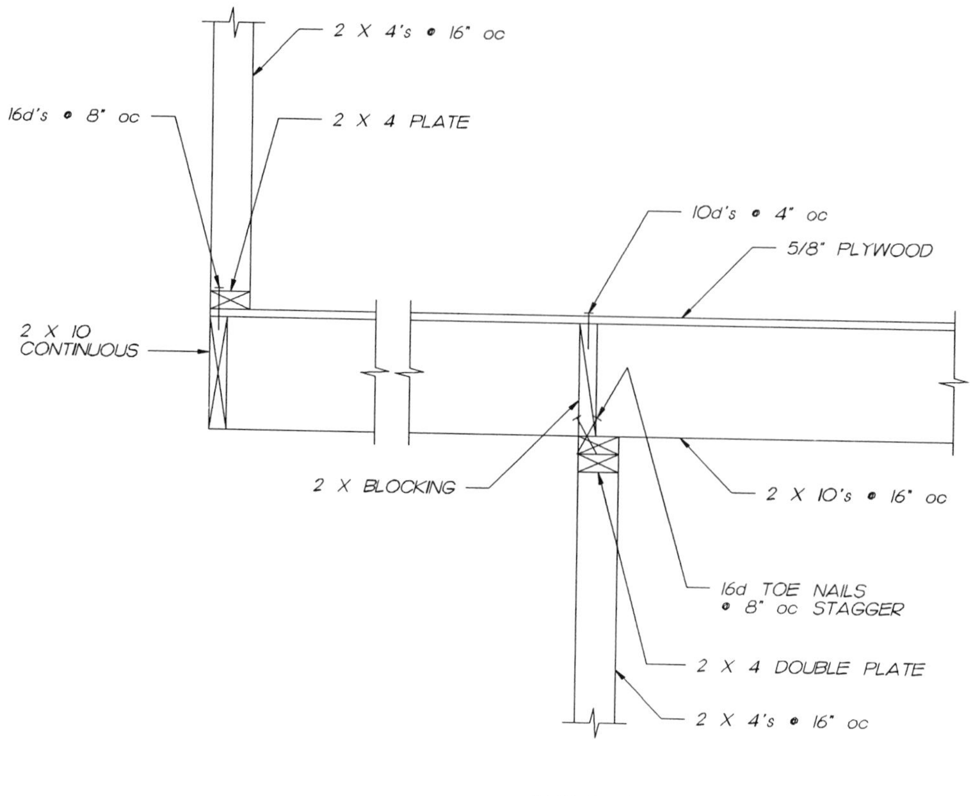

W3342

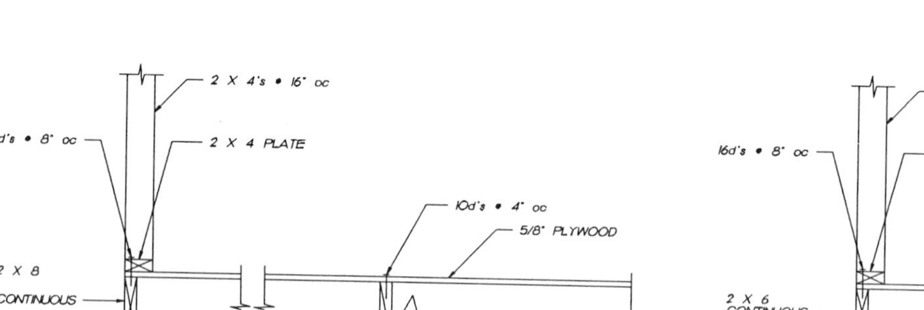

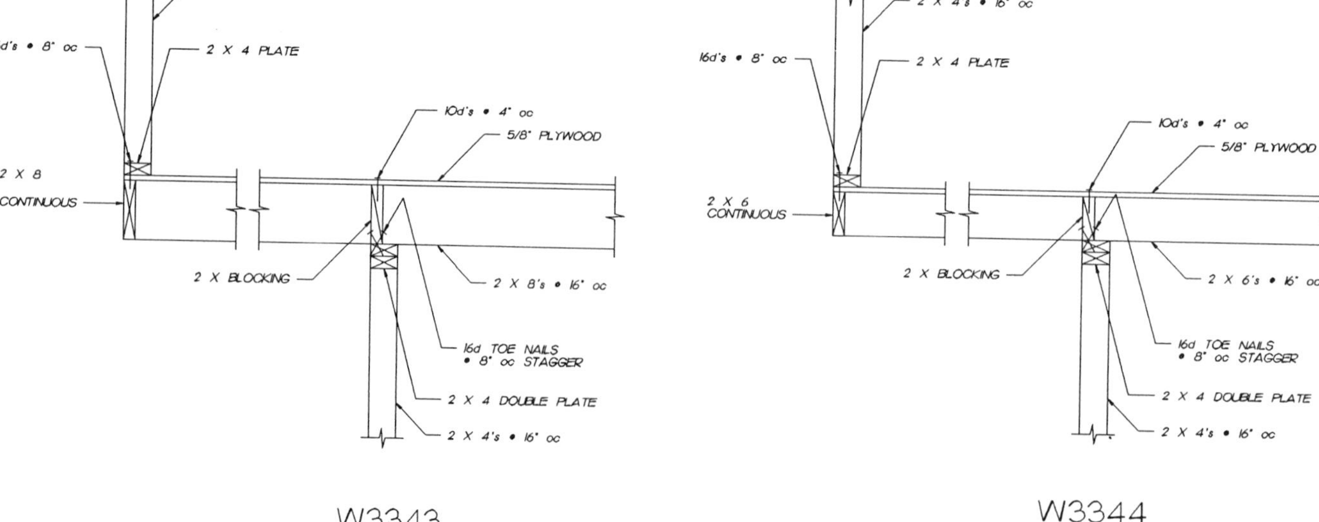

W3343

W3344

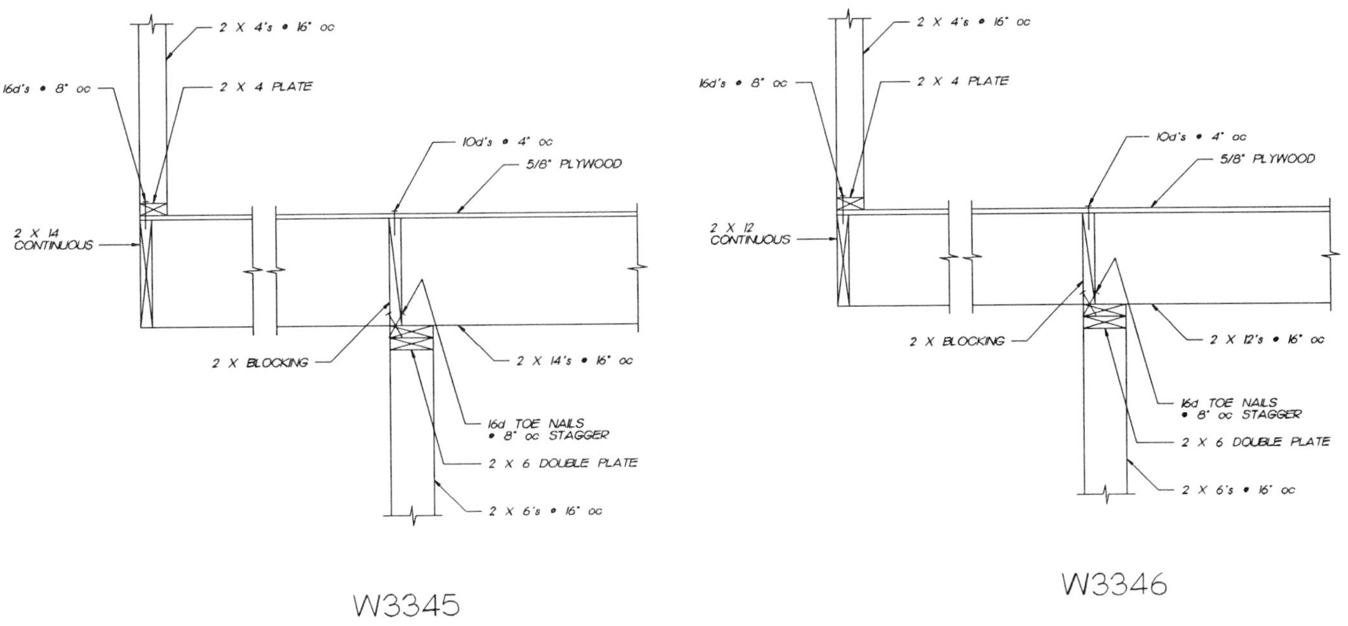

W3345

W3346

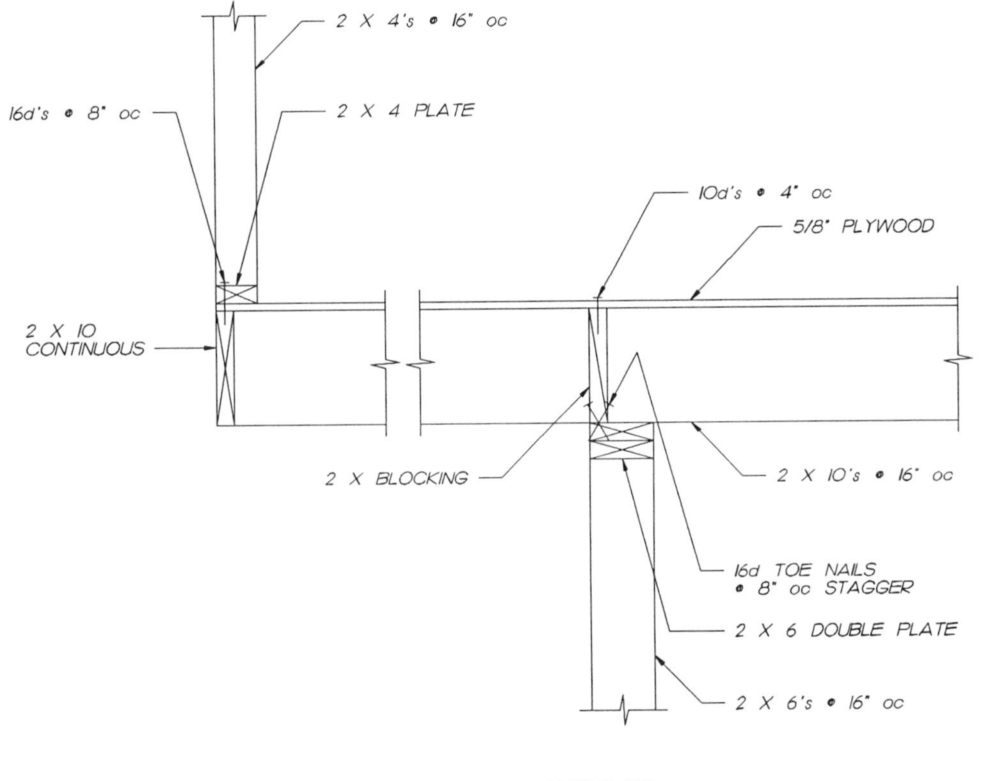

W3347

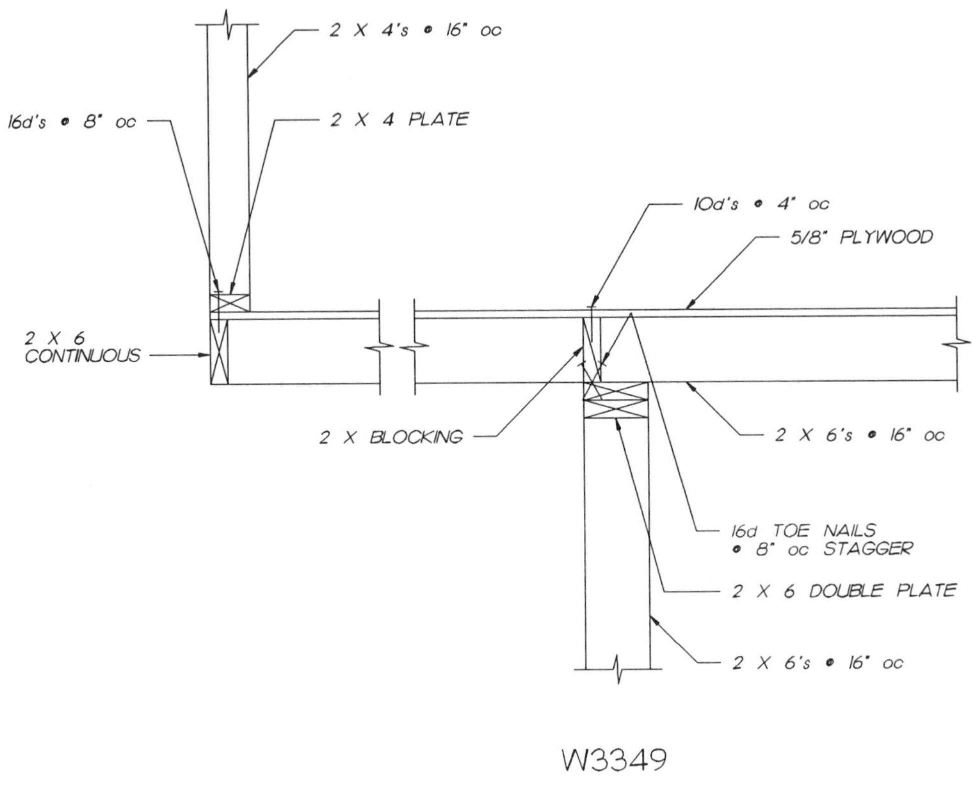
2 X 4's @ 16" oc
16d's @ 8" oc
2 X 4 PLATE
10d's @ 4" oc
5/8" PLYWOOD
2 X 6
CONTINUOUS
2 X BLOCKING
2 X 6's @ 16" oc
16d TOE NAILS
@ 8" oc STAGGER
2 X 6 DOUBLE PLATE
2 X 6's @ 16" oc

W3349

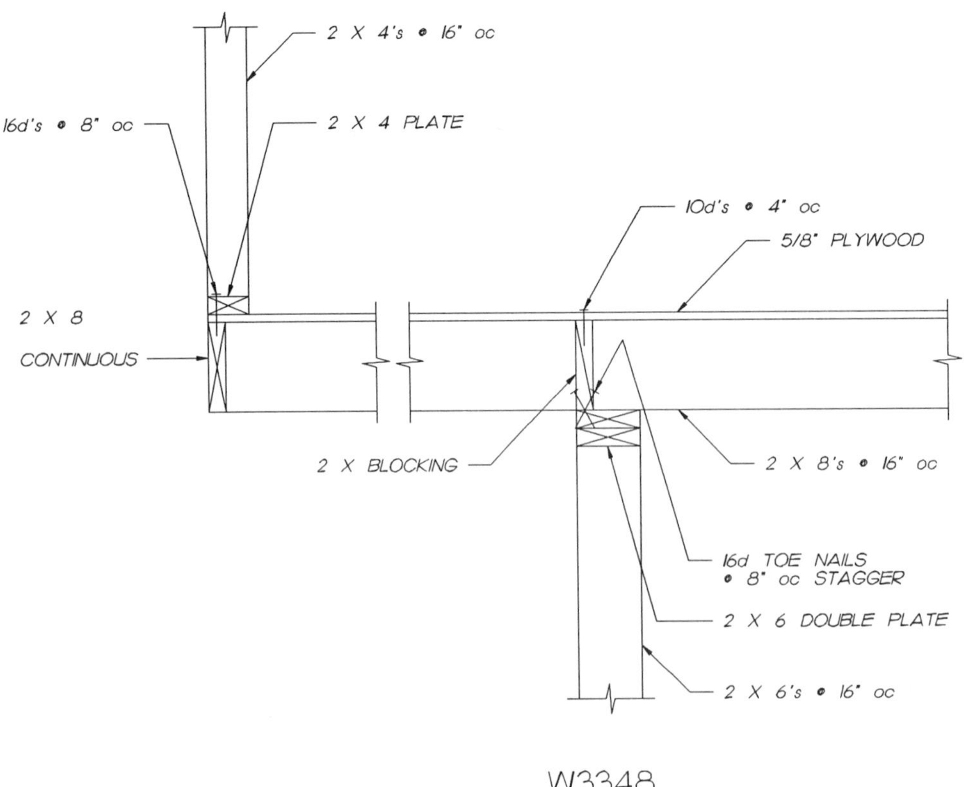
2 X 4's @ 16" oc
16d's @ 8" oc
2 X 4 PLATE
10d's @ 4" oc
5/8" PLYWOOD
2 X 8
CONTINUOUS
2 X BLOCKING
2 X 8's @ 16" oc
16d TOE NAILS
@ 8" oc STAGGER
2 X 6 DOUBLE PLATE
2 X 6's @ 16" oc

W3348

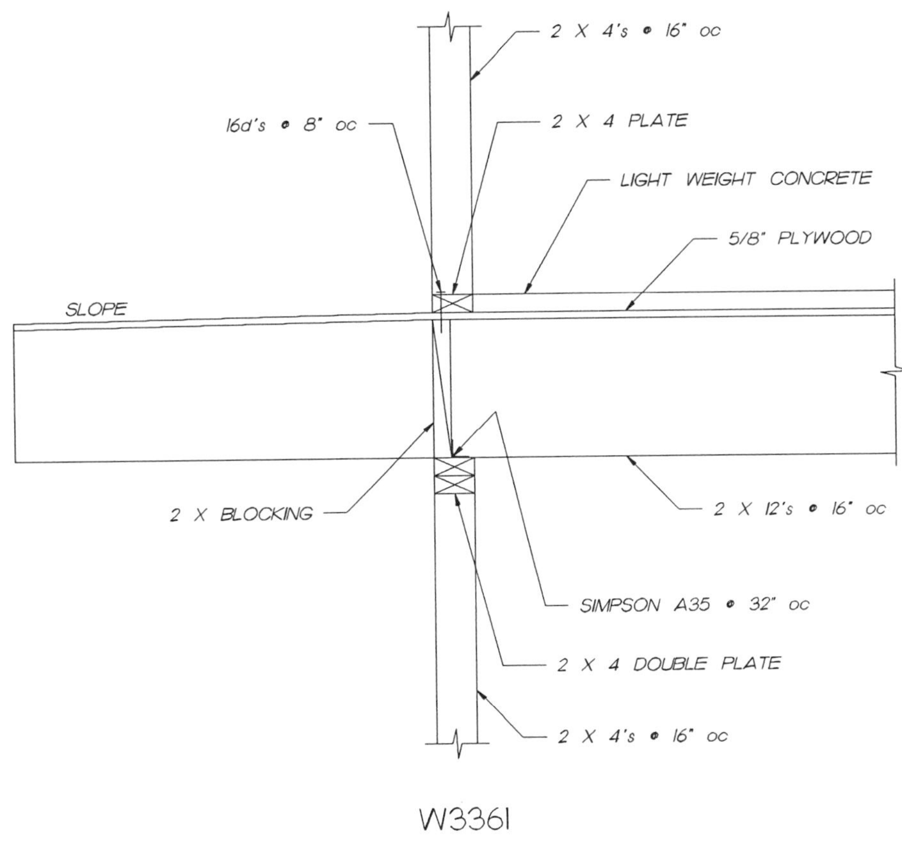
2 X 4's @ 16" oc
16d's @ 8" oc
2 X 4 PLATE
LIGHT WEIGHT CONCRETE
5/8" PLYWOOD
SLOPE
2 X BLOCKING
2 X 12's @ 16" oc
SIMPSON A35 @ 32" oc
2 X 4 DOUBLE PLATE
2 X 4's @ 16" oc

W3361

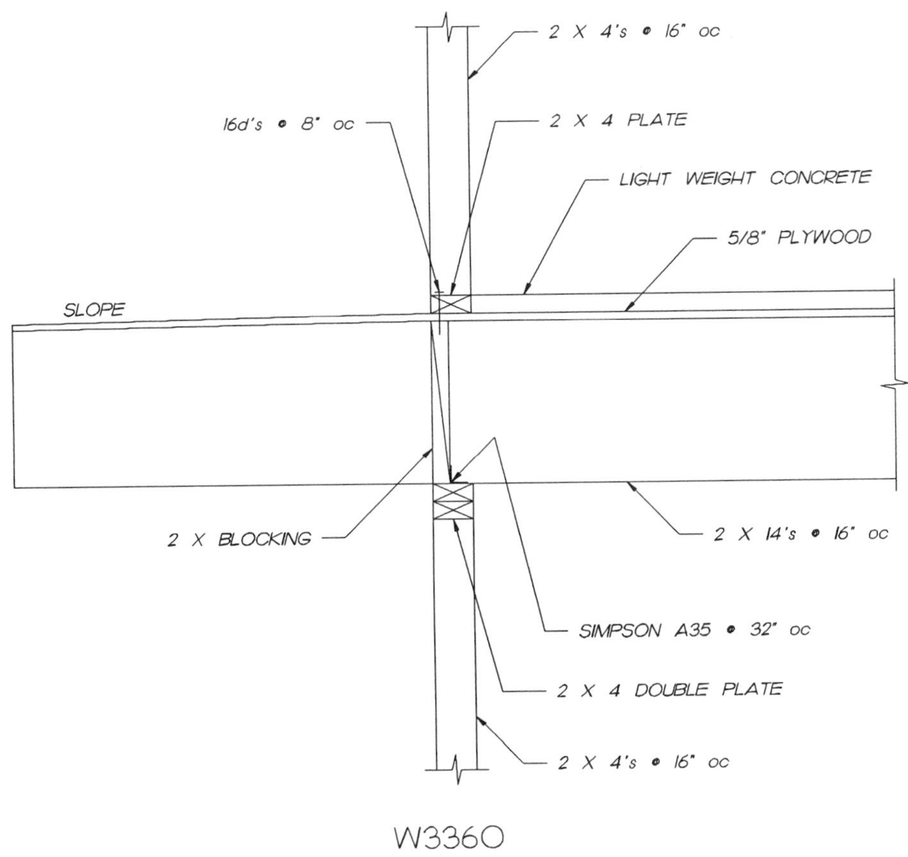
2 X 4's @ 16" oc
16d's @ 8" oc
2 X 4 PLATE
LIGHT WEIGHT CONCRETE
5/8" PLYWOOD
SLOPE
2 X BLOCKING
2 X 14's @ 16" oc
SIMPSON A35 @ 32" oc
2 X 4 DOUBLE PLATE
2 X 4's @ 16" oc

W3360

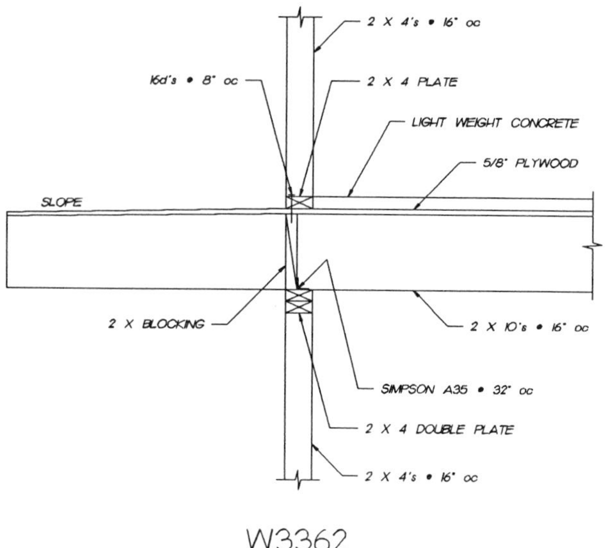

W3362

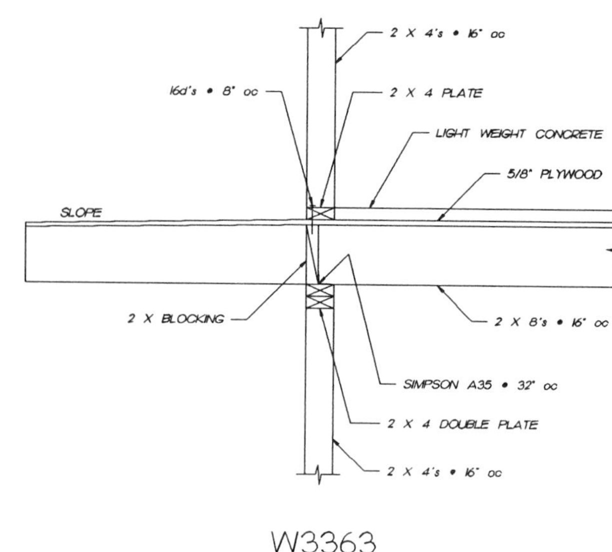

W3363

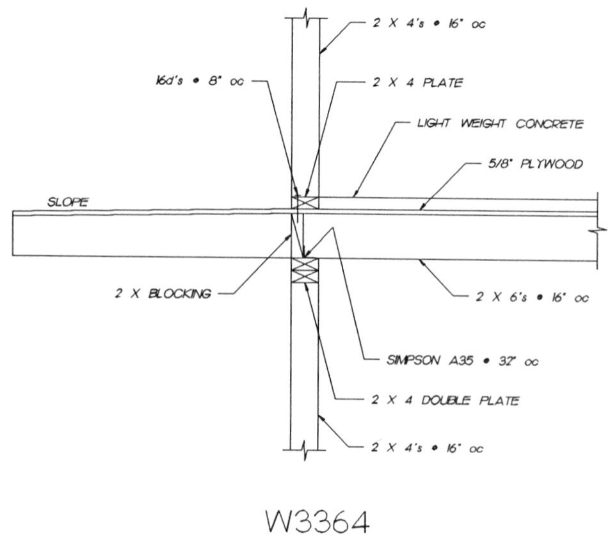

W3364

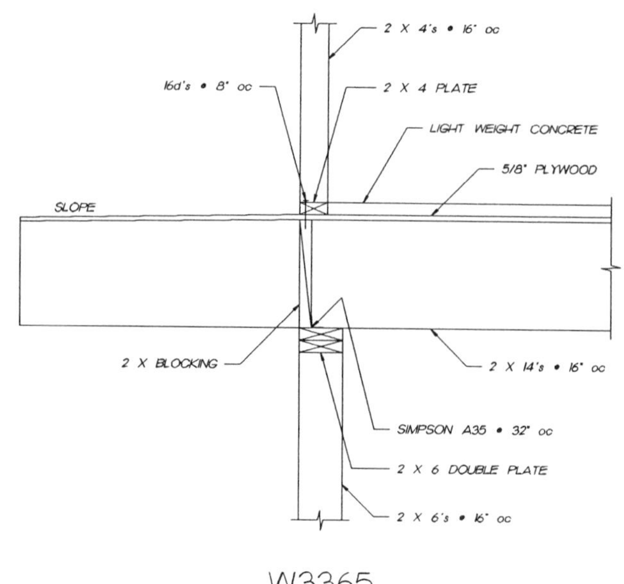

W3365

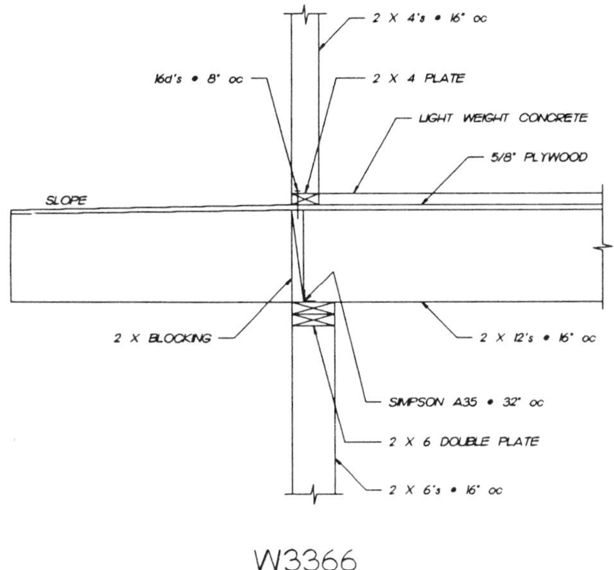

W3366

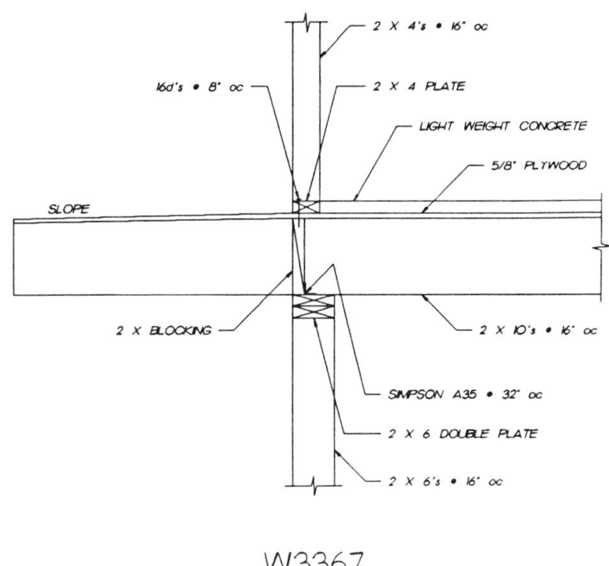

W3367

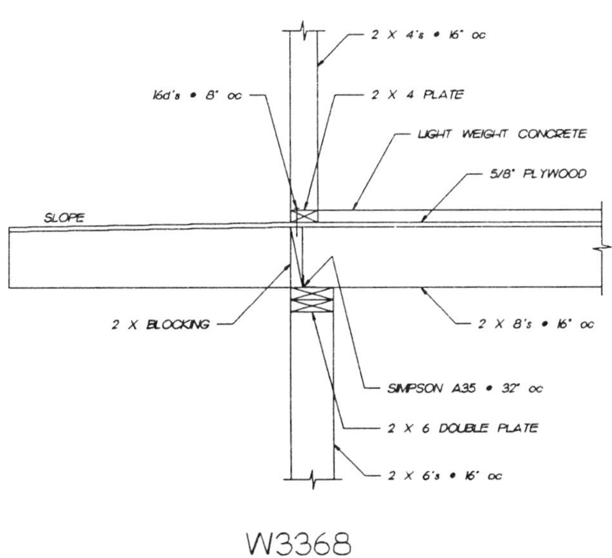

W3368

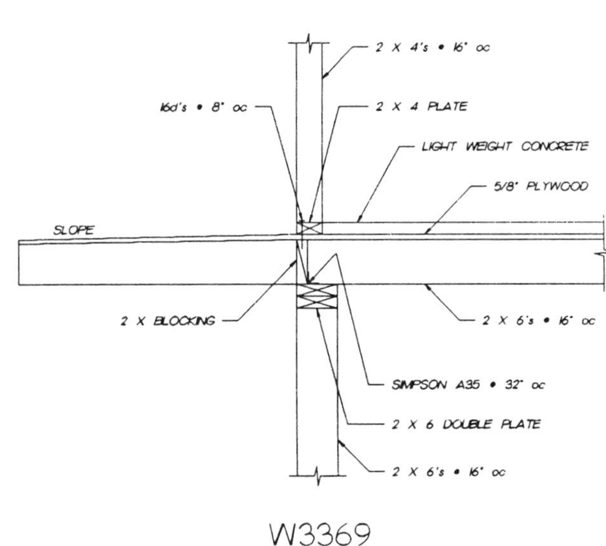

W3369

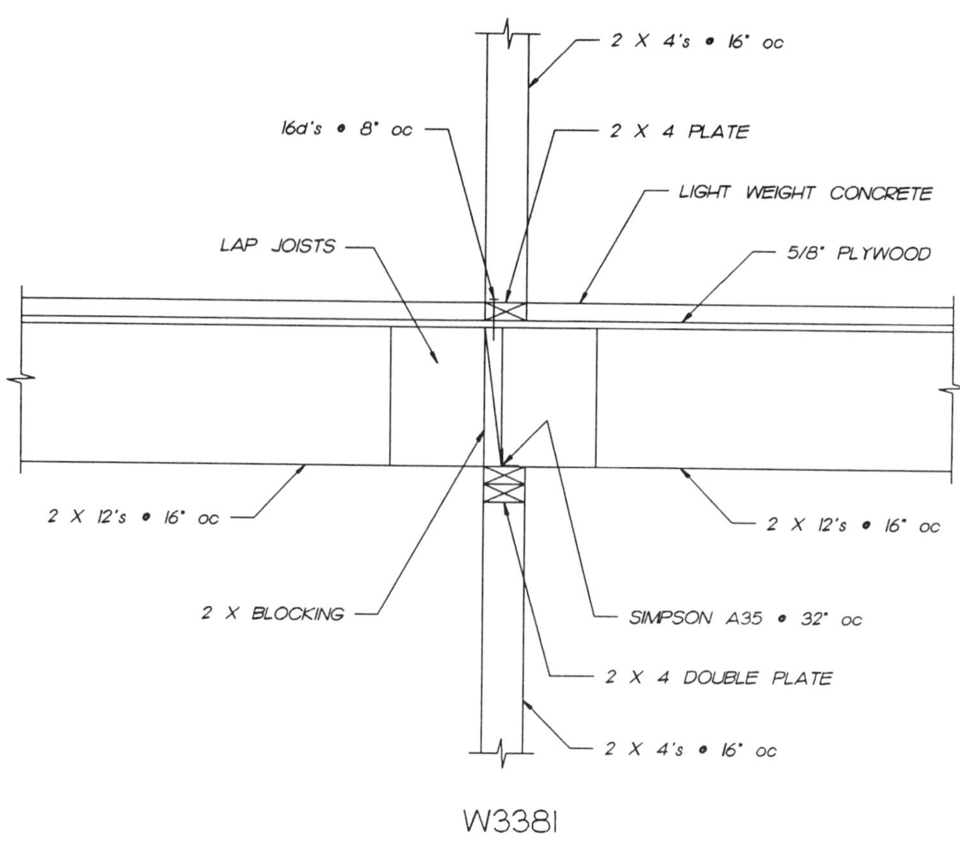
2 X 4's @ 16" oc
16d's @ 8" oc
2 X 4 PLATE
LIGHT WEIGHT CONCRETE
LAP JOISTS
5/8" PLYWOOD
2 X 12's @ 16" oc
2 X 12's @ 16" oc
2 X BLOCKING
SIMPSON A35 @ 32" oc
2 X 4 DOUBLE PLATE
2 X 4's @ 16" oc

W3381

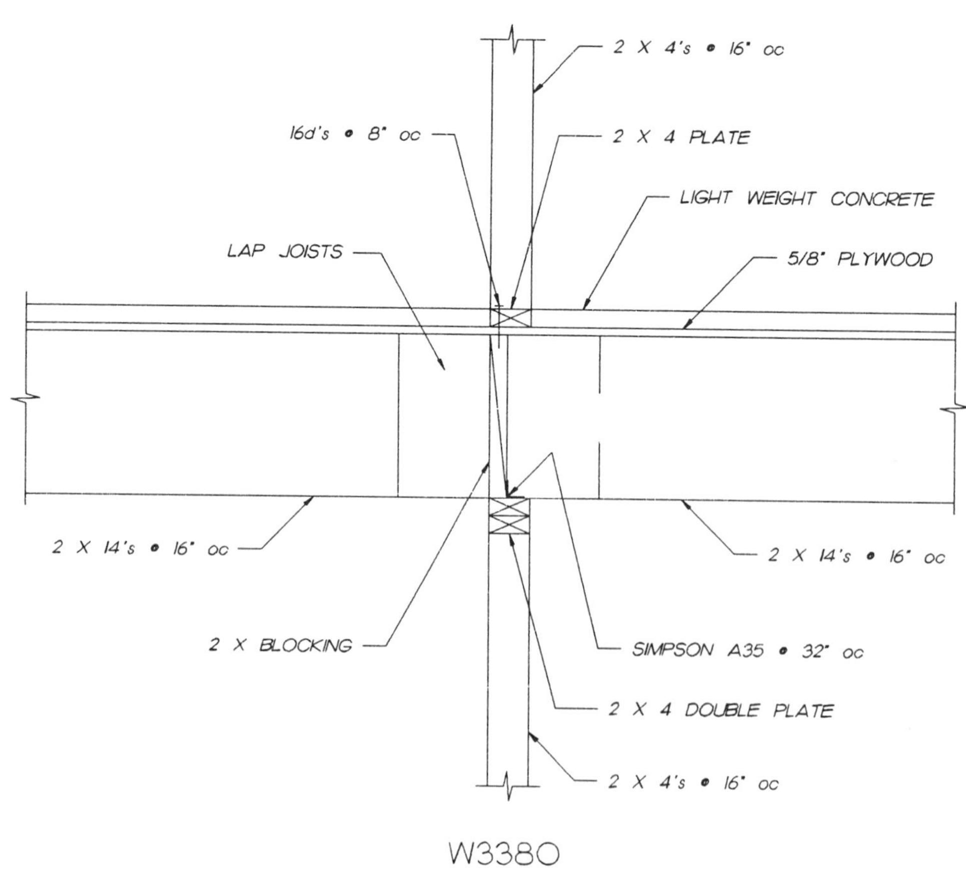
2 X 4's @ 16" oc
16d's @ 8" oc
2 X 4 PLATE
LIGHT WEIGHT CONCRETE
LAP JOISTS
5/8" PLYWOOD
2 X 14's @ 16" oc
2 X 14's @ 16" oc
2 X BLOCKING
SIMPSON A35 @ 32" oc
2 X 4 DOUBLE PLATE
2 X 4's @ 16" oc

W3380

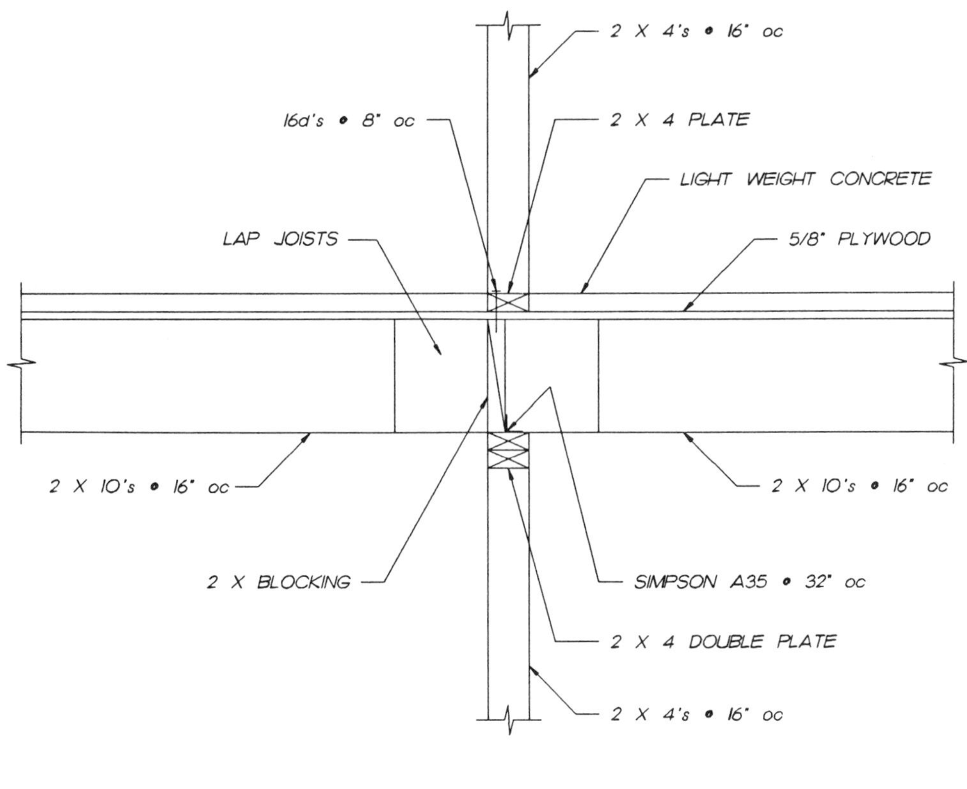
2 X 4's @ 16" oc
16d's @ 8" oc
2 X 4 PLATE
LIGHT WEIGHT CONCRETE
LAP JOISTS
5/8" PLYWOOD
2 X 10's @ 16" oc
2 X 10's @ 16" oc
2 X BLOCKING
SIMPSON A35 @ 32" oc
2 X 4 DOUBLE PLATE
2 X 4's @ 16" oc

W3382

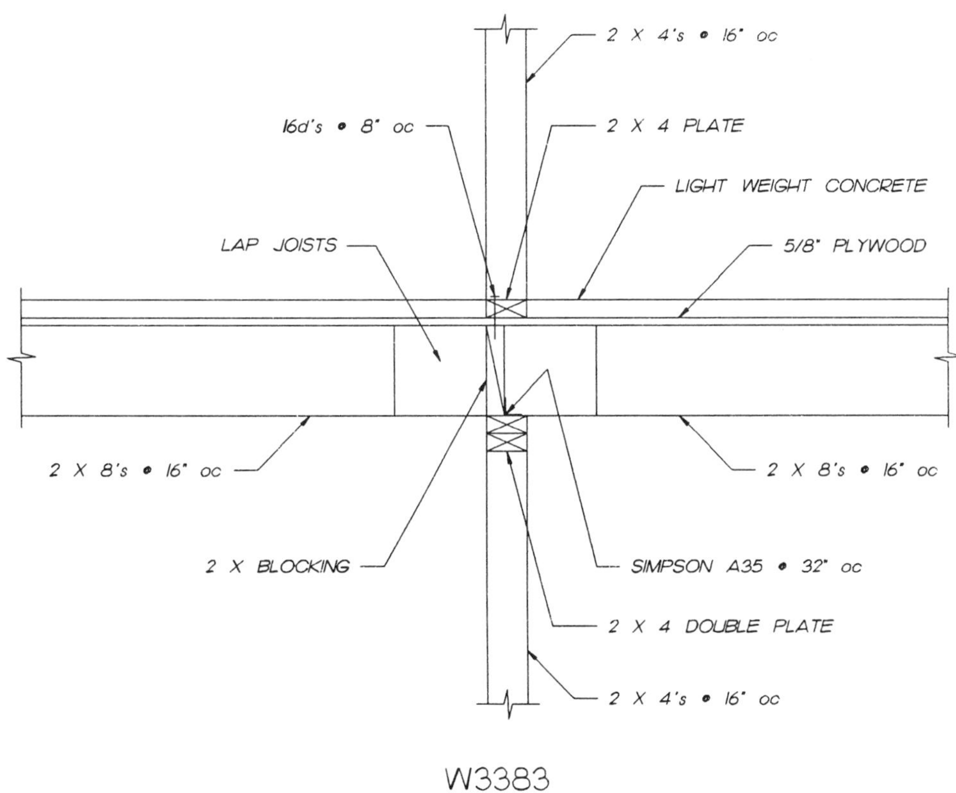
2 X 4's @ 16" oc
16d's @ 8" oc
2 X 4 PLATE
LIGHT WEIGHT CONCRETE
LAP JOISTS
5/8" PLYWOOD
2 X 8's @ 16" oc
2 X 8's @ 16" oc
2 X BLOCKING
SIMPSON A35 @ 32" oc
2 X 4 DOUBLE PLATE
2 X 4's @ 16" oc

W3383

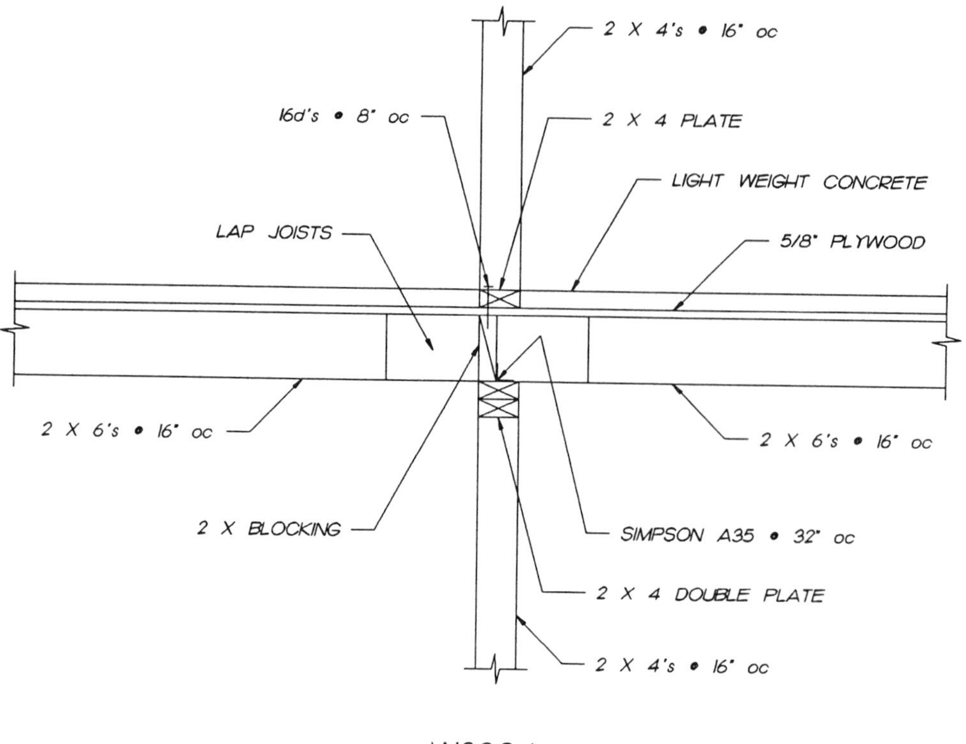
2 X 4's @ 16" oc
16d's @ 8" oc
2 X 4 PLATE
LIGHT WEIGHT CONCRETE
LAP JOISTS
5/8" PLYWOOD
2 X 6's @ 16" oc
2 X 6's @ 16" oc
2 X BLOCKING
SIMPSON A35 @ 32" oc
2 X 4 DOUBLE PLATE
2 X 4's @ 16" oc

W3384

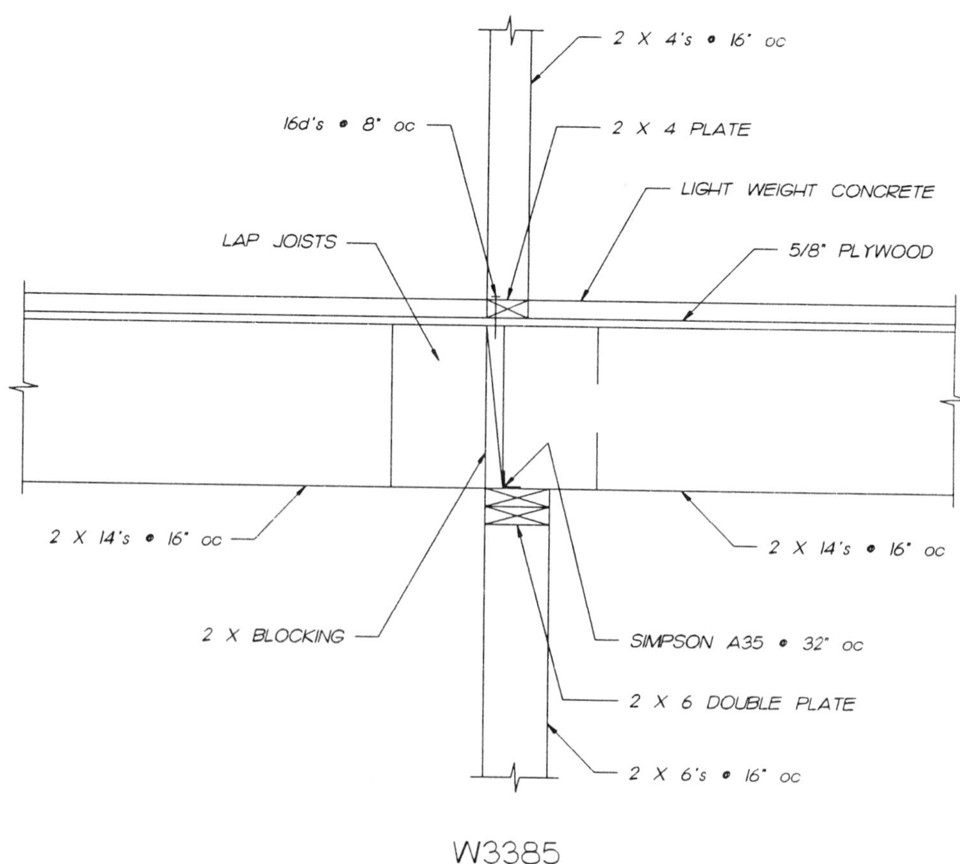
2 X 4's @ 16" oc
16d's @ 8" oc
2 X 4 PLATE
LIGHT WEIGHT CONCRETE
LAP JOISTS
5/8" PLYWOOD
2 X 14's @ 16" oc
2 X 14's @ 16" oc
2 X BLOCKING
SIMPSON A35 @ 32" oc
2 X 6 DOUBLE PLATE
2 X 6's @ 16" oc

W3385

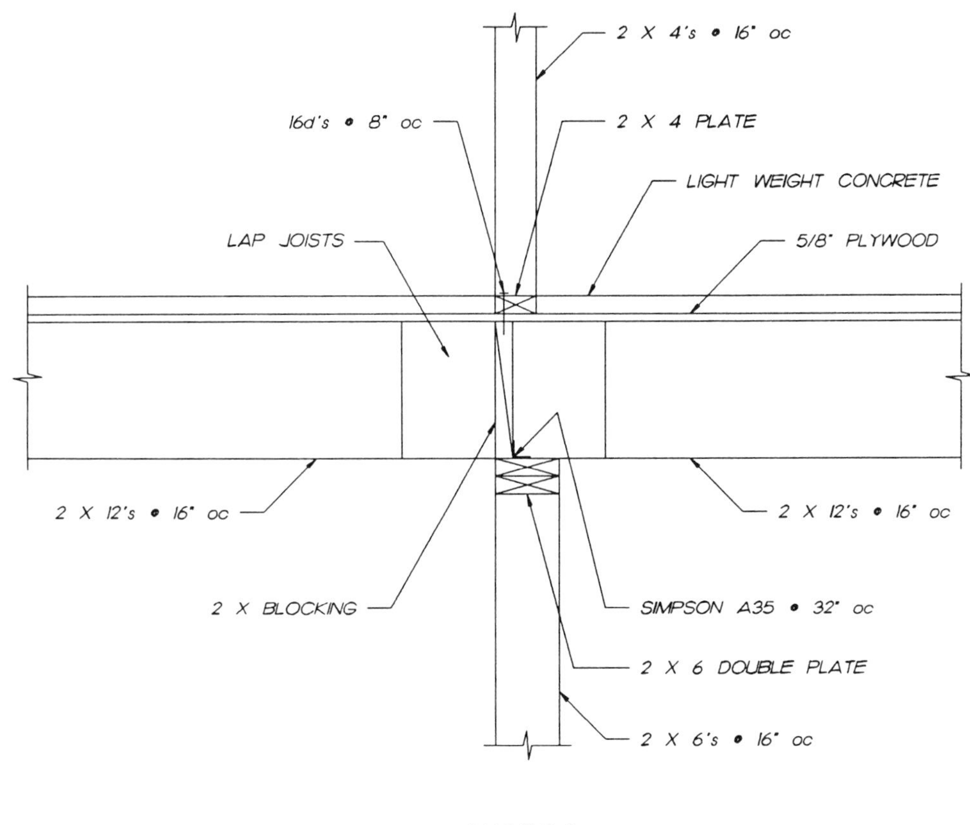
2 X 4's @ 16" oc
16d's @ 8" oc
2 X 4 PLATE
LIGHT WEIGHT CONCRETE
LAP JOISTS
5/8" PLYWOOD
2 X 12's @ 16" oc
2 X 12's @ 16" oc
2 X BLOCKING
SIMPSON A35 @ 32" oc
2 X 6 DOUBLE PLATE
2 X 6's @ 16" oc

W3386

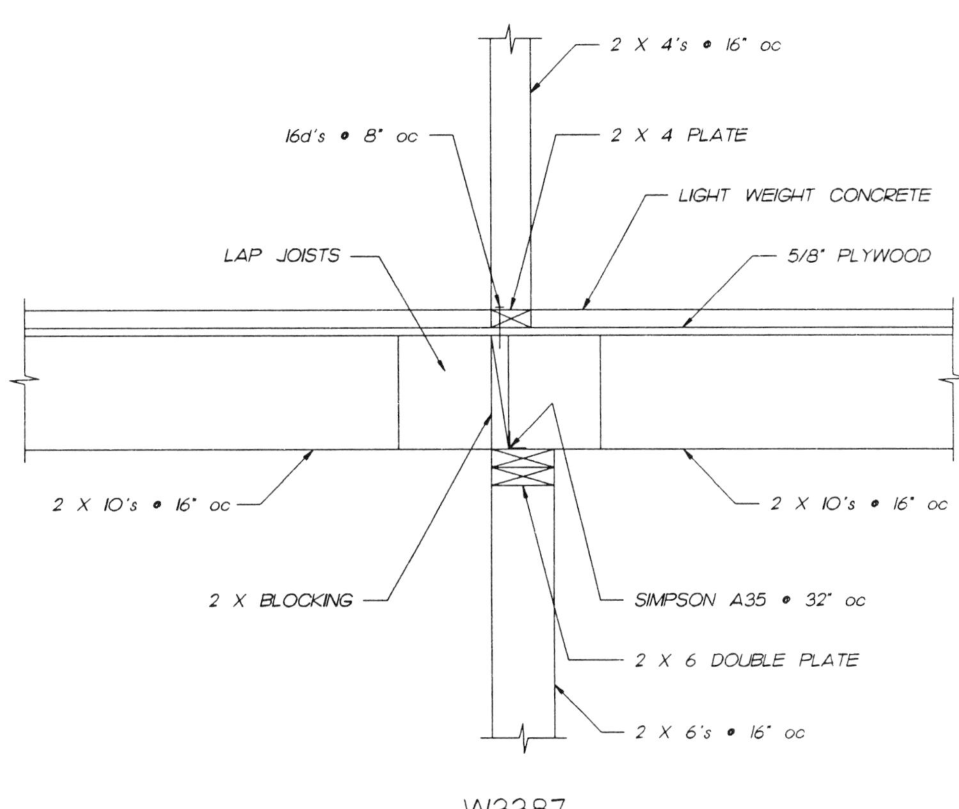
2 X 4's @ 16" oc
16d's @ 8" oc
2 X 4 PLATE
LIGHT WEIGHT CONCRETE
LAP JOISTS
5/8" PLYWOOD
2 X 10's @ 16" oc
2 X 10's @ 16" oc
2 X BLOCKING
SIMPSON A35 @ 32" oc
2 X 6 DOUBLE PLATE
2 X 6's @ 16" oc

W3387

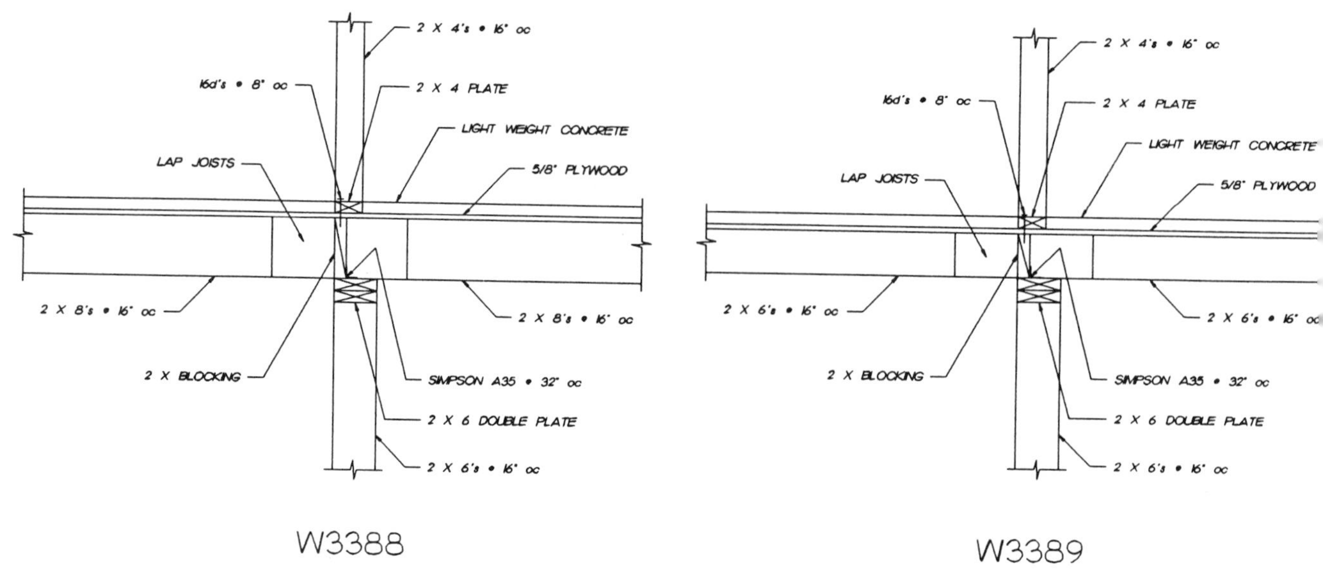
2 X 4's @ 16" oc
16d's @ 8" oc
2 X 4 PLATE
LIGHT WEIGHT CONCRETE
LAP JOISTS
5/8" PLYWOOD
2 X 8's @ 16" oc
2 X 8's @ 16" oc
2 X BLOCKING
SIMPSON A35 @ 32" oc
2 X 6 DOUBLE PLATE
2 X 6's @ 16" oc
W3388
2 X 4's @ 16" oc
16d's @ 8" oc
2 X 4 PLATE
LIGHT WEIGHT CONCRETE
LAP JOISTS
5/8" PLYWOOD
2 X 6's @ 16" oc
2 X 6's @ 16" oc
2 X BLOCKING
SIMPSON A35 @ 32" oc
2 X 6 DOUBLE PLATE
2 X 6's @ 16" oc
W3389

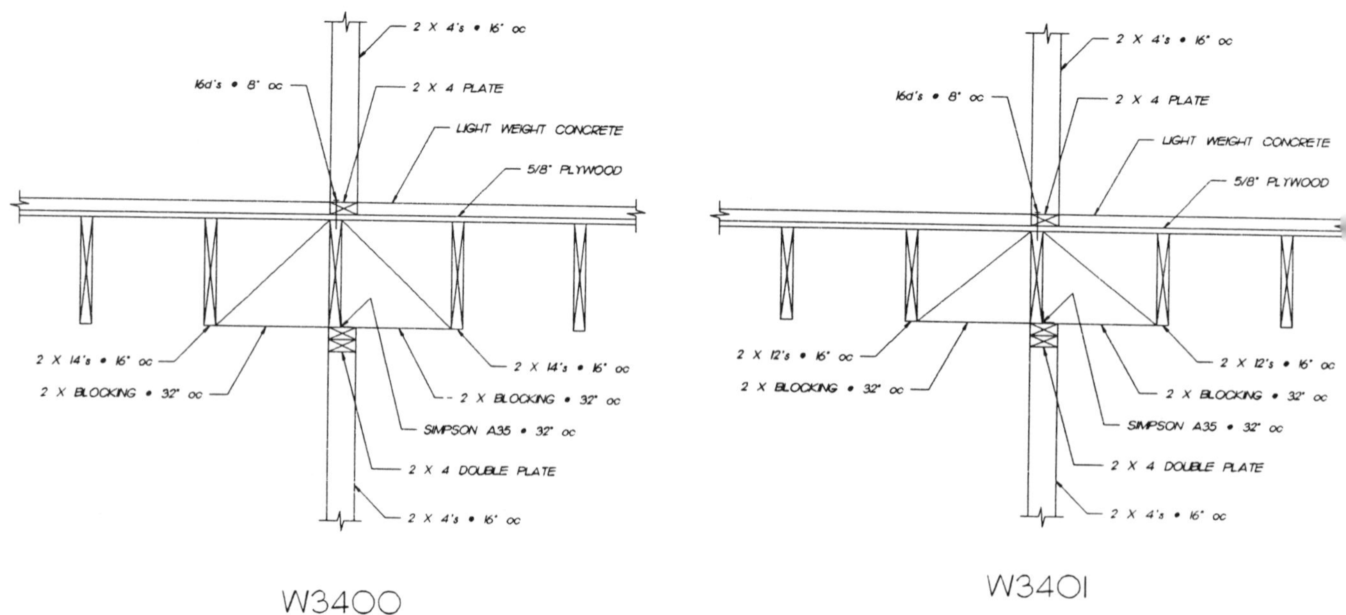
2 X 4's @ 16" oc
16d's @ 8" oc
2 X 4 PLATE
LIGHT WEIGHT CONCRETE
5/8" PLYWOOD
2 X 14's @ 16" oc
2 X BLOCKING @ 32" oc
2 X 14's @ 16" oc
2 X BLOCKING @ 32" oc
SIMPSON A35 @ 32" oc
2 X 4 DOUBLE PLATE
2 X 4's @ 16" oc
W3400
2 X 4's @ 16" oc
16d's @ 8" oc
2 X 4 PLATE
LIGHT WEIGHT CONCRETE
5/8" PLYWOOD
2 X 12's @ 16" oc
2 X BLOCKING @ 32" oc
2 X 12's @ 16" oc
2 X BLOCKING @ 32" oc
SIMPSON A35 @ 32" oc
2 X 4 DOUBLE PLATE
2 X 4's @ 16" oc
W3401

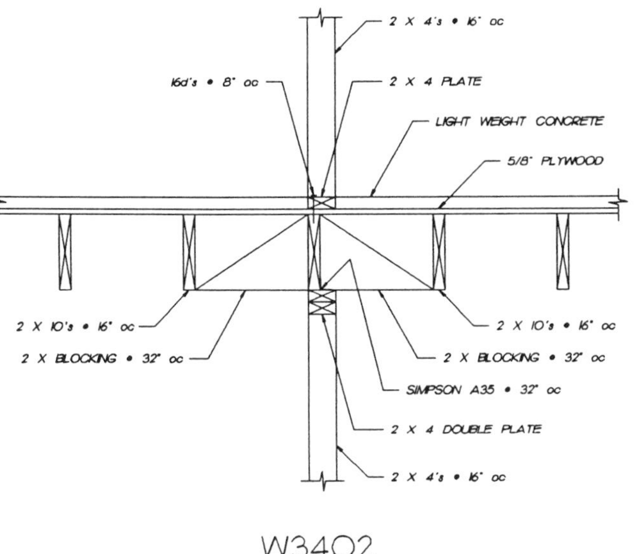

W3402

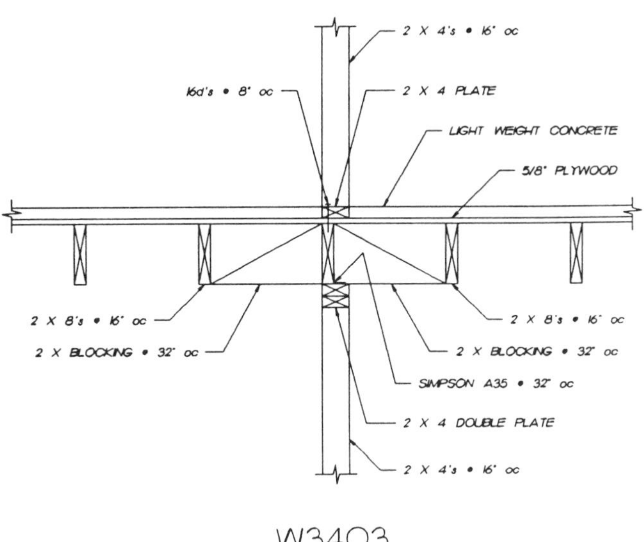

W3403

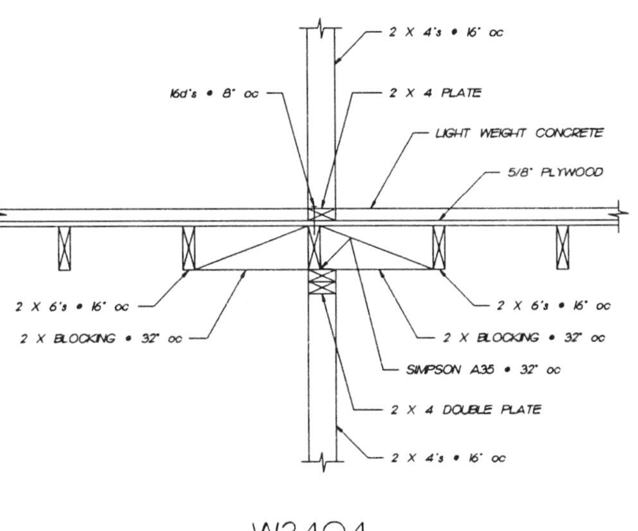

W3404

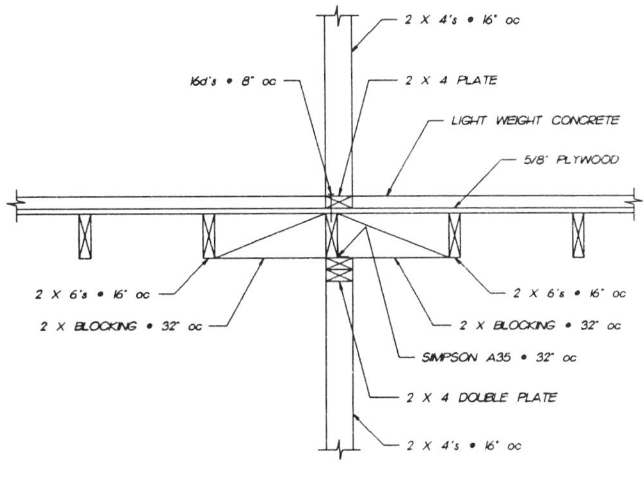

W3405

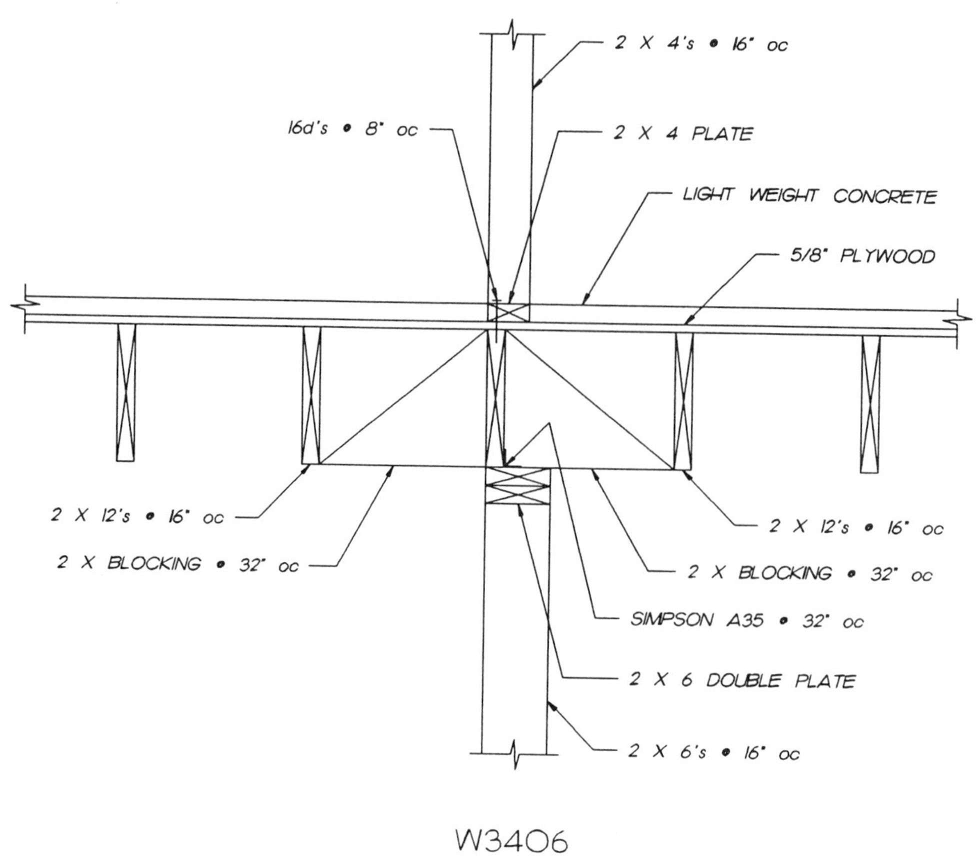

W3406

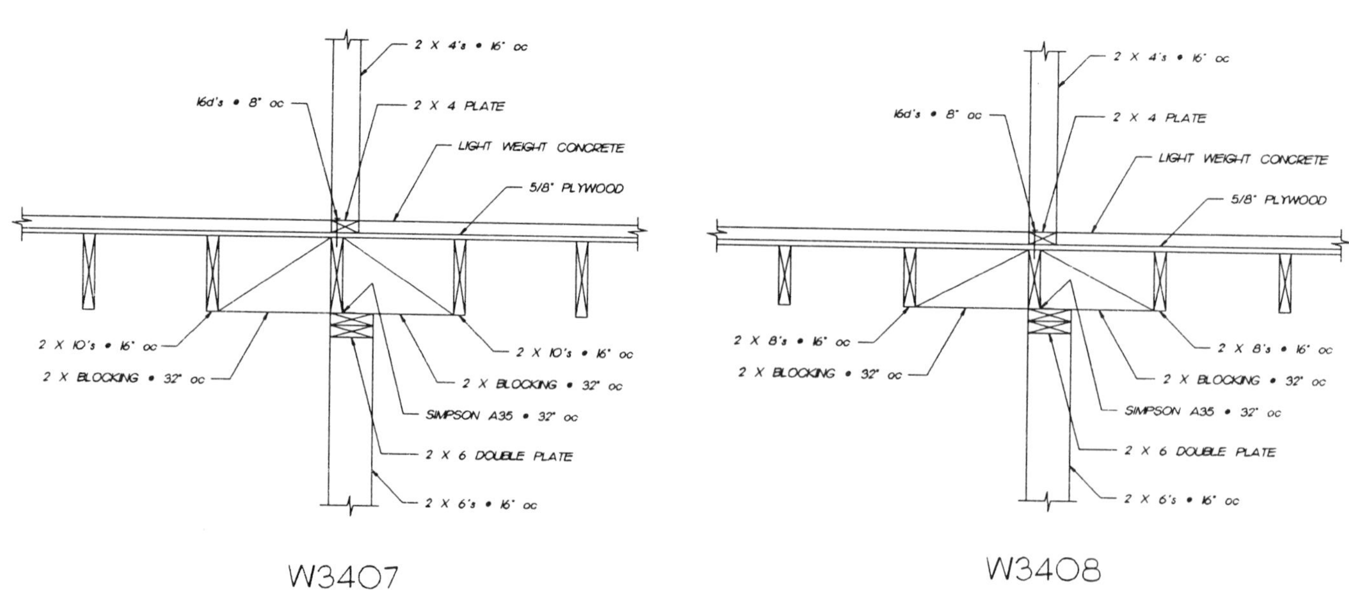

W3407

W3408

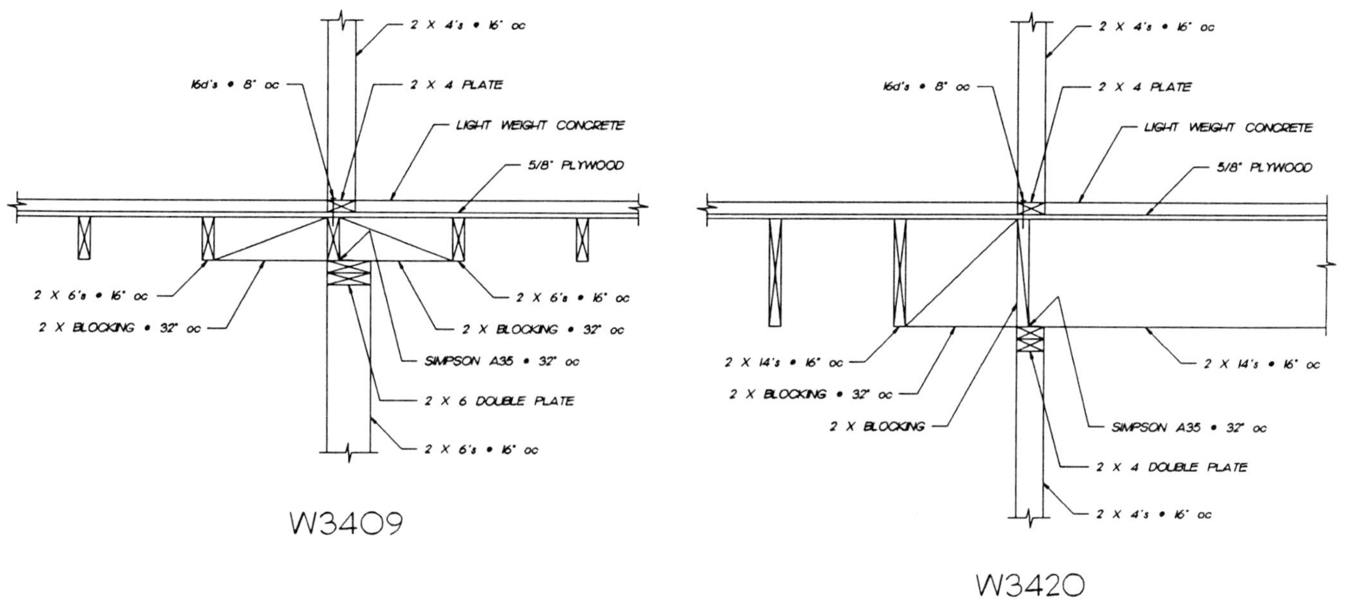
2 X 4's • 16" oc
16d's • 8" oc
2 X 4 PLATE
LIGHT WEIGHT CONCRETE
5/8" PLYWOOD
2 X 6's • 16" oc
2 X BLOCKING • 32" oc
2 X 6's • 16" oc
2 X BLOCKING • 32" oc
SIMPSON A35 • 32" oc
2 X 6 DOUBLE PLATE
2 X 6's • 16" oc
W3409
2 X 4's • 16" oc
16d's • 8" oc
2 X 4 PLATE
LIGHT WEIGHT CONCRETE
5/8" PLYWOOD
2 X 14's • 16" oc
2 X BLOCKING • 32" oc
2 X BLOCKING
2 X 14's • 16" oc
SIMPSON A35 • 32" oc
2 X 4 DOUBLE PLATE
2 X 4's • 16" oc
W3420

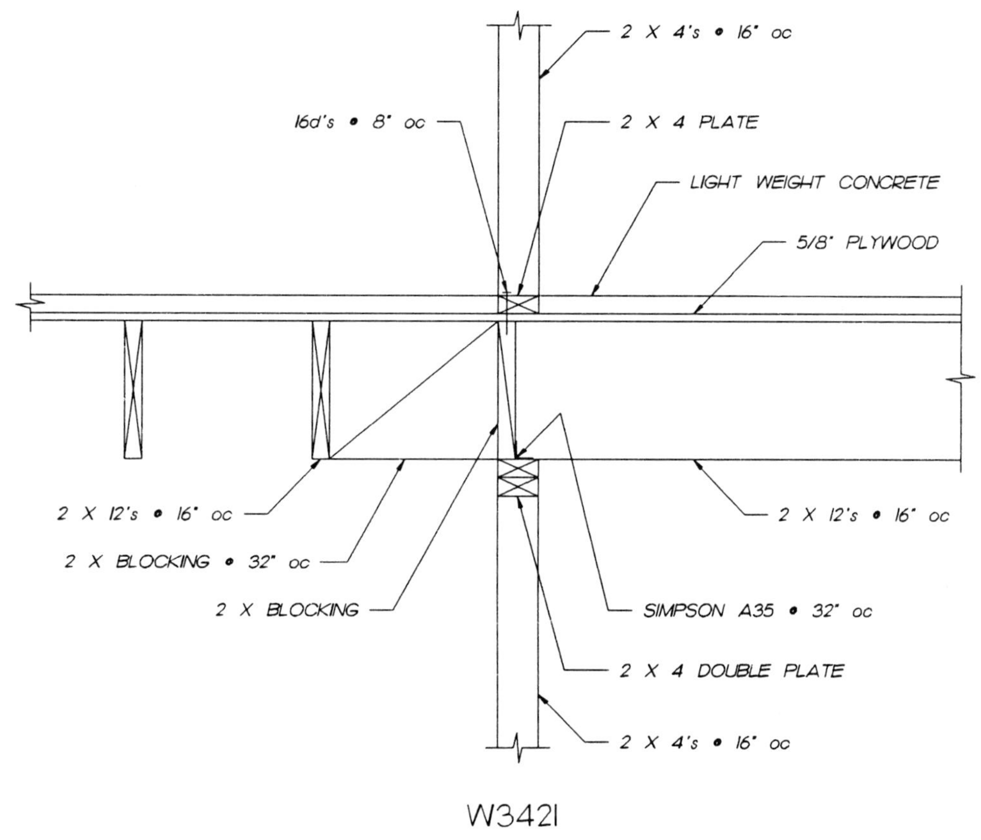
2 X 4's • 16" oc
16d's • 8" oc
2 X 4 PLATE
LIGHT WEIGHT CONCRETE
5/8" PLYWOOD
2 X 12's • 16" oc
2 X BLOCKING • 32" oc
2 X BLOCKING
2 X 12's • 16" oc
SIMPSON A35 • 32" oc
2 X 4 DOUBLE PLATE
2 X 4's • 16" oc
W3421

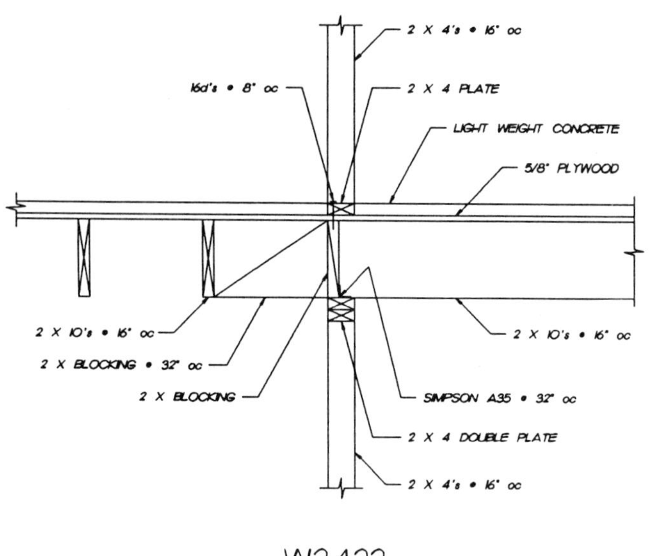

W3422

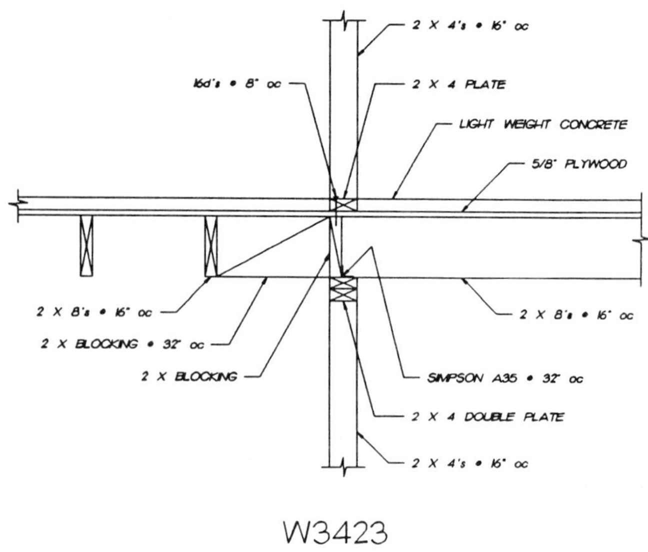

W3423

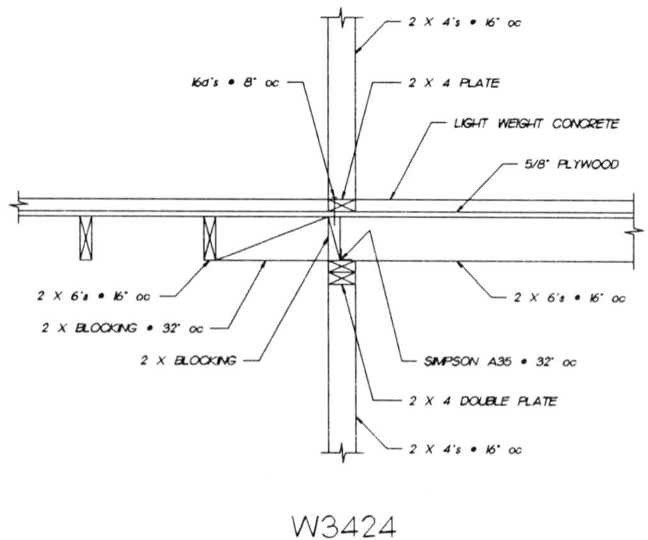

W3424

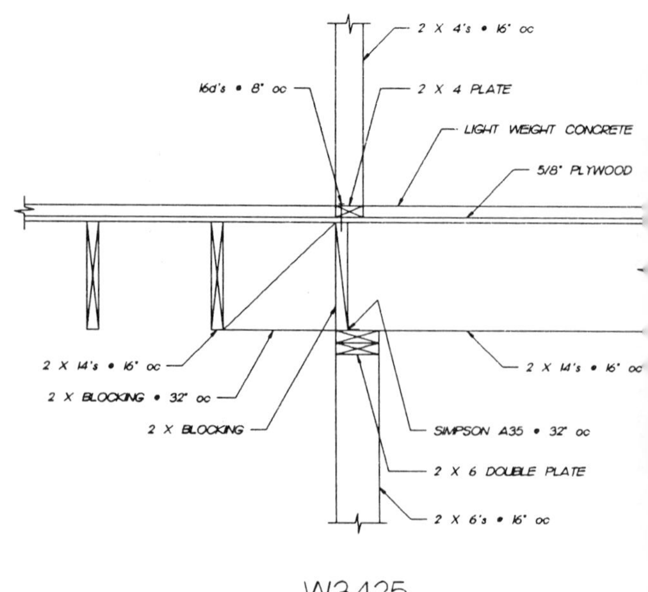

W3425

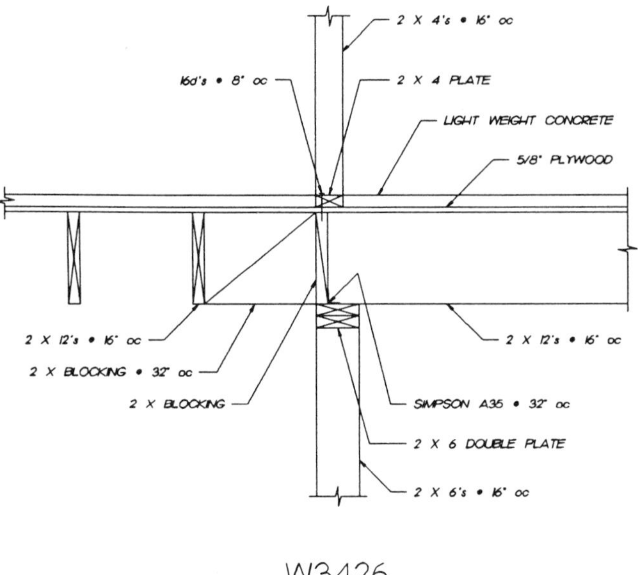

W3426

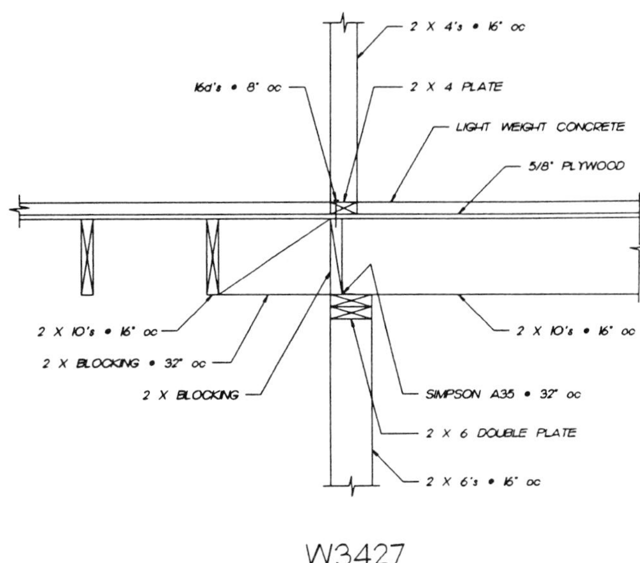

W3427

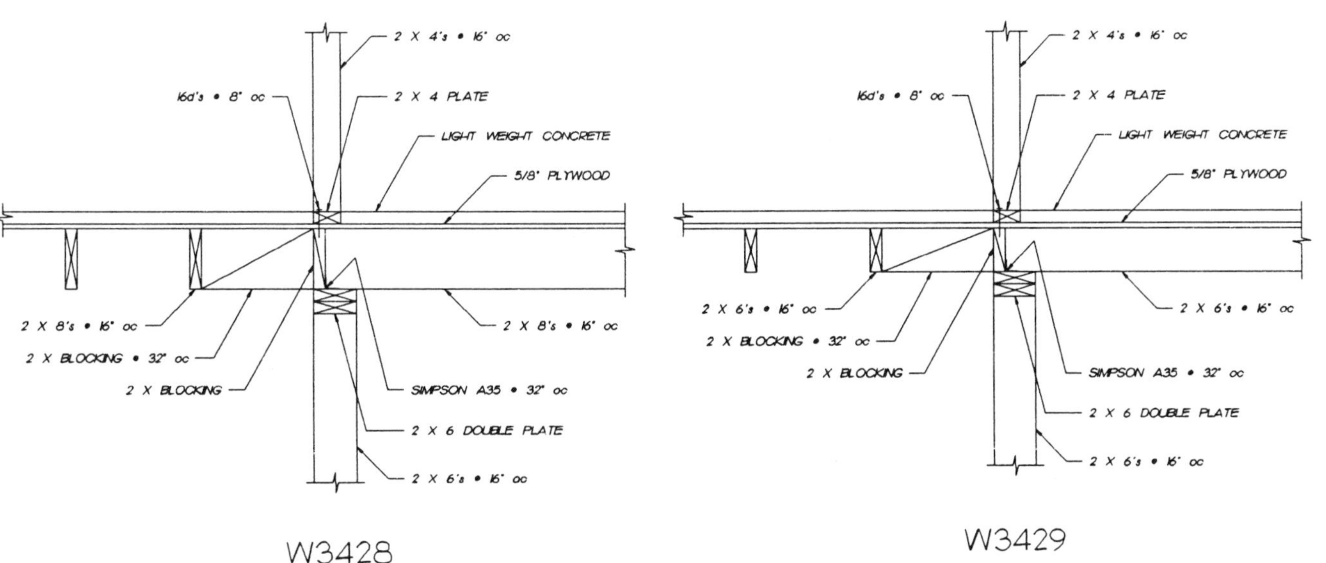

W3428

W3429

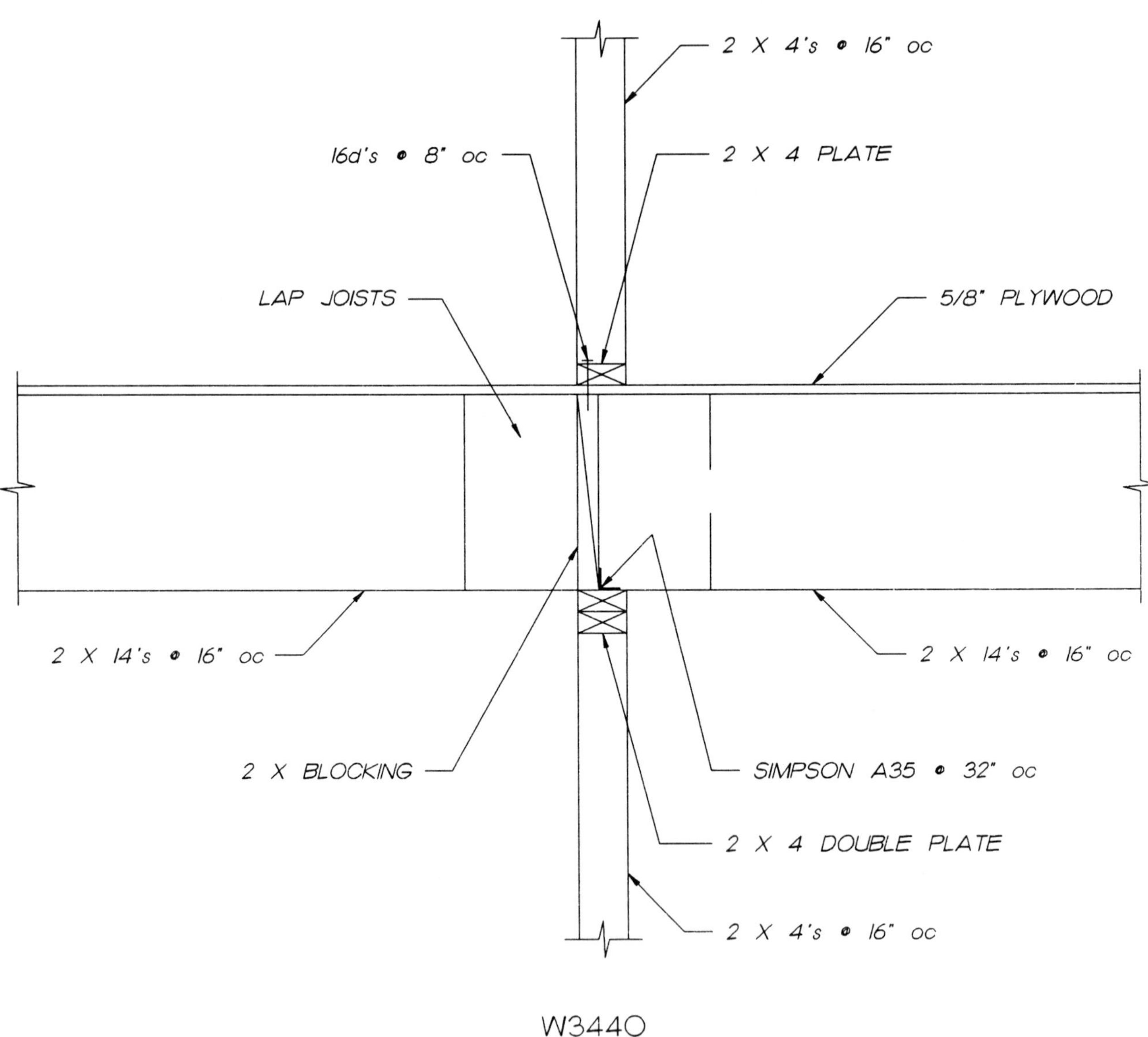
2 X 4's @ 16" oc
16d's @ 8" oc
2 X 4 PLATE
LAP JOISTS
5/8" PLYWOOD
2 X 14's @ 16" oc
2 X 14's @ 16" oc
2 X BLOCKING
SIMPSON A35 @ 32" oc
2 X 4 DOUBLE PLATE
2 X 4's @ 16" oc

W3440

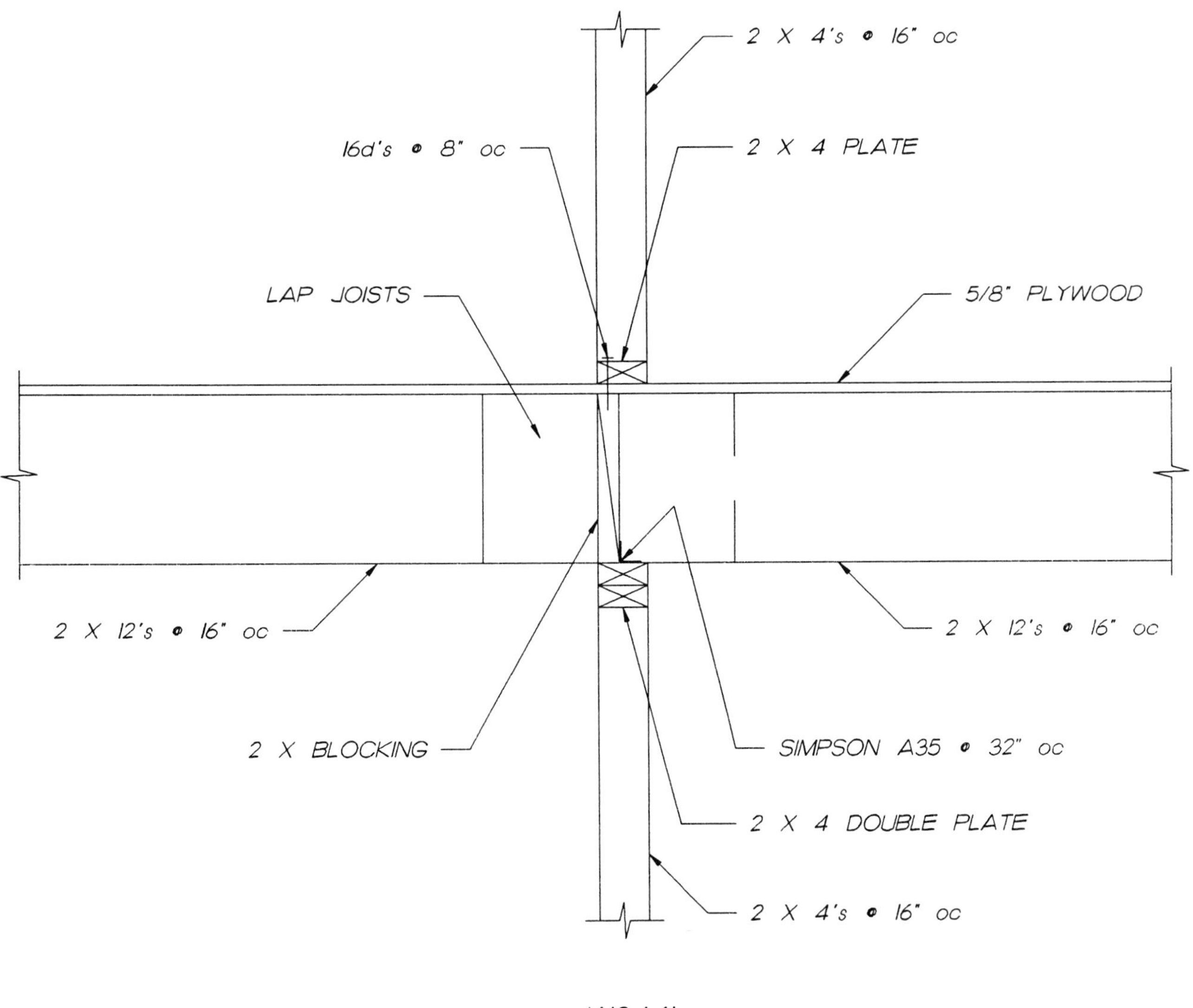
2 X 4's @ 16" oc
16d's @ 8" oc
2 X 4 PLATE
LAP JOISTS
5/8" PLYWOOD
2 X 12's @ 16" oc
2 X 12's @ 16" oc
2 X BLOCKING
SIMPSON A35 @ 32" oc
2 X 4 DOUBLE PLATE
2 X 4's @ 16" oc

W3441

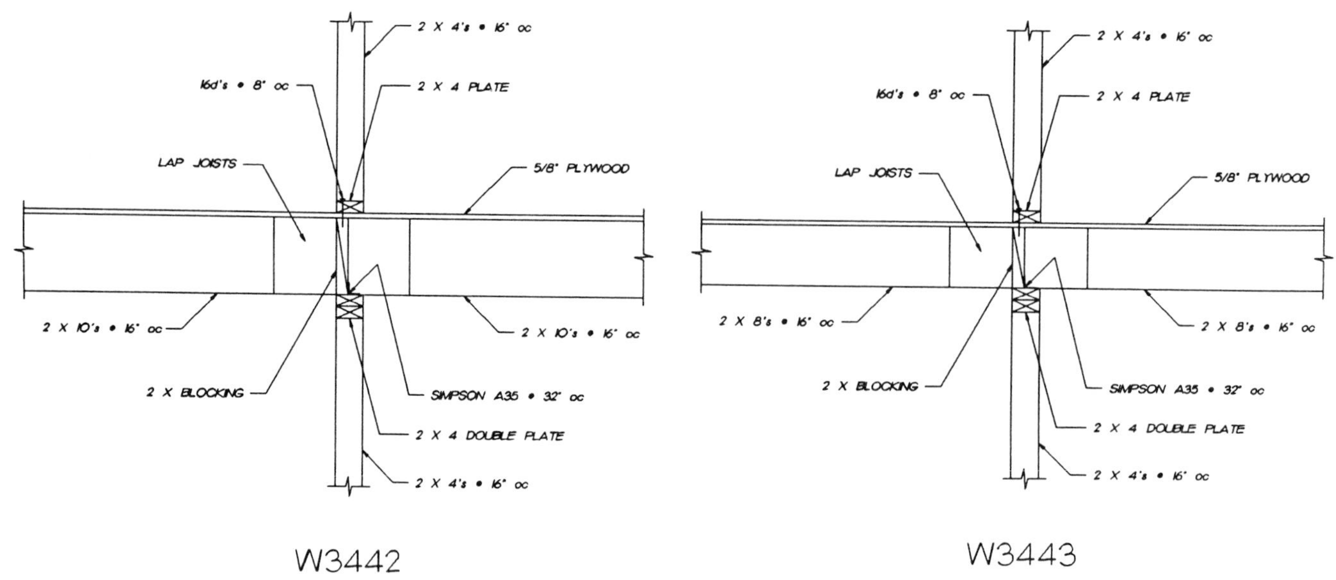

W3442

W3443

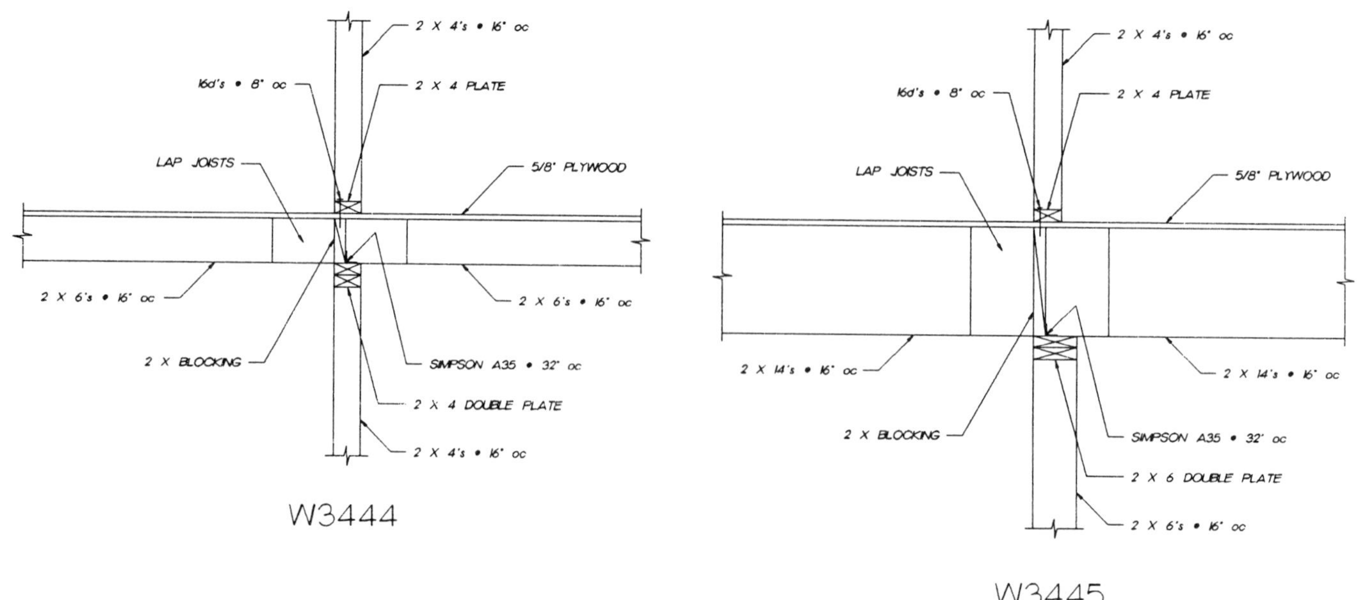

W3444

W3445

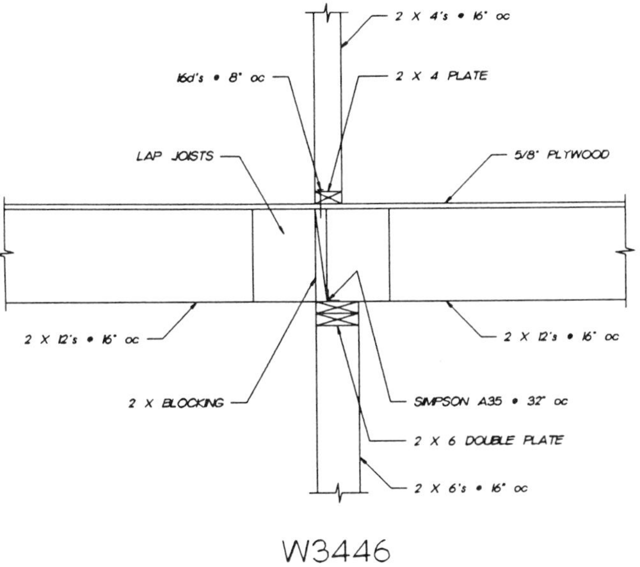

W3446

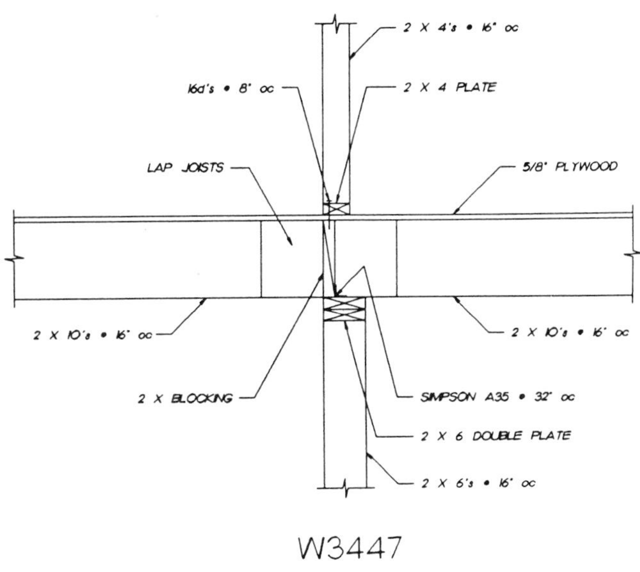

W3447

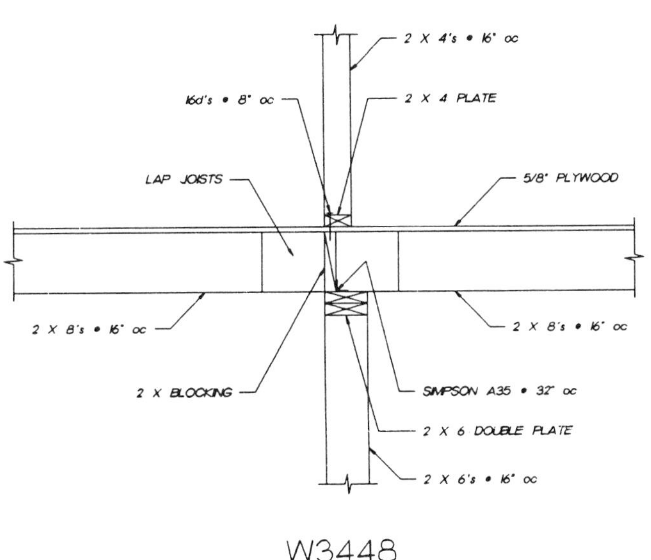

W3448

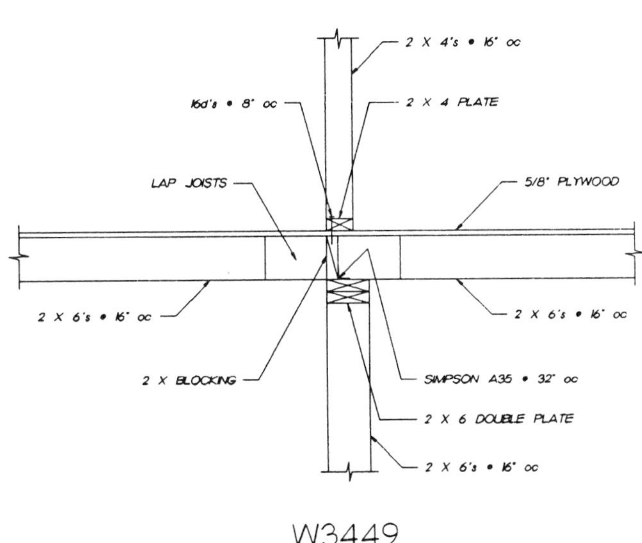

W3449

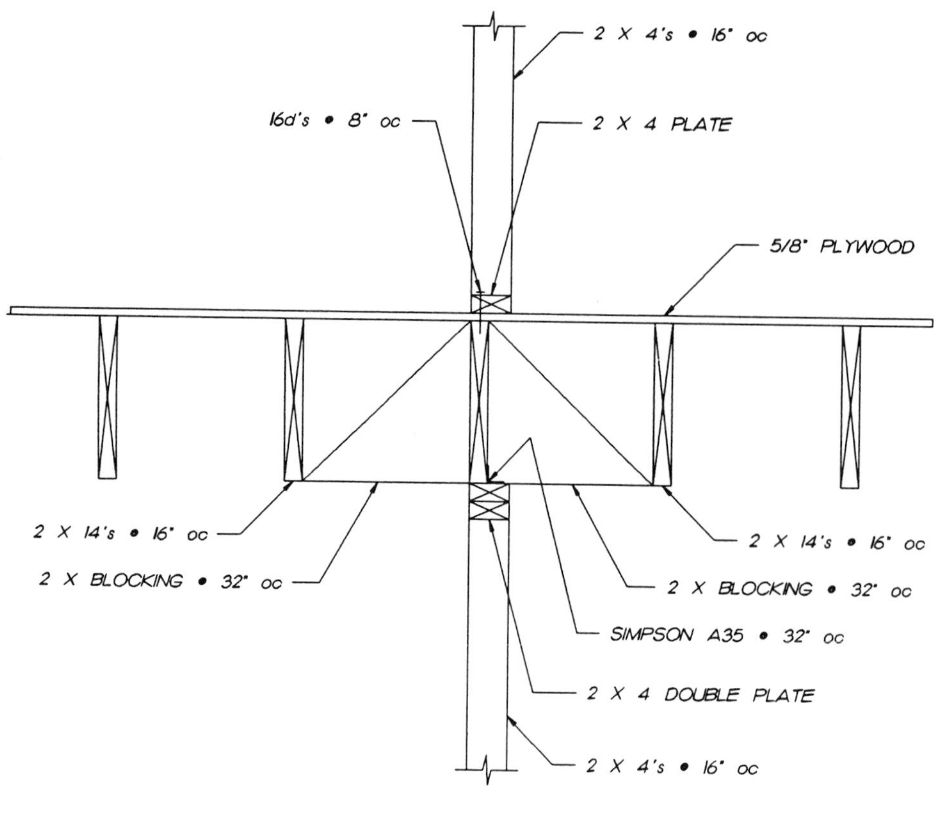
2 X 4's • 16" oc
16d's • 8" oc
2 X 4 PLATE
5/8" PLYWOOD
2 X 14's • 16" oc
2 X BLOCKING • 32" oc
2 X 14's • 16" oc
2 X BLOCKING • 32" oc
SIMPSON A35 • 32" oc
2 X 4 DOUBLE PLATE
2 X 4's • 16" oc

W3460

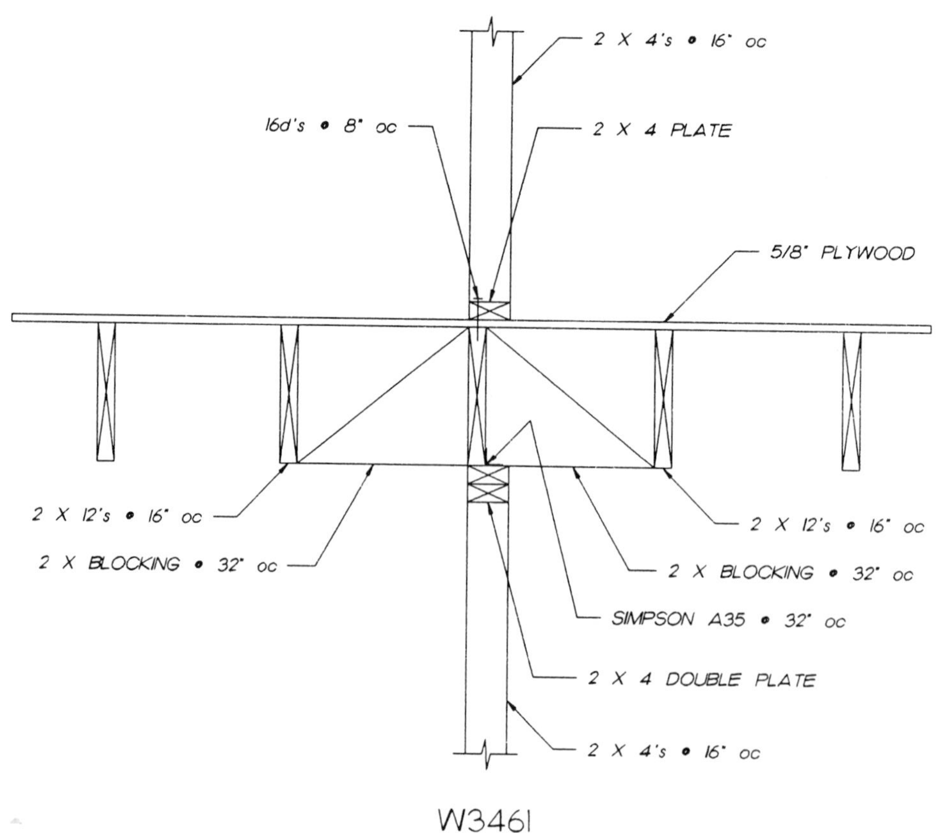
2 X 4's • 16" oc
16d's • 8" oc
2 X 4 PLATE
5/8" PLYWOOD
2 X 12's • 16" oc
2 X BLOCKING • 32" oc
2 X 12's • 16" oc
2 X BLOCKING • 32" oc
SIMPSON A35 • 32" oc
2 X 4 DOUBLE PLATE
2 X 4's • 16" oc

W3461

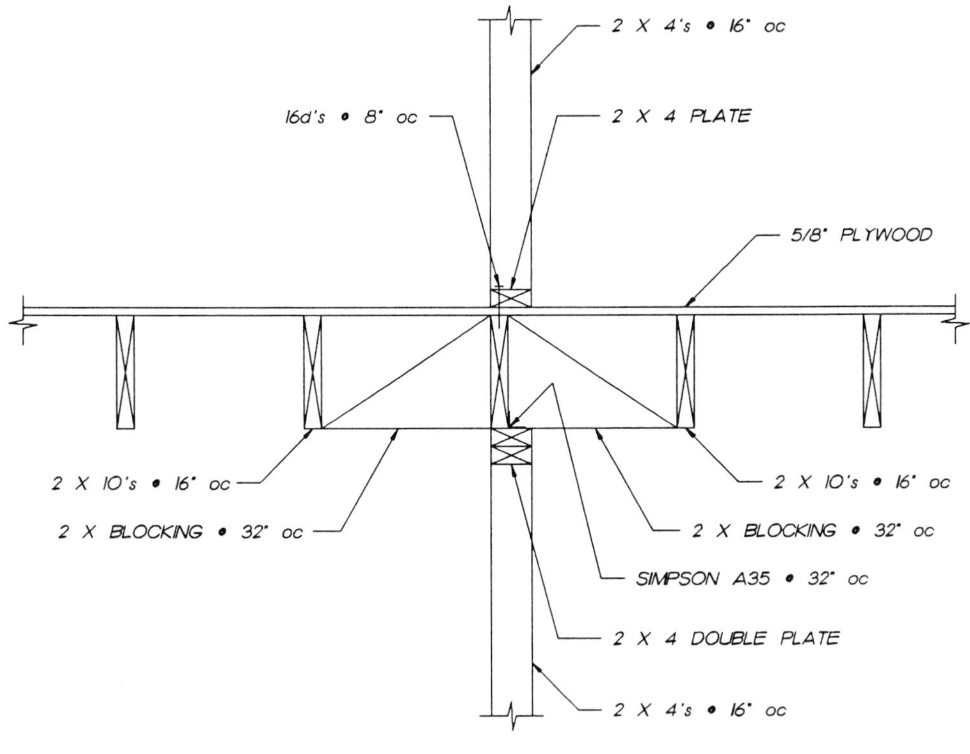
2 X 4's @ 16" oc
16d's @ 8" oc
2 X 4 PLATE
5/8" PLYWOOD
2 X 10's @ 16" oc
2 X 10's @ 16" oc
2 X BLOCKING @ 32" oc
2 X BLOCKING @ 32" oc
SIMPSON A35 @ 32" oc
2 X 4 DOUBLE PLATE
2 X 4's @ 16" oc

W3462

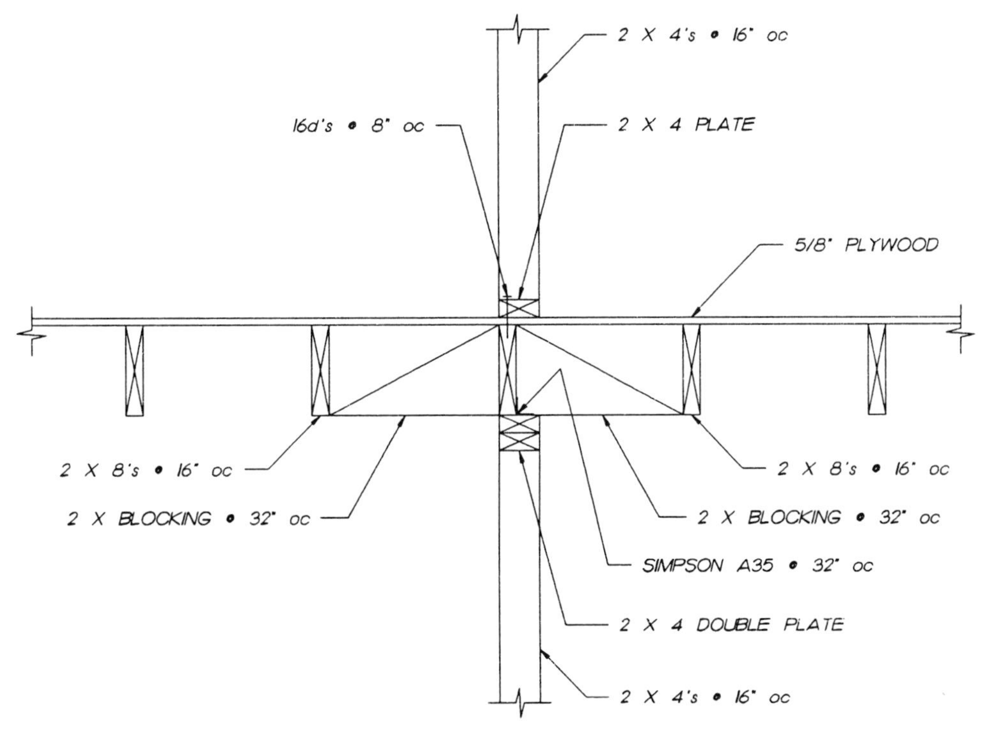
2 X 4's @ 16" oc
16d's @ 8" oc
2 X 4 PLATE
5/8" PLYWOOD
2 X 8's @ 16" oc
2 X 8's @ 16" oc
2 X BLOCKING @ 32" oc
2 X BLOCKING @ 32" oc
SIMPSON A35 @ 32" oc
2 X 4 DOUBLE PLATE
2 X 4's @ 16" oc

W3463

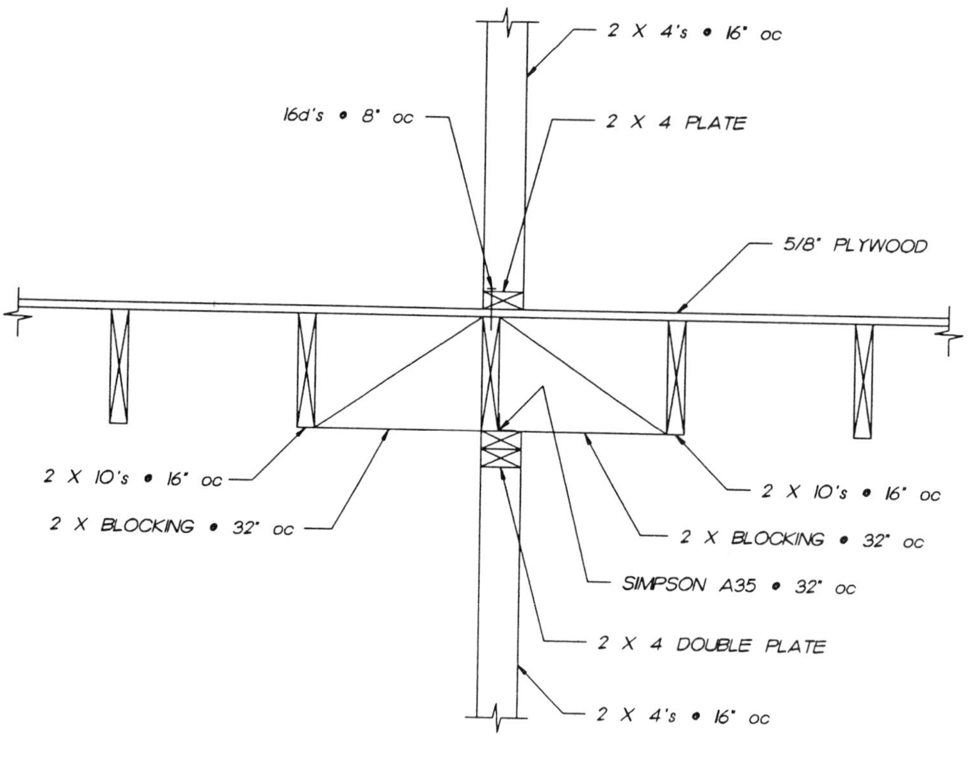
2 X 4's @ 16" oc
16d's @ 8" oc
2 X 4 PLATE
5/8" PLYWOOD
2 X 10's @ 16" oc
2 X BLOCKING @ 32" oc
2 X 10's @ 16" oc
2 X BLOCKING @ 32" oc
SIMPSON A35 @ 32" oc
2 X 4 DOUBLE PLATE
2 X 4's @ 16" oc

W3462

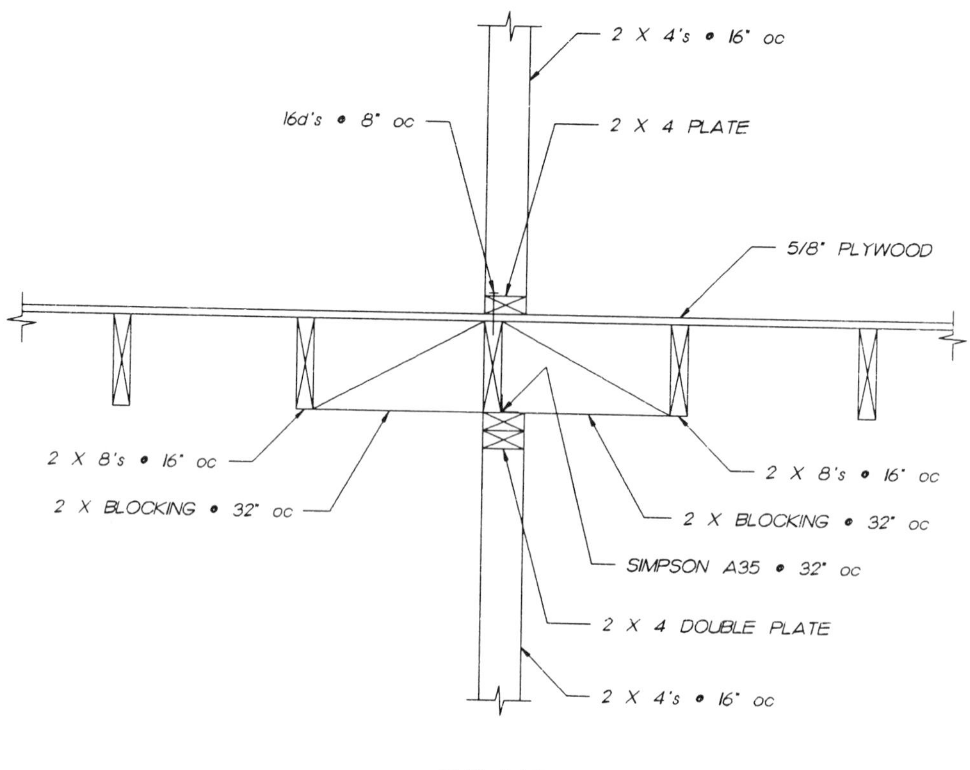
2 X 4's @ 16" oc
16d's @ 8" oc
2 X 4 PLATE
5/8" PLYWOOD
2 X 8's @ 16" oc
2 X BLOCKING @ 32" oc
2 X 8's @ 16" oc
2 X BLOCKING @ 32" oc
SIMPSON A35 @ 32" oc
2 X 4 DOUBLE PLATE
2 X 4's @ 16" oc

W3463

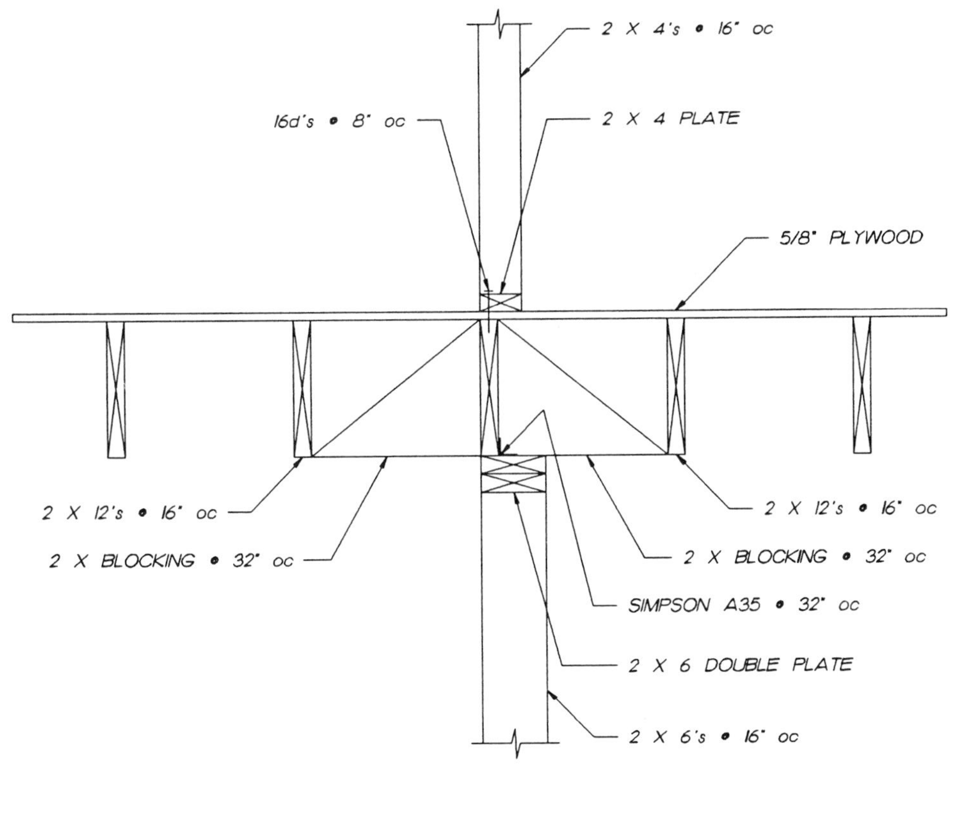
2 X 4's @ 16" oc
16d's @ 8" oc
2 X 4 PLATE
5/8" PLYWOOD
2 X 12's @ 16" oc
2 X 12's @ 16" oc
2 X BLOCKING @ 32" oc
2 X BLOCKING @ 32" oc
SIMPSON A35 @ 32" oc
2 X 6 DOUBLE PLATE
2 X 6's @ 16" oc

W3466

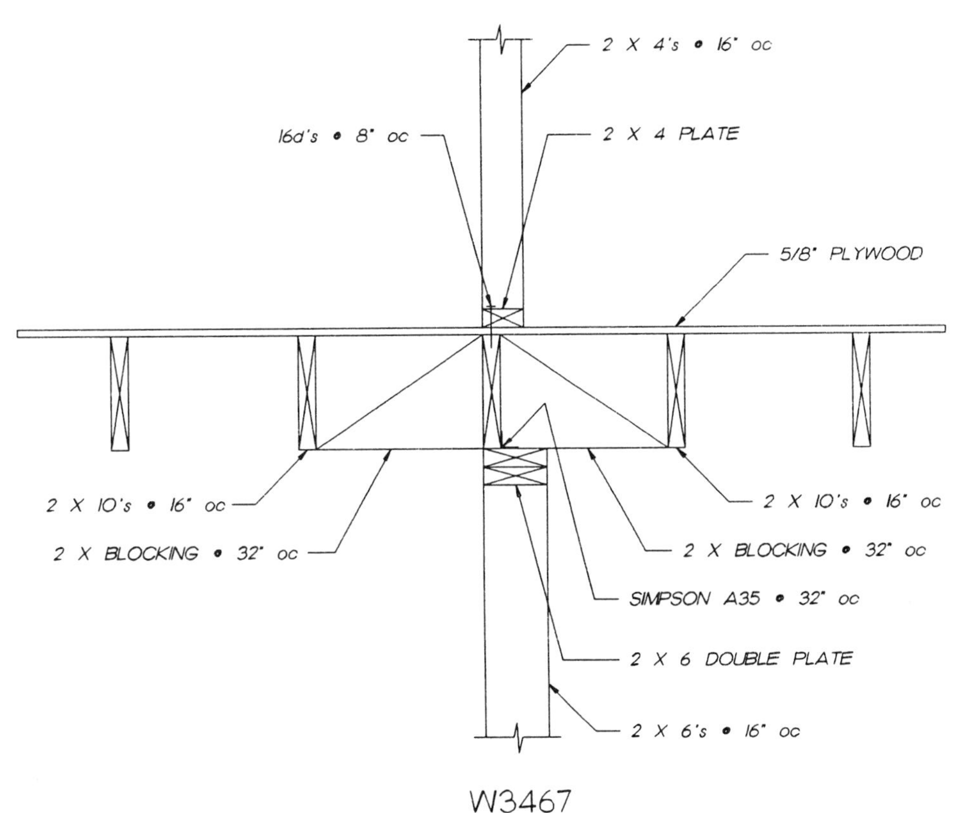
2 X 4's @ 16" oc
16d's @ 8" oc
2 X 4 PLATE
5/8" PLYWOOD
2 X 10's @ 16" oc
2 X 10's @ 16" oc
2 X BLOCKING @ 32" oc
2 X BLOCKING @ 32" oc
SIMPSON A35 @ 32" oc
2 X 6 DOUBLE PLATE
2 X 6's @ 16" oc

W3467

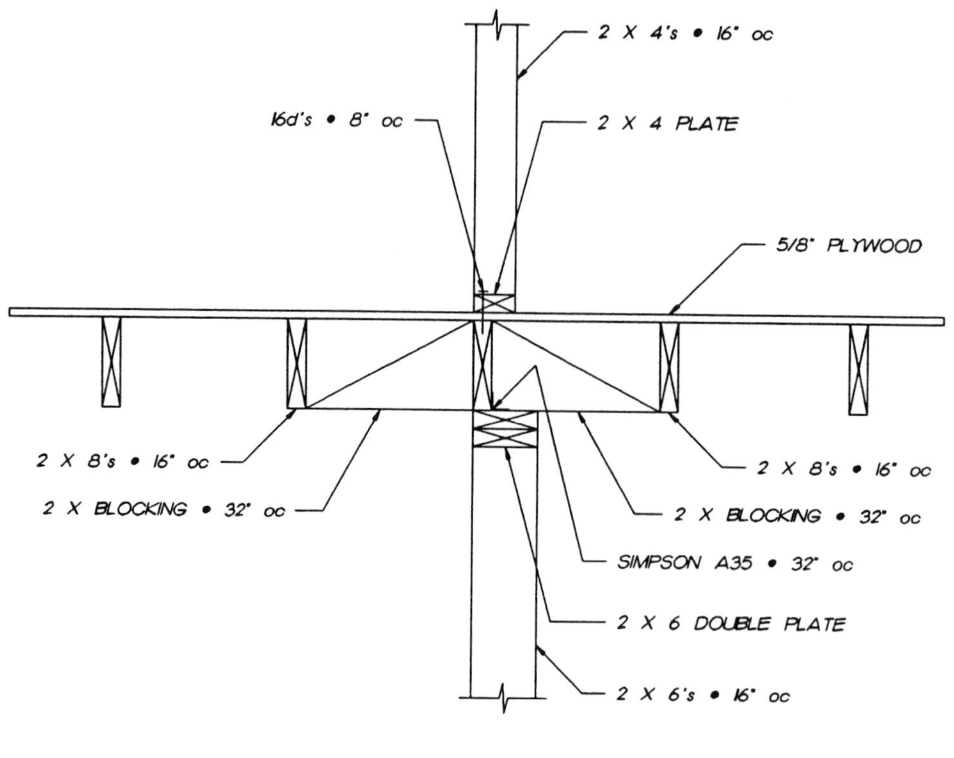

W3468

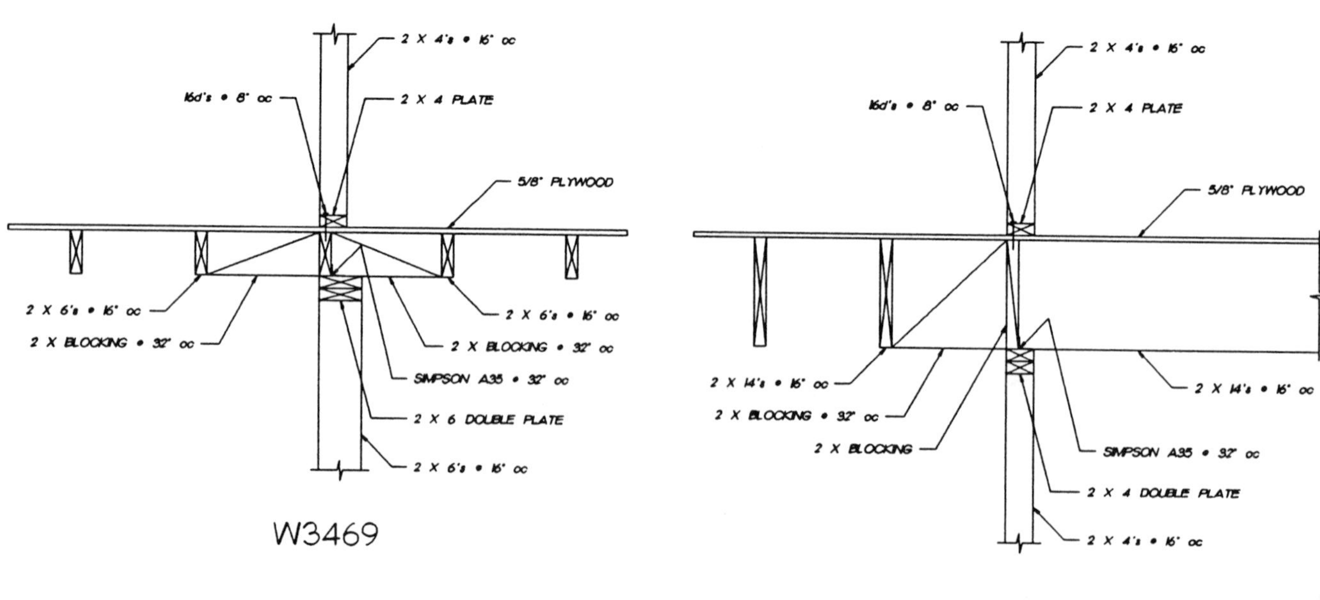

W3469

W3480

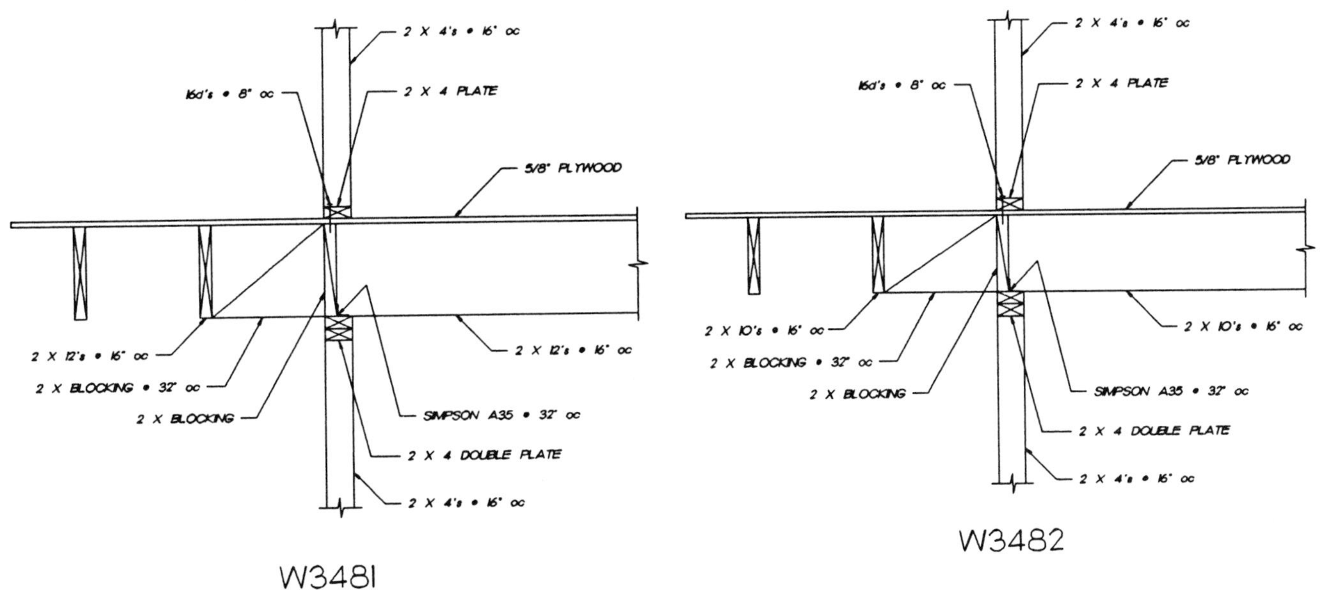

W3481

W3482

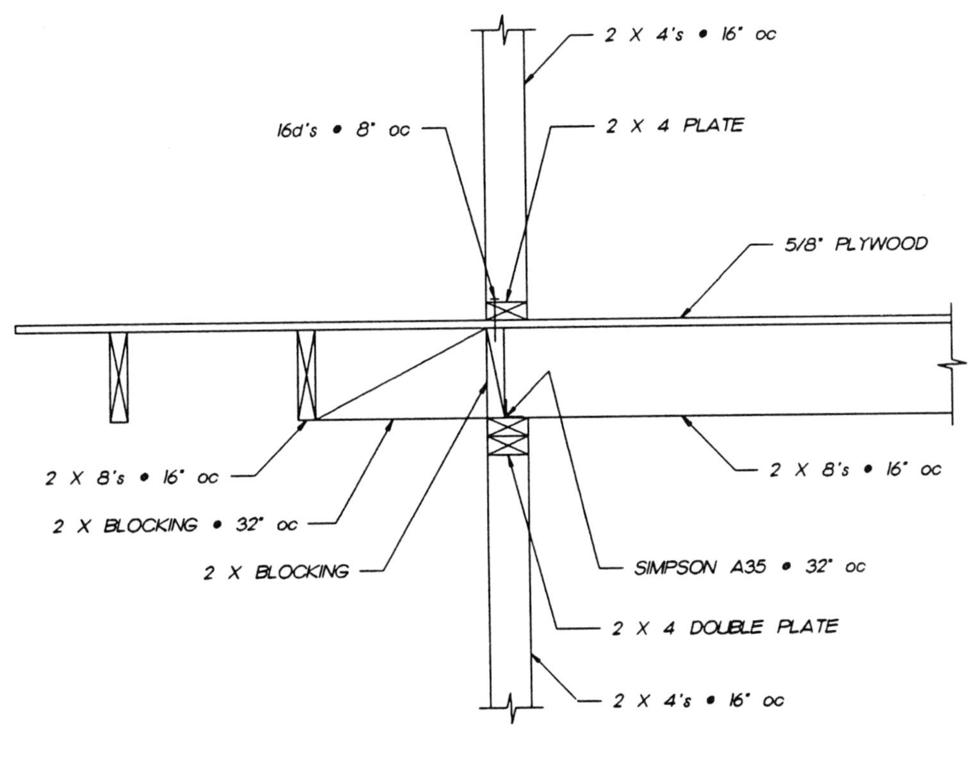

W3483

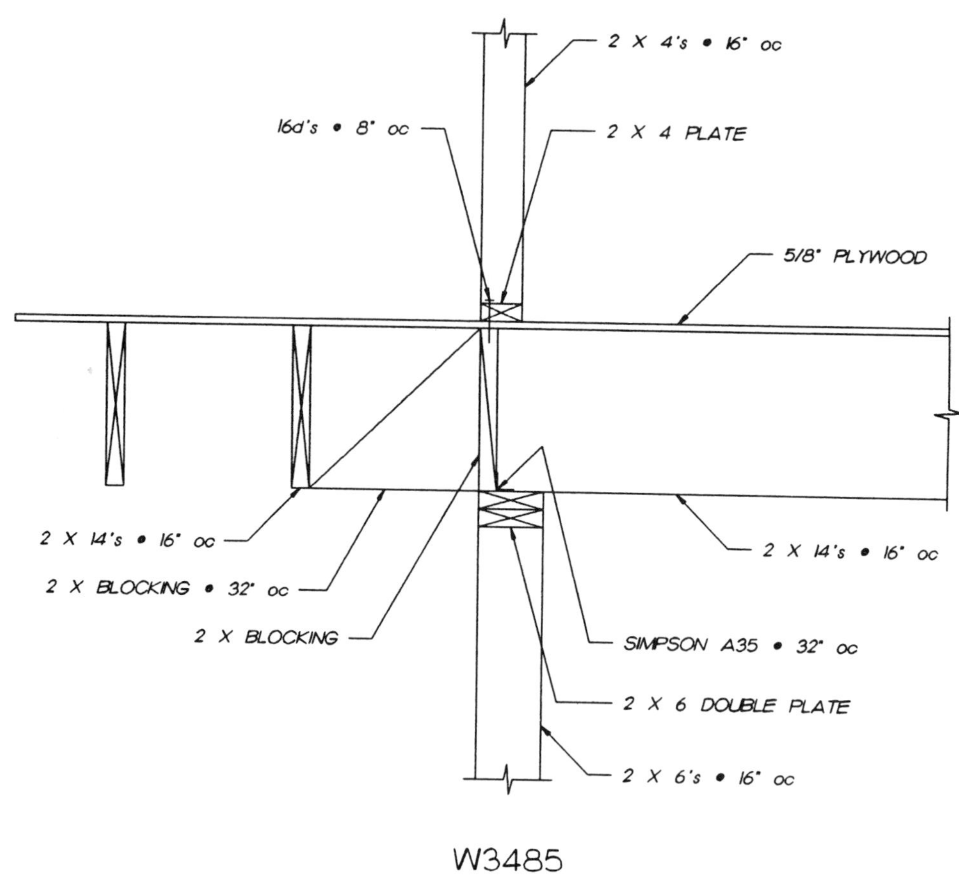
2 X 4's • 16" oc
16d's • 8" oc
2 X 4 PLATE
5/8" PLYWOOD
2 X 14's • 16" oc
2 X BLOCKING • 32" oc
2 X BLOCKING
2 X 14's • 16" oc
SIMPSON A35 • 32" oc
2 X 6 DOUBLE PLATE
2 X 6's • 16" oc

W3485

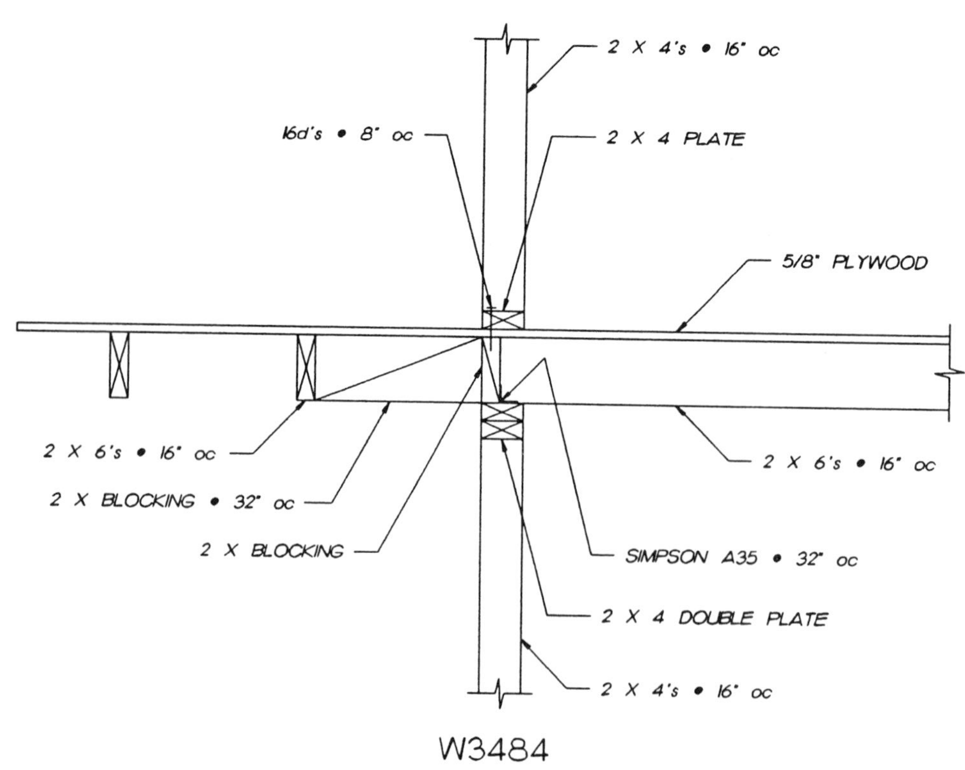
2 X 4's • 16" oc
16d's • 8" oc
2 X 4 PLATE
5/8" PLYWOOD
2 X 6's • 16" oc
2 X BLOCKING • 32" oc
2 X BLOCKING
2 X 6's • 16" oc
SIMPSON A35 • 32" oc
2 X 4 DOUBLE PLATE
2 X 4's • 16" oc

W3484

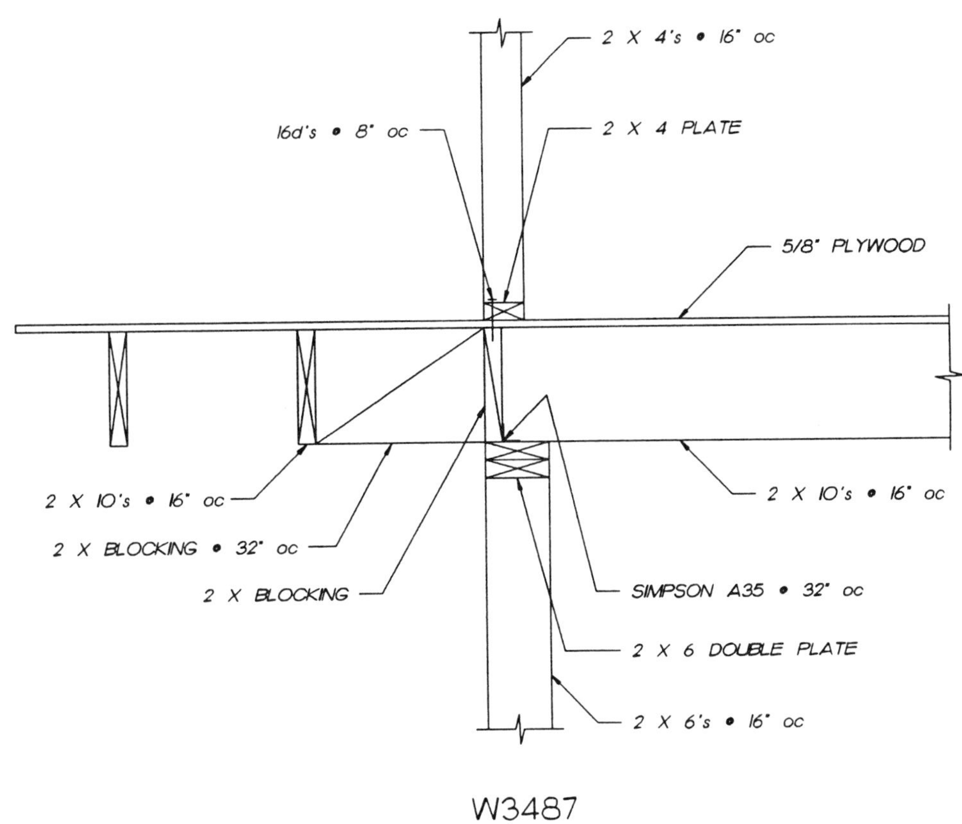
2 X 4's • 16" oc
16d's • 8" oc
2 X 4 PLATE
5/8" PLYWOOD
2 X 10's • 16" oc
2 X BLOCKING • 32" oc
2 X BLOCKING
2 X 10's • 16" oc
SIMPSON A35 • 32" oc
2 X 6 DOUBLE PLATE
2 X 6's • 16" oc

W3487

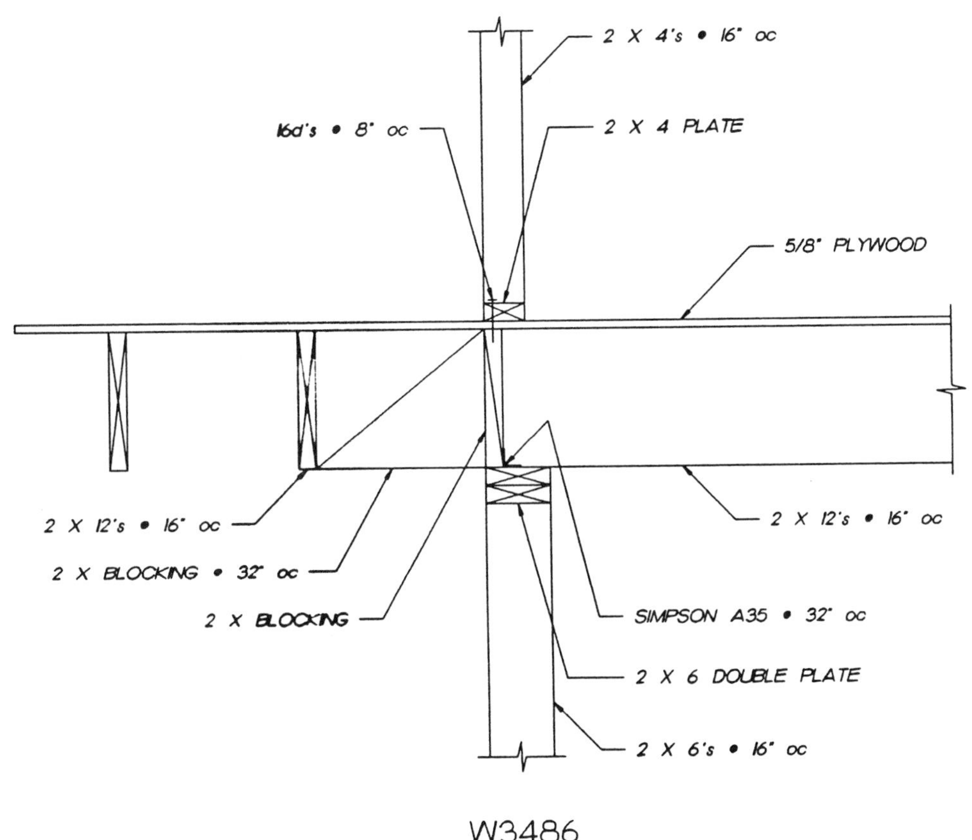
2 X 4's • 16" oc
16d's • 8" oc
2 X 4 PLATE
5/8" PLYWOOD
2 X 12's • 16" oc
2 X BLOCKING • 32" oc
2 X BLOCKING
2 X 12's • 16" oc
SIMPSON A35 • 32" oc
2 X 6 DOUBLE PLATE
2 X 6's • 16" oc

W3486

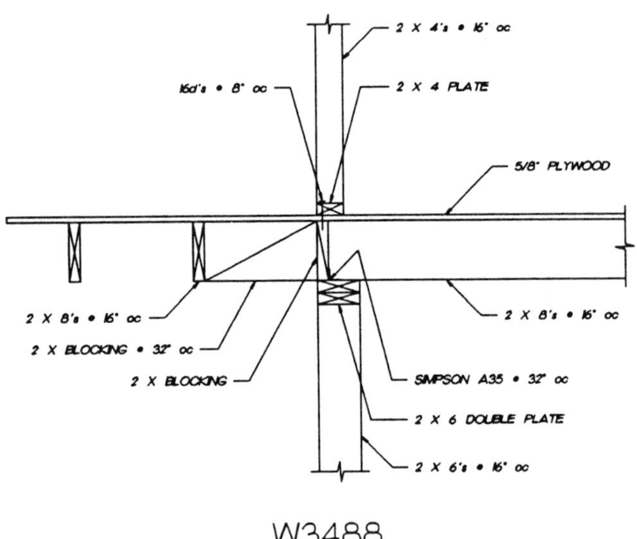

W3488

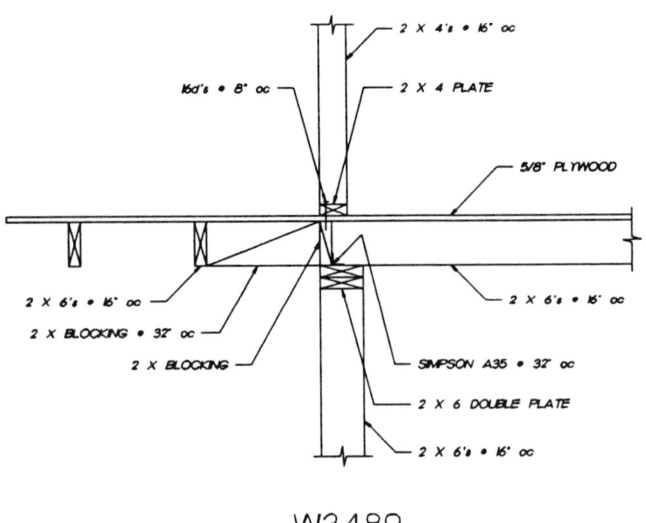

W3489

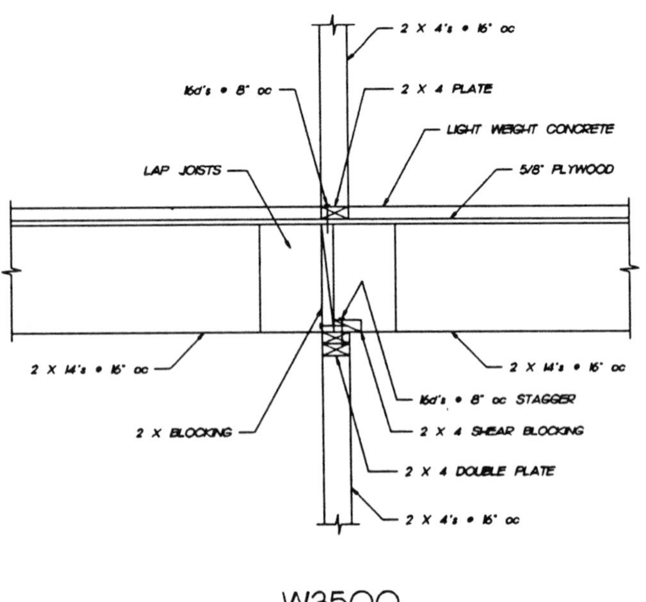

W3500

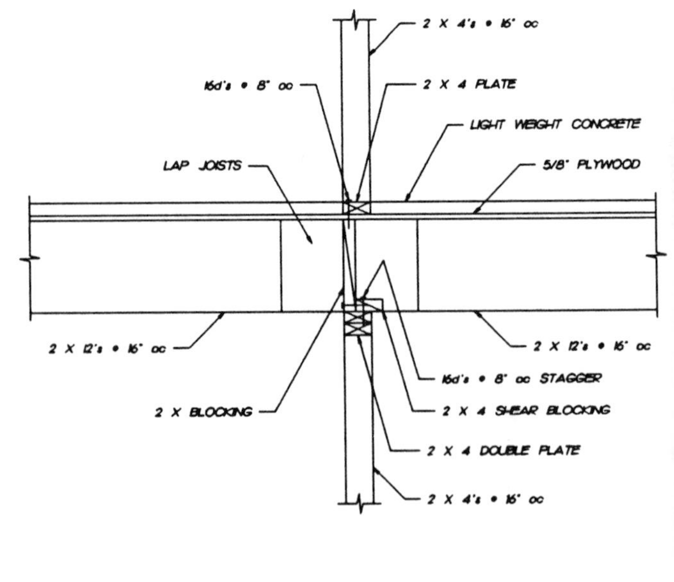

W3501

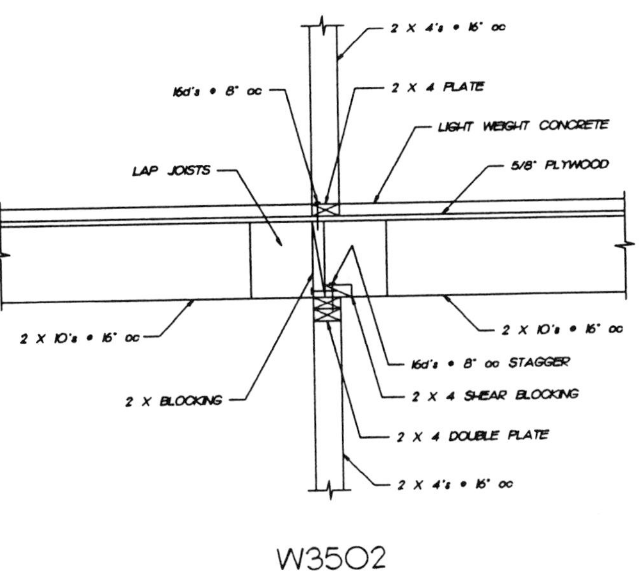

W3502

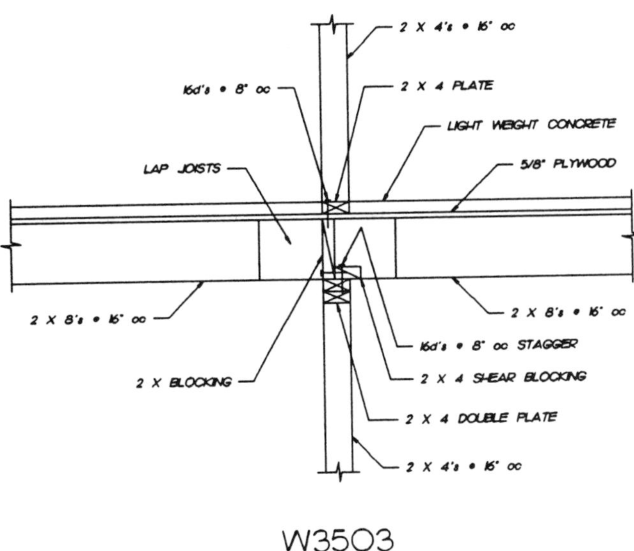

W3503

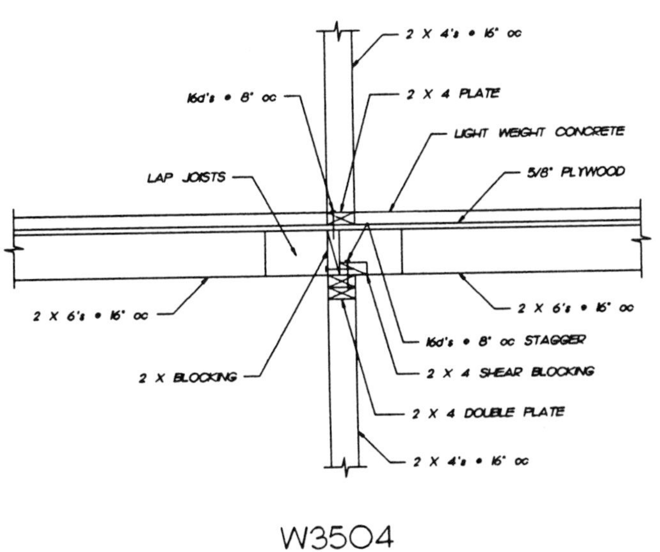

W3504

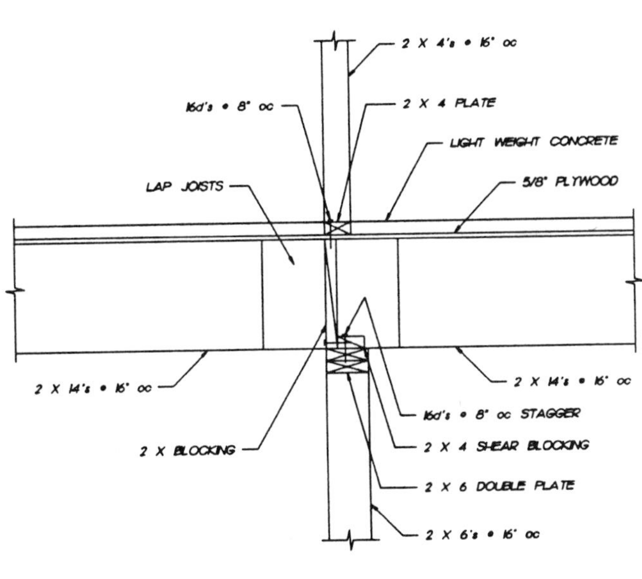

W3505

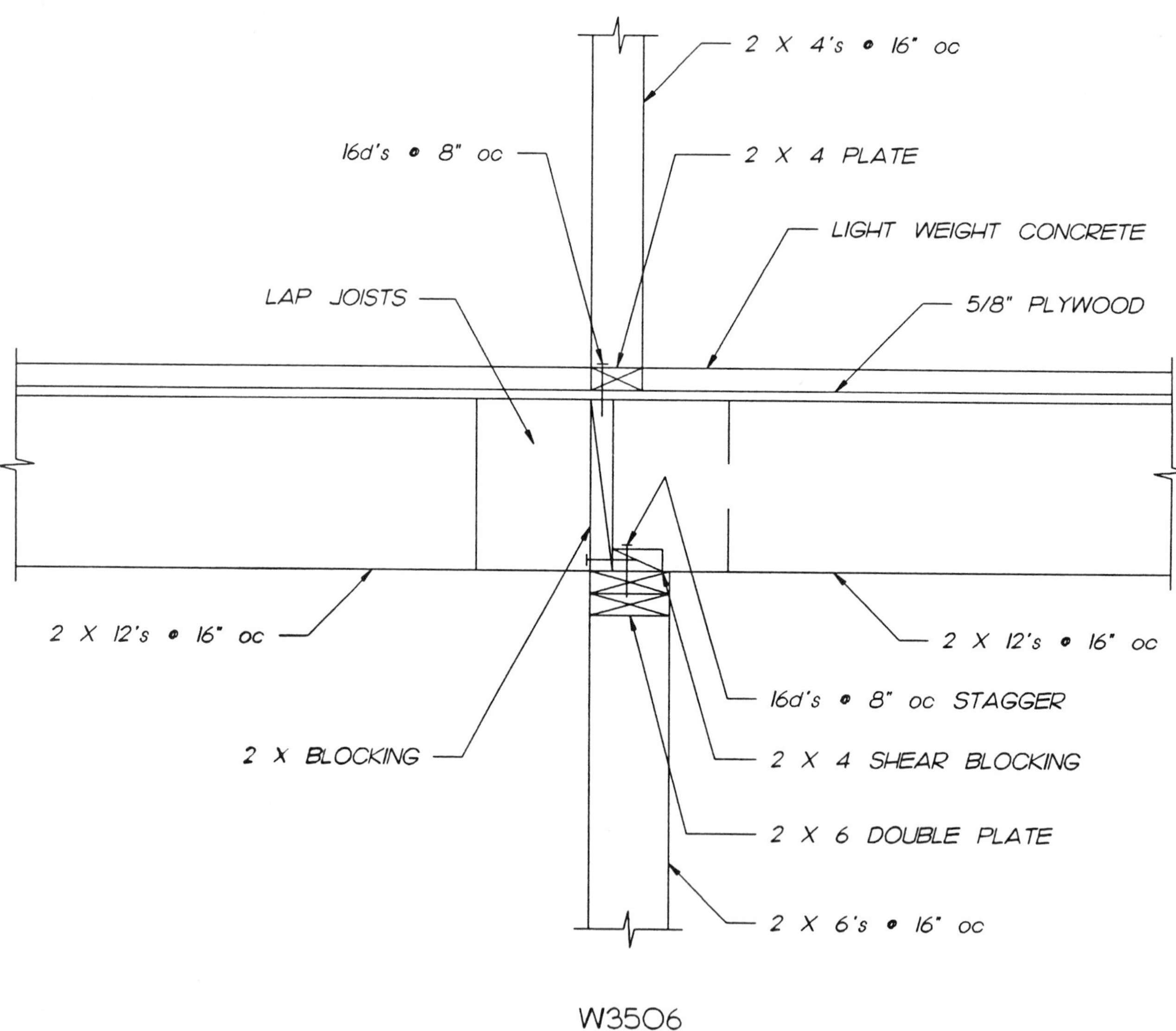
2 X 4's @ 16" oc
16d's @ 8" oc
2 X 4 PLATE
LIGHT WEIGHT CONCRETE
LAP JOISTS
5/8" PLYWOOD
2 X 12's @ 16" oc
2 X 12's @ 16" oc
16d's @ 8" oc STAGGER
2 X BLOCKING
2 X 4 SHEAR BLOCKING
2 X 6 DOUBLE PLATE
2 X 6's @ 16" oc

W3506

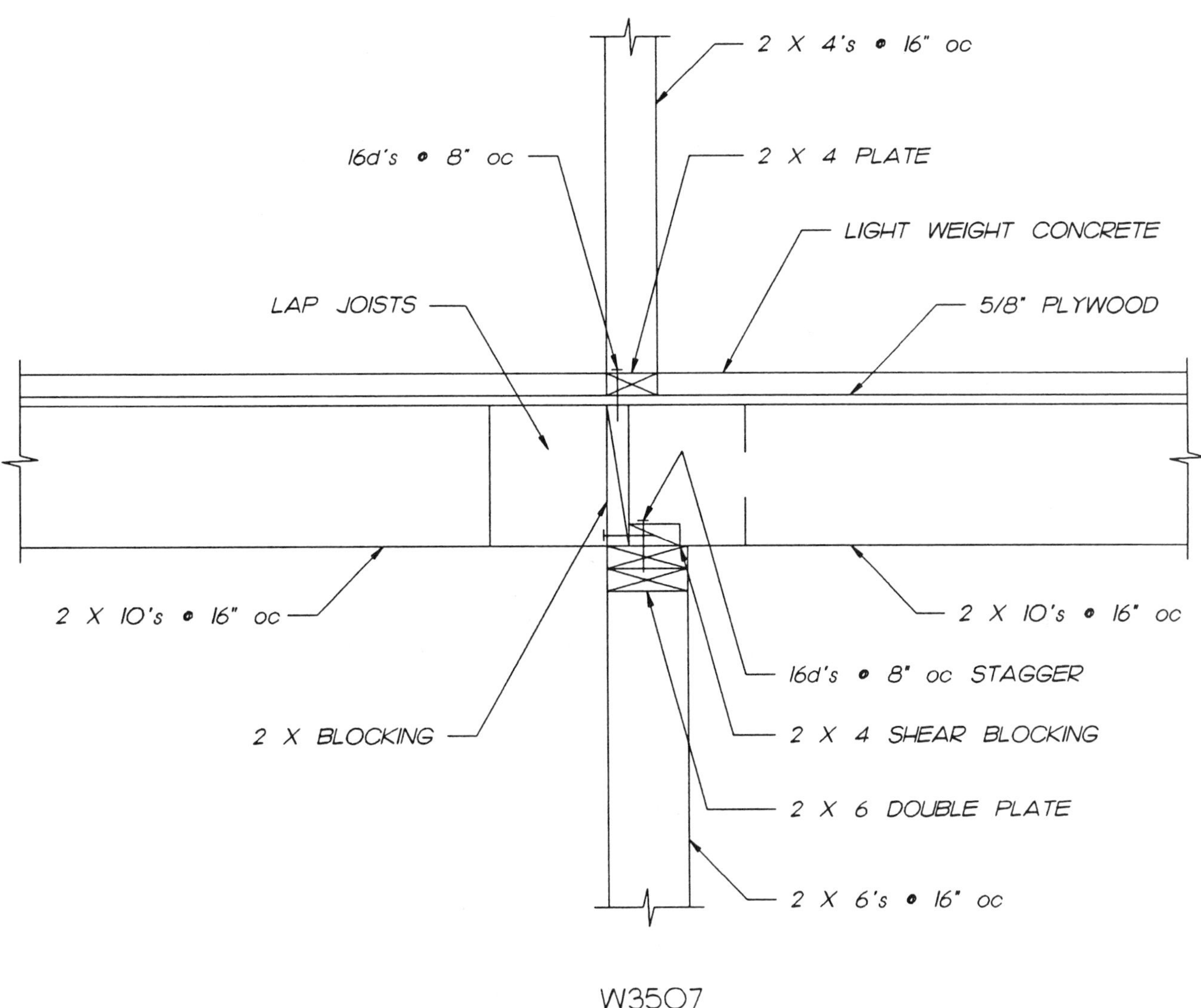
2 X 4's @ 16" oc
16d's @ 8" oc
2 X 4 PLATE
LIGHT WEIGHT CONCRETE
LAP JOISTS
5/8" PLYWOOD
2 X 10's @ 16" oc
2 X 10's @ 16" oc
16d's @ 8" oc STAGGER
2 X BLOCKING
2 X 4 SHEAR BLOCKING
2 X 6 DOUBLE PLATE
2 X 6's @ 16" oc

W3507

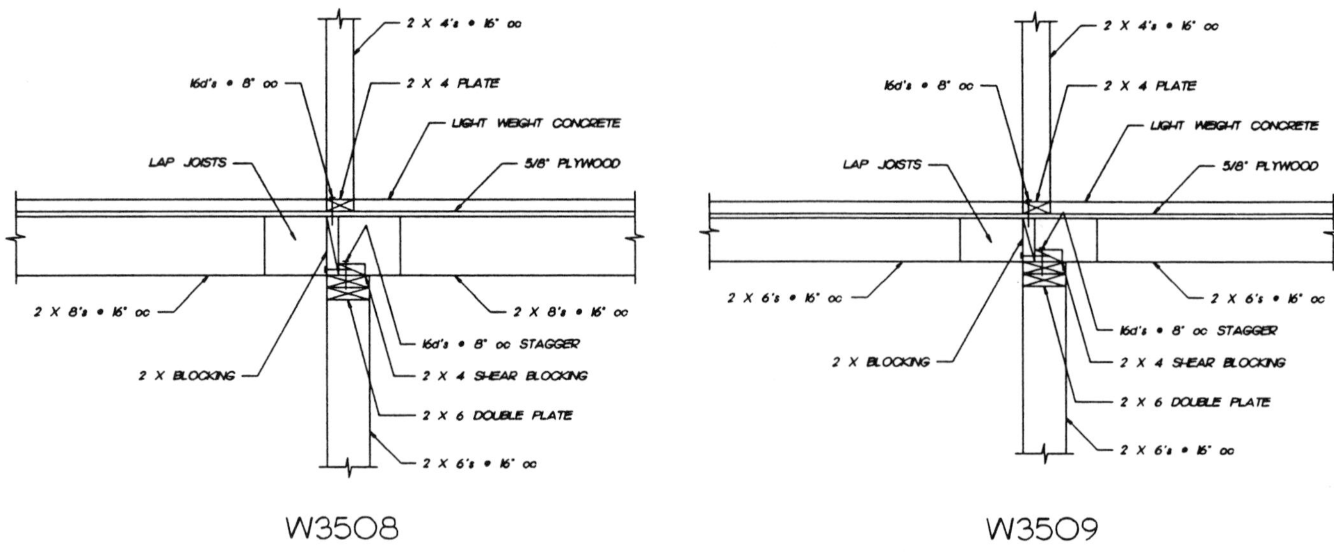
2 X 4's • 16" oc
16d's • 8" oc
2 X 4 PLATE
LIGHT WEIGHT CONCRETE
LAP JOISTS
5/8" PLYWOOD
2 X 8's • 16" oc
2 X 8's • 16" oc
16d's • 8" oc STAGGER
2 X BLOCKING
2 X 4 SHEAR BLOCKING
2 X 6 DOUBLE PLATE
2 X 6's • 16" oc
W3508
2 X 4's • 16" oc
16d's • 8" oc
2 X 4 PLATE
LIGHT WEIGHT CONCRETE
LAP JOISTS
5/8" PLYWOOD
2 X 6's • 16" oc
2 X 6's • 16" oc
16d's • 8" oc STAGGER
2 X BLOCKING
2 X 4 SHEAR BLOCKING
2 X 6 DOUBLE PLATE
2 X 6's • 16" oc
W3509

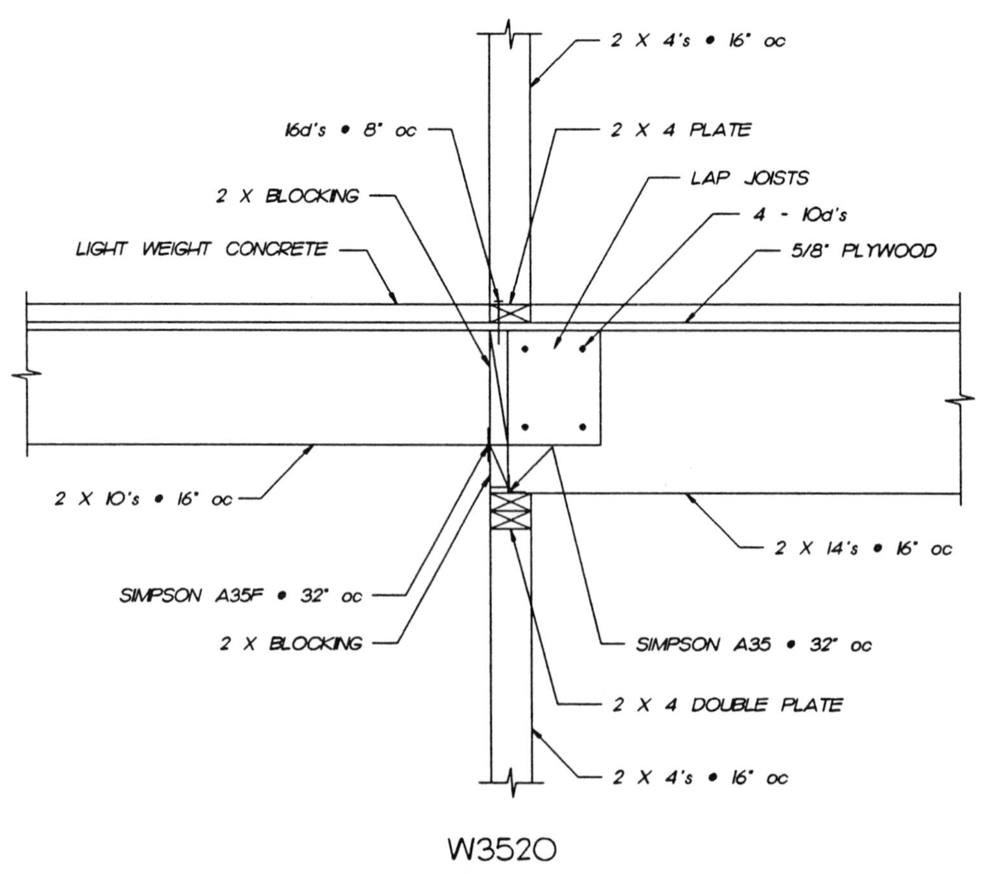
2 X 4's • 16" oc
16d's • 8" oc
2 X 4 PLATE
LAP JOISTS
2 X BLOCKING
4 - 10d's
LIGHT WEIGHT CONCRETE
5/8" PLYWOOD
2 X 10's • 16" oc
2 X 14's • 16" oc
SIMPSON A35F • 32" oc
2 X BLOCKING
SIMPSON A35 • 32" oc
2 X 4 DOUBLE PLATE
2 X 4's • 16" oc
W3520

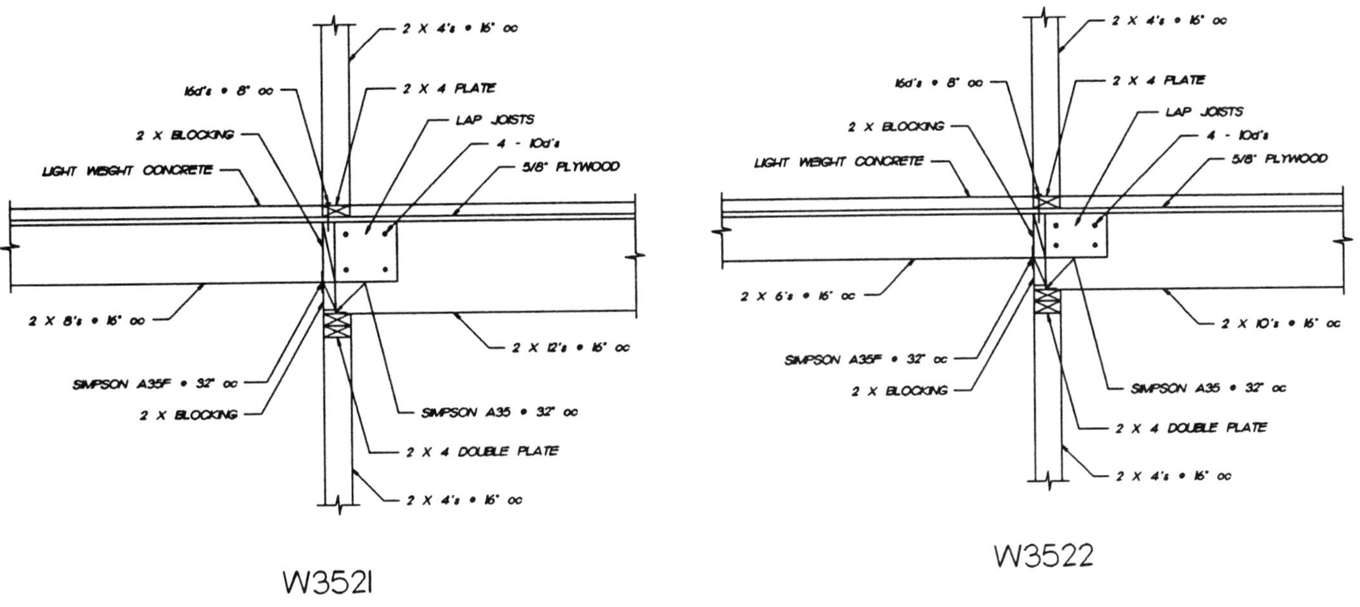

W3521

W3522

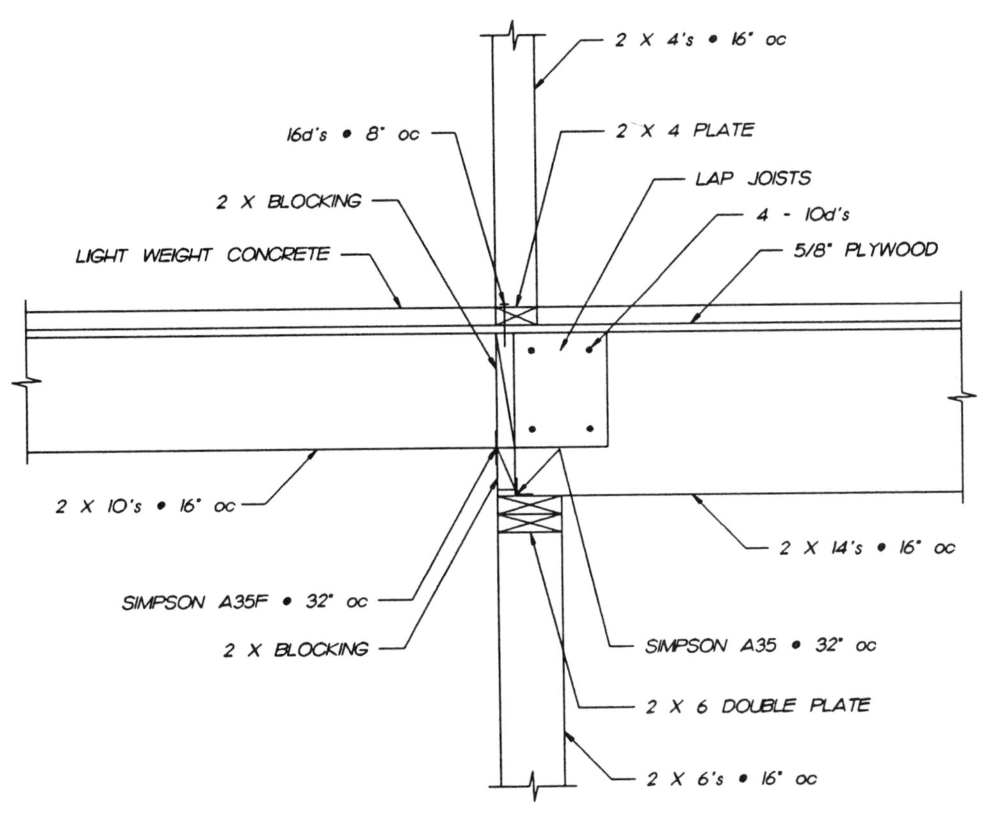

W3523

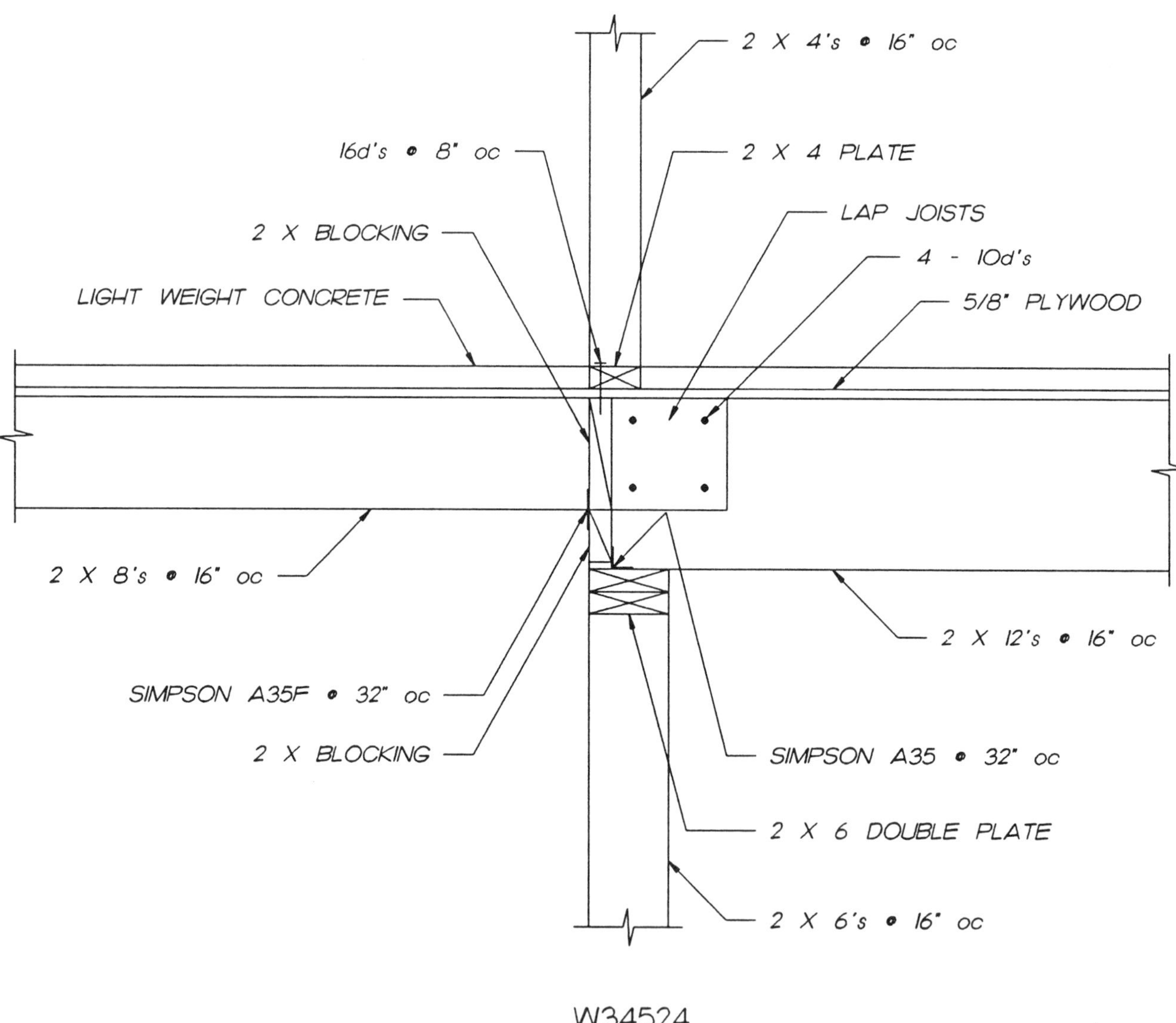
2 X 4's @ 16" oc
16d's @ 8" oc
2 X 4 PLATE
LAP JOISTS
2 X BLOCKING
4 - 10d's
LIGHT WEIGHT CONCRETE
5/8" PLYWOOD
2 X 8's @ 16" oc
2 X 12's @ 16" oc
SIMPSON A35F @ 32" oc
2 X BLOCKING
SIMPSON A35 @ 32" oc
2 X 6 DOUBLE PLATE
2 X 6's @ 16" oc

W34524

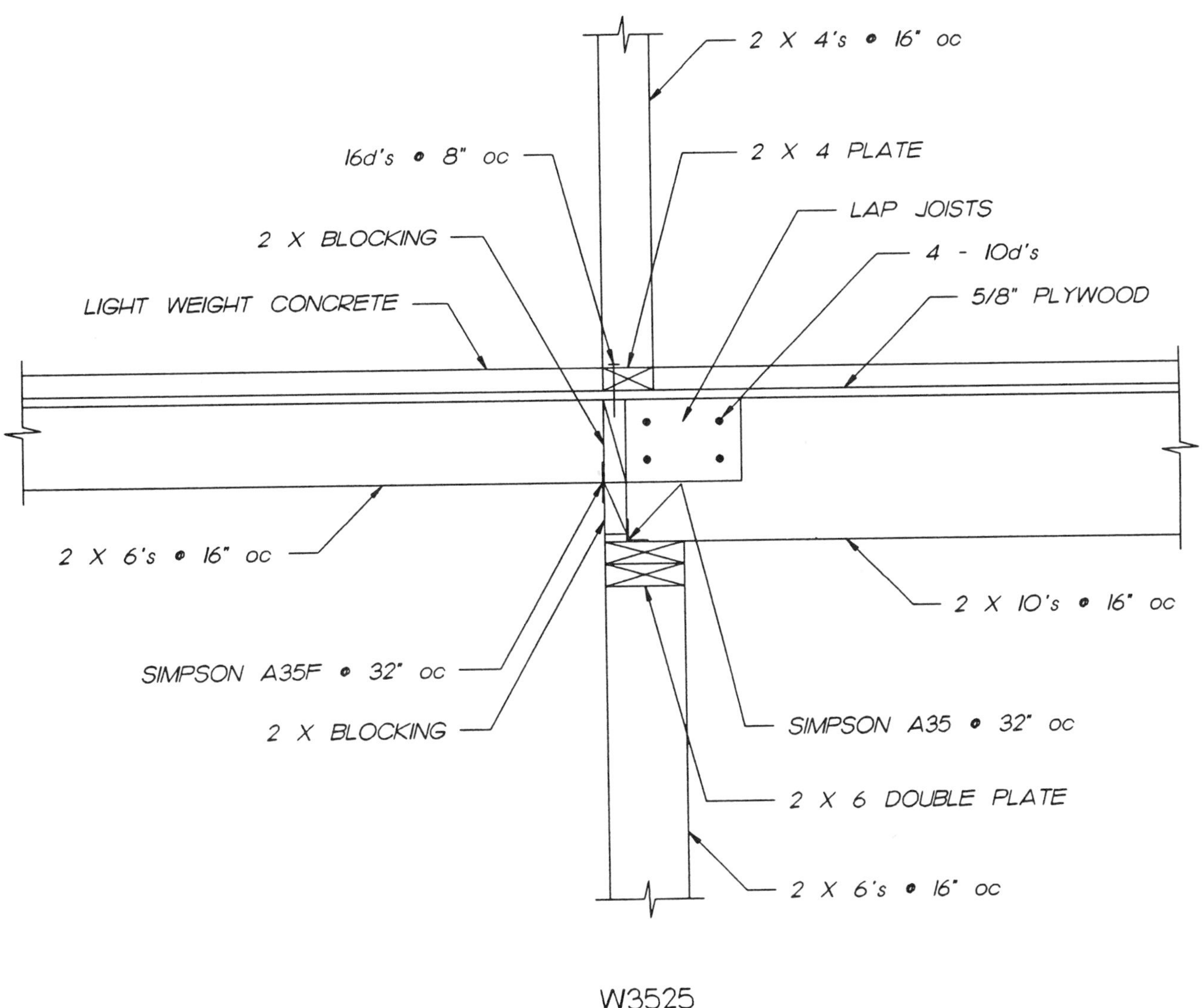
2 X 4's @ 16" oc
16d's @ 8" oc
2 X 4 PLATE
LAP JOISTS
2 X BLOCKING
4 - 10d's
LIGHT WEIGHT CONCRETE
5/8" PLYWOOD
2 X 6's @ 16" oc
2 X 10's @ 16" oc
SIMPSON A35F @ 32" oc
2 X BLOCKING
SIMPSON A35 @ 32" oc
2 X 6 DOUBLE PLATE
2 X 6's @ 16" oc

W3525

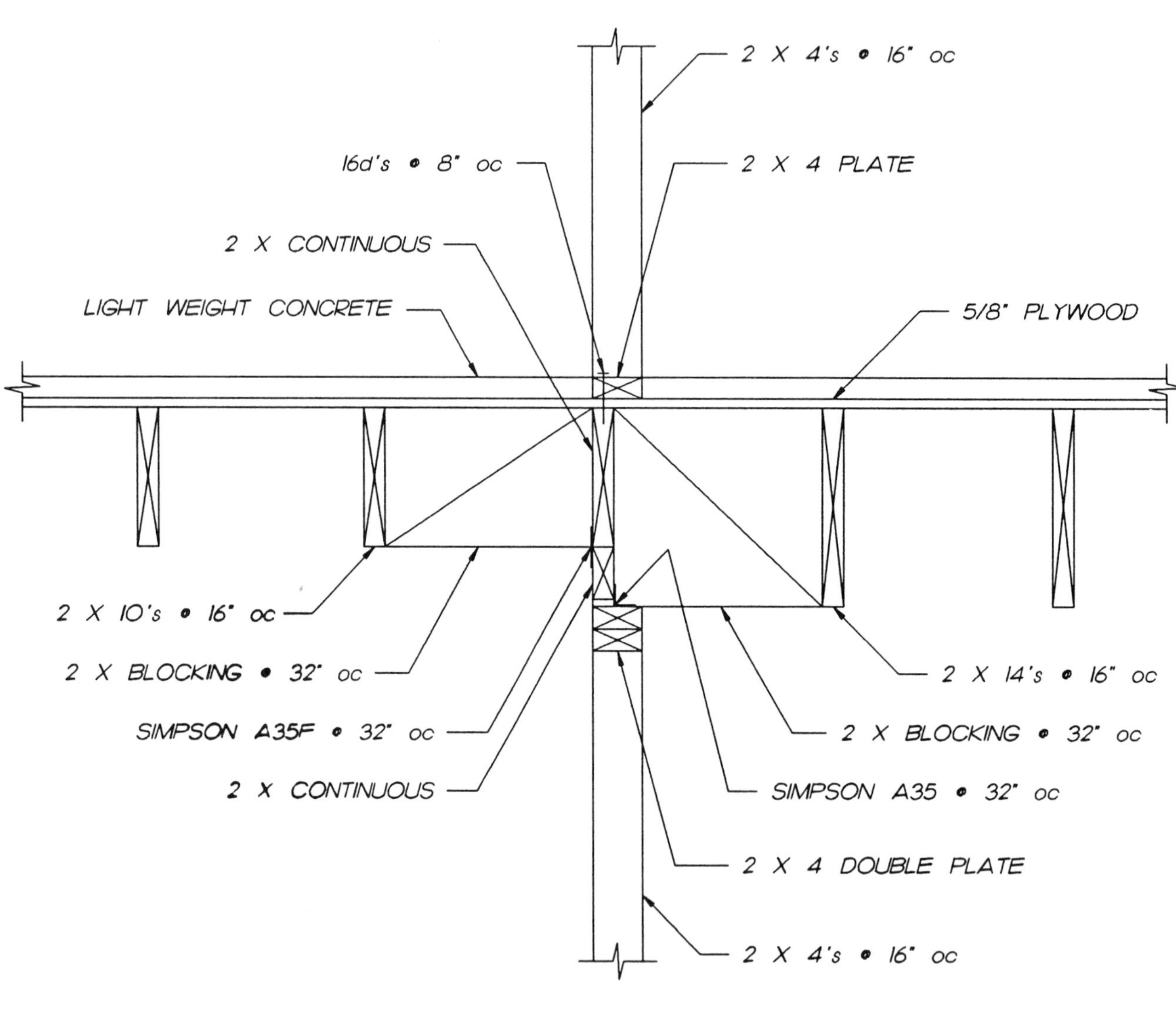

2 X 4's @ 16" oc
16d's @ 8" oc
2 X 4 PLATE
2 X CONTINUOUS
LIGHT WEIGHT CONCRETE
5/8" PLYWOOD
2 X 10's @ 16" oc
2 X BLOCKING @ 32" oc
SIMPSON A35F @ 32" oc
2 X CONTINUOUS
2 X 14's @ 16" oc
2 X BLOCKING @ 32" oc
SIMPSON A35 @ 32" oc
2 X 4 DOUBLE PLATE
2 X 4's @ 16" oc

W3540

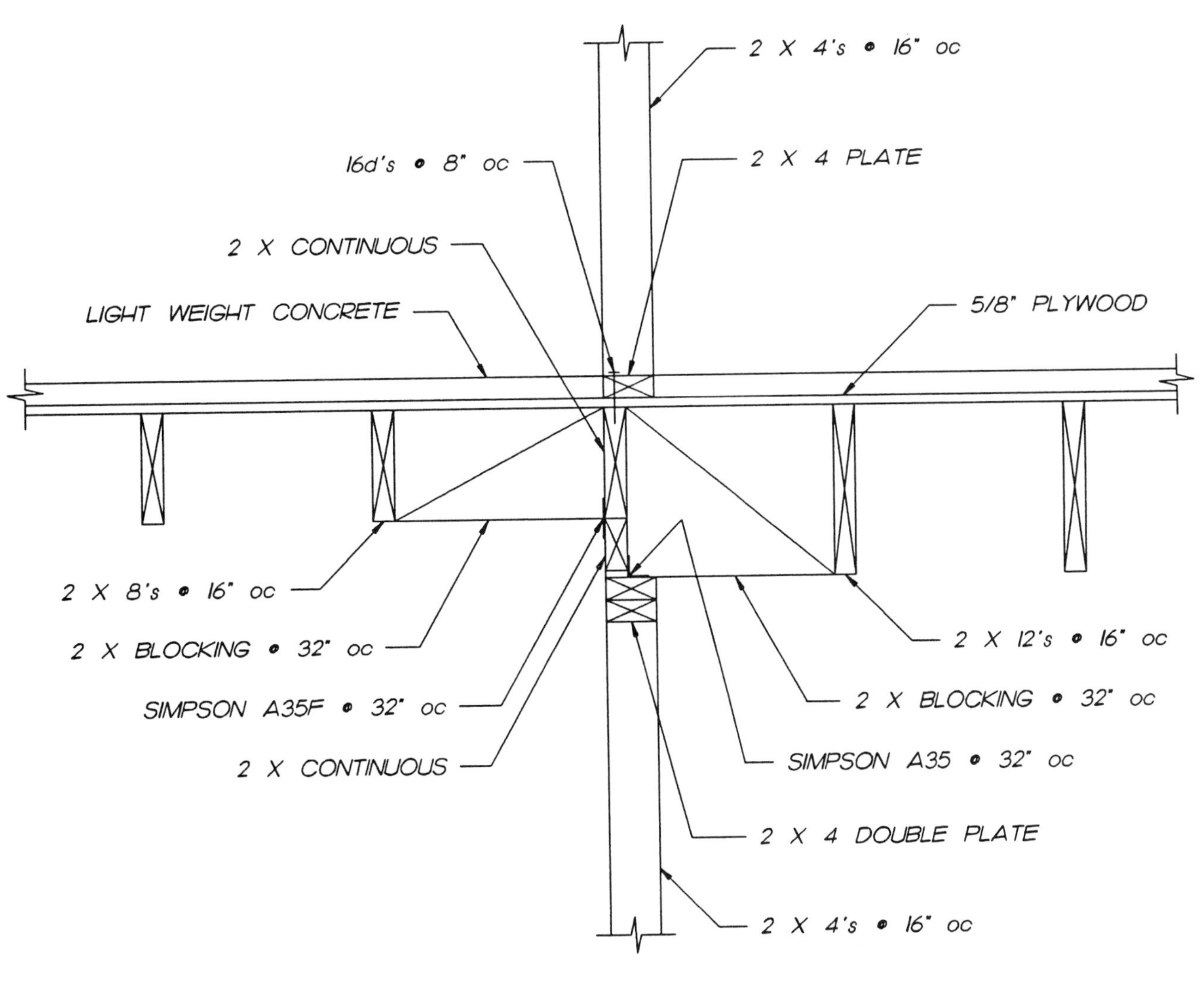
2 X 4's @ 16" oc
16d's @ 8" oc
2 X 4 PLATE
2 X CONTINUOUS
LIGHT WEIGHT CONCRETE
5/8" PLYWOOD
2 X 8's @ 16" oc
2 X BLOCKING @ 32" oc
2 X 12's @ 16" oc
SIMPSON A35F @ 32" oc
2 X BLOCKING @ 32" oc
2 X CONTINUOUS
SIMPSON A35 @ 32" oc
2 X 4 DOUBLE PLATE
2 X 4's @ 16" oc

W3541

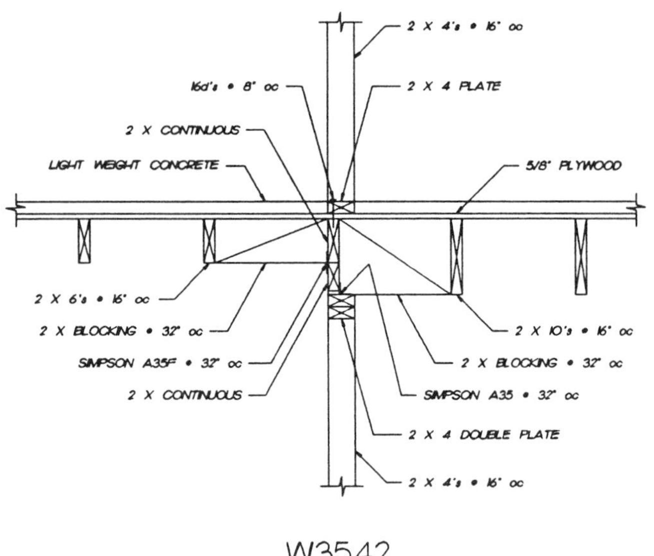

W3542

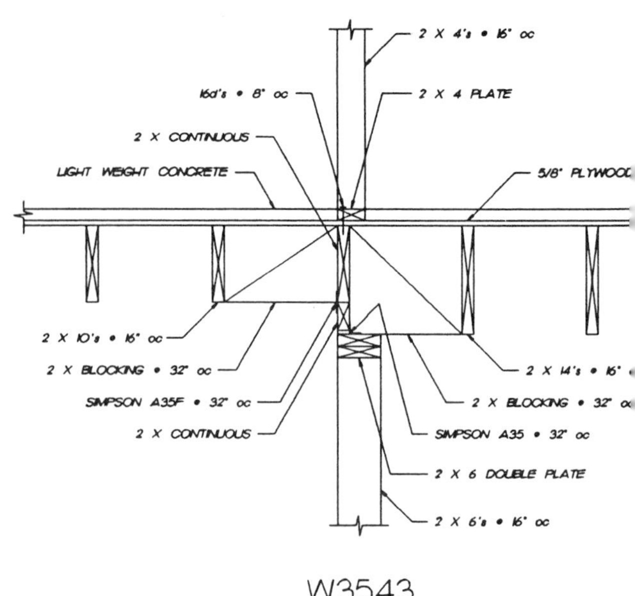

W3543

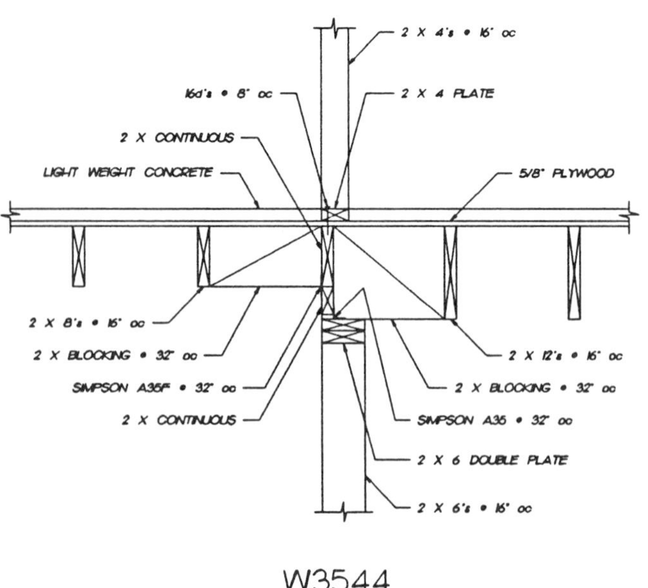

W3544

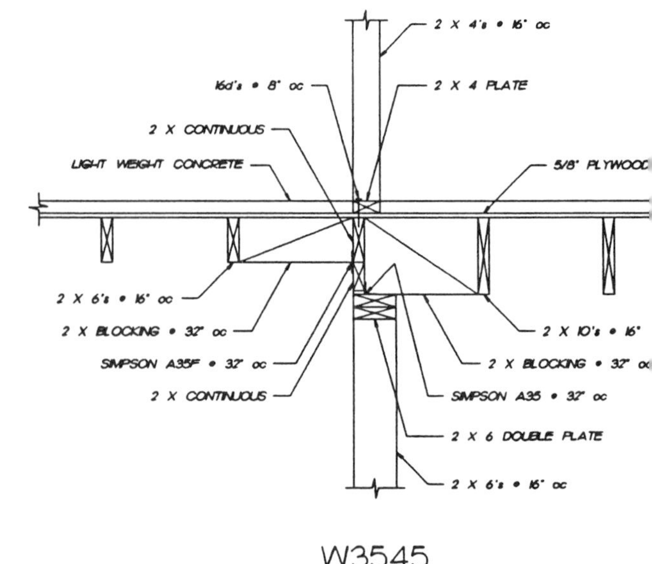

W3545

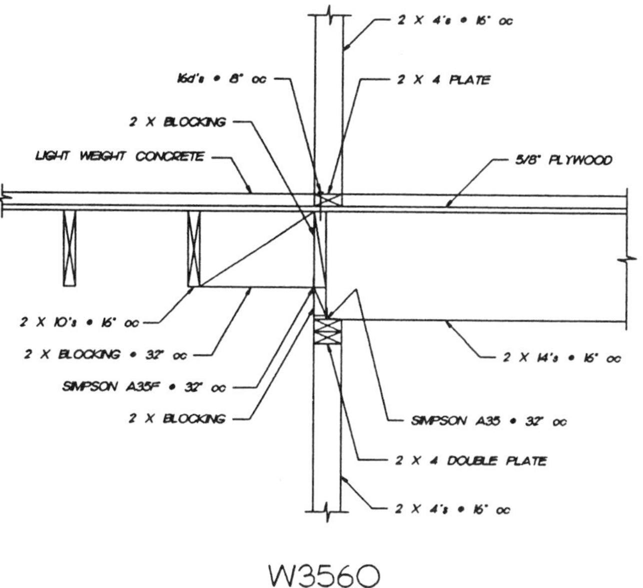

W3560

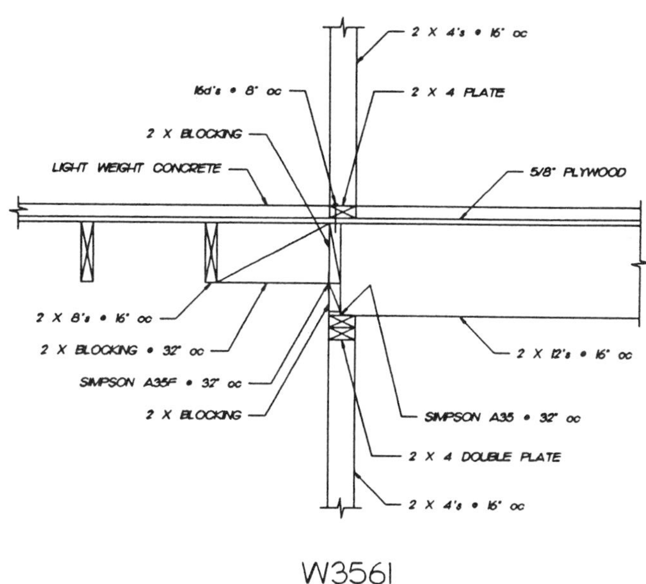

W3561

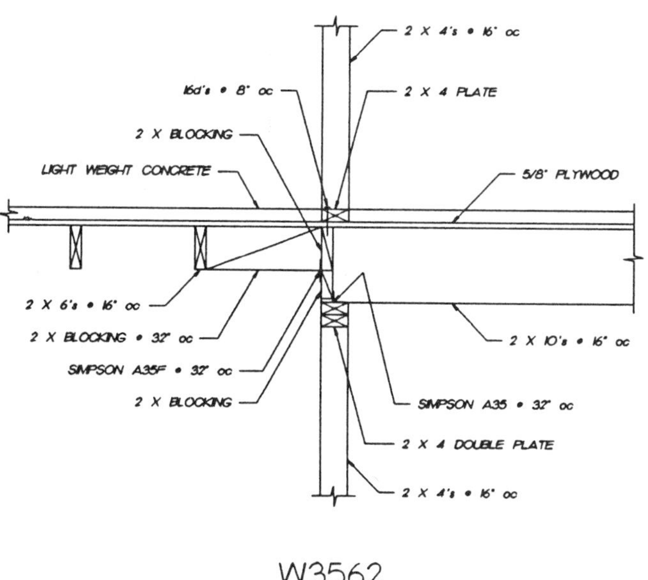

W3562

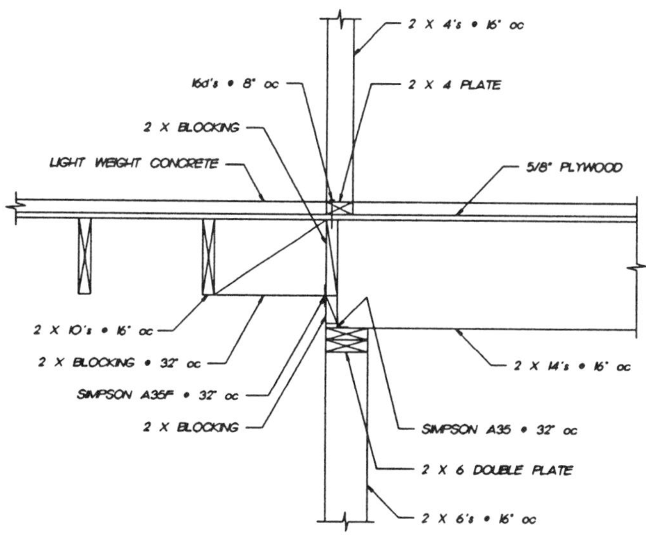

W3563

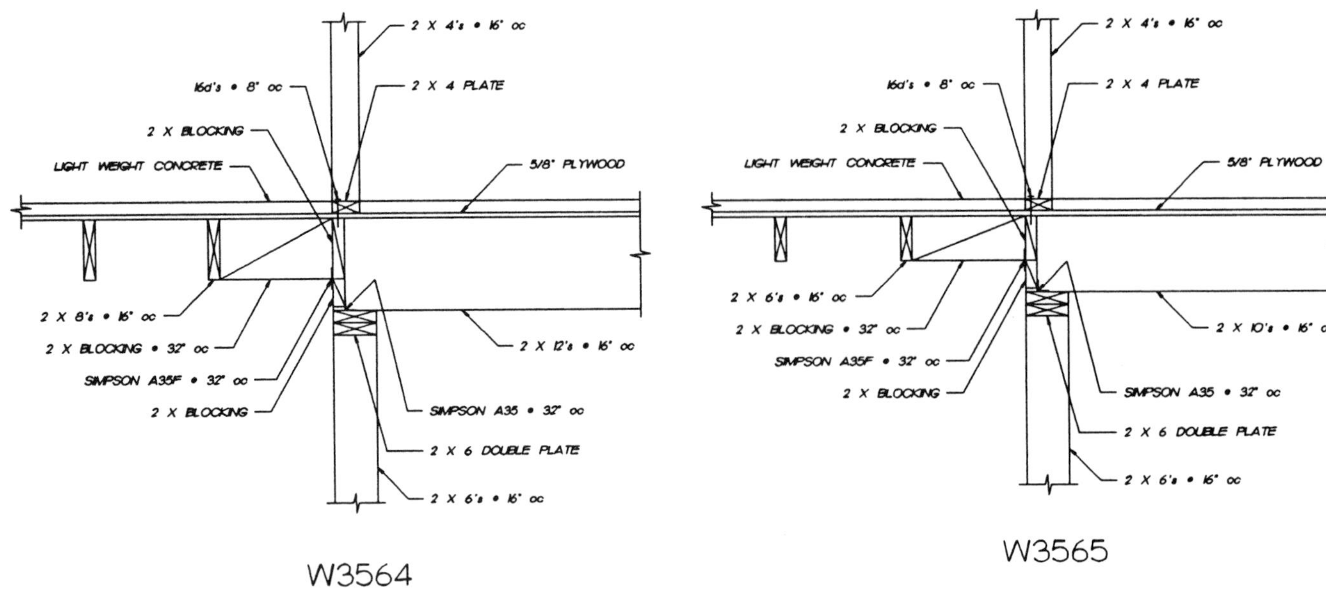

W3564

W3565

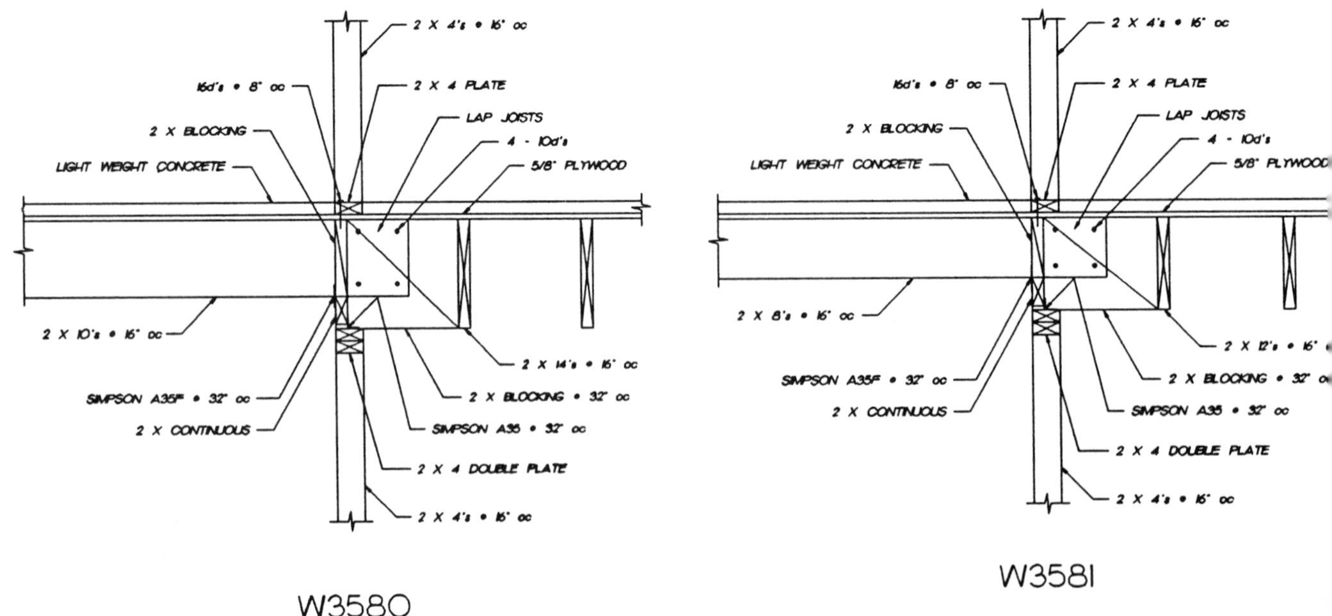

W3580

W3581

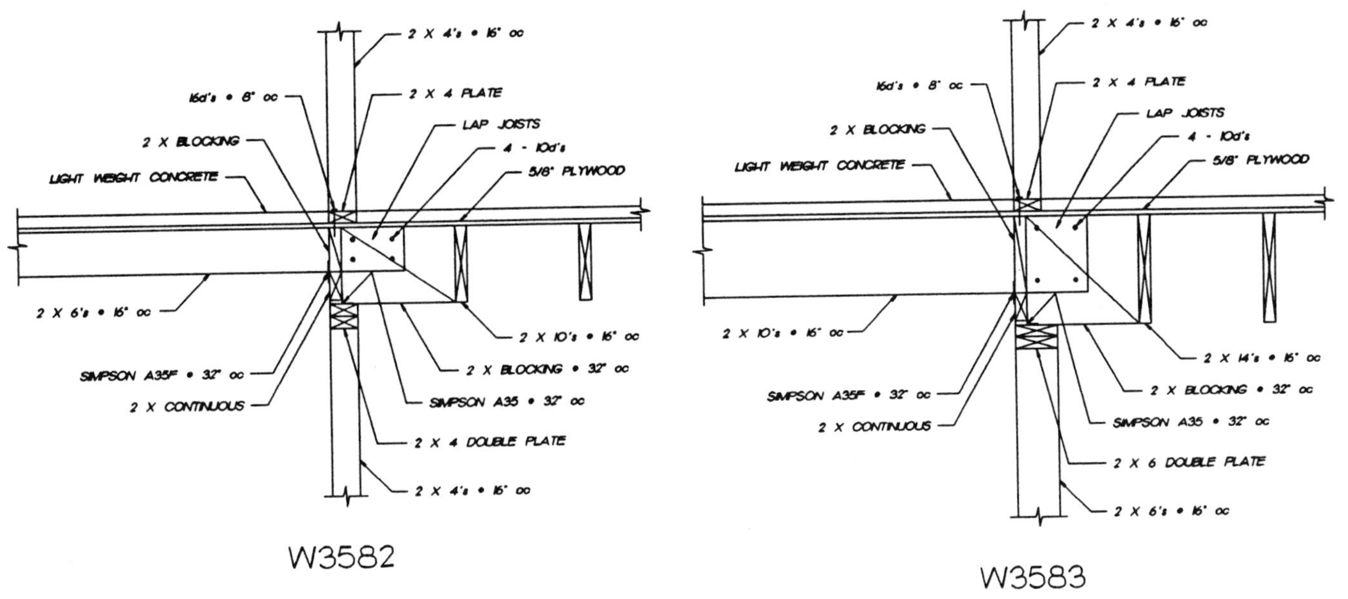

W3582

W3583

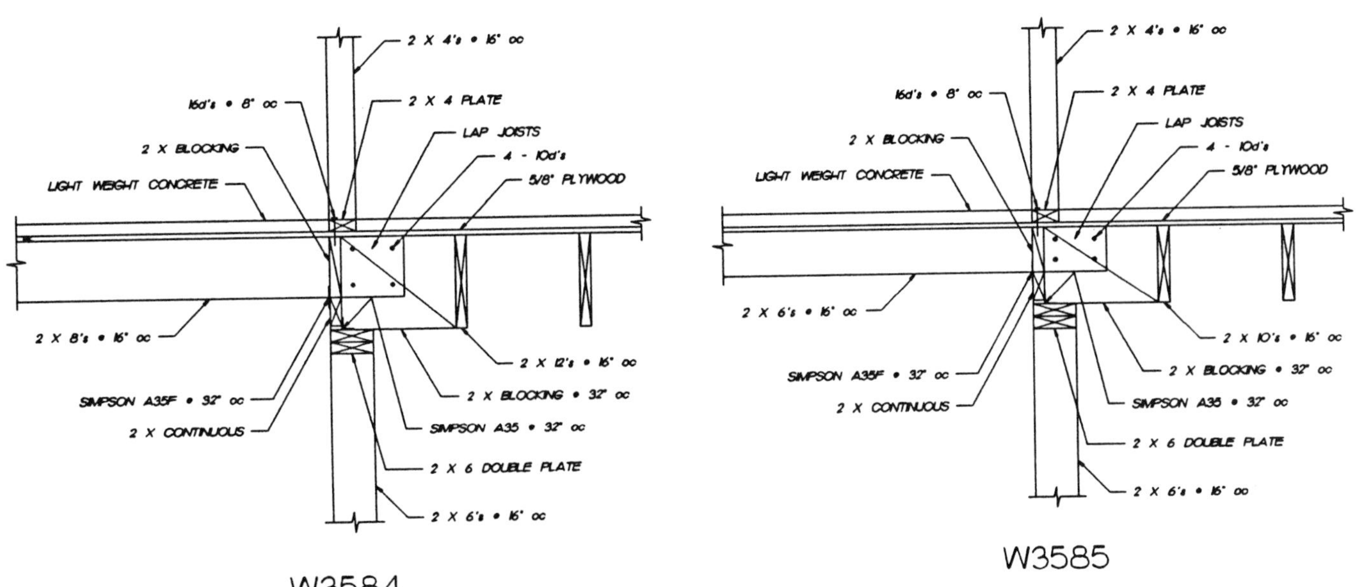

W3584

W3585

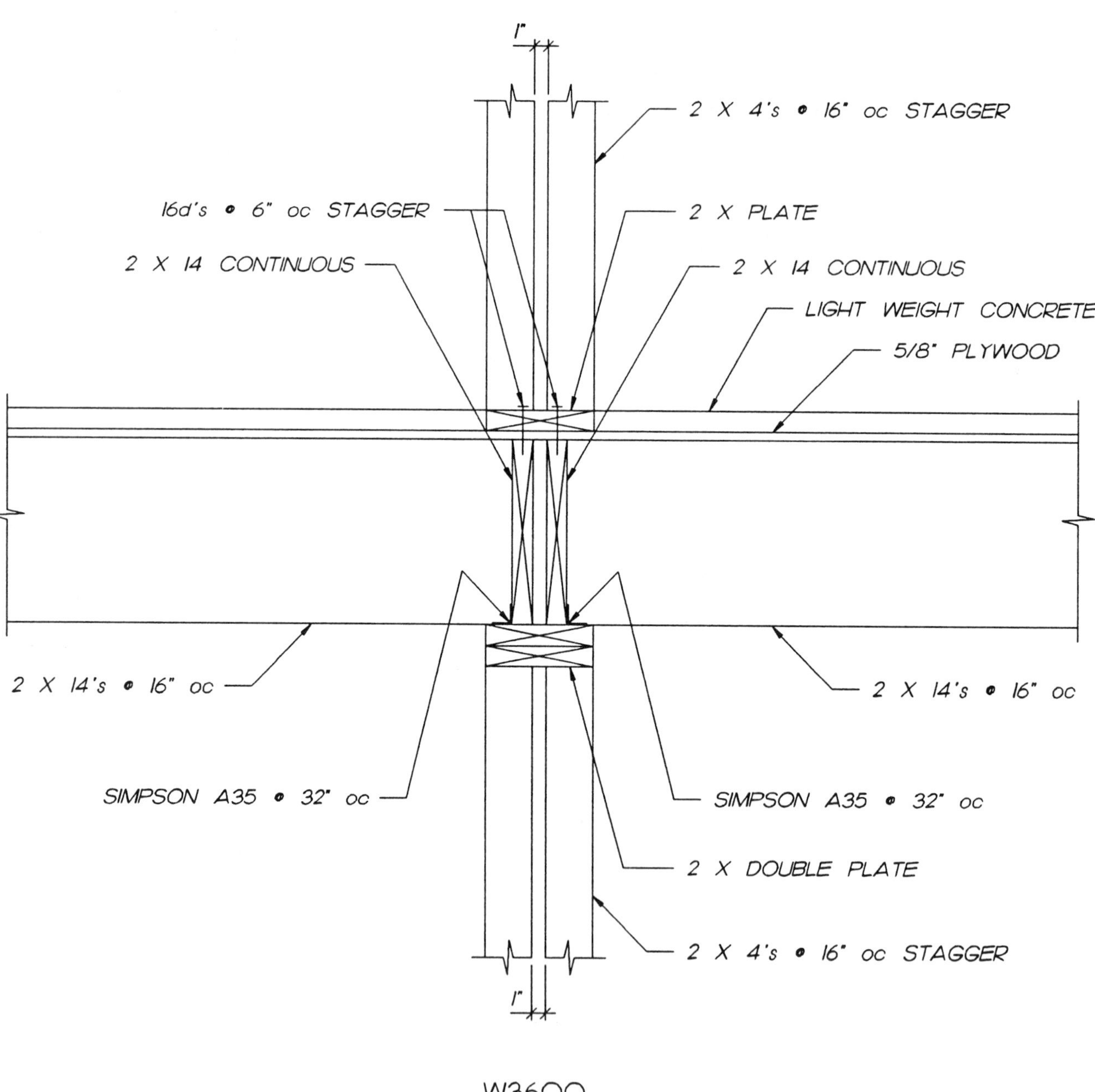
1"
2 X 4's @ 16" oc STAGGER
16d's @ 6" oc STAGGER
2 X PLATE
2 X 14 CONTINUOUS
2 X 14 CONTINUOUS
LIGHT WEIGHT CONCRETE
5/8" PLYWOOD
2 X 14's @ 16" oc
2 X 14's @ 16" oc
SIMPSON A35 @ 32" oc
SIMPSON A35 @ 32" oc
2 X DOUBLE PLATE
2 X 4's @ 16" oc STAGGER
1"

W3600

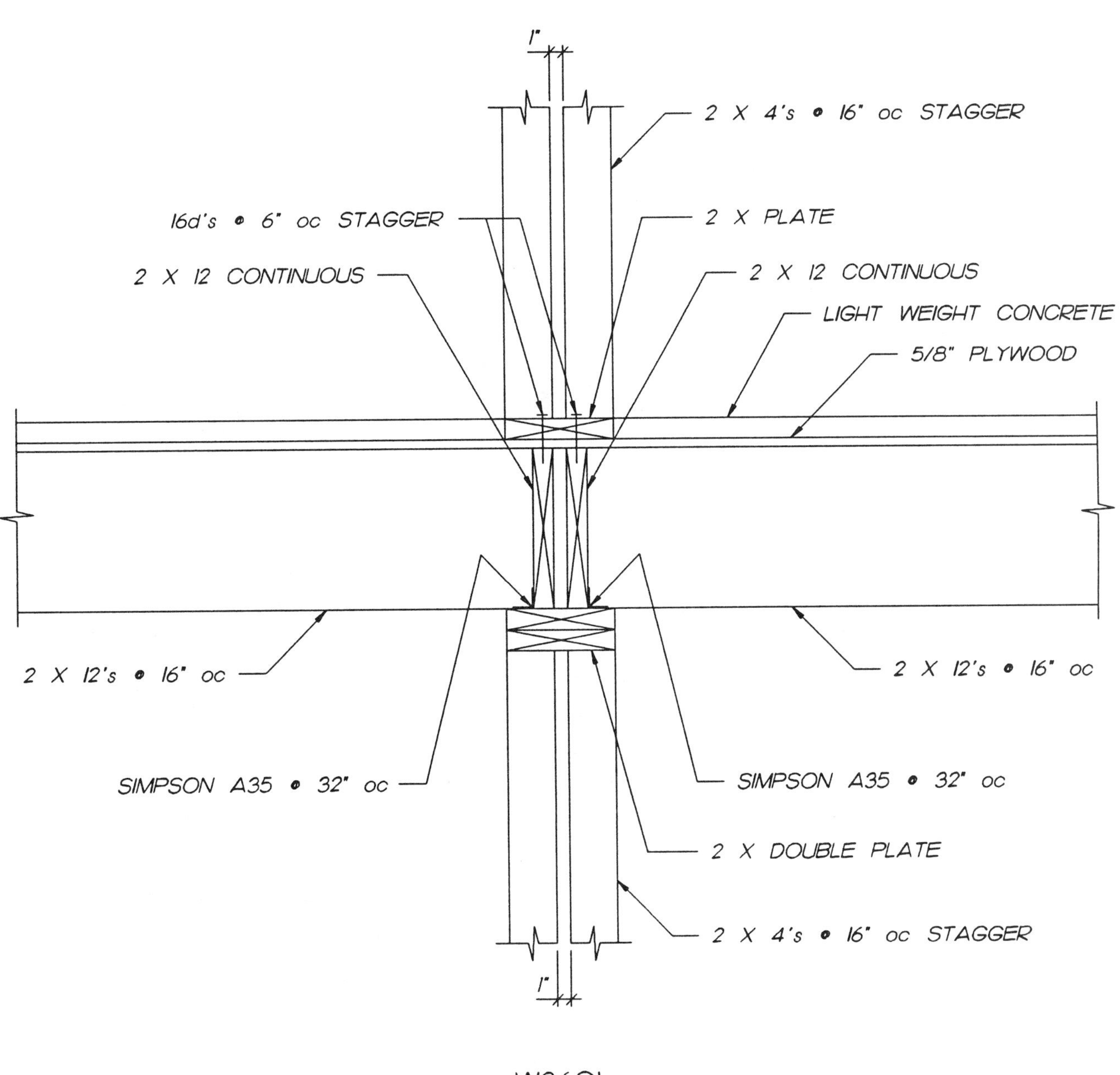
1"
2 X 4's @ 16" oc STAGGER
16d's @ 6" oc STAGGER
2 X PLATE
2 X 12 CONTINUOUS
2 X 12 CONTINUOUS
LIGHT WEIGHT CONCRETE
5/8" PLYWOOD
2 X 12's @ 16" oc
2 X 12's @ 16" oc
SIMPSON A35 @ 32" oc
SIMPSON A35 @ 32" oc
2 X DOUBLE PLATE
2 X 4's @ 16" oc STAGGER
1"

W3601

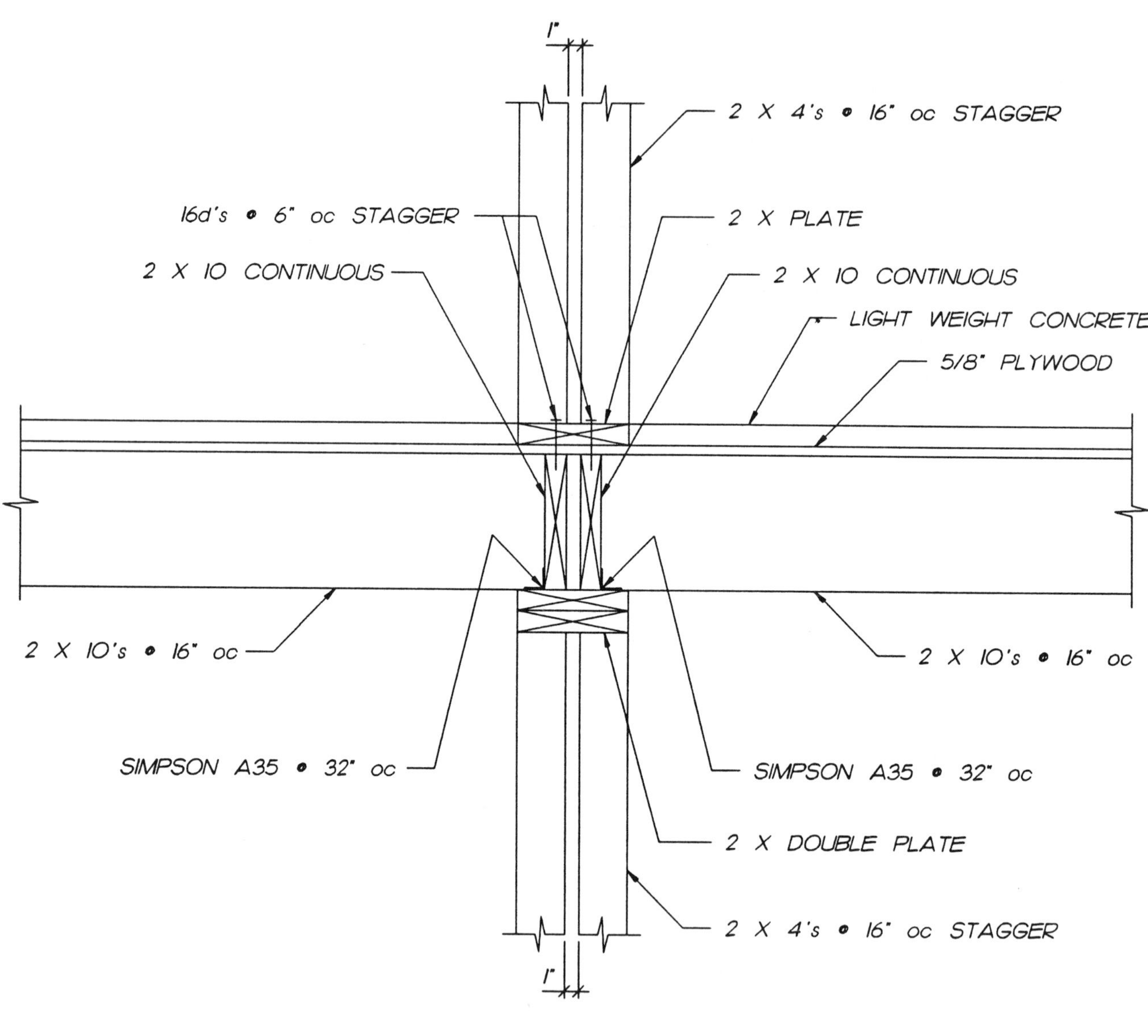

1"
2 X 4's @ 16" oc STAGGER
16d's @ 6" oc STAGGER
2 X PLATE
2 X 10 CONTINUOUS
2 X 10 CONTINUOUS
LIGHT WEIGHT CONCRETE
5/8" PLYWOOD
2 X 10's @ 16" oc
2 X 10's @ 16" oc
SIMPSON A35 @ 32" oc
SIMPSON A35 @ 32" oc
2 X DOUBLE PLATE
2 X 4's @ 16" oc STAGGER
1"

W3602

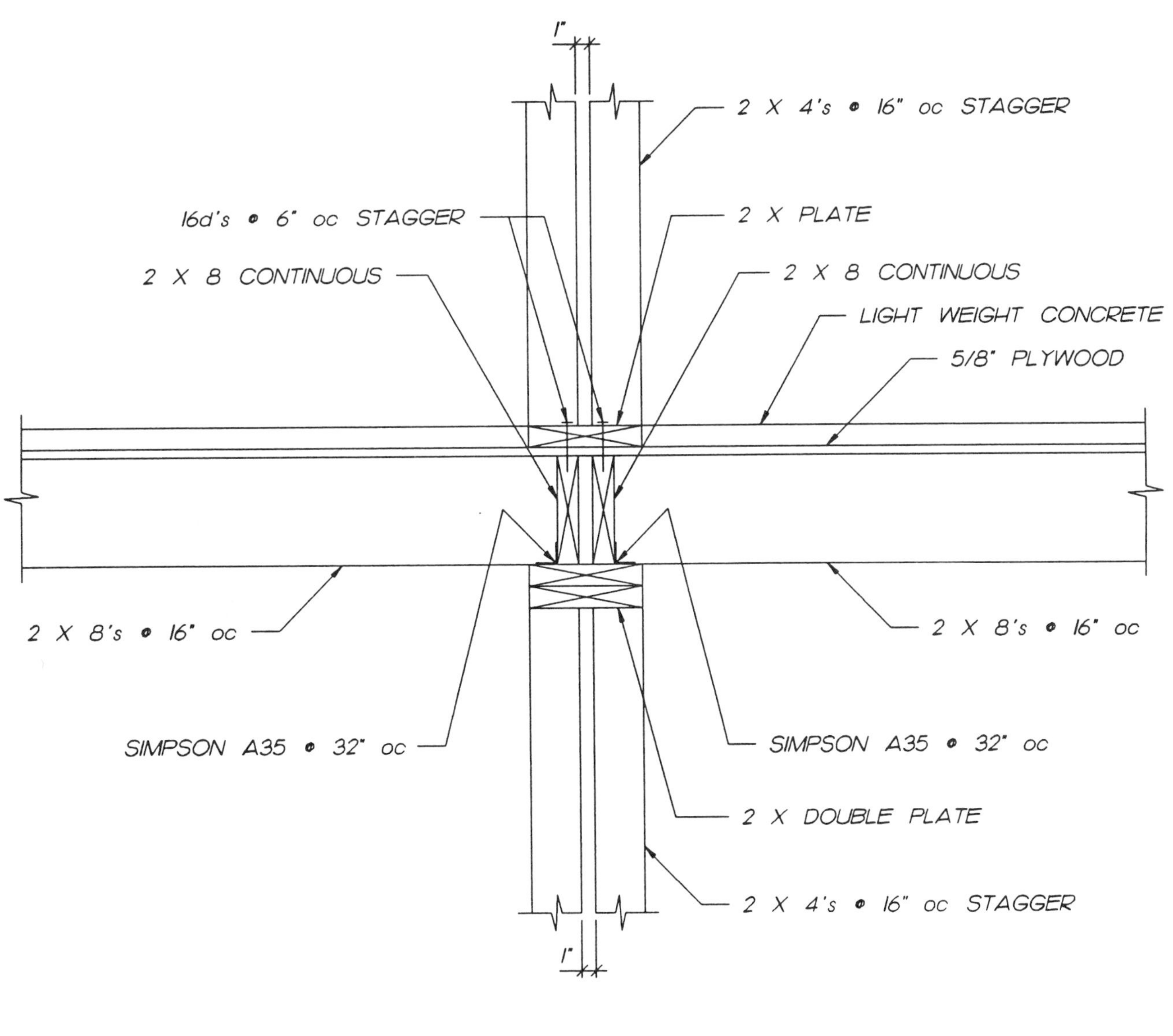
1"
2 X 4's @ 16" oc STAGGER
16d's @ 6" oc STAGGER
2 X PLATE
2 X 8 CONTINUOUS
2 X 8 CONTINUOUS
LIGHT WEIGHT CONCRETE
5/8" PLYWOOD
2 X 8's @ 16" oc
2 X 8's @ 16" oc
SIMPSON A35 @ 32" oc
SIMPSON A35 @ 32" oc
2 X DOUBLE PLATE
2 X 4's @ 16" oc STAGGER
1"

W3603

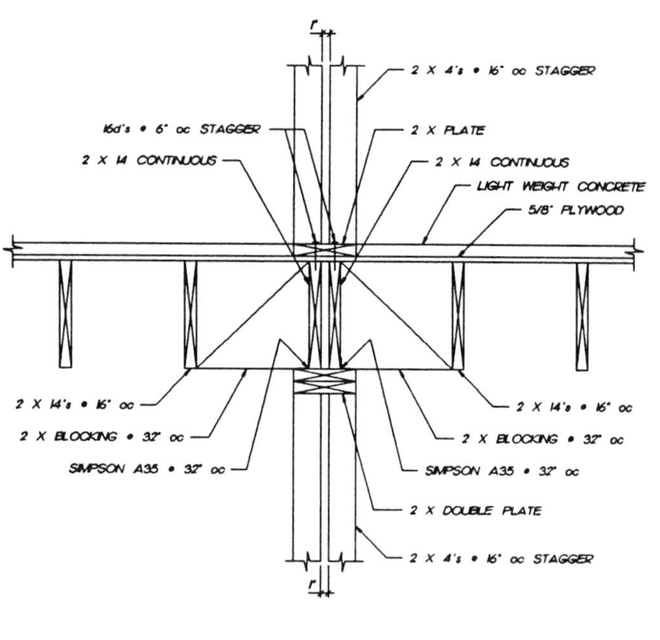

W3620

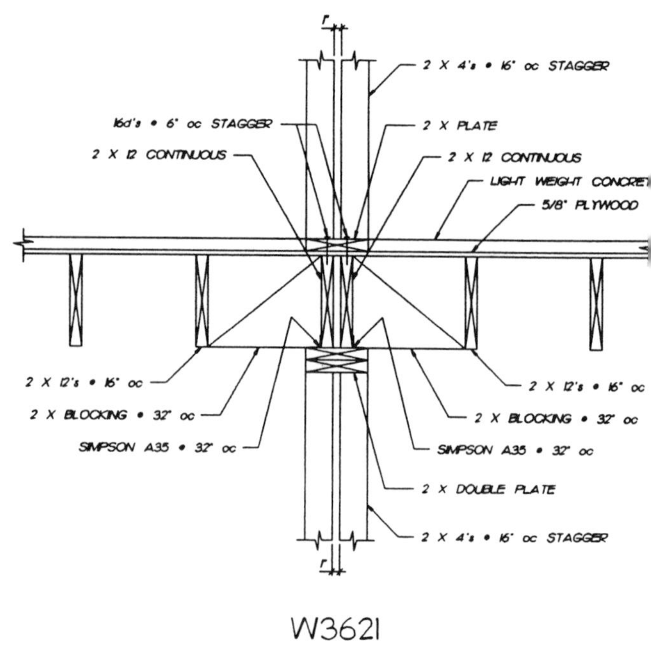

W3621

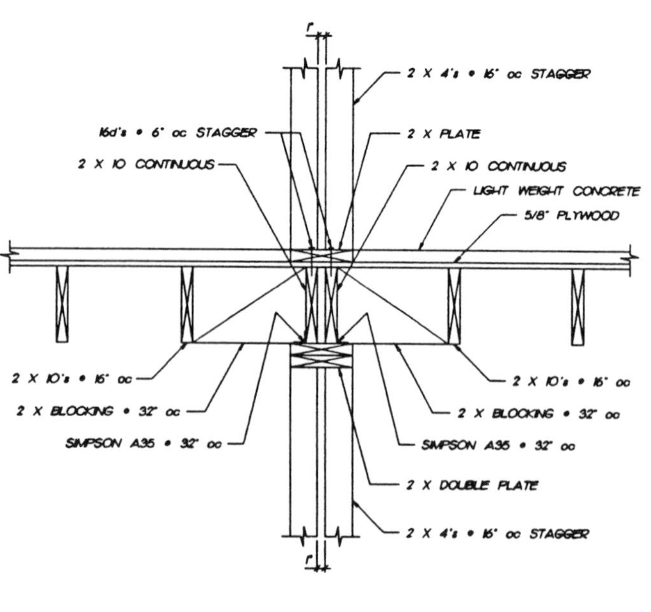

W3622

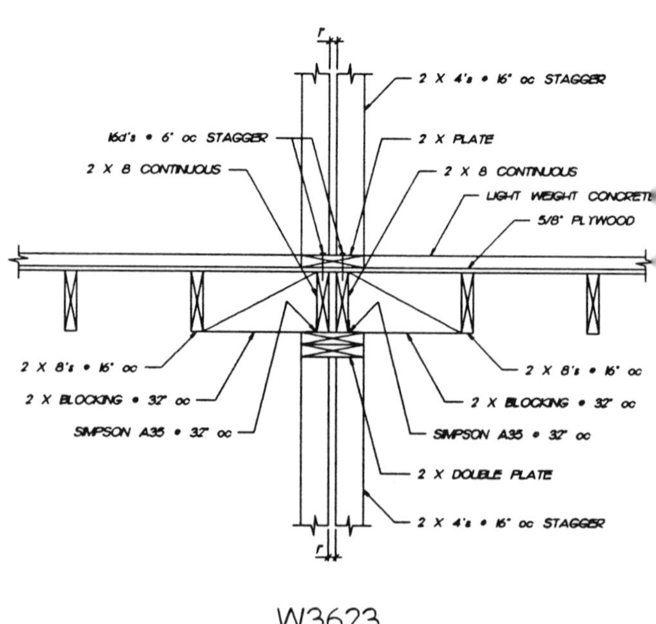

W3623

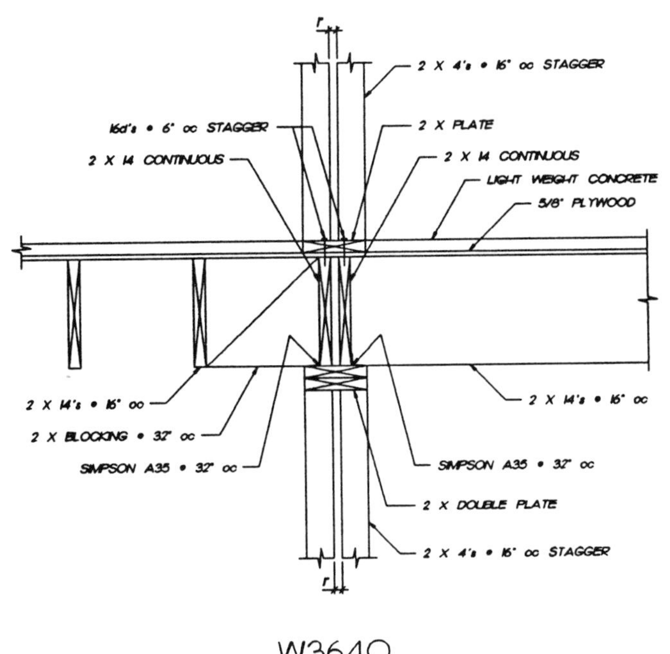

W3640

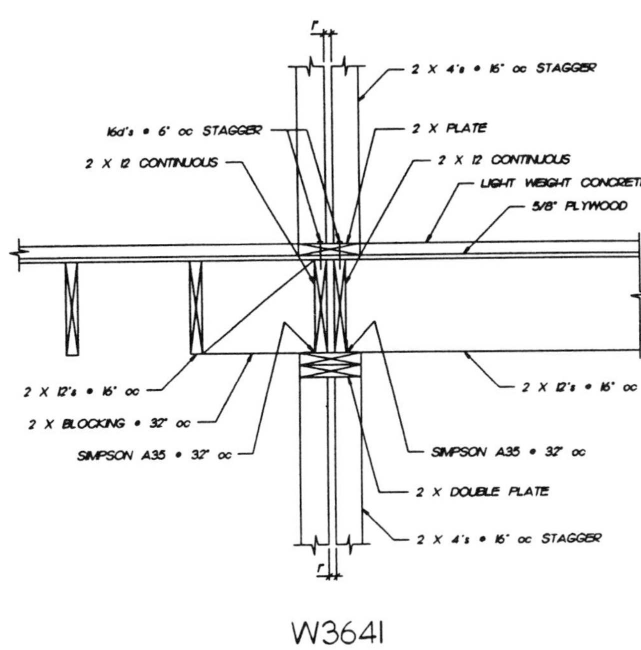

W3641

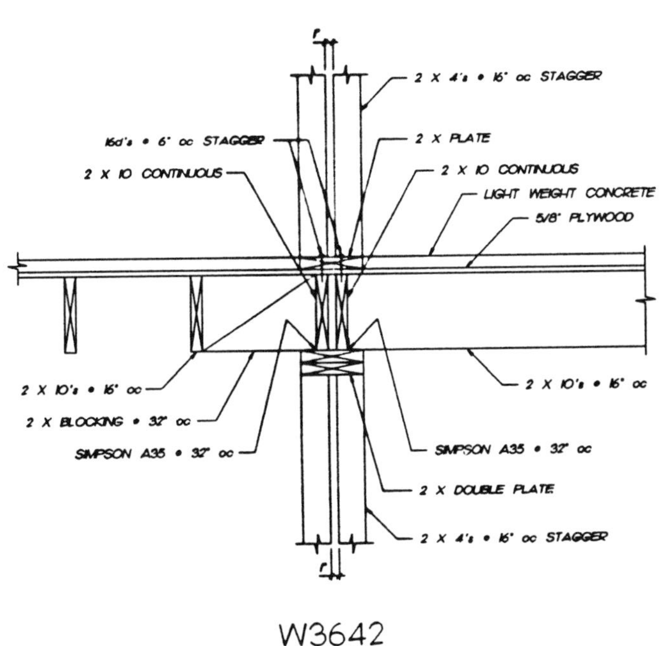

W3642

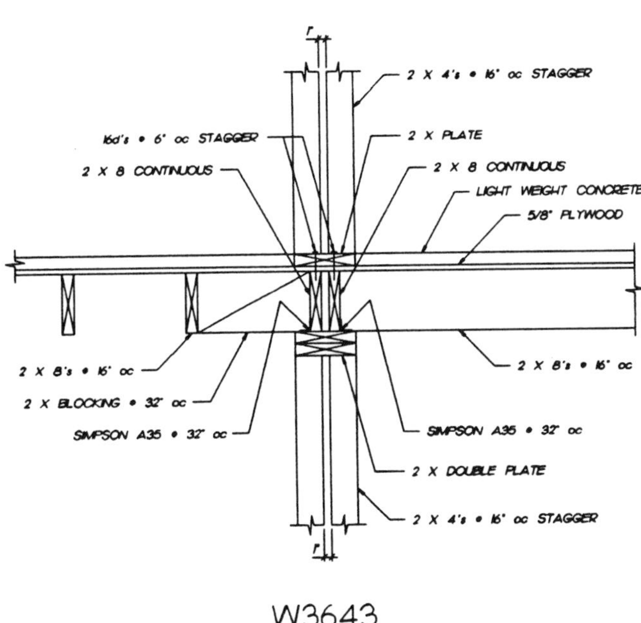

W3643

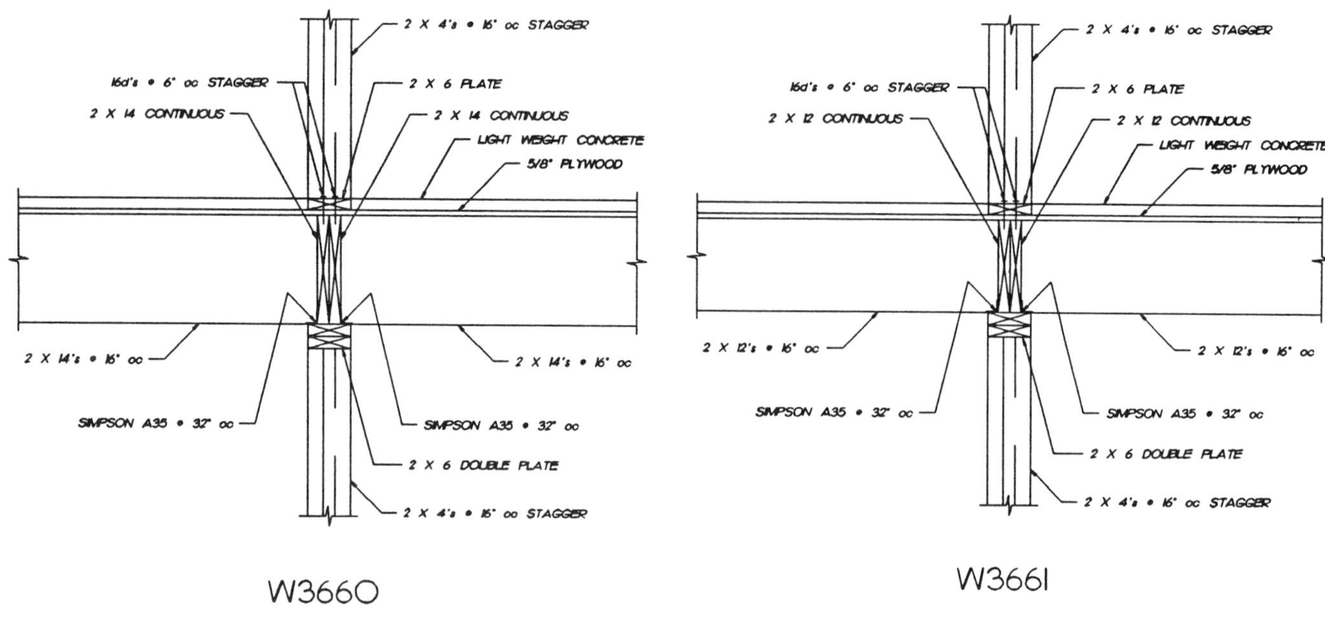

W3660

W3661

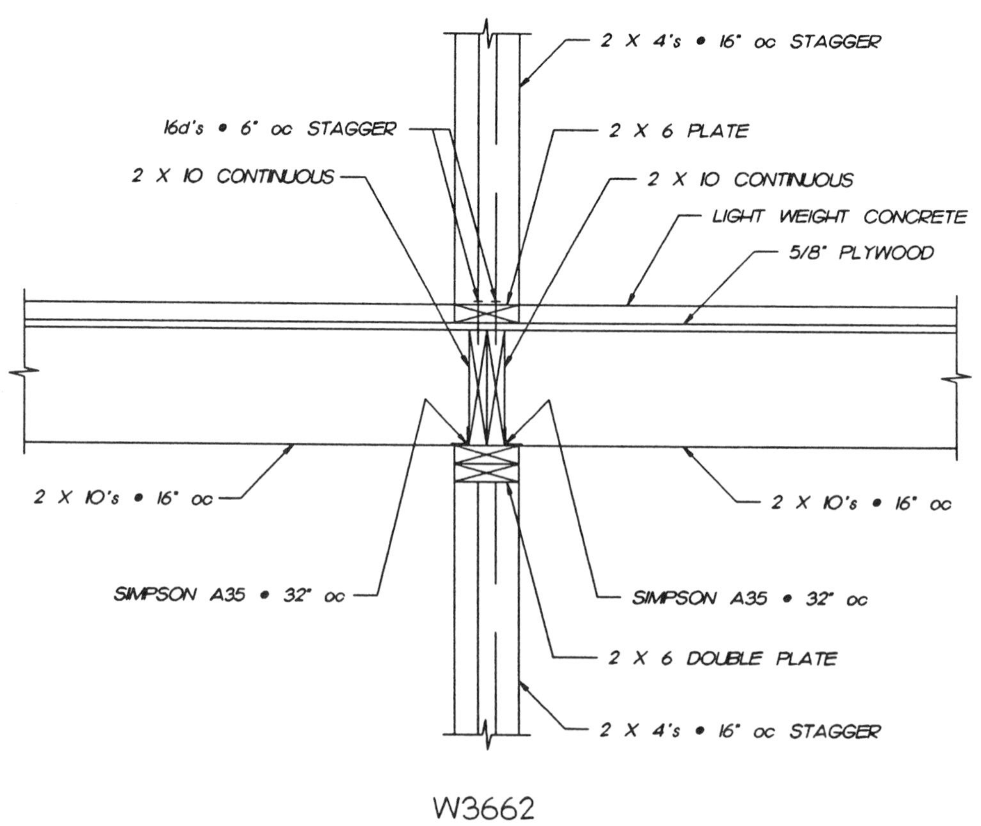

W3662

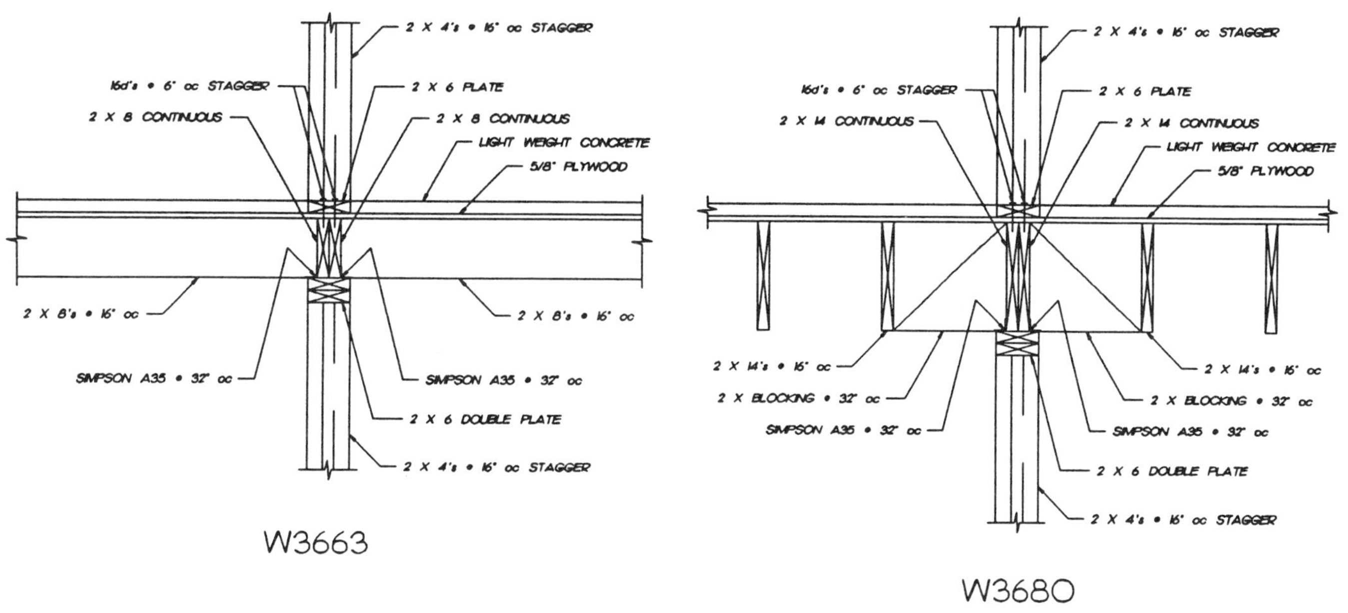

W3663

W3680

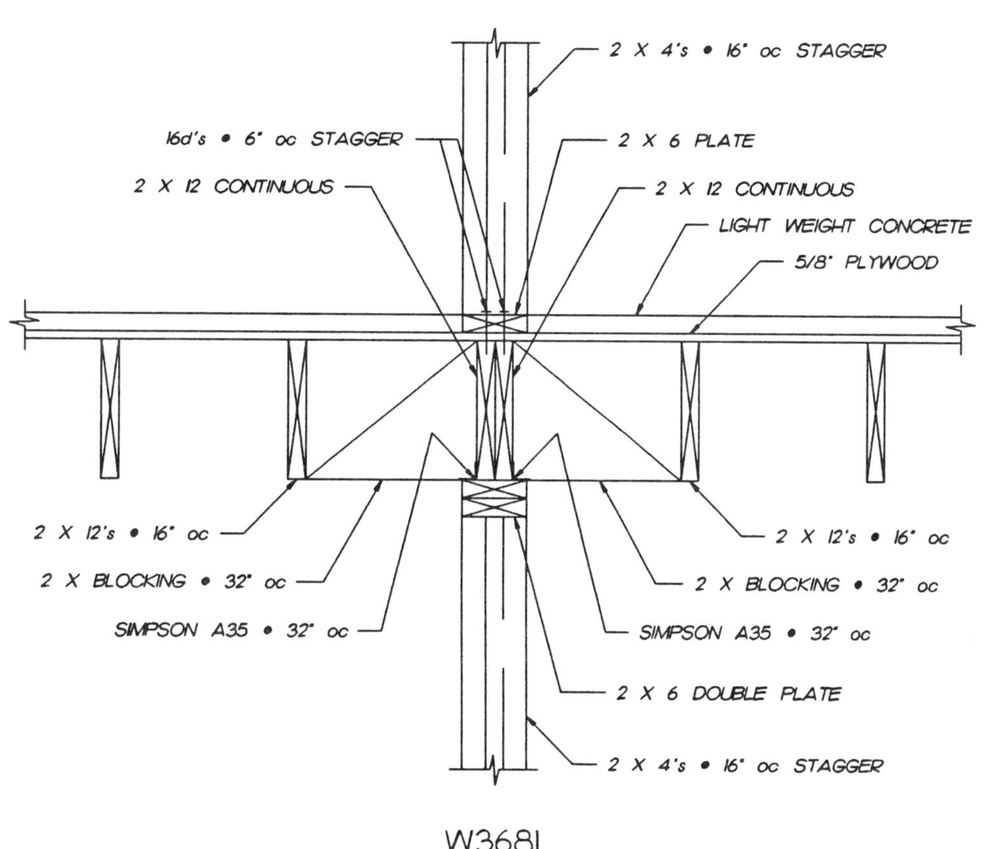

W3681

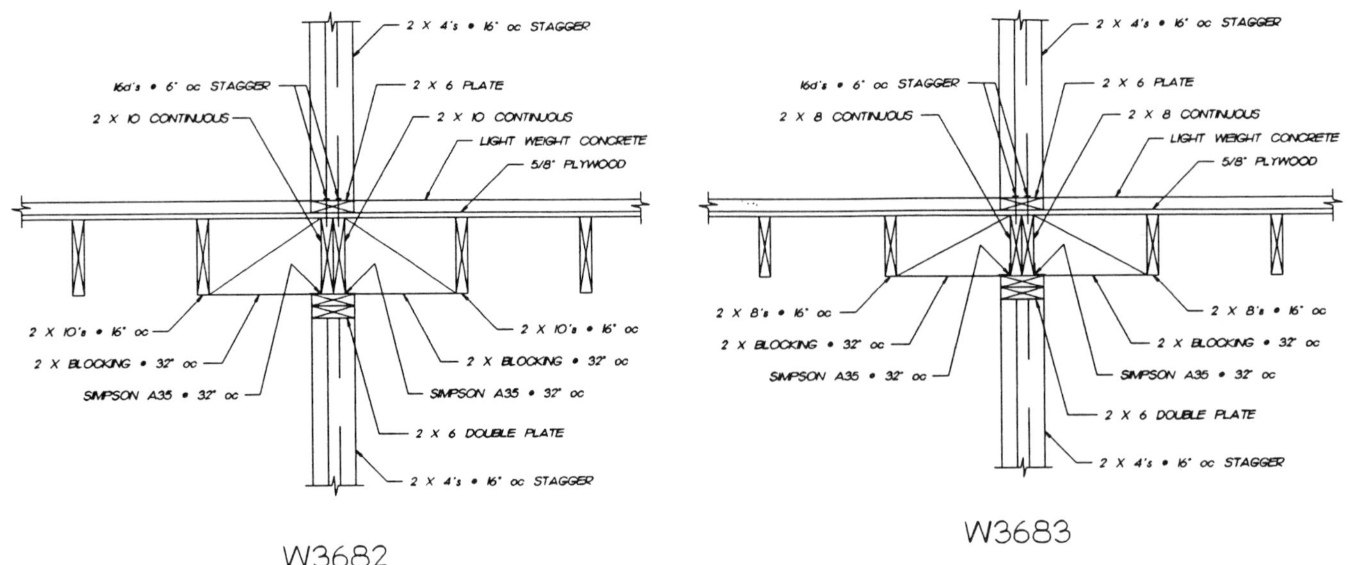

W3682

W3683

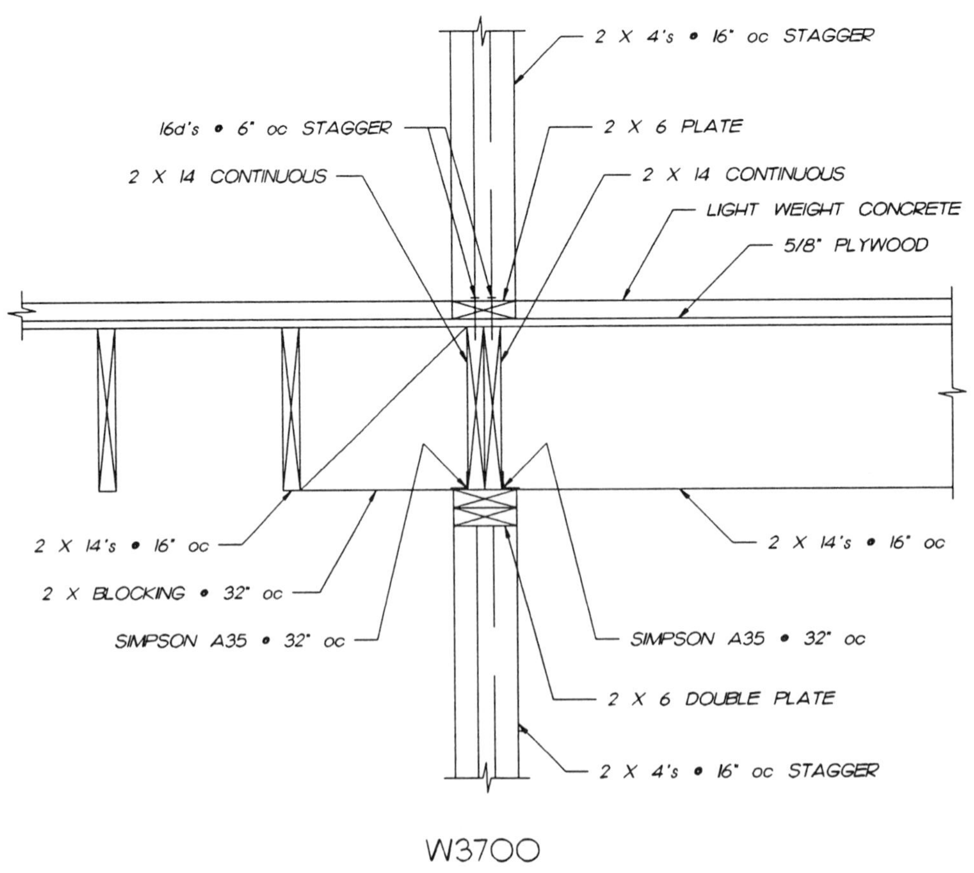

W3700

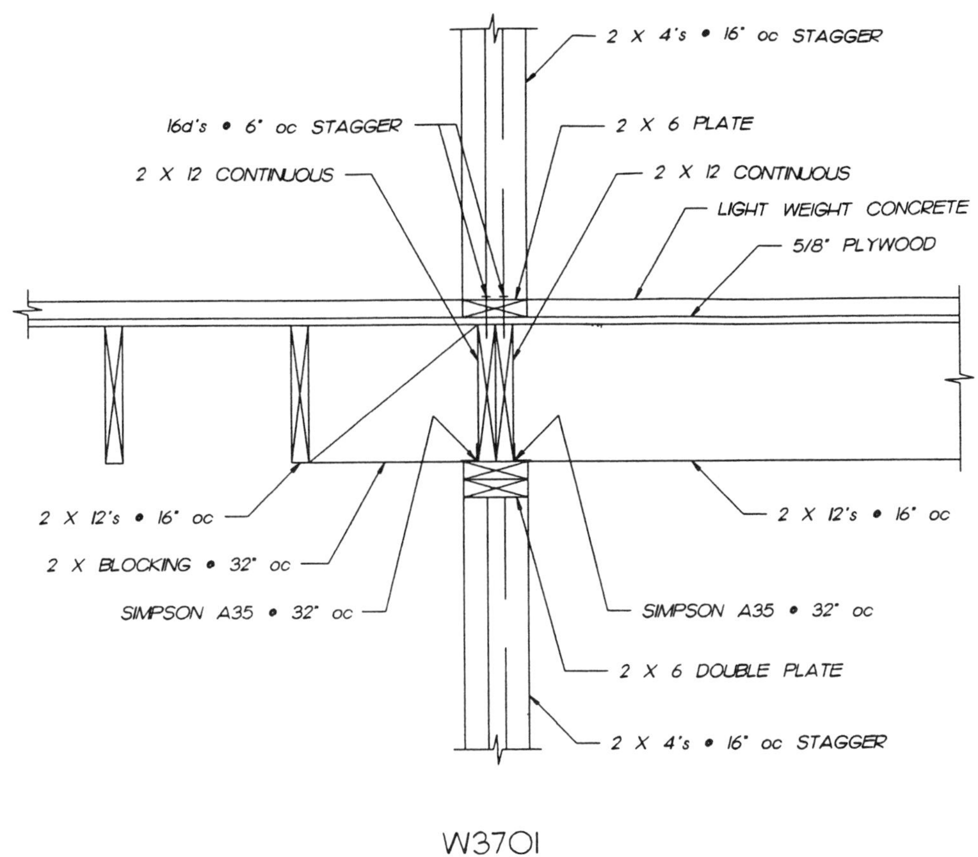

W3701

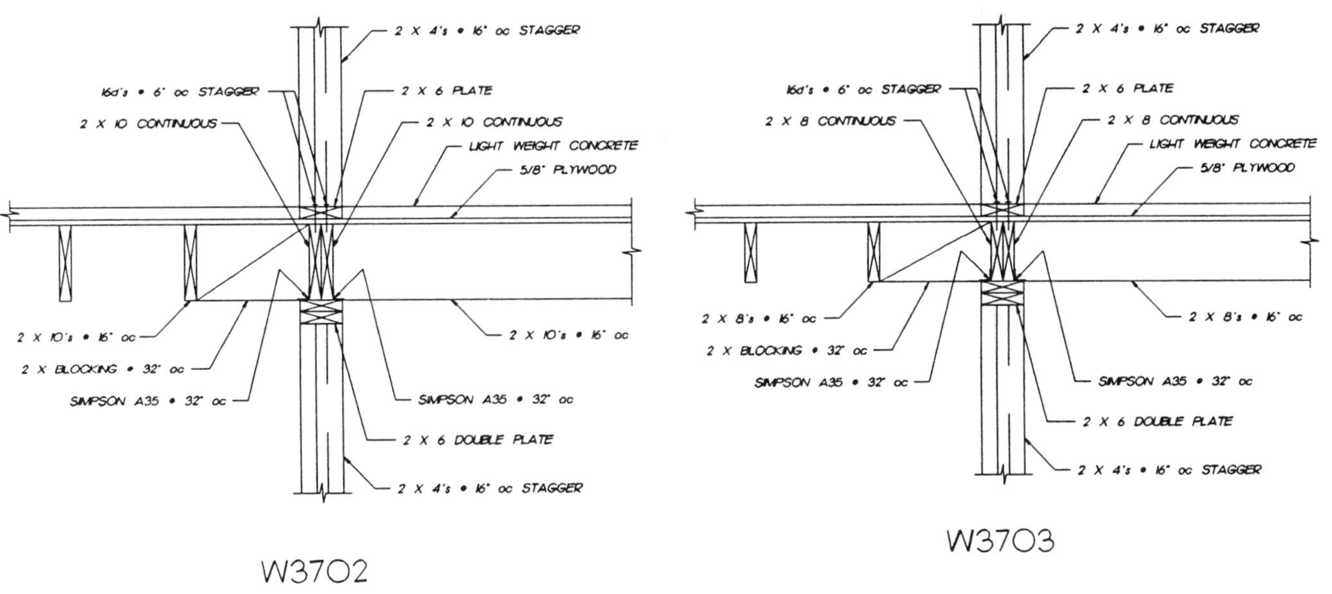

W3702

W3703

CHAPTER 1 WOOD	SECTION 4	WALLS & ROOF
EXTERIOR WALL RAFTERS PERP. TO WALL FLAT ROOF METAL SHEAR RESIST. CLIPS	————————	W4020 - W4023
EXTERIOR WALL RAFTERS PARALELL TO WALL FLAT ROOF METAL SHEAR RESIST. CLIPS	————————	W4040 - W4043
EXTERIOR WALL RAFTERS PERP. TO WALL FLAT ROOF METAL SHEAR RESIST. CLIPS	WITH A PARAPET	W4060 - W4063
EXTERIOR WALL RAFTERS PARALELL TO WALL FLAT ROOF METAL SHEAR RESIST. CLIPS	WITH A PARAPET	W4080 - W4083
EXTERIOR WALL RAFTERS PERP. TO WALL FLAT ROOF METAL SHEAR RESIST. CLIPS	WITH A ROOF OVERHANG	W4100 - W4103
EXTERIOR WALL RAFTERS PARALELL TO WALL FLAT ROOF METAL SHEAR RESIST. CLIPS	WITH A ROOF OVERHANG	W4120 - W4123
EXTERIOR WALL RAFTERS PERP. TO WALL FLAT ROOF METAL SHEAR RESIST. CLIPS	WITH A ROOF OVERHANG WITH A PARAPET	W4140 - W4143
EXTERIOR WALL RAFTERS PARALELL TO WALL FLAT ROOF METAL SHEAR RESIST. CLIPS	WITH A ROOF OVERHANG WITH A PARAPET	W4160 - W4163
EXTERIOR WALL RAFTERS PERP. TO WALL SCARF SLOPE ROOF TO WALL METAL SHEAR RESIST. CLIPS	————————	W4180 - W4183
EXTERIOR WALL RAFTERS PARALELL TO WALL SCARF SLOPE ROOF TO WALL METAL SHEAR RESIST. CLIPS	————————	W4200 - W4203
EXTERIOR WALL RAFTERS PERP. TO WALL SCARF SLOPE ROOF TO WALL METAL SHEAR RESIST. CLIPS	WITH A PARAPET	W4220 - W4223
EXTERIOR WALL RAFTERS PARALELL TO WALL SCARF SLOPE ROOF TO WALL METAL SHEAR RESIST. CLIPS	WITH A PARAPET	W4240 - W4243
EXTERIOR WALL RAFTERS PERP. TO WALL SCARF SLOPE AWAY FROM WALL METAL SHEAR RESIST. CLIPS	————————	W4260 - W4263
EXTERIOR WALL RAFTERS PARALELL TO WALL SCARF SLOPE AWAY FROM WALL METAL SHEAR RESIST. CLIPS	————————	W4280 - W4283
EXTERIOR WALL RAFTERS PERP. TO WALL SCARF SLOPE AWAY FROM WALL METAL SHEAR RESIST. CLIPS	WITH A PARAPET	W4300 - W4303
EXTERIOR WALL RAFTERS PARALELL TO WALL SCARF SLOPE AWAY FROM WALL METAL SHEAR RESIST. CLIPS	WITH A PARAPET	W4320 - W4323
EXTERIOR WALL - OVERHANG RAFTERS PARALELL TO WALL SCARF SLOPE ROOF TO WALL METAL SHEAR RESIST. CLIPS	————————	W4340 - W4343

CHAPTER 1 WOOD	SECTION 4	WALLS & ROOF
EXTERIOR WALL - OVERHANG RAFTERS PARALELL TO WALL SCARF SLOPE ROOF TO WALL METAL SHEAR RESIST. CLIPS	————	W4360 - W4363
EXTERIOR WALL - OVERHANG RAFTERS PERP. TO WALL SCARF SLOPE ROOF TO WALL METAL SHEAR RESIST. CLIPS	WITH A PARAPET	W4380 - W4383
EXTERIOR WALL - OVERHANG RAFTERS PARALELL TO WALL SCARF SLOPE ROOF TO WALL METAL SHEAR RESIST. CLIPS	WITH A PARAPET	W4400 - W4403
EXTERIOR WALL - OVERHANG RAFTERS PERP. TO WALL 3 TO 1 ROOF SLOPE METAL SHEAR RESIST. CLIPS	NO CEILING JOISTS	W4420 - W4423
EXTERIOR WALL - OVERHANG RAFTERS PERP. TO WALL 3 TO 1 ROOF SLOPE TOE NAIL SHEAR RESISTANCE	NO CEILING JOISTS	W4440 - W4443
EXTERIOR WALL - OVERHANG RAFTERS PERP. TO WALL 3 TO 1 ROOF SLOPE TOE NAIL SHEAR RESISTANCE	WITH CEILING JOIST PERP. TO WALL	W4460 - W4463
EXTERIOR WALL - OVERHANG RAFTERS PERP. TO WALL 3 TO 1 ROOF SLOPE TOE NAIL SHEAR RESISTANCE	WITH CEILING JOIST PARALELL TO WALL	W4480 - W4483
EXTERIOR WALL - OVERHANG RAFTERS PERP. TO WALL 2 TO 1 ROOF SLOPE TOE NAIL SHEAR RESISTANCE	NO CEILING JOISTS	W4500 - W4503
EXTERIOR WALL - OVERHANG RAFTERS PERP. TO WALL 2 TO 1 ROOF SLOPE TOE NAIL SHEAR RESISTANCE	WITH CEILING JOIST PERP. TO WALL	W4520 - W4523
EXTERIOR WALL - OVERHANG RAFTERS PERP. TO WALL 2 TO 1 ROOF SLOPE TOE NAIL SHEAR RESISTANCE	WITH CEILING JOIST PARALELL TO WALL	W4540 - W4543
INTERIOR WALL RAFTERS PERP. TO WALL FLAT ROOF METAL SHEAR RESIST. CLIPS	————	W4560 - W4563
INTERIOR WALL RAFTERS PARALELL TO WALL FLAT ROOF METAL SHEAR RESIST. CLIPS	————	W4580 - W4583
INTERIOR WALL RFTRS PERP & PARALELL ALT SIDES FLAT ROOF METAL SHEAR RESIST. CLIPS	————	W4600 - W4603
INTERIOR WALL RAFTERS PERP. TO WALL FLAT ROOF METAL SHEAR RESIST. CLIPS	DROPPED CEILING CEILING JOISTS PERP. TO WALL	W4620 - W4623
INTERIOR WALL RAFTERS PARALELL TO WALL FLAT ROOF METAL SHEAR RESIST. CLIPS	DROPPED CEILING CEILING JOISTS PERP. TO WALL	W4640 - W4643
INTERIOR WALL RFTRS PERP & PARALELL ALT SIDES FLAT ROOF METAL SHEAR RESIST. CLIPS	DROPPED CEILING CEILING JOISTS PERP. TO WALL	W4660 - W4663
INTERIOR WALL RAFTERS PARALELL TO WALL SCARF SLOPED ROOF METAL SHEAR RESIST. CLIPS	————	W4680 - W4683

CHAPTER 1 WOOD SECTION 4 WALLS & ROOF

NTERIOR WALL RAFTERS PARALELL TO WALL SCARF SLOPED ROOF METAL SHEAR RESTANT CLIPS	————	W4700 - W4703
NTERIOR WALL FTRS PERP & PARALELL ALT SIDES SCARF SLOPED ROOF METAL SHEAR RESTANT CLIPS	————	W4720 - W4723
NTERIOR DIVISION WALL RAFTERS PERP. TO WALL FLAT ROOF METAL SHEAR RESIST. CLIPS	NO CEILING JOISTS SEPARATED STUDS	W4740 - W4743
INTERIOR DIVISION WALL RAFTERS PARALELL TO WALL FLAT ROOF METAL SHEAR RESIST. CLIPS	NO CEILING JOISTS SEPARATED STUDS	W4760 - W4763
INTERIOR DIVISION WALL RFTRS PERP & PARALELL ALT SIDES FLAT ROOF METAL SHEAR RESIST. CLIPS	NO CEILING JOISTS SEPARATED STUDS	W4780 - W4783
INTERIOR DIVISION WALL RAFTERS PERP. TO WALL FLAT ROOF METAL SHEAR RESIST. CLIPS	NO CEILING JOISTS STAGGERED STUDS	W4800 - W4803
INTERIOR DIVISION WALL RAFTERS PARALELL TO WALL FLAT ROOF METAL SHEAR RESIST. CLIPS	NO CEILING JOISTS STAGGERED STUDS	W4820 - W4823
INTERIOR DIVISION WALL RFTRS PERP & PARALELL ALT SIDES FLAT ROOF METAL SHEAR RESIST. CLIPS	NO CEILING JOISTS STAGGERED STUDS	W4840 - W4843
RIDGE BEAM & HIP BEAM INTERSECTION	PLAN	W4860
	ELEVATION	W4861
REDGE BEAM SECTION	KING POST SUPPORT	W4880
	NO KING POST	W4881
ROOF SECTION OF RIDGE CEILING & WALL	DIAGONAL ROOF SUPPORTS TO WALL	W4900
	DIAGONAL & VERT. SUPPORT TO WALL	W4901
	DIAGONAL & VERT. SUPPORT TO WALL WITH RAFTER TIES	W4902
	RIDGE ALIGNS WITH WALL	W4903
ROOF MECHANICAL EQUIPMENT PLATFORM	————	W4920 - W4921
ROOF SCARF CONNECTION TO RAFTERS	USING METAL CLIPS - ELEV. & SECT.	W4940 & W4942
	USING TOE NAILS - - ELEV. & SECT.	W4941 & W4943
PARTITION CONNECTION TO ROOF RAFTERS	RAFTERS PERP. TO PARTITION	W4960
	RAFTERS PARALELL TO PARTITION	W4961

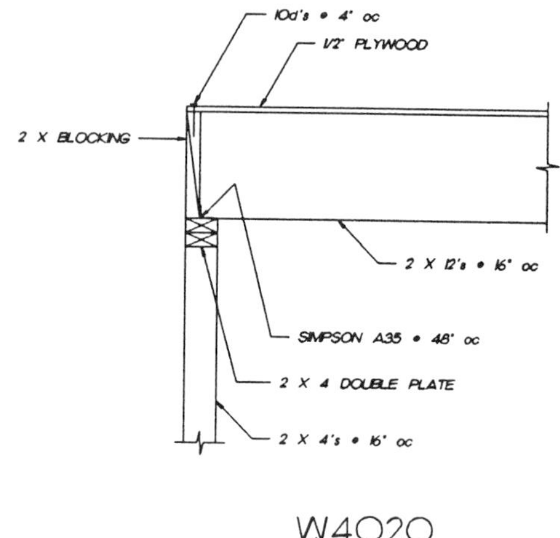

W4020

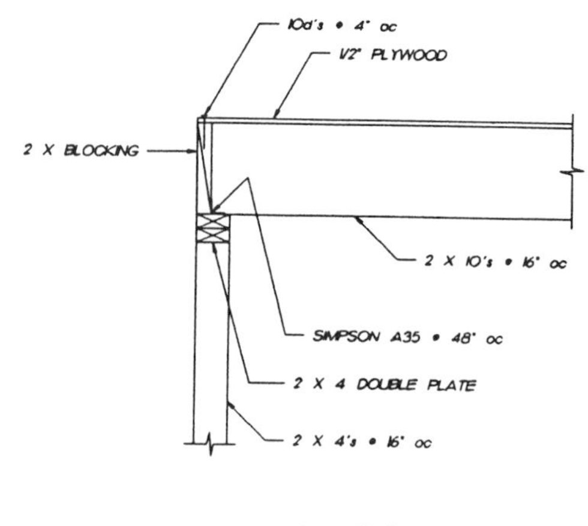

W4021

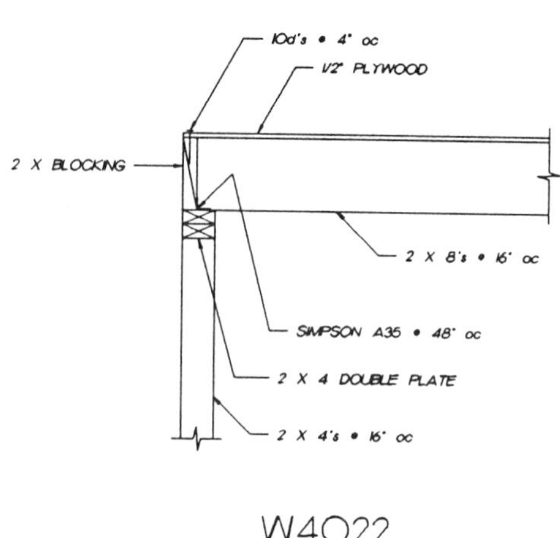

W4022

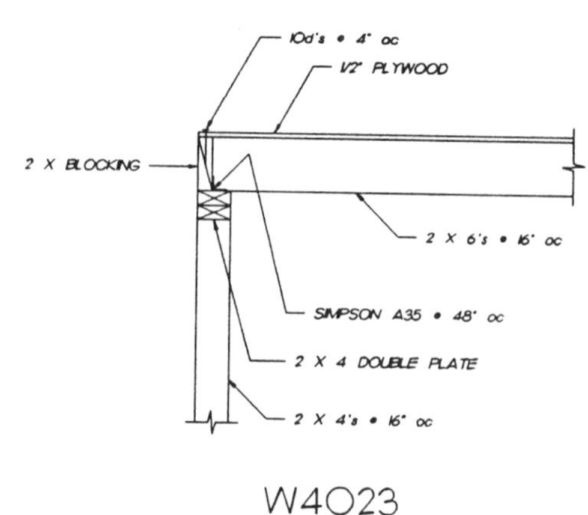

W4023

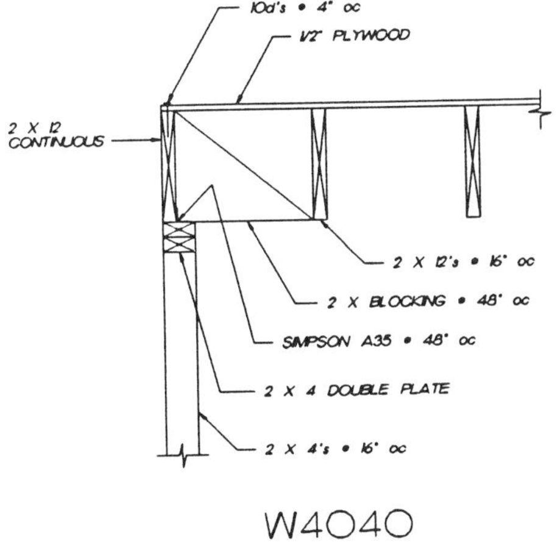

W4040

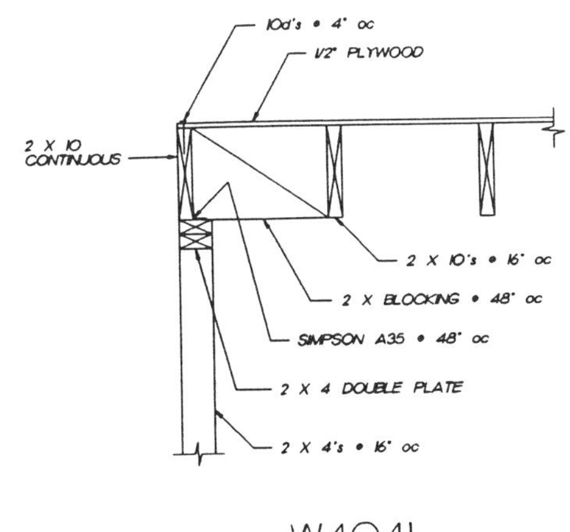

W4041

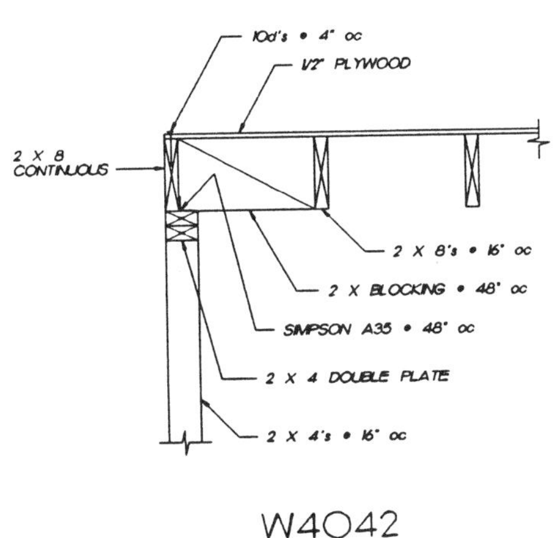

W4042

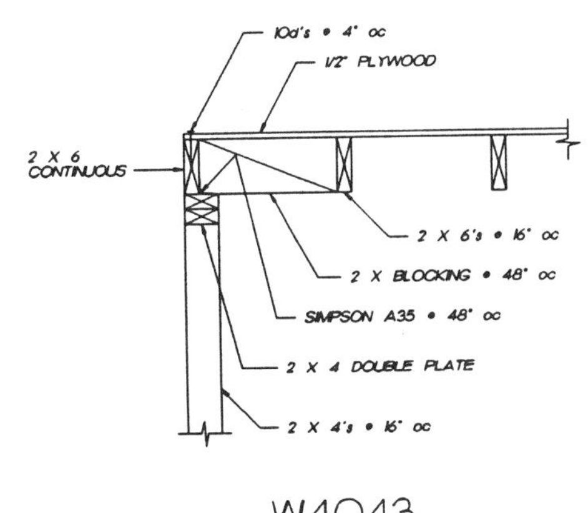

W4043

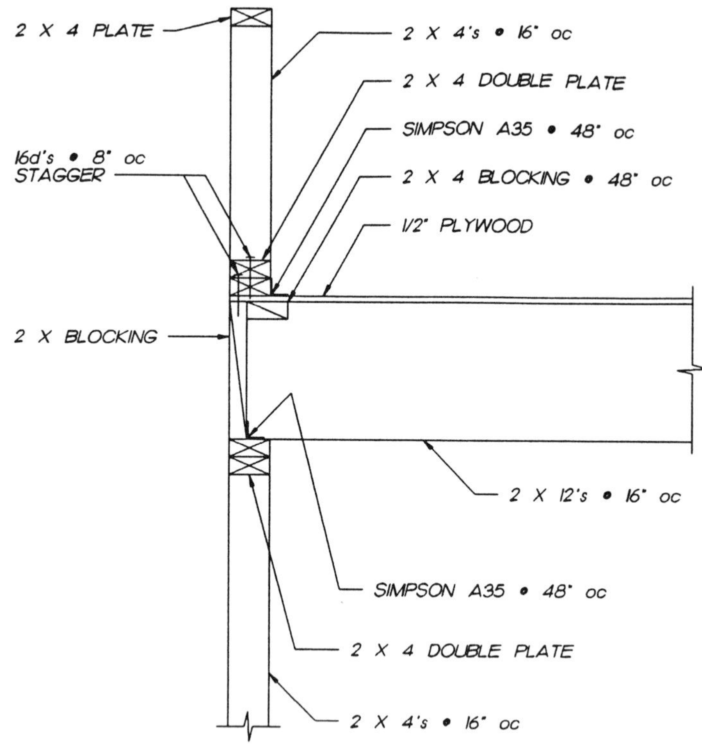

2 X 4 PLATE
2 X 4's • 16" oc
2 X 4 DOUBLE PLATE
SIMPSON A35 • 48" oc
16d's • 8" oc
STAGGER
2 X 4 BLOCKING • 48" oc
1/2" PLYWOOD
2 X BLOCKING
2 X 12's • 16" oc
SIMPSON A35 • 48" oc
2 X 4 DOUBLE PLATE
2 X 4's • 16" oc

W4060

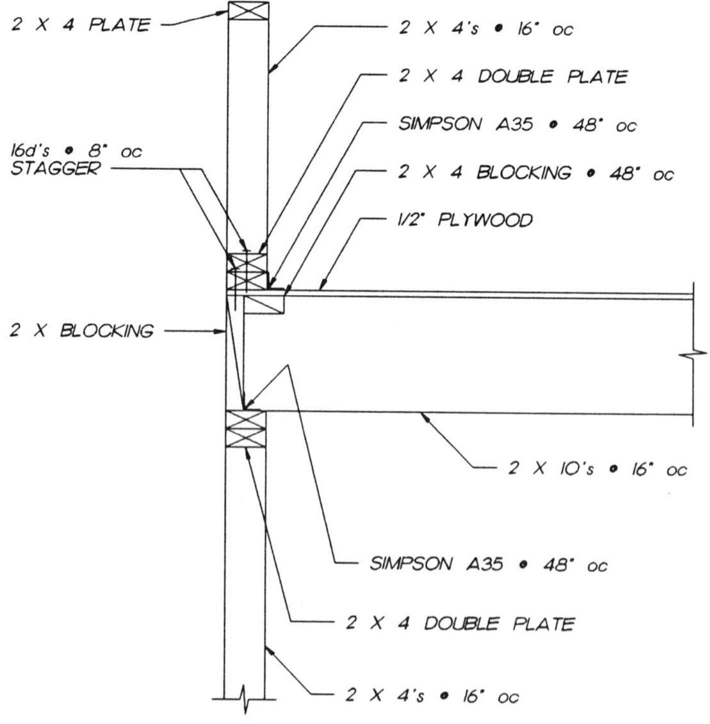

2 X 4 PLATE
2 X 4's • 16" oc
2 X 4 DOUBLE PLATE
SIMPSON A35 • 48" oc
16d's • 8" oc
STAGGER
2 X 4 BLOCKING • 48" oc
1/2" PLYWOOD
2 X BLOCKING
2 X 10's • 16" oc
SIMPSON A35 • 48" oc
2 X 4 DOUBLE PLATE
2 X 4's • 16" oc

W4061

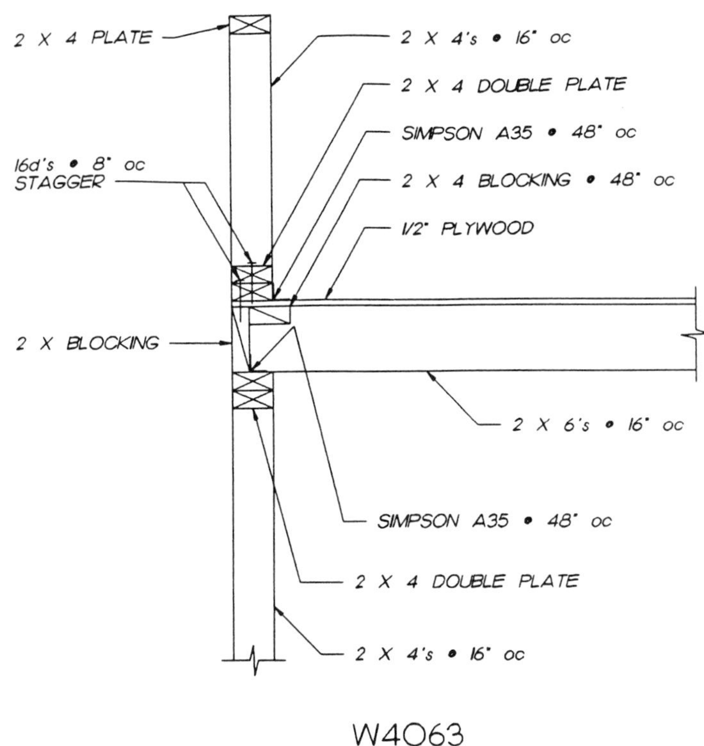
2 X 4 PLATE
2 X 4's • 16" oc
2 X 4 DOUBLE PLATE
SIMPSON A35 • 48" oc
16d's • 8" oc
STAGGER
2 X 4 BLOCKING • 48" oc
1/2" PLYWOOD
2 X BLOCKING
2 X 6's • 16" oc
SIMPSON A35 • 48" oc
2 X 4 DOUBLE PLATE
2 X 4's • 16" oc

W4063

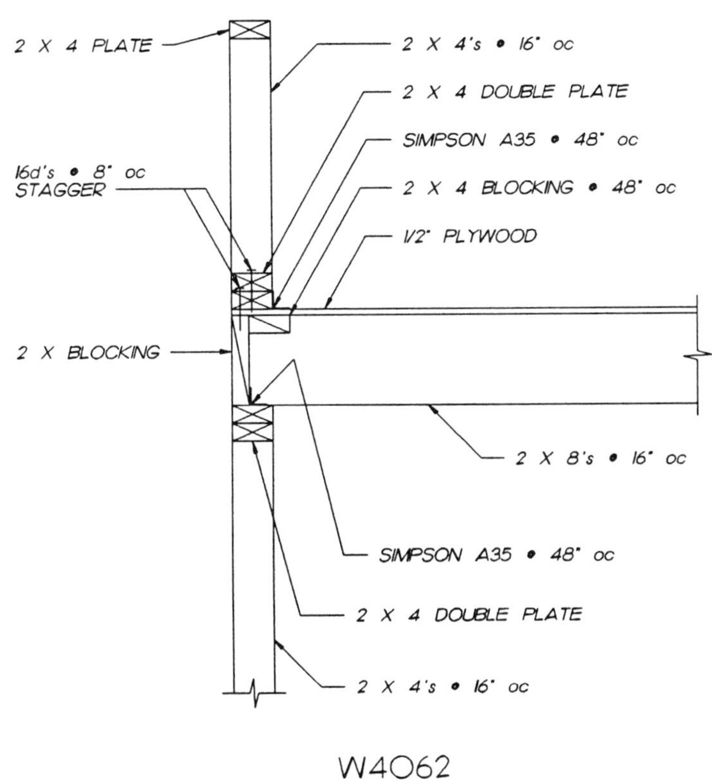
2 X 4 PLATE
2 X 4's • 16" oc
2 X 4 DOUBLE PLATE
SIMPSON A35 • 48" oc
16d's • 8" oc
STAGGER
2 X 4 BLOCKING • 48" oc
1/2" PLYWOOD
2 X BLOCKING
2 X 8's • 16" oc
SIMPSON A35 • 48" oc
2 X 4 DOUBLE PLATE
2 X 4's • 16" oc

W4062

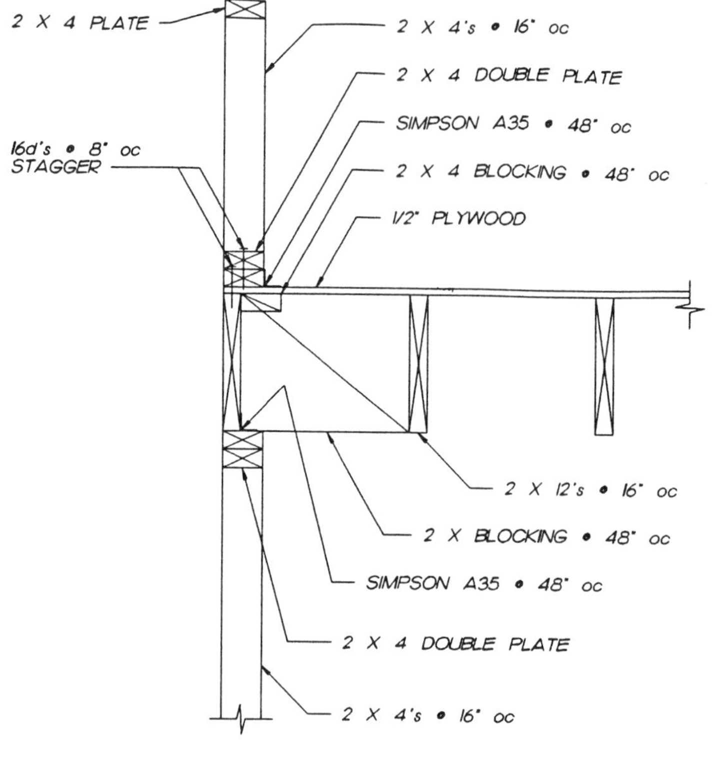

W4080

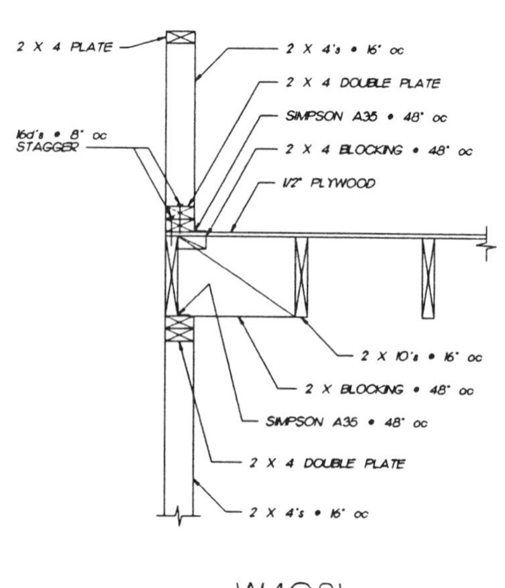

W4081

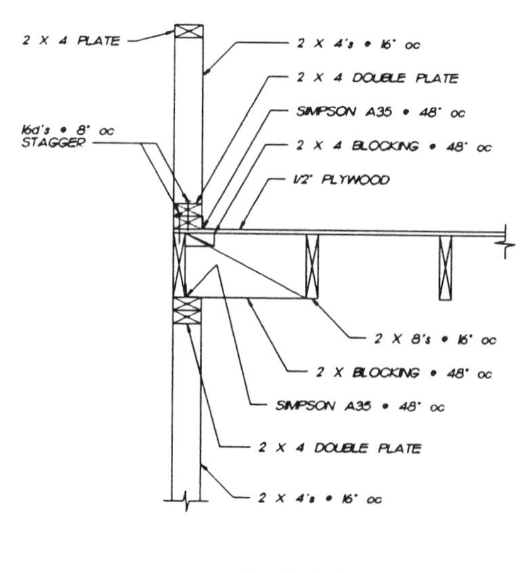

W4082

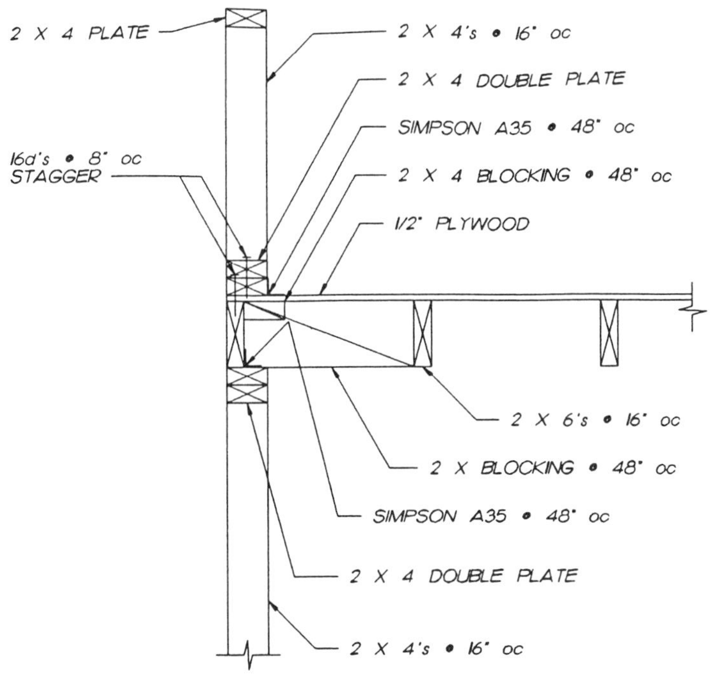

W4083

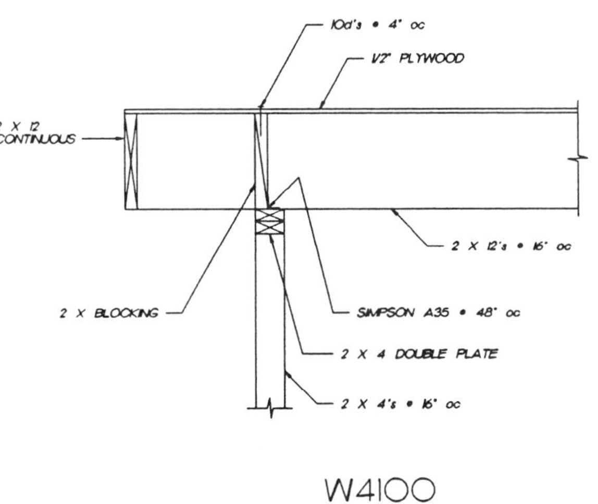

W4100

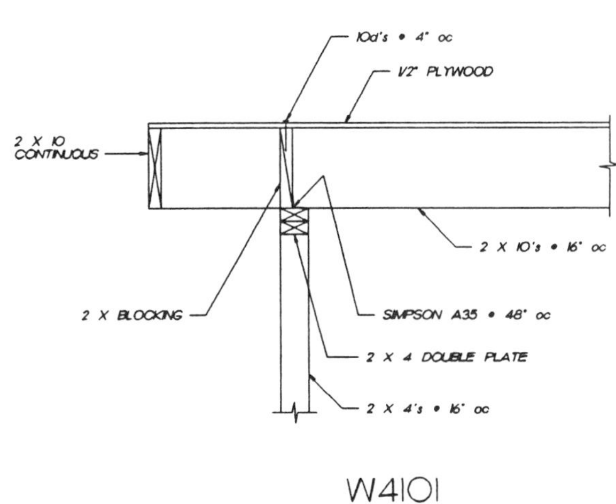

W4101

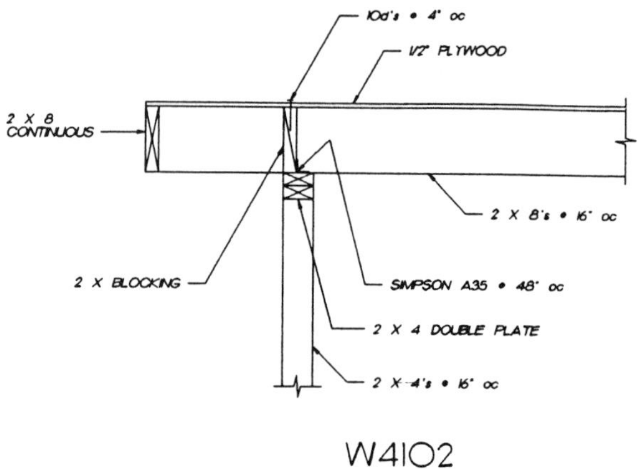

W4102

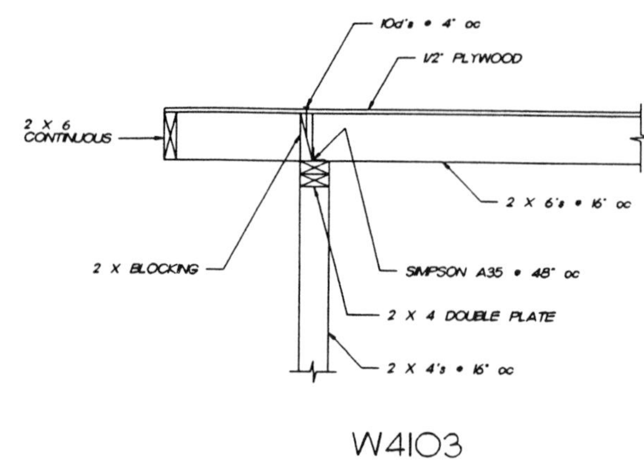

W4103

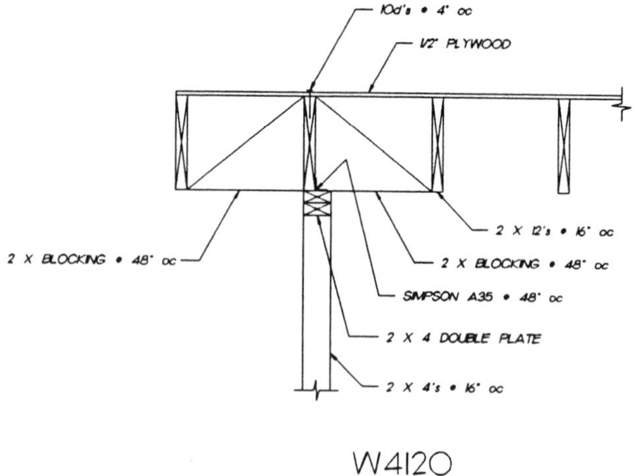

W4120

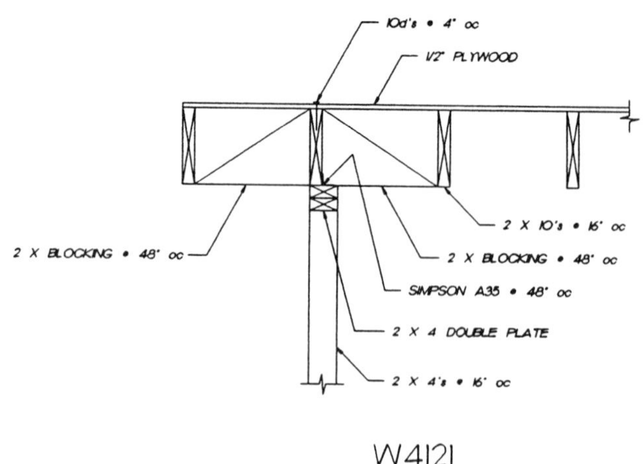

W4121

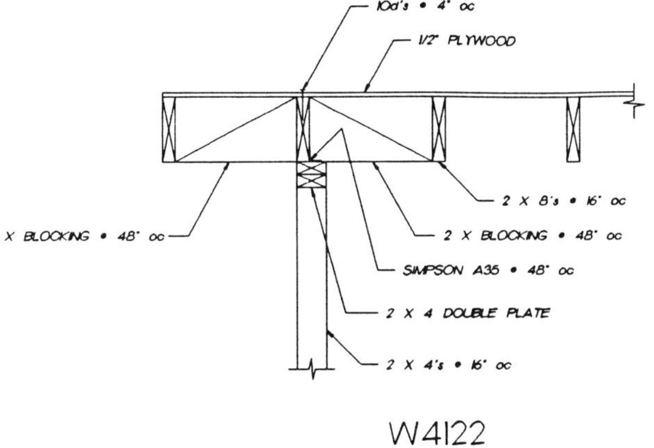

W4122

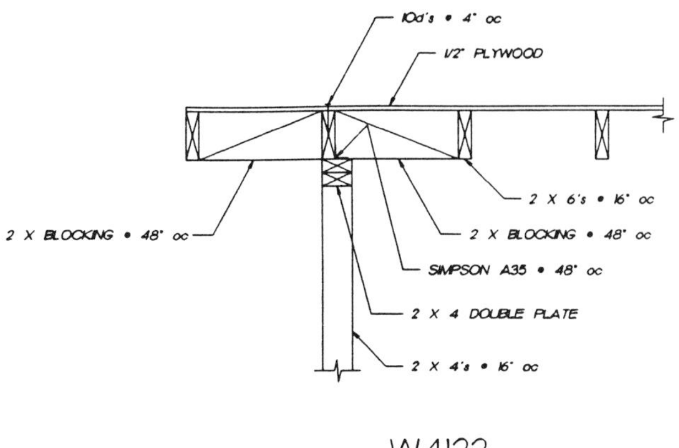

W4123

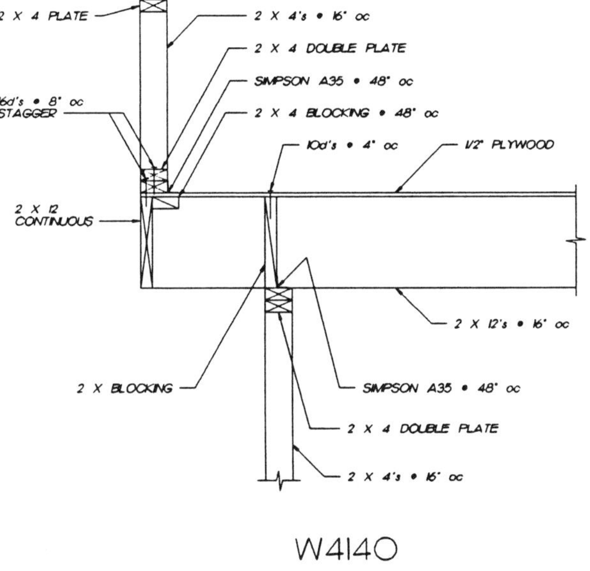

W4140

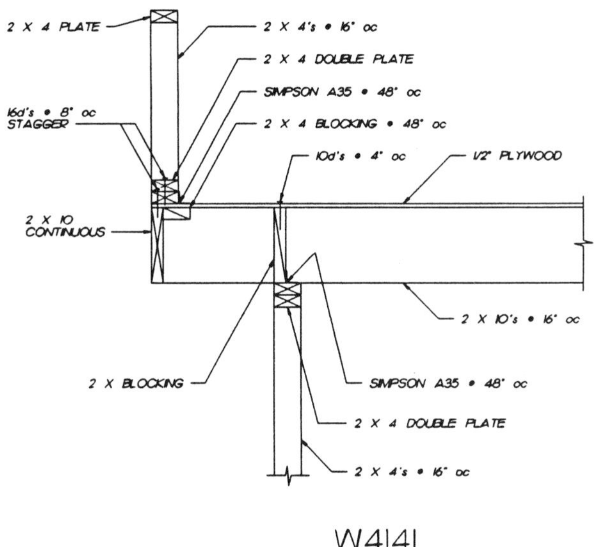

W4141

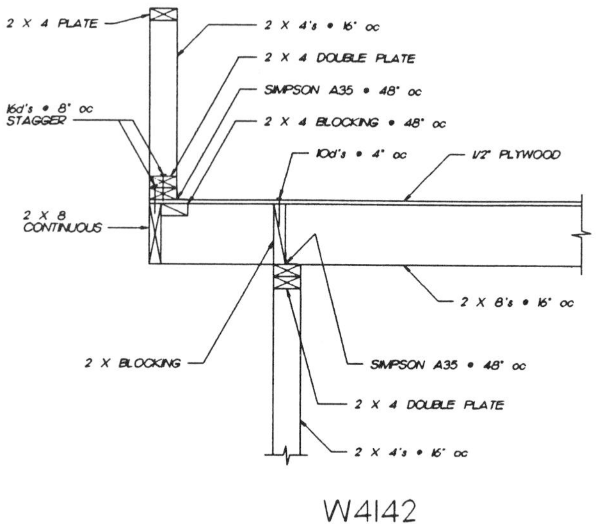

W4142

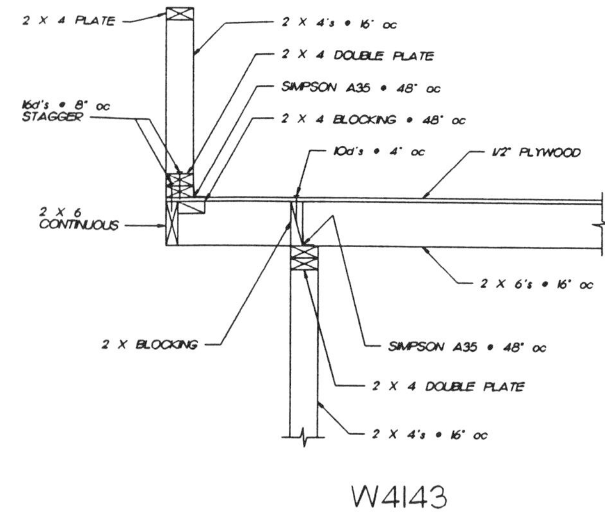

W4143

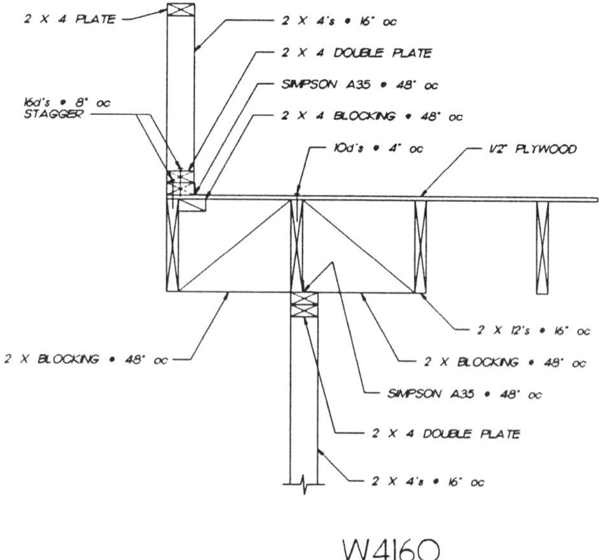

W4160

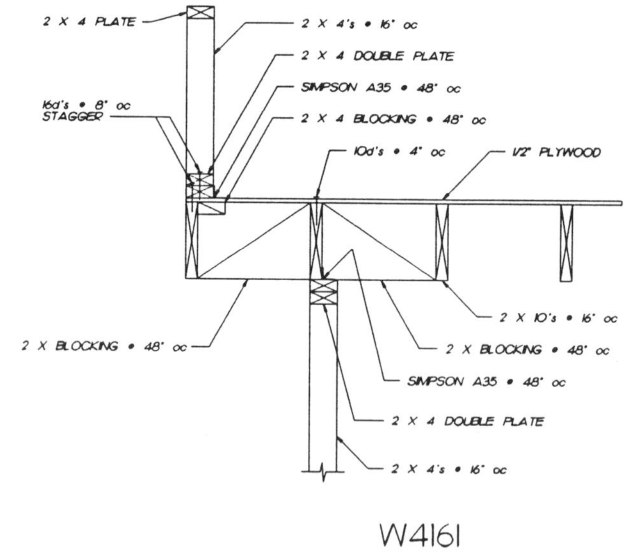

W4161

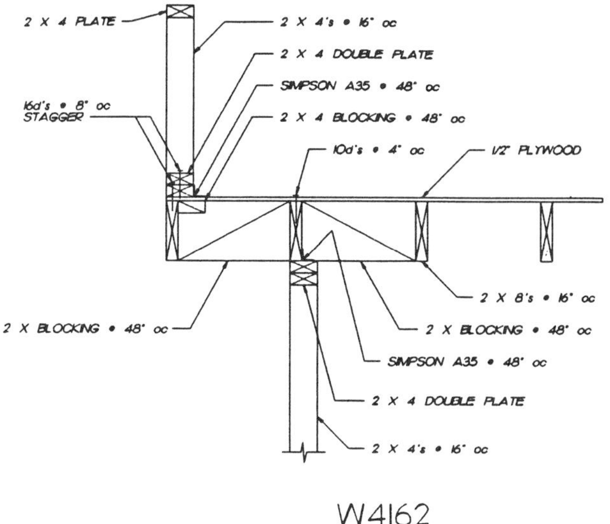

W4162

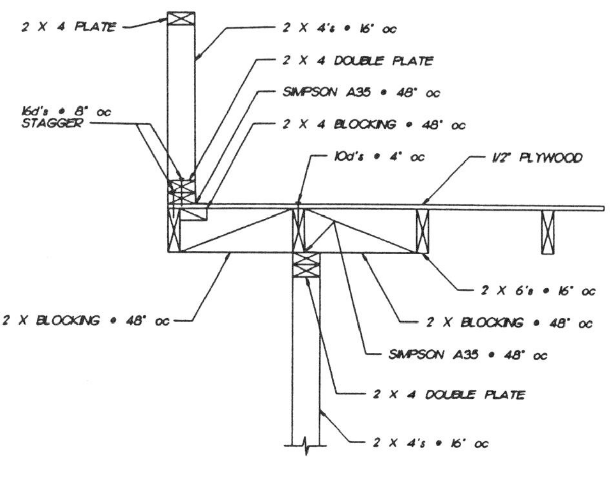

W4163

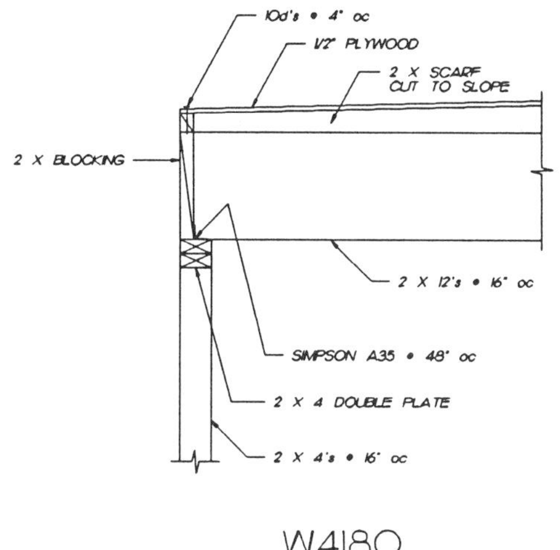

W4180

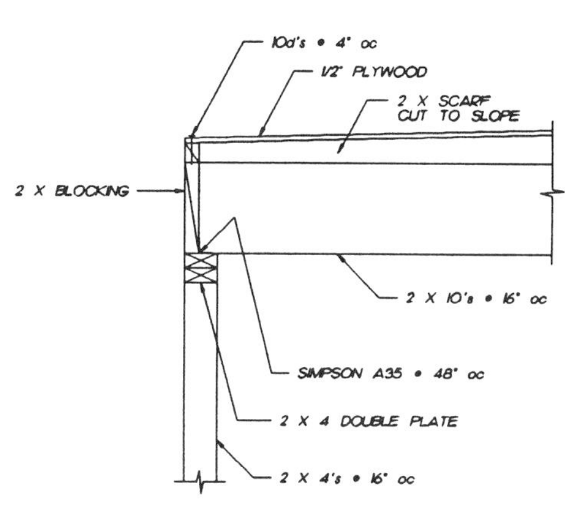

W4180

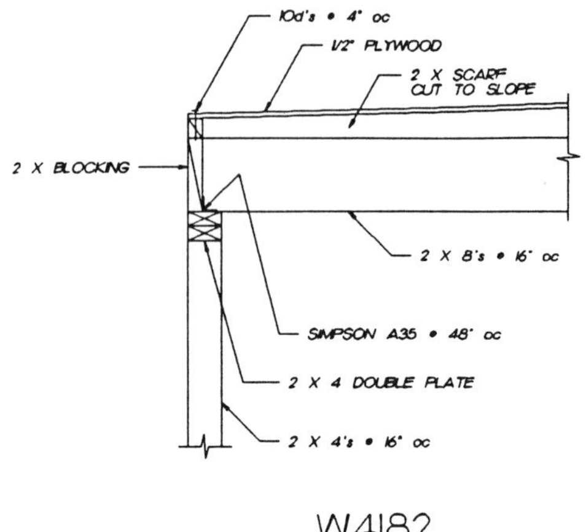

W4182

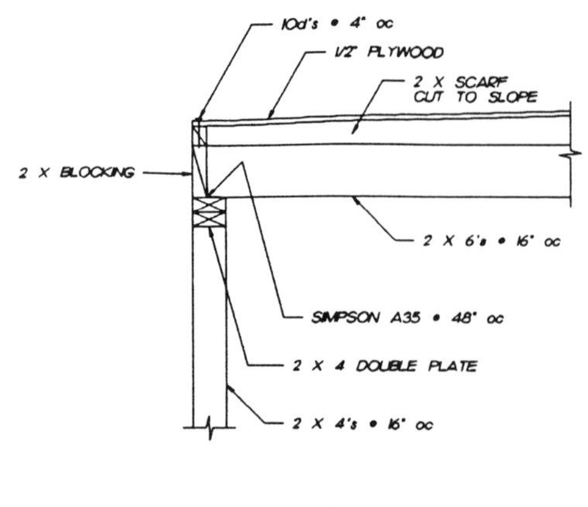

W4183

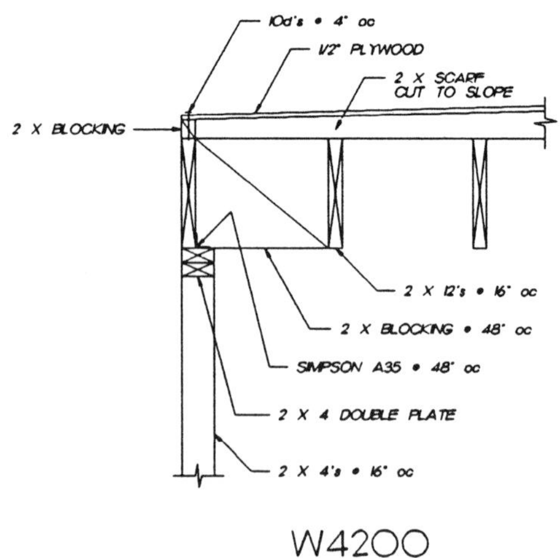

W4200

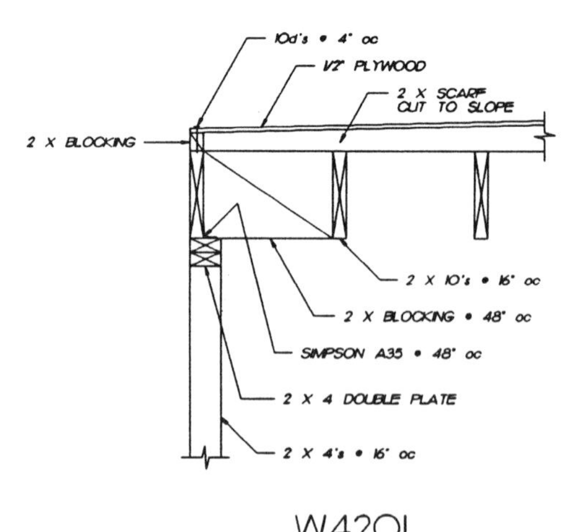

W4201

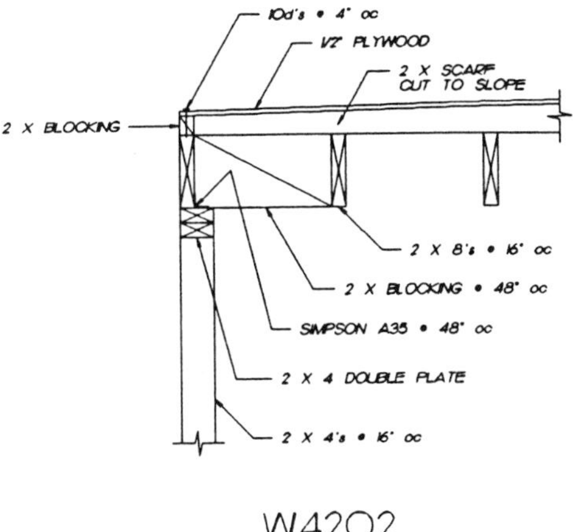

W4202

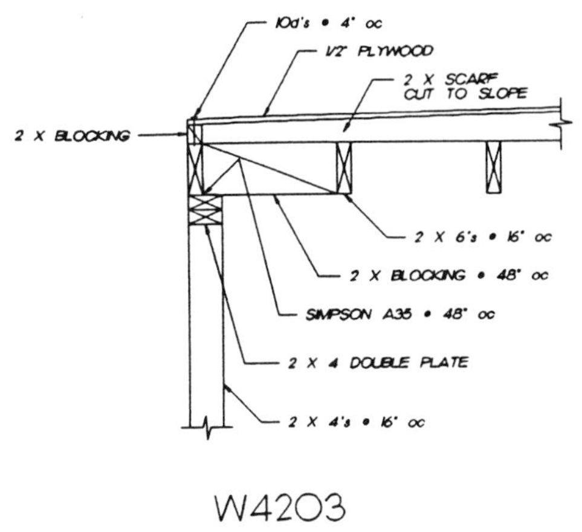

W4203

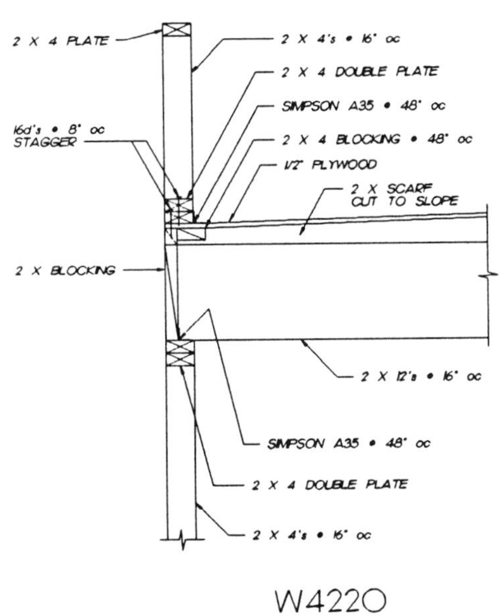

W4220

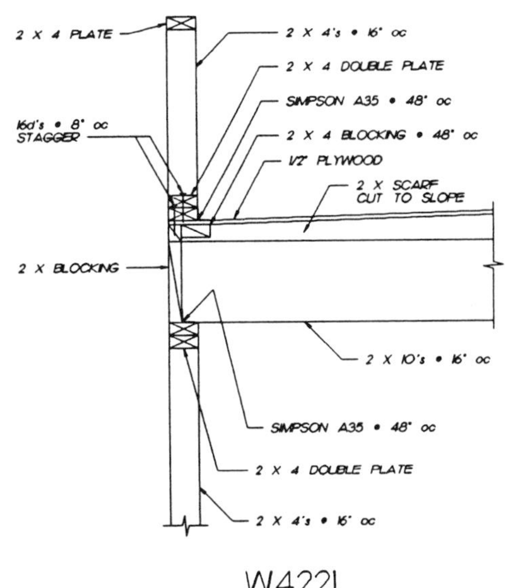

W4221

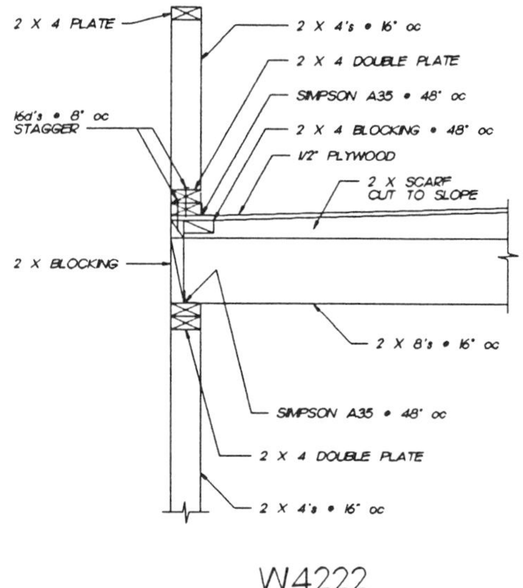

W4222

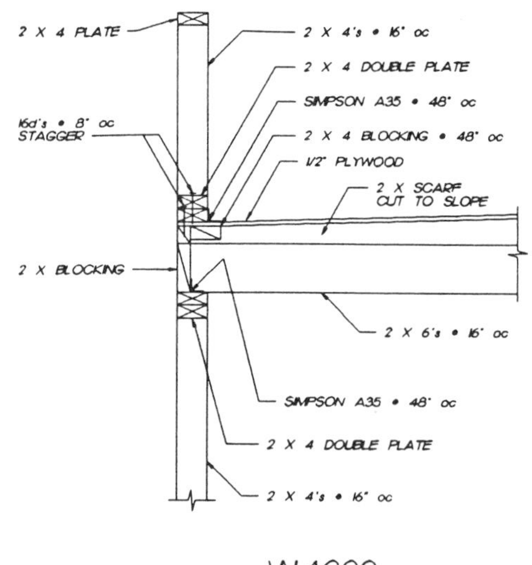

W4223

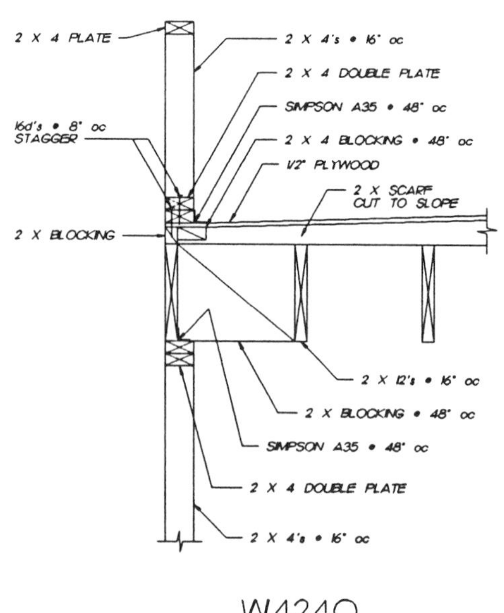

W4240

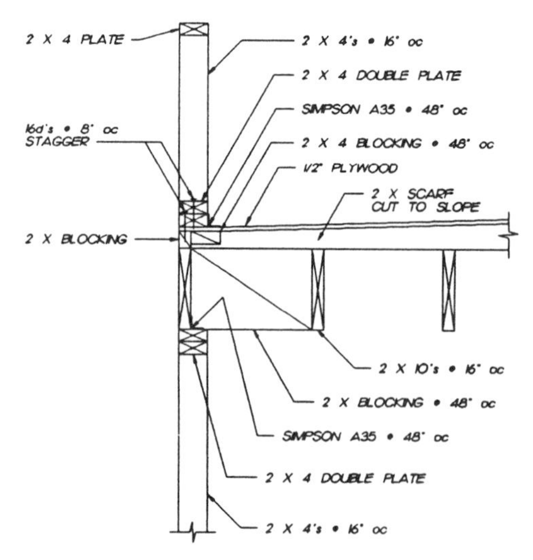

W4241

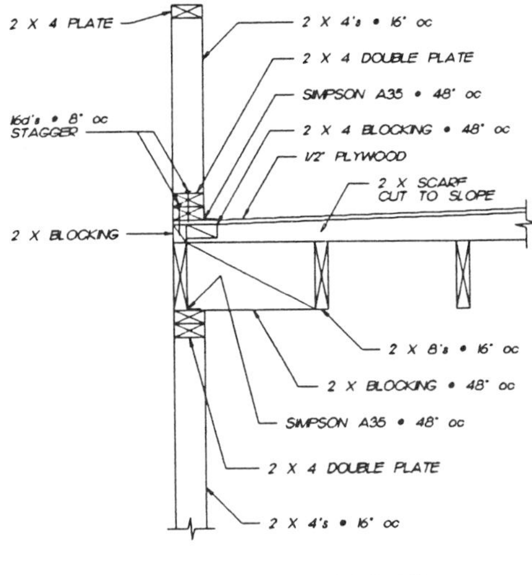

W4242

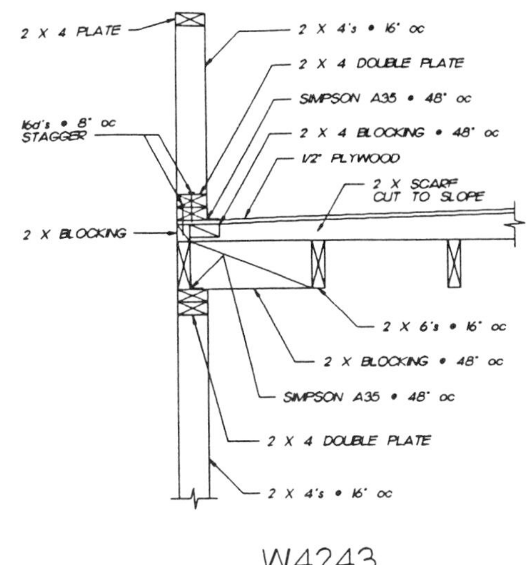

W4243

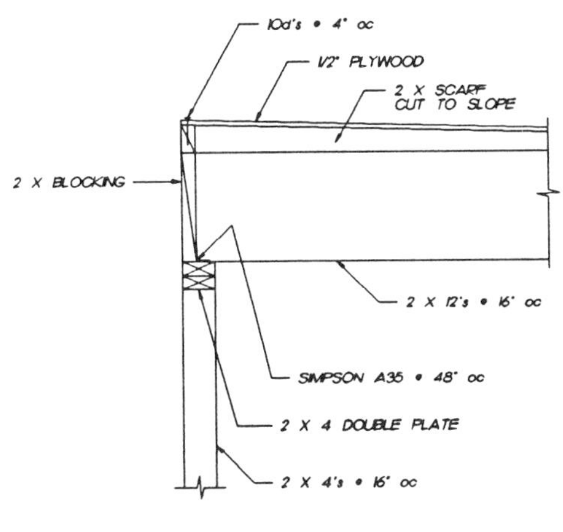

W4260

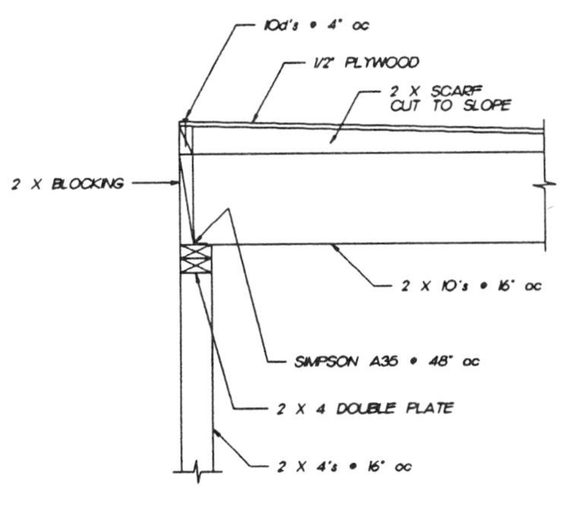

W4261

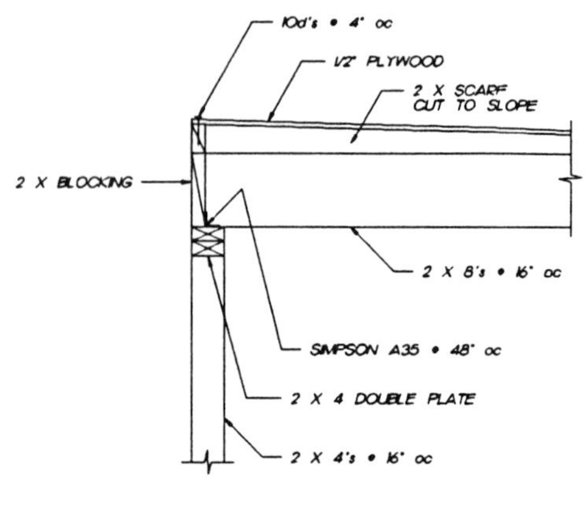

W4262

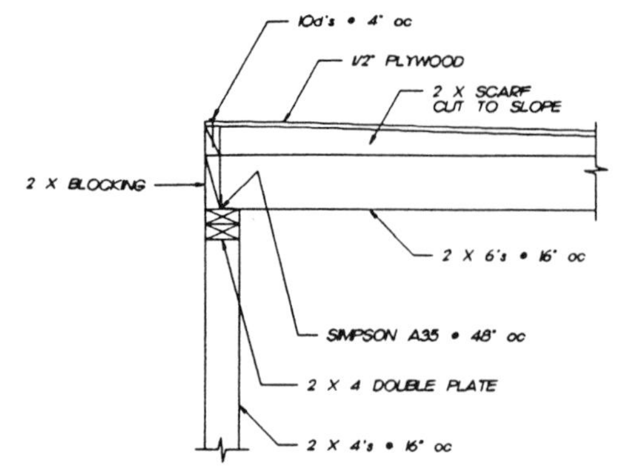

W4263

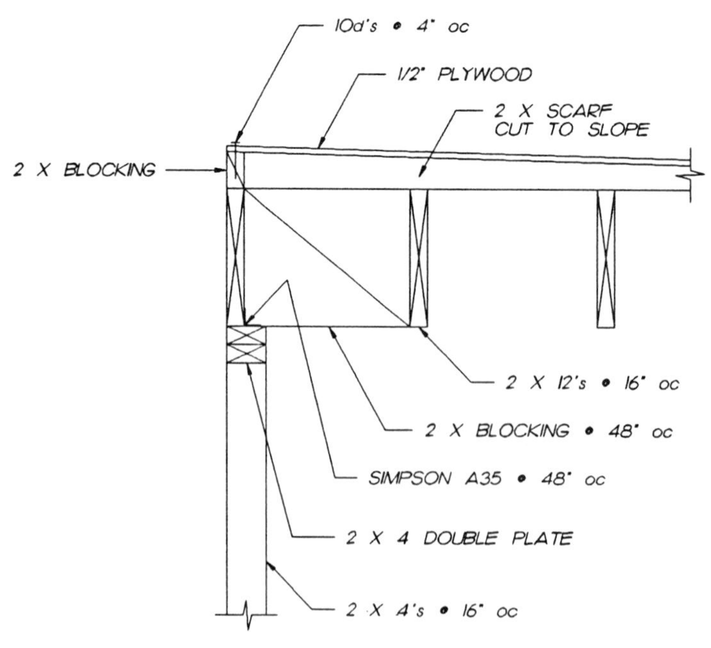

W4280

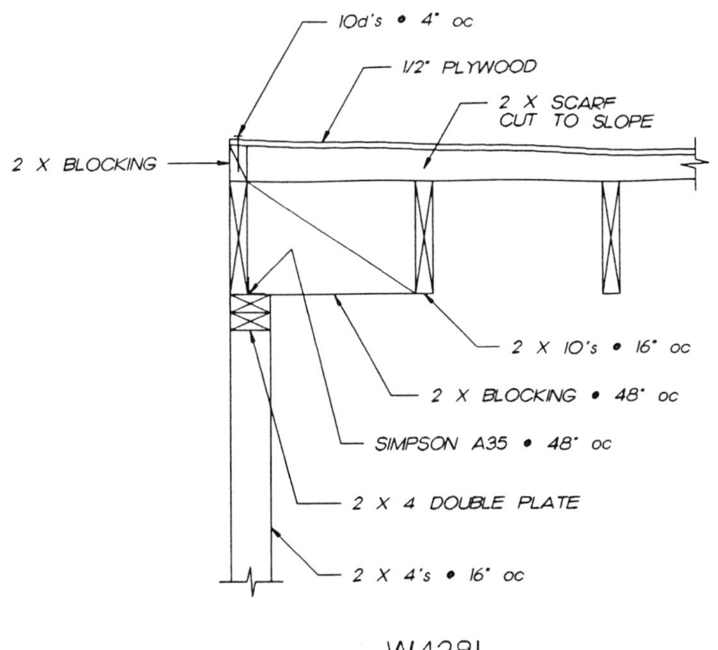

W4281

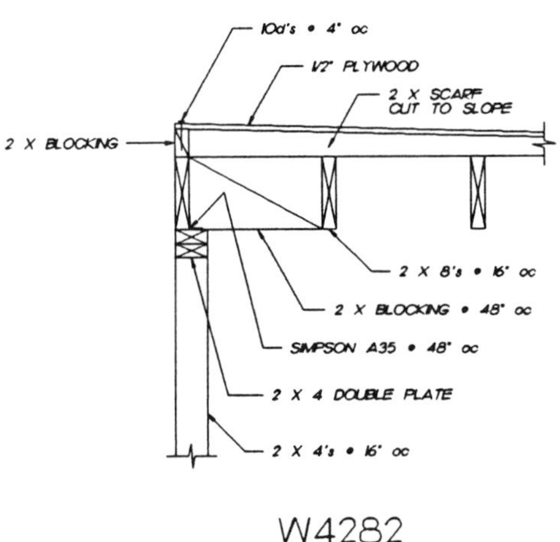

W4282

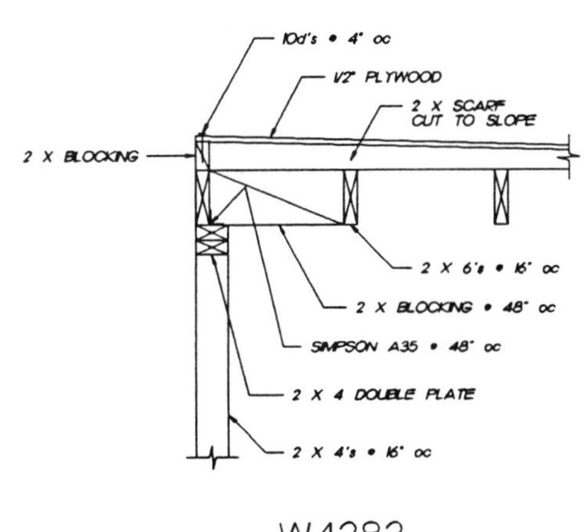

W4283

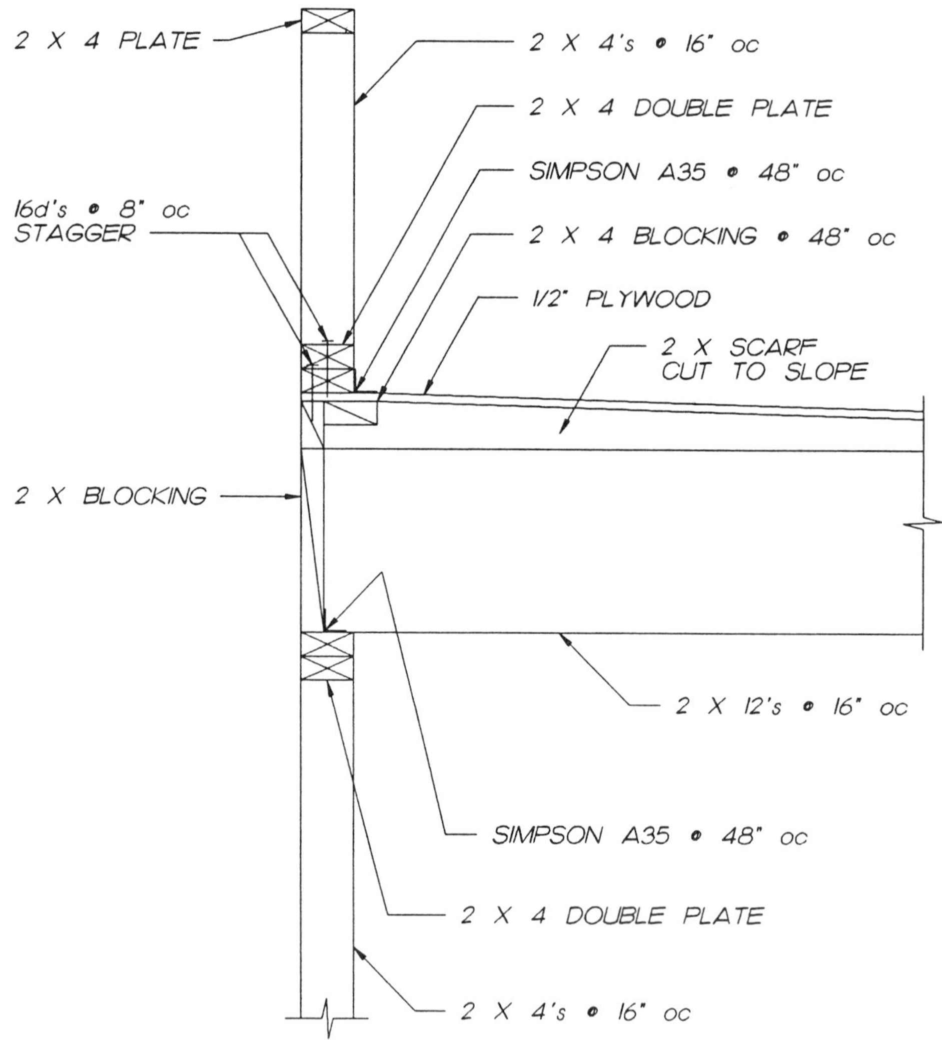
2 X 4 PLATE
2 X 4's @ 16" oc
2 X 4 DOUBLE PLATE
SIMPSON A35 @ 48" oc
16d's @ 8" oc
STAGGER
2 X 4 BLOCKING @ 48" oc
1/2" PLYWOOD
2 X SCARF
CUT TO SLOPE
2 X BLOCKING
2 X 12's @ 16" oc
SIMPSON A35 @ 48" oc
2 X 4 DOUBLE PLATE
2 X 4's @ 16" oc

W4300

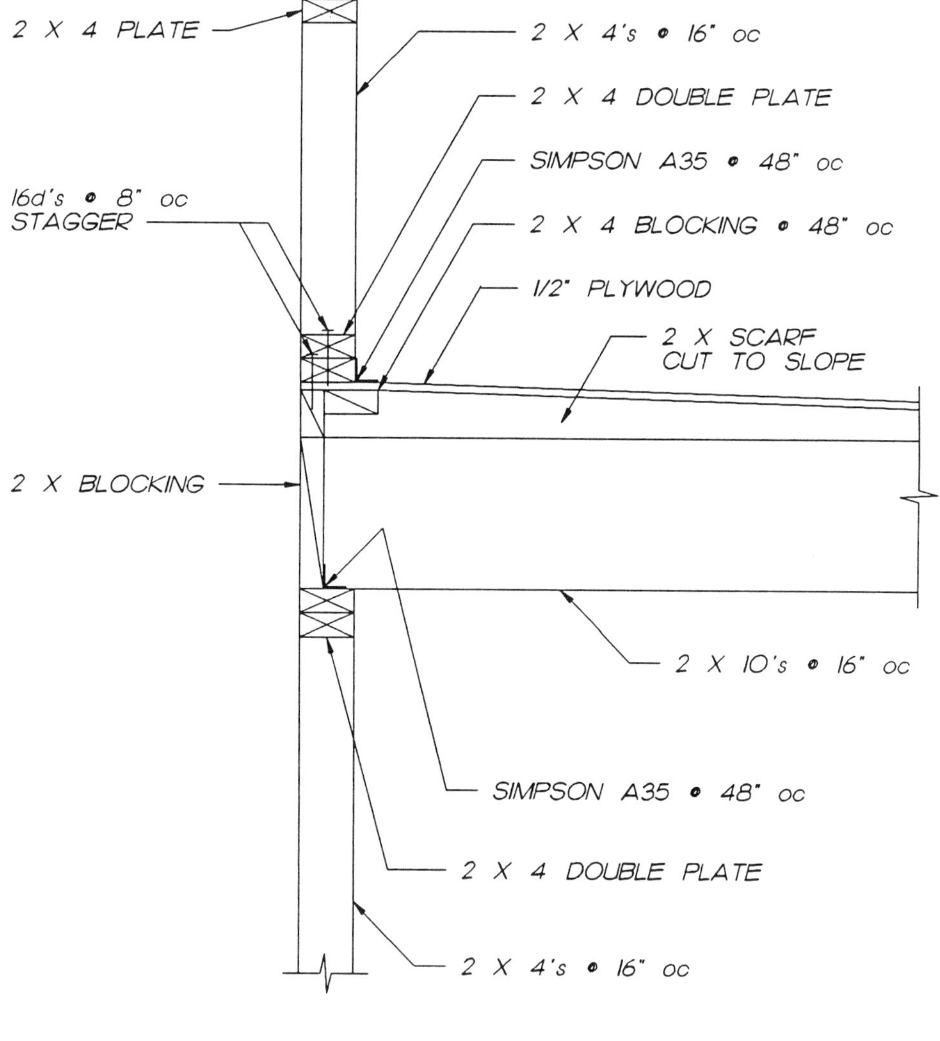
2 X 4 PLATE
2 X 4's @ 16" oc
2 X 4 DOUBLE PLATE
SIMPSON A35 @ 48" oc
16d's @ 8" oc
STAGGER
2 X 4 BLOCKING @ 48" oc
1/2" PLYWOOD
2 X SCARF
CUT TO SLOPE
2 X BLOCKING
2 X 10's @ 16" oc
SIMPSON A35 @ 48" oc
2 X 4 DOUBLE PLATE
2 X 4's @ 16" oc

W4301

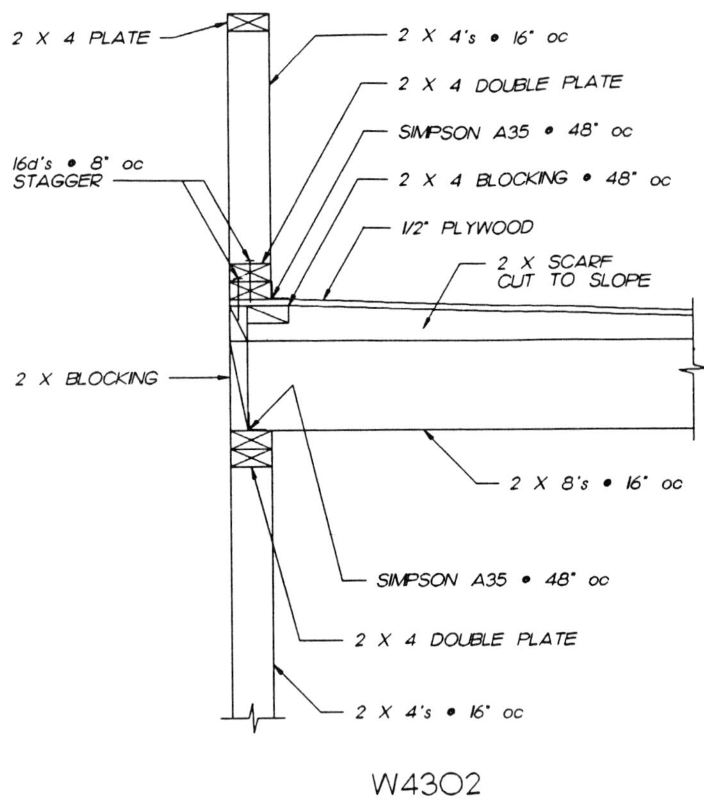
2 X 4 PLATE
2 X 4's • 16" oc
2 X 4 DOUBLE PLATE
SIMPSON A35 • 48" oc
16d's • 8" oc
STAGGER
2 X 4 BLOCKING • 48" oc
1/2" PLYWOOD
2 X SCARF
CUT TO SLOPE
2 X BLOCKING
2 X 8's • 16" oc
SIMPSON A35 • 48" oc
2 X 4 DOUBLE PLATE
2 X 4's • 16" oc

W4302

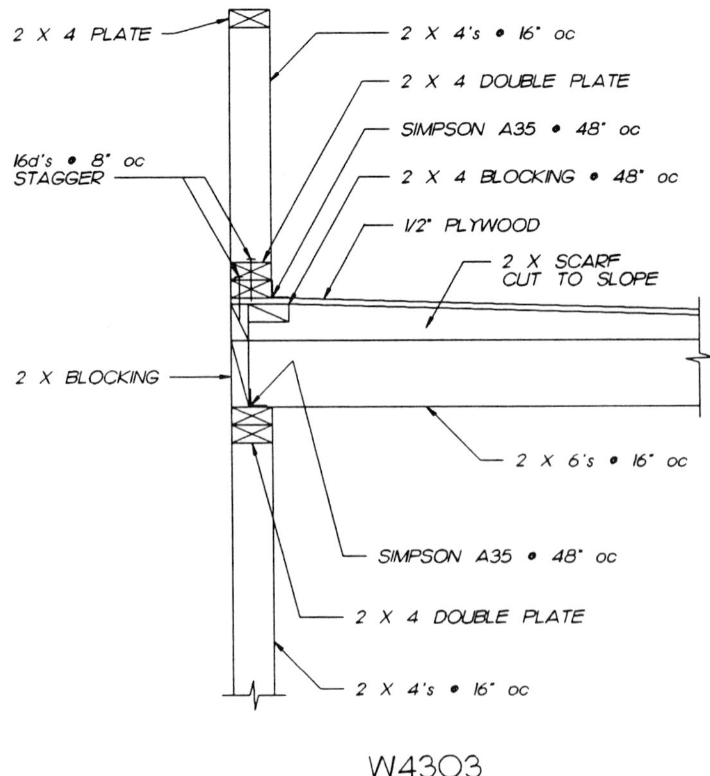
2 X 4 PLATE
2 X 4's • 16" oc
2 X 4 DOUBLE PLATE
SIMPSON A35 • 48" oc
16d's • 8" oc
STAGGER
2 X 4 BLOCKING • 48" oc
1/2" PLYWOOD
2 X SCARF
CUT TO SLOPE
2 X BLOCKING
2 X 6's • 16" oc
SIMPSON A35 • 48" oc
2 X 4 DOUBLE PLATE
2 X 4's • 16" oc

W4303

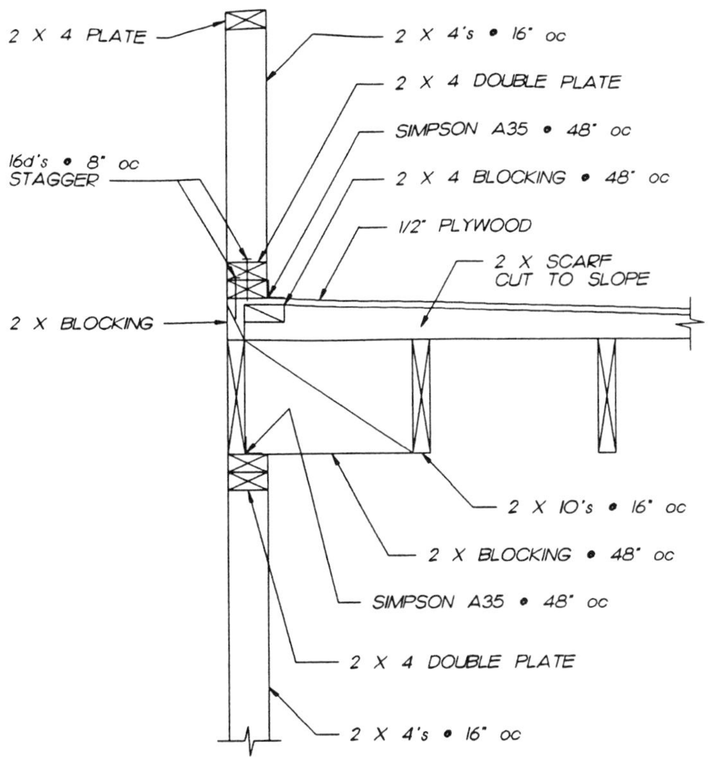
2 X 4 PLATE
2 X 4's • 16" oc
2 X 4 DOUBLE PLATE
SIMPSON A35 • 48" oc
16d's • 8" oc
STAGGER
2 X 4 BLOCKING • 48" oc
1/2" PLYWOOD
2 X SCARF
CUT TO SLOPE
2 X BLOCKING
2 X 10's • 16" oc
2 X BLOCKING • 48" oc
SIMPSON A35 • 48" oc
2 X 4 DOUBLE PLATE
2 X 4's • 16" oc

W4321

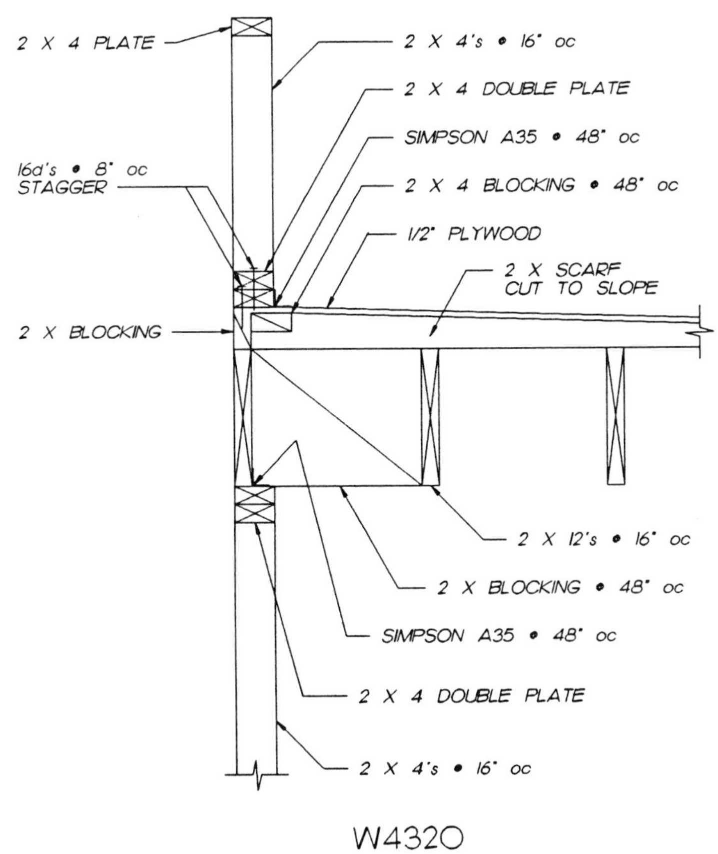
2 X 4 PLATE
2 X 4's • 16" oc
2 X 4 DOUBLE PLATE
SIMPSON A35 • 48" oc
16d's • 8" oc
STAGGER
2 X 4 BLOCKING • 48" oc
1/2" PLYWOOD
2 X SCARF
CUT TO SLOPE
2 X BLOCKING
2 X 12's • 16" oc
2 X BLOCKING • 48" oc
SIMPSON A35 • 48" oc
2 X 4 DOUBLE PLATE
2 X 4's • 16" oc

W4320

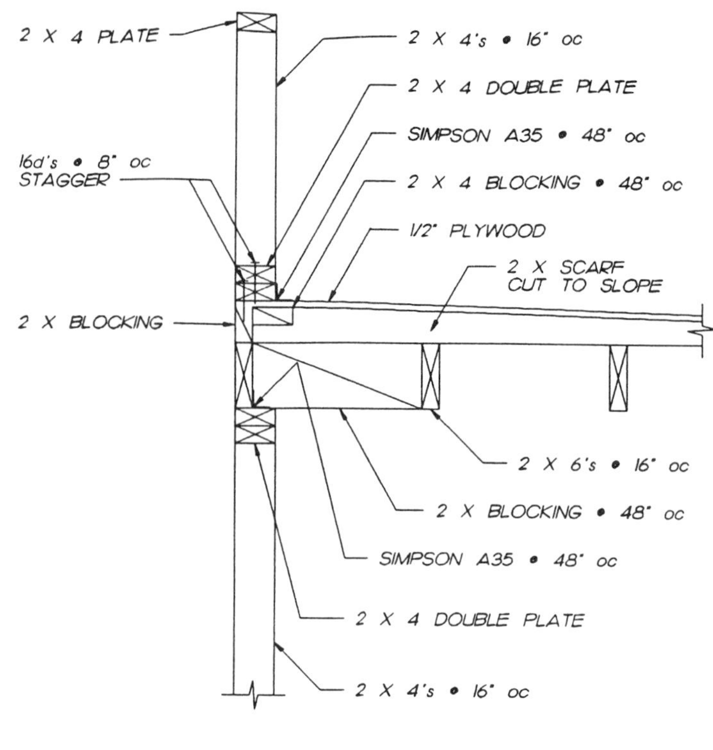
2 X 4 PLATE
2 X 4's • 16" oc
2 X 4 DOUBLE PLATE
SIMPSON A35 • 48" oc
16d's • 8" oc
STAGGER
2 X 4 BLOCKING • 48" oc
1/2" PLYWOOD
2 X SCARF
CUT TO SLOPE
2 X BLOCKING
2 X 6's • 16" oc
2 X BLOCKING • 48" oc
SIMPSON A35 • 48" oc
2 X 4 DOUBLE PLATE
2 X 4's • 16" oc

W4323

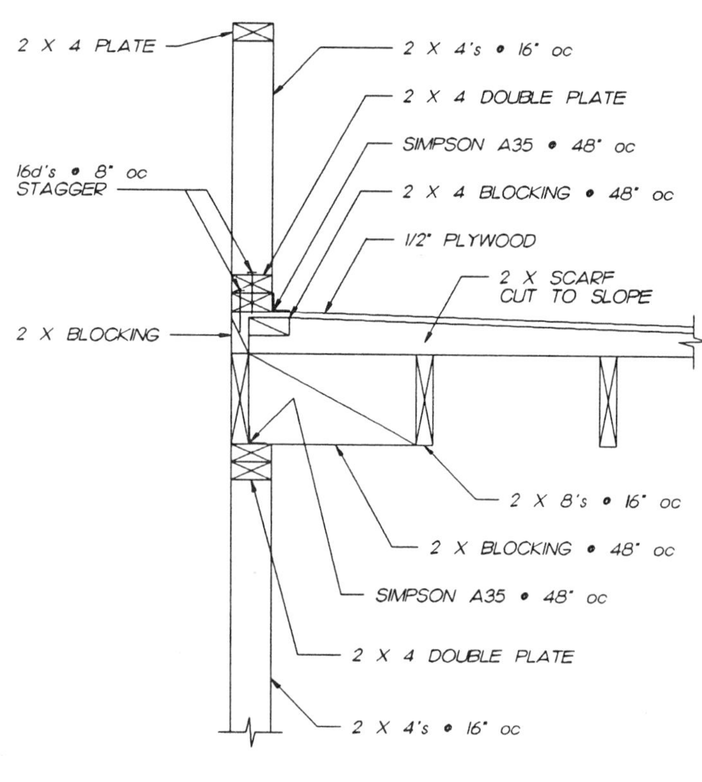
2 X 4 PLATE
2 X 4's • 16" oc
2 X 4 DOUBLE PLATE
SIMPSON A35 • 48" oc
16d's • 8" oc
STAGGER
2 X 4 BLOCKING • 48" oc
1/2" PLYWOOD
2 X SCARF
CUT TO SLOPE
2 X BLOCKING
2 X 8's • 16" oc
2 X BLOCKING • 48" oc
SIMPSON A35 • 48" oc
2 X 4 DOUBLE PLATE
2 X 4's • 16" oc

W4322

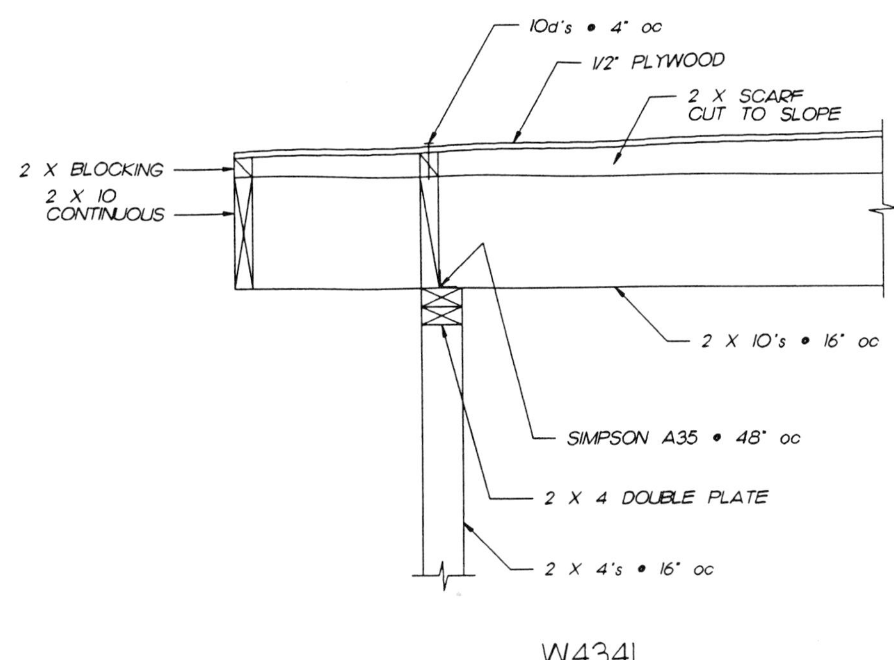
10d's • 4" oc
1/2" PLYWOOD
2 X SCARF
CUT TO SLOPE
2 X BLOCKING
2 X 10
CONTINUOUS
2 X 10's • 16" oc
SIMPSON A35 • 48" oc
2 X 4 DOUBLE PLATE
2 X 4's • 16" oc

W4341

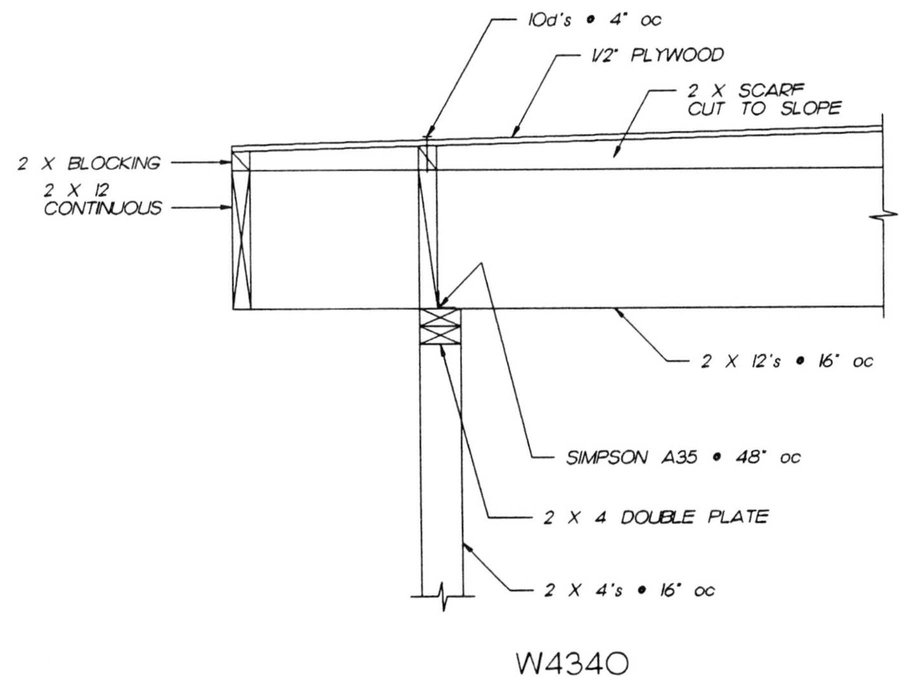
10d's • 4" oc
1/2" PLYWOOD
2 X SCARF
CUT TO SLOPE
2 X BLOCKING
2 X 12
CONTINUOUS
2 X 12's • 16" oc
SIMPSON A35 • 48" oc
2 X 4 DOUBLE PLATE
2 X 4's • 16" oc

W4340

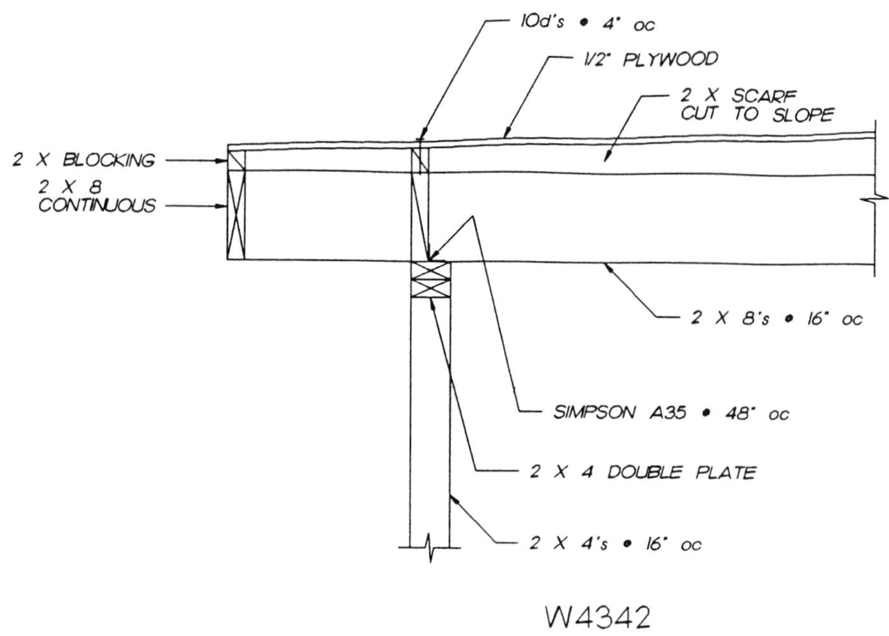
10d's • 4" oc
1/2" PLYWOOD
2 X SCARF
CUT TO SLOPE
2 X BLOCKING
2 X 8
CONTINUOUS
2 X 8's • 16" oc
SIMPSON A35 • 48" oc
2 X 4 DOUBLE PLATE
2 X 4's • 16" oc

W4342

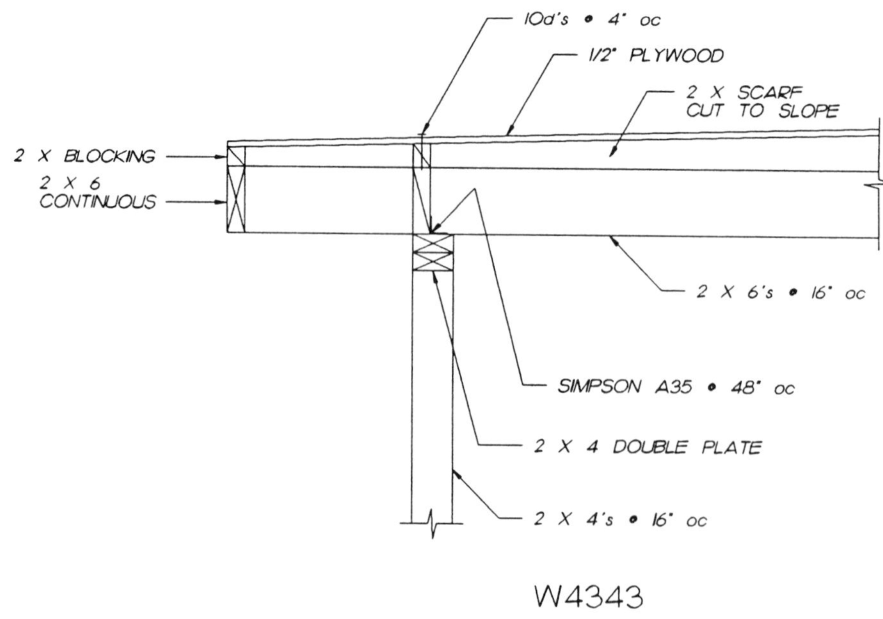
10d's • 4" oc
1/2" PLYWOOD
2 X SCARF
CUT TO SLOPE
2 X BLOCKING
2 X 6
CONTINUOUS
2 X 6's • 16" oc
SIMPSON A35 • 48" oc
2 X 4 DOUBLE PLATE
2 X 4's • 16" oc

W4343

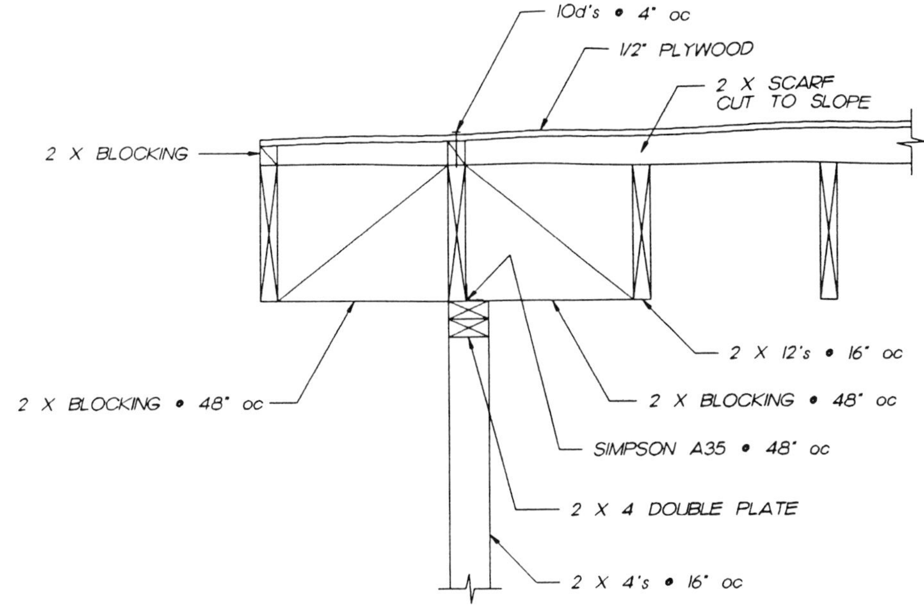
10d's @ 4" oc
1/2" PLYWOOD
2 X SCARF
CUT TO SLOPE
2 X BLOCKING
2 X 12's @ 16" oc
2 X BLOCKING @ 48" oc
2 X BLOCKING @ 48" oc
SIMPSON A35 @ 48" oc
2 X 4 DOUBLE PLATE
2 X 4's @ 16" oc

W4360

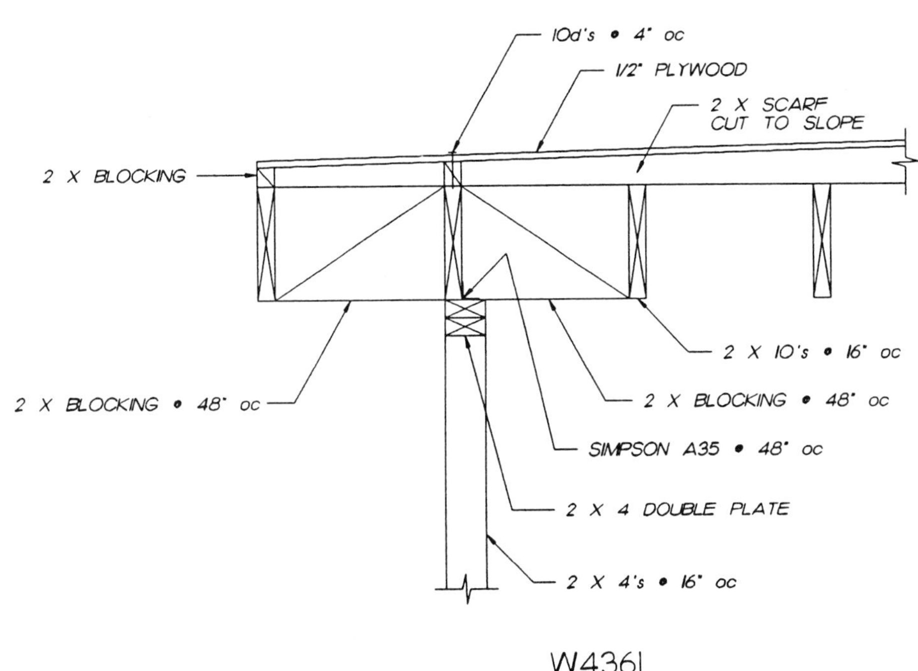
10d's @ 4" oc
1/2" PLYWOOD
2 X SCARF
CUT TO SLOPE
2 X BLOCKING
2 X 10's @ 16" oc
2 X BLOCKING @ 48" oc
2 X BLOCKING @ 48" oc
SIMPSON A35 @ 48" oc
2 X 4 DOUBLE PLATE
2 X 4's @ 16" oc

W4361

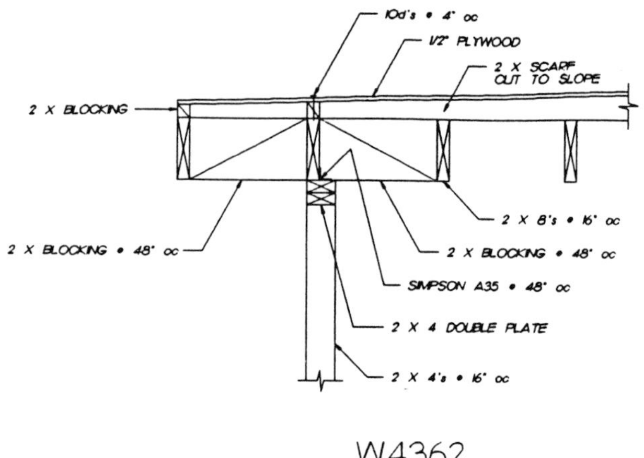

W4362

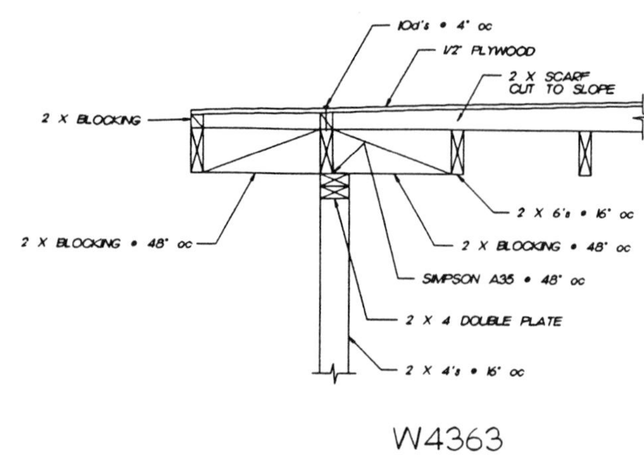

W4363

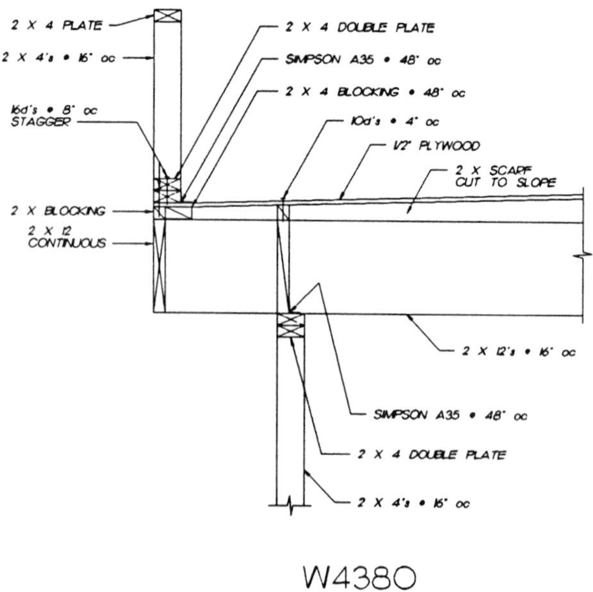

W4380

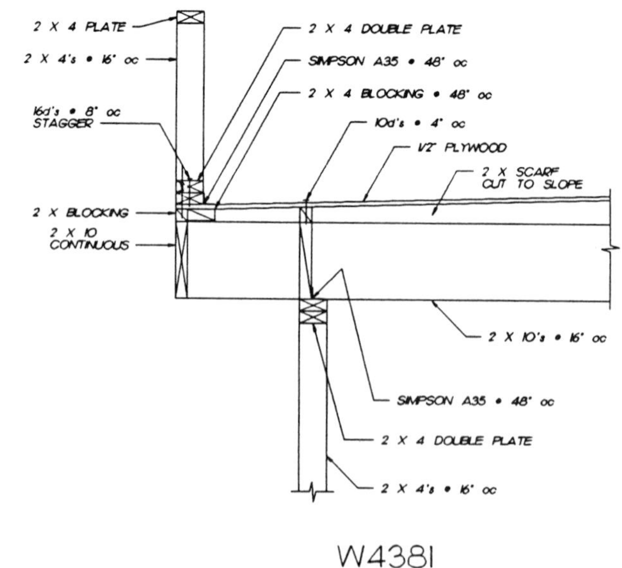

W4381

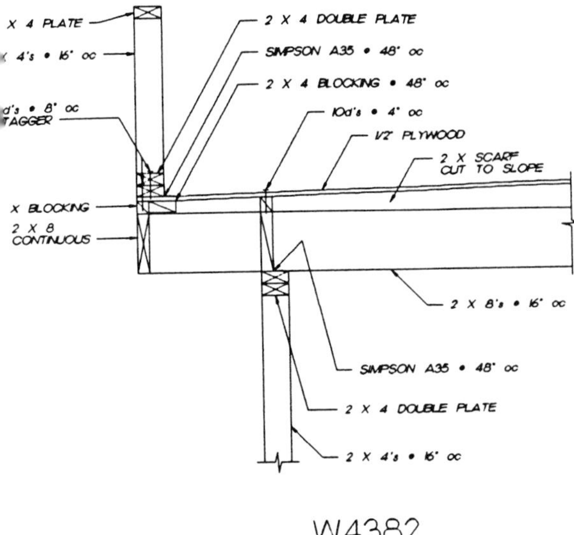

W4382

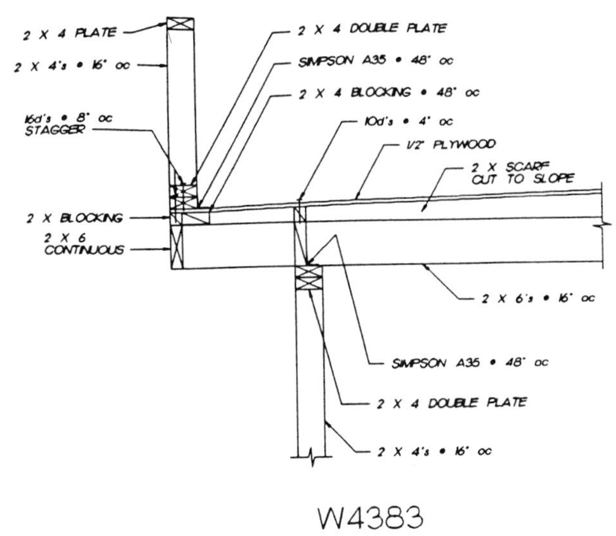

W4383

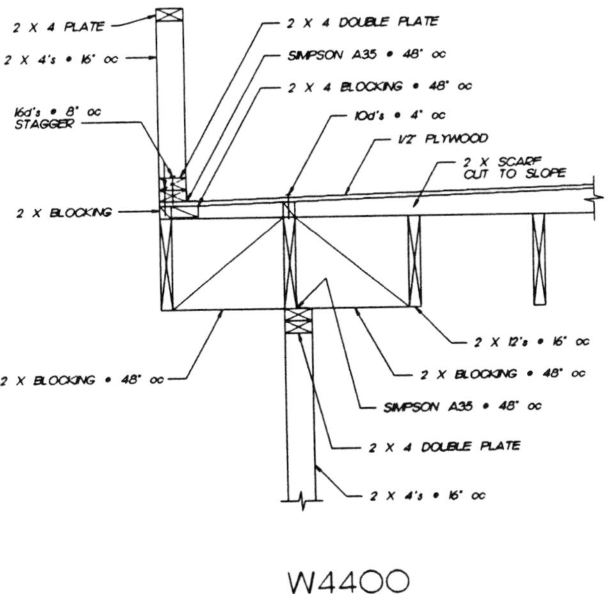

W4400

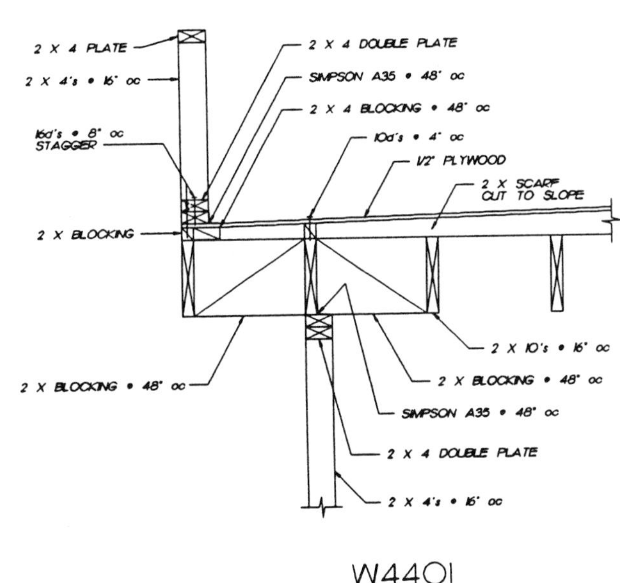

W4401

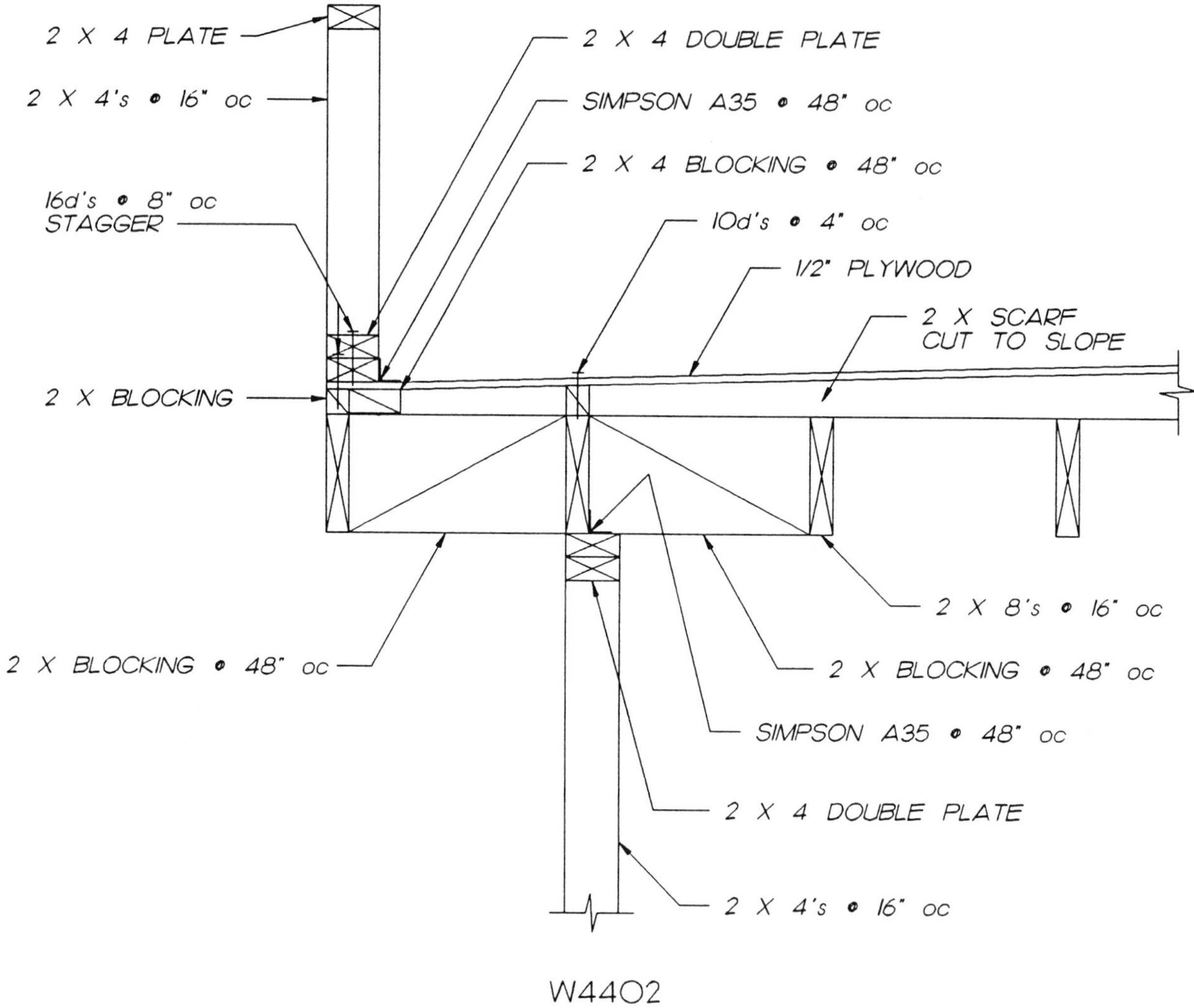
2 X 4 PLATE
2 X 4'S @ 16" OC
16d'S @ 8" OC
STAGGER
2 X BLOCKING
2 X 4 DOUBLE PLATE
SIMPSON A35 @ 48" OC
2 X 4 BLOCKING @ 48" OC
10d'S @ 4" OC
1/2" PLYWOOD
2 X SCARF
CUT TO SLOPE
2 X 8'S @ 16" OC
2 X BLOCKING @ 48" OC
2 X BLOCKING @ 48" OC
SIMPSON A35 @ 48" OC
2 X 4 DOUBLE PLATE
2 X 4'S @ 16" OC

W4402

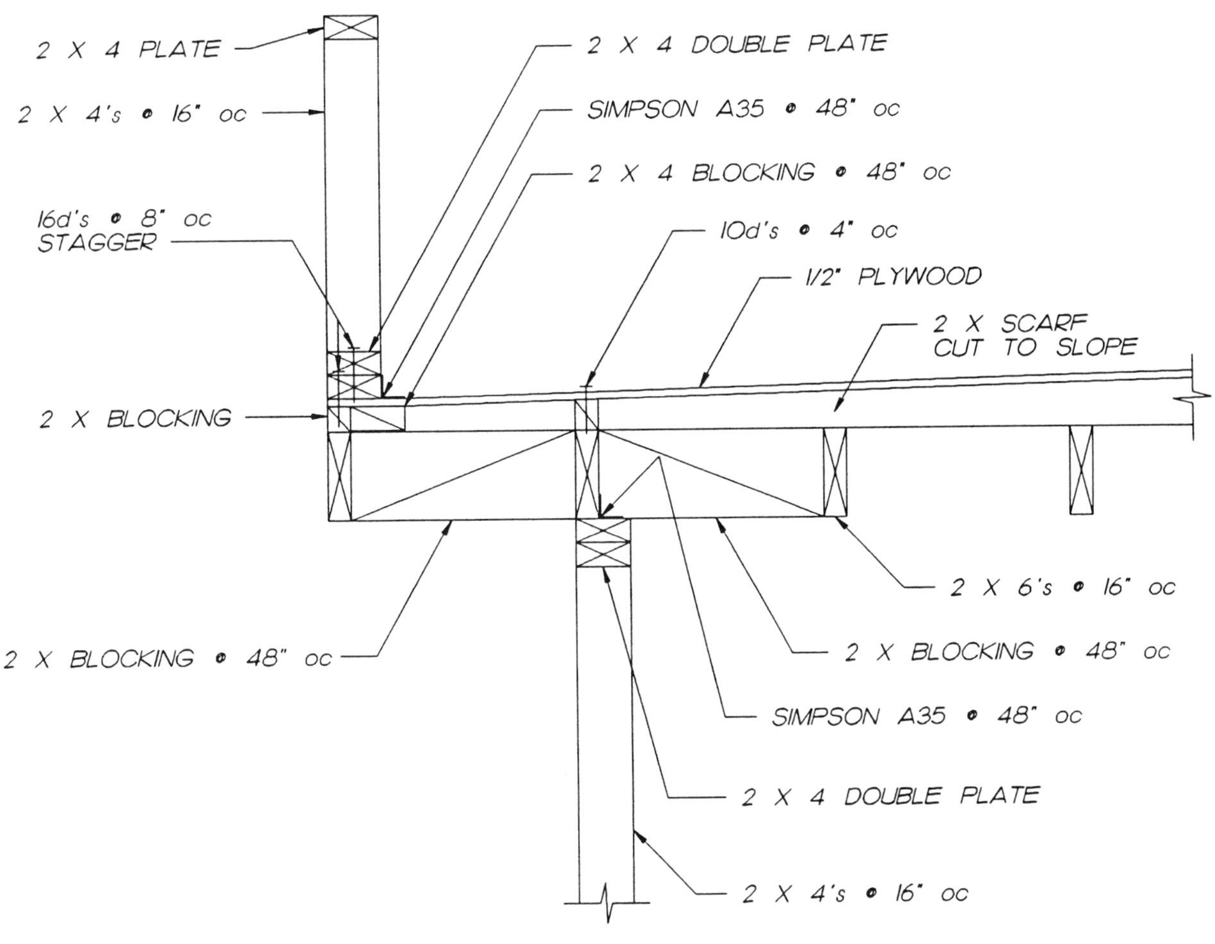

2 X 4 PLATE
2 X 4's @ 16" oc
16d's @ 8" oc
STAGGER
2 X BLOCKING
2 X 4 DOUBLE PLATE
SIMPSON A35 @ 48" oc
2 X 4 BLOCKING @ 48" oc
10d's @ 4" oc
1/2" PLYWOOD
2 X SCARF
CUT TO SLOPE
2 X 6's @ 16" oc
2 X BLOCKING @ 48" oc
2 X BLOCKING @ 48" oc
SIMPSON A35 @ 48" oc
2 X 4 DOUBLE PLATE
2 X 4's @ 16" oc

W4403

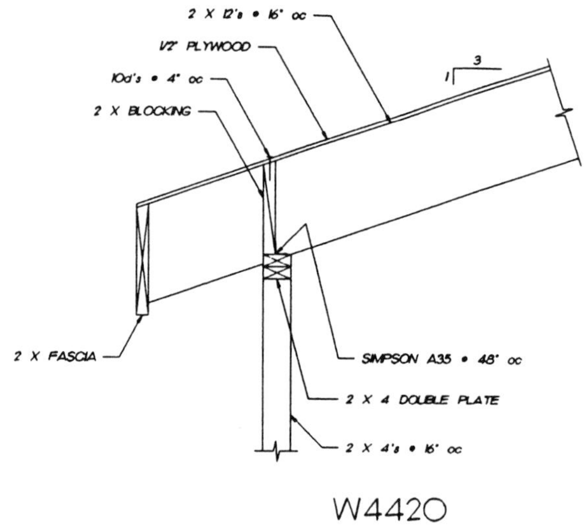

W4420

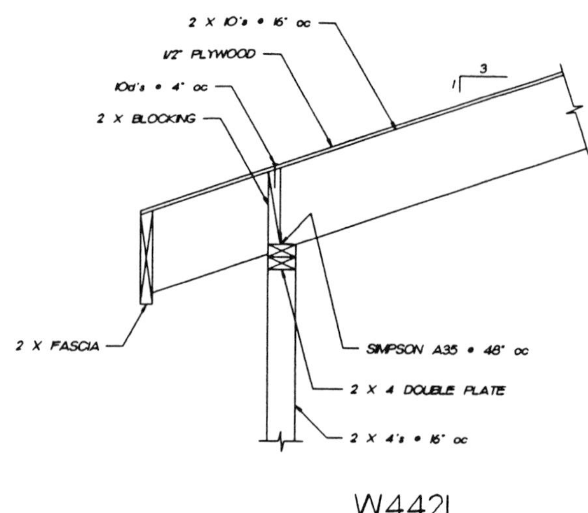

W4421

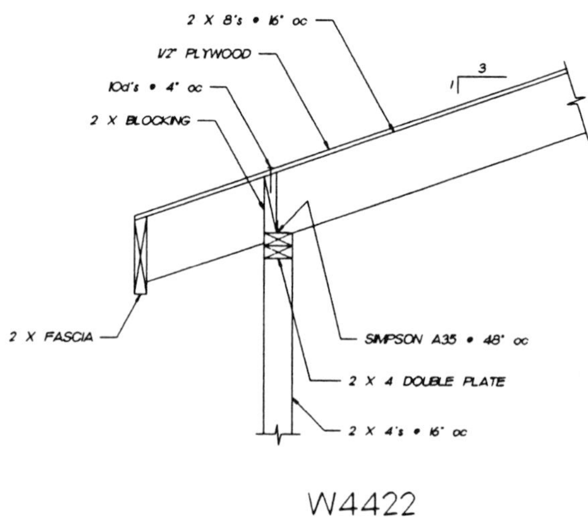

W4422

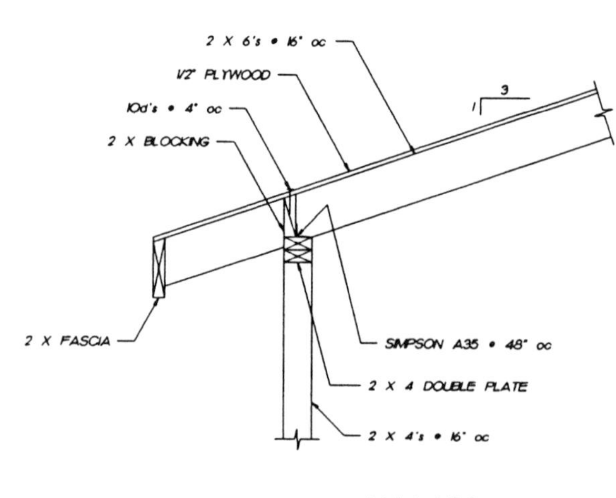

W4423

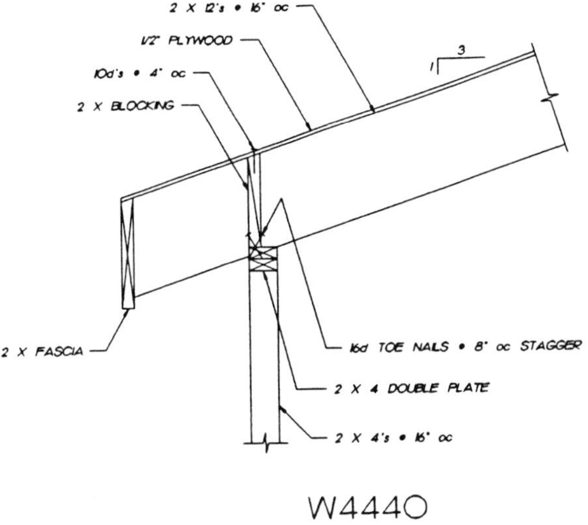

W4440

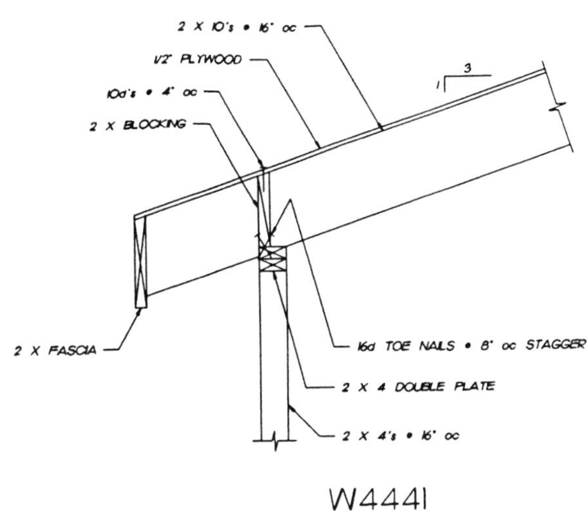

W4441

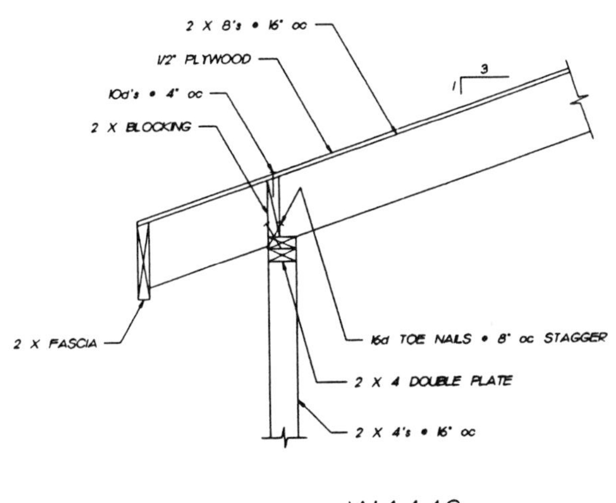

W4442

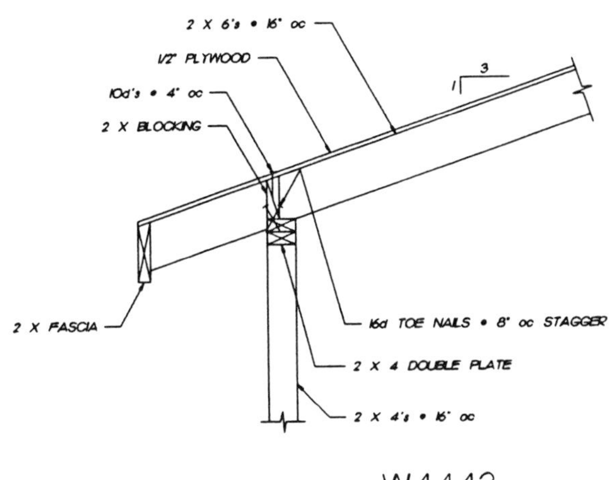

W4443

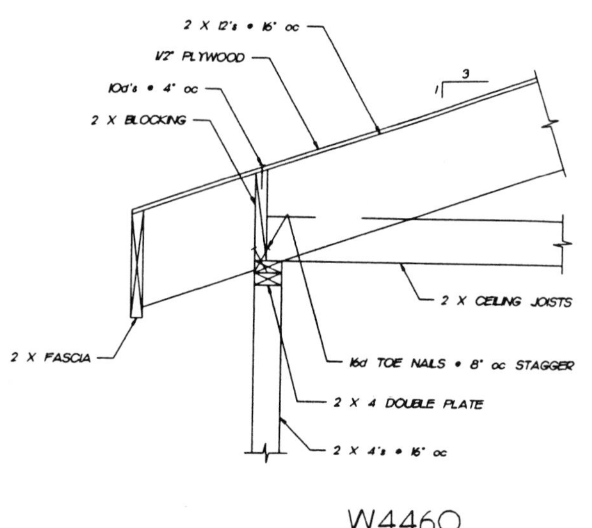

W4460

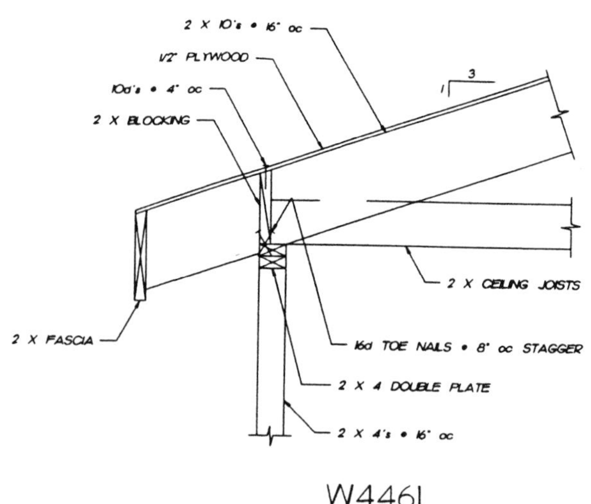

W4461

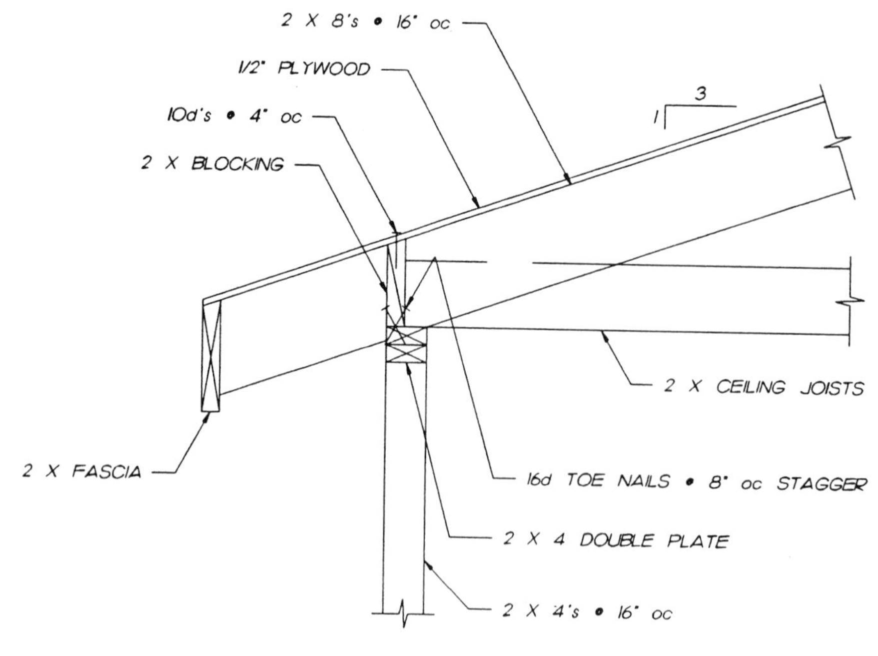

W4462

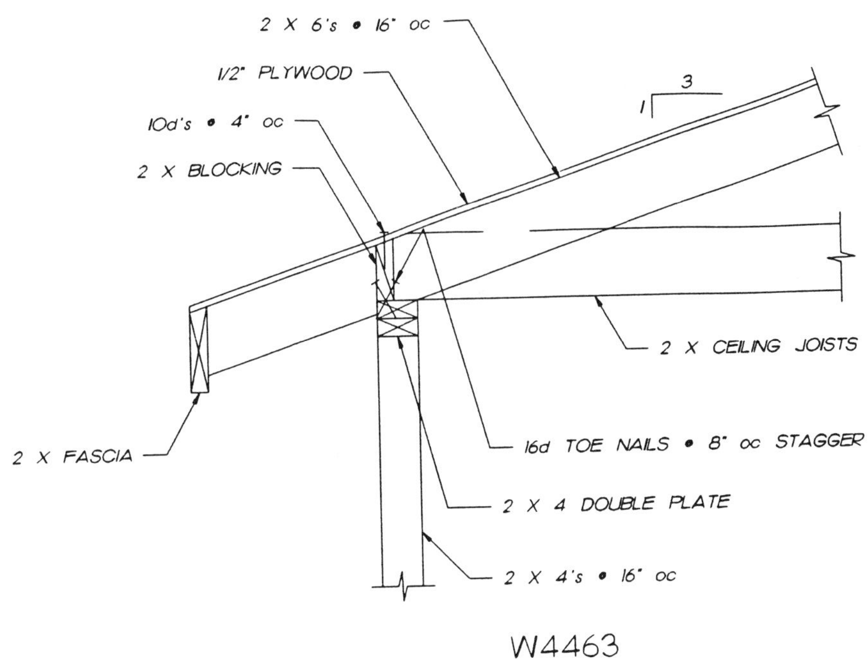

W4463

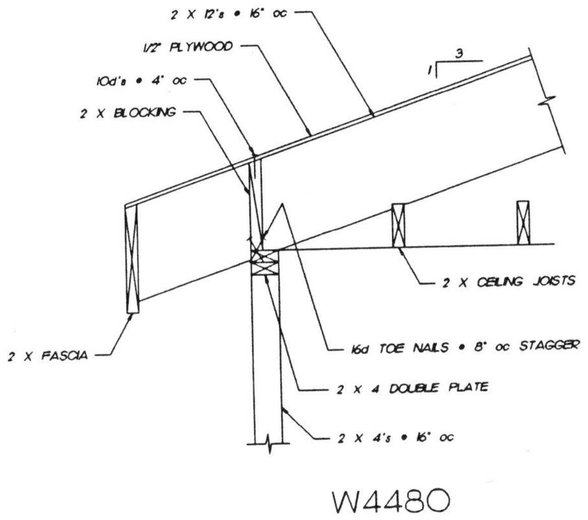

W4480

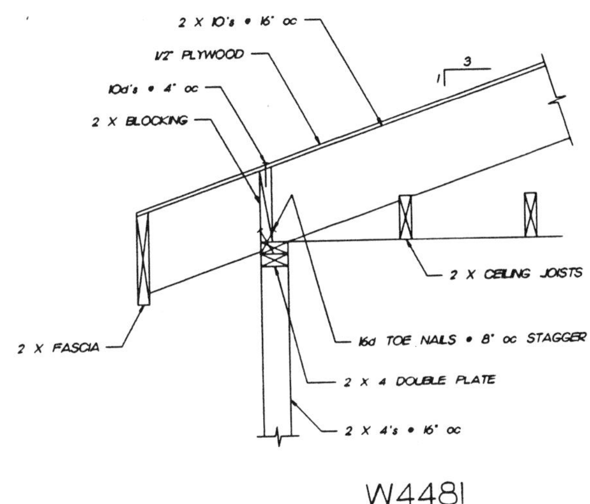

W4481

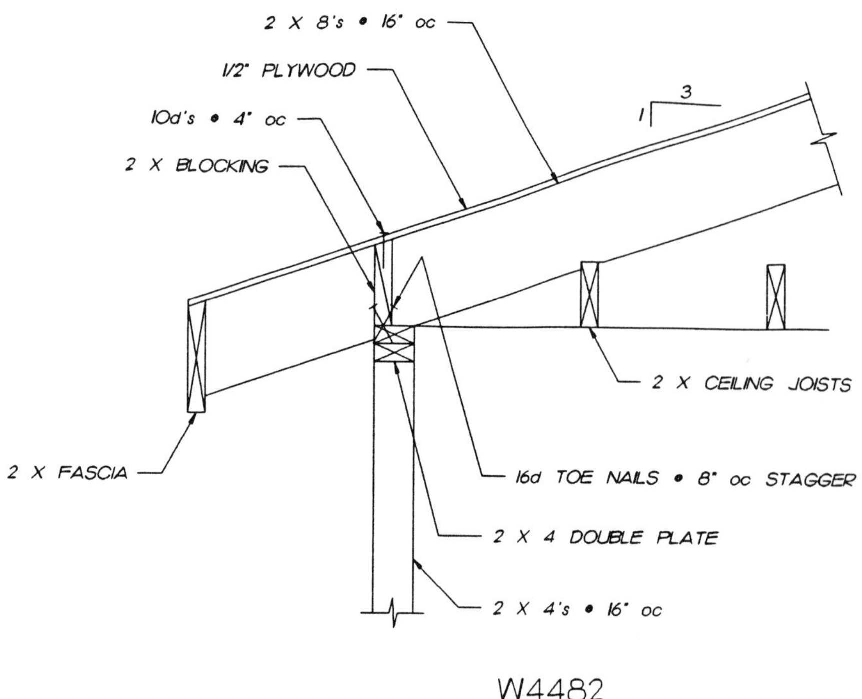
2 X 8's • 16" oc
1/2" PLYWOOD
10d's • 4" oc
2 X BLOCKING
1
3
2 X CEILING JOISTS
2 X FASCIA
16d TOE NAILS • 8" oc STAGGER
2 X 4 DOUBLE PLATE
2 X 4's • 16" oc

W4482

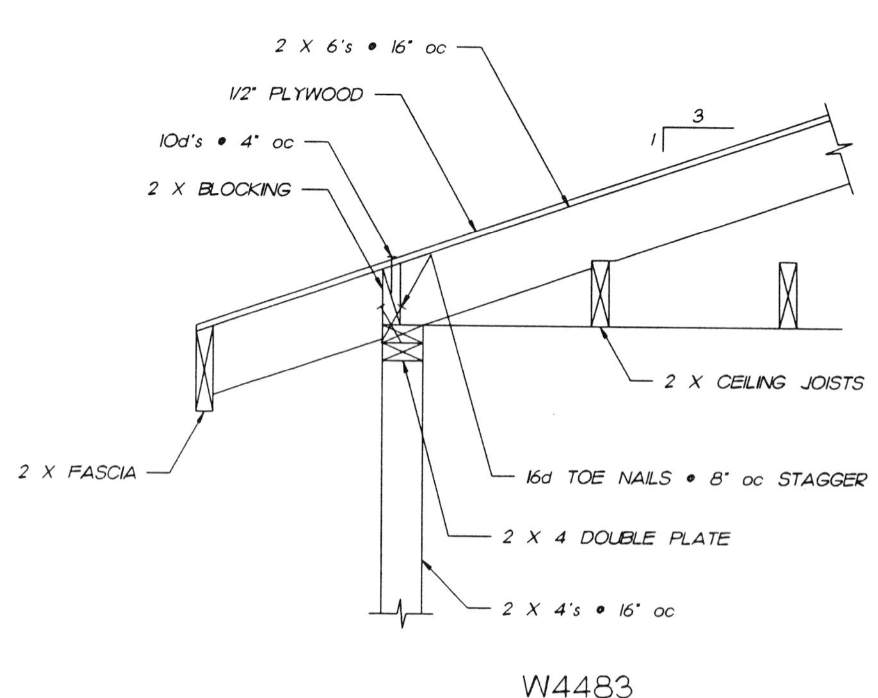
2 X 6's • 16" oc
1/2" PLYWOOD
10d's • 4" oc
2 X BLOCKING
1
3
2 X CEILING JOISTS
2 X FASCIA
16d TOE NAILS • 8" oc STAGGER
2 X 4 DOUBLE PLATE
2 X 4's • 16" oc

W4483

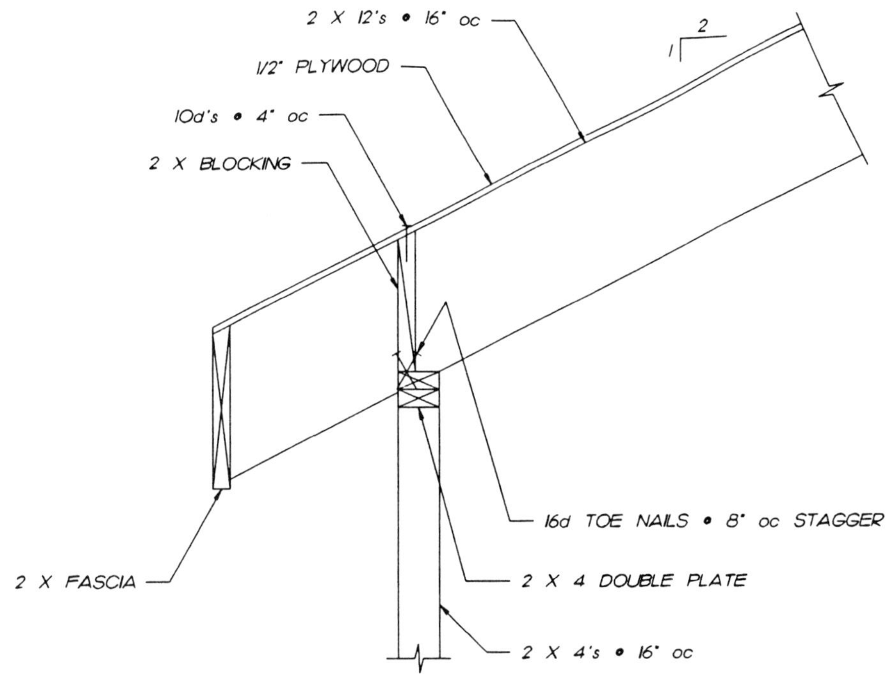
2 X 12's • 16" oc
1/2" PLYWOOD
10d's • 4" oc
2 X BLOCKING
1
2
16d TOE NAILS • 8" oc STAGGER
2 X FASCIA
2 X 4 DOUBLE PLATE
2 X 4's • 16" oc

W4500

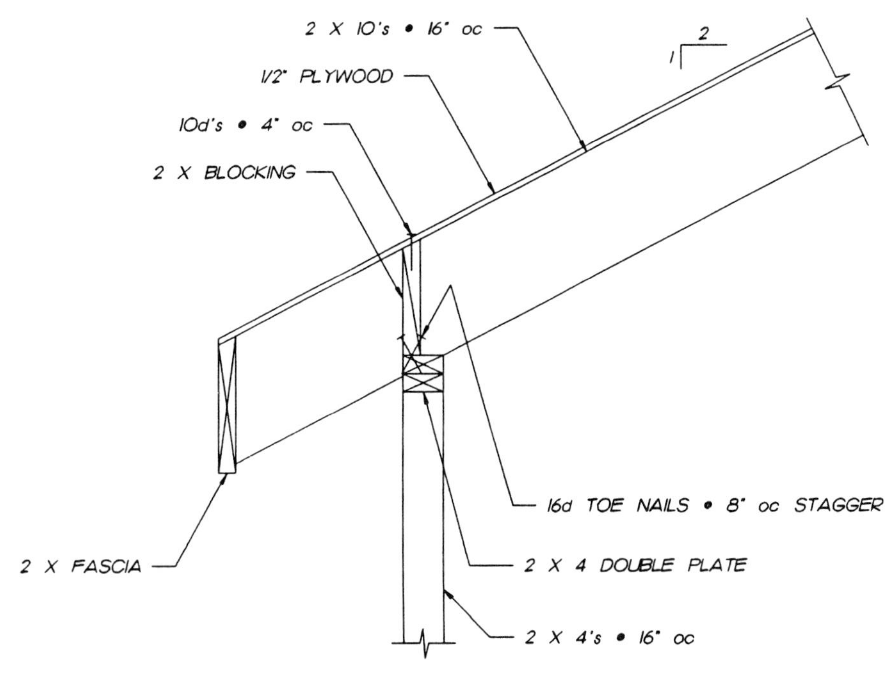
2 X 10's • 16" oc
1/2" PLYWOOD
10d's • 4" oc
2 X BLOCKING
1
2
16d TOE NAILS • 8" oc STAGGER
2 X FASCIA
2 X 4 DOUBLE PLATE
2 X 4's • 16" oc

W4501

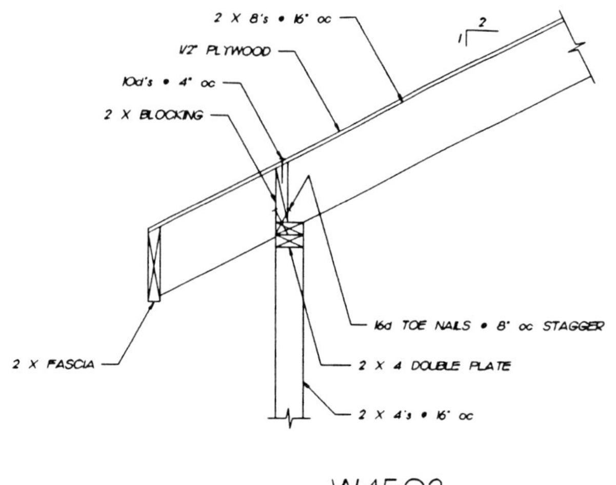

W4502

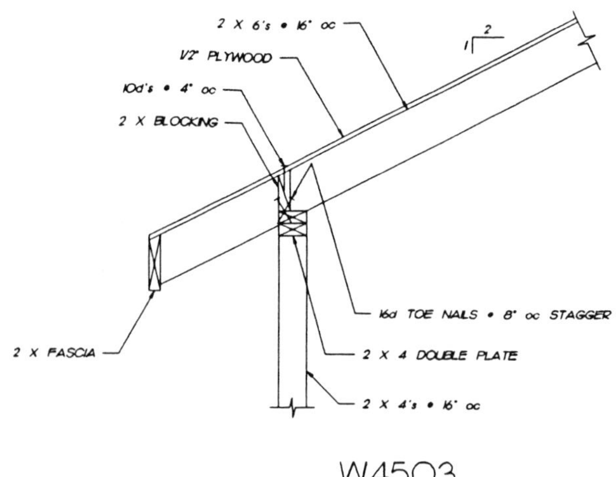

W4503

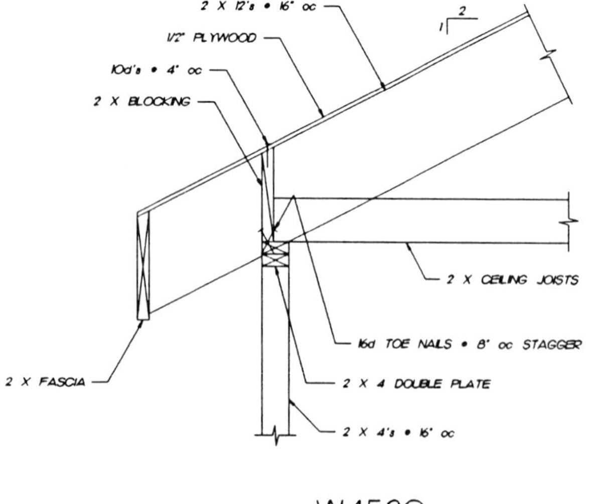

W4520

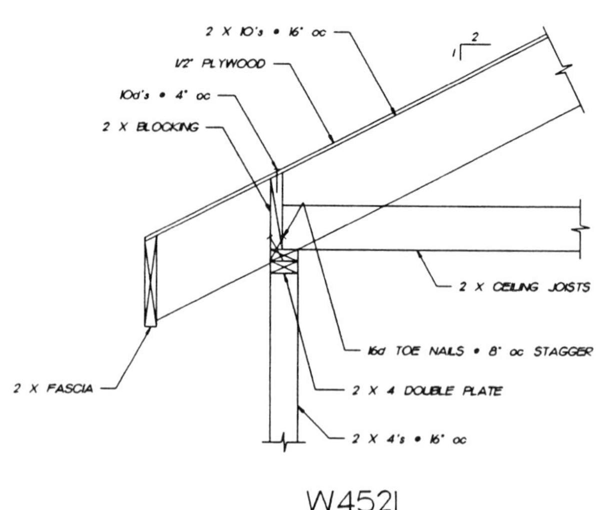

W4521

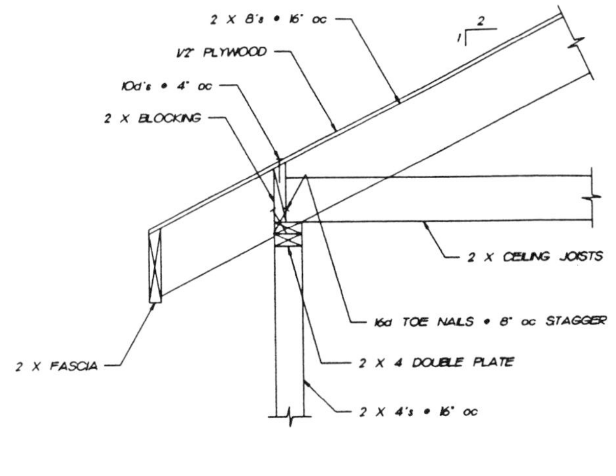

W4522

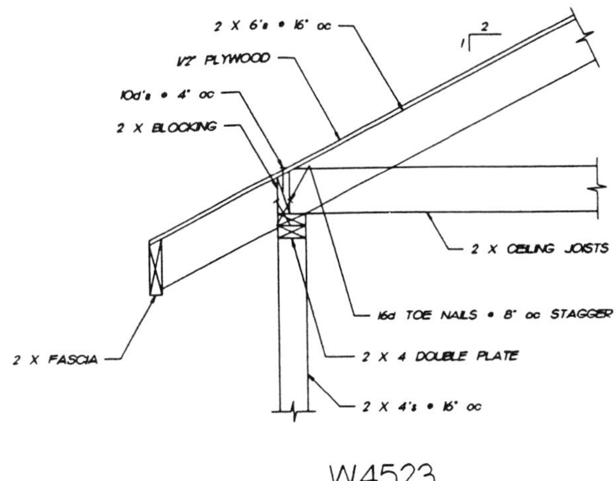

W4523

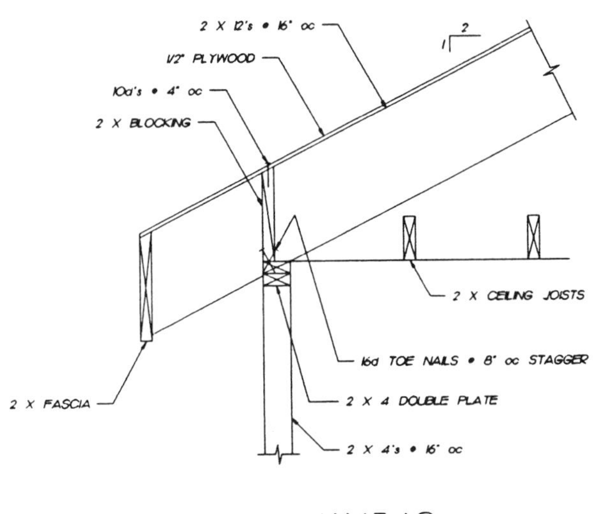

W4540

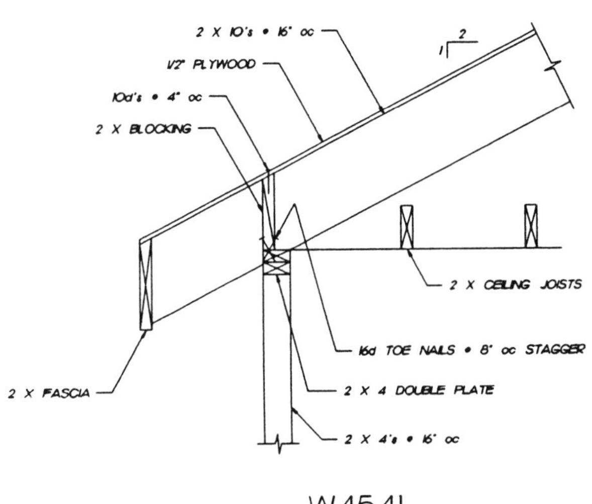

W4541

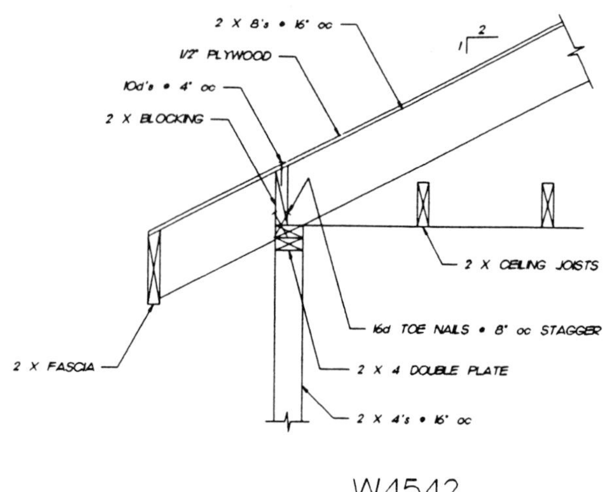

W4542

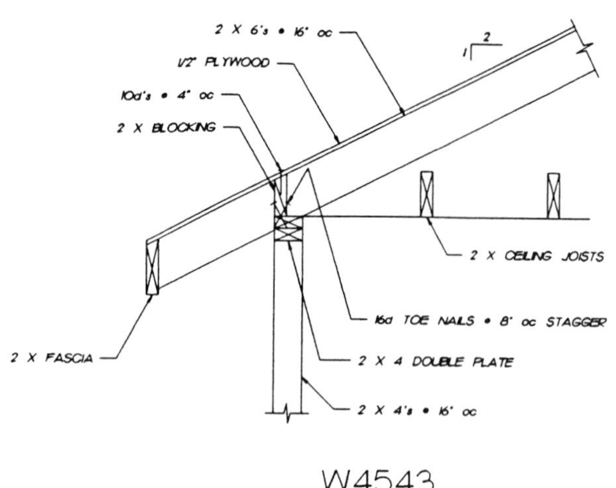

W4543

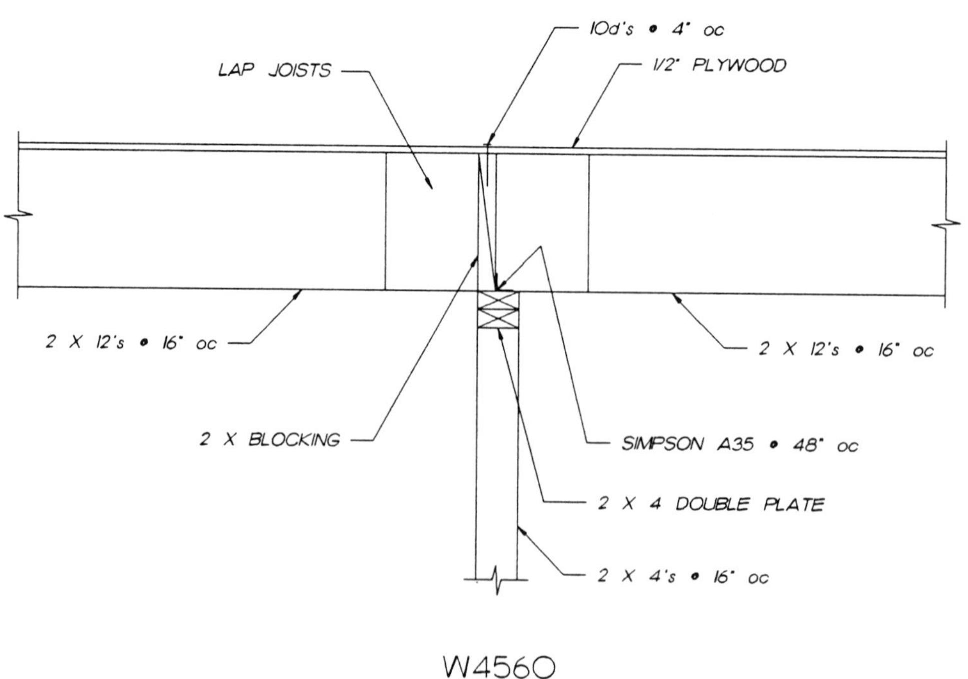

W4560

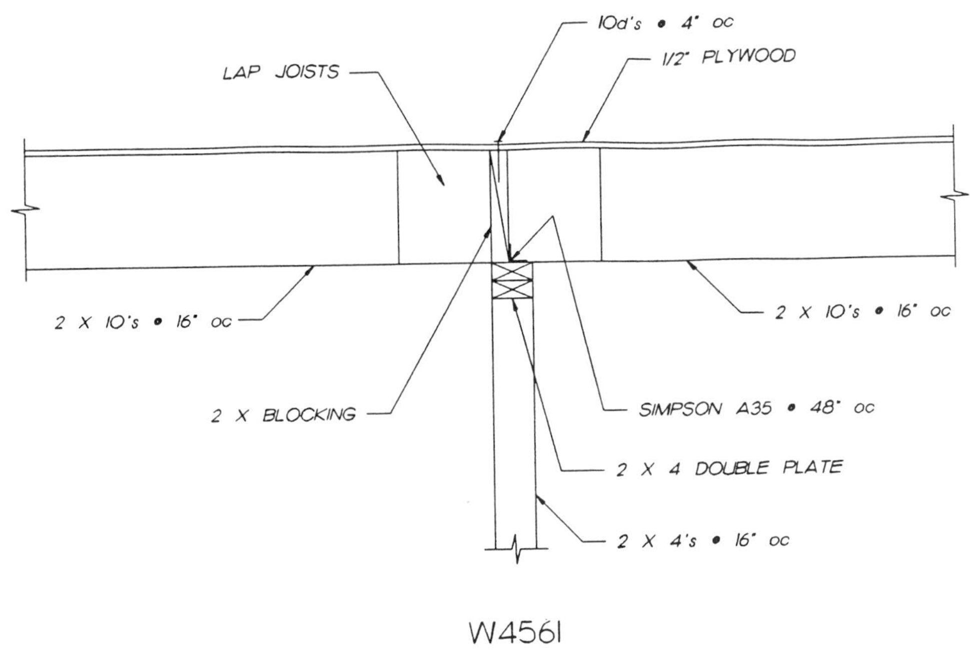

W4561

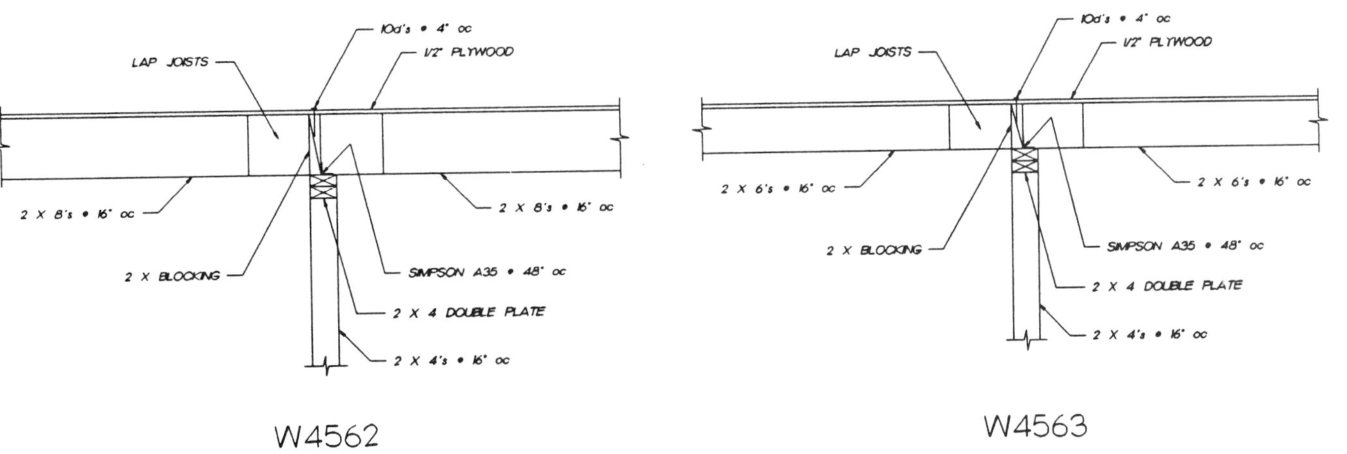

W4562

W4563

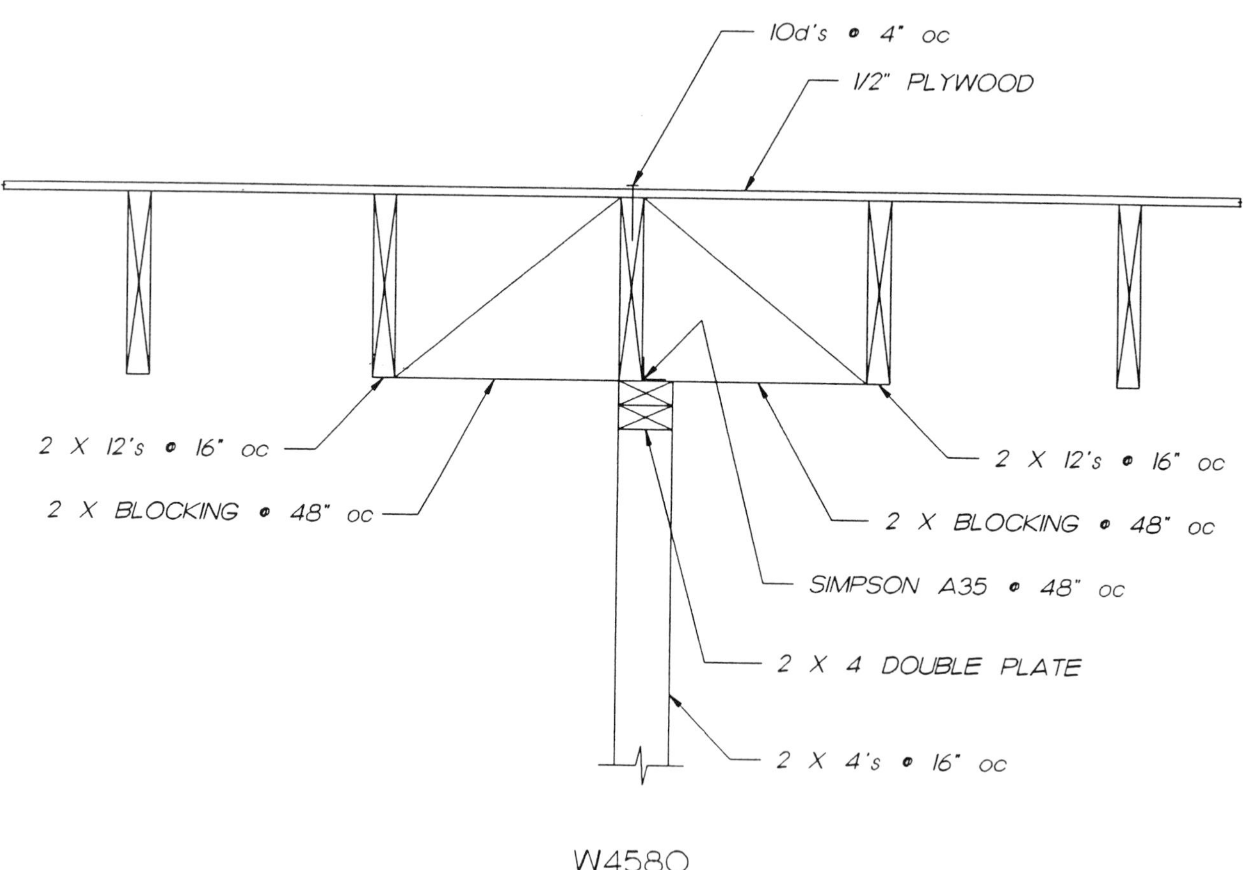
10d's @ 4" oc
1/2" PLYWOOD
2 X 12's @ 16" oc
2 X BLOCKING @ 48" oc
2 X 12's @ 16" oc
2 X BLOCKING @ 48" oc
SIMPSON A35 @ 48" oc
2 X 4 DOUBLE PLATE
2 X 4's @ 16" oc

W4580

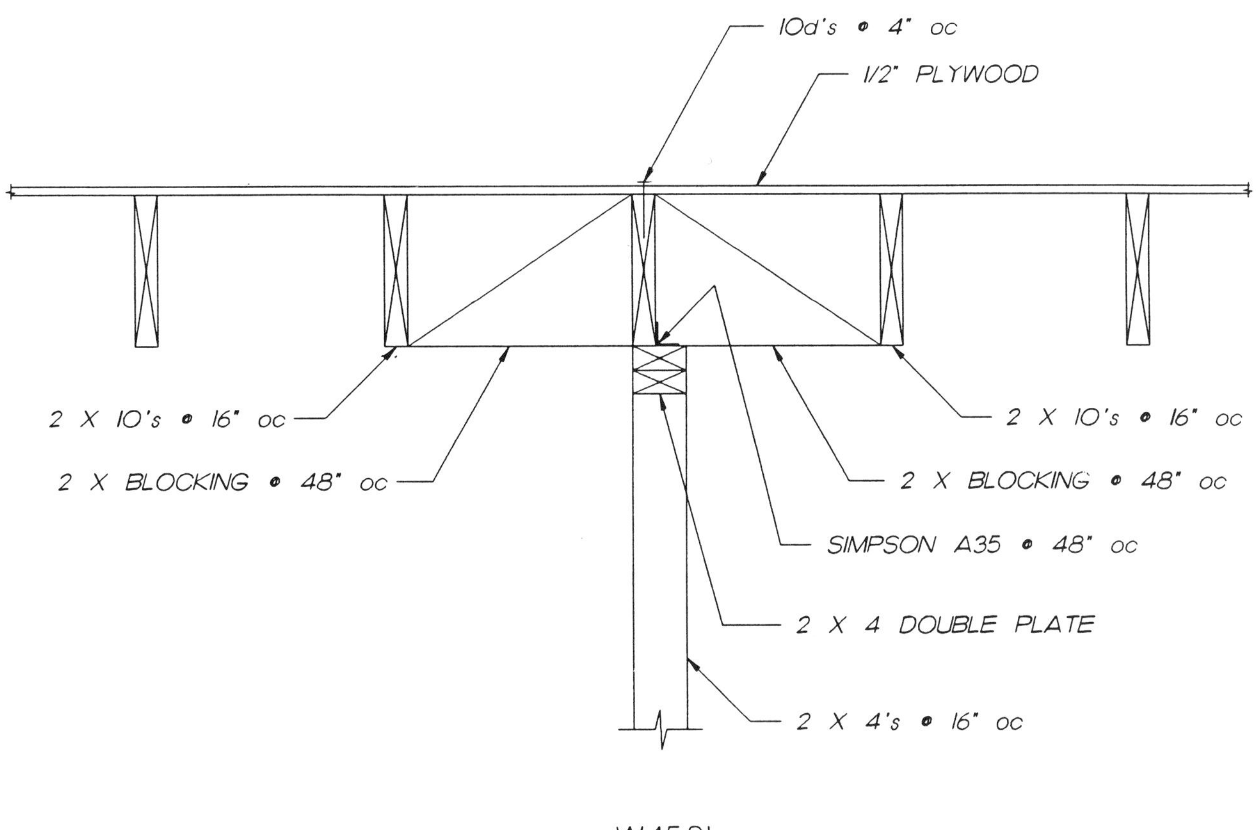
10d's @ 4" oc
1/2" PLYWOOD
2 X 10's @ 16" oc
2 X BLOCKING @ 48" oc
2 X 10's @ 16" oc
2 X BLOCKING @ 48" oc
SIMPSON A35 @ 48" oc
2 X 4 DOUBLE PLATE
2 X 4's @ 16" oc

W4581

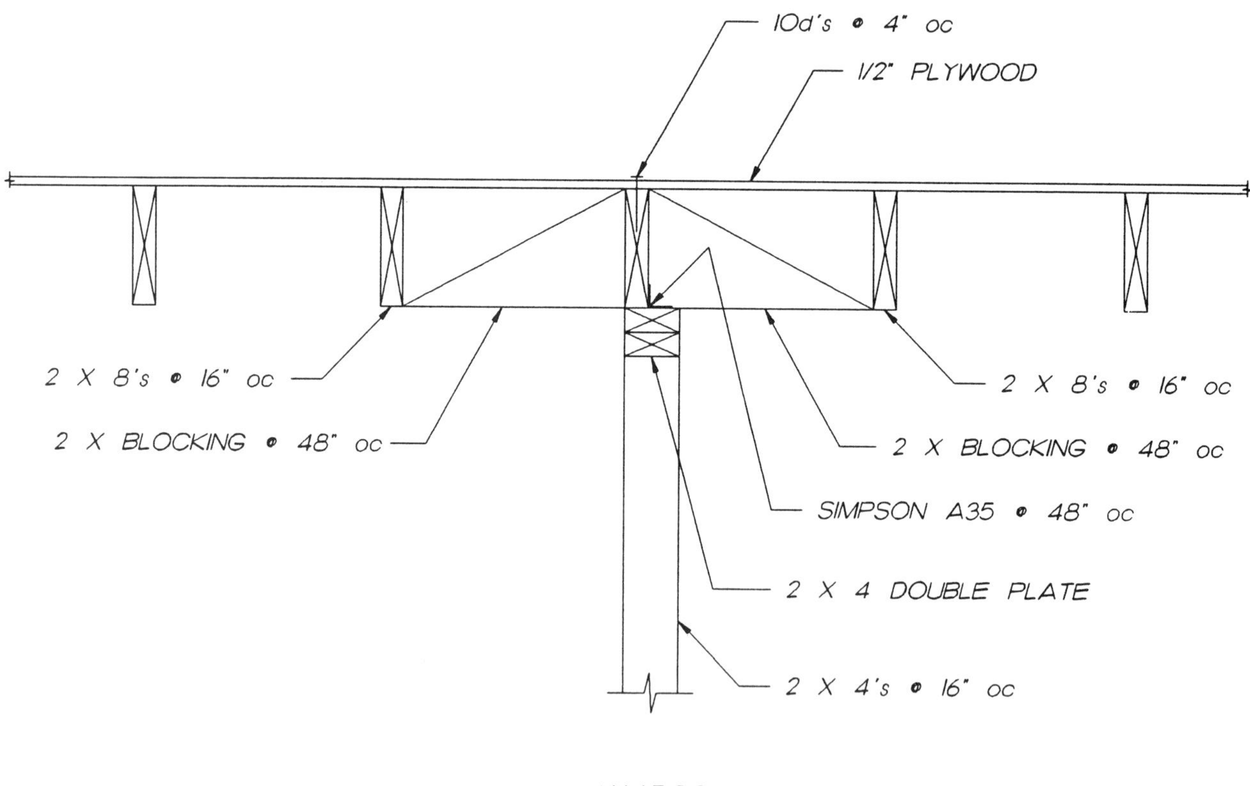
10d's @ 4" oc
1/2" PLYWOOD
2 X 8's @ 16" oc
2 X BLOCKING @ 48" oc
2 X 8's @ 16" oc
2 X BLOCKING @ 48" oc
SIMPSON A35 @ 48" oc
2 X 4 DOUBLE PLATE
2 X 4's @ 16" oc

W4582

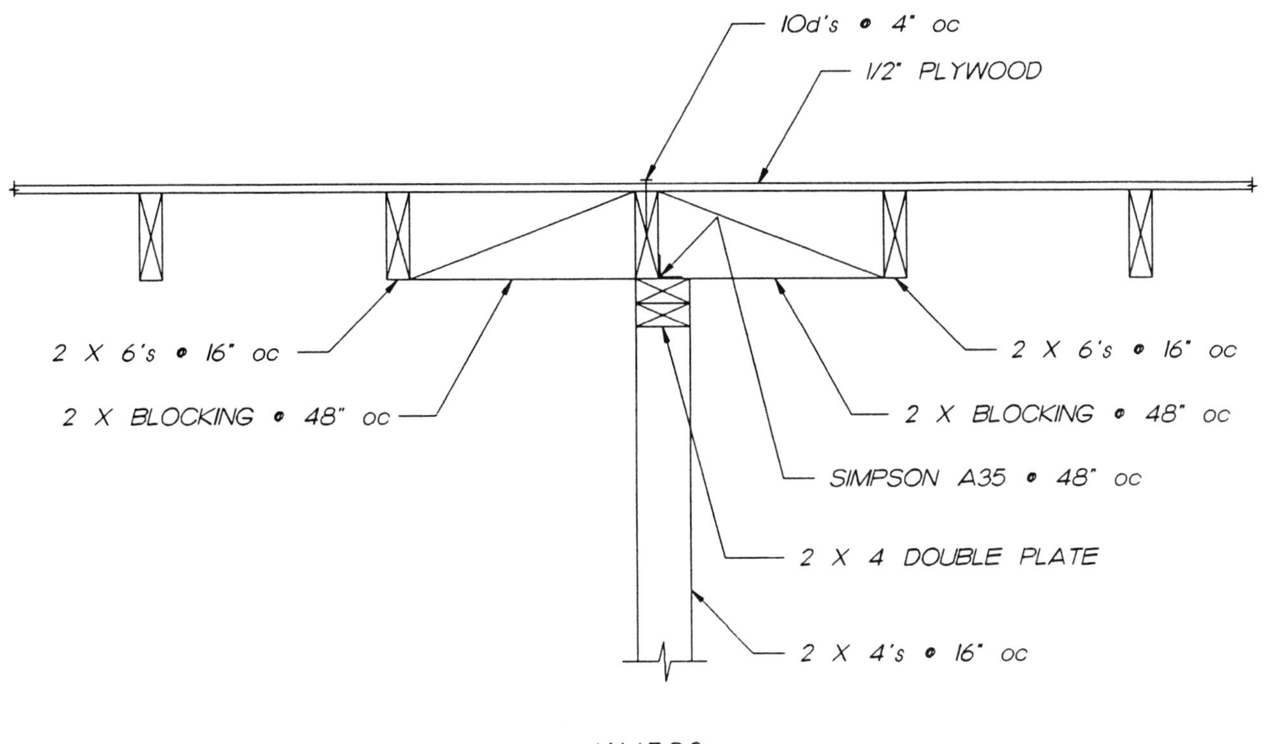
10d's @ 4" oc
1/2" PLYWOOD
2 X 6's @ 16" oc
2 X BLOCKING @ 48" oc
2 X 6's @ 16" oc
2 X BLOCKING @ 48" oc
SIMPSON A35 @ 48" oc
2 X 4 DOUBLE PLATE
2 X 4's @ 16" oc

W4583

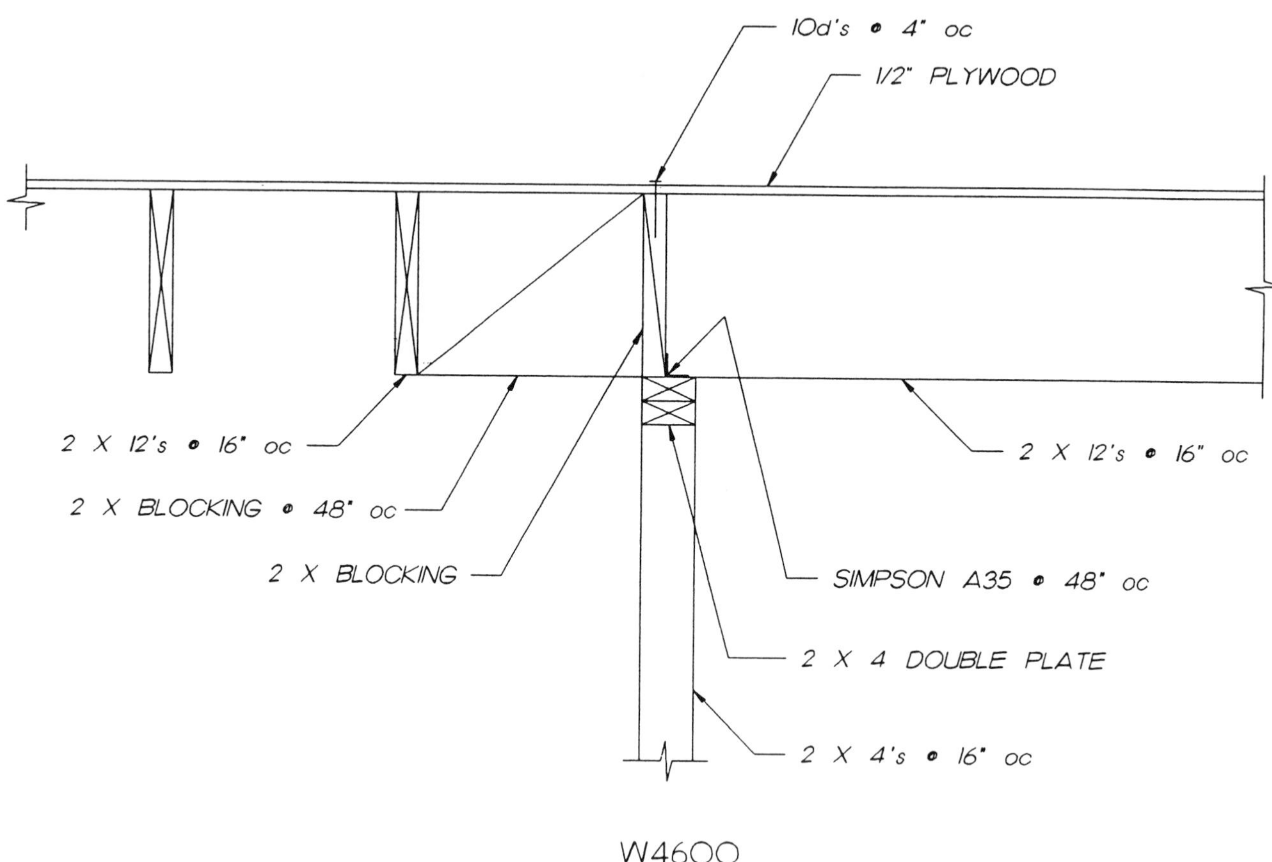
10d's @ 4" oc
1/2" PLYWOOD
2 X 12's @ 16" oc
2 X 12's @ 16" oc
2 X BLOCKING @ 48" oc
2 X BLOCKING
SIMPSON A35 @ 48" oc
2 X 4 DOUBLE PLATE
2 X 4's @ 16" oc

W4600

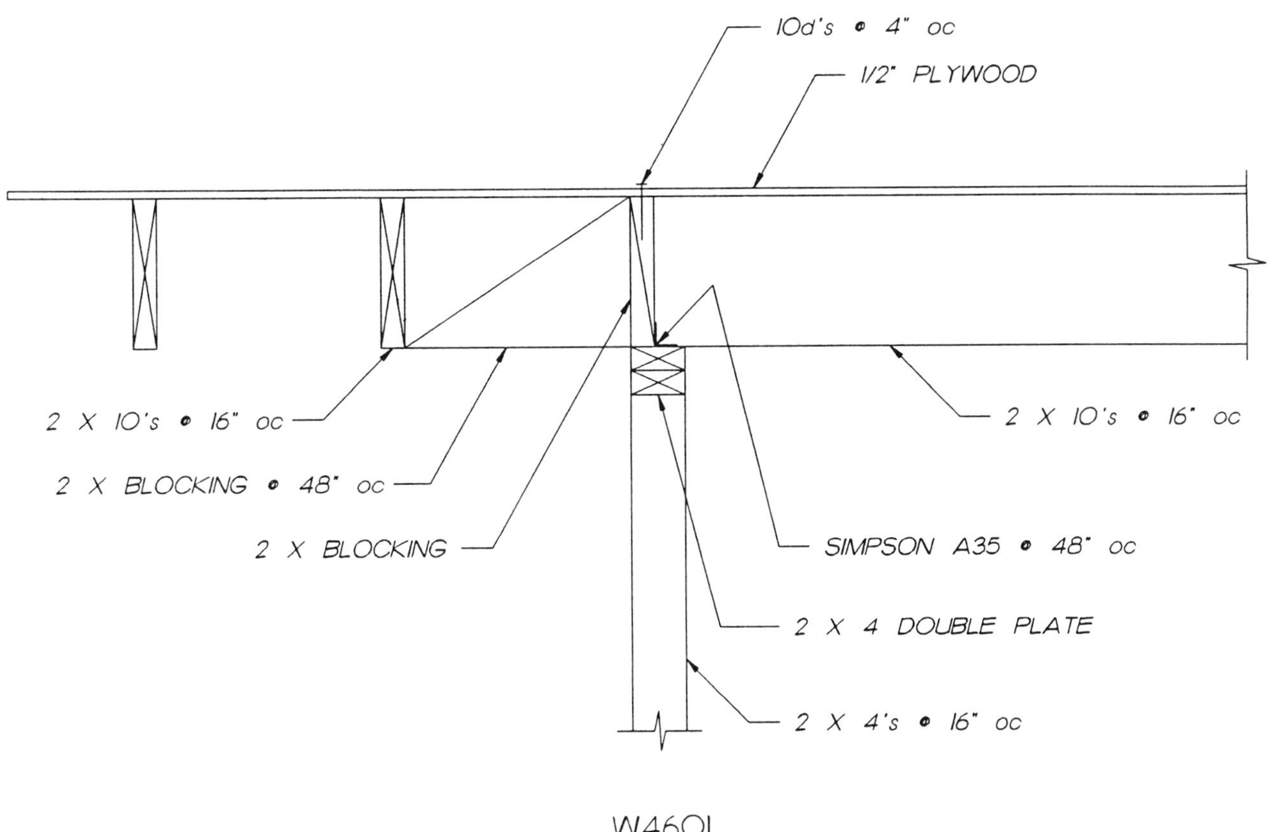
10d's @ 4" oc
1/2" PLYWOOD
2 X 10's @ 16" oc
2 X BLOCKING @ 48" oc
2 X BLOCKING
2 X 10's @ 16" oc
SIMPSON A35 @ 48" oc
2 X 4 DOUBLE PLATE
2 X 4's @ 16" oc

W4601

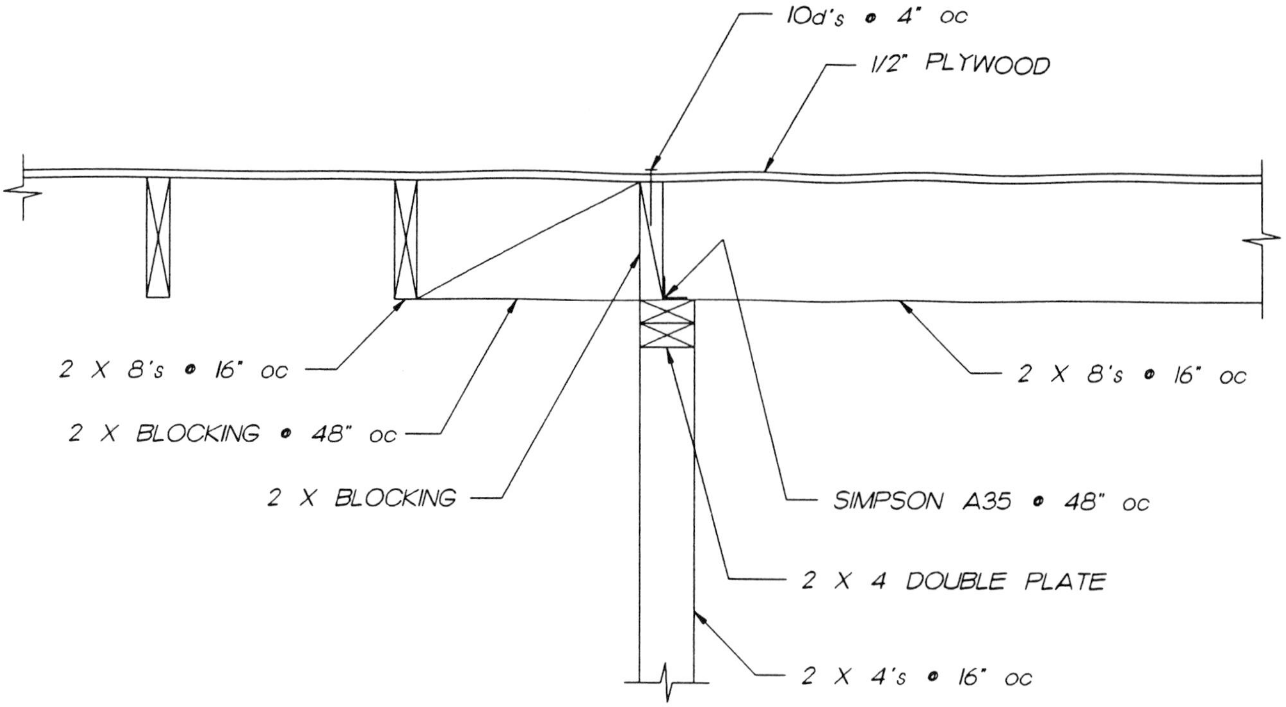

10d's @ 4" oc
1/2" PLYWOOD
2 X 8's @ 16" oc
2 X BLOCKING @ 48" oc
2 X BLOCKING
2 X 8's @ 16" oc
SIMPSON A35 @ 48" oc
2 X 4 DOUBLE PLATE
2 X 4's @ 16" oc

W4602

10d's @ 4" oc
1/2" PLYWOOD
2 X 6's @ 16" oc
2 X BLOCKING @ 48" oc
2 X BLOCKING
2 X 6's @ 16" oc
SIMPSON A35 @ 48" oc
2 X 4 DOUBLE PLATE
2 X 4's @ 16" oc

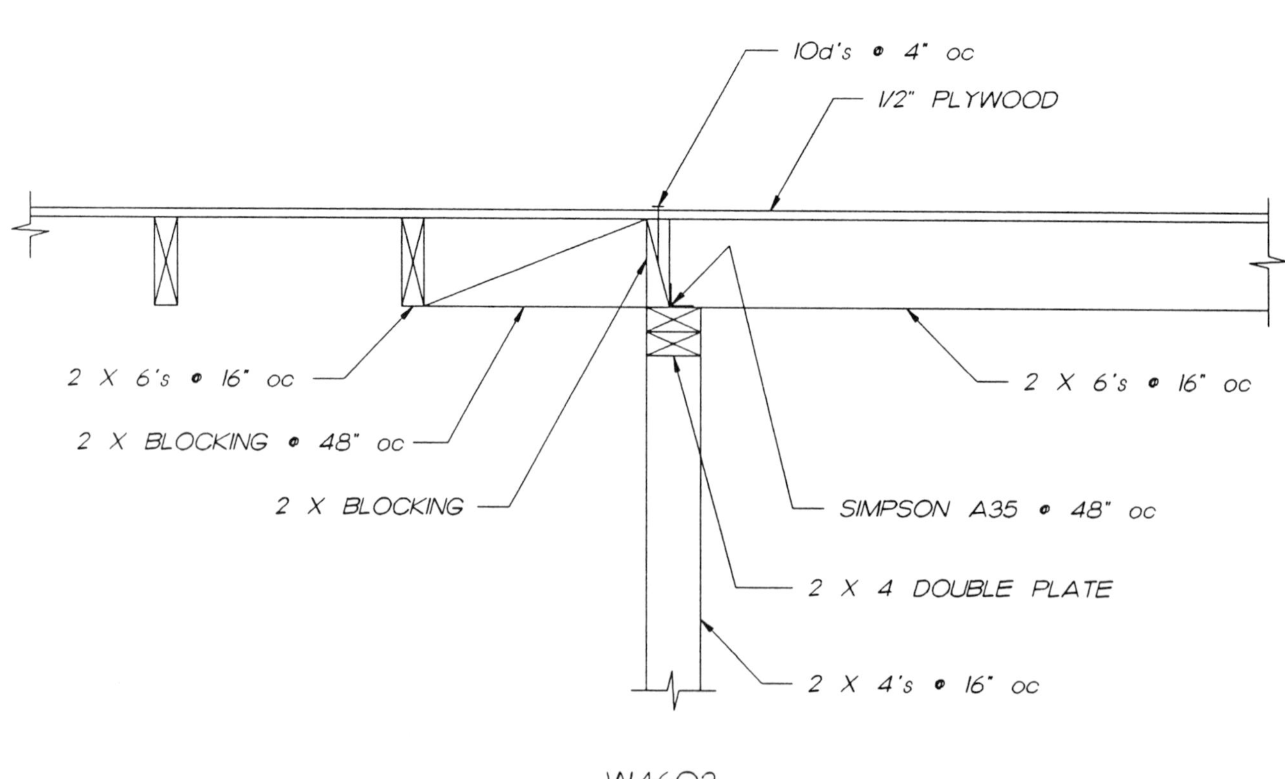

W4603

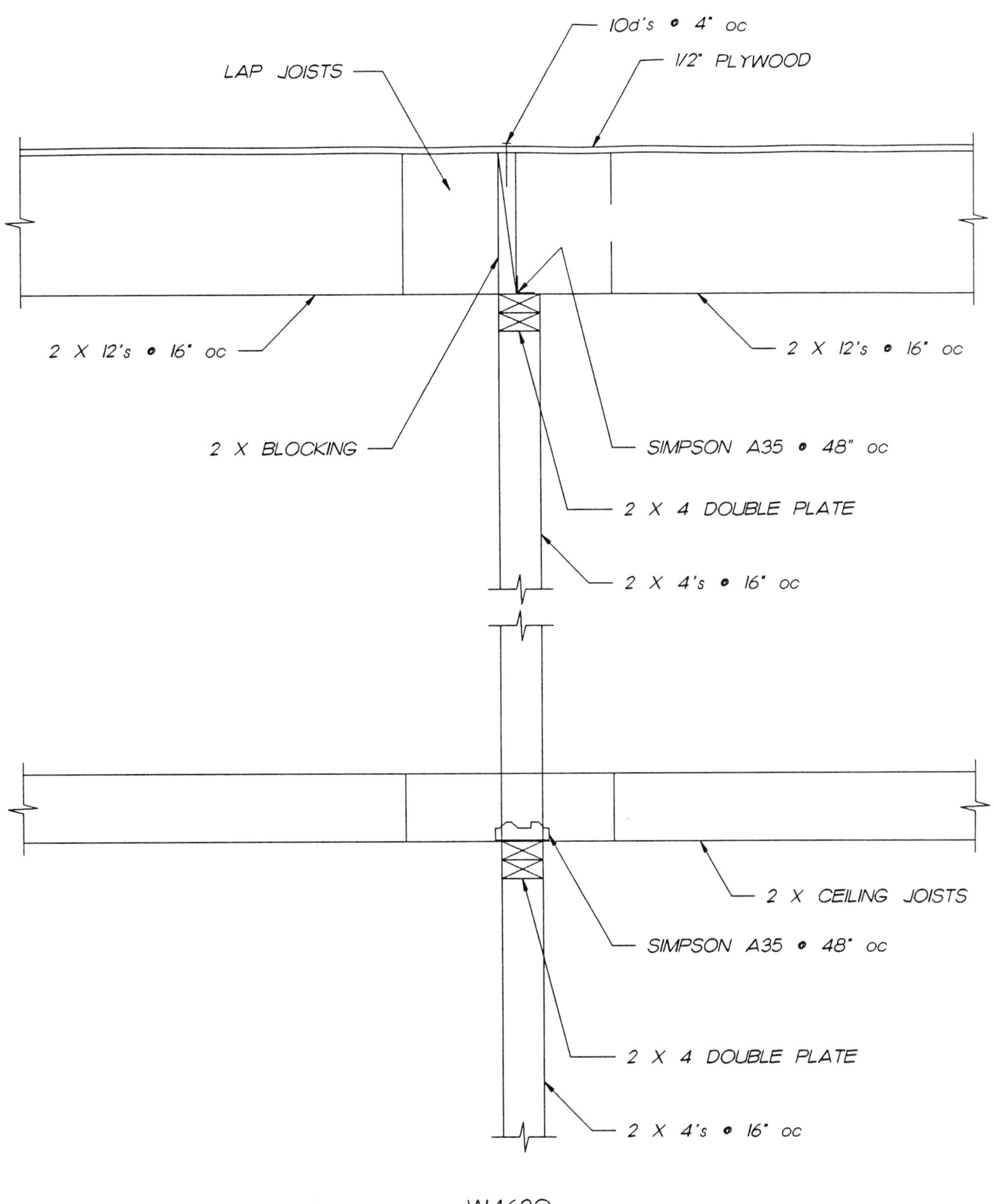
10d's @ 4" oc
1/2" PLYWOOD
LAP JOISTS
2 X 12's @ 16" oc
2 X 12's @ 16" oc
2 X BLOCKING
SIMPSON A35 @ 48" oc
2 X 4 DOUBLE PLATE
2 X 4's @ 16" oc
2 X CEILING JOISTS
SIMPSON A35 @ 48" oc
2 X 4 DOUBLE PLATE
2 X 4's @ 16" oc

W4620

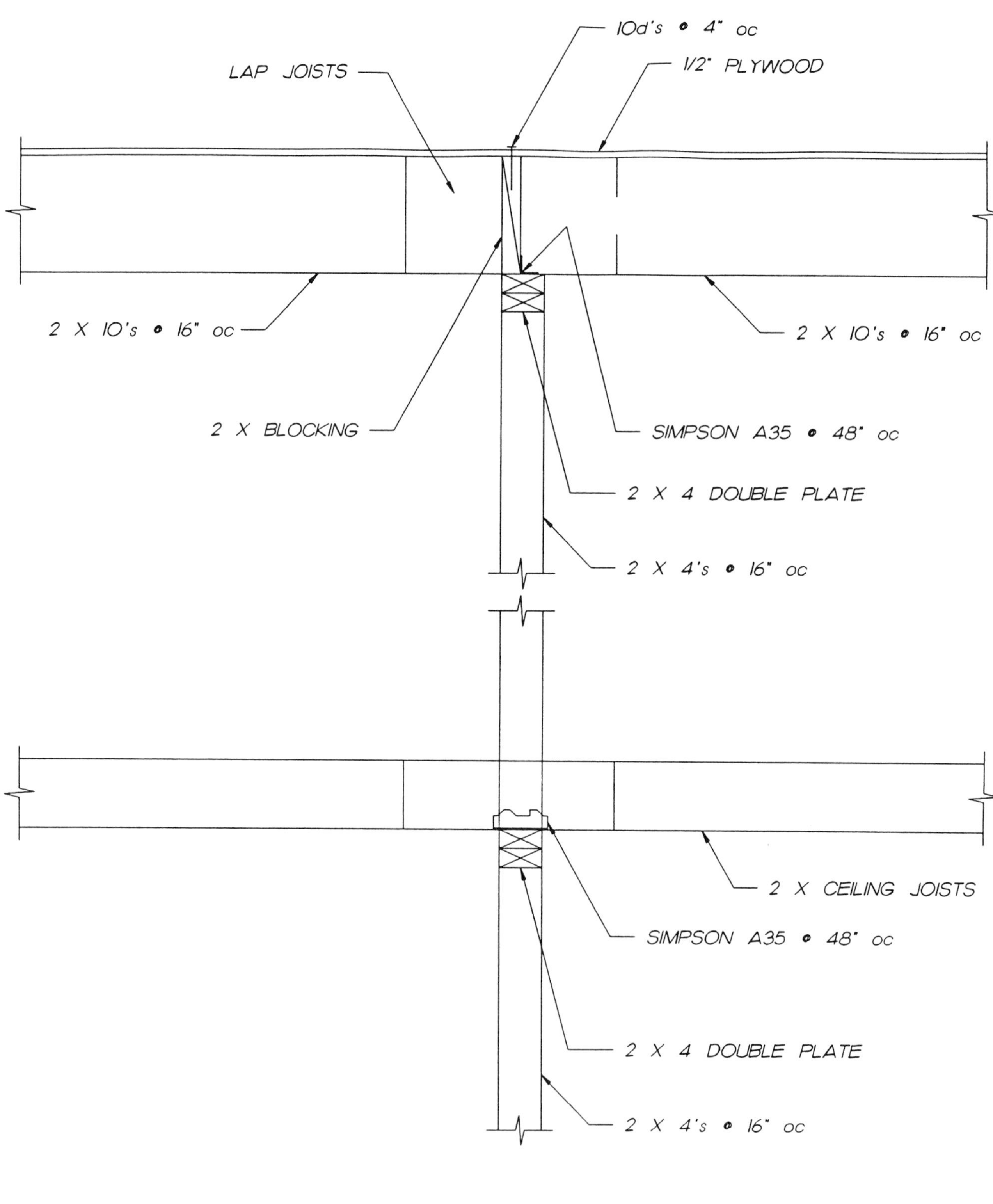
10d's @ 4" oc
LAP JOISTS
1/2" PLYWOOD
2 X 10's @ 16" oc
2 X 10's @ 16" oc
2 X BLOCKING
SIMPSON A35 @ 48" oc
2 X 4 DOUBLE PLATE
2 X 4's @ 16" oc
2 X CEILING JOISTS
SIMPSON A35 @ 48" oc
2 X 4 DOUBLE PLATE
2 X 4's @ 16" oc

W4621

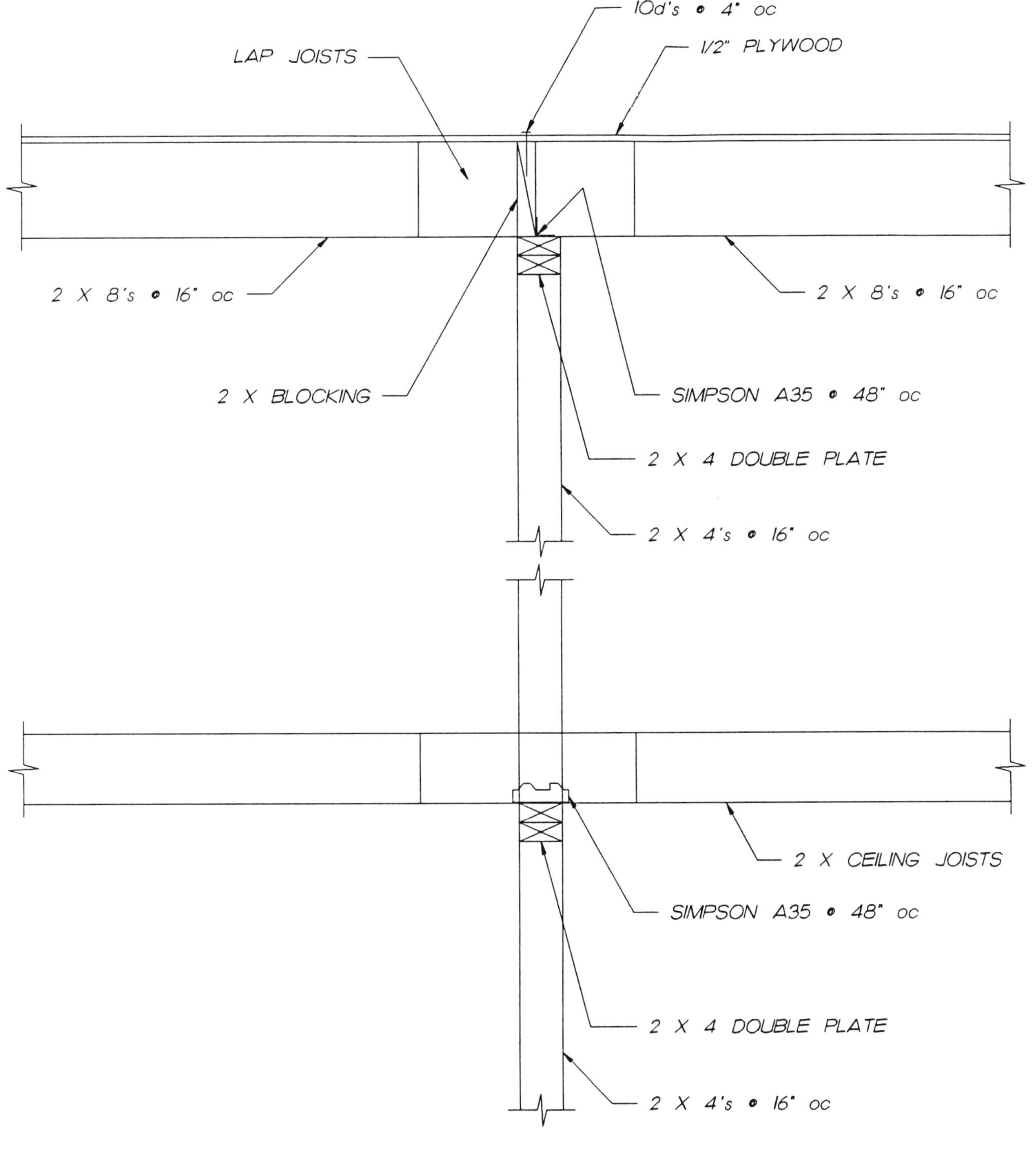
10d's @ 4" oc
1/2" PLYWOOD
LAP JOISTS
2 X 8's @ 16" oc
2 X 8's @ 16" oc
2 X BLOCKING
SIMPSON A35 @ 48" oc
2 X 4 DOUBLE PLATE
2 X 4's @ 16" oc
2 X CEILING JOISTS
SIMPSON A35 @ 48" oc
2 X 4 DOUBLE PLATE
2 X 4's @ 16" oc

W4622

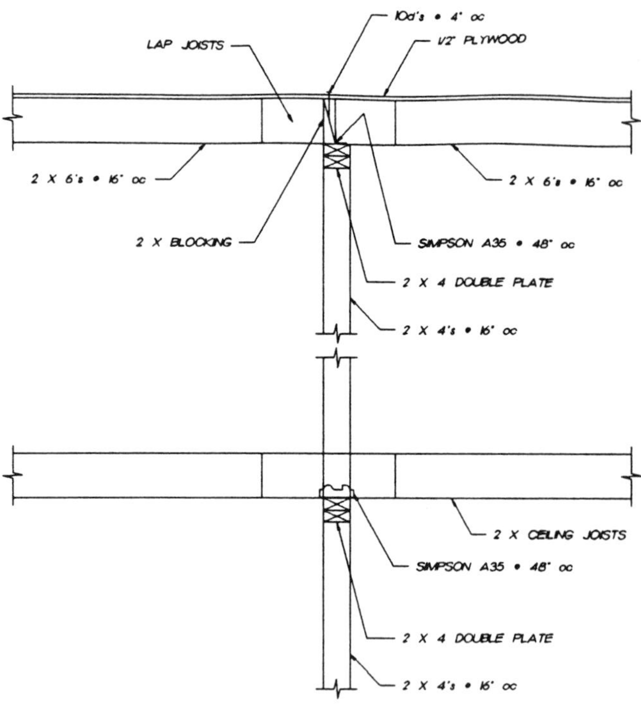
10d's • 4" oc
LAP JOISTS
1/2" PLYWOOD
2 X 6's • 16" oc
2 X 6's • 16" oc
2 X BLOCKING
SIMPSON A35 • 48" oc
2 X 4 DOUBLE PLATE
2 X 4's • 16" oc
2 X CEILING JOISTS
SIMPSON A35 • 48" oc
2 X 4 DOUBLE PLATE
2 X 4's • 16" oc

W4623

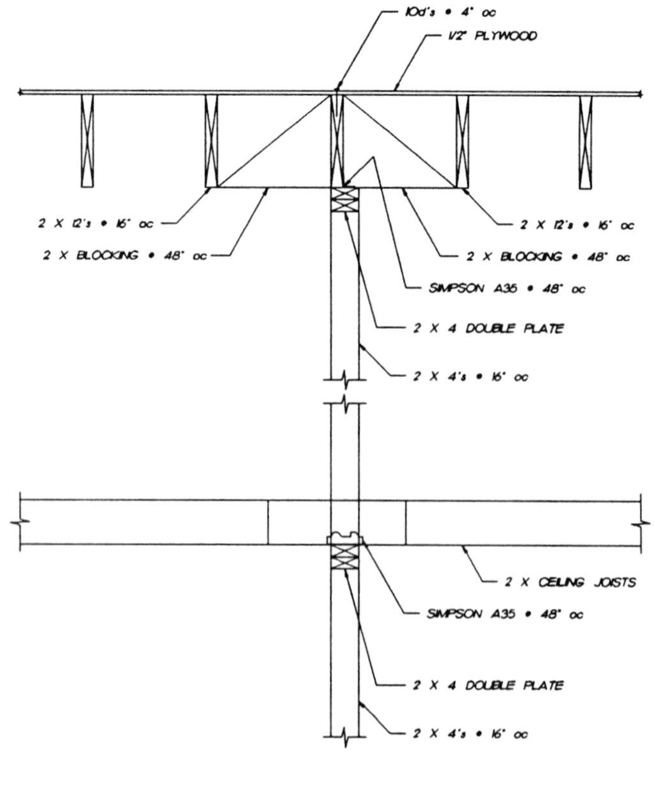
10d's • 4" oc
1/2" PLYWOOD
2 X 12's • 16" oc
2 X BLOCKING • 48" oc
2 X 12's • 16" oc
2 X BLOCKING • 48" oc
SIMPSON A35 • 48" oc
2 X 4 DOUBLE PLATE
2 X 4's • 16" oc
2 X CEILING JOISTS
SIMPSON A35 • 48" oc
2 X 4 DOUBLE PLATE
2 X 4's • 16" oc

W4640

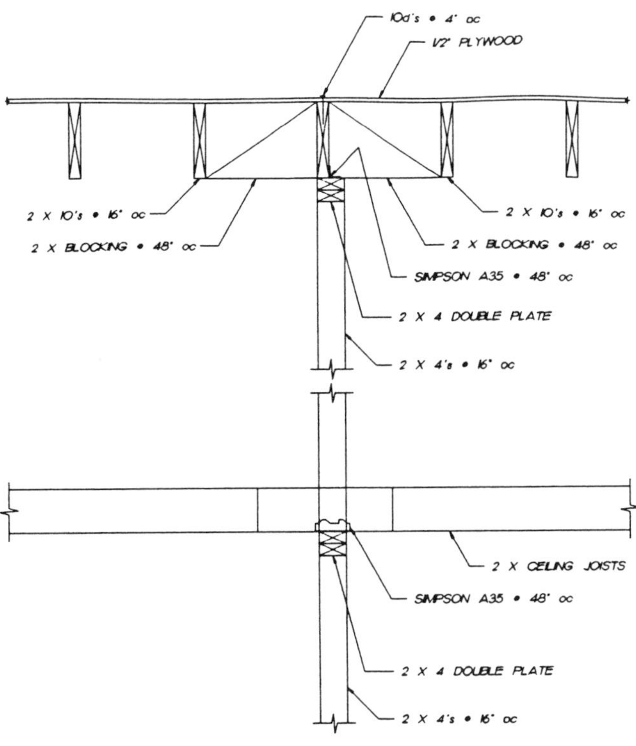
10d's • 4" oc
1/2" PLYWOOD
2 X 10's • 16" oc
2 X BLOCKING • 48" oc
2 X 10's • 16" oc
2 X BLOCKING • 48" oc
SIMPSON A35 • 48" oc
2 X 4 DOUBLE PLATE
2 X 4's • 16" oc
2 X CEILING JOISTS
SIMPSON A35 • 48" oc
2 X 4 DOUBLE PLATE
2 X 4's • 16" oc

W4641

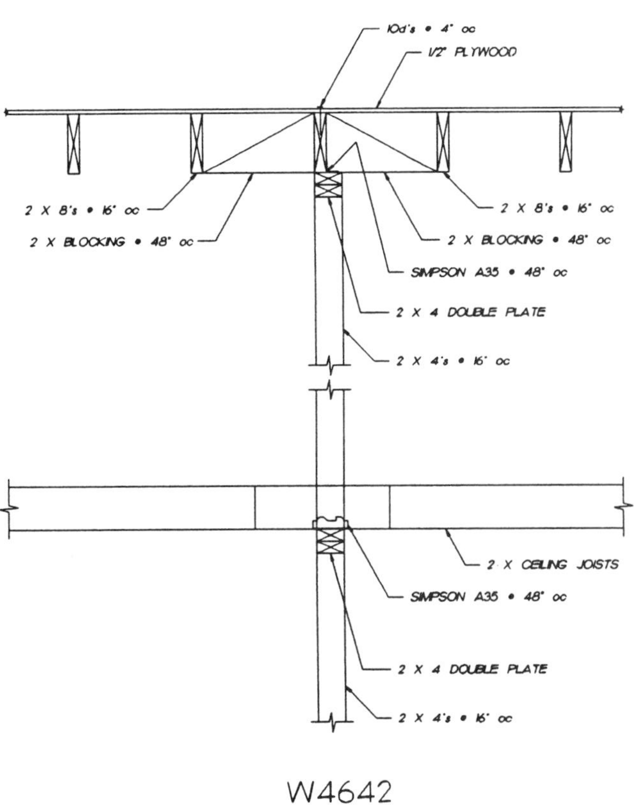
10d's • 4" oc
1/2" PLYWOOD
2 X 8's • 16" oc
2 X BLOCKING • 48" oc
2 X 8's • 16" oc
2 X BLOCKING • 48" oc
SIMPSON A35 • 48" oc
2 X 4 DOUBLE PLATE
2 X 4's • 16" oc
2 X CEILING JOISTS
SIMPSON A35 • 48" oc
2 X 4 DOUBLE PLATE
2 X 4's • 16" oc

W4642

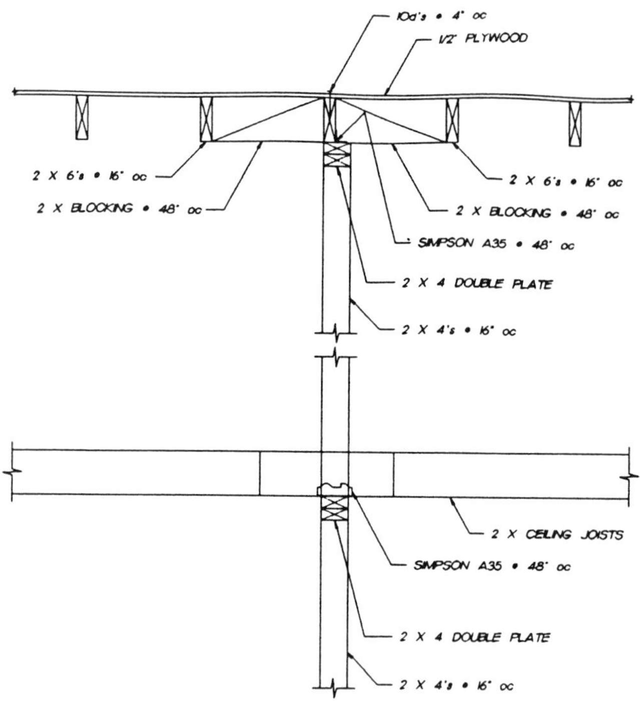
10d's • 4" oc
1/2" PLYWOOD
2 X 6's • 16" oc
2 X BLOCKING • 48" oc
2 X 6's • 16" oc
2 X BLOCKING • 48" oc
SIMPSON A35 • 48" oc
2 X 4 DOUBLE PLATE
2 X 4's • 16" oc
2 X CEILING JOISTS
SIMPSON A35 • 48" oc
2 X 4 DOUBLE PLATE
2 X 4's • 16" oc

W4643

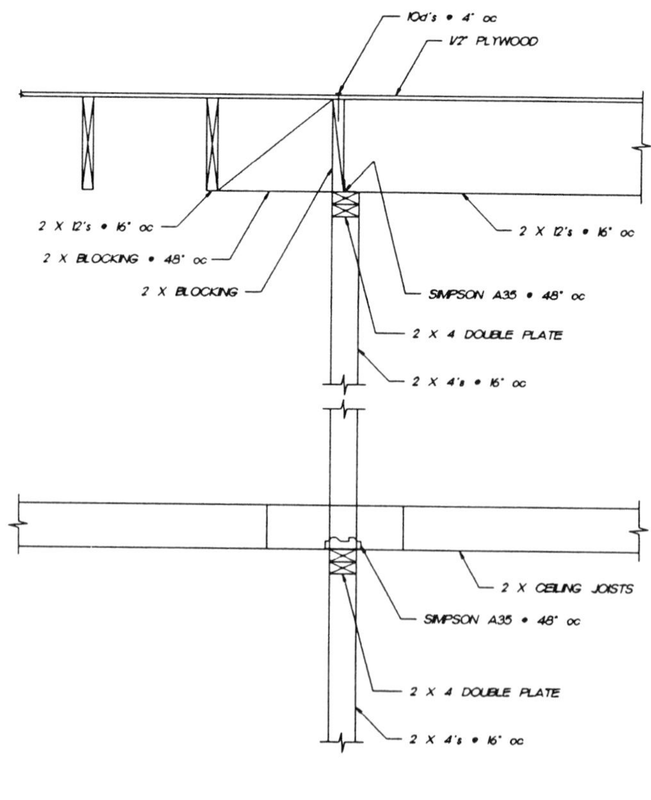
10d's • 4" oc
1/2" PLYWOOD
2 X 12's • 16" oc
2 X BLOCKING • 48" oc
2 X BLOCKING
2 X 12's • 16" oc
SIMPSON A35 • 48" oc
2 X 4 DOUBLE PLATE
2 X 4's • 16" oc
2 X CEILING JOISTS
SIMPSON A35 • 48" oc
2 X 4 DOUBLE PLATE
2 X 4's • 16" oc

W4660

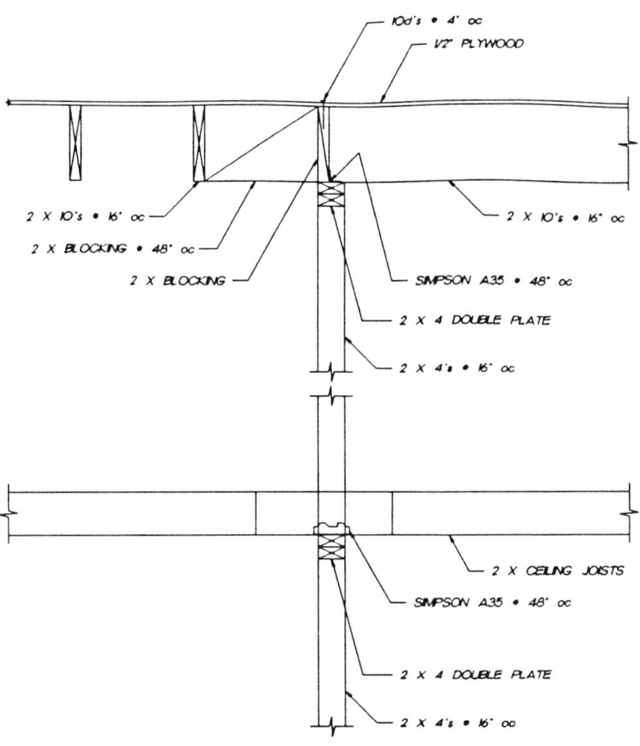
10d's @ 4" oc
1/2" PLYWOOD
2 X 10's @ 16" oc
2 X BLOCKING @ 48" oc
2 X BLOCKING
2 X 10's @ 16" oc
SIMPSON A35 @ 48" oc
2 X 4 DOUBLE PLATE
2 X 4's @ 16" oc
2 X CEILING JOISTS
SIMPSON A35 @ 48" oc
2 X 4 DOUBLE PLATE
2 X 4's @ 16" oc

W4661

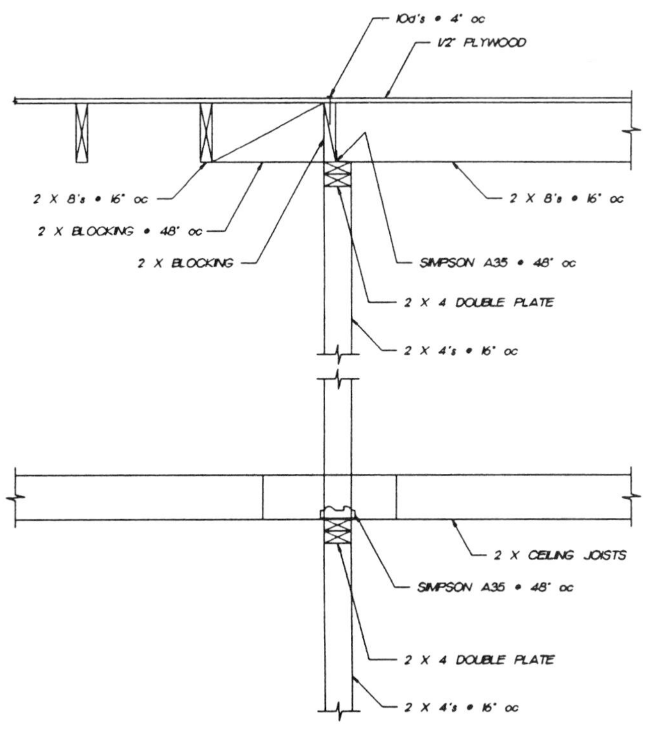
10d's @ 4" oc
1/2" PLYWOOD
2 X 8's @ 16" oc
2 X BLOCKING @ 48" oc
2 X BLOCKING
2 X 8's @ 16" oc
SIMPSON A35 @ 48" oc
2 X 4 DOUBLE PLATE
2 X 4's @ 16" oc
2 X CEILING JOISTS
SIMPSON A35 @ 48" oc
2 X 4 DOUBLE PLATE
2 X 4's @ 16" oc

W4662

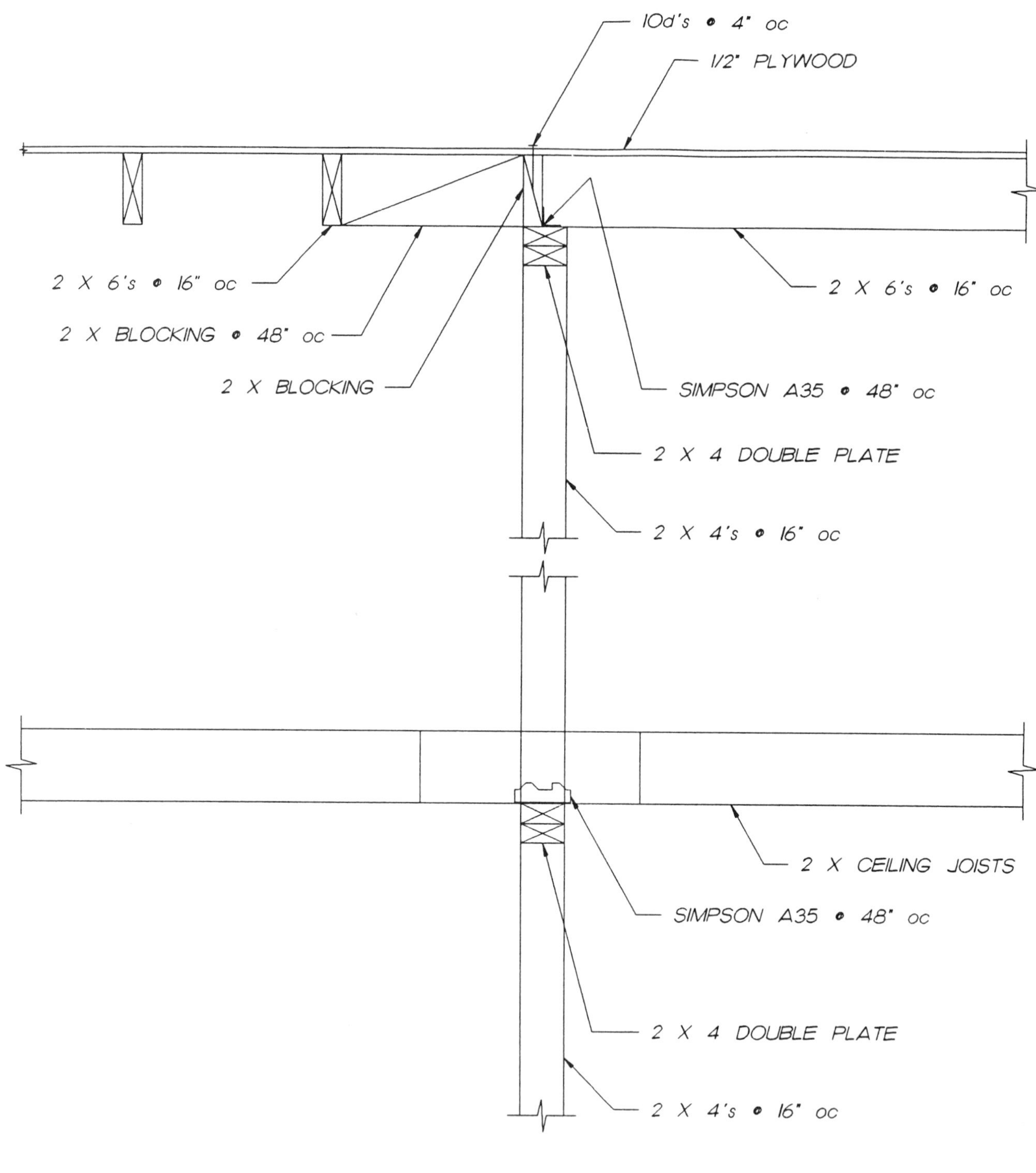
10d's @ 4" oc
1/2" PLYWOOD
2 X 6's @ 16" oc
2 X BLOCKING @ 48" oc
2 X BLOCKING
2 X 6's @ 16" oc
SIMPSON A35 @ 48" oc
2 X 4 DOUBLE PLATE
2 X 4's @ 16" oc
2 X CEILING JOISTS
SIMPSON A35 @ 48" oc
2 X 4 DOUBLE PLATE
2 X 4's @ 16" oc

W4663

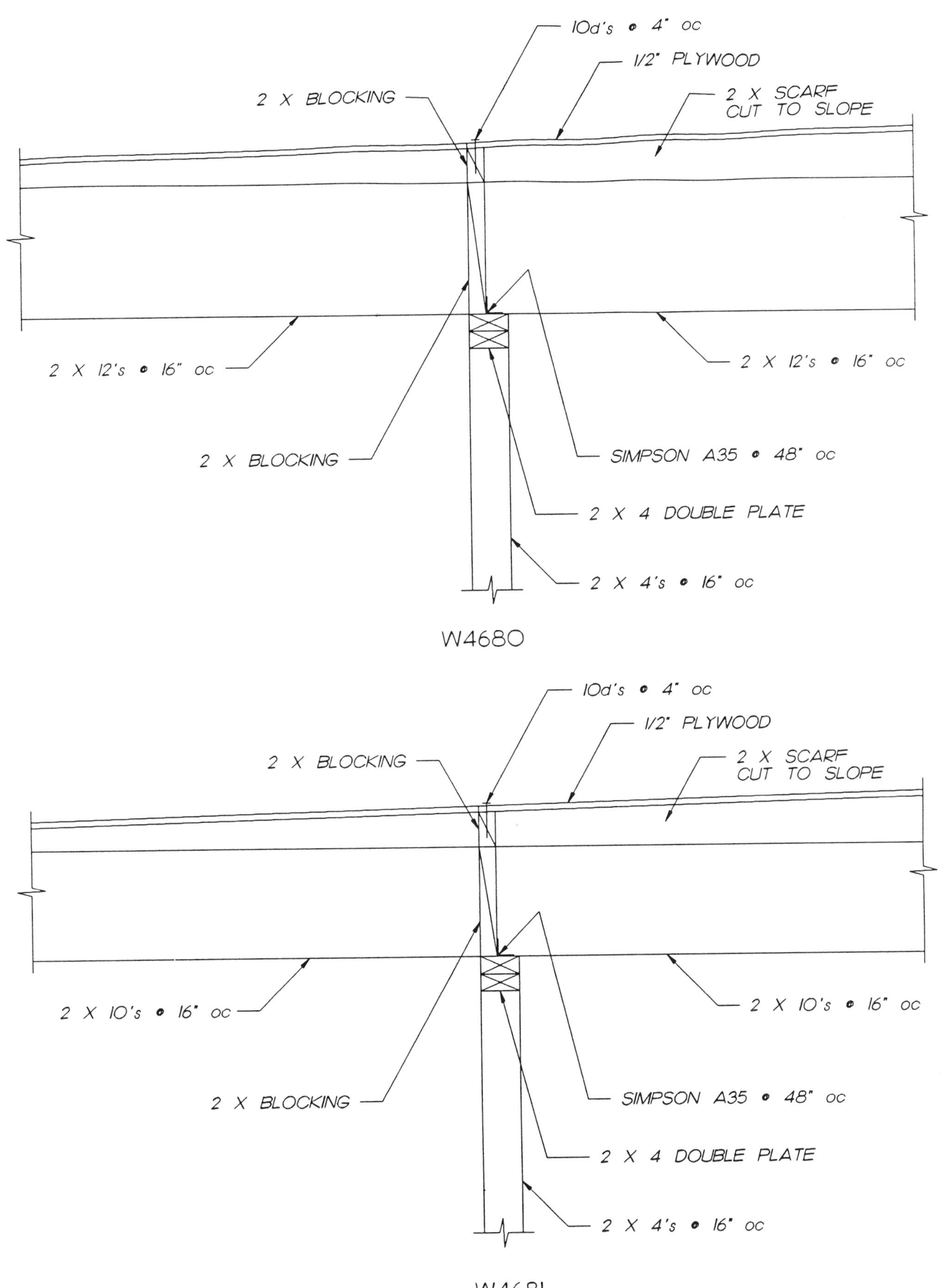
10d's @ 4" oc
1/2" PLYWOOD
2 X BLOCKING
2 X SCARF
CUT TO SLOPE
2 X 12's @ 16" oc
2 X 12's @ 16" oc
2 X BLOCKING
SIMPSON A35 @ 48" oc
2 X 4 DOUBLE PLATE
2 X 4's @ 16" oc
W4680
10d's @ 4" oc
1/2" PLYWOOD
2 X BLOCKING
2 X SCARF
CUT TO SLOPE
2 X 10's @ 16" oc
2 X 10's @ 16" oc
2 X BLOCKING
SIMPSON A35 @ 48" oc
2 X 4 DOUBLE PLATE
2 X 4's @ 16" oc
W4681

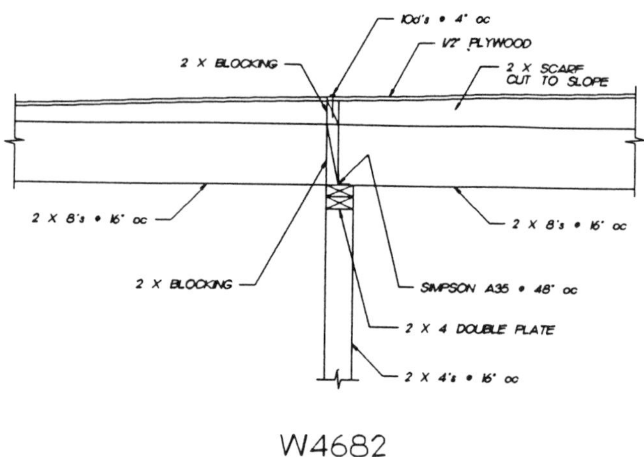

W4682

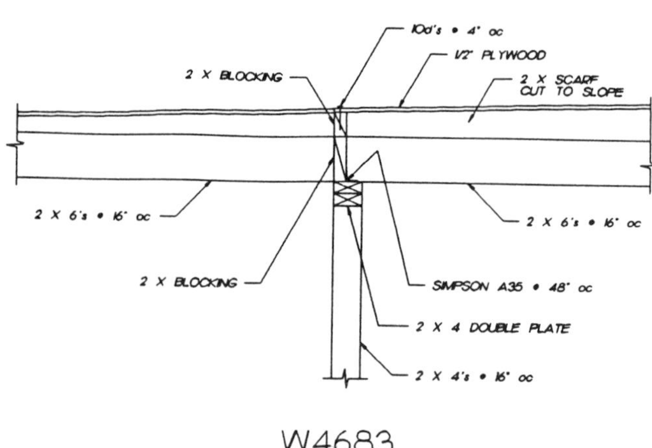

W4683

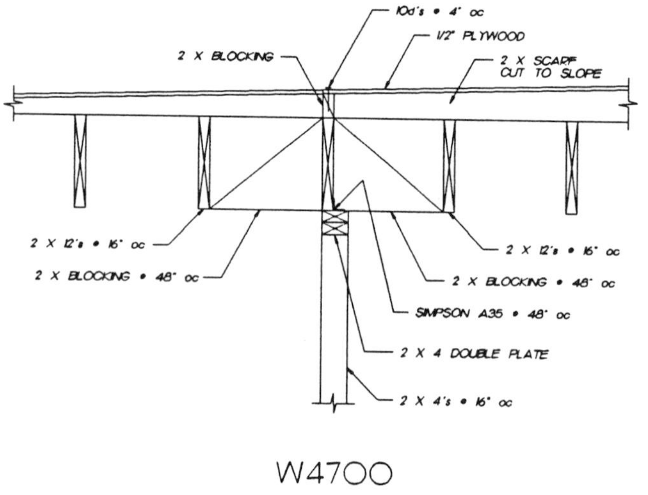

W4700

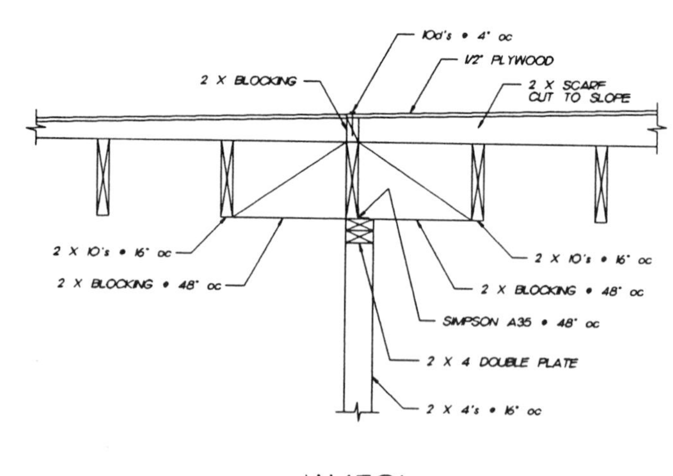

W4701

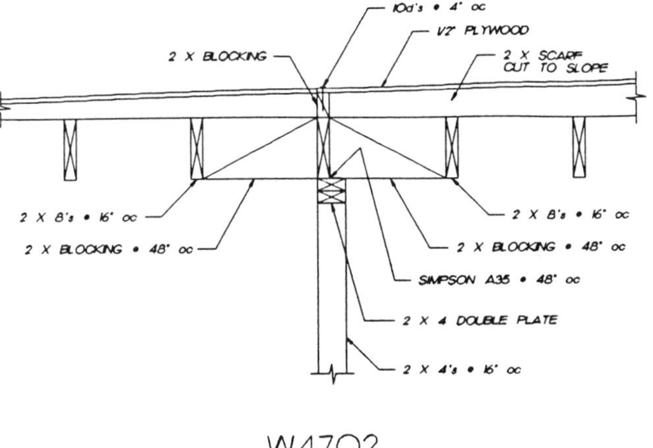

W4702

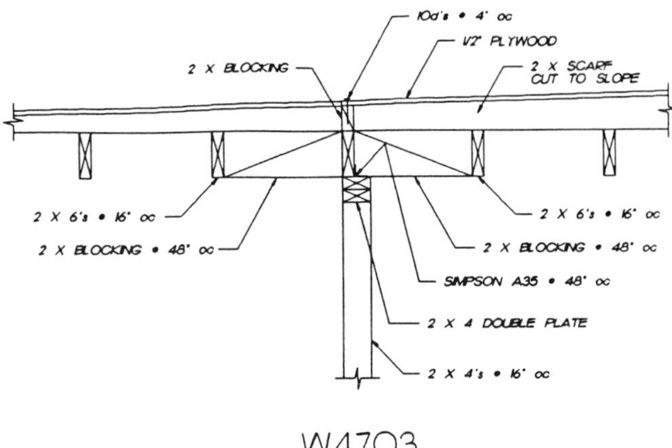

W4703

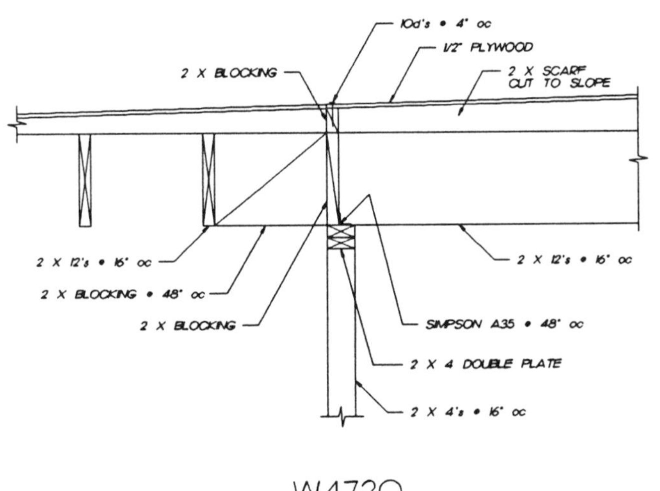

W4720

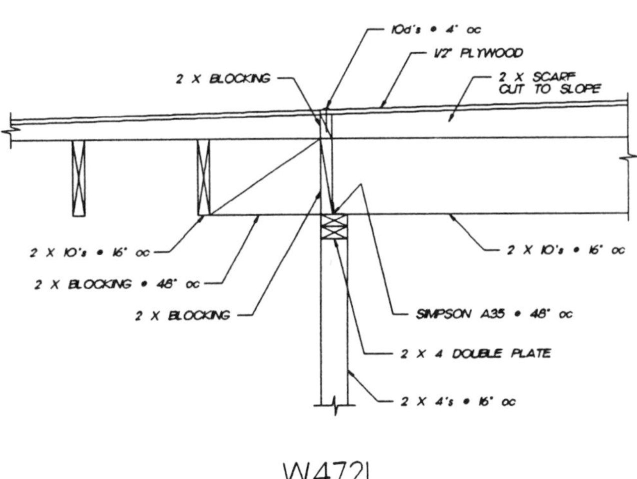

W4721

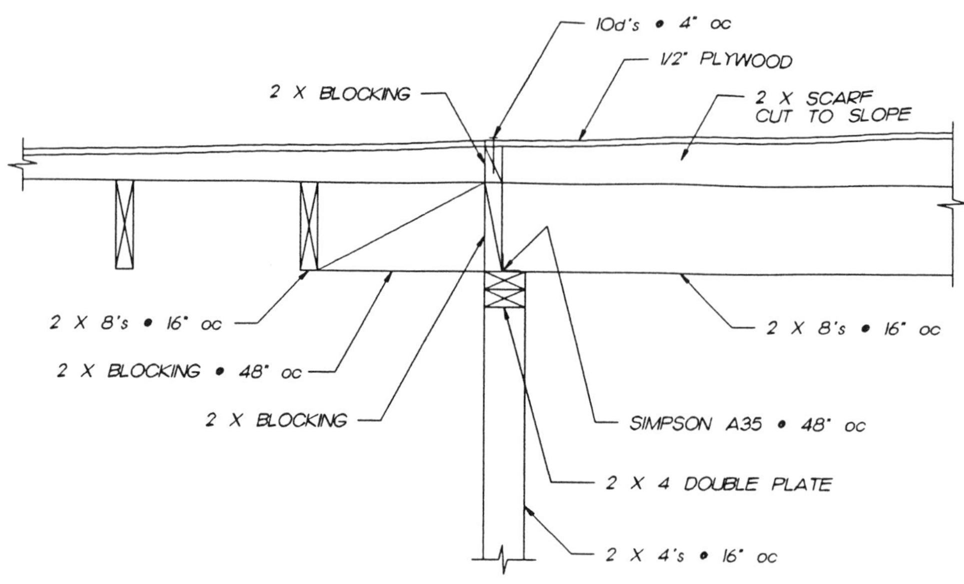
10d's • 4" oc
1/2" PLYWOOD
2 X BLOCKING
2 X SCARF
CUT TO SLOPE
2 X 8's • 16" oc
2 X BLOCKING • 48" oc
2 X BLOCKING
2 X 8's • 16" oc
SIMPSON A35 • 48" oc
2 X 4 DOUBLE PLATE
2 X 4's • 16" oc

W4722

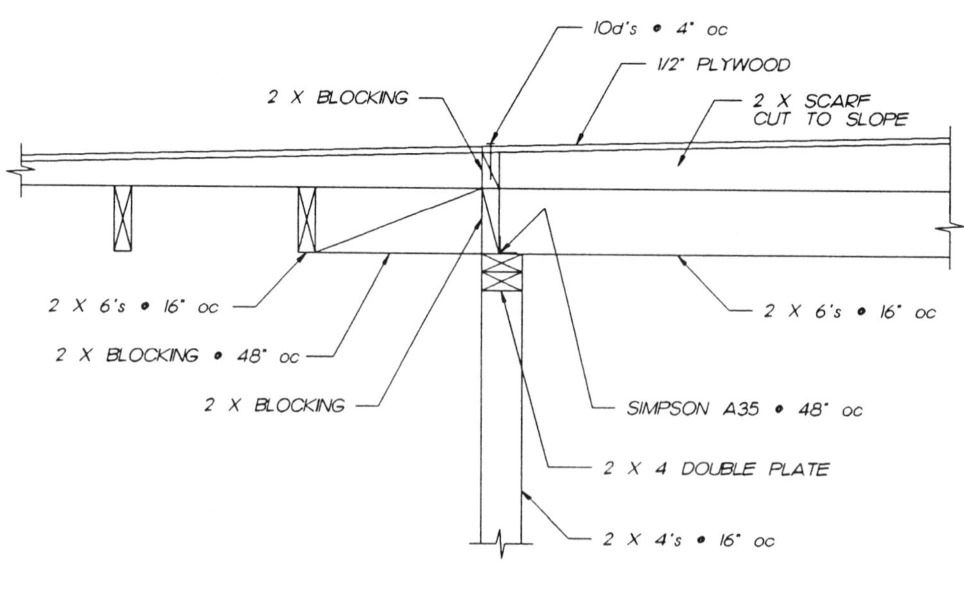
10d's • 4" oc
1/2" PLYWOOD
2 X BLOCKING
2 X SCARF
CUT TO SLOPE
2 X 6's • 16" oc
2 X BLOCKING • 48" oc
2 X BLOCKING
2 X 6's • 16" oc
SIMPSON A35 • 48" oc
2 X 4 DOUBLE PLATE
2 X 4's • 16" oc

W4723

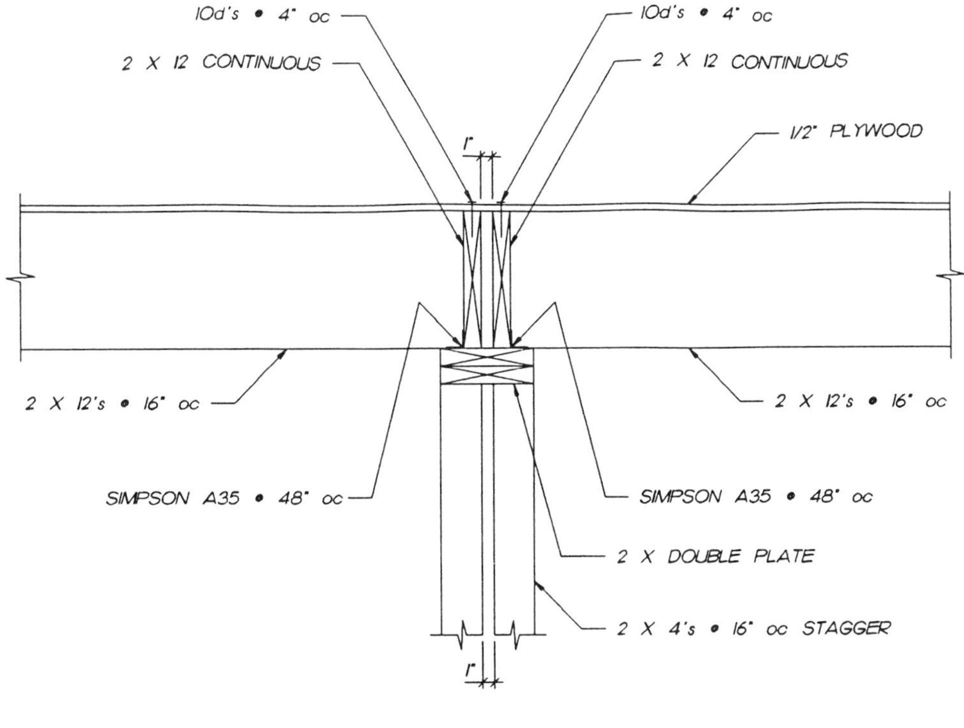
10d's • 4" oc
10d's • 4" oc
2 X 12 CONTINUOUS
2 X 12 CONTINUOUS
1/2" PLYWOOD
1"
2 X 12's • 16" oc
2 X 12's • 16" oc
SIMPSON A35 • 48" oc
SIMPSON A35 • 48" oc
2 X DOUBLE PLATE
2 X 4's • 16" oc STAGGER
1"

W4740

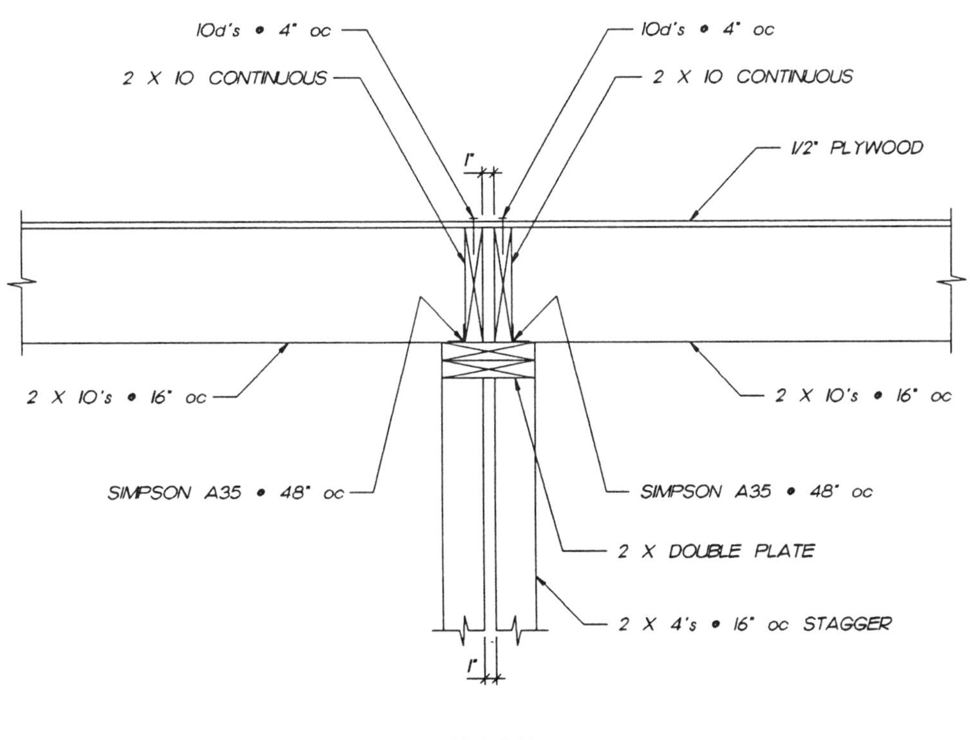
10d's • 4" oc
10d's • 4" oc
2 X 10 CONTINUOUS
2 X 10 CONTINUOUS
1/2" PLYWOOD
1"
2 X 10's • 16" oc
2 X 10's • 16" oc
SIMPSON A35 • 48" oc
SIMPSON A35 • 48" oc
2 X DOUBLE PLATE
2 X 4's • 16" oc STAGGER
1"

W4741

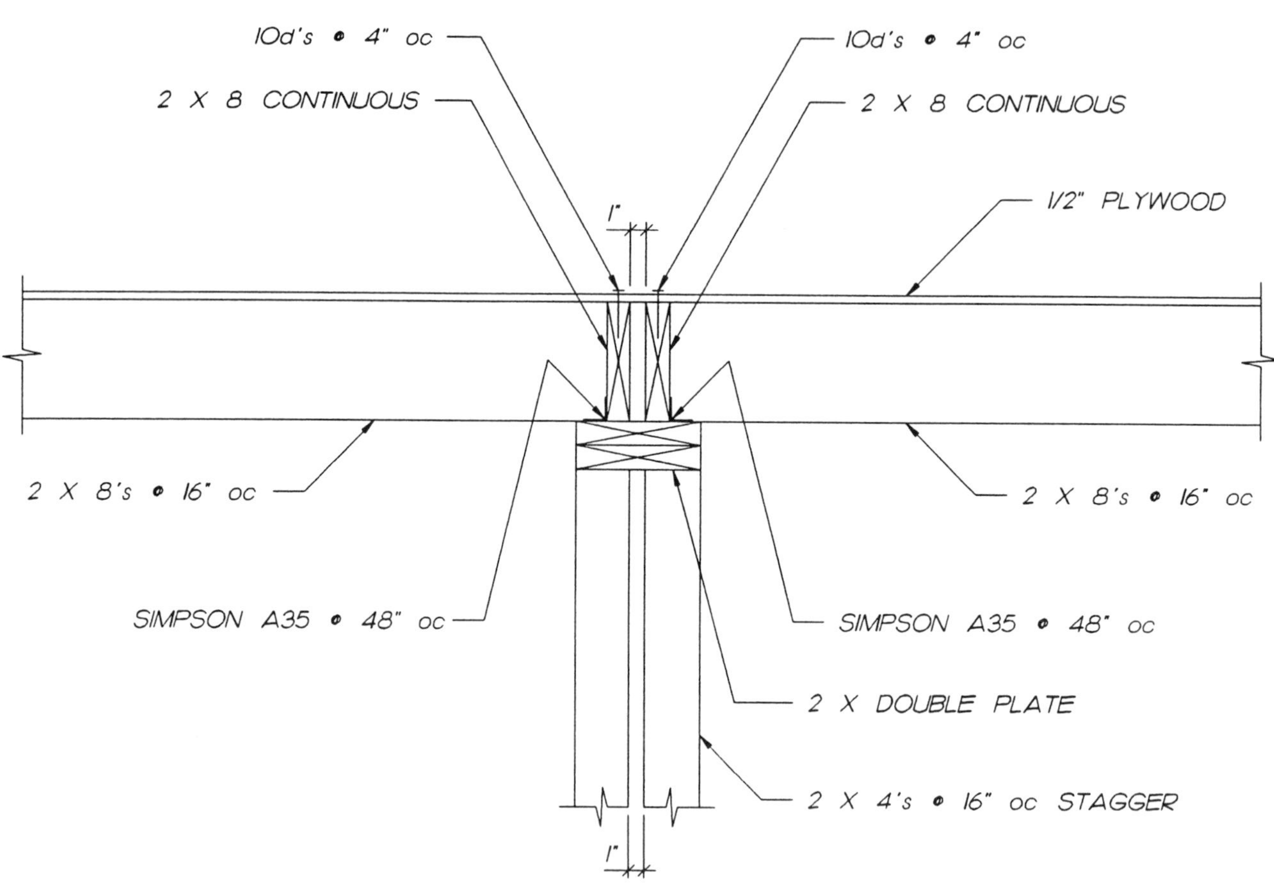

10d's @ 4" oc
2 X 8 CONTINUOUS
10d's @ 4" oc
2 X 8 CONTINUOUS
1/2" PLYWOOD
1"
2 X 8's @ 16" oc
2 X 8's @ 16" oc
SIMPSON A35 @ 48" oc
SIMPSON A35 @ 48" oc
2 X DOUBLE PLATE
2 X 4's @ 16" oc STAGGER
1"

W4742

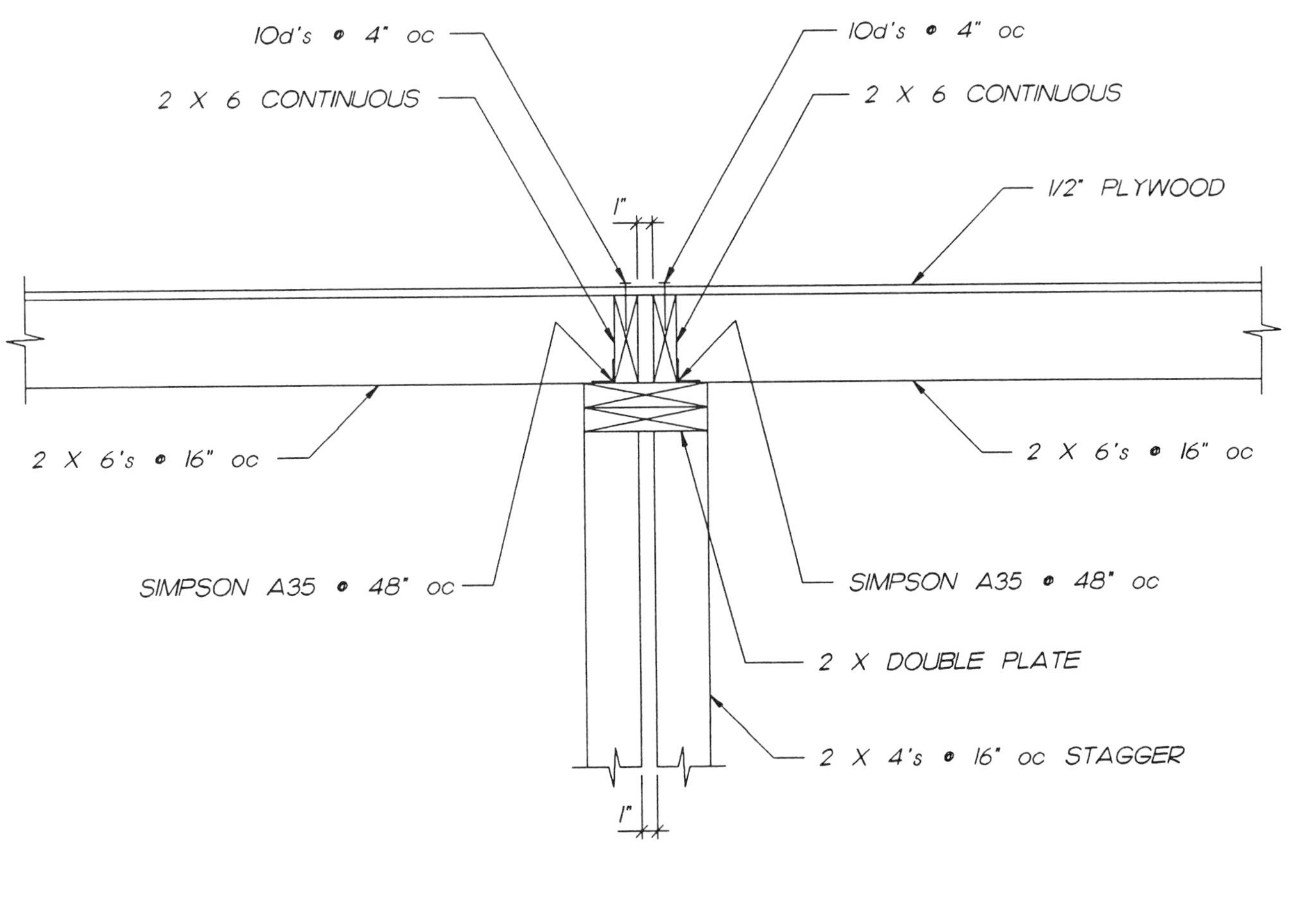

10d's @ 4" oc
10d's @ 4" oc
2 X 6 CONTINUOUS
2 X 6 CONTINUOUS
1/2" PLYWOOD
1"
2 X 6's @ 16" oc
2 X 6's @ 16" oc
SIMPSON A35 @ 48" oc
SIMPSON A35 @ 48" oc
2 X DOUBLE PLATE
2 X 4's @ 16" oc STAGGER
1"

W4743

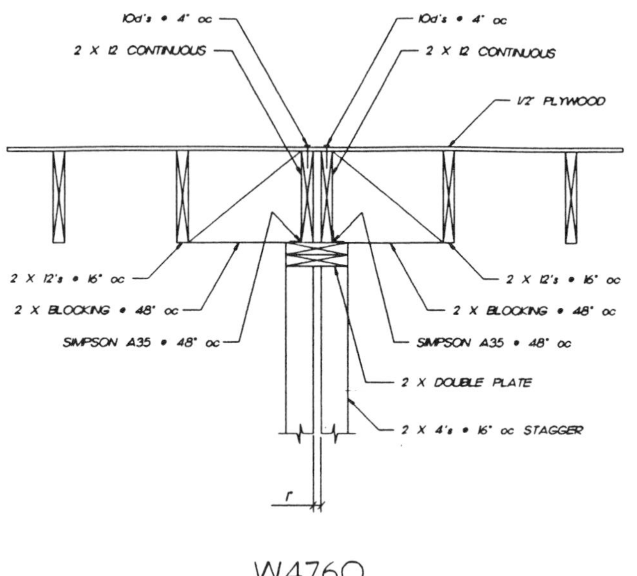

W4760

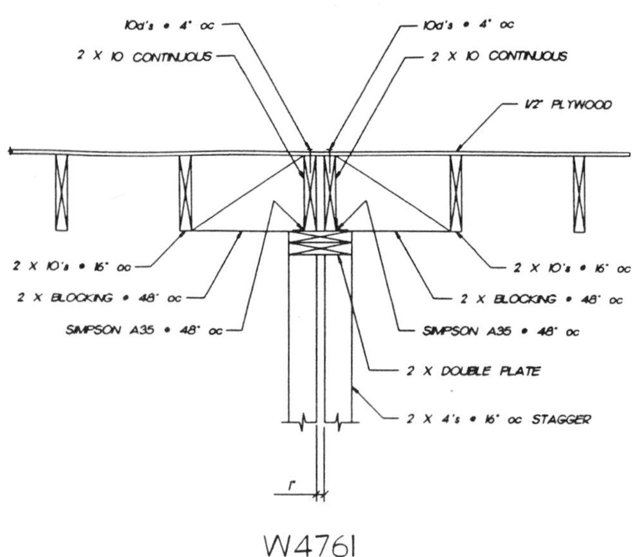

W4761

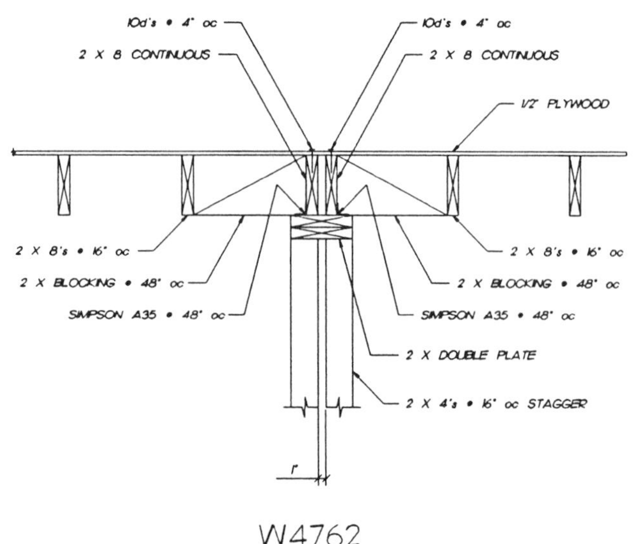

W4762

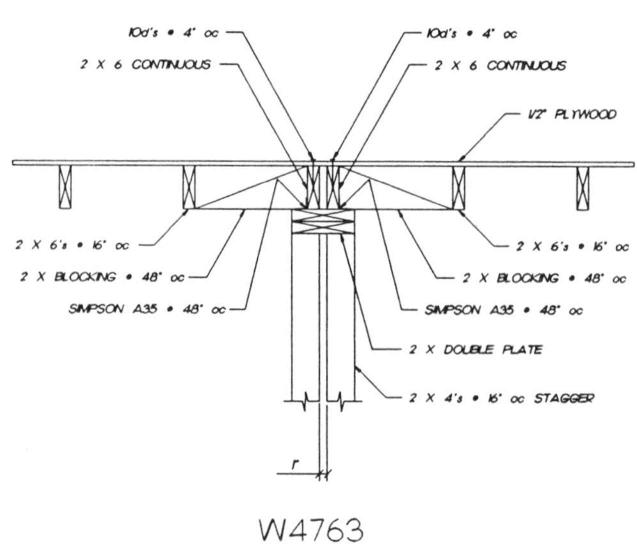

W4763

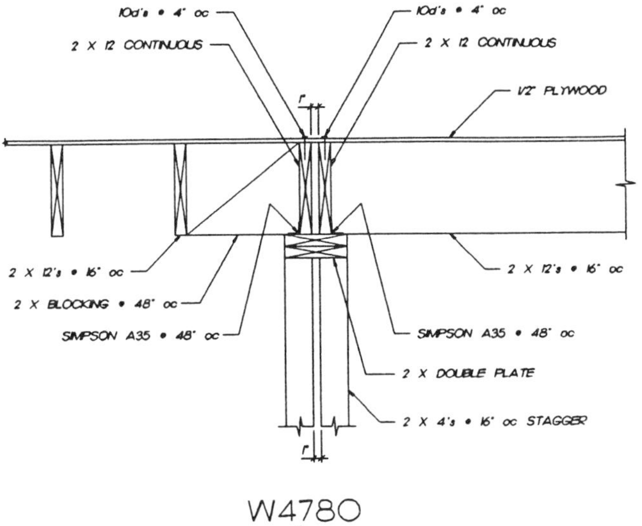

W4780

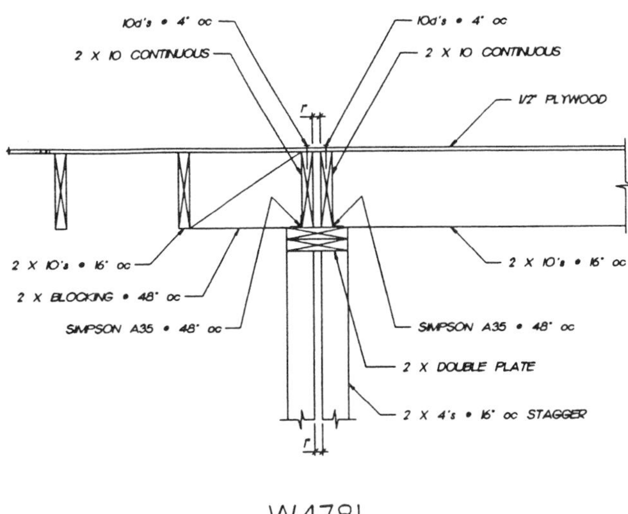

W4781

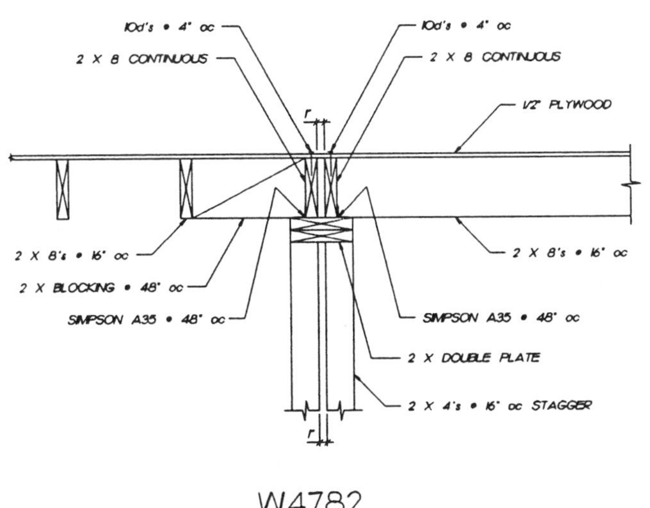

W4782

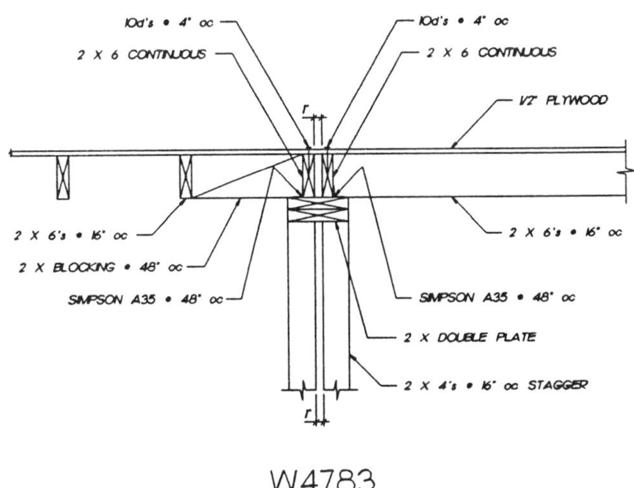

W4783

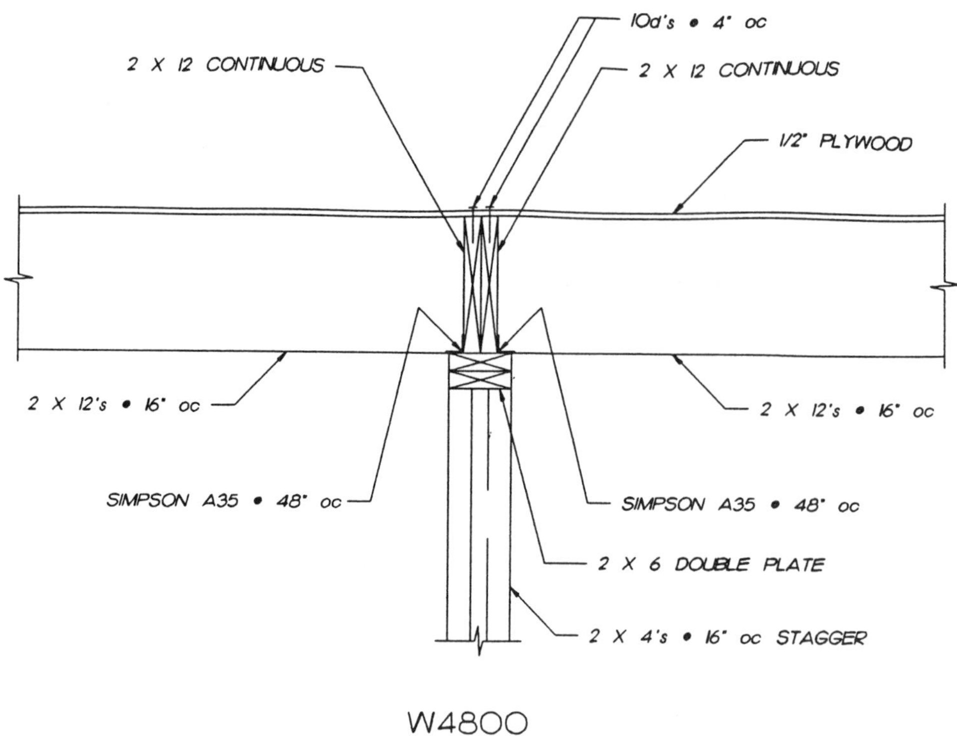
10d's • 4" oc
2 X 12 CONTINUOUS
2 X 12 CONTINUOUS
1/2" PLYWOOD
2 X 12's • 16" oc
2 X 12's • 16" oc
SIMPSON A35 • 48" oc
SIMPSON A35 • 48" oc
2 X 6 DOUBLE PLATE
2 X 4's • 16" oc STAGGER

W4800

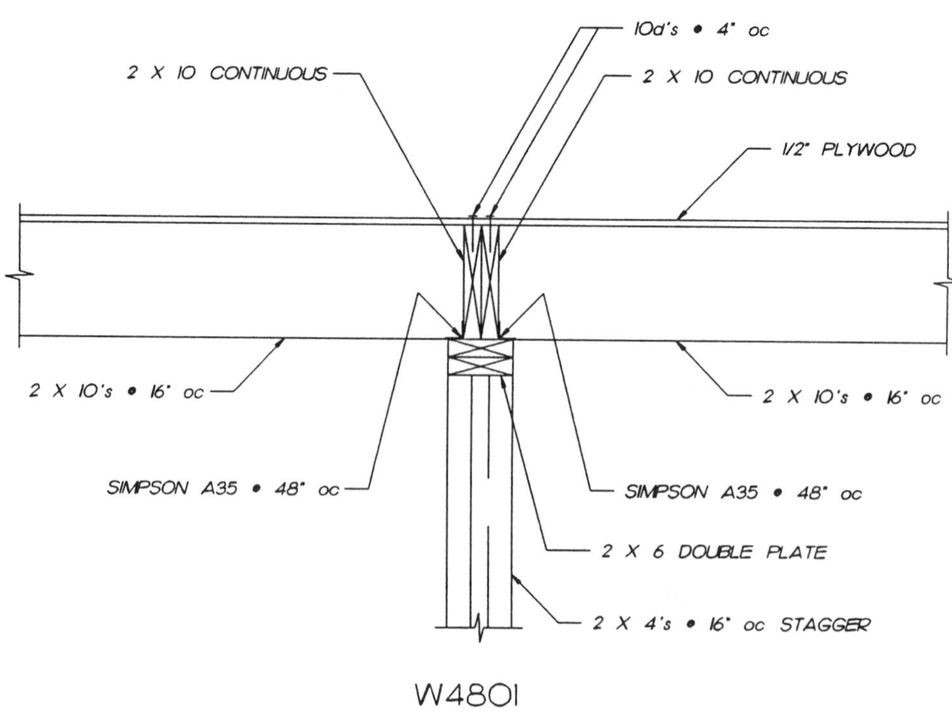
10d's • 4" oc
2 X 10 CONTINUOUS
2 X 10 CONTINUOUS
1/2" PLYWOOD
2 X 10's • 16" oc
2 X 10's • 16" oc
SIMPSON A35 • 48" oc
SIMPSON A35 • 48" oc
2 X 6 DOUBLE PLATE
2 X 4's • 16" oc STAGGER

W4801

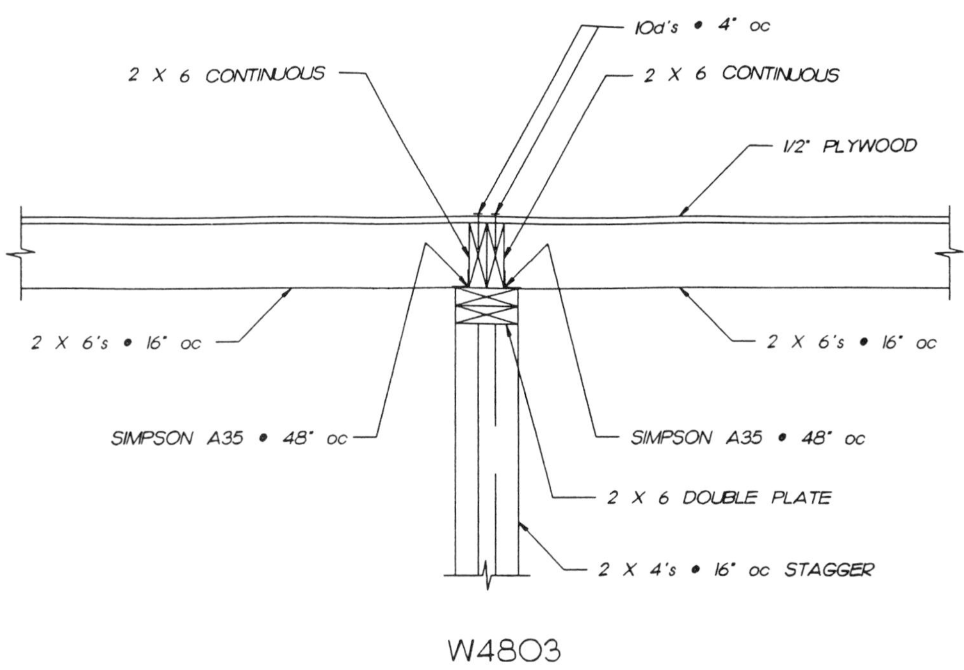
10d's • 4" oc
2 X 6 CONTINUOUS
2 X 6 CONTINUOUS
1/2" PLYWOOD
2 X 6's • 16" oc
2 X 6's • 16" oc
SIMPSON A35 • 48" oc
SIMPSON A35 • 48" oc
2 X 6 DOUBLE PLATE
2 X 4's • 16" oc STAGGER

W4803

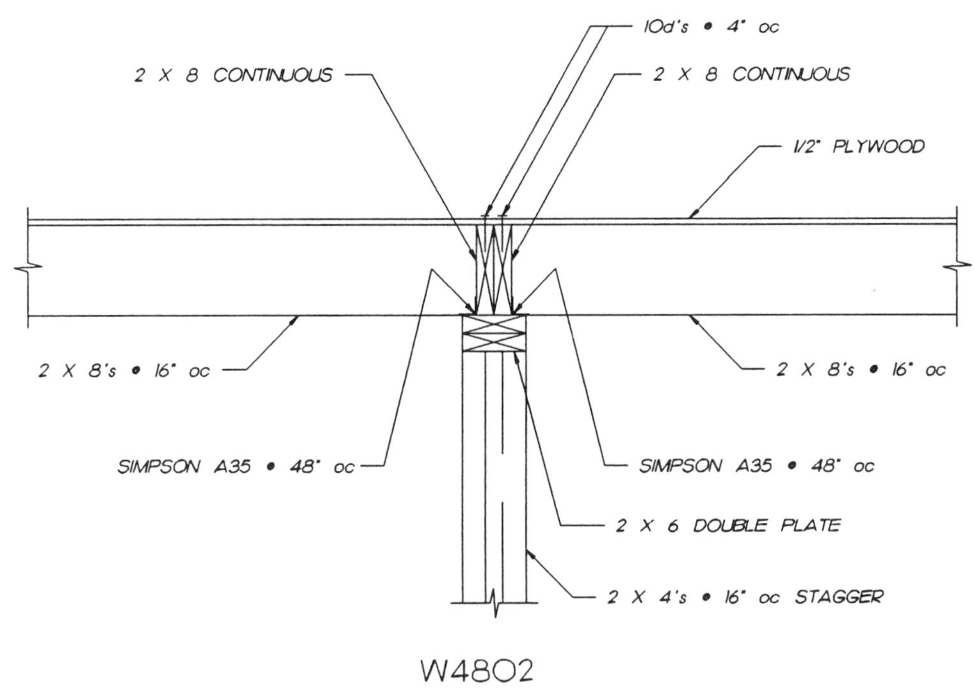
10d's • 4" oc
2 X 8 CONTINUOUS
2 X 8 CONTINUOUS
1/2" PLYWOOD
2 X 8's • 16" oc
2 X 8's • 16" oc
SIMPSON A35 • 48" oc
SIMPSON A35 • 48" oc
2 X 6 DOUBLE PLATE
2 X 4's • 16" oc STAGGER

W4802

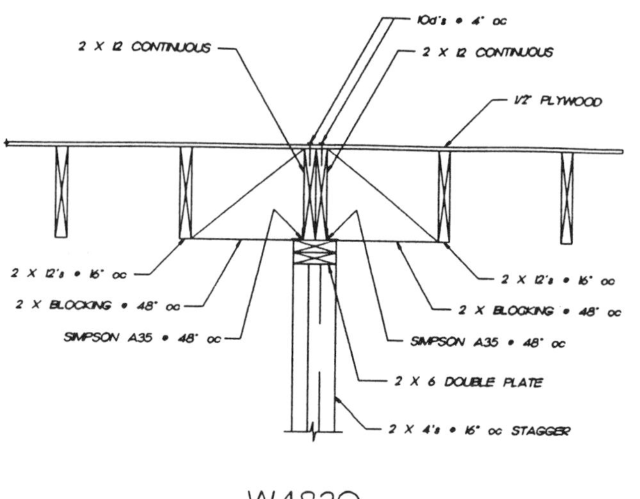

W4820

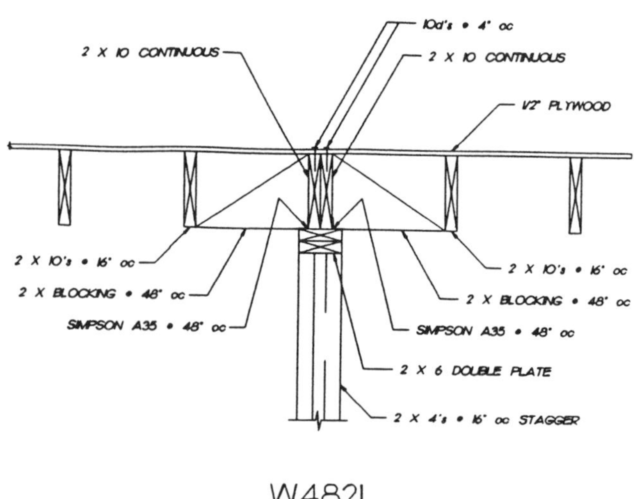

W4821

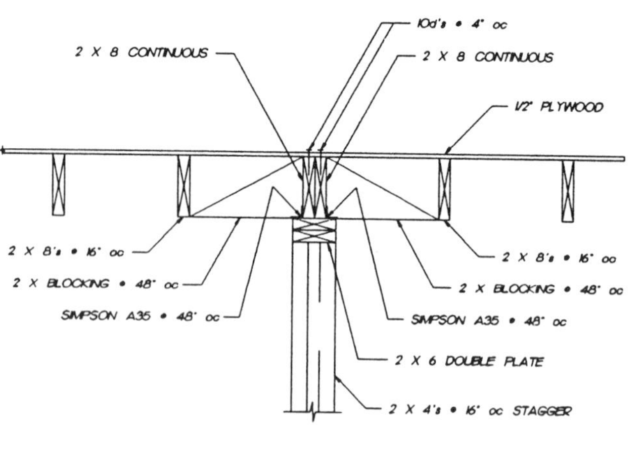

W4822

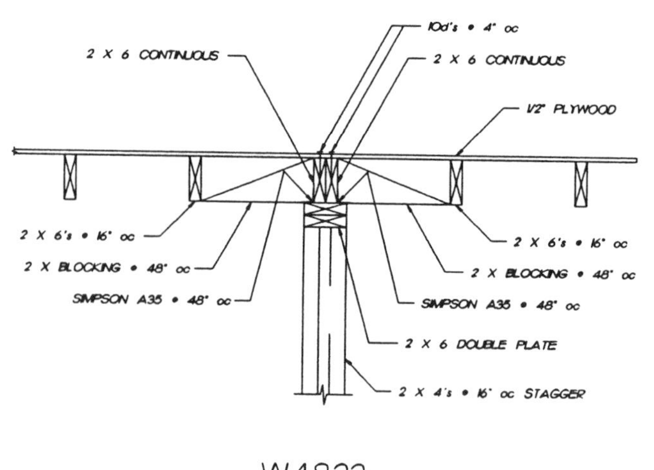

W4823

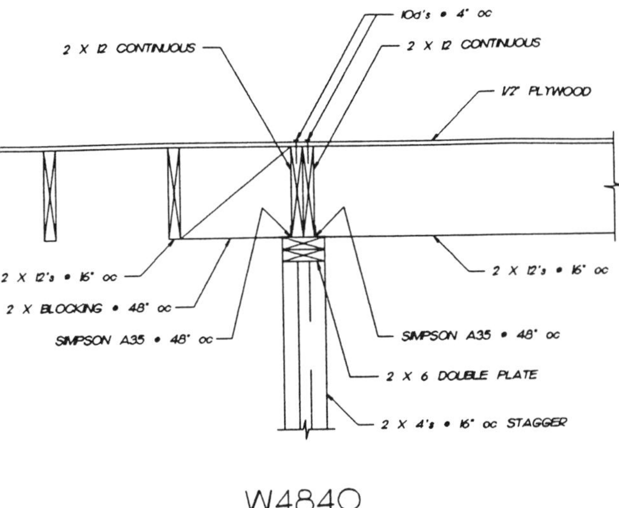

W4840

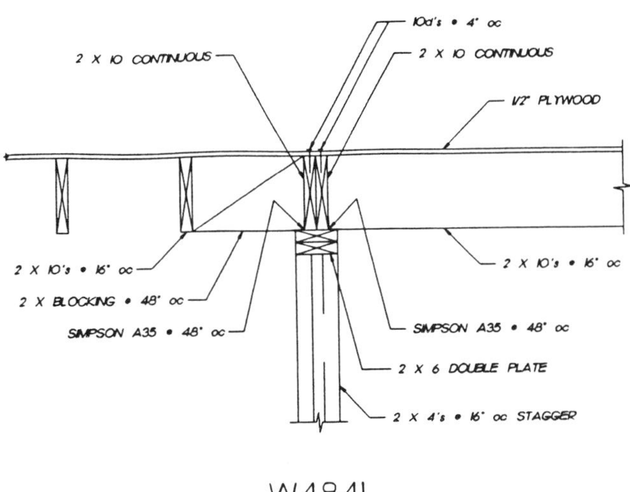

W4841

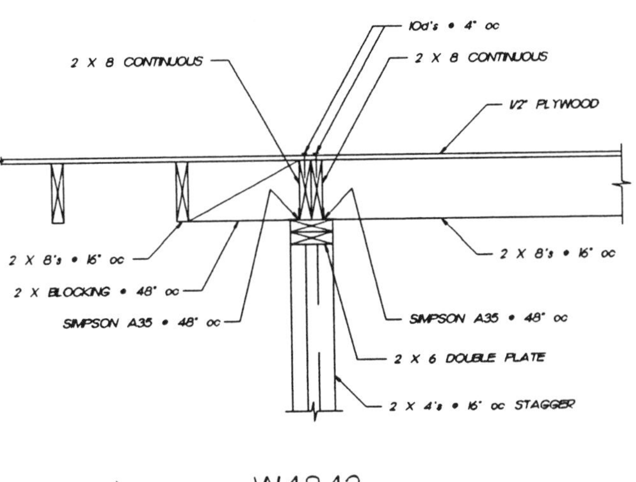

W4842

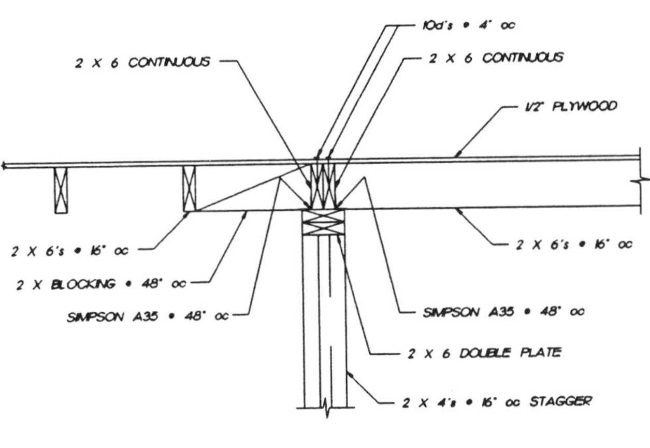

W4843

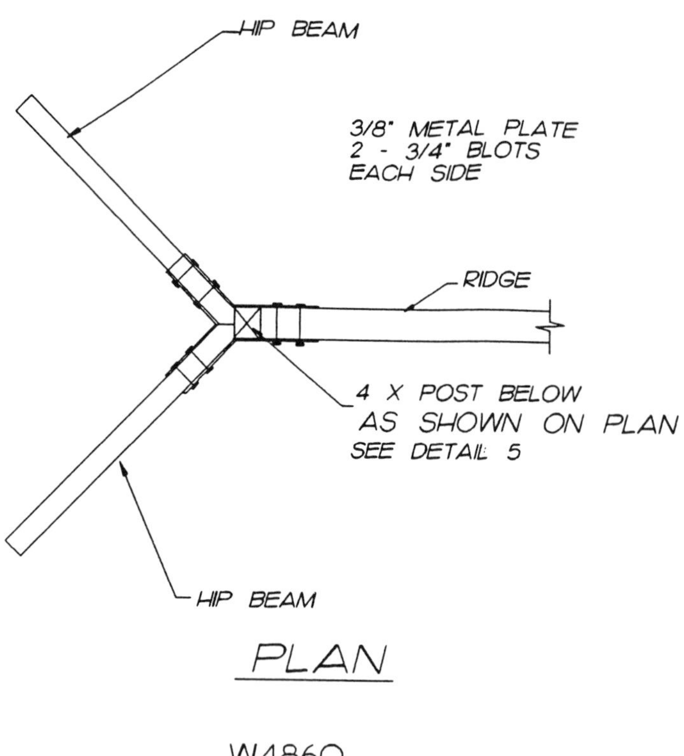
HIP BEAM
3/8" METAL PLATE
2 - 3/4" BLOTS
EACH SIDE
RIDGE
4 X POST BELOW
AS SHOWN ON PLAN
SEE DETAIL 5
HIP BEAM
PLAN

W4860

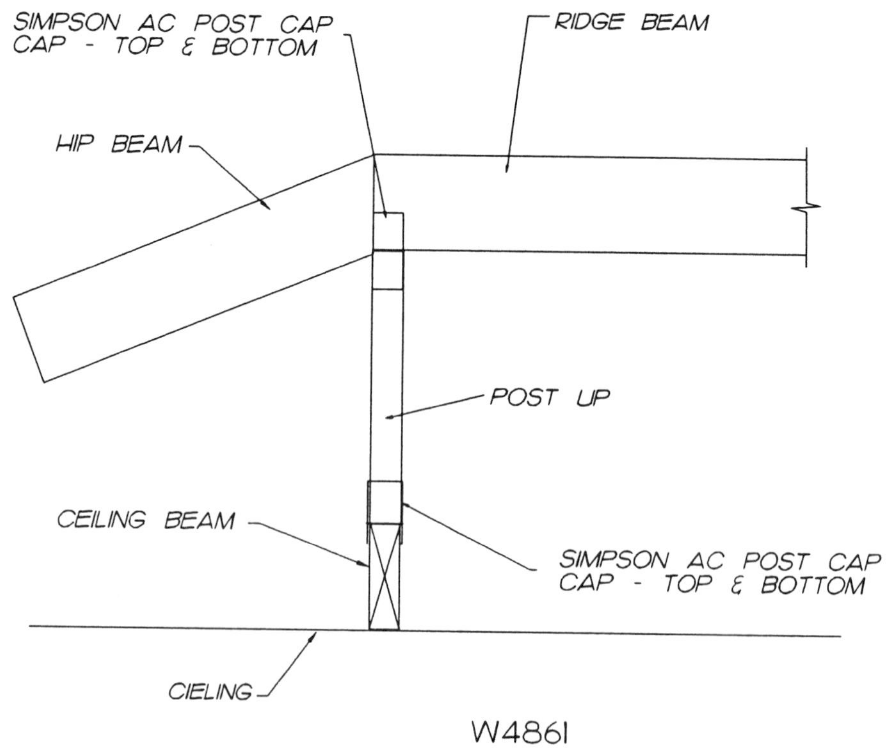
SIMPSON AC POST CAP
CAP - TOP & BOTTOM
RIDGE BEAM
HIP BEAM
POST UP
CEILING BEAM
SIMPSON AC POST CAP
CAP - TOP & BOTTOM
CIELING

W4861

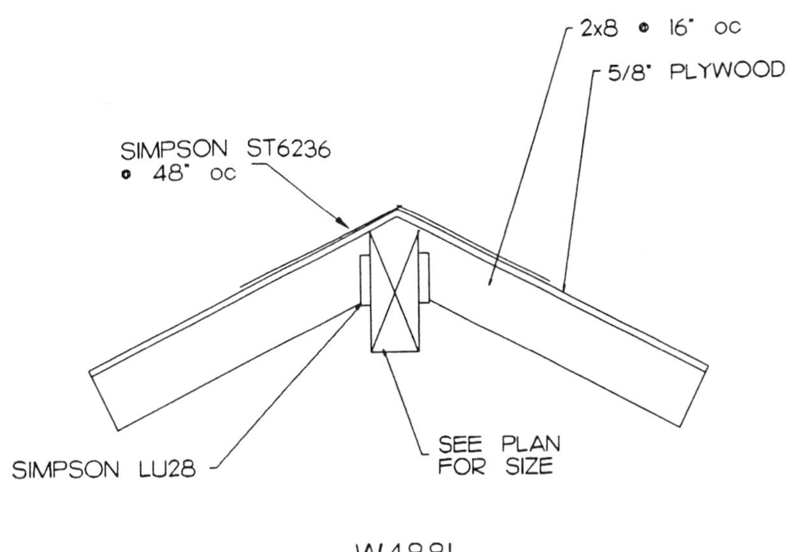
2x8 @ 16" OC
5/8" PLYWOOD
SIMPSON ST6236
@ 48" OC
SIMPSON LU28
SEE PLAN
FOR SIZE

W4881

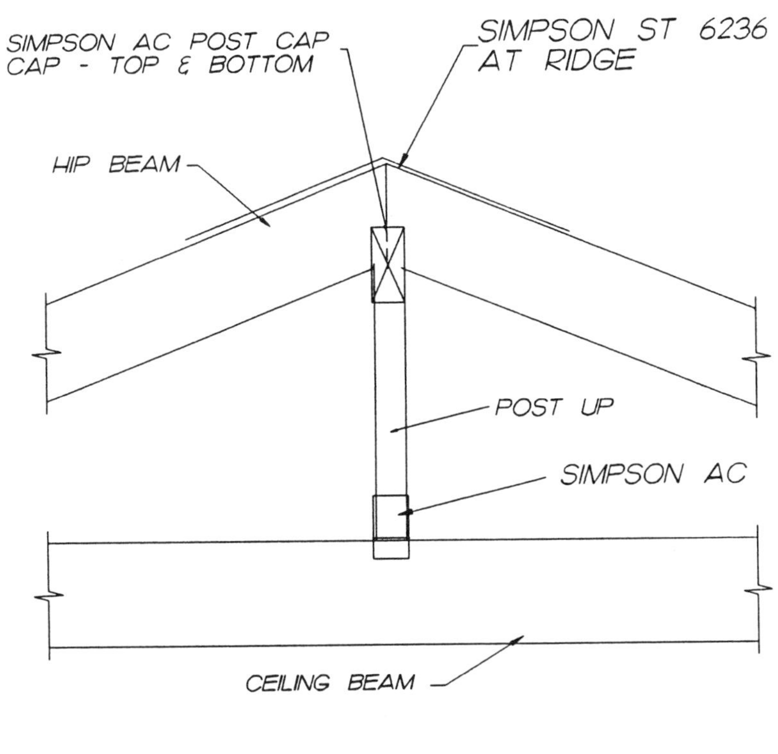
SIMPSON AC POST CAP
CAP - TOP & BOTTOM
SIMPSON ST 6236
AT RIDGE
HIP BEAM
POST UP
SIMPSON AC
CEILING BEAM

W4880

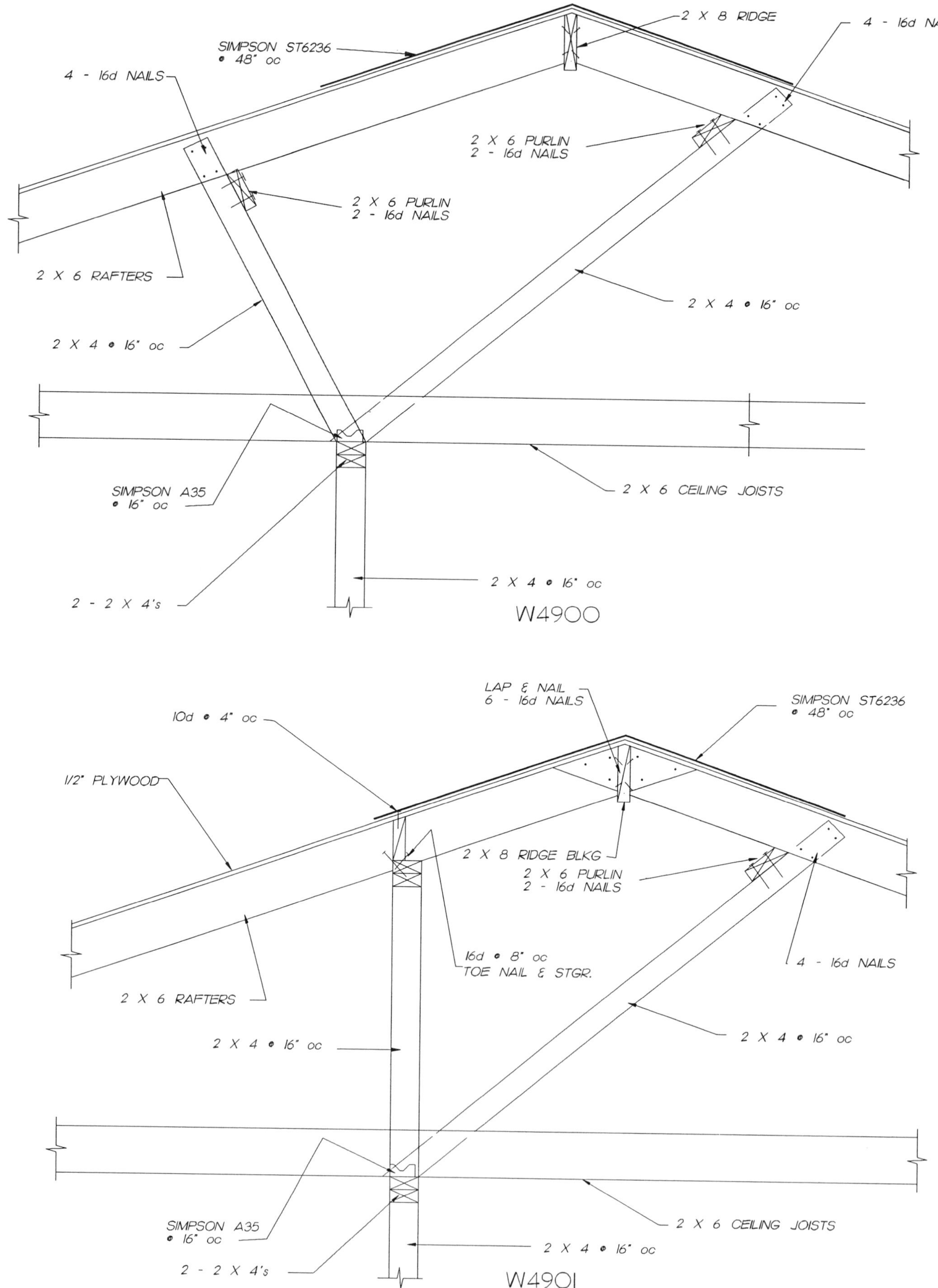

SIMPSON ST6236
@ 48" oc
2 X 8 RIDGE
4 - 16d NAIL
4 - 16d NAILS
2 X 6 PURLIN
2 - 16d NAILS
2 X 6 PURLIN
2 - 16d NAILS
2 X 6 RAFTERS
2 X 4 @ 16" oc
2 X 4 @ 16" oc
SIMPSON A35
@ 16" oc
2 X 6 CEILING JOISTS
2 X 4 @ 16" oc
2 - 2 X 4's
W4900
LAP & NAIL
6 - 16d NAILS
SIMPSON ST6236
@ 48" oc
10d @ 4" oc
1/2" PLYWOOD
2 X 8 RIDGE BLKG
2 X 6 PURLIN
2 - 16d NAILS
16d @ 8" oc
TOE NAIL & STGR.
4 - 16d NAILS
2 X 6 RAFTERS
2 X 4 @ 16" oc
2 X 4 @ 16" oc
SIMPSON A35
@ 16" oc
2 X 6 CEILING JOISTS
2 X 4 @ 16" oc
2 - 2 X 4's
W4901

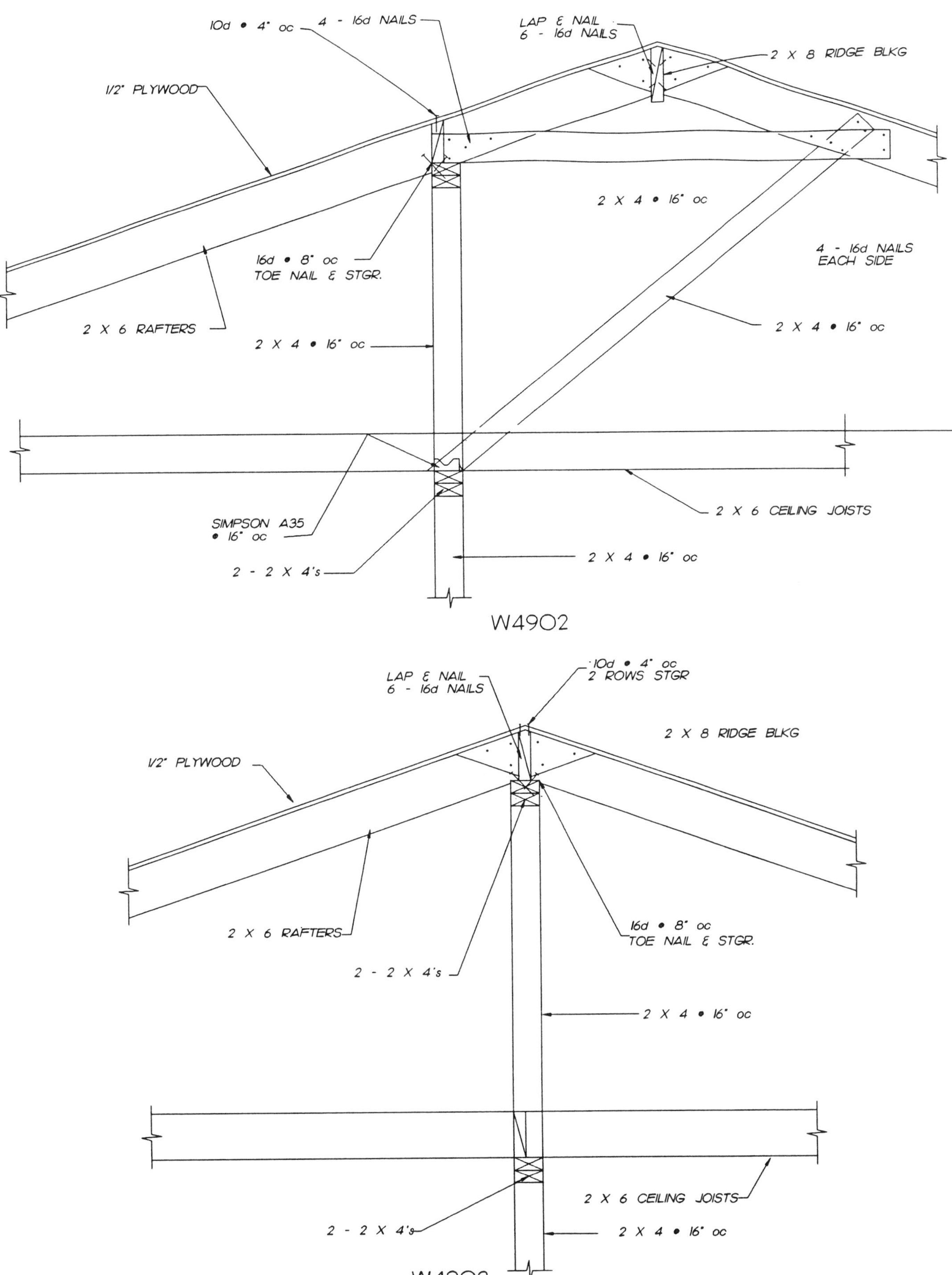
10d • 4" oc
4 - 16d NAILS
LAP & NAIL
6 - 16d NAILS
2 X 8 RIDGE BLKG
1/2" PLYWOOD
2 X 4 • 16" oc
4 - 16d NAILS
EACH SIDE
16d • 8" oc
TOE NAIL & STGR.
2 X 6 RAFTERS
2 X 4 • 16" oc
2 X 4 • 16" oc
SIMPSON A35
• 16" oc
2 X 6 CEILING JOISTS
2 X 4 • 16" oc
2 - 2 X 4's
W4902
10d • 4" oc
2 ROWS STGR
LAP & NAIL
6 - 16d NAILS
2 X 8 RIDGE BLKG
1/2" PLYWOOD
2 X 6 RAFTERS
16d • 8" oc
TOE NAIL & STGR.
2 - 2 X 4's
2 X 4 • 16" oc
2 X 6 CEILING JOISTS
2 - 2 X 4's
2 X 4 • 16" oc
W4903

2 X PLANKS
SIMPSON A35
SIMPSON 35 • 24" oc
CONCRETE FILL
2 JOISTS • 12" oc
4 X PERIMETER BEAMS
6 X SUPPORT BEAMS
2 Ls 6" X 4" X 5/16"
3/4" DIA. M. BOLTS
• 24" oc 4X4 SHIM
2 X 4 STIFFENERS
2 X JOISTS

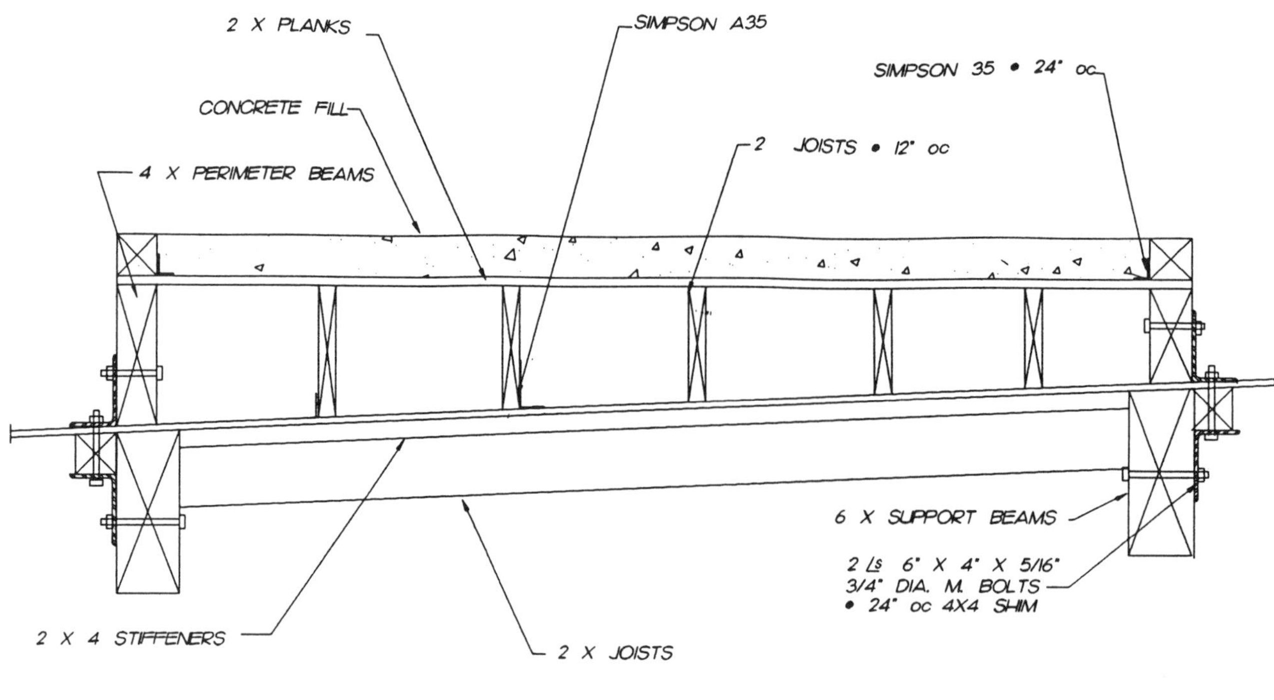

W4921

2 X PLANKS
SIMPSON A35
2 Ls 6" X 4" X 5/16"
3/4" DIA. M. BOLTS
• 24" oc 4X4 SHIM
CONCRETE FILL
2 X SHAPED
JOISTS • 12" oc
4 X PERIMETER BEAMS
SIMPSON A35 • 24" oc
6 X SUPPORT BEAMS
2 X JOISTS
2 X 4 STIFFENERS

W4920

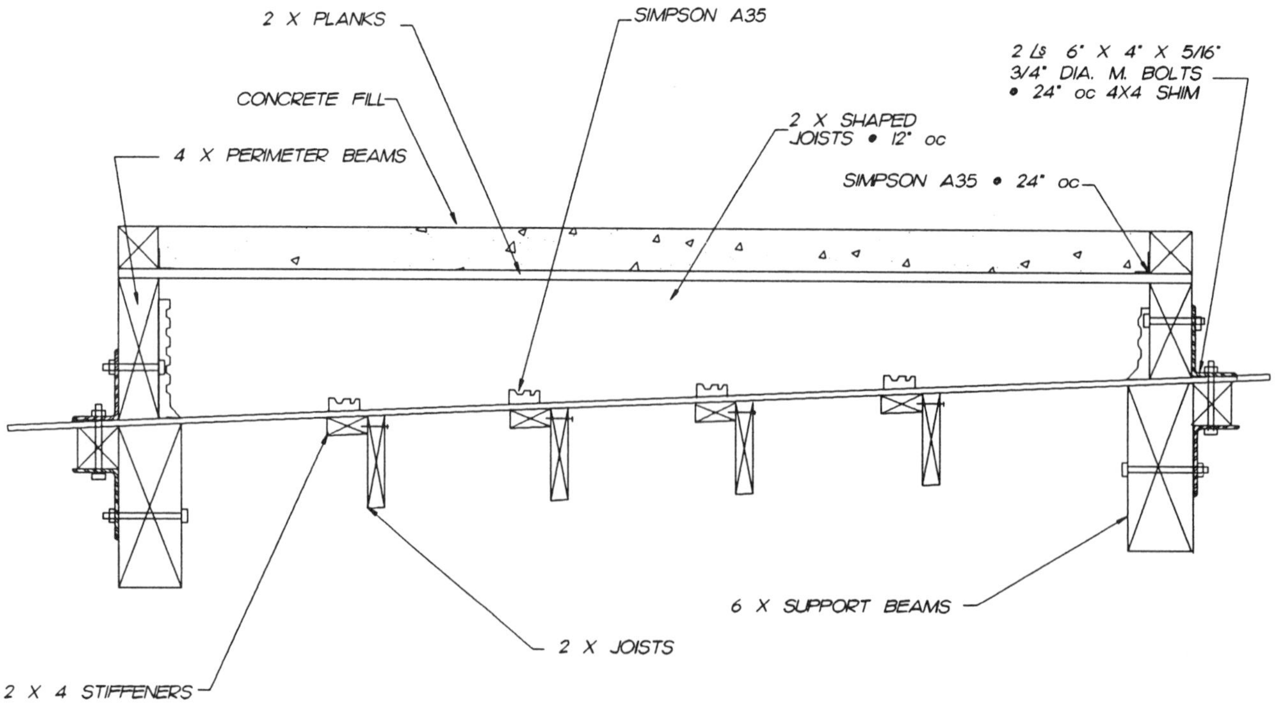

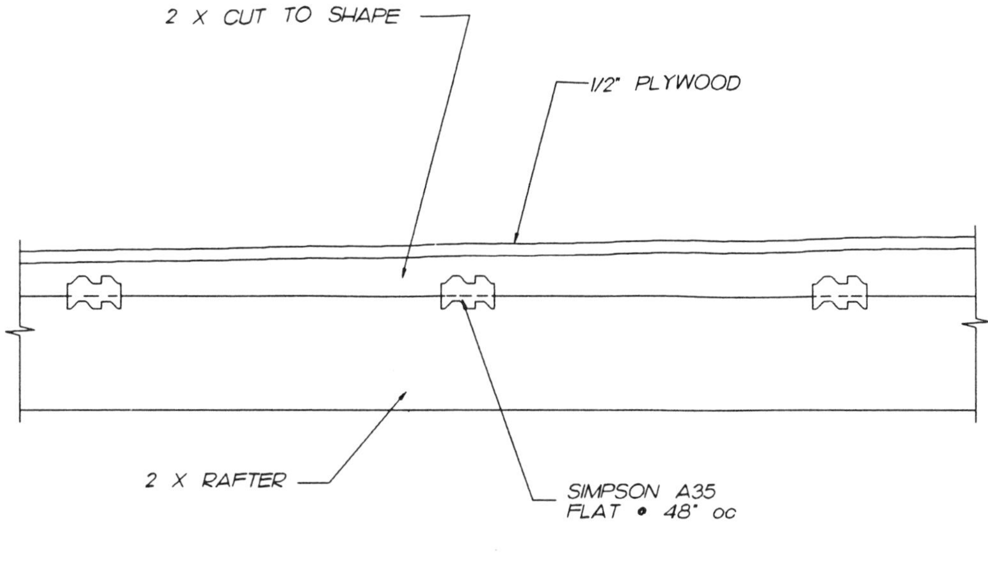
2 X CUT TO SHAPE
1/2" PLYWOOD
2 X RAFTER
SIMPSON A35
FLAT @ 48" oc

W4940

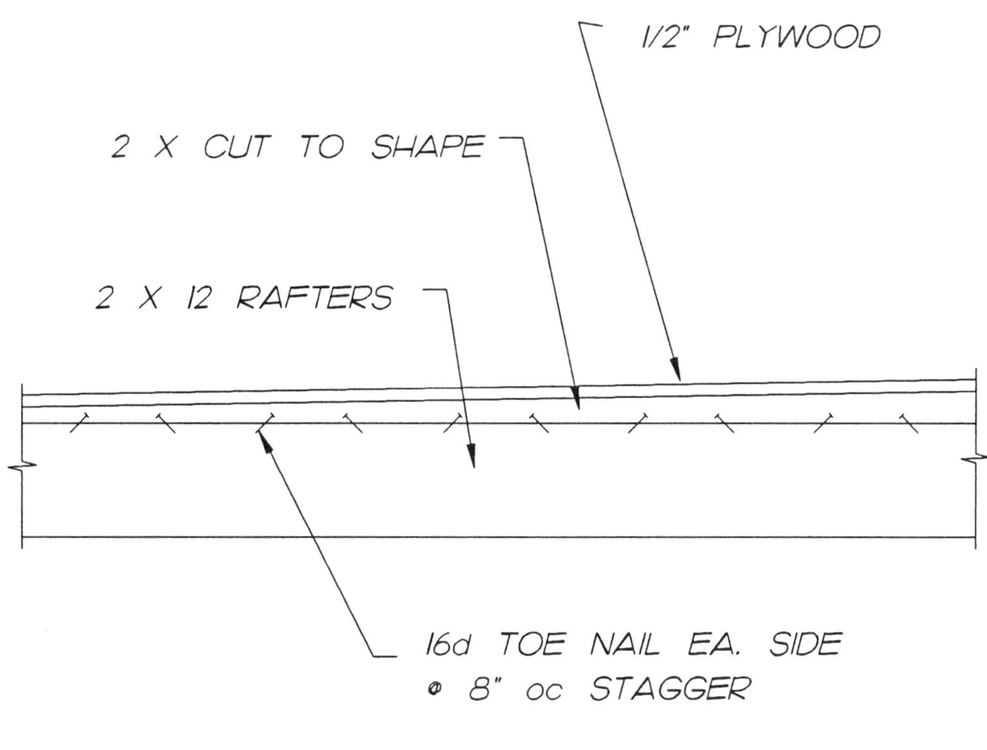
1/2" PLYWOOD
2 X CUT TO SHAPE
2 X 12 RAFTERS
16d TOE NAIL EA. SIDE
@ 8" oc STAGGER

W4941

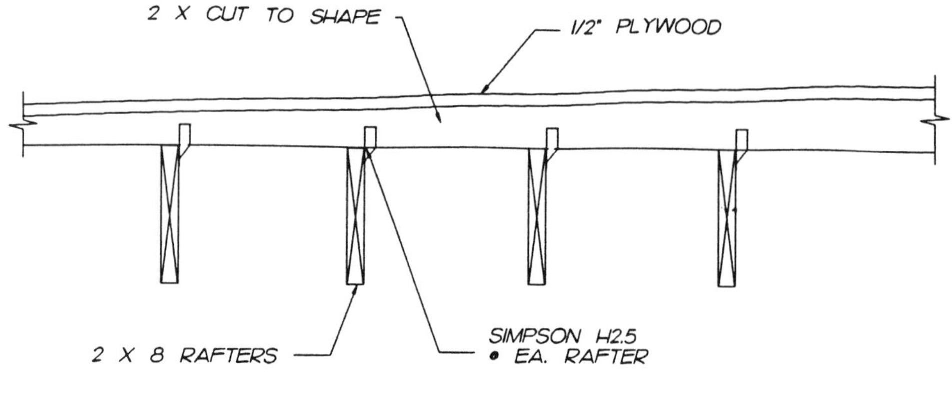
2 X CUT TO SHAPE
1/2" PLYWOOD
2 X 8 RAFTERS
SIMPSON H2.5
@ EA. RAFTER

W4942

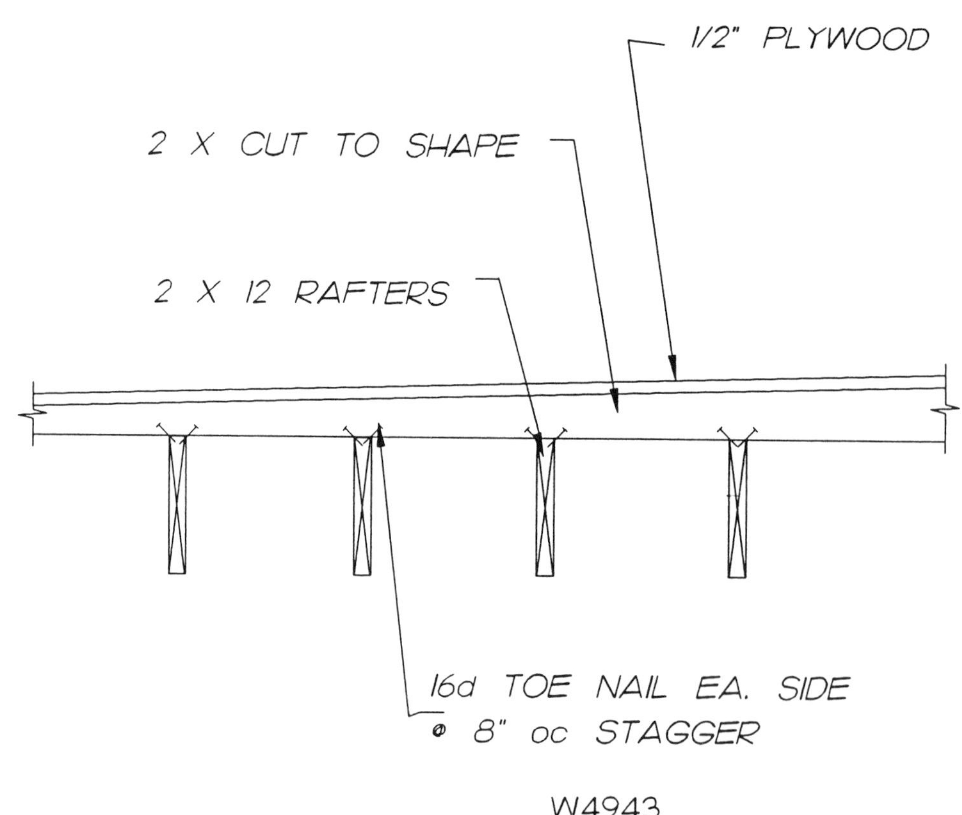
1/2" PLYWOOD
2 X CUT TO SHAPE
2 X 12 RAFTERS
16d TOE NAIL EA. SIDE
@ 8" oc STAGGER

W4943

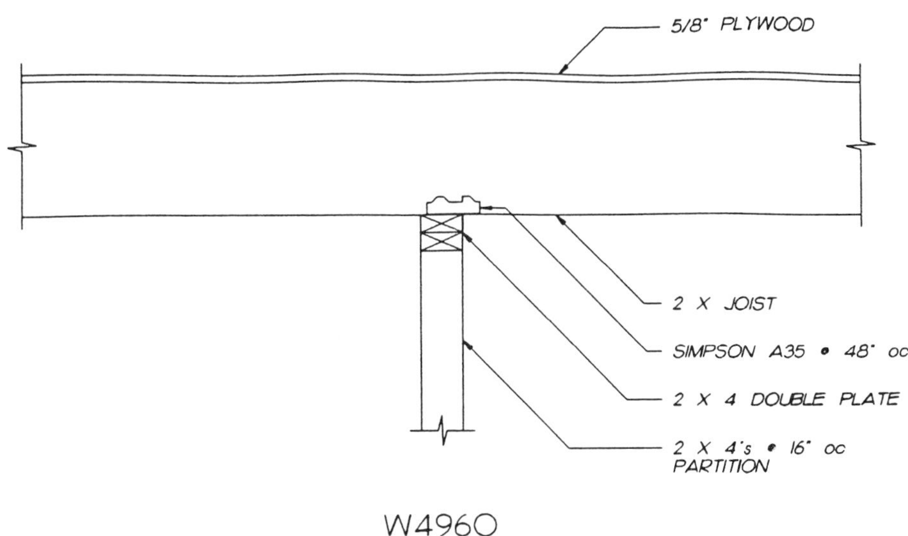
5/8" PLYWOOD
2 X JOIST
SIMPSON A35 @ 48" oc
2 X 4 DOUBLE PLATE
2 X 4's @ 16" oc
PARTITION

W4960

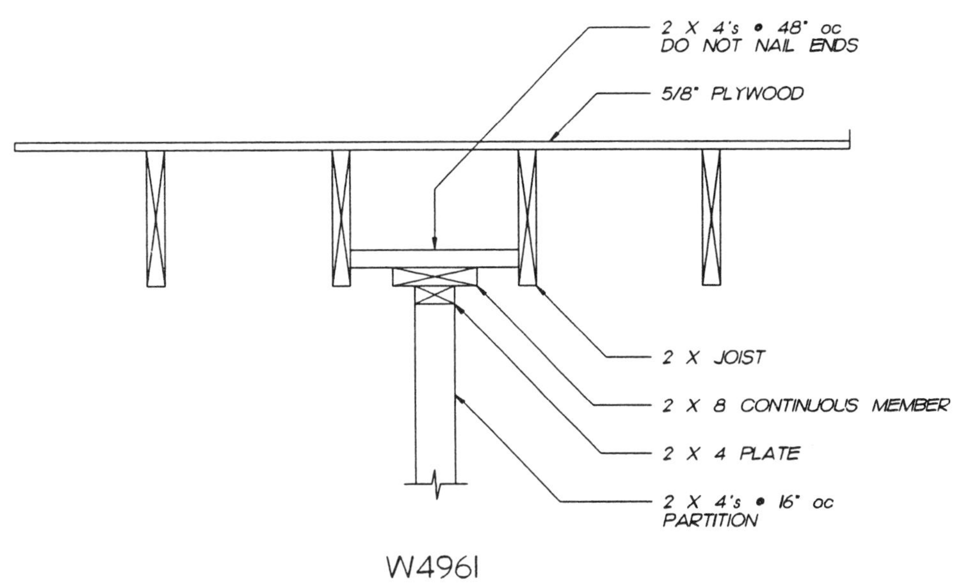
2 X 4's @ 48" oc
DO NOT NAIL ENDS
5/8" PLYWOOD
2 X JOIST
2 X 8 CONTINUOUS MEMBER
2 X 4 PLATE
2 X 4's @ 16" oc
PARTITION

W4961

CHAPTER I WOOD	SECTION 7	GLUE LAMINATED
BEAM TO GIRDER CONNECTION	BEAM ON ONE SIDE	W7020
BEAM TO GIRDER CONNECTION	EQUAL BEAMS EACH SIDE	W7040
BEAM TO GIRDER CONNECTION	UNEQUAL BEAMS EACH SIDE	W7060
BEAM SUPPORT AT MASONRY PILASTER	————————	W7080
BEAM SUPPORT AT CONCRETE PILASTER	————————	W7100
BEAM SECT. & ROOF PURLINS		W7120
HINGE CONNECTION EQUAL SIZE BEAM	NORMAL LOAD SUPPORT	W7140
HINGE CONNECTION UNEQUAL SIZE BEAM	NORMAL LOAD SUPPORT	W7160
HINGE CONNECTION EQUAL SIZE BEAM	HEAVY LOAD SUPPORT	W7180
HINGE CONNECTION UNEQUAL SIZE BEAM	HEAVY LOAD SUPPORT	W7200
BEAM SUPPORTED BY PIPE COL	CANTERLEVER BEAM	W7220
BEAM SUPPORTED BY PIPE COL	SIMPLE SUPPORT	W7240
BEAM SUPPORTED BY PIPE COL	INTERSECTION OF 4 BEAMS	W7260
BEAM SUPPORTED BY PIPE COL	INTERSECTION OF 4 BEAMS PLAN & SECTION	W7280
BEAM SUPPORTED BY PIPE COL ABOVE & BELOW	————————	W7300
BEAM SUPPORTED BY PIPE COL	CANTILEVER BEAM HINGE SUPPORT	W7320

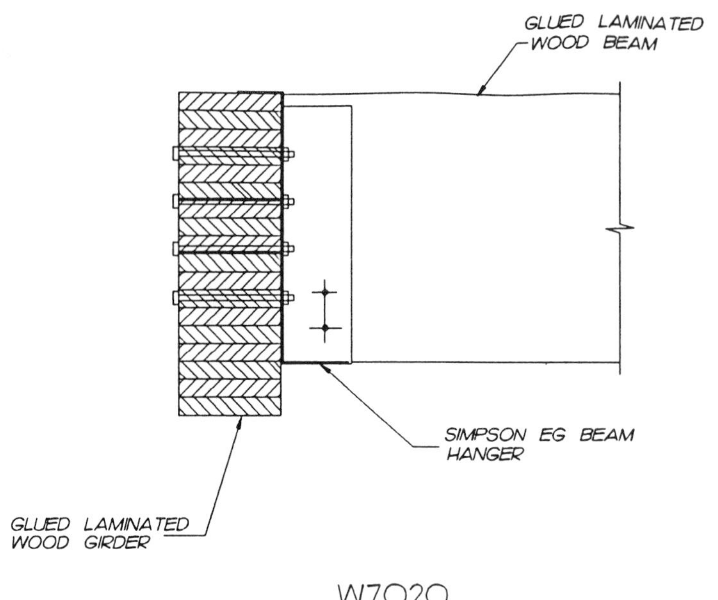
GLUED LAMINATED
WOOD BEAM
SIMPSON EG BEAM
HANGER
GLUED LAMINATED
WOOD GIRDER

W7020

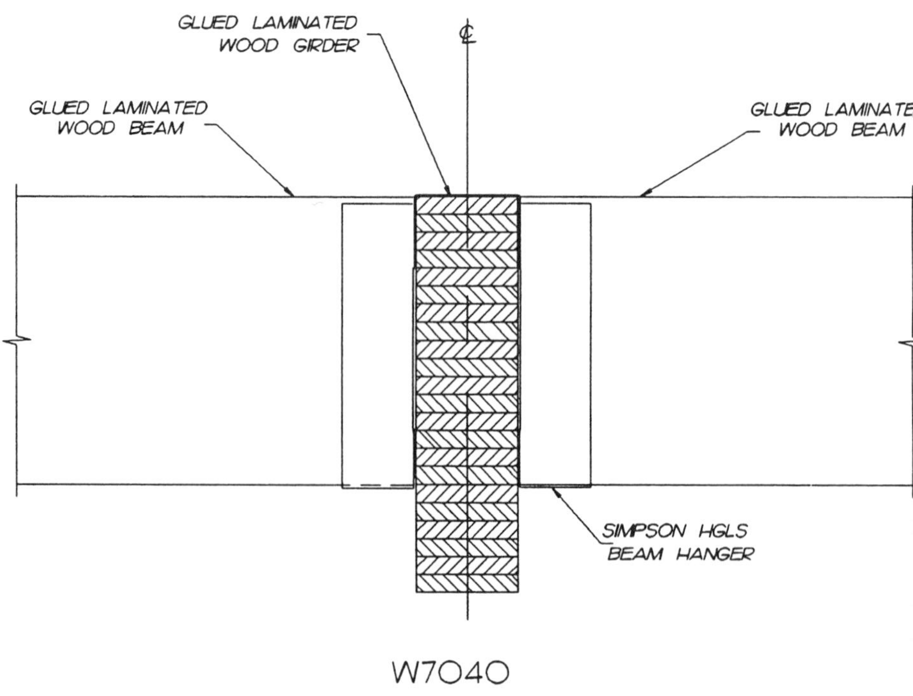
GLUED LAMINATED
WOOD GIRDER
℄
GLUED LAMINATED
WOOD BEAM
GLUED LAMINATE
WOOD BEAM
SIMPSON HGLS
BEAM HANGER

W7040

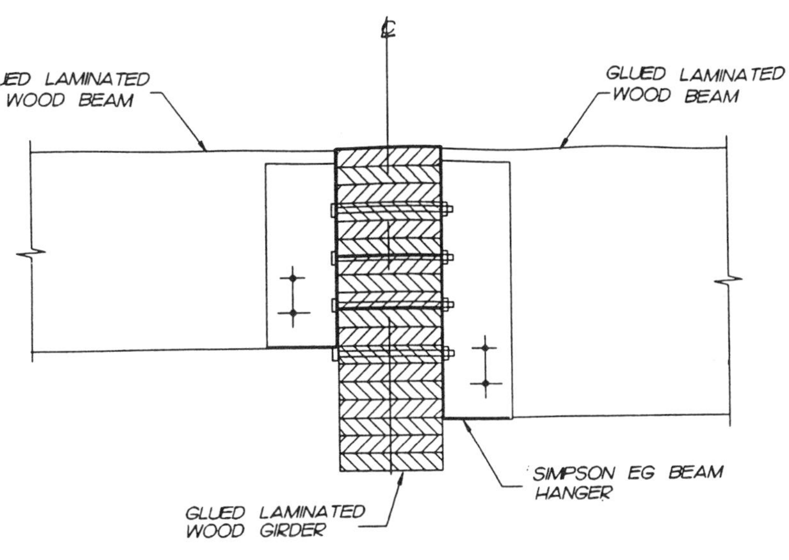
ED LAMINATED
WOOD BEAM
GLUED LAMINATED
WOOD BEAM
SIMPSON EG BEAM
HANGER
GLUED LAMINATED
WOOD GIRDER

W7060

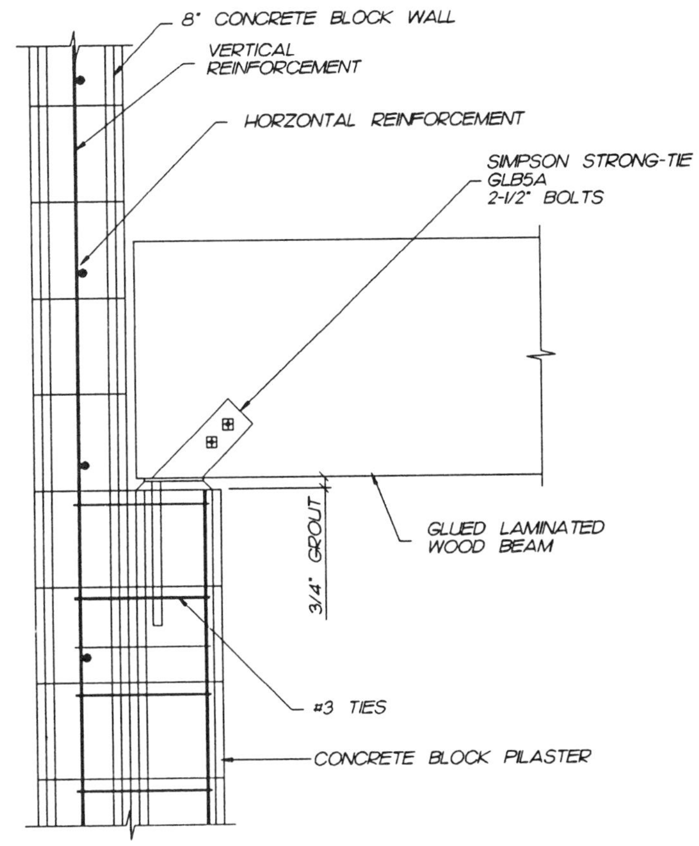
8" CONCRETE BLOCK WALL
VERTICAL
REINFORCEMENT
HORZONTAL REINFORCEMENT
SIMPSON STRONG-TIE
GLB5A
2-1/2" BOLTS
3/4" GROUT
GLUED LAMINATED
WOOD BEAM
#3 TIES
CONCRETE BLOCK PILASTER

W7080

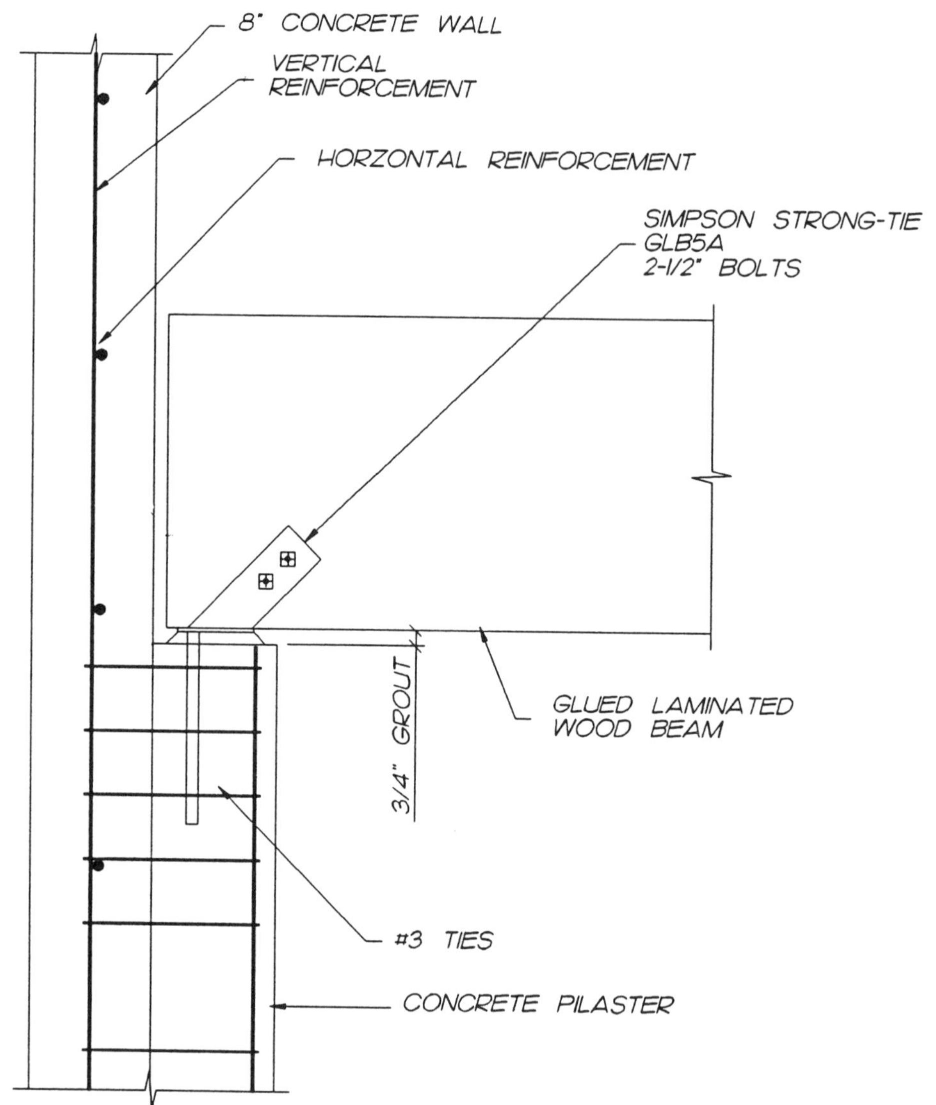
8" CONCRETE WALL
VERTICAL
REINFORCEMENT
HORZONTAL REINFORCEMENT
SIMPSON STRONG-TIE
GLB5A
2-1/2" BOLTS
GLUED LAMINATED
WOOD BEAM
3/4" GROUT
#3 TIES
CONCRETE PILASTER

W7100

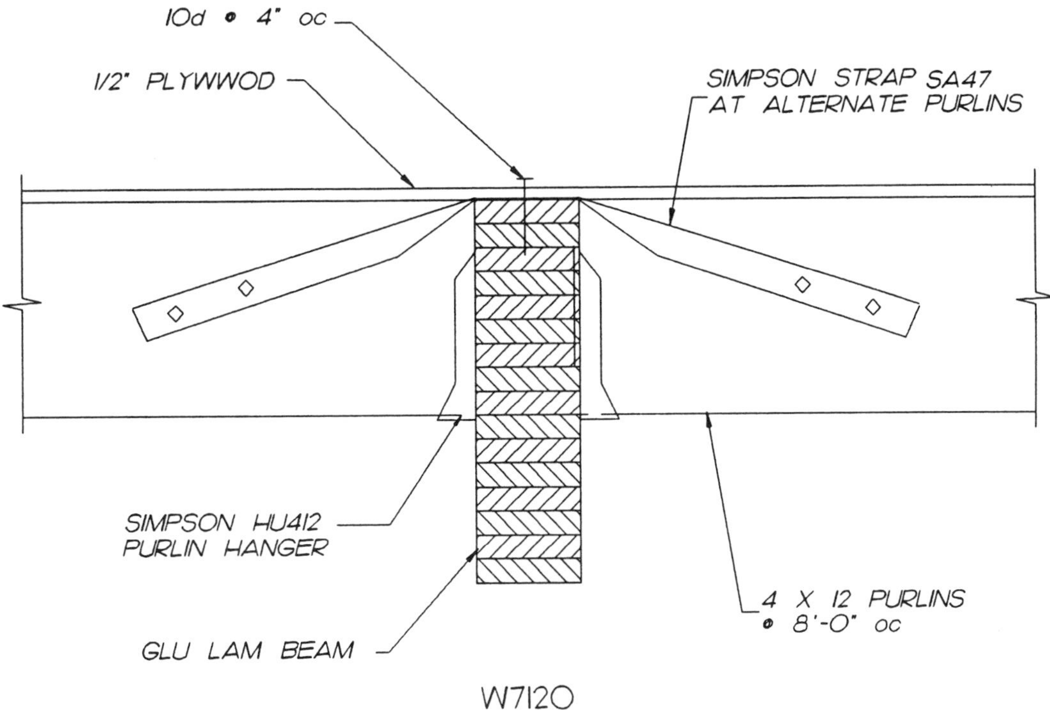
10d @ 4" oc
1/2" PLYWWOD
SIMPSON STRAP SA47
AT ALTERNATE PURLINS
SIMPSON HU412
PURLIN HANGER
GLU LAM BEAM
4 X 12 PURLINS
@ 8'-0" oc

W7120

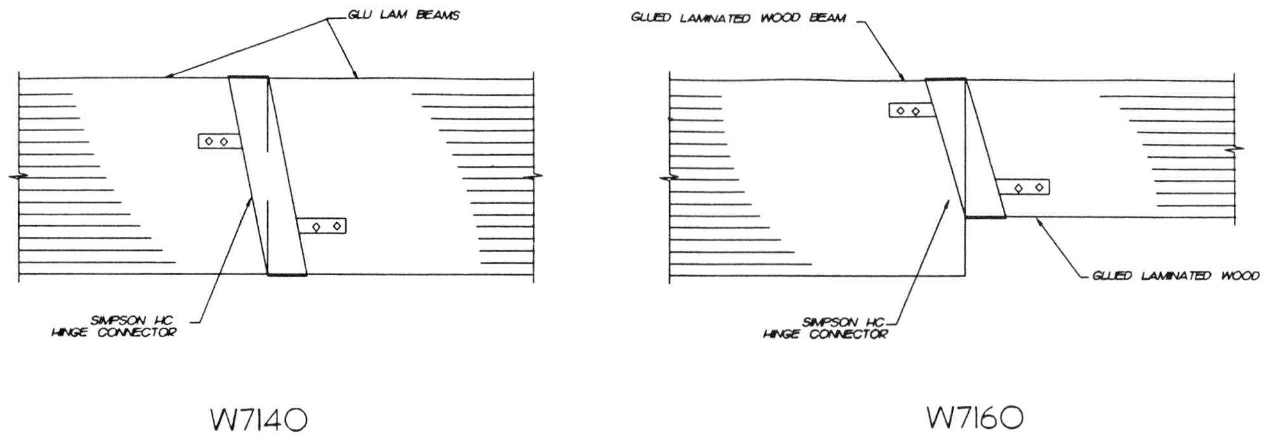

W7140

W7160

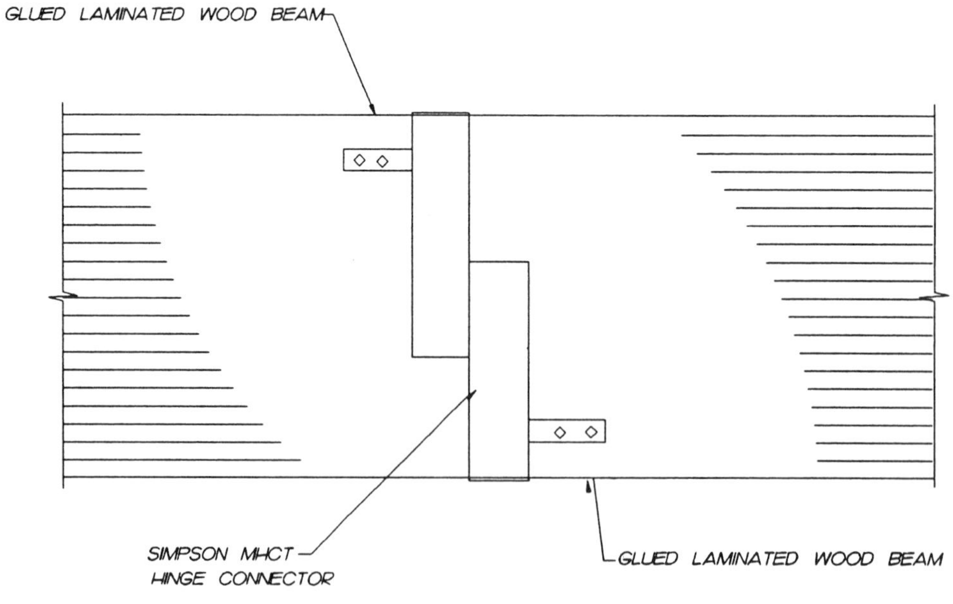

W7180

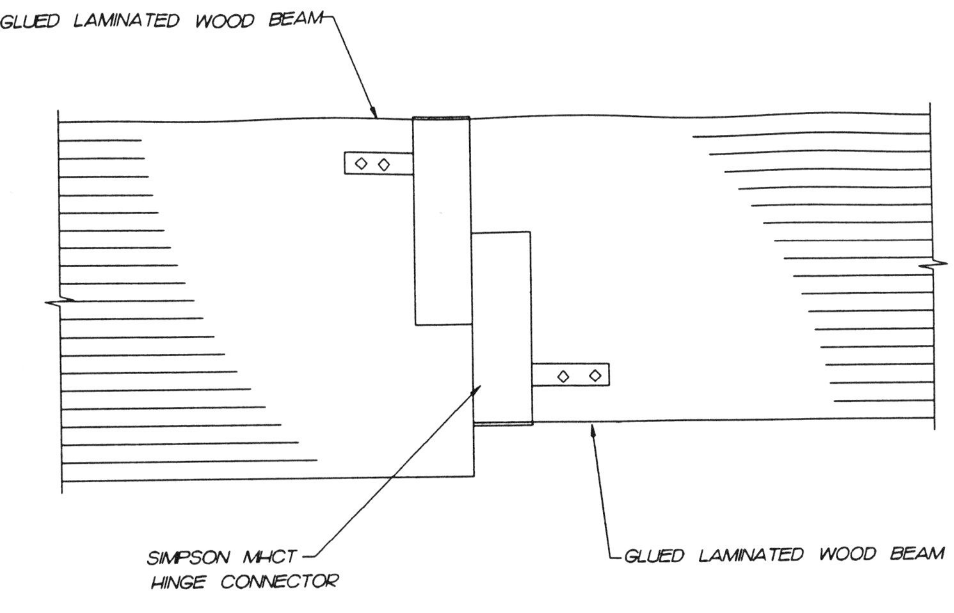

W7200

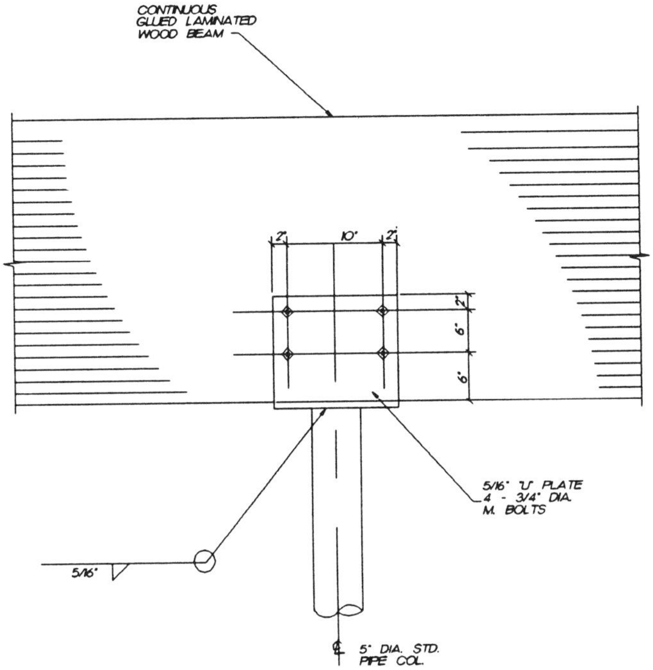

W7220

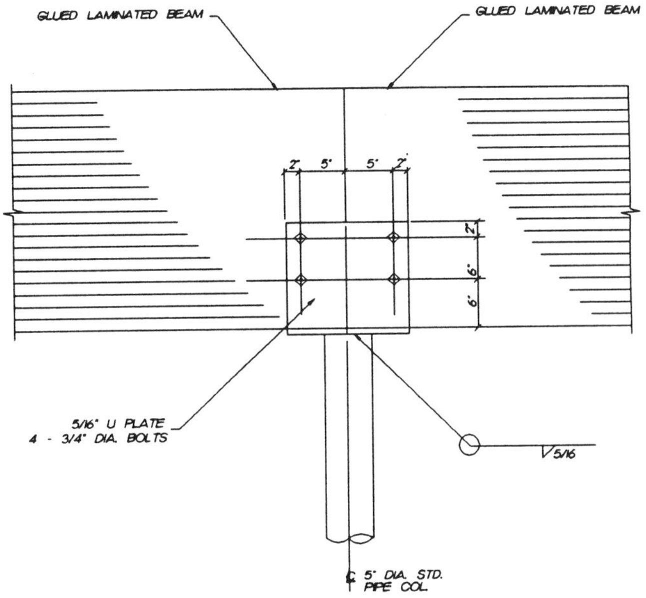

W7240

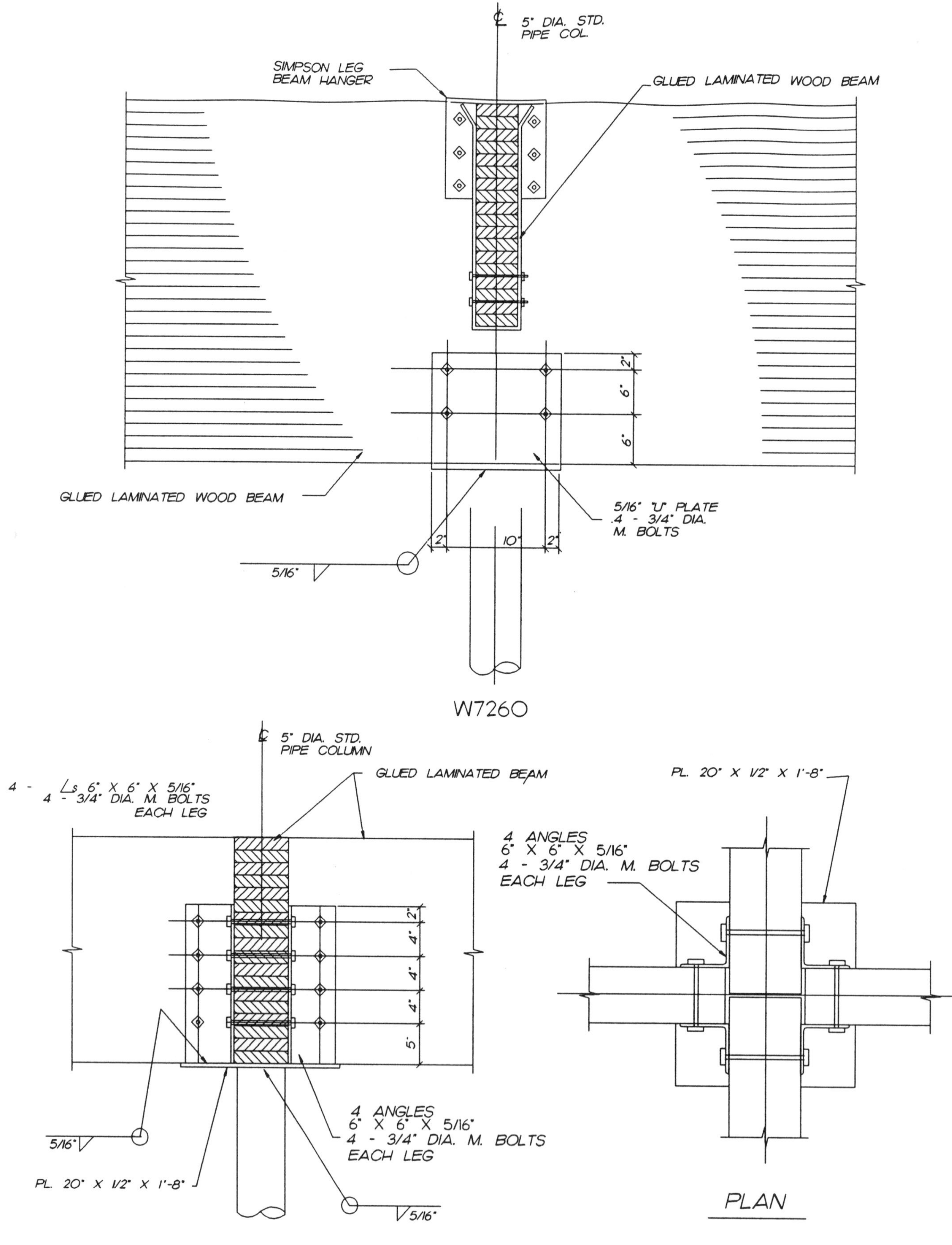

℄ 5" DIA. STD. PIPE COL.
SIMPSON LEG BEAM HANGER
GLUED LAMINATED WOOD BEAM
2"
6"
6"
GLUED LAMINATED WOOD BEAM
5/16" "U" PLATE
.4 - 3/4" DIA. M. BOLTS
2"
10"
2"
5/16"
W7260
℄ 5" DIA. STD. PIPE COLUMN
4 - Ls 6" X 6" X 5/16"
4 - 3/4" DIA. M. BOLTS EACH LEG
GLUED LAMINATED BEAM
PL. 20" X 1/2" X 1'-8"
4 ANGLES 6" X 6" X 5/16" 4 - 3/4" DIA. M. BOLTS EACH LEG
2"
4"
4"
4"
5"
5/16"
4 ANGLES 6" X 6" X 5/16" 4 - 3/4" DIA. M. BOLTS EACH LEG
PL. 20" X 1/2" X 1'-8"
5/16"
PLAN
W7280

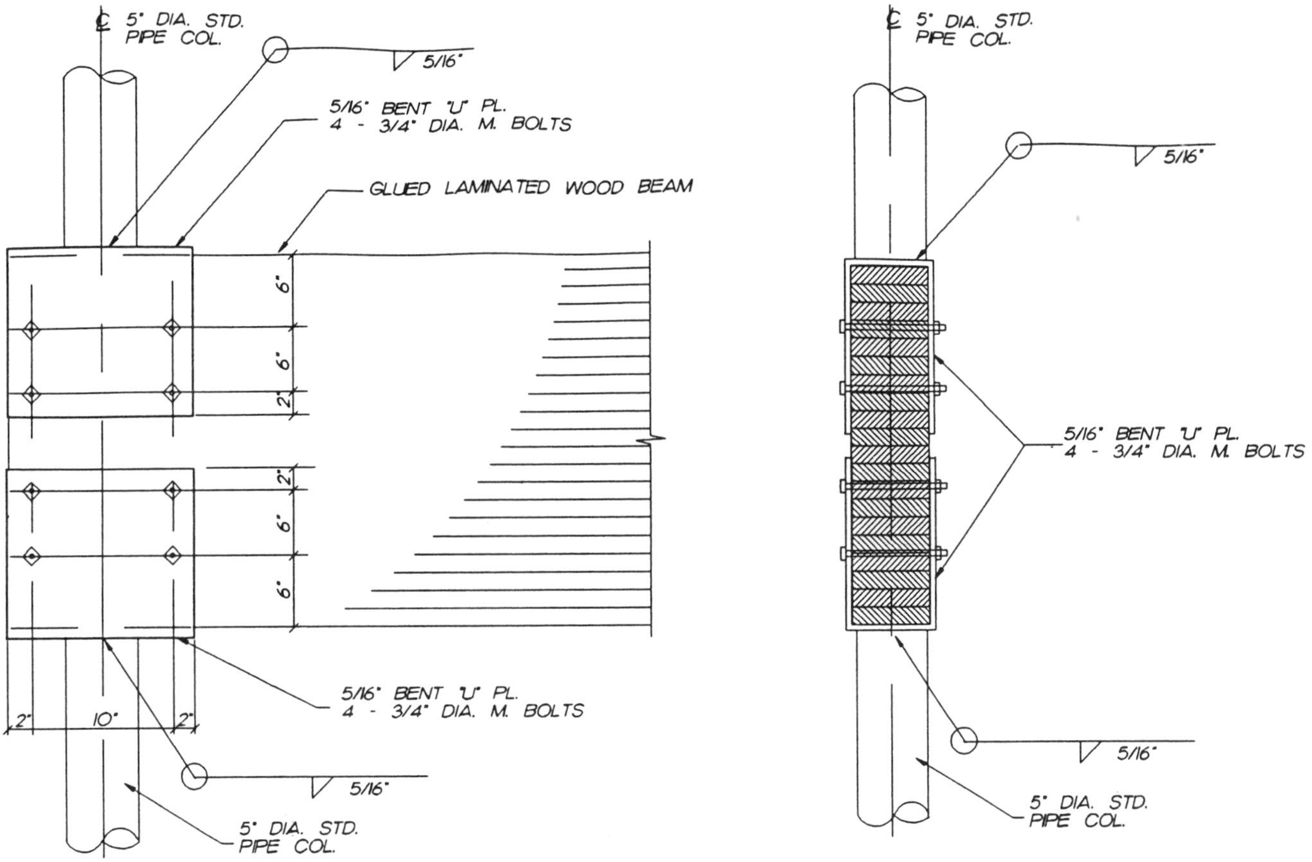

5" DIA. STD. PIPE COL.

5/16"

5/16" BENT "U" PL.
4 - 3/4" DIA. M. BOLTS

GLUED LAMINATED WOOD BEAM

6" 6" 2" 2" 6" 6"

2" 10" 2"

5/16" BENT "U" PL.
4 - 3/4" DIA. M. BOLTS

5/16"

5" DIA. STD. PIPE COL.

5" DIA. STD. PIPE COL.

5/16"

5/16" BENT "U" PL.
4 - 3/4" DIA. M. BOLTS

5/16"

5" DIA. STD. PIPE COL.

W7300

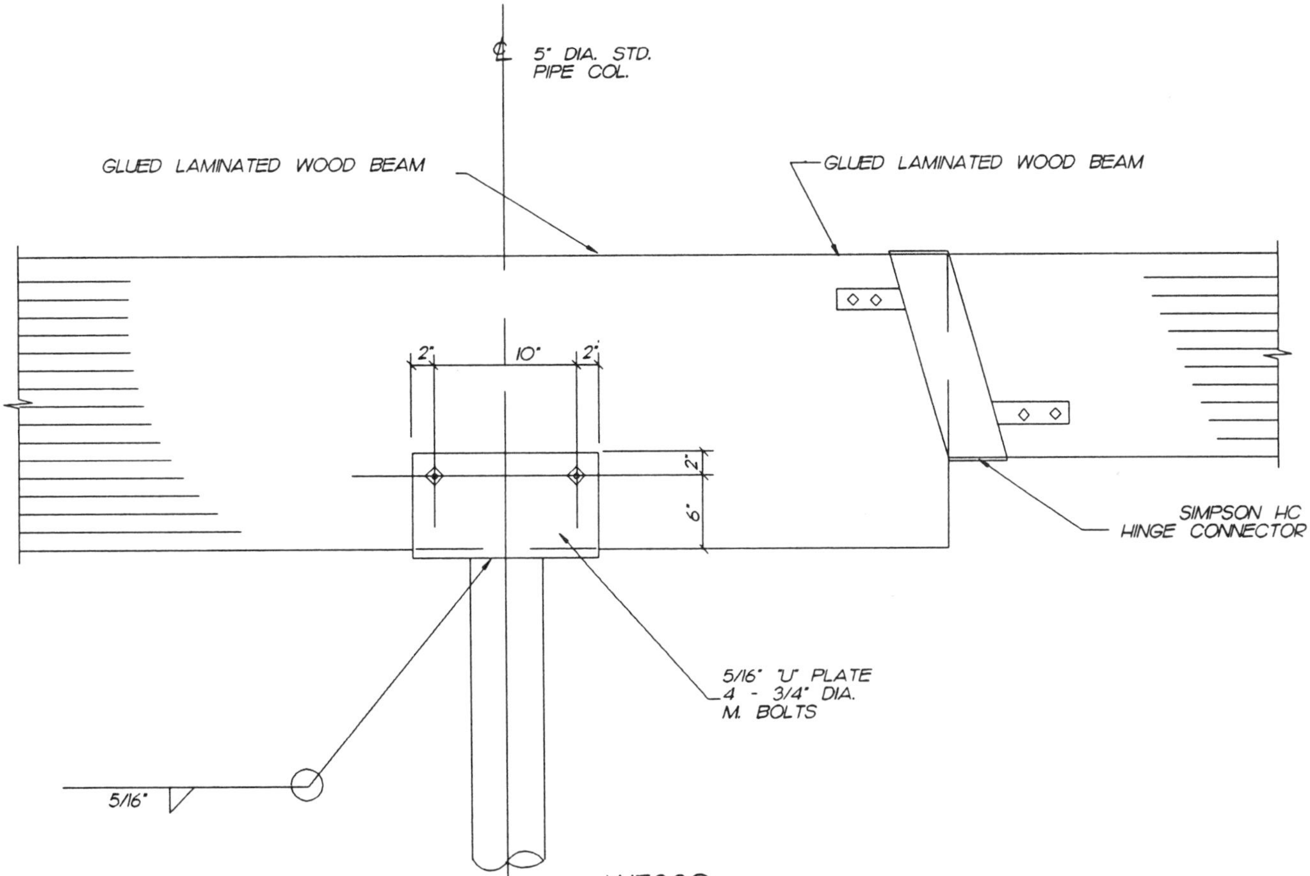

5" DIA. STD. PIPE COL.

GLUED LAMINATED WOOD BEAM

GLUED LAMINATED WOOD BEAM

2" 10" 2"

2" 6"

SIMPSON HC HINGE CONNECTOR

5/16" "U" PLATE
4 - 3/4" DIA.
M. BOLTS

5/16"

W7320

CHAPTER 1	WOOD SECTION 8		CONNECTIONS
BEAM SUPPORT FLOOR & WALL ONE SIDE	JOISTS PERP. TO BEAM		W8020 - W8023
	JOISTS PARALELL TO BEAM		W8040 - W8043
BEAM SUPPORT FLOOR & WALL BOTH SIDES	JOISTS PERP. TO BEAM		W8060 - W8063
	JOISTS PARALELL TO BEAM		W8080 - W8083
BEAM SUPPORT FLOOR & WALL BOTH SIDES	JOISTS PERP. & PARALELL ALT. SIDES		W8100 - W8103

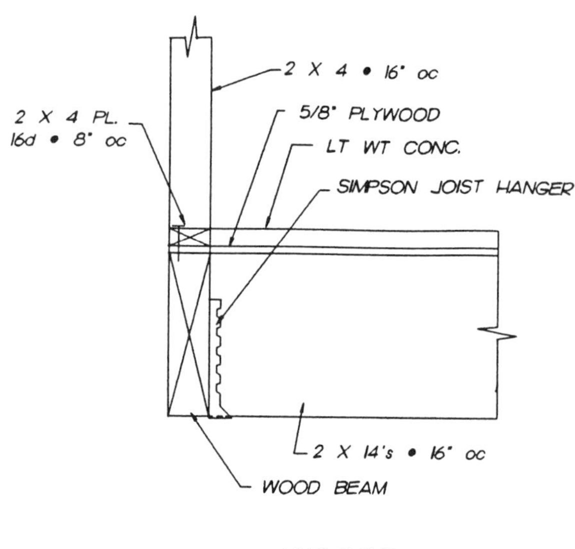

W8020

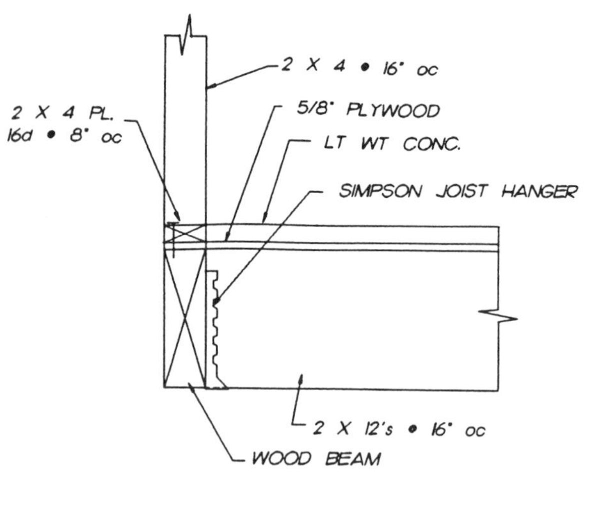

W8021

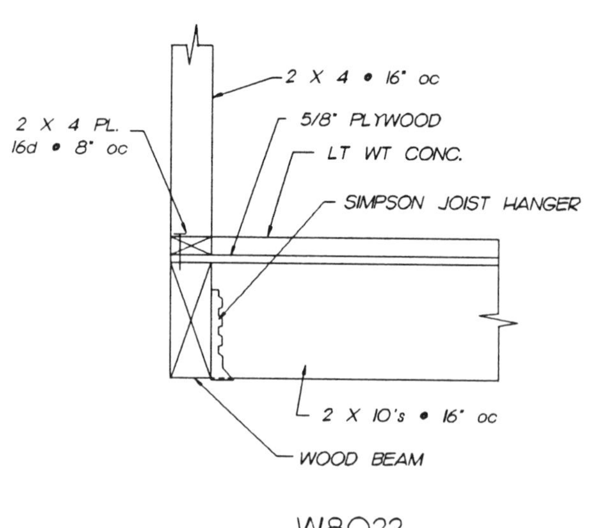

W8022

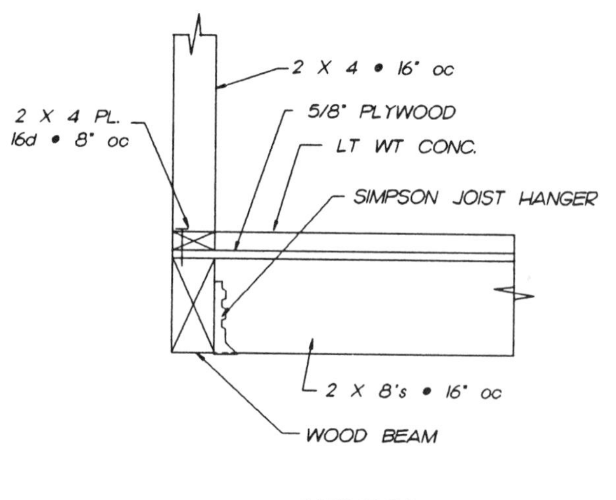

W8023

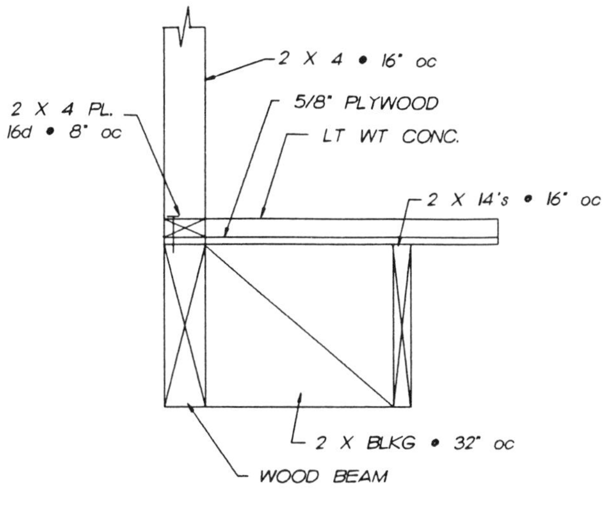

W8040

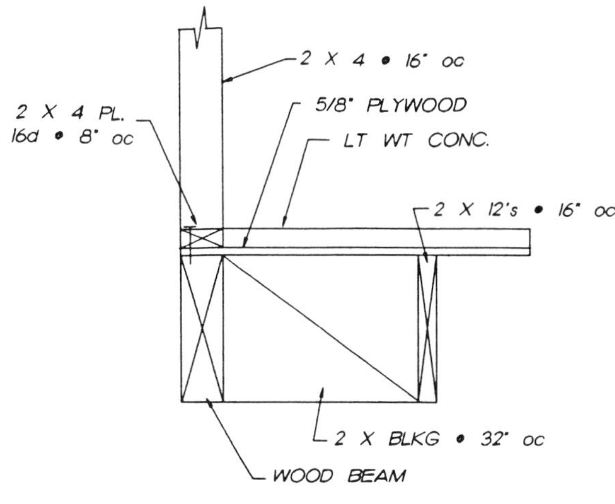

W8041

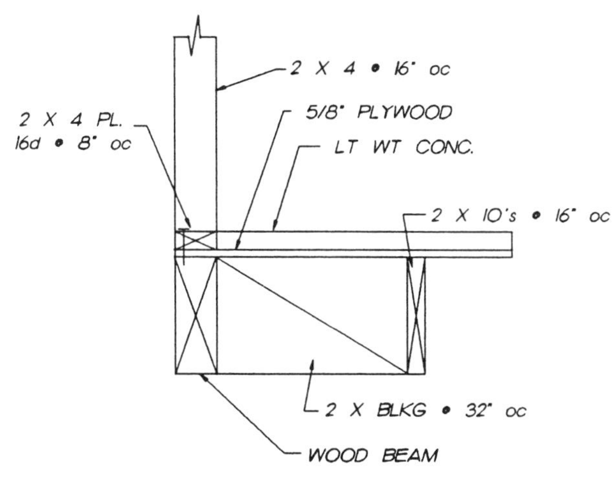

W8042

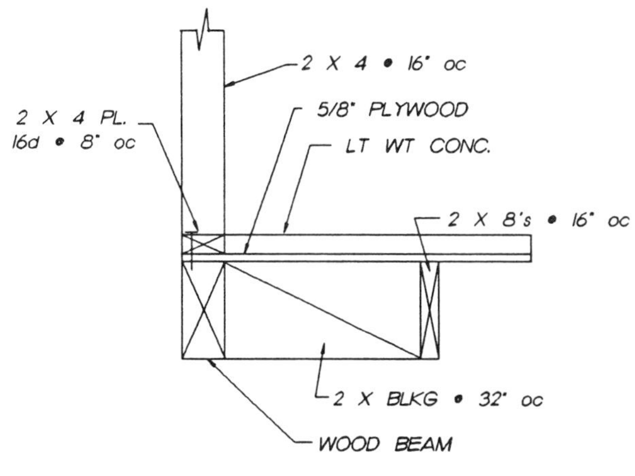

W8043

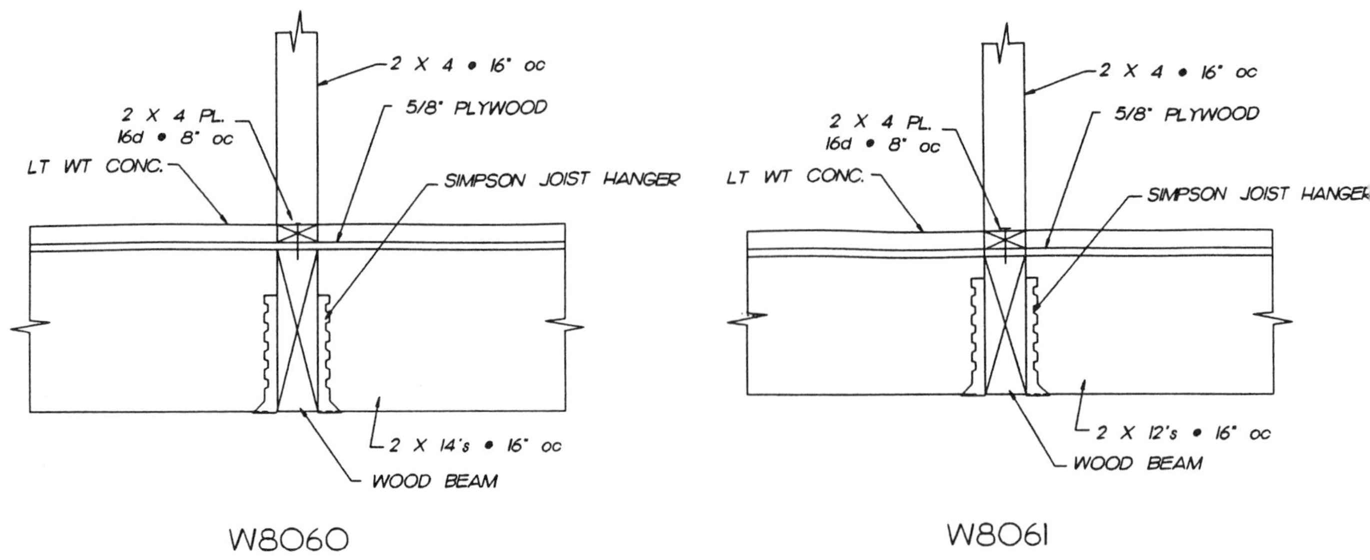
2 X 4 • 16" oc
2 X 4 PL.
16d • 8" oc
5/8" PLYWOOD
LT WT CONC.
SIMPSON JOIST HANGER
2 X 14's • 16" oc
WOOD BEAM
W8060
2 X 4 • 16" oc
2 X 4 PL.
16d • 8" oc
5/8" PLYWOOD
LT WT CONC.
SIMPSON JOIST HANGER
2 X 12's • 16" oc
WOOD BEAM
W8061

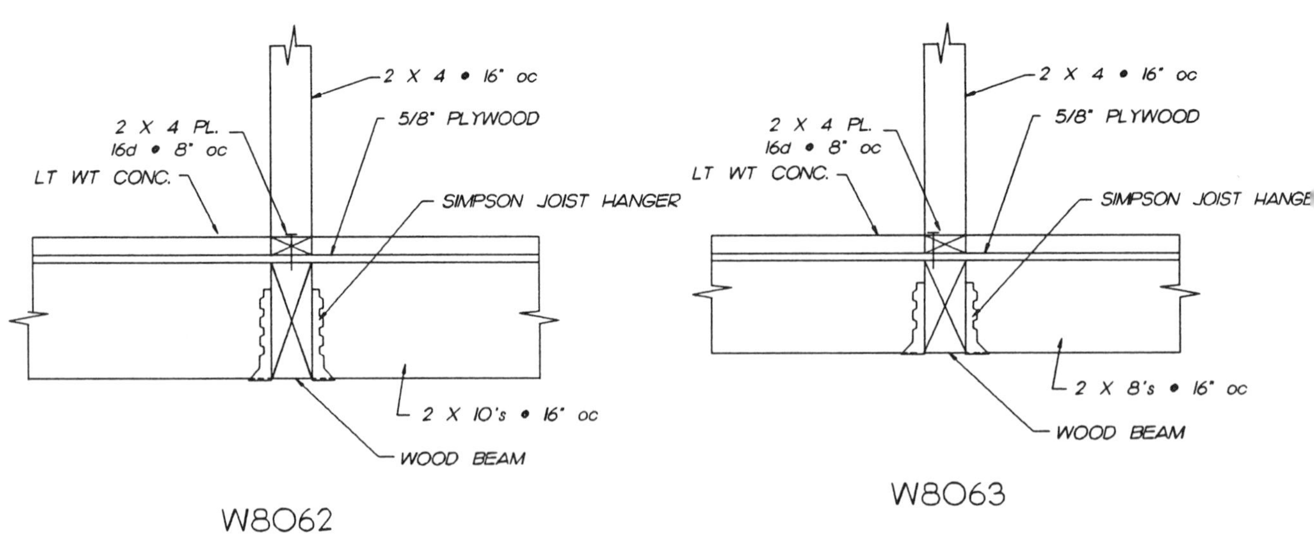
2 X 4 • 16" oc
2 X 4 PL.
16d • 8" oc
5/8" PLYWOOD
LT WT CONC.
SIMPSON JOIST HANGER
2 X 10's • 16" oc
WOOD BEAM
W8062
2 X 4 • 16" oc
2 X 4 PL.
16d • 8" oc
5/8" PLYWOOD
LT WT CONC.
2 X 8's • 16" oc
WOOD BEAM
W8063

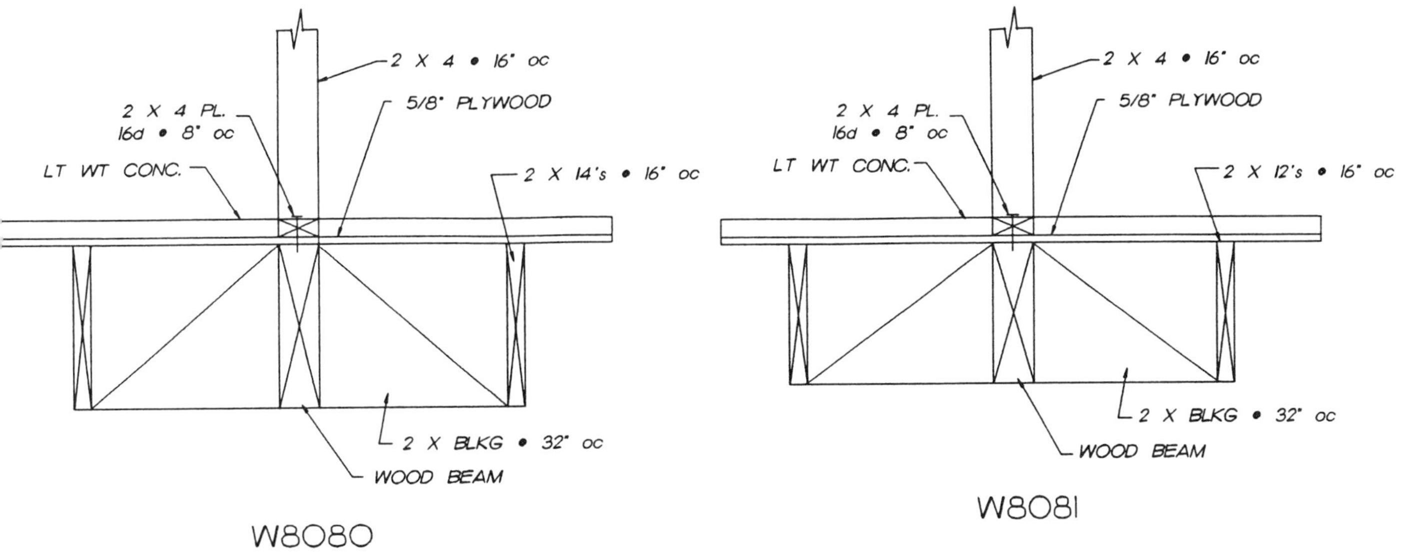

W8080

W8081

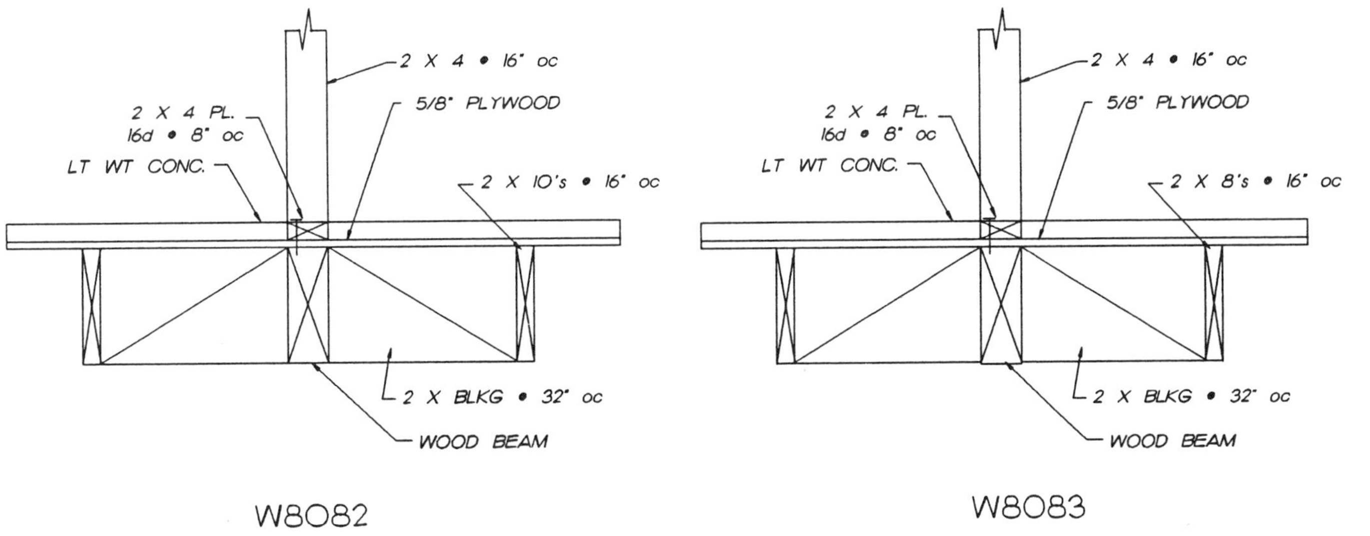

W8082

W8083

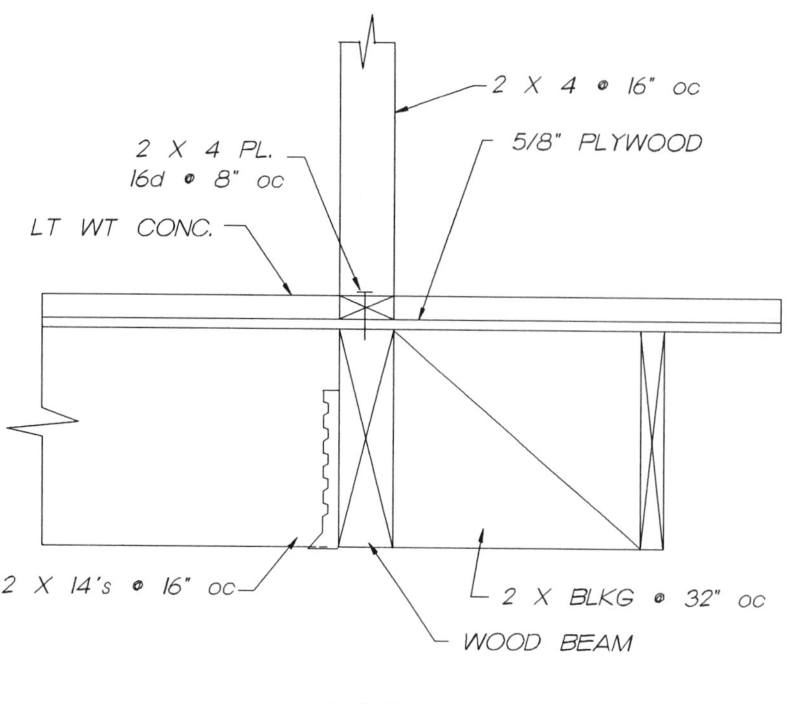
2 X 4 @ 16" oc
5/8" PLYWOOD
2 X 4 PL.
16d @ 8" oc
LT WT CONC.
2 X 14's @ 16" oc
2 X BLKG @ 32" oc
WOOD BEAM

W8100

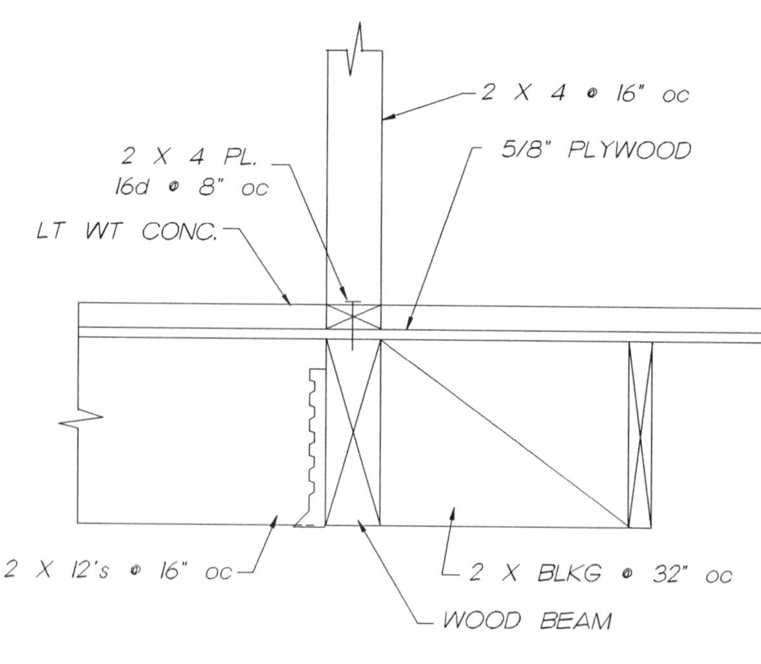
2 X 4 @ 16" oc
5/8" PLYWOOD
2 X 4 PL.
16d @ 8" oc
LT WT CONC.
2 X 12's @ 16" oc
2 X BLKG @ 32" oc
WOOD BEAM

W8101

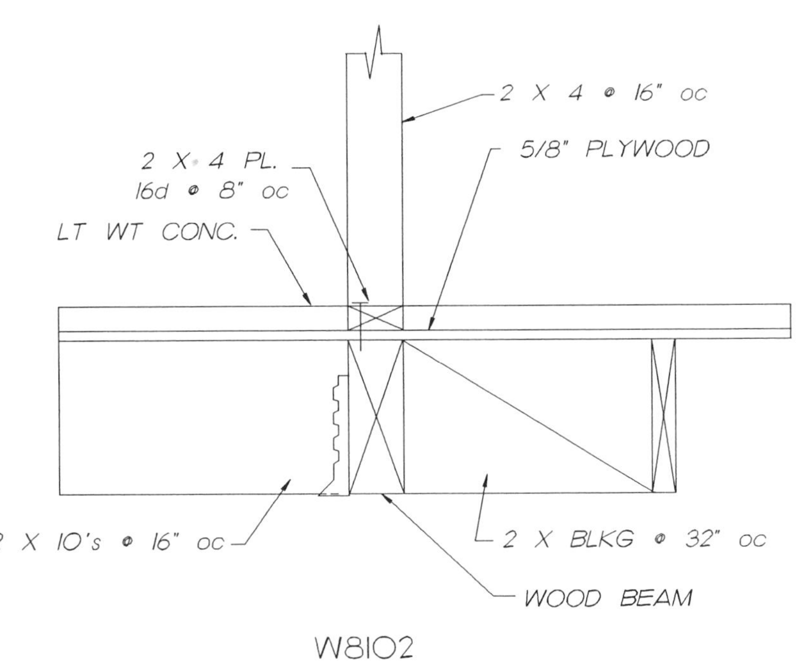
2 X 4 @ 16" oc
5/8" PLYWOOD
2 X 4 PL.
16d @ 8" oc
LT WT CONC.
2 X 10's @ 16" oc
2 X BLKG @ 32" oc
WOOD BEAM

W8102

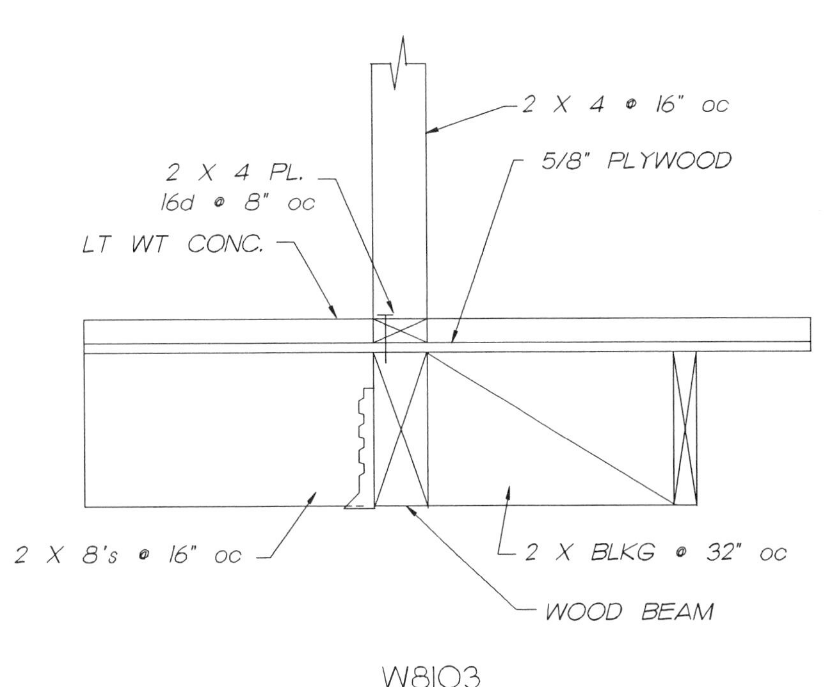
2 X 4 @ 16" oc
5/8" PLYWOOD
2 X 4 PL.
16d @ 8" oc
LT WT CONC.
2 X 8's @ 16" oc
2 X BLKG @ 32" oc
WOOD BEAM

W8103

CHAPTER I	WOOD SECTION 9	MISCELLANEOUS
STUD WALL DOUBLE PL. SPLICE	————	W9020
DIAGONAL SHEATHING PLAN	————	W9040
SEISMIC HOLD DOWN	————	W9060 - W9064
HORIZONTAL SEISMIC STRAPS WALL TO BEAM	————	W9080 - W9084
VERTICAL SEISMIC STRAP FLOOR TO FLOOR	————	W9100
WALL DOUBLE PL. STRAP ACROSS VERTICAL PIPE COL.	————	W9120
STAIR SUPPORTS	AT FOUNDATION	W9140
	AT LANDINGS	W9141 - W9142
JOIST BRIDGING	————	W9160
BOLTING OF LATERALLY LAMINATED 2 X MEMBERS	————	W9180
SUPPORT OF VENEER OVER WINDOWS & DOORS	————	W9200 - W9201

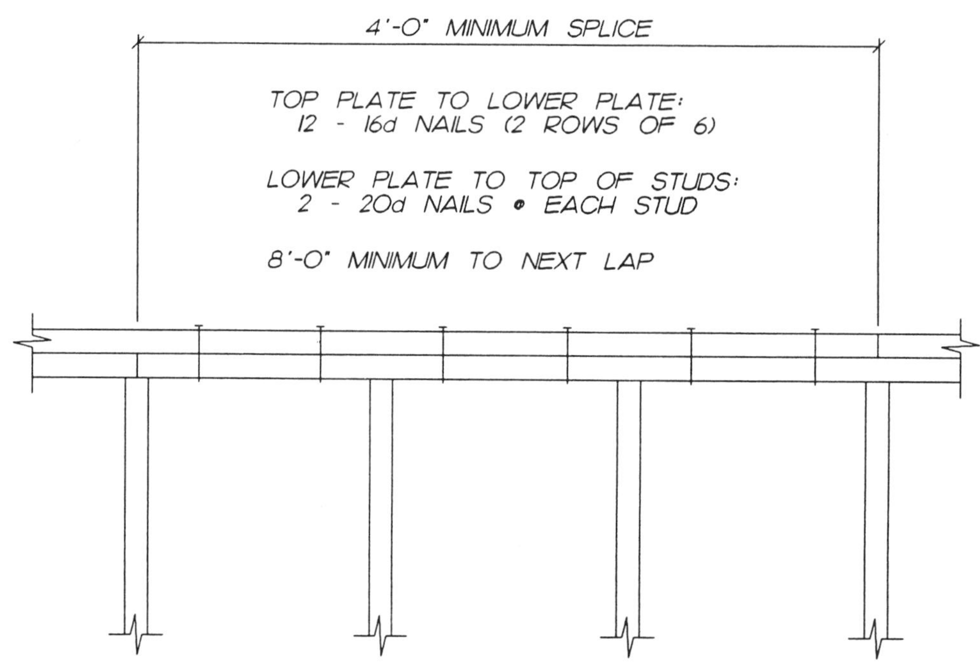
4'-0" MINIMUM SPLICE
TOP PLATE TO LOWER PLATE:
12 - 16d NAILS (2 ROWS OF 6)
LOWER PLATE TO TOP OF STUDS:
2 - 20d NAILS @ EACH STUD
8'-0" MINIMUM TO NEXT LAP
STUDS @ 16" oc UNLESS OTHERWISE NOTED

W9020

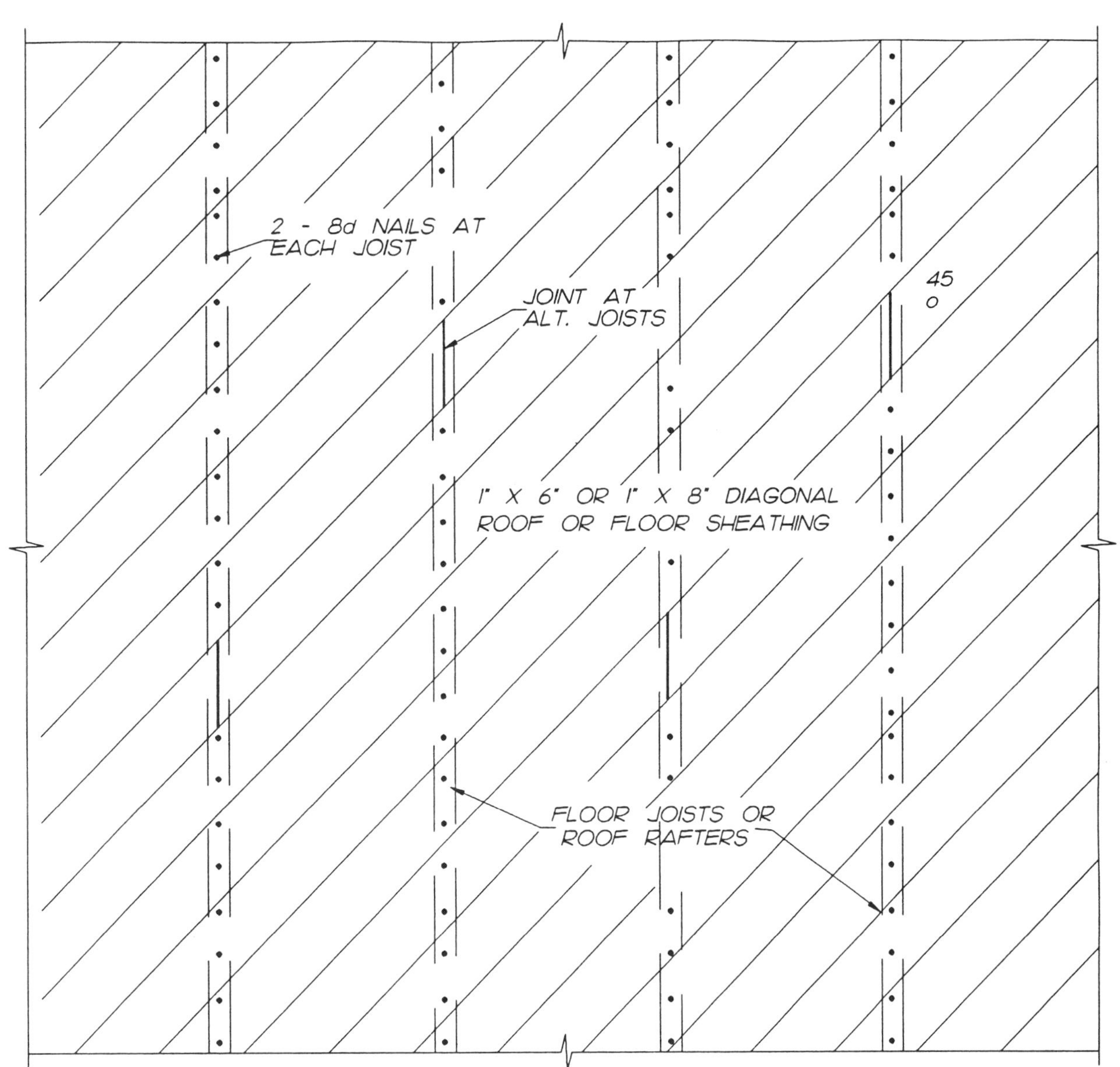

TYPICAL DIAGONAL SHEATHING PLAN

W9040

4 X 6 POST
TOP OF CONCRETE
SIMPSON HD5A
2 X 4 MUDSILL

W9060

4 X 6 POST
SIMPSON HD9
TOP OF CONCRETE
THREADED ROD
MUDSILL

W9061

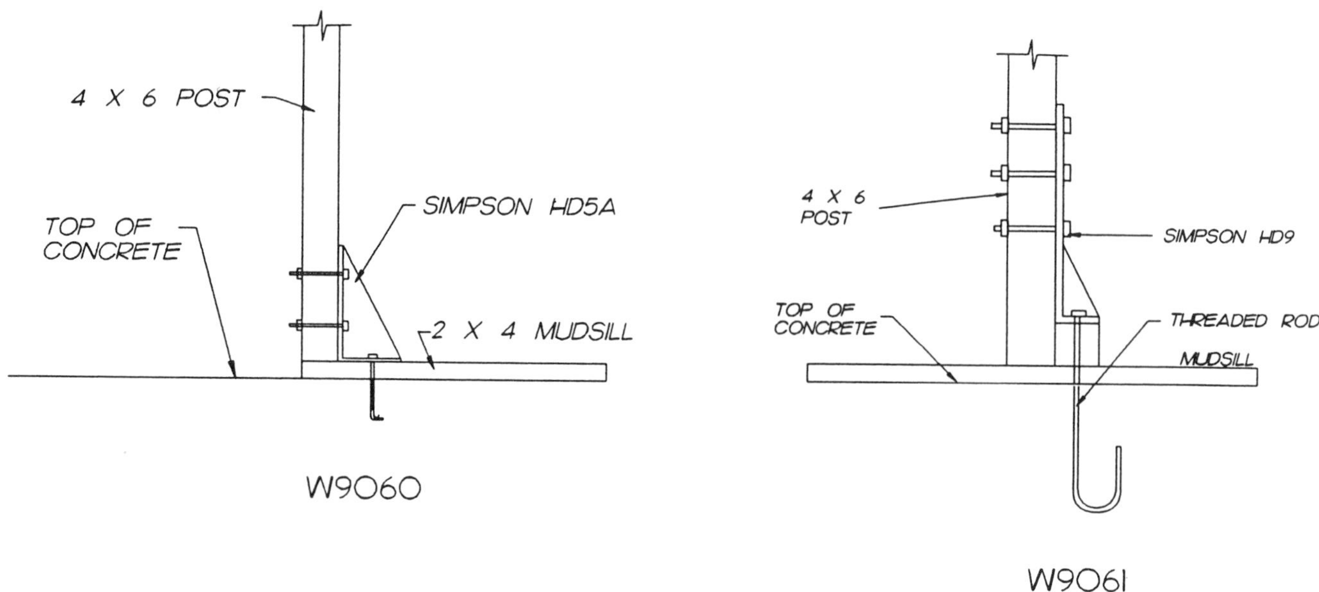

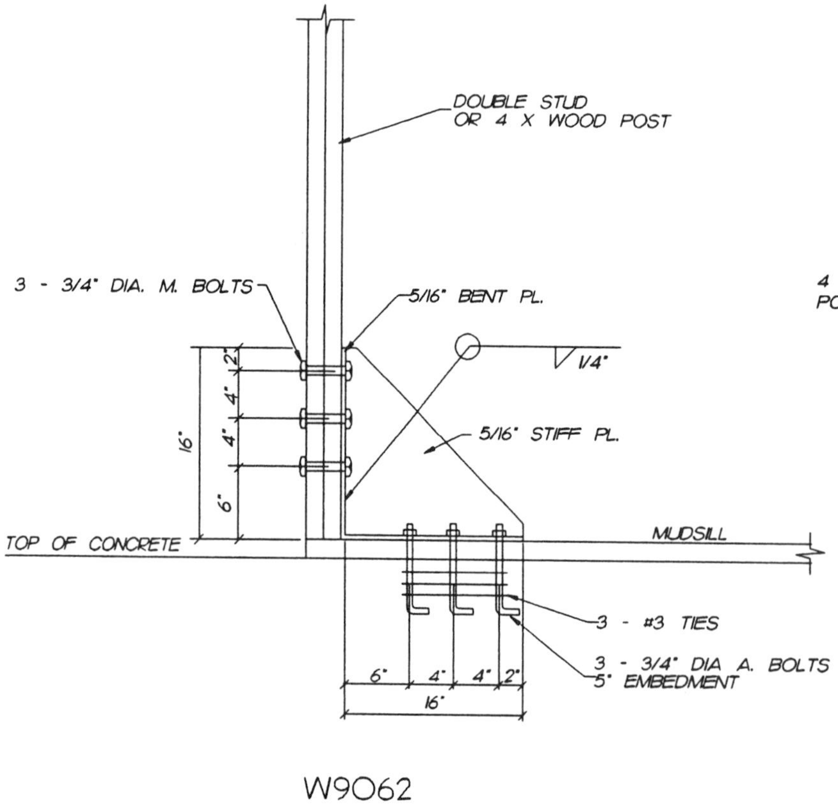

DOUBLE STUD OR 4 X WOOD POST
3 - 3/4" DIA. M. BOLTS
5/16" BENT PL.
1/4"
5/16" STIFF PL.
16"
2"
4"
4"
6"
TOP OF CONCRETE
MUDSILL
3 - #3 TIES
3 - 3/4" DIA A. BOLTS
5" EMBEDMENT
6"
4"
4"
2"
16"

W9062

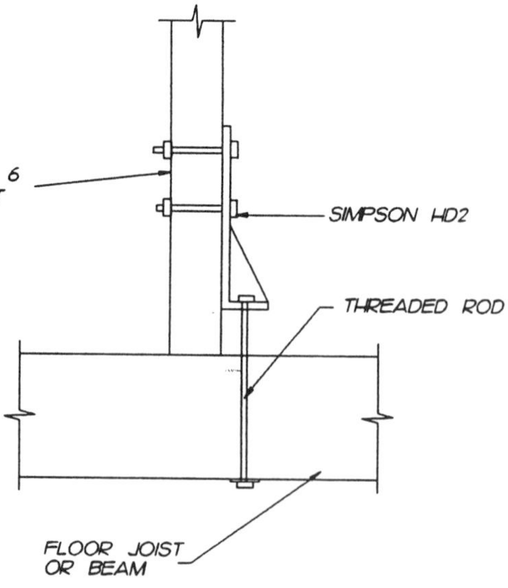

4 X 6 POST
SIMPSON HD2
THREADED ROD
FLOOR JOIST OR BEAM

W9063

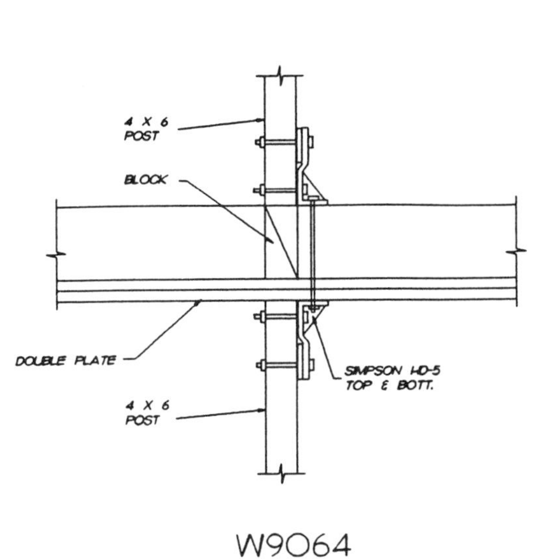

W9064

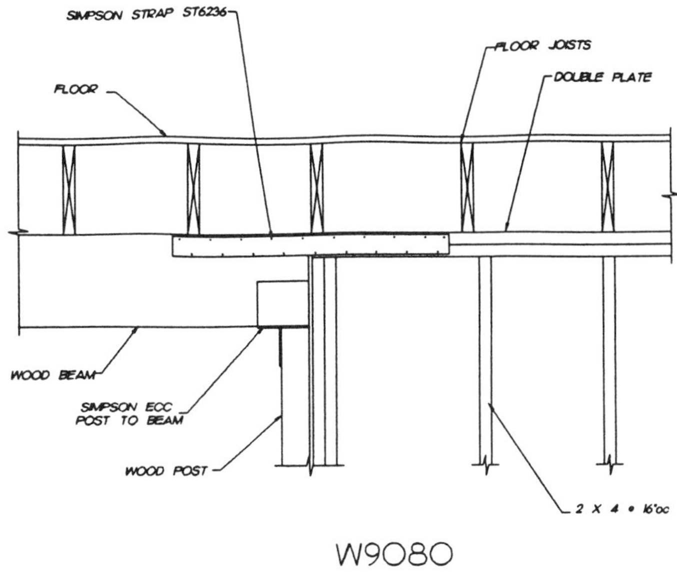

W9080

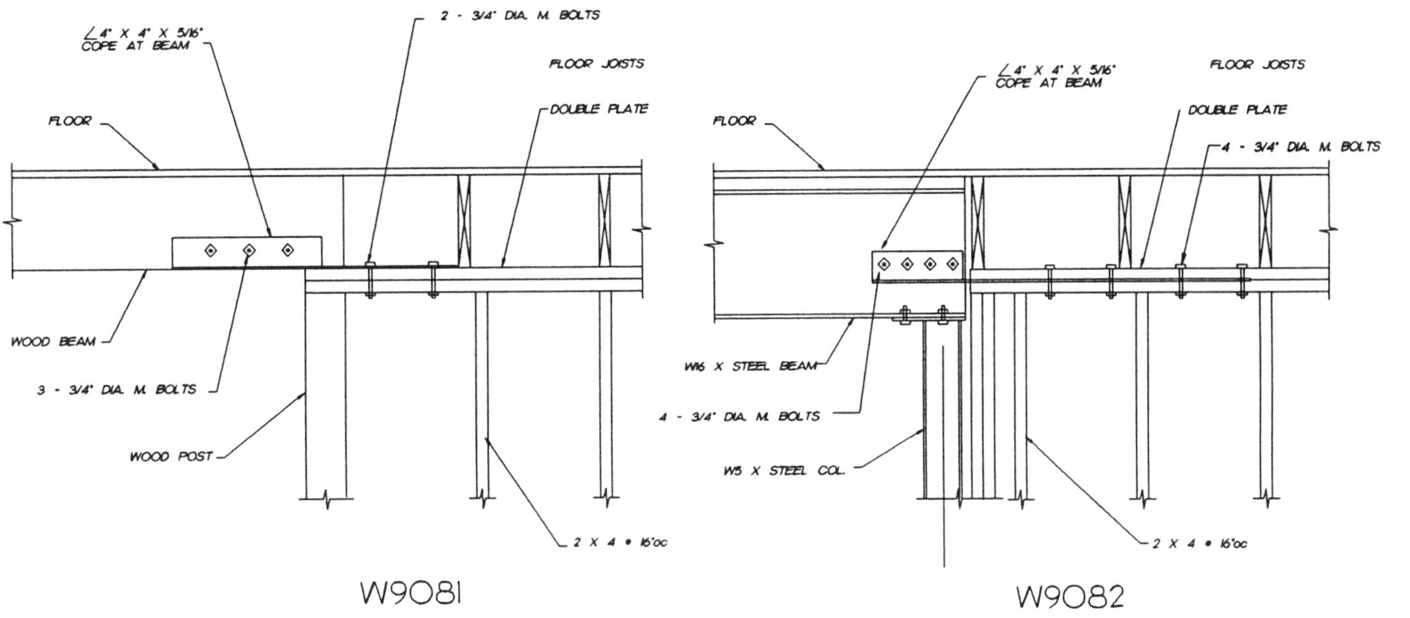

W9081

W9082

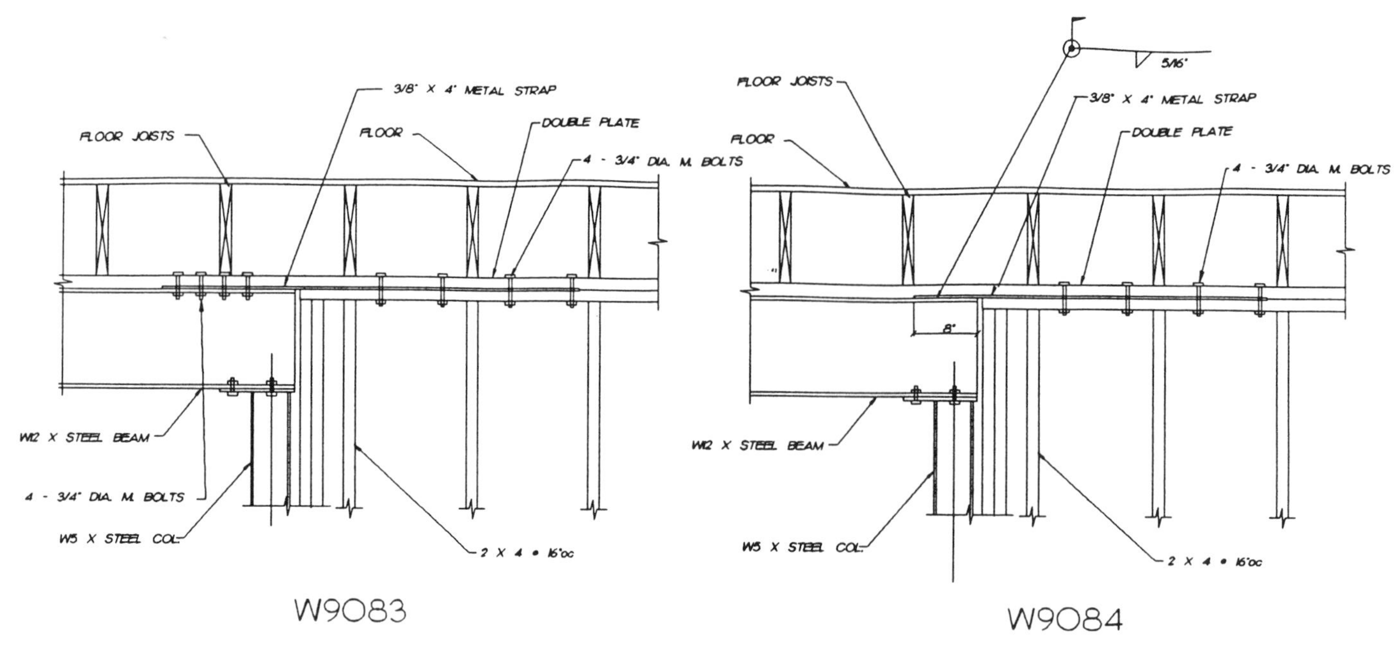
3/8" X 4" METAL STRAP
FLOOR JOISTS
FLOOR
DOUBLE PLATE
4 - 3/4" DIA. M. BOLTS
W12 X STEEL BEAM
4 - 3/4" DIA. M. BOLTS
W5 X STEEL COL.
2 X 4 • 16"oc
W9083
5/16"
FLOOR JOISTS
3/8" X 4" METAL STRAP
FLOOR
DOUBLE PLATE
4 - 3/4" DIA. M. BOLTS
8"
W12 X STEEL BEAM
W5 X STEEL COL.
2 X 4 • 16"oc
W9084

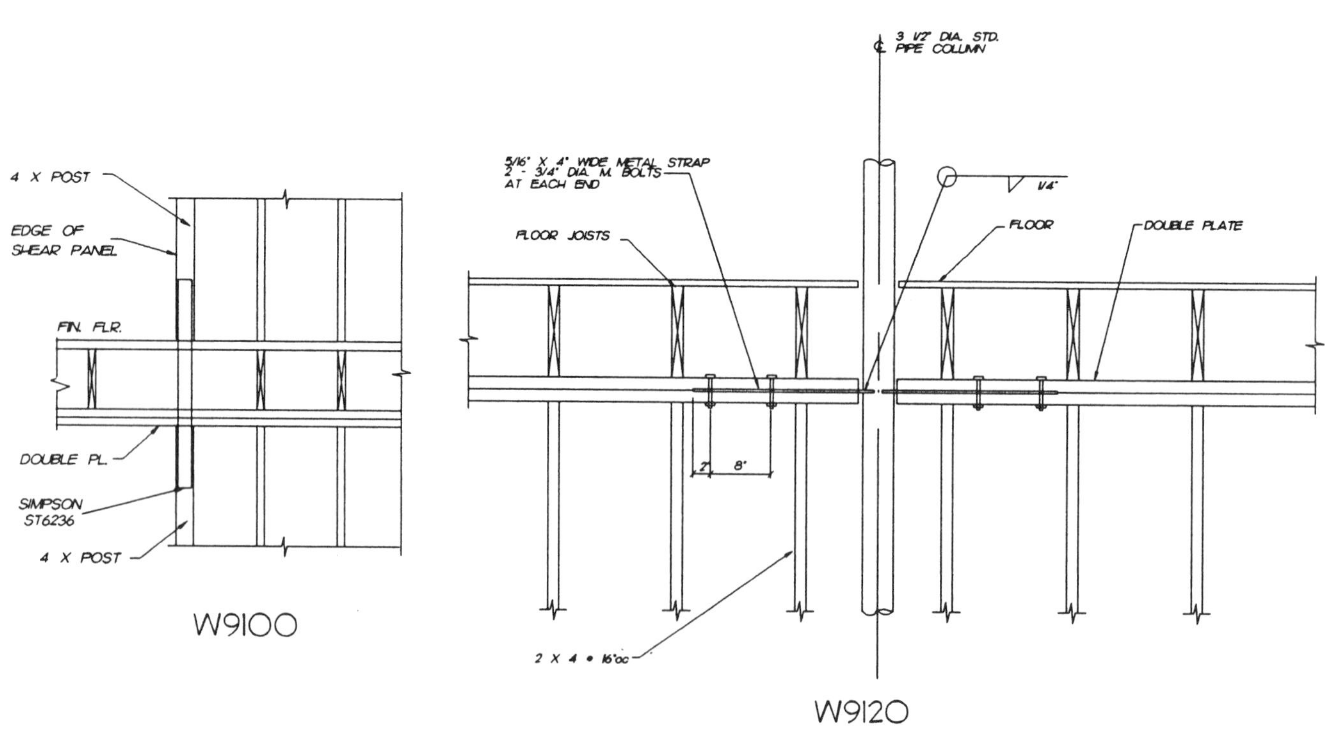
4 X POST
EDGE OF SHEAR PANEL
FIN. FLR.
DOUBLE PL.
SIMPSON ST6236
4 X POST
W9100
3 1/2" DIA. STD. PIPE COLUMN
5/16" X 4" WIDE METAL STRAP
2 - 3/4" DIA. M. BOLTS AT EACH END
1/4"
FLOOR JOISTS
FLOOR
DOUBLE PLATE
2"
8"
2 X 4 • 16"oc
W9120

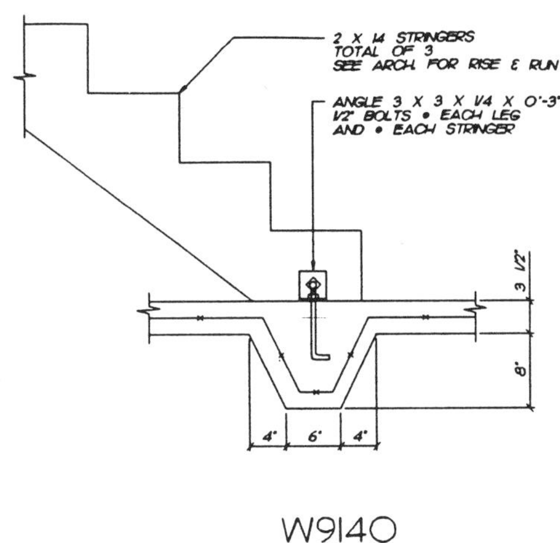

W9140

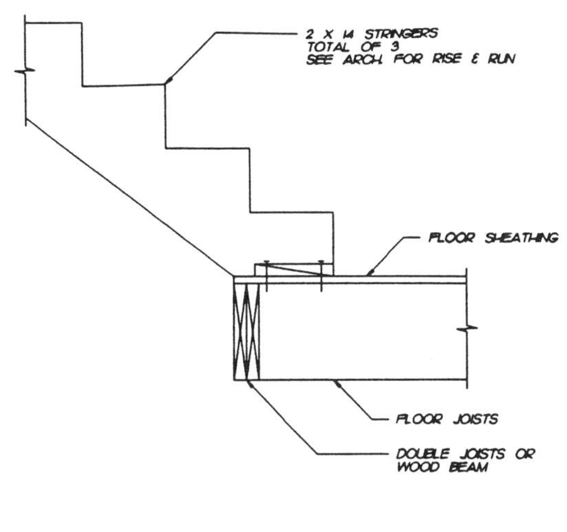

W9141

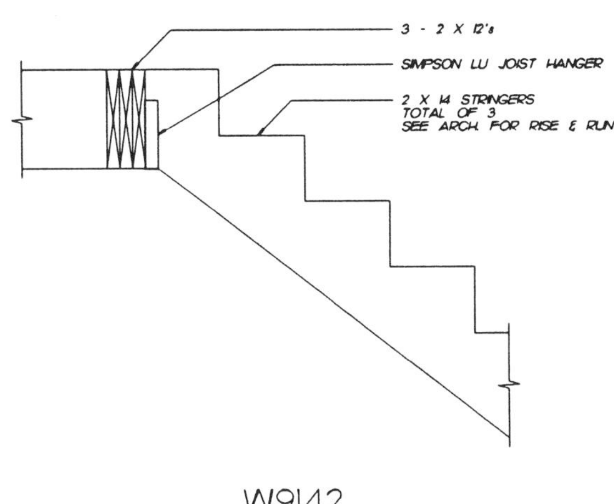

W9142

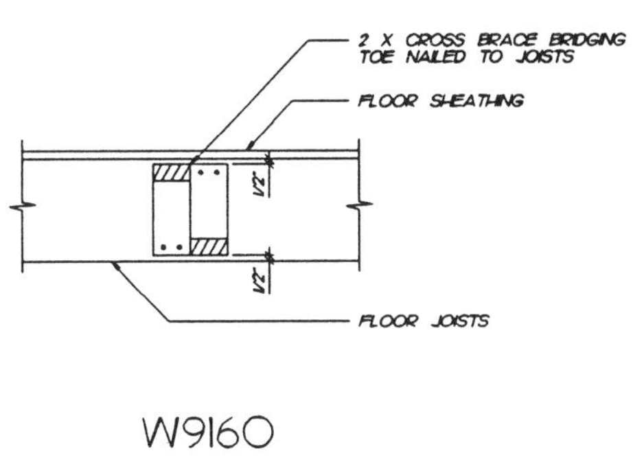

W9160

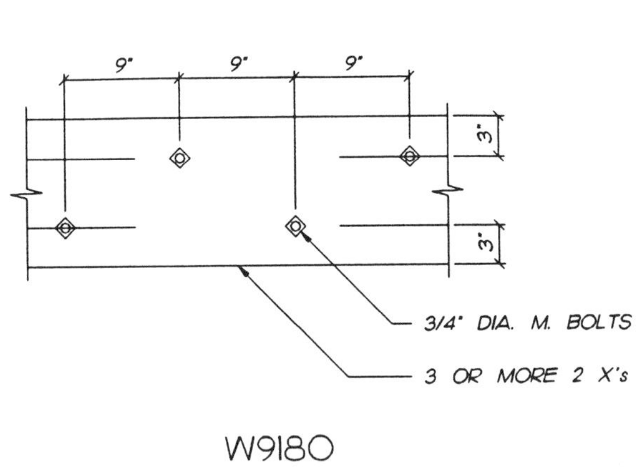

W9180

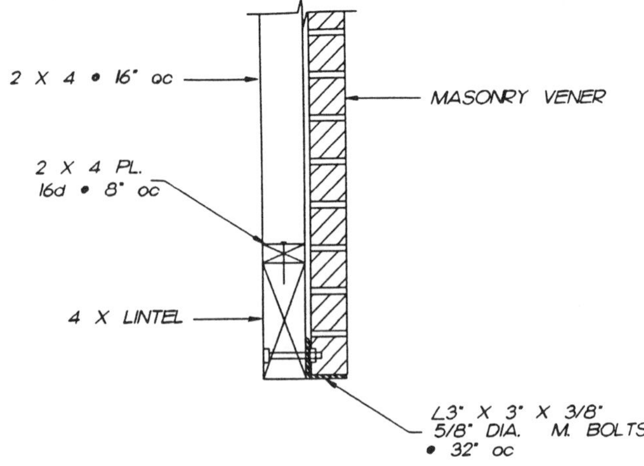

W9200

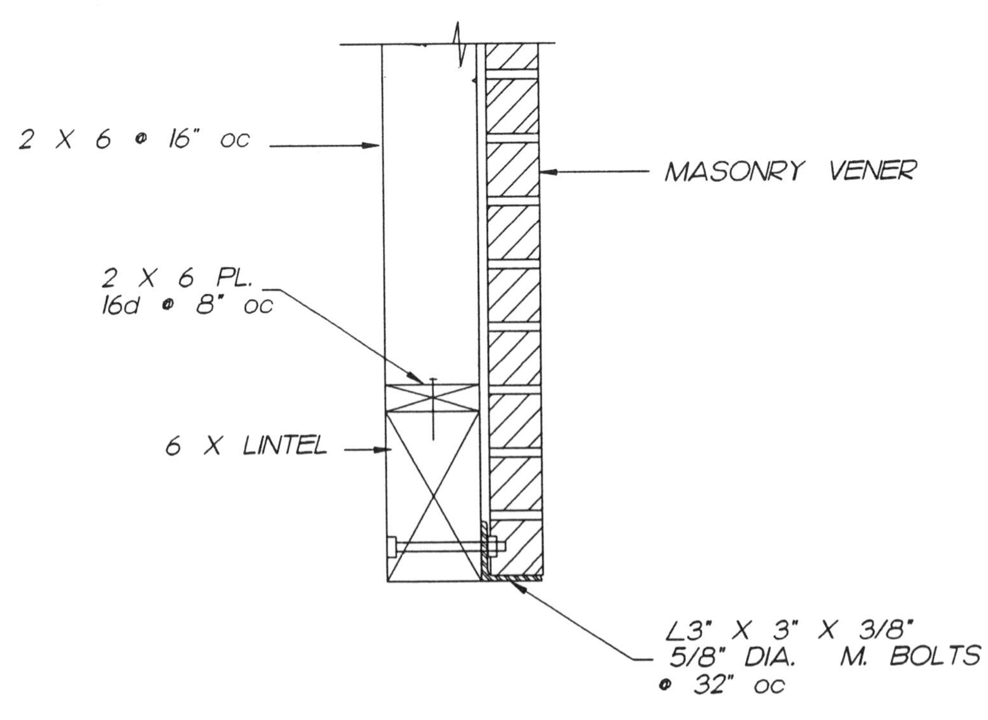

W9201

Masonry Structural Details

ne details and assemblies presented in this chapter apply to reinforced oncrete block and brick masonry. A certain amount of masonry is used details in other chapters; however, it is not their predominant material. ost of the details and assemblies presented specify dimensions of the oncrete blocks and bricks, and the reinforcement sizes and spacing. This formation is included to make the detail a more helpful example of the se of masonry as a construction material. Details are drawing symbols nd, as such, should not be construed as actual design drawings. The equired design data must be derived from design calculations. Standard ructural details should be used with an informed concern for the strength f the design.

The structural strength of masonry compares favorably with the lower alues of the f'_c of reinforced concrete, and so masonry is often used in eu of concrete. The use of masonry can resolve many fire rating problems nd also eliminate the cost of building the wood formwork required for oncrete. The most usual applications of masonry in construction are for alls, lintels, jambs, pilasters, and columns; it is rarely used for floor or oof spans. The masonry details are also categorized by their type and anufactured size. These categories of use, type, and size have resulted the creation of a large number of detail standards, which are shown in is chapter.

The accessibility of the information in the chapter is facilitated by arnging the material in a logical, relevant sequence so that the reader may eadily locate a particular detail. The basic concept of this arrangement places the drawings in an assigned hierarchy which starts with the m general condition, then progresses to a discrete number of associa explicit conditions.

Each drawing is identified by a coded designator starting with a mate identification character and followed by a four-digit number. Since t chapter is concerned with masonry details, the first character of the de designator is the letter M. The detail numbers which follow are coded represent a construction function section and the detail number in t particular section. The last digit of the detail number represents the nu ber of the variation of the original detail. This arrangement allows some degree of parallelism between the different chapters. For examp the numbers for the masonry columns are similar to those for the concr columns. However, an exact comparison between chapters will reveal t the drawings are not identical.

The details are organized in a sequence of sections, each of wh pertains to a particular function in the construction process. The first d of the detail number is used to describe the particular construction p pose or function of the detail. This chapter consists of nine function s tions, which are defined as follows:

Foundations	M1020 to M1xxx
Columns	M2020 to M2xxx
Walls and wood floors	M3020 to M3xxx
Walls and wood roofs	M4020 to M4xxx
Walls and concrete	M5020 to M5xxx
Walls and steel floors	M6020 to M6xxx
Retaining walls	M7020 to M7xxx
Connections	M8020 to M8xxx
Miscellaneous	M9020 to M9xxx

The following is an example of the number coding system:

Given No. M3142

The number 3 indicates that the detail is in the brick masonry walls a wood floors section; the number 14 indicates that it is the fourteer drawing in this section; and the number 2 indicates that it is the seco variation of the original configuration. A look at Detail M3142 shows brick masonry wall supporting a wood floor. The catalog charts prior the drawings are presented to assist the reader in locating details. T construction function section numbers at the top of the charts are arrang in numerical order, as outlined above. The detail descriptions and the respective particulars are stated in the left-hand columns, and the des nation numbers are stated in the right-hand column of the chart.

The two dominant types of structural masonry units used in mode construction are hollow-unit block and solid brick. Each of these types additionally defined by the material composition and basic configuratic Hollow masonry units are generally known as concrete blocks, since th are shell units, and the material used in their manufacture consists of sar cement, and natural crushed rock. The substance composition of concre blocks often depends on the availability or the natural source of the a gregate. Lightweight hollow masonry units use aggregates composed coal cinders, slag, shale, or volcanic ash. The sizes and shapes of the ma standard concrete blocks are shown in the drawings prior to the det drawings. These configurations are in accordance with the requiremer of the Concrete Masonry Association of California. Table 2-1 gives the AST designation numbers for the various types of concrete blocks.

ble 2-1. *ASTM Specifications for Concrete Block Masonry*

Masonry unit	Grade of concrete	ASTM Spec. No.
Hollow load bearing	A	C90
Solid load bearing	A	C145
Hollow nonbearing		C129
Concrete building brick	A	C53

Concrete block masonry is used to construct vertical structural members ch as walls, beams, pilasters, columns, and foundation walls. The blocks stacked vertically and joined to the adjacent blocks' horizontal and rtical surfaces by mortar joints. Since the actual dimensions of the units ⅜ in less than the nominal dimensions, the mortar joints are made so at the finished construction will equal an even inch. Except for three-arter-length units, the nominal dimensions of the lengths of concrete ock masonry units are in multiples of 8 in. It is recommended that the sign of concrete block construction utilize the vertical and horizontal minal modular dimensions of the units. The width of concrete block asonry members is determined by the actual size of the block; however, is designated by its nominal dimension. It is common practice to des-nate concrete block units in terms of their nominal dimensions in the quence of width, height, and length.

Hollow-unit masonry walls are constructed by lapping head joints of e units in alternate successive vertical courses by one-half the unit length provide a vertical unobstructed core within the wall. This arrangement the blocks, known as common bond, is the method most often used in e construction of masonry walls. A stack bond arrangement is constructed nen the blocks of each successive vertical course are placed directly over ch other and all the vertical cells and head joints are aligned. Stack bond nstruction requires that the wall be tied together with horizontal rein-rcement or that open-end blocks be used to obtain a bond between the ocks in the mortar head joints. The vertical core formed by the aligned lls of the masonry units provides a space for the reinforcing steel. Vertical res that contain reinforcement should be not less than 2 × 3 in in plan mensions and must be filled with cement grout.

Horizontal and vertical mortar joints of hollow-unit masonry construc-on are ⅜ in thick. The horizontal surface of a concrete block bed joint ould have a mortar cover on the exterior shell and cross web to form vertical core to contain the reinforcement. Any mortar that may overflow to the vertical core should be removed to allow the cement grout to ond to the interior surface of the core and to permit proper reinforcement earance. The vertical mortar joints, or the head joints, as they are called, ould be filled with mortar for a distance in from the exterior face of e block equal to, but not less than, the thickness of the face shell of the ock. The mortar joints between concrete blocks should be made straight d with a uniform thickness. Excess mortar that is squeezed from a joint the result of positioning a block should be removed with a trowel. oncrete block is usually constructed with either flush or concave bed ints (horizontal) and head joints (vertical). Concave joints are made by oling the surface of the joint so that the mortar will be compressed into e joint rather than being removed from the surface; however, mortar ints should be tooled only when the mortar is partially set and still plastic. Also, mortar joints should be made flush with the shell surface the block whenever the surface is to be covered with a coat of plaster

Horizontal and vertical cells of concrete blocks that contain reinforci bars, ledger bolts, and other inserts must be filled solid with cement grc to secure them in position. Grout pours should not be made in lifts high than 4′-0″, and the grout should be well consolidated in place. Cleanc holes should be provided at the bottom of all cores that are to be fille and a period of 1 hour should elapse between successive 4′-0″-high grc pours. Grout pours should be stopped 1½ in below the top of the cell a course so that a key for the succeeding pour can be formed. The grouti of masonry beams within walls should be performed in one operatic and the tops of the masonry cells directly below the beam that are remain unfilled with grout should be covered with metal lath to preve grout leakage. Masonry beams and lintels should be supported in pla with shores. It is recommended that these shores remain in place for period of not less than seven days, or at least until the member is capab of supporting its own weight and any construction load that may occu

Many of the concrete hollow-unit masonry details that are presented this chapter are also presented as details constructed of reinforced groute solid brick masonry. This type of construction consists of two or mo wythes of bricks which are bonded together by mortar joints and by solid vertical core of cement grout between the wythes. The grout, tl mortar, and the bricks are bonded to each other to such a degree th they will react as a monolithic material. The structural capacity of a rei forced grouted brick wall depends on the area of reinforcement, the quali of the mortar and cement grout, the durability of the brick, the size ar compressive strength of the brick, and the water absorption factor. Bric are manufactured by molding mixtures of clay and shale into oblo shapes, which are then hardened by a kiln burning process. Each man facturer produces its own particular size of bricks, which are genera around 3.5 in square × 12 in long. Although the materials used in tl manufacture of bricks may vary, depending on the local source of suppl they must conform to the requirements of ASTM Specification C62. Als bricks are produced with a variety of surface textures, colors, and finishe Table 2-2 lists the standard nominal modular sizes of brick masonry uni

The mortar joints of reinforced brick masonry should be straight ar have a uniform thickness, the bricks should be laid in full head and be joints, the joints should be not less than ½ in thick, and any excess mort should be removed from the surface of the bricks after they are place Bricks should be dampened at the time they are laid to prevent the initi suction of the surface from removing too much water from the mortar

Table 2-2. *Standard Nominal Modular Sizes of Brick*

	Face dimensions in the wall	
Thickness, in	**Height, in**	**Length, in**
4	2	12
4	2⅔	8
4	2⅔	12
4	4	8
4	4	12
4	5⅓	8
4	5⅓	12

out. The suction of the brick is a prime factor in creating the bond etween the mortar and the brick; a mortar mixture that contains a larger nount of water will have a higher bond strength. When a brick member r wall is constructed of more than two wythes, the interior brick should e floated into place with not less than ¾ in of cement grout surrounding . Brick walls are constructed with the same common bond or stack bond rangement as that described for hollow-unit masonry. The method of oling mortar joints and the time required for soffit shoring previously escribed for hollow-unit masonry, also apply to reinforced grouted brick nasonry.

Detail M9065 shows the minimum clearance dimensions required for einforcing steel in grouted brick masonry walls. The grout space should ot be less than the sum of the diameters of the vertical and horizontal einforcing bars plus ¼ in clear on each side of the bars; note that point ontact between the vertical and horizontal bars is permitted. It is recmmended that the grout space be not less than 2 in thick for reinforced routed masonry and not less than ¾ in thick for unreinforced grouted nasonry. Reinforcing steel serves the same function in grouted masonry s in reinforced concrete, that is, to resist tension or compression caused y external forces. The methods and principles of working stress design or grouted masonry are the same as those for reinforced concrete, except hat the allowable working stresses are adjusted for the masonry materials sed. The reinforcing bars are embedded in the grout space between the rick tiers or hollow-unit shells and are spaced and covered as in Table -3. Grout pour height requirements for brick walls vary in different building codes; however, it is generally recommended that low-lift grout pours ot exceed 12 in in height for vertical core widths less than 2 in. It is also ecommended that high-lift grout pours may be made when the vertical ore is 2 in or more in width and the grout lift height does not exceed 8 times the core width for mortar-type grout, or 64 times the core width or pea gravel grout, or a maximum height of 12′-0″. High-lift grout pours lso require that the exterior tiers of the wall be tied together with recangular ties of No. 9 gauge wire, 4 in wide by 2 in long and spaced 24 n o.c. horizontally and 16 in o.c. vertically, for walls using a common ond arrangement of the bricks. When a stacked bond arrangement of the ricks is used, the wire ties between the exterior tiers should be spaced

able 2-3. *Minimum Reinforcing Steel Spacing and Cover in Grouted Masonry*

ocation of reinforcement in masonry	Distance
laximum spacing between reinforcing bars in walls	48″
linimum spacing between parallel bars	1 bar diameter or not less than 1″
linimum cover of reinforcing bars at the bottom of foundations	3″
linimum cover of reinforcing bars in ertical members exposed to weather or soil	2″
linimum cover of reinforcing bars in olumns and at the bottoms or sides of irders or beams	1½″
linimum cover of reinforcing bars in nterior walls	¾″

Note: Reinforcing bars that are perpendicular to each other are permitted point contact t their intersection.

24 in o.c. horizontally and 12 in o.c. vertically. The grout pours for rei forced masonry should be terminated 1½ in below the top of the brick to form a key for the succeeding pour. The Uniform Building Code of th International Conference of Building Officials recommends that high-li grout pours not exceed 4′-0″. Also, grout should be consolidated by v brating or puddling to achieve a bond with the reinforcing bars an masonry.

Grout and mortar are a mixture of water, portland cement, sand ag gregate, and lime putty or hydrated lime. The proportions of the mixtur are specified by the volume ratio of the ingredients. Mortar and grout ar mixed in the ratio of 1 part portland cement, ¼ to ½ part lime putty o hydrated lime, and 2½ to 3 parts damp, loose sand. When the grout is t be poured in a space that is 3 in or more in width, the mix may be 1 par portland cement, 2 to 3 parts damp, loose sand, and 2 parts pea gravel o ⅜-in aggregate. Grout should have a fluid consistency when it is pumpe into place; however, it should not be so fluid that the constituent aggregate of the mixture will segregate. Mortar should be used within 1½ hour after it is mixed, but it may be retempered with water during that time t maintain a workable plasticity. Each building code specifies requirement for mortar and grout mixtures; however, the general requirements shoul conform to ASTM Specification C270 for mortar and ASTM Specificatio C476 for grout. Since mortar and grout are composed of water and cemen the temperature conditions at the time they are placed can be critical. Th temperature of masonry should be maintained at 50°F for 24 hours fo mixes composed of high-early-strength cement and 72 hours for mixe composed of regular types of cement.

Acceptable construction of hollow-unit masonry and grouted brick ma sonry depends a great deal on the quality of the workmanship in the fiel Problems will often occur involving the placement of reinforcing steel the locations and the details for embedded conduits or pipes, the placin and aligning of construction joints, and the waterproofing of wall surfaces The engineer should try to anticipate these conditions so that the con struction can proceed without delays or extra costs. The size and locatio of reinforcing bars should be accurately dimensioned on the drawings vertical reinforcement should be secured in place at the top and the botton and at an distance not exceeding 192 bar diameters apart; horizontal rein forcing bars should be placed in bond beam or channel blocks in hollow unit masonry and within grout widths in brick; and reinforcing bars shoul be lapped 30 bar diameters or not less than 24 in. Possible problem caused by locating pipes or conduits that pass through a masonry wall o are embedded within a wall can be avoided by coordinating the structura drawings with the requirements of the mechanical and electrical drawings The location and the method of providing for the pipe or conduit in the wall should be shown on the structural drawings; all pipes that pass throug masonry walls should be sleeved with a standard wrought iron pipe to prevent the pipe from bonding with the masonry. Pipes and conduits i cores that are not filled with cement grout are not considered as embedde in the wall. However, when a pipe or conduit is embedded in masonry construction, care should be taken to ensure that it is not located at a poin of high shear or flexure stress so that it does not reduce the structura value of the wall.

Masonry walls expand or contract, depending on the climate conditions and the composition of the material. The coefficient of thermal expansior of hollow masonry and brick masonry will vary in relation to the materials used in the manufacture; however, the engineer should consider the ne essity for vertical construction joints in long walls to allow for thermal rowth, particularly in climates where the extreme temperature values will ary over a wide range in a relatively short time.

There are many recommended types and spacings of vertical wall joints. n general, a vertical wall joint should be of sufficient width and flexibility o permit the free horizontal movement of the wall and also be capable f maintaining the structural capacity of the wall. This can be accomplished y aligning a full-height vertical joint through the wall cross section. The oint space is filled with a compressible bituminous or waterproof material, nd the horizontal reinforcing bars are lapped across the joint but are not onded to the grout on one side. The absence of expansion joints in nasonry walls can be a source of severe cracking and can therefore reduce he structural value of the wall.

Masonry is a comparatively porous material; if it is not waterproofed nd is exposed to normal weather conditions, it will allow a certain amount f moisture to permeate it, especially if the walls are approximately 8 in hick. Moisture that penetrates a wall from the exterior to the interior of he building will leave a white chalklike stain on the interior surface of he masonry. This phenomenon is caused by efflorescence, which occurs when the moisture that passes through the wall evaporates from the interior surface, leaving a white, dry, water-soluble salt residue that is derived rom the masonry material. Efflorescence can be prevented by waterproofing the exterior surface of the wall and by sealing any cracks in the mortar joints by retooling them. Other sources of water intrusion in masonry walls can be traced to insert openings or cracks, to exposed grout surfaces at door and window openings, and to the tops of parapet walls. There are many commerical products available for waterproofing masonry walls. The recommended methods and frequency of application of these materials are usually specified in their guarantee of performance. All masonry walls that are constructed below grade should be waterproofed with two layers of either asphalt or coal tar pitch material. Materials specifically manufactured for this purpose are also available and should be applied o the exterior surface of the wall in accordance with the manufacturer's ecommendations. A definite method of removing large amounts of water hat may collect adjacent to walls below grade should also be provided. This can be done either by installing perforated ceramic drain tiles adjacent o the bottom of the wall or by backfilling the area adjacent to the bottom of the wall with a continuous pocket of crushed rock or coarse aggregate. The tiles or crushed rock backfill should be sloped to permit the water o flow away from the wall. Exposed grout surfaces of masonry walls at windows, doors, and parapets should be covered with galvanized sheet metal flashing. Jamb, sill, and header frames at doors and windows should be pressure-caulked to the masonry surface. The layers of the roofing material should extend above the roof onto the surface of the parapet wall and should be either flashed into the wall or extended over the top of the wall and covered by the parapet sheet metal flashing. Improper moisture protection of masonry walls can be a source of much damage within a building. Small leaks around windows and doors or through parapet copings may cause damage to large areas of finished plaster walls and ceilings and expensive floor covering materials; therefore, it is quite important that positive steps be taken in the construction of a building to eliminate the possibility of water leakage.

Reinforced grouted masonry is designed by the same basic methods and principles used to design reinforced concrete. It is also assumed that grouted masonry will react to externally applied loads in the same manner as reinforced concrete will react, that is, that the masonry is not capab of resisting tensile stress, that the tension in a masonry member is resiste only by the reinforcing steel, and that the cement grout and the reinforcin steel are bonded together and will react as an analogous monolithic m terial. The primary difference between reinforced concrete and reinforce grouted masonry is the working stress values assigned to the groute masonry by the various building codes or design criteria. These valu are determined by compressive tests on the masonry and are specified f'_m. The values of f'_m depend on the type of masonry used and on th quality of field workmanship during construction. The values of the a lowable unit stresses on masonry acting in compression, shear, and bon are listed in the building codes for masonry construction performed wit continuous inspection and without continuous inspection. Since the qualit of field workmanship is an important factor in the strength of reinforce grouted masonry, the allowable unit working stresses with continuou inspection are double the allowable stresses that do not require continuou inspection. Valid continuous inspection requires that a registered deput building inspector be on the job during the execution of all masonr construction and that he or she inspect the work and verify that it is bein performed in accordance with the building code requirements and th it complies with the engineer's structural design. The building code is nc a design manual; it is a specification of the minimum safe design an construction requirements. Engineers, contractors, and field inspector should bear this fact in mind when they use masonry as a constructio material. This is not to be construed as meaning that masonry is not good construction material; rather, it is to stress that the quality of th workmanship is a determining factor in the structural strength of masonr used as a structural material. No construction is better than the least tech nically qualified person on the job, regardless of whether that person i designing, drafting, supervising, or mixing mortar.

Section 1: Foundations. Details M1020 to M1125 show a series o concrete continuous footings for 8-in-thick and 12-in-thick masonry walls Details M1140 to M1234 show a series of concrete continuous footings fo brick masonry walls. The footing configurations are rectangular, L-shaped or inverted-T-shaped. The balance of the details in this section are con cerned with concrete subgrade stairs and walls.

Section 2: Columns. Details M2020 to M2024 show a series of plan sections of the arrangement of reinforcement in rectangular hollow-uni masonry columns. Details M2040 to M2042 show a series of plan section of the arrangement of reinforcement in rectangular brick masonry col umns. Details M2060 to M2084 show the plan sections of steel column connected to a masonry wall.

Section 3: Masonry Walls and Wood Floors. Details M3020 tc M3109 show a series of wood floor connections to either 8-in- or 12-in thick concrete block walls. Details M3120 and M3121 show sections of full height concrete block walls supporting wood floors. Details M3140 tc M3224 show a series of wood floor connections to brick masonry walls.

Section 4: Masonry Walls and Wood Roofs. Details M4020 tc M4203 show a series of wood roof connections to 8-in-thick concrete block walls. Details M4200 and M4201 show sections of full-height concrete block walls supporting wood roofs. Details M4240 to M4423 show a series of wood roof connections to brick masonry walls.

ction 5: Masonry Walls. Details M5020 to M5060 show a series of ıe-story concrete block walls supporting concrete slabs. Details M5080 M5083 show a series of two-story concrete block walls supporting con-ete slabs.

ction 6: Masonry Walls and Steel Floors. Details M6020 to 6221 show a series of details of concrete block walls supporting metal ecking and concrete floors. Details M6240 to M6440 show a series of etails of brick masonry walls supporting metal decking and concrete oors. Floors are supported by either steel beams, metal light gage joists, bar joists.

ction 7: Cantilever Retaining Walls. Details M7020 to M7061 ow a series of concrete block retaining walls for a variety of backfill opes with the retained soil not over the footing. Details M7080 to M7121 ow a series of concrete block retaining walls for a variety of backfill opes with the retained soil over the footing.

ction 8: Masonry Wall Connections. Details M8020 to M8201 ow the corner and intersection connection of masonry walls.

ction 9: Miscellaneous. This section contains various details that ay be needed but do not fit into any of the previous sections. The section ows details of wall openings, jambs and lintels, wall and slab connections, ıd stairs.

CHAPTER 2	MASONRY	SECTION I	FOUNDATIONS
CONTINUOUS FOOTING EXTERIOR WALL SLAB ON GRADE	RECTANGULAR SHAPE	8" CONCRETE BLOCK	M1020 - M1023
		12" CONCRETE BLOCK	M1024 - M1027
		BRICK MASONRY	M1140 - M1143
CONTINUOUS FOOTING EXTERIOR WALL SLAB ON GRADE WITH A CURB	RECTANGULAR SHAPE	8" CONCRETE BLOCK	M1040 - M1043
		12" CONCRETE BLOCK	M1044 - M1047
		BRICK MASONRY	M1160 - M1163
CONTINUOUS FOOTING EXTERIOR WALL SLAB ON GRADE	L SHAPE	8" CONCRETE BLOCK	M1060 - M1063
		12" CONCRETE BLOCK	M1064 - M1066
		BRICK MASONRY	M1180 - M1183
CONTINUOUS FOOTING EXTERIOR WALL SLAB ON GRADE	T SHAPE	8" CONCRETE BLOCK	M1080 - M1083
		12" CONCRETE BLOCK	M1084
		BRICK MASONRY	M1200 - M1203
CONTINUOUS FOOTING INTERIOR WALL SLAB ON GRADE	RECTANGULAR SHAPE	8" CONCRETE BLOCK	M1100 - M1103
		12" CONCRETE BLOCK	M1104 - M1106
		BRICK MASONRY	M1220 - M1223
CONTINUOUS FOOTING INTERIOR WALL SLAB ON GRADE	T SHAPE	8" CONCRETE BLOCK	M1120 - M1123
		12" CONCRETE BLOCK	M1124 - M1125
		BRICK MASONRY BLOCK	M1240 - M1243
CONCRETE STEPS ON GRADE MASONRY SIDE WALLS	————	8" CONCRETE	M1260 -M1261
		BRICK MASONRY	M1262
ELEVATOR PIT	————	————	M1280 - M1282
LOADING DOCK WALL	————	————	M1300
MASONRY WALL & WOOD FLOOR CONNECTION AT GRADE	————	JOISTS PERP.	M1340
		JOISTS PARALELL	M1341

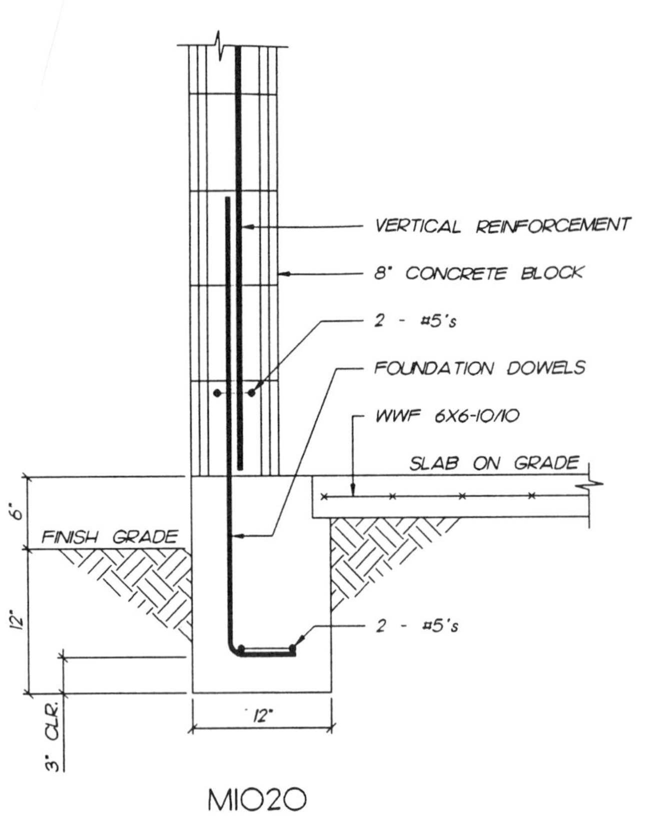
VERTICAL REINFORCEMENT
8" CONCRETE BLOCK
2 - #5's
FOUNDATION DOWELS
WWF 6X6-10/10
SLAB ON GRADE
6"
FINISH GRADE
12"
2 - #5's
3" CLR.
12"

M1020

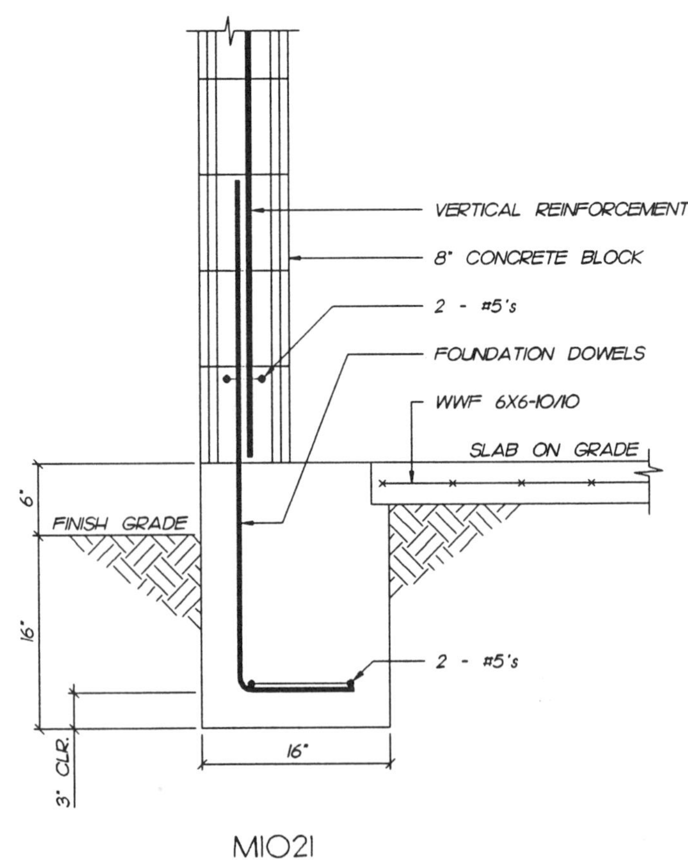
VERTICAL REINFORCEMENT
8" CONCRETE BLOCK
2 - #5's
FOUNDATION DOWELS
WWF 6X6-10/10
SLAB ON GRADE
6"
FINISH GRADE
16"
2 - #5's
3" CLR.
16"

M1021

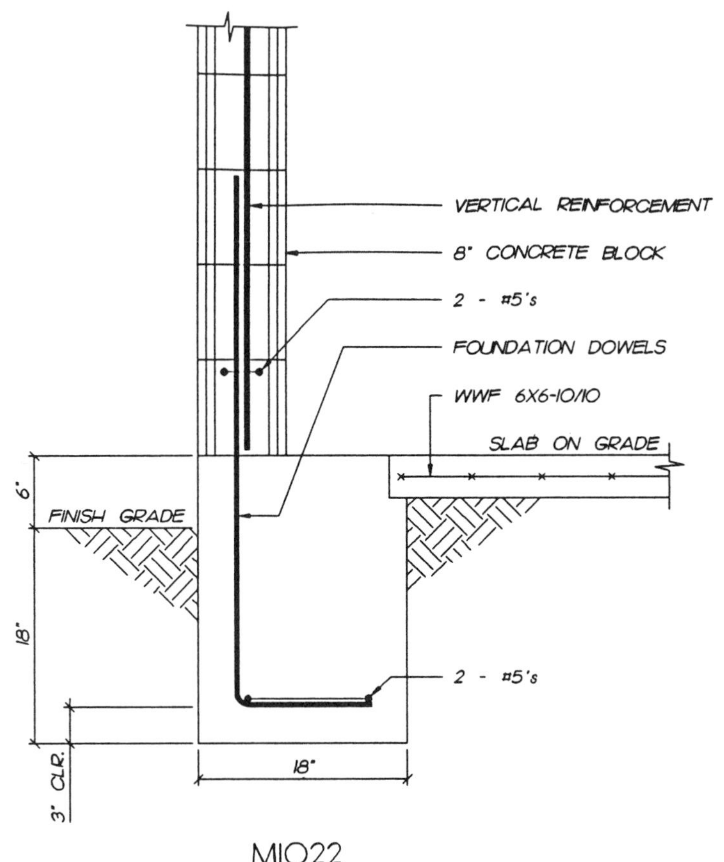
VERTICAL REINFORCEMENT
8" CONCRETE BLOCK
2 - #5's
FOUNDATION DOWELS
WWF 6X6-10/10
SLAB ON GRADE
6"
FINISH GRADE
18"
2 - #5's
3" CLR.
18"

MIO22

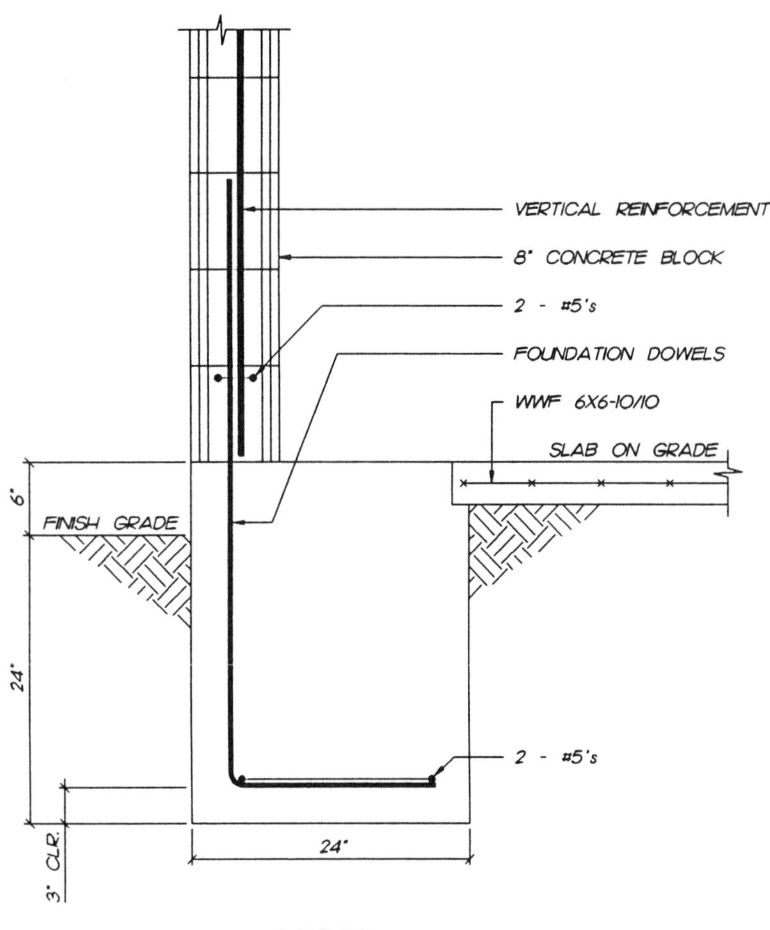
VERTICAL REINFORCEMENT
8" CONCRETE BLOCK
2 - #5's
FOUNDATION DOWELS
WWF 6X6-10/10
SLAB ON GRADE
6"
FINISH GRADE
24"
2 - #5's
3" CLR.
24"

MIO23

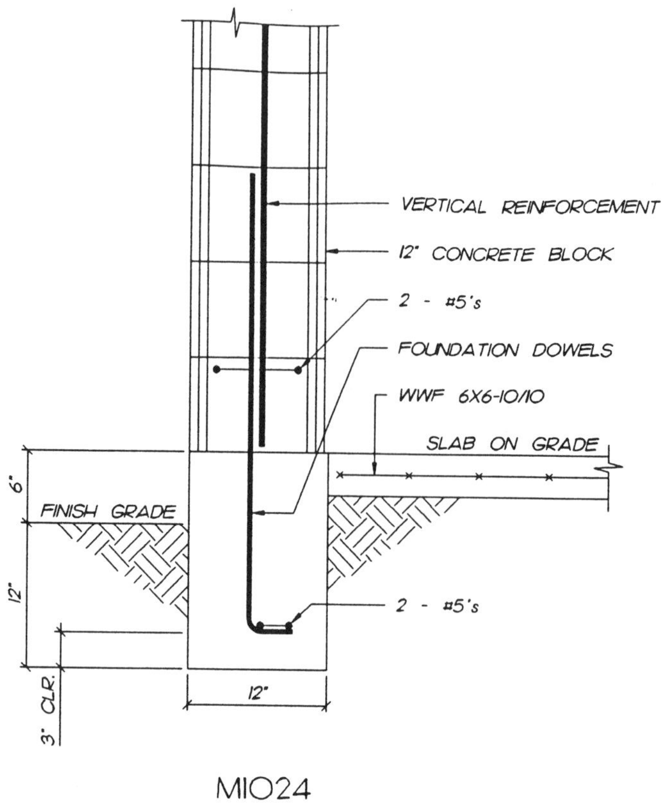

MIO24

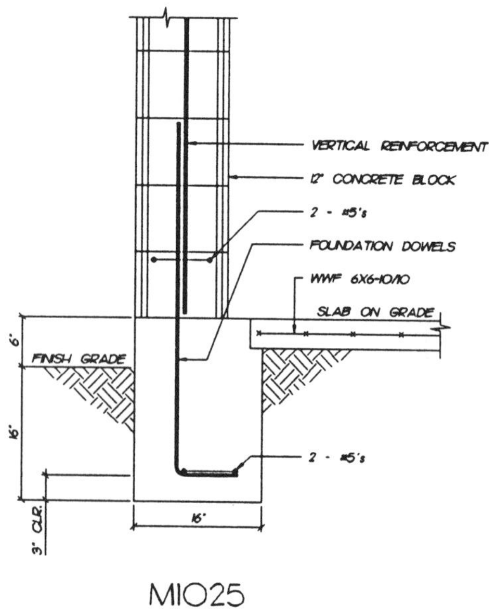

MIO25

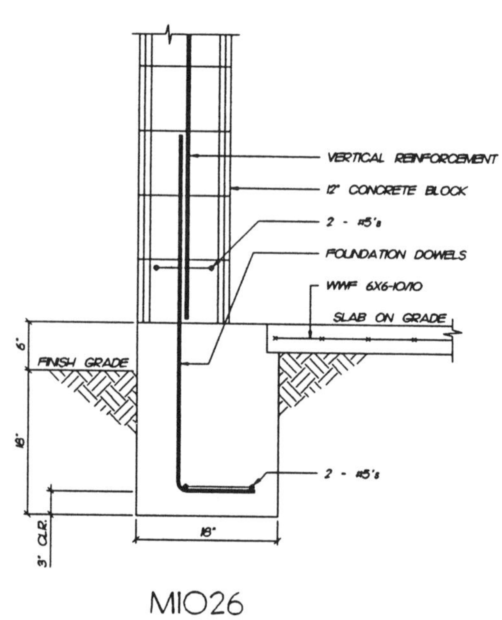

MIO26

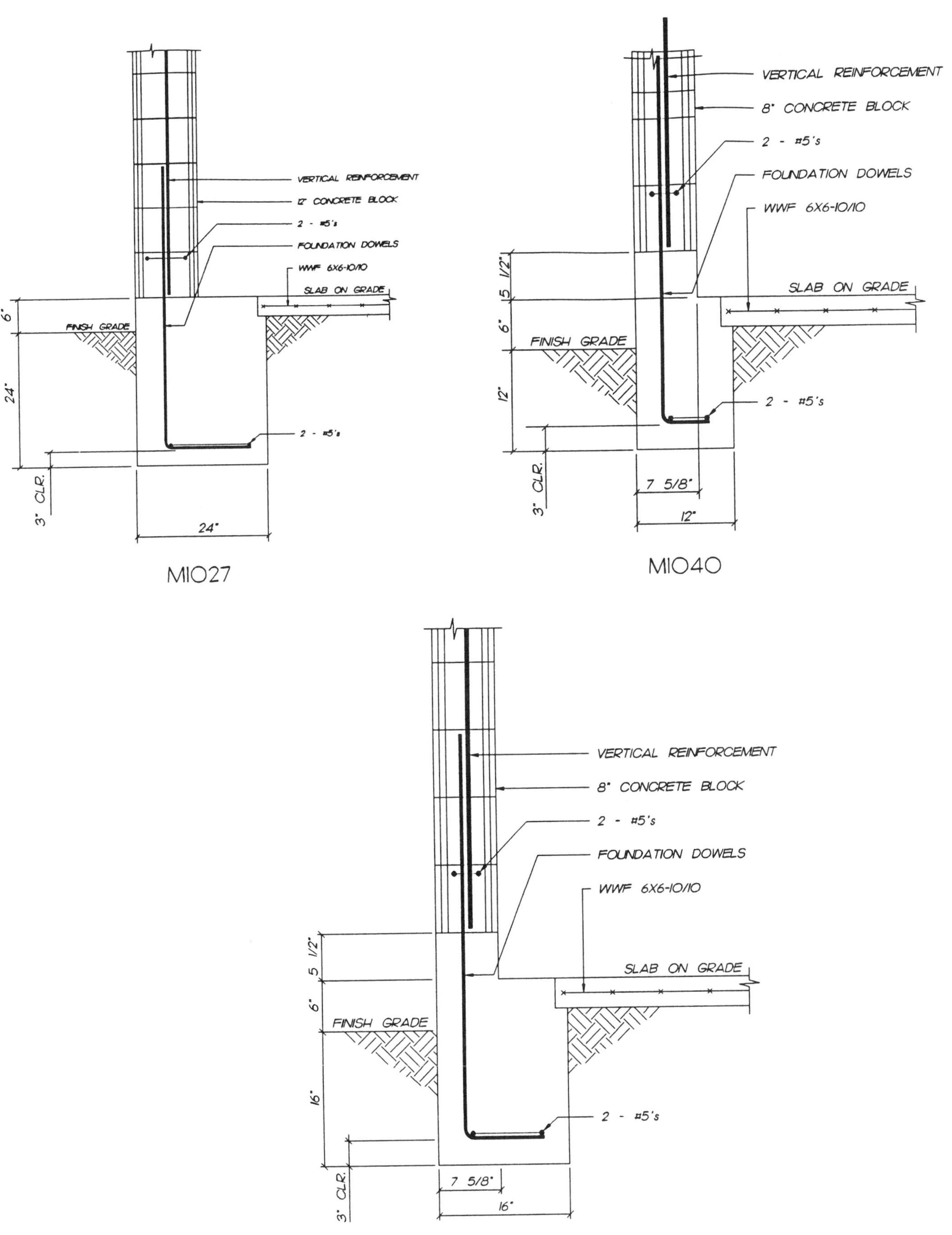
VERTICAL REINFORCEMENT
12" CONCRETE BLOCK
2 - #5's
FOUNDATION DOWELS
WWF 6X6-10/10
SLAB ON GRADE
FINISH GRADE
6"
24"
3" CLR.
2 - #5's
24"
M1027
VERTICAL REINFORCEMENT
8" CONCRETE BLOCK
2 - #5's
FOUNDATION DOWELS
WWF 6X6-10/10
SLAB ON GRADE
FINISH GRADE
5 1/2"
6"
12"
3" CLR.
2 - #5's
7 5/8"
12"
M1040
VERTICAL REINFORCEMENT
8" CONCRETE BLOCK
2 - #5's
FOUNDATION DOWELS
WWF 6X6-10/10
SLAB ON GRADE
FINISH GRADE
5 1/2"
6"
16"
3" CLR.
2 - #5's
7 5/8"
16"
M1041

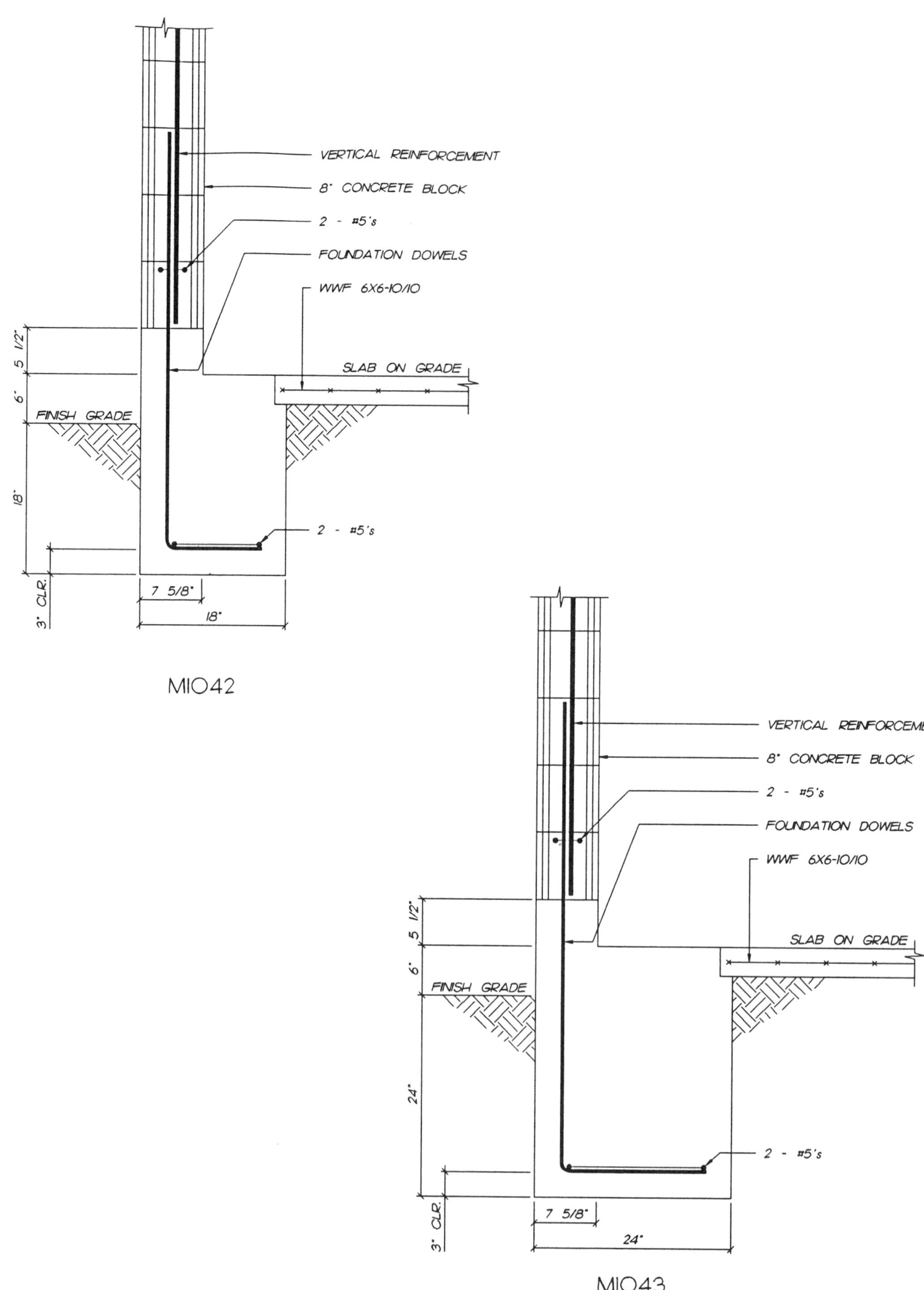
VERTICAL REINFORCEMENT
8" CONCRETE BLOCK
2 - #5's
FOUNDATION DOWELS
WWF 6X6-10/10
SLAB ON GRADE
5 1/2"
6"
FINISH GRADE
18"
2 - #5's
3" CLR.
7 5/8"
18"
MIO42
VERTICAL REINFORCEME
8" CONCRETE BLOCK
2 - #5's
FOUNDATION DOWELS
WWF 6X6-10/10
SLAB ON GRADE
5 1/2"
6"
FINISH GRADE
24"
2 - #5's
3" CLR.
7 5/8"
24"
MIO43

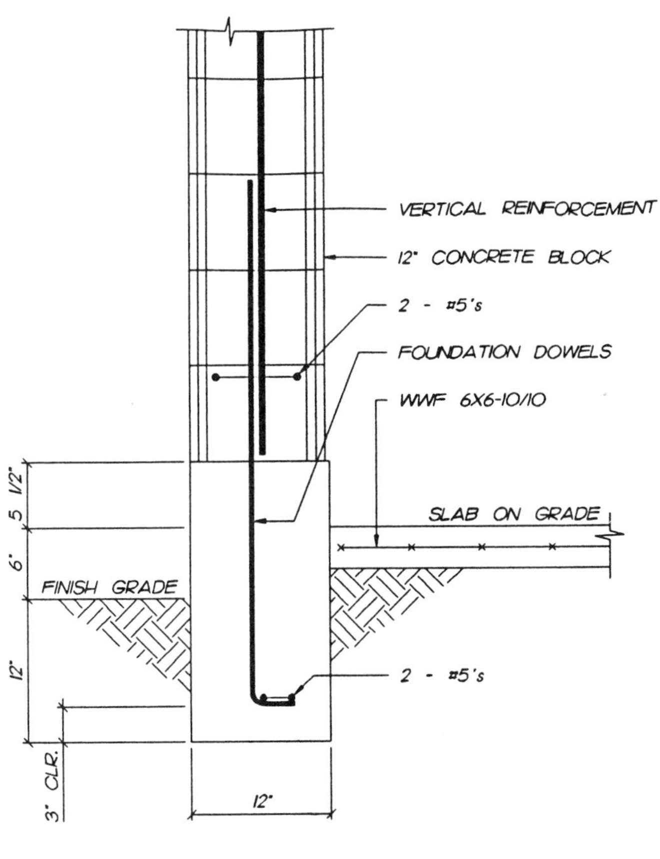
VERTICAL REINFORCEMENT
12" CONCRETE BLOCK
2 - #5's
FOUNDATION DOWELS
WWF 6X6-10/10
SLAB ON GRADE
FINISH GRADE
2 - #5's
5 1/2"
6"
12"
3" CLR.
12"

MIO44

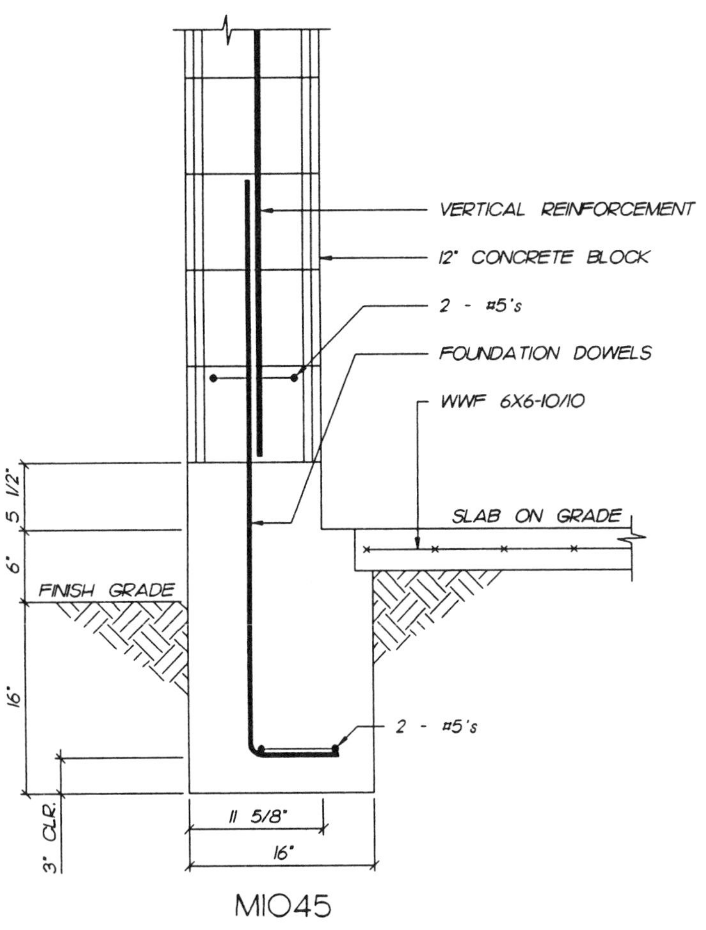
VERTICAL REINFORCEMENT
12" CONCRETE BLOCK
2 - #5's
FOUNDATION DOWELS
WWF 6X6-10/10
SLAB ON GRADE
FINISH GRADE
2 - #5's
5 1/2"
6"
16"
3" CLR.
11 5/8"
16"

MIO45

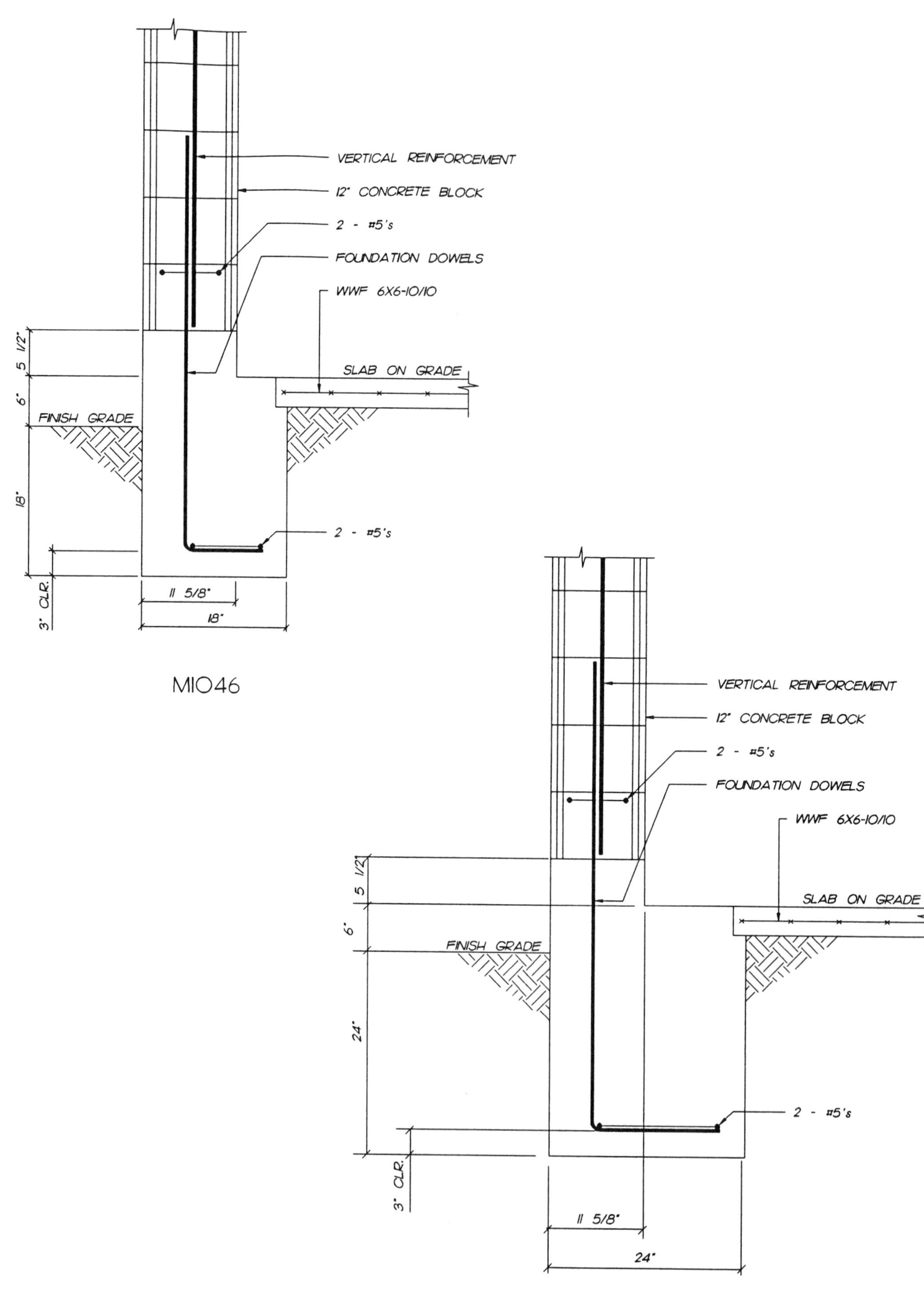
VERTICAL REINFORCEMENT
12" CONCRETE BLOCK
2 - #5's
FOUNDATION DOWELS
WWF 6X6-10/10
SLAB ON GRADE
FINISH GRADE
5 1/2"
6"
18"
2 - #5's
3" CLR.
11 5/8"
18"
MIO46
VERTICAL REINFORCEMENT
12" CONCRETE BLOCK
2 - #5's
FOUNDATION DOWELS
WWF 6X6-10/10
SLAB ON GRADE
FINISH GRADE
5 1/2"
6"
24"
2 - #5's
3" CLR.
11 5/8"
24"
MIO47

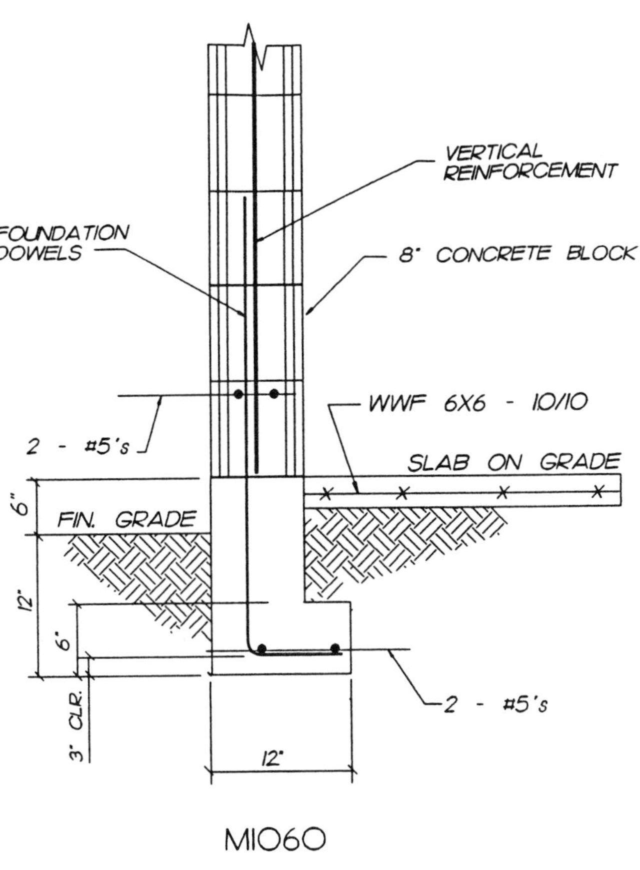
VERTICAL
REINFORCEMENT
FOUNDATION
DOWELS
8" CONCRETE BLOCK
WWF 6X6 - 10/10
2 - #5's
SLAB ON GRADE
6"
FIN. GRADE
12"
6"
2 - #5's
3" CLR.
12"

M1060

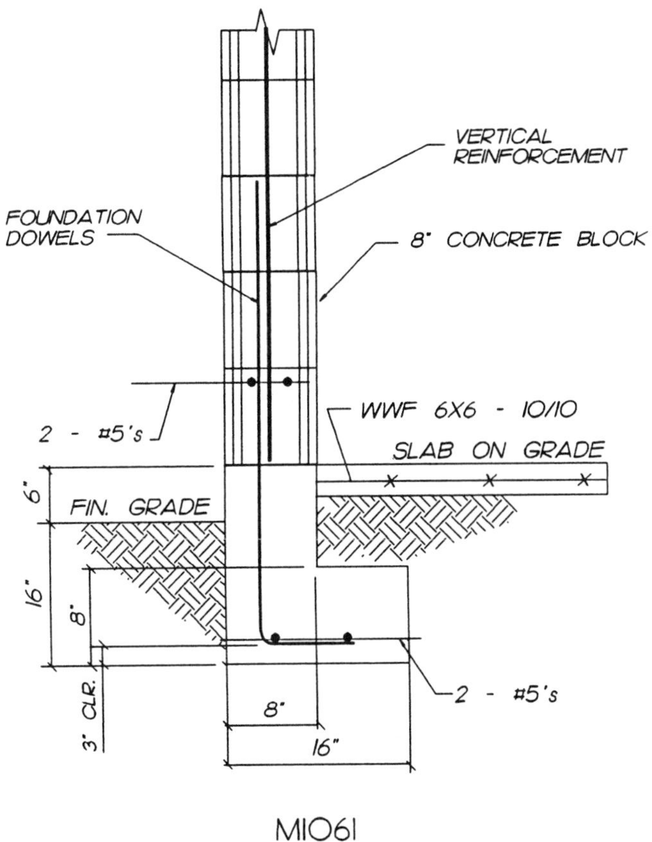
VERTICAL
REINFORCEMENT
FOUNDATION
DOWELS
8" CONCRETE BLOCK
WWF 6X6 - 10/10
2 - #5's
SLAB ON GRADE
6"
FIN. GRADE
16"
8"
2 - #5's
3" CLR.
8"
16"

M1061

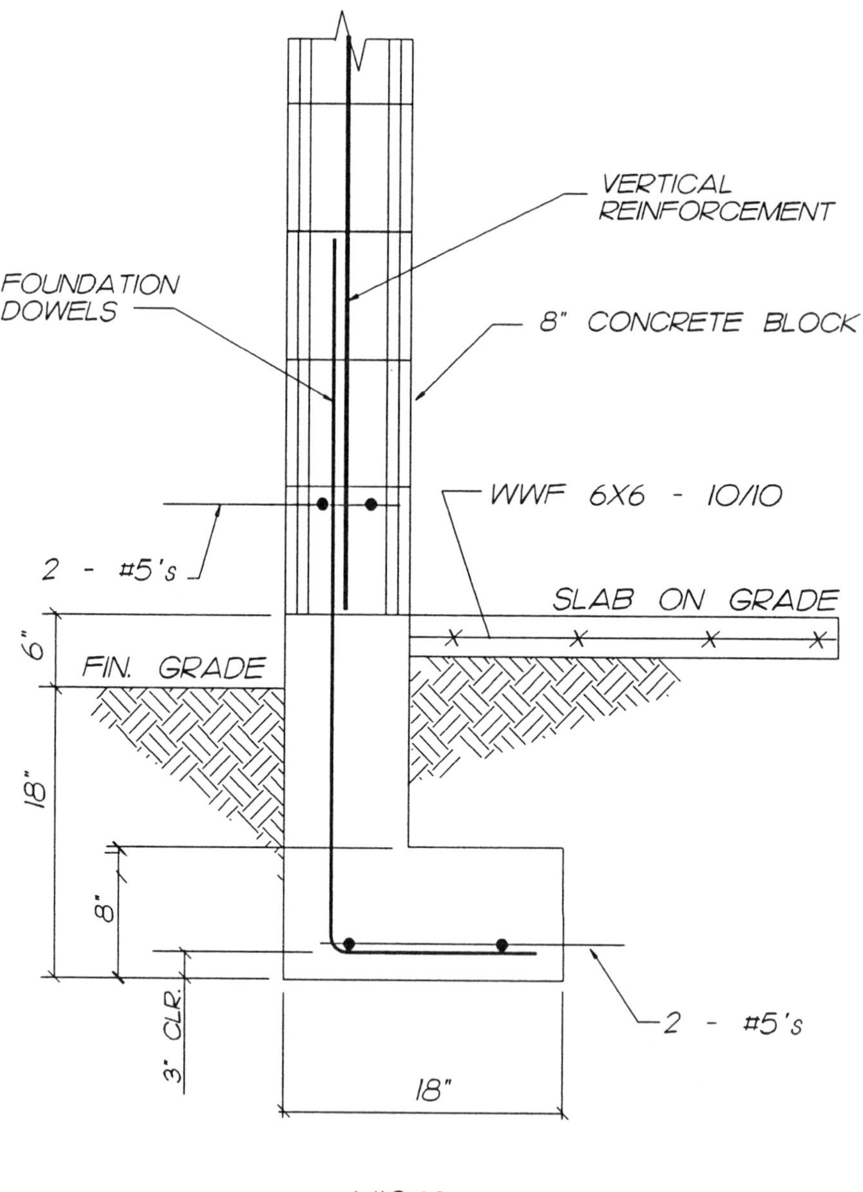

VERTICAL
REINFORCEMENT
FOUNDATION
DOWELS
8" CONCRETE BLOCK
WWF 6X6 - 10/10
2 - #5's
SLAB ON GRADE
6"
FIN. GRADE
18"
8"
3" CLR.
2 - #5's
18"

MIO62

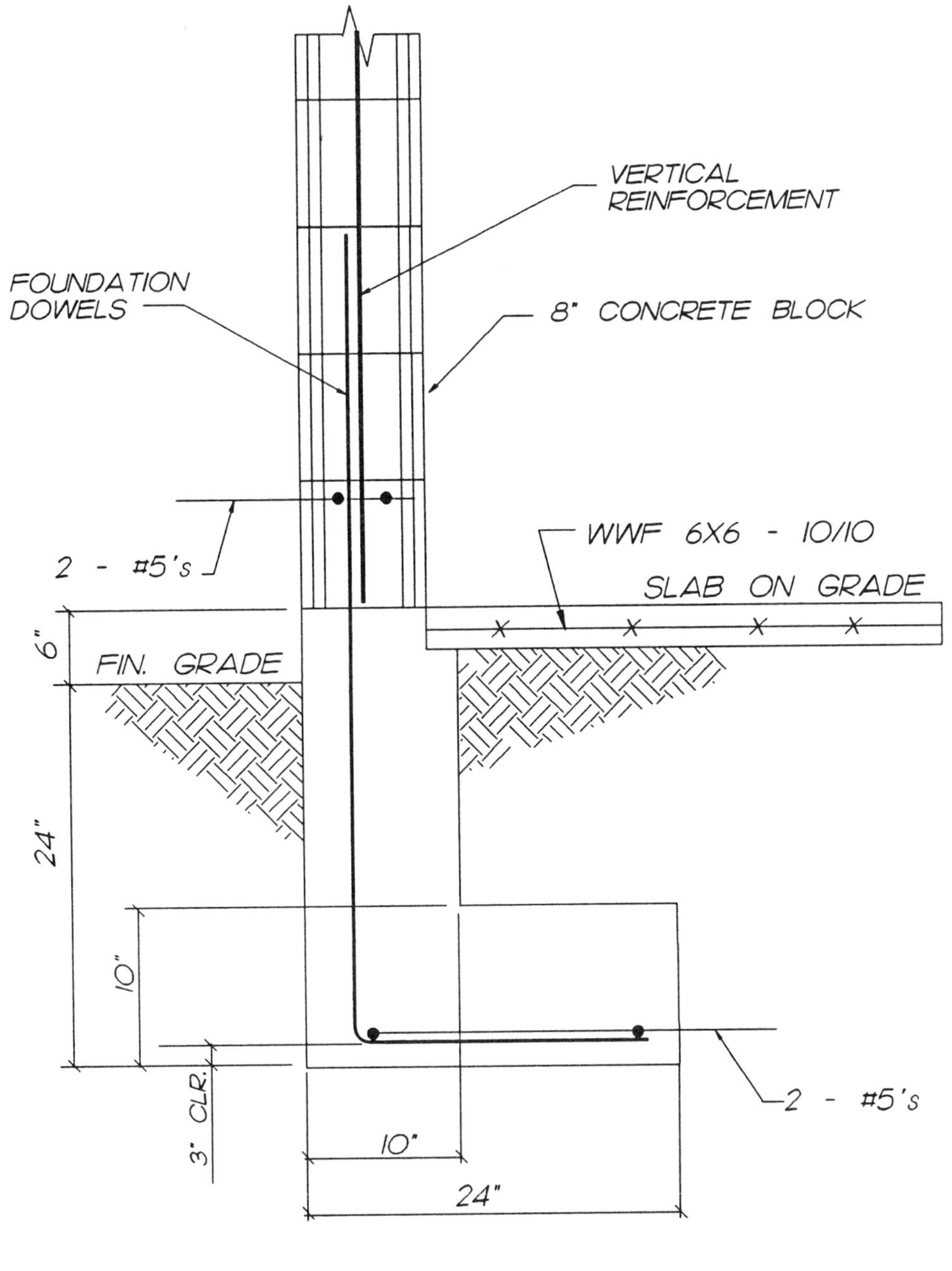
VERTICAL REINFORCEMENT
FOUNDATION DOWELS
8" CONCRETE BLOCK
2 - #5's
WWF 6X6 - 10/10
SLAB ON GRADE
6"
FIN. GRADE
24"
10"
3" CLR.
2 - #5's
10"
24"

M1063

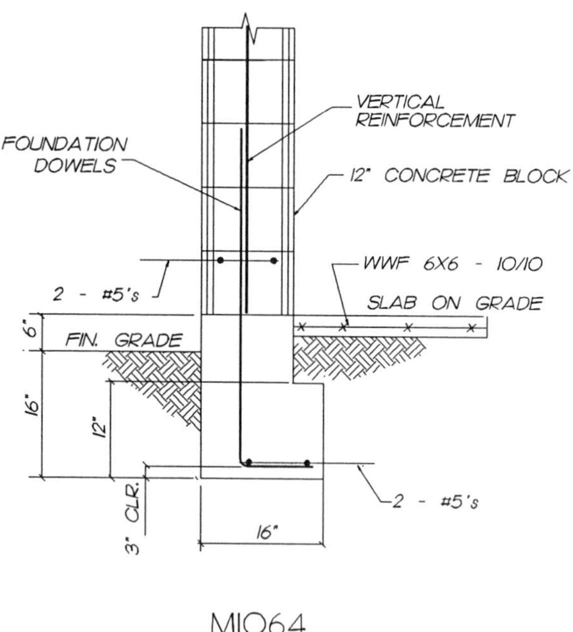

M1064

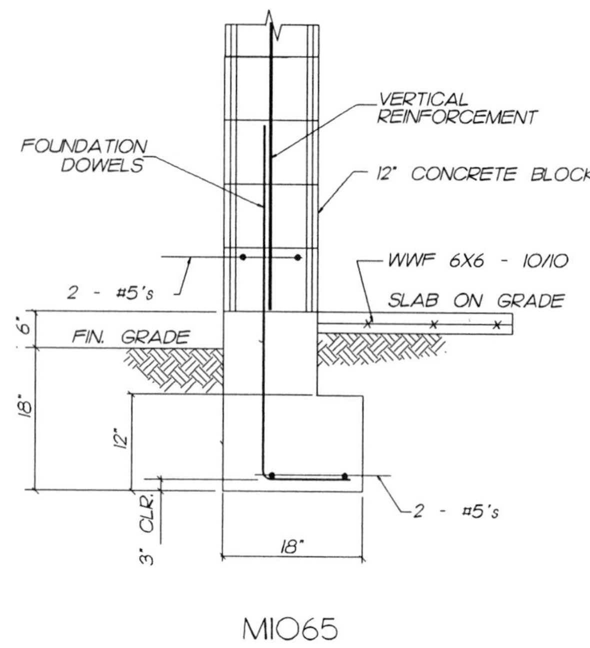

M1065

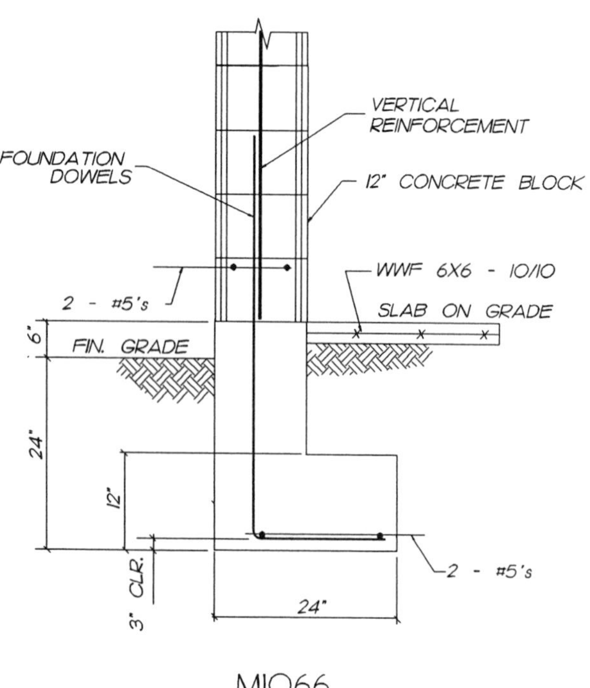

M1066

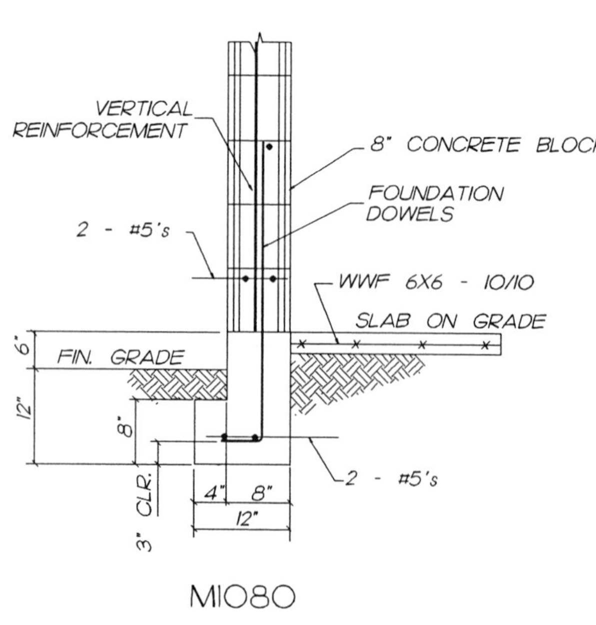

M1080

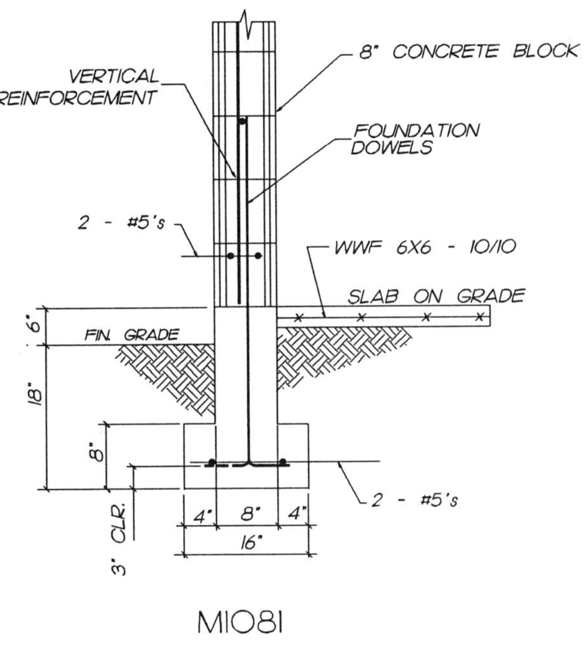

M1081

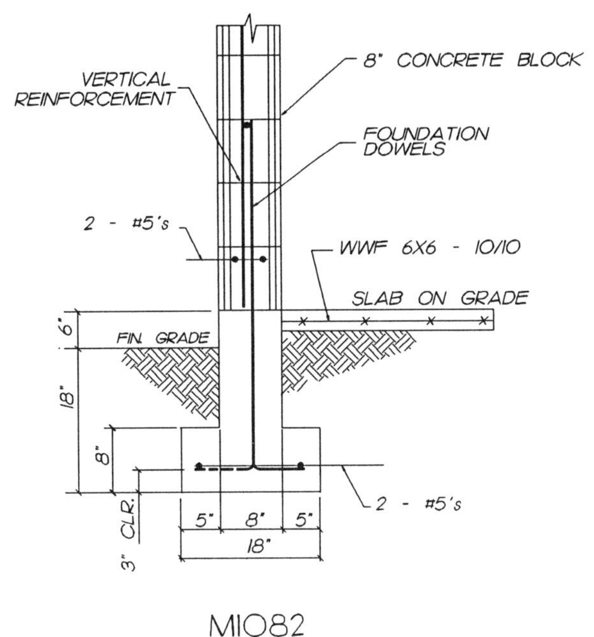

M1082

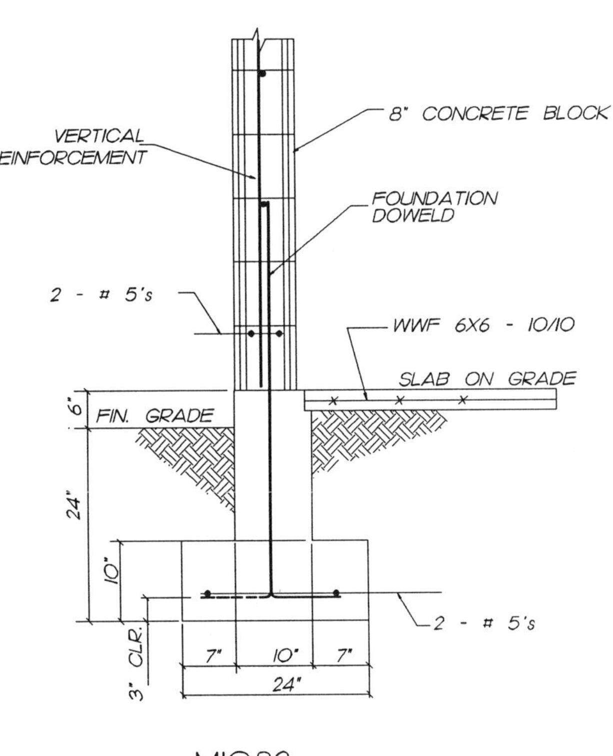

M1083

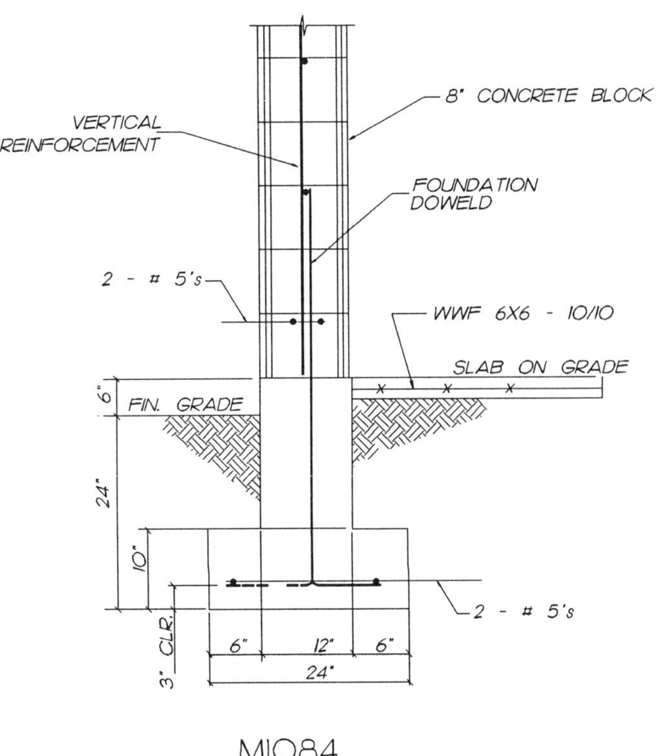

M1084

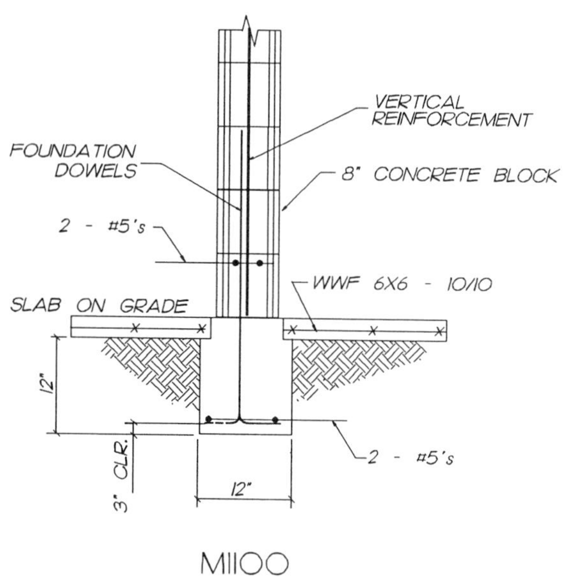

MI100

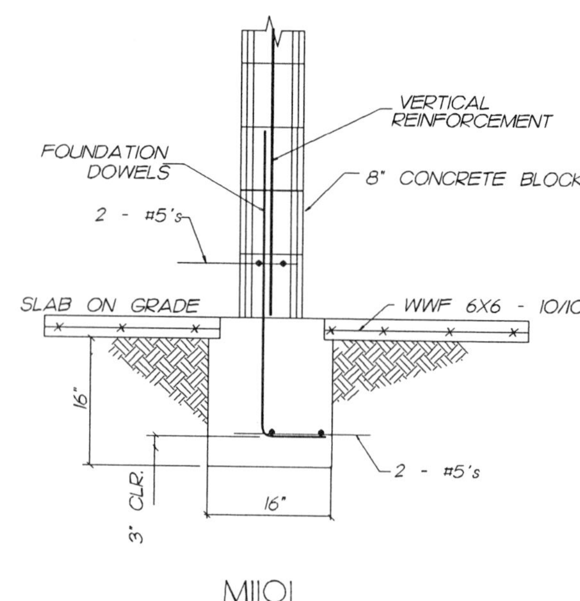

MI101

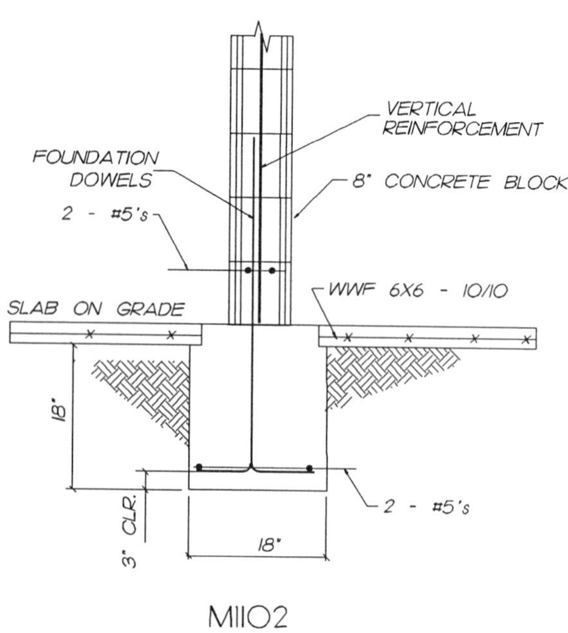

MI102

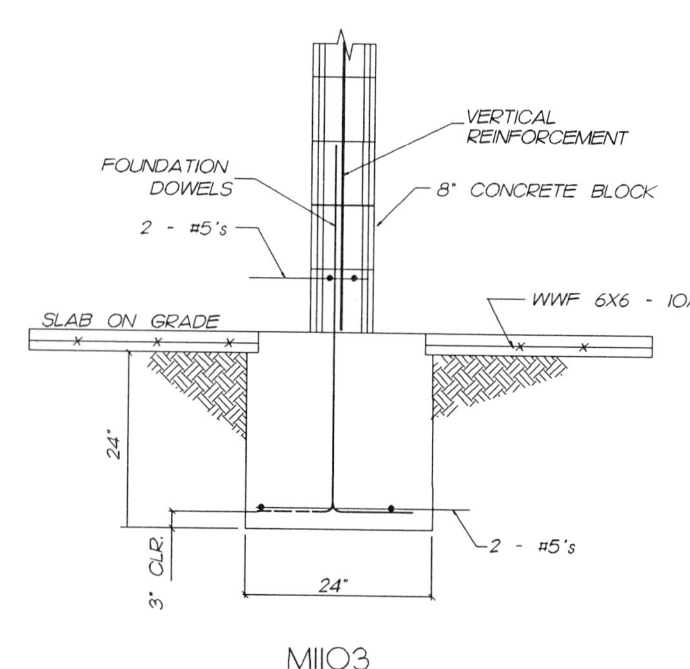

MI103

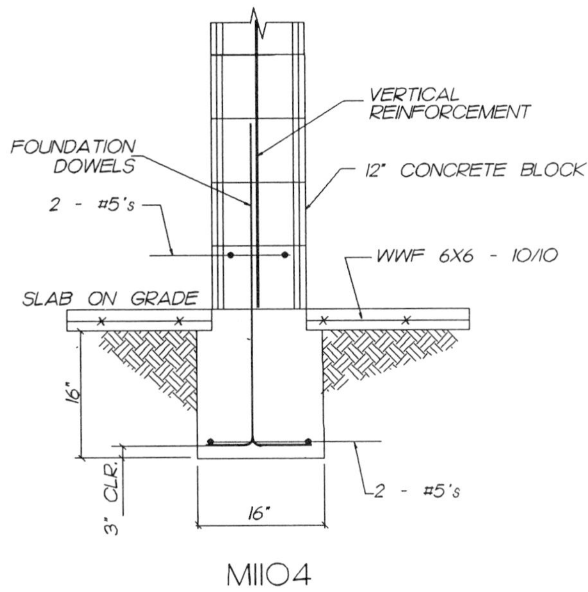

M1104

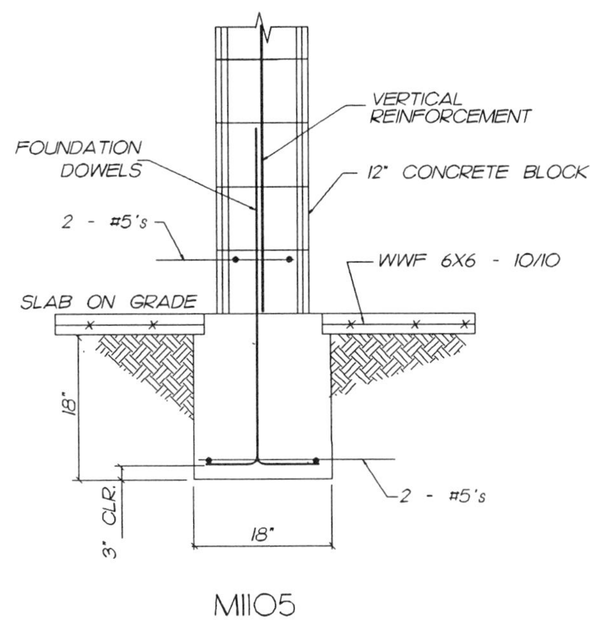

M1105

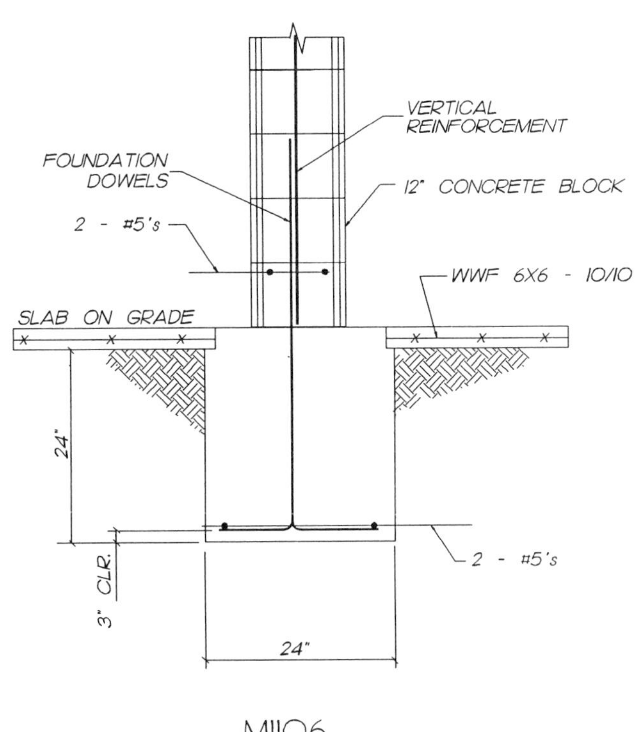

M1106

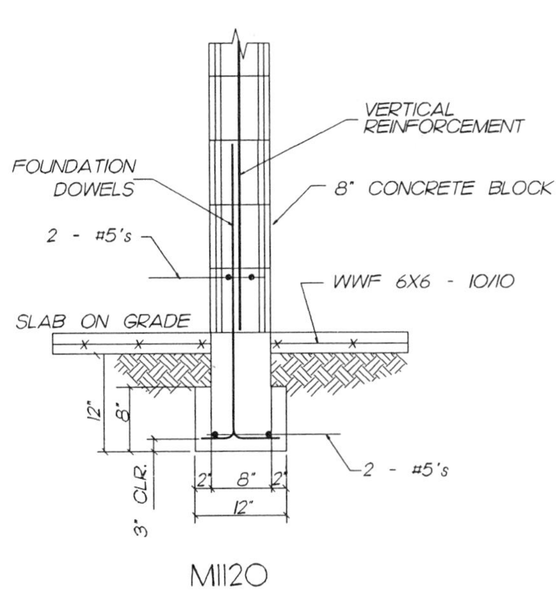

M1120

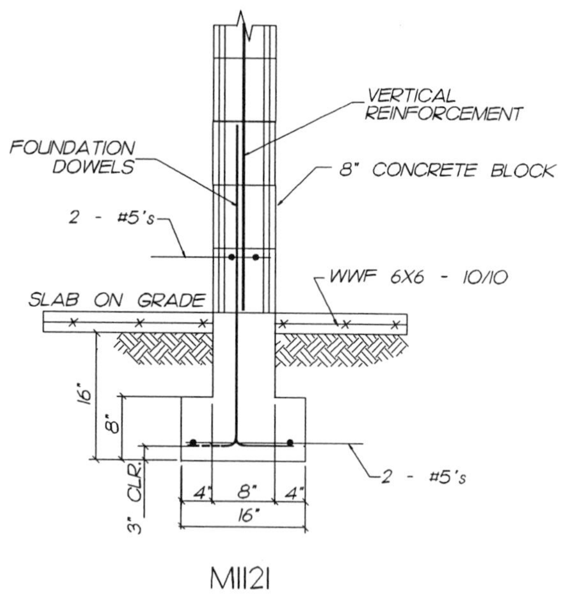
VERTICAL
REINFORCEMENT
FOUNDATION
DOWELS
8" CONCRETE BLOCK
2 - #5's
WWF 6X6 - 10/10
SLAB ON GRADE
16"
8"
3" CLR.
4"
8"
4"
16"
2 - #5's
M1121

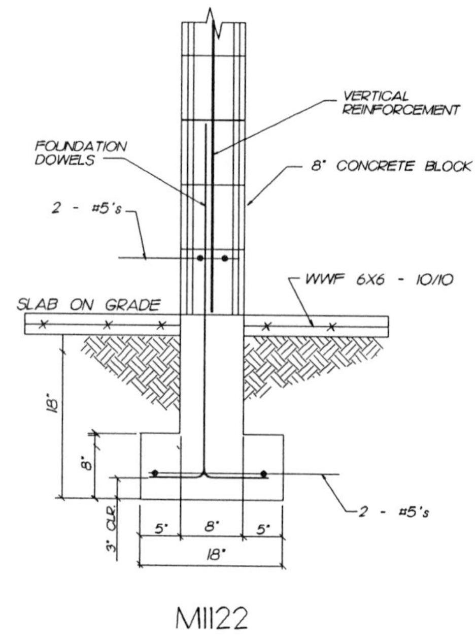
VERTICAL
REINFORCEMENT
FOUNDATION
DOWELS
8" CONCRETE BLOCK
2 - #5's
WWF 6X6 - 10/10
SLAB ON GRADE
18"
8"
3" CLR.
5"
8"
5"
18"
2 - #5's
M1122

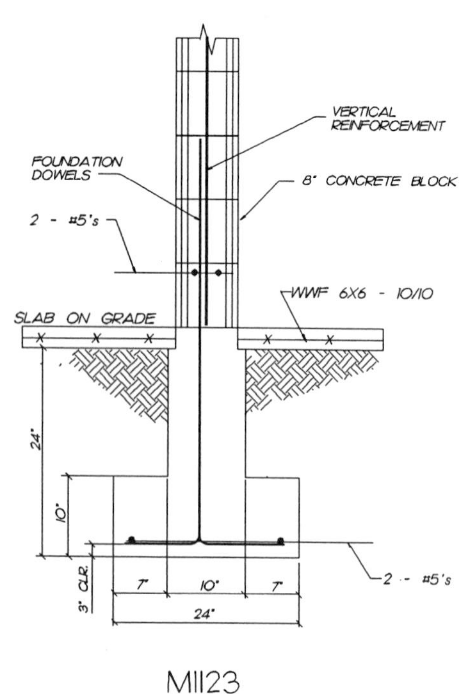
VERTICAL
REINFORCEMENT
FOUNDATION
DOWELS
8" CONCRETE BLOCK
2 - #5's
WWF 6X6 - 10/10
SLAB ON GRADE
24"
10"
3" CLR.
7"
10"
7"
24"
2 - #5's
M1123

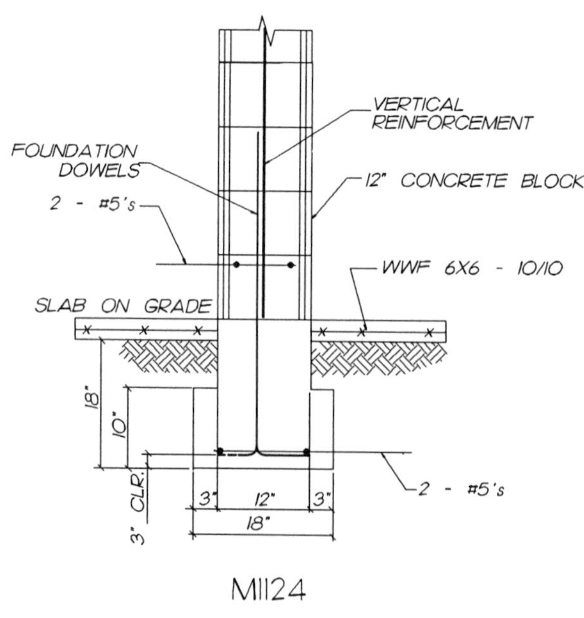
VERTICAL
REINFORCEMENT
FOUNDATION
DOWELS
12" CONCRETE BLOCK
2 - #5's
WWF 6X6 - 10/10
SLAB ON GRADE
18"
10"
3" CLR.
3"
12"
3"
18"
2 - #5's
M1124

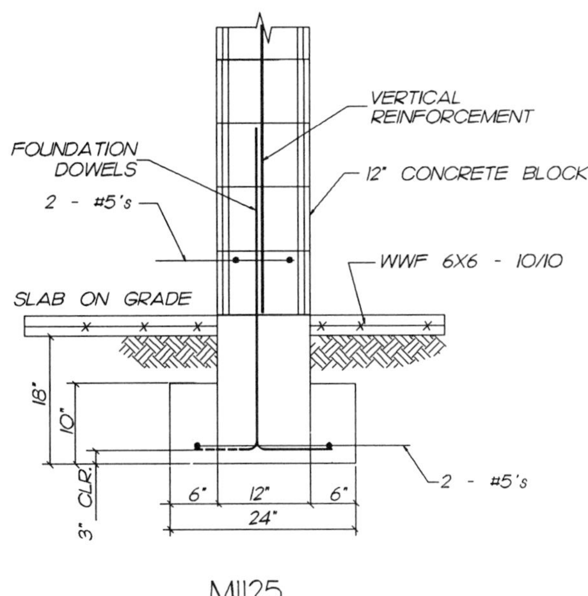

M1125

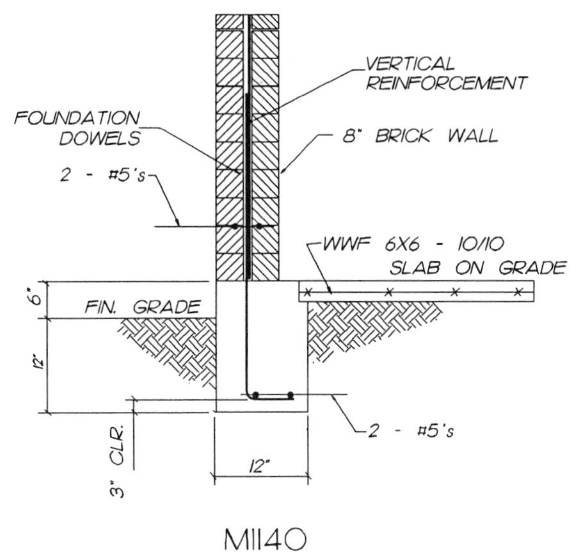

M1140

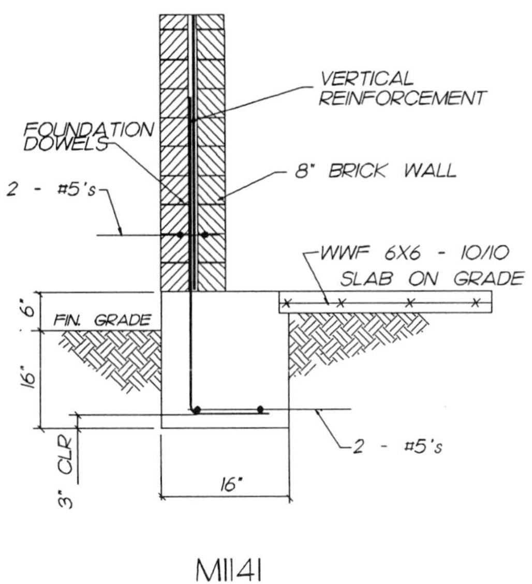

M1141

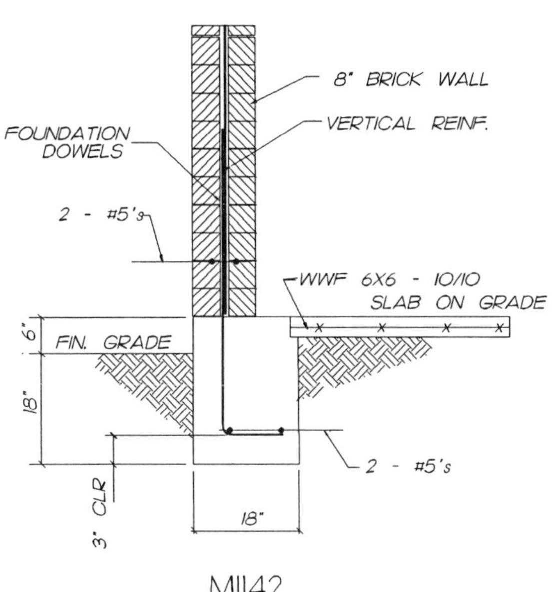

M1142

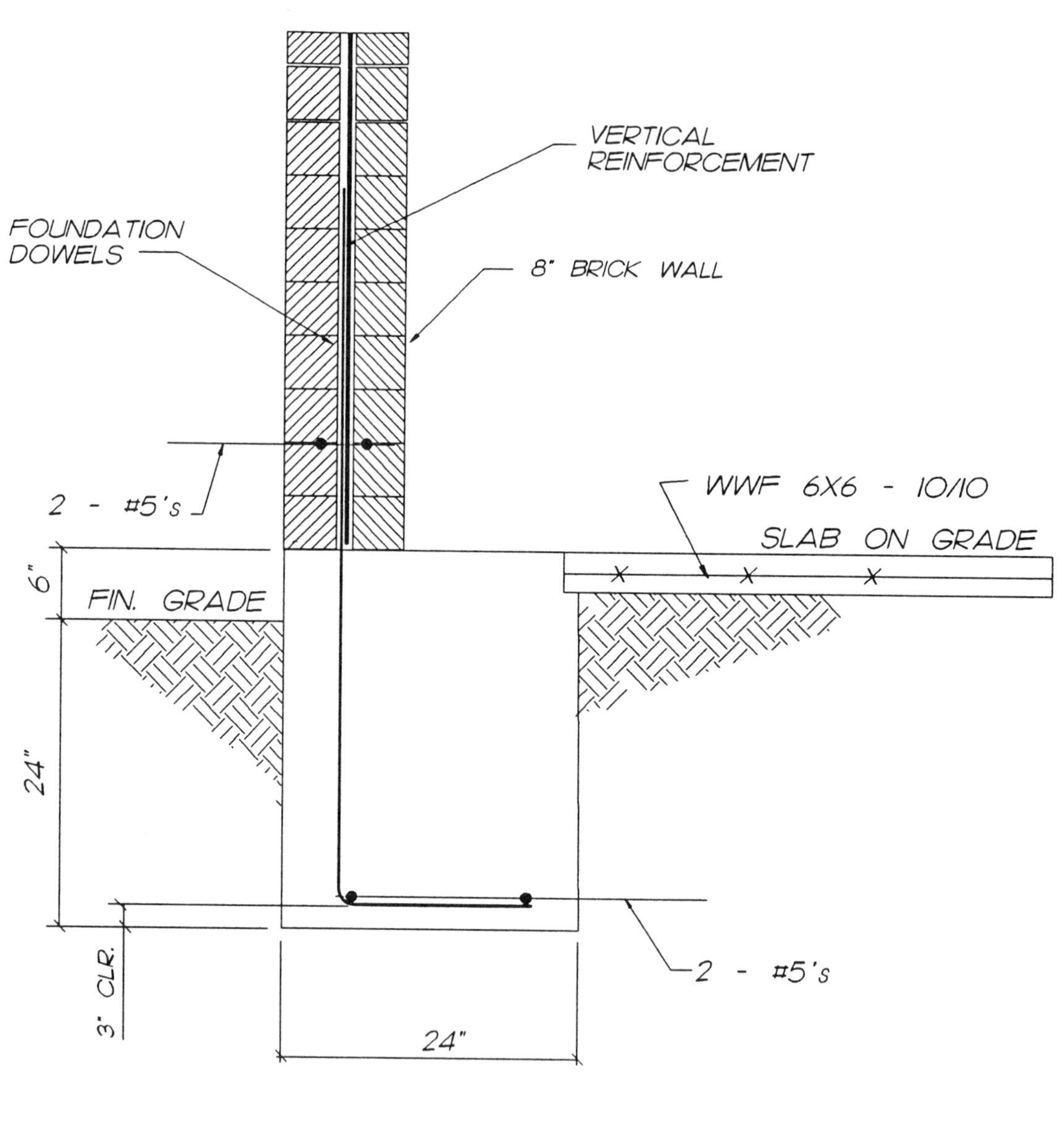

VERTICAL
REINFORCEMENT
FOUNDATION
DOWELS
8" BRICK WALL
2 - #5's
WWF 6X6 - 10/10
SLAB ON GRADE
6"
FIN. GRADE
24"
3" CLR.
2 - #5's
24"

M1143

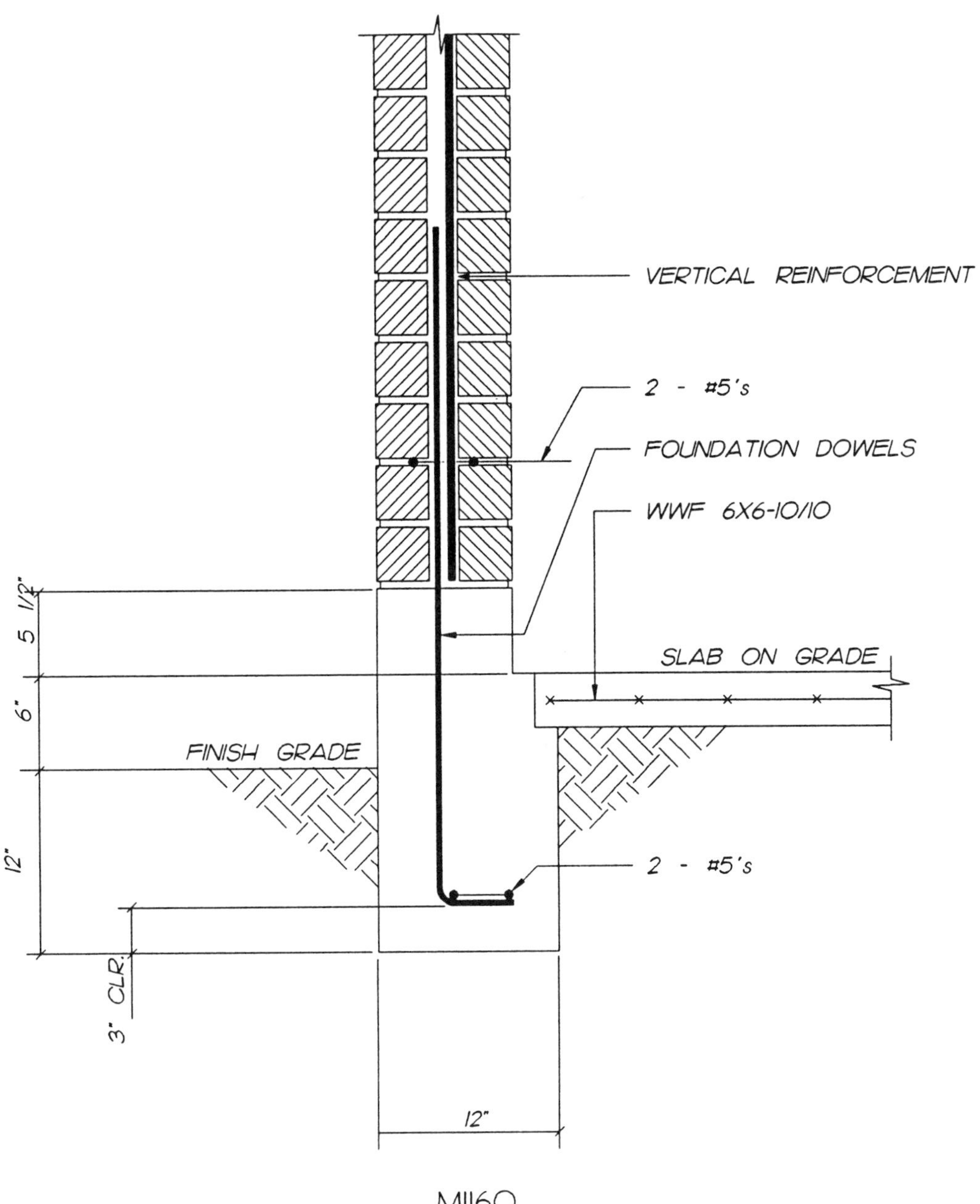
VERTICAL REINFORCEMENT
2 - #5's
FOUNDATION DOWELS
WWF 6X6-10/10
SLAB ON GRADE
5 1/2"
6"
FINISH GRADE
12"
2 - #5's
3" CLR.
12"

M1160

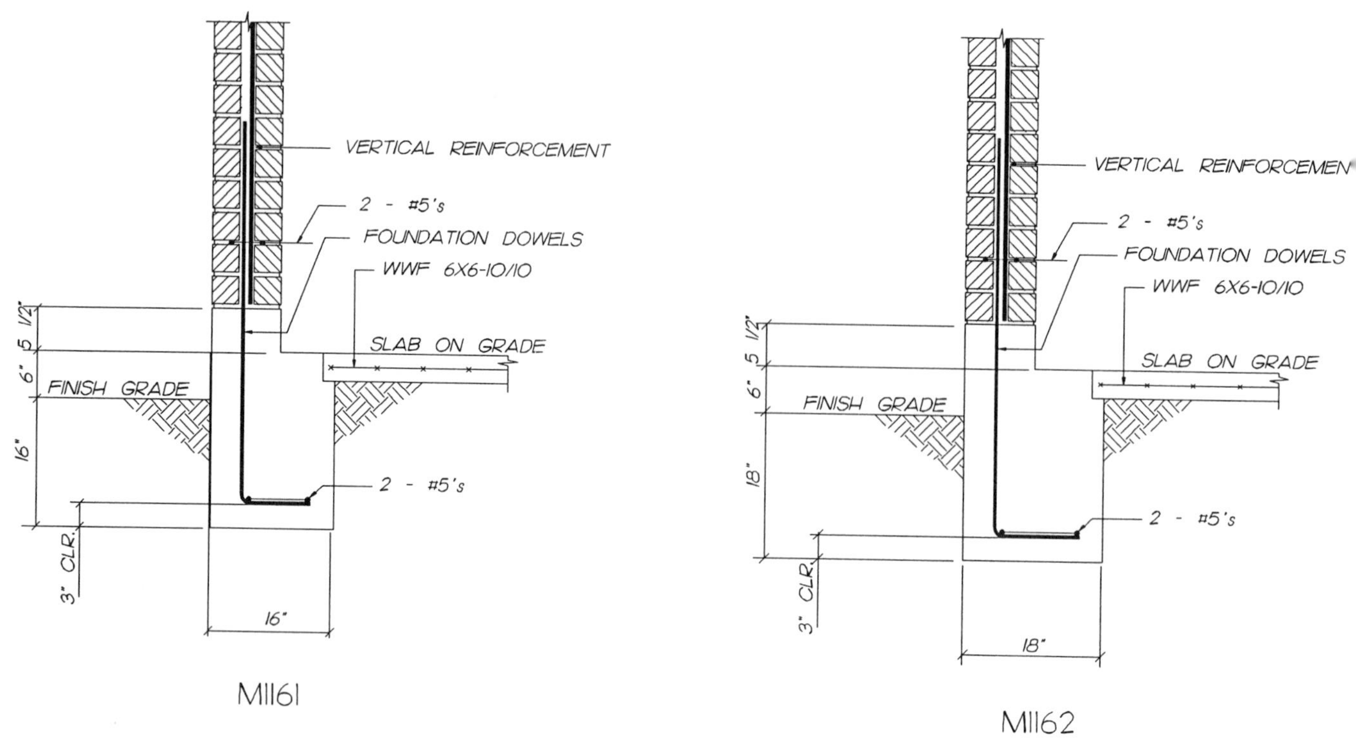
VERTICAL REINFORCEMENT
2 - #5's
FOUNDATION DOWELS
WWF 6X6-10/10
SLAB ON GRADE
FINISH GRADE
5 1/2"
6"
16"
2 - #5's
3" CLR.
16"
M1161
VERTICAL REINFORCEMEN
2 - #5's
FOUNDATION DOWELS
WWF 6X6-10/10
SLAB ON GRADE
FINISH GRADE
5 1/2"
6"
18"
2 - #5's
3" CLR.
18"
M1162

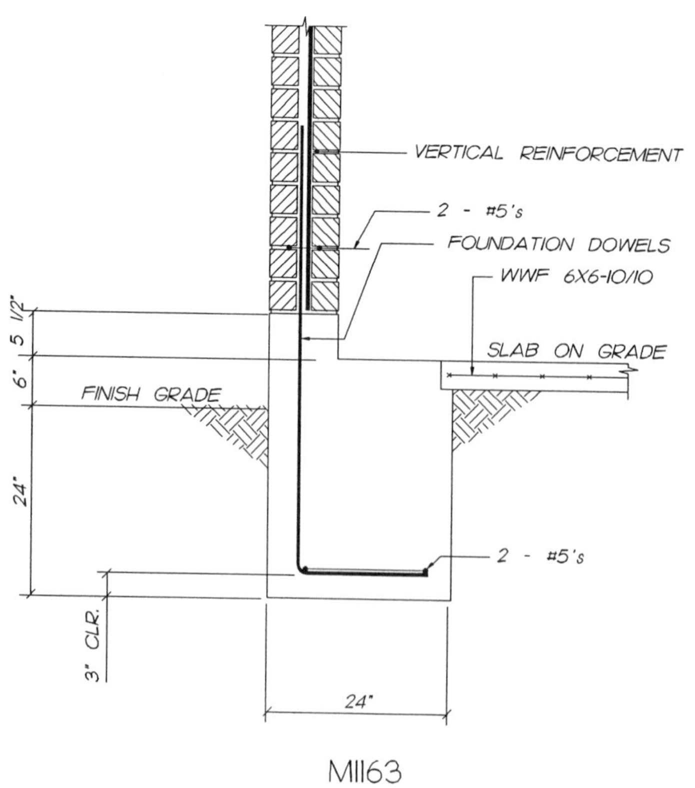
VERTICAL REINFORCEMENT
2 - #5's
FOUNDATION DOWELS
WWF 6X6-10/10
SLAB ON GRADE
FINISH GRADE
5 1/2"
6"
24"
2 - #5's
3" CLR.
24"
M1163

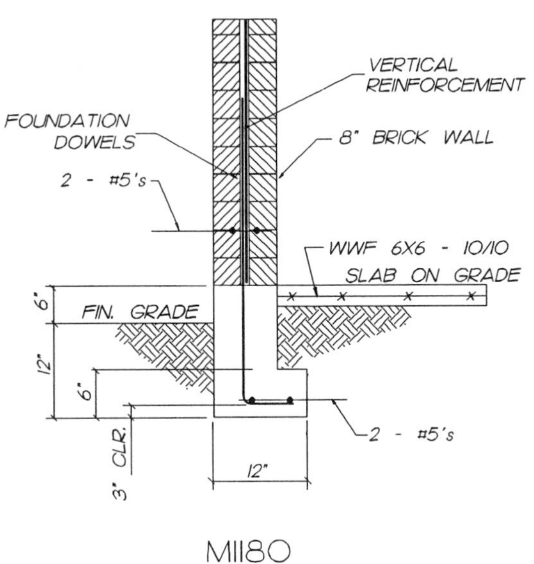

M1180

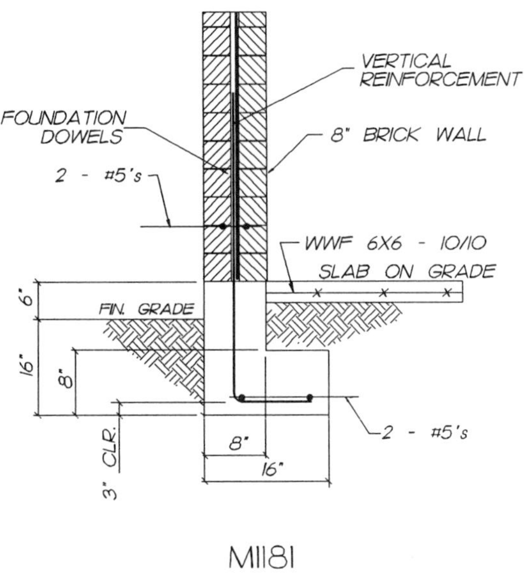

M1181

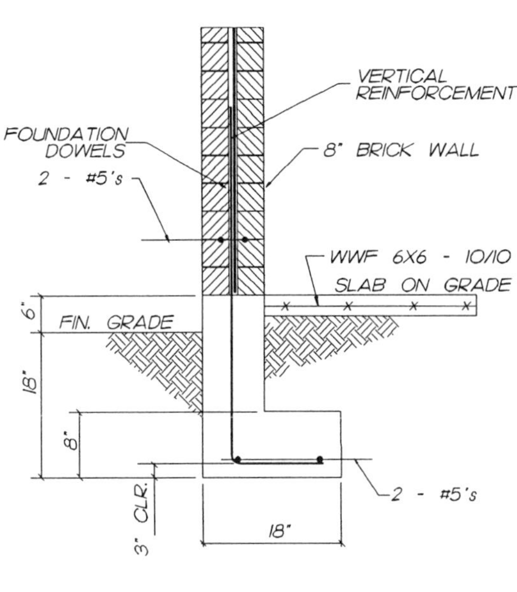

M1182

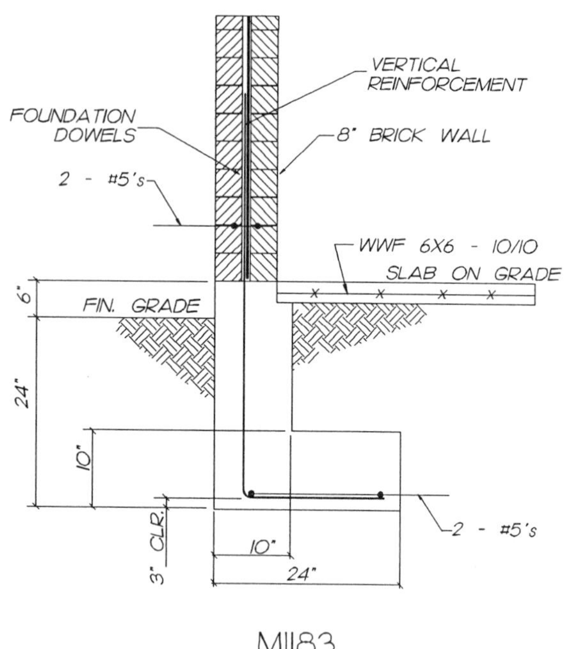

M1183

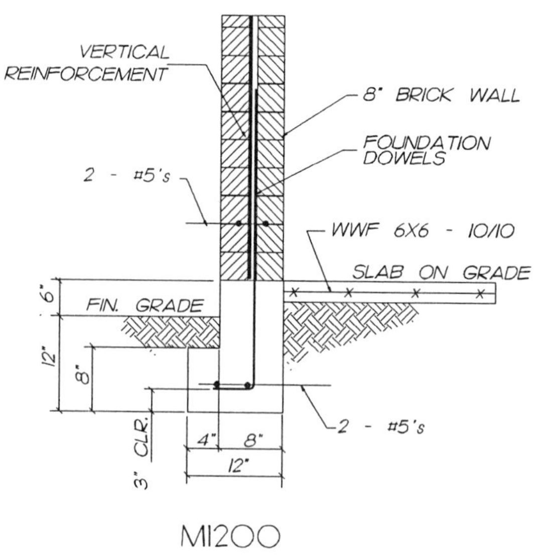

M1200

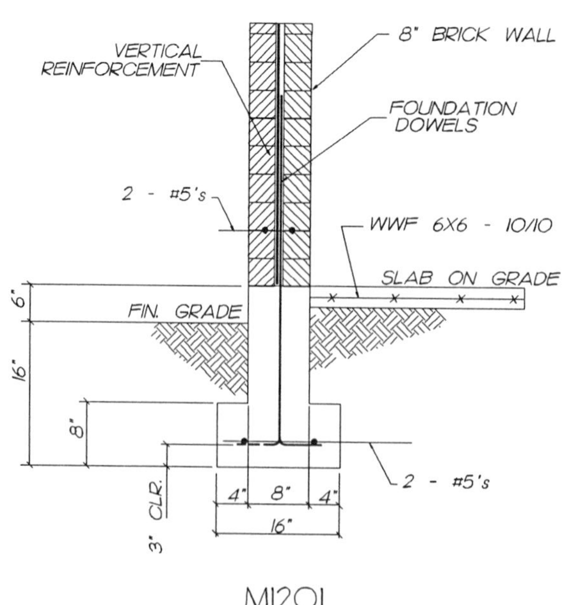

M1201

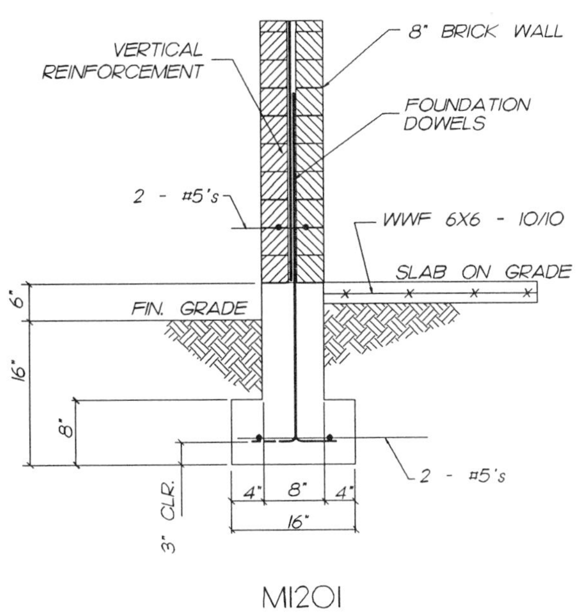

M1201

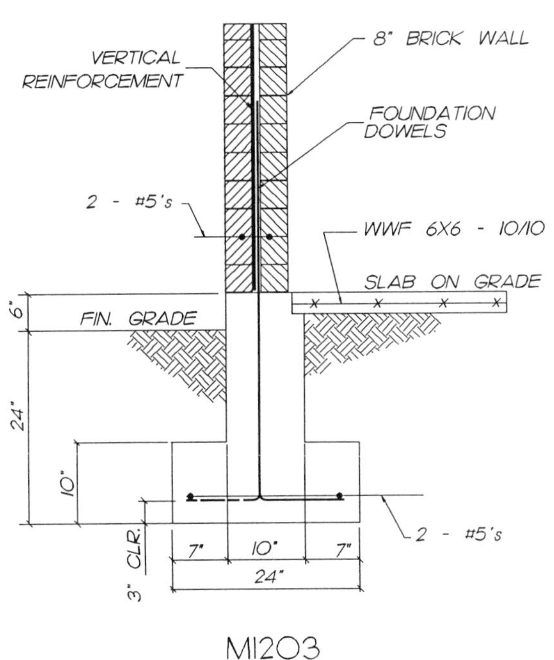

M1203

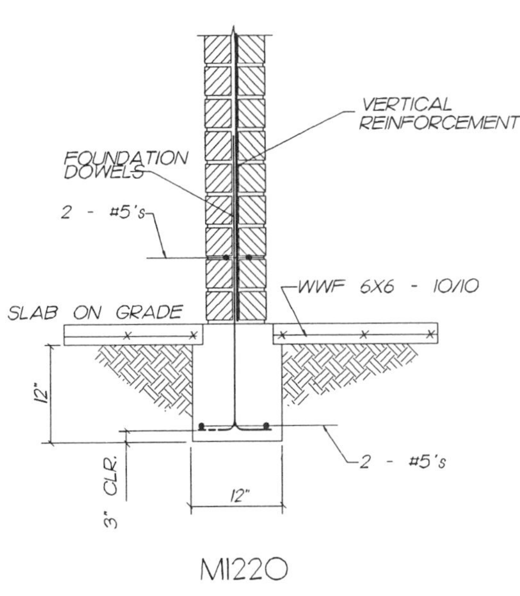

M1220

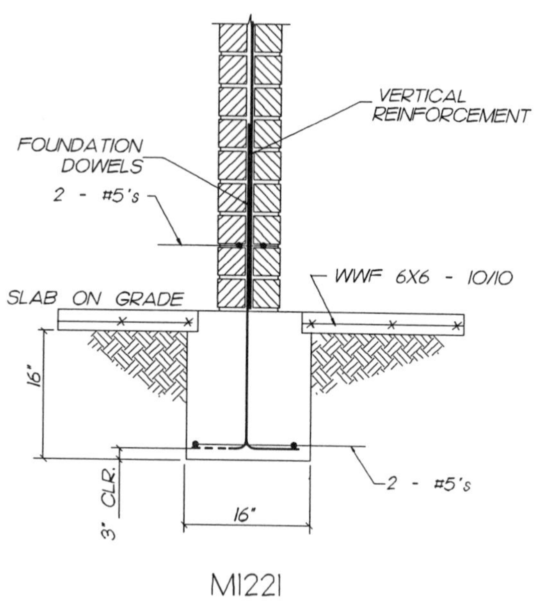

M1221

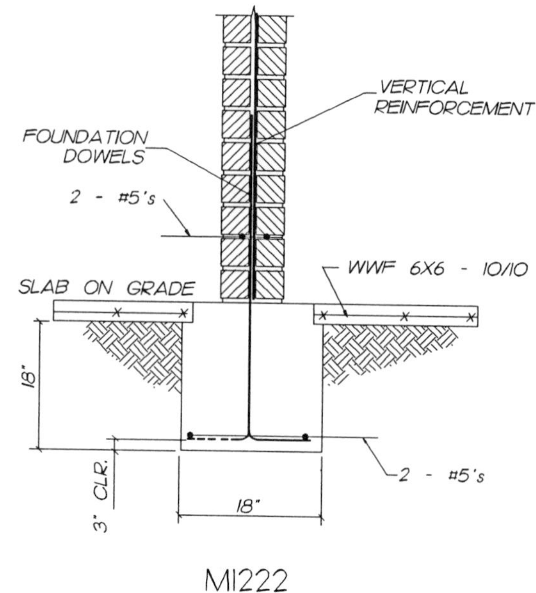

M1222

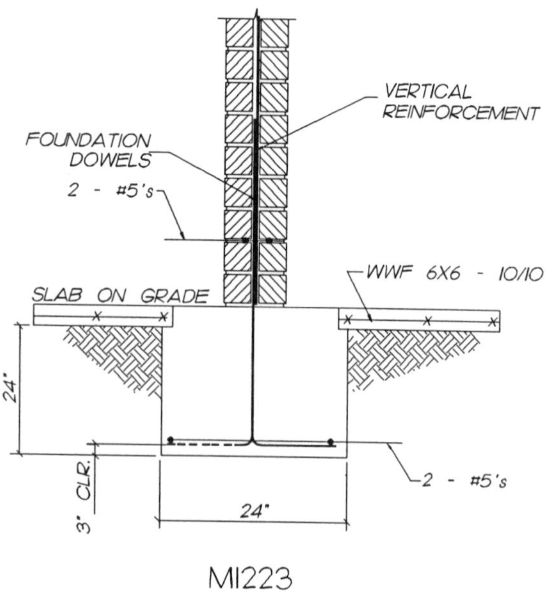

M1223

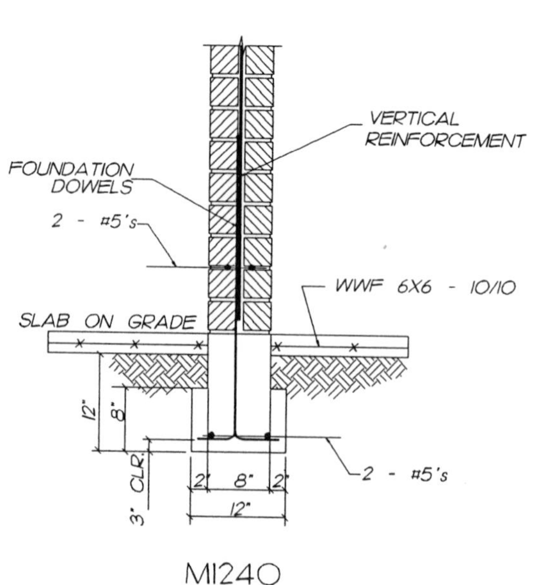

M1240

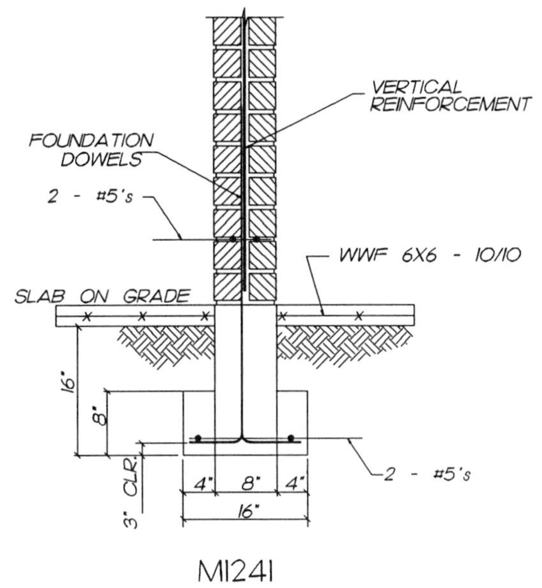

M1241

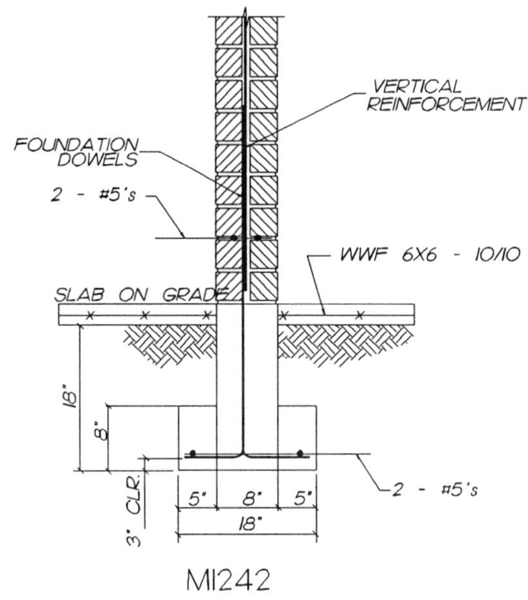

M1242

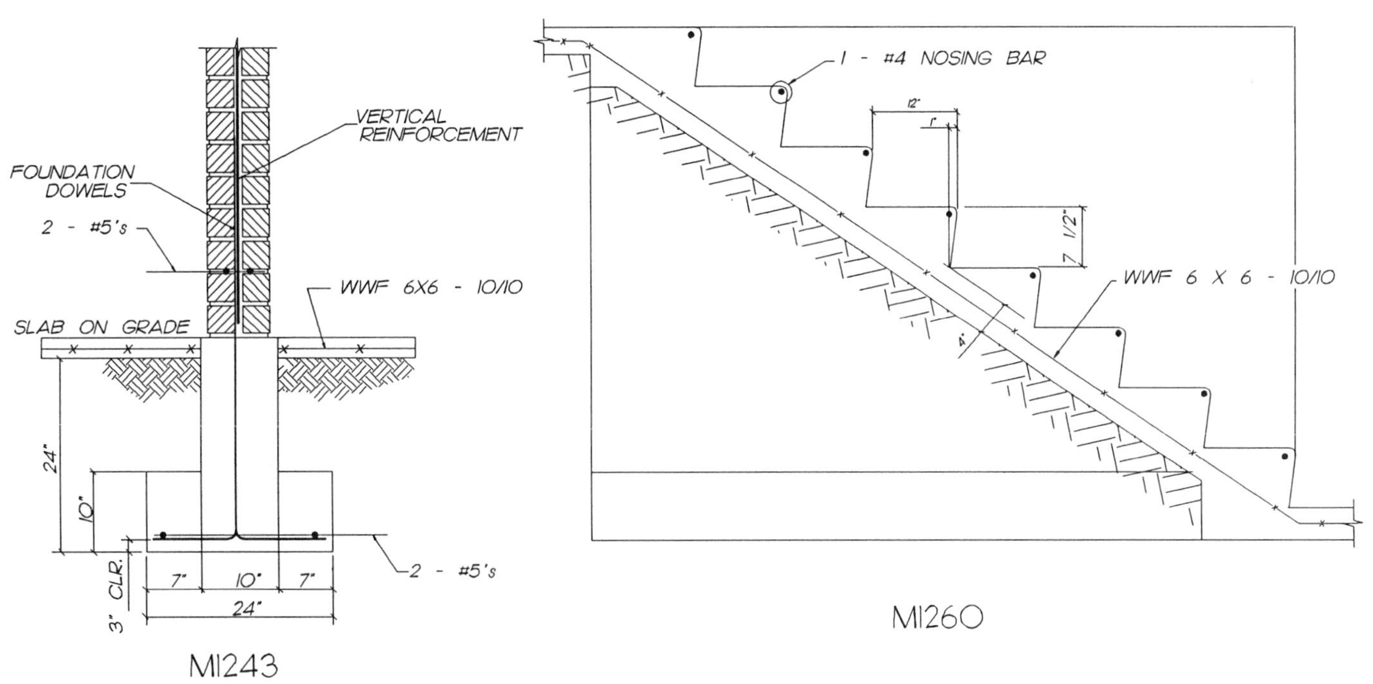

M1243

M1260

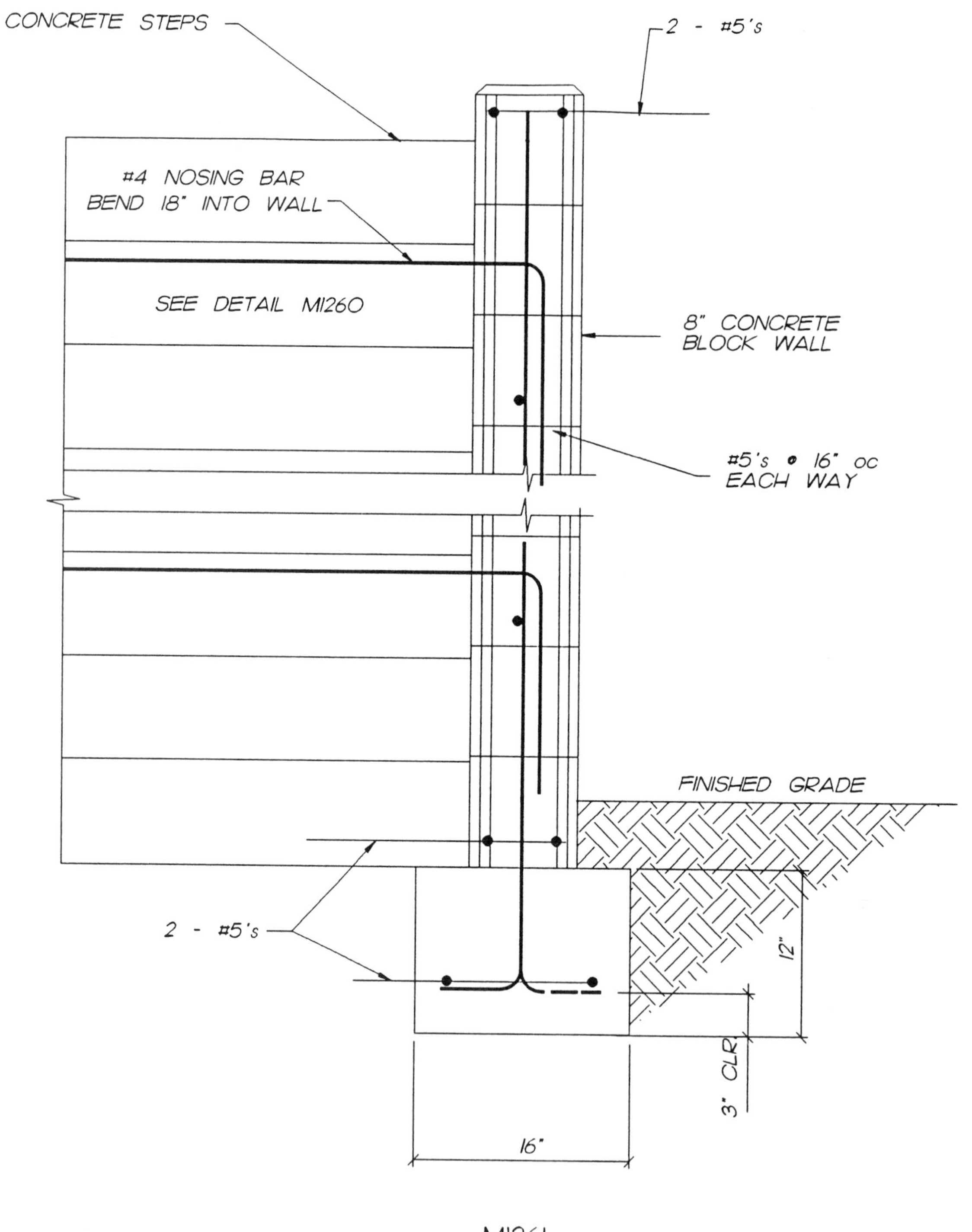
CONCRETE STEPS
2 - #5's
#4 NOSING BAR
BEND 18" INTO WALL
SEE DETAIL M1260
8" CONCRETE
BLOCK WALL
#5's @ 16" oc
EACH WAY
FINISHED GRADE
2 - #5's
12"
3" CLR.
16"

M1261

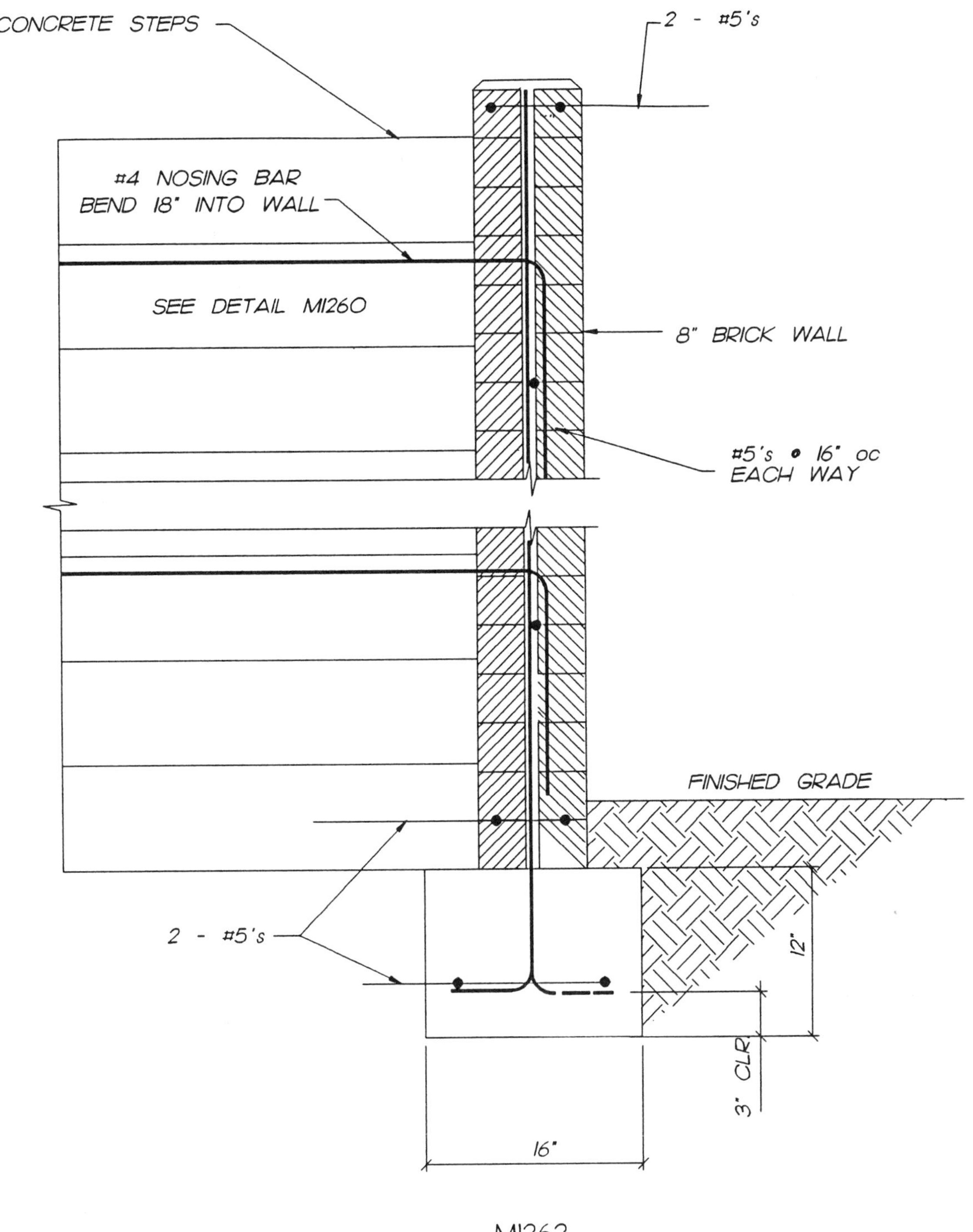
CONCRETE STEPS
2 - #5's
#4 NOSING BAR
BEND 18" INTO WALL
SEE DETAIL M1260
8" BRICK WALL
#5's @ 16" OC
EACH WAY
FINISHED GRADE
2 - #5's
12"
3" CLR
16"

M1262

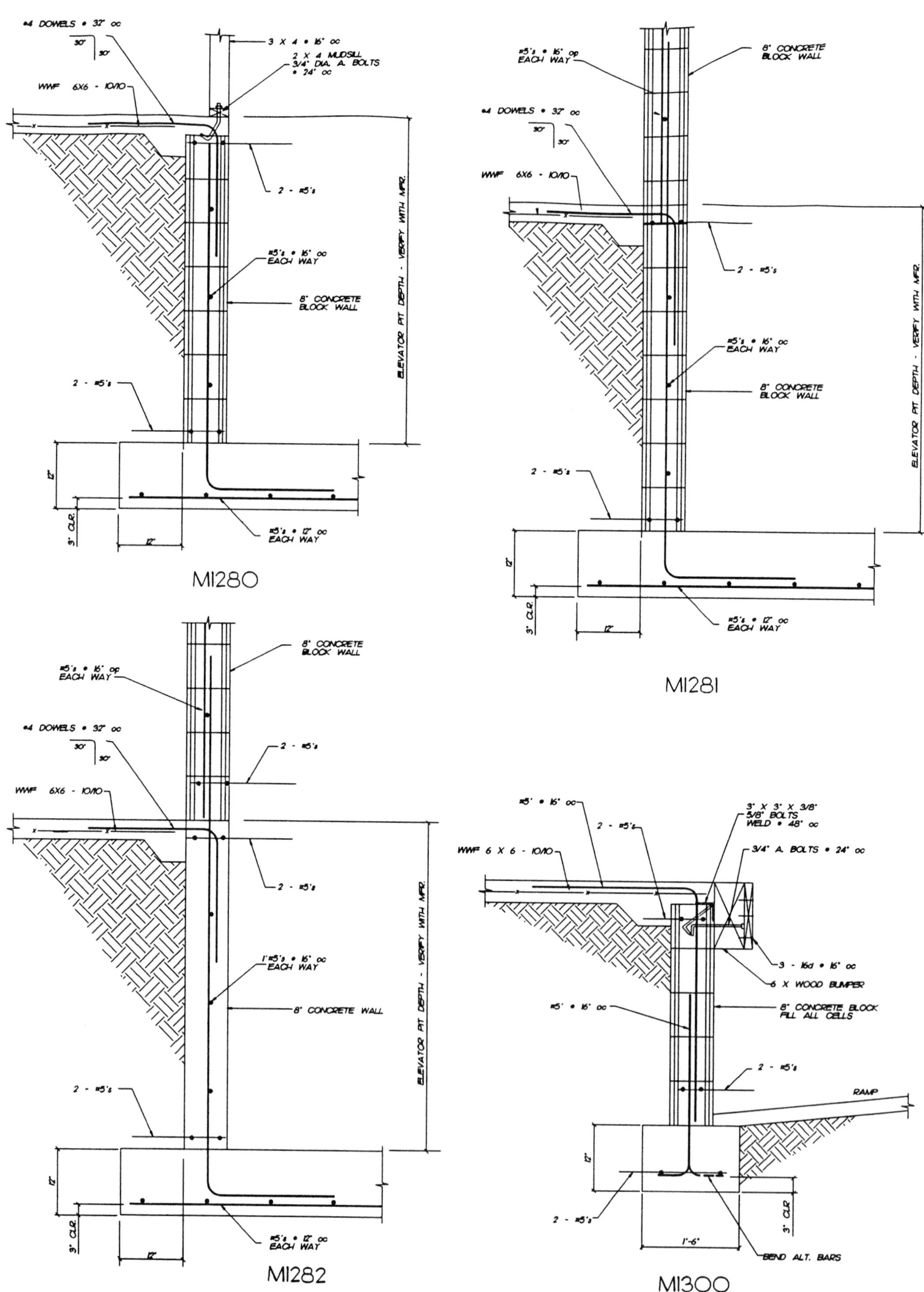
#4 DOWELS • 32" oc
30"
30"
3 X 4 • 16" oc
2 X 4 MUDSILL
3/4" DIA. A. BOLTS
• 24" oc
WWF 6X6 - 10/10
2 - #5's
#5's • 16" oc
EACH WAY
8" CONCRETE
BLOCK WALL
ELEVATOR PIT DEPTH - VERIFY WITH MFR.
2 - #5's
12"
3" CLR.
12"
#5's • 12" oc
EACH WAY
M1280
#5's • 16" op
EACH WAY
8" CONCRETE
BLOCK WALL
#4 DOWELS • 32" oc
30"
30"
WWF 6X6 - 10/10
2 - #5's
#5's • 16" oc
EACH WAY
8" CONCRETE
BLOCK WALL
ELEVATOR PIT DEPTH - VERIFY WITH MFR.
2 - #5's
12"
3" CLR.
12"
#5's • 12" oc
EACH WAY
M1281
8" CONCRETE
BLOCK WALL
#5's • 16" op
EACH WAY
#4 DOWELS • 32" oc
30"
30"
2 - #5's
WWF 6X6 - 10/10
2 - #5's
1"#5's • 16" oc
EACH WAY
8" CONCRETE WALL
ELEVATOR PIT DEPTH - VERIFY WITH MFR.
2 - #5's
12"
3" CLR.
12"
#5's • 12" oc
EACH WAY
M1282
#5' • 16" oc
2 - #5's
3" X 3" X 3/8"
5/8" BOLTS
WELD • 48" oc
3/4" A. BOLTS • 24" oc
WWF 6 X 6 - 10/10
3 - 16d • 16" oc
6 X WOOD BUMPER
#5' • 16" oc
8" CONCRETE BLOCK
FILL ALL CELLS
2 - #5's
RAMP
12"
3" CLR.
2 - #5's
1'-6"
BEND ALT. BARS
M1300

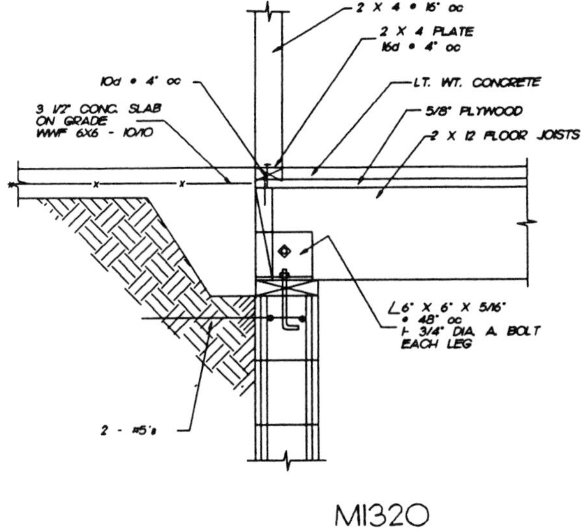

M1320

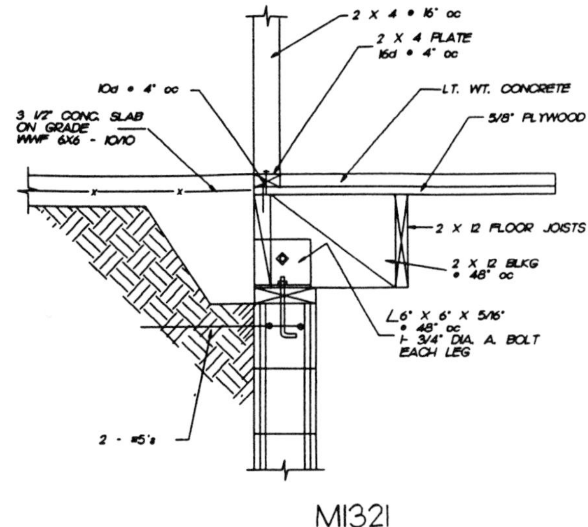

M1321

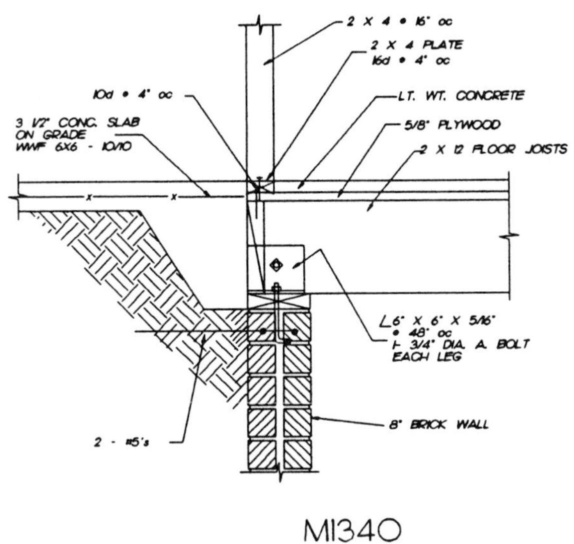

M1340

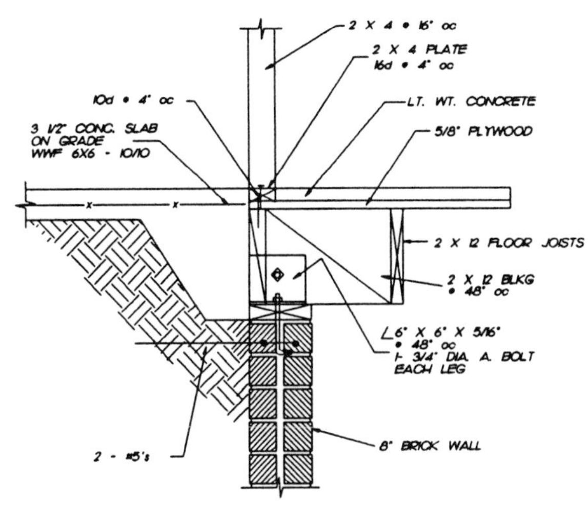

M1341

CHAPTER 2 MASONRY SECTION 2			COLUMNS
CONCRETE BLOCK	PLAN SECTIONS	————	M2020 - M2024
BRICK MASONRY	PLAN SECTIONS	————	M2040 - M2042
CONCRETE BLOCK WALL STEEL & CONC. COLUMNS	PLAN SECTIONS	————	M2060 - M2064
BRICK MASONRY WALLS STEEL & CONC. COLUMNS	PLAN SECTIONS	————	M2080 - M2084

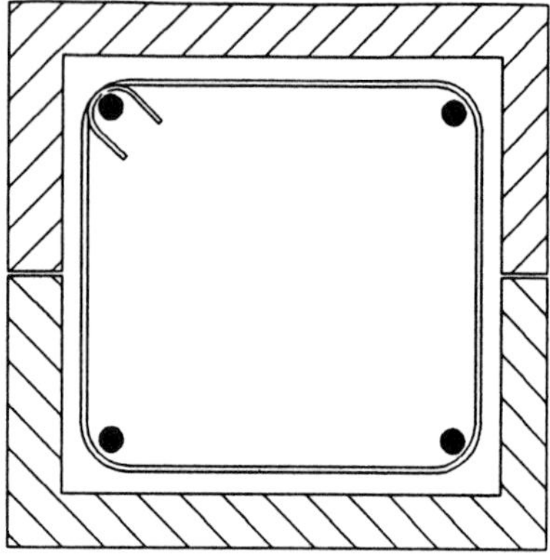

M2020

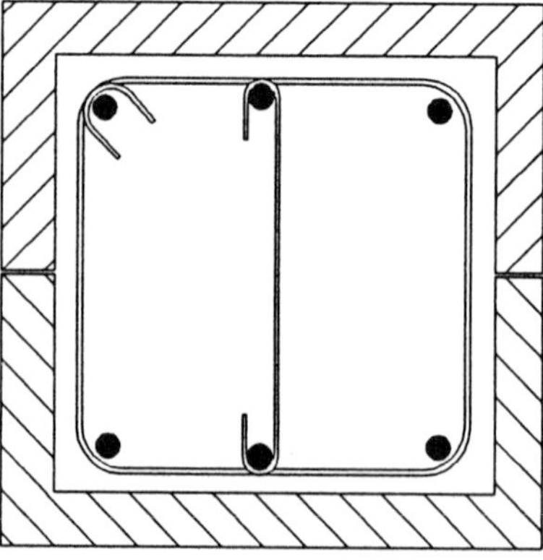

M2021

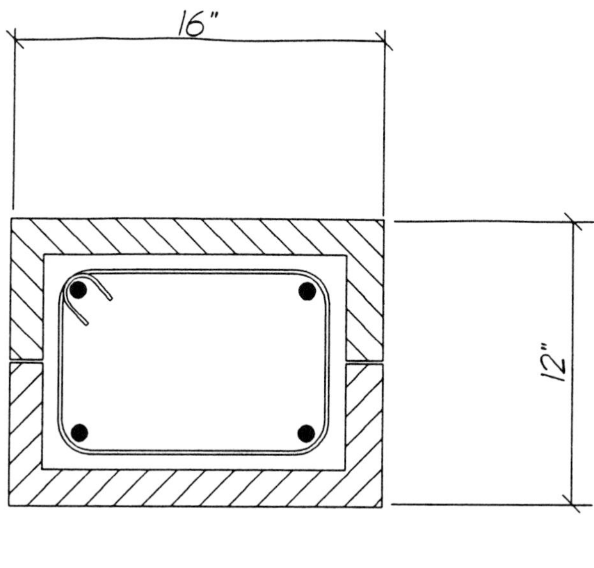

M2022

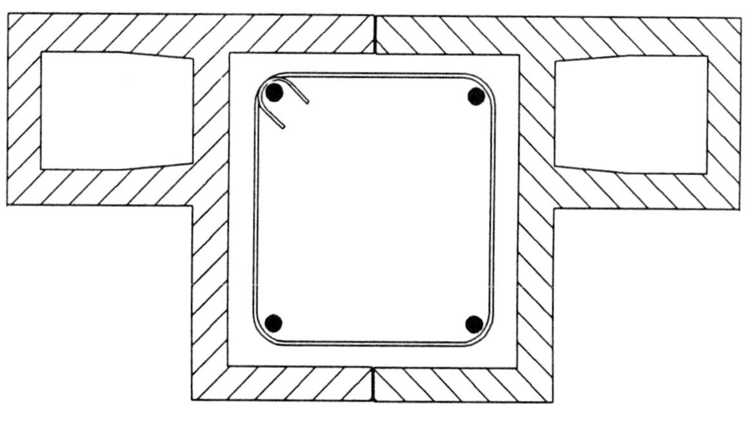

M2023

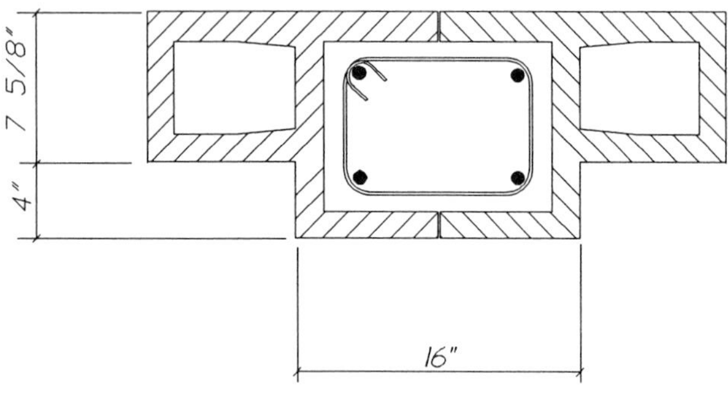
7 5/8"
4"
16"

M2024

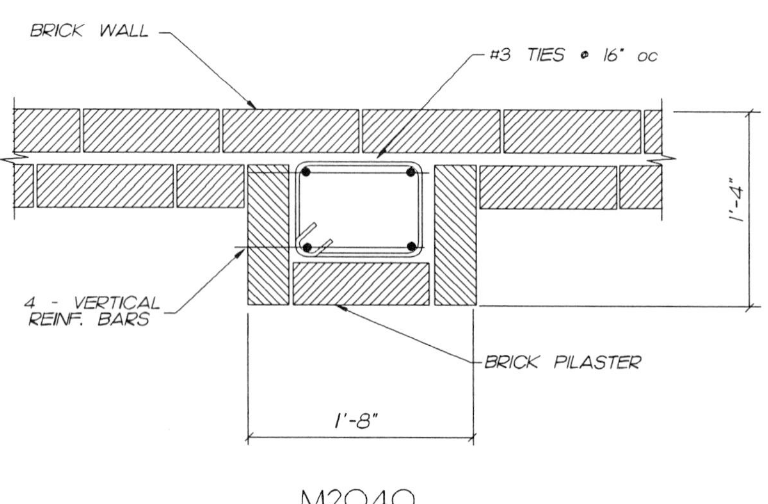
BRICK WALL
#3 TIES @ 16" oc
1'-4"
4 - VERTICAL
REINF. BARS
BRICK PILASTER
1'-8"

M2040

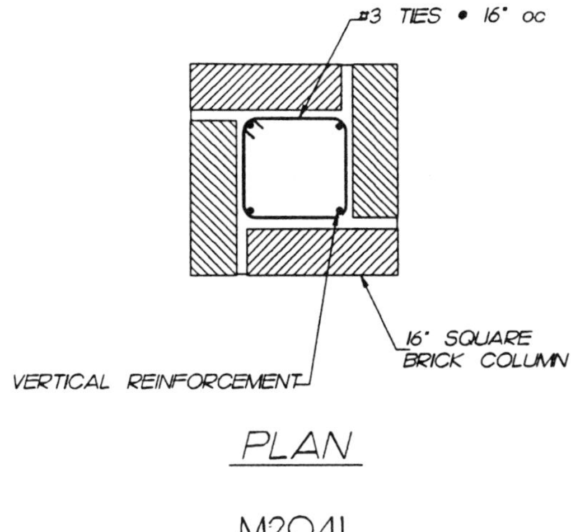
#3 TIES • 16" oc
16" SQUARE
BRICK COLUMN
VERTICAL REINFORCEMENT
PLAN

M2041

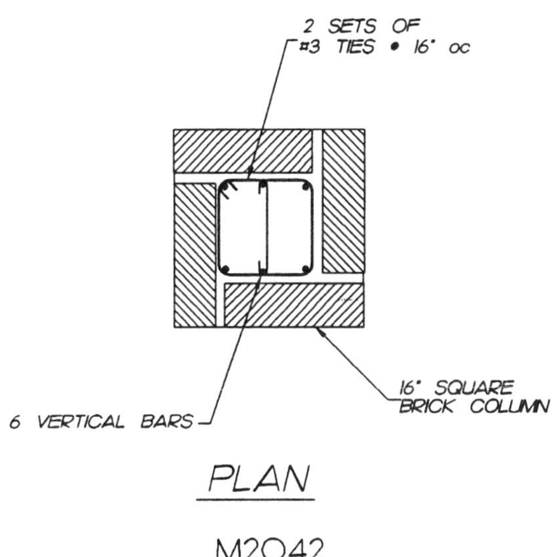
2 SETS OF
#3 TIES • 16" oc
16" SQUARE
BRICK COLUMN
6 VERTICAL BARS
PLAN

M2042

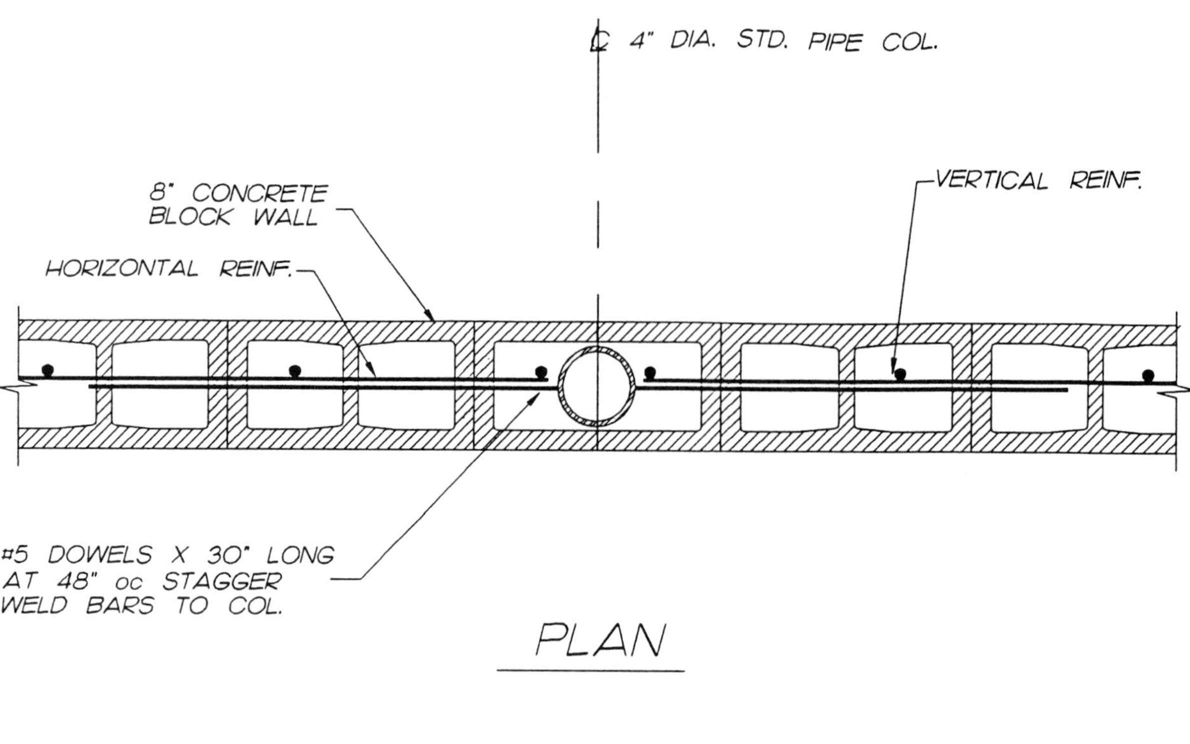

PLAN

M2060

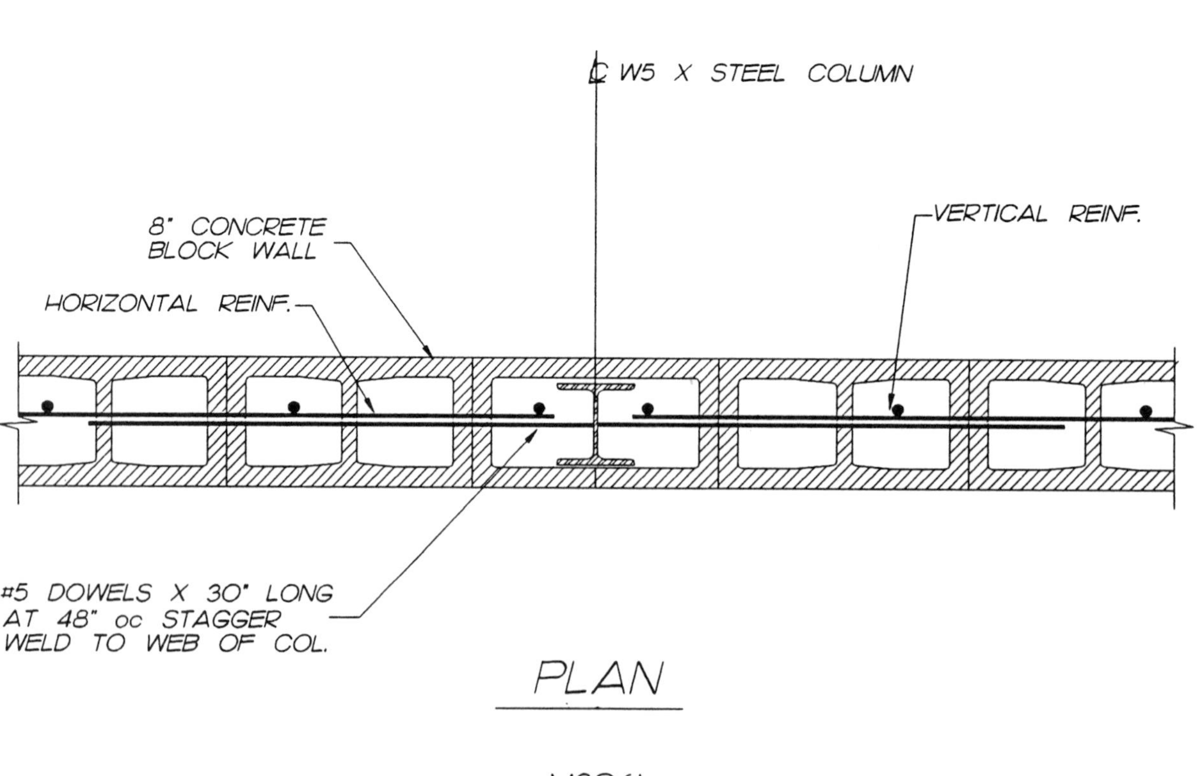

PLAN

M2061

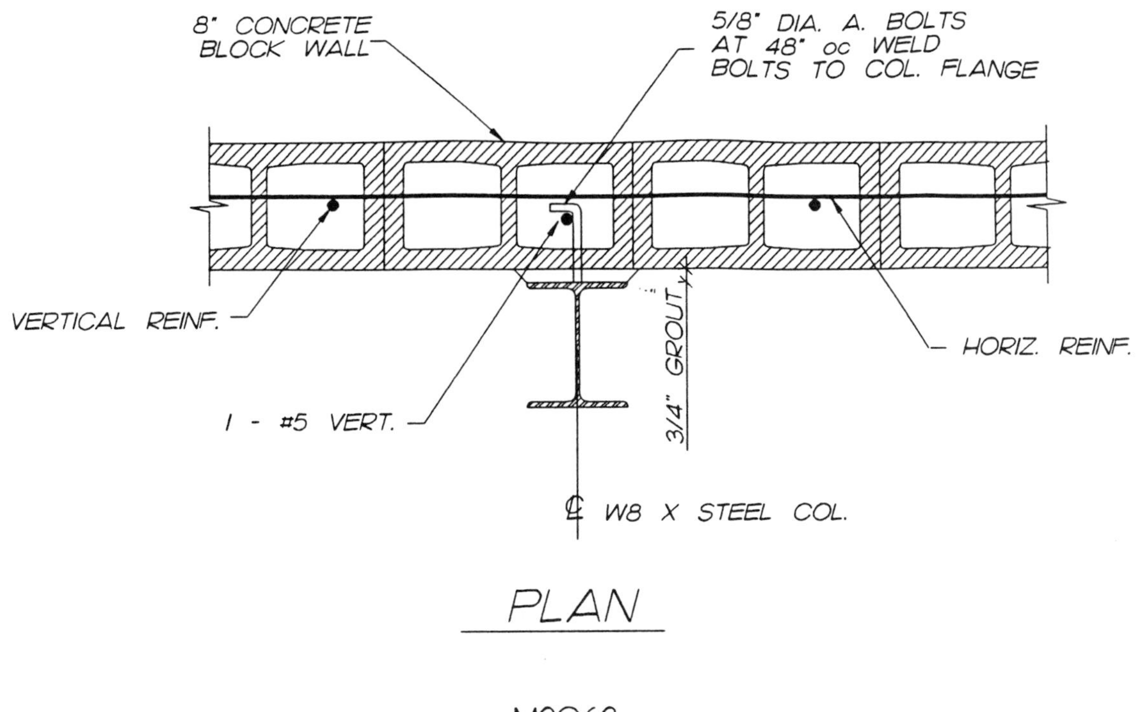
8" CONCRETE
BLOCK WALL
5/8" DIA. A. BOLTS
AT 48" oc WELD
BOLTS TO COL. FLANGE
VERTICAL REINF.
HORIZ. REINF.
3/4" GROUT
1 - #5 VERT.
℄ W8 X STEEL COL.
PLAN

M2062

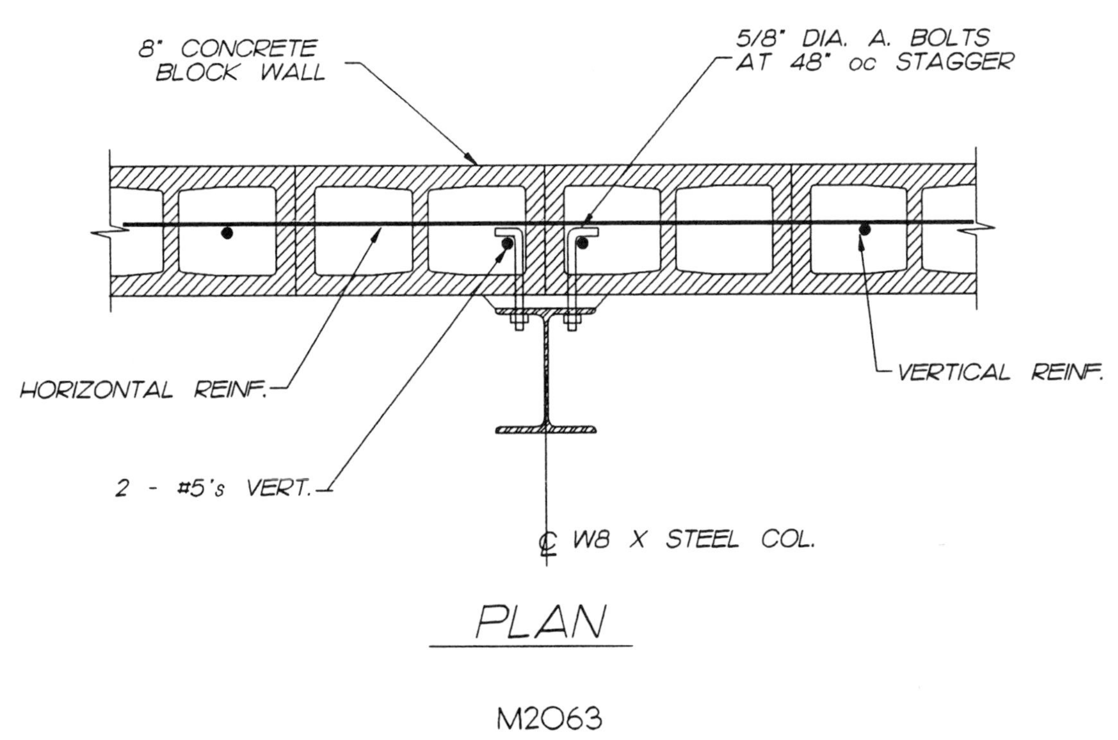
8" CONCRETE
BLOCK WALL
5/8" DIA. A. BOLTS
AT 48" oc STAGGER
HORIZONTAL REINF.
VERTICAL REINF.
2 - #5's VERT.
℄ W8 X STEEL COL.
PLAN

M2063

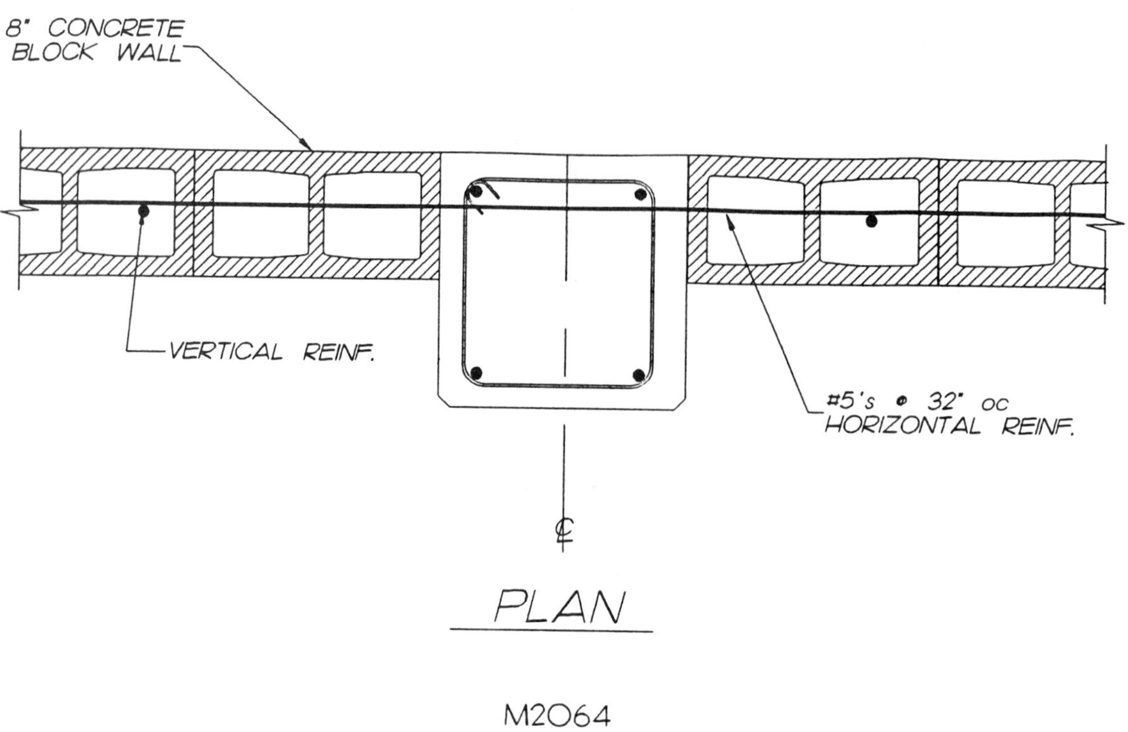

M2064

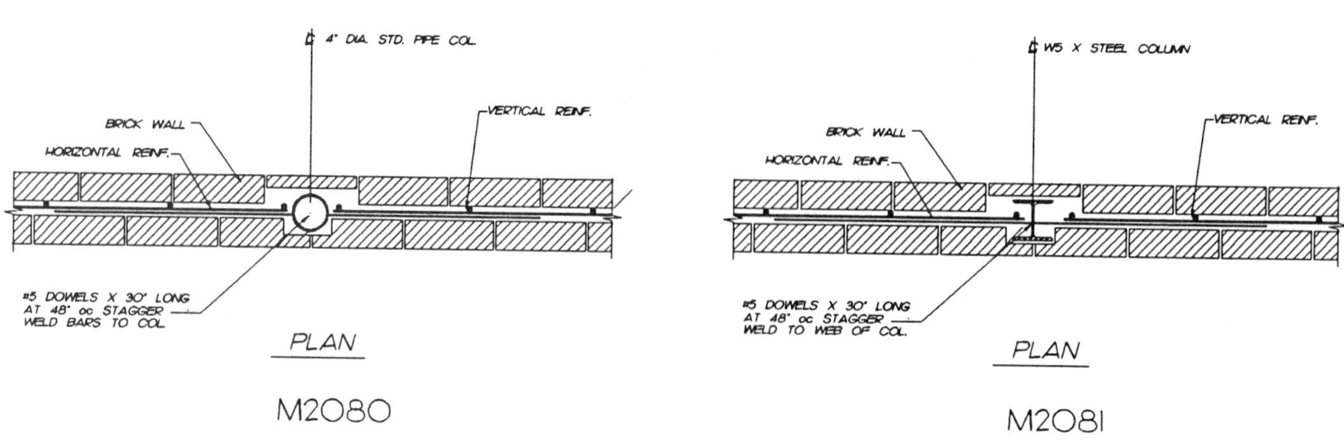

M2080

M2081

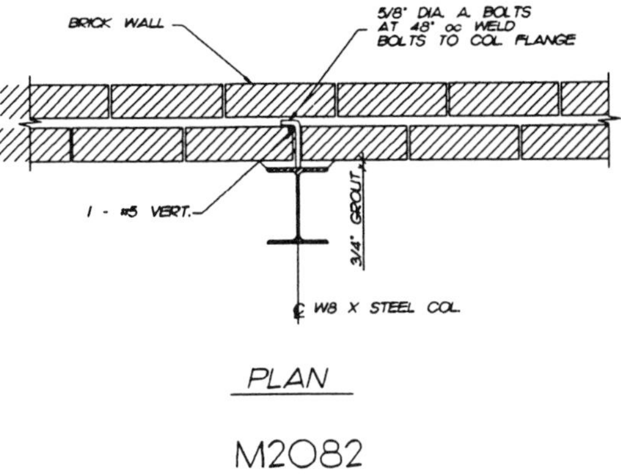

PLAN

M2082

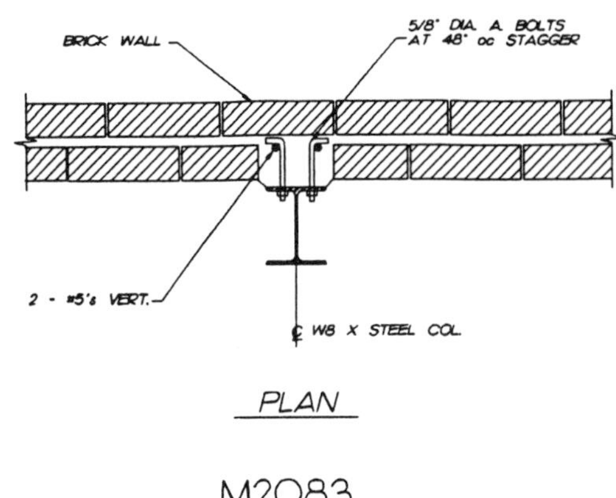

PLAN

M2083

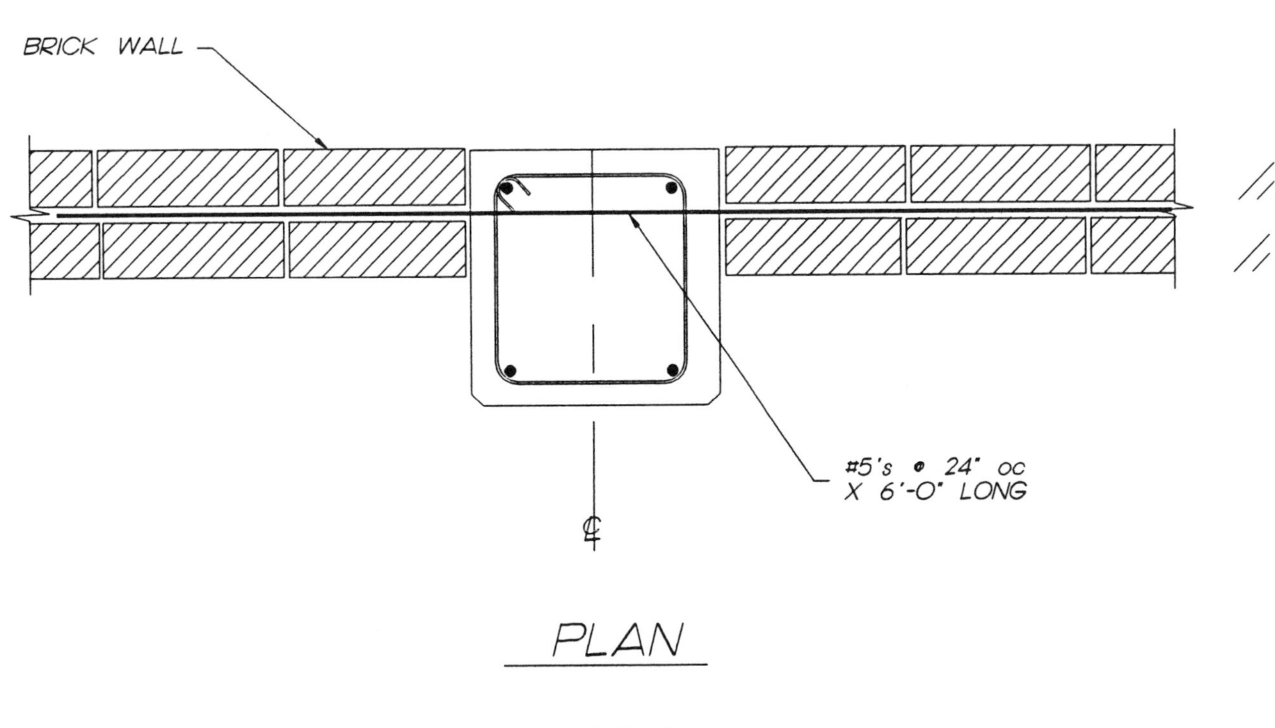

PLAN

M2084

CHAPTER 2 MASONRY SECTION 3 WALL - WOOD FLOOR			
EXTERIOR WALL JOISTS PERP. TO WALL 4 X LEDGER METAL TIE STRAPS NO LT. WT. CONCRETE	———	8" CONCRETE BLOCK	M3020 - M3024
		12" CONCRETE BLOCK	M3025 - M3029
		BRICK MASONRY	M3140 - M3144
EXTERIOR WALL JOISTS PARALELL TO WALL 4 X LEDGER METAL TIE STRAPS NO LT. WT. CONCRETE	———	8" CONCRETE BLOCK	M3040 - M3044
		12" CONCRETE BLOCK	M3045 - M3049
		BRICK MASONRY	M3160 - M3164
INTERIOR WALL JOISTS PERP. TO WALL 4 X LEDGER METAL TIE STRAPS NO LT. WT. CONCRETE	———	8" CONCRETE BLOCK	M3060 - M3064
		12" CONCRETE BLOCK	M3065 - M3069
		BRICK MASONRY	M3180 - M3184
INTERIOR WALL JOISTS PARALELL TO WALL 4 X LEDGER METAL TIE STRAPS NO LT. WT. CONCRETE	———	8" CONCRETE BLOCK	M3080 - M3084
		12" CONCRETE BLOCK	M3085 - M3089
		BRICK MASONRY	M3200 - M3204
INTERIOR WALL JSTS PERP & PARALELL ALT SIDES 4 X LEDGER METAL TIE STRAPS NO LT. WT. CONCRETE	———	8" CONCRETE BLOCK	M3100 - M3104
		12" CONCRETE BLOCK	M3105 - M3109
		BRICK MASONRY	M3220 - M3224
BASEMENT WALL WOOD FLOOR OVER	JOISTS PERP. TO WALL	8" CONCRETE BLOCK	M3120
	JOISTS PARALELL TO WALL	8" CONCRETE BLOCK	M3121

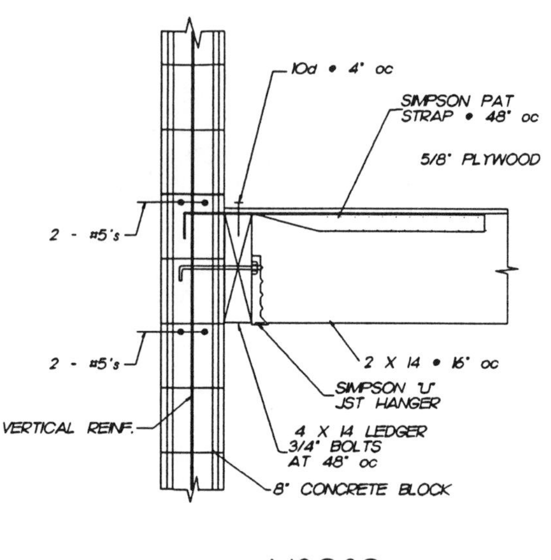

M3020

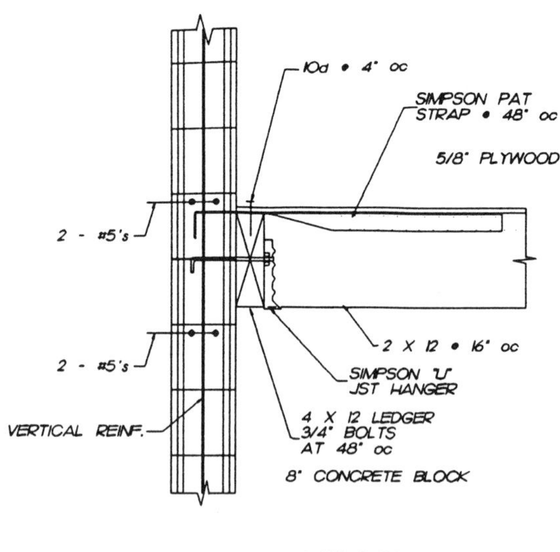

M3021

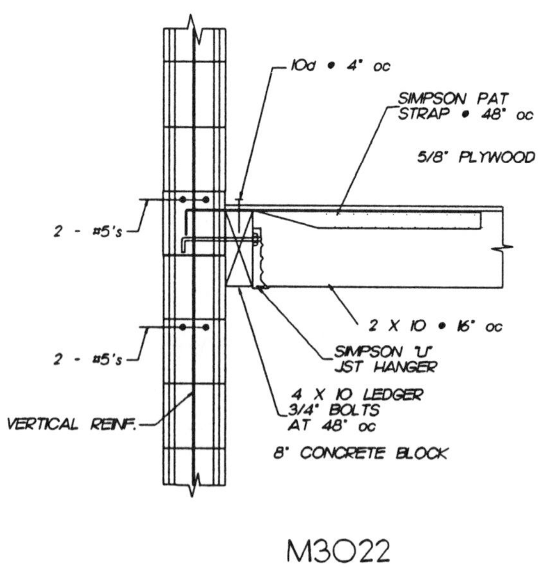

M3022

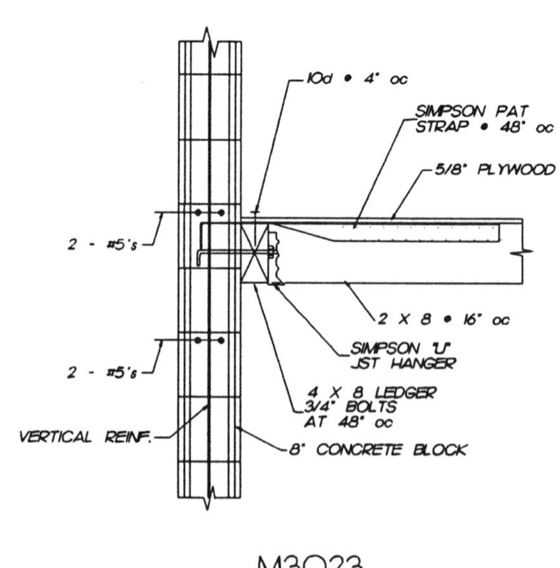

M3023

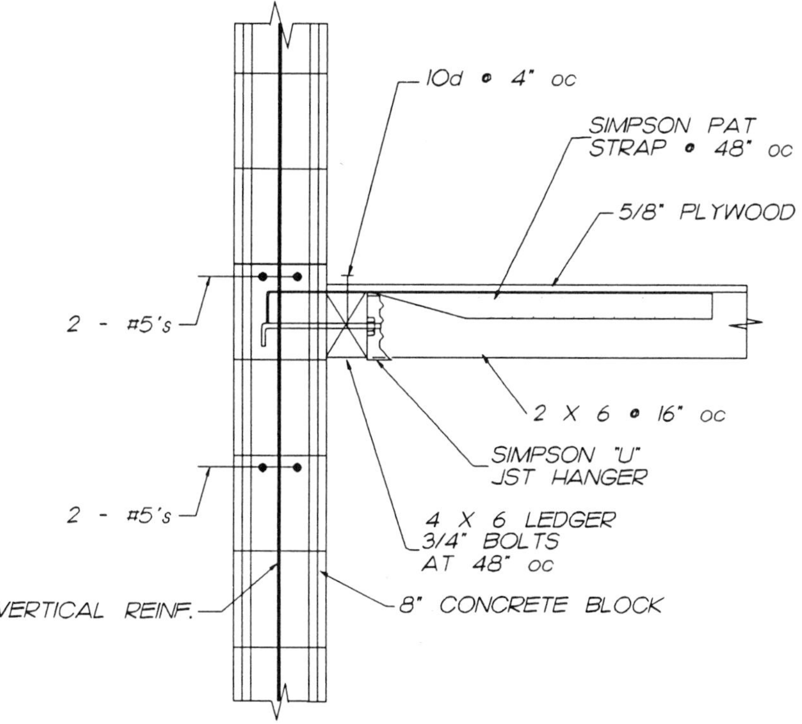

M3024

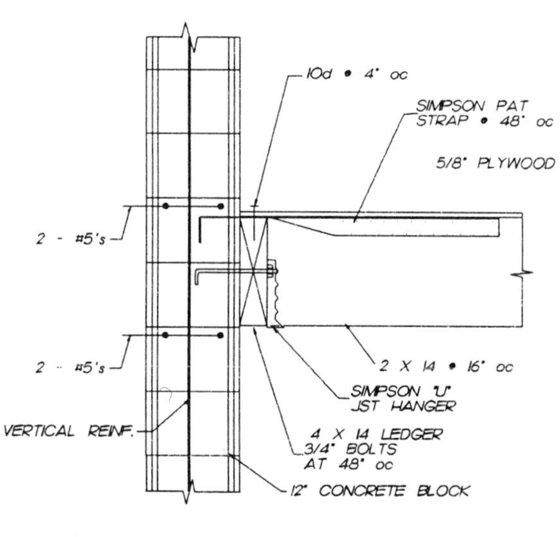

M3025

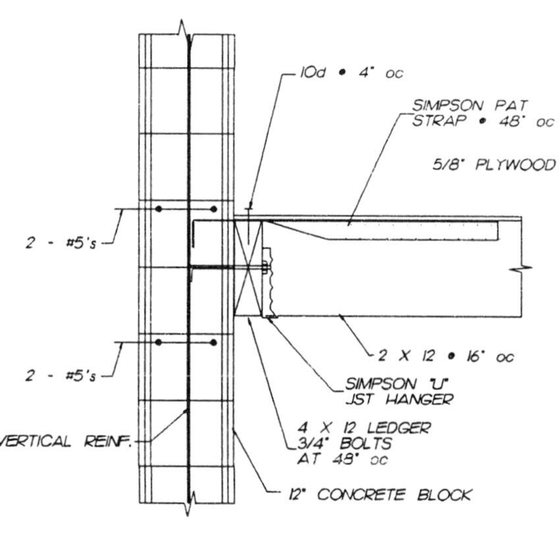

M3026

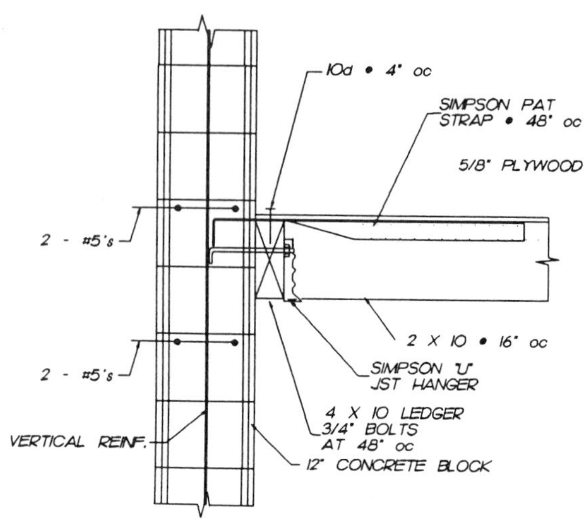

M3027

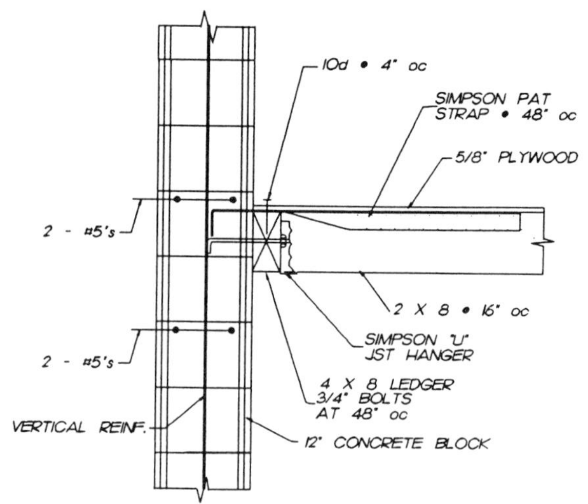

M3028

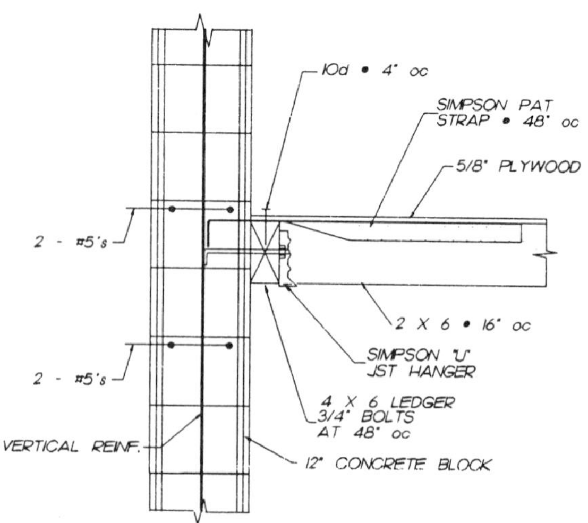

M3029

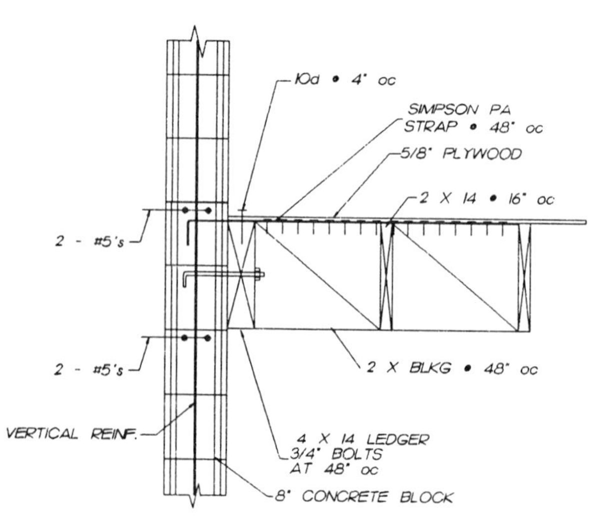

M3040

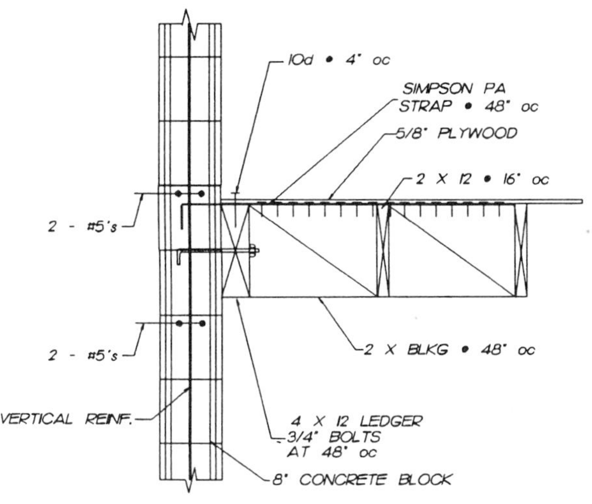

M3041

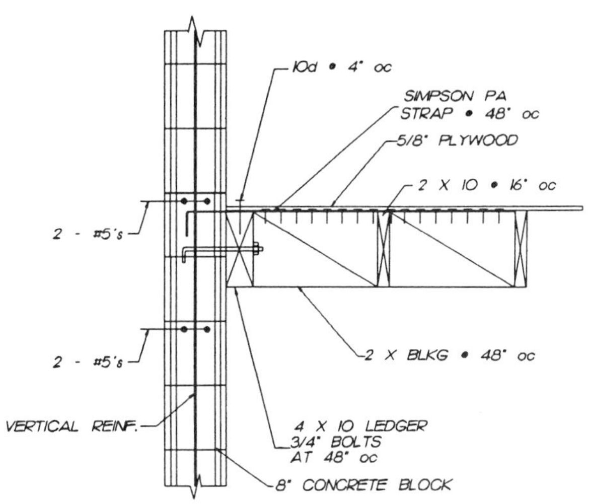

M3042

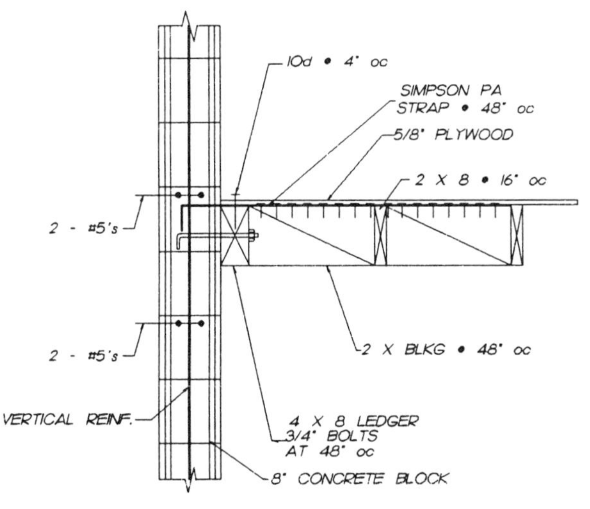

M3043

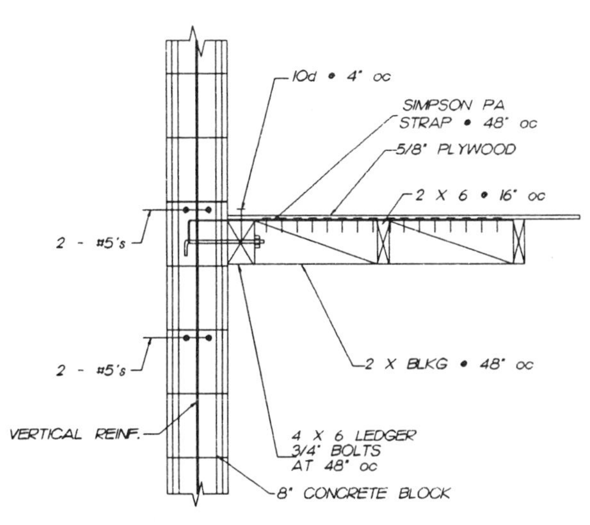

M3044

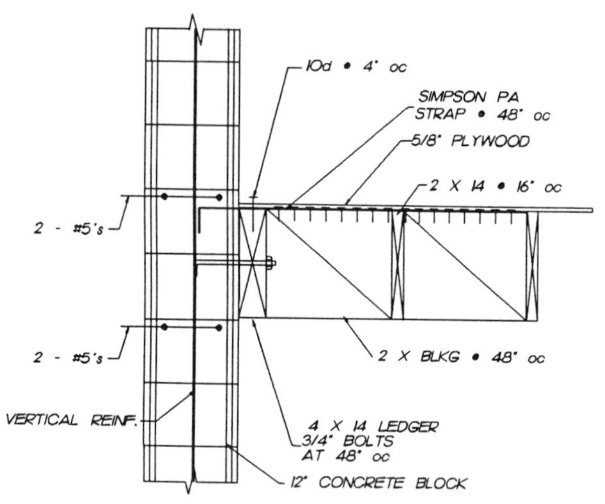

M3045

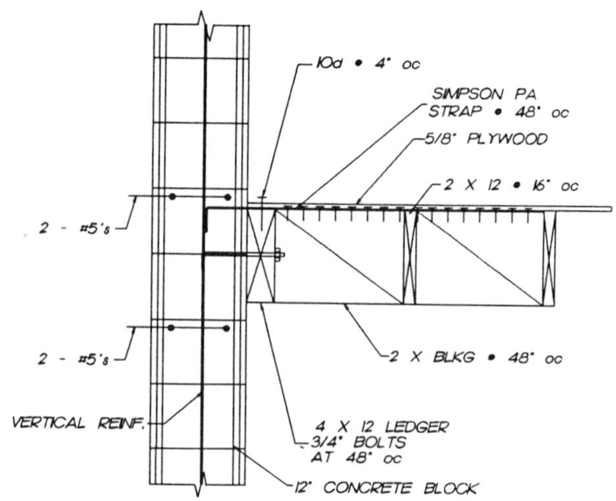

M3046

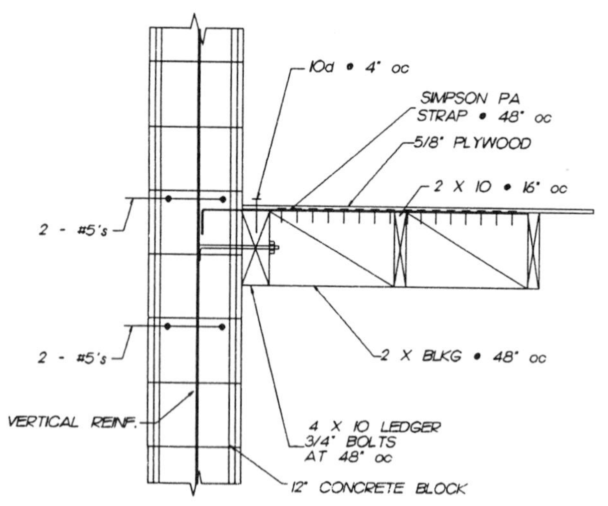

M3047

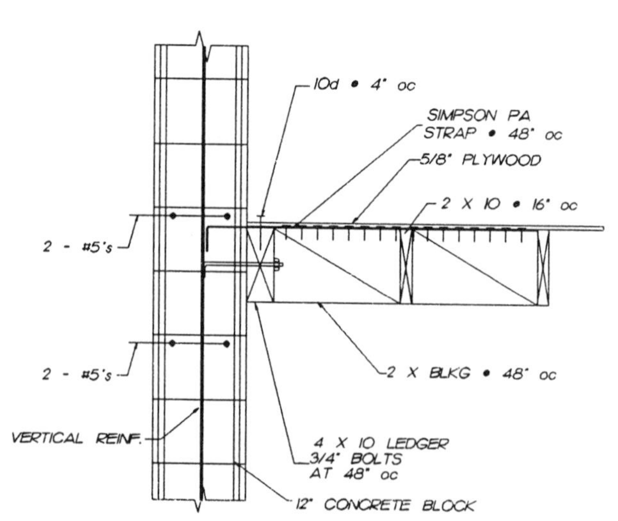

M3048

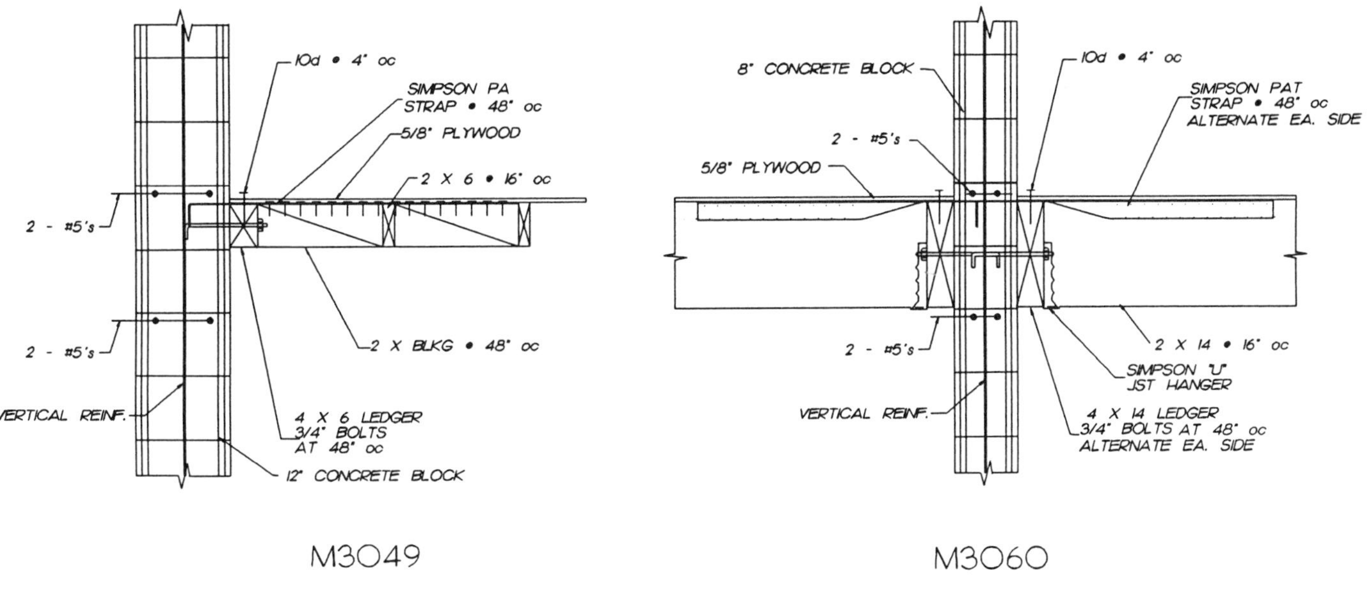

M3049

M3060

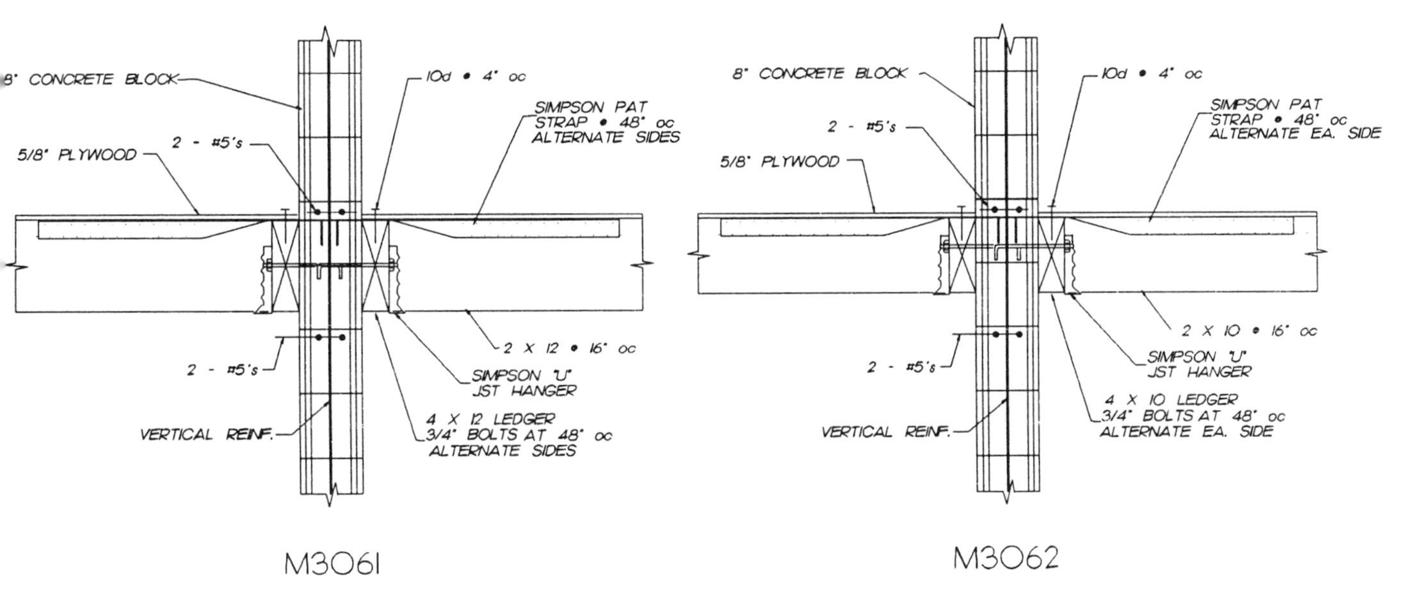

M3061

M3062

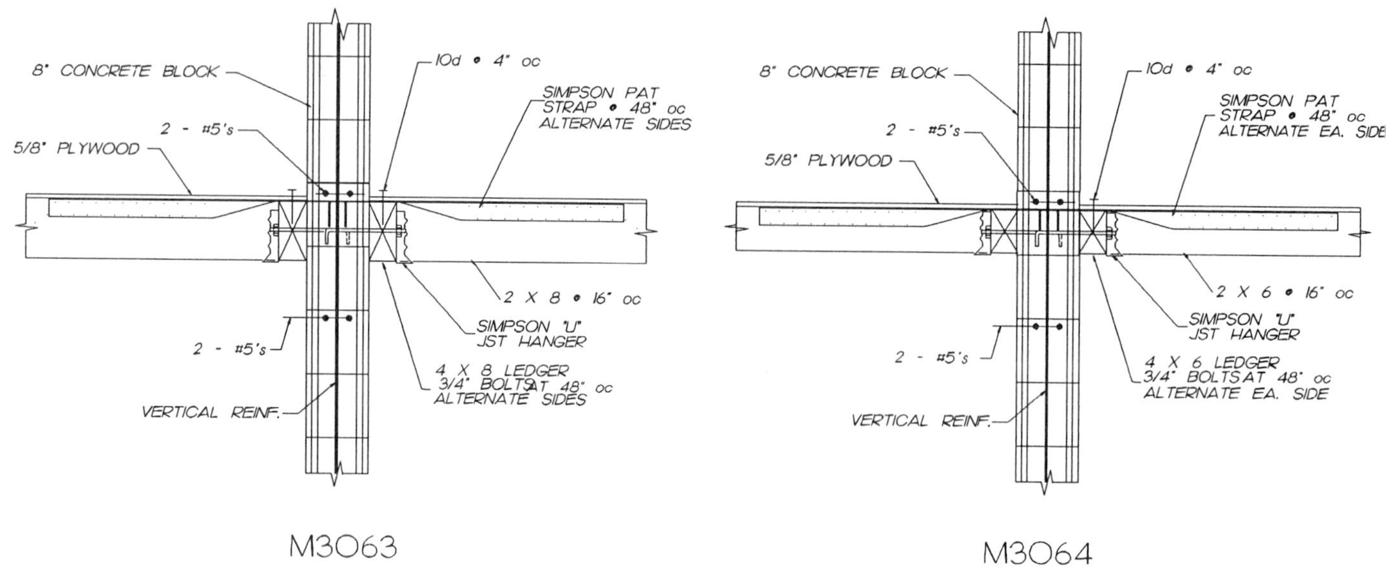

M3063

M3064

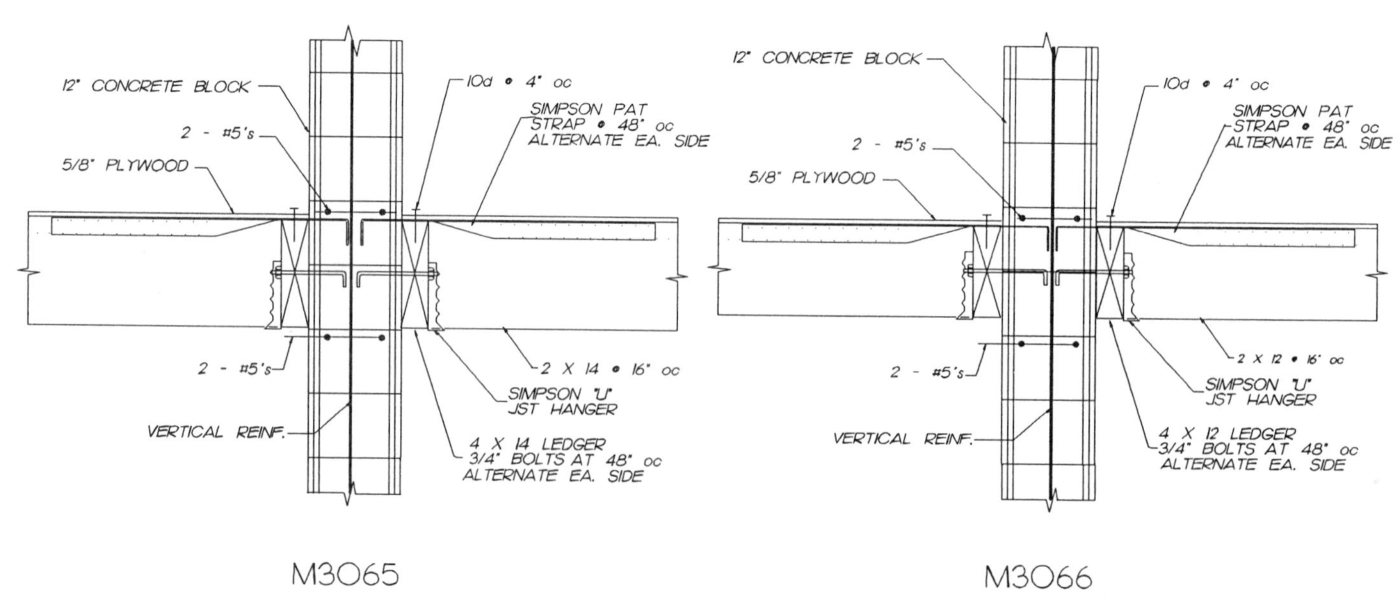

M3065

M3066

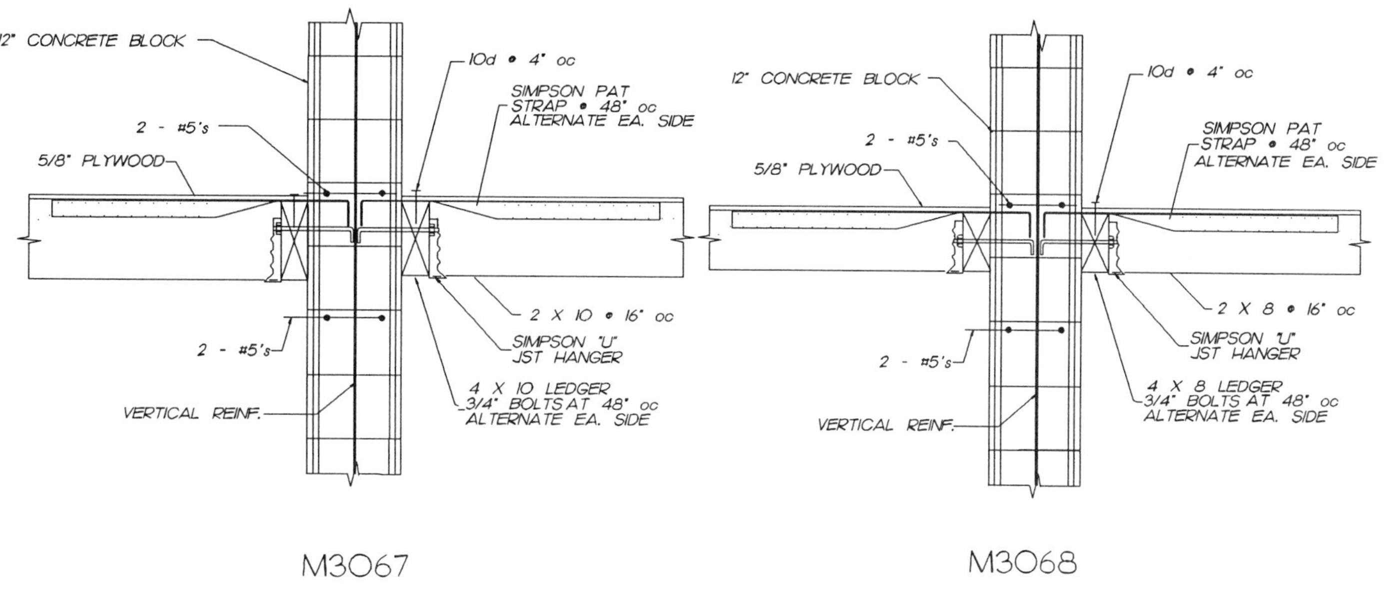
12" CONCRETE BLOCK
10d @ 4" oc
SIMPSON PAT STRAP @ 48" oc ALTERNATE EA. SIDE
2 - #5's
5/8" PLYWOOD
2 X 10 @ 16" oc
SIMPSON "U" JST HANGER
2 - #5's
4 X 10 LEDGER 3/4" BOLTS AT 48" oc ALTERNATE EA. SIDE
VERTICAL REINF.
M3067
12" CONCRETE BLOCK
10d @ 4" oc
SIMPSON PAT STRAP @ 48" oc ALTERNATE EA. SIDE
2 - #5's
5/8" PLYWOOD
2 X 8 @ 16" oc
SIMPSON "U" JST HANGER
2 - #5's
4 X 8 LEDGER 3/4" BOLTS AT 48" oc ALTERNATE EA. SIDE
VERTICAL REINF.
M3068

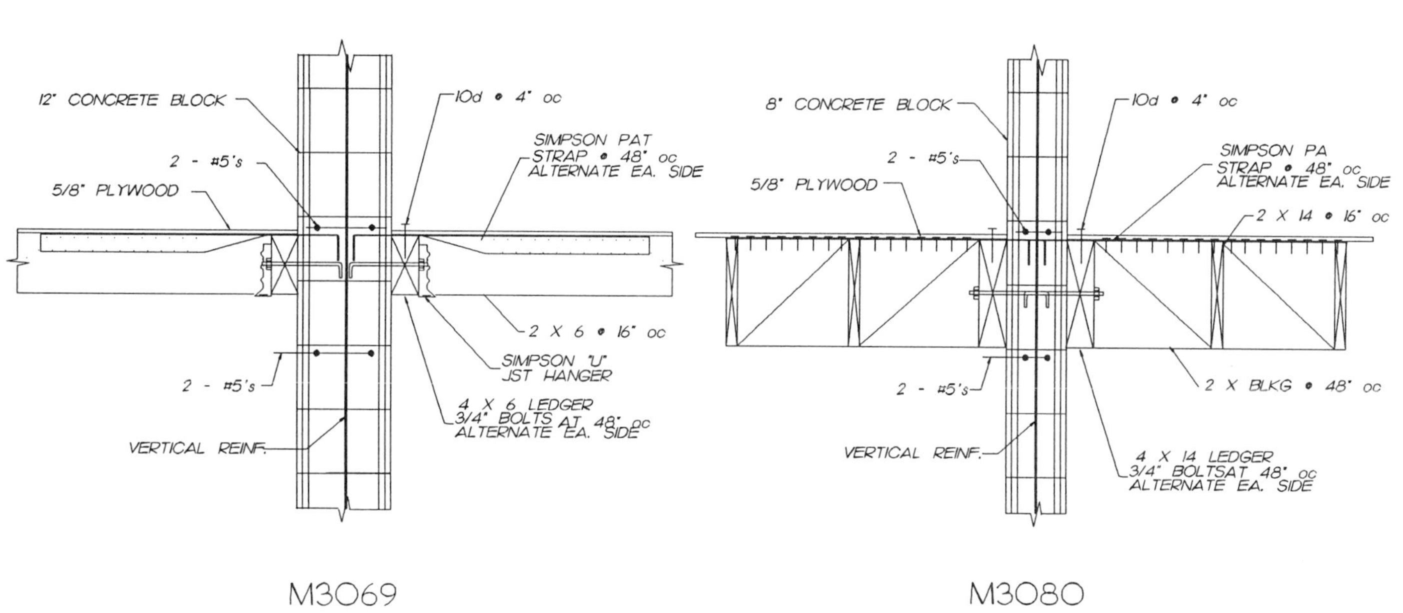
12" CONCRETE BLOCK
10d @ 4" oc
SIMPSON PAT STRAP @ 48" oc ALTERNATE EA. SIDE
2 - #5's
5/8" PLYWOOD
2 X 6 @ 16" oc
SIMPSON "U" JST HANGER
2 - #5's
4 X 6 LEDGER 3/4" BOLTS AT 48" oc ALTERNATE EA. SIDE
VERTICAL REINF.
M3069
8" CONCRETE BLOCK
10d @ 4" oc
SIMPSON PA STRAP @ 48" oc ALTERNATE EA. SIDE
2 - #5's
5/8" PLYWOOD
2 X 14 @ 16" oc
2 X BLKG @ 48" oc
2 - #5's
VERTICAL REINF.
4 X 14 LEDGER 3/4" BOLTSAT 48" oc ALTERNATE EA. SIDE
M3080

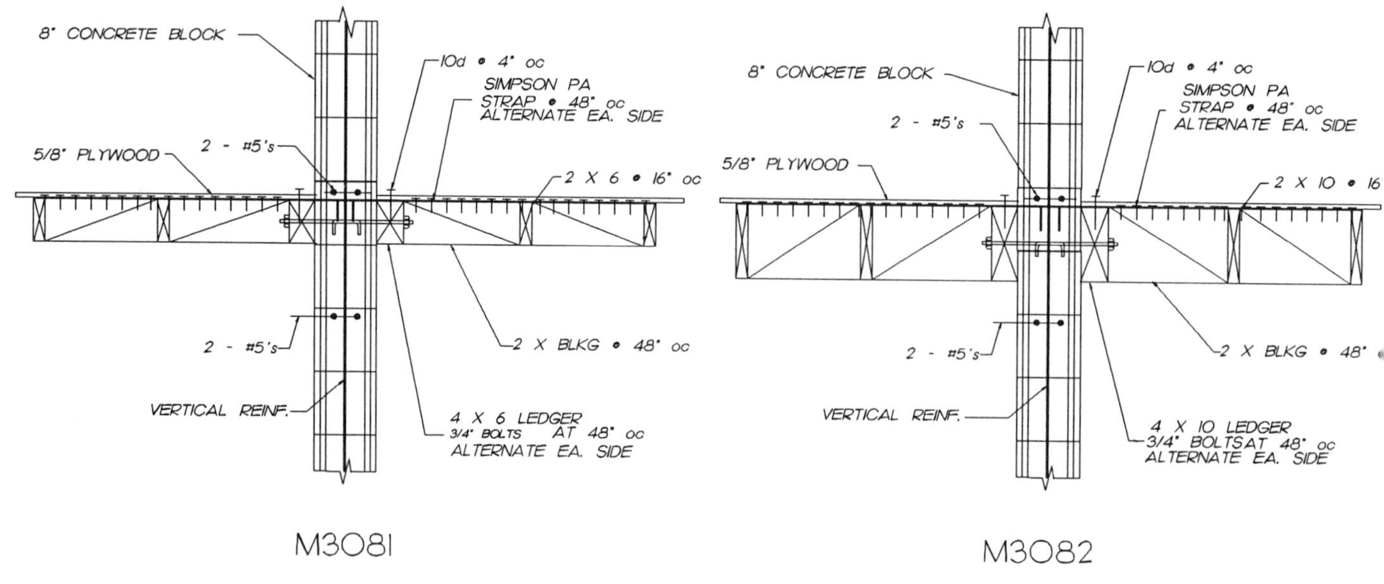
8" CONCRETE BLOCK
10d @ 4" oc
SIMPSON PA
STRAP @ 48" oc
ALTERNATE EA. SIDE
5/8" PLYWOOD
2 - #5's
2 X 6 @ 16" oc
2 - #5's
2 X BLKG @ 48" oc
VERTICAL REINF.
4 X 6 LEDGER
3/4" BOLTS AT 48" oc
ALTERNATE EA. SIDE
M3081
8" CONCRETE BLOCK
10d @ 4" oc
SIMPSON PA
STRAP @ 48" oc
ALTERNATE EA. SIDE
2 - #5's
5/8" PLYWOOD
2 X 10 @ 16
2 X BLKG @ 48"
2 - #5's
VERTICAL REINF.
4 X 10 LEDGER
3/4" BOLTS AT 48" oc
ALTERNATE EA. SIDE
M3082

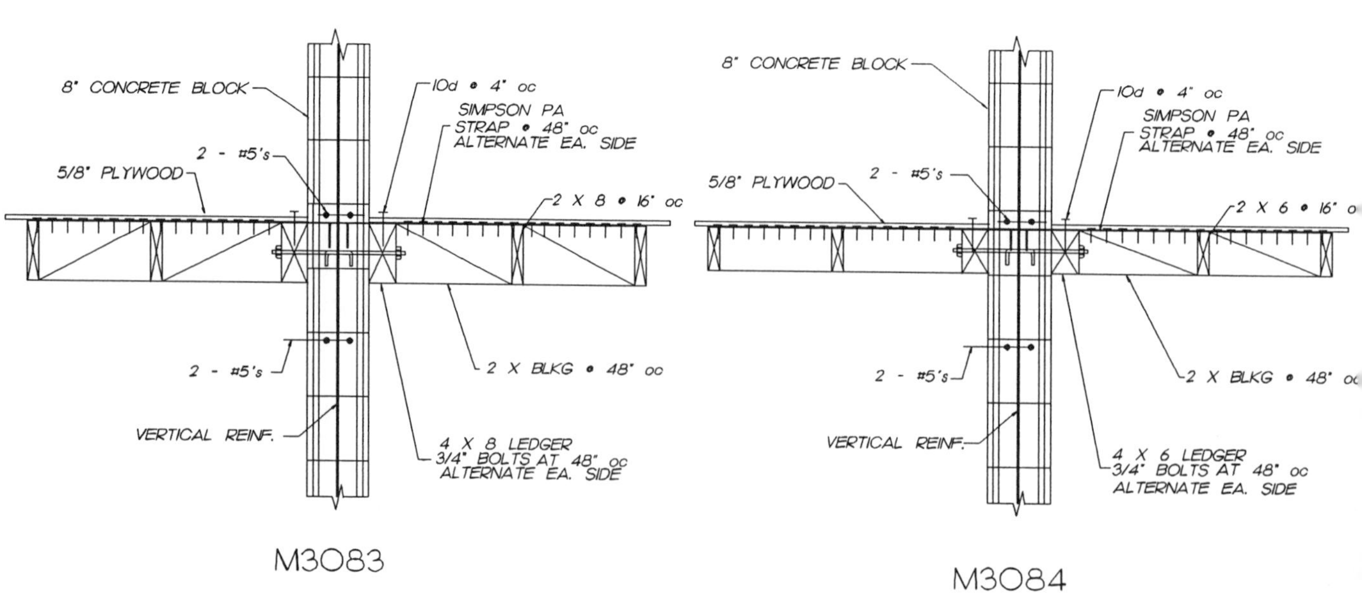
8" CONCRETE BLOCK
10d @ 4" oc
SIMPSON PA
STRAP @ 48" oc
ALTERNATE EA. SIDE
2 - #5's
5/8" PLYWOOD
2 X 8 @ 16" oc
2 - #5's
2 X BLKG @ 48" oc
VERTICAL REINF.
4 X 8 LEDGER
3/4" BOLTS AT 48" oc
ALTERNATE EA. SIDE
M3083
8" CONCRETE BLOCK
10d @ 4" oc
SIMPSON PA
STRAP @ 48" oc
ALTERNATE EA. SIDE
5/8" PLYWOOD
2 - #5's
2 X 6 @ 16" o
2 - #5's
2 X BLKG @ 48" oc
VERTICAL REINF.
4 X 6 LEDGER
3/4" BOLTS AT 48" oc
ALTERNATE EA. SIDE
M3084

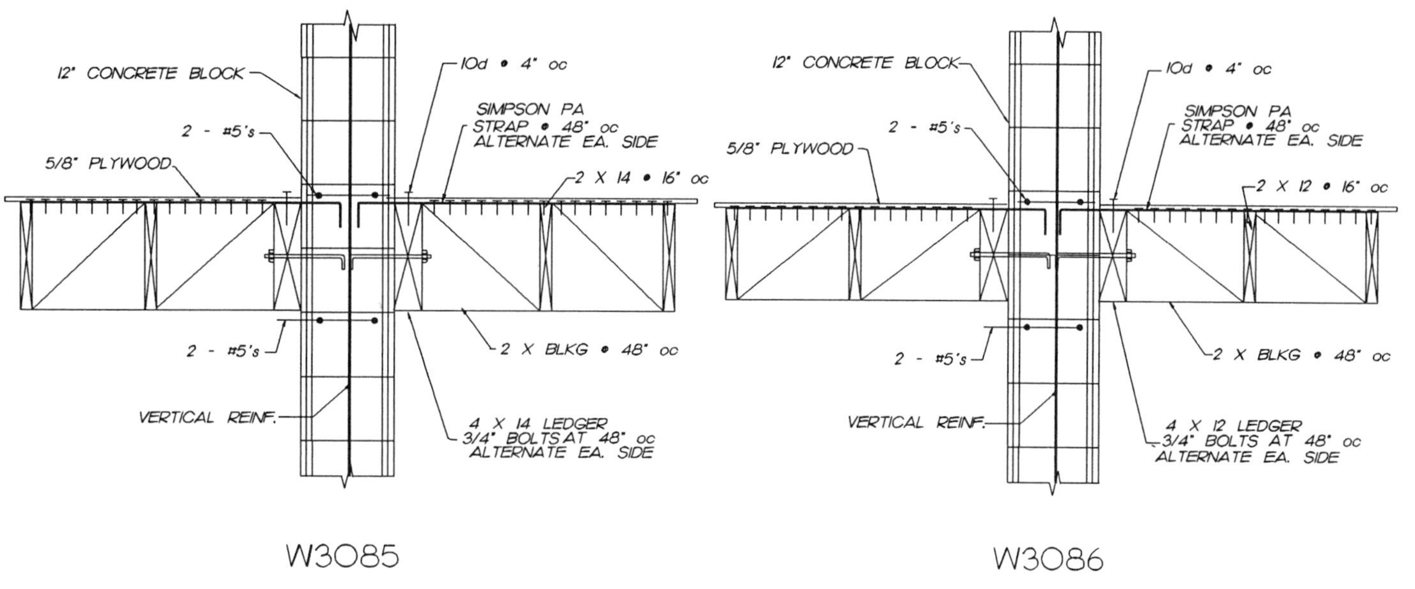
12" CONCRETE BLOCK
10d @ 4" oc
SIMPSON PA STRAP @ 48" oc ALTERNATE EA. SIDE
2 - #5's
5/8" PLYWOOD
2 X 14 @ 16" oc
2 - #5's
2 X BLKG @ 48" oc
VERTICAL REINF.
4 X 14 LEDGER 3/4" BOLTS AT 48" oc ALTERNATE EA. SIDE
W3085
12" CONCRETE BLOCK
10d @ 4" oc
SIMPSON PA STRAP @ 48" oc ALTERNATE EA. SIDE
2 - #5's
5/8" PLYWOOD
2 X 12 @ 16" oc
2 X BLKG @ 48" oc
2 - #5's
VERTICAL REINF.
4 X 12 LEDGER 3/4" BOLTS AT 48" oc ALTERNATE EA. SIDE
W3086

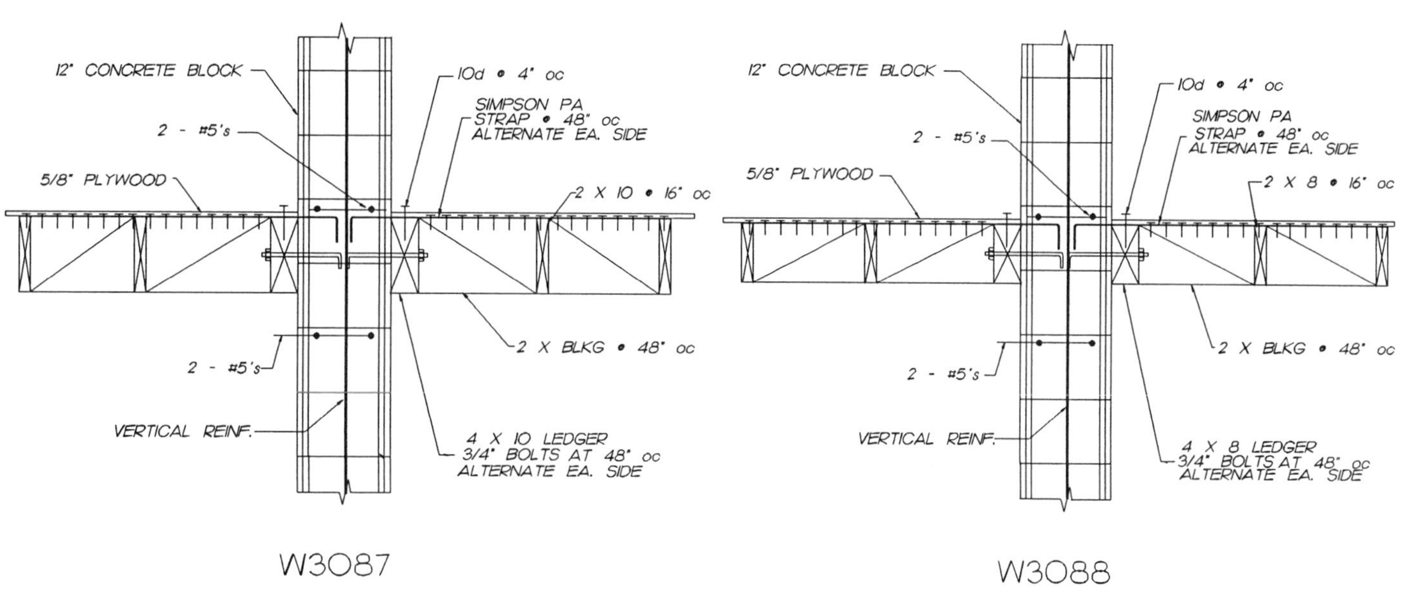
12" CONCRETE BLOCK
10d @ 4" oc
SIMPSON PA STRAP @ 48" oc ALTERNATE EA. SIDE
2 - #5's
5/8" PLYWOOD
2 X 10 @ 16" oc
2 X BLKG @ 48" oc
2 - #5's
VERTICAL REINF.
4 X 10 LEDGER 3/4" BOLTS AT 48" oc ALTERNATE EA. SIDE
W3087
12" CONCRETE BLOCK
10d @ 4" oc
SIMPSON PA STRAP @ 48" oc ALTERNATE EA. SIDE
2 - #5's
5/8" PLYWOOD
2 X 8 @ 16" oc
2 X BLKG @ 48" oc
2 - #5's
VERTICAL REINF.
4 X 8 LEDGER 3/4" BOLTS AT 48" oc ALTERNATE EA. SIDE
W3088

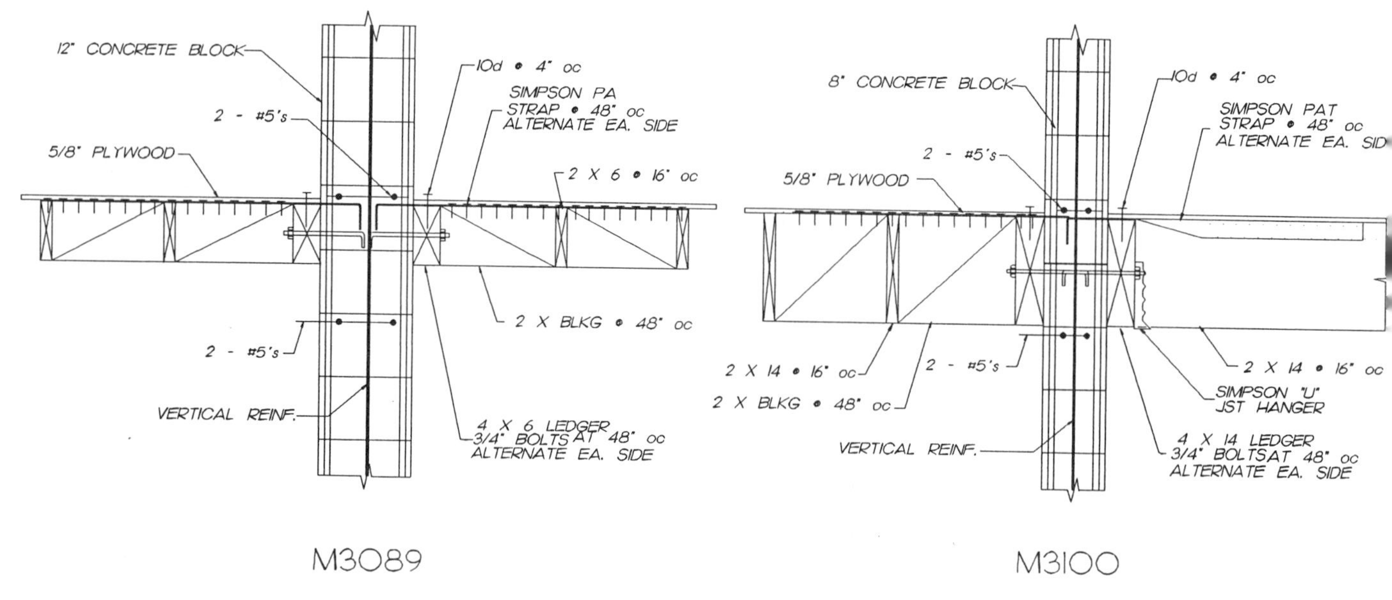
12" CONCRETE BLOCK
2 - #5's
5/8" PLYWOOD
10d @ 4" oc
SIMPSON PA STRAP @ 48" oc ALTERNATE EA. SIDE
2 X 6 @ 16" oc
2 X BLKG @ 48" oc
2 - #5's
VERTICAL REINF.
4 X 6 LEDGER 3/4" BOLTS AT 48" oc ALTERNATE EA. SIDE
M3089
8" CONCRETE BLOCK
10d @ 4" oc
SIMPSON PAT STRAP @ 48" oc ALTERNATE EA. SID
2 - #5's
5/8" PLYWOOD
2 X 14 @ 16" oc
2 X BLKG @ 48" oc
2 - #5's
2 X 14 @ 16" oc
SIMPSON "U" JST HANGER
VERTICAL REINF.
4 X 14 LEDGER 3/4" BOLTSAT 48" oc ALTERNATE EA. SIDE
M3100

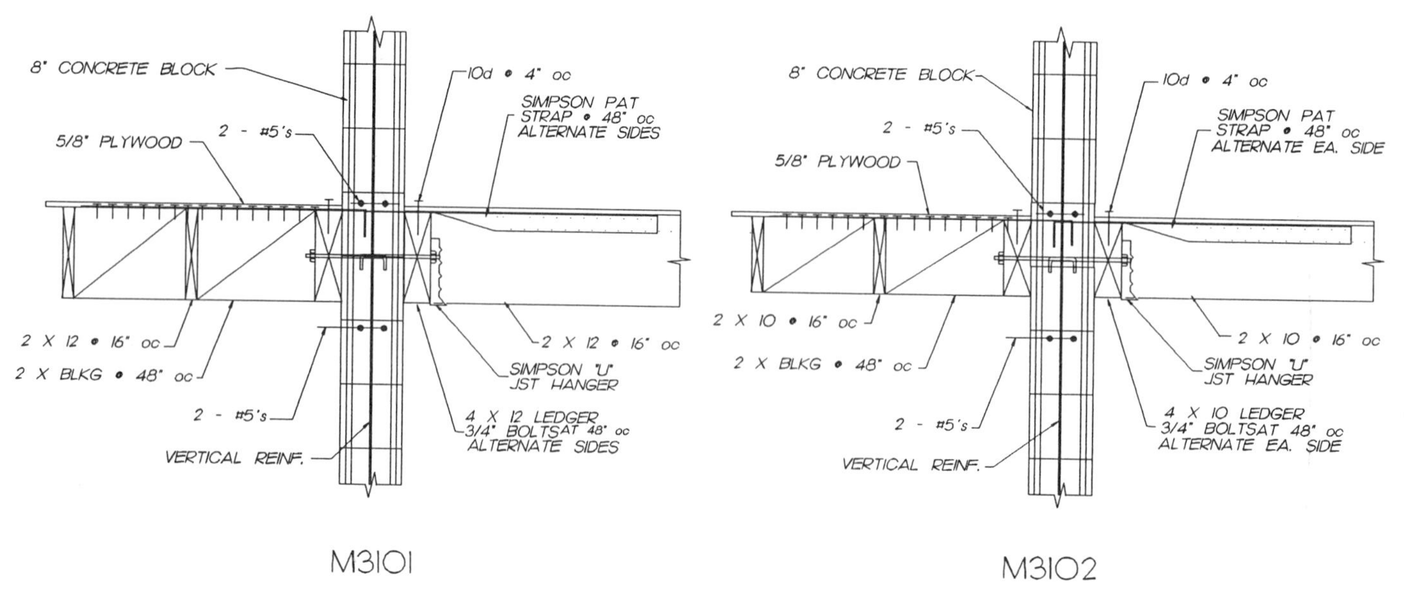
8" CONCRETE BLOCK
10d @ 4" oc
SIMPSON PAT STRAP @ 48" oc ALTERNATE SIDES
5/8" PLYWOOD
2 - #5's
2 X 12 @ 16" oc
2 X BLKG @ 48" oc
2 X 12 @ 16" oc
SIMPSON "U" JST HANGER
2 - #5's
4 X 12 LEDGER 3/4" BOLTSAT 48" oc ALTERNATE SIDES
VERTICAL REINF.
M3101
8" CONCRETE BLOCK
10d @ 4" oc
SIMPSON PAT STRAP @ 48" oc ALTERNATE EA. SIDE
2 - #5's
5/8" PLYWOOD
2 X 10 @ 16" oc
2 X BLKG @ 48" oc
2 X 10 @ 16" oc
SIMPSON "U" JST HANGER
2 - #5's
4 X 10 LEDGER 3/4" BOLTSAT 48" oc ALTERNATE EA. SIDE
VERTICAL REINF.
M3102

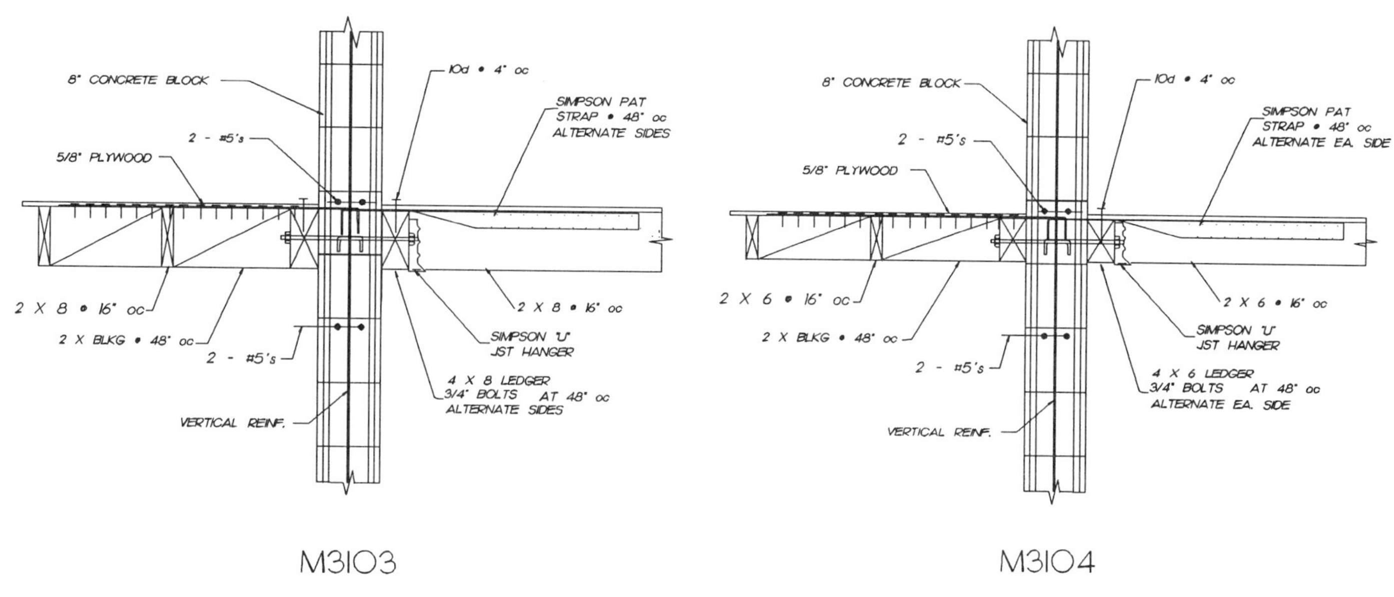

M3103

M3104

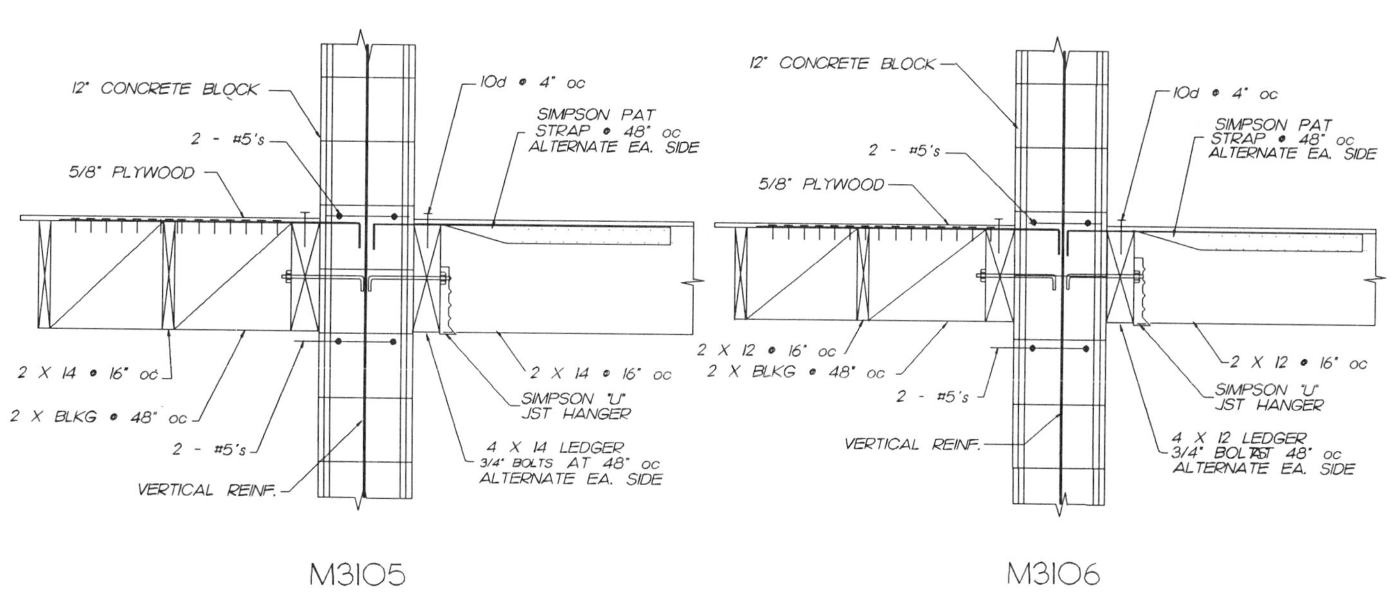

M3105

M3106

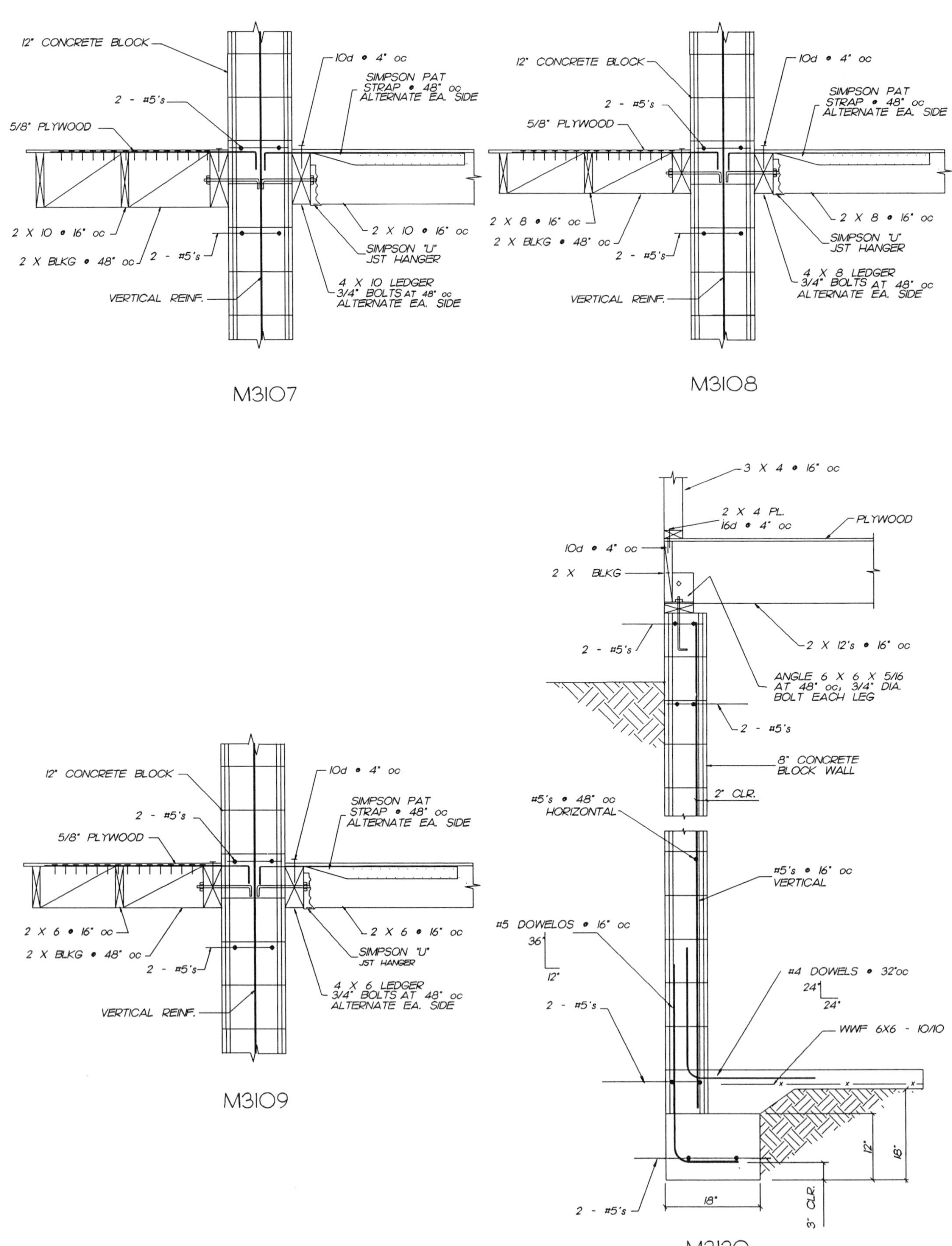
12" CONCRETE BLOCK
10d @ 4" oc
SIMPSON PAT STRAP @ 48" oc ALTERNATE EA. SIDE
2 - #5's
5/8" PLYWOOD
2 X 10 @ 16" oc
2 X BLKG @ 48" oc
2 - #5's
2 X 10 @ 16" oc
SIMPSON "U" JST HANGER
4 X 10 LEDGER 3/4" BOLTS AT 48" oc ALTERNATE EA. SIDE
VERTICAL REINF.
M3107
12" CONCRETE BLOCK
10d @ 4" oc
SIMPSON PAT STRAP @ 48" oc ALTERNATE EA. SIDE
2 - #5's
5/8" PLYWOOD
2 X 8 @ 16" oc
2 X BLKG @ 48" oc
2 - #5's
2 X 8 @ 16" oc
SIMPSON "U" JST HANGER
4 X 8 LEDGER 3/4" BOLTS AT 48" oc ALTERNATE EA. SIDE
VERTICAL REINF.
M3108
3 X 4 @ 16" oc
2 X 4 PL. 16d @ 4" oc
PLYWOOD
10d @ 4" oc
2 X BLKG
2 - #5's
2 X 12's @ 16" oc
ANGLE 6 X 6 X 5/16 AT 48" oc, 3/4" DIA. BOLT EACH LEG
2 - #5's
8" CONCRETE BLOCK WALL
2" CLR.
#5's @ 48" oc HORIZONTAL
#5's @ 16" oc VERTICAL
#5 DOWELOS @ 16" oc
36"
12"
#4 DOWELS @ 32"oc
24"
24"
2 - #5's
WWF 6X6 - 10/10
12"
18"
18"
3" CLR.
2 - #5's
M3120
12" CONCRETE BLOCK
10d @ 4" oc
SIMPSON PAT STRAP @ 48" oc ALTERNATE EA. SIDE
2 - #5's
5/8" PLYWOOD
2 X 6 @ 16" oc
2 X BLKG @ 48" oc
2 - #5's
2 X 6 @ 16" oc
SIMPSON "U" JST HANGER
4 X 6 LEDGER 3/4" BOLTS AT 48" oc ALTERNATE EA. SIDE
VERTICAL REINF.
M3109

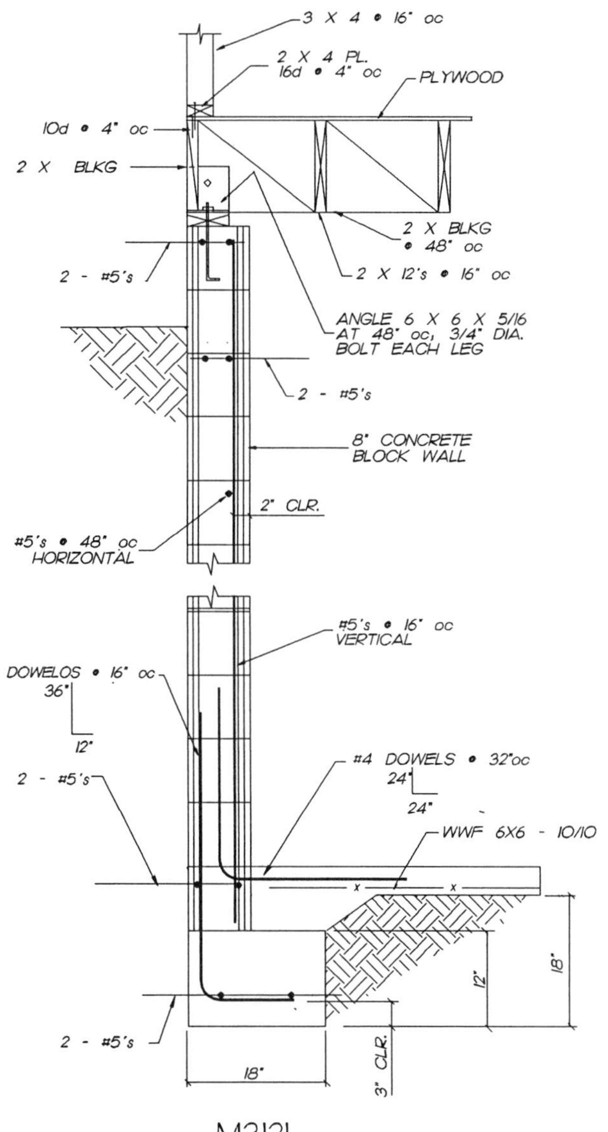

M3121

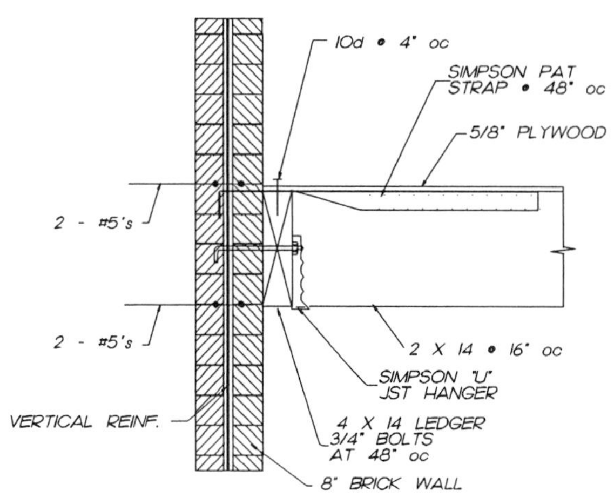

M3140

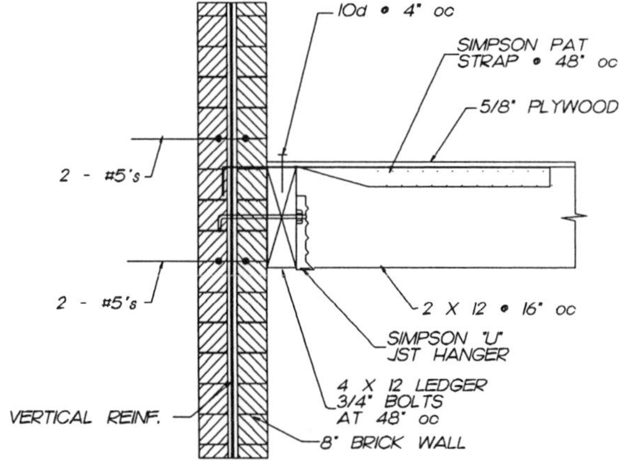

M3141

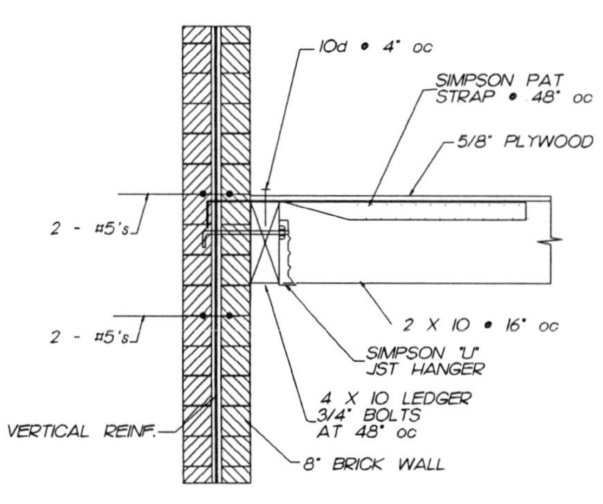

M3142

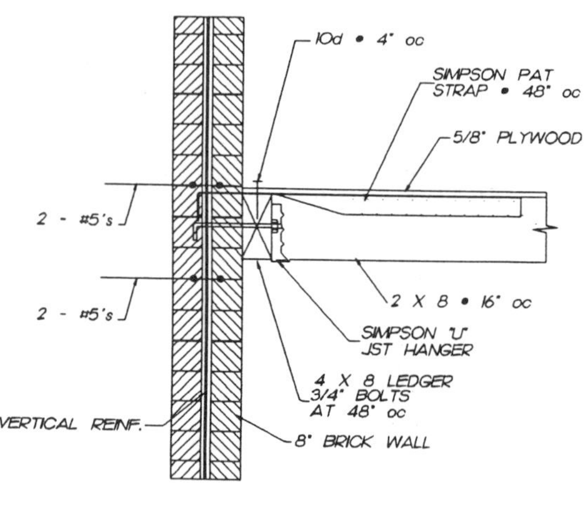

M3143

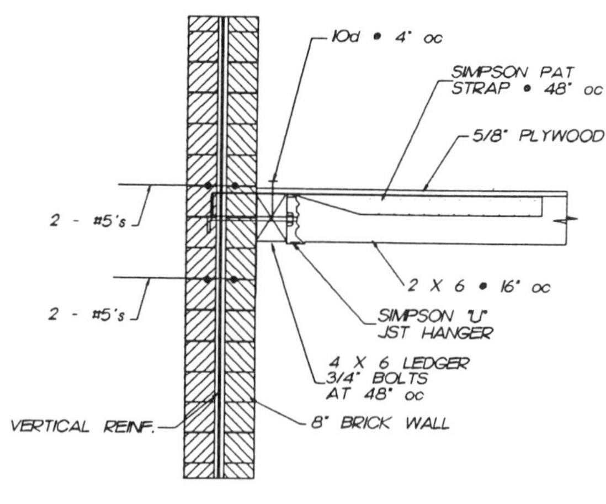

M3144

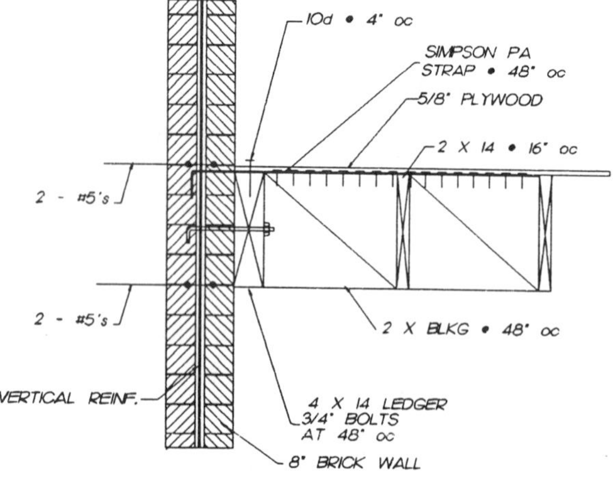

M3160

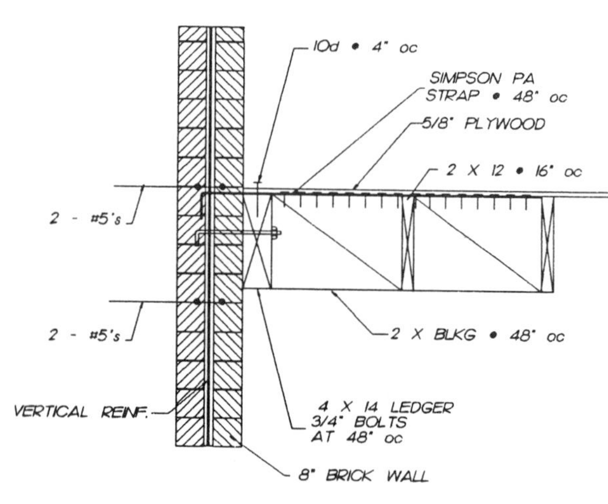

M3161

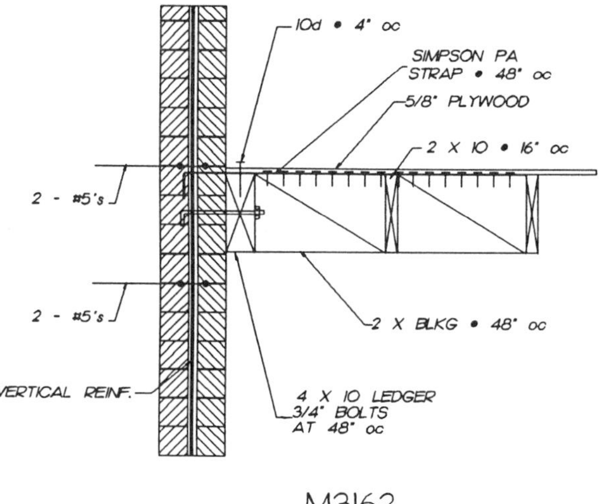

M3162

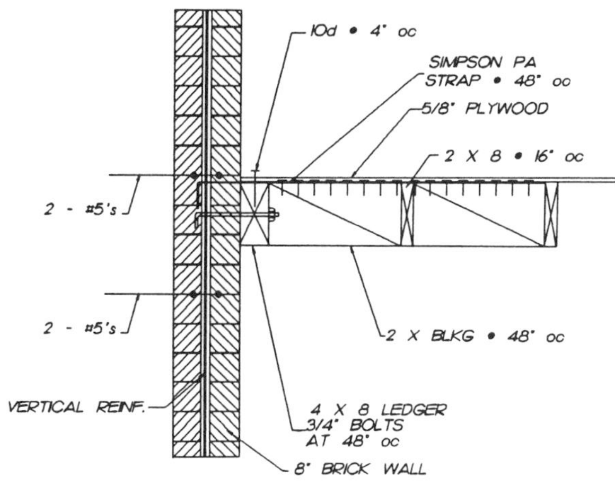

M3163

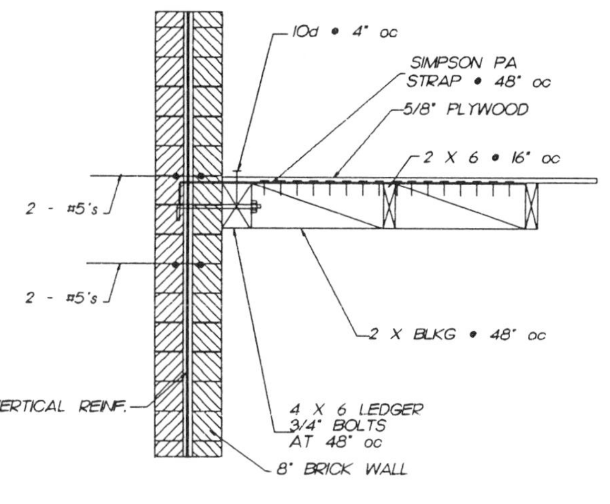

M3164

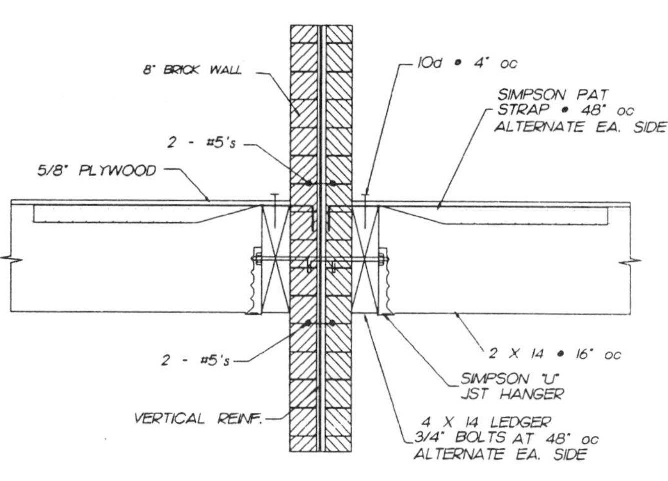

M3180

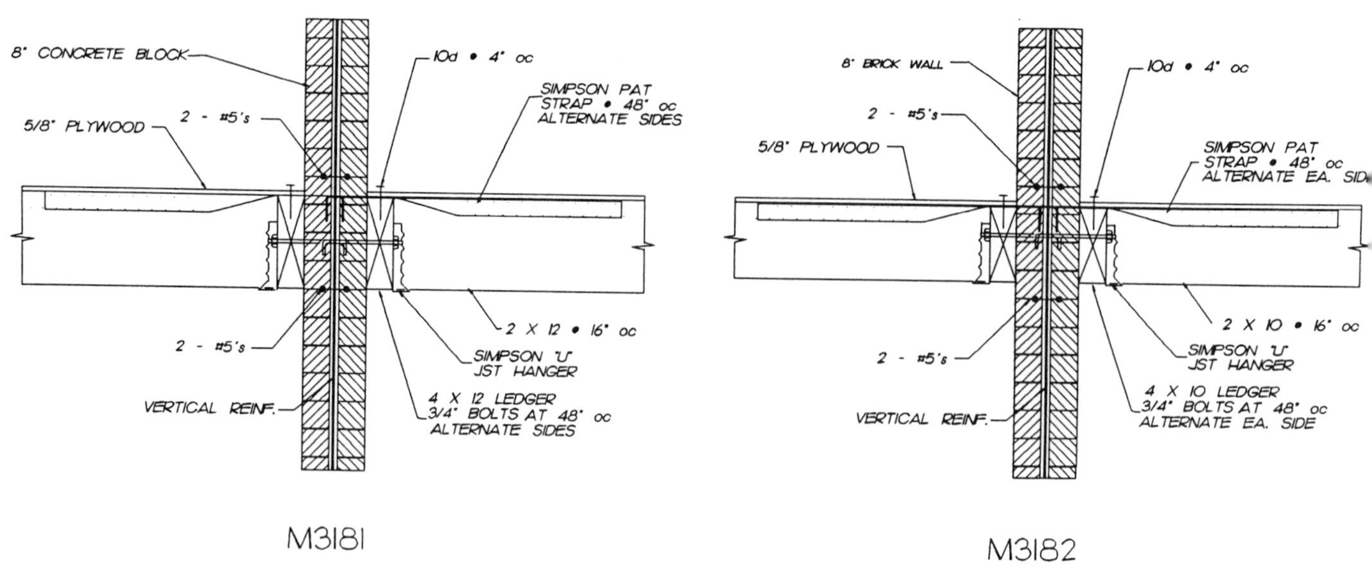
8" CONCRETE BLOCK
10d • 4" oc
SIMPSON PAT STRAP • 48" oc ALTERNATE SIDES
5/8" PLYWOOD
2 - #5's
2 X 12 • 16" oc
SIMPSON "U" JST HANGER
4 X 12 LEDGER 3/4" BOLTS AT 48" oc ALTERNATE SIDES
VERTICAL REINF.
M3181
8" BRICK WALL
10d • 4" oc
2 - #5's
SIMPSON PAT STRAP • 48" oc ALTERNATE EA. SIDE
5/8" PLYWOOD
2 X 10 • 16" oc
SIMPSON "U" JST HANGER
4 X 10 LEDGER 3/4" BOLTS AT 48" oc ALTERNATE EA. SIDE
VERTICAL REINF.
M3182

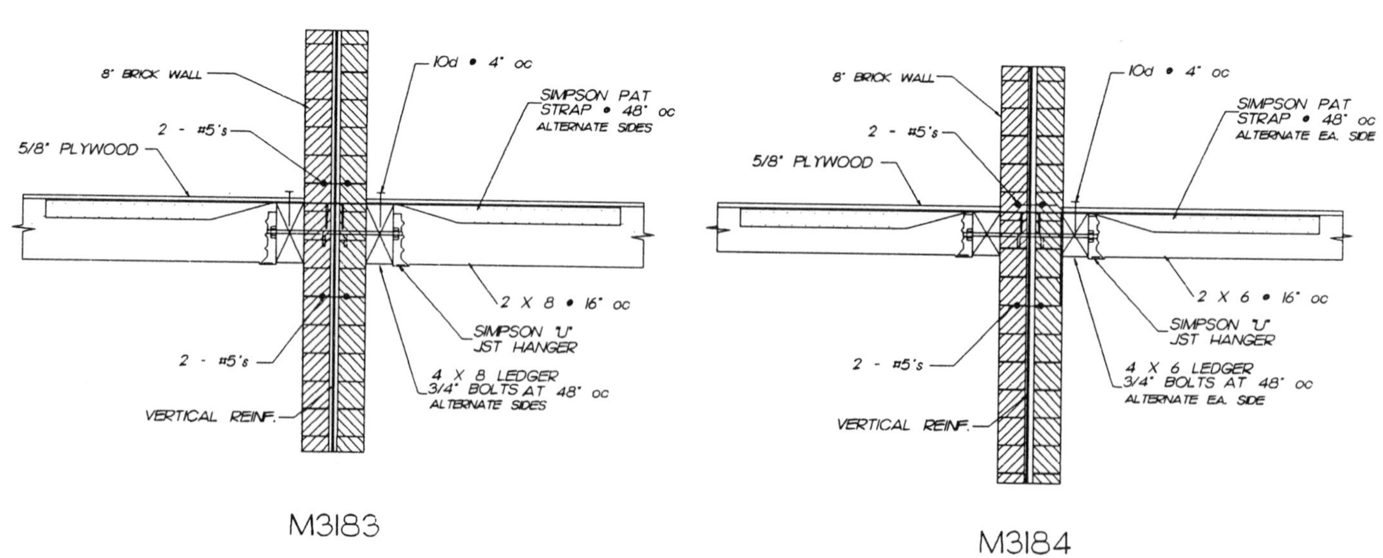
8" BRICK WALL
10d • 4" oc
SIMPSON PAT STRAP • 48" oc ALTERNATE SIDES
2 - #5's
5/8" PLYWOOD
2 X 8 • 16" oc
SIMPSON "U" JST HANGER
4 X 8 LEDGER 3/4" BOLTS AT 48" oc ALTERNATE SIDES
VERTICAL REINF.
M3183
8" BRICK WALL
10d • 4" oc
SIMPSON PAT STRAP • 48" oc ALTERNATE EA. SIDE
2 - #5's
5/8" PLYWOOD
2 X 6 • 16" oc
SIMPSON "U" JST HANGER
4 X 6 LEDGER 3/4" BOLTS AT 48" oc ALTERNATE EA. SIDE
VERTICAL REINF.
M3184

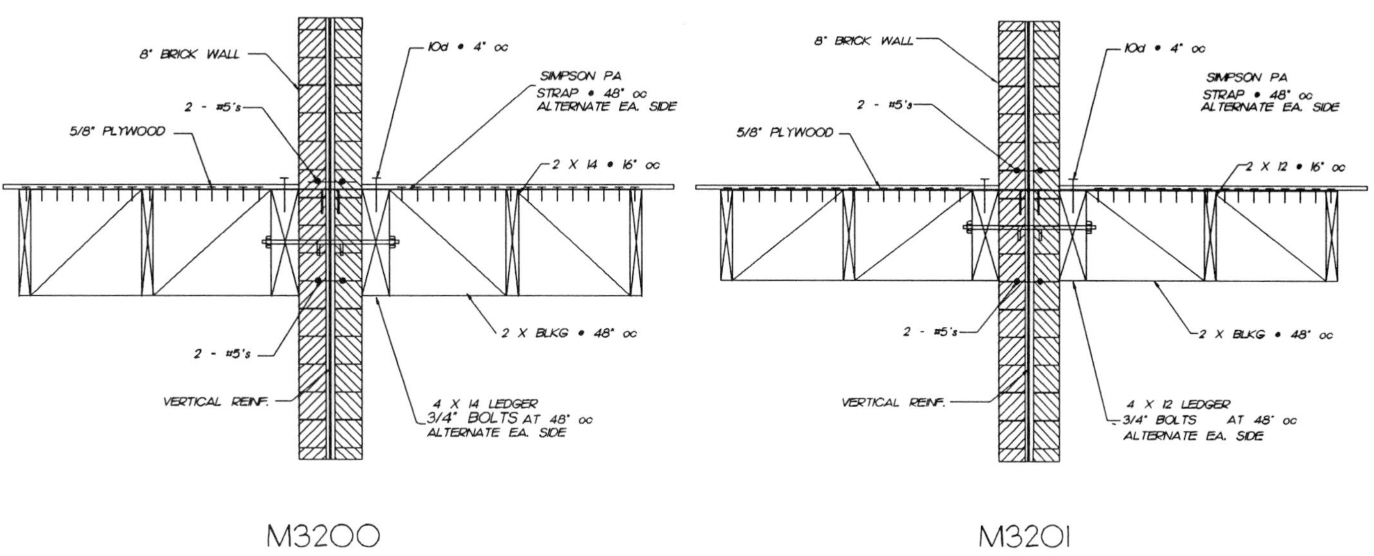
8" BRICK WALL
2 - #5's
5/8" PLYWOOD
10d • 4" oc
SIMPSON PA STRAP • 48" oc ALTERNATE EA. SIDE
2 X 14 • 16" oc
2 X BLKG • 48" oc
2 - #5's
VERTICAL REINF.
4 X 14 LEDGER
3/4" BOLTS AT 48" oc
ALTERNATE EA. SIDE
M3200
8" BRICK WALL
2 - #5's
5/8" PLYWOOD
10d • 4" oc
SIMPSON PA STRAP • 48" oc ALTERNATE EA. SIDE
2 X 12 • 16" oc
2 - #5's
2 X BLKG • 48" oc
VERTICAL REINF.
4 X 12 LEDGER
3/4" BOLTS AT 48" oc
ALTERNATE EA. SIDE
M3201

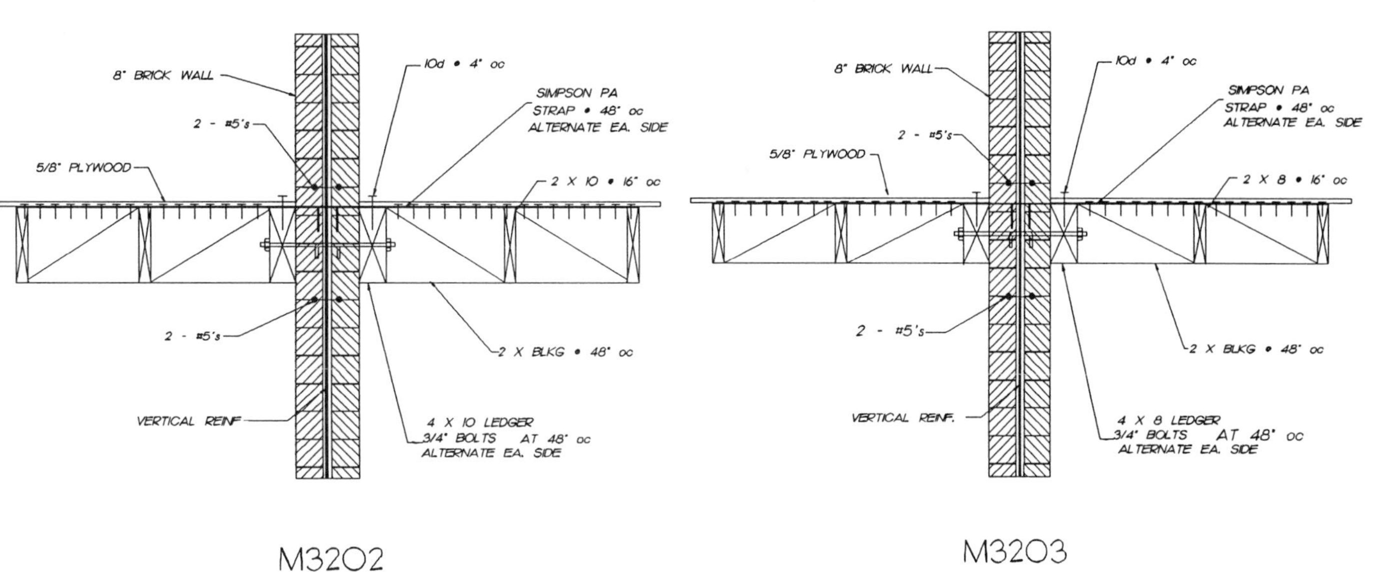
8" BRICK WALL
2 - #5's
5/8" PLYWOOD
10d • 4" oc
SIMPSON PA STRAP • 48" oc ALTERNATE EA. SIDE
2 X 10 • 16" oc
2 - #5's
2 X BLKG • 48" oc
VERTICAL REINF
4 X 10 LEDGER
3/4" BOLTS AT 48" oc
ALTERNATE EA. SIDE
M3202
8" BRICK WALL
2 - #5's
5/8" PLYWOOD
10d • 4" oc
SIMPSON PA STRAP • 48" oc ALTERNATE EA. SIDE
2 X 8 • 16" oc
2 - #5's
2 X BLKG • 48" oc
VERTICAL REINF.
4 X 8 LEDGER
3/4" BOLTS AT 48" oc
ALTERNATE EA. SIDE
M3203

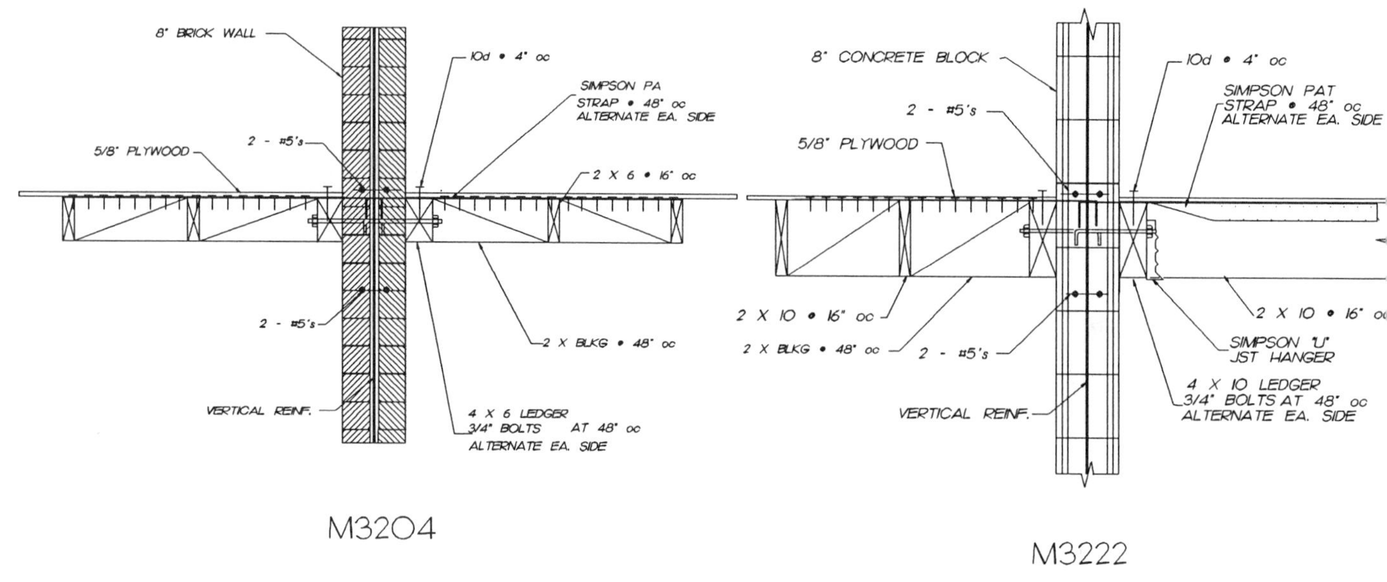

M3204

M3222

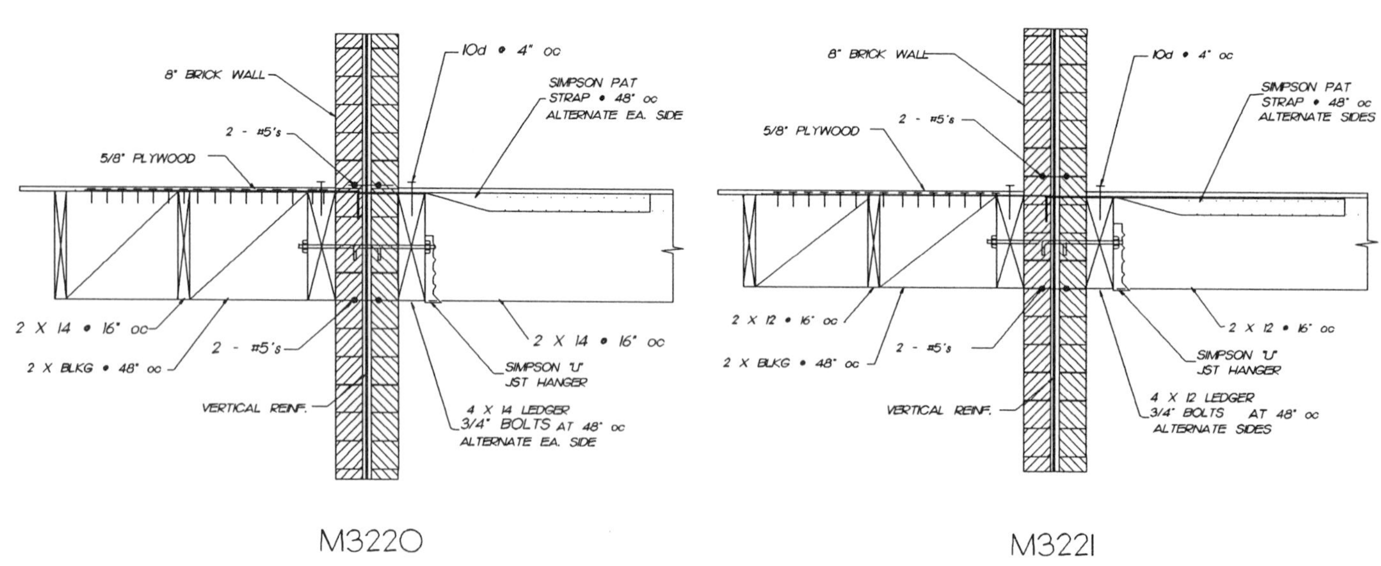

M3220

M3221

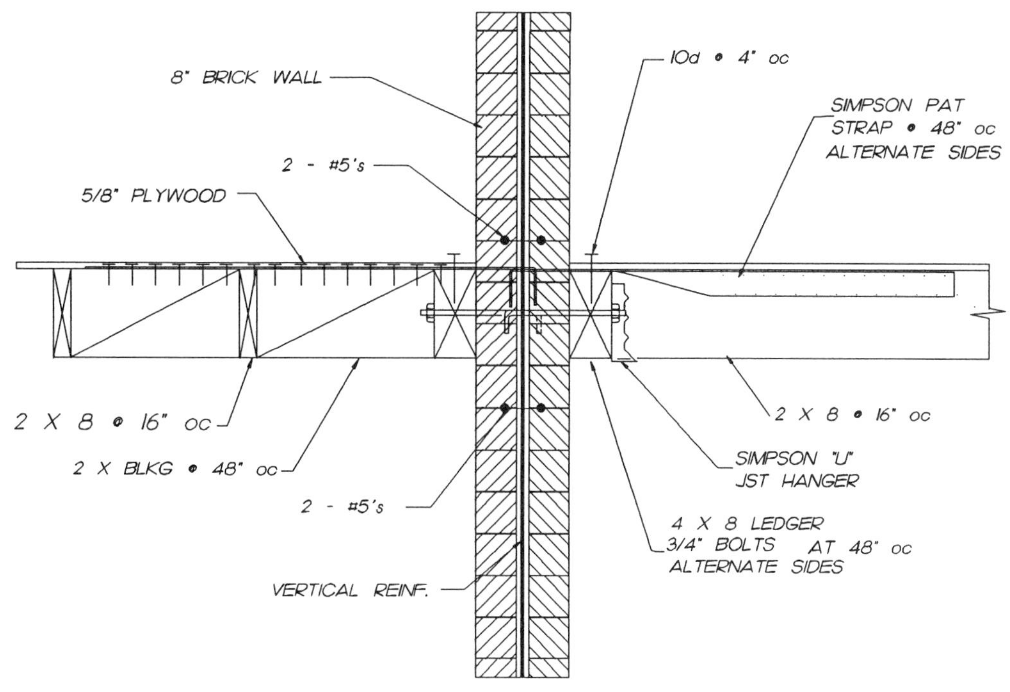
8" BRICK WALL
10d @ 4" oc
SIMPSON PAT
STRAP @ 48" oc
ALTERNATE SIDES
2 - #5's
5/8" PLYWOOD
2 X 8 @ 16" oc
2 X BLKG @ 48" oc
2 - #5's
2 X 8 @ 16" oc
SIMPSON "U"
JST HANGER
4 X 8 LEDGER
3/4" BOLTS AT 48" oc
ALTERNATE SIDES
VERTICAL REINF.

M3223

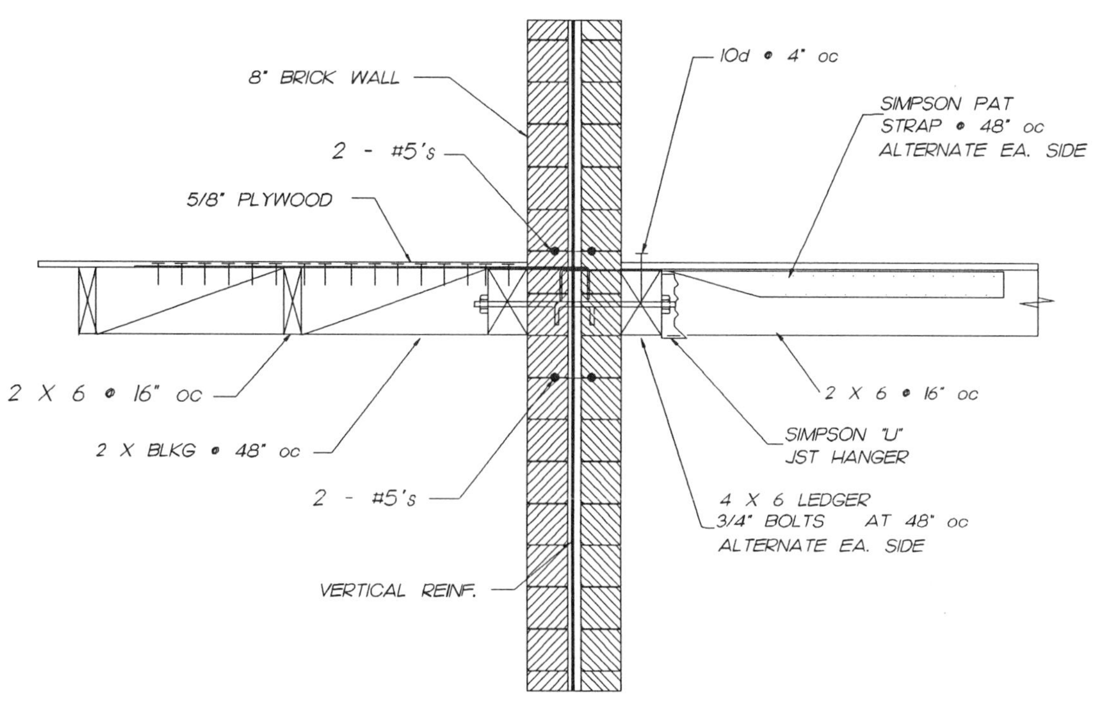
8" BRICK WALL
10d @ 4" oc
SIMPSON PAT
STRAP @ 48" oc
ALTERNATE EA. SIDE
2 - #5's
5/8" PLYWOOD
2 X 6 @ 16" oc
2 X BLKG @ 48" oc
2 - #5's
2 X 6 @ 16" oc
SIMPSON "U"
JST HANGER
4 X 6 LEDGER
3/4" BOLTS AT 48" oc
ALTERNATE EA. SIDE
VERTICAL REINF.

M3224

CHAPTER 2 MASONRY	SECTION 4	WALL - WOOD ROOF	
EXTERIOR WALL RAFTERS PERP. TO WALL CLIP ANGLE RAFTERS TO WALL	————	8" CONCRETE BLOCK	M4020 - M4023
		BRICK MASONRY	M4240 - M4243
EXTERIOR WALL RAFTERS PARALELL TO WALL CLIP ANGLE RAFTERS TO WALL	————	8" CONCRETE BLOCK	M4040 - M4043
		BRICK MASONRY	M4260 - M4263
EXTERIOR WALL RAFTERS PERP. TO WALL ROOF OVERHANG CLIP ANGLE RAFTERS TO WALL	————	8" CONCRETE BLOCK	M4060 - M4063
		BRICK MASONRY	M4280 - M4283
EXTERIOR WALL RAFTERS PARALELL TO WALL ROOF OVERHANG CLIP ANGLE RAFTERS TO WALL	————	8" CONCRETE BLOCK	M4080 - M4083
		BRICK MASONRY	M4300 - M4303
EXTERIOR WALL RAFTERS PERP. TO WALL SLOPED ROOF OVERHANG CLIP ANGLE RAFTERS TO WALL	————	8" CONCRETE BLOCK	M4100 - M4103
		BRICK MASONRY	M4320 - M4323
EXTERIOR WALL & PARAPET RAFTERS PERP. TO WALL SLOPED ROOF NO OVERHANG LEDGER SUPPORT	————	8" CONCRETE BLOCK	M4120 - M4123
		BRICK MASONRY	M4340 - M4343
EXTERIOR WALL & PARAPET RAFTERS PARALELL TO WALL SLOPED ROOF NO OVERHANG LEDGER SUPPORT	————	8" CONCRETE BLOCK	M4140 - M4143
		BRICK MASONRY	M4360 - M4363
INTERIOR WALL RAFTERS PERP. TO WALL CLIP ANGLE RAFTERS TO WALL	————	8" CONCRETE BLOCK	M4160 - M4163
		BRICK MASONRY	M4380 - M4383
INTERIOR WALL RAFTERS PARALELL TO WALL CLIP ANGLE RAFTERS TO WALL	————	8" CONCRETE BLOCK	M4180 - M4183
		BRICK MASONRY	M4400 - M4403
INTERIOR WALL RTRS PERP & PARALELL ALT SIDES CLIP ANGLE RAFTERS TO WALL	————	8" CONCRETE BLOCK	M4200 - M4203
		BRICK MASONRY	M4420 - M4423
ONE STORY CONCRETE BLOCK WALL	WOOD ROOF	RAFTERS PERP. TO WALL	M4220
		RAFTERS PARALELL TO WALL	M4221

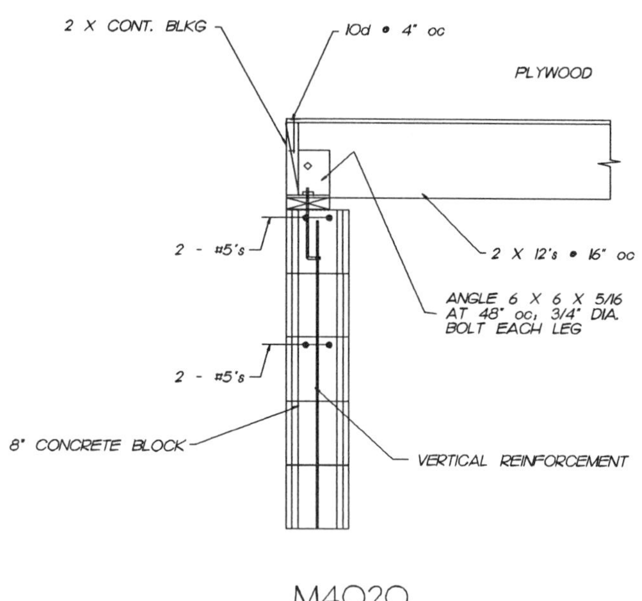

M4020

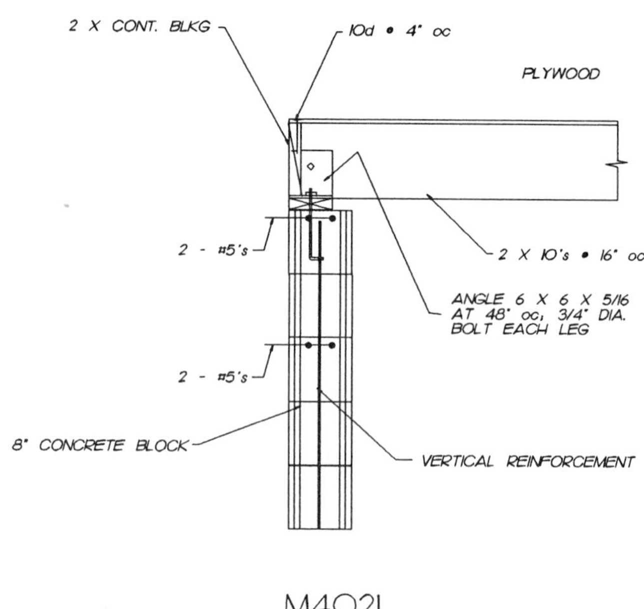

M4021

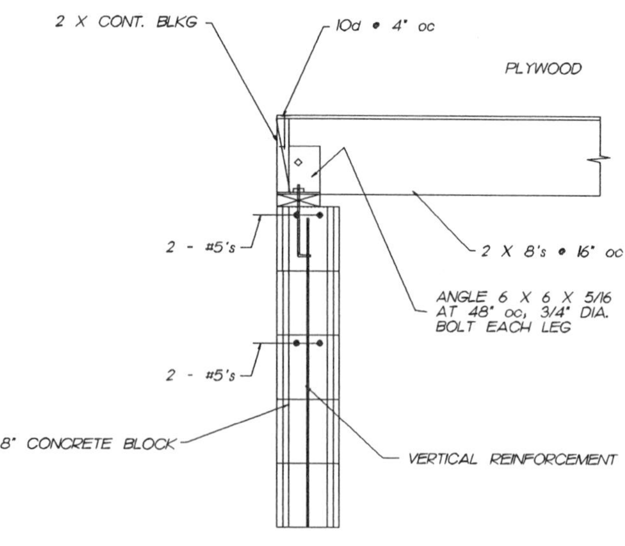

M4022

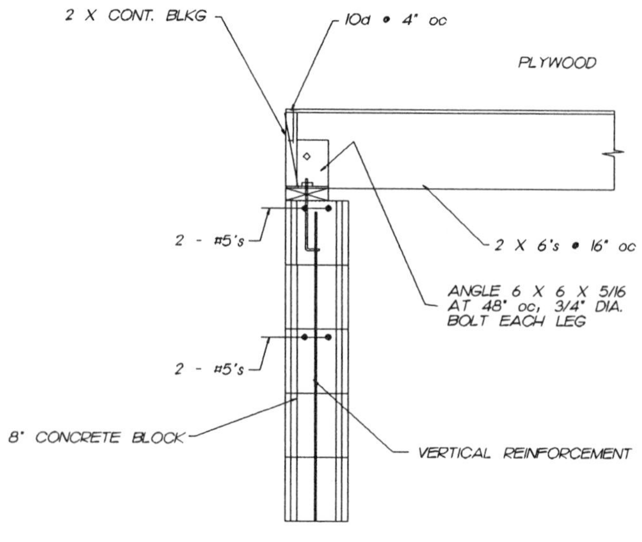

M4023

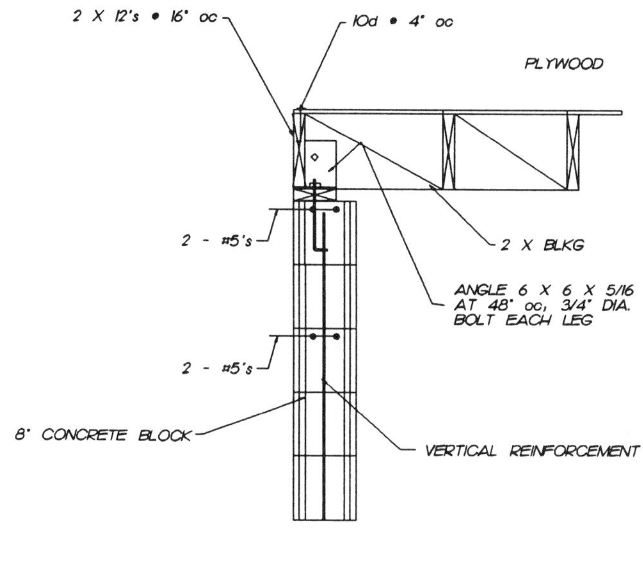

M4040

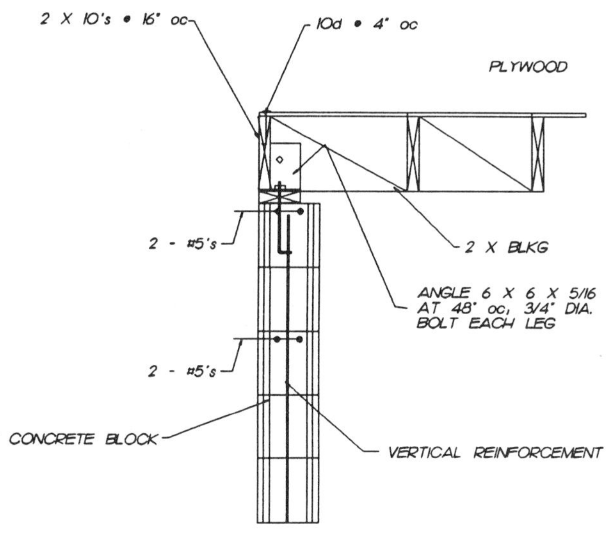

M4041

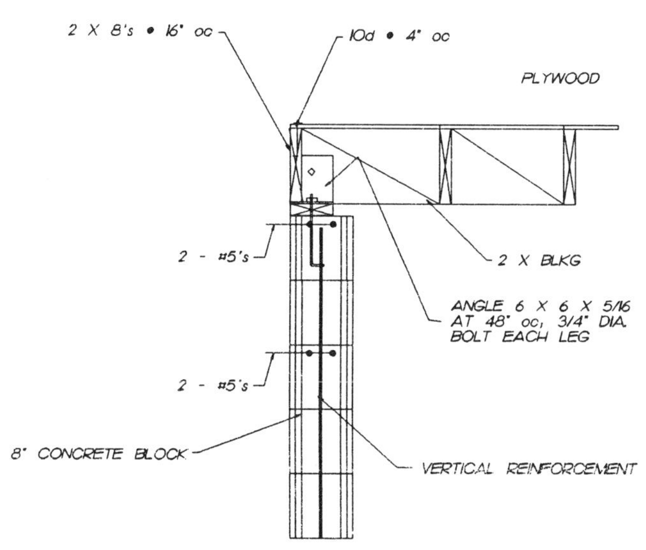

M4042

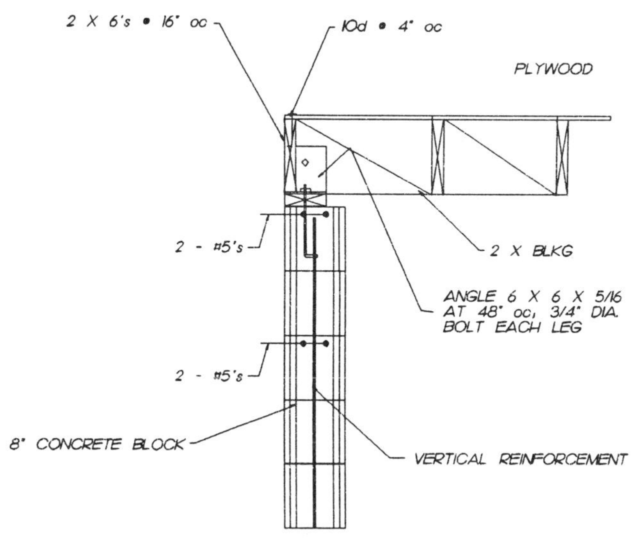

M4043

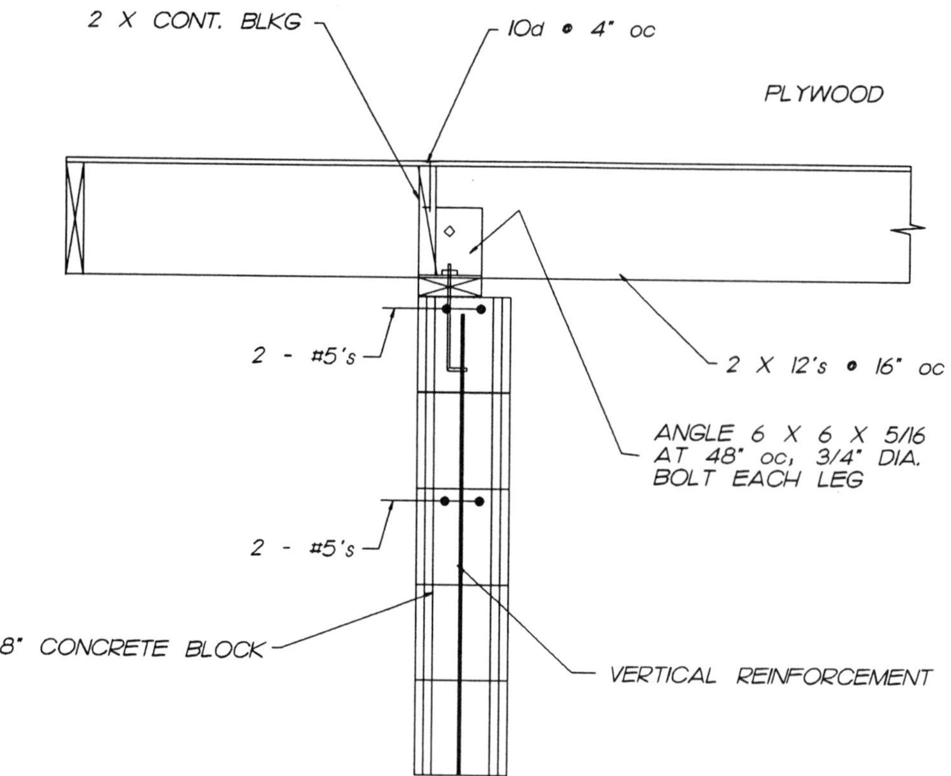
2 X CONT. BLKG
10d @ 4" oc
PLYWOOD
2 - #5's
2 X 12's @ 16" oc
ANGLE 6 X 6 X 5/16
AT 48" oc, 3/4" DIA.
BOLT EACH LEG
2 - #5's
8" CONCRETE BLOCK
VERTICAL REINFORCEMENT

M4060

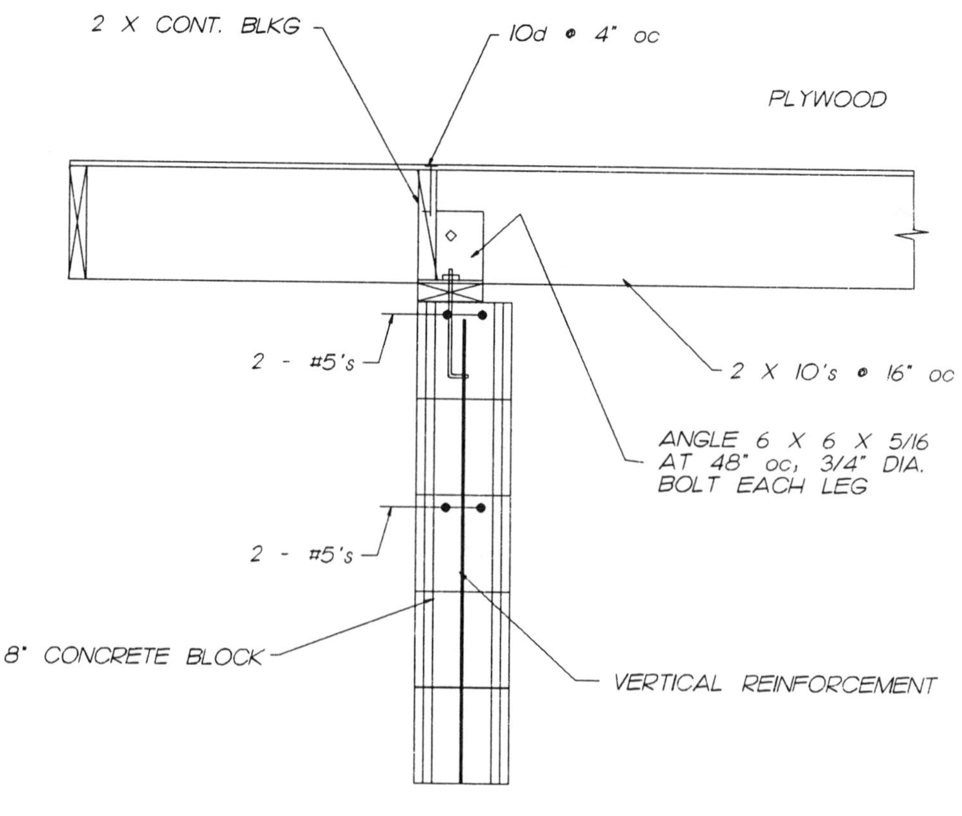
2 X CONT. BLKG
10d @ 4" oc
PLYWOOD
2 - #5's
2 X 10's @ 16" oc
ANGLE 6 X 6 X 5/16
AT 48" oc, 3/4" DIA.
BOLT EACH LEG
2 - #5's
8" CONCRETE BLOCK
VERTICAL REINFORCEMENT

M4061

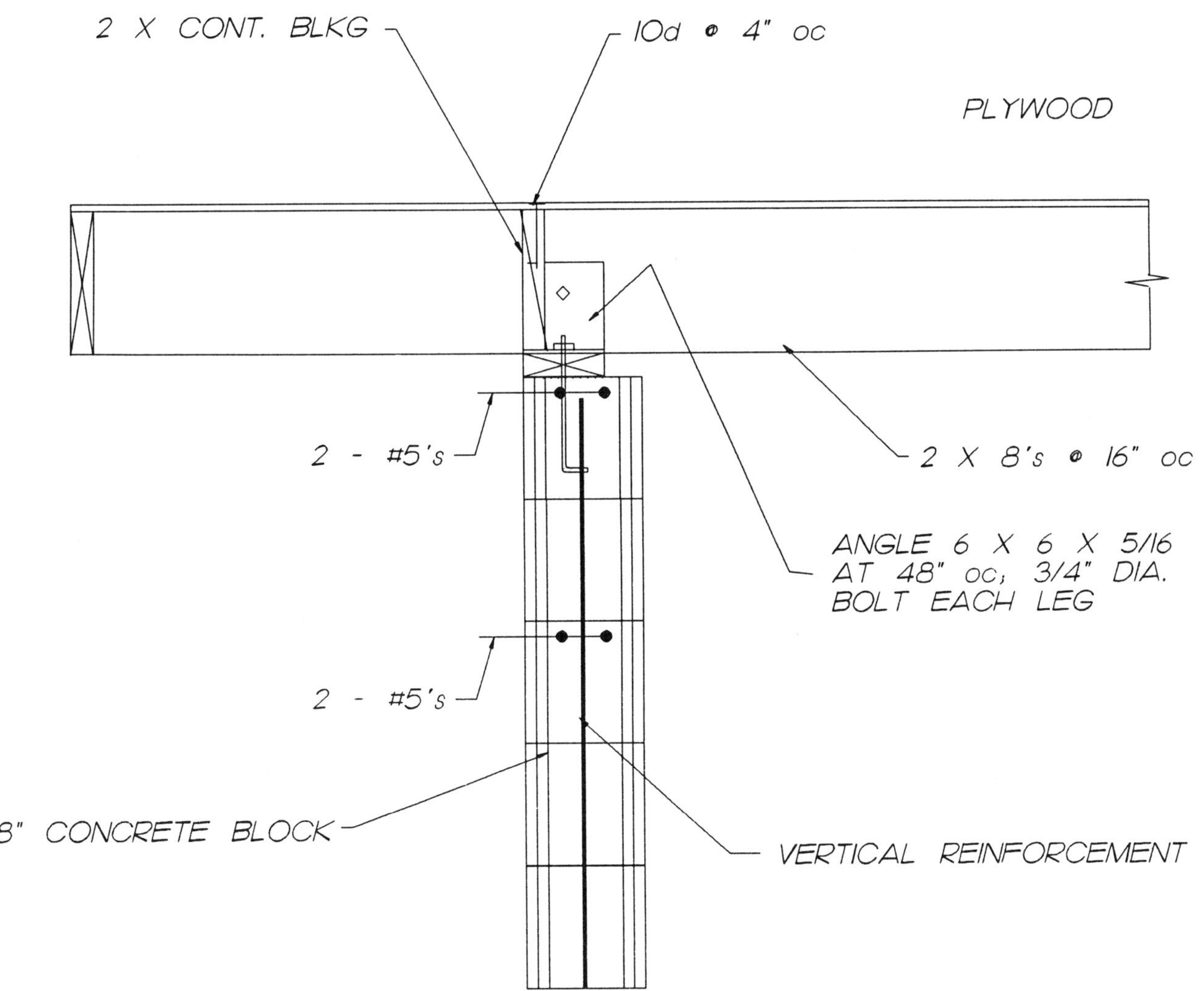

2 X CONT. BLKG
10d @ 4" oc
PLYWOOD
2 - #5's
2 X 8's @ 16" oc
ANGLE 6 X 6 X 5/16
AT 48" oc; 3/4" DIA.
BOLT EACH LEG
2 - #5's
8" CONCRETE BLOCK
VERTICAL REINFORCEMENT

M4062

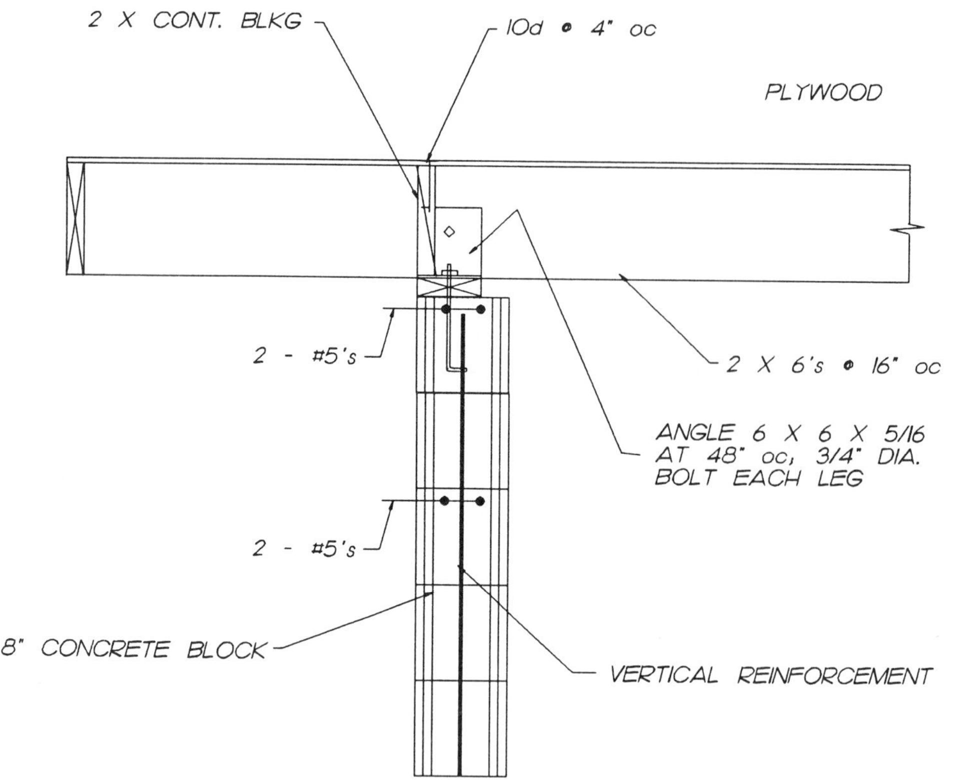

2 X CONT. BLKG
10d @ 4" oc
PLYWOOD
2 - #5's
2 X 6's @ 16" oc
ANGLE 6 X 6 X 5/16
AT 48" oc, 3/4" DIA.
BOLT EACH LEG
2 - #5's
8" CONCRETE BLOCK
VERTICAL REINFORCEMENT

M4063

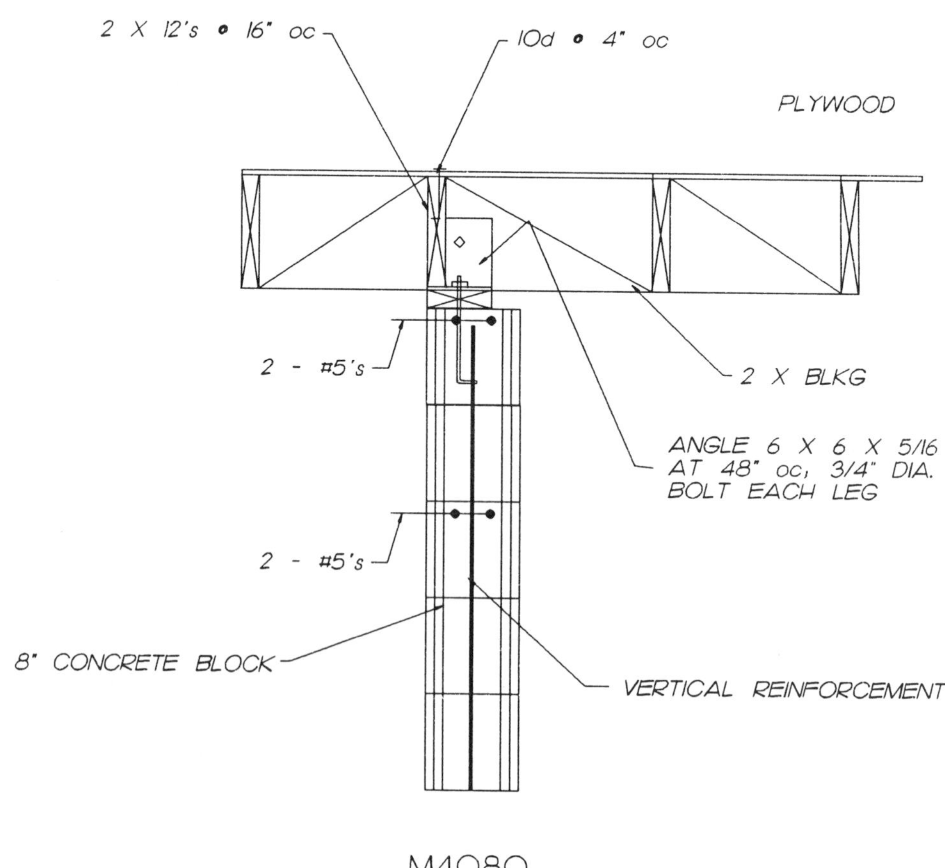

2 X 12's @ 16" oc
10d @ 4" oc
PLYWOOD
2 - #5's
2 X BLKG
ANGLE 6 X 6 X 5/16
AT 48" oc, 3/4" DIA.
BOLT EACH LEG
2 - #5's
8" CONCRETE BLOCK
VERTICAL REINFORCEMENT

M4080

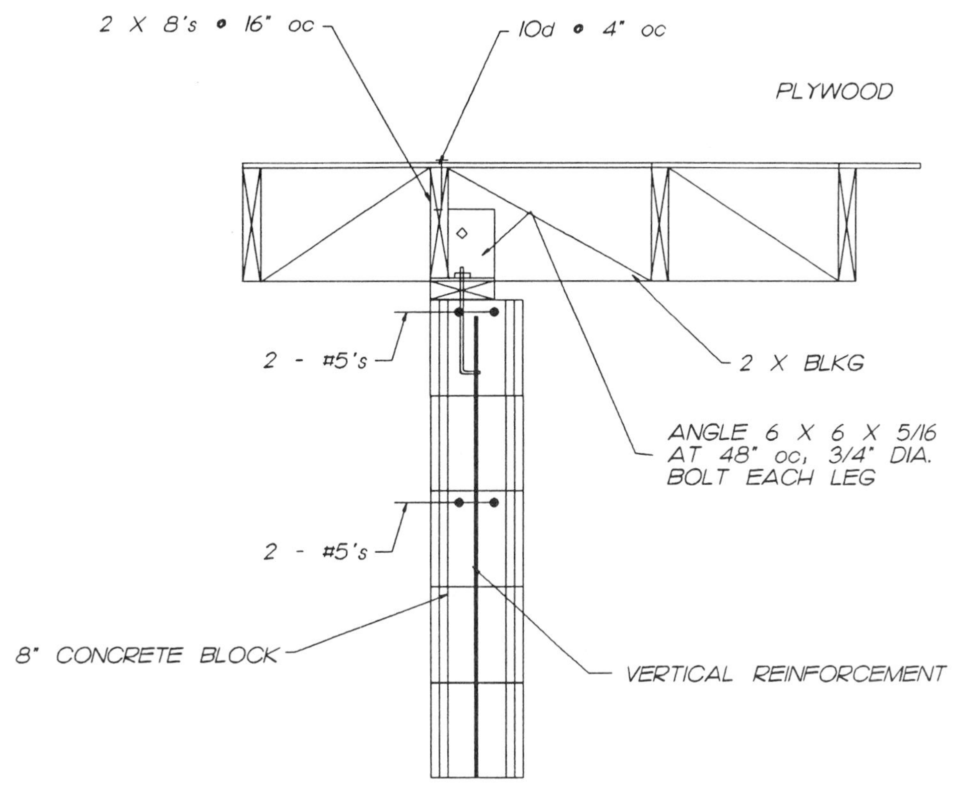
2 X 8's @ 16" oc
10d @ 4" oc
PLYWOOD
2 - #5's
2 X BLKG
ANGLE 6 X 6 X 5/16
AT 48" oc, 3/4" DIA.
BOLT EACH LEG
2 - #5's
8" CONCRETE BLOCK
VERTICAL REINFORCEMENT

M4082

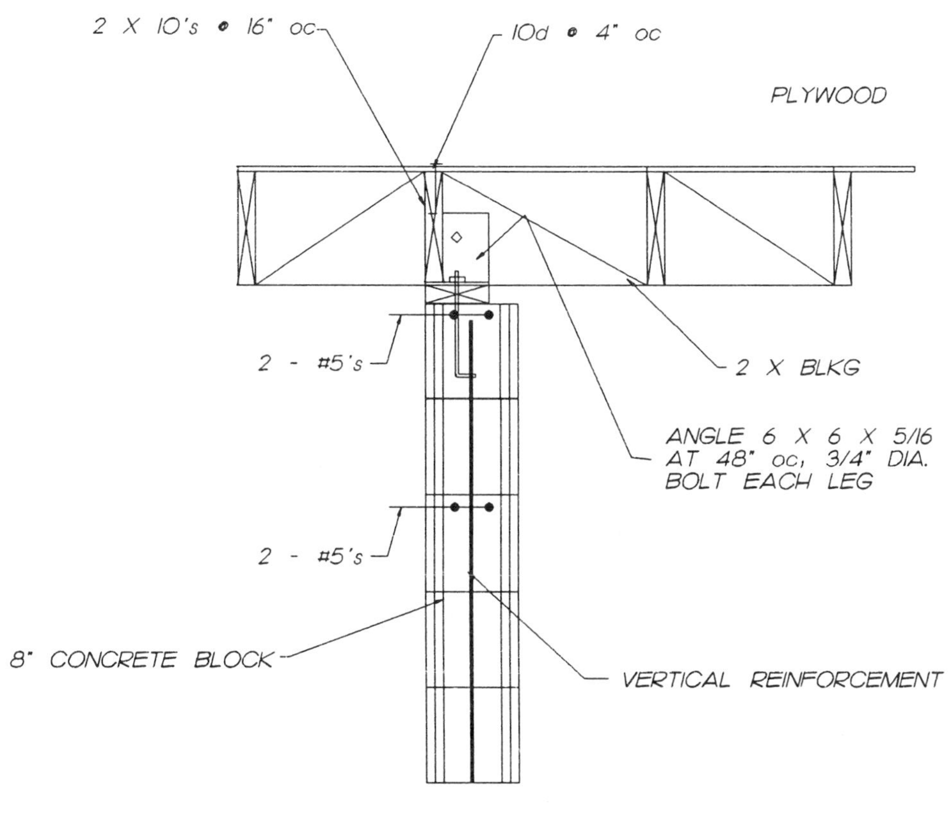
2 X 10's @ 16" oc
10d @ 4" oc
PLYWOOD
2 - #5's
2 X BLKG
ANGLE 6 X 6 X 5/16
AT 48" oc, 3/4" DIA.
BOLT EACH LEG
2 - #5's
8" CONCRETE BLOCK
VERTICAL REINFORCEMENT

M4081

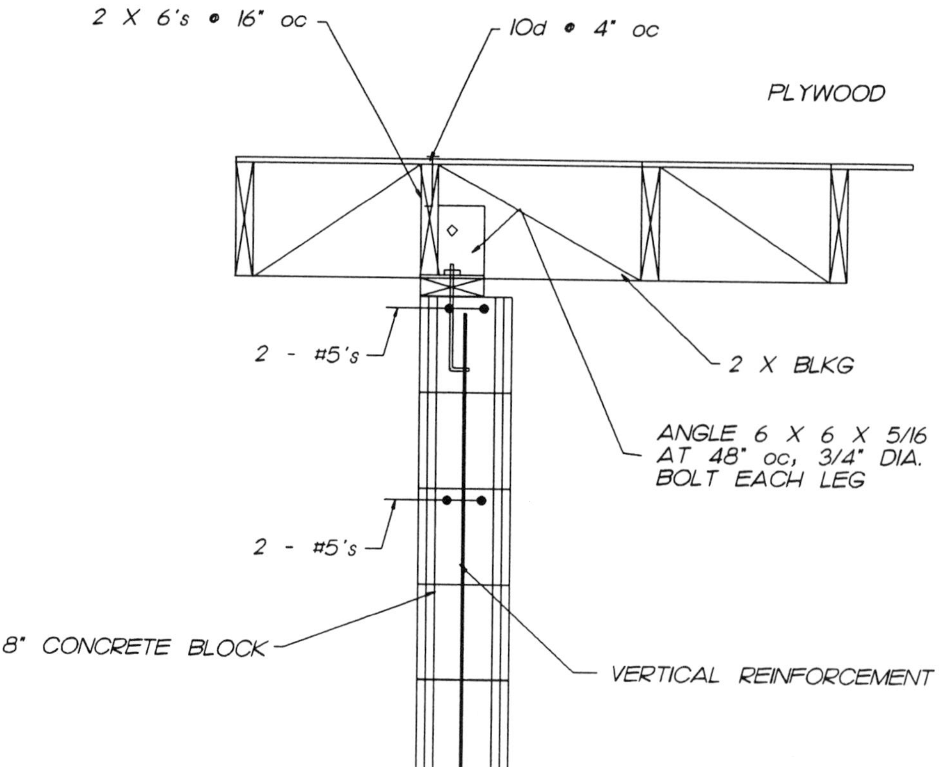
2 X 6's @ 16" oc
10d @ 4" oc
PLYWOOD
2 - #5's
2 X BLKG
ANGLE 6 X 6 X 5/16
AT 48" oc, 3/4" DIA.
BOLT EACH LEG
2 - #5's
8" CONCRETE BLOCK
VERTICAL REINFORCEMENT

M4083

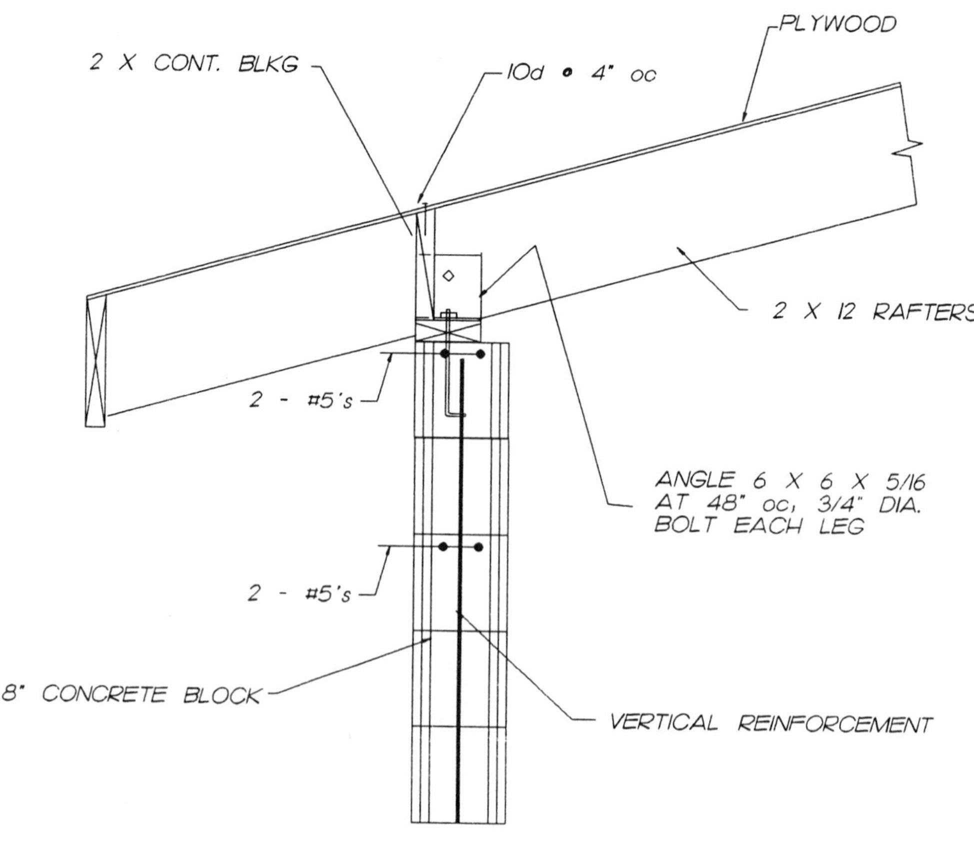
PLYWOOD
2 X CONT. BLKG
10d @ 4" oc
2 X 12 RAFTERS
2 - #5's
ANGLE 6 X 6 X 5/16
AT 48" oc, 3/4" DIA.
BOLT EACH LEG
2 - #5's
8" CONCRETE BLOCK
VERTICAL REINFORCEMENT

M4100

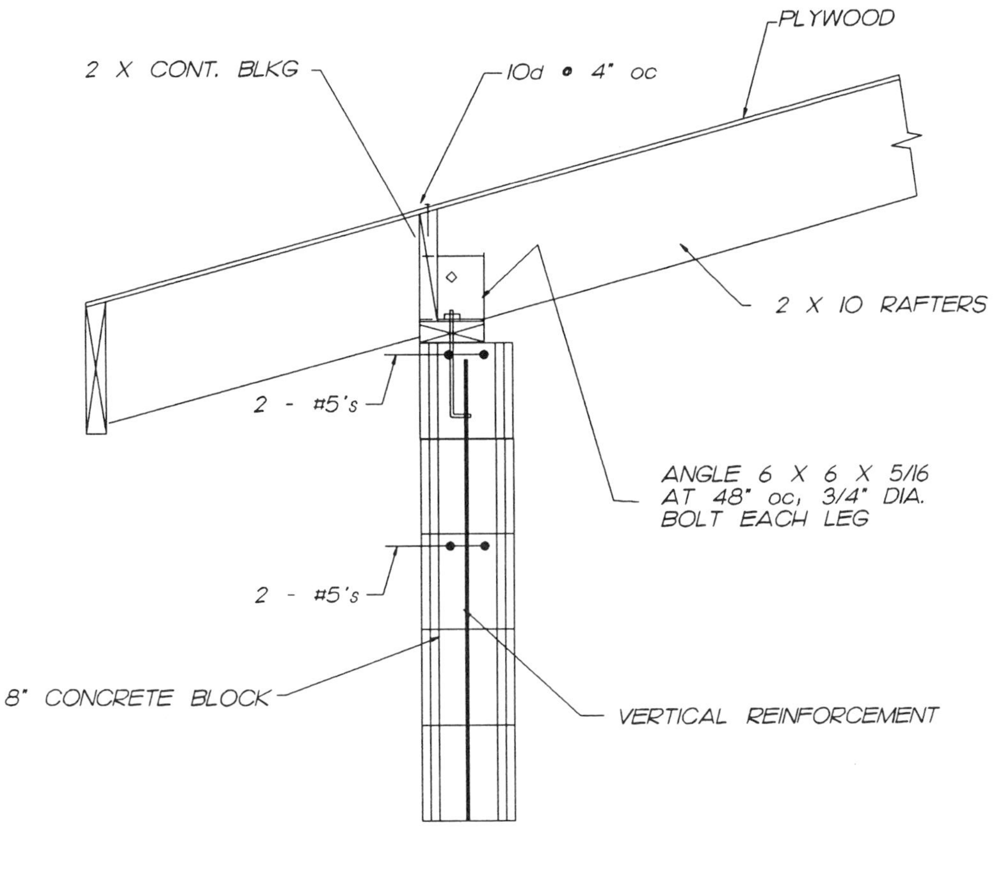
PLYWOOD
2 X CONT. BLKG
10d @ 4" oc
2 X 10 RAFTERS
2 - #5's
ANGLE 6 X 6 X 5/16
AT 48" oc, 3/4" DIA.
BOLT EACH LEG
2 - #5's
8" CONCRETE BLOCK
VERTICAL REINFORCEMENT

M4101

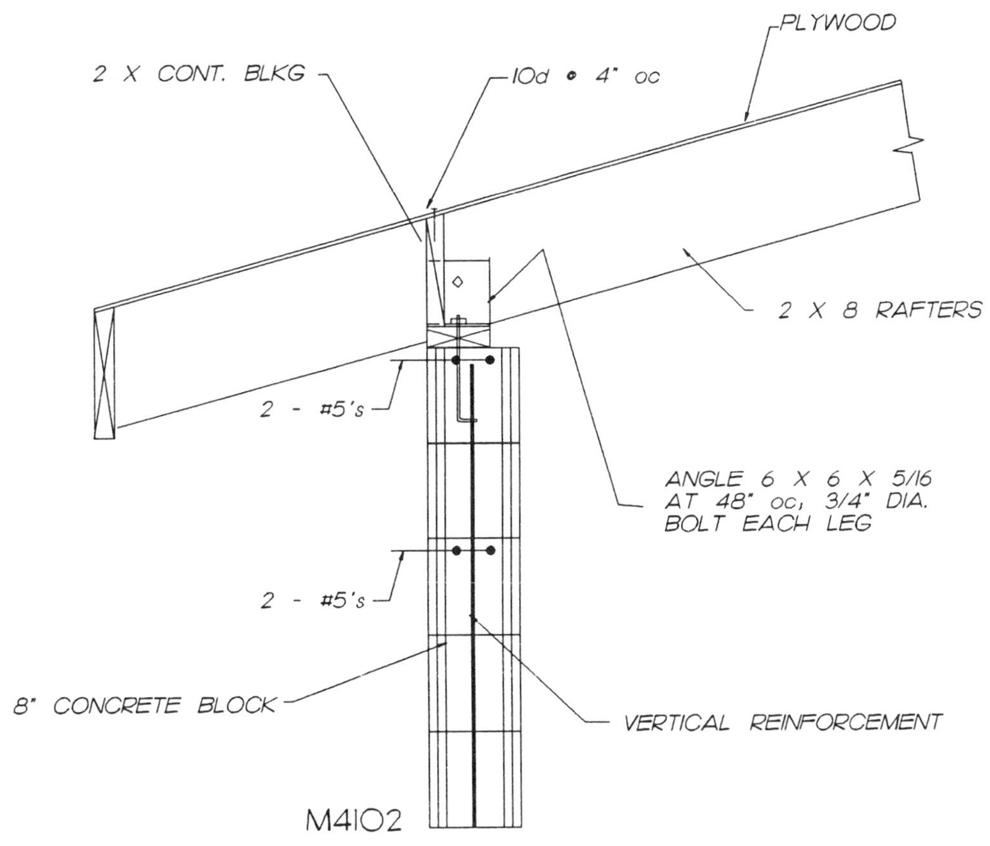
PLYWOOD
2 X CONT. BLKG
10d @ 4" oc
2 X 8 RAFTERS
2 - #5's
ANGLE 6 X 6 X 5/16
AT 48" oc, 3/4" DIA.
BOLT EACH LEG
2 - #5's
8" CONCRETE BLOCK
VERTICAL REINFORCEMENT

M4102

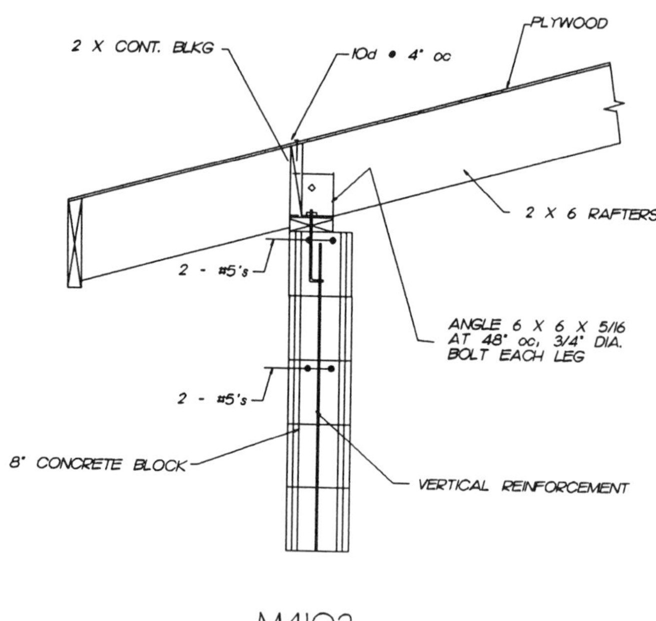

M4103

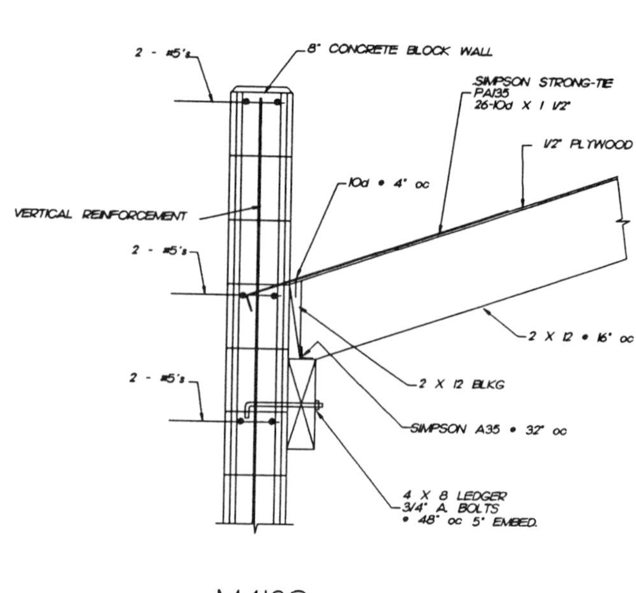

M4120

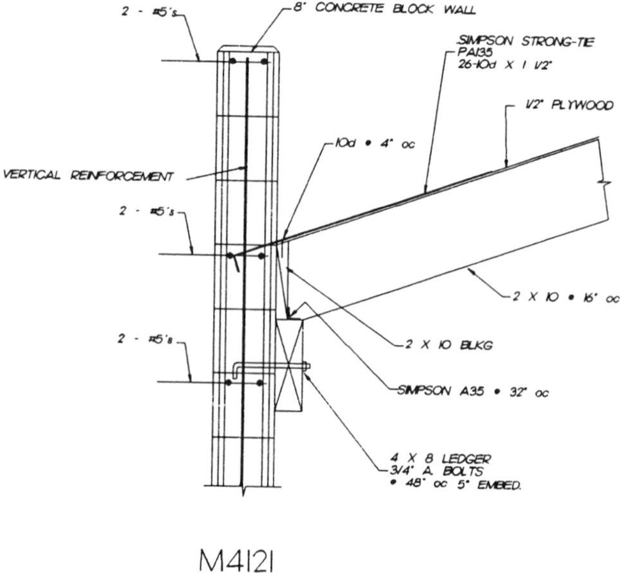

M4121

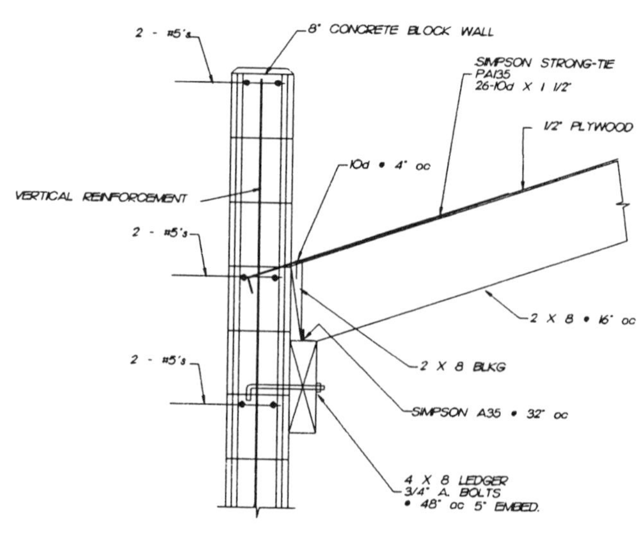

M4122

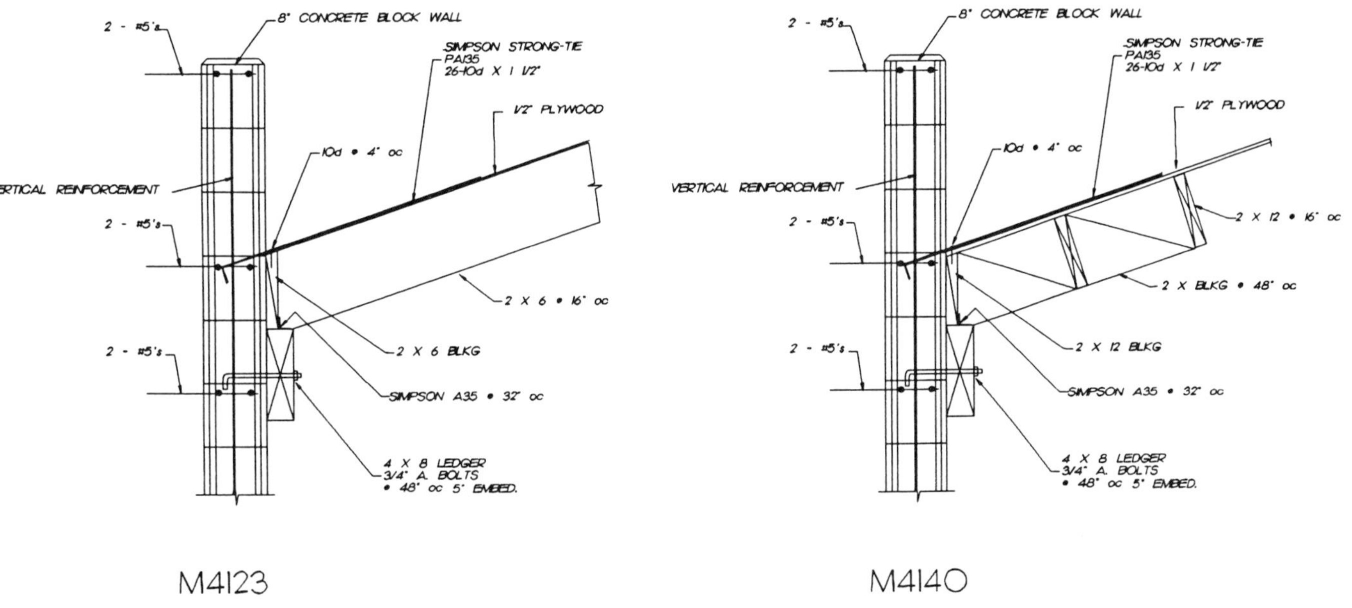

M4123

M4140

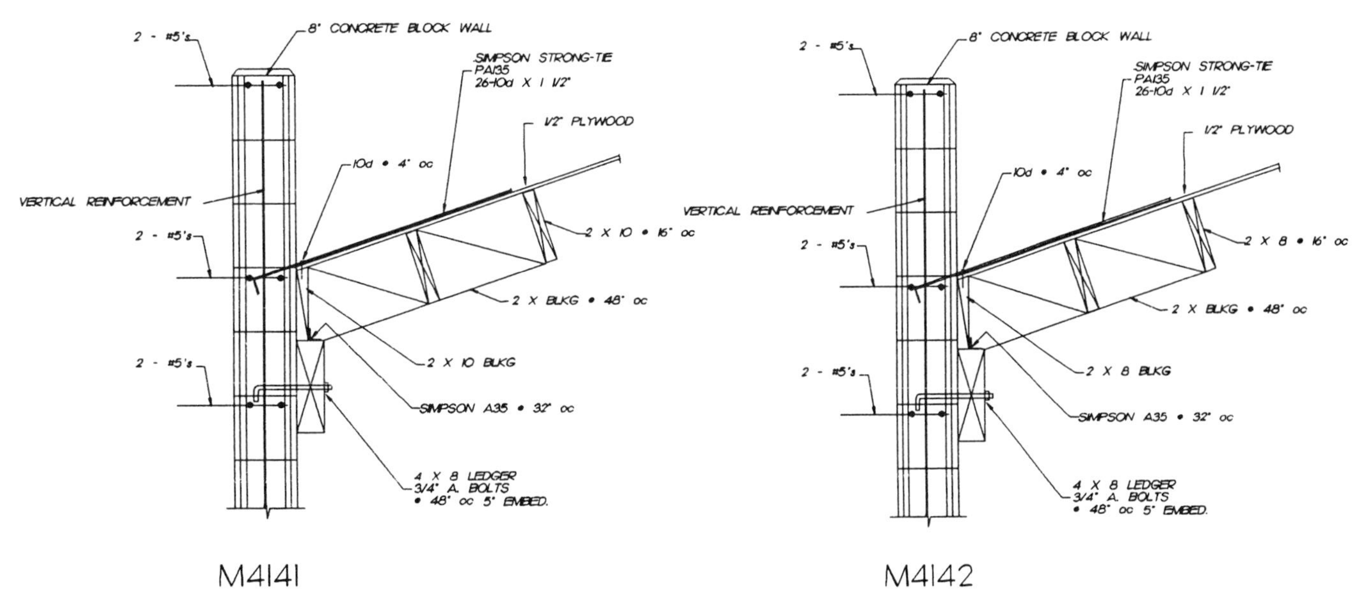

M4141

M4142

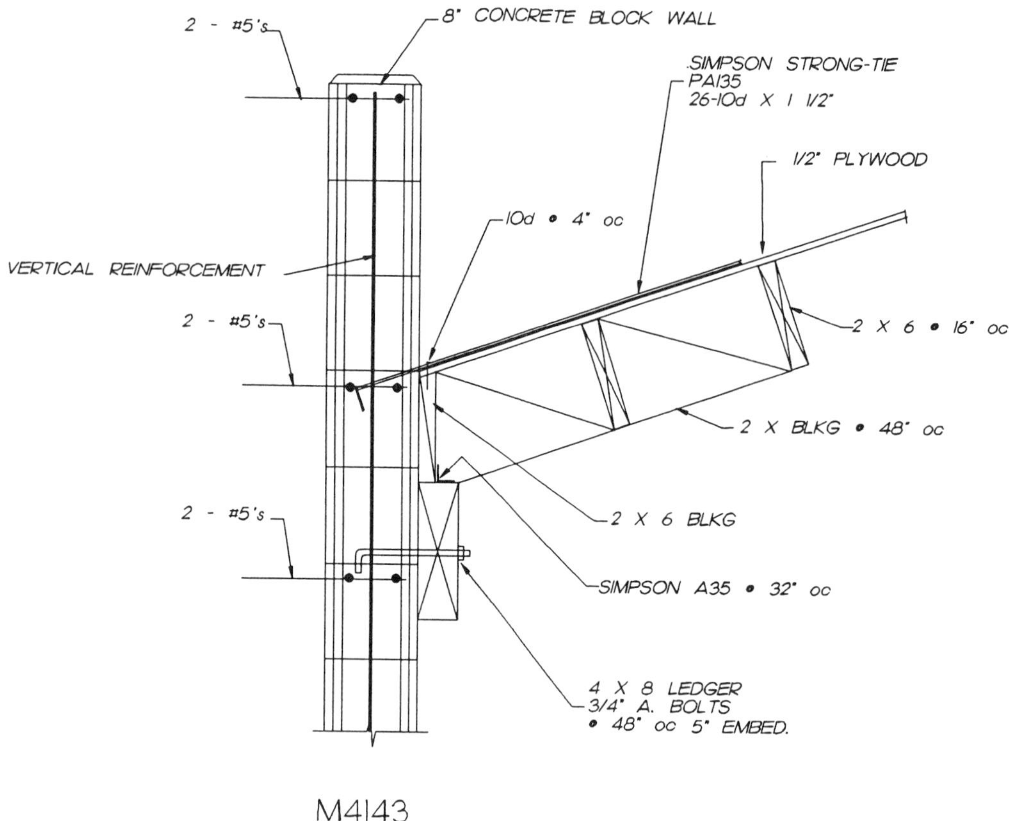
2 - #5's
8" CONCRETE BLOCK WALL
SIMPSON STRONG-TIE
PA135
26-10d X 1 1/2"
1/2" PLYWOOD
10d @ 4" oc
VERTICAL REINFORCEMENT
2 - #5's
2 X 6 @ 16" oc
2 X BLKG @ 48" oc
2 X 6 BLKG
2 - #5's
SIMPSON A35 @ 32" oc
4 X 8 LEDGER
3/4" A. BOLTS
@ 48" oc 5" EMBED.

M4143

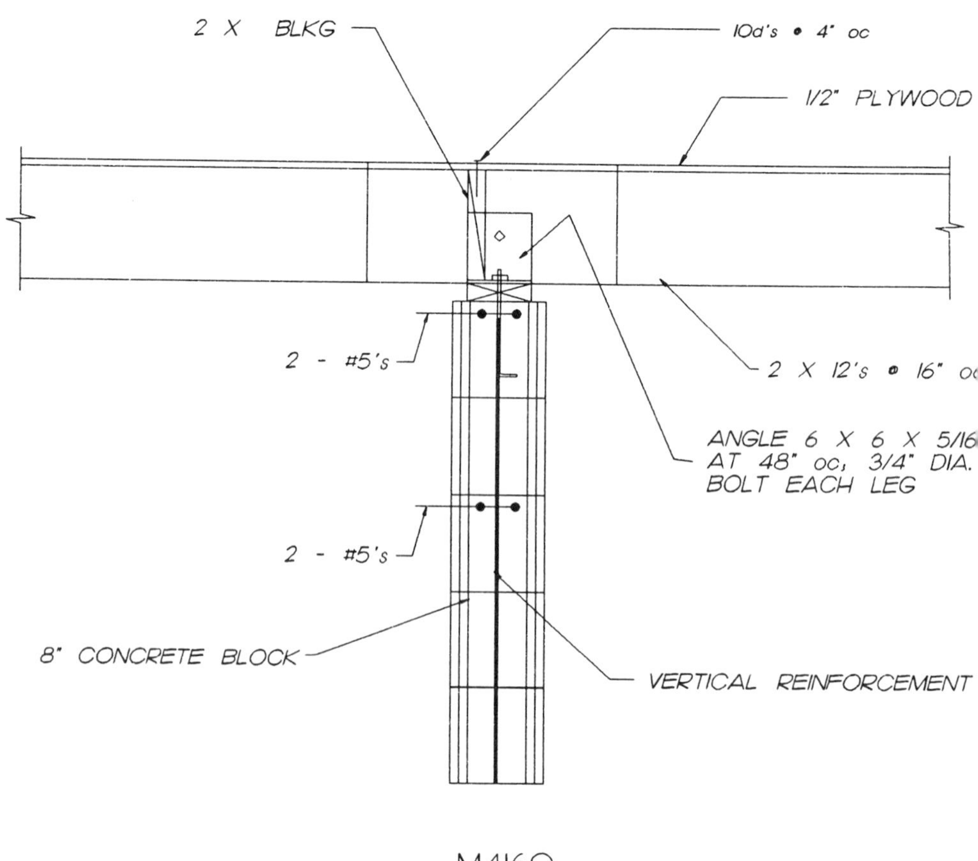
2 X BLKG
10d's @ 4" oc
1/2" PLYWOOD
2 - #5's
2 X 12's @ 16" oc
ANGLE 6 X 6 X 5/16
AT 48" oc, 3/4" DIA.
BOLT EACH LEG
2 - #5's
8" CONCRETE BLOCK
VERTICAL REINFORCEMENT

M4160

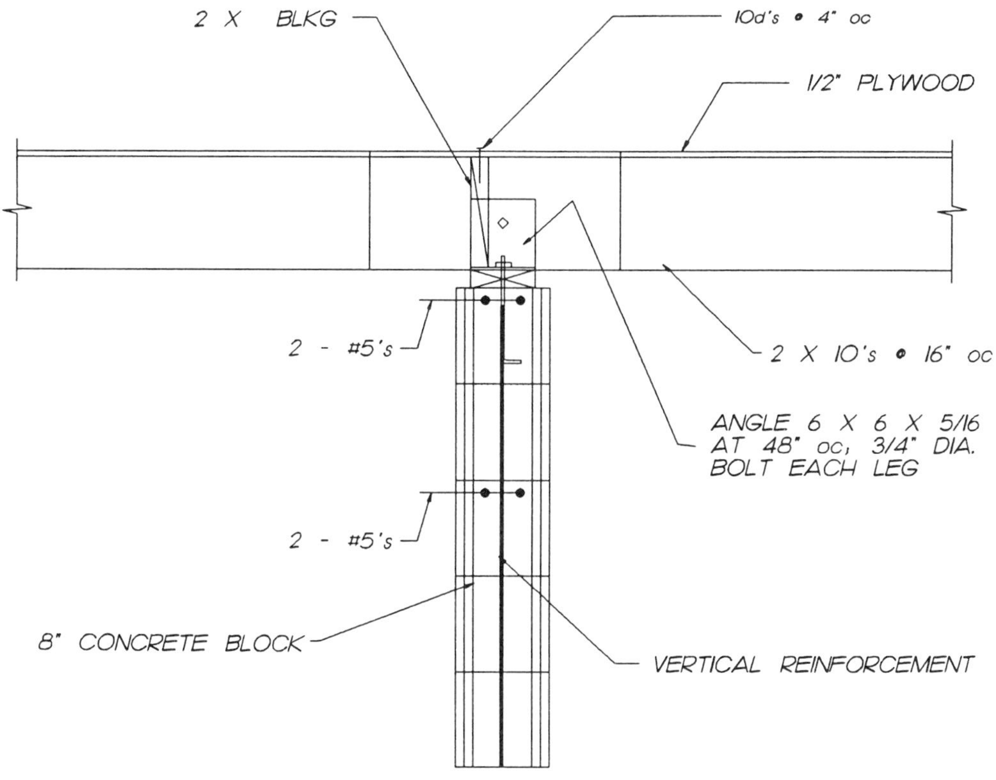
2 X BLKG
10d's @ 4" oc
1/2" PLYWOOD
2 - #5's
2 X 10's @ 16" oc
ANGLE 6 X 6 X 5/16
AT 48" oc, 3/4" DIA.
BOLT EACH LEG
2 - #5's
8" CONCRETE BLOCK
VERTICAL REINFORCEMENT

M4161

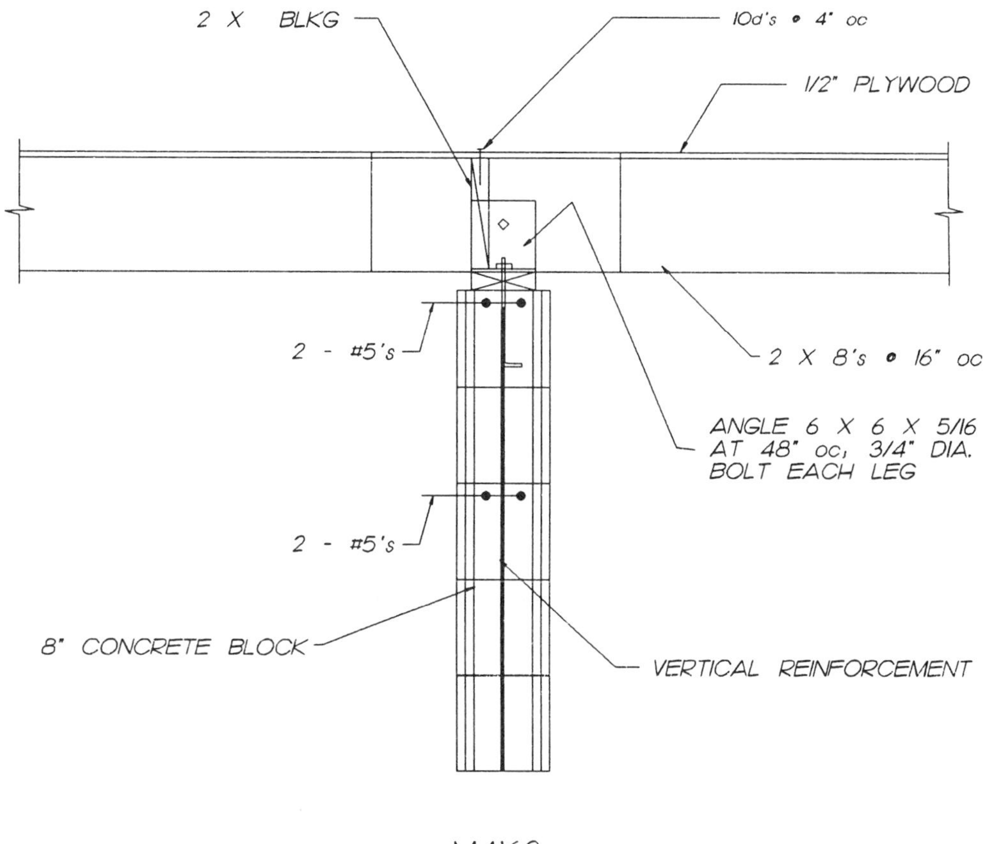
2 X BLKG
10d's @ 4" oc
1/2" PLYWOOD
2 - #5's
2 X 8's @ 16" oc
ANGLE 6 X 6 X 5/16
AT 48" oc, 3/4" DIA.
BOLT EACH LEG
2 - #5's
8" CONCRETE BLOCK
VERTICAL REINFORCEMENT

M4162

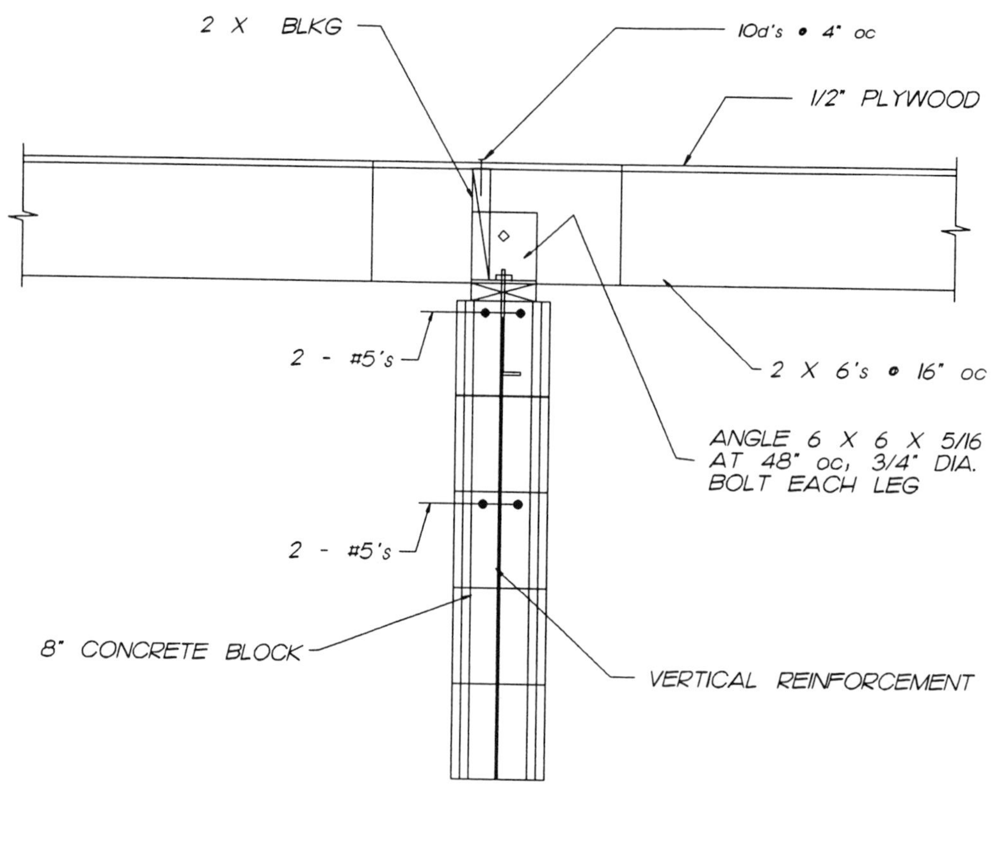

M4163

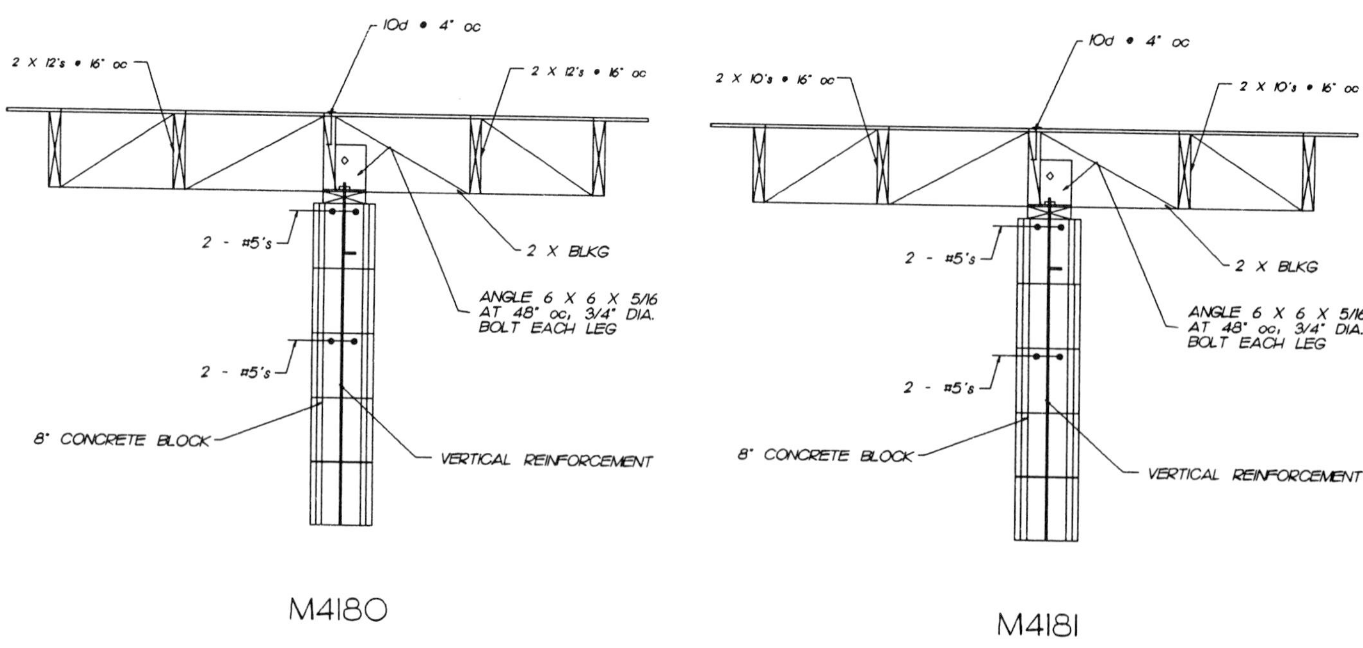

M4180

M4181

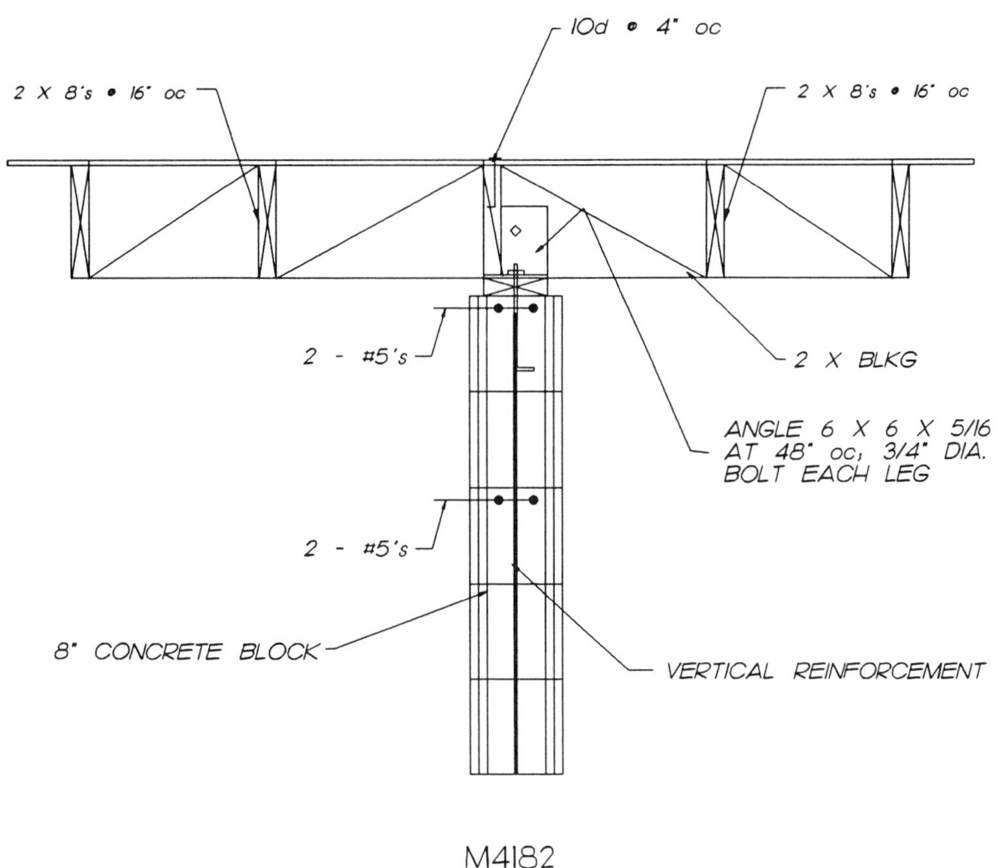

M4182

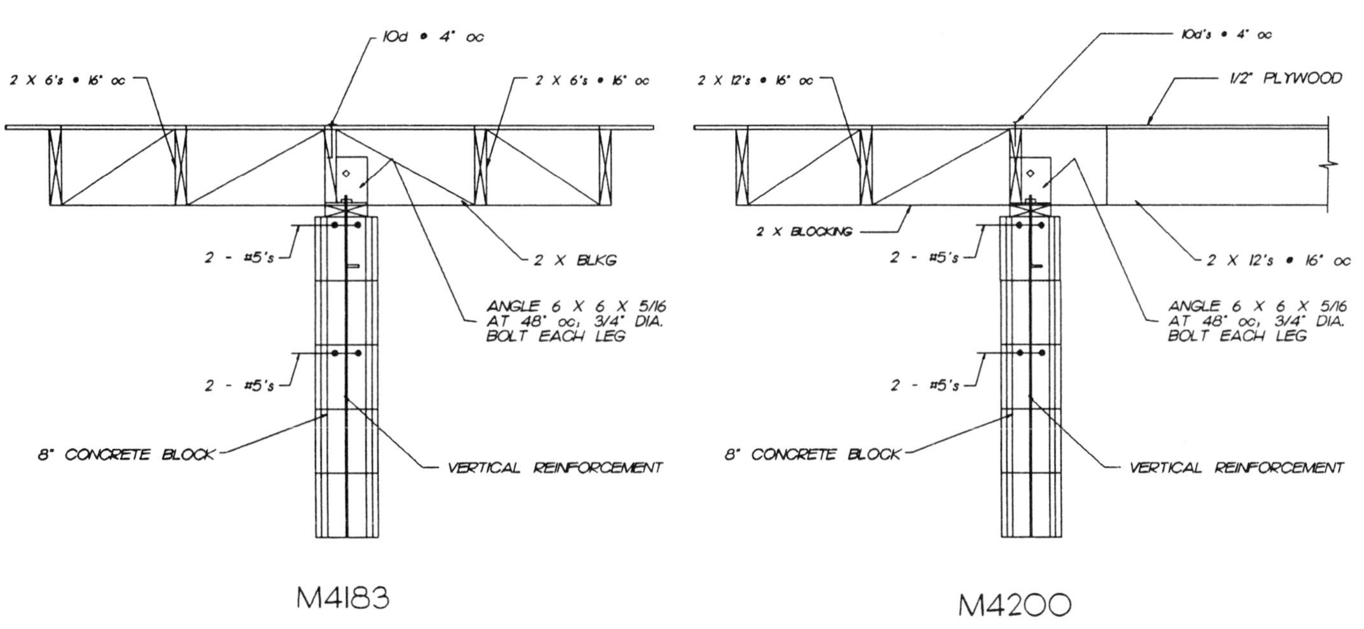

M4183

M4200

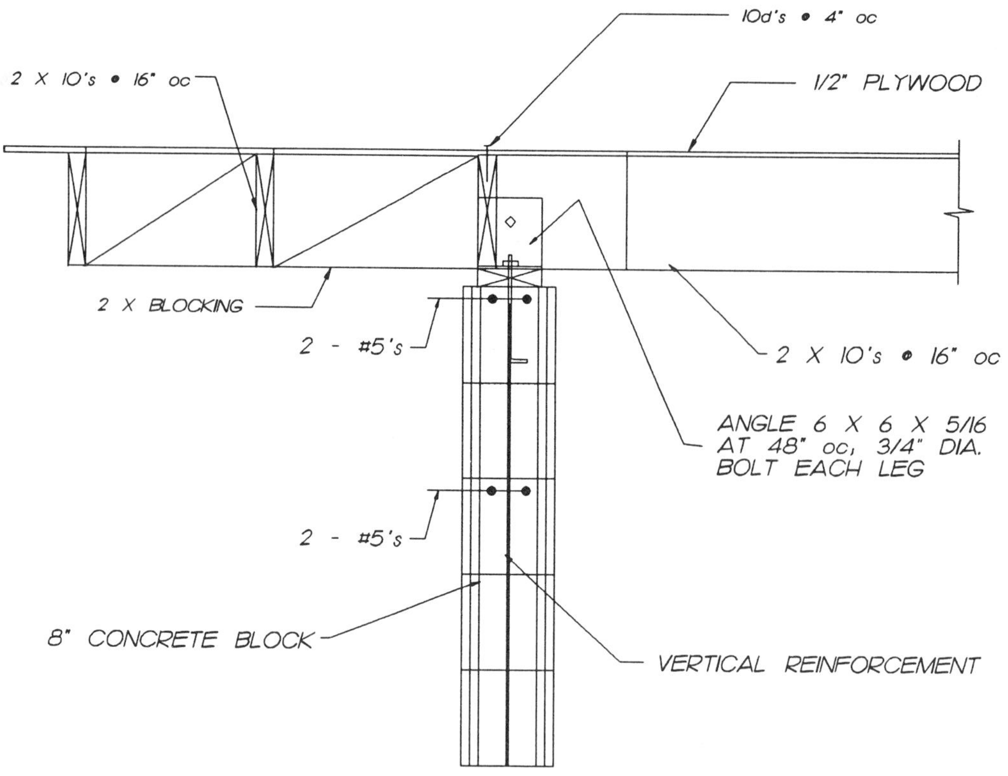
10d's • 4" oc
2 X 10's • 16" oc
1/2" PLYWOOD
2 X BLOCKING
2 - #5's
2 X 10's • 16" oc
ANGLE 6 X 6 X 5/16
AT 48" oc, 3/4" DIA.
BOLT EACH LEG
2 - #5's
8" CONCRETE BLOCK
VERTICAL REINFORCEMENT

M4201

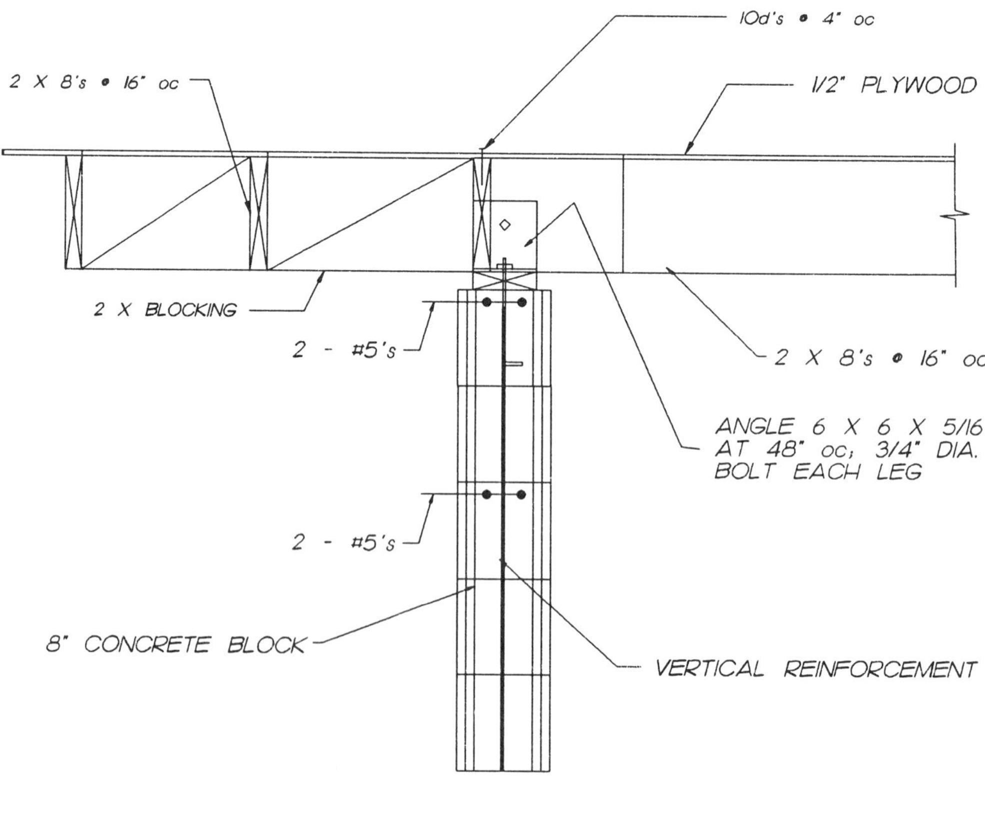
10d's • 4" oc
2 X 8's • 16" oc
1/2" PLYWOOD
2 X BLOCKING
2 - #5's
2 X 8's • 16" oc
ANGLE 6 X 6 X 5/16
AT 48" oc, 3/4" DIA.
BOLT EACH LEG
2 - #5's
8" CONCRETE BLOCK
VERTICAL REINFORCEMENT

M4202

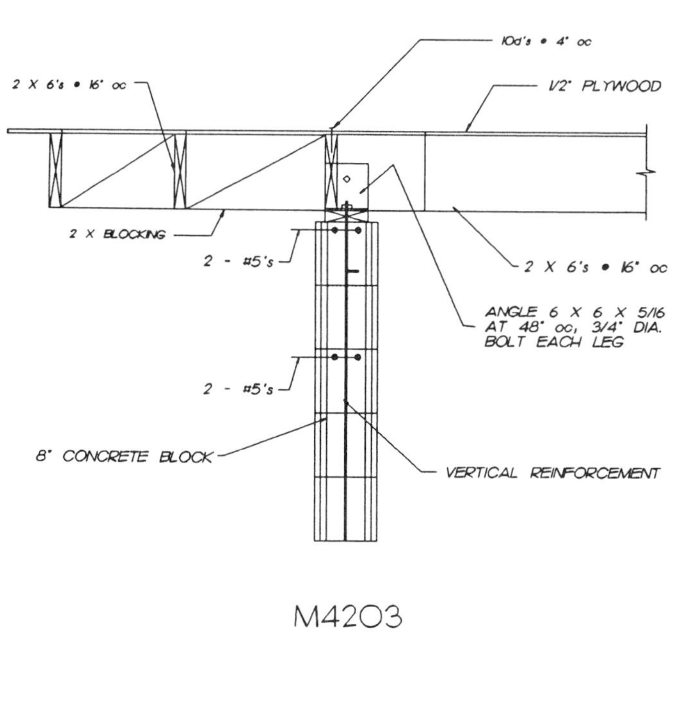
10d's @ 4" oc
2 X 6's @ 16" oc
1/2" PLYWOOD
2 X BLOCKING
2 - #5's
2 X 6's @ 16" oc
ANGLE 6 X 6 X 5/16
AT 48" oc, 3/4" DIA.
BOLT EACH LEG
2 - #5's
8" CONCRETE BLOCK
VERTICAL REINFORCEMENT

M4203

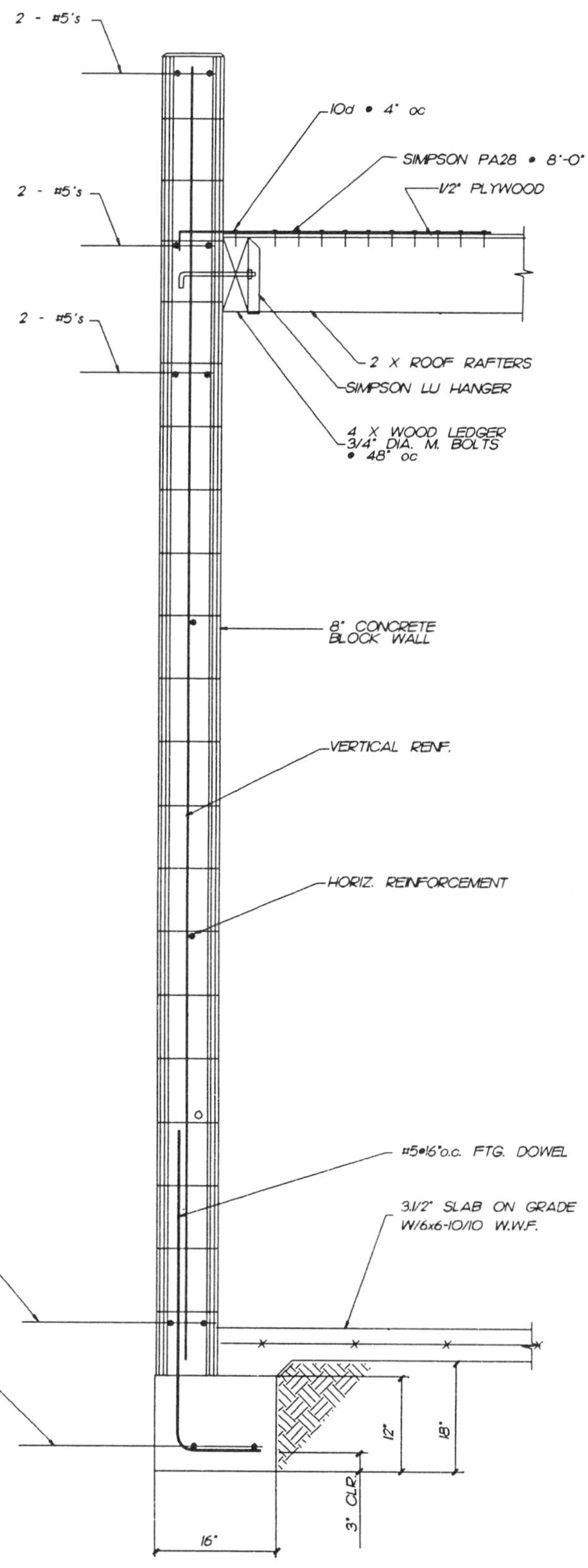
2 - #5's
10d @ 4" oc
SIMPSON PA28 @ 8'-0" oc
1/2" PLYWOOD
2 - #5's
2 - #5's
2 X ROOF RAFTERS
SIMPSON LU HANGER
4 X WOOD LEDGER
3/4" DIA. M. BOLTS
@ 48" oc
8" CONCRETE
BLOCK WALL
VERTICAL REINF.
HORIZ. REINFORCEMENT
#5@16"o.c. FTG. DOWEL
3 1/2" SLAB ON GRADE
W/6x6-10/10 W.W.F.
2 - #5's
2 - #5's
12"
18"
3" CLR.
16"

M4220

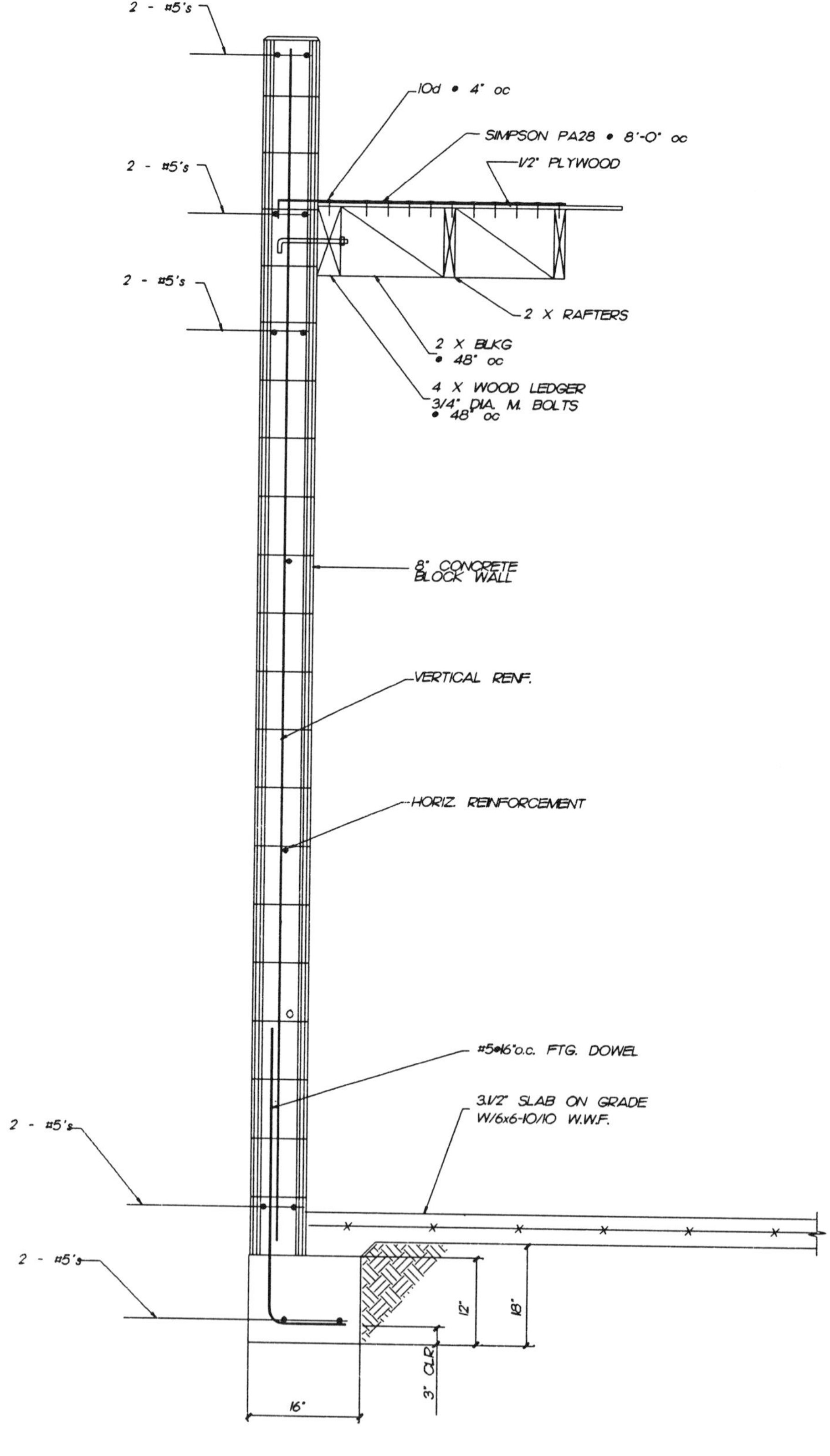
2 - #5's
10d • 4" oc
SIMPSON PA28 • 8'-0" oc
1/2" PLYWOOD
2 - #5's
2 - #5's
2 X RAFTERS
2 X BLKG
• 48" oc
4 X WOOD LEDGER
3/4" DIA. M. BOLTS
• 48" oc
8" CONCRETE
BLOCK WALL
VERTICAL REINF.
HORIZ. REINFORCEMENT
#5•16"o.c. FTG. DOWEL
3 1/2" SLAB ON GRADE
W/6x6-10/10 W.W.F.
2 - #5's
2 - #5's
12"
18"
3" CLR.
16"

M4221

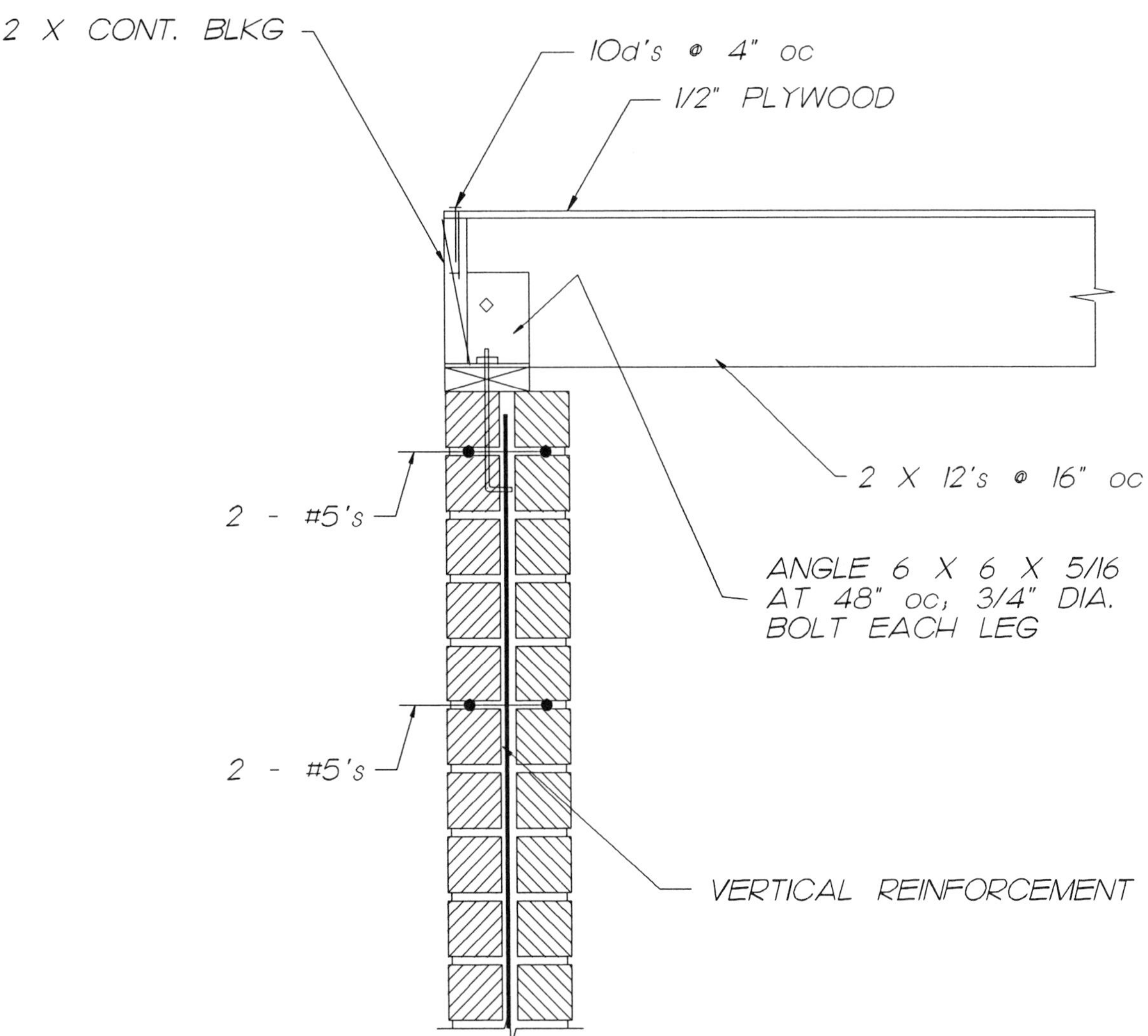
2 X CONT. BLKG
10d's @ 4" oc
1/2" PLYWOOD
2 X 12's @ 16" oc
2 - #5's
ANGLE 6 X 6 X 5/16
AT 48" oc, 3/4" DIA.
BOLT EACH LEG
2 - #5's
VERTICAL REINFORCEMENT

M4240

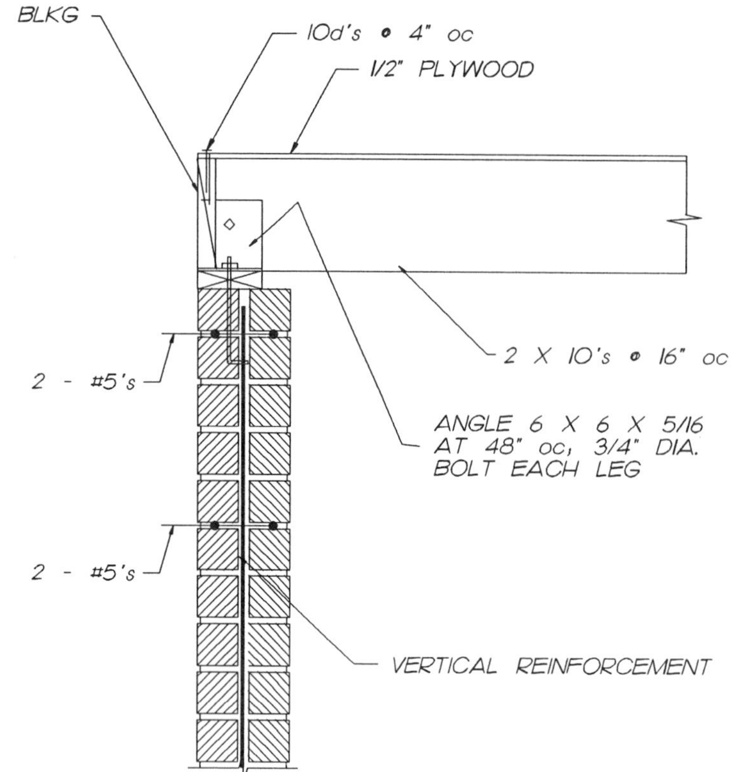
2 X CONT. BLKG
10d's @ 4" oc
1/2" PLYWOOD
2 - #5's
2 X 10's @ 16" oc
ANGLE 6 X 6 X 5/16
AT 48" oc, 3/4" DIA.
BOLT EACH LEG
2 - #5's
VERTICAL REINFORCEMENT

M4241

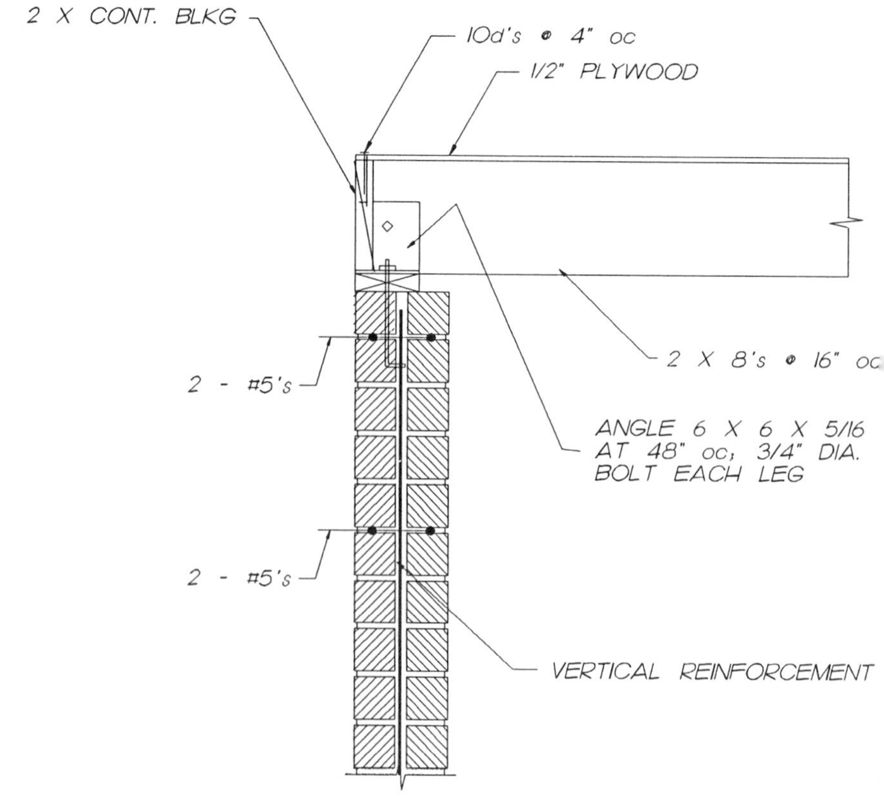
2 X CONT. BLKG
10d's @ 4" oc
1/2" PLYWOOD
2 - #5's
2 X 8's @ 16" oc
ANGLE 6 X 6 X 5/16
AT 48" oc, 3/4" DIA.
BOLT EACH LEG
2 - #5's
VERTICAL REINFORCEMENT

M4242

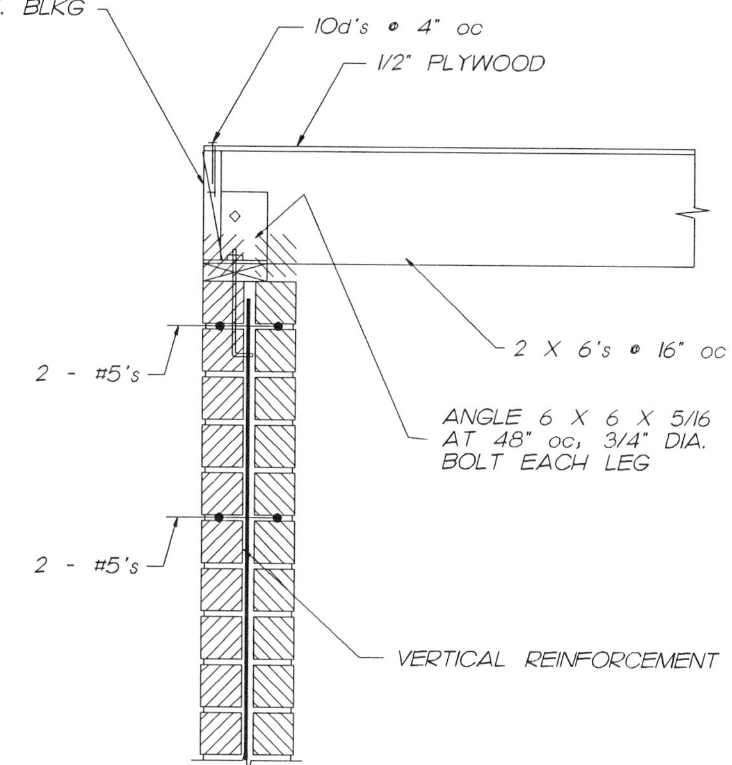
2 X CONT. BLKG
10d's @ 4" oc
1/2" PLYWOOD
2 X 6's @ 16" oc
2 - #5's
ANGLE 6 X 6 X 5/16
AT 48" oc, 3/4" DIA.
BOLT EACH LEG
2 - #5's
VERTICAL REINFORCEMENT

M4243

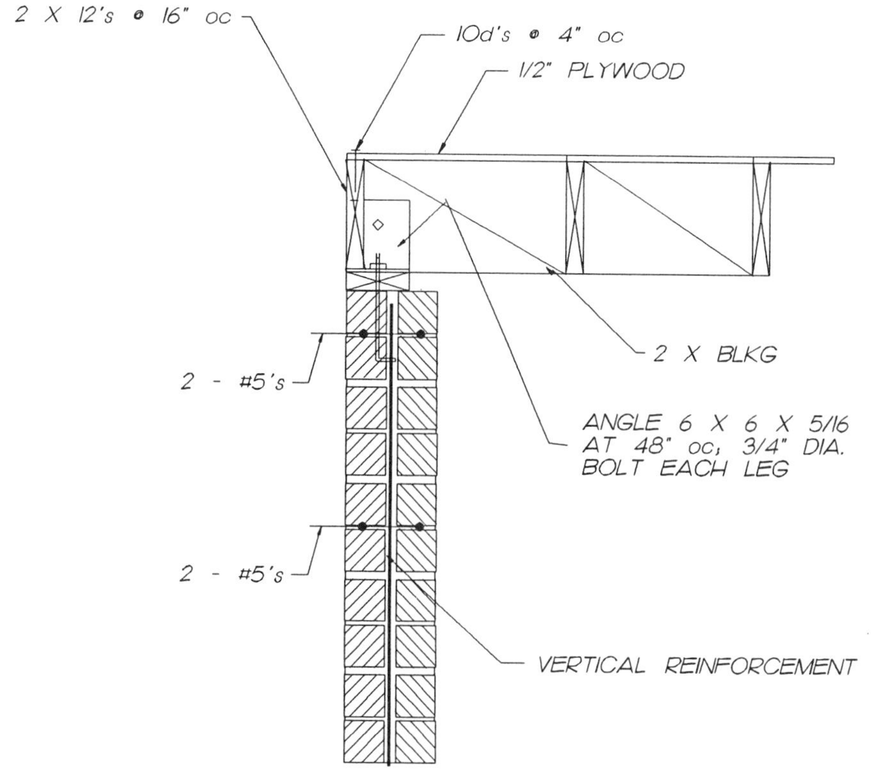
2 X 12's @ 16" oc
10d's @ 4" oc
1/2" PLYWOOD
2 X BLKG
2 - #5's
ANGLE 6 X 6 X 5/16
AT 48" oc, 3/4" DIA.
BOLT EACH LEG
2 - #5's
VERTICAL REINFORCEMENT

M4260

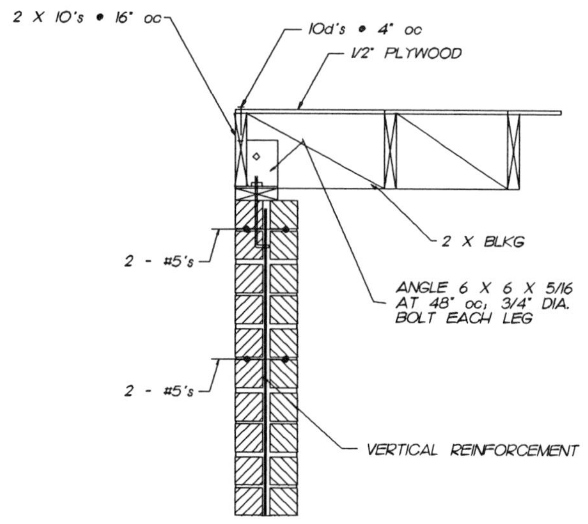

M4261

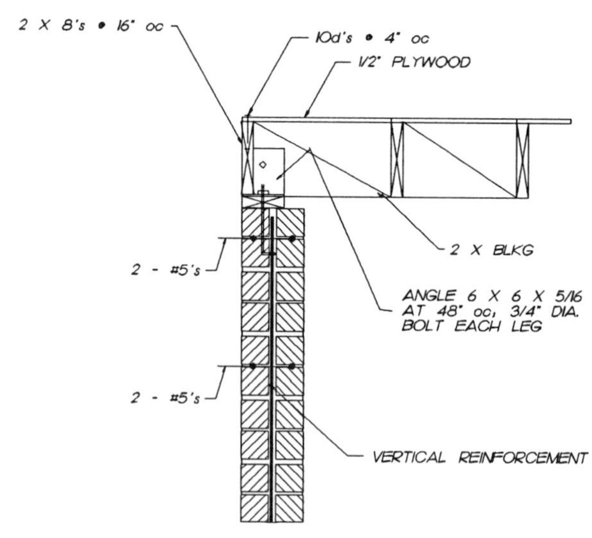

M4262

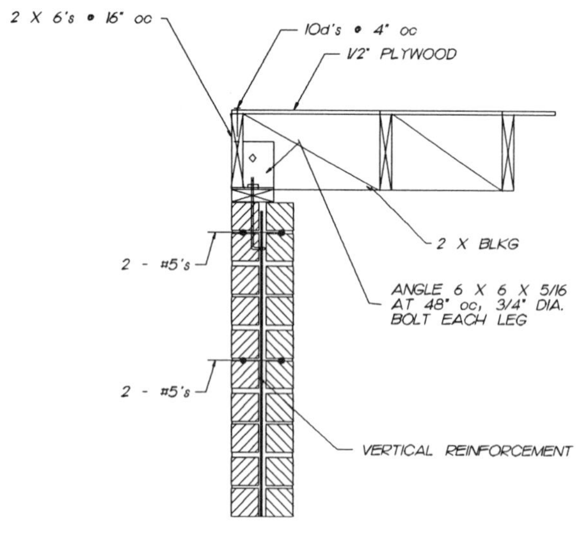

M4263

2 X CONT. BLKG
10d's • 4" oc
1/2" PLYWOOD
2 X 12's • 16" oc
ANGLE 6 X 6 X 5/16
AT 48" oc, 3/4" DIA.
BOLT EACH LEG
2 - #5's
2 - #5's
VERTICAL REINFORCEMENT

M4280

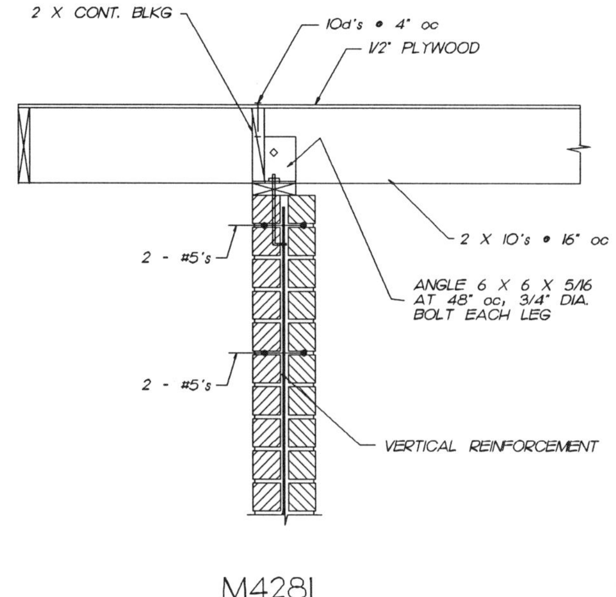

M4281

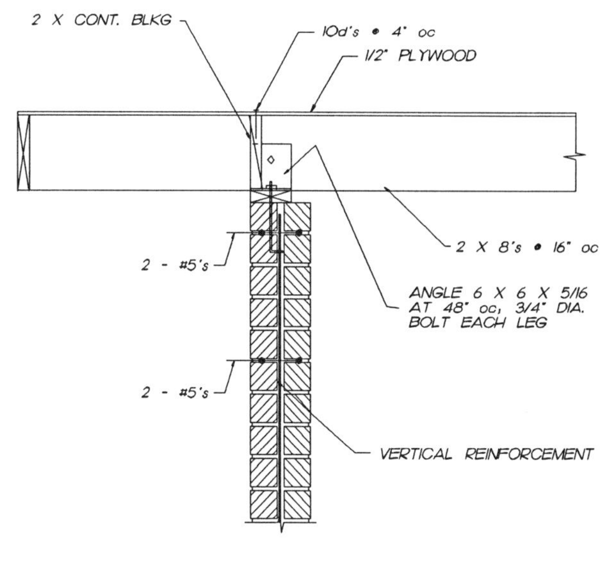

M4282

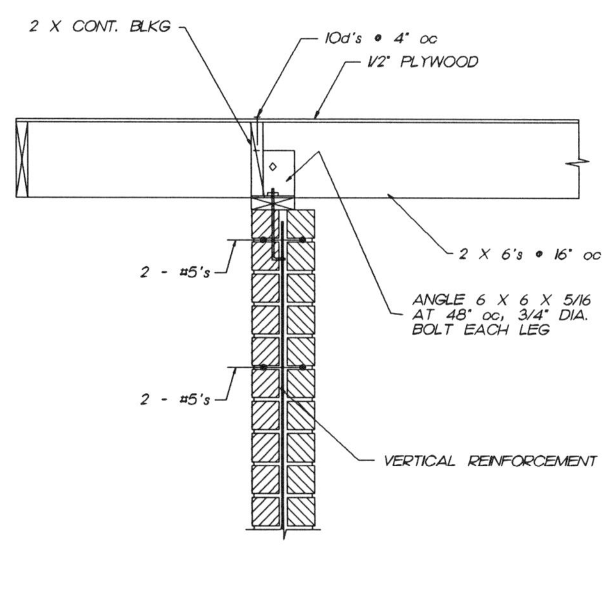

M4300

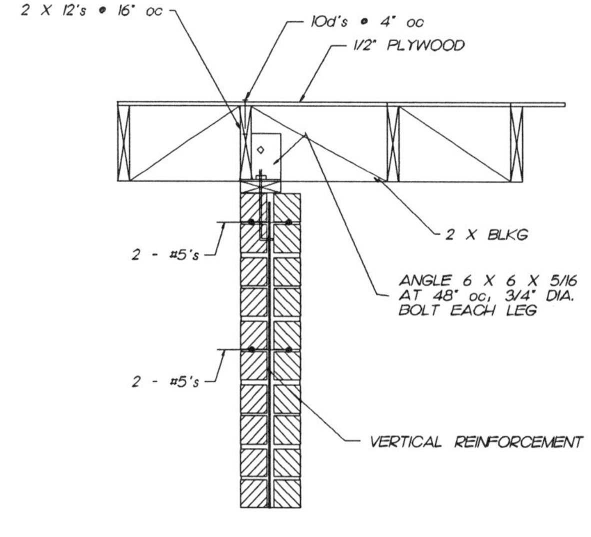

M4283

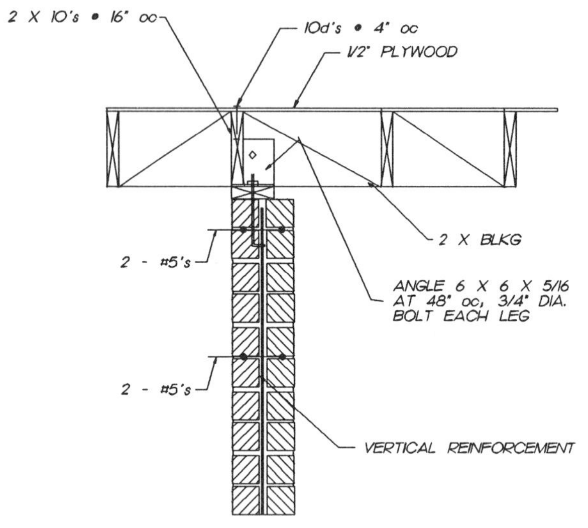

M4301

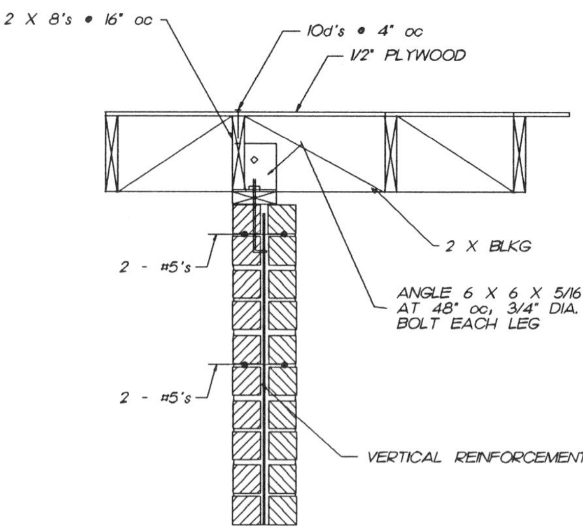

M4302

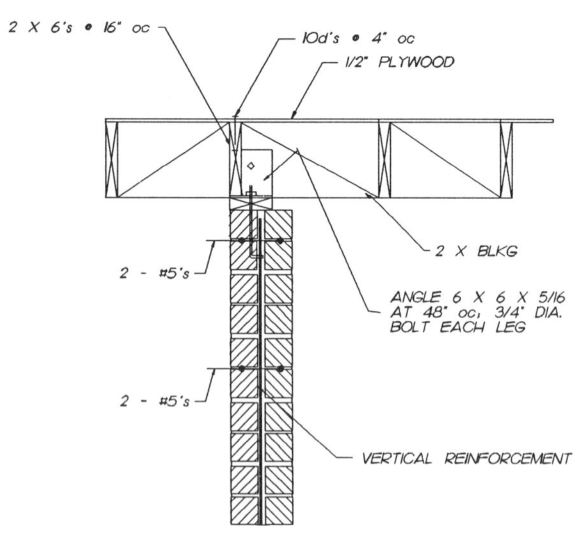

M4303

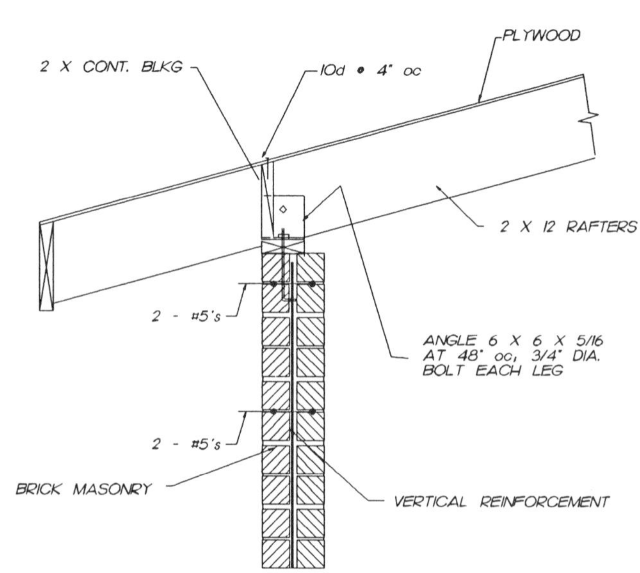

M4320

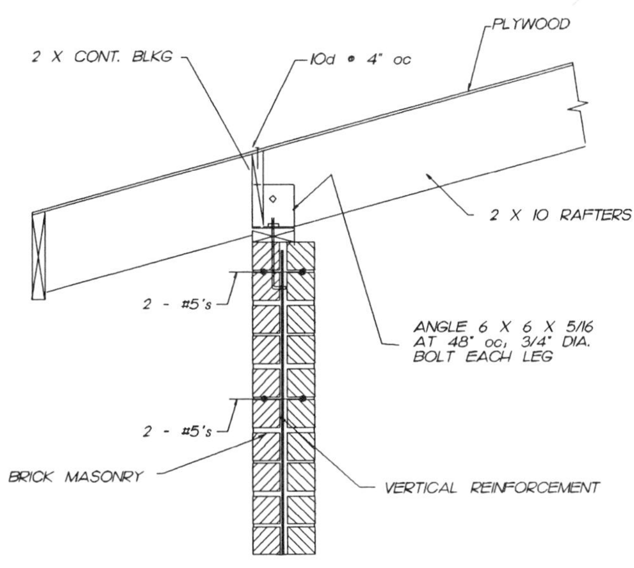

M4321

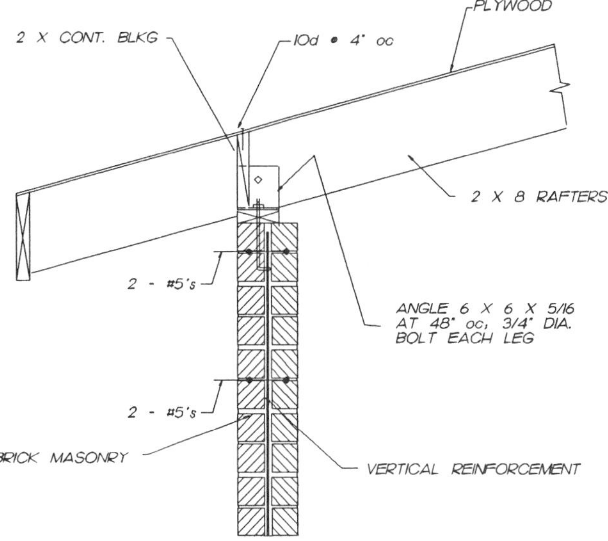

M4322

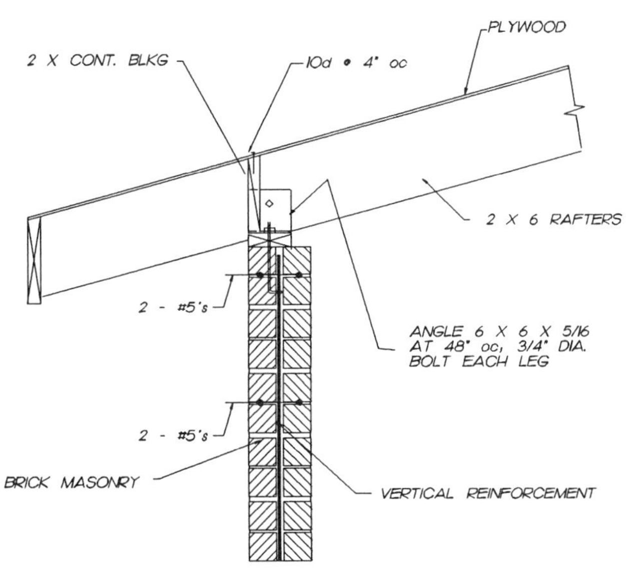

M4323

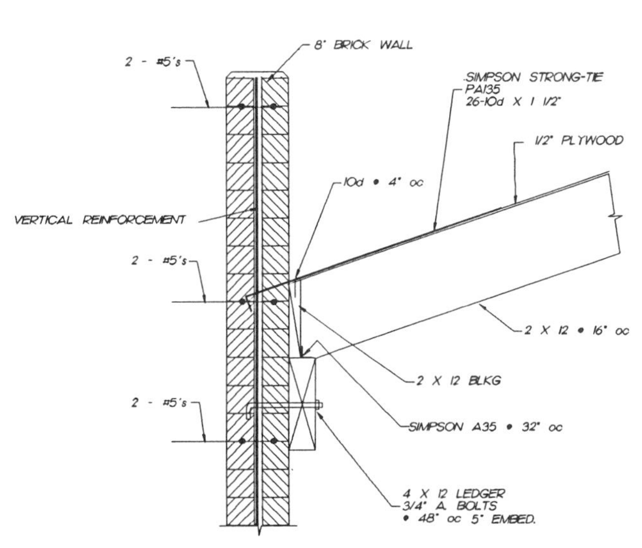

M4340

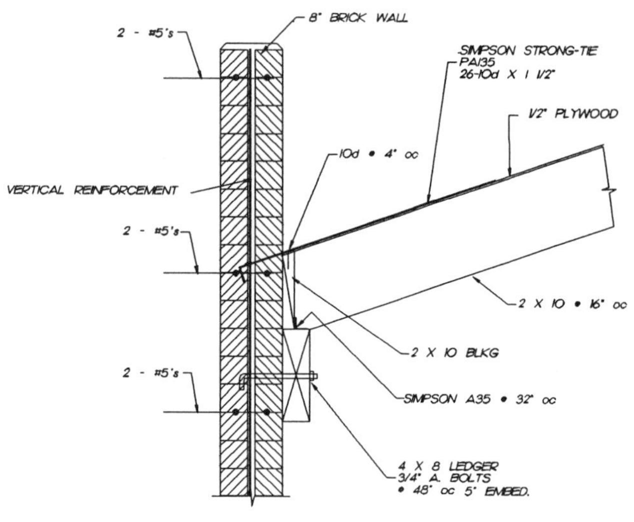

M4341

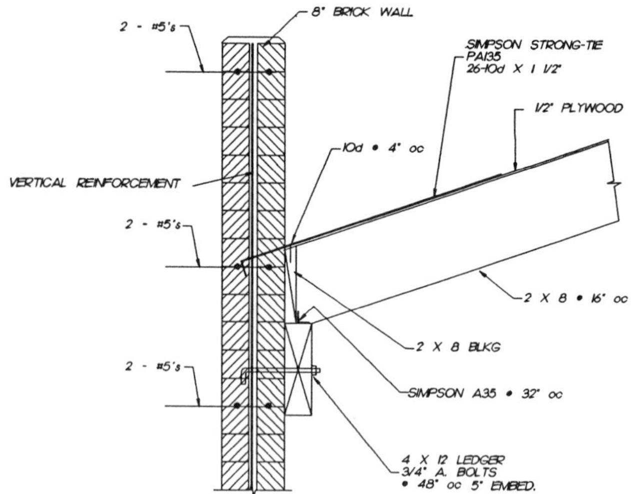

M4342

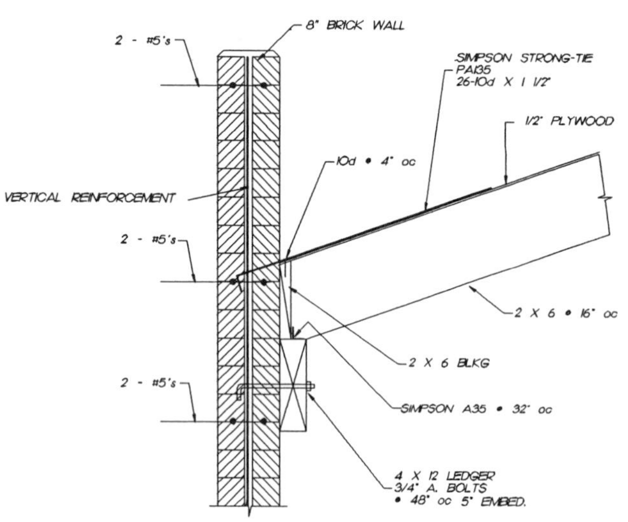

M4343

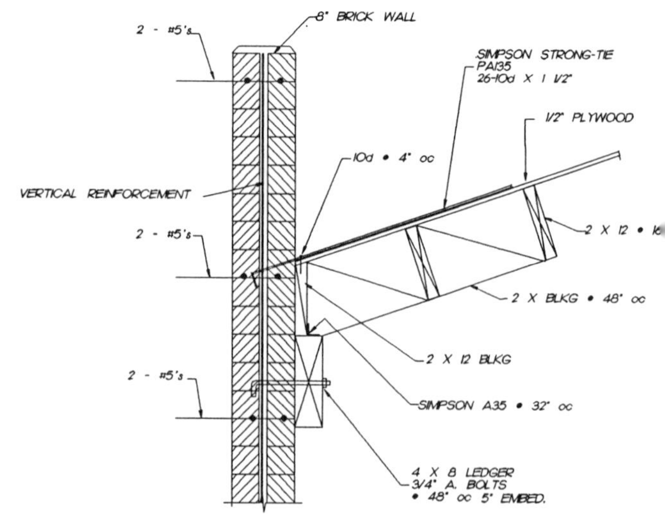

M4360

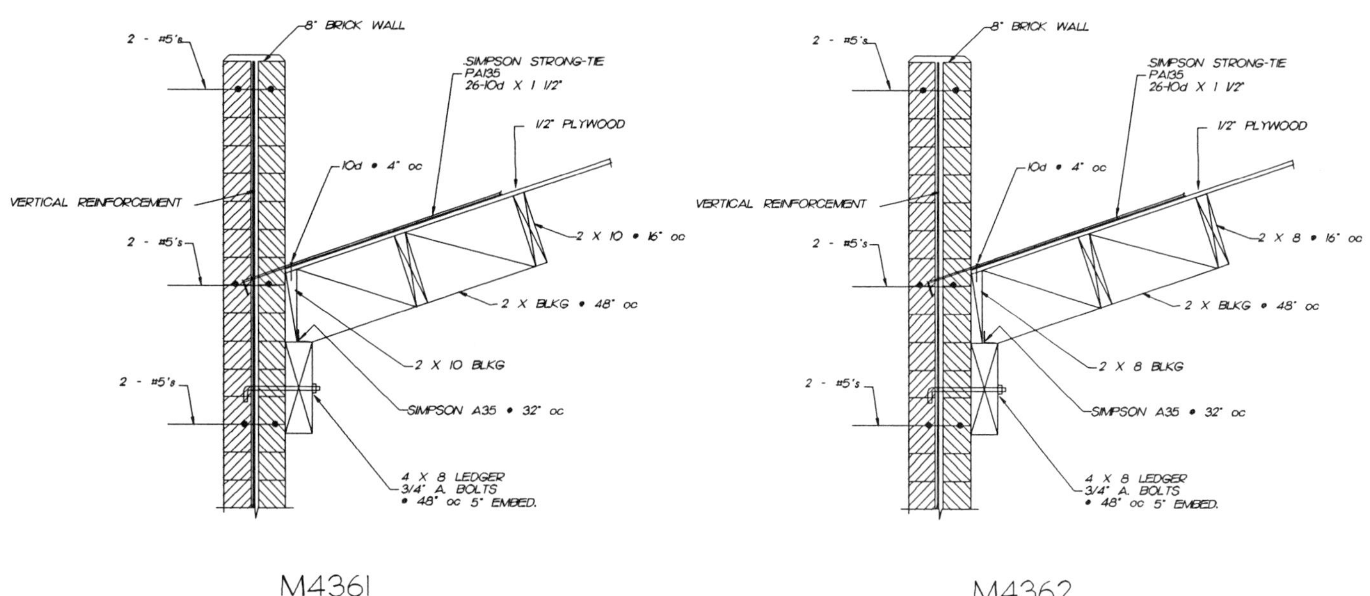
2 - #5's
8" BRICK WALL
SIMPSON STRONG-TIE
PA135
26-10d X 1 1/2"
1/2" PLYWOOD
10d • 4" oc
VERTICAL REINFORCEMENT
2 X 10 • 16" oc
2 - #5's
2 X BLKG • 48" oc
2 X 10 BLKG
2 - #5's
SIMPSON A35 • 32" oc
4 X 8 LEDGER
3/4" A. BOLTS
• 48" oc 5" EMBED.
M4361
2 - #5's
8" BRICK WALL
SIMPSON STRONG-TIE
PA135
26-10d X 1 1/2"
1/2" PLYWOOD
10d • 4" oc
VERTICAL REINFORCEMENT
2 X 8 • 16" oc
2 - #5's
2 X BLKG • 48" oc
2 X 8 BLKG
2 - #5's
SIMPSON A35 • 32" oc
4 X 8 LEDGER
3/4" A. BOLTS
• 48" oc 5" EMBED.
M4362

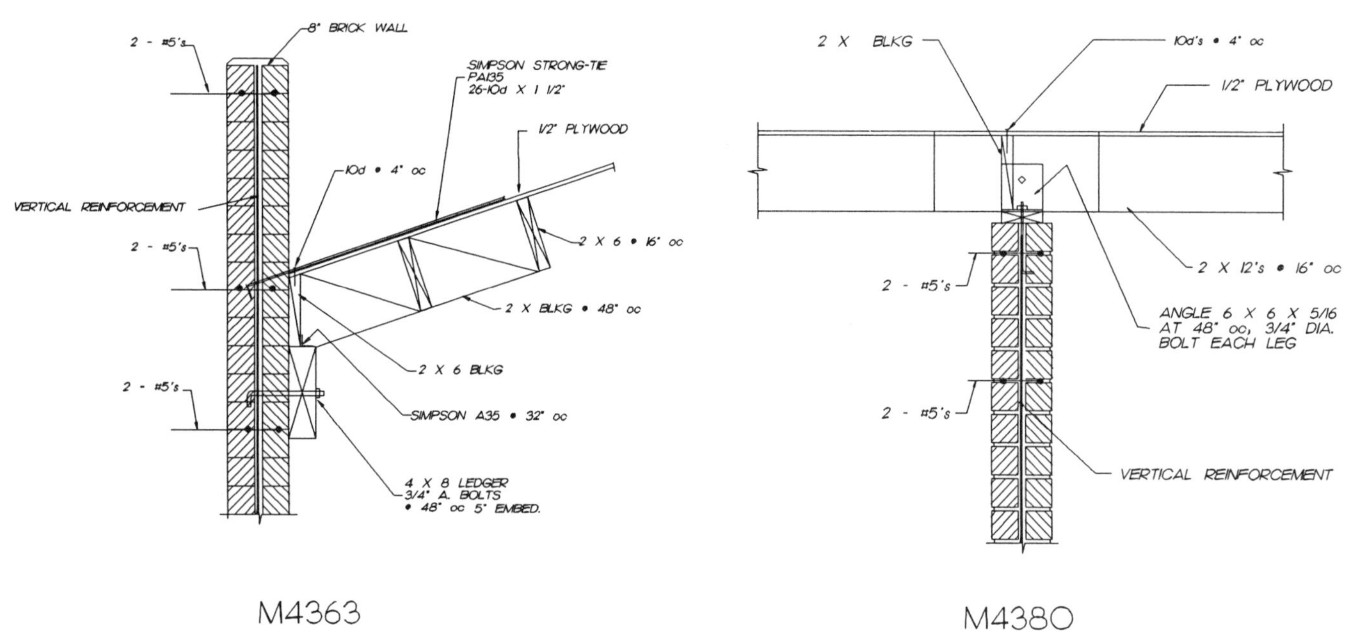
2 - #5's
8" BRICK WALL
SIMPSON STRONG-TIE
PA135
26-10d X 1 1/2"
1/2" PLYWOOD
10d • 4" oc
VERTICAL REINFORCEMENT
2 X 6 • 16" oc
2 - #5's
2 X BLKG • 48" oc
2 X 6 BLKG
2 - #5's
SIMPSON A35 • 32" oc
4 X 8 LEDGER
3/4" A. BOLTS
• 48" oc 5" EMBED.
M4363
2 X BLKG
10d's • 4" oc
1/2" PLYWOOD
2 X 12's • 16" oc
2 - #5's
ANGLE 6 X 6 X 5/16
AT 48" oc, 3/4" DIA.
BOLT EACH LEG
2 - #5's
VERTICAL REINFORCEMENT
M4380

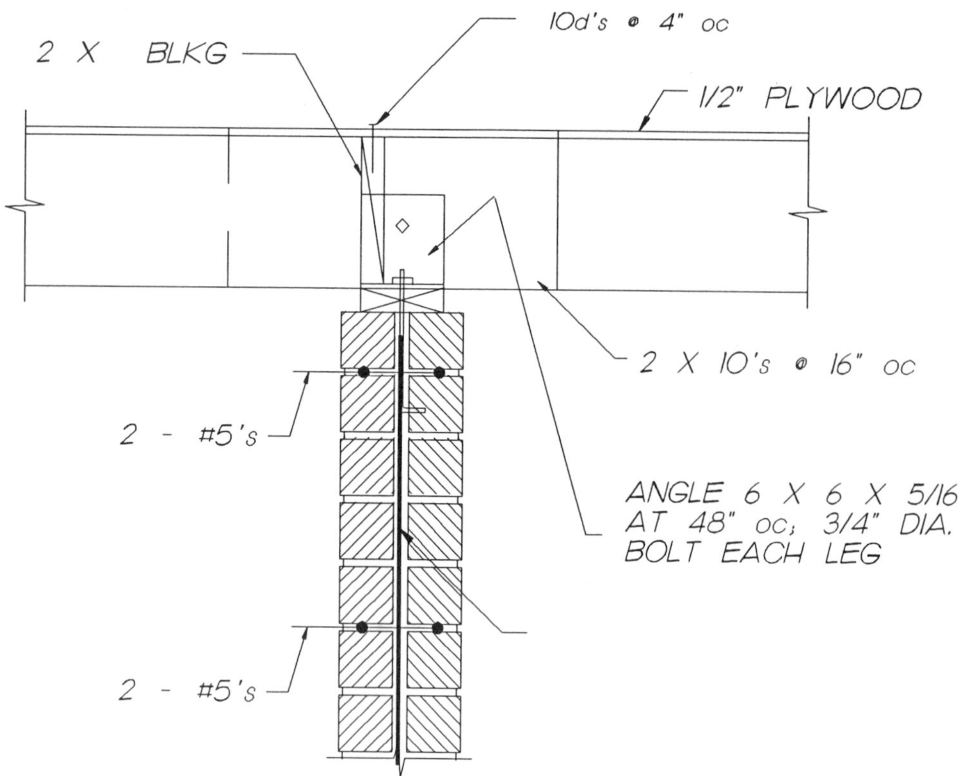
10d's @ 4" oc
2 X BLKG
1/2" PLYWOOD
2 X 10's @ 16" oc
2 - #5's
ANGLE 6 X 6 X 5/16
AT 48" oc; 3/4" DIA.
BOLT EACH LEG
2 - #5's

M4381

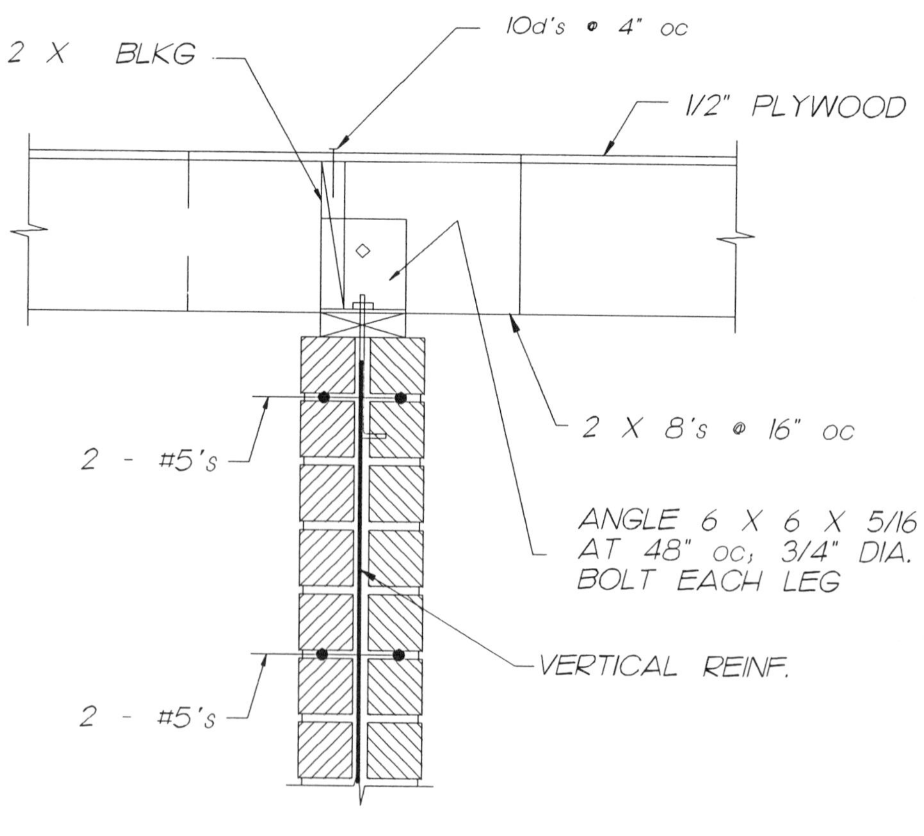
10d's @ 4" oc
2 X BLKG
1/2" PLYWOOD
2 X 8's @ 16" oc
2 - #5's
ANGLE 6 X 6 X 5/16
AT 48" oc; 3/4" DIA.
BOLT EACH LEG
VERTICAL REINF.
2 - #5's

M4382

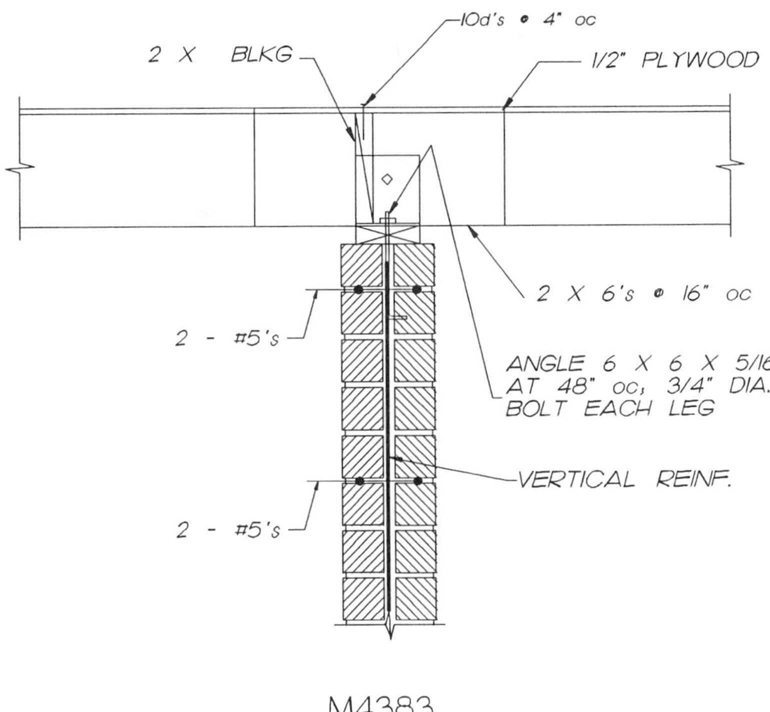
10d's @ 4" oc
2 X BLKG
1/2" PLYWOOD
2 X 6's @ 16" oc
2 - #5's
ANGLE 6 X 6 X 5/16
AT 48" oc, 3/4" DIA.
BOLT EACH LEG
VERTICAL REINF.
2 - #5's

M4383

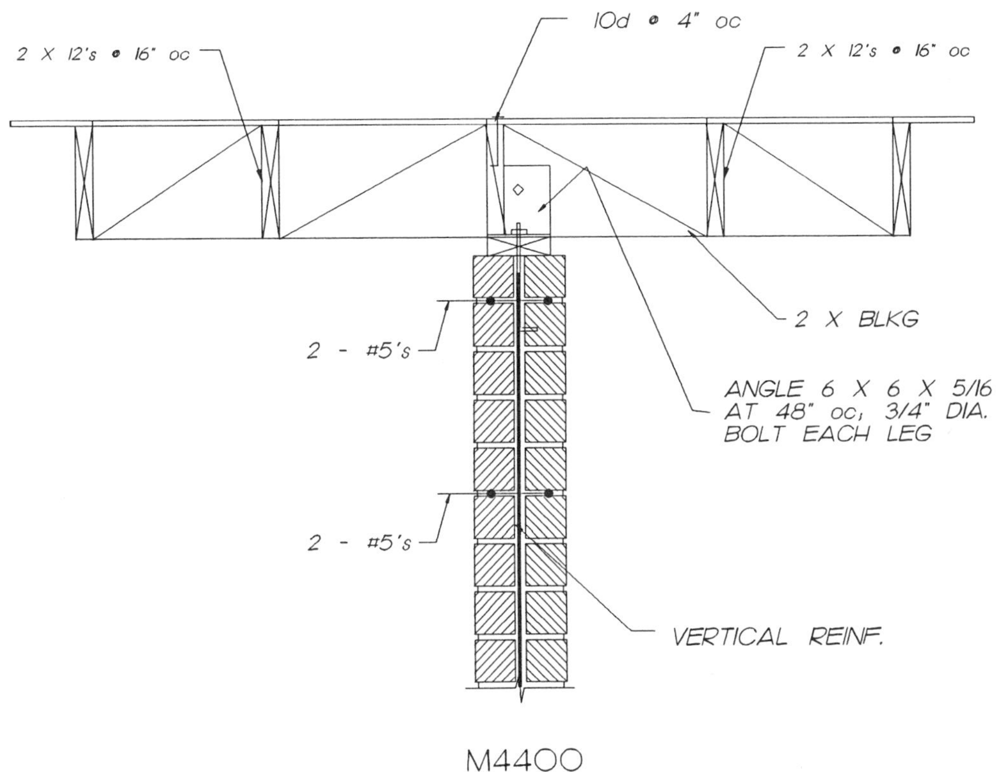
10d @ 4" oc
2 X 12's @ 16" oc
2 X 12's @ 16" oc
2 X BLKG
2 - #5's
ANGLE 6 X 6 X 5/16
AT 48" oc, 3/4" DIA.
BOLT EACH LEG
2 - #5's
VERTICAL REINF.

M4400

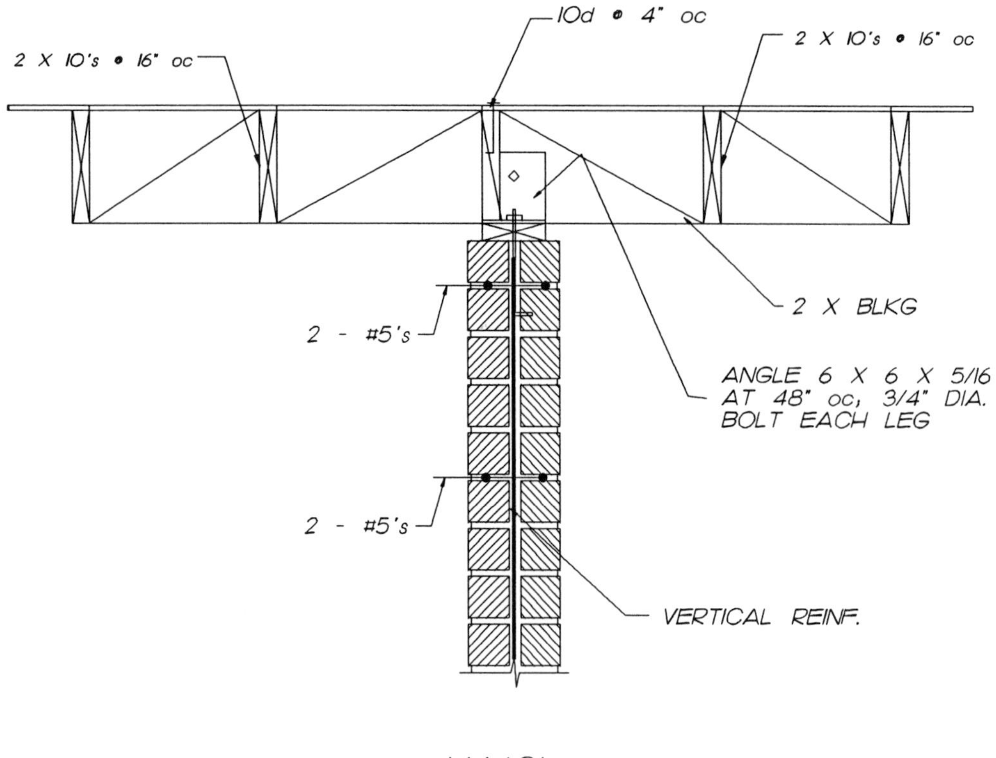
10d @ 4" oc
2 X 10's @ 16" oc
2 X 10's @ 16" oc
2 X BLKG
2 - #5's
ANGLE 6 X 6 X 5/16
AT 48" oc, 3/4" DIA.
BOLT EACH LEG
2 - #5's
VERTICAL REINF.

M4401

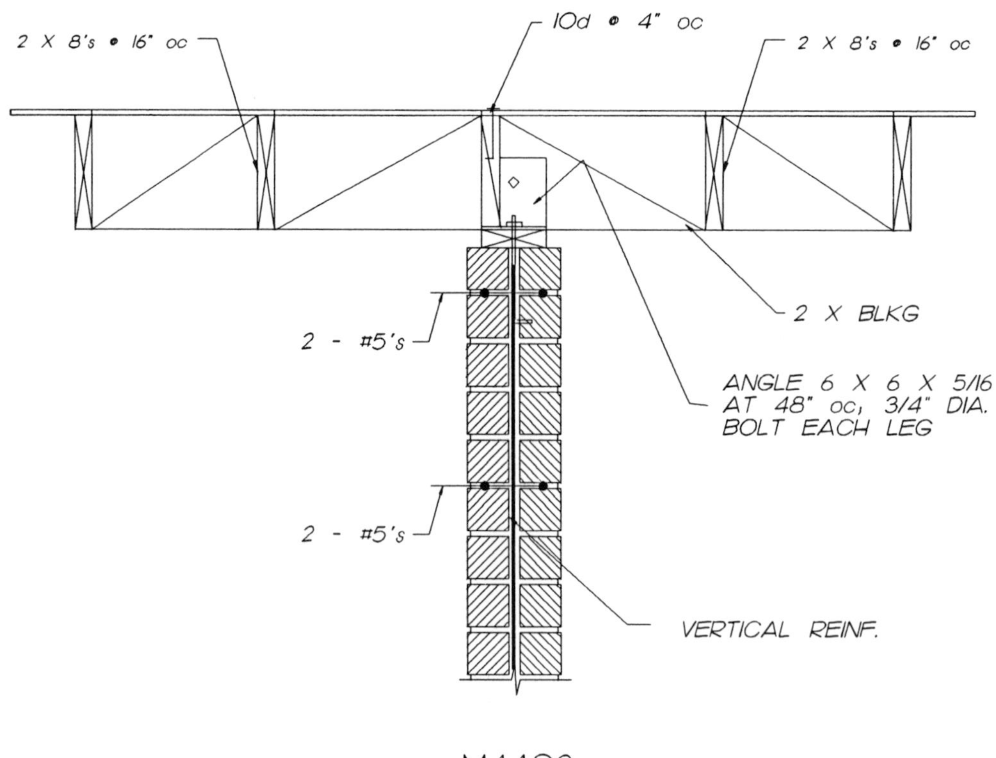
10d @ 4" oc
2 X 8's @ 16" oc
2 X 8's @ 16" oc
2 X BLKG
2 - #5's
ANGLE 6 X 6 X 5/16
AT 48" oc, 3/4" DIA.
BOLT EACH LEG
2 - #5's
VERTICAL REINF.

M4402

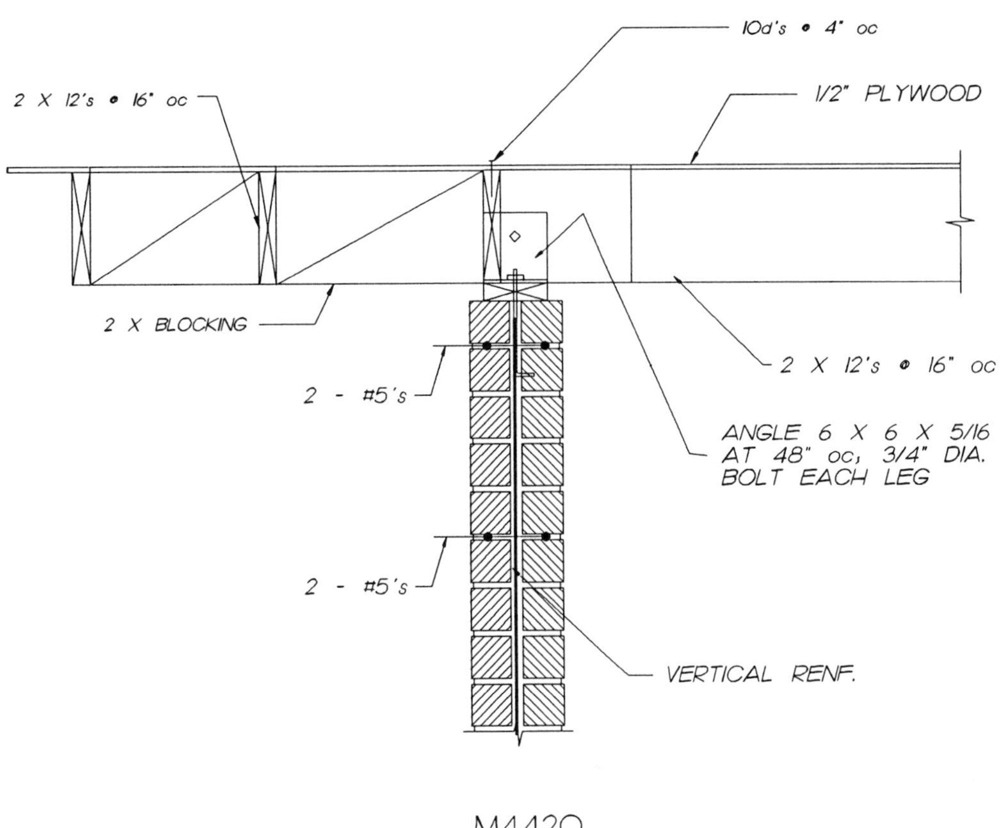
10d's @ 4" oc
2 X 12's @ 16" oc
1/2" PLYWOOD
2 X BLOCKING
2 X 12's @ 16" oc
2 - #5's
ANGLE 6 X 6 X 5/16
AT 48" oc, 3/4" DIA.
BOLT EACH LEG
2 - #5's
VERTICAL RENF.

M4420

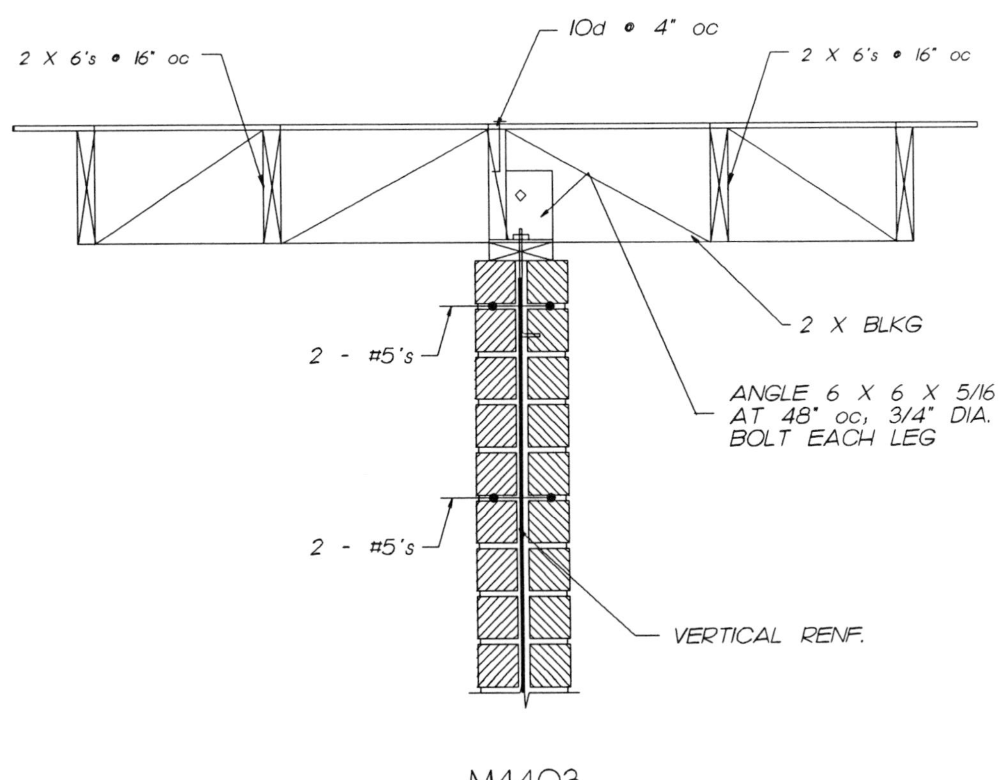
10d @ 4" oc
2 X 6's @ 16" oc
2 X 6's @ 16" oc
2 X BLKG
2 - #5's
ANGLE 6 X 6 X 5/16
AT 48" oc, 3/4" DIA.
BOLT EACH LEG
2 - #5's
VERTICAL RENF.

M4403

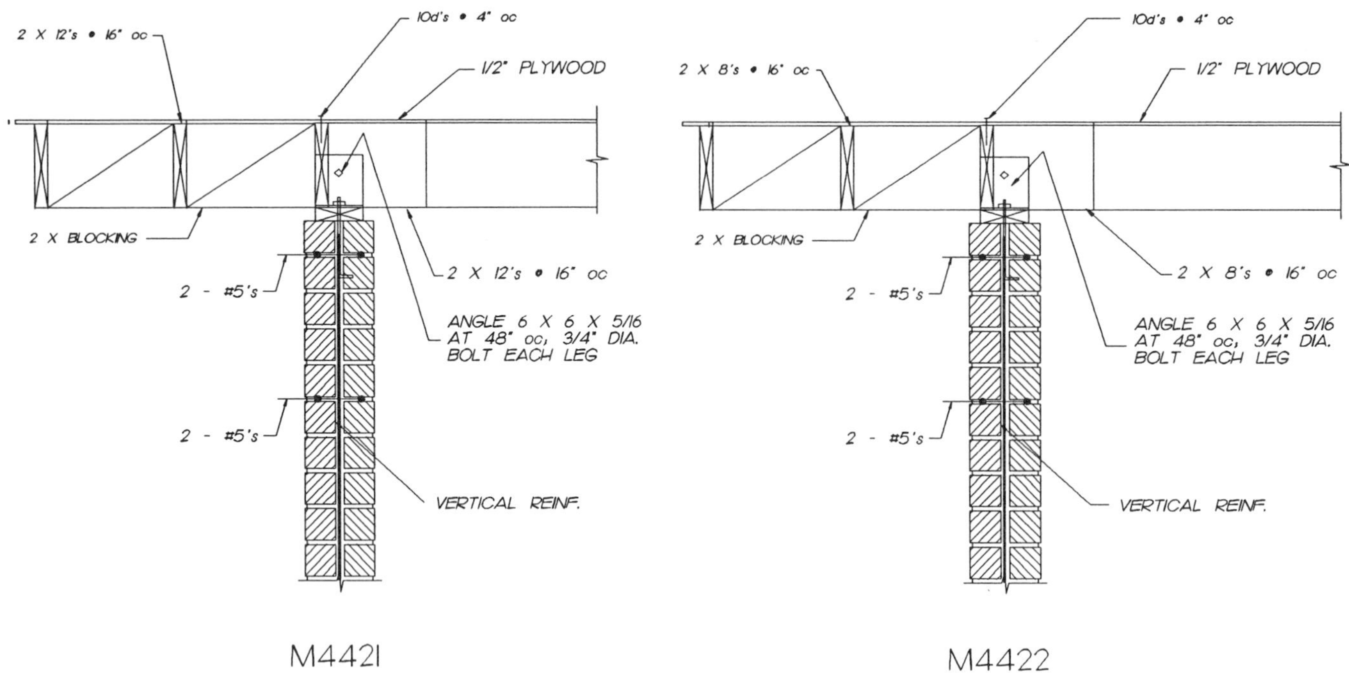
10d's • 4" oc
2 X 12's • 16" oc
1/2" PLYWOOD
2 X BLOCKING
2 - #5's
2 X 12's • 16" oc
ANGLE 6 X 6 X 5/16
AT 48" oc, 3/4" DIA.
BOLT EACH LEG
2 - #5's
VERTICAL REINF.
M4421
10d's • 4" oc
2 X 8's • 16" oc
1/2" PLYWOOD
2 X BLOCKING
2 - #5's
2 X 8's • 16" oc
ANGLE 6 X 6 X 5/16
AT 48" oc, 3/4" DIA.
BOLT EACH LEG
2 - #5's
VERTICAL REINF.
M4422

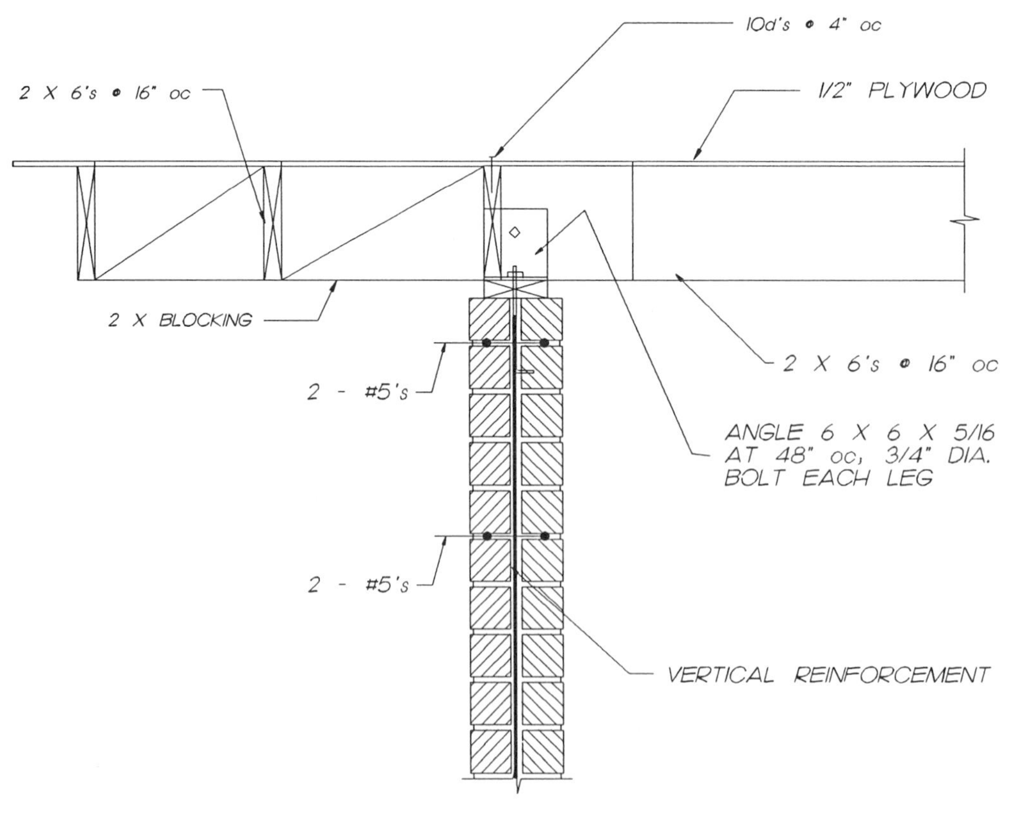
10d's • 4" oc
2 X 6's • 16" oc
1/2" PLYWOOD
2 X BLOCKING
2 - #5's
2 X 6's • 16" oc
ANGLE 6 X 6 X 5/16
AT 48" oc, 3/4" DIA.
BOLT EACH LEG
2 - #5's
VERTICAL REINFORCEMENT

M4423

CHAPTER 2 MASONRY		SECTION 5	WALL - CONCRETE
EXTERIOR BASEMENT WALL CONCRETE SLAB		8" CONCRETE BLOCK	M5020 - M5025
WALL SUPPORTED BY A SLAB WITH A SLAB ON TOP WALL		8" CONCRETE BLOCK	M5040 - M5041
INTERIOR WALL STRUCTURAL SLAB ON TOP SLAB ON GRADE AT BOTTOM		8" CONCRETE BLOCK	M5060
EXTERIOR BASEMENT WALL TWO LEVEL CONCRETE SLAB		8" CONCRETE BLOCK	M5080 - M5080
INTERIOR BASEMENT WALL TWO LEVEL CONCRETE SLAB		8" CONCRETE BLOCK	M5081
EXTERIOR BASEMENT WALL ONE LEVEL CONCRETE SLAB WITH RAMP SLAB ON GRADE		8" CONCRETE BLOCK	M5082
EXTERIOR BASEMENT WALL TWO LEVEL CONCRETE SLAB WITH RAMP SLAB ON GRADE		8" CONCRETE BLOCK	M5083

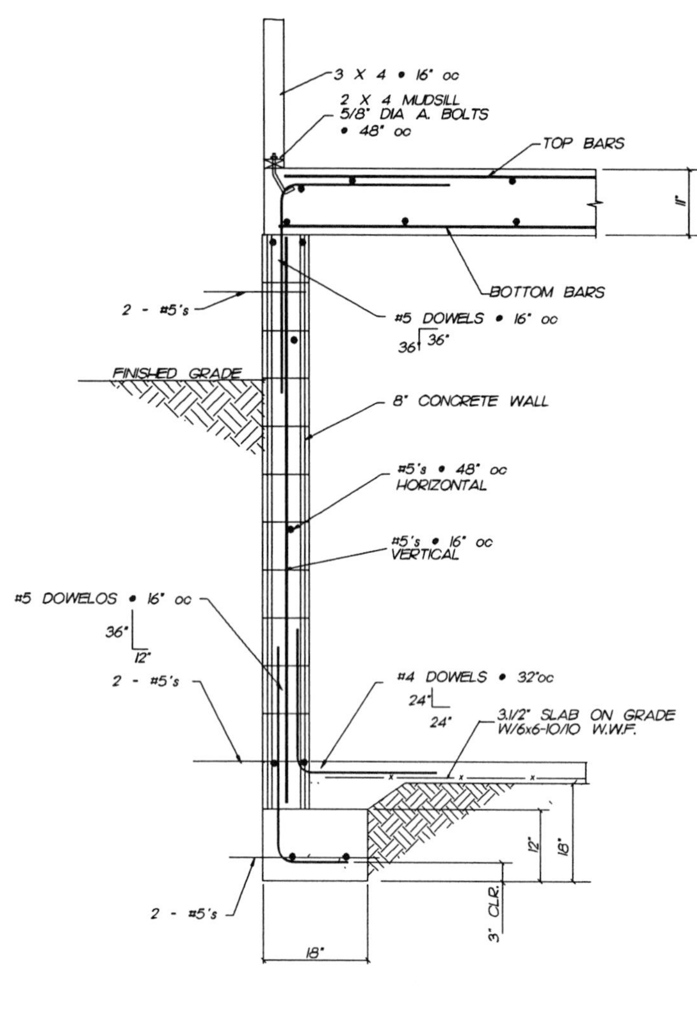

M5020

3 X 4 • 16" oc
2 X 4 MUDSILL
5/8" A. BOLTS
• 48" .oc U.O.N.
TOP BARS
#5's • 16" oc
11"
3 - #5's
BOTTOM BARS
8" CONCRETE BLOCK
BEYOND
2-#5
FIN. GRADE
#5's • 16" oc
VERTICAL
#5's • 48" oc
HORIZONTAL
#5 DOWELS • 16" oc 42" 12"
#4 DOWELS • 32" oc 32" 32"
2-#5
3.1/2" SLAB ON GRADE
W/6x6-10/10 W.W.F.
2-#5
12"
3" CLR.
18"

M5021

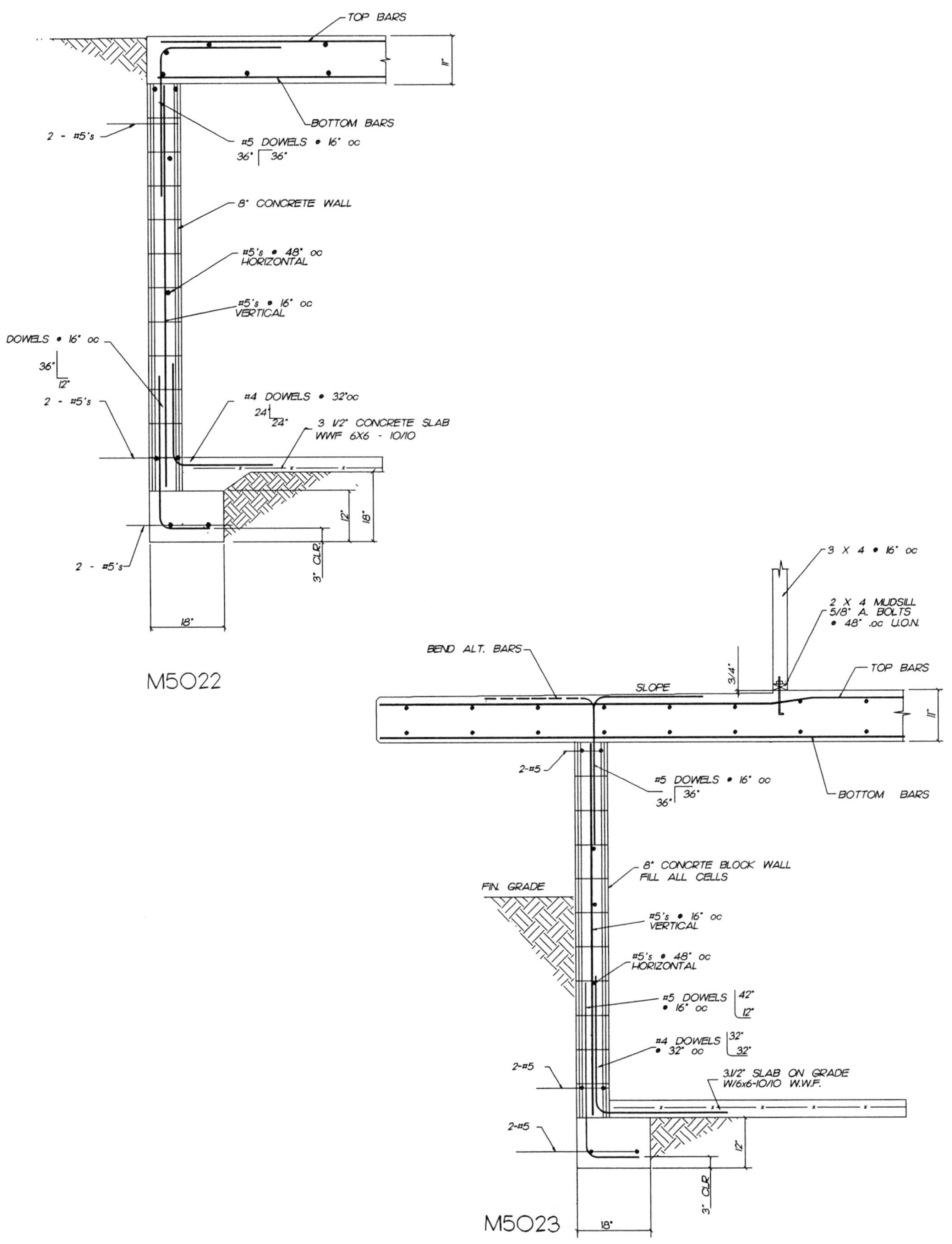
TOP BARS
11"
BOTTOM BARS
2 - #5's
#5 DOWELS • 16" oc
36" 36"
8" CONCRETE WALL
#5's • 48" oc
HORIZONTAL
#5's • 16" oc
VERTICAL
5 DOWELS • 16" oc
36"
12"
2 - #5's
#4 DOWELS • 32"oc
24
24"
3 1/2" CONCRETE SLAB
WWF 6X6 - 10/10
12"
18"
2 - #5's
3" CLR.
18"
M5022
3 X 4 • 16" oc
2 X 4 MUDSILL
5/8" A. BOLTS
• 48" .oc U.O.N.
BEND ALT. BARS
3/4"
SLOPE
TOP BARS
11"
2-#5
#5 DOWELS • 16" oc
36" 36"
BOTTOM BARS
8" CONCRTE BLOCK WALL
FILL ALL CELLS
FIN. GRADE
#5's • 16" oc
VERTICAL
#5's • 48" oc
HORIZONTAL
#5 DOWELS • 16" oc
42"
12"
#4 DOWELS • 32" oc
32"
32"
2-#5
3 1/2" SLAB ON GRADE
W/6x6-10/10 W.W.F.
2-#5
12"
3" CLR.
18"
M5023

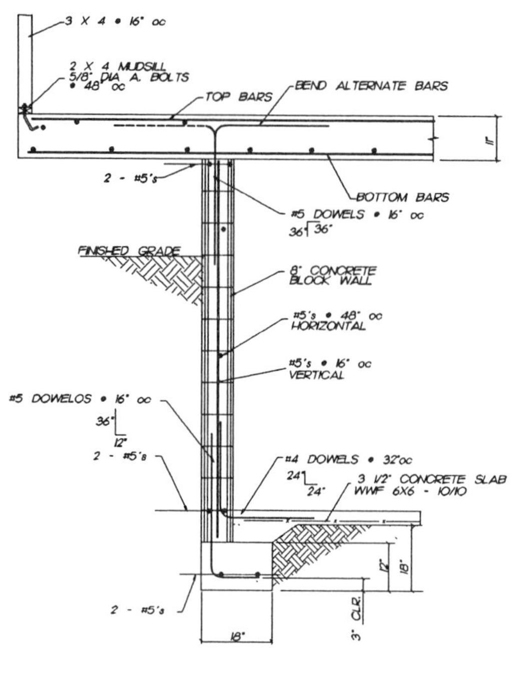

M5024

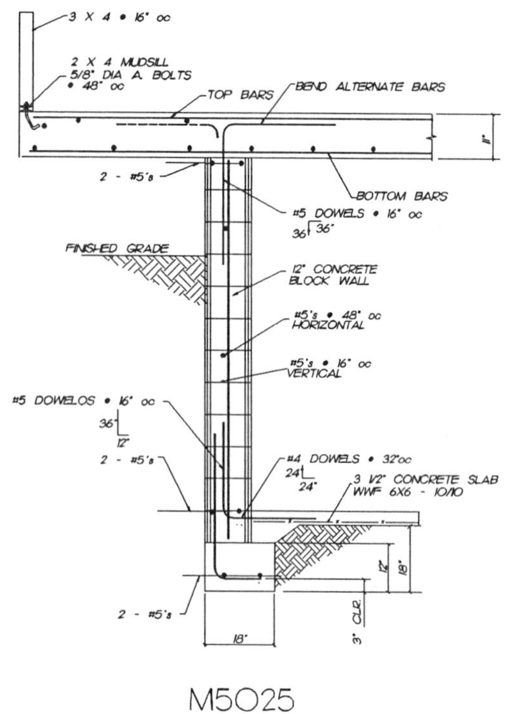

M5025

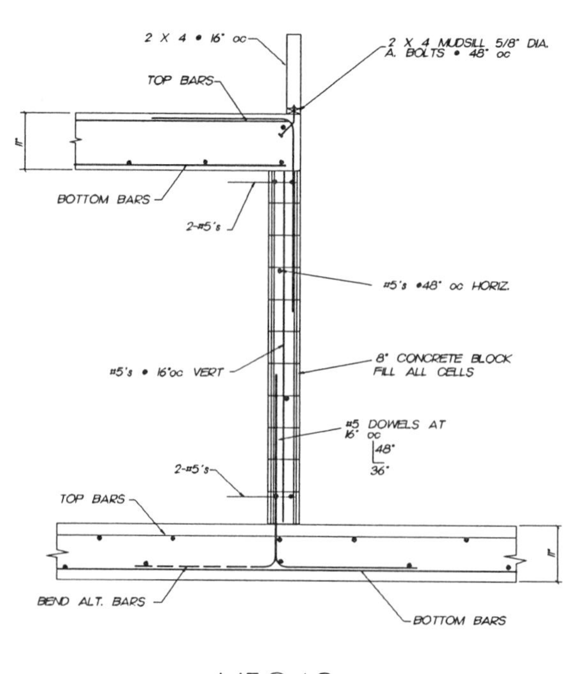

M5040

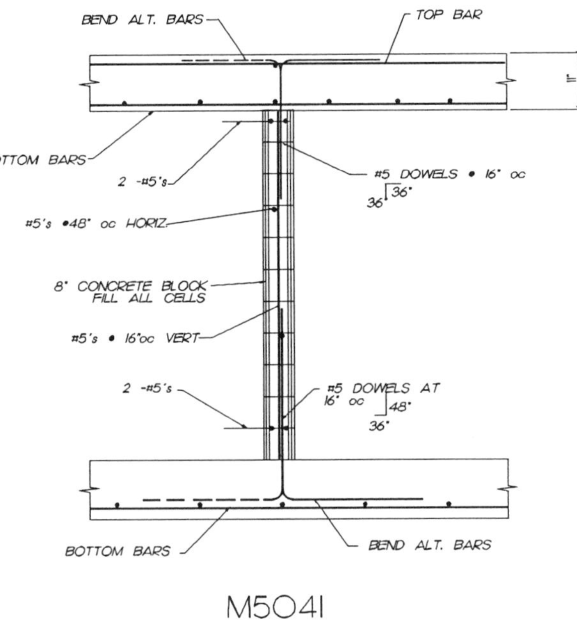

M5041

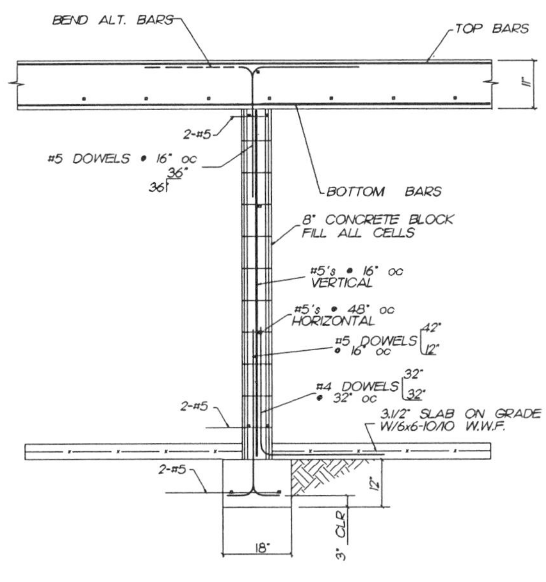

M5060

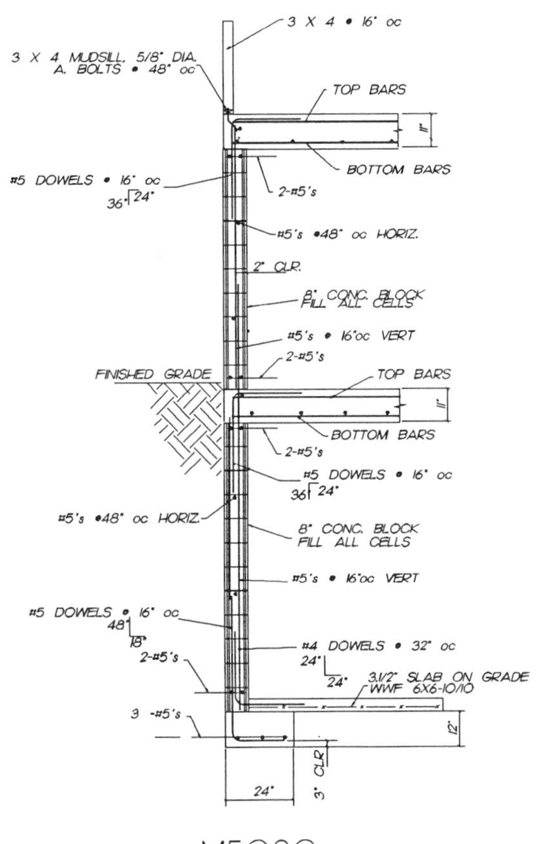

M5080

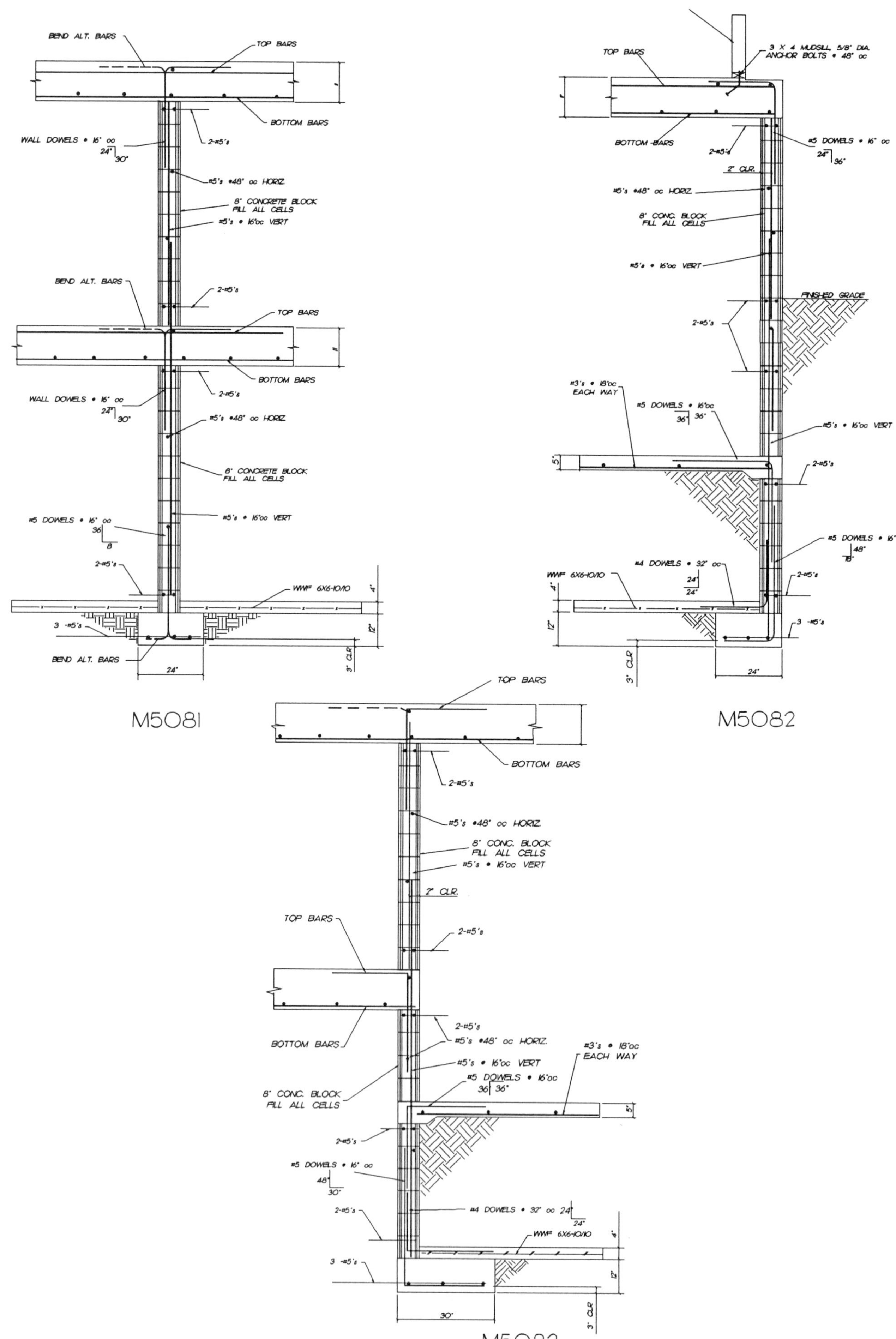
M5081
M5082
M5083

CHAPTER 2 MASONRY SECTION 6 WALL - STEEL FLOOR		
EXTERIOR WALL METAL DECKING PARALELL TO WALL DECKING SUPPORTED BY STEEL BMS. BM. CONNECTED WITH CLIP ANGLES USING ANCHOR BOLTS & MACHINE BOLTS	8" CONCRETE BLOCK	M6020 - M6024
	12" CONCRETE BLOCK	M6025 - M6029
	BRICK MASONRY	M6240 - M6244
EXTERIOR WALL METAL DECKING PERP. TO WALL DECKING SUPPORTED BY STEEL BMS. BM. CONNECTED WITH CLIP ANGLES USING ANCHOR BOLTS & MACHINE BOLTS	8" CONCRETE BLOCK	M6040 - M6044
	12" CONCRETE BLOCK	M6045 - M6049
	BRICK MASONRY	M6260 - M6264
EXTERIOR WALL METAL DECKING PARALELL TO WALL DECKING SUPPORTED BY LIGHT METAL JOISTS JOISTS CONNECTED WITH SEAT ANGLES USING WELDS & ANCHOR BOLTS	8" CONCRETE BLOCK	M6060
	12" CONCRETE BLOCK	M6061
	BRICK MASONRY	M6280
EXTERIOR WALL METAL DECKING PERP. TO WALL DECKING SUPPORTED BY LIGHT METAL JOISTS METAL DECKING PERP. TO WALL DECKING CONNECTED WITH SEAT ANGLES USING WELDS & ANCHOR BOLTS	8" CONCRETE BLOCK	M6080
	12" CONCRETE BLOCK	M6081
	BRICK MASONRY	M6300
EXTERIOR WALL METAL DECKING PERP. TO WALL DECKING SUPPORTED BY LIGHT METAL JOISTS METAL DECKING PERP. TO WALL DECKING CONNECTED WITH SEAT ANGLES USING WELDS & ANCHOR BOLTS	8" CONCRETE BLOCK	M6100
	BRICK MASONRY	M6320
EXTERIOR WALL METAL DECKING PARALELL TO WALL DECKING SUPPORTED BY METAL BAR JOISTS BAR JOISTS CONNECTED WITH SEAT ANGLES USING WELDS & ANCHOR BOLTS	8" CONCRETE BLOCK	M6120
	12" CONCRETE BLOCK	M6121
	BRICK MASONRY	M6340
EXTERIOR WALL METAL DECKING PERP. TO WALL DECKING SUPPORTED BY METAL BAR JOISTS BAR JOISTS CONNECTED WITH SEAT ANGLES USING WELDS & ANCHOR BOLTS	8" CONCRETE BLOCK	M6140
	12" CONCRETE BLOCK	M6141
	BRICK MASONRY	M6360
EXTERIOR WALL METAL DECKING PARALELL TO WALL METAL DECKING CONNECTED WITH SEAT ANGLES USING WELDS & ANCHOR BOLTS	8" CONCRETE BLOCK	M6160
	12" CONCRETE BLOCK	M6161
	BRICK MASONRY	M6380
INTERIOR WALL METAL DECKING PARALELL TO WALL METAL DECKING CONNECTED WITH SEAT ANGLES USING WELDS & ANCHOR BOLTS	8" CONCRETE BLOCK	M6180
	12" CONCRETE BLOCK	M6181
	BRICK MASONRY	M6400
INTERIOR WALL METAL DECKING PARALELL TO WALL DECKING SUPPORTED BY METAL BAR JOISTS BAR JOISTS CONNECTED WITH SEAT ANGLES USING WELDS & ANCHOR BOLTS	8" CONCRETE BLOCK	M6200
	12" CONCRETE BLOCK	M6201
	BRICK MASONRY	M6420
INTERIOR WALL METAL DECKING PERP. TO WALL DECKING SUPPORTED BY METAL BAR JOISTS BAR JOISTS CONNECTED WITH SEAT ANGLES USING WELDS & ANCHOR BOLTS	8" CONCRETE BLOCK	M6220
	12" CONCRETE BLOCK	M6221
	BRICK MASONRY	M6440

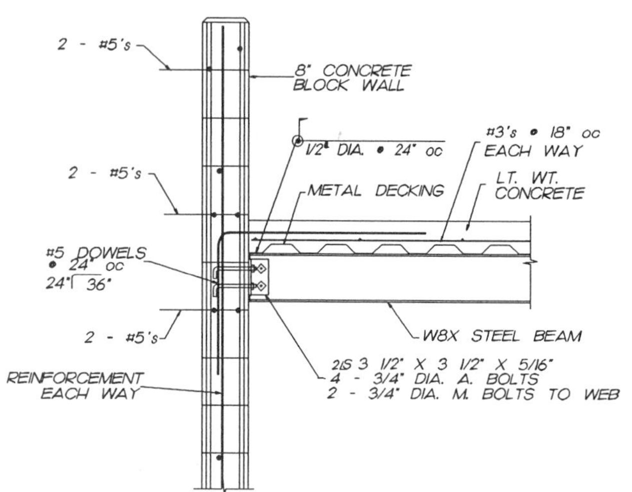

M6020

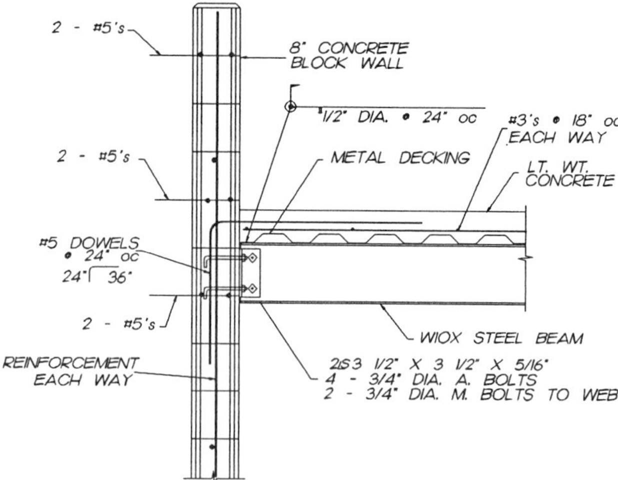

M6021

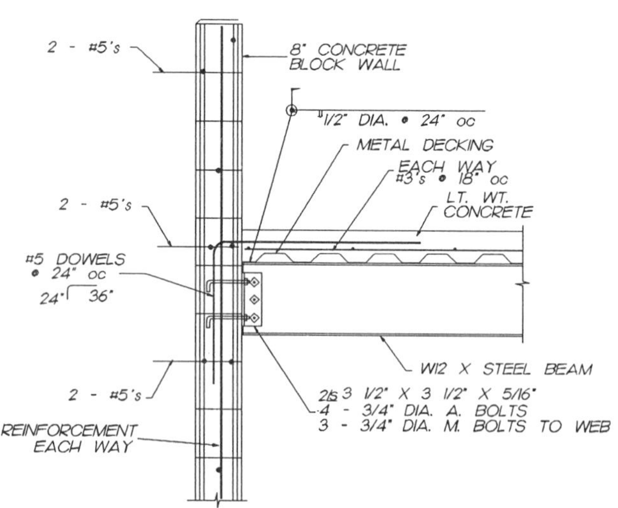

M6022

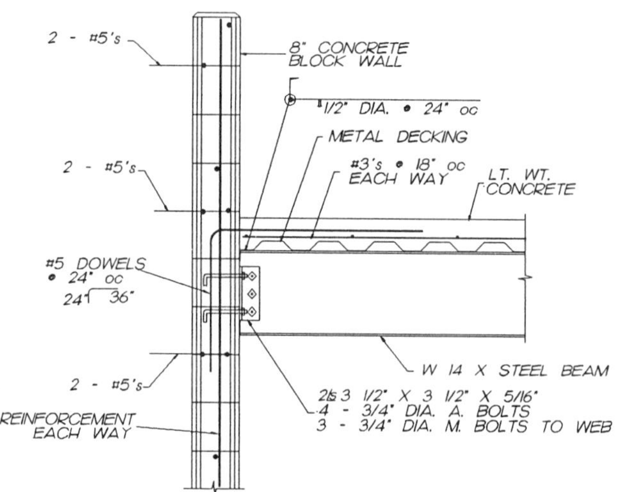

M6023

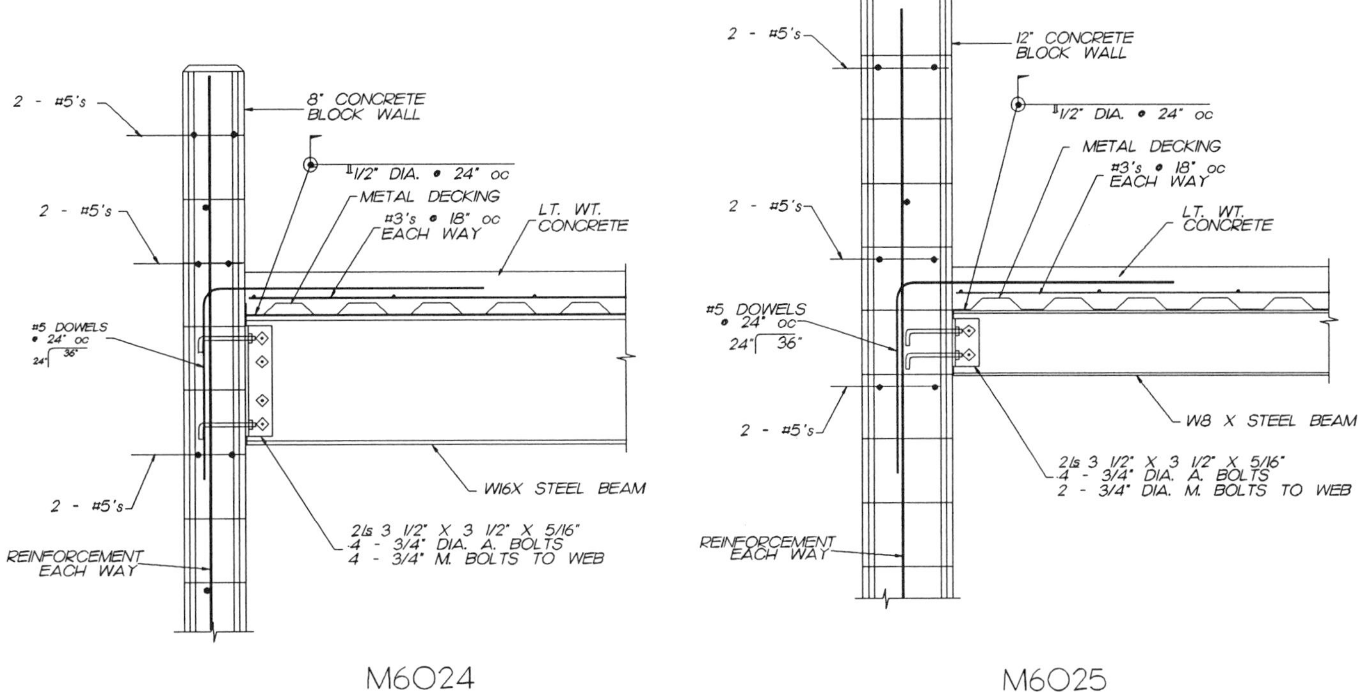

M6024

M6025

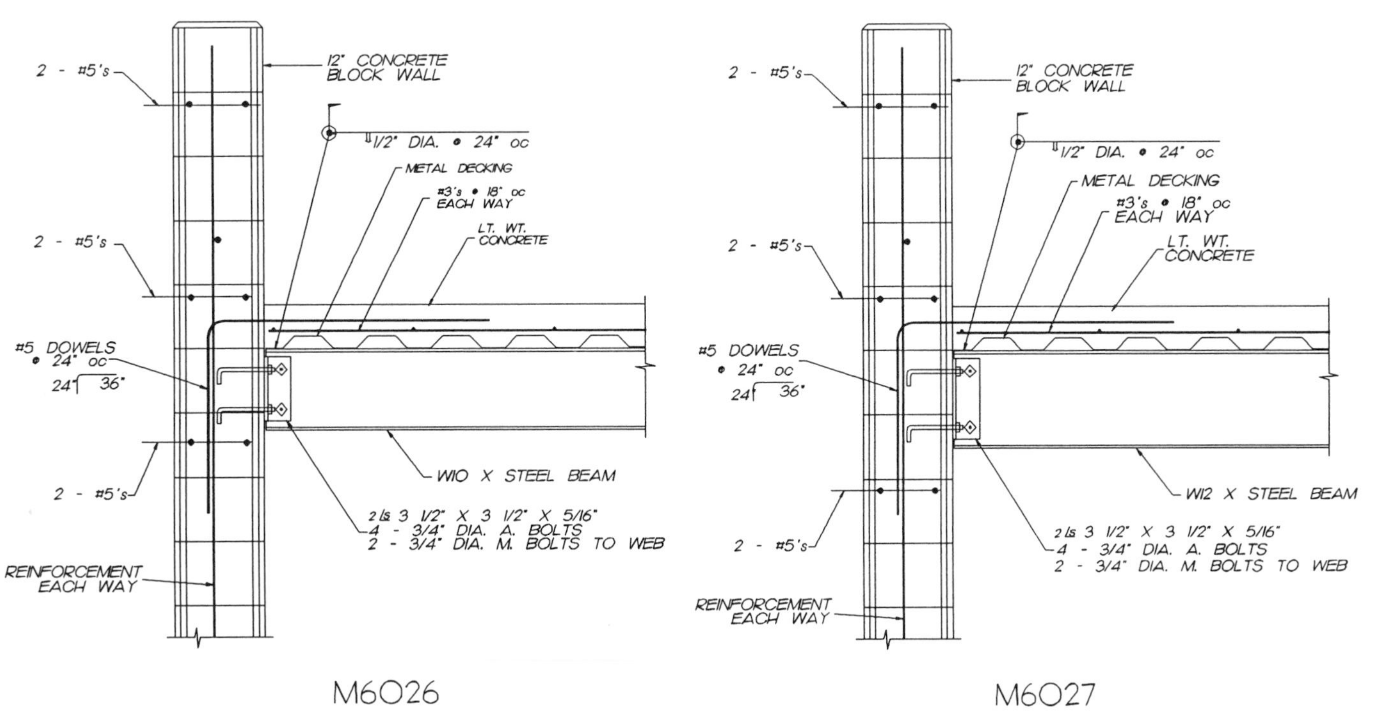

M6026

M6027

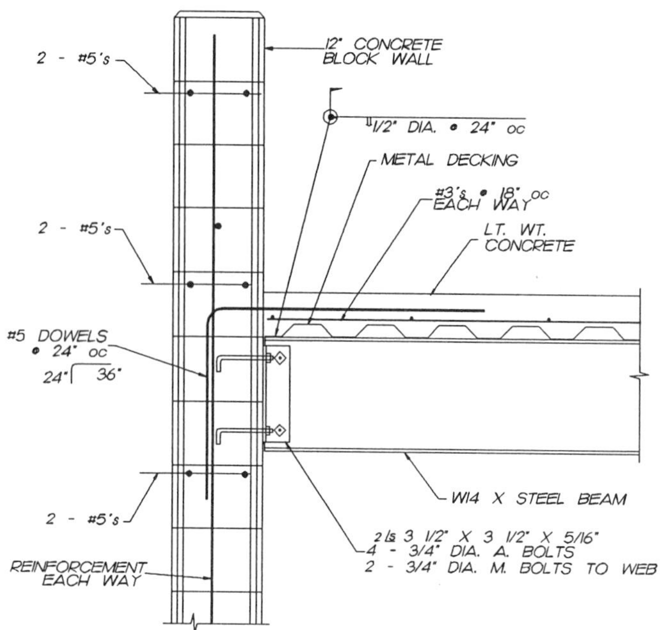
2 - #5's
12" CONCRETE BLOCK WALL
1/2" DIA. @ 24" oc
METAL DECKING
#3's @ 18" oc EACH WAY
2 - #5's
LT. WT. CONCRETE
#5 DOWELS @ 24" oc
24" 36"
W14 X STEEL BEAM
2 - #5's
2 Ls 3 1/2" X 3 1/2" X 5/16"
4 - 3/4" DIA. A. BOLTS
2 - 3/4" DIA. M. BOLTS TO WEB
REINFORCEMENT EACH WAY

M6028

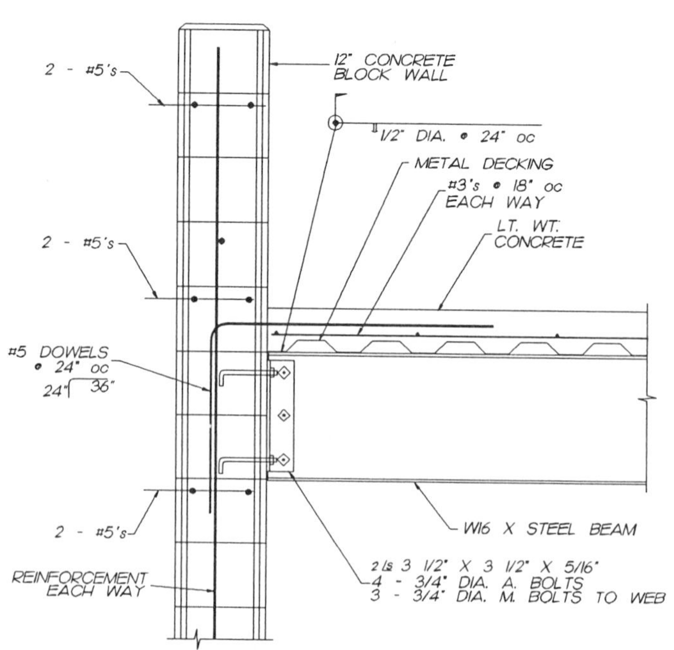
2 - #5's
12" CONCRETE BLOCK WALL
1/2" DIA. @ 24" oc
METAL DECKING
#3's @ 18" oc EACH WAY
LT. WT. CONCRETE
2 - #5's
#5 DOWELS @ 24" oc
24" 36"
W16 X STEEL BEAM
2 - #5's
2 Ls 3 1/2" X 3 1/2" X 5/16"
4 - 3/4" DIA. A. BOLTS
3 - 3/4" DIA. M. BOLTS TO WEB
REINFORCEMENT EACH WAY

M6029

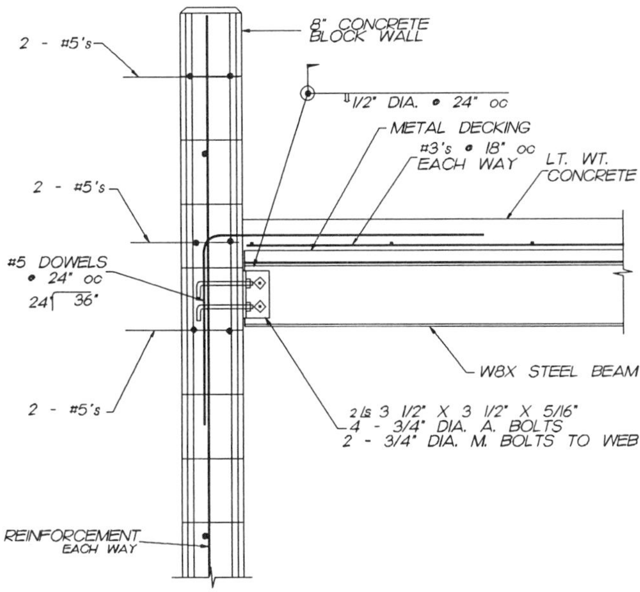
8" CONCRETE BLOCK WALL
2 - #5's
1/2" DIA. @ 24" oc
METAL DECKING
#3's @ 18" oc EACH WAY
LT. WT. CONCRETE
2 - #5's
#5 DOWELS @ 24" oc
24 36"
W8X STEEL BEAM
2 - #5's
2 Ls 3 1/2" X 3 1/2" X 5/16"
4 - 3/4" DIA. A. BOLTS
2 - 3/4" DIA. M. BOLTS TO WEB
REINFORCEMENT EACH WAY

M6040

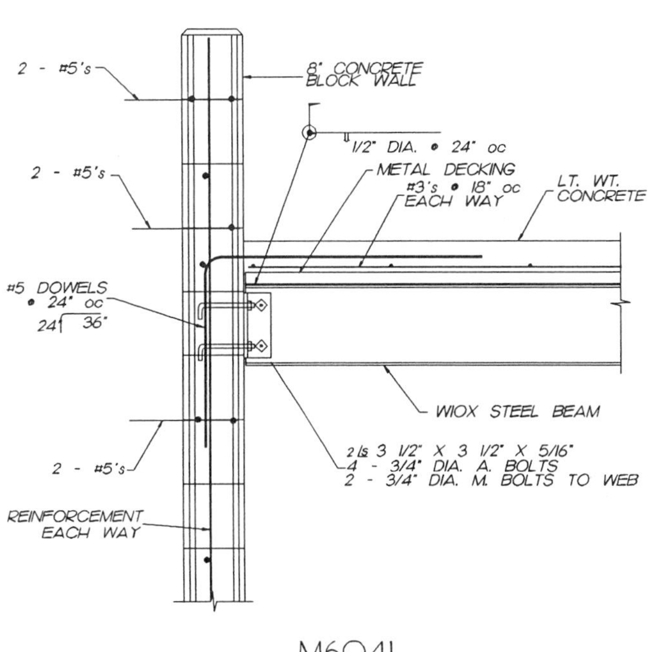
8" CONCRETE BLOCK WALL
2 - #5's
1/2" DIA. @ 24" oc
METAL DECKING
#3's @ 18" oc EACH WAY
LT. WT. CONCRETE
2 - #5's
#5 DOWELS @ 24" oc
24 36"
W10X STEEL BEAM
2 - #5's
2 Ls 3 1/2" X 3 1/2" X 5/16"
4 - 3/4" DIA. A. BOLTS
2 - 3/4" DIA. M. BOLTS TO WEB
REINFORCEMENT EACH WAY

M6041

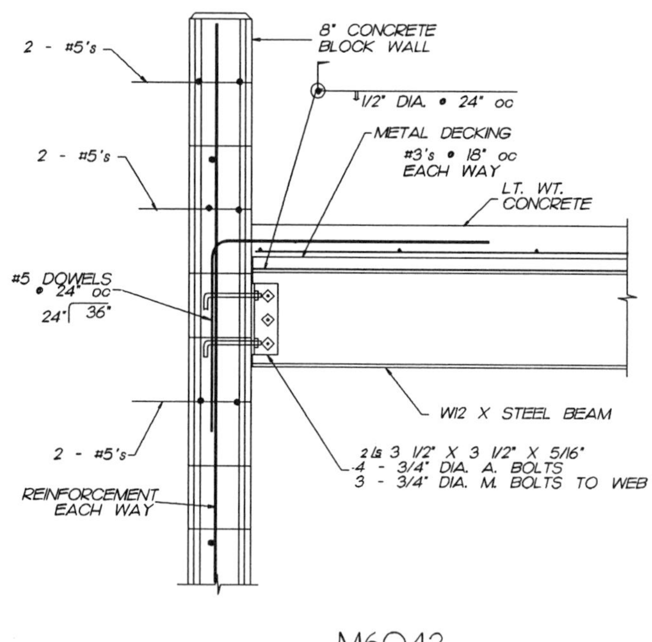

M6042

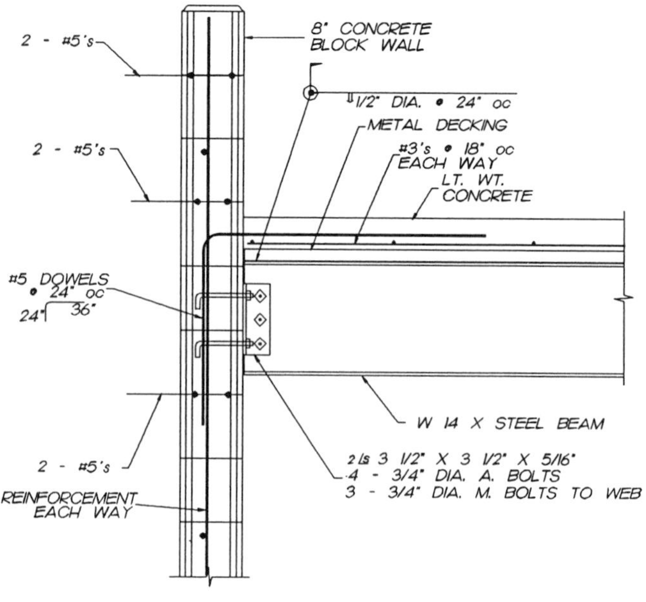

M6043

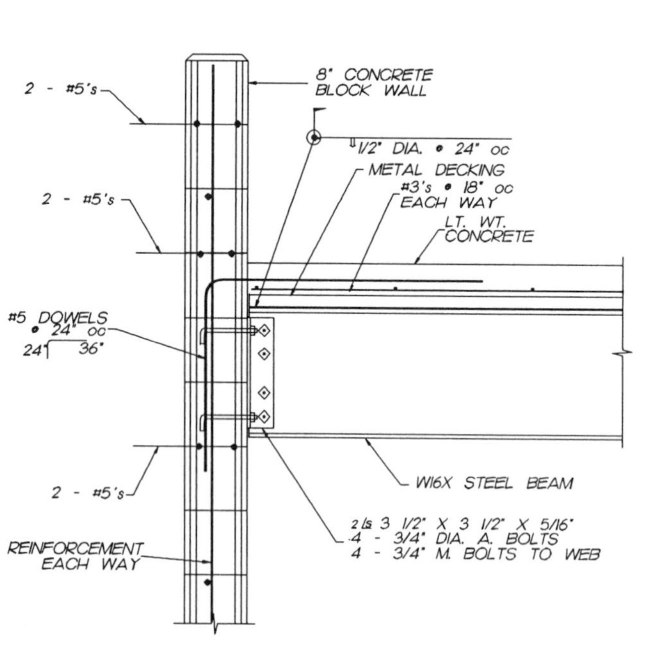

M6044

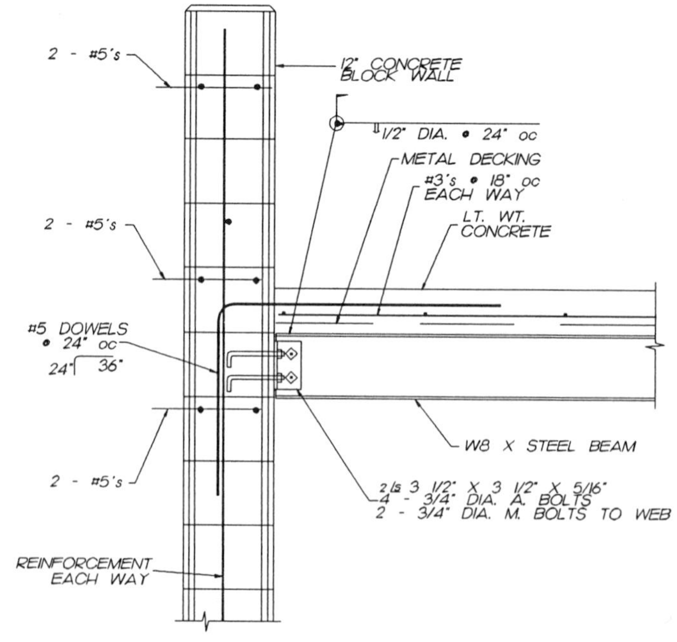

M6045

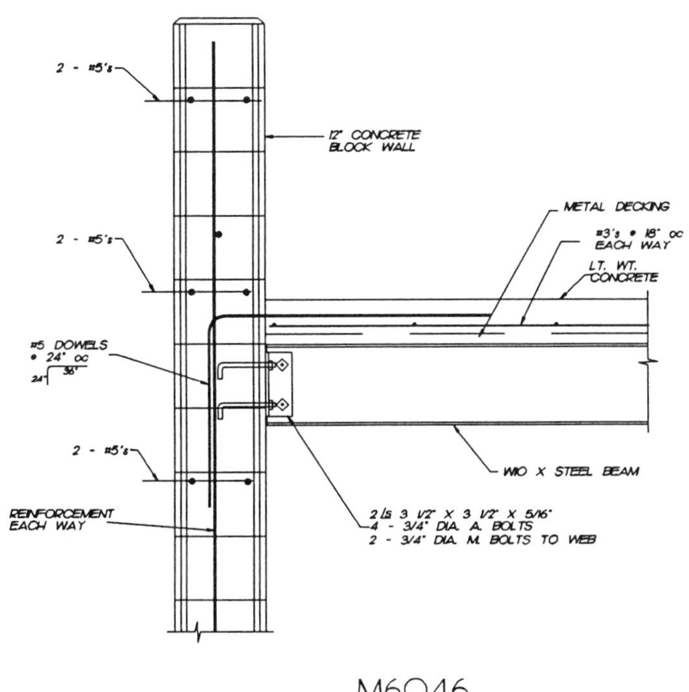

M6046

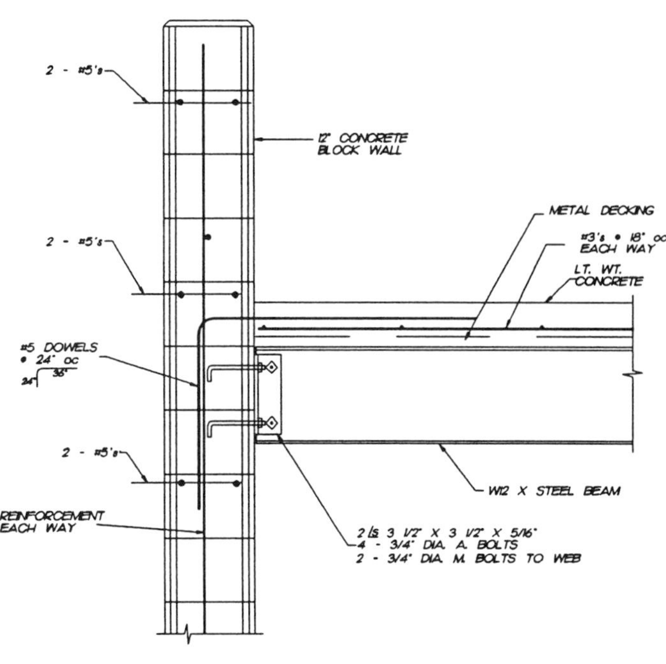

M6047

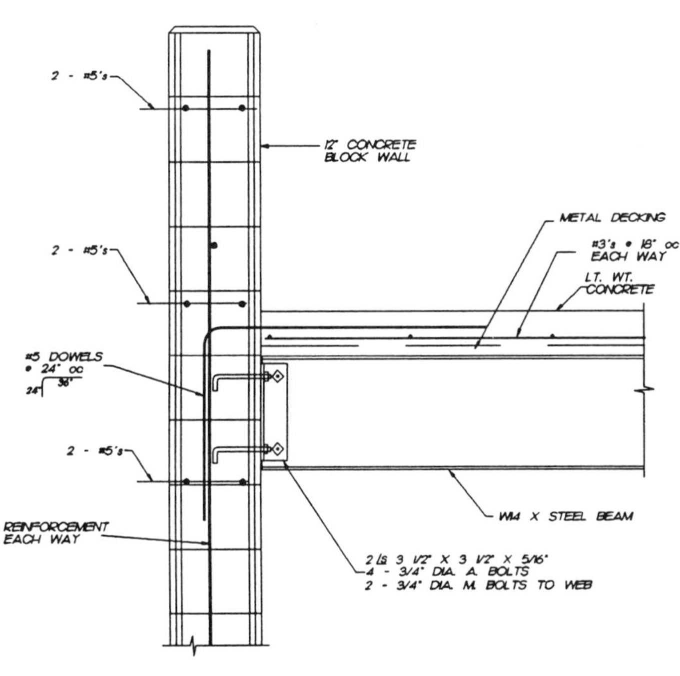

M6048

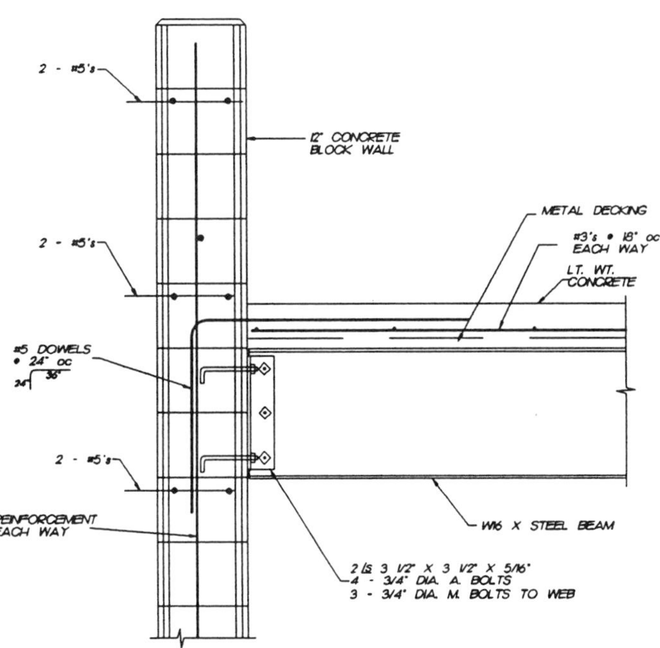

M6049

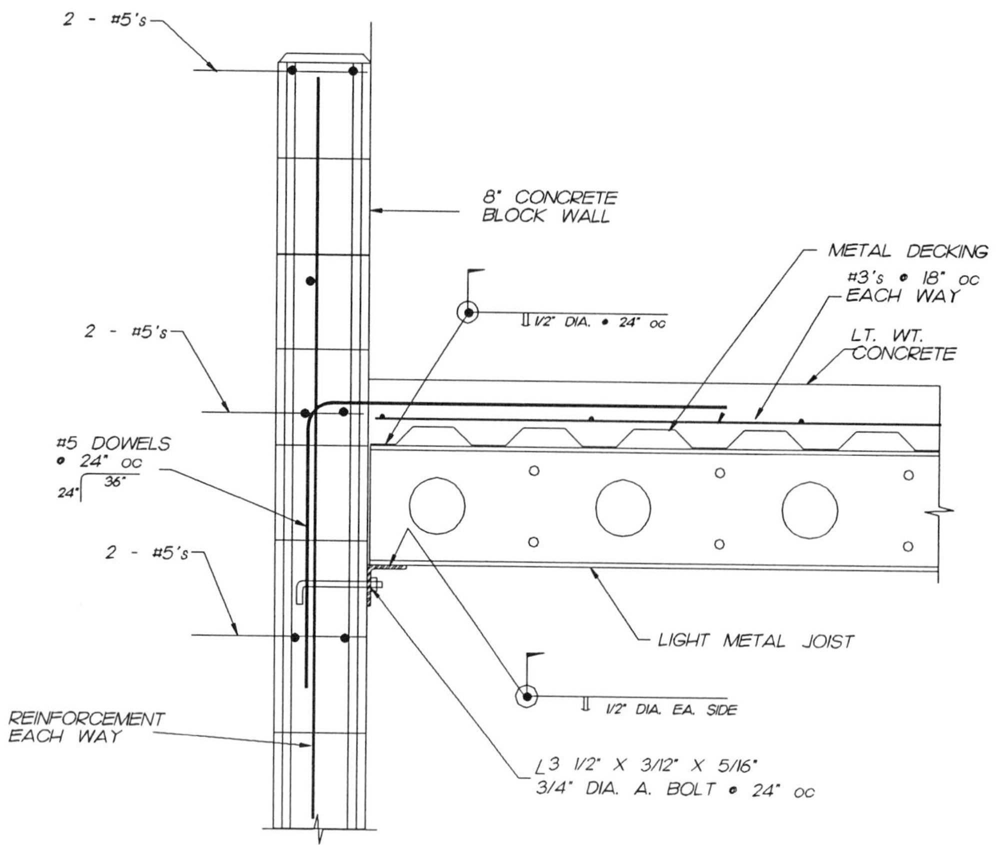

M6060

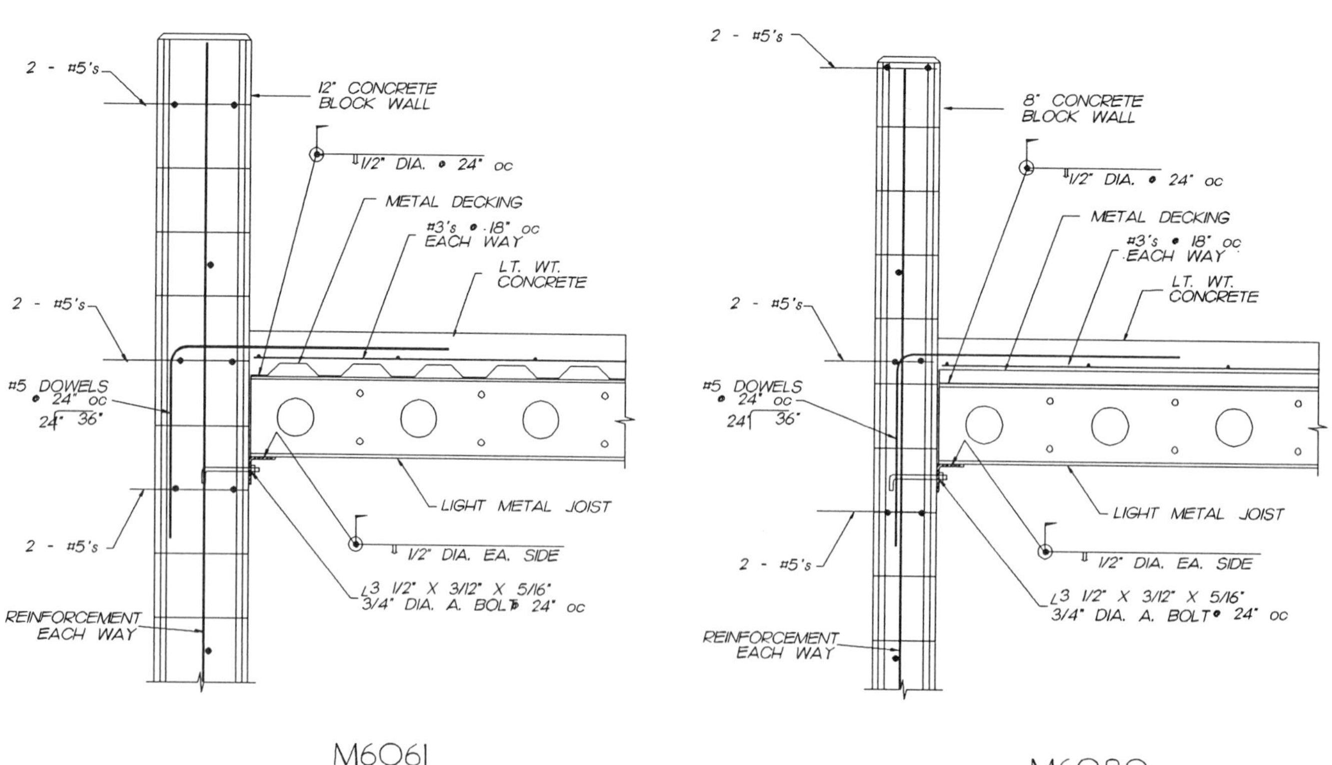

M6061

M6080

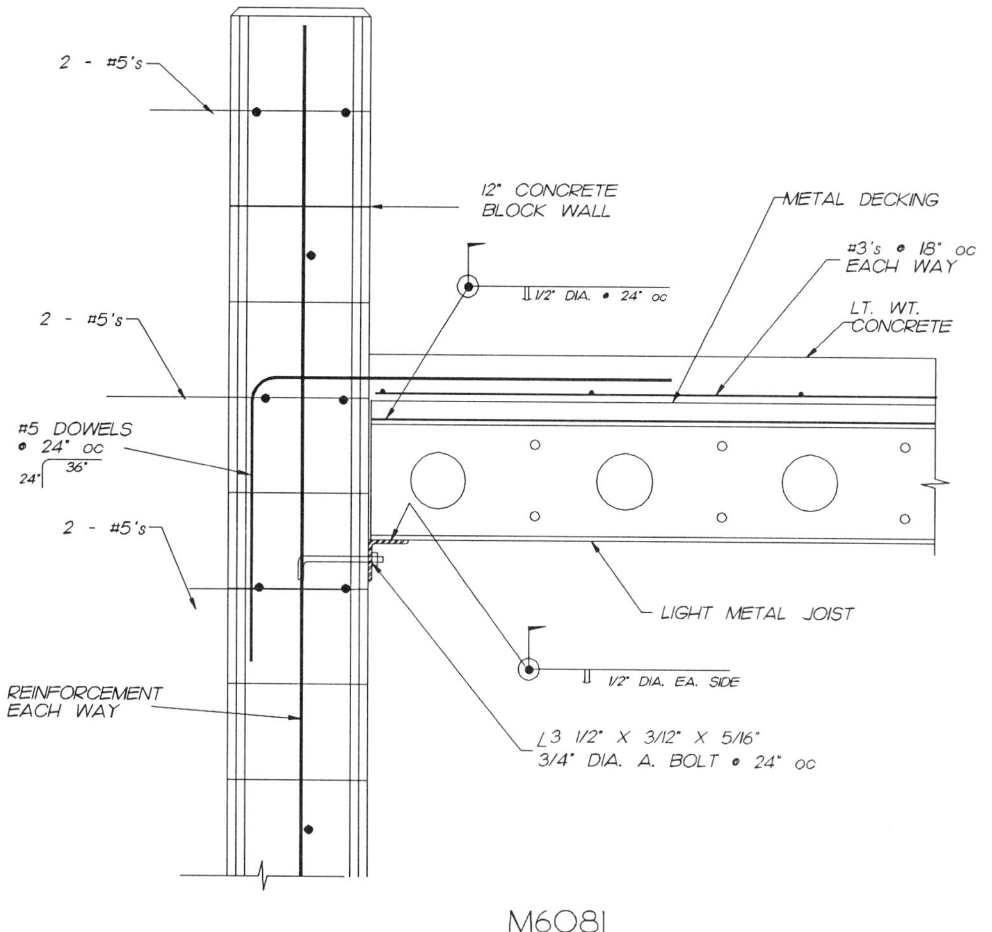

M6081

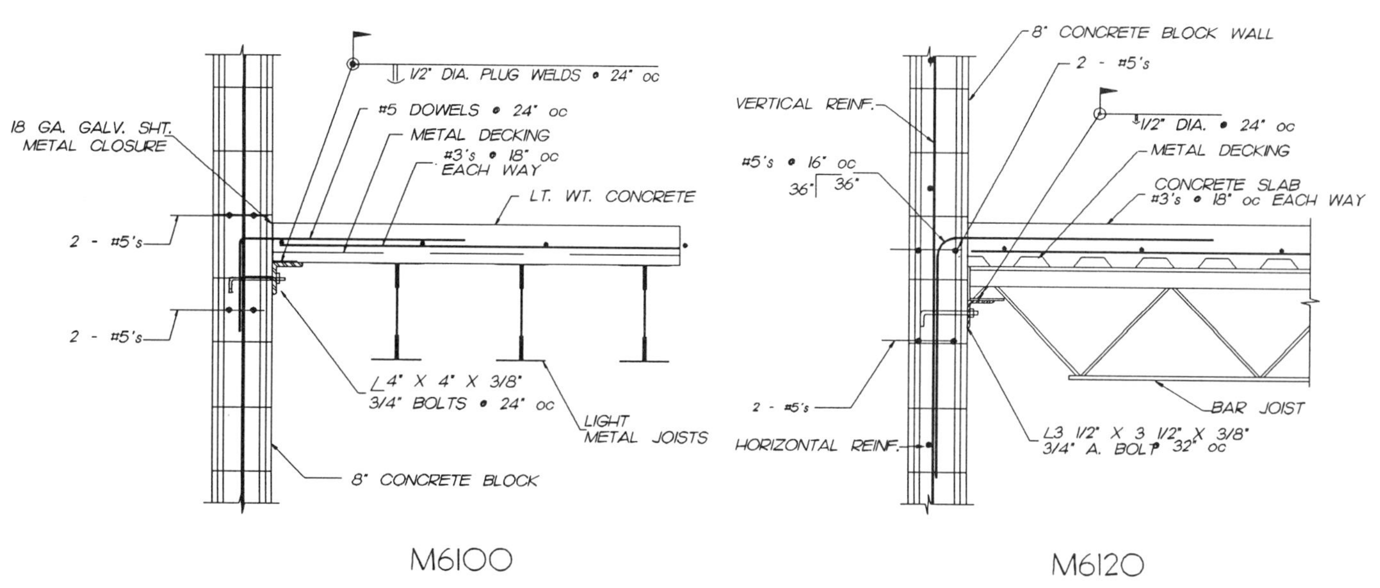

M6100

M6120

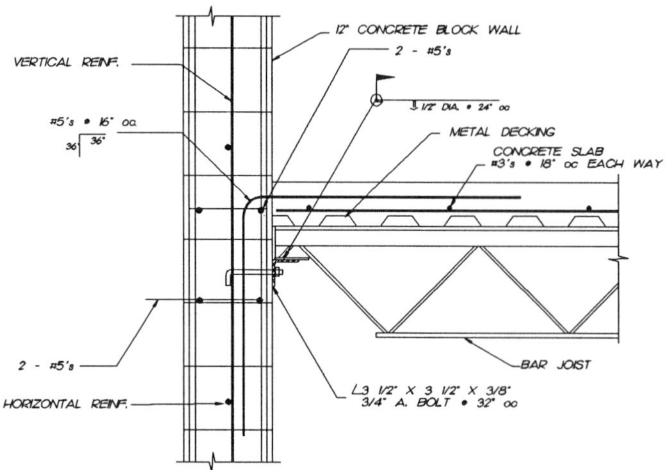

M6121

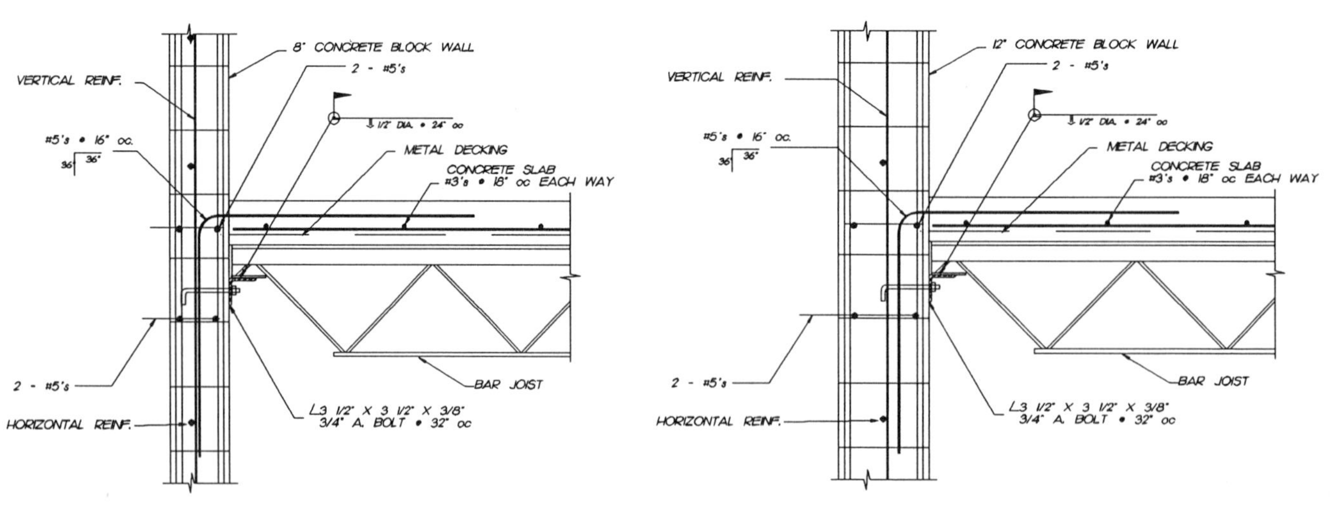

M6140

M6141

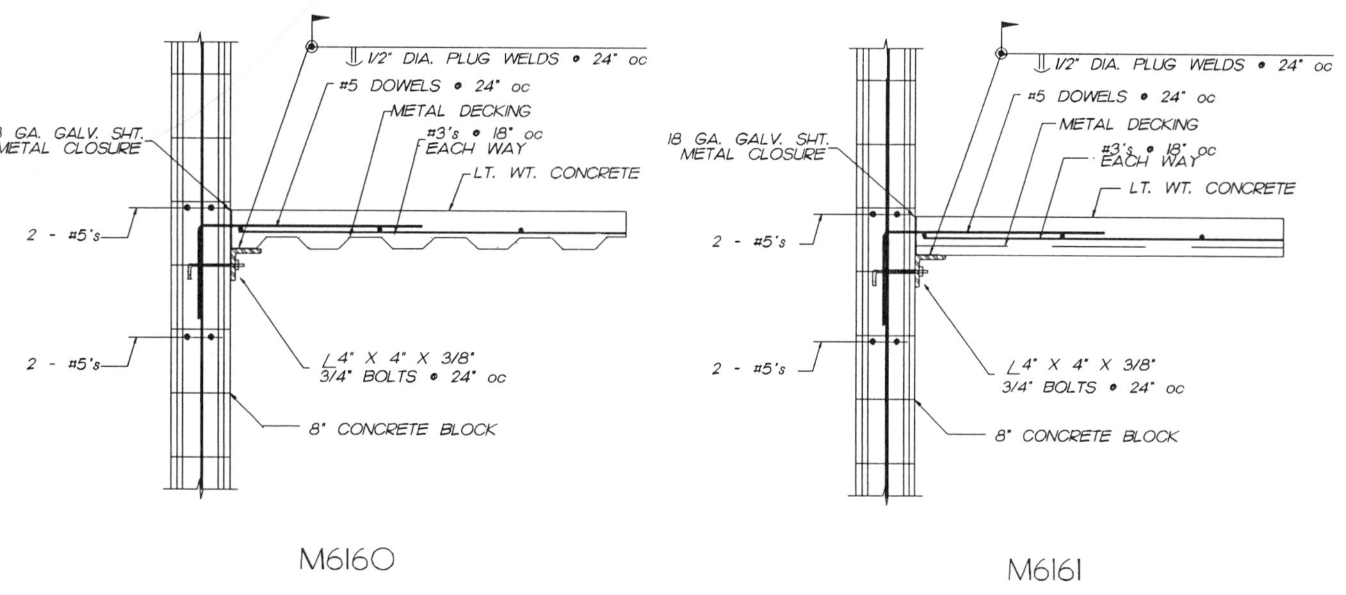
1/2" DIA. PLUG WELDS @ 24" oc
#5 DOWELS @ 24" oc
METAL DECKING
#3's @ 18" oc
EACH WAY
LT. WT. CONCRETE
18 GA. GALV. SHT.
METAL CLOSURE
2 - #5's
2 - #5's
∠4" X 4" X 3/8"
3/4" BOLTS @ 24" oc
8" CONCRETE BLOCK
M6160
1/2" DIA. PLUG WELDS @ 24" oc
#5 DOWELS @ 24" oc
METAL DECKING
#3's @ 18" oc
EACH WAY
LT. WT. CONCRETE
18 GA. GALV. SHT.
METAL CLOSURE
2 - #5's
2 - #5's
∠4" X 4" X 3/8"
3/4" BOLTS @ 24" oc
8" CONCRETE BLOCK
M6161

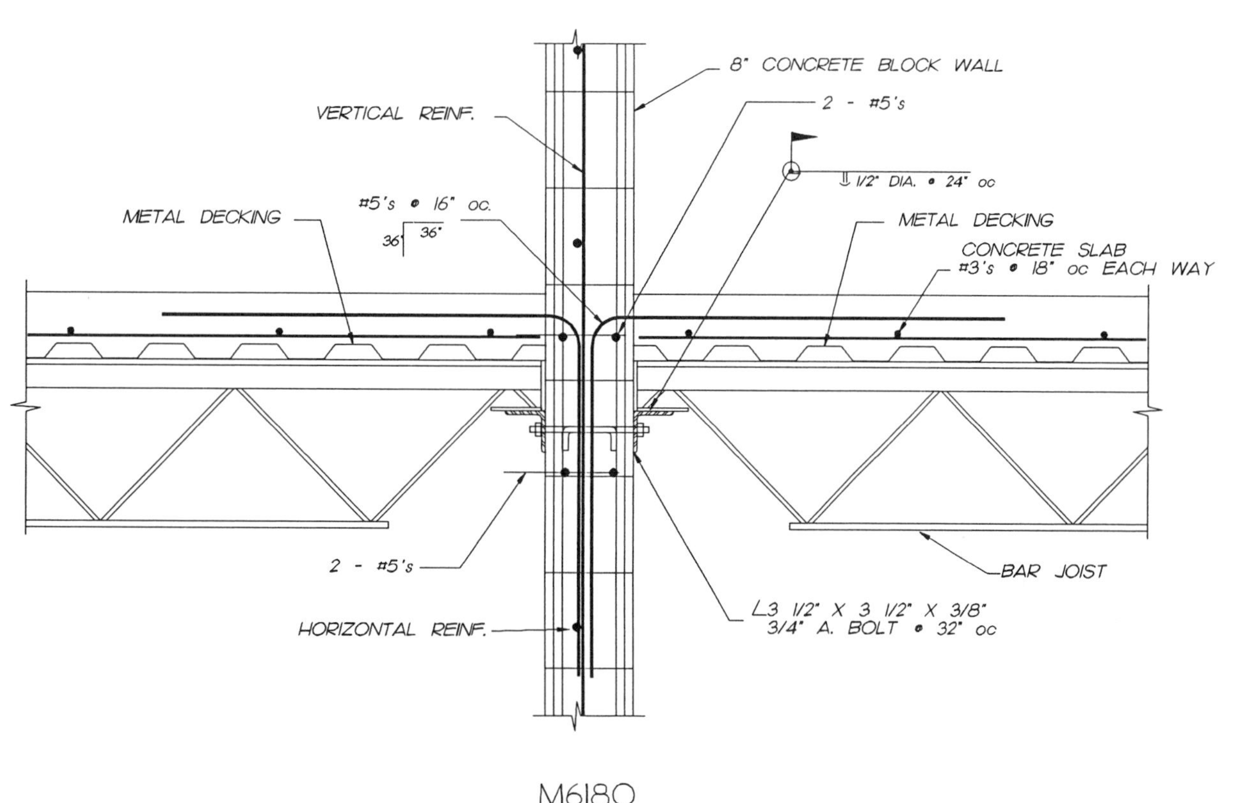
8" CONCRETE BLOCK WALL
2 - #5's
VERTICAL REINF.
1/2" DIA. @ 24" oc
METAL DECKING
#5's @ 16" oc.
36"
36"
METAL DECKING
CONCRETE SLAB
#3's @ 18" oc EACH WAY
2 - #5's
BAR JOIST
∠3 1/2" X 3 1/2" X 3/8"
3/4" A. BOLT @ 32" oc
HORIZONTAL REINF.
M6180

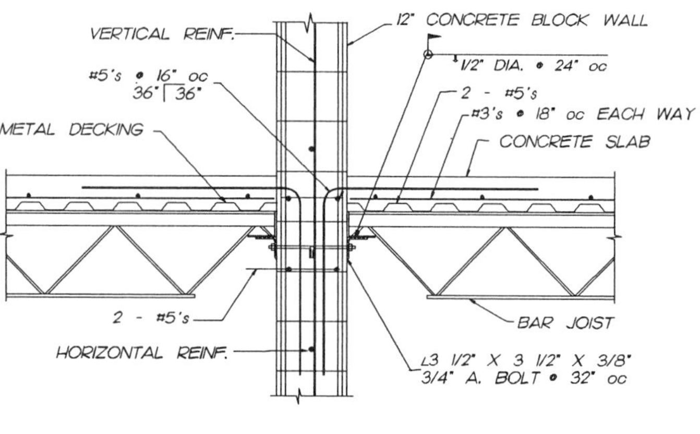

M6181

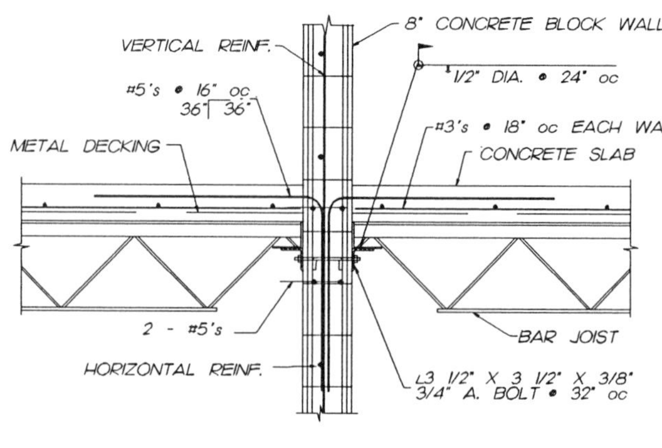

M6200

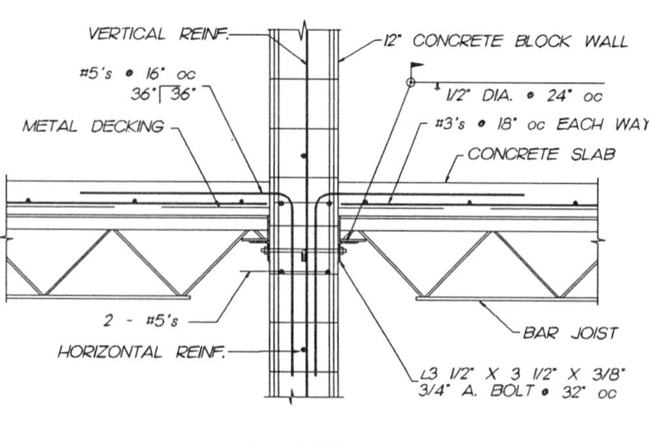

M6201

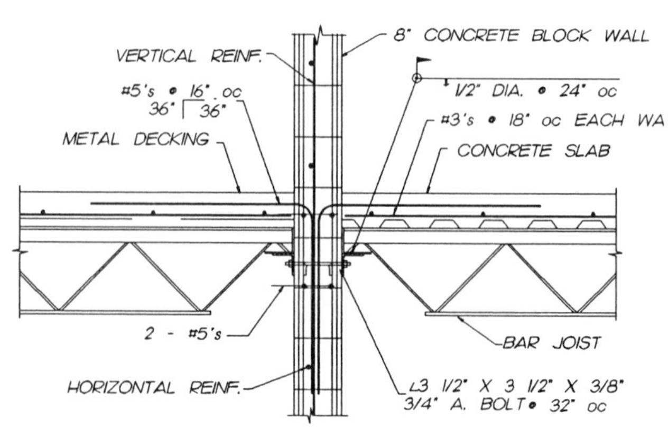

M6220

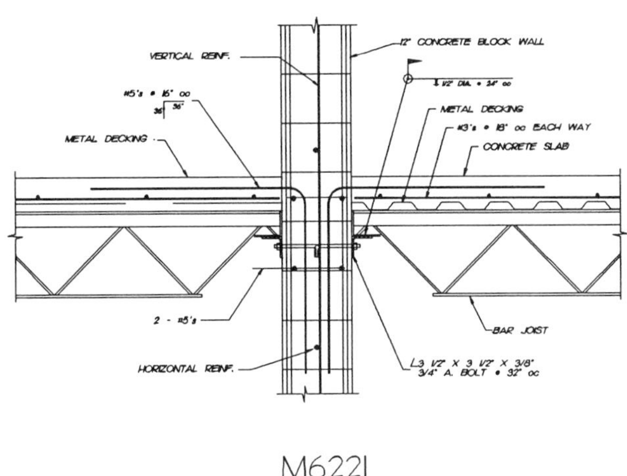

M6221

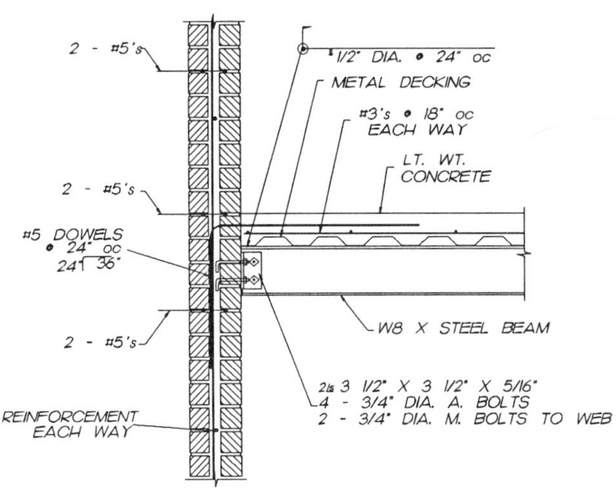

M6240

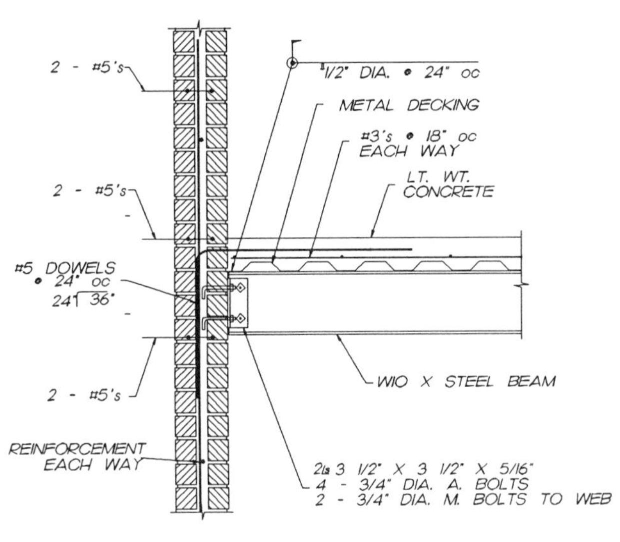

M6241

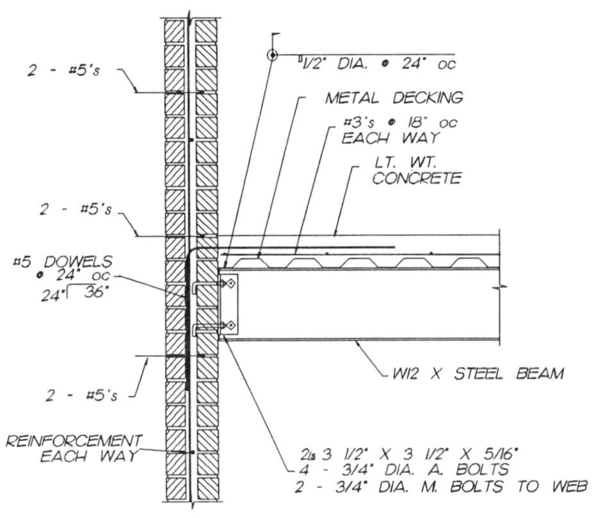

M6242

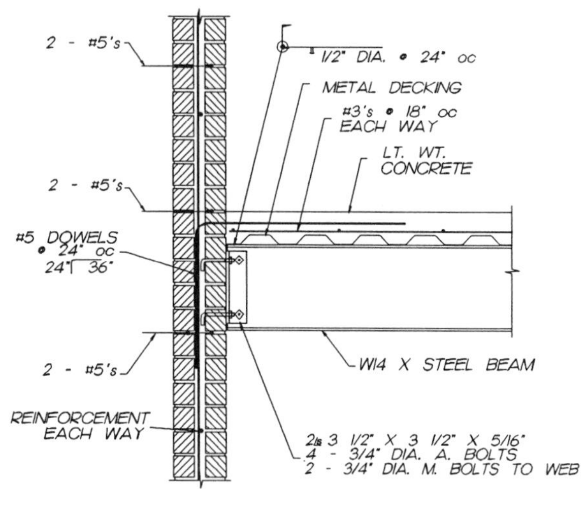

M6243

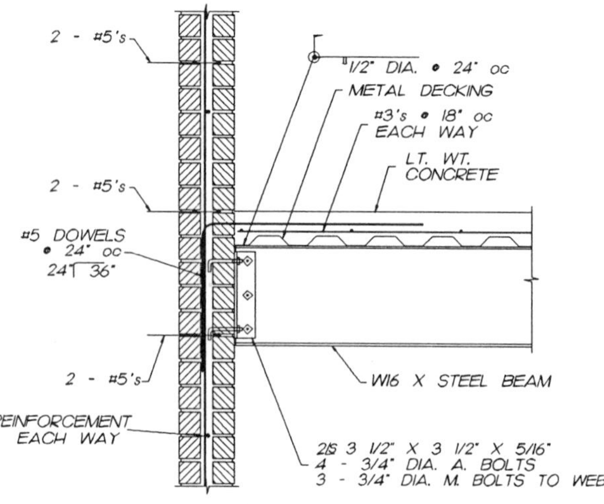

M6244

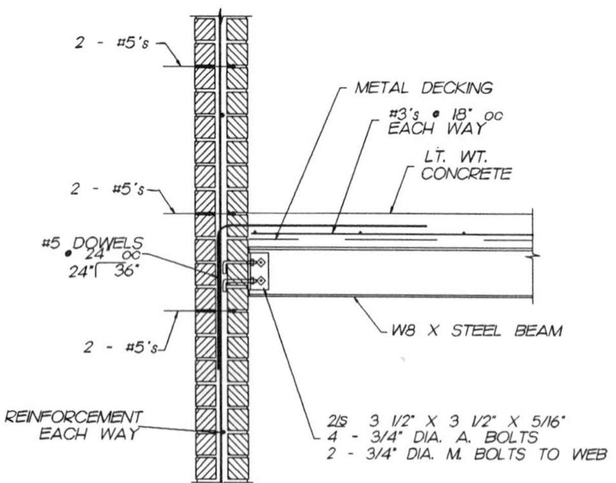

M6260

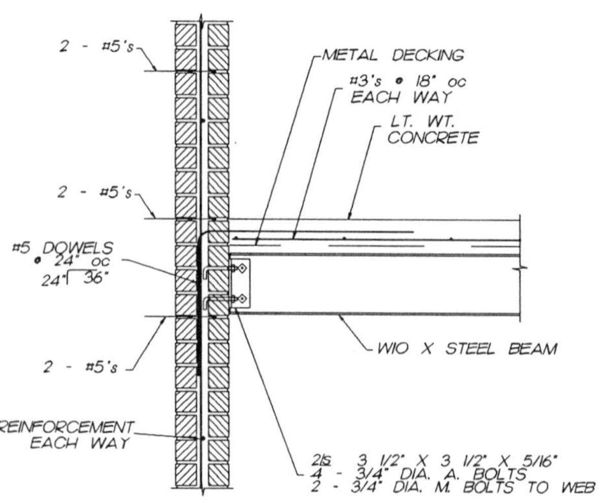

M6261

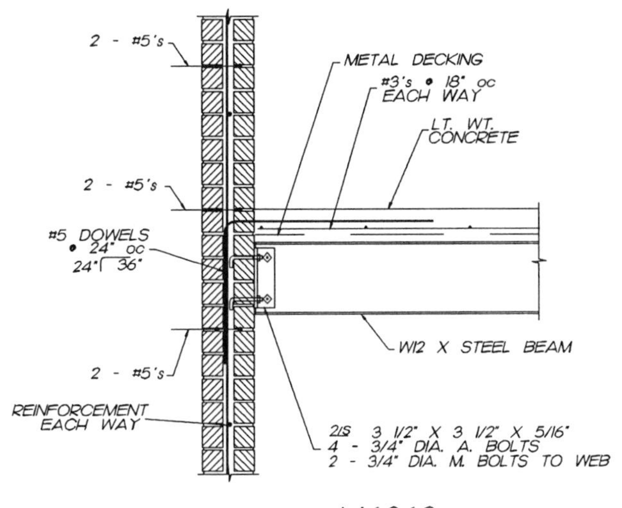

M6262

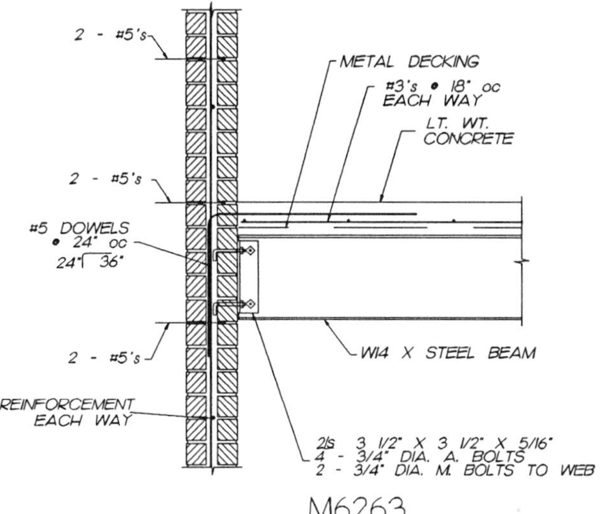

M6263

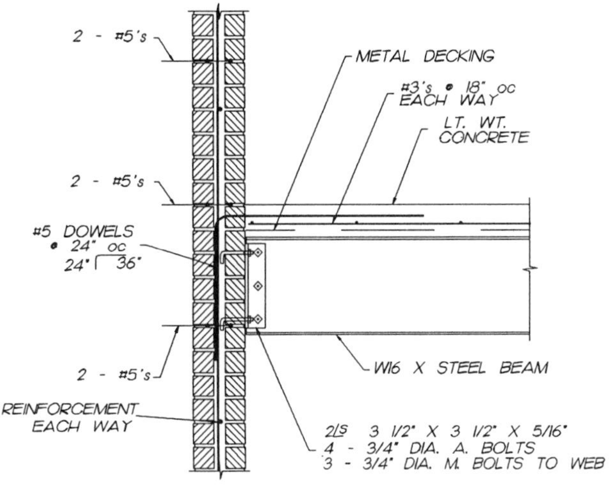

M6264

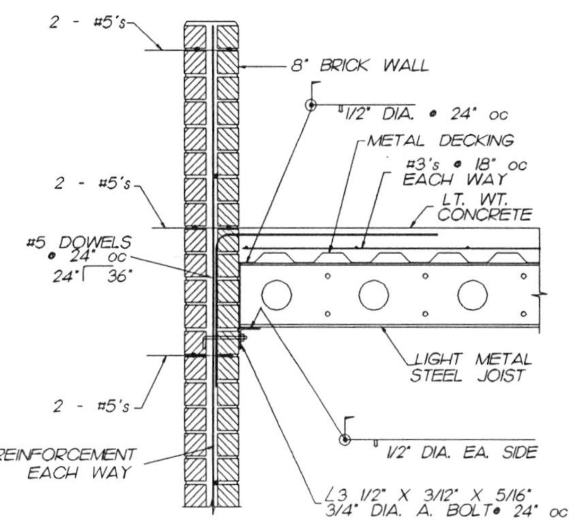

M6280

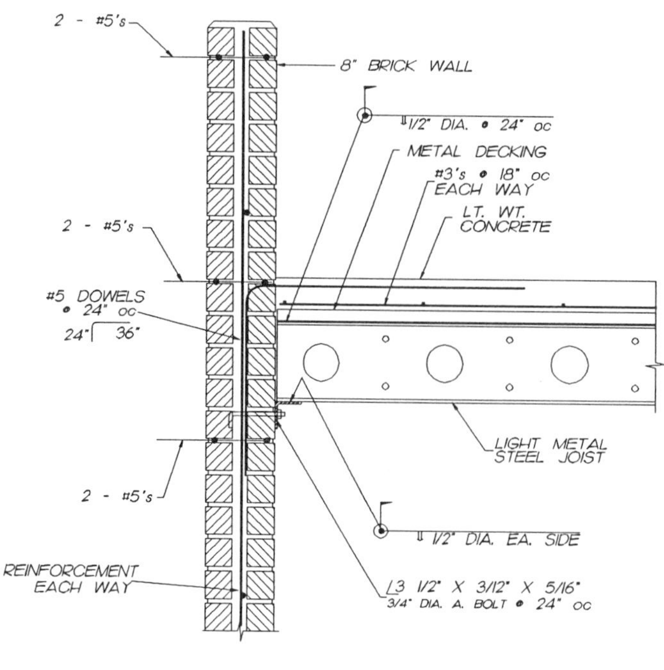

M6300

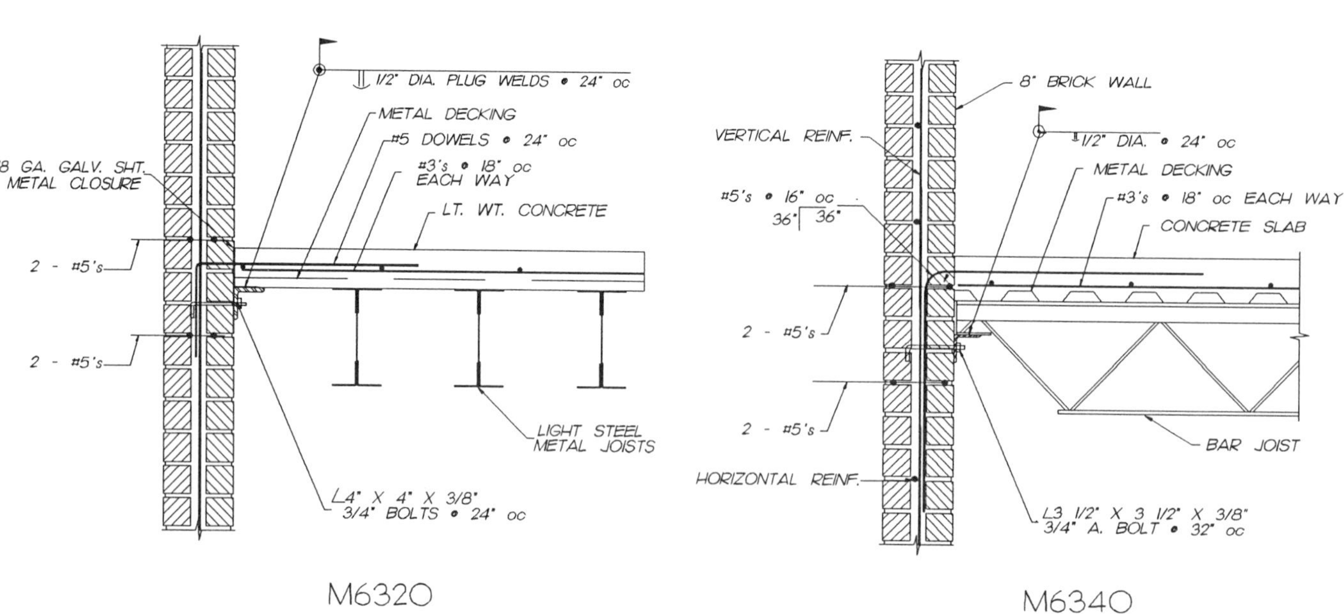

M6320

M6340

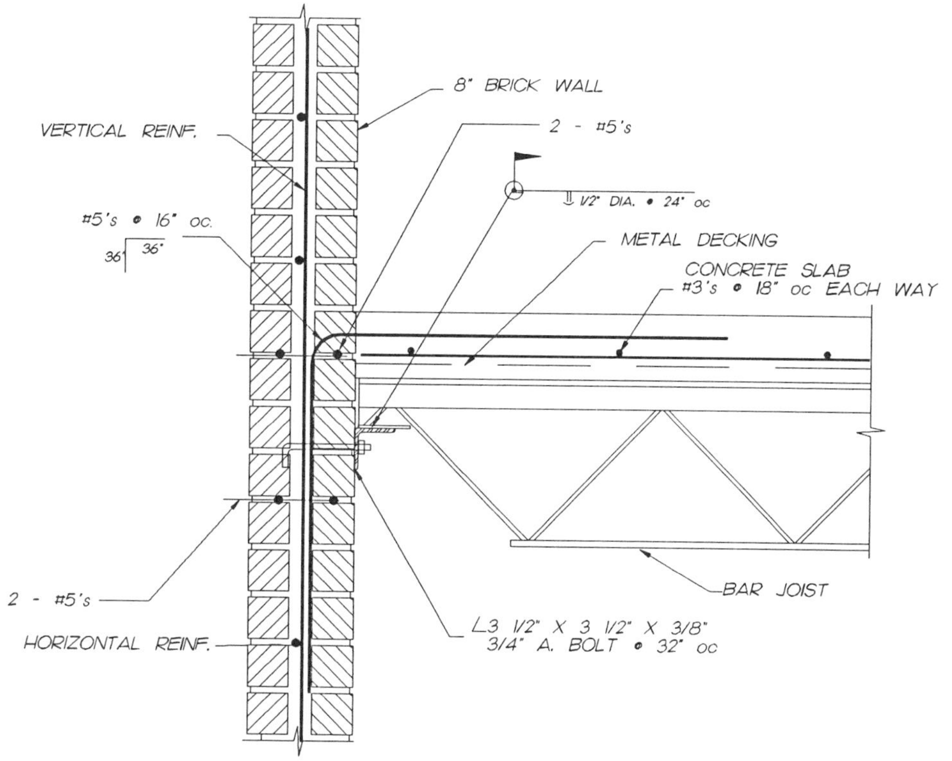

M6360

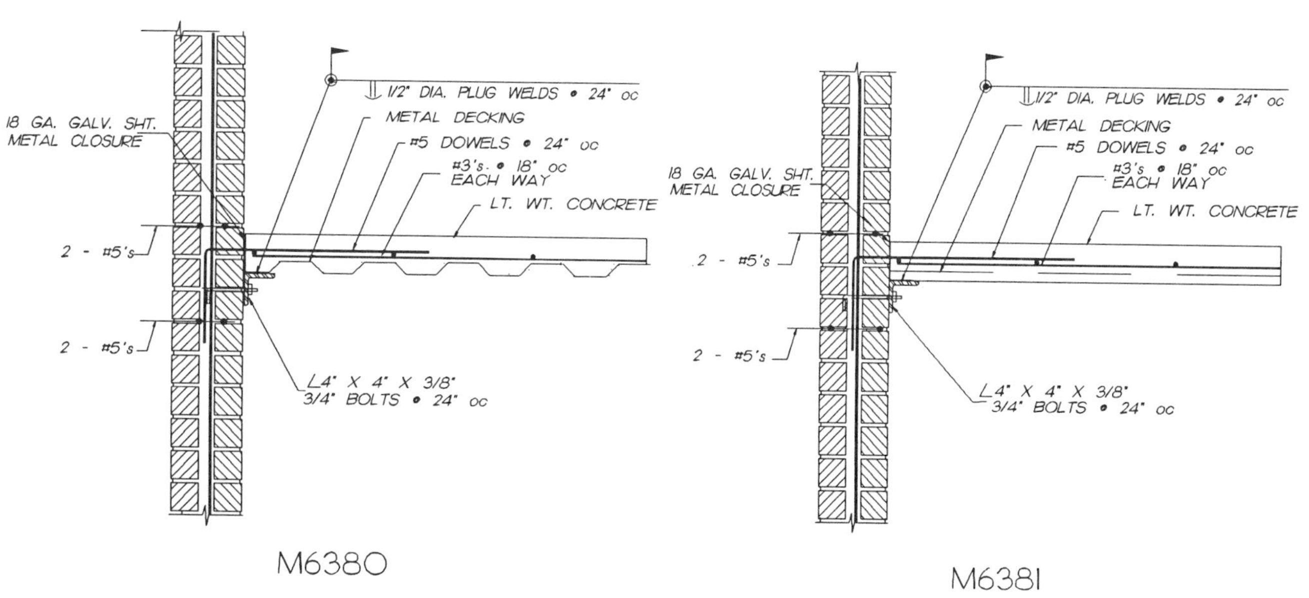

M6380

M6381

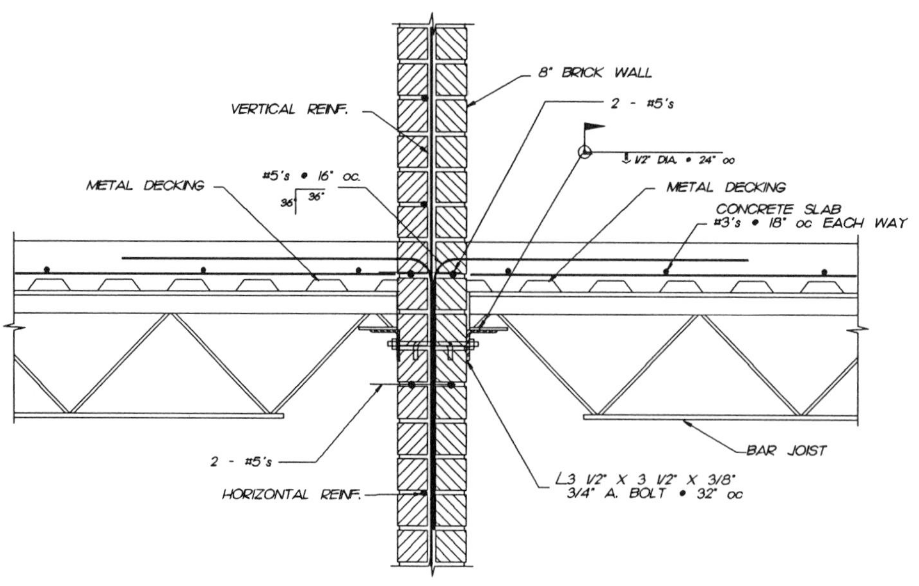

M6400

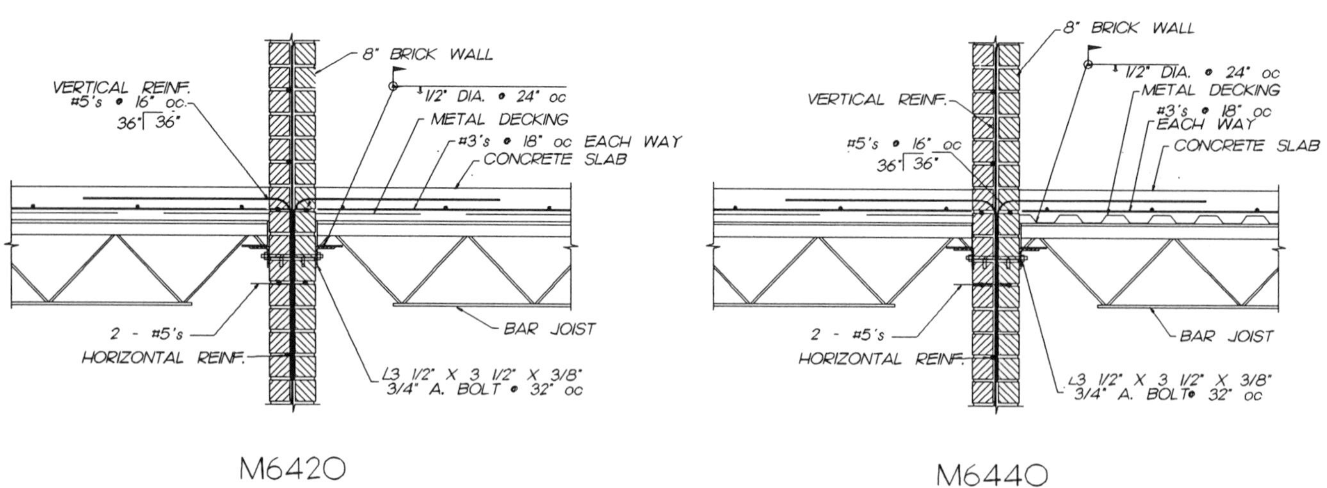

M6420

M6440

Description	Drawings
RETAINING WALL SOIL IS NOT OVER THE FOOTING RETAINED SLOPE - FLAT STEM HEIGHTS - 4'-0" to 8'-0"	M7020 - M7023
RETAINING WALL SOIL IS NOT OVER THE FOOTING RETAINED SLOPE - 2 HORIZ. to 1 VERT. STEM HEIGHTS - 4'-0" to 8'-0"	M7040 - M7043
RETAINING WALL SOIL IS NOT OVER THE FOOTING RETAINED SLOPE - 1 HORIZ. to 1 VERT. STEM HEIGHTS - 4'-0" to 5'-4"	M7060 - M7061
RETAINING WALL SOIL IS OVER THE FOOTING RETAINED SLOPE - FLAT STEM HEIGHTS - 4'-0" to 8'-0"	M7080 - M7083
RETAINING WALL SOIL IS OVER THE FOOTING RETAINED SLOPE - 2 HORIZ. to 1 VERT. STEM HEIGHTS - 4'-0" to 6'-0"	M7100 - M7103
RETAINING WALL SOIL IS OVER THE FOOTING RETAINED SLOPE - 1 HORIZ. to 1 VERT. STEM HEIGHTS - 4'-0" to 4'-8"	M7120 - M7121

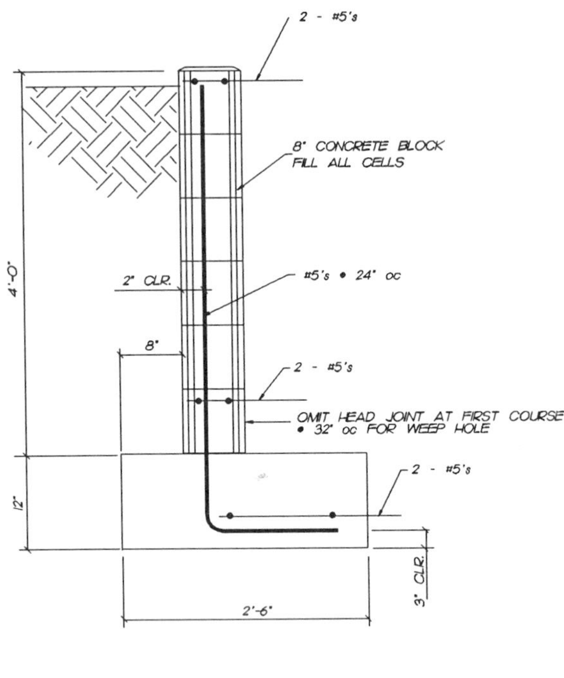
2 - #5's
8" CONCRETE BLOCK
FILL ALL CELLS
#5's • 24" oc
2" CLR.
4'-0"
8"
2 - #5's
OMIT HEAD JOINT AT FIRST COURSE
• 32" oc FOR WEEP HOLE
2 - #5's
12"
3" CLR.
2'-6"

M7020

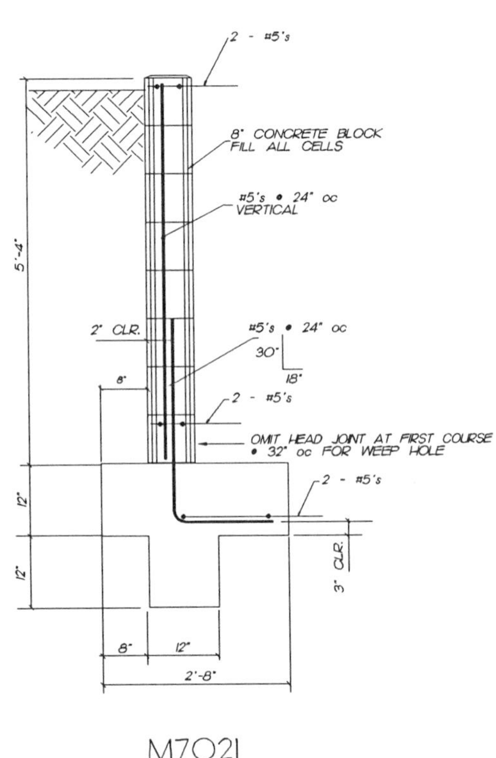
2 - #5's
8" CONCRETE BLOCK
FILL ALL CELLS
#5's • 24" oc
VERTICAL
5'-4"
2" CLR.
#5's • 24" oc
30"
18"
8"
2 - #5's
OMIT HEAD JOINT AT FIRST COURSE
• 32" oc FOR WEEP HOLE
2 - #5's
12"
12"
3" CLR.
8"
12"
2'-8"

M7021

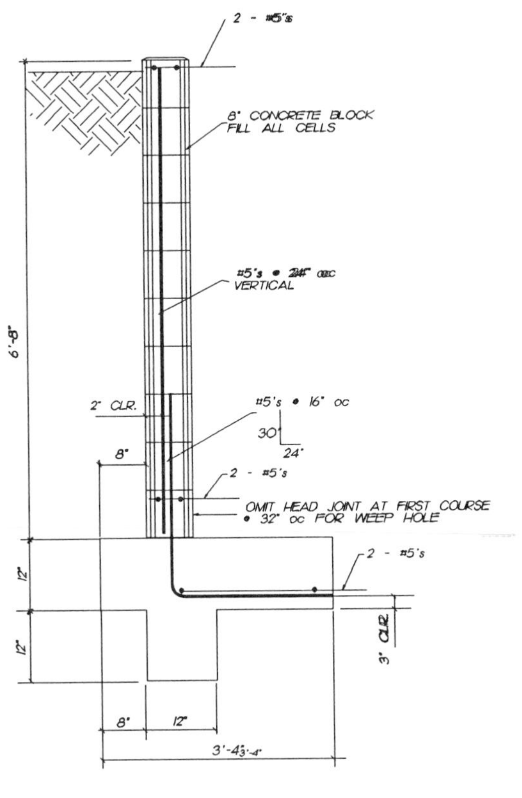
2 - #5's
8" CONCRETE BLOCK
FILL ALL CELLS
#5's @ 24" oc
VERTICAL
6'-8"
2" CLR.
#5's @ 16" oc
30"
24"
8"
2 - #5's
OMIT HEAD JOINT AT FIRST COURSE
@ 32" oc FOR WEEP HOLE
2 - #5's
12"
12"
3" CLR.
8"
12"
3'-4"

M7022

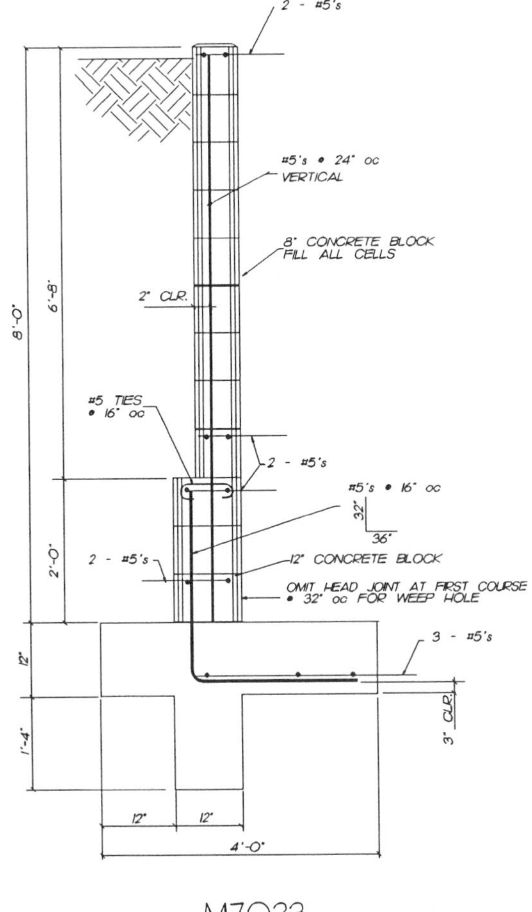
2 - #5's
#5's @ 24" oc
VERTICAL
8" CONCRETE BLOCK
FILL ALL CELLS
2" CLR.
8'-0"
6'-8"
#5 TIES
@ 16" oc
2 - #5's
#5's @ 16" oc
32"
36"
2 - #5's
12" CONCRETE BLOCK
2'-0"
OMIT HEAD JOINT AT FIRST COURSE
@ 32" oc FOR WEEP HOLE
3 - #5's
12"
1'-4"
3" CLR.
12"
12"
4'-0"

M7023

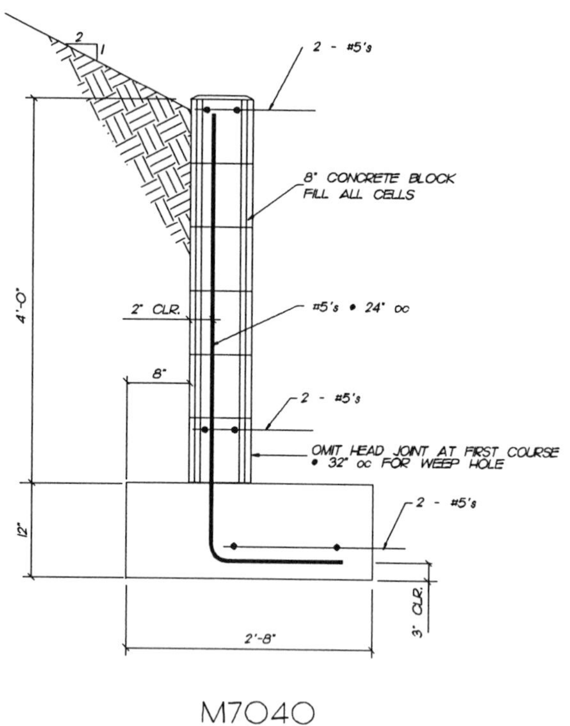

M7040

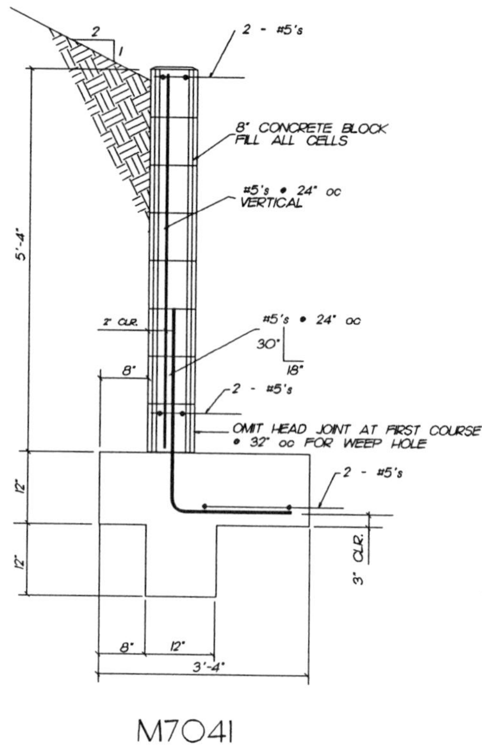

M7041

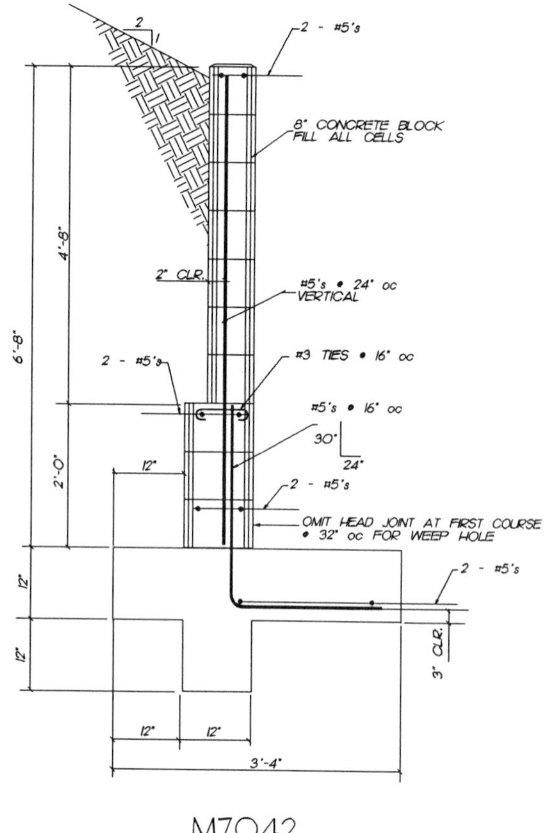

M7042

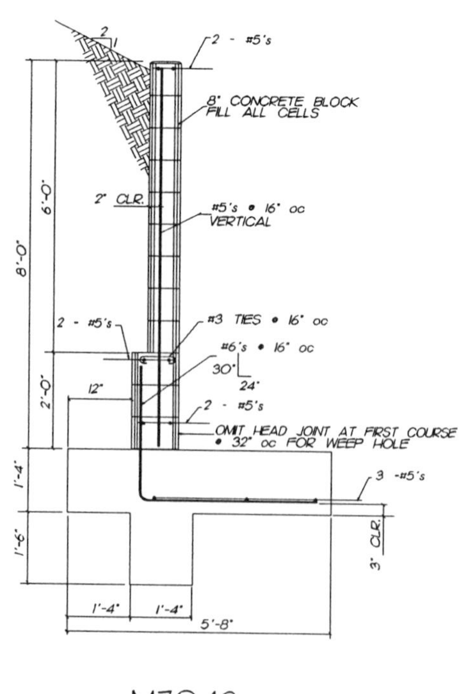

M7043

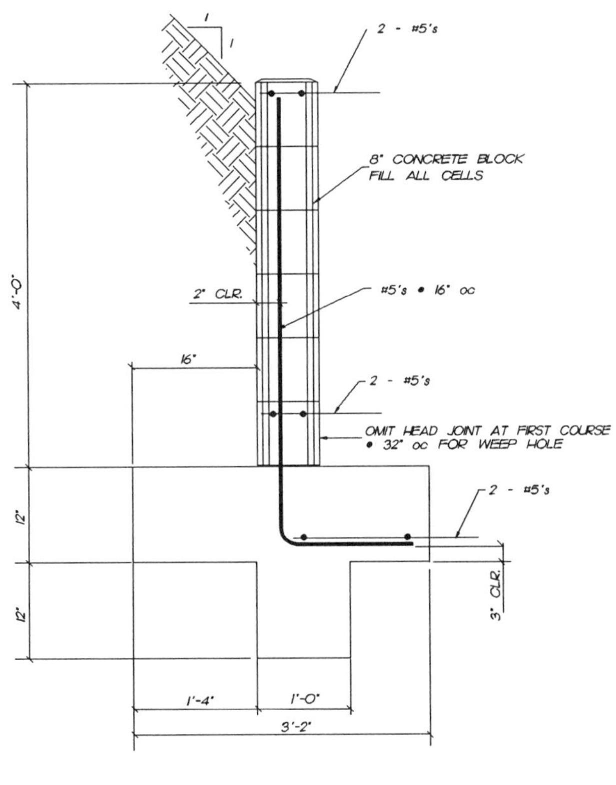

M7060

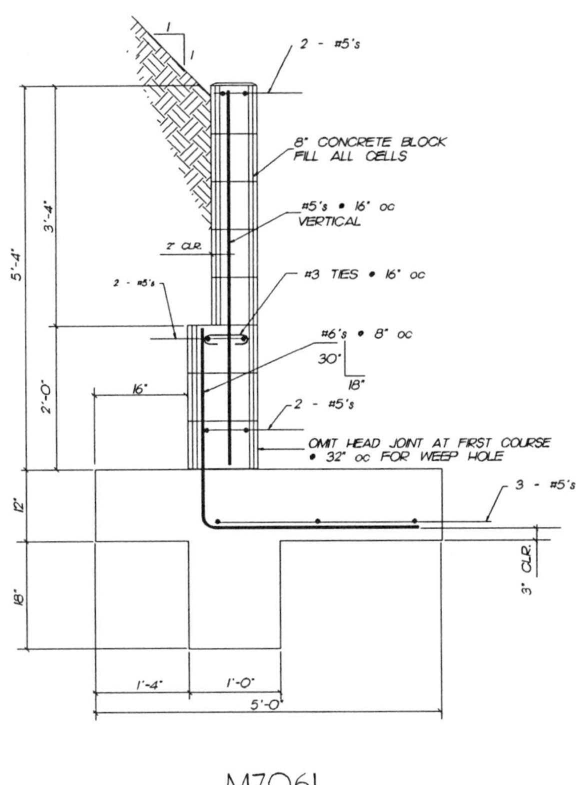

M7061

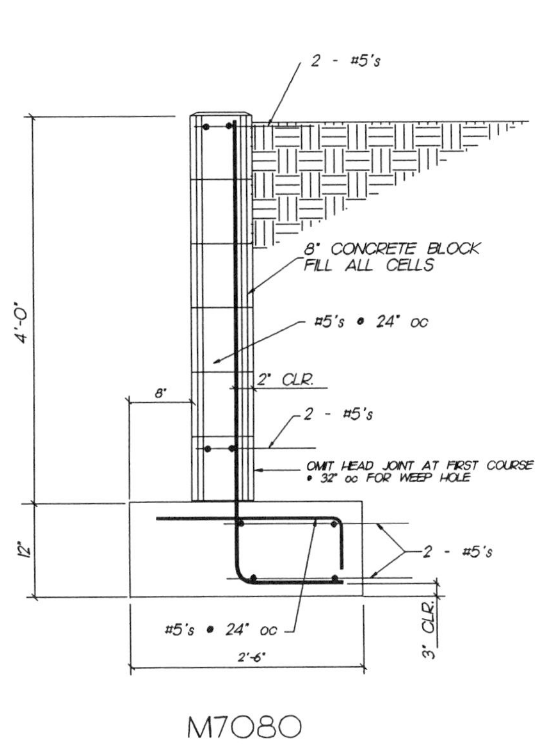

M7080

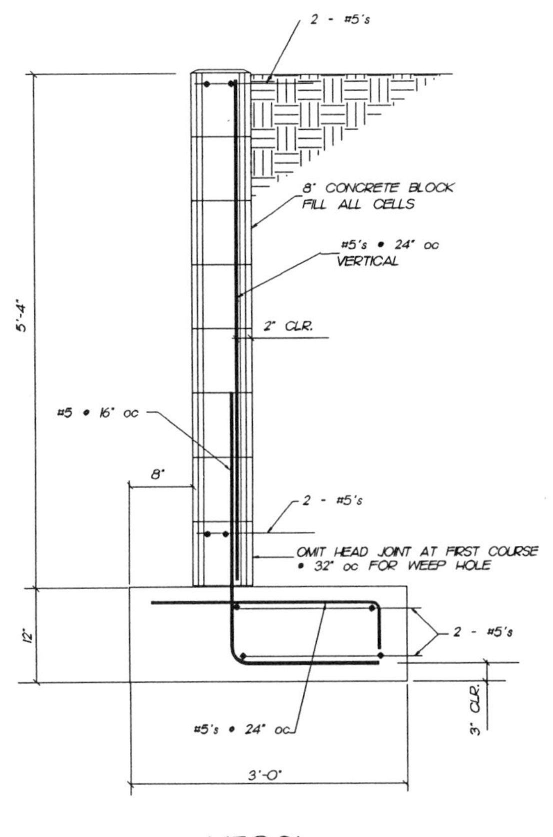

M7081

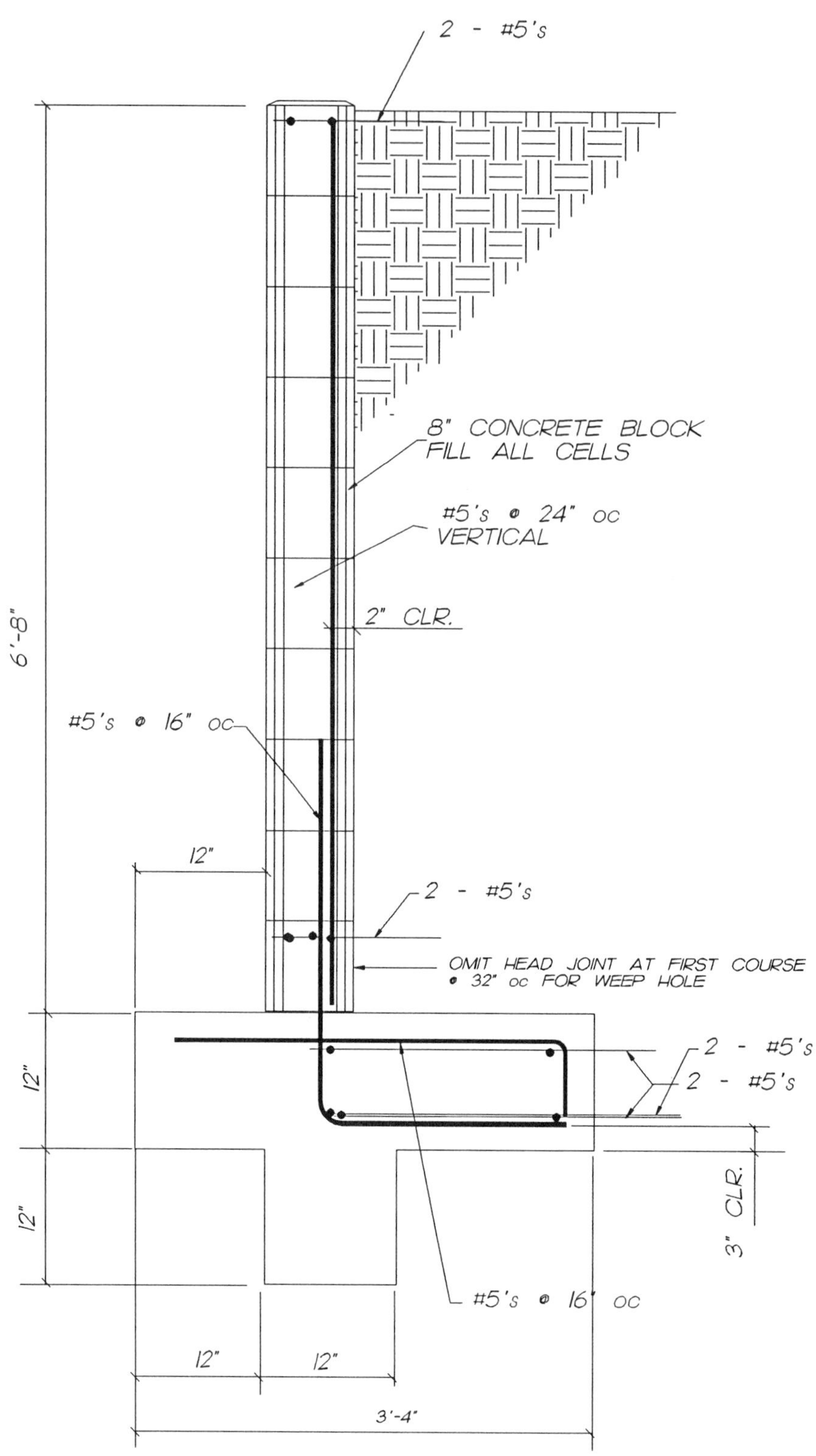
2 - #5's
8" CONCRETE BLOCK
FILL ALL CELLS
#5's @ 24" oc
VERTICAL
2" CLR.
6'-8"
#5's @ 16" oc
12"
2 - #5's
OMIT HEAD JOINT AT FIRST COURSE
@ 32" oc FOR WEEP HOLE
2 - #5's
2 - #5's
12"
12"
3" CLR.
#5's @ 16" oc
12"
12"
3'-4"

M7082

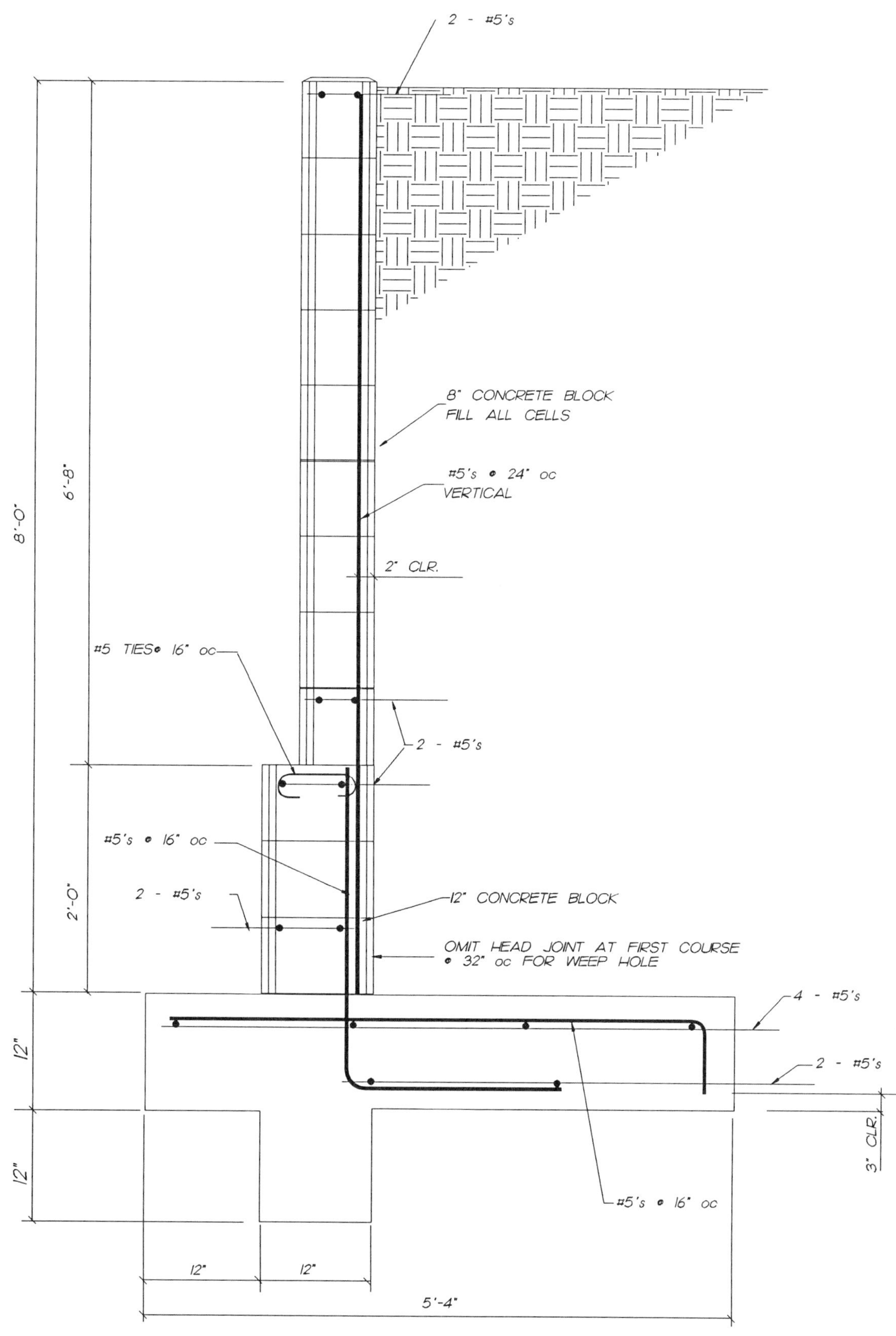
2 - #5's
8" CONCRETE BLOCK
FILL ALL CELLS
#5's @ 24" oc
VERTICAL
2" CLR.
#5 TIES @ 16" oc
2 - #5's
#5's @ 16" oc
2 - #5's
12" CONCRETE BLOCK
OMIT HEAD JOINT AT FIRST COURSE
@ 32" oc FOR WEEP HOLE
4 - #5's
2 - #5's
3" CLR.
#5's @ 16" oc
8'-0"
6'-8"
2'-0"
12"
12"
12"
12"
5'-4"

M7083

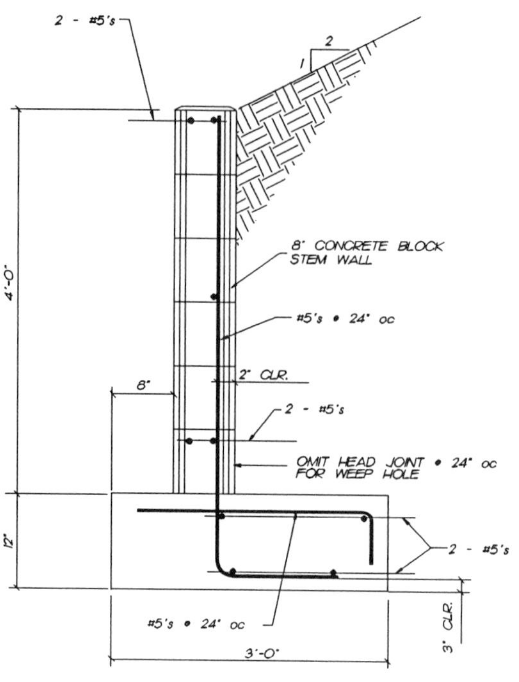

M7100

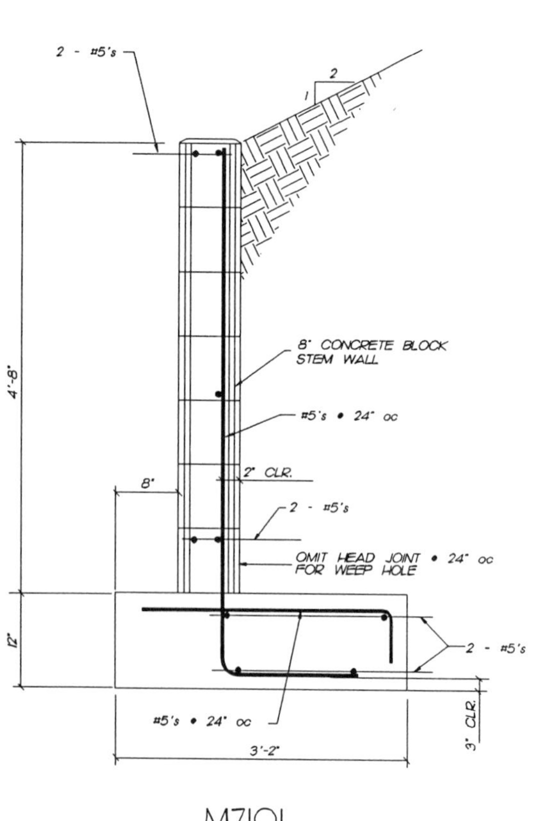

M7101

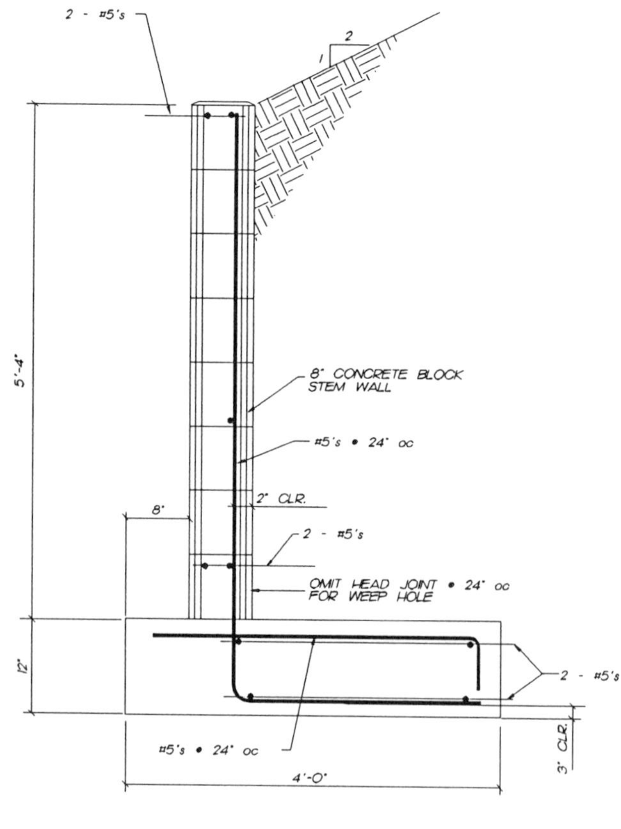

M7102

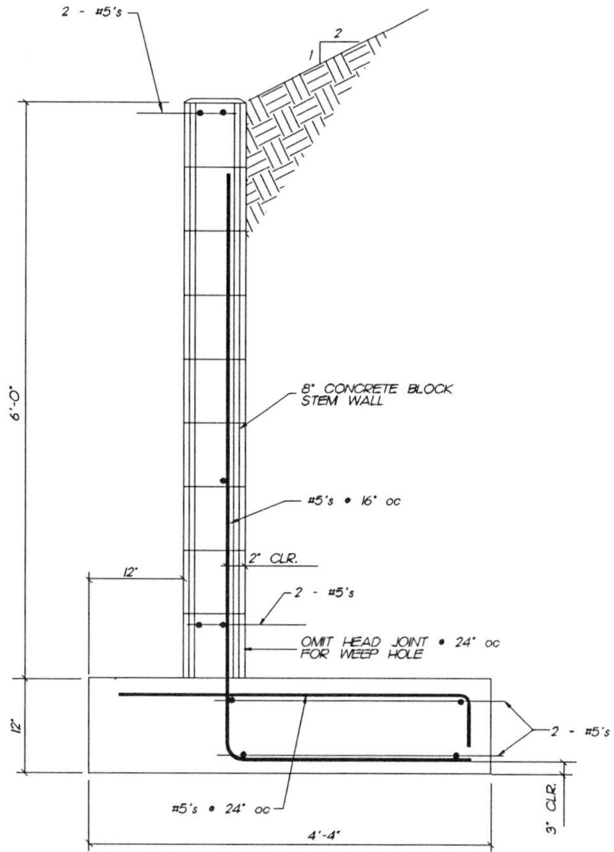

M7103

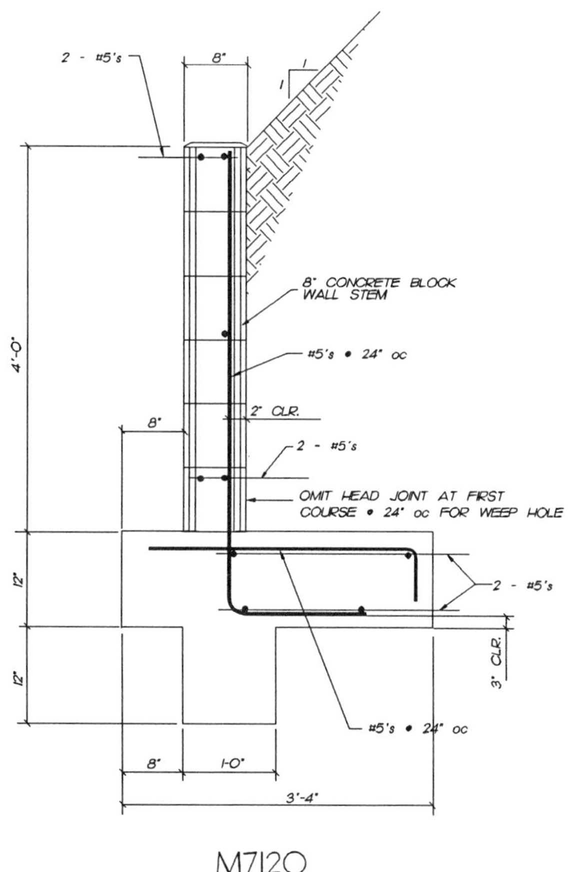

M7120

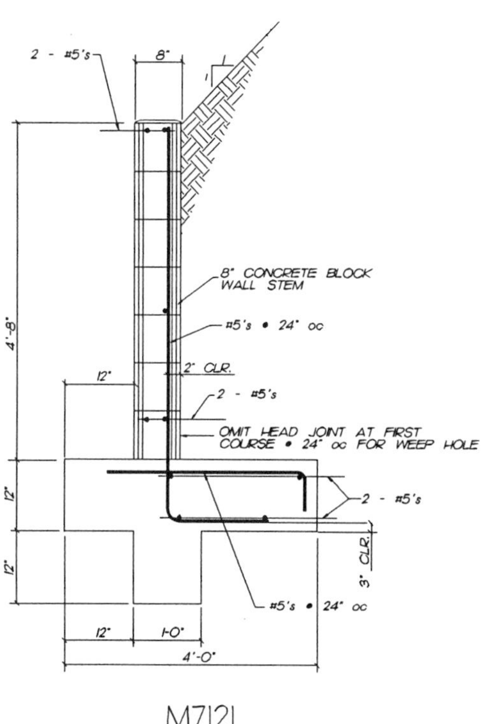

M7121

CHAPTER 2 MASONRY SECTION 8	CONNECTIONS
CONCRETE BLOCK WALL & WOOD STUD WALL INTERSECTION	M8020
8" CONCRETE END WALL & CONCRETE FLOOR SLAB	M8040
CONCRETE BLOCK BOND BEAM CORNER	M8041
12" CONCRETE END WALL & CONCRETE FLOOR SLAB	M8042
8" CONCRETE INTERIOR WALL & CONCRETE FLOOR SLAB	M8060
12" CONCRETE INTERIOR WALL & CONCRETE FLOOR SLAB	M8061
8" CONCRETE END WALL & STEEL ANGLES & PLATES TO CONC. SLAB	M8080
8" CONCRETE INTERIOR WALL & STEEL ANGLES & PLATES TO CONC. SLAB	M8081
CONCRETE BLOCK BOND BEAM INTERSECTION	M8100
CONCRETE BLOCK WALL INTERSECTION	M8101
BRICK WALL & WOOD STUD WALL INTERSECT.	M8120
BRICK WALL & CONCRETE FLOOR SLAB	M8140
BRICK WALL & BOND BEAM CORNER	M8141
BRICK WALL INTERIOR SUPPORT OF A CONCRETE SLAB	M8160
BRICK WALL END CONNECT - STEEL ANGLES & PLATES TO CONC. SLAB	M8180
BRICK WALL INT. CONNECT - STEEL ANGLES & PLATES TO CONC. SLAB	M8181
BRICK WALL BOND BEAM INTERSECTION	M8200
BRICK WALL INTERSECTION	M8201

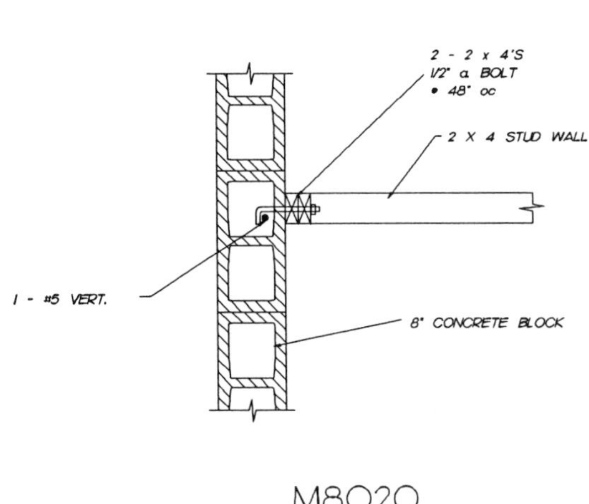

M8020

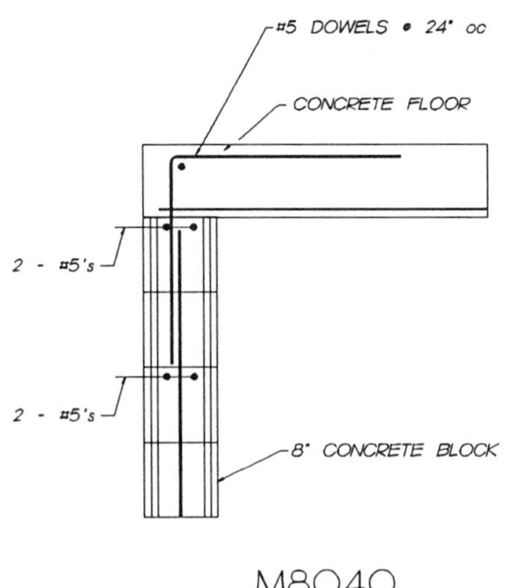

M8040

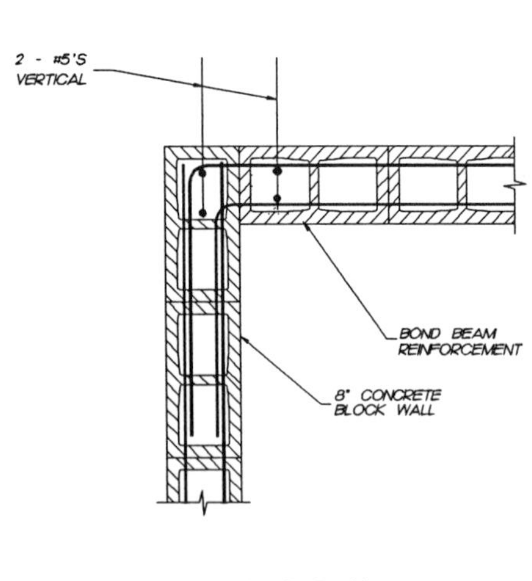

M8041

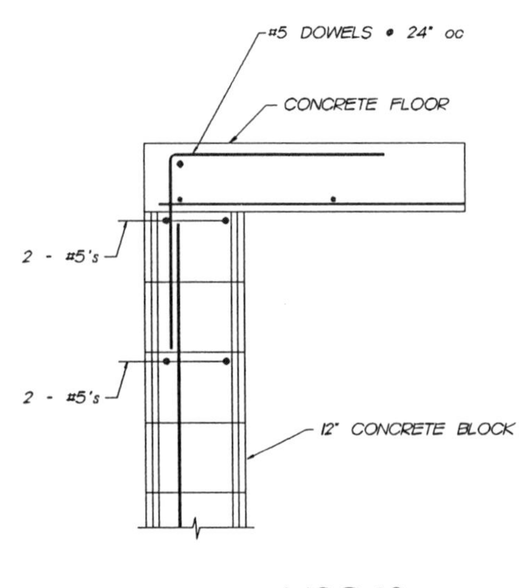

M8042

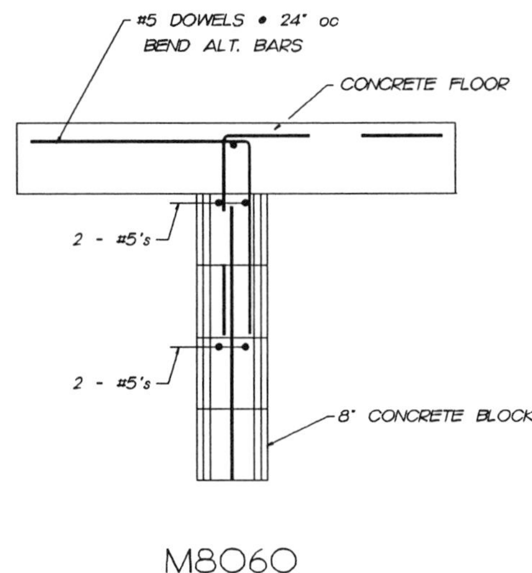

M8060

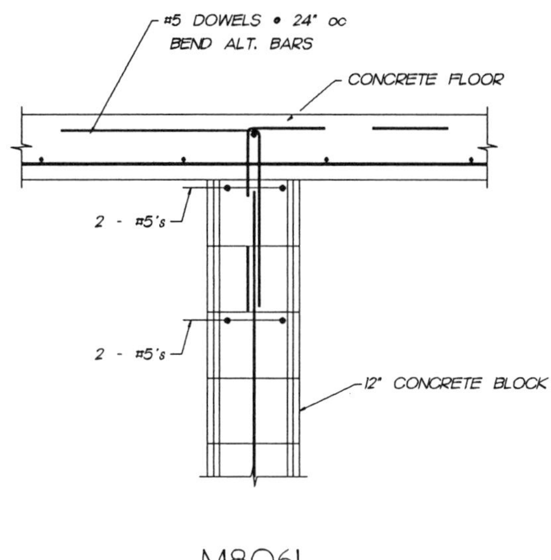

M8061

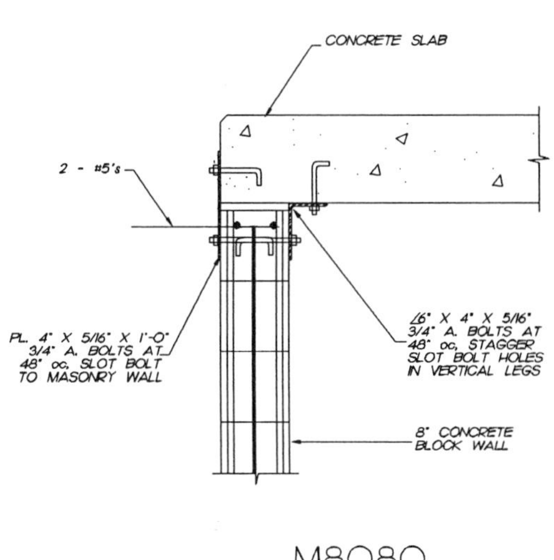

M8080

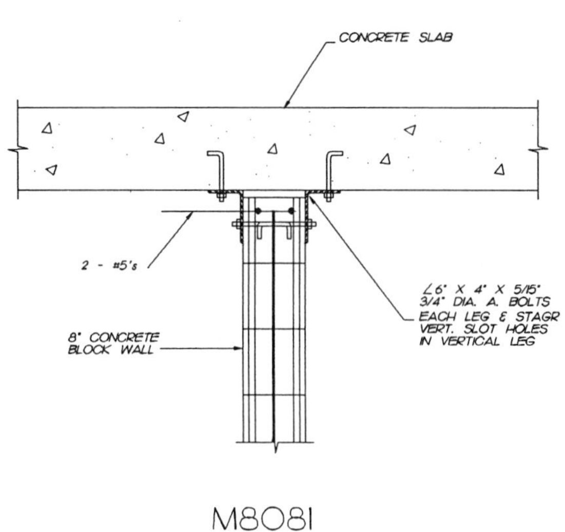

M8081

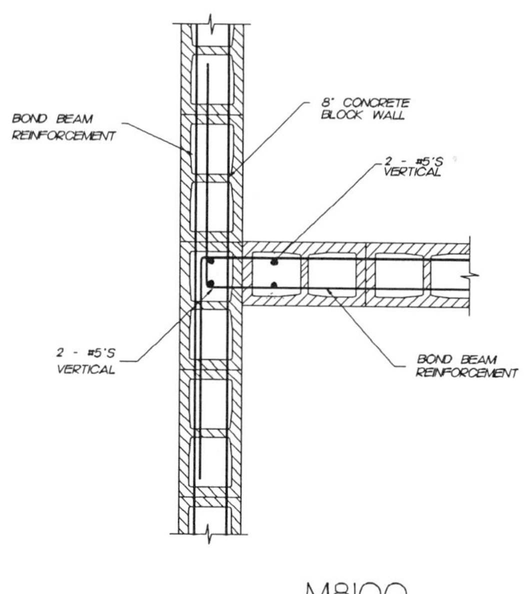

M8100

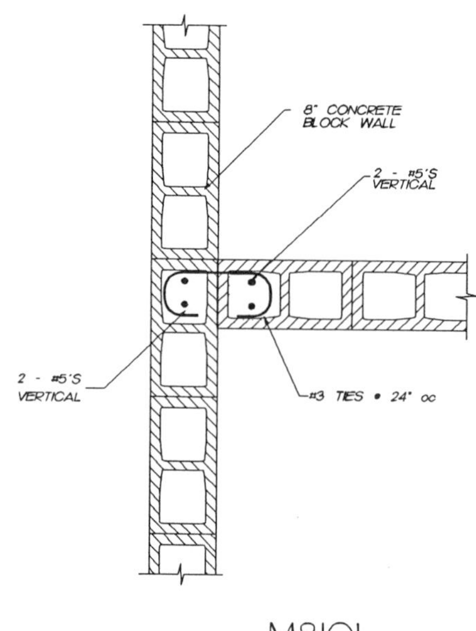

M8101

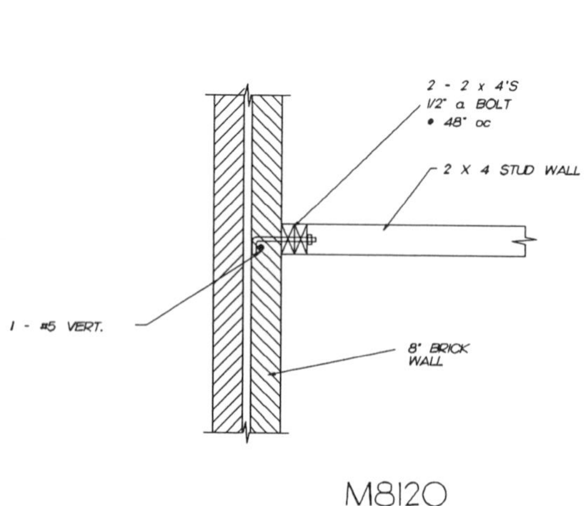

M8120

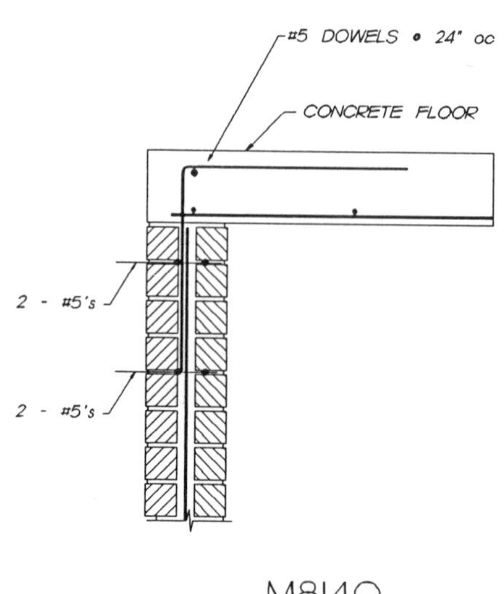

M8140

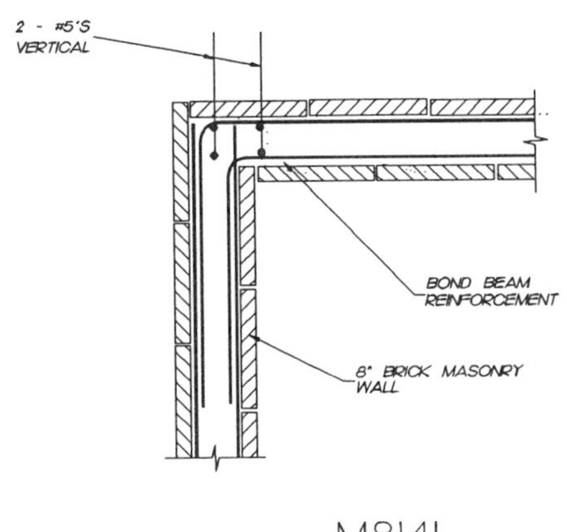

M8141

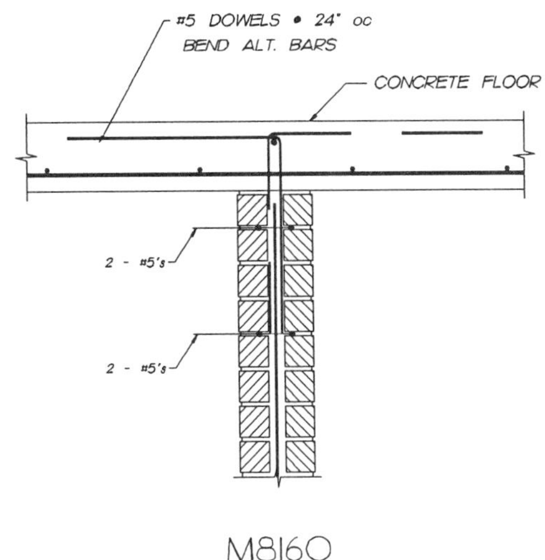

M8160

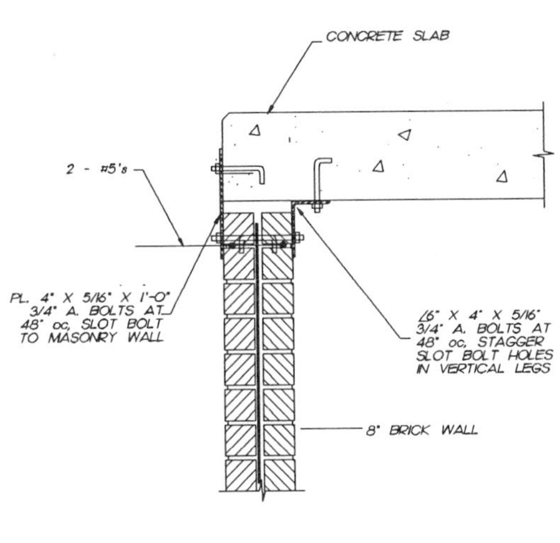

M8180

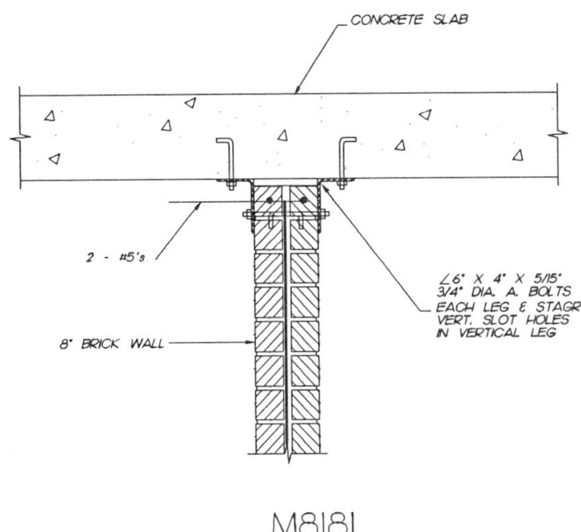

M8181

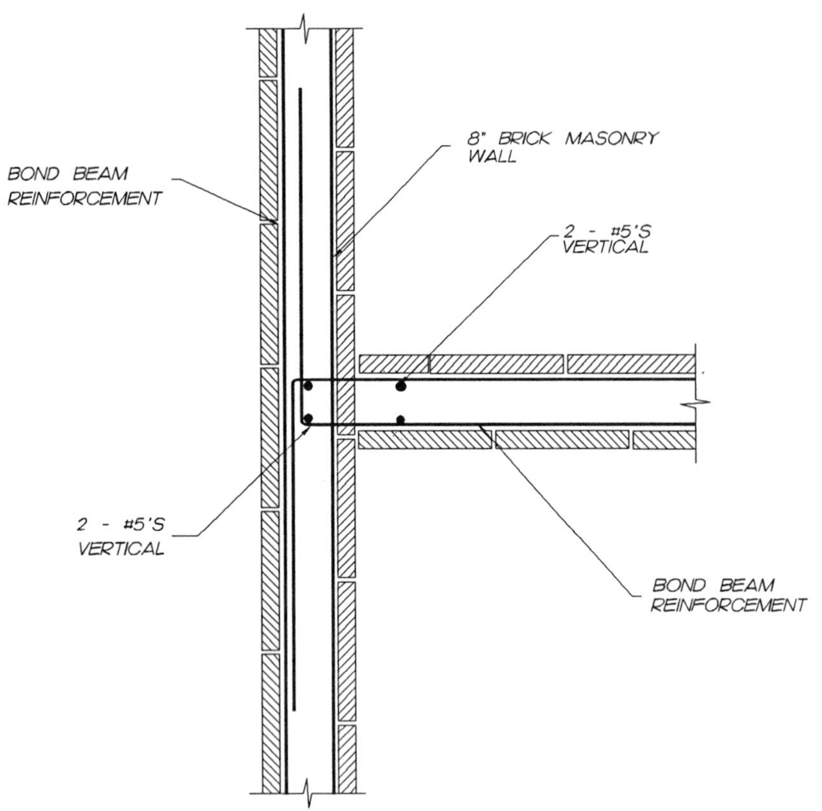
8" BRICK MASONRY WALL
BOND BEAM REINFORCEMENT
2 - #5'S VERTICAL
2 - #5'S VERTICAL
BOND BEAM REINFORCEMENT

M8200

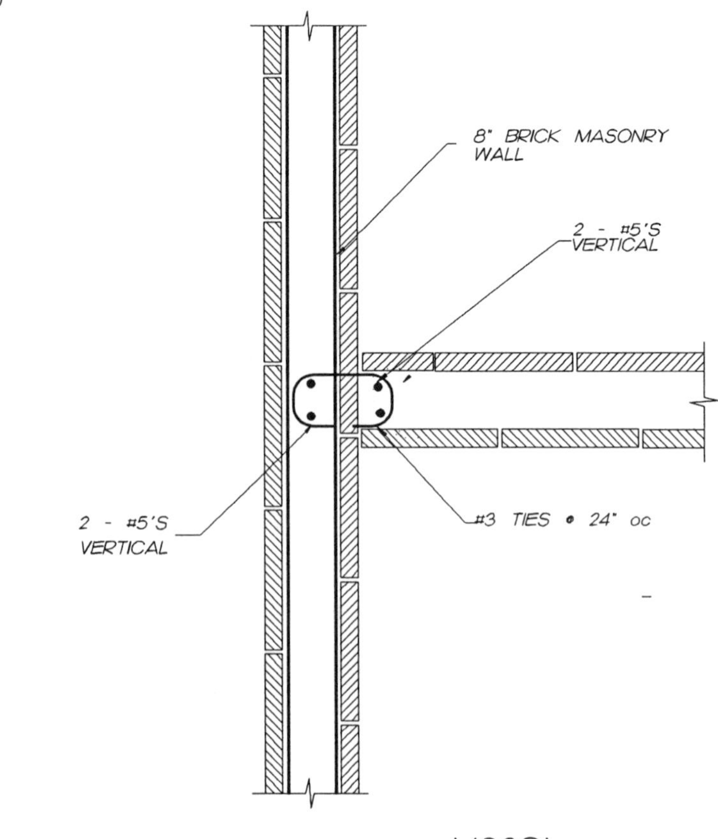
8" BRICK MASONRY WALL
2 - #5'S VERTICAL
2 - #5'S VERTICAL
#3 TIES @ 24" oc

M8201

CHAPTER 2 MASONRY SECTION 9	MISCELLANEOUS
WALL DOOR OPENING	M9020
WALL WINDOW OPENING	M9021
8" CONCRETE BLOCK JAMB & SILL SECTION	M9022
8" CONCRETE BLOCK LINTEL SECTION	M9023
WALL DOOR OPENING WITH STEEL CHANNEL JAMB & LINTEL	M9040
8" CONCRETE BLOCK STEEL CHANNEL JAMB & LINTEL	M9041 - M9042
BRICK WALL JAMB & LINTEL	M9060 - M9063
SECTION OF 3 WYTHE BRICK WALL	M9064
TYPICAL PLACEMENT OF REINFORCEMENT IN BRICK WALL	M9065
VERITCAL REINFORCEMENT IN A CONCRETE BEAM SUPPORT	M9080
STEEL BEAMS IN MASONRY WALLS	M9100 - M9101
EXTERIOR BRICK WALL SUPPORT OF A CONCRETE SLAB	M9120 - M9121
CANTILEVERED STEEL CHANNEL CONNECTION TO A MASONRY WALL	M9140
MASONRY WALLS & CONCRETE STAIRS	M9160 - M9161
5'-0" HIGH FREE STANDING WALL	M9180
6'-0" HIGH FREE STANDING WALL	M9181

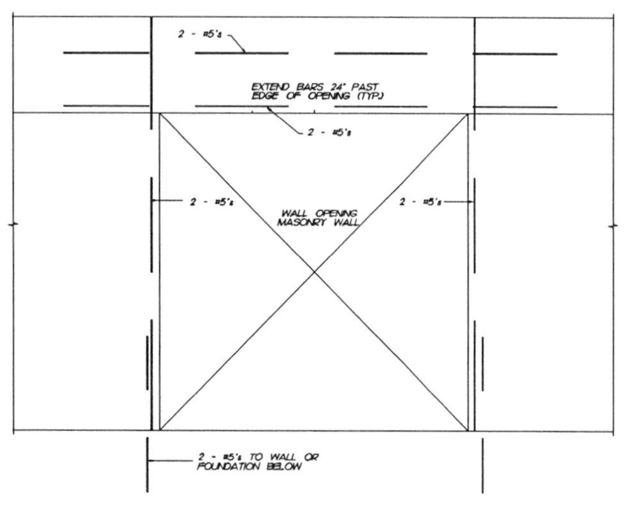

M9020

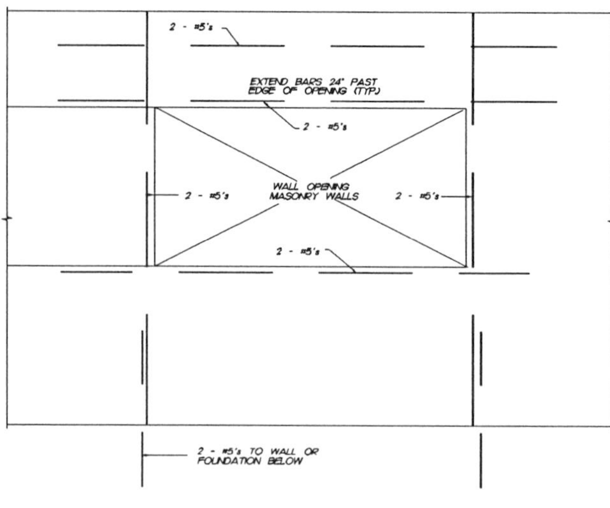

M9021

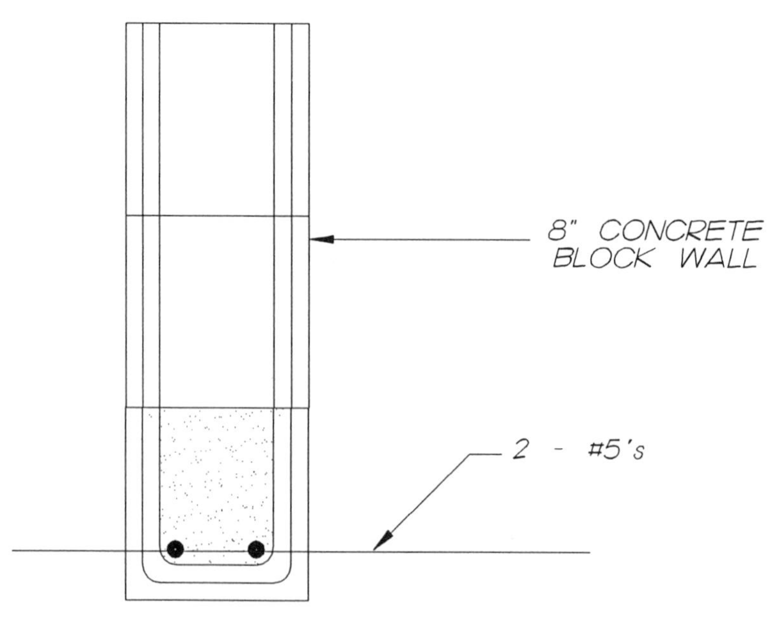

M9022

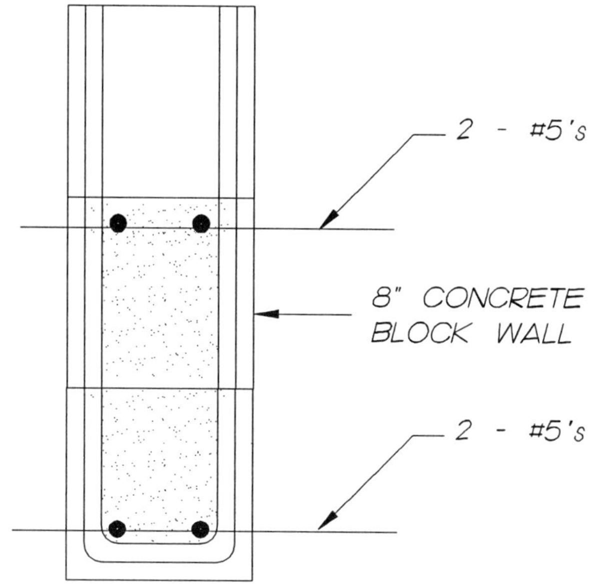

M9023

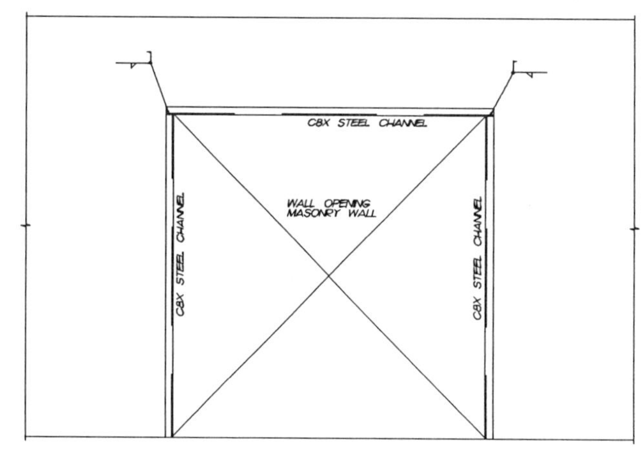

M9040

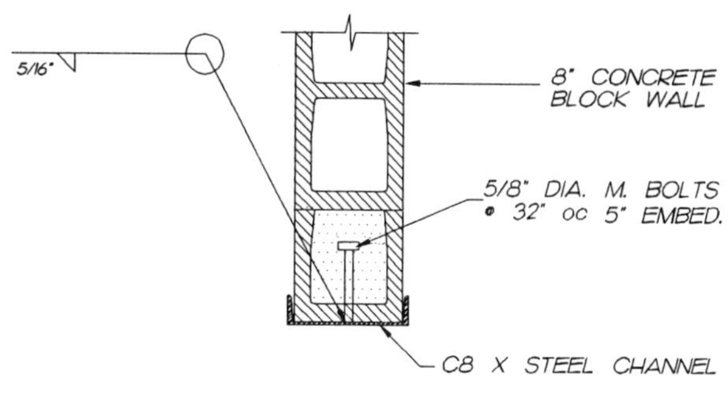

M9041

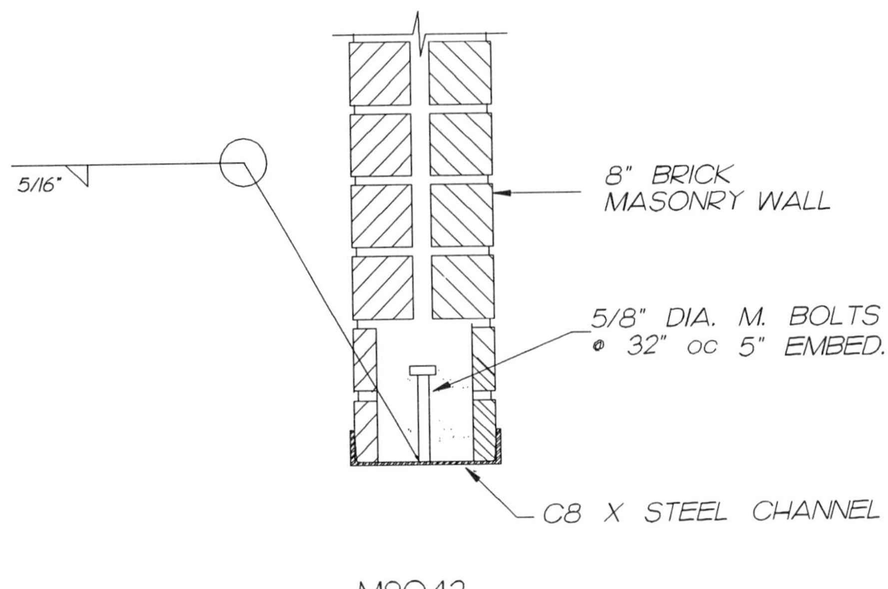

M9042

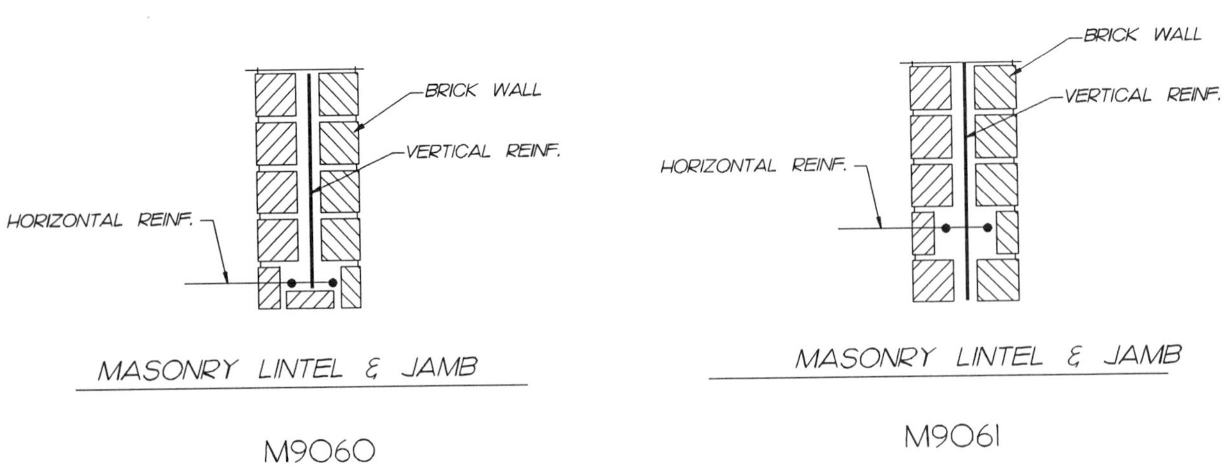

M9060

M9061

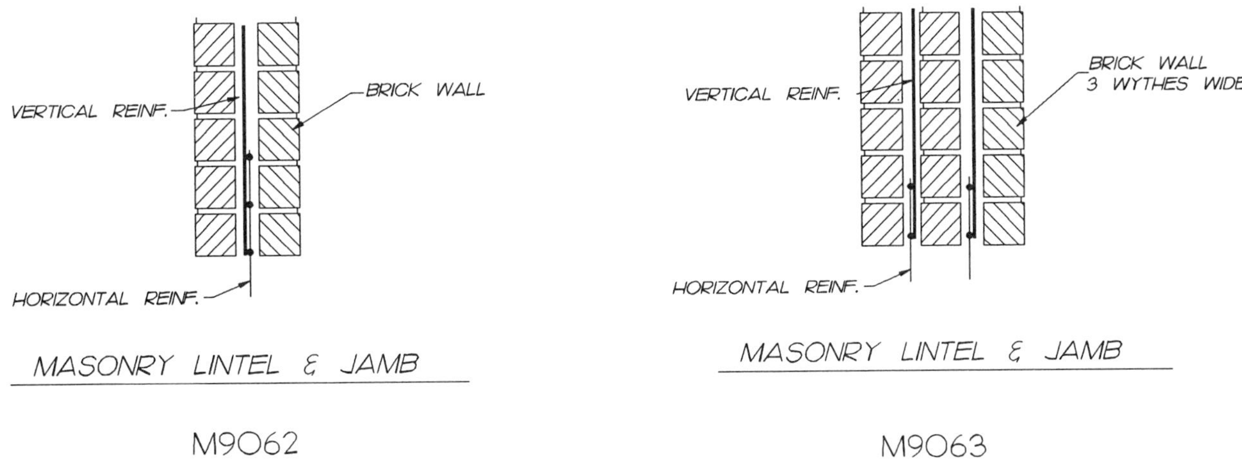

MASONRY LINTEL & JAMB

M9062

MASONRY LINTEL & JAMB

M9063

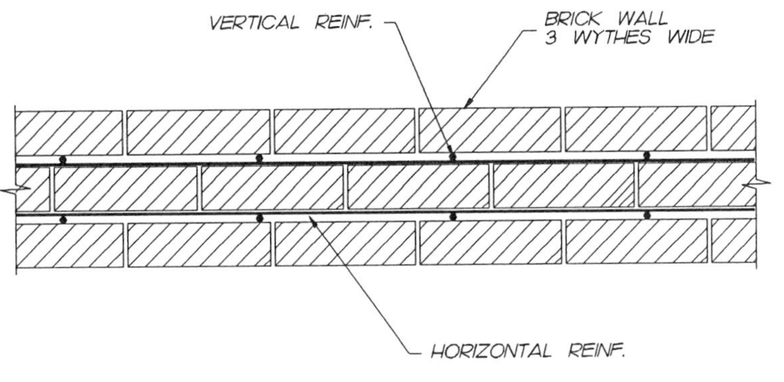

PLAN SECTION - TYPICAL 3 WYTHE BRICK WALL

M9064

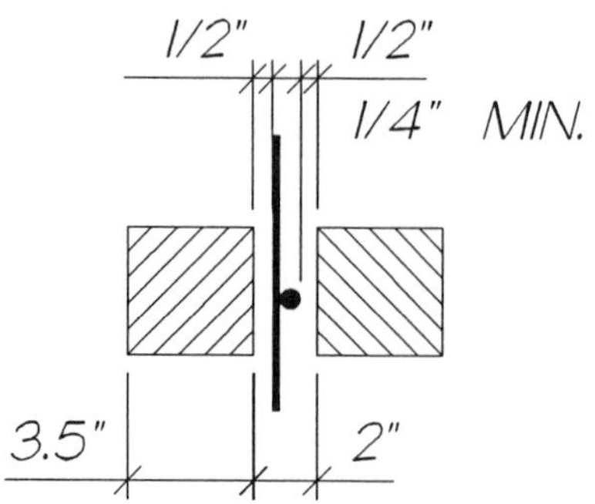

REINFORCEMENT IN BRICK WALLS

M9065

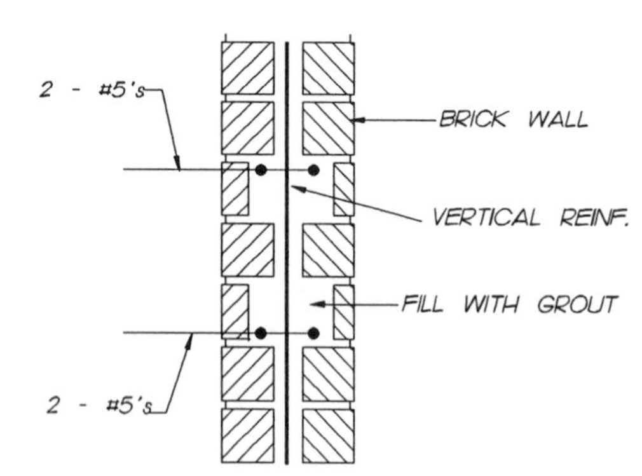

BOND BEAM SECTION

M9080

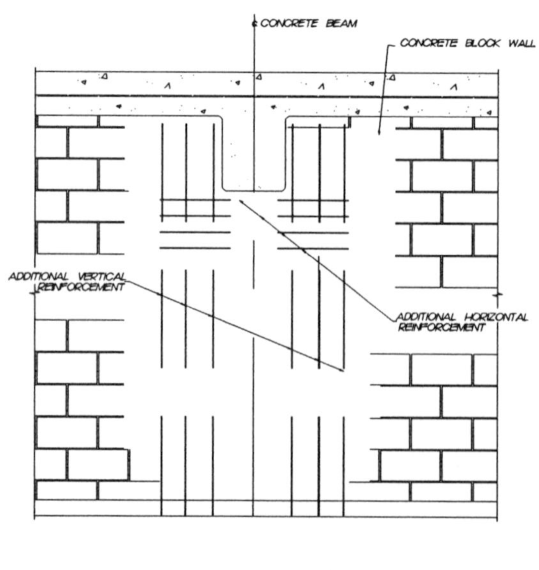

M9066

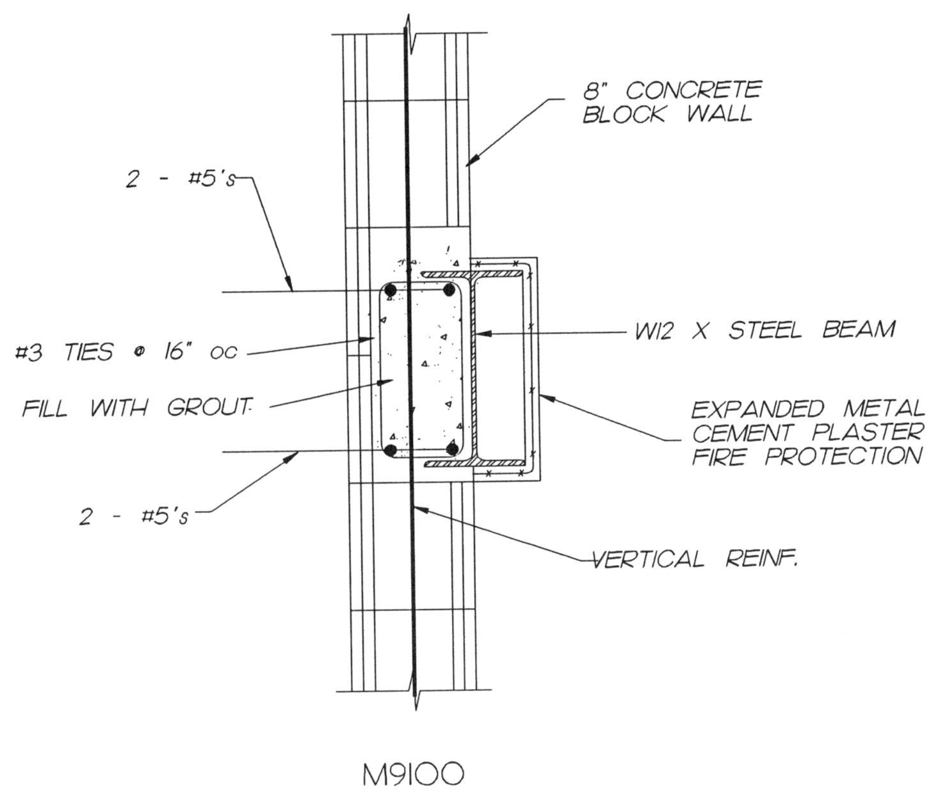

M9100

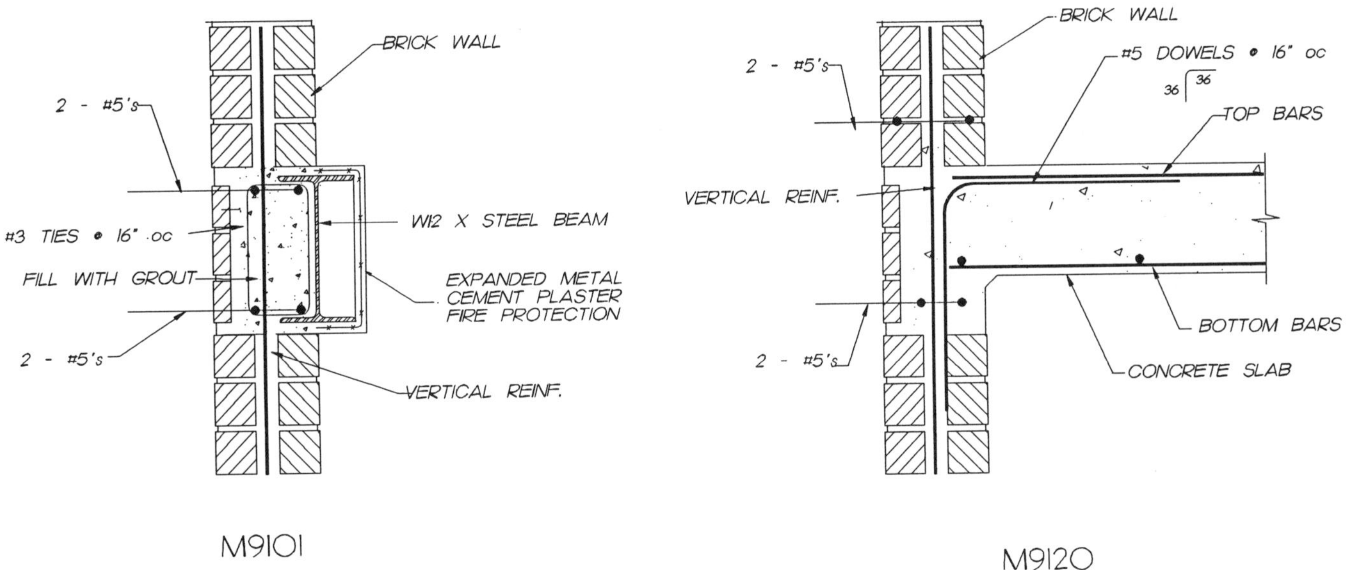

M9101

M9120

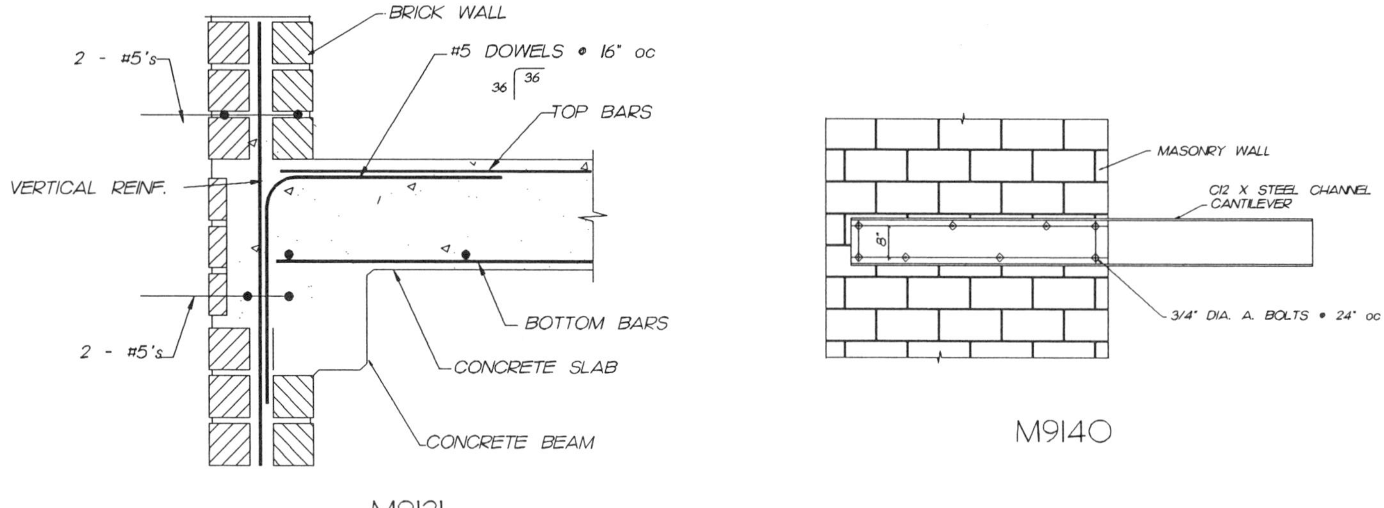

M9121

M9140

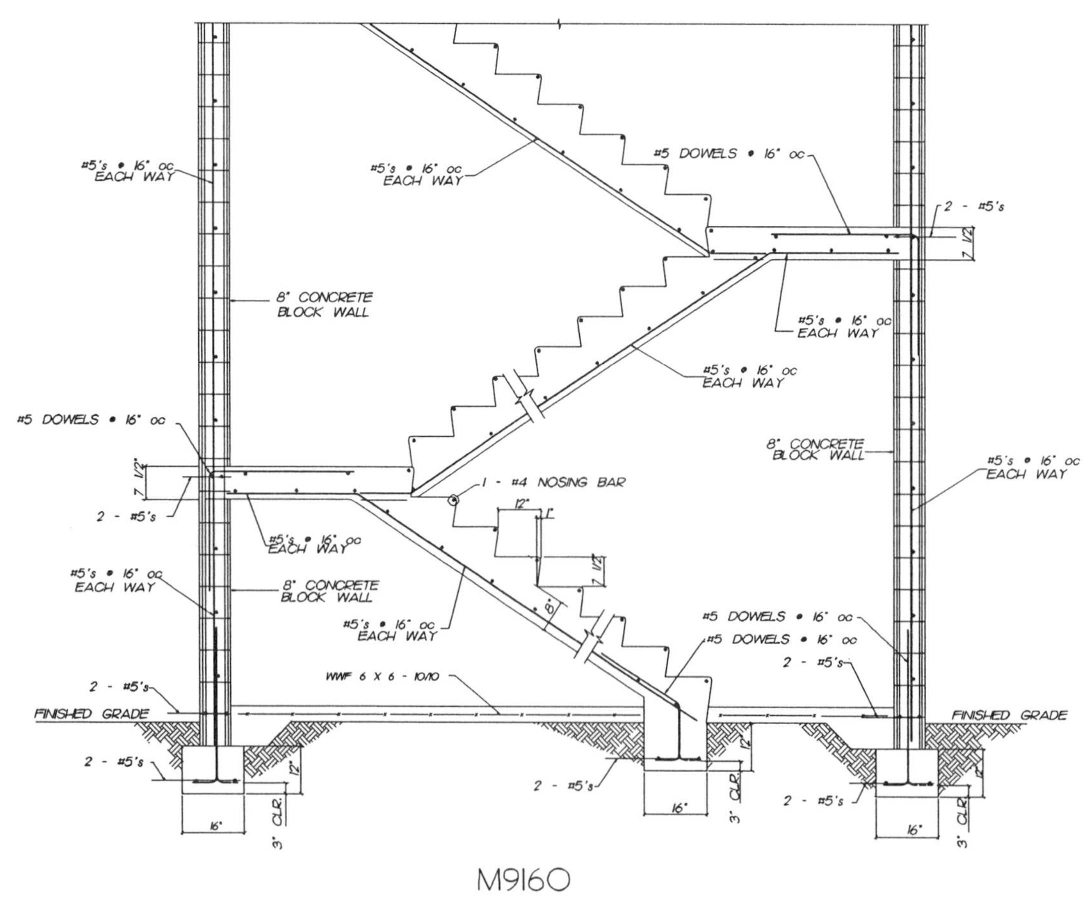

M9160

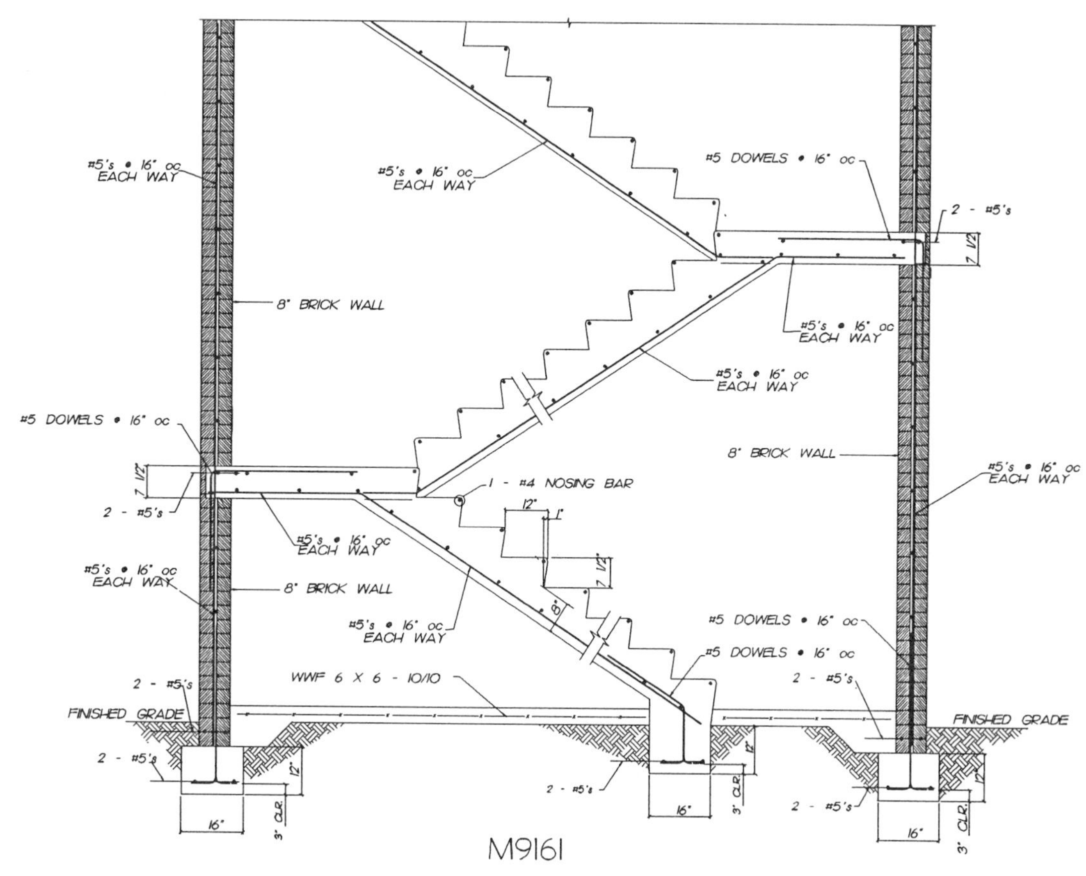

M9161

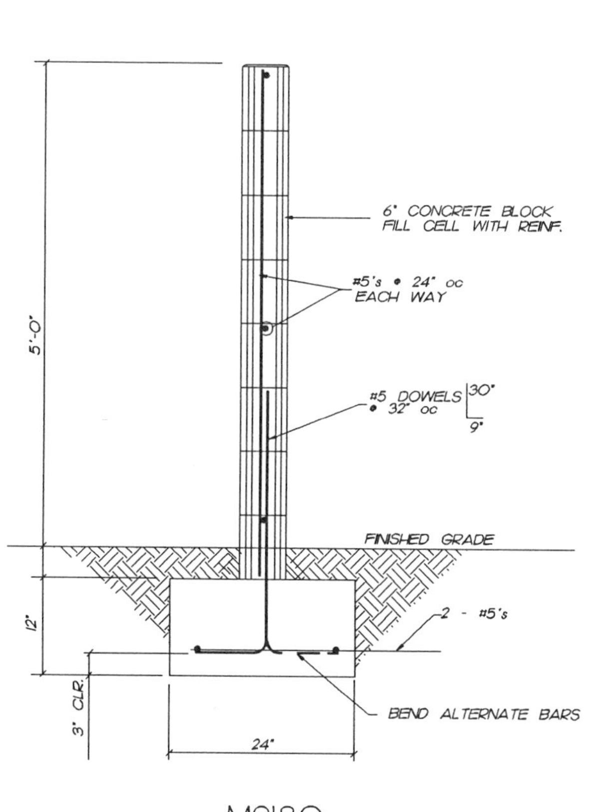

M9180

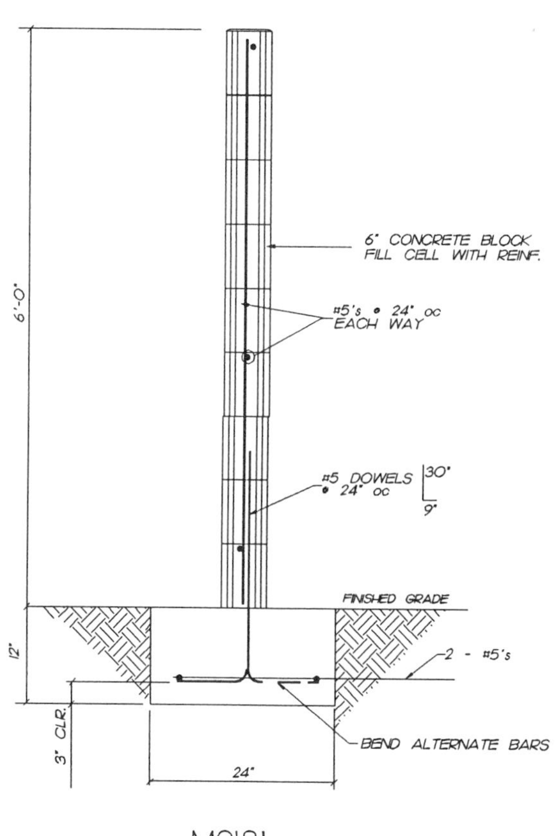

M9181

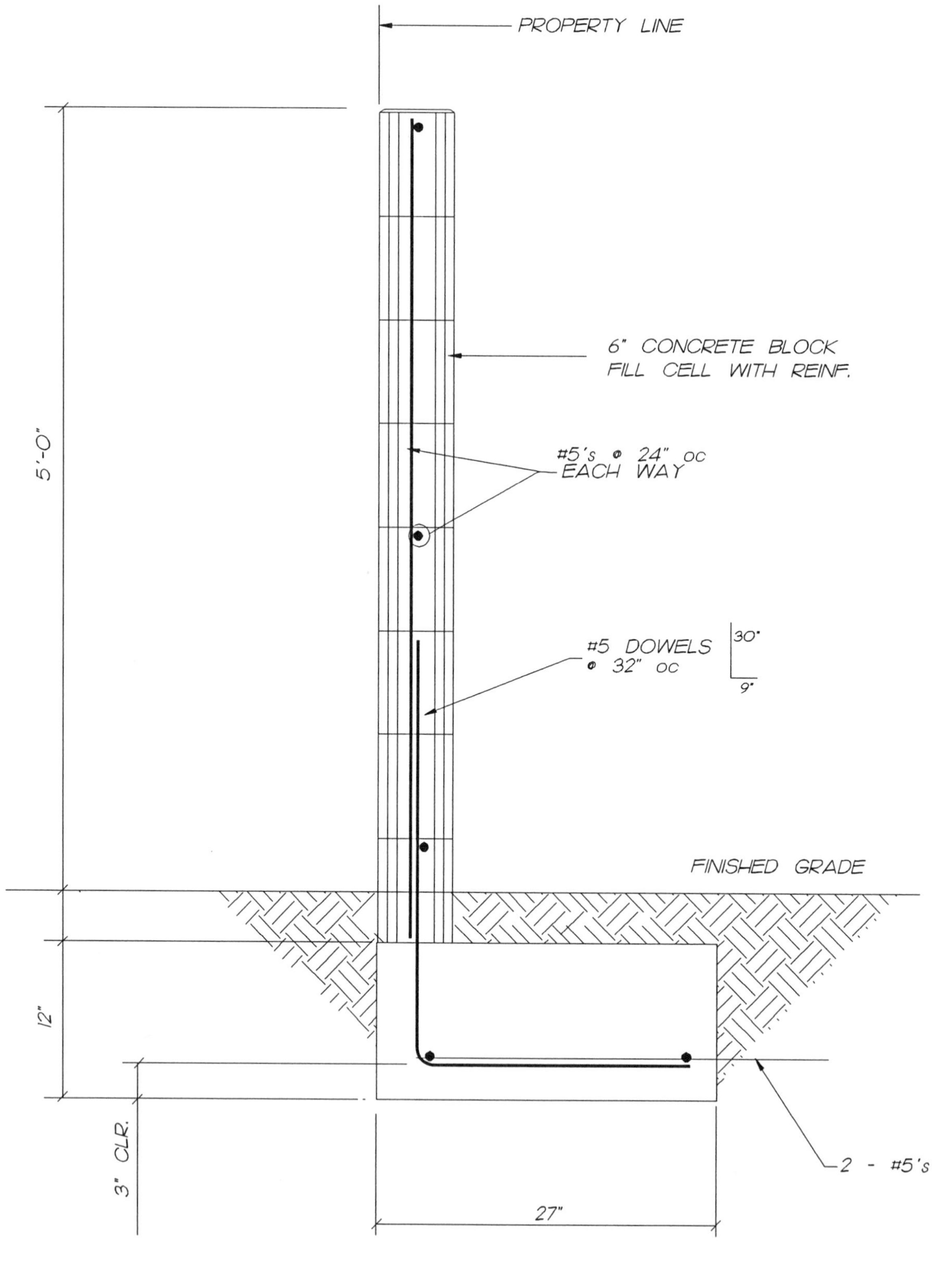
PROPERTY LINE
6" CONCRETE BLOCK
FILL CELL WITH REINF.
#5's @ 24" oc
EACH WAY
#5 DOWELS
@ 32" oc
30"
9"
FINISHED GRADE
5'-0"
12"
3" CLR.
27"
2 - #5's

M9182

Concrete Structural Details

e details and assemblies presented in this chapter primarily apply to
inforced concrete construction. There are details in other chapters of
is book that incorporate a certain amount of concrete; however, they
e not included in this chapter because concrete is not their predominant
aterial. Most of the details and assembles presented in this chapter specify
mensions and reinforcement requirements. This information is included
make the detail a more helpful example of the use of concrete as a
nstruction material. Details should not be construed as actual design
awings; dimensions and reinforcement can be determined only by de-
gn calculations. Standard structural details should be used with an in-
rmed concern for the strength of the design. The arbitrary application
these details may not satisfy the requirements determined by the struc-
ral design. Also, several of the details are presented in their most basic
ate, that is, without dimensions or reinforcement. These types of details
e used to demonstrate the most elementary configurations.

Reinforced concrete has a wide spectrum of uses in construction. Ap-
ications of the material include such components as flat slab, columns,
eams, foundations, tilt-ups (precast walls), and many more. The variety
uses of reinforced concrete has resulted in the creation of a large
umber of detail standards, which are shown in this chapter.

The accessibility of the information in the chapter is facilitated by arnging the material in a logical, relevant sequence so that the reader may
adily locate a particular detail. The basic concept of this arrangement
aces the drawings in an assigned hierarchy which starts with the most

general condition, then progresses to a discrete number of associat explicit conditions.

Each drawing is identified by a coded designator starting with a mater identification character and followed by a four-digit number. Since th chapter is concerned with concrete details, the first character of the det designator is the letter C. The detail numbers which follow are coded represent a construction function section and the detail number in th particular section. The last digit of the detail number represents the nu ber of the variation of the original detail. This arrangement allows f some degree of parallelism between the different chapters. For examp the numbers for the concrete columns are similar to those for the ste or masonry columns. However, an exact comparison between chapte will reveal that the drawings are not identical.

The details are organized in a sequence of sections, each of whi pertains to a particular function in the construction process. The first di of the detail number is used to describe the particular construction pu pose or function of the detail. This chapter consists of nine function se tions, which are defined as follows:

Foundations	C1020 to C1xxx
Columns	C2020 to C2xxx
Walls and wood floors	C3020 to C3xxx
Walls and wood roofs	C4020 to C4xxx
Walls and concrete slabs	C5020 to C5xxx
Walls and steel floors	C6020 to C6xxx
Retaining walls	C7020 to C7xxx
Precast and connections	C8020 to C8xxx
Miscellaneous	C9020 to C9xxx

The following is an example of the number coding system:

Given No. C4161

The number 4 indicates that the detail is in the concrete walls and wo roofs section; the number 16 indicates that it is the sixteenth drawing this section; and the number 1 indicates that it is the first variation of t original configuration. A look at Detail C4161 shows a concrete wall su porting a wood roof framing. The catalog charts prior to the drawings a presented to assist the reader in locating details. The construction functi section numbers at the top of the charts are arranged in numerical ord as outlined above. The detail descriptions and their respective particula are stated in the left-hand columns, and the designation numbers are stat in the right-hand column of the chart.

The type and the strength of the concrete required for the constructi of the details in this chapter are not specified. The requirements for stru tural concrete for a building depend upon the concrete's compressi strength, durability, watertightness, and resistance to deterioration, a upon the economics of the construction. The strength of concrete reinforcement that may be required to resist the design loads on a structu was not considered as a governing factor in the presentation of the details; however, strength of materials is a primary factor in the structu design process. The strength of concrete is measured by its capacity resist compression, and is expressed as a unit compressive stress in ter of pounds per square inch. The compressive stress used to specify t strength of a concrete mixture 28 days after it is placed or molded f ıboratory test samples is designated as f'_c. The values of f'_c for the various 'ater and cement ratio mixtures are verified by testing the compressive trength of a representative number of standard-size specimens. The Amer-:an Society for Testing and Materials (ASTM) specifies definite rigid re-uirements for the testing procedures for concrete specimens. The pecimen samples are taken from fresh concrete batches and are molded ıto cylinders 6 in in diameter and 12 in high, or are taken from cores of ıe existing concrete. These test cylinders are subjected to laboratory ɔading to determine their ultimate compressive strength at 7 days and at 8 days after the sample is molded and cured. Tests of a number of pecimens of a particular mixture can yield a comparatively wide range f values. In order to obtain a valid value of the f'_c, it is necessary to test everal specimens of a mixture and correlate the results mathematically. roper quality control of concrete mixtures should produce concrete with n average compressive strength approximately 15 percent greater than ıe f'_c for working stress designs and 25 percent greater for ultimate stress esigns. The design, testing, sampling, and evaluating of concrete mixtures re performed by a responsible certified laboratory using definite ASTM tandards of procedure.

Structural concrete strength is generally specified by the water-cement atio of the mixture, or by the value of f'_c, and the weight or volume roportions of the mixture. Concrete is composed of a mixture of cement, rater, coarse aggregate, fine aggregate, and, in certain cases, chemical dmixtures. The type and quality of cement used in commercial structural oncrete is usually Type I or Type II portland cement, which is produced ı accordance with ASTM standards. Other types of commerical cements re available; however, Type I or Type II cement is most commonly used ı building construction. Fine aggregate and coarse aggregate are two ifferent materials in the concrete mixture; however, both are used as hemically inert fillers to increase the volume of the mixture. Coarse ggregates consist of particles of crushed rock greater than ¼ in but usually ot larger than 1½ in. Fine aggregates consist of graded natural sand and articles of crushed rock less than ¼ in. Various types of aggregate can e employed in concrete to achieve a specific quality that may be required s a condition of the use of the concrete; for example, concrete used for ireproofing or thermal insulation requires a lightweight mixture. Many ommercial chemical admixtures are also available for use in structural oncrete. These admixtures are used to produce a specific quality in the nished concrete, such as watertightness, hard exterior surfaces, color, nd high early strength immediately after the concrete is placed.

The water-cement ratio of a concrete mixture is the primary factor in etermining the compressive strength of the hardened concrete. The water nd cement combine chemically to create a physical paste which adheres ɔ and binds the aggregates. The result of this chemical and physical henomenon is a concrete mixture that is capable of acting as a monolithic naterial after it hardens. The water-cement ratio is usually specified as the umber of gallons of water per sack of cement. This ratio can also be xpressed in terms of weight or as the number of cubic feet of water and he number of cubic feet of cement. One cubic foot of water equals 7.48 allons; one standard sack of cement equals 1 cubic foot. These values are iven to demonstrate arithmetically that a water-cement ratio of 7.5 gallons f water per sack of cement will contain approximately 1 cubic foot of rater per cubic foot of cement. The compressive strength of concrete is nversely proportional to the water-cement ratio; that is, as the water-cement ratio increases, the f'_c value decreases. The f'_c values of structural concrete range between 2000 and 5000 psi and are usually designated in 500-psi increments, except for $f'_c = 3750$ psi.

Efficient methods for the proper placing of concrete depend upon the plasticity or workability of the wet mixture. The degree of plasticity of wet concrete is determined by a slump test. This test is performed by measuring the subsidence or vertical displacement of a sample of the wet concrete mixture. The sample is formed in a metal truncated cone 12 in high with top and bottom diameters equal to 4 in and 8 in, respectively. The results of the test are given as the number of inches of vertical displacement that is measured immediately after the cone form is removed. The amount of slump should vary between 2 in and 6 in. The degree of plasticity required depends on the clearance between reinforcing bars, which is related to the size of the coarse aggregate, the method of delivering the concrete to the forms, and the intricacy of the formwork. The American Concrete Institute Building Code (ACI) recommends a maximum slump of 4 in and a minimum slump of 1 in for the various structural elements of the building. The minimum slump value should be maintained for construction of concrete ramps, sloping walls, and slabs. When the slump is too high, the wet concrete mixture will be loose and will thus permit the coarse aggregate to separate from the cement paste; when the slump is too low, the wet concrete mixture will be stiff and will create internal void spaces in the hardened concrete. Very often the wet concrete must be mechanically vibrated or rodded as it is placed to prevent the formation of internal voids and to ensure that the reinforcing bars are completely surrounded by concrete. Mechanical vibrating should be performed with care so that the position of the reinforcing bars is not altered. Designs using reinforced concrete make the assumption that the concrete material is homogeneous and well consolidated when it is placed; if this condition is met, the concrete will react as a monolithic material when it hardens. A positive physical bond must exist between the reinforcement and the concrete to permit it to react as a monolithic material. Internal voids caused by poor consolidation will seriously reduce the structural bonding of the two different materials. Also, reinforcement surfaces should be free of excessive rust and mill scale, paint, soil, or other substances that will reduce the bonding capacity. To ensure good consolidation, concrete should be placed in continuous layers of uniform depth, it should not be dropped into place from a height greater than 5′-0″, and it should not be chuted into place at a steep angle for long distances. Concrete mixes that are composed of small-size aggregates are often used when the clearance dimensions between the reinforcing bars and the forms are relatively small. As the concrete is being placed, it should be continuously moved away from the sides of the forms to allow the smaller-size aggregates to settle near the exposed surfaces, thus giving the structure a smooth finish and eliminating surface voids.

After concrete is placed, it must be cured to obtain a uniform hardness. The curing process consists of maintaining an approximately even external and internal temperature condition. Retaining the exposed surfaces' moisture will be favorable to uniform hardening of concrete. The time rate of hardening is not constant; it proceeds quite rapidly immediately after the concrete is placed, and the concrete will reach approximately 75 percent of its f'_c value within the first 7 days after it is placed. The concrete will continue to harden at a progressively slower rate for a long period of time; however, it should attain at least the f'_c value within 28 days after it is in place. Cold weather, that is, temperatures less than 40°F, will retard the water and cement chemical reaction required to produce the cement paste in the mixture. Hot weather, that is, temperatures greater than 90°F, will cause the exposed surface moisture to evaporate too rapidly, resulting in uneven hardening and surface cracks. The method and length of time required to cure a particular concrete construction should be determined with regard to the temperature and moisture conditions at the job site at the time the concrete is placed. The concrete should be kept moist for at least one day immediately after it is placed, and it should be prevented from rapid drying for a period of 7 days after it is placed. High-early-strength concrete mixtures require only a 3-day curing period.

The formwork for a concrete structure is constructed to reflect the architectural and engineering design configuration. Although concrete forms are generally designed and constructed by the contractor, the resident engineer or building inspector should verify that the forms satisfy two main conditions: (1) the form structure is safe, and (2) the forms meet the architectural and engineering design requirements as they are shown on the working drawings. Since the formwork is often used as a working deck during construction, it should be designed and constructed to resist vertical and lateral forces. Lateral bracing is required to plumb the structure and to resist forces from the construction work. Most form failures are the result of improper or insufficient lateral bracing. The contractor must provide sufficient vertical support shoring of the form structure to prevent excessive deflection or the overstressing of any form members. The sides of beams, columns, and walls should be braced against the lateral hydrostatic pressure exerted by the wet concrete mixture. These side forms may be removed as soon as the concrete attains its initial hardness; however, the vertical supporting members should not be removed until the concrete attains a strength capable of supporting the dead load of the structure and any construction live loads that may occur. Early removal of the vertical form supports can result in excessive deflections, since the material has not reached a modulus of elasticity. The contact surfaces of the forms are coated with a heavy oil to prevent the concrete from adhering or bonding to the forms and to permit the forms to be removed without damaging the exposed concrete surfaces. Also, the sharp corners of rectangular columns, walls, and beams are formed with a 45° chamfer strip to eliminate the possibility of sharp edge spalling of the concrete when the forms are removed. Damaged concrete surfaces, surface voids, and form tie holes are patched with a patching mortar that is composed of the same proportions of sand and cement as the poured concrete. The surface finish of poured concrete depends on the degree of exposure of the members and the architectural quality of the building. ACI Standard 301-84 defines the various methods and criteria for concrete surface finishes.

Structural reinforced concrete members are used to resist external loads from bending, compression, or a combination of both. The ability of plain concrete to resist tensile stress is quite low and undependable. In the analysis and design of reinforced concrete, it is assumed that the concrete material offers no resistance to tensile stress and that tension is resisted only by the reinforcing steel. Reinforcing steel is also used to resist compression in a member acting in flexure or when the unit compressive stress exceeds the permissible compressive stress of the concrete.

The structural analysis is mathematically performed by the "transformed section method." This method is based on the assumptions that (1) the concrete resists compression, (2) the reinforcing steel resists tension and compression, (3) the concrete cannot resist tension, (4) the reinforcing steel and the concrete are bonded together and therefore react togethe and (5) the reinforcing steel and the concrete are individually elastic. Whe a reinforced concrete member deflects or deforms from an externall applied load, the unit stress in the reinforcing steel is directly proportiona to the unit stress in the concrete. The proportional relationship betwee the two materials is denoted as the value n. This value n for a concret mixture is defined as the ratio of the modulus of elasticity of the reinforc ing steel E_s and the modulus of elasticity of the concrete E_c, or n = $30{,}000/f'_c$. It can be seen in the last equation that n is inversely proportiona to the value of f'_c. The values of n vary from 15.0 for $f'_c = 2000$ psi to n = 6.0 for $f'_c = 5000$ psi. The reinforcing steel area of a concrete bear or column can be converted or transformed into an equivalent cross sectional area of concrete by multiplying the cross-sectional area of th reinforcement, designated by A_s, by the value of n. After the reinforce concrete is transformed into an analogous homogeneous concrete mem ber, it can be structurally analyzed using the basic rules of mechanics an equilibrium.

The working stress design method (WSD), as applied to reinforce concrete members, uses maximum allowable working stresses of the ma terials, the actual dead load, and a building-code-specified live load. Th various building codes and design criteria specify the maximum allowabl design stress for concrete and reinforcing steel. This stress is determine by dividing the yield point stress of the material by a factor of safety Reinforced concrete structures designed by the working stress desig method do not utilize the full stress capacity of the materials because th factor of safety is generally applied both to the working stresses of eac material and to the loads.

An alternative method of reinforced concrete structural design is calle the ultimate strength design method (USD). This method is based upo designing the members to their yield point stress and individually increas ing the dead load and the live load by a required load factor instead c using a single general safety factor. The structural safety factor is calculate by the dead load and the live factors, which are based upon a mathematica probability of occurrence. The ultimate strength design method has bee adopted by the ACI, which prescribes the guidelines for design criteri and methods. The strength of materials used in USD is usually greate than that of materials used in the WSD method, because this method allow the designer to design to a value comparatively close to the yield strengt of the material. For this reason, it is not unusual to use concrete wit $f'_c = 4000$ psi or greater, and hard-grade reinforcement (ASTM A615-60) The USD method produces significant economies in the use of concret and reinforcing steel and also creates a structure that is more realistic i the use of materials.

Table 3-1 shows the various reinforcing steel strength values and thei respective ASTM specification titles and numbers. Reinforcing steel bar are manufactured in standard sizes, which are designated by a numbe indicating the number of eighths of an inch of nominal diameter of deformed reinforcing bar; for example, a #6 bar is 6/8 (3/4) in in diamete Plain reinforcing bars, that is, bars that are not surface deformed, ar designated by the diameter in inches, for example, 1/2 in or 5/8 in. Rein forcing bars that are 1/4 in in diameter are never deformed and are alway designated as #2 bars. Reinforcing bars that are manufactured with a raise surface pattern are referred to as deformed bars. These surface projection serve to increase the contact surface area between the concrete and th reinforcing steel, and they also provide a mechanical device to increas

able 3-1. *Concrete Reinforcing Steel for ASTM Specifications*

inimum yield-point strength, psi	Grade of steel	ASTM specification title	ASTM Spec. No.
40,000	Intermediate	Specifications for billet-steel bars for concrete reinforcement	A15
	Intermediate	Specifications for axle-steel bars for concrete reinforcement	A160
	Intermediate	Specifications for special large size deformed billet-steel bars for concrete reinforcement	A408
50,000	Hard	Specifications for billet-steel bars for concrete reinforcement	A15
	Regular	Specifications for rail-steel bars for concrete reinforcement	A16
	Hard	Specifications for axle-steel bars for concrete reinforcement	A160
	Hard	Specifications for special large size deformed billet-steel bars for concrete reinforcement	A408
60,000		Specifications for deformed billet-steel bars for concrete reinforcement with 60,000 psi minimum yield strength	A432
		Specifications for deformed bail steel bars for concrete reinforcement with 60,000 psi minimum yield strength	A61
70,000		Specifications for cold-drawn steel wire for concrete reinforcement	A82
75,000		Specificationss for high-strength deformed billet steel bars for concrete reinforcement with 75,000 psi minimum yield strength	A431
Bars and rod mats		Specifications for fabricated steel bar or rod mats for concrete reinforcement	A184
Welded wire fabric		Specifications for welded steel wire fabric for concrete reinforcement	A185

the bar bond capacity. Welded wire fabric consists of cold-drawn wi arranged in a rectangular pattern and welded at the points of intersectio The size and spacing of the wire can be varied in either direction. Th welded wire fabric shown in the following drawings is 6 × 6–10/10; th 6 × 6 indicates that the wires are 6 in o.c. in each direction, and the 1 10 indicates that the wires are 10 gauge in each direction. Cold-draw wire or plain reinforcing rod that is used for round column spiral rei forcement is designated by the rod diameter in inches. Table 3-2 show the reinforcing bar designation numbers and their dimensions.

Reinforcing bars should be accurately placed to ensure that the com pleted construction will reflect the engineer's design. Small inaccuracie may appreciably increase the stress in a member and thus cause seriou surface cracks; therefore, the engineer must rely a great deal on the i telligence and integrity of the workers who are constructing the design Shop drawings prepared from the information given on the engineer working drawings will help to coordinate the field work with the structur design. The engineer responsible for the structural design should chec the shop drawings for dimensions, reinforcing bar sizes, and details placement of the reinforcing bars. It is also a good practice for the enginee to visit the job site during the construction to inspect the accuracy of th placement of the reinforcing steel. The ACI *Specification for Structur Concrete for Buildings*, No. 301-84, recommends allowable fabrication an placing tolerances for reinforcement. The bars should be wired togethe and secured in place to prevent any movement that might be caused b placing the concrete. Galvanized metal or concrete block bar chairs ar used to support the reinforcement in the forms to obtain the require concrete cover. In no instance should these chairs be used to suppo form boards or construction loads other than the weight of the reinforcin bars. Table 3-3 gives the recommended concrete cover for reinforcing bar in the various structural members. Table 3-4 gives the recommended clea distances between reinforcing bars.

Reinforcing bars can be connected either by a lapped splice or by butt-welded connection. Splices should not be made at points of critic stress in the member or at points which are not specified on the workin drawings, without the authorization of the design engineer. Lapped splice acting in tension or compression should not be made with reinforcin bars larger in size than #11. If deformed reinforcing bars are used an

Table 3-2. *Concrete Reinforcing Steel Bar Sizes and Dimensions*

Deformed-bar designation	Weight, lb/ft	Diameter, in	Cross-section area, in[a]	Perimeter, in	Max. outsid dia., in
#2	0.167	0.250	0.55	0.786	
#3	0.376	.375	0.11	1.178	7/16
#4	0.668	0.500	0.20	1.571	9/16
#5	1.043	0.625	0.31	1.963	11/16
#6	1.502	0.750	0.44	2.356	7/8
#7	2.044	0.875	0.60	2.749	1
#8	2.670	1.000	0.79	3.142	1 1/8
#9	3.400	1.128	1.00	3.544	1 1/4
#10	4.303	1.270	1.27	3.990	1 7/16
#11	5.313	1.410	1.56	4.430	1 5/8
#14S	7.65	1.693	2.25	5.32	1 15/16
#18S	13.60	2.257	4.00	7.09	2 1/2

ıble 3-3. *Minimum Clear Cover of Concrete for Reinforcing Steel*

Location of reinforcement in concrete	Clear distance
·inforcement in footings and other structural members in which e concrete is poured directly against the ground	3″
ɔrmed concrete surfaces to be exposed to weather or in contact ith the ground for bar sizes greater than #5	2″
ɔrmed concrete surfaces to be exposed to weather or in contact ith the ground for bar size #5 or less	1½″
abs and walls not exposed to weather or in contact with the ground	¾″
·ams and girders not exposed to weather or in contact with the ound	1½
oor joists with a maximum clear spacing of 30″	¾″
ɔlumn spirals or ties (not less than 1½ times the maximum size the coarse aggregate or . . .)	1½″

Note: Except for concrete slabs or joists, the concrete cover protectuion shall not be less an the nominal diameter of the reinforcing bar.

= 3000 psi, the recommended length of lap in a splice is as follows: nsion splices, not less than 24, 30, and 36 bar diameters for specified eld strengths of 40,000, 50,000, and 60,000 psi, respectively; compression ɔlices, not less than 24, 30, and 36 bar diameters for specified yield rengths of 40,000, 50,000, and 60,000 psi, respectively. These bar diameter ·ngths should be doubled for plain reinforcing bars and increased by ne-third for f'_c less than 3000 psi. The minimum length of lap should not e less than 1′-0″. Welded splices are usually made for large-size rein- ɔrcing bars. This is done by butt welding the square cut ends together ɔ that the connection is capable of resisting 125 percent of the specified ield strength of the bar in tension. Various mechanical bar connectors re available; however, their use would depend upon the approval of the ngineer and their acceptability under the design criteria. In the case of ıp splicing, it is always good practice to extend the length of the bars to ıe next even half foot, since the value of the bond strength of the concrete ıay not be 100 percent for the first few inches of insertion.

able 3-4. *Minimum Clear Spacing Distance between Reinforcing Bars*

Space between bars	Clear distance
lear distance between parallel bars xcept in columns and layers of bars ı beams and girders	Not less than the nominal diameter of the bars, or 1⅓ times the size of the coarse aggregate, or not less than 1″
lear distance between layers of rein- ɔrcement in beams or girders; the bars ı each layer shall be directly above nd below the bars in the adjacent layer	Not less than 1″
lear distance between bars in walls nd slab	Not more than three times the wall or slab thickness or more than 18″
lear distance between bars in spirally einforced and tied columns	1½ times the nominal bar diameter, 1½ times the maximum size of the coarse aggregate, or not less than 1½″

Note: The clear distances above also apply for the clear distances between contact splices nd adjacent splices of reinforcing bars.

The preceding is only a brief discussion of some aspects of using rei forced concrete as a construction material. The design, construction, ar quality control of reinforced concrete are covered by a set of criteria. Th material is presented as a guide for the use of the structural details in th book. The American Concrete Institute and the Portland Cement Assoc ation have compiled a large amount of information on the many aspec of reinforced concrete construction. These standards are subject to revisic whenever the studies of the committees responsible indicate that deve opments in concrete design and construction warrant a change.

Working drawings are made to communicate the design requiremen and configurations to the contractor. These drawings should be suppl mented by shop drawings of the individual reinforcing bars and the method of placement. The accuracy of design, the quality of workmanshi and the quality control of the construction materials are important facto in reinforced concrete design and construction. Once the reinforcing ste and the concrete are placed, a total commitment is made by the enginee the job-site inspector, the laboratory testing the materials, and the co tractor. The owner of the building must be able to rely on this commitmer

Section 1: Foundations. Details C1020 to C1105 are a series of co crete continuous footings for 8-in-thick and 10-in-thick concrete concre walls. The footing configurations are rectangular, L-shaped, or inverted- shaped. Details C1120 are series of spread footings supporting wood, stee or concrete columns. Details C1127 to C1129 use a concrete grade bea to restrain the bottom of the column in bending, thus making a fixed en connection. Details C1140 and C1141 show concrete columns at the sprea footing with either column tie reinforcement or spiral reinforcemen Details C1160 to C1165 show pile caps supporting round concrete columr and square concrete piles. Details C1180 to C1185 show a series of detai of pile caps supporting steel wide-flange columns and steel piles. Th balance of the details in this section are concerned with concrete subgrad stairs and walls.

Section 2: Columns. Details C2020 to C2064 show a series of pla sections of the arrangement of reinforcement in square or round concre columns. Details C2080 to C2105 show a series of details of concre column connections to concrete slabs and concrete column splices. Detai C2120 to C2124 show the plan sections of steel columns connected to concrete wall.

Section 3: Concrete Walls and Wood Floors. Details C3020 to C310 show a series of wood floor connections to either 8-in or 10-in-thic concrete walls. Details C3120 and C3121 show sections of full-height cor crete walls supporting wood floors.

Section 4: Concrete Walls and Wood Roofs. Details C4020 to C420 show a series of wood roof connectors to 8-in-thick concrete walls. Detai C4200 and C4201 show sections of full-height concrete walls supportin wood roofs.

Section 5: Concrete Walls and Slabs. Details C5020 to C5060 shov a series of one-story concrete walls supporting concrete slabs. Detail C5080 to C5083 show a series of two-story concrete walls supportin concrete slabs.

Section 6: Concrete Walls and Steel Floors. Details C6020 to C622 show a series of details of a concrete wall supporting metal decking anc concrete floor. Floors are supported by either steel beams, metal ligh gage joists, or bar joists.

Section 7: Cantilever Retaining Walls. Details C7020 to C7061 show a series of concrete retaining walls for a variety of backfill slopes with the retained soil not over the footing. Details C7080 to C7120 show a series of concrete retaining walls for a variety of backfill slopes with the retained soil over the footing.

Section 8: Precast Concrete Walls and Concrete Connections. Details C8020 to C8161 show precast or tilt-up wall sections, connections, and ledger details. Details C8180 to C8261 show corner and intersection connections of concrete walls.

Section 9: Miscellaneous. This section contains various details that may be needed but do not fit into any of the previous sections. The section shows details of reinforcement bar bends, wall openings, jambs and lintels, stairs, and free-standing garden walls.

CHAPTER 3 CONCRETE SECTION 1 FOUNDATIONS			
CONTINUOUS FOOTING EXTERIOR WALL SLAB ON GRADE	RECTANGULAR SHAPE	8" CONCRETE WALL	C1020 - C1023
		10" CONCRETE WALL	C1024 - C1027
CONTINUOUS FOOTING EXTERIOR WALL SLAB ON GRADE WITH A CURB	L SHAPE	8" CONCRETE WALL	C1040 - C1043
		10" CONCRETE WALL	M1044 - C1046
CONTINUOUS FOOTING EXTERIOR WALL SLAB ON GRADE		8" CONCRETE WALL	C1060 - C1063
		10" CONCRETE WALL	C1064 - C1066
CONTINUOUS FOOTING INTERIOR WALL SLAB ON GRADE	RECTANGULAR SHAPE	8" CONCRETE WALL	C1080 - C1083
		10" CONCRETE WALL	C1080 - C1086
CONTINUOUS FOOTING INTERIOR WALL SLAB ON GRADE	T SHAPE	8" CONCRETE WALL	C1100 - C1103
		10" CONCRETE WALL	C1104 - C1106
SPREAD FOOTINGS	WOOD POST		C1120 - C1121
	PIPE COLUMN		C1122 - C114
	WIDE FLANGE COLUMN		C1125
	PIPE COLUMN & PEDESTAL		C1126
	WIDE FLANGE COLUMN & GRADE BEAM		C1127 - C1129
SPREAD FOOTINGS	CONCRETE COLUMNS WITH TIE REINF.		C1140
	CONCRETE COLUMNS WITH SPIRALS		C1141
PILE CAPS ROUND CONCRETE COLUMNS SPIRAL REINFORCEMENT 12" SQUARE PRECAST PILES	————	2 PILES	C1160
	————	3 PILES	C1161
	————	4 PILES	C1162
	————	5 PILES	C1163
	————	6 PILES	C1164
	————	7 PILES	C1165
PILE CAPS STEEL WIDE FLANGE COLUMNS 12" HP STEEL PILES	————	2 PILES	C1180
	————	3 PILES	C1181
	————	4 PILES	C1182
	————	5 PILES	C1183
	————	6 PILES	C1184
	————	7 PILES	C1185
CONCRETE STEPS ON GRADE CONCRETE SIDE WALLS	————	————	C1200 - C1201
STEPPED CONTINUOUS FOOTING	————	————	C1220
MASONRY WALL & WOOD FLOOR CONNECTION AT GRADE	————	————	C1240 - C1241
ELEVATOR PIT	————	————	C1260 - C1261
BASEMENT WALLS WITH WOOD FLOORS ABOVE	————	————	C1280 - C1281

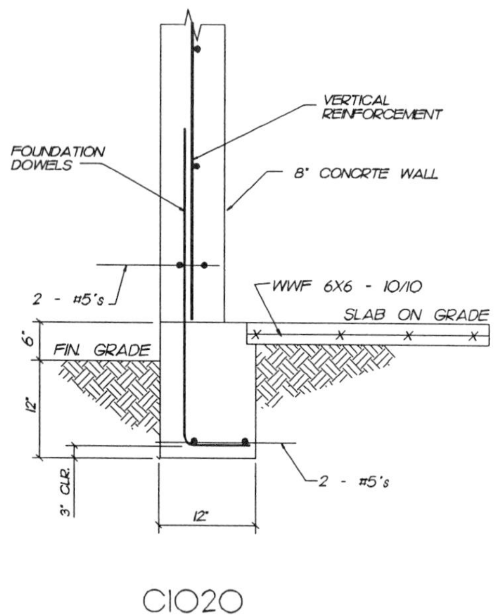

C1020

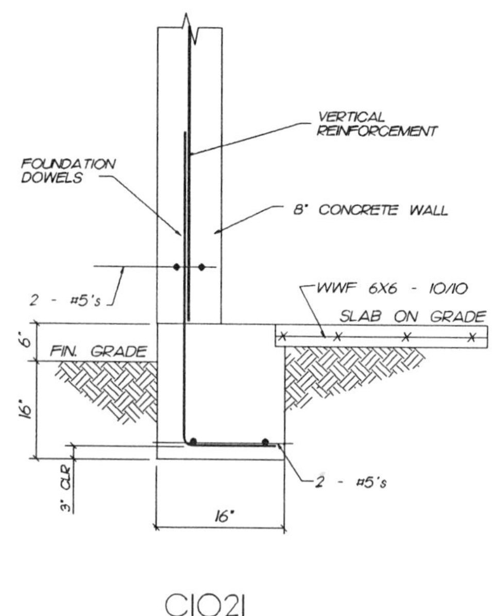

C1021

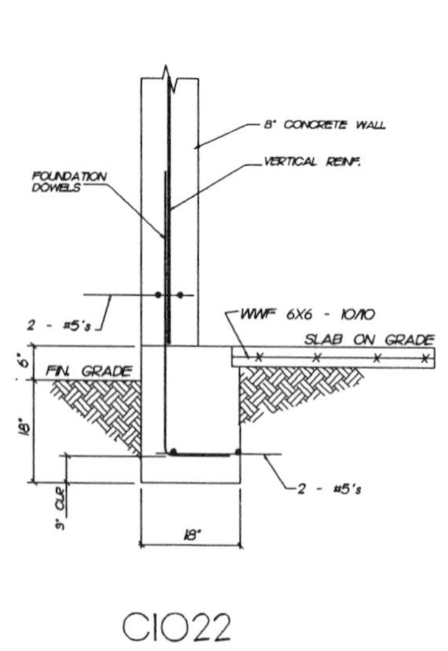

C1022

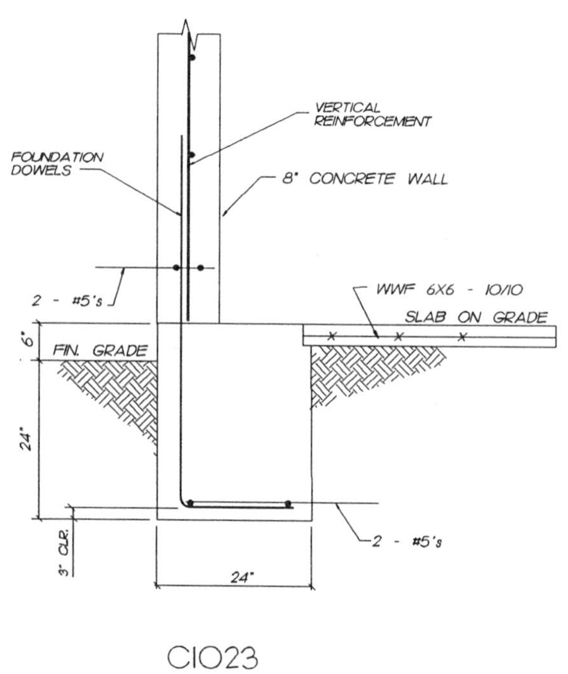

C1023

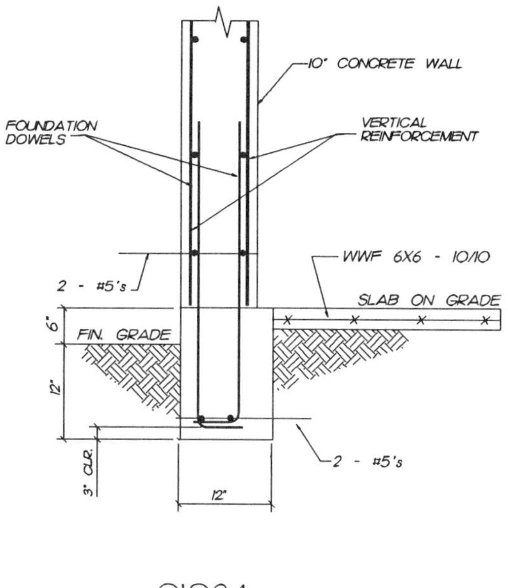

CIO24

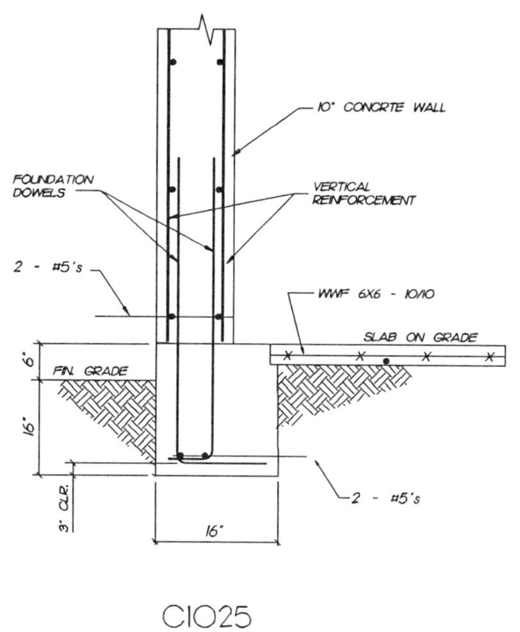

CIO25

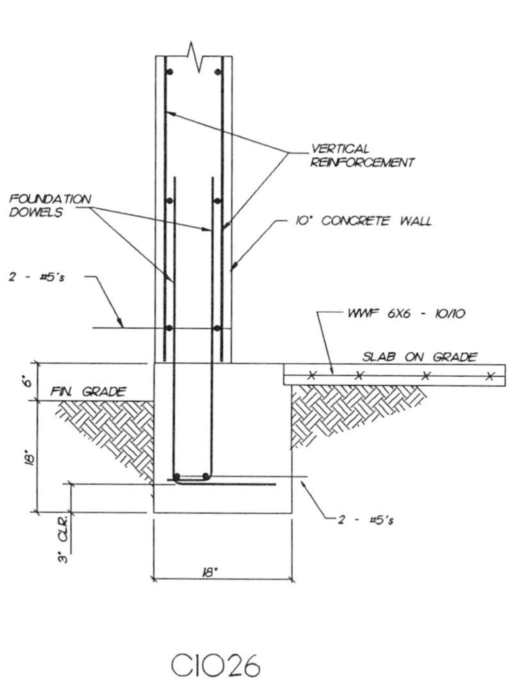

CIO26

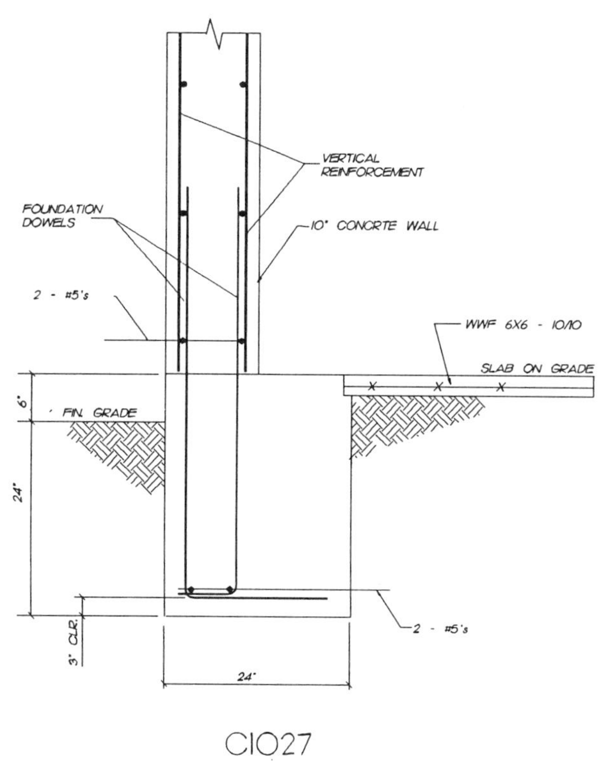

CIO27

VERTICAL REINFORCEMENT
FOUNDATION DOWELS
8" CONCRETE WALL
2 - #5's
WWF 6X6 - 10/10
SLAB ON GRADE
6"
FIN. GRADE
12"
6"
3" CLR.
2 - #5's
12"

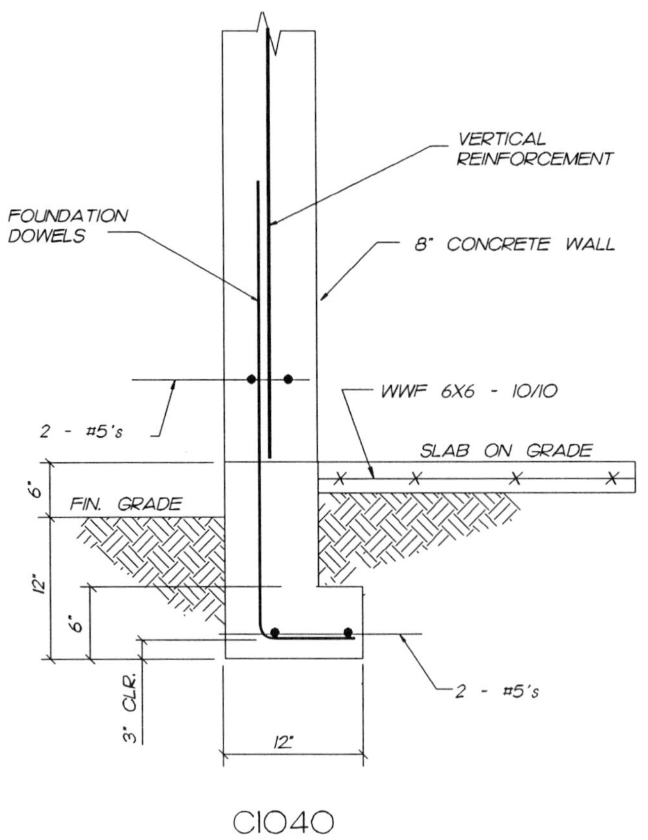

C1040

VERTICAL REINFORCEMENT
FOUNDATION DOWELS
8" CONCRETE WALL
2 - #5's
WWF 6X6 - 10/10
SLAB ON GRADE
6"
FIN. GRADE
16"
8"
3" CLR.
2 - #5's
8"
16"

C1041

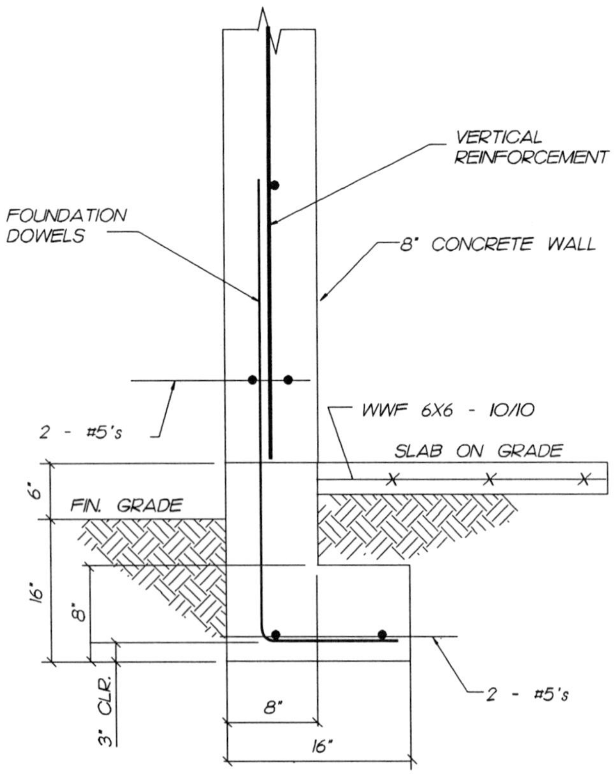

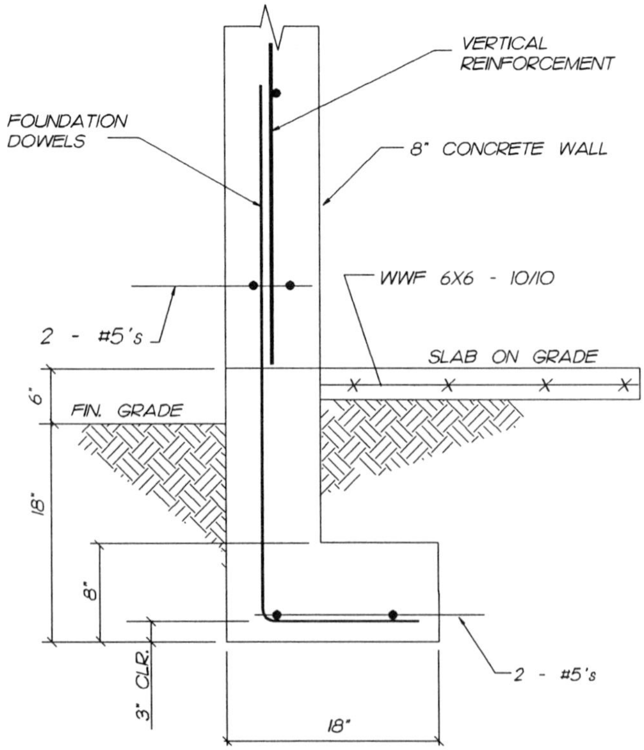

VERTICAL REINFORCEMENT
FOUNDATION DOWELS
8" CONCRETE WALL
WWF 6X6 - 10/10
2 - #5's
SLAB ON GRADE
6"
FIN. GRADE
18"
8"
3" CLR.
2 - #5's
18"

C1042

VERTICAL
REINFORCEMENT
FOUNDATION
DOWELS
8" CONCRETE WALL
2 - #5's
WWF 6X6 - 10/10
SLAB ON GRADE
6"
FIN. GRADE
24"
10"
3" CLR.
10"
24"
2 - #5's

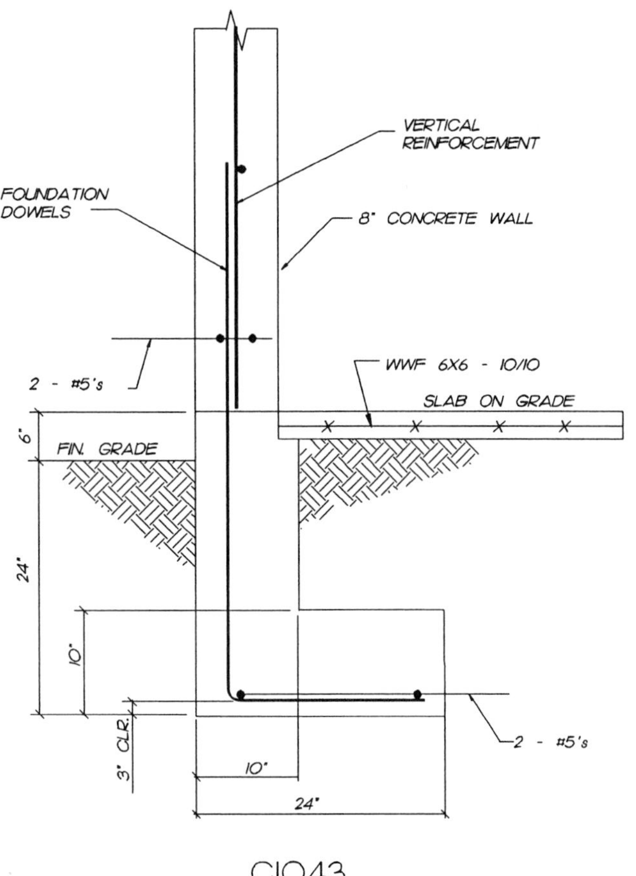

C1043

VERTICAL
REINFORCEMENT
FOUNDATION
DOWELS
10" CONCRTE WALL
2 - #5's
WWF 6X6 - 10/10
SLAB ON GRADE
6"
FIN. GRADE
16"
12"
3" CLR.
16"
2 - #5's

C1044

VERTICAL
REINFORCEMENT
FOUNDATION
DOWELS
10" CONCRETE WALL
2 - #5's
WWF 6X6 - 10/10
SLAB ON GRADE
6"
FIN. GRADE
18"
12"
3" CLR.
18"
2 - #5's

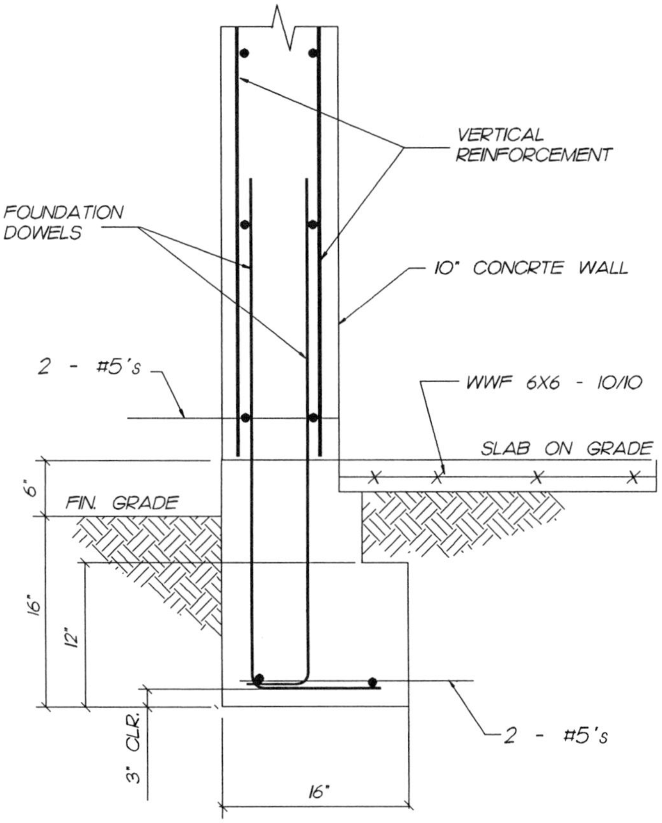

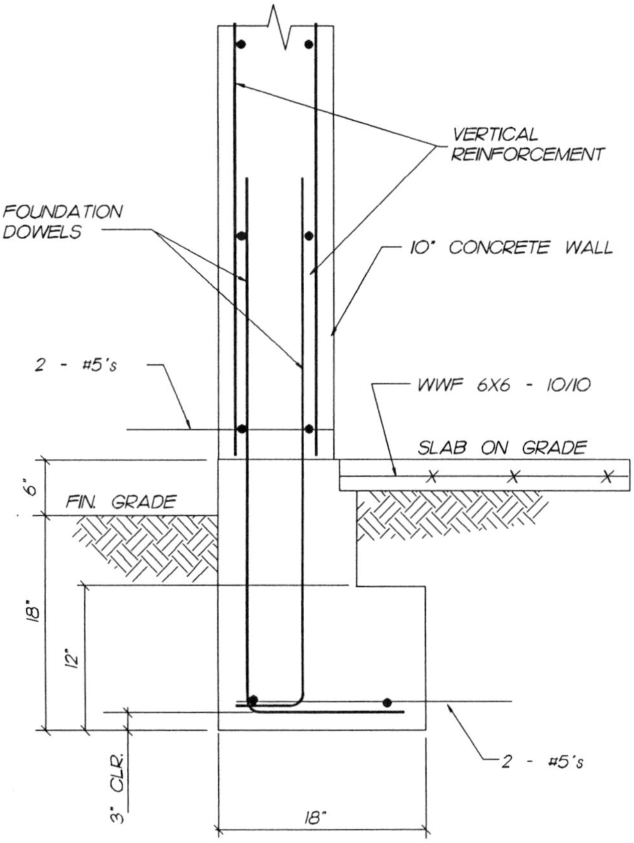

C1045

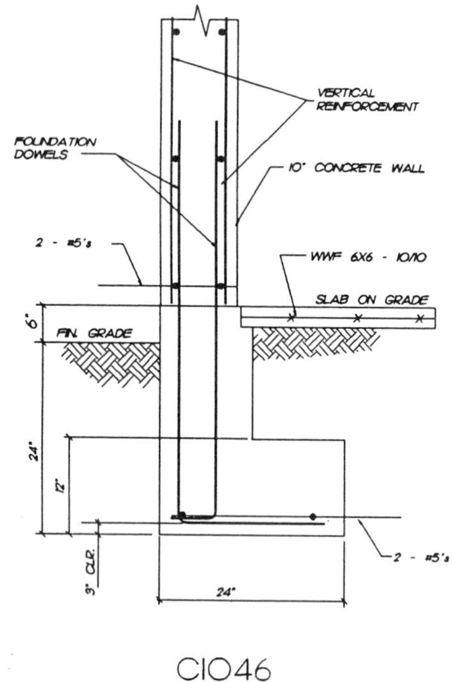

C1046

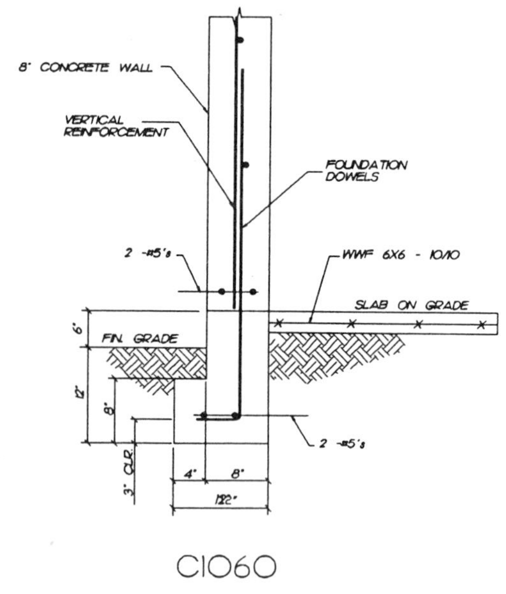

C1060

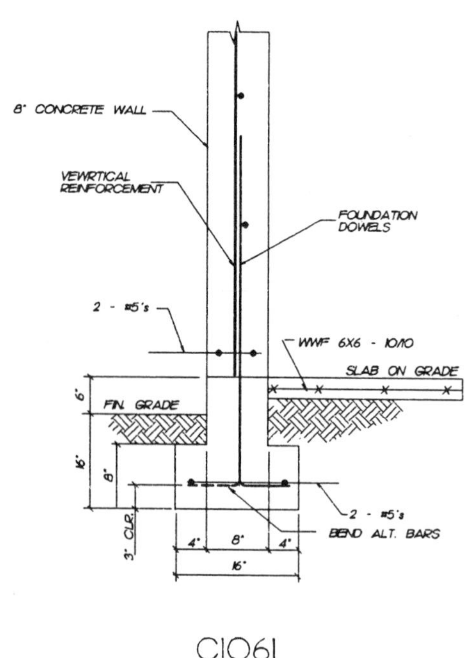

C1061

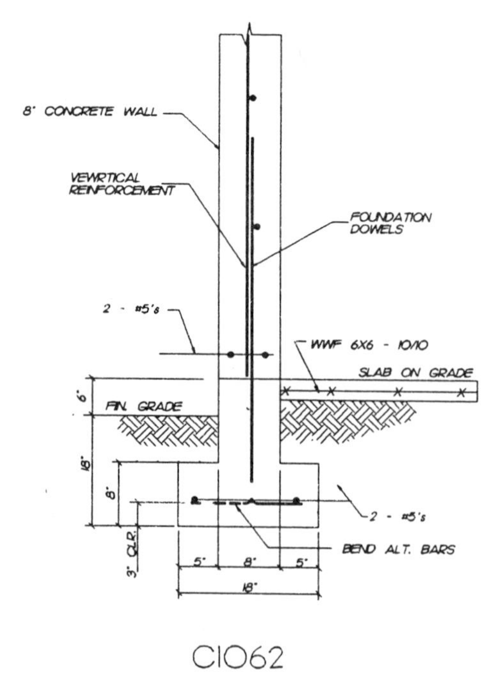

C1062

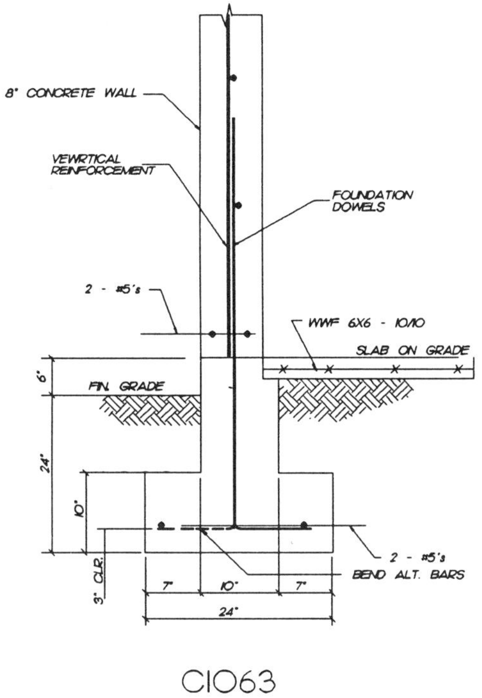

C1063

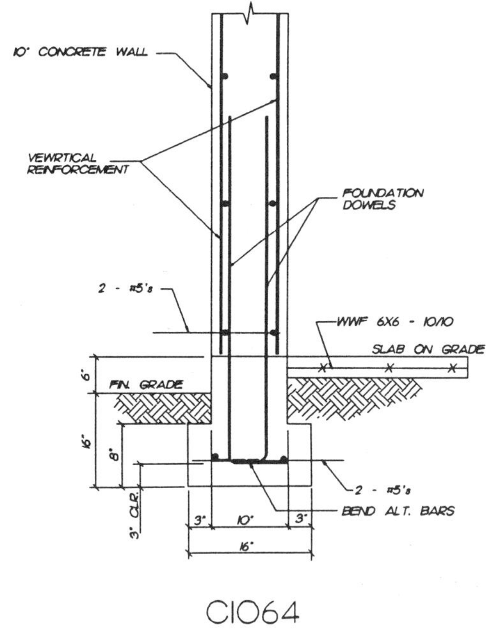

C1064

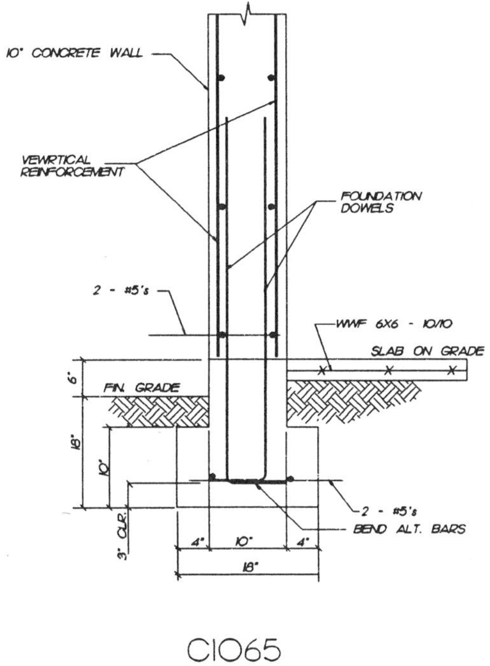

C1065

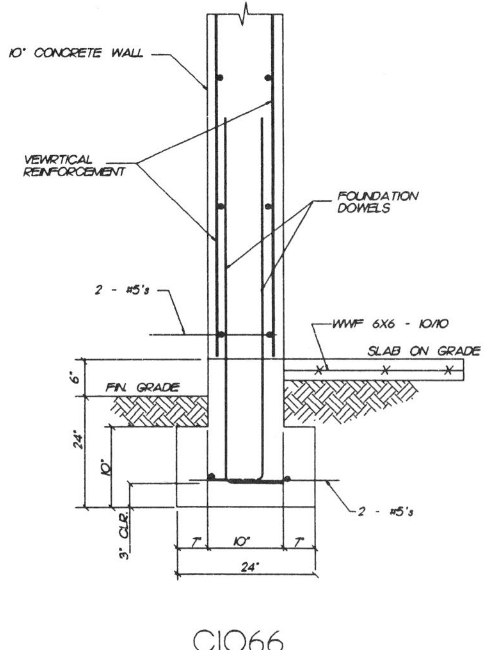

C1066

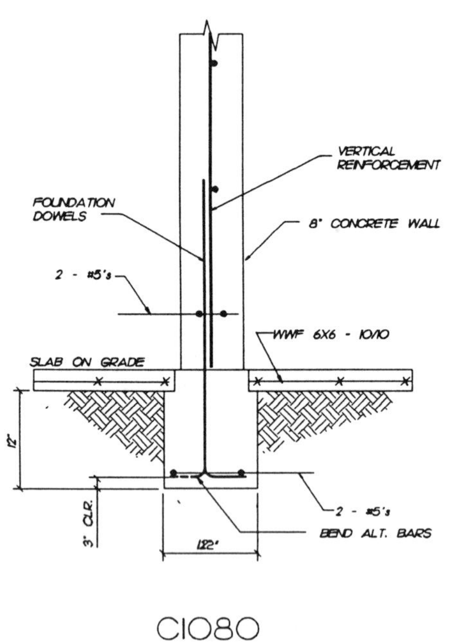

C1080

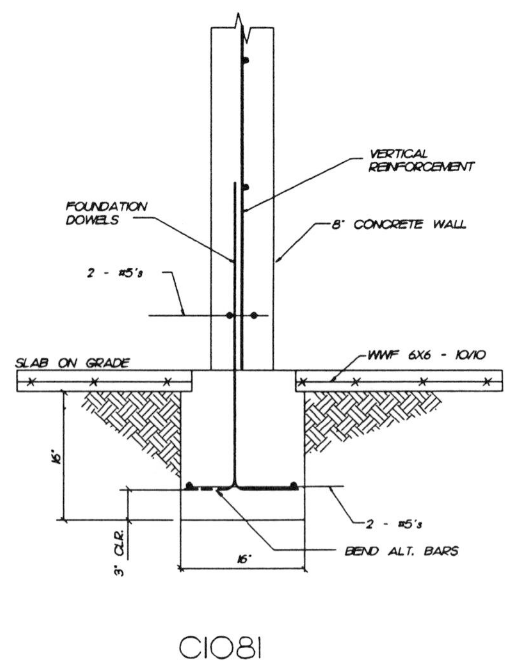

C1081

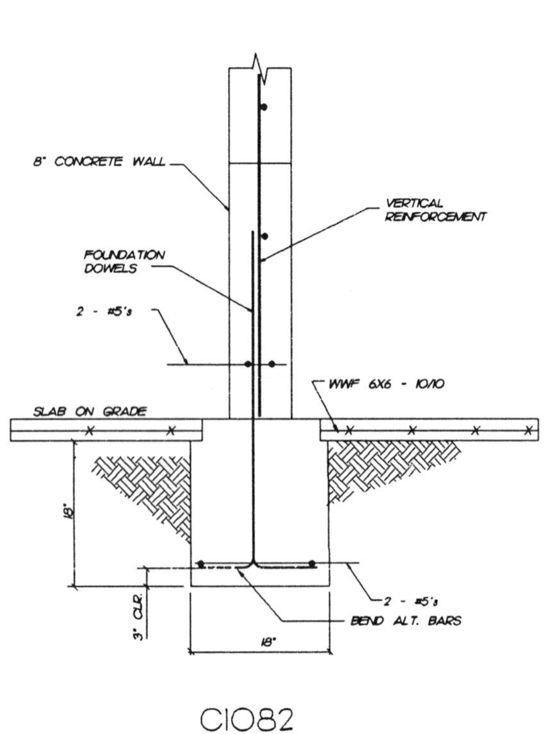

C1082

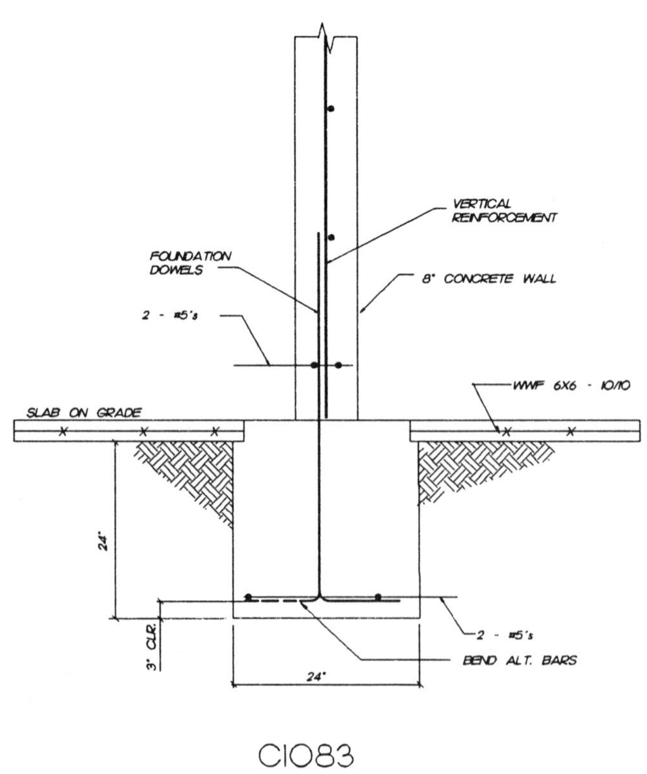

C1083

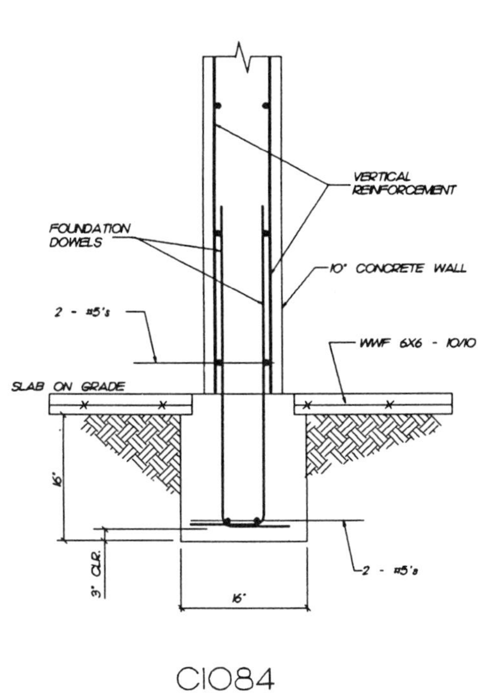

C1084

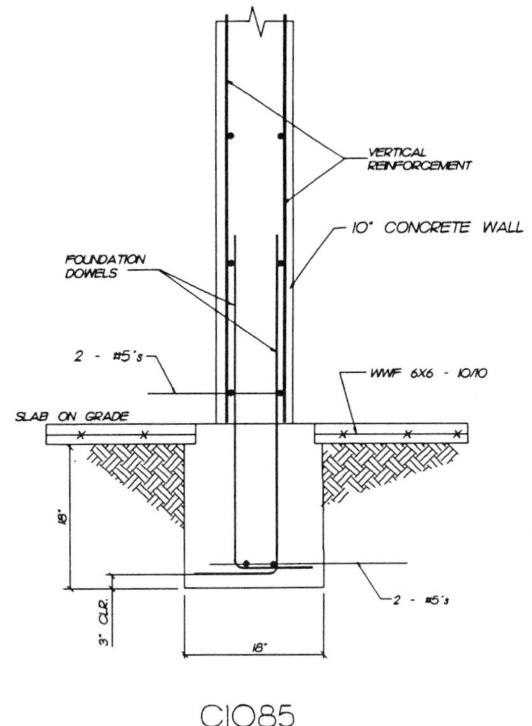

C1085

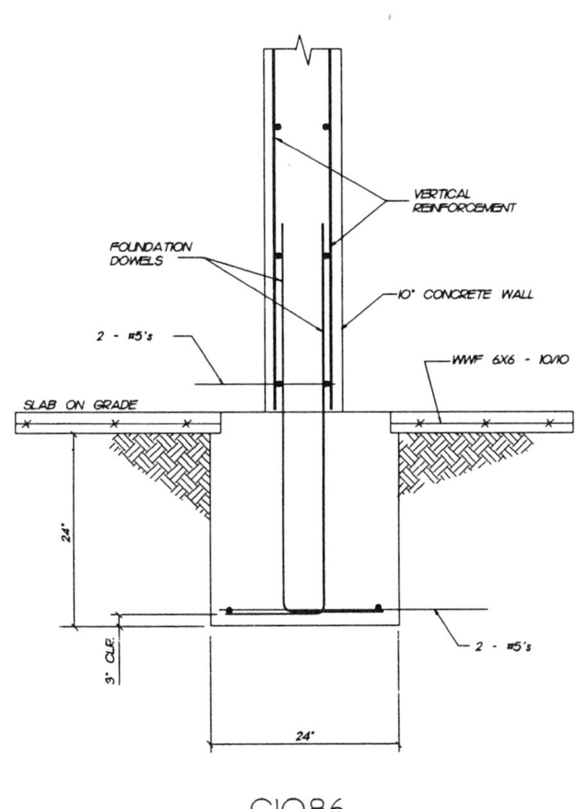

C1086

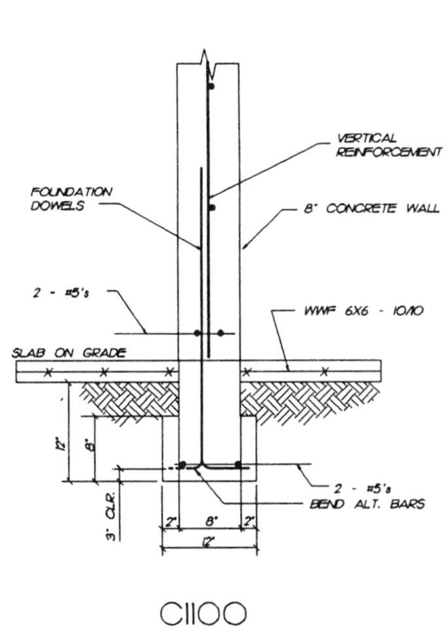

C1100

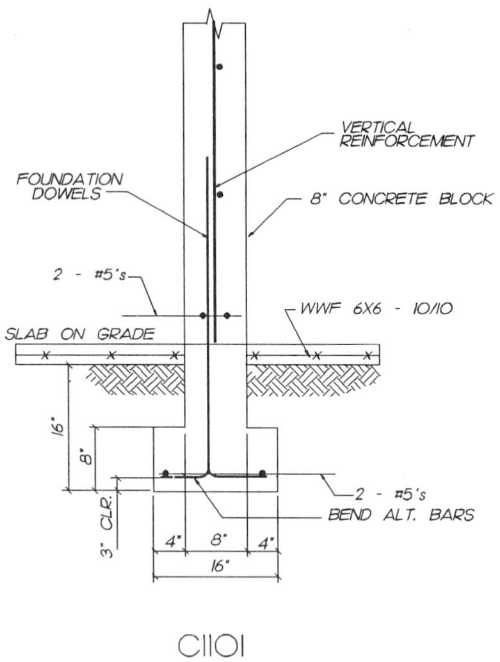

C1101

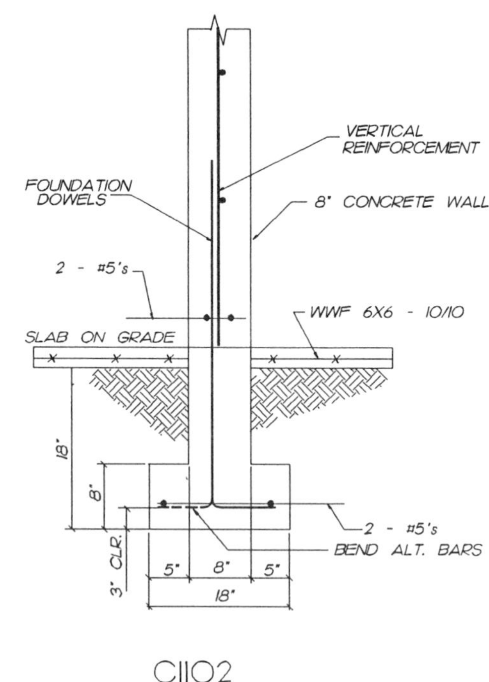

C1102

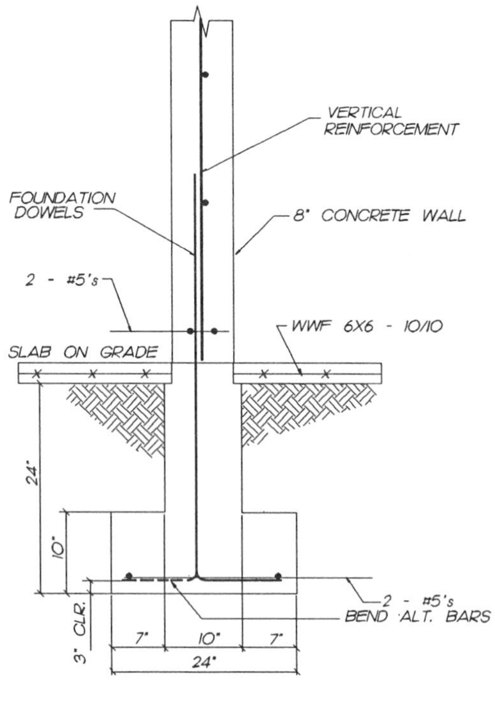

C1103

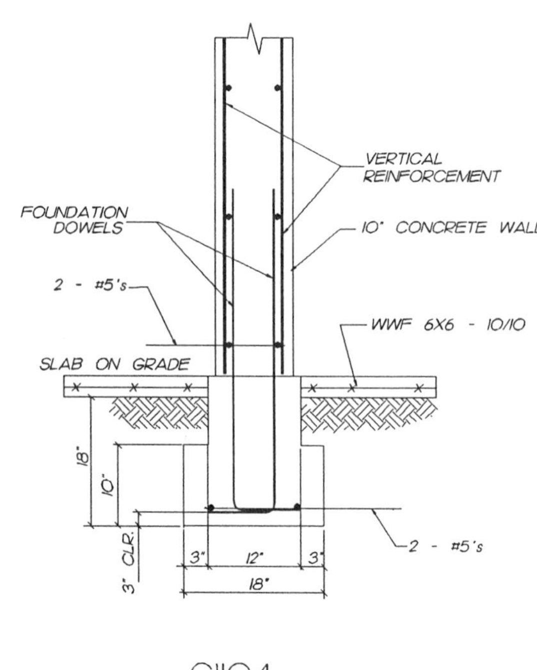

C1104

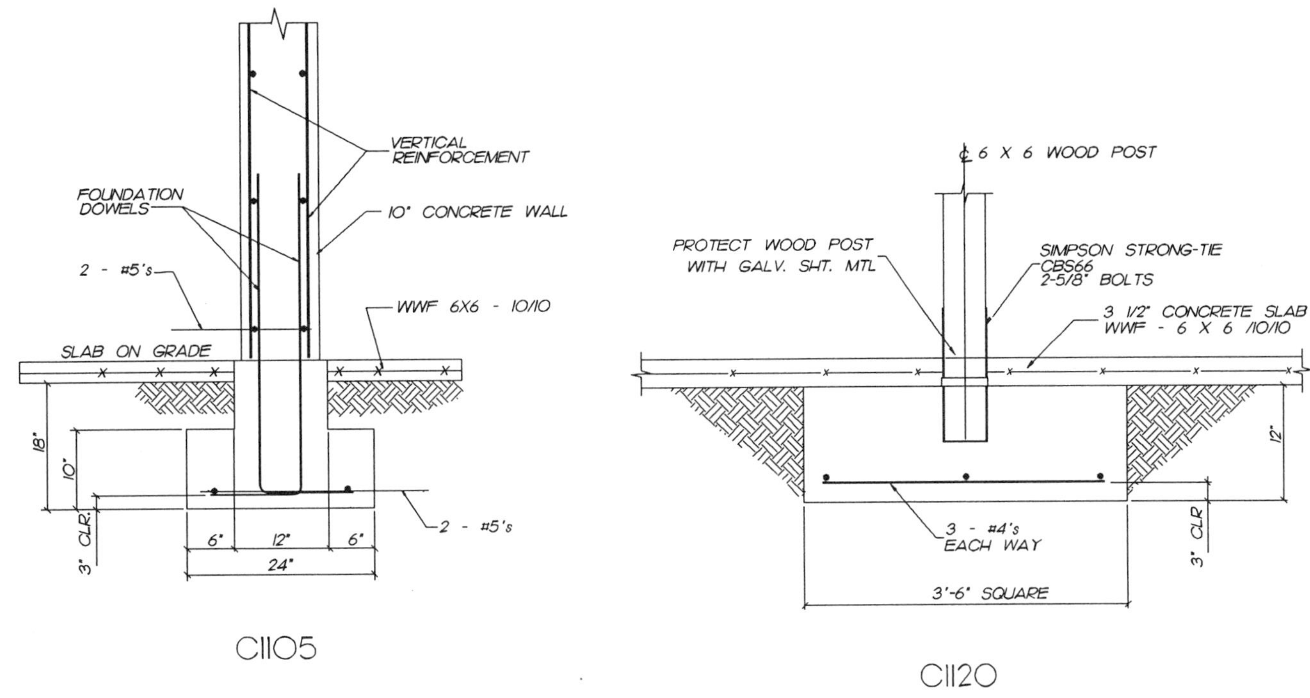

C1105

C1120

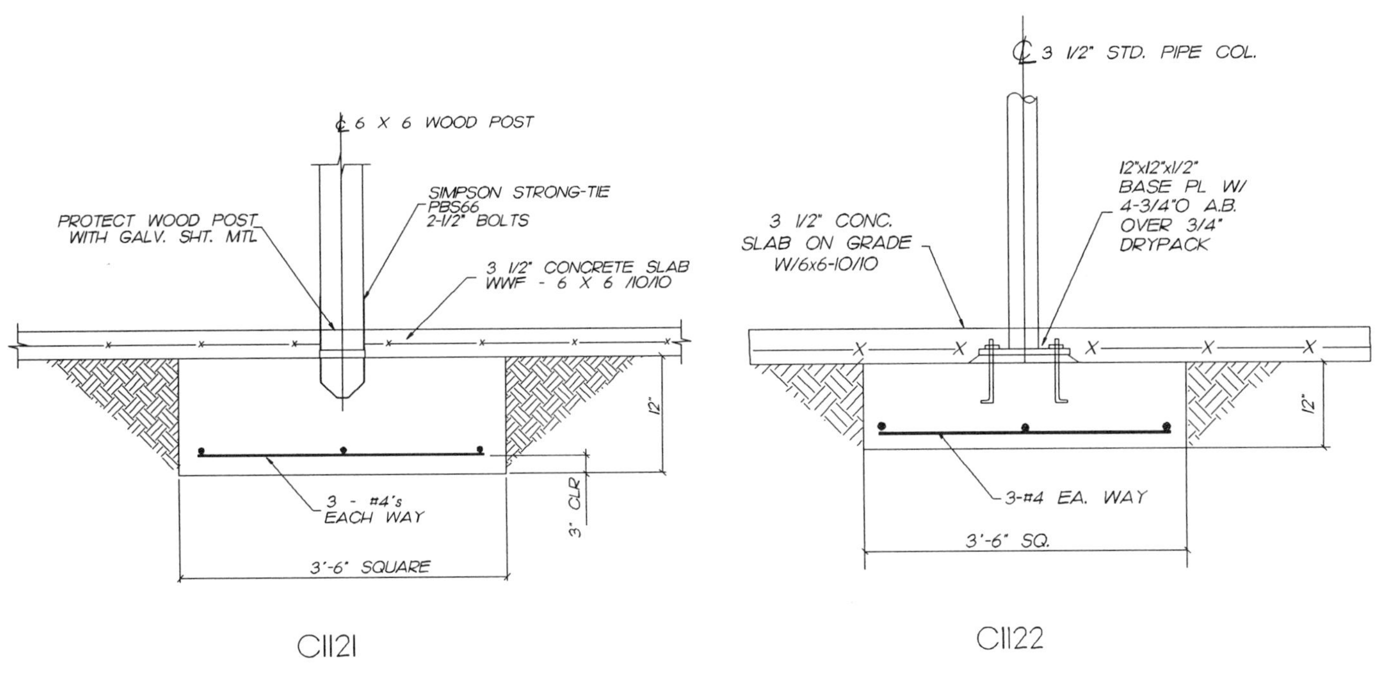

C1121

C1122

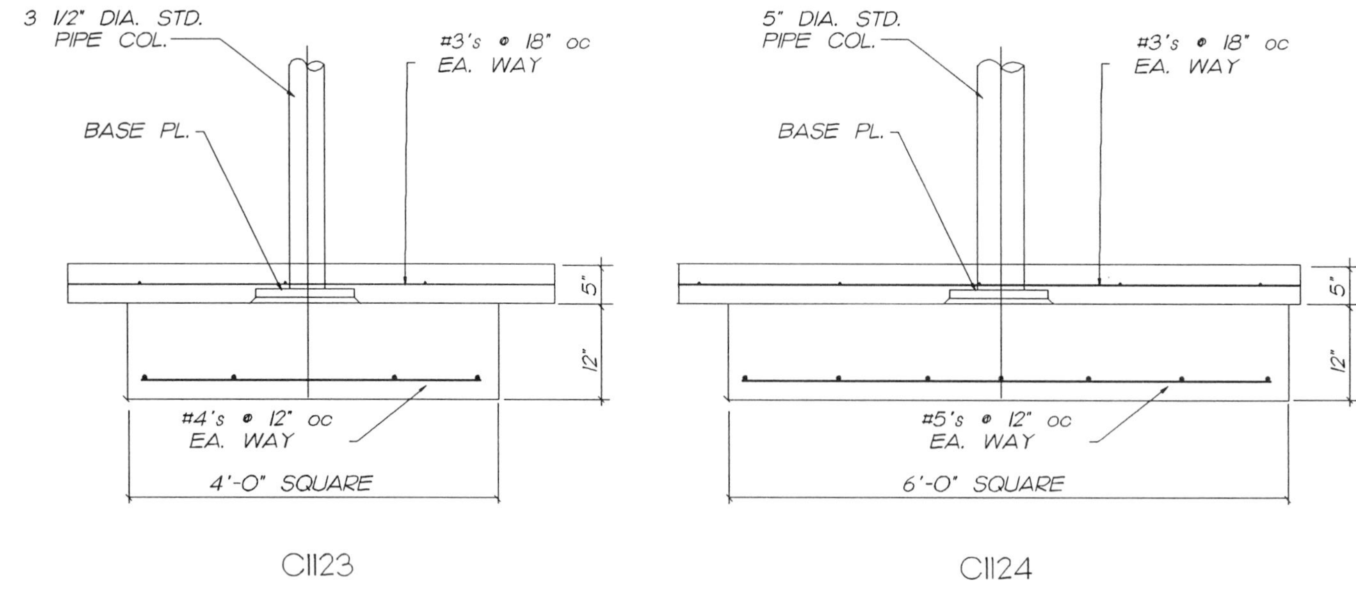

3 1/2" DIA. STD.
PIPE COL.
#3's @ 18" oc
EA. WAY
BASE PL.
5"
12"
#4's @ 12" oc
EA. WAY
4'-0" SQUARE

C1123

5" DIA. STD.
PIPE COL.
#3's @ 18" oc
EA. WAY
BASE PL.
5"
12"
#5's @ 12" oc
EA. WAY
6'-0" SQUARE

C1124

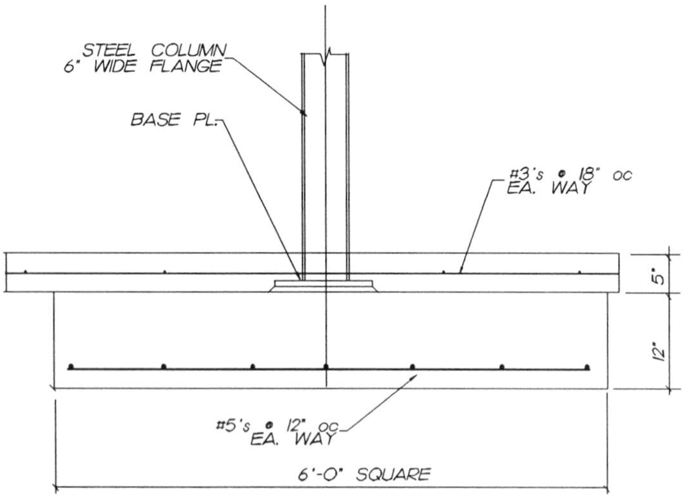

STEEL COLUMN
6" WIDE FLANGE
BASE PL.
#3's @ 18" oc
EA. WAY
5"
12"
#5's @ 12" oc
EA. WAY
6'-0" SQUARE

C1125

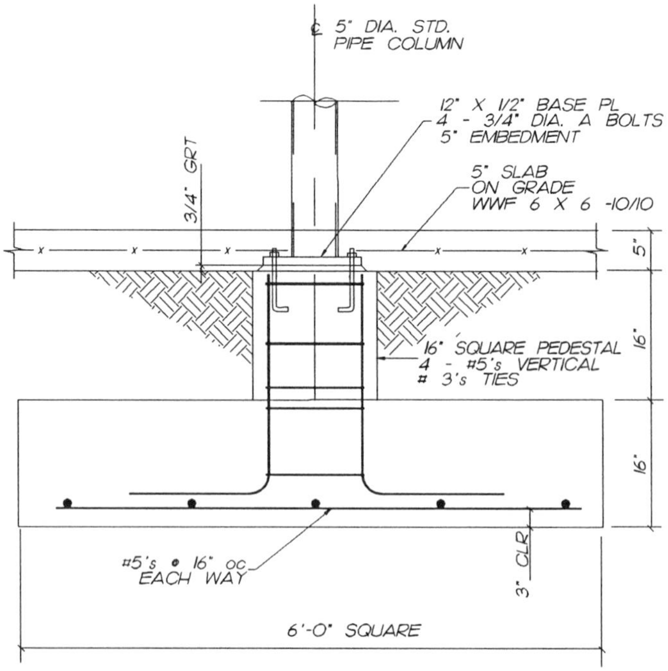

℄ 5" DIA. STD.
PIPE COLUMN
12" X 1/2" BASE PL
4 - 3/4" DIA. A BOLTS
5" EMBEDMENT
3/4" GRT
5" SLAB
ON GRADE
WWF 6 X 6 -10/10
5"
16" SQUARE PEDESTAL
4 - #5's VERTICAL
3's TIES
16"
16"
#5's @ 16" oc
EACH WAY
3" CLR
6'-0" SQUARE

C1126

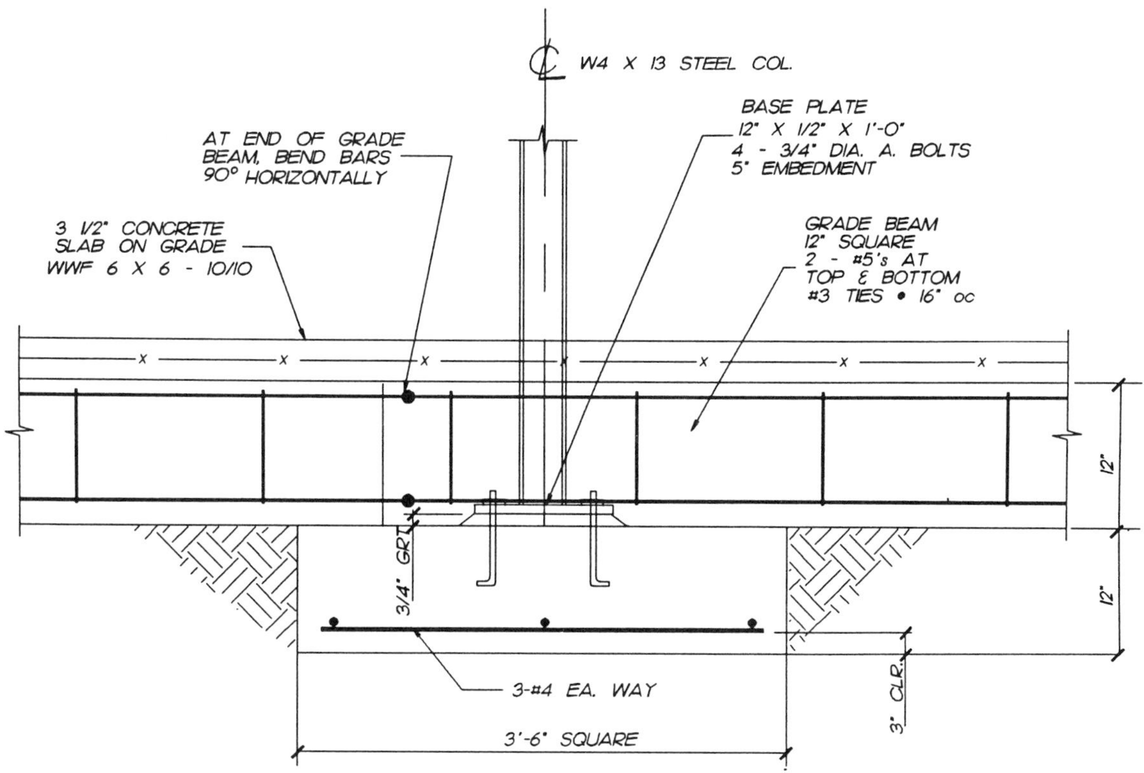
W4 X 13 STEEL COL.
BASE PLATE
12" X 1/2" X 1'-0"
4 - 3/4" DIA. A. BOLTS
5" EMBEDMENT
AT END OF GRADE
BEAM, BEND BARS
90° HORIZONTALLY
3 1/2" CONCRETE
SLAB ON GRADE
WWF 6 X 6 - 10/10
GRADE BEAM
12" SQUARE
2 - #5's AT
TOP & BOTTOM
#3 TIES @ 16" oc
12"
3/4" GRT.
12"
3-#4 EA. WAY
3" CLR.
3'-6" SQUARE

C1127

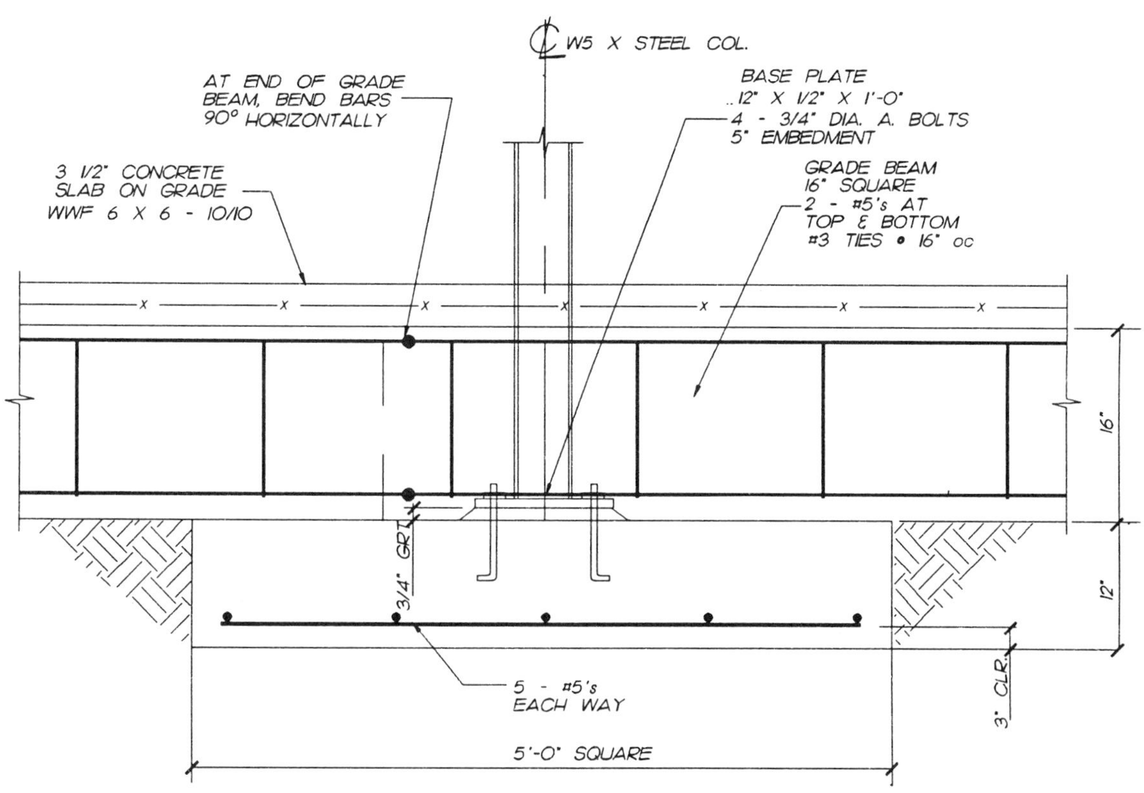
W5 X STEEL COL.
AT END OF GRADE
BEAM, BEND BARS
90° HORIZONTALLY
BASE PLATE
12" X 1/2" X 1'-0"
4 - 3/4" DIA. A. BOLTS
5" EMBEDMENT
3 1/2" CONCRETE
SLAB ON GRADE
WWF 6 X 6 - 10/10
GRADE BEAM
16" SQUARE
2 - #5's AT
TOP & BOTTOM
#3 TIES @ 16" oc
16"
3/4" GRT.
12"
5 - #5's
EACH WAY
3" CLR.
5'-0" SQUARE

C1128

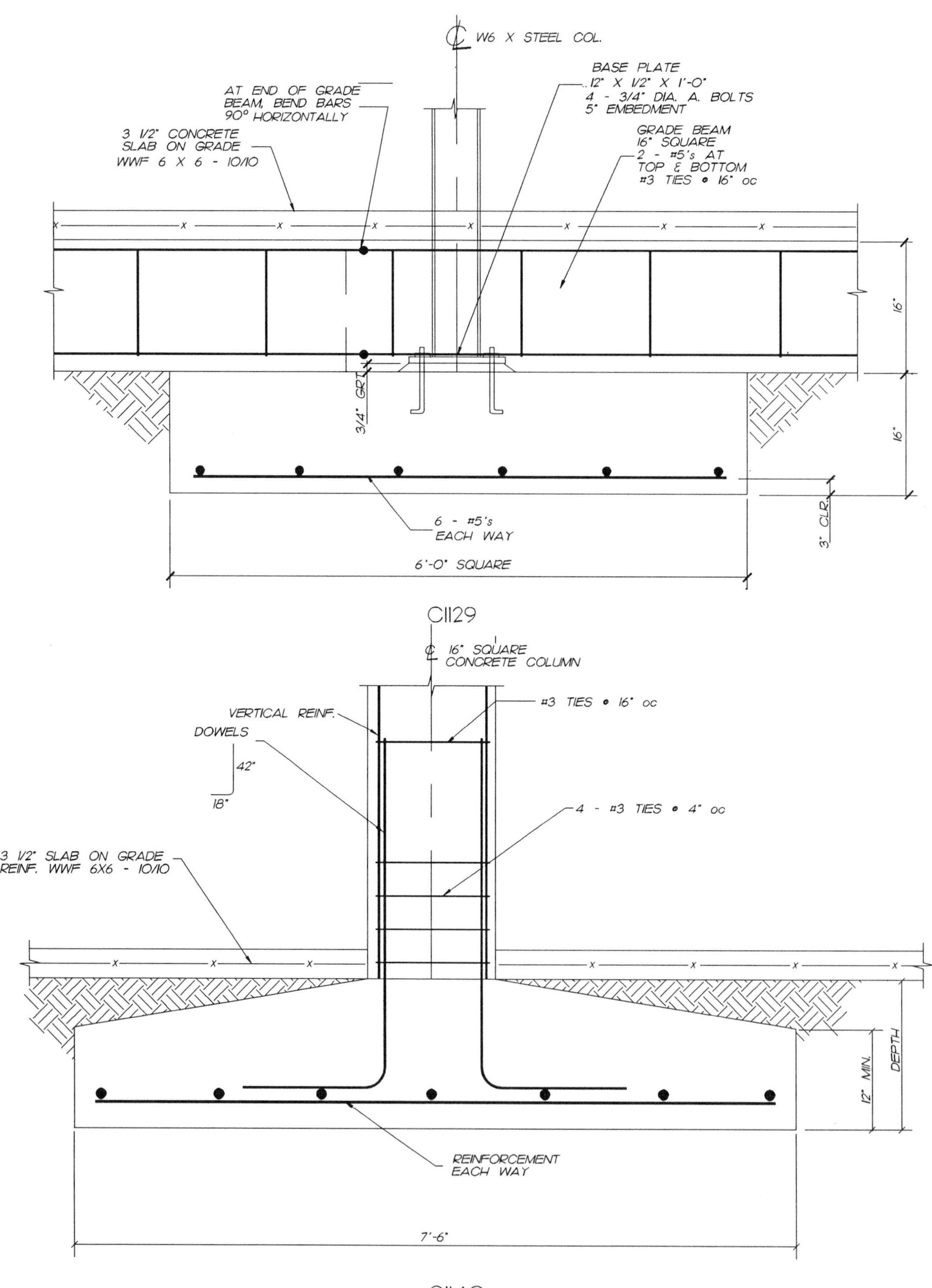
W6 X STEEL COL.
BASE PLATE
12" X 1/2" X 1'-0"
4 - 3/4" DIA. A. BOLTS
5" EMBEDMENT
AT END OF GRADE
BEAM, BEND BARS
90° HORIZONTALLY
3 1/2" CONCRETE
SLAB ON GRADE
WWF 6 X 6 - 10/10
GRADE BEAM
16" SQUARE
2 - #5's AT
TOP & BOTTOM
#3 TIES @ 16" oc
16"
16"
3/4" GRT.
6 - #5's
EACH WAY
3" CLR.
6'-0" SQUARE
C1129

16" SQUARE
CONCRETE COLUMN
VERTICAL REINF.
DOWELS
42"
18"
#3 TIES @ 16" oc
4 - #3 TIES @ 4" oc
3 1/2" SLAB ON GRADE
REINF. WWF 6X6 - 10/10
12" MIN.
DEPTH
REINFORCEMENT
EACH WAY
7'-6"
C1140

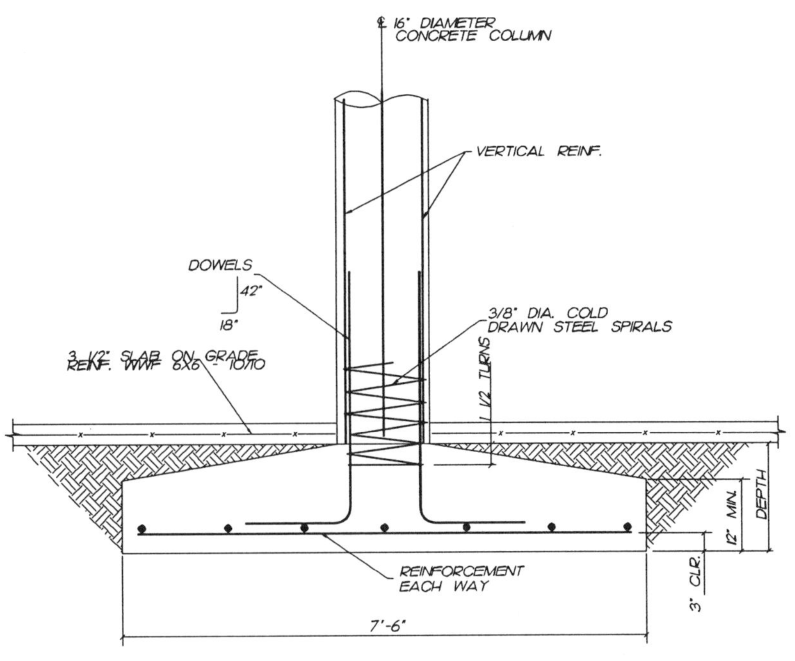
℄ 16" DIAMETER
CONCRETE COLUMN
VERTICAL REINF.
DOWELS
42"
18"
3/8" DIA. COLD
DRAWN STEEL SPIRALS
3 1/2" SLAB ON GRADE
REINF. WWF 6X6 - 10/10
1 1/2 TURNS
12" MIN.
DEPTH
REINFORCEMENT
EACH WAY
3" CLR.
7'-6"

C1141

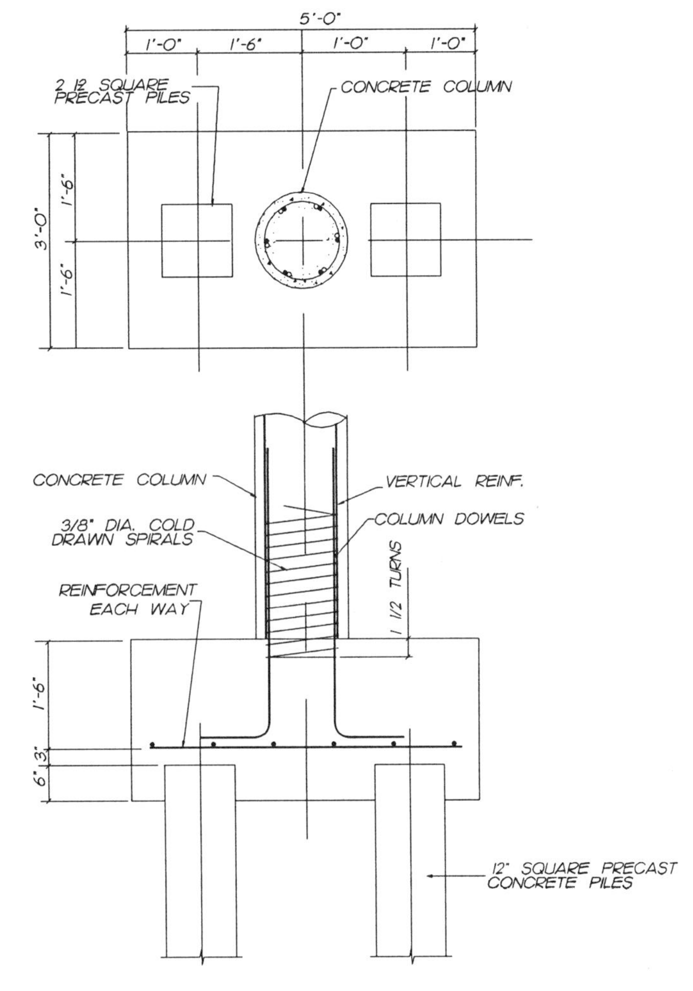
5'-0"
1'-0"
1'-6"
1'-0"
1'-0"
2 12 SQUARE
PRECAST PILES
CONCRETE COLUMN
3'-0"
1'-6"
1'-6"
CONCRETE COLUMN
VERTICAL REINF.
3/8" DIA. COLD
DRAWN SPIRALS
COLUMN DOWELS
REINFORCEMENT
EACH WAY
1 1/2 TURNS
1'-6"
3"
6"
12" SQUARE PRECAST
CONCRETE PILES

C1160

6'-6"
3 - 12 SQUARE PRECAST PILES
CONCRETE COLUMN
1'-0"
2'-7 1/2"
1'-0"
1'-6"
1'-6"
CONCRETE COLUMN
VERTICAL REINF.
COLUMN DOWELS
3/8" DIA. COLD DRAWN SPIRALS
REINFORCEMENT EACH WAY
1 1/2 TURNS
1'-6"
3"
6"
12" SQUARE PRECAST CONCRETE PILES

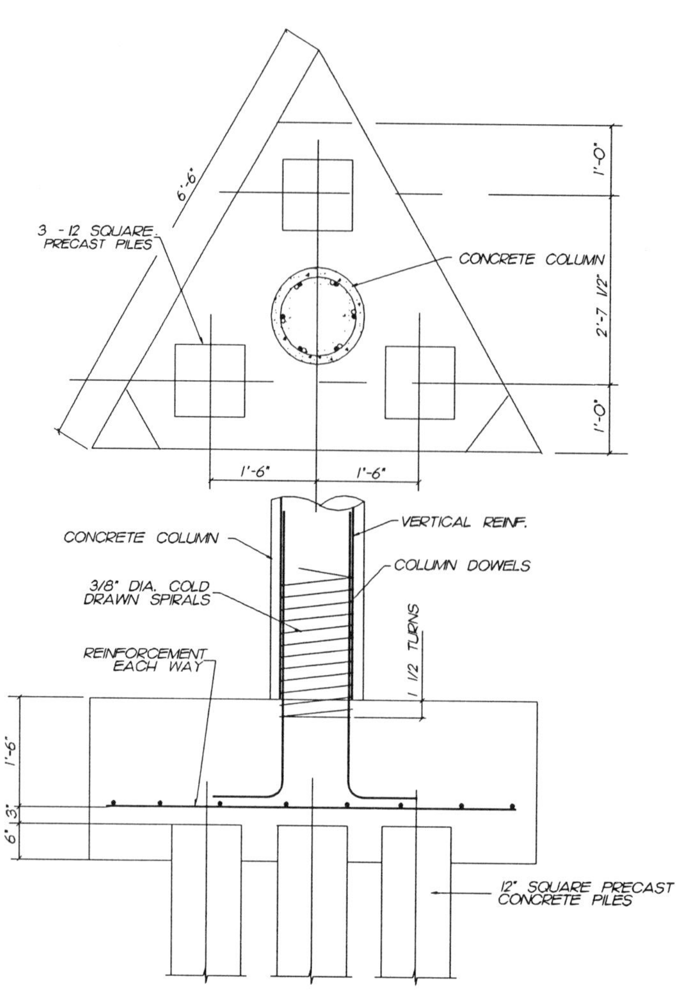

C1161

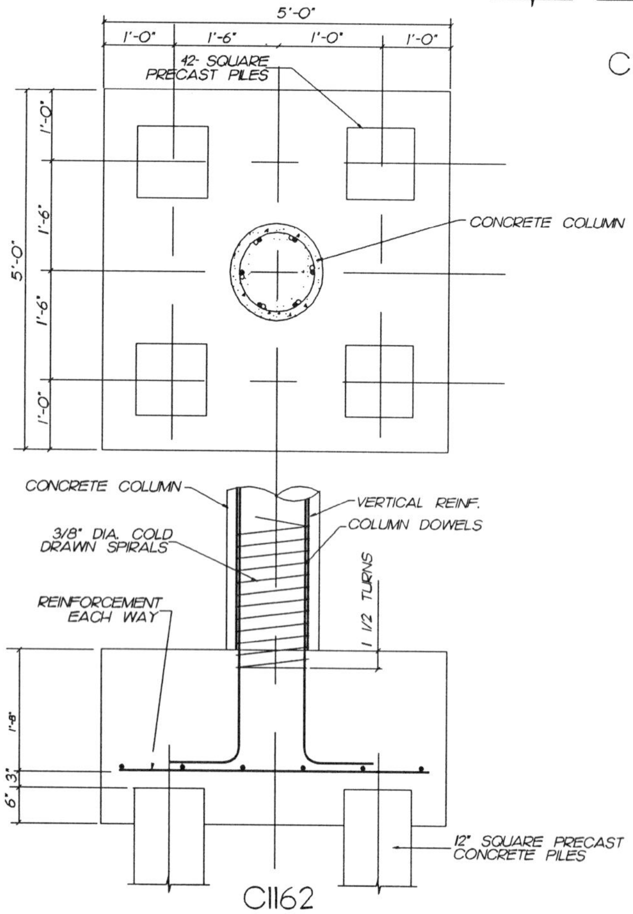

5'-0"
1'-0"
1'-6"
1'-0"
1'-0"
42- SQUARE PRECAST PILES
CONCRETE COLUMN
1'-0"
1'-6"
5'-0"
1'-6"
1'-0"
CONCRETE COLUMN
VERTICAL REINF.
COLUMN DOWELS
3/8" DIA. COLD DRAWN SPIRALS
REINFORCEMENT EACH WAY
1 1/2 TURNS
1'-8"
3"
6"
12" SQUARE PRECAST CONCRETE PILES

C1162

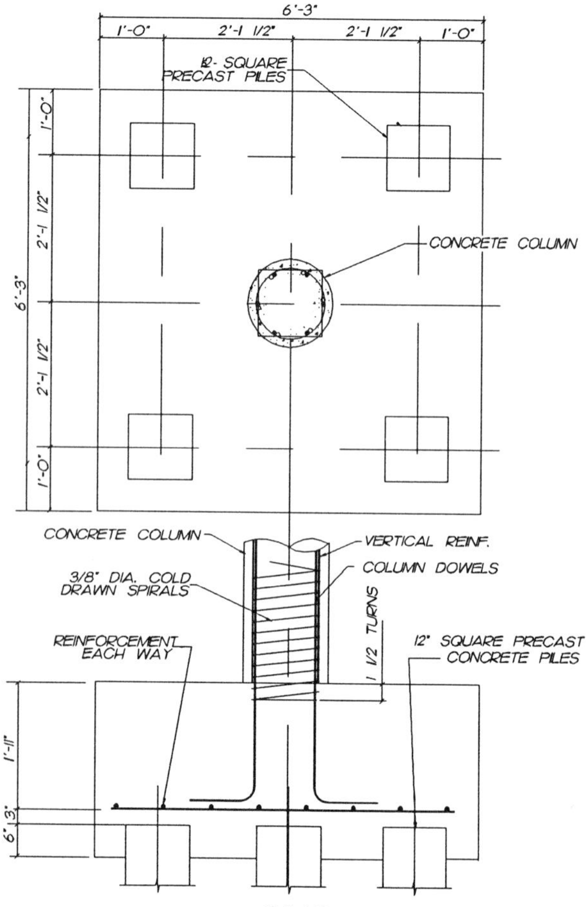

6'-3"
1'-0"
2'-1 1/2"
2'-1 1/2"
1'-0"
12- SQUARE PRECAST PILES
CONCRETE COLUMN
1'-0"
2'-1 1/2"
6'-3"
2'-1 1/2"
1'-0"
CONCRETE COLUMN
VERTICAL REINF.
COLUMN DOWELS
3/8" DIA. COLD DRAWN SPIRALS
REINFORCEMENT EACH WAY
1 1/2 TURNS
12" SQUARE PRECAST CONCRETE PILES
1'-11"
3"
6"

C1163

8'-0"
1'-0"
3'-0"
3'-0"
1'-0"
6 - 12" SQUARE PRECAST PILES
CONCRETE COLUMN
5'-0"
1'-0"
1'-6"
1'-0"
CONCRETE COLUMN
VERTICAL REINF.
COLUMN DOWELS
3/8" DIA. COLD DRAWN SPIRALS
REINFORCEMENT EACH WAY
1 1/2 TURNS
2'-11"
2'-2"
6"
3"
6 -12" SQUARE PRECAST CONCRETE PILES

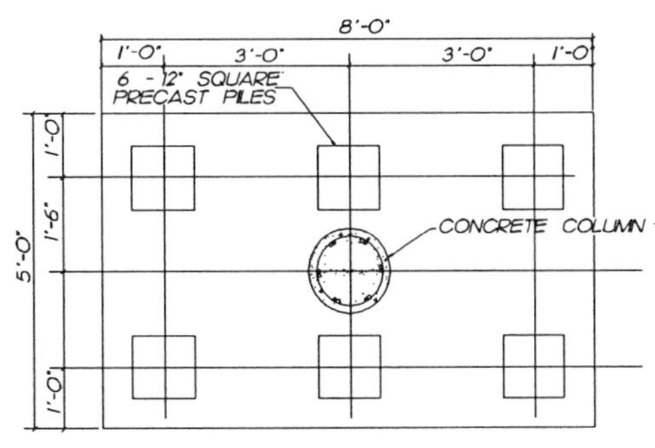

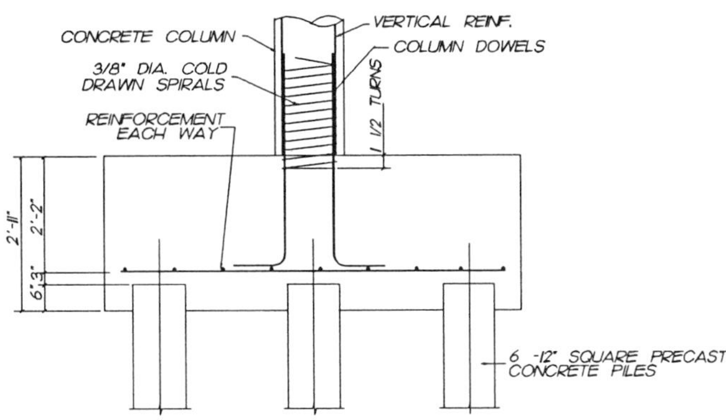

C1164

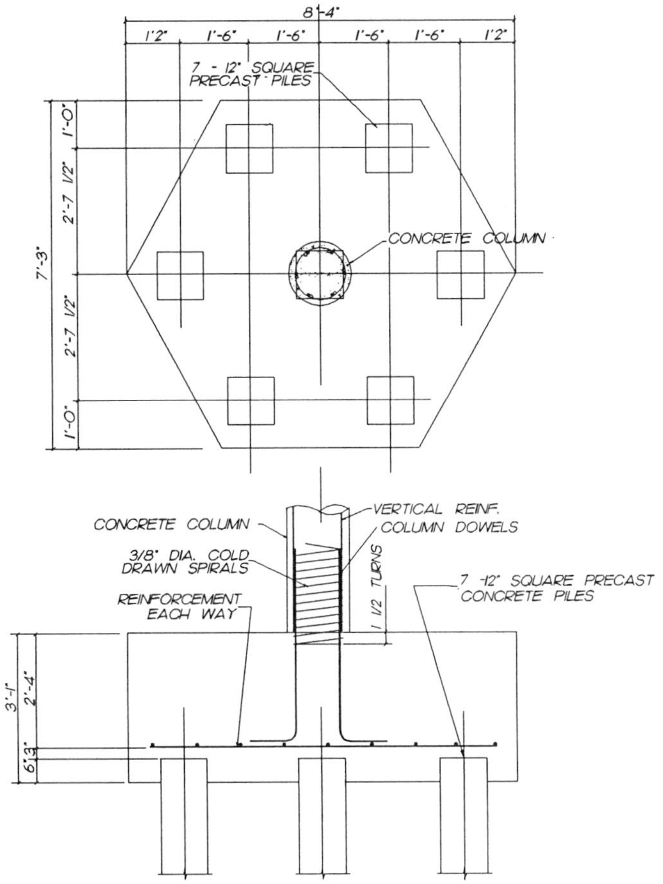

8'-4"
1'2"
1'-6"
1'-6"
1'-6"
1'-6"
1'2"
7 - 12" SQUARE PRECAST PILES
CONCRETE COLUMN
7'-3"
1'-0"
2'-7 1/2"
2'-7 1/2"
1'-0"
CONCRETE COLUMN
VERTICAL REINF.
COLUMN DOWELS
3/8" DIA. COLD DRAWN SPIRALS
REINFORCEMENT EACH WAY
1 1/2 TURNS
7 -12" SQUARE PRECAST CONCRETE PILES
3'-1"
2'-4"
6"
3"

C1165

5'-0"
1'-0"
1'-6"
1'-0"
1'-0"
℄ 14" WIDE FLANGE COLUMN
3'-0"
1'-6"
1'-6"
4 - 3/4" DIA. ANCHOR BOLTS 5" EMBEDMENT
℄ 14" WIDE FLANGE COLUMN
REINFORCEMENT EACH WAY
3/4" GROUT
1'-6"
6"
3"
BEARING PLs. 14" X 3/4" X 1'2"
2 - HP12X53 STEEL PILES

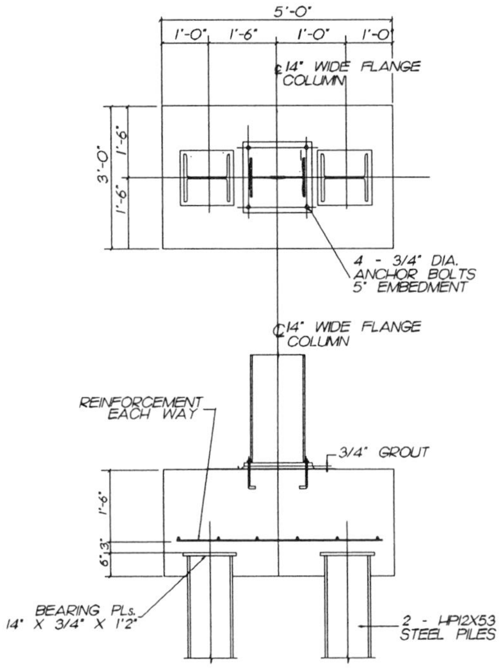

C1180

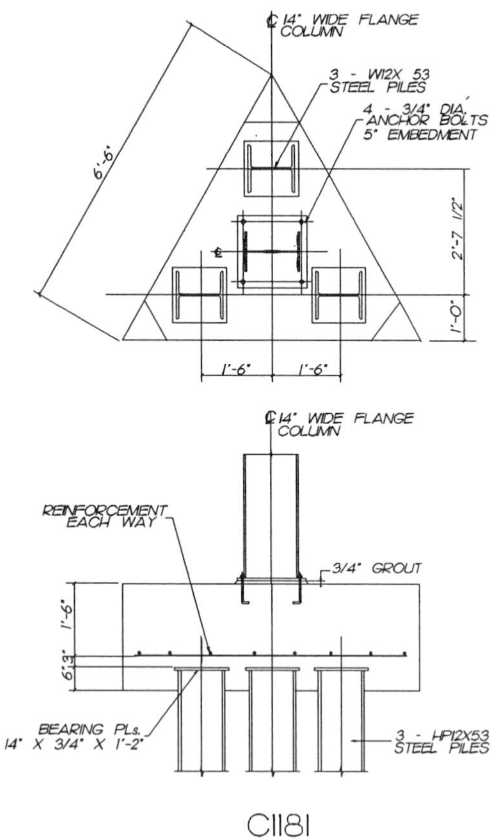

C1181

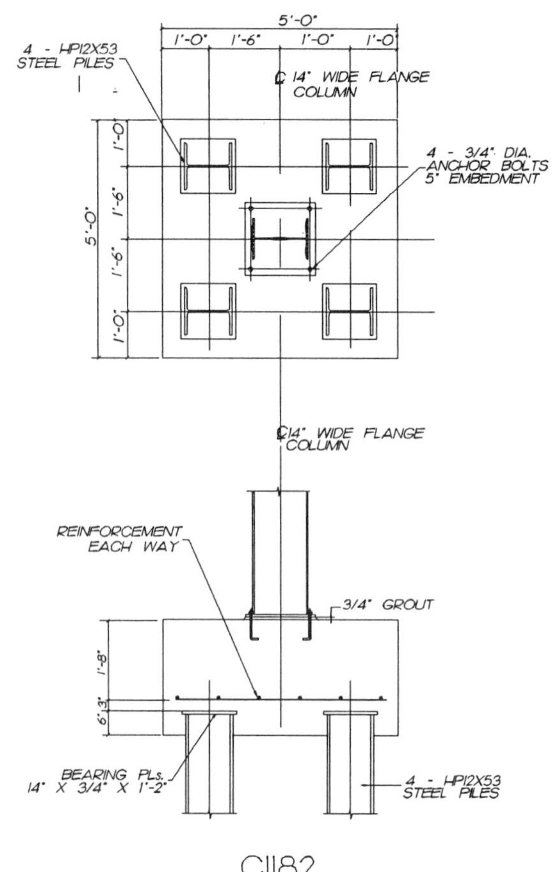

C1182

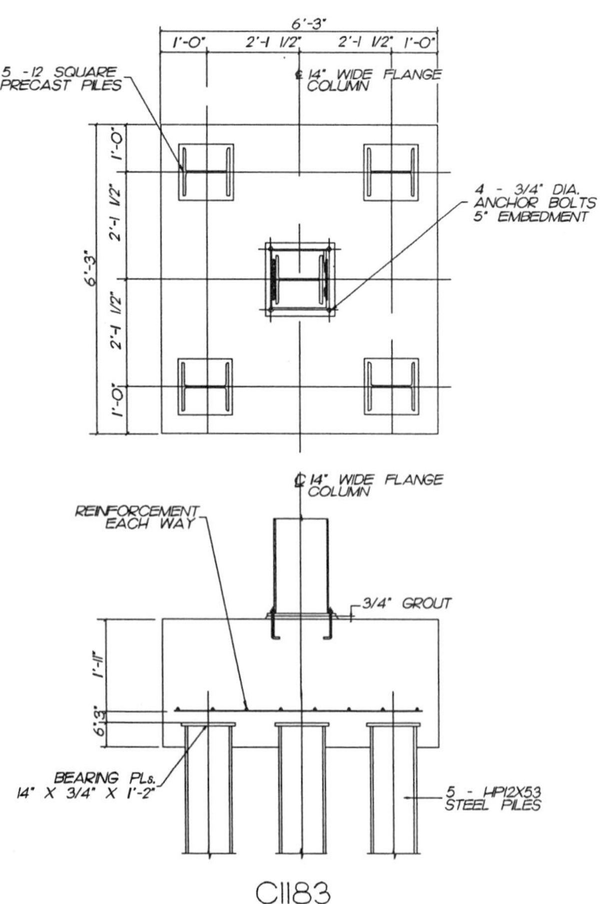

C1183

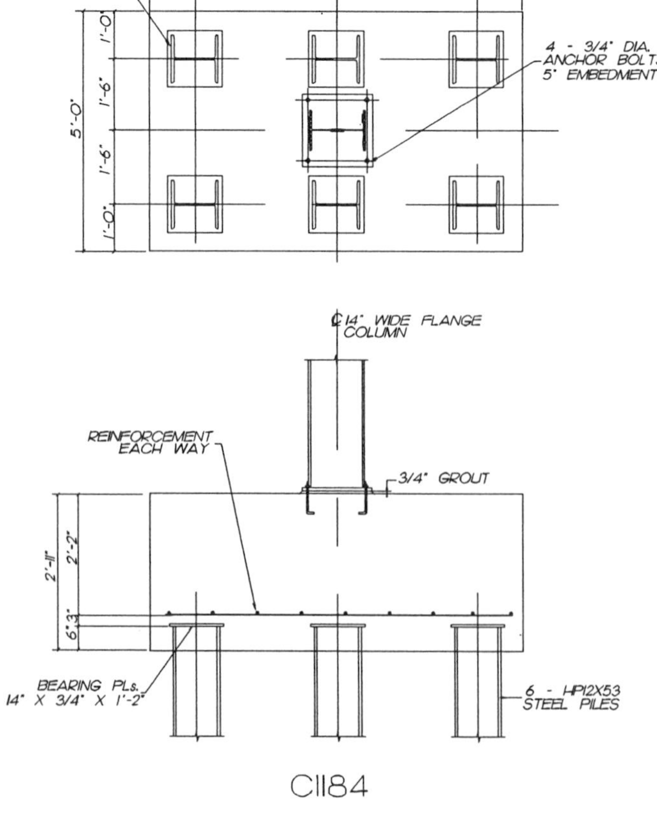

C1184

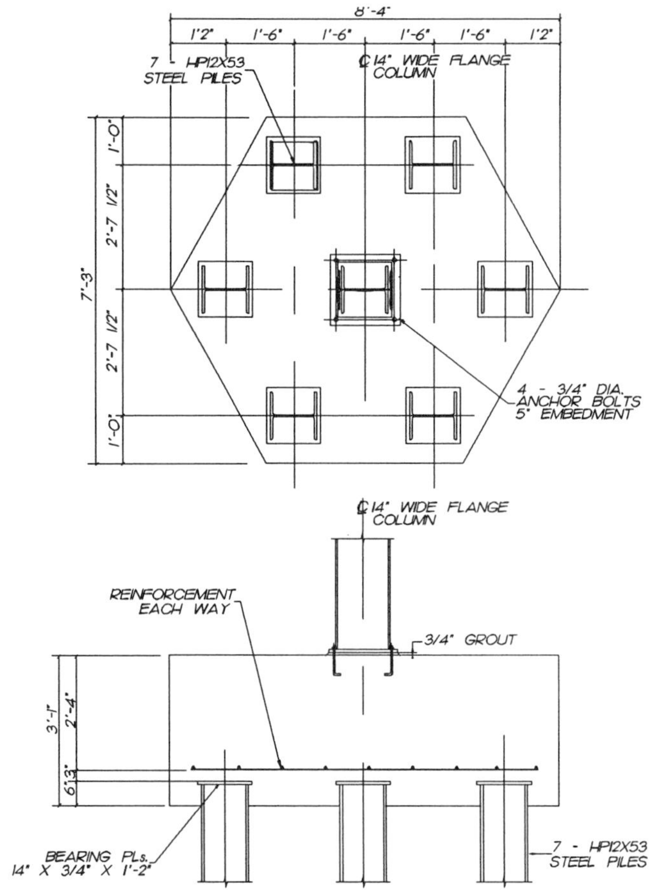

8'-4"
1'2"
1'-6"
1'-6"
1'-6"
1'-6"
1'2"
7 - HP12X53
STEEL PILES
℄ 14" WIDE FLANGE
COLUMN
1'-0"
2'-7 1/2"
7'-3"
2'-7 1/2"
1'-0"
4 - 3/4" DIA.
ANCHOR BOLTS
5" EMBEDMENT
℄ 14" WIDE FLANGE
COLUMN
REINFORCEMENT
EACH WAY
3/4" GROUT
3'-1"
2'-4"
6"
3"
BEARING PLs.
14" X 3/4" X 1'-2"
7 - HP12X53
STEEL PILES

C1185

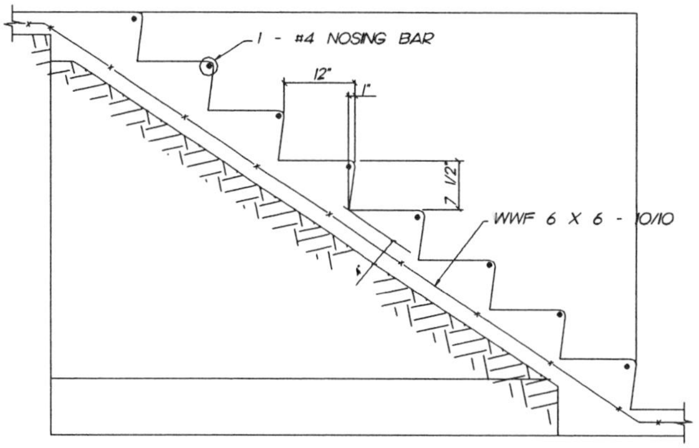

1 - #4 NOSING BAR
12"
1"
7 1/2"
WWF 6 X 6 - 10/10

C1200

8"
CONCRETE STEPS
2 - #5's
#4 NOSING BAR
BEND 18" INTO WALL
#5's @ 16" oc
EACH WAY
FINISHED GRADE
2 - #5's
12"
BEND ALT. BARS
16"
3" CLR.

C1201

12"
12"
1.5 X "H"
FTG. DEPTH
"H"
12"

TYPICAL STEP OF A CONTINUOUS WALL FOOTING

C1220

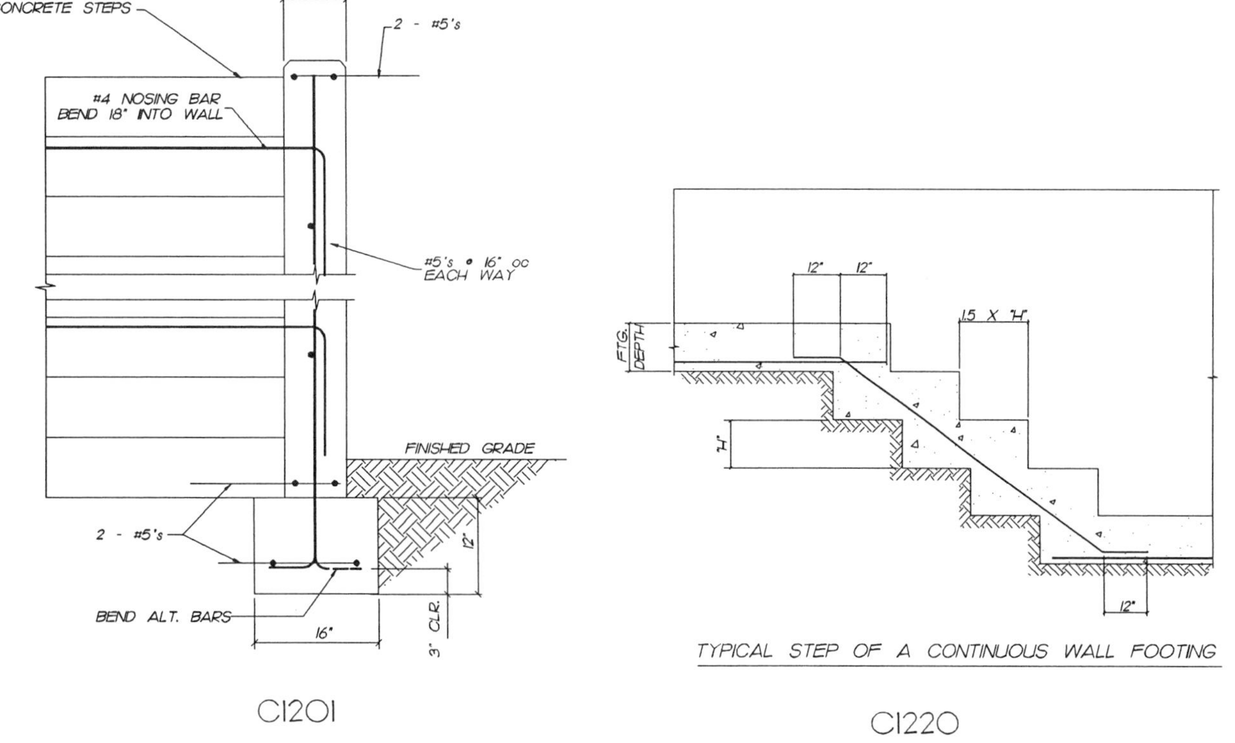

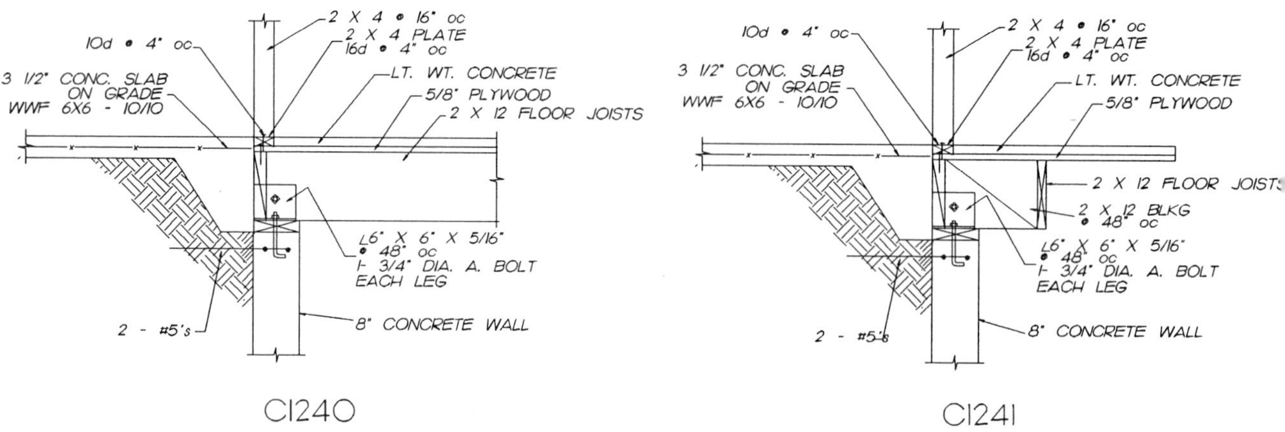

C1240

C1241

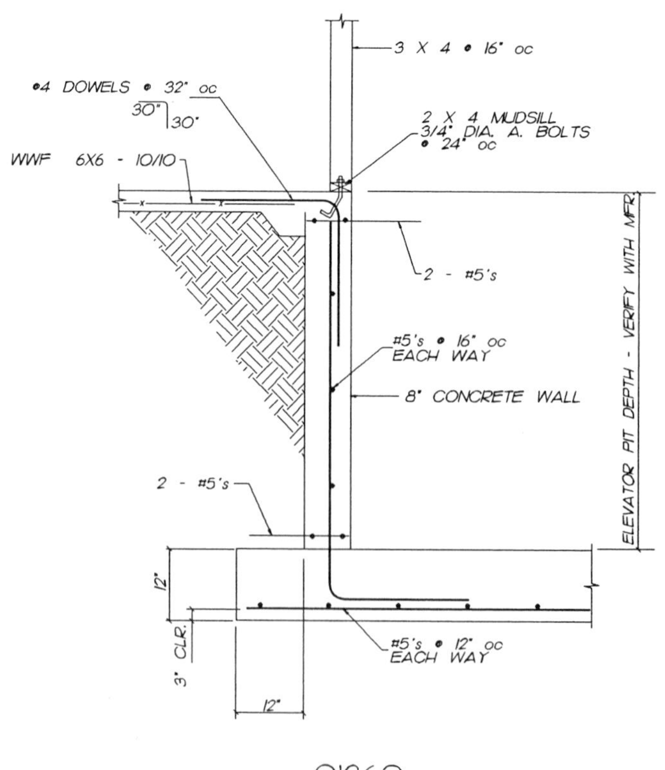

C1260

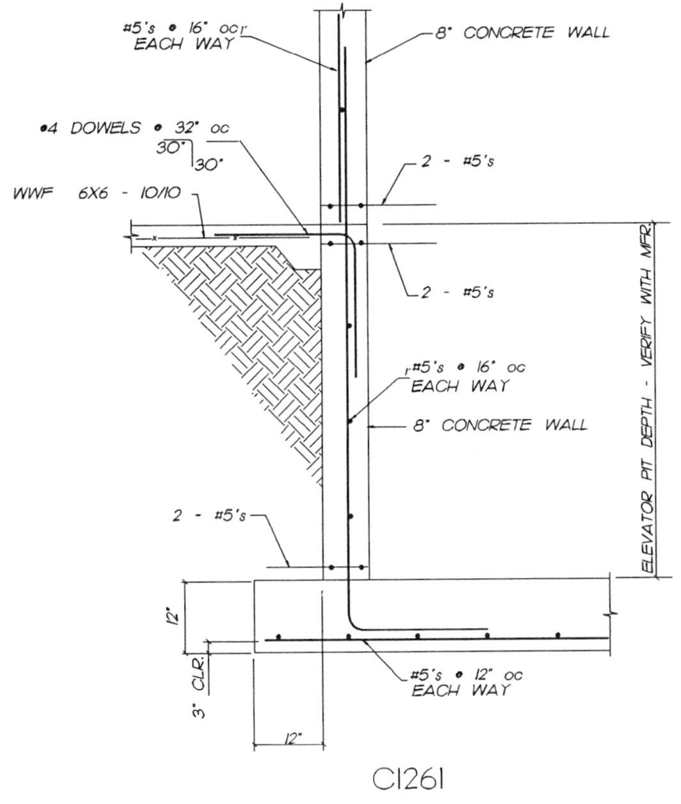

C1261

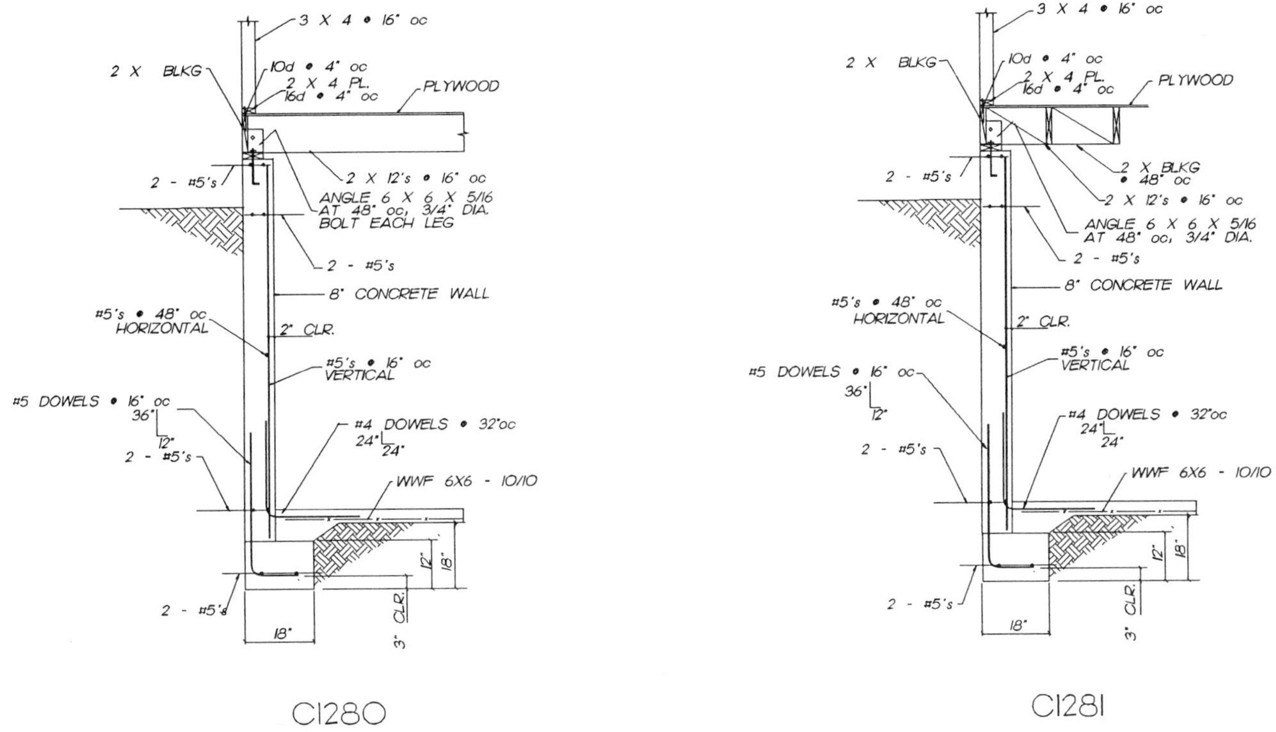

C1280

C1281

CHAPTER 3 CONCRETE SECTION 2			COLUMNS
SQUARE COLUMNS	PLAN SECTIONS	————	C2020 - C2024
ROUND COLUMNS WITH SQUARE REINFORCEMENT	PLAN SECTIONS	————	C2040 - C2043
ROUND COLUMNS SPIRAL REINFORCEMENT	PLAN SECTIONS	————	C2060 - C2064
CONCRETE COLUMN SUPPORTING A SLAB WITH COLUMN CAPITAL AND ON A SPREAD FOOTING	SQUARE COLUMN WITH TIES		C2080
	ROUND COLUMN WITH SPIRALS		C2081
CONCRETE COLUMN SUPPORTING A SLAB WITH COLUMN DROP PANEL AND ON A SPREAD FOOTING	SQUARE COLUMN WITH TIES		C2082
	ROUND COLUMN WITH SPIRALS		C2083
CONCRETE COLUMN SPLICE AT BEAM & SLAB	SQUARE COLUMN WITH TIES		C2100
	ROUND COLUMN WITH SPIRALS		C2101
CONCRETE COLUMN SPLICE AT COLUMN CAPITAL OR DROP PANEL	SQUARE COLUMN WITH TIES		C2102
	ROUND COLUMN WITH SPIRALS		C2103
TOP OF CONCRETE COLUMN AT COLUMN CAPITAL OR DROP PANEL	SQUARE COLUMN WITH TIES		C2104
	ROUND COLUMN WITH SPIRALS		C2105
STEEL COLUMNS IN WALLS	PIPE COLUMN		C2120
	WIDE FLANGE COLUMN		C2121
COLUMNS ATTACHED TO WALL	WIDE FLANGE COLUMN		C2122 - C1223
	SQUARE COLUMN WITH TIES		C2124

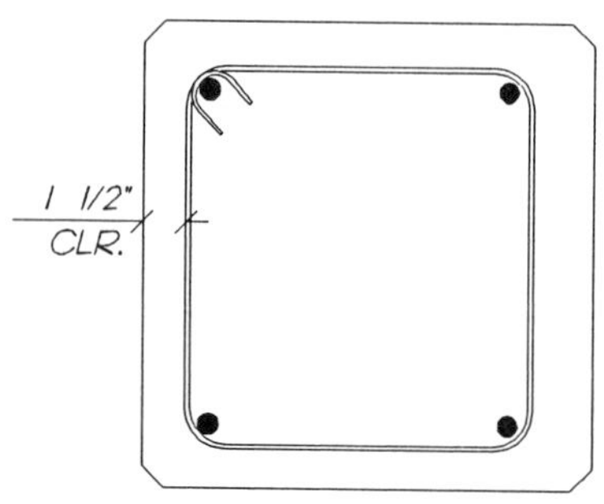

SQUARE CONCRETE COLUMN
4 VERTICAL BARS - SINGLE TIES

C2020

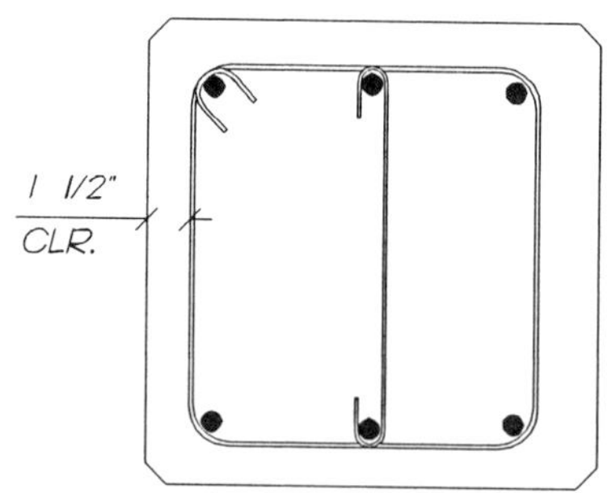

SQUARE CONCRETE COLUMN
6 VERTICAL BARS- 2 SETS OF TIES

C2021

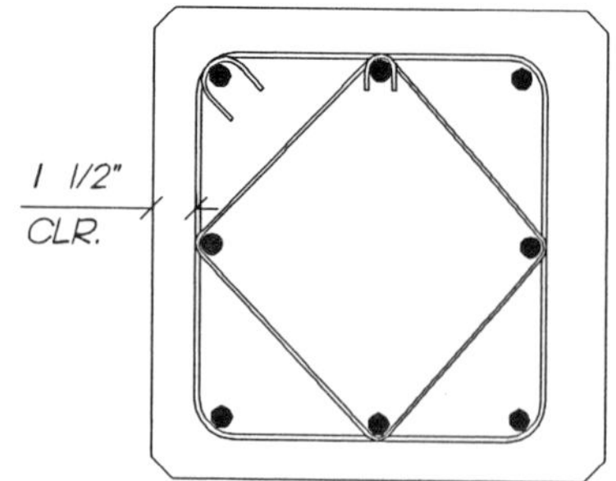

SQUARE CONCRETE COLUMN
8 VERTICAL BARS- 2 SETS OFTIES

C2022

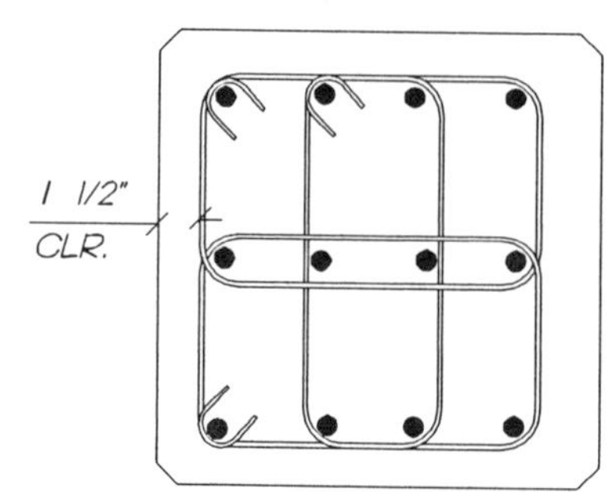

SQUARE CONCRETE COLUMN
10 VERTICAL BARS-3 SETS OF TIES

C2023

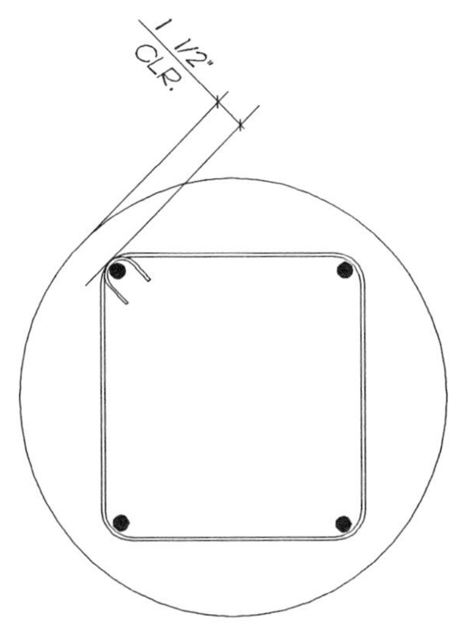

ROUND CONCRETE COLUMN
4 VERTICAL BARS - SINGLE TIES

C2040

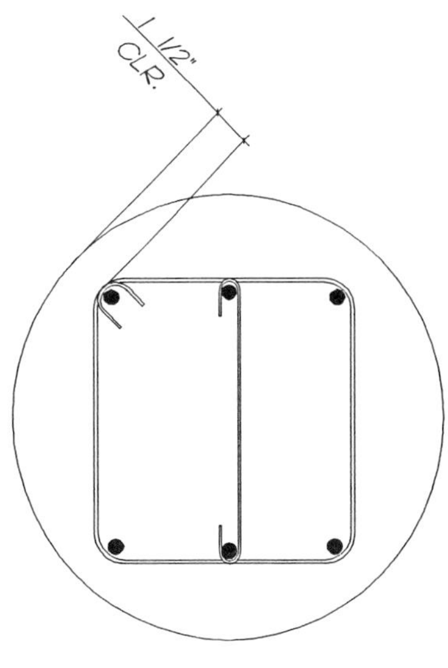

ROUND CONCRETE COLUMN
6 VERTICAL BARS - 2 SETS OF TIES

C2041

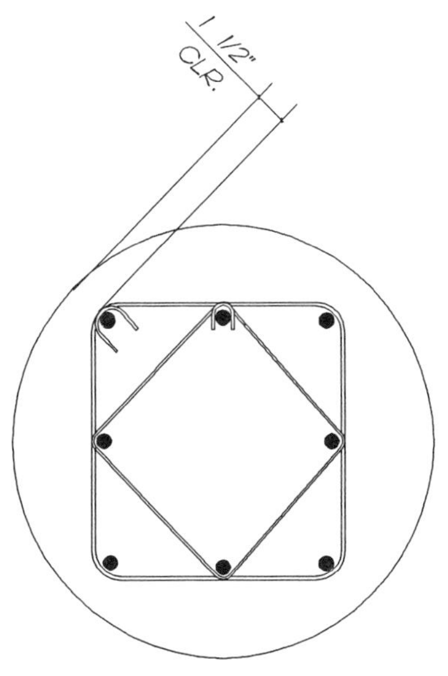

ROUND CONCRETE COLUMN
8 VERTICAL BARS - 2 SETS OF TIES

C2042

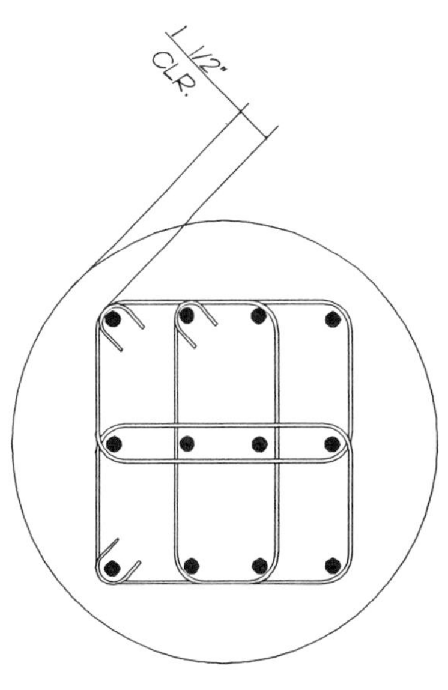

ROUND CONCRETE COLUMN
10 VERTICAL BARS - 3 SETS OF TIES

C2043

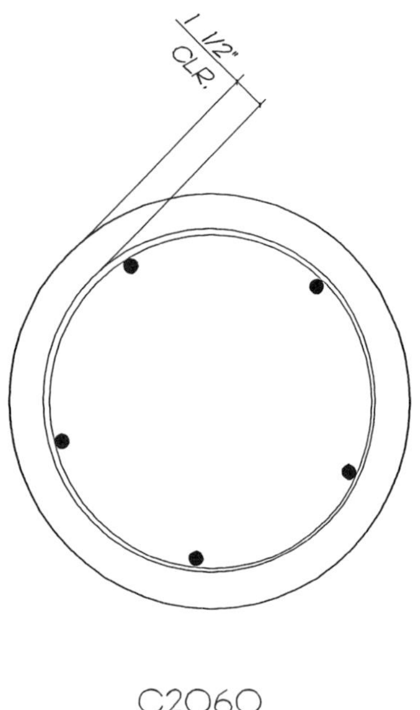

C2060

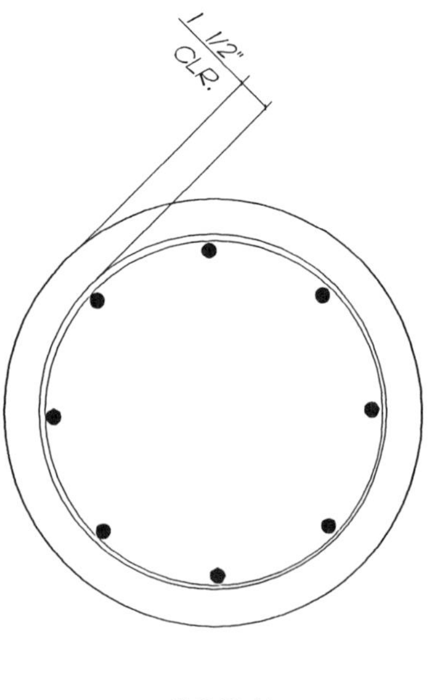

C2061

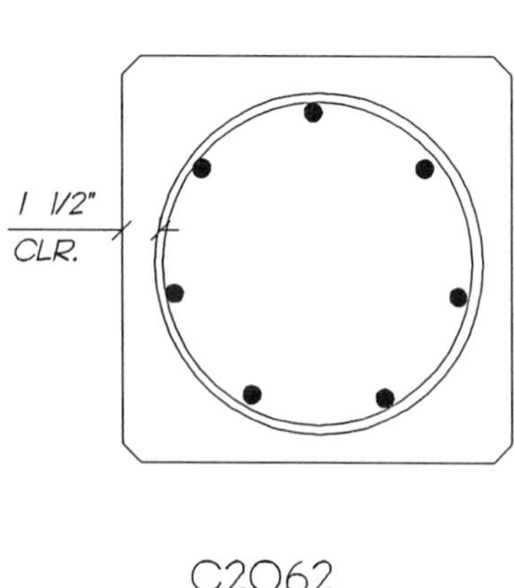

C2062

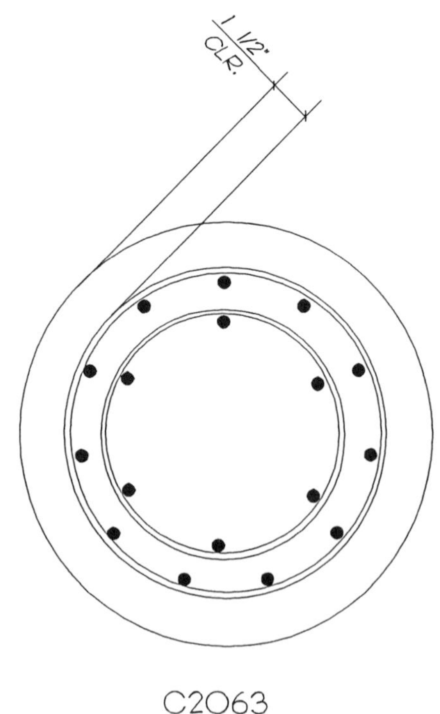

C2063

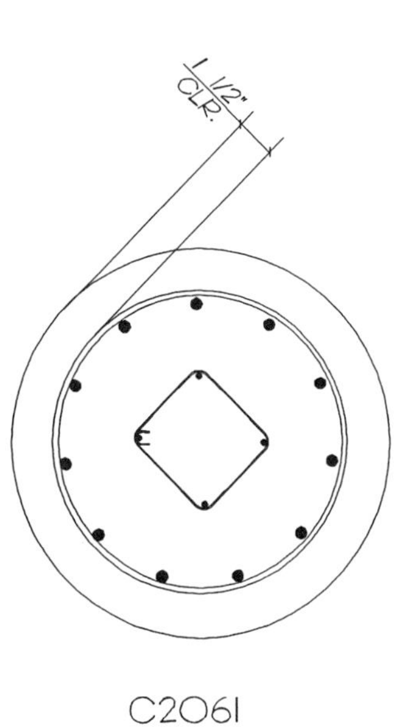

C2061

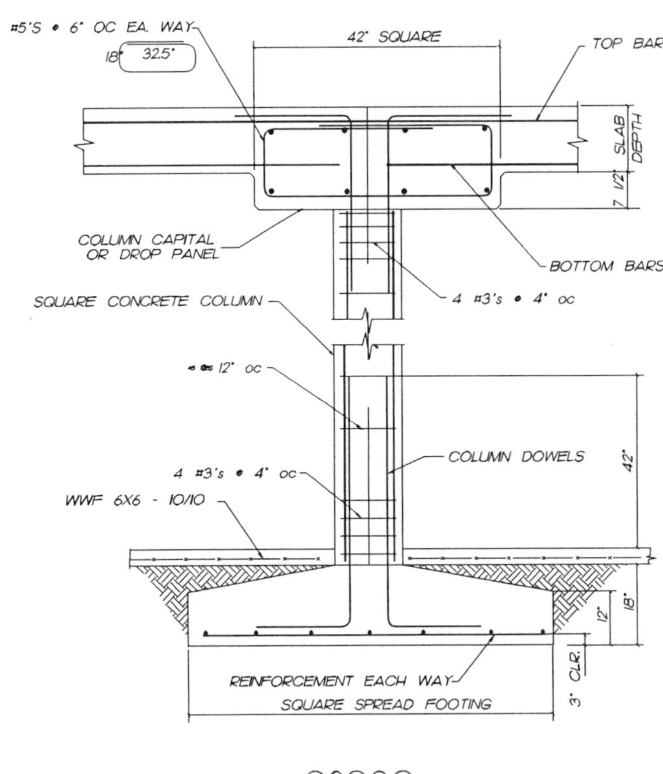

C2080

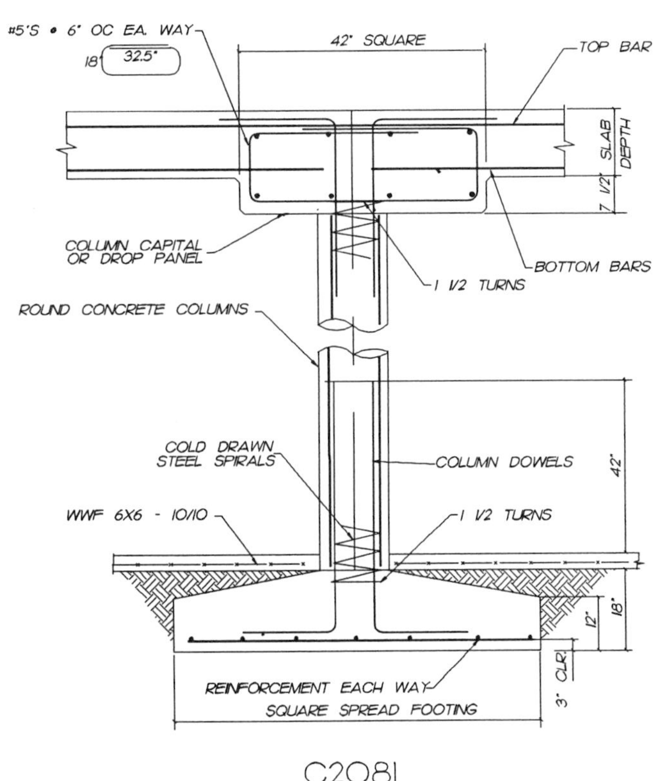

C2081

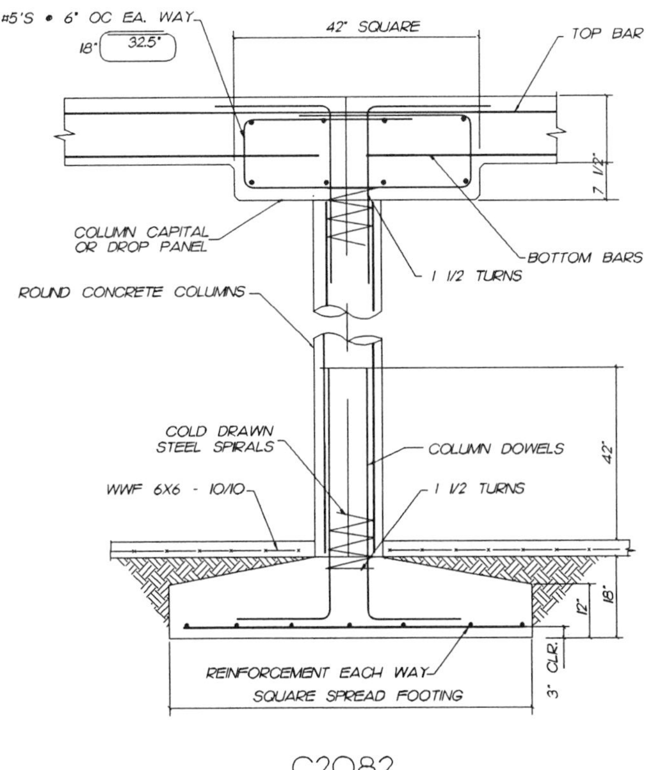

C2082

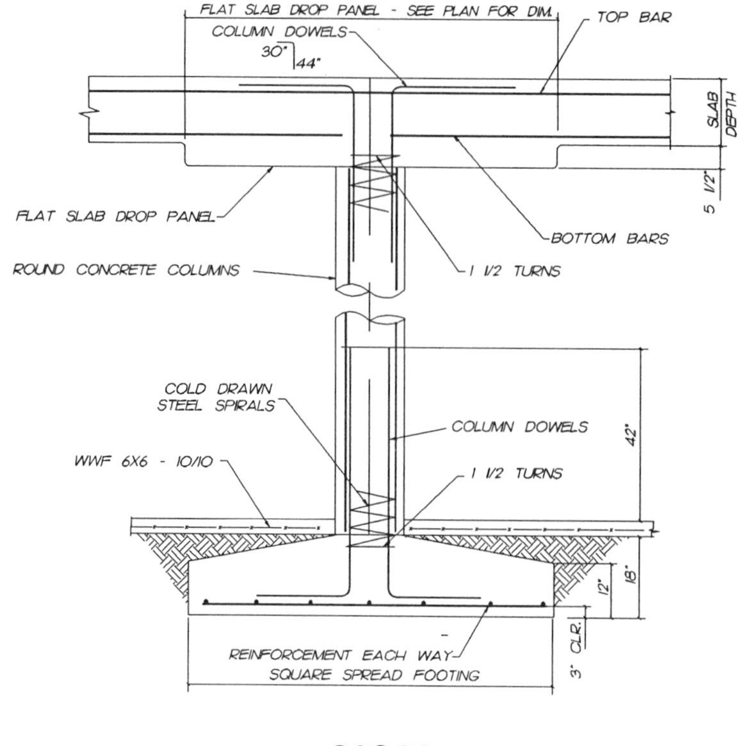

C2083

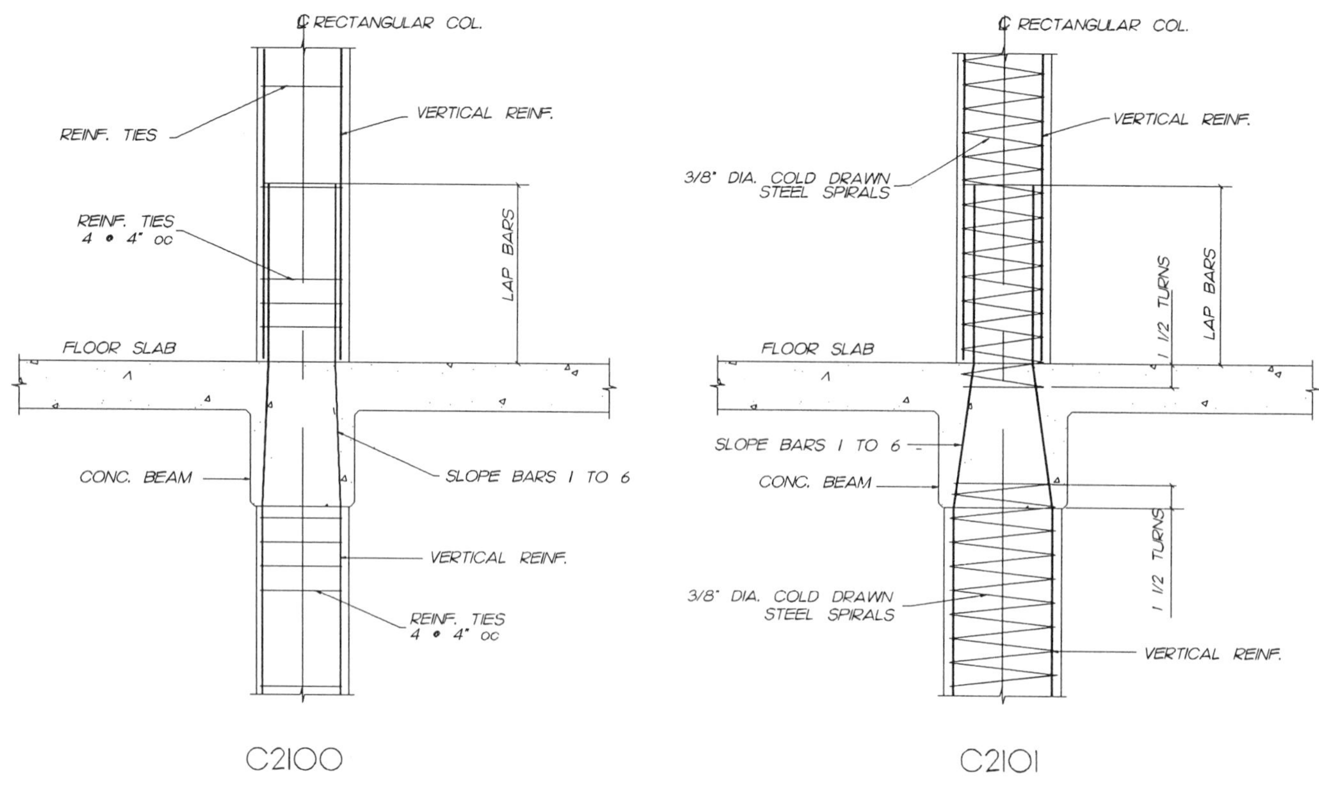

C2100

C2101

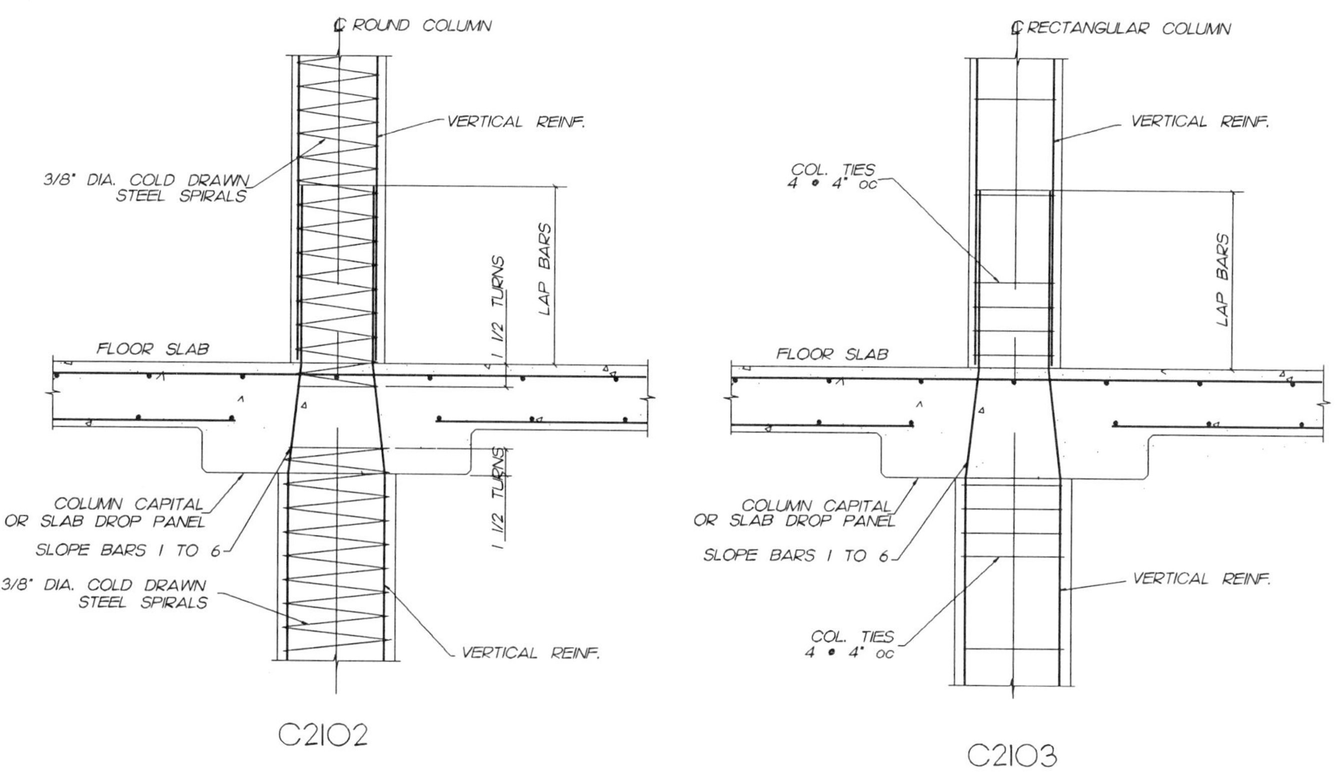

℄ ROUND COLUMN
VERTICAL REINF.
3/8" DIA. COLD DRAWN STEEL SPIRALS
LAP BARS
1 1/2 TURNS
FLOOR SLAB
1 1/2 TURNS
COLUMN CAPITAL OR SLAB DROP PANEL
SLOPE BARS 1 TO 6
3/8" DIA. COLD DRAWN STEEL SPIRALS
VERTICAL REINF.
C2102
℄ RECTANGULAR COLUMN
VERTICAL REINF.
COL. TIES 4 @ 4" oc
LAP BARS
FLOOR SLAB
COLUMN CAPITAL OR SLAB DROP PANEL
SLOPE BARS 1 TO 6
VERTICAL REINF.
COL. TIES 4 @ 4" oc
C2103

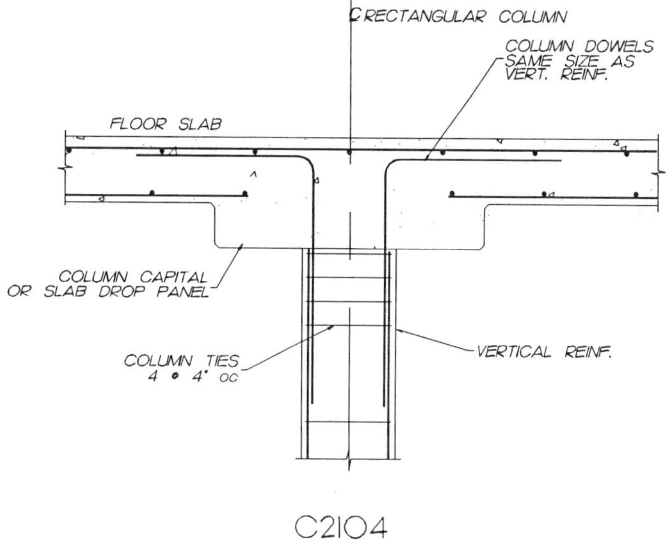

℄ RECTANGULAR COLUMN
COLUMN DOWELS SAME SIZE AS VERT. REINF.
FLOOR SLAB
COLUMN CAPITAL OR SLAB DROP PANEL
COLUMN TIES 4 @ 4" oc
VERTICAL REINF.
C2104

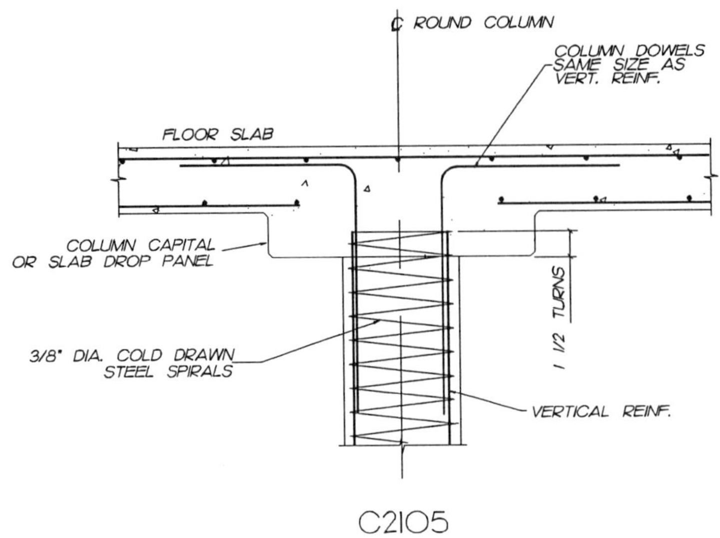

C2105

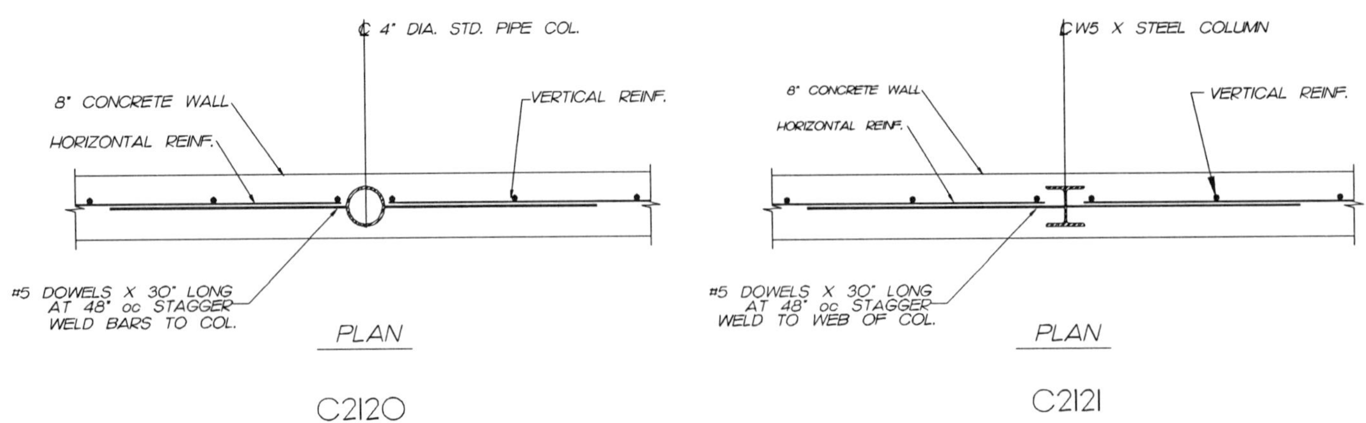

C2120

C2121

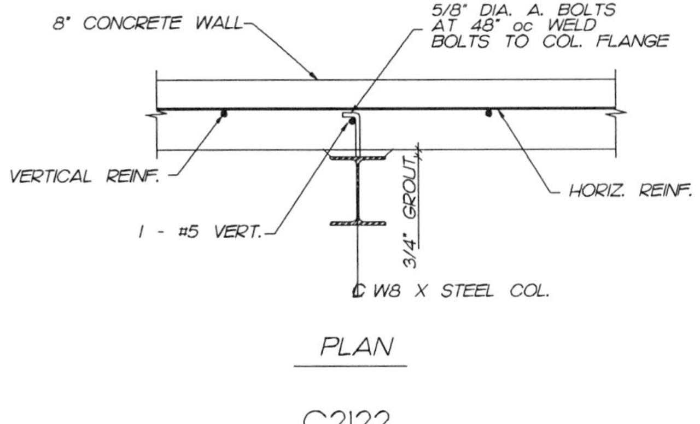

C2122

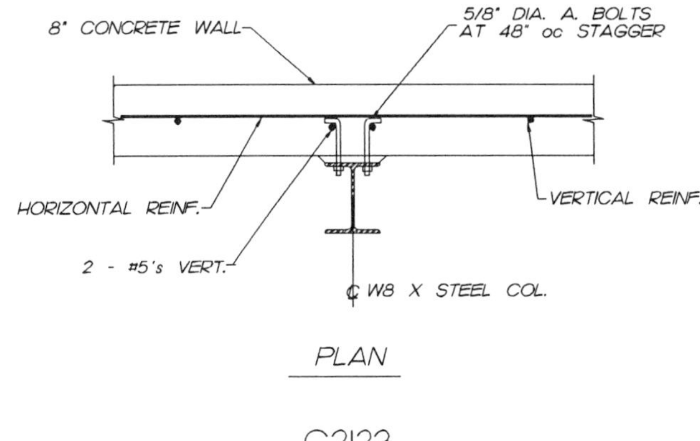

C2123

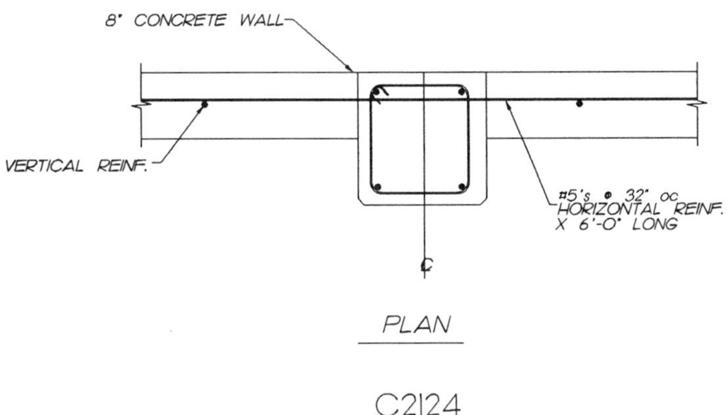

C2124

CHAPTER 3 CONCRETE SECTION 3 WALL - WOOD FLOOR

EXTERIOR WALL JOISTS PERP. TO WALL 4 X LEDGER METAL TIE STRAPS NO LT. WT. CONCRETE	————	8" CONCRETE WALL	C3020 - C3024
		10" CONCRETE WALL	C3025 - C3029
		————	————
EXTERIOR WALL JOISTS PARALELL TO WALL 4 X LEDGER METAL TIE STRAPS NO LT. WT. CONCRETE	————	8" CONCRETE WALL	C3040 - C3044
		10" CONCRETE WALL	C3045 - C3049
		————	————
INTERIOR WALL JOISTS PERP. TO WALL 4 X LEDGER METAL TIE STRAPS NO LT. WT. CONCRETE	————	8" CONCRETE WALL	C3060 - C3064
		10" CONCRETE WALL	C3065 - C3069
		————	————
INTERIOR WALL JOISTS PARALELL TO WALL 4 X LEDGER METAL TIE STRAPS NO LT. WT. CONCRETE	————	8" CONCRETE WALL	C3080 - C3084
		10" CONCRETE WALL	C3085 - C3089
		————	————
INTERIOR WALL JSTS PERP & PARALELL ALT SIDES 4 X LEDGER METAL TIE STRAPS NO LT. WT. CONCRETE	————	8" CONCRETE WALL	C3100 - C3104
		10" CONCRETE WALL	C3105 - C3109
		————	————
BASEMENT WALL WOOD FLOOR OVER	JOISTS PERP TO WALL	8" CONCRETE WALL	C3120
	JOISTS PARALELL TO WALL	8" CONCRETE WALL	C3121

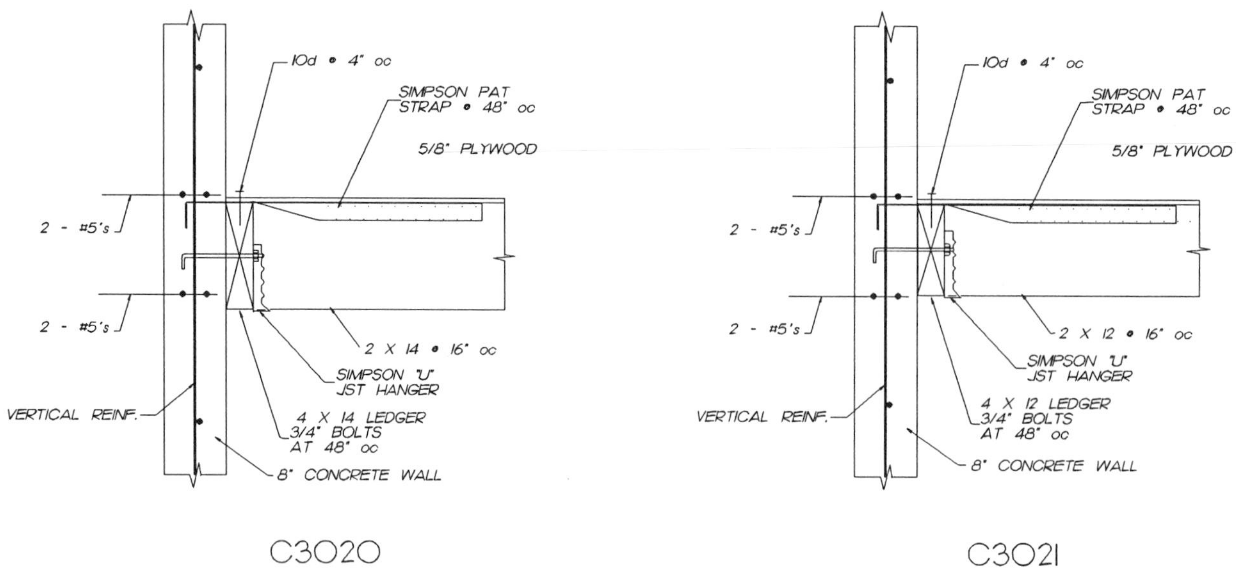

C3020

C3021

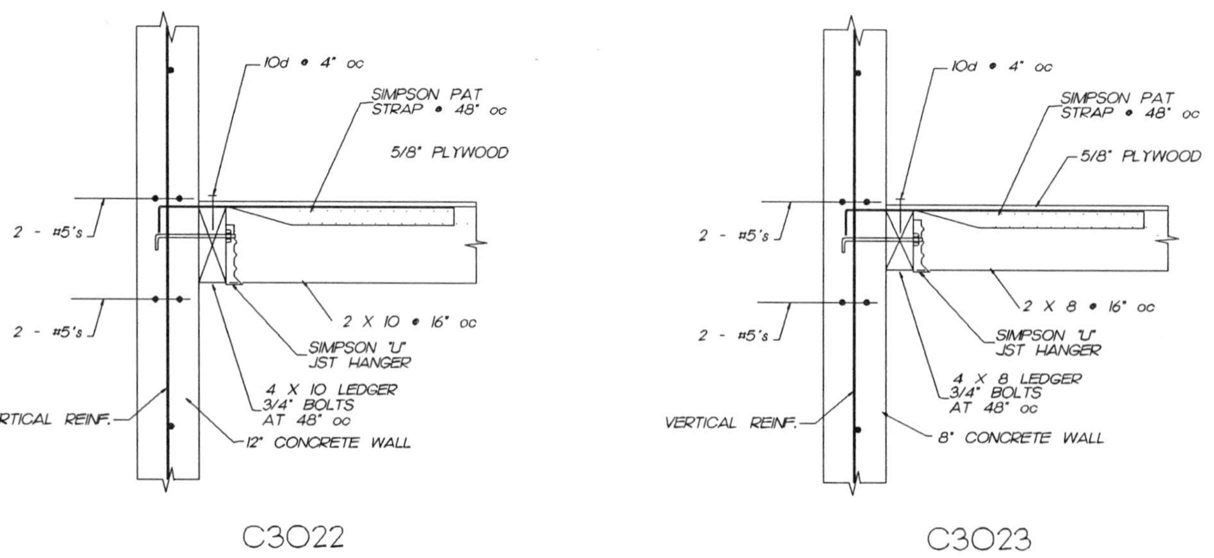

C3022

C3023

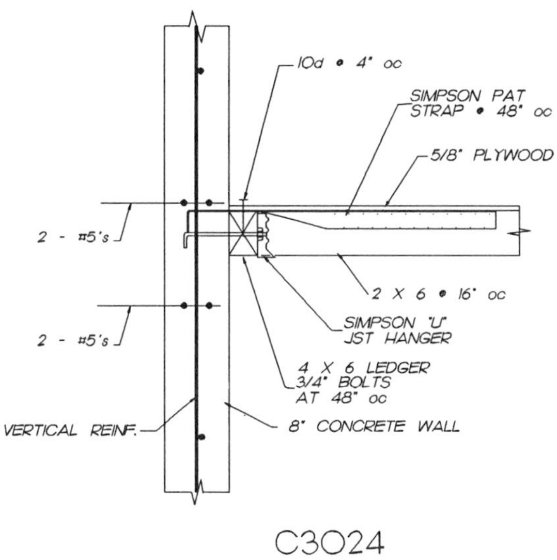

C3024

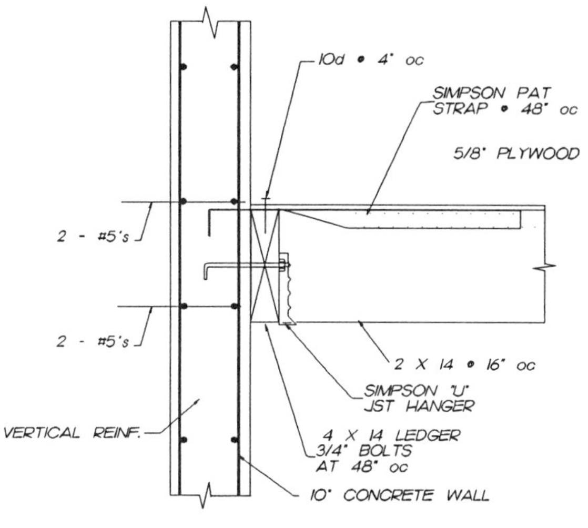

C3025

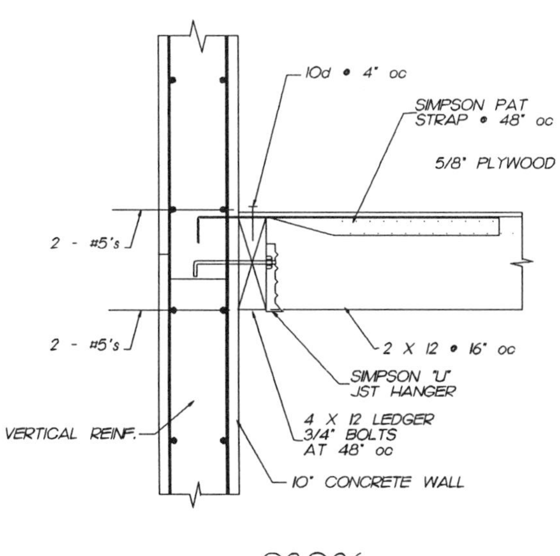

C3026

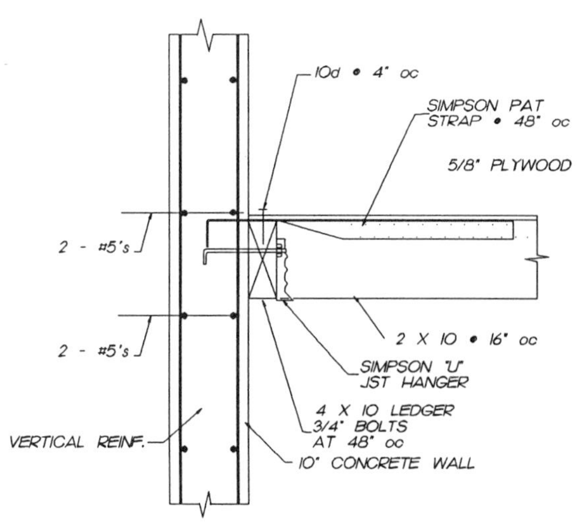

C3027

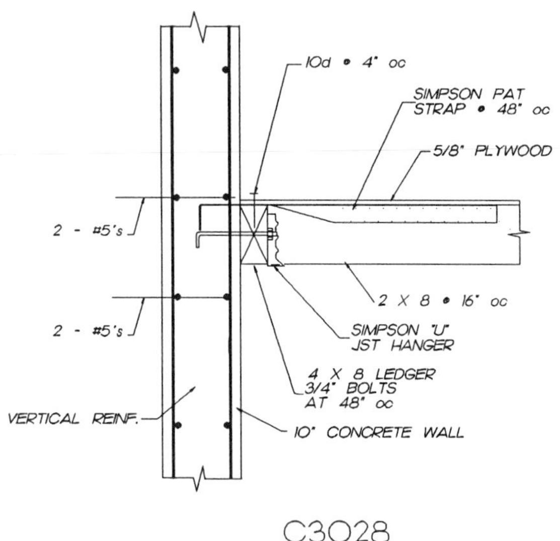

C3028

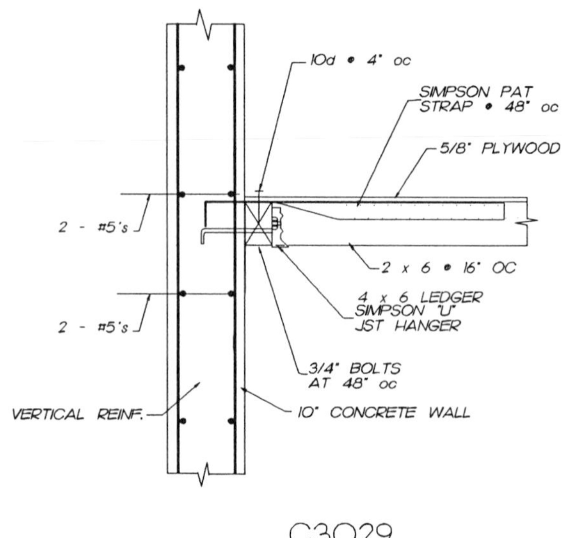

C3029

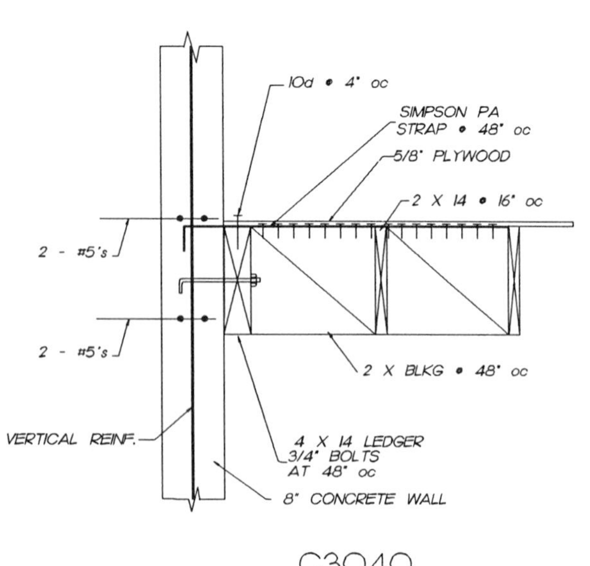

C3040

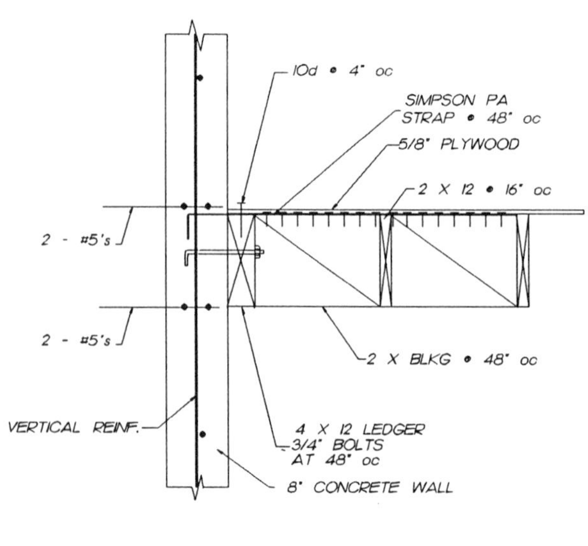

C3041

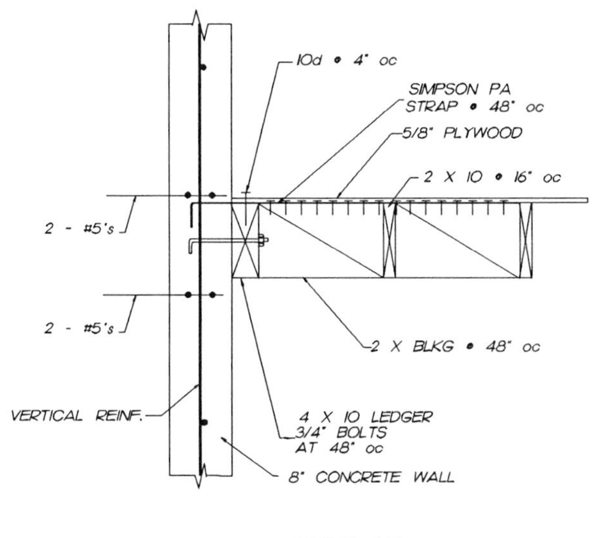

C3042

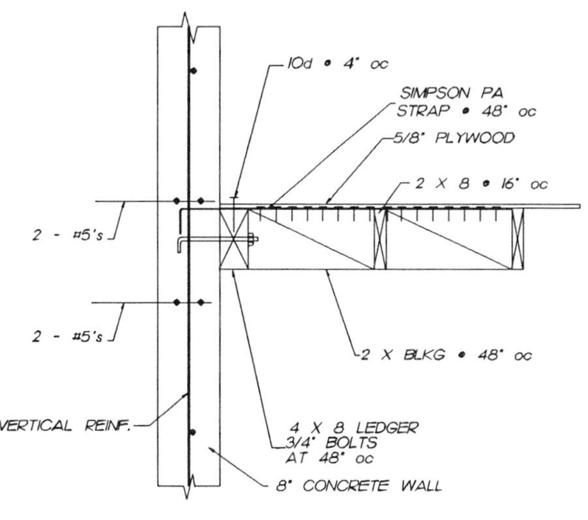

C3043

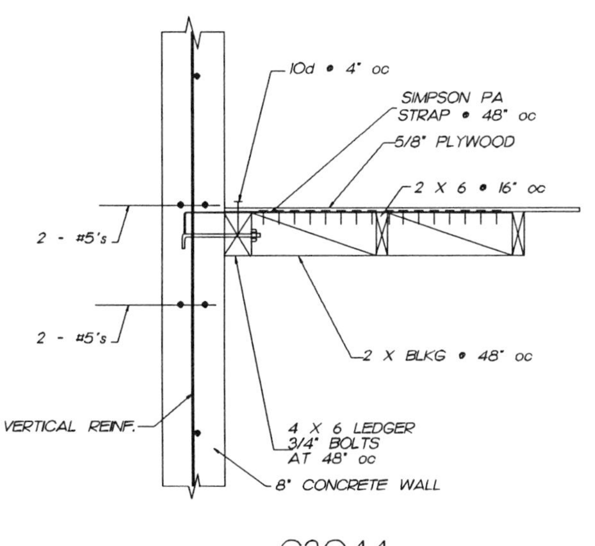

C3044

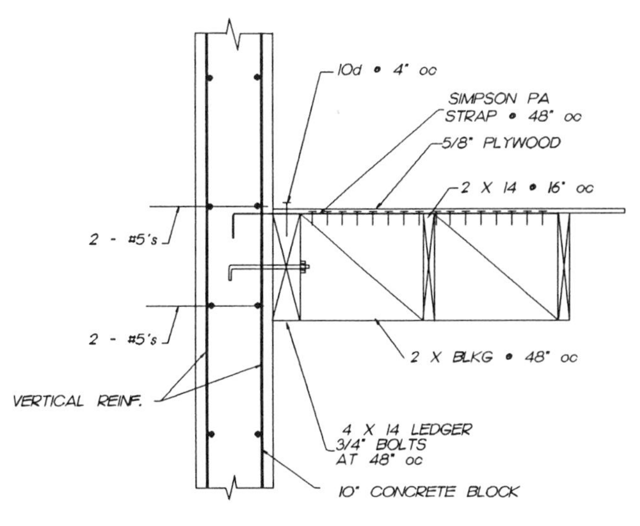

C3045

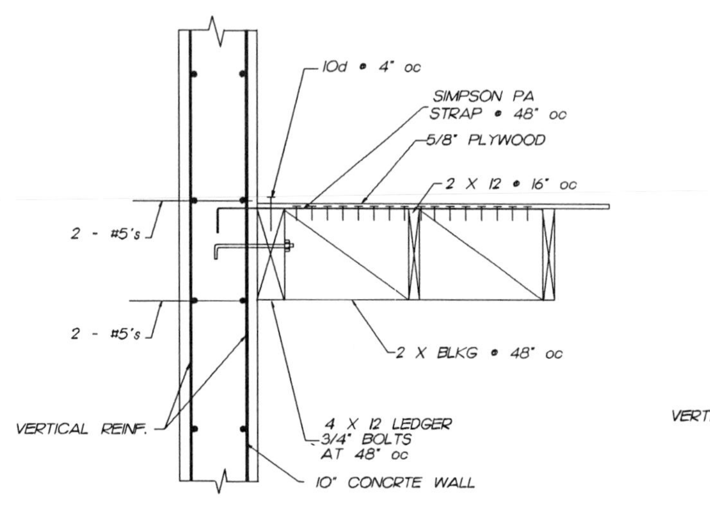

C3046

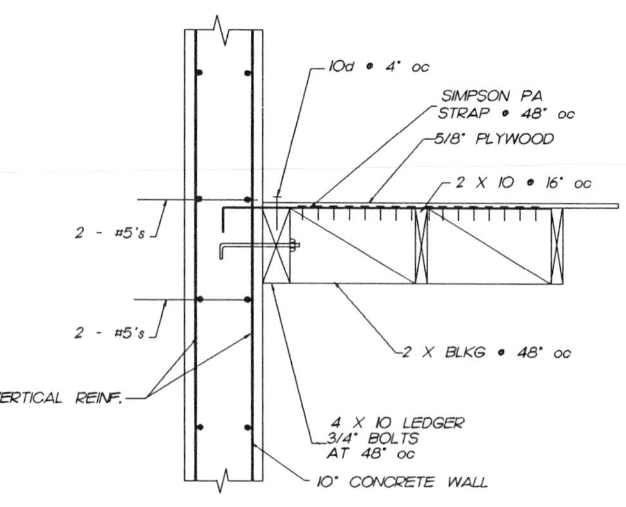

C3047

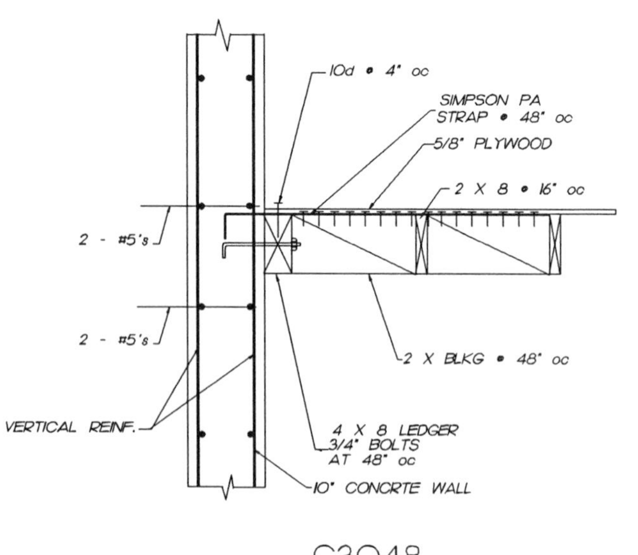

C3048

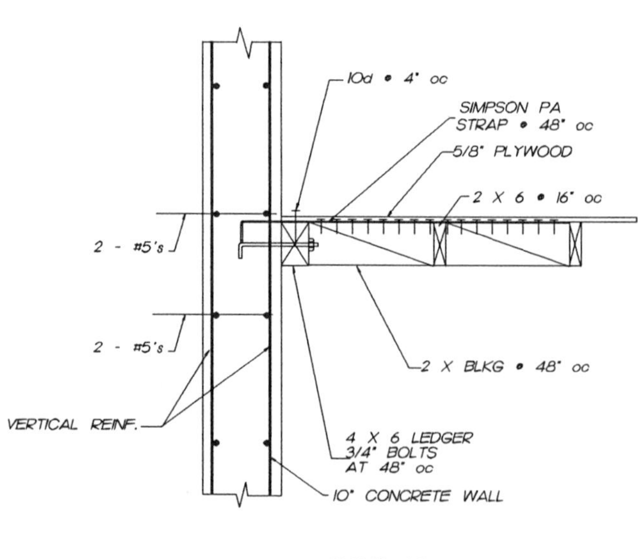

C3049

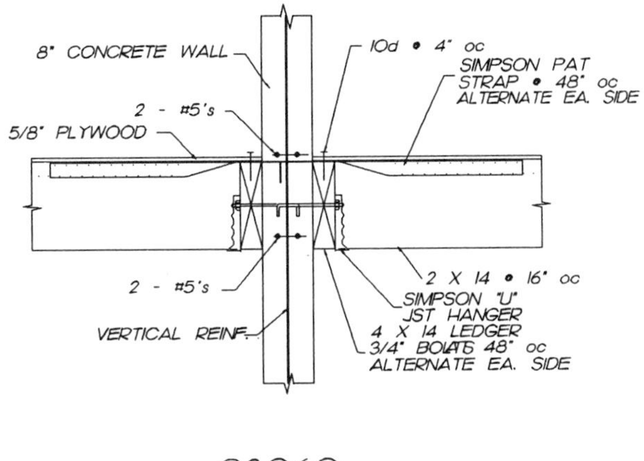

C3060

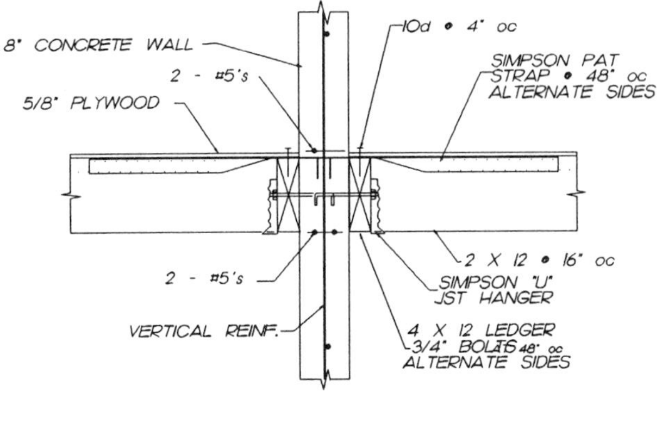

C3061

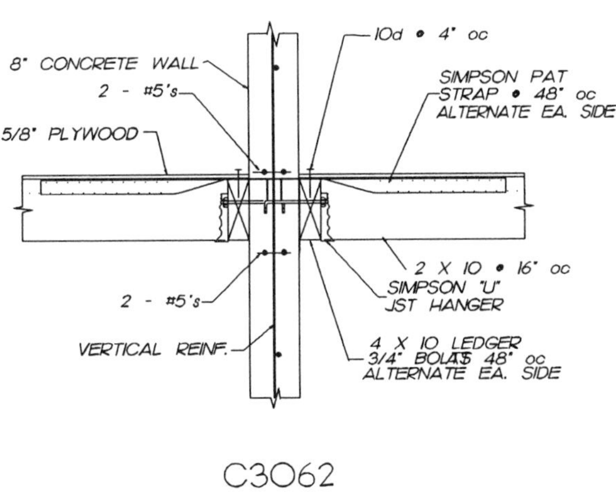

C3062

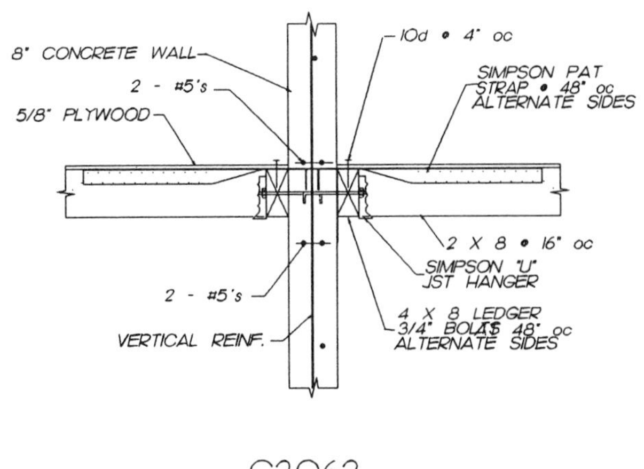

C3063

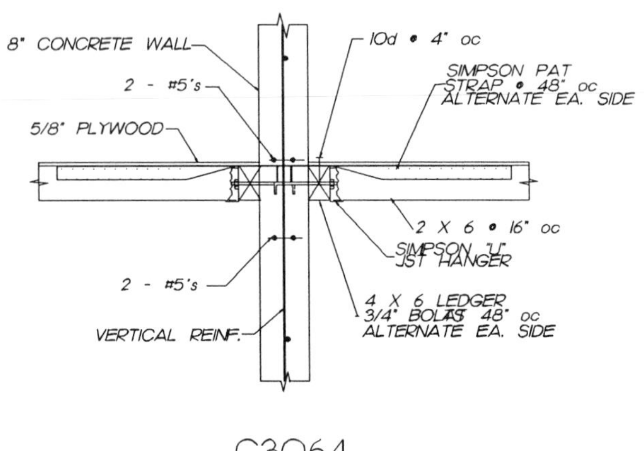

C3064

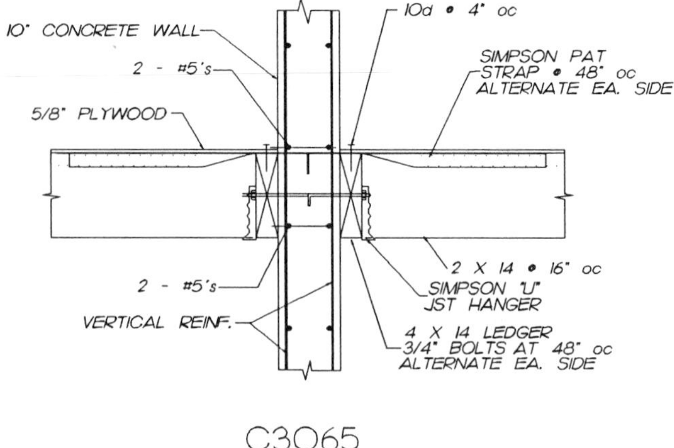

C3065

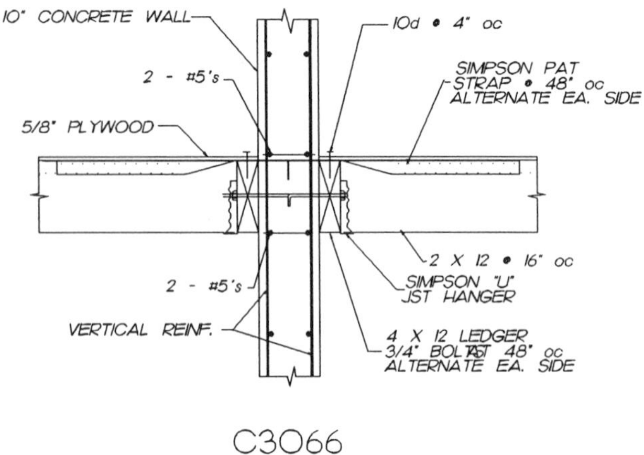

C3066

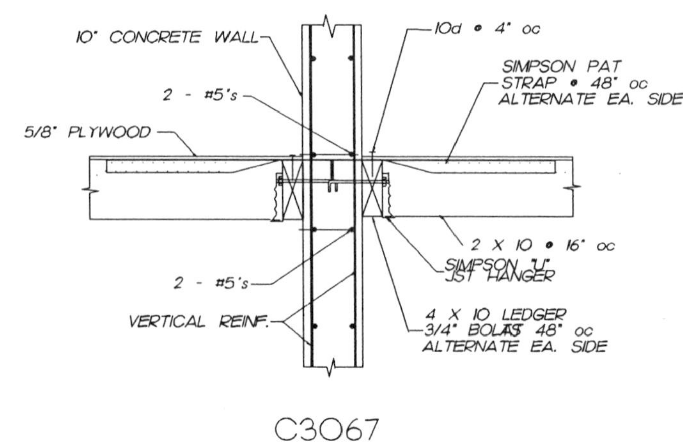

C3067

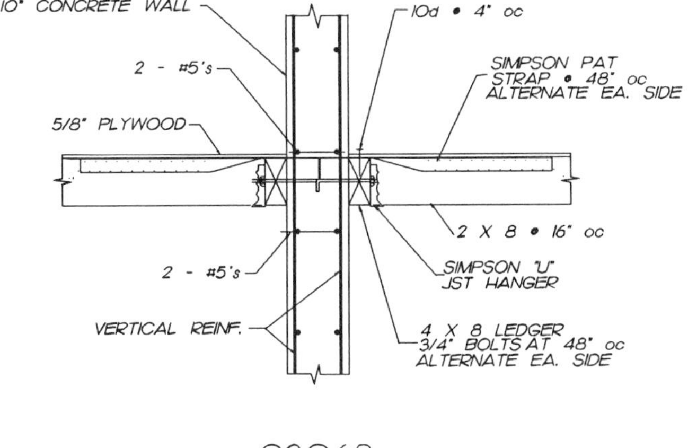

C3068

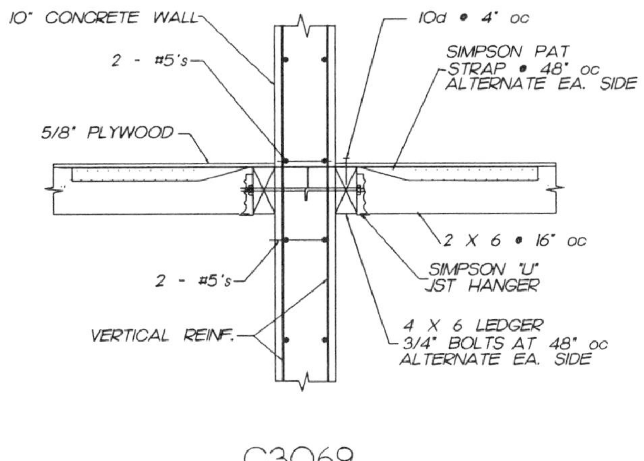

C3069

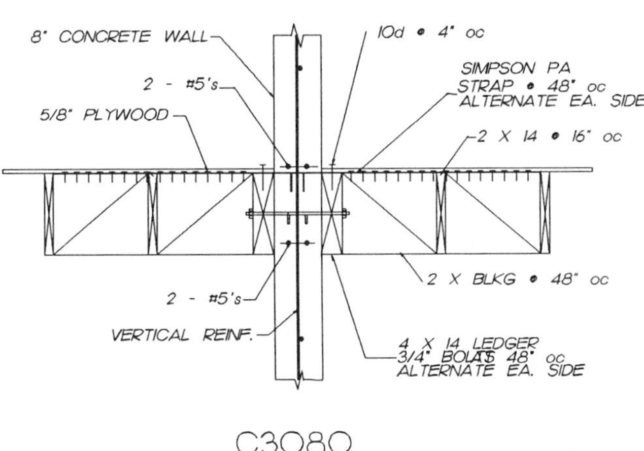

C3080

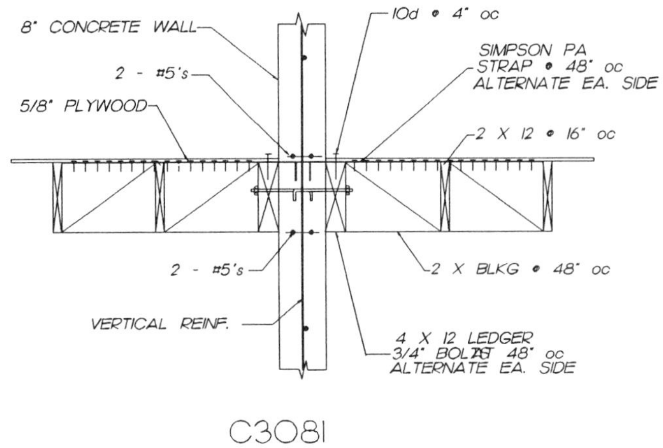

C3081

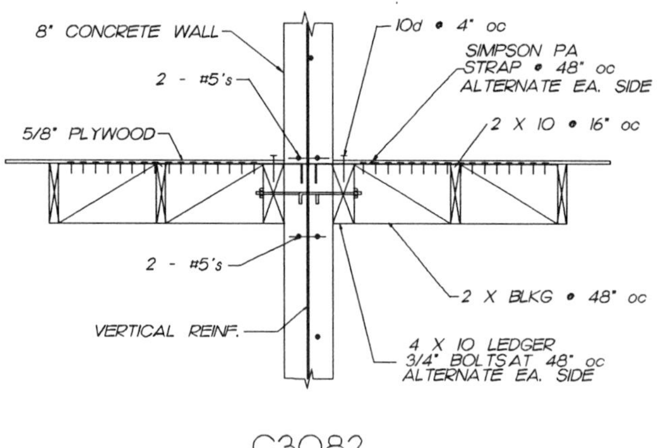

C3082

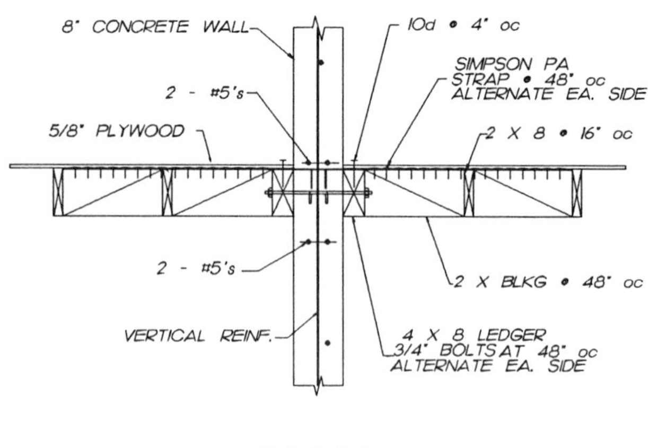

C3083

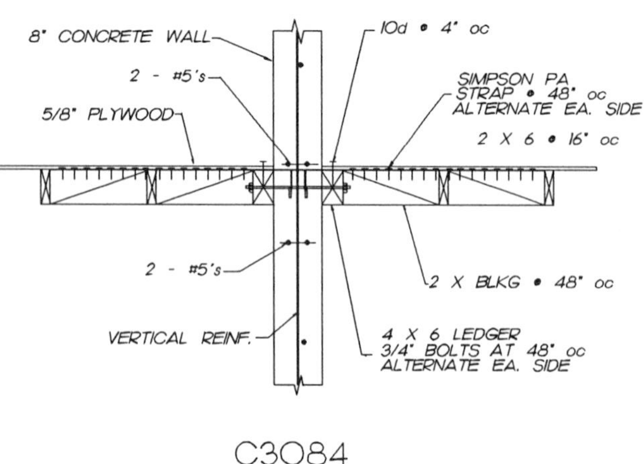

C3084

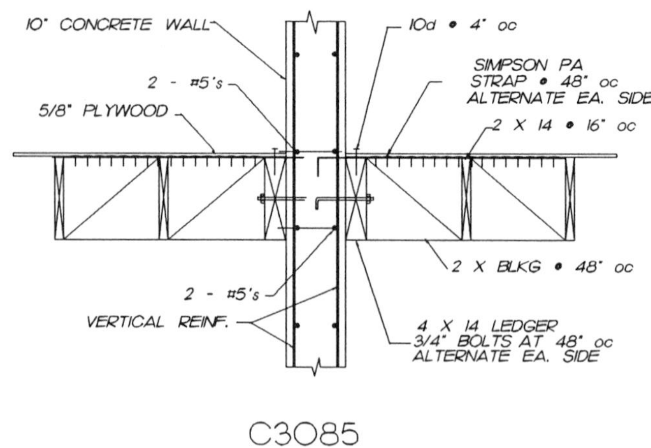

C3085

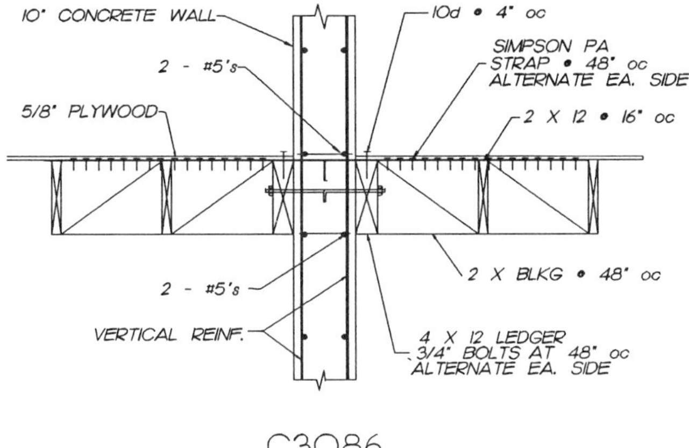

C3086

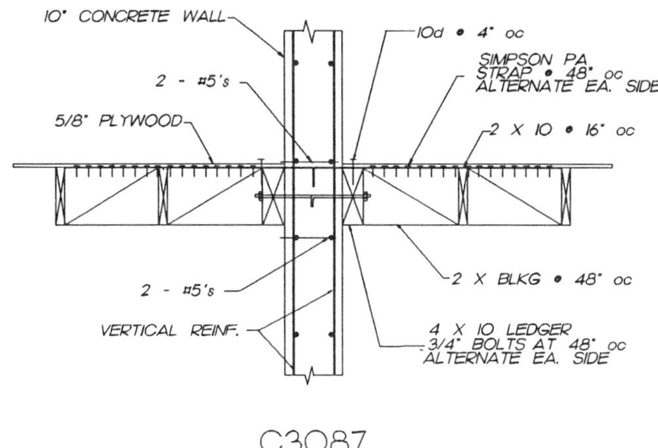

C3087

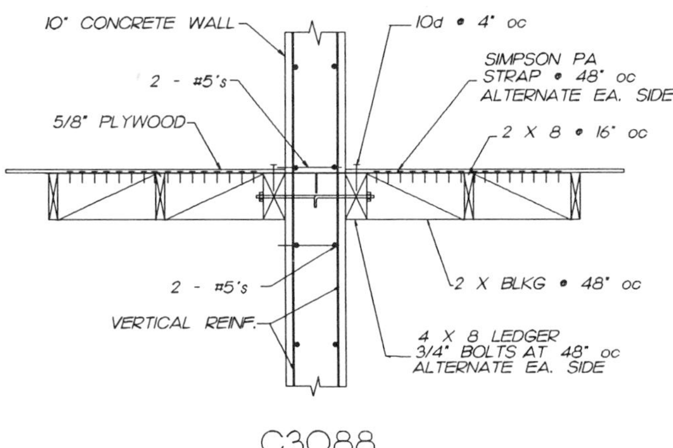

C3088

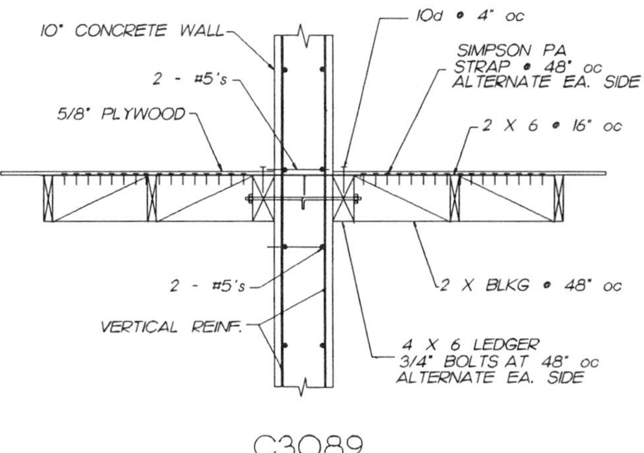

C3089

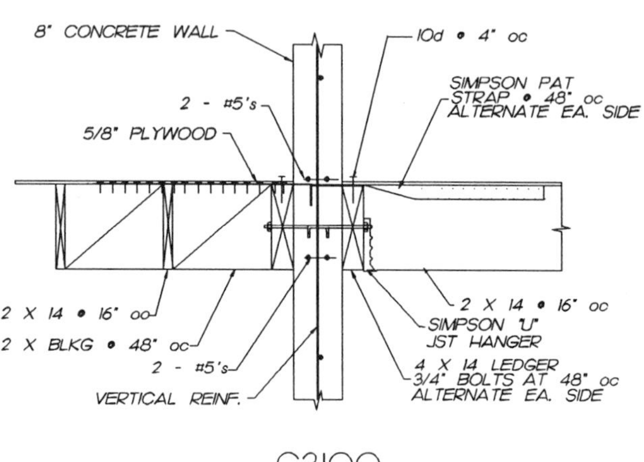

C3100

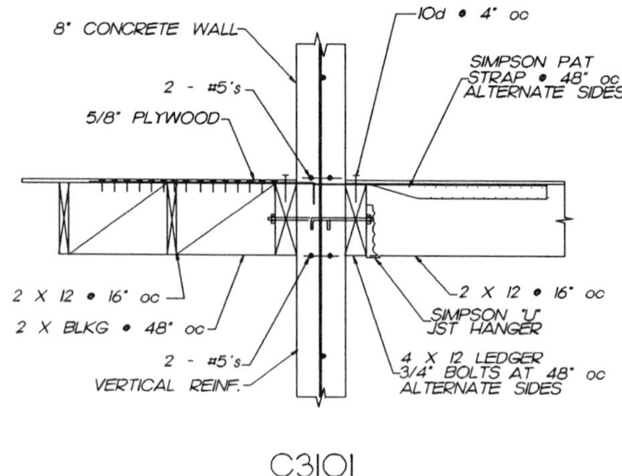

C3101

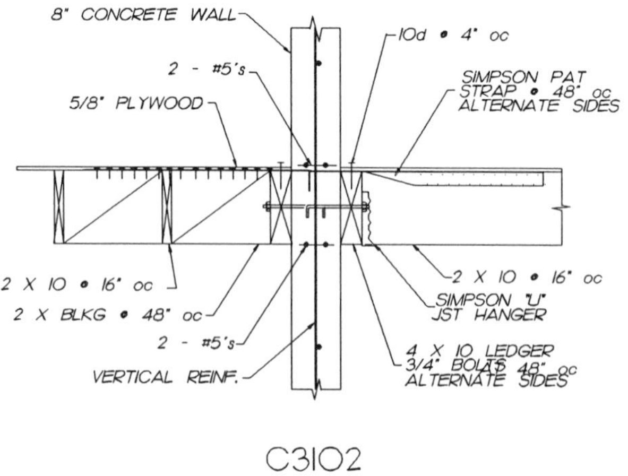

C3102

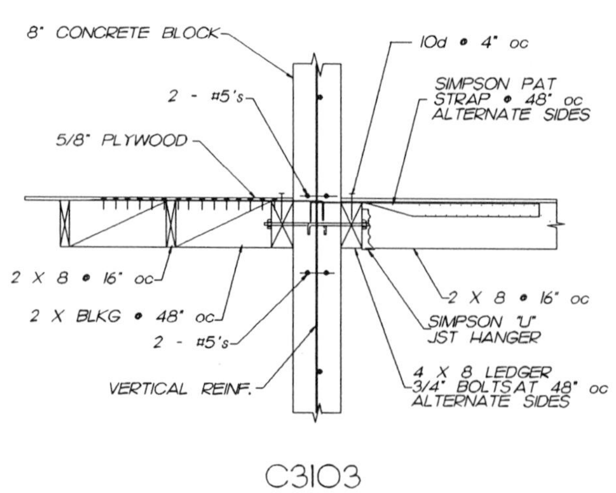

C3103

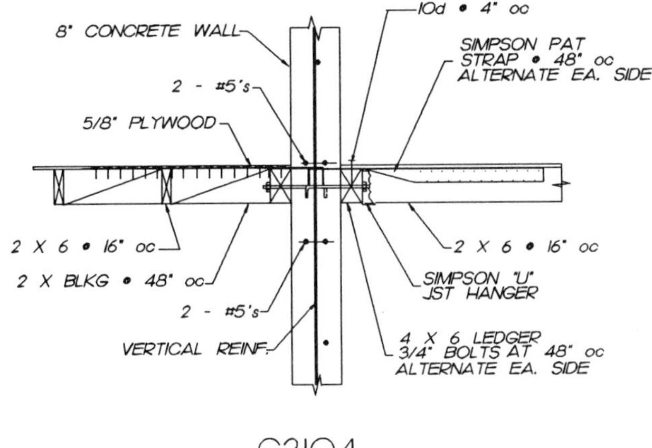

C3104

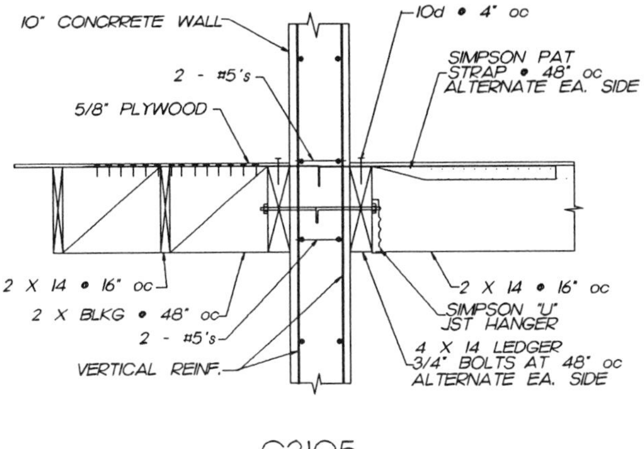

C3105

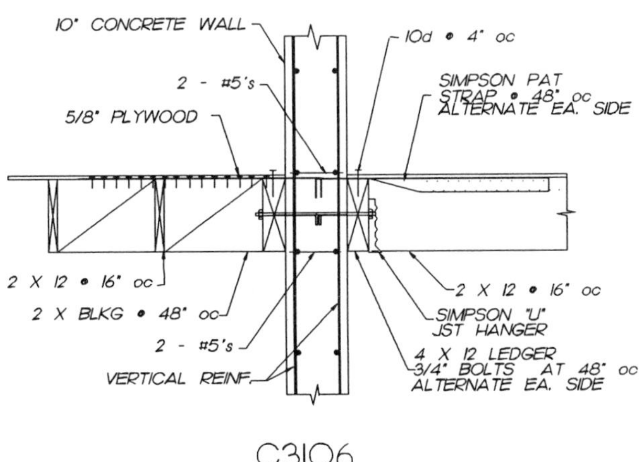

C3106

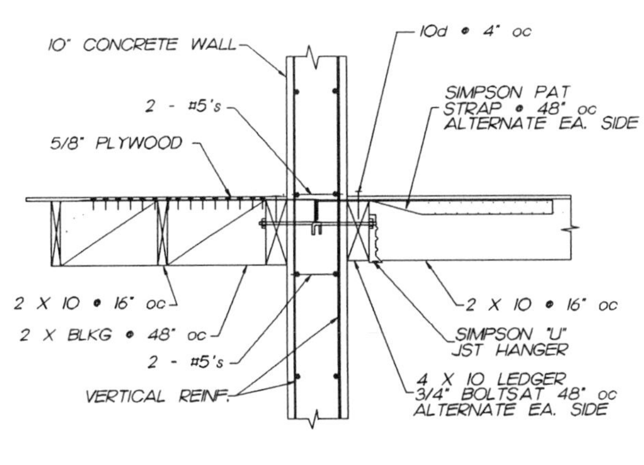

C3107

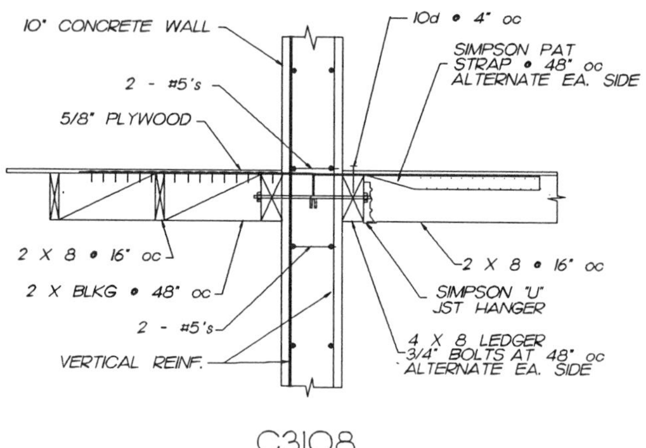

C3108

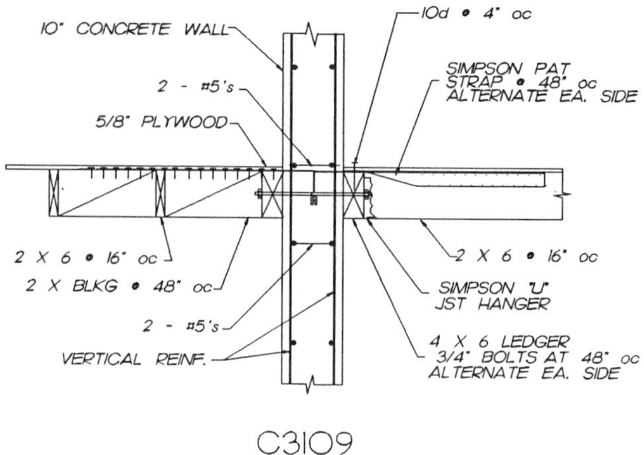

C3109

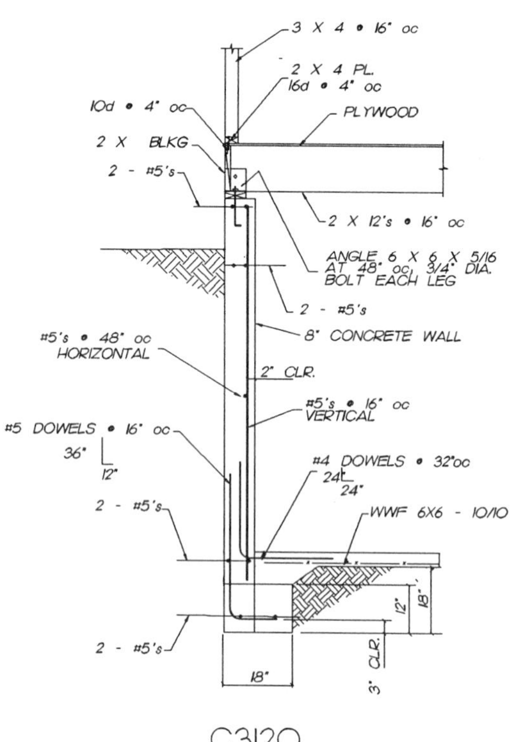

C3120

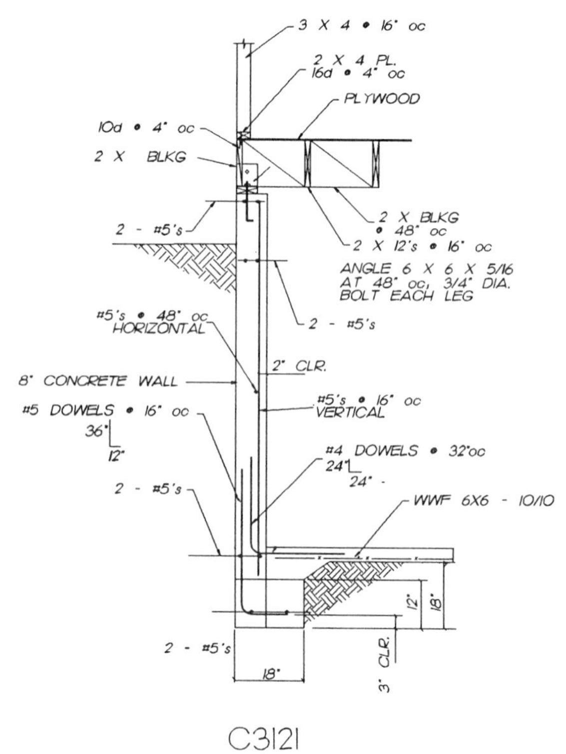

C3121

CHAPTER 3 CONCRETE SECTION 4 WALL - WOOD ROOF			
EXTERIOR WALL RAFTERS PERP. TO WALL CLIP ANGLE RAFTERS TO WALL	________	8" CONCRETE WALL	C4020 - C4023
EXTERIOR WALL RAFTERS PARALELL TO WALL CLIP ANGLE RAFTERS TO WALL	________	8" CONCRETE WALL	C4040 - C4043
EXTERIOR WALL RAFTERS PERP. TO WALL ROOF OVERHANG CLIP ANGLE RAFTERS TO WALL	________	8" CONCRETE WALL	C4060 - C4063
EXTERIOR WALL RAFTERS PARALELL TO WALL ROOF OVERHANG CLIP ANGLE RAFTERS TO WALL	________	8" CONCRETE WALL	C4080 - C4083
EXTERIOR WALL RAFTERS PERP. TO WALL SLOPED ROOF OVERHANG CLIP ANGLE RAFTERS TO WALL	________	8" CONCRETE WALL	C4100 - C4103
EXTERIOR WALL & PARAPET RAFTERS PERP. TO WALL SLOPED ROOF NO OVERHANG LEDGER SUPPORT	________	8" CONCRETE WALL	C4120 - C4123
EXTERIOR WALL & PARAPET RAFTERS PARALELL TO WALL SLOPED ROOF NO OVERHANG LEDGER SUPPORT	________	8" CONCRETE WALL	C4140 - C4143
INTERIOR WALL RAFTERS PERP. TO WALL CLIP ANGLE RAFTERS TO WALL	________	8" CONCRETE WALL	C4160 - C4163
INTERIOR WALL RAFTERS PARALELL TO WALL CLIP ANGLE RAFTERS TO WALL	________	8" CONCRETE WALL	C4180 - C4183
INTERIOR WALL RTRS PERP & PARALELL ALT SIDES CLIP ANGLE RAFTERS TO WALL	________	8" CONCRETE WALL	C4200 - C4203
ONE STORY CONCRETE WALL	WOOD ROOF	RAFTERS PERP. TO WALL	C4220
		RAFTERS PARALELL TO WALL	C4221

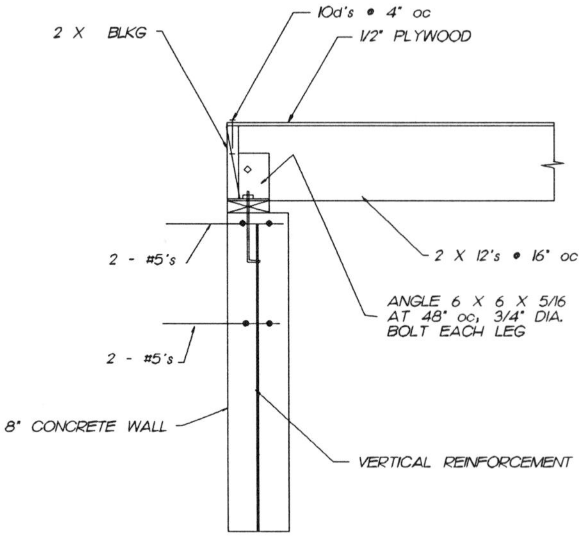

C4020

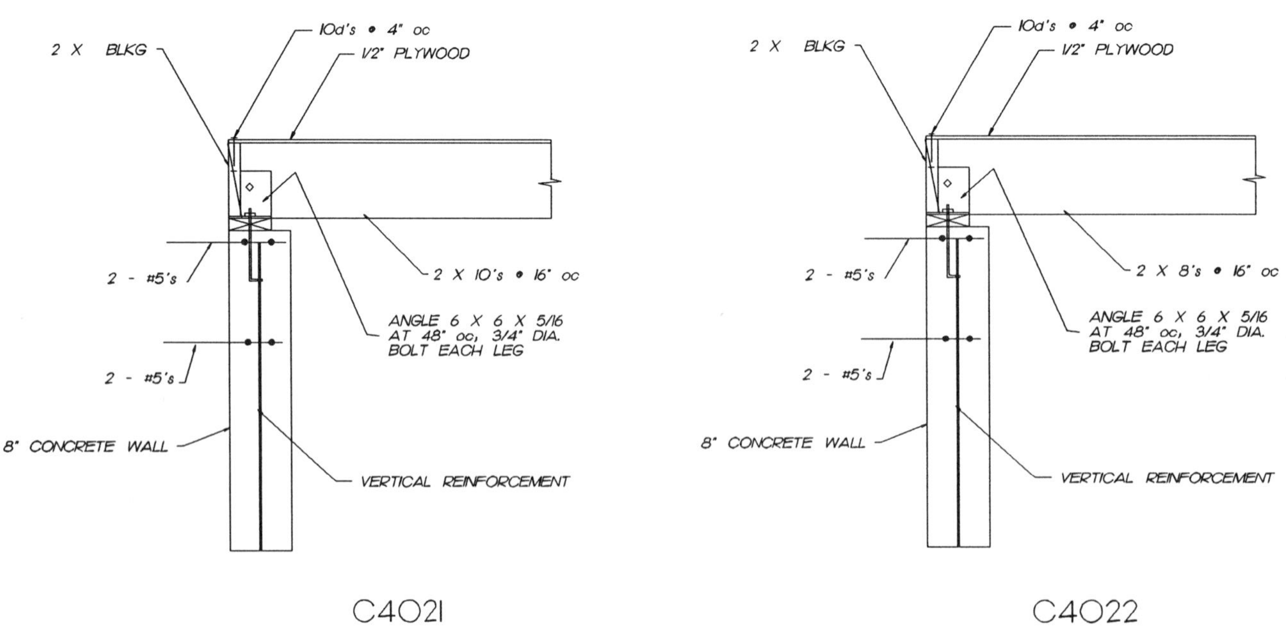

C4021

C4022

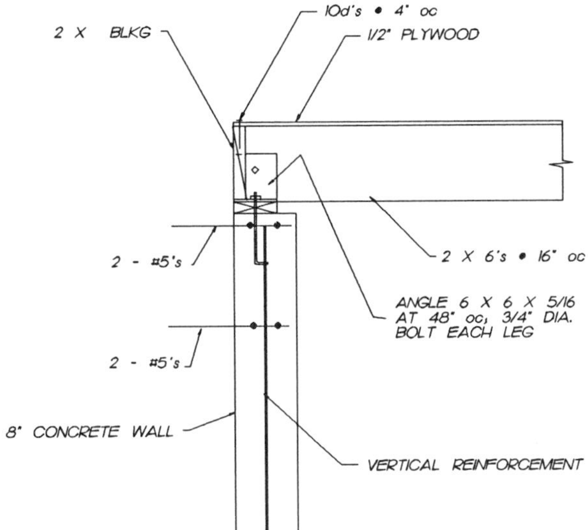
10d's • 4" oc
2 X BLKG
1/2" PLYWOOD
2 - #5's
2 X 6's • 16" oc
ANGLE 6 X 6 X 5/16
AT 48" oc, 3/4" DIA.
BOLT EACH LEG
2 - #5's
8" CONCRETE WALL
VERTICAL REINFORCEMENT

C4023

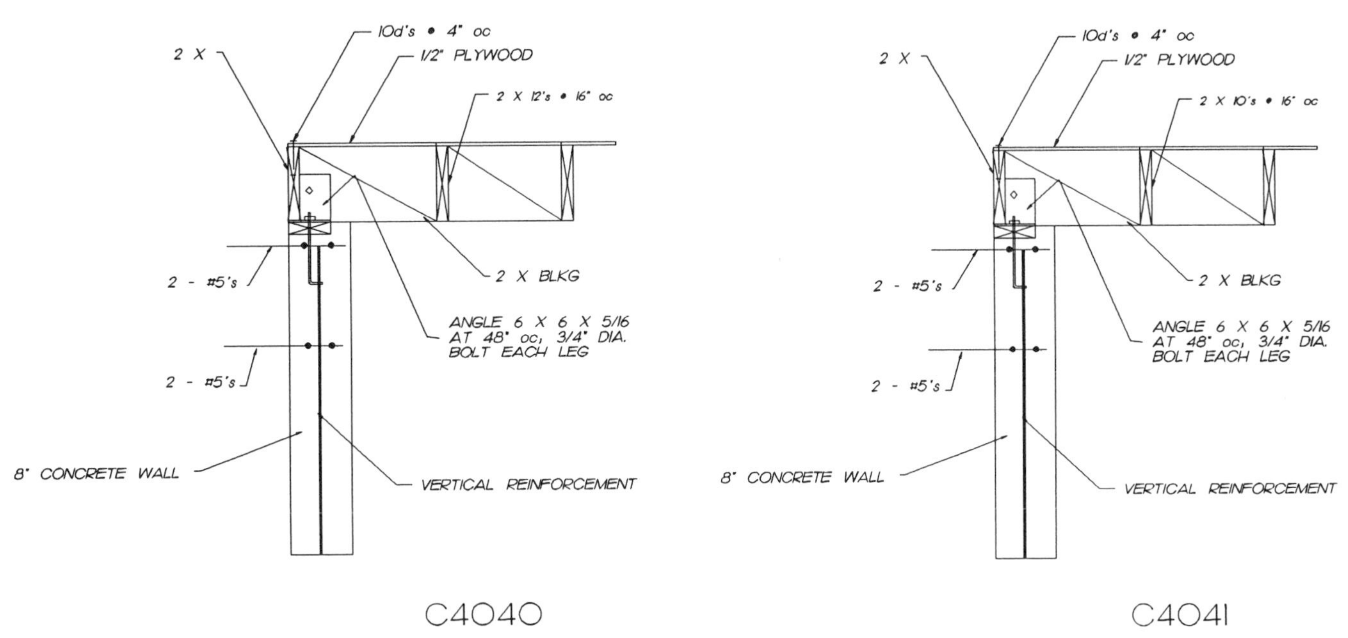
10d's • 4" oc
2 X
1/2" PLYWOOD
2 X 12's • 16" oc
2 - #5's
2 X BLKG
ANGLE 6 X 6 X 5/16
AT 48" oc, 3/4" DIA.
BOLT EACH LEG
2 - #5's
8" CONCRETE WALL
VERTICAL REINFORCEMENT
C4040
10d's • 4" oc
2 X
1/2" PLYWOOD
2 X 10's • 16" oc
2 - #5's
2 X BLKG
ANGLE 6 X 6 X 5/16
AT 48" oc, 3/4" DIA.
BOLT EACH LEG
2 - #5's
8" CONCRETE WALL
VERTICAL REINFORCEMENT
C4041

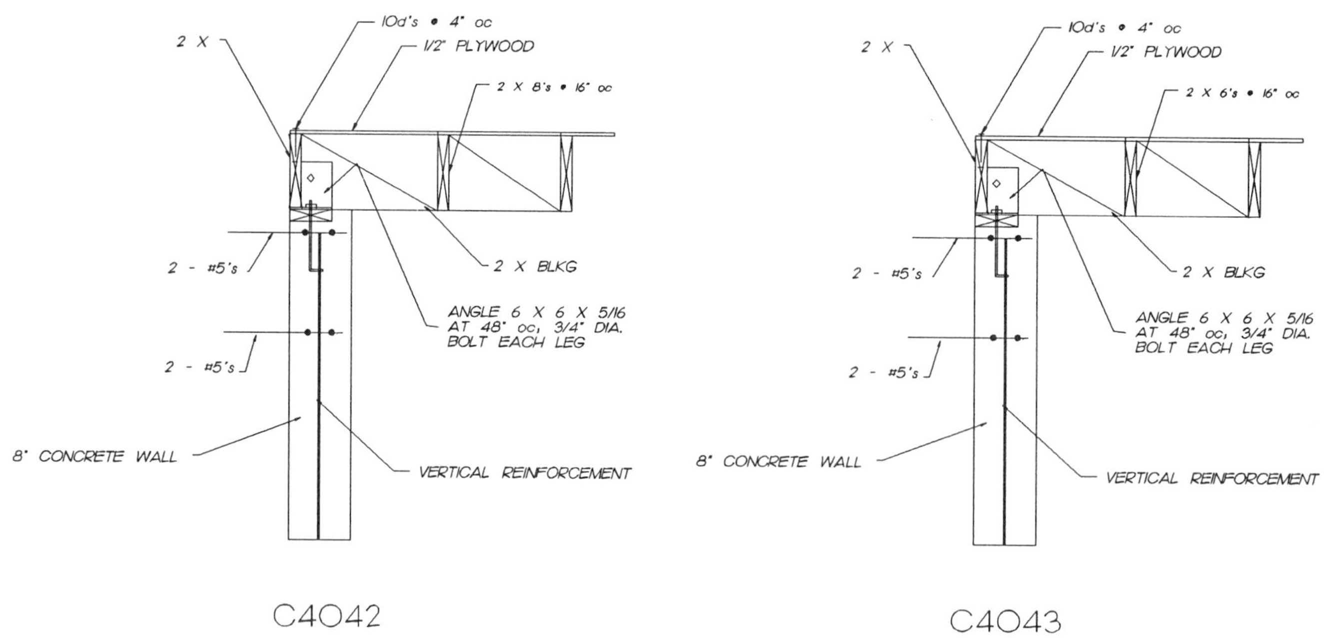
2 X
10d's @ 4" oc
1/2" PLYWOOD
2 X 8's @ 16" oc
2 - #5's
2 X BLKG
ANGLE 6 X 6 X 5/16
AT 48" oc, 3/4" DIA.
BOLT EACH LEG
2 - #5's
8" CONCRETE WALL
VERTICAL REINFORCEMENT
C4042
2 X
10d's @ 4" oc
1/2" PLYWOOD
2 X 6's @ 16" oc
2 - #5's
2 X BLKG
ANGLE 6 X 6 X 5/16
AT 48" oc, 3/4" DIA.
BOLT EACH LEG
2 - #5's
8" CONCRETE WALL
VERTICAL REINFORCEMENT
C4043

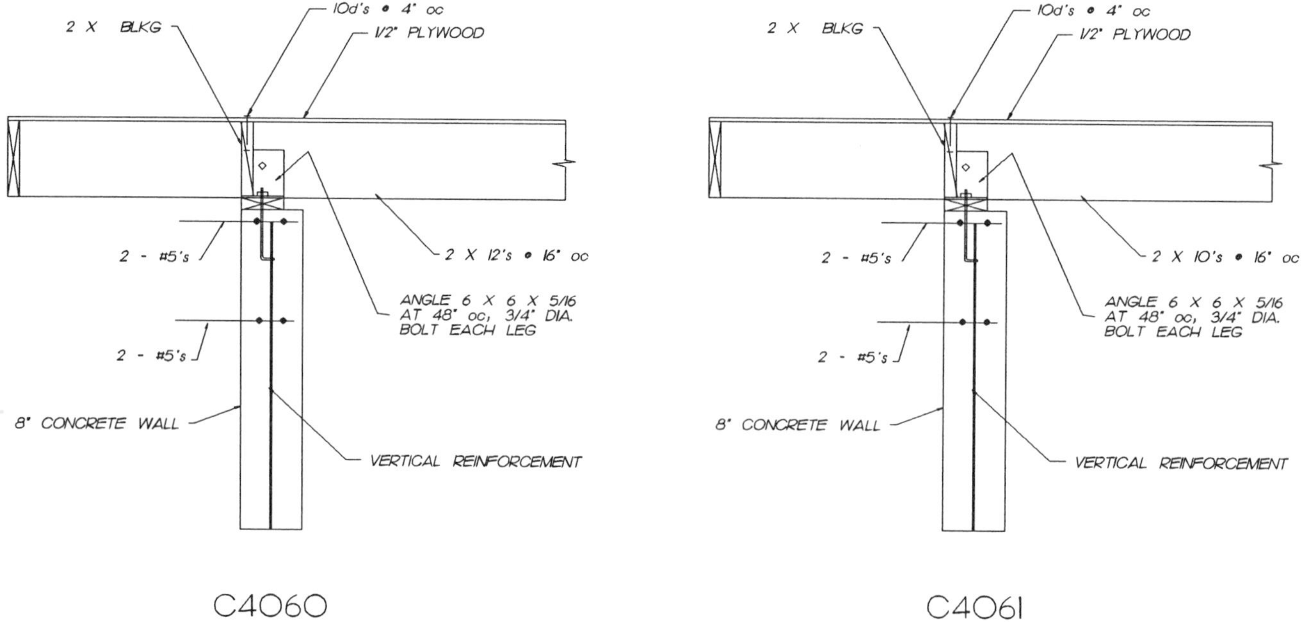
2 X BLKG
10d's @ 4" oc
1/2" PLYWOOD
2 - #5's
2 X 12's @ 16" oc
ANGLE 6 X 6 X 5/16
AT 48" oc, 3/4" DIA.
BOLT EACH LEG
2 - #5's
8" CONCRETE WALL
VERTICAL REINFORCEMENT
C4060
2 X BLKG
10d's @ 4" oc
1/2" PLYWOOD
2 - #5's
2 X 10's @ 16" oc
ANGLE 6 X 6 X 5/16
AT 48" oc, 3/4" DIA.
BOLT EACH LEG
2 - #5's
8" CONCRETE WALL
VERTICAL REINFORCEMENT
C4061

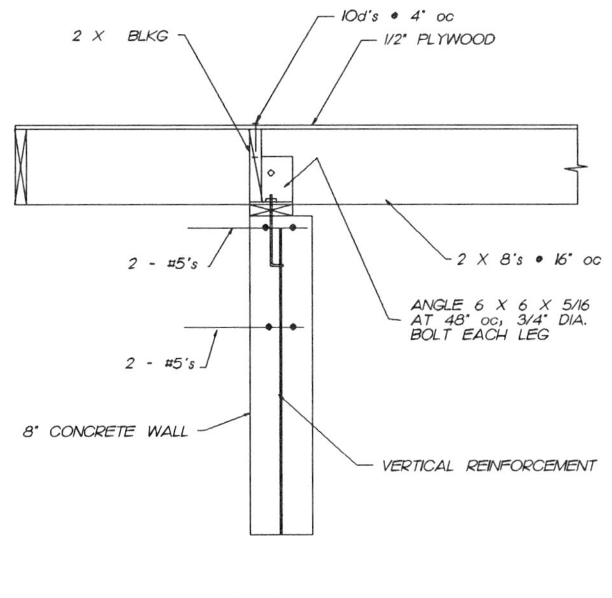
10d's @ 4" oc
2 X BLKG
1/2" PLYWOOD
2 X 8's @ 16" oc
2 - #5's
ANGLE 6 X 6 X 5/16
AT 48" oc, 3/4" DIA.
BOLT EACH LEG
2 - #5's
8" CONCRETE WALL
VERTICAL REINFORCEMENT

C4062

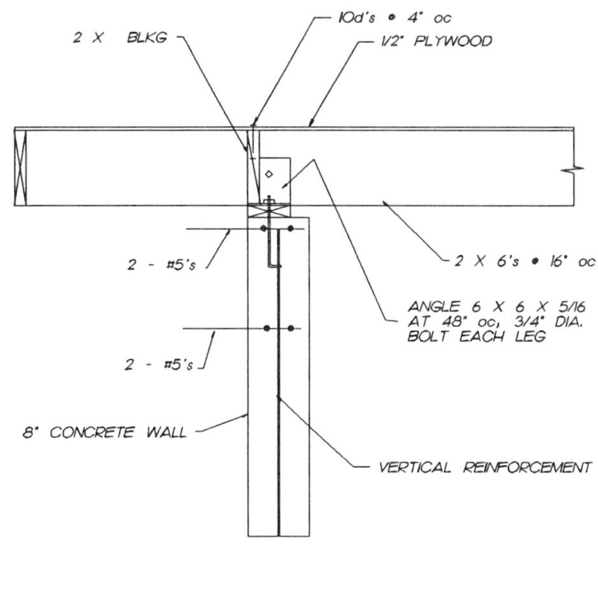
10d's @ 4" oc
2 X BLKG
1/2" PLYWOOD
2 X 6's @ 16" oc
2 - #5's
ANGLE 6 X 6 X 5/16
AT 48" oc, 3/4" DIA.
BOLT EACH LEG
2 - #5's
8" CONCRETE WALL
VERTICAL REINFORCEMENT

C4063

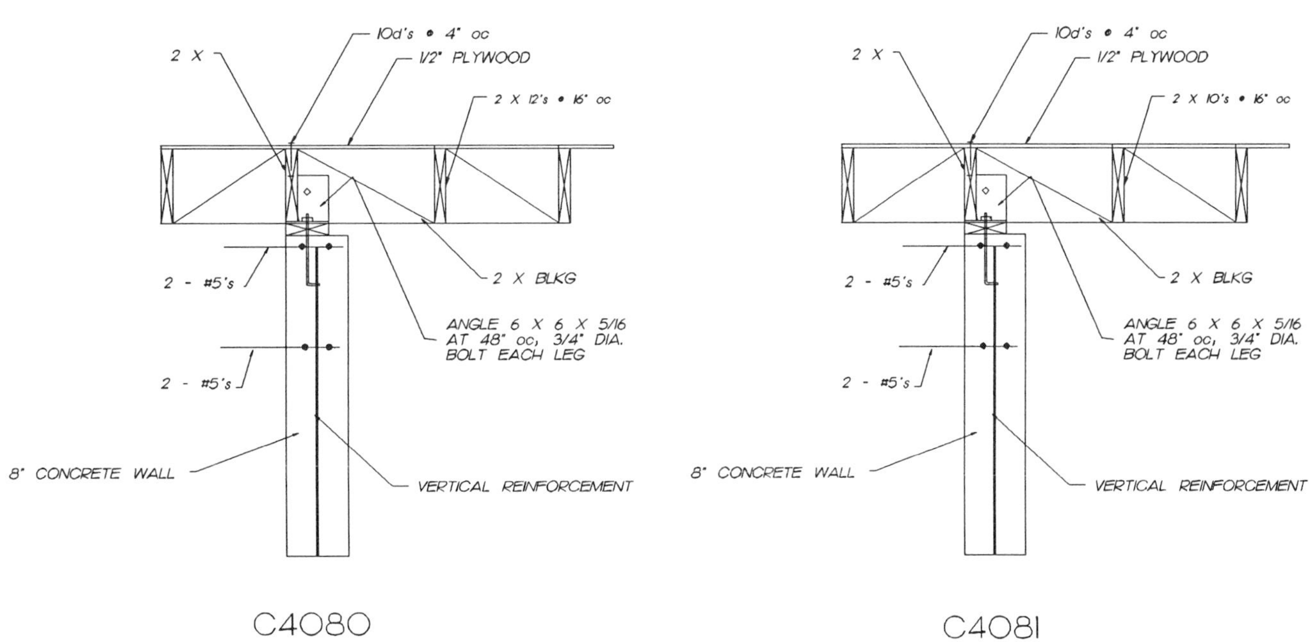
10d's @ 4" oc
2 X
1/2" PLYWOOD
2 X 12's @ 16" oc
2 - #5's
2 X BLKG
ANGLE 6 X 6 X 5/16
AT 48" oc, 3/4" DIA.
BOLT EACH LEG
2 - #5's
8" CONCRETE WALL
VERTICAL REINFORCEMENT
C4080
10d's @ 4" oc
2 X
1/2" PLYWOOD
2 X 10's @ 16" oc
2 - #5's
2 X BLKG
ANGLE 6 X 6 X 5/16
AT 48" oc, 3/4" DIA.
BOLT EACH LEG
2 - #5's
8" CONCRETE WALL
VERTICAL REINFORCEMENT
C4081

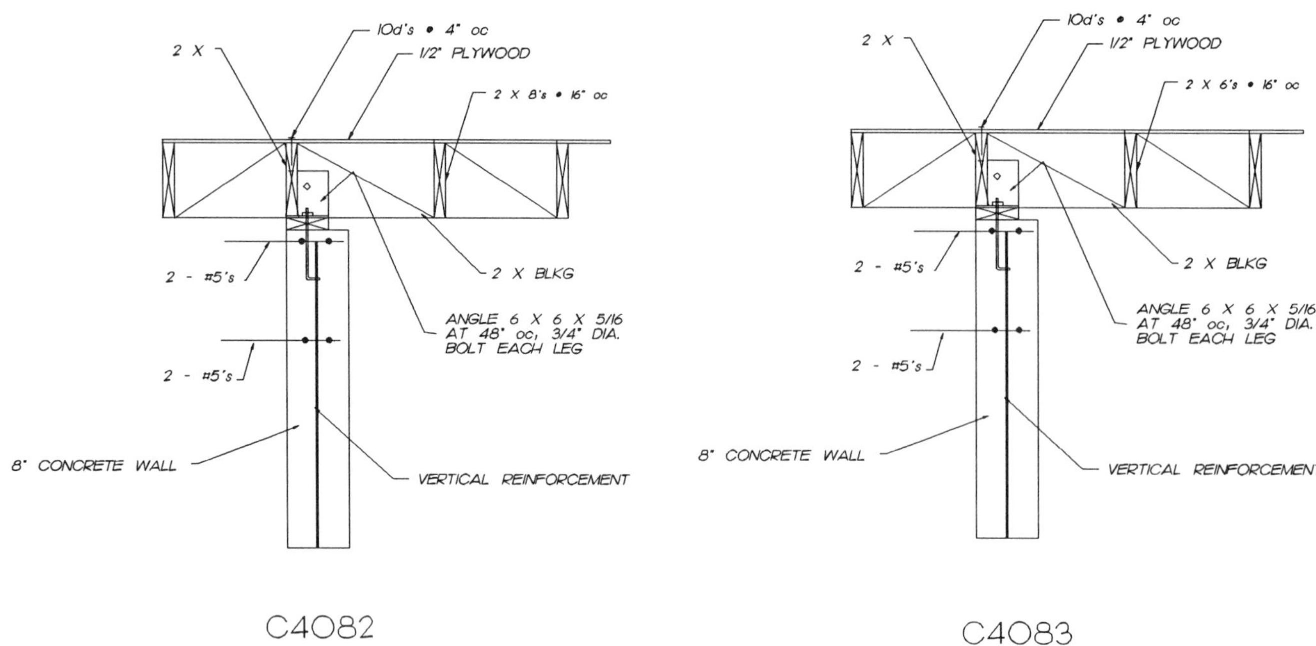

C4082

C4083

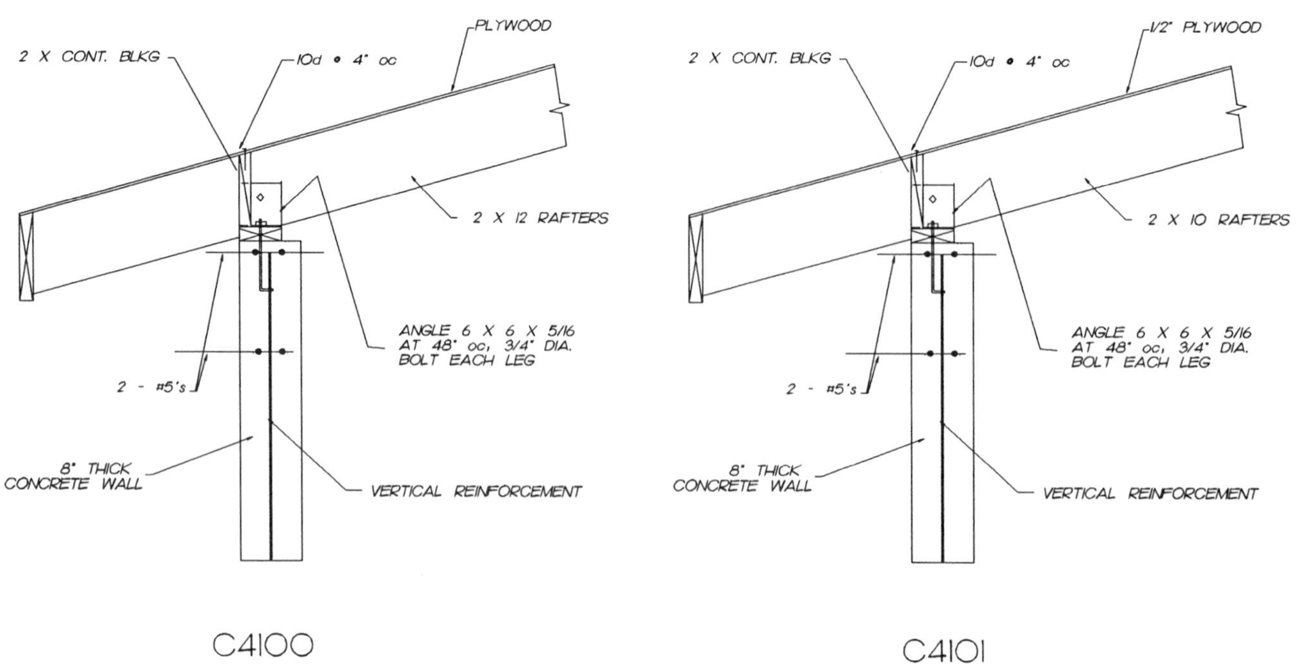

C4100

C4101

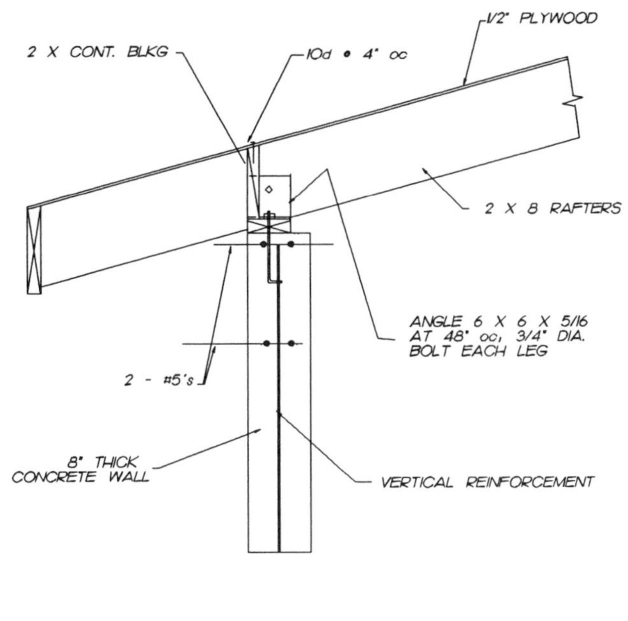

C4102

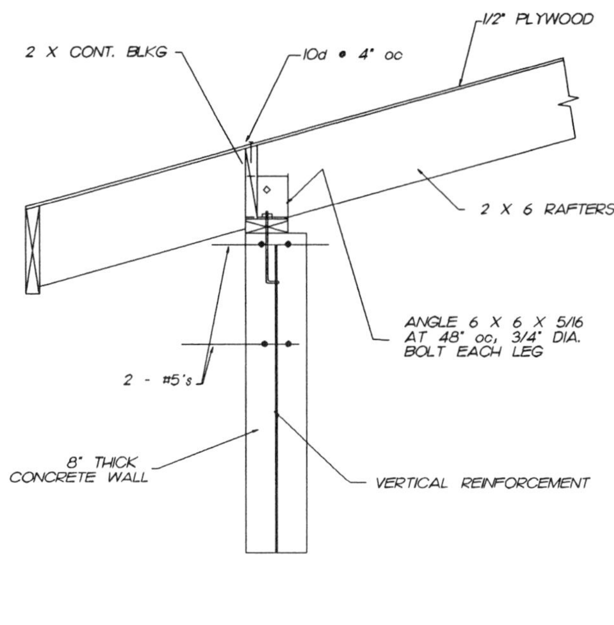

C4103

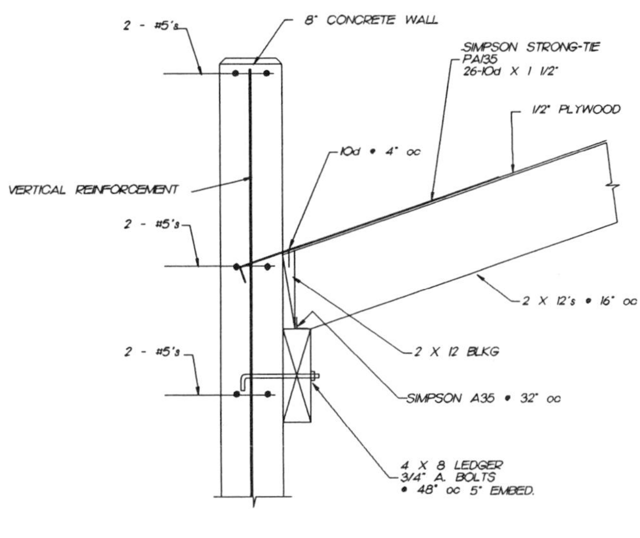

C4120

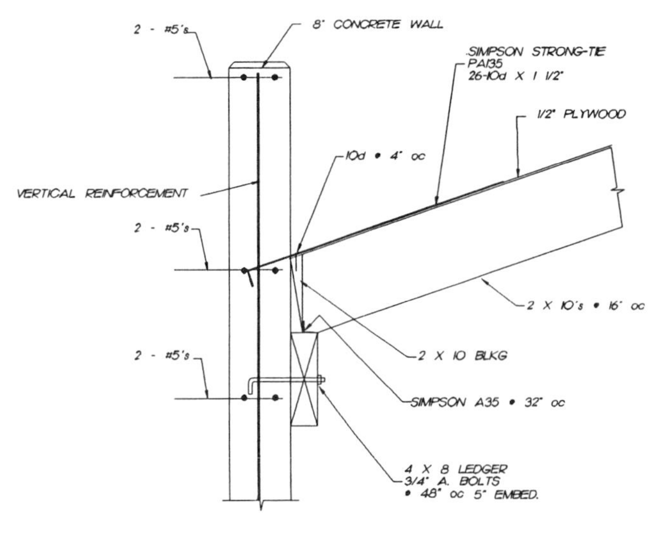

C4121

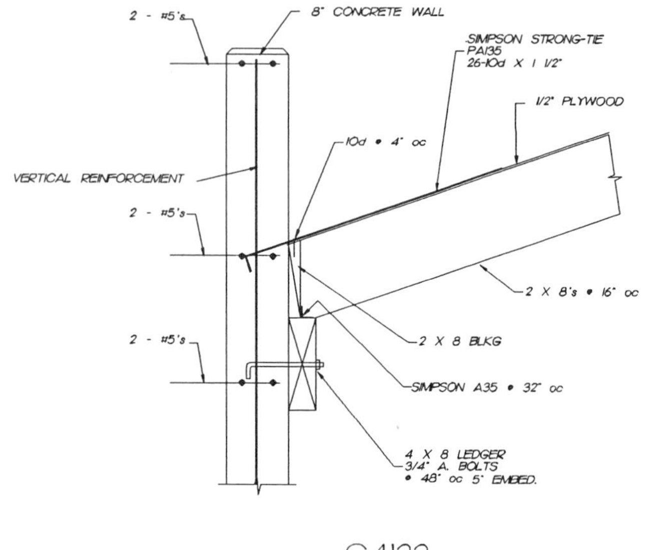

C4122

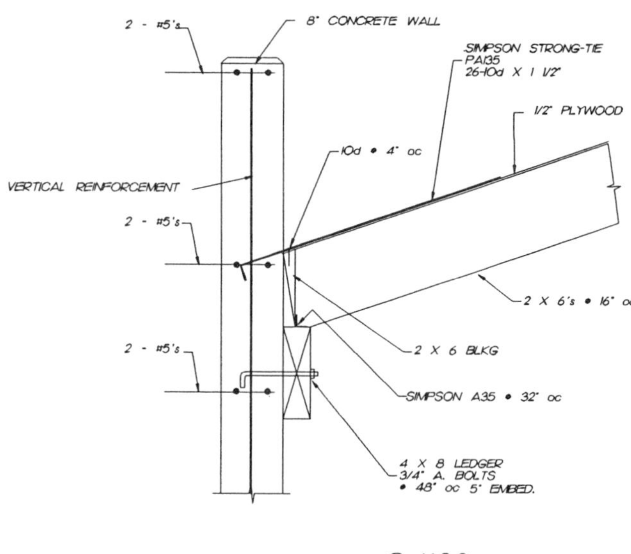

C4123

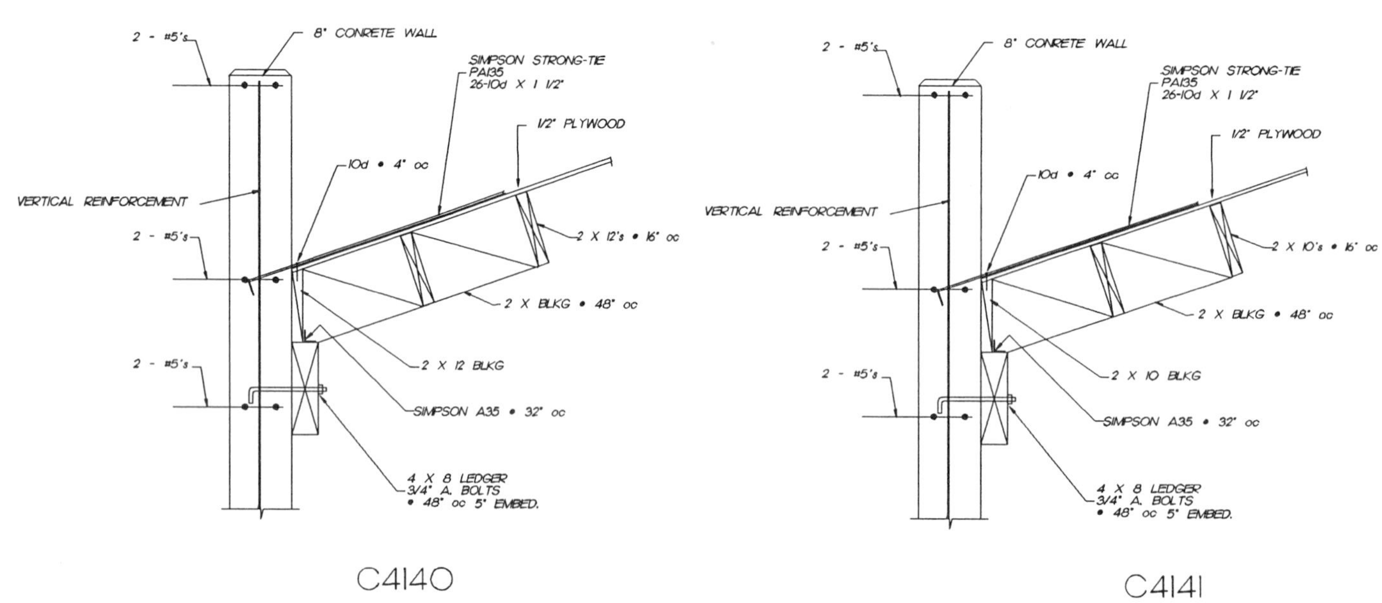

C4140

C4141

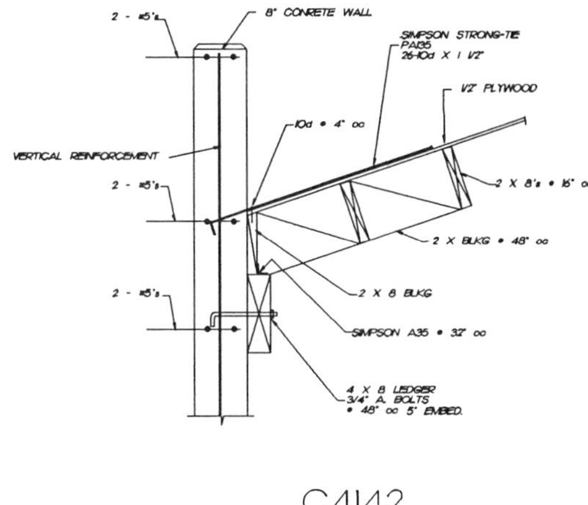

C4142

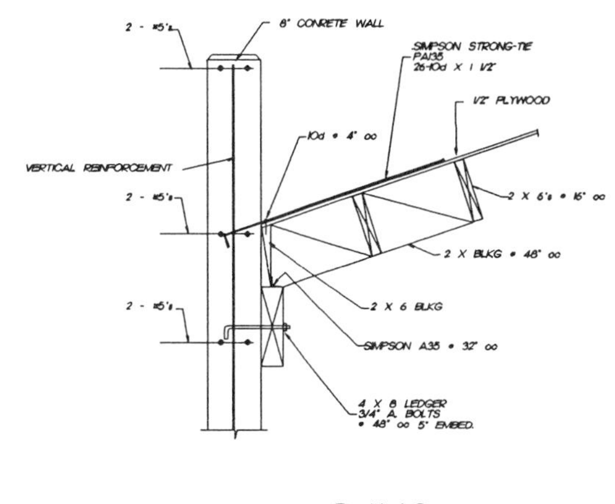

C4143

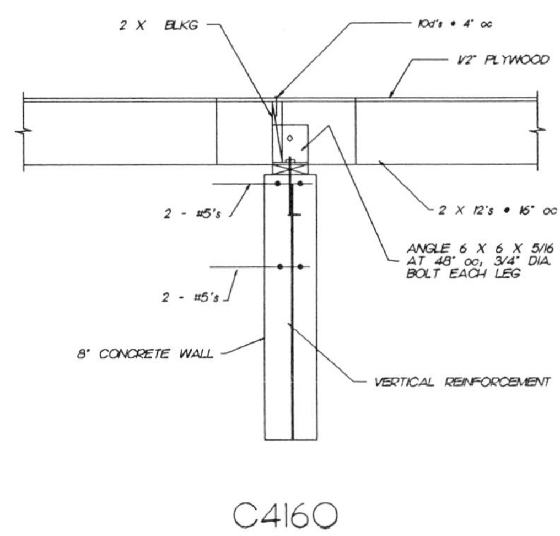

C4160

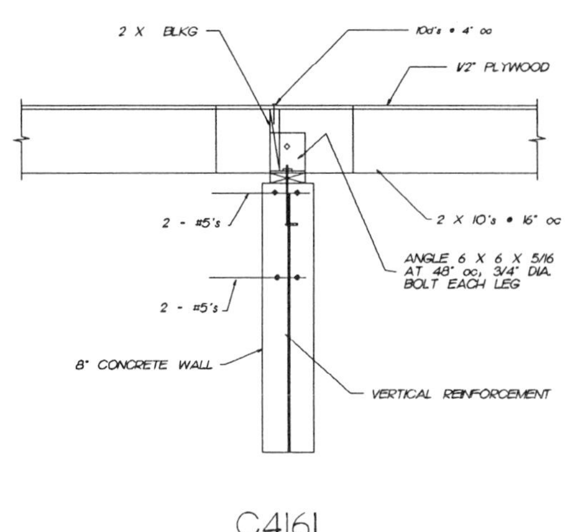

C4161

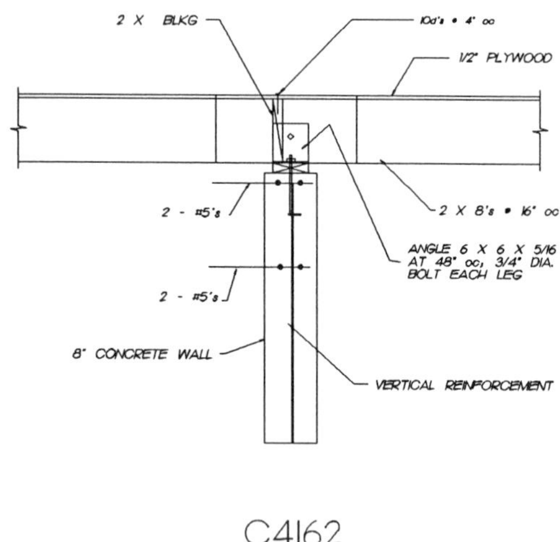

C4162

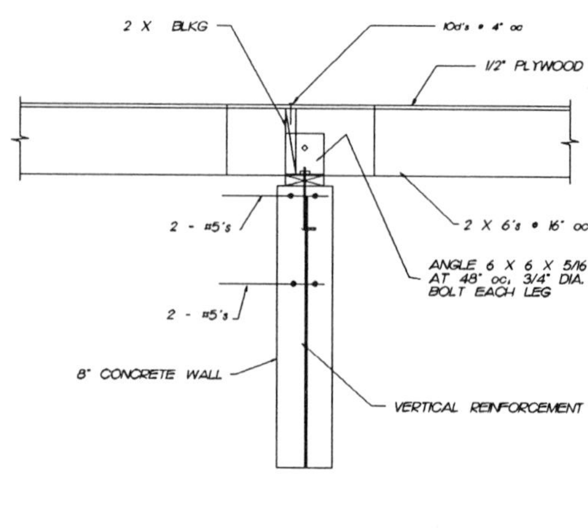

C4163

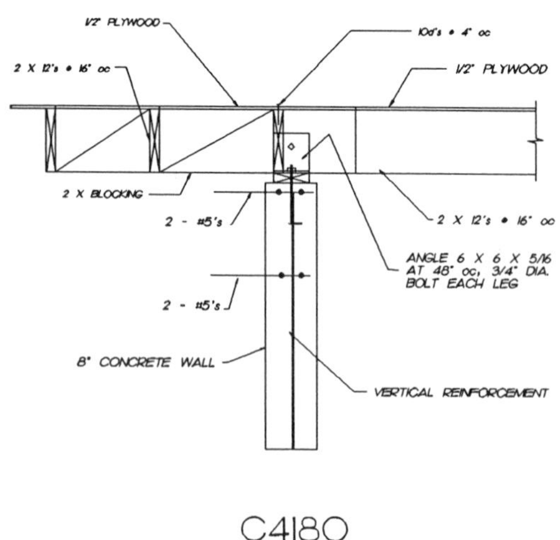

C4180

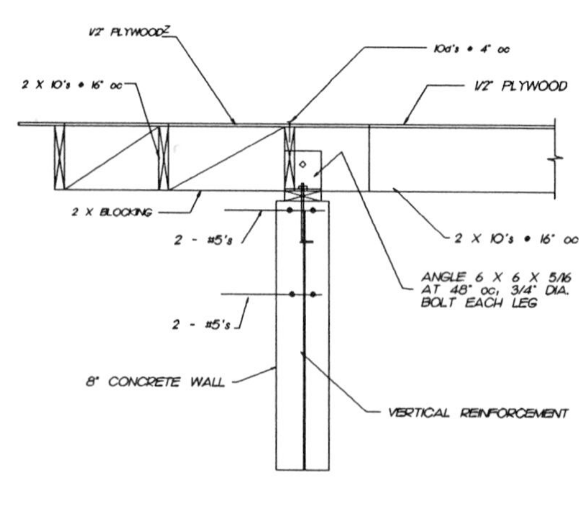

C4181

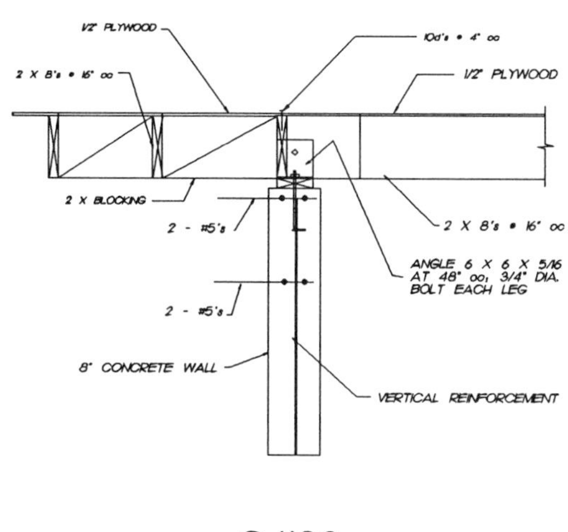

C4182

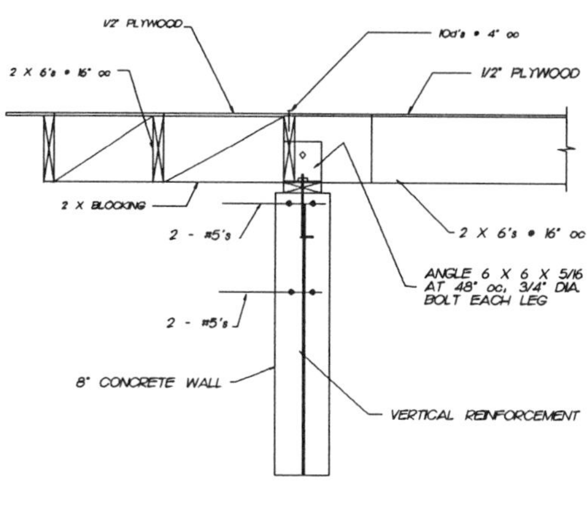

C4183

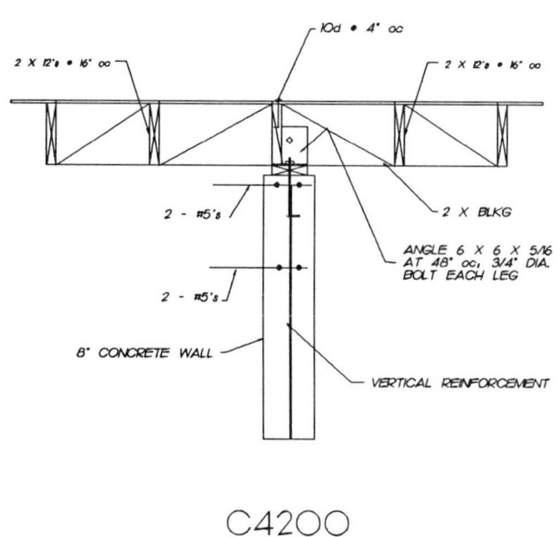

C4200

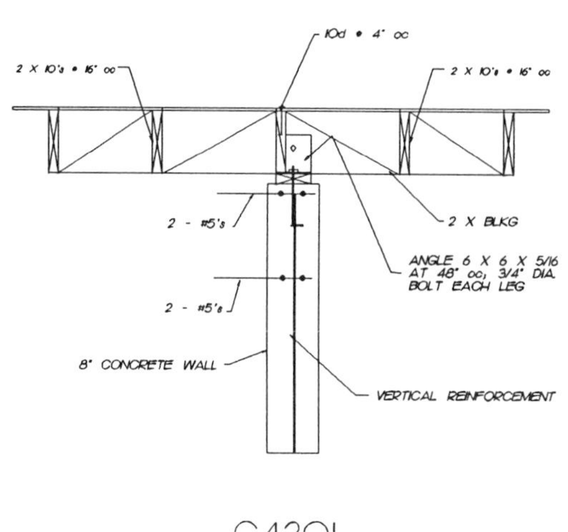

C4201

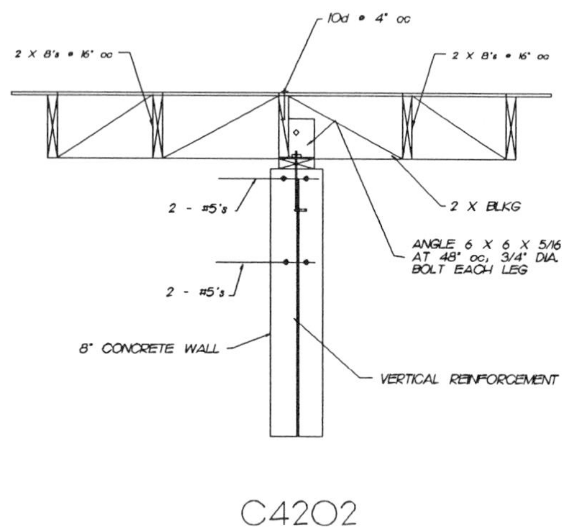

C4202

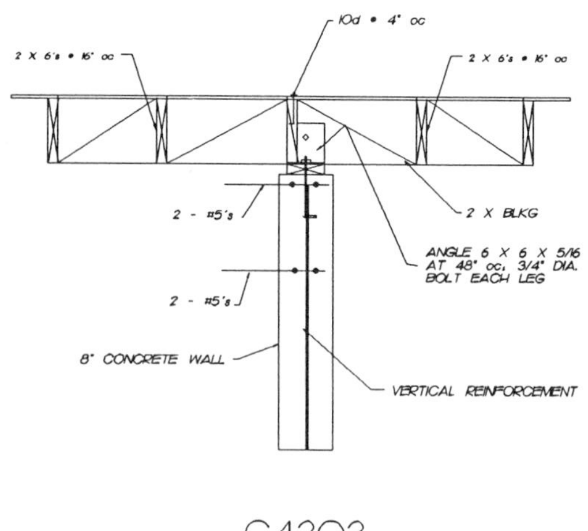

C4203

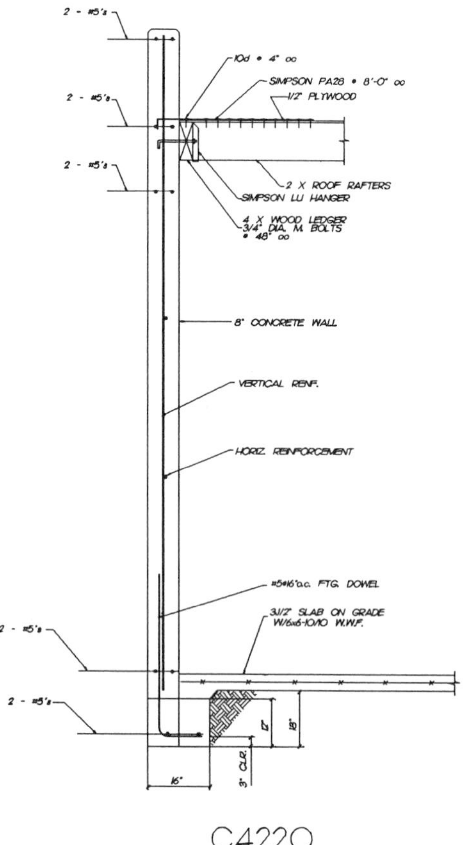

C4220

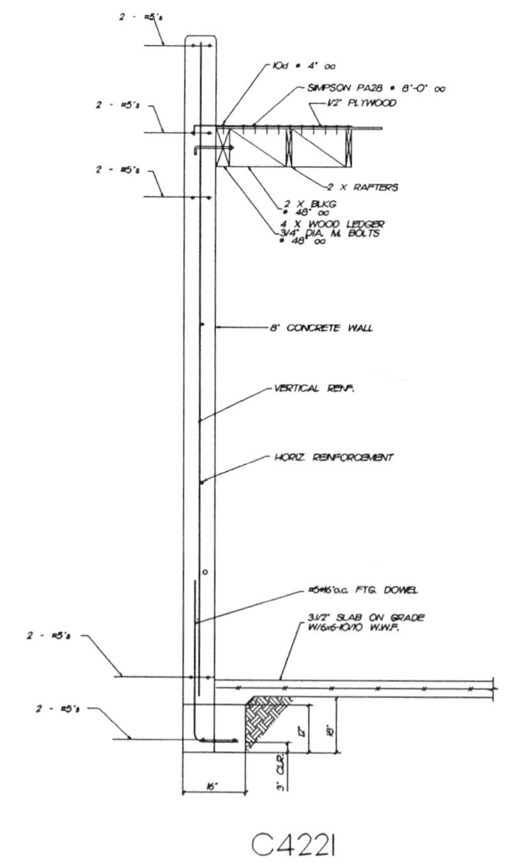

C4221

CHAPTER 3 CONCRETE SECTION 5 WALL - CONCRETE

EXTERIOR BASEMENT WALL CONCRETE SLAB	______	8" CONCRETE WALL	C5020 - C5025
WALL SUPPORTED BY A SLAB WITH A SLAB ON TOP WALL	______	8" CONCRETE WALL	C5040 - C5041
INTERIOR WALL STRUCTURAL SLAB ON TOP SLAB ON GRADE AT BOTTOM	______	8" CONCRETE WALL	C5060
EXTERIOR BASEMENT WALL TWO LEVEL CONCRETE SLAB	______	8" CONCRETE WALL	C5080 - C5080
INTERIOR BASEMENT WALL TWO LEVEL CONCRETE SLAB	______	8" CONCRETE WALL	C5081
EXTERIOR BASEMENT WALL ONE LEVEL CONCRETE SLAB WITH RAMP SLAB ON GRADE	______	8" CONCRETE WALL	C5082
EXTERIOR BASEMENT WALL TWO LEVEL CONCRETE SLAB WITH RAMP SLAB ON GRADE	______	8" CONCRETE WALL	C5083

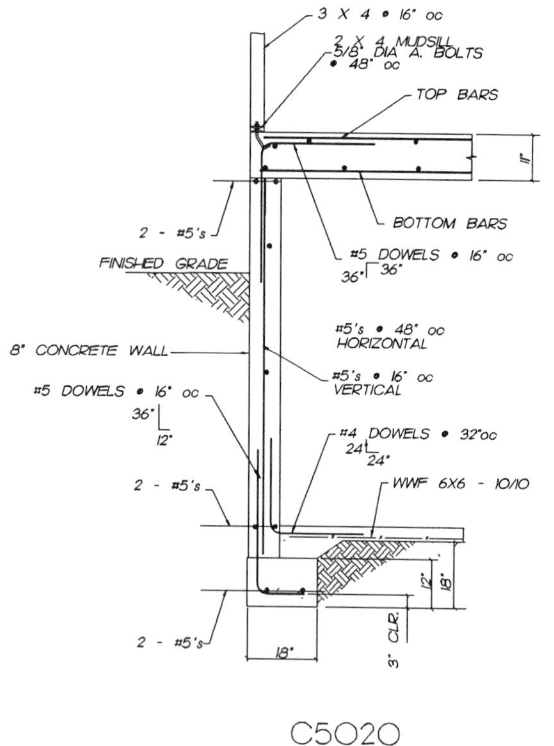

C5020

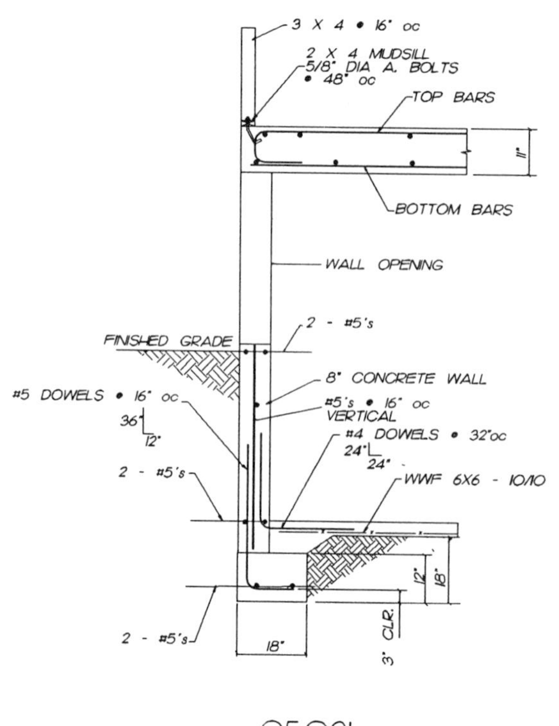

C5021

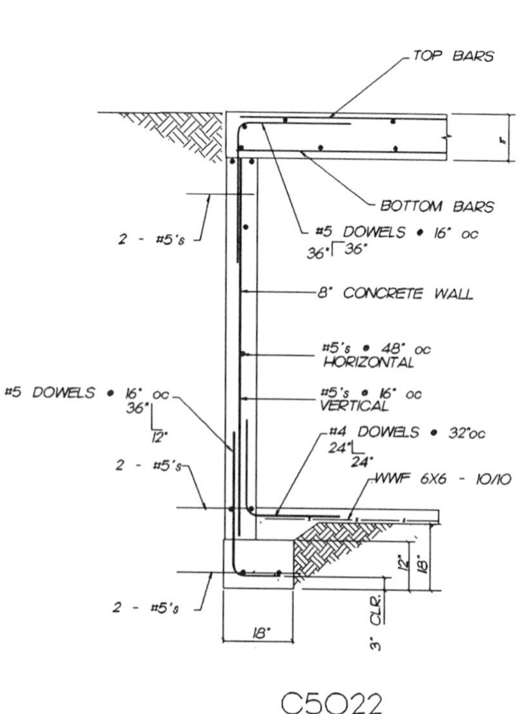

C5022

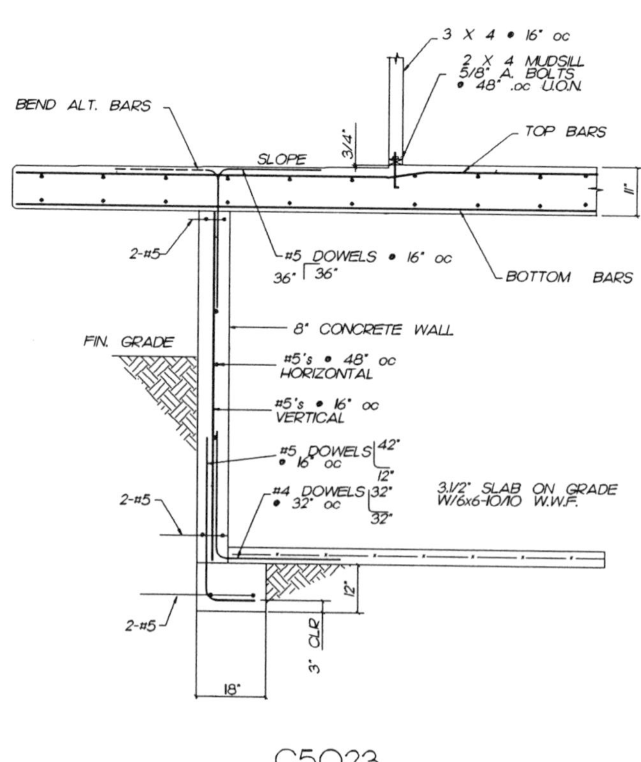

C5023

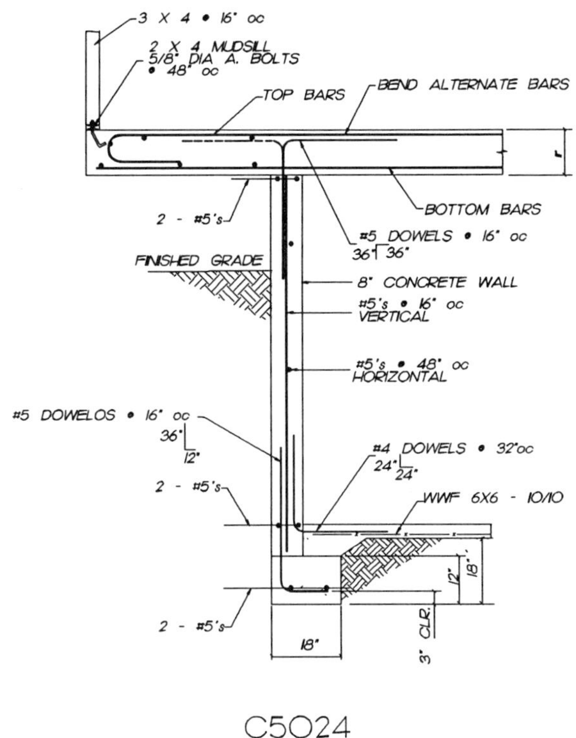

C5024

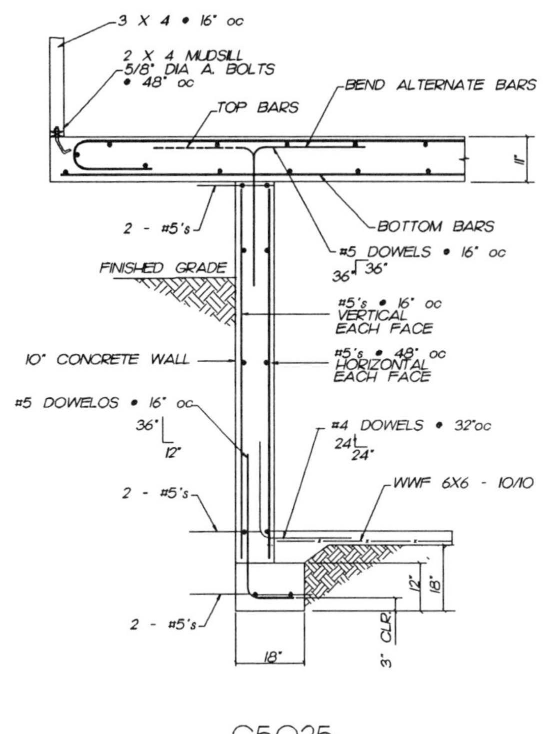

C5025

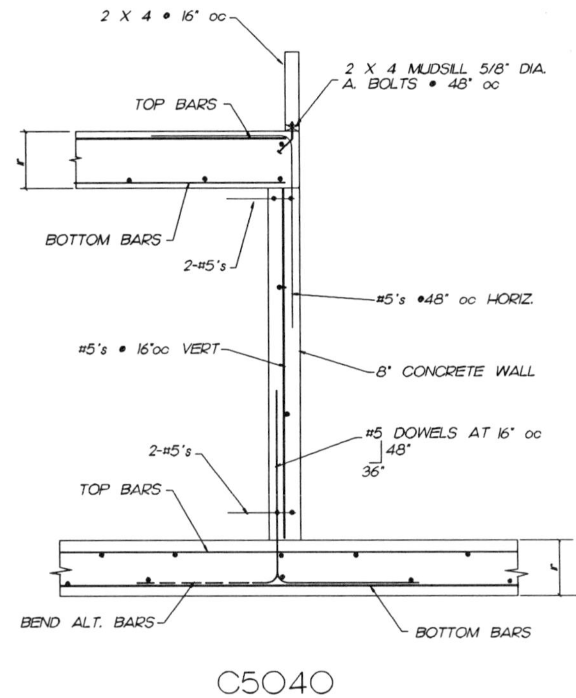

C5040

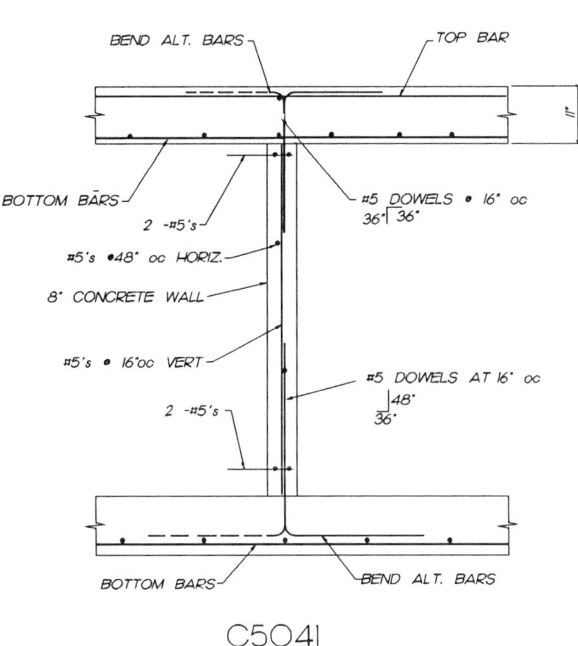

C5041

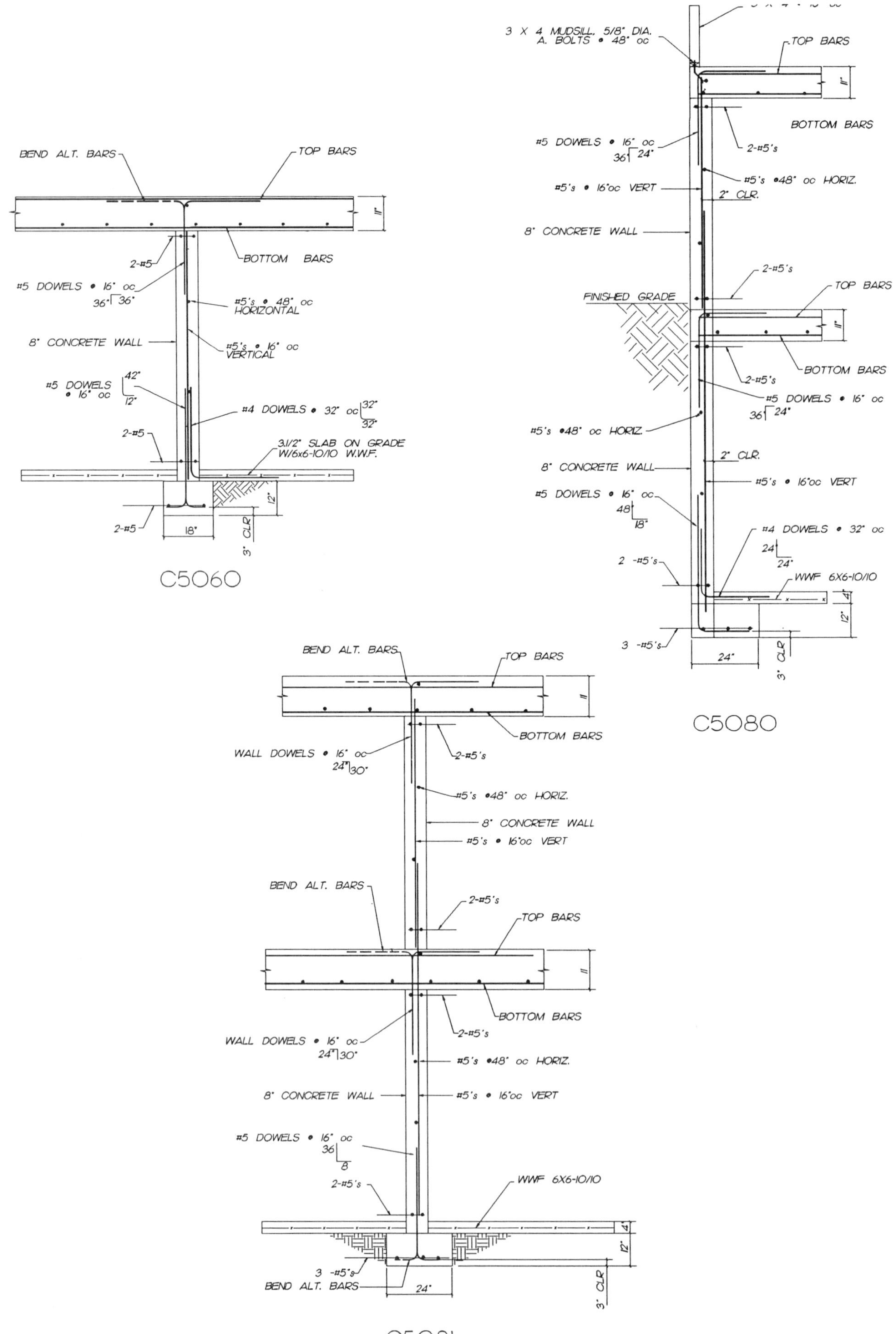

BEND ALT. BARS
TOP BARS
11"
2-#5
BOTTOM BARS
#5 DOWELS • 16" oc
36" 36"
#5's • 48" oc HORIZONTAL
8" CONCRETE WALL
#5's • 16" oc VERTICAL
#5 DOWELS • 16" oc
42"
12"
#4 DOWELS • 32" oc
32"
32"
2-#5
3 1/2" SLAB ON GRADE W/6x6-10/10 W.W.F.
12"
2-#5
18"
3" CLR
C5060
3 X 4 MUDSILL, 5/8" DIA. A. BOLTS • 48" oc
TOP BARS
11"
BOTTOM BARS
#5 DOWELS • 16" oc
36 24"
2-#5's
#5's • 48" oc HORIZ.
#5's • 16"oc VERT
2" CLR.
8" CONCRETE WALL
2-#5's
FINISHED GRADE
TOP BARS
11"
BOTTOM BARS
2-#5's
#5 DOWELS • 16" oc
36 24"
#5's • 48" oc HORIZ.
2" CLR.
8" CONCRETE WALL
#5's • 16"oc VERT
#5 DOWELS • 16" oc
48
18"
#4 DOWELS • 32" oc
24
24"
2 -#5's
WWF 6X6-10/10
4"
12"
3 -#5's
24"
3" CLR
C5080
BEND ALT. BARS
TOP BARS
11
BOTTOM BARS
WALL DOWELS • 16" oc
24" 30"
2-#5's
#5's • 48" oc HORIZ.
8" CONCRETE WALL
#5's • 16"oc VERT
BEND ALT. BARS
2-#5's
TOP BARS
11
BOTTOM BARS
WALL DOWELS • 16" oc
24" 30"
2-#5's
#5's • 48" oc HORIZ.
8" CONCRETE WALL
#5's • 16"oc VERT
#5 DOWELS • 16" oc
36
8
2-#5's
WWF 6X6-10/10
4
12"
3 -#5's
BEND ALT. BARS
24"
3" CLR
C5081

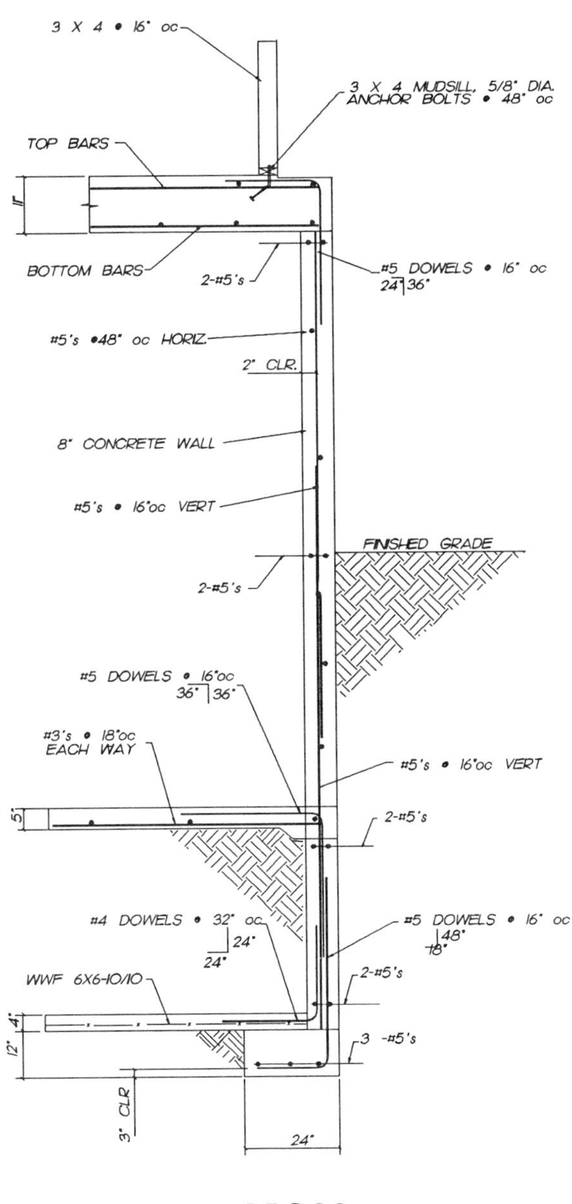
3 X 4 • 16" oc
3 X 4 MUDSILL, 5/8" DIA. ANCHOR BOLTS • 48" oc
TOP BARS
11"
BOTTOM BARS
2-#5's
#5 DOWELS • 16" oc
24" 36"
#5's •48" oc HORIZ.
2" CLR.
8" CONCRETE WALL
#5's • 16"oc VERT
FINISHED GRADE
2-#5's
#5 DOWELS • 16"oc
36" 36"
#3's • 18"oc EACH WAY
#5's • 16"oc VERT
5"
2-#5's
#4 DOWELS • 32" oc
24"
24"
#5 DOWELS • 16" oc
48"
18"
2-#5's
WWF 6X6-10/10
4"
12"
3 -#5's
3" CLR
24"

C5082

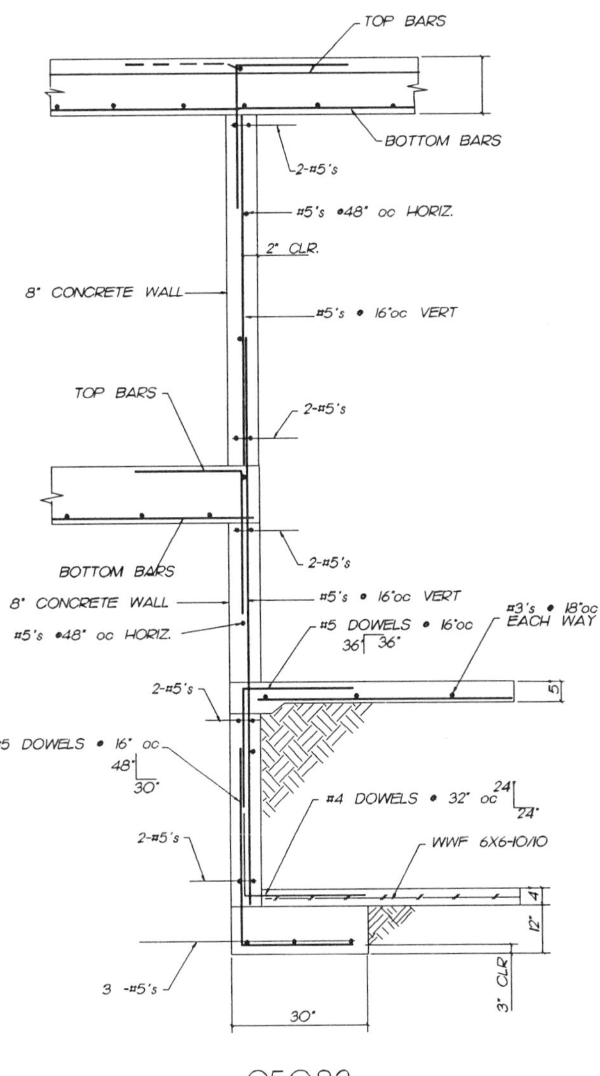
TOP BARS
BOTTOM BARS
2-#5's
#5's •48" oc HORIZ.
2" CLR.
8" CONCRETE WALL
#5's • 16"oc VERT
TOP BARS
2-#5's
BOTTOM BARS
2-#5's
#5's • 16"oc VERT
8" CONCRETE WALL
#5's •48" oc HORIZ.
#5 DOWELS • 16"oc
36 36"
#3's • 18"oc EACH WAY
5"
2-#5's
#5 DOWELS • 16" oc
48"
30"
#4 DOWELS • 32" oc
24
24"
2-#5's
WWF 6X6-10/10
4"
12"
3 -#5's
3" CLR
30"

C5083

CHAPTER 3 CONCRETE SECTION 6 WALL - STEEL FLOOR

Description	Wall	Detail
EXTERIOR WALL METAL DECKING PARALELL TO WALL DECKING SUPPORTED BY STEEL BMS. BM. CONNECTED WITH CLIP ANGLES USING ANCHOR BOLTS & MACHINE BOLTS	8" CONCRETE WALL	C6020 - C6024
	10" CONCRETE WALL	C6025 - C6029
	———	———
EXTERIOR WALL METAL DECKING PERP. TO WALL DECKING SUPPORTED BY STEEL BMS. BM. CONNECTED WITH CLIP ANGLES USING ANCHOR BOLTS & MACHINE BOLTS	8" CONCRETE WALL	C6040 - C6044
	10" CONCRETE WALL	C6045 - C6049
	———	———
EXTERIOR WALL METAL DECKING PARALELL TO WALL DECKING SUPPORTED BY LIGHT METAL JOISTS JOISTS CONNECTED WITH SEAT ANGLES USING WELDS & ANCHOR BOLTS	8" CONCRETE WALL	C6060
	10" CONCRETE WALL	C6061
	———	———
EXTERIOR WALL METAL DECKING PERP. TO WALL DECKING SUPPORTED BY LIGHT METAL JOISTS METAL DECKING PERP. TO WALL DECKING CONNECTED WITH SEAT ANGLES USING WELDS & ANCHOR BOLTS	8" CONCRETE WALL	C6080
	10" CONCRETE WALL	C6081
	———	———
EXTERIOR WALL METAL DECKING PERP. TO WALL DECKING SUPPORTED BY LIGHT METAL JOISTS METAL DECKING PERP. TO WALL DECKING CONNECTED WITH SEAT ANGLES USING WELDS & ANCHOR BOLTS	8" CONCRETE BLOCK	C6100
	———	———
	———	———
EXTERIOR WALL METAL DECKING PARALELL TO WALL DECKING SUPPORTED BY METAL BAR JOISTS BAR JOISTS CONNECTED WITH SEAT ANGLES USING WELDS & ANCHOR BOLTS	8" CONCRETE WALL	C6120
	10" CONCRETE WALL	C6121
	———	———
EXTERIOR WALL METAL DECKING PERP. TO WALL DECKING SUPPORTED BY METAL BAR JOISTS BAR JOISTS CONNECTED WITH SEAT ANGLES USING WELDS & ANCHOR BOLTS	8" CONCRETE WALL	C6140
	10" CONCRETE WALL	C6141
	———	———
EXTERIOR WALL METAL DECKING PARALELL TO WALL METAL DECKING CONNECTED WITH SEAT ANGLES USING WELDS & ANCHOR BOLTS	8" CONCRETE WALL	C6160
	10" CONCRETE WALL	C6161
	———	———
INTERIOR WALL METAL DECKING PARALELL TO WALL METAL DECKING CONNECTED WITH SEAT ANGLES USING WELDS & ANCHOR BOLTS	8" CONCRETE WALL	C6180
	10" CONCRETE WALL	C6181
	———	———
INTERIOR WALL METAL DECKING PARALELL TO WALL DECKING SUPPORTED BY METAL BAR JOISTS BAR JOISTS CONNECTED WITH SEAT ANGLES USING WELDS & ANCHOR BOLTS	8" CONCRETE WALL	C6200
	10" CONCRETE WALL	C6201
	———	———
INTERIOR WALL METAL DECKING PERP. TO WALL DECKING SUPPORTED BY METAL BAR JOISTS BAR JOISTS CONNECTED WITH SEAT ANGLES USING WELDS & ANCHOR BOLTS	8" CONCRETE WALL	C6220
	10" CONCRETE WALL	C6221
	———	———

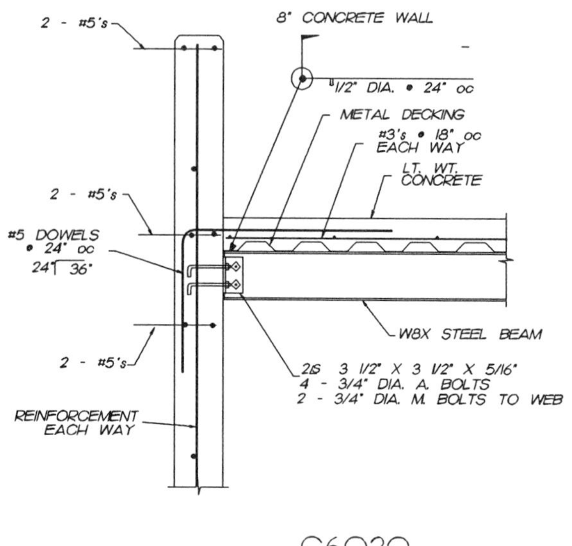

C6020

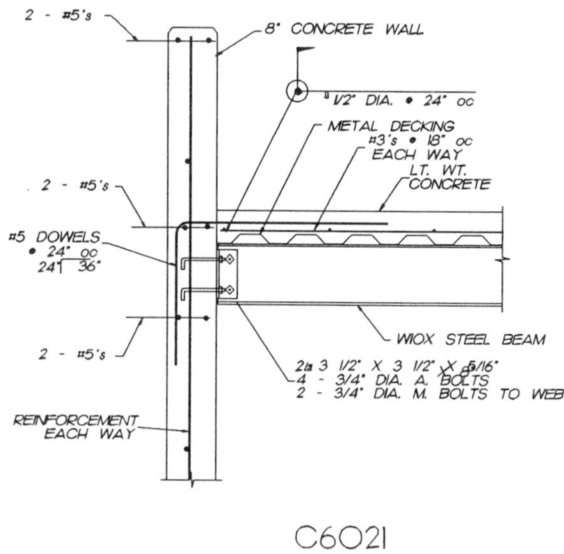

C6021

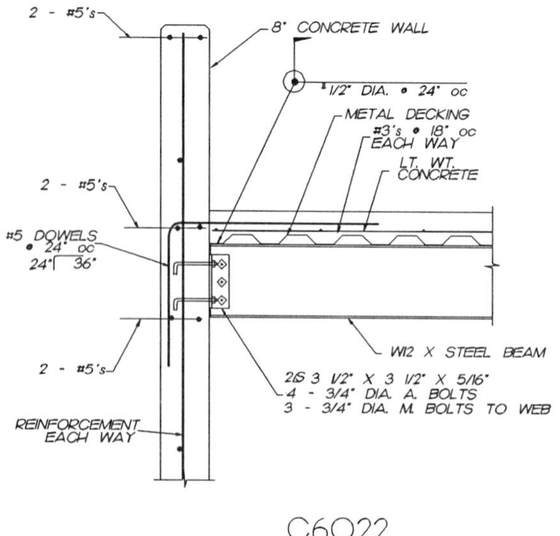

C6022

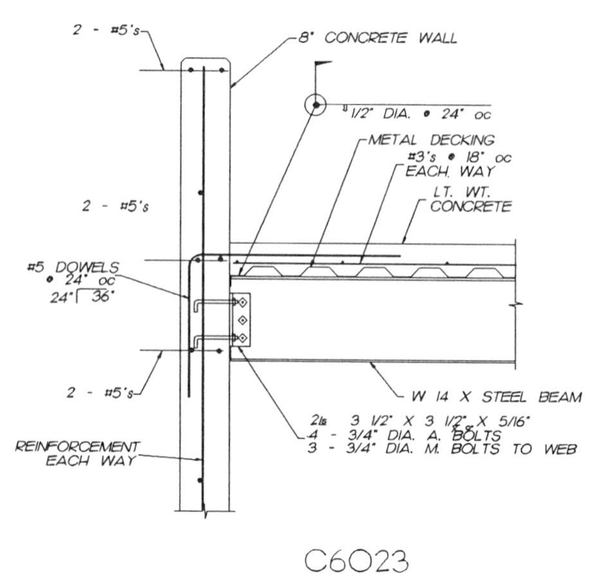

C6023

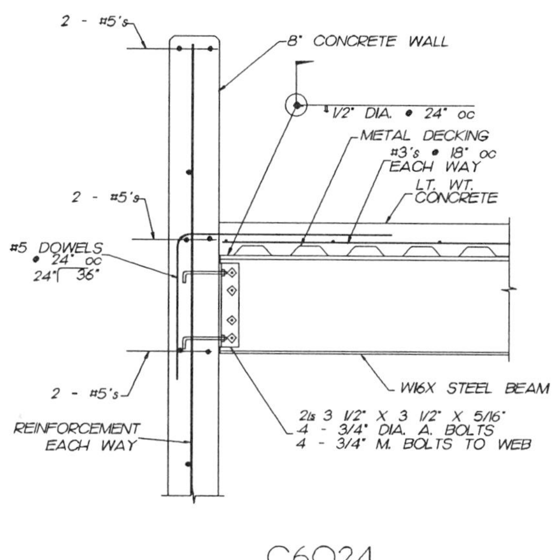

C6024

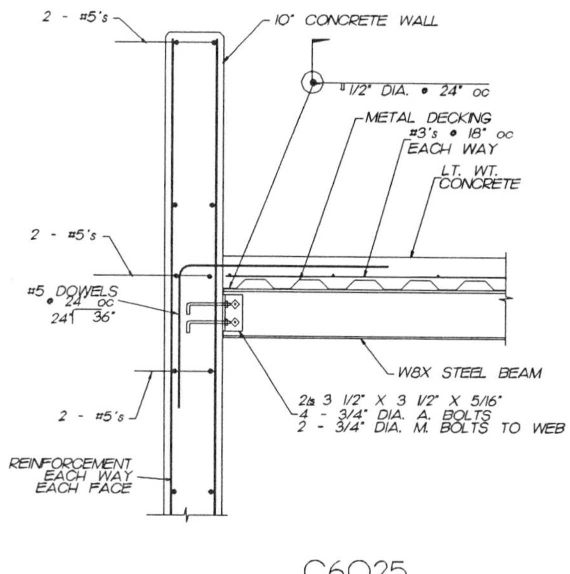

C6025

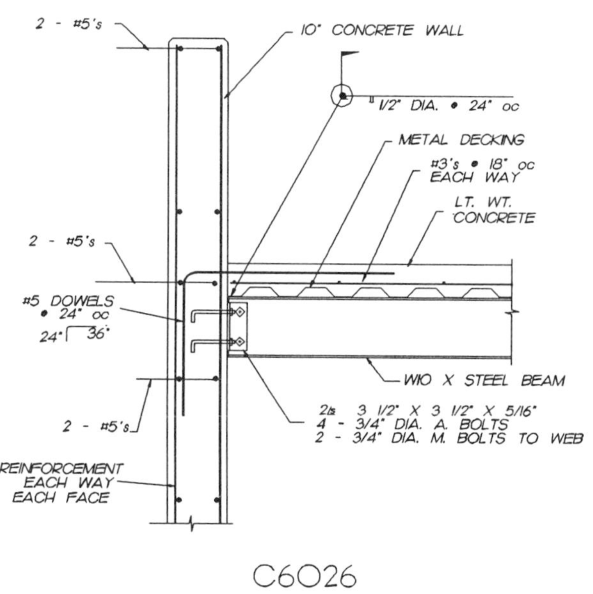

C6026

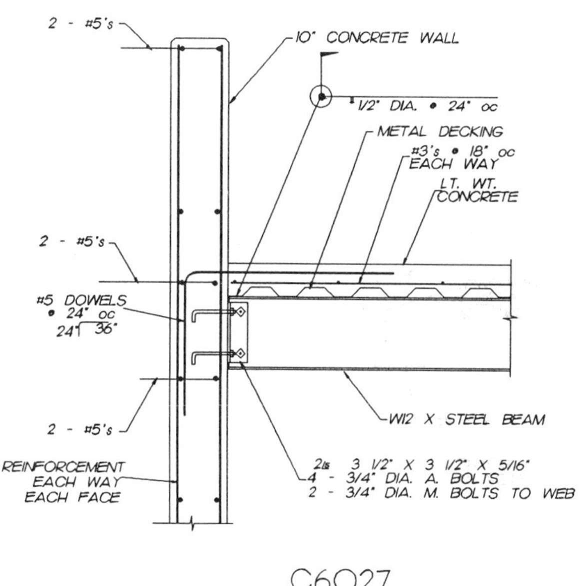

C6027

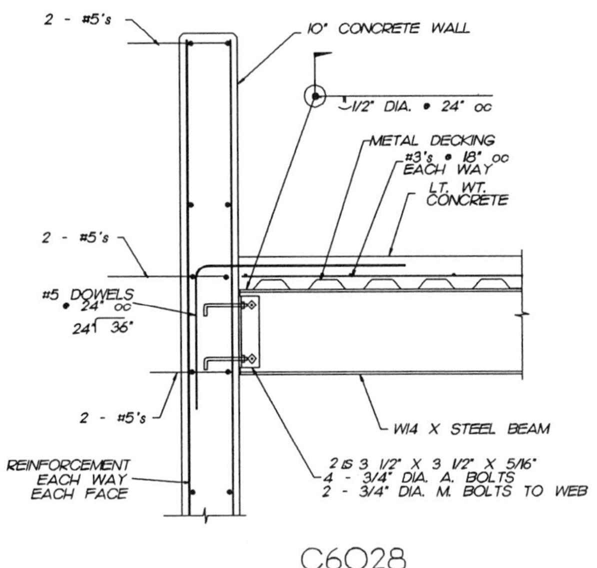

C6028

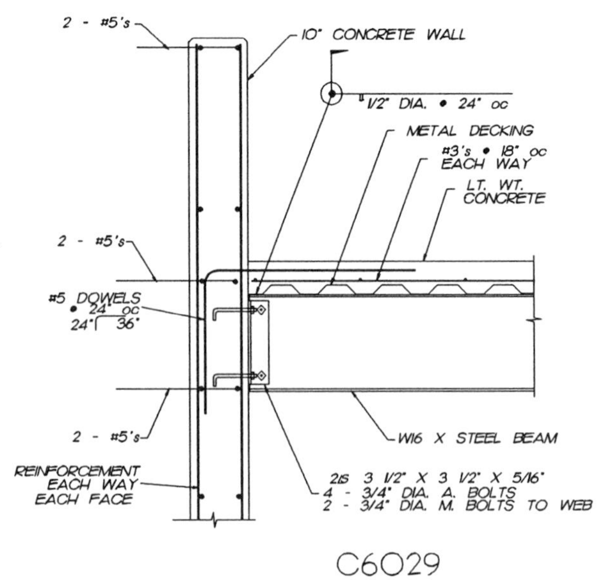

C6029

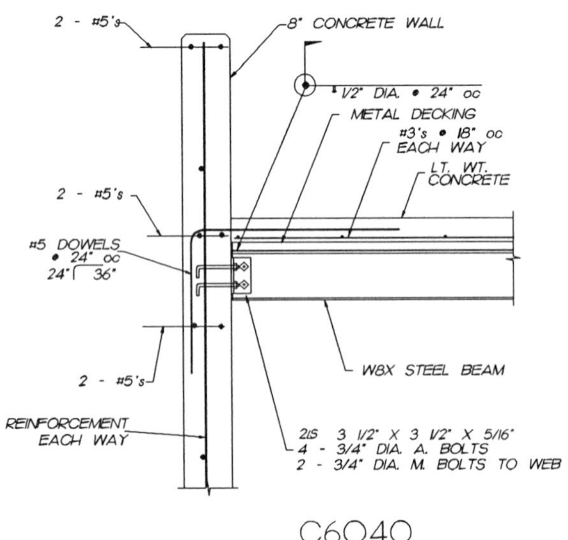

C6040

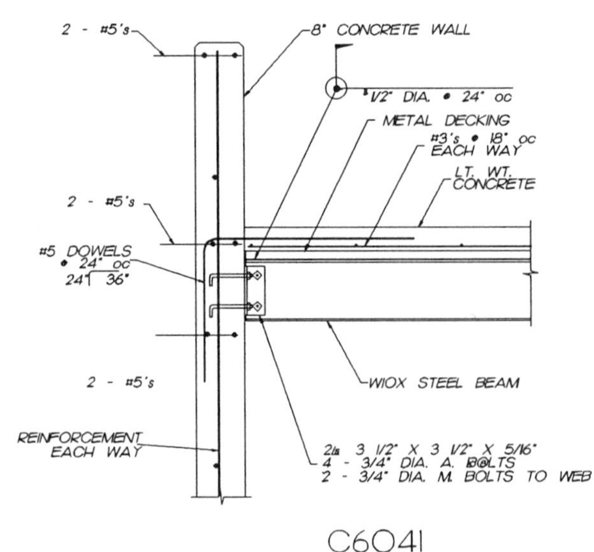

C6041

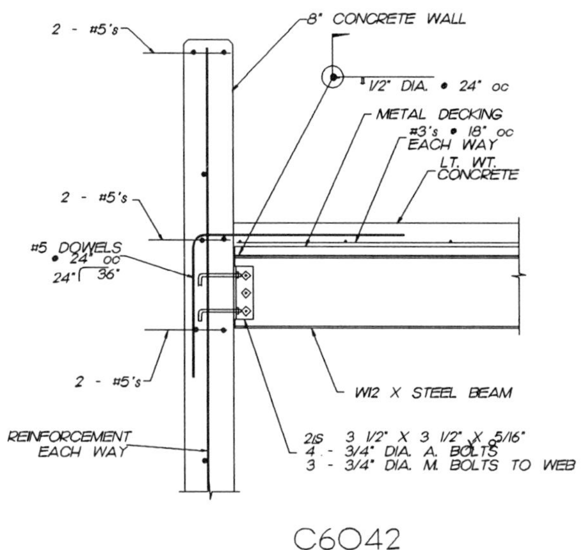

C6042

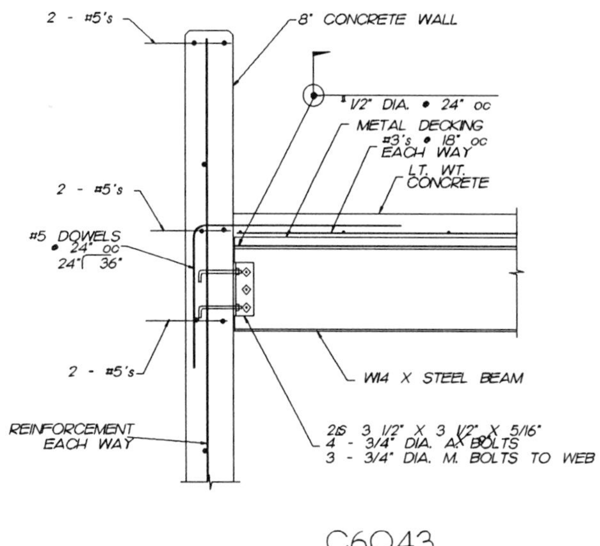

C6043

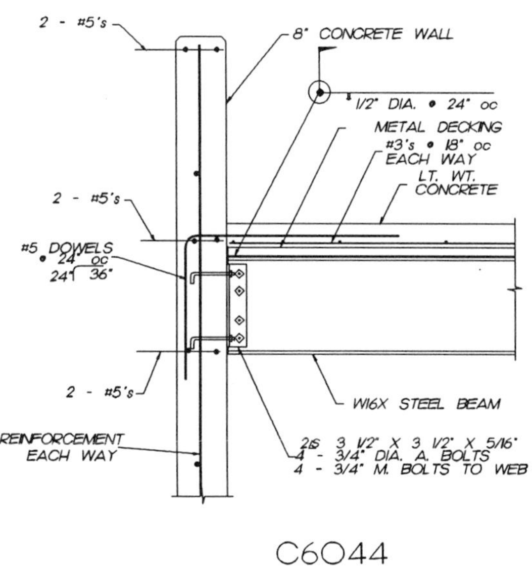

C6044

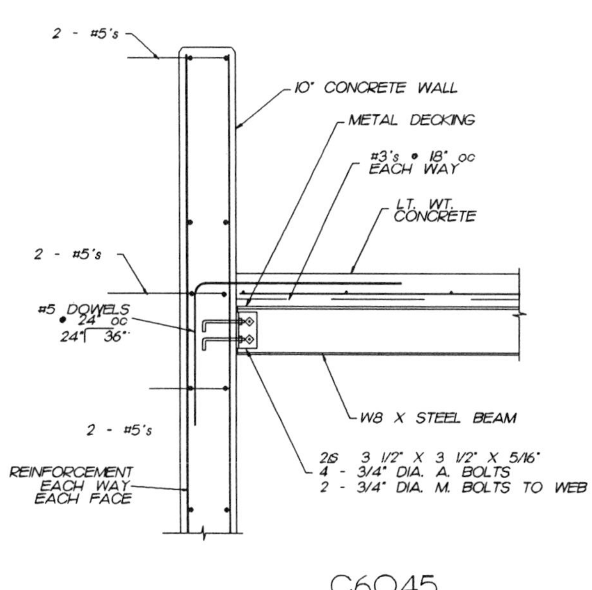

C6045

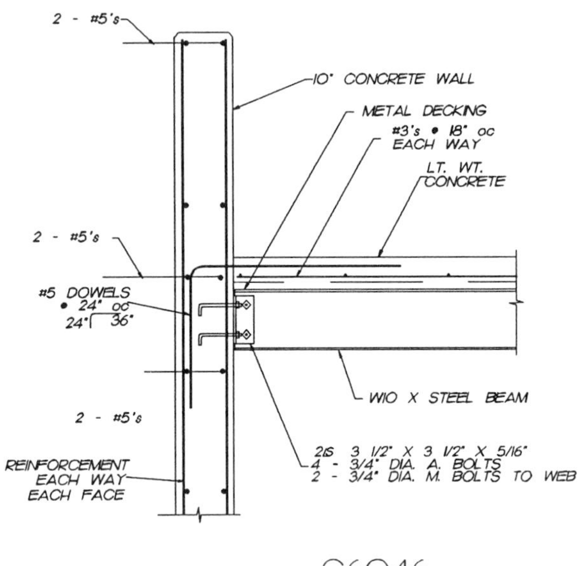

C6046

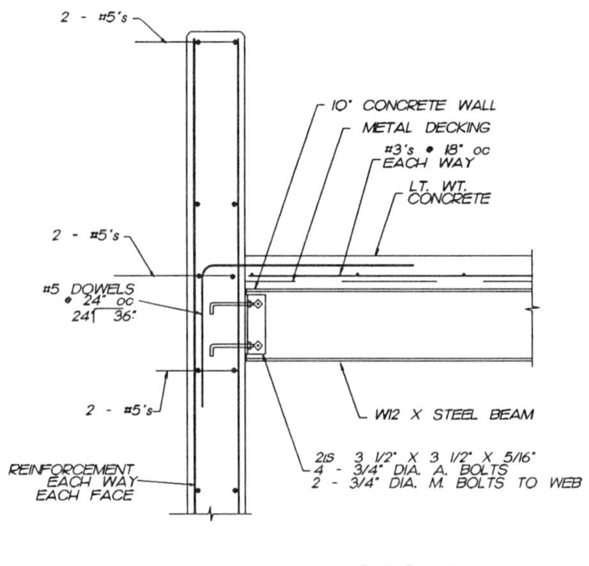

C6047

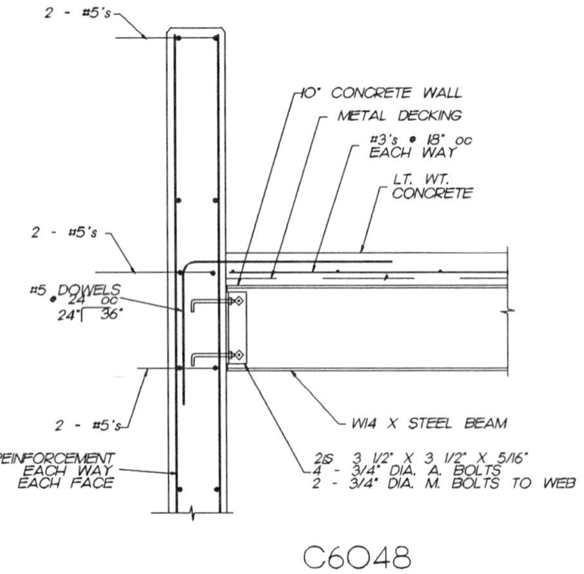

C6048

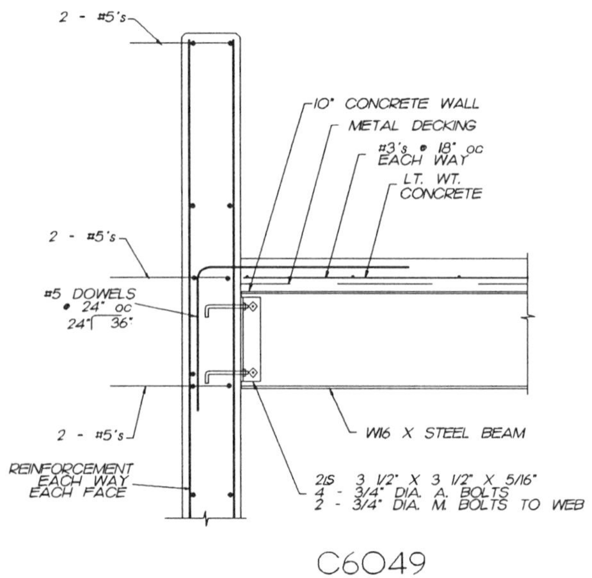

C6049

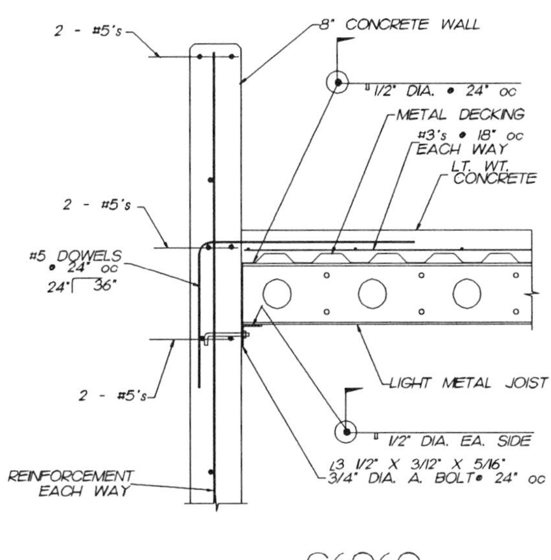

C6060

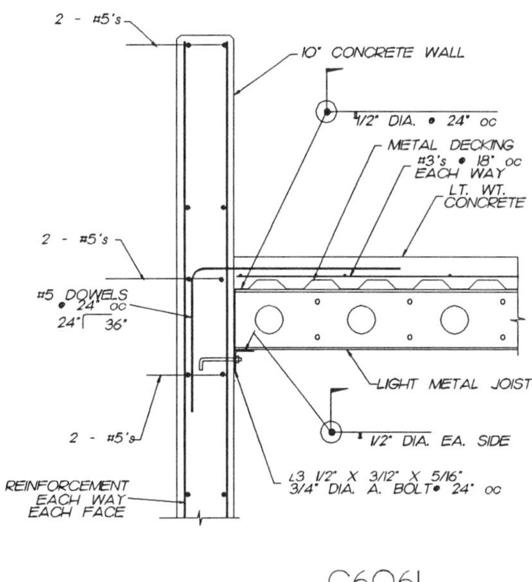

C6061

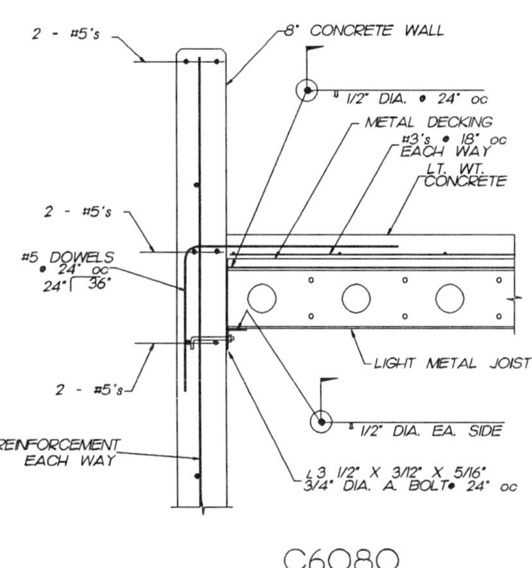

C6080

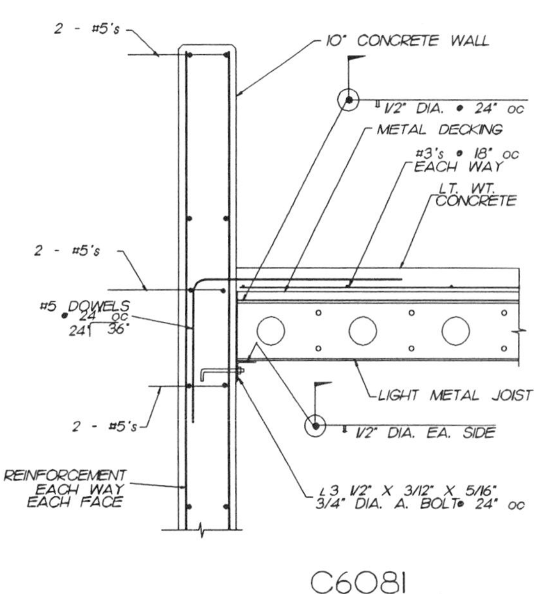

C6081

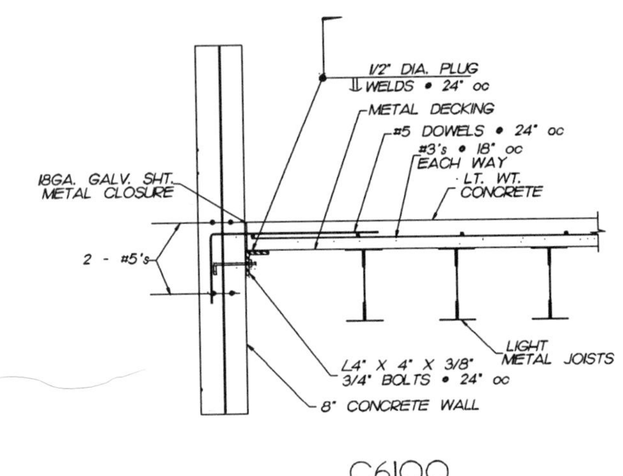

C6100

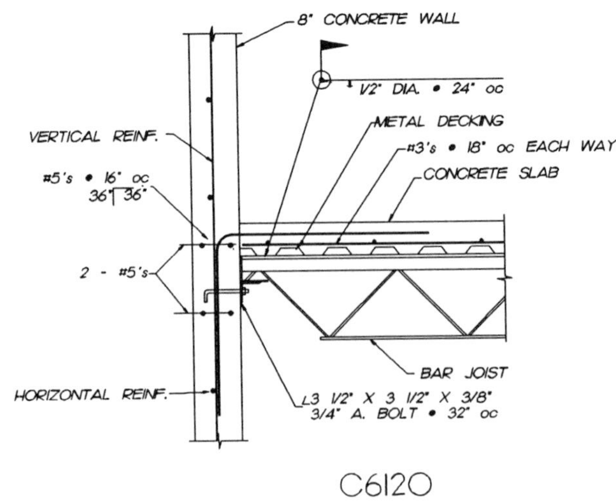

C6120

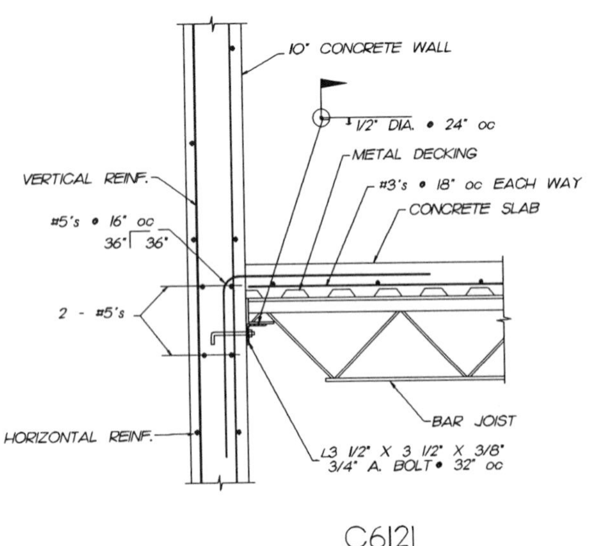

C6121

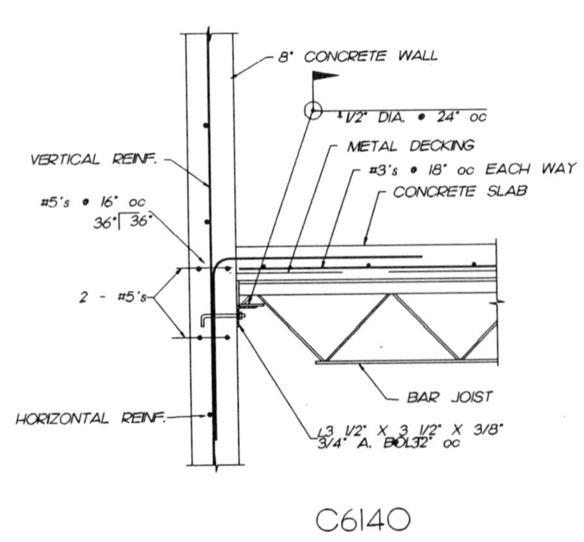

C6140

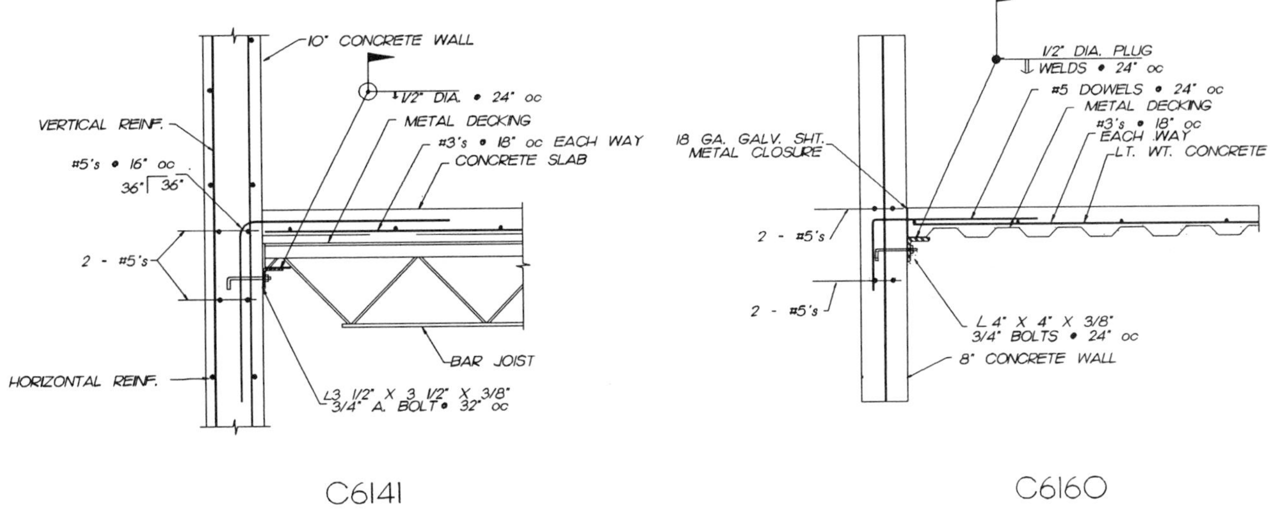

C6141

C6160

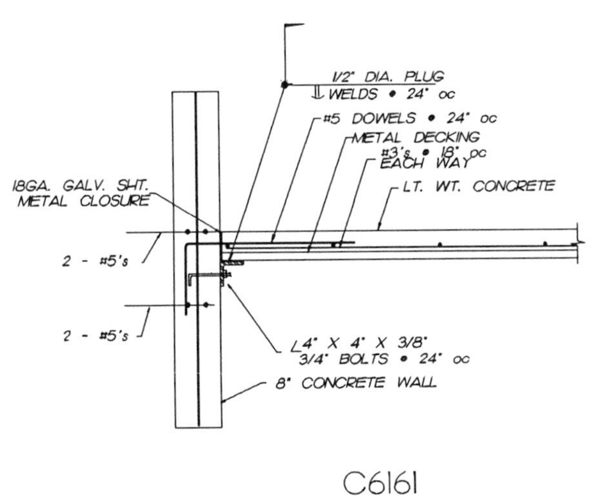

C6161

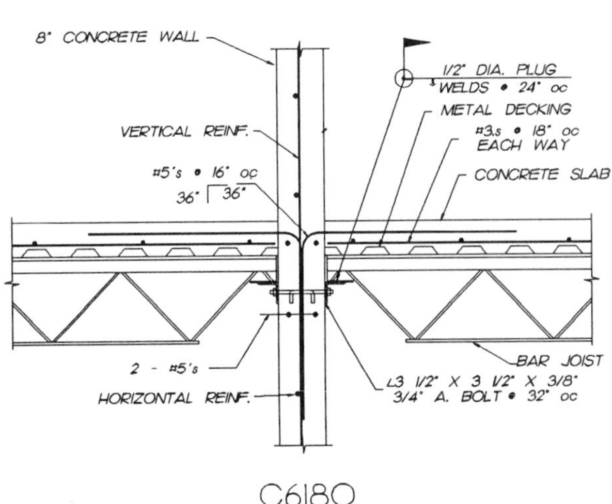

C6180

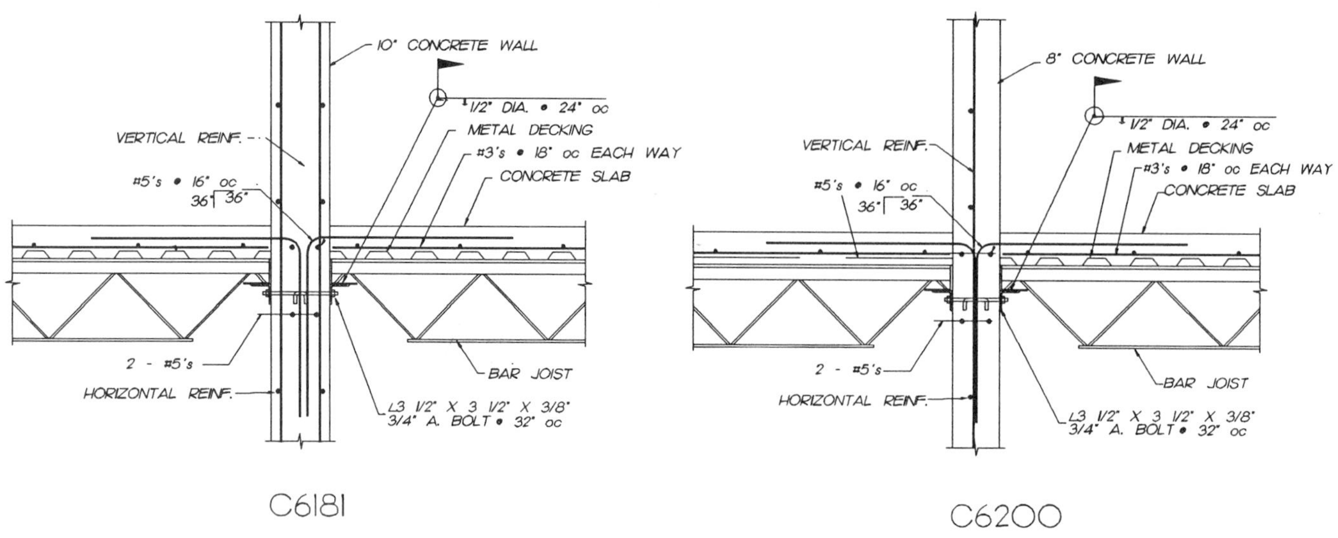

C6181

C6200

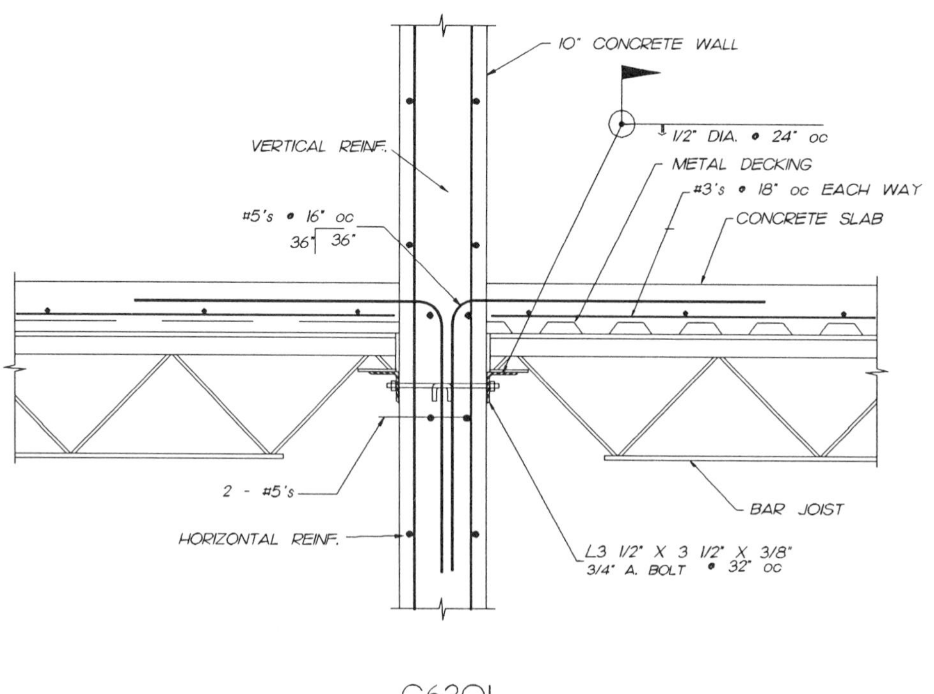

C6201

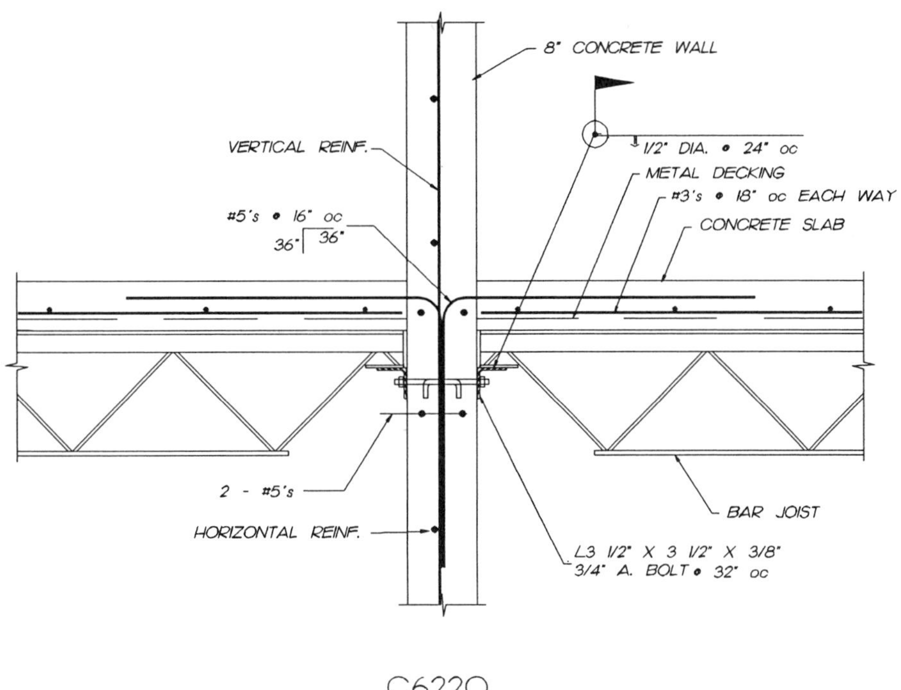
8" CONCRETE WALL
VERTICAL REINF.
1/2" DIA. @ 24" oc
METAL DECKING
#3's @ 18" oc EACH WAY
#5's @ 16" oc
36" 36"
CONCRETE SLAB
2 - #5's
BAR JOIST
HORIZONTAL REINF.
L3 1/2" X 3 1/2" X 3/8"
3/4" A. BOLT @ 32" oc

C6220

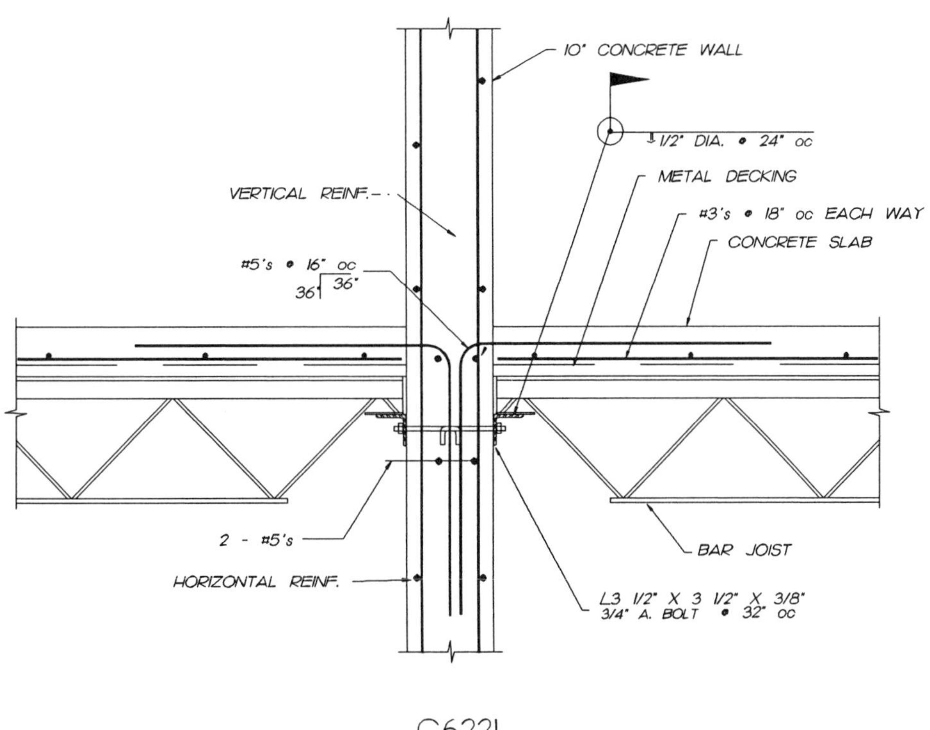
10" CONCRETE WALL
1/2" DIA. @ 24" oc
METAL DECKING
VERTICAL REINF.
#3's @ 18" oc EACH WAY
CONCRETE SLAB
#5's @ 16" oc
36 36"
2 - #5's
BAR JOIST
HORIZONTAL REINF.
L3 1/2" X 3 1/2" X 3/8"
3/4" A. BOLT @ 32" oc

C6221

RETAINING WALL SOIL IS NOT OVER THE FOOTING RETAINED SLOPE - FLAT STEM HEIGHTS - 4'-0" to 8'-0"	C7020 - C7023
RETAINING WALL SOIL IS NOT OVER THE FOOTING RETAINED SLOPE - 2 HORIZ. to I VERT. STEM HEIGHTS - 4'-0" to 8'-0"	C7040 - C7043
RETAINING WALL SOIL IS NOT OVER THE FOOTING RETAINED SLOPE - I HORIZ. to I VERT. STEM HEIGHTS - 4'-0" to 5'-4"	C7060 - C7061
RETAINING WALL SOIL IS OVER THE FOOTING RETAINED SLOPE - FLAT STEM HEIGHTS - 4'-0" to 8'-0"	C7080 - C7083
RETAINING WALL SOIL IS OVER THE FOOTING RETAINED SLOPE - 2 HORIZ. to I VERT. STEM HEIGHTS - 4'-0" to 8'-0"'	C7100 - C7103
RETAINING WALL SOIL IS OVER THE FOOTING RETAINED SLOPE - I HORIZ. to I VERT. STEM HEIGHTS - 4'-0" to 5'-0"	C7120 - C7121

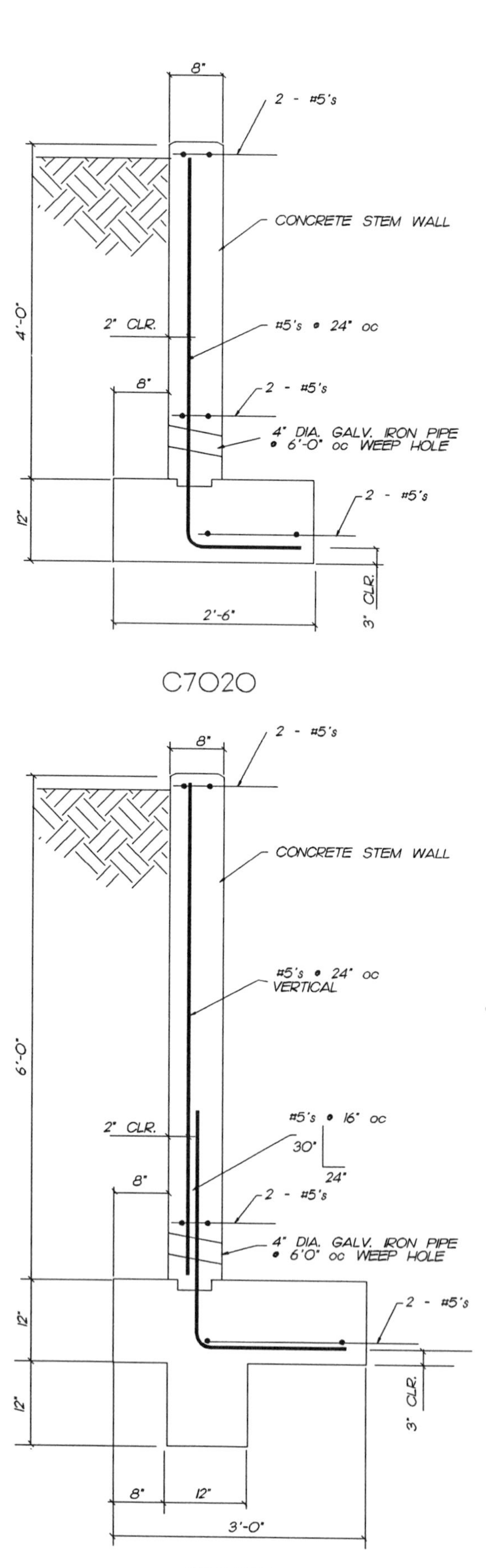

C7020

C7022

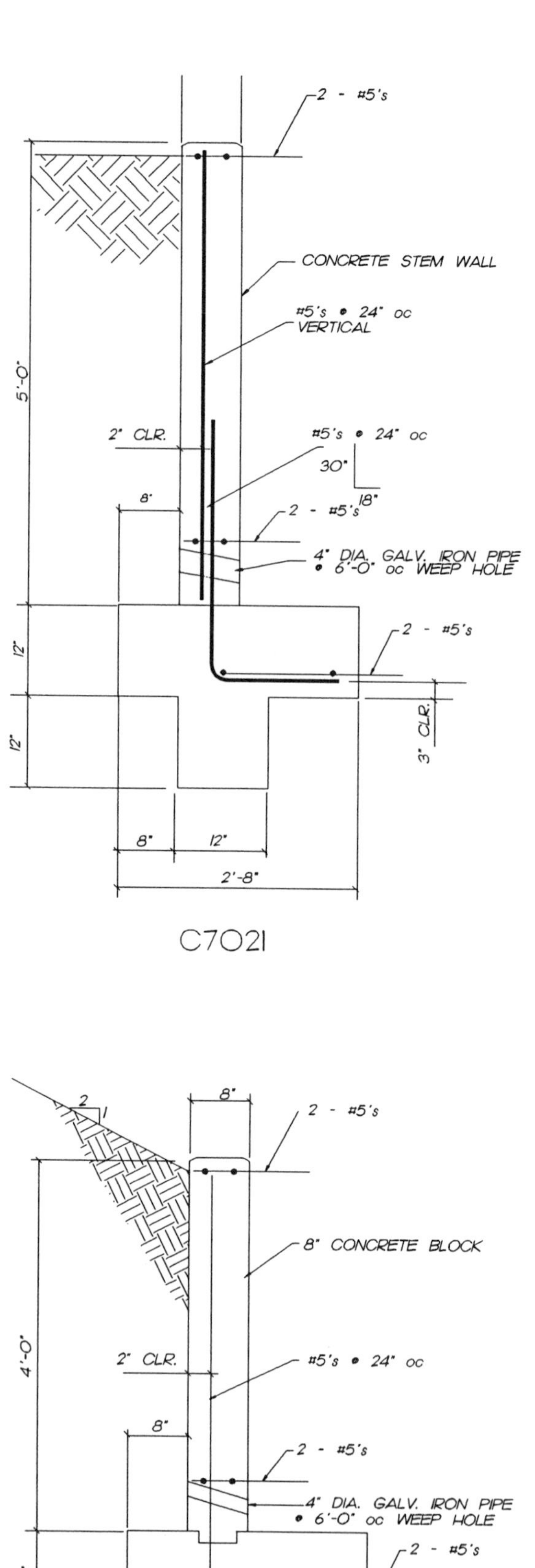

C7021

C7023

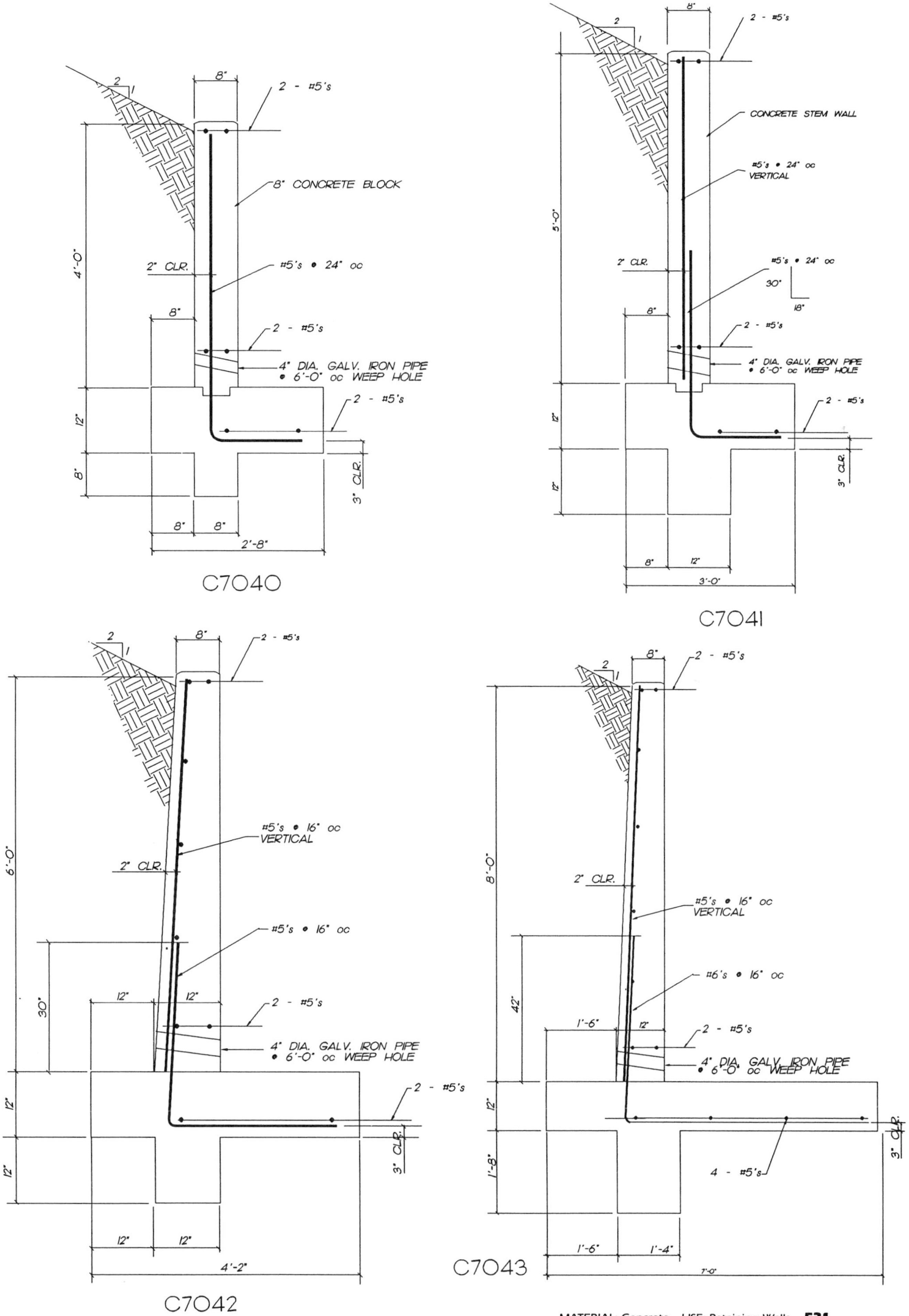
2
1
8"
2 - #5's
8" CONCRETE BLOCK
4'-0"
2" CLR.
#5's • 24" oc
8"
2 - #5's
4" DIA. GALV. IRON PIPE
• 6'-0" oc WEEP HOLE
2 - #5's
12"
8"
3" CLR.
8"
8"
2'-8"
C7040
2
1
8"
2 - #5's
CONCRETE STEM WALL
#5's • 24" oc
VERTICAL
5'-0"
2" CLR.
#5's • 24" oc
30"
18"
8"
2 - #5's
4" DIA. GALV. IRON PIPE
• 6'-0" oc WEEP HOLE
2 - #5's
12"
12"
3" CLR.
8"
12"
3'-0"
C7041
2
1
8"
2 - #5's
#5's • 16" oc
VERTICAL
6'-0"
2" CLR.
#5's • 16" oc
30"
12"
12"
2 - #5's
4" DIA. GALV. IRON PIPE
• 6'-0" oc WEEP HOLE
2 - #5's
12"
12"
3" CLR.
12"
12"
4'-2"
C7042
2
1
8"
2 - #5's
8'-0"
2" CLR.
#5's • 16" oc
VERTICAL
#6's • 16" oc
42"
1'-6"
12"
2 - #5's
4" DIA. GALV. IRON PIPE
• 6'-0" oc WEEP HOLE
12"
3" CLR.
1'-8"
4 - #5's
1'-6"
1'-4"
7'-0"
C7043

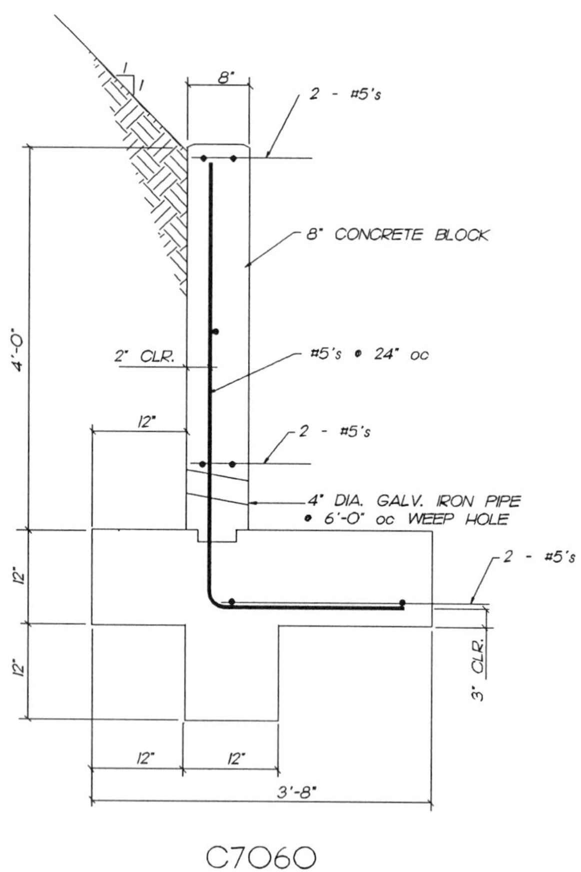

C7060

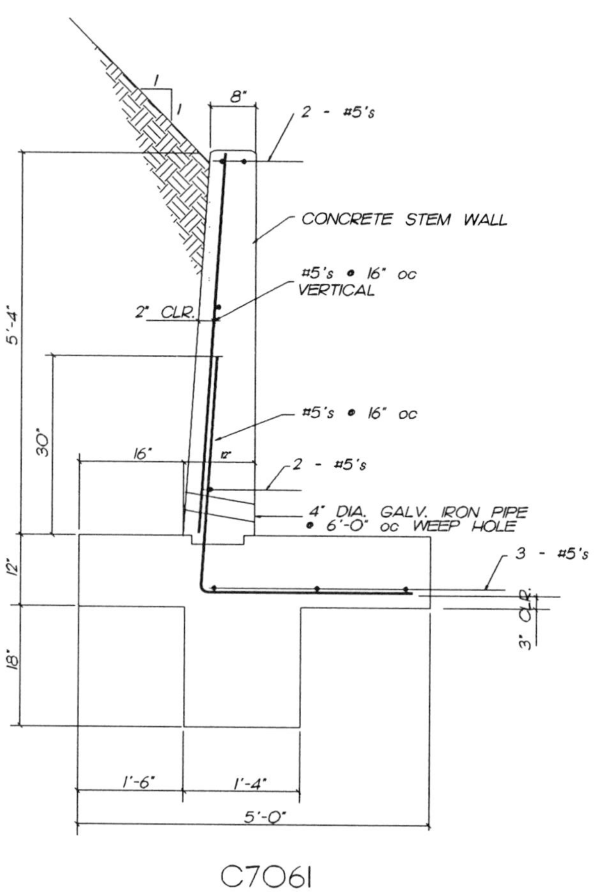

C7061

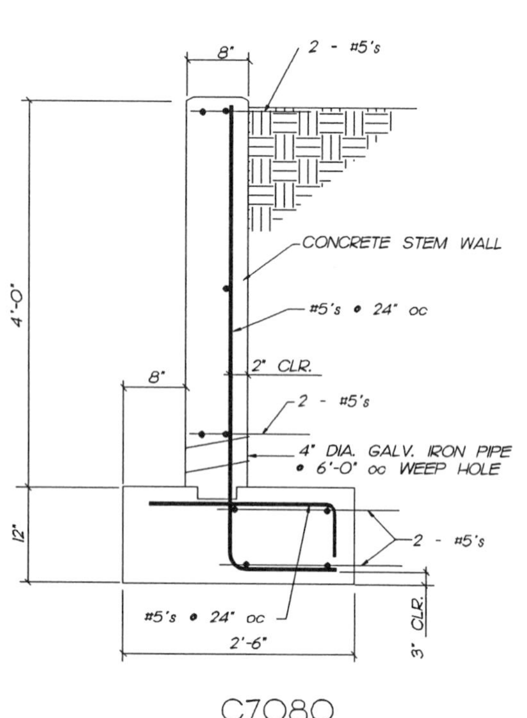

C7080

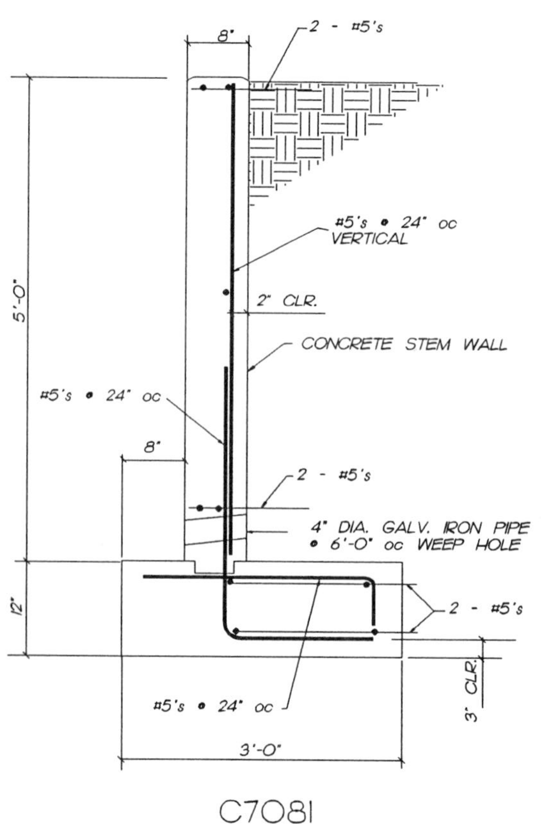

C7081

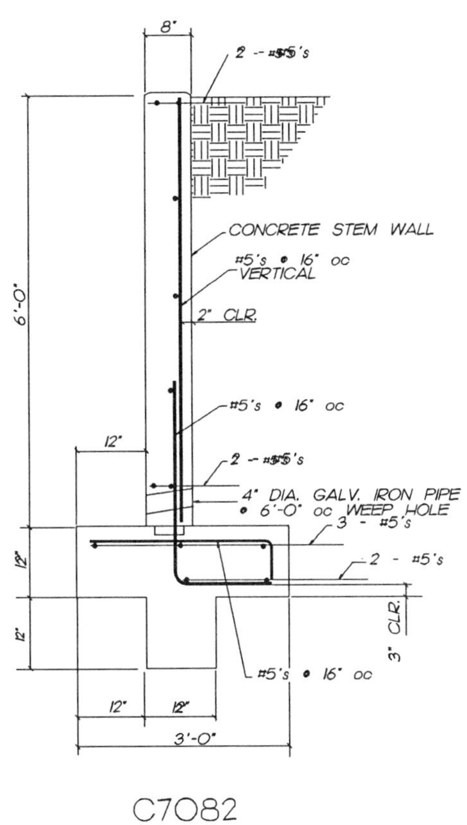

C7082

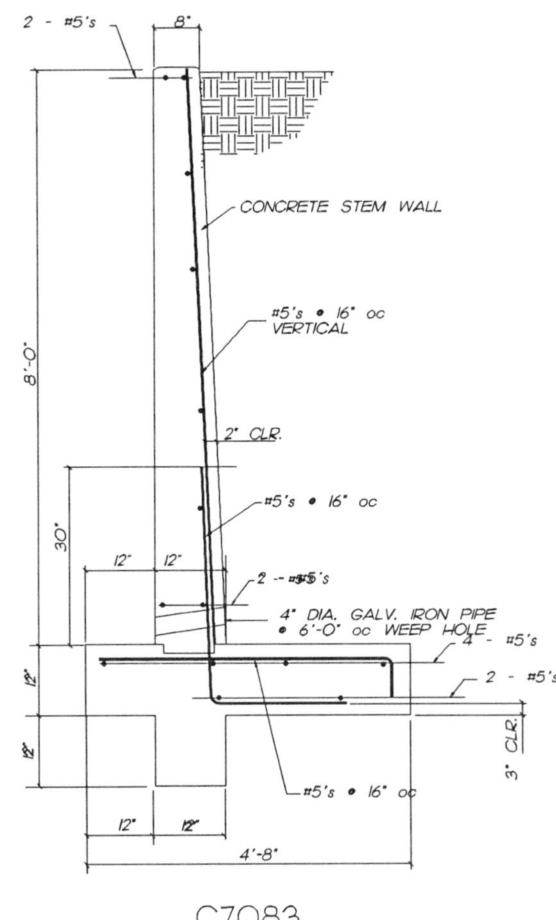

C7083

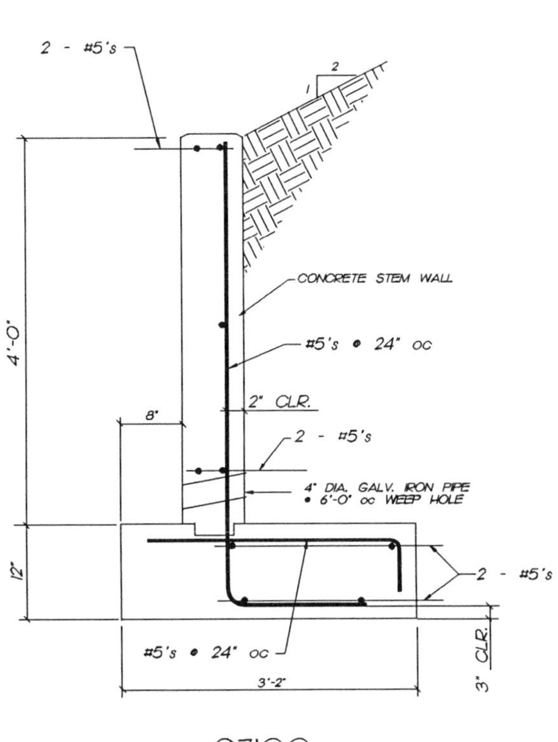

C7100

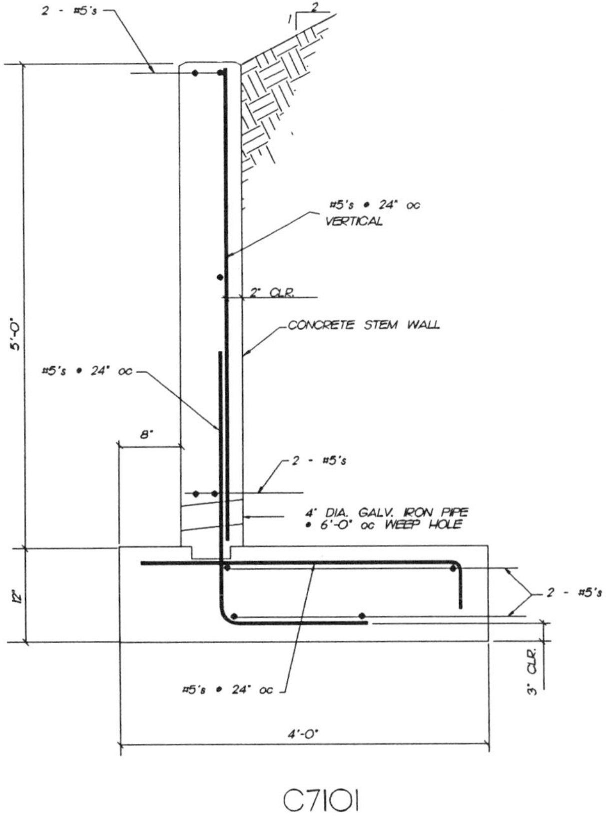

C7101

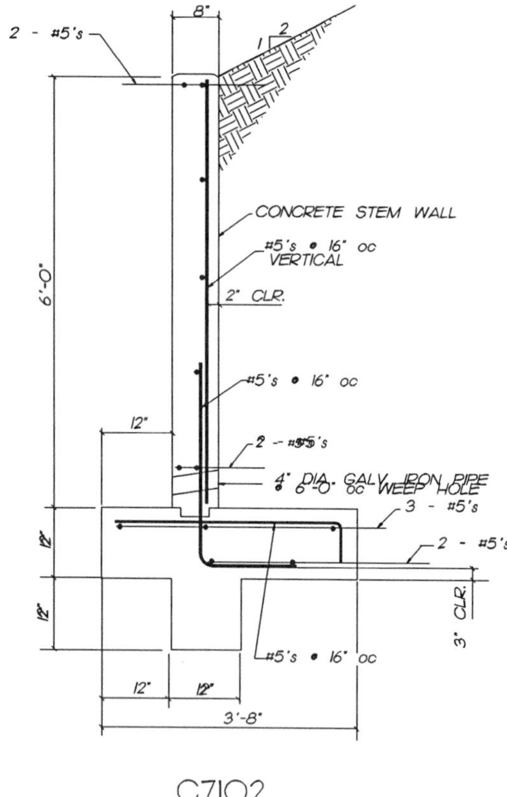
8"
2 - #5's
1
2
CONCRETE STEM WALL
#5's @ 16" oc
VERTICAL
2" CLR.
6'-0"
#5's @ 16" oc
12"
2 - #5's
4" DIA. GALV. IRON PIPE
@ 6'-0" oc WEEP HOLE
3 - #5's
2 - #5's
12"
12"
3" CLR.
#5's @ 16" oc
12"
12"
3'-8"

C7102

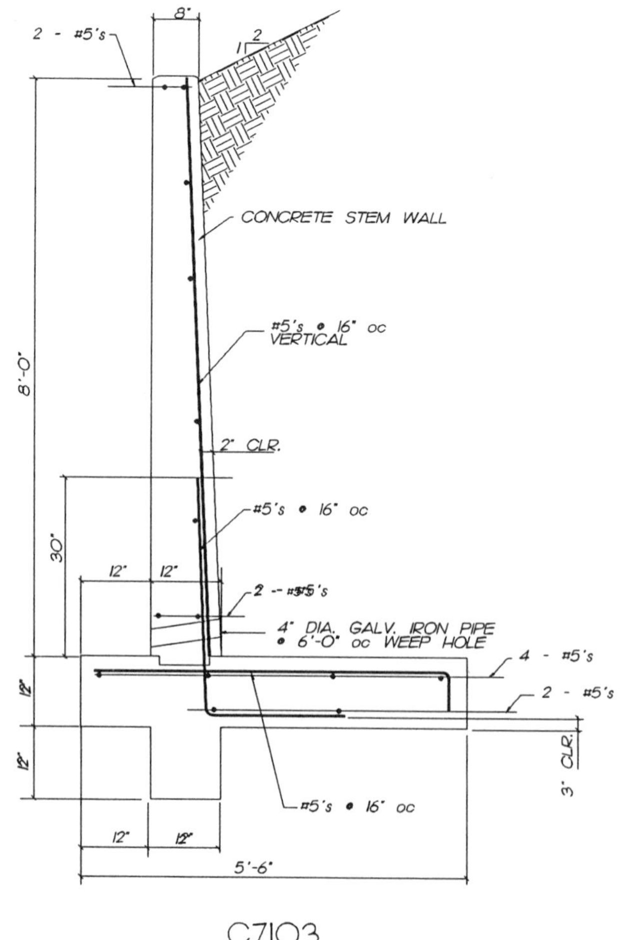
8"
2 - #5's
1
2
CONCRETE STEM WALL
#5's @ 16" oc
VERTICAL
8'-0"
2" CLR.
#5's @ 16" oc
30"
12"
12"
2 - #5's
4" DIA. GALV. IRON PIPE
@ 6'-0" oc WEEP HOLE
4 - #5's
2 - #5's
12"
12"
3" CLR.
#5's @ 16" oc
12"
12"
5'-6"

C7103

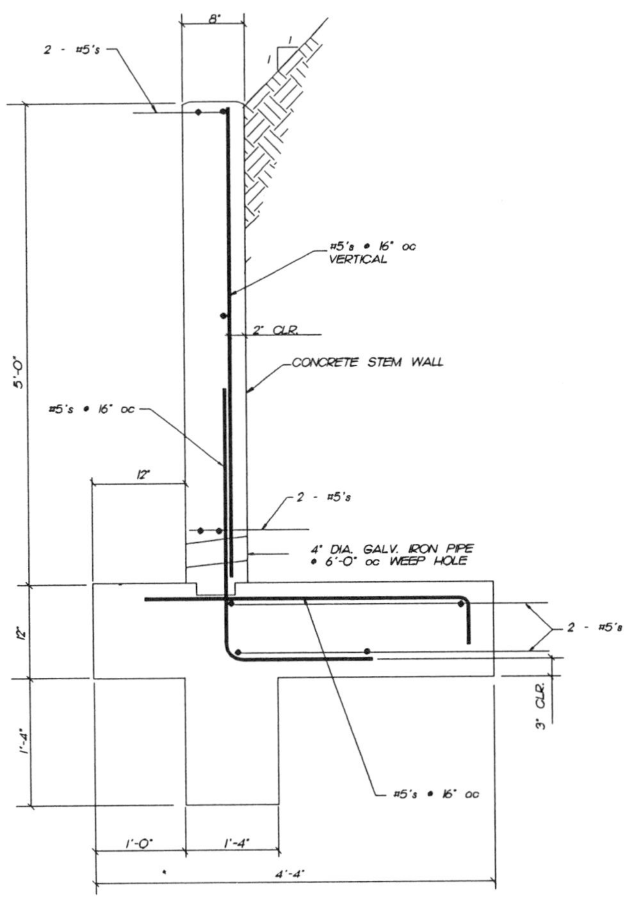
8"
2 - #5's
1
1
#5's @ 16" oc
VERTICAL
2" CLR.
CONCRETE STEM WALL
5'-0"
#5's @ 16" oc
12"
2 - #5's
4" DIA. GALV. IRON PIPE
@ 6'-0" oc WEEP HOLE
2 - #5's
12"
3" CLR.
1'-4"
#5's @ 16" oc
1'-0"
1'-4"
4'-4"

C7121

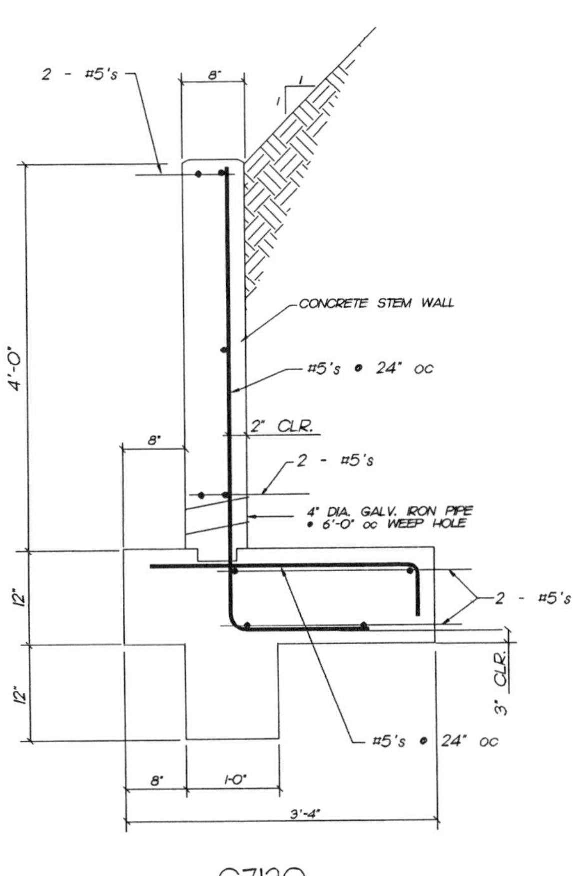
2 - #5's
8"
1
1
CONCRETE STEM WALL
4'-0"
#5's @ 24" oc
2" CLR.
8"
2 - #5's
4" DIA. GALV. IRON PIPE
@ 6'-0" oc WEEP HOLE
2 - #5's
12"
3" CLR.
12"
#5's @ 24" oc
8"
1-0"
3'-4"

C7120

CHAPTER 3 CONCRETE SECTION 8	PRECAST CONCRETE WALLS AND CONCRETE CONNECTIONS
PRECAST CONCRETE WALL - PURLINS PERP. TO WALL	C8020
PRECAST CONCRETE WALL - STIFFENERS PERP. TO WALL	C8021
PRECAST CONCRETE WALL - JOIST & RAFTERS PERP. TO WALL	C8022
PRECAST WALL - TYP. PURLIN LEDGER	C8044 - 8042
PRECAST WALL - TYP. STIFFENER LEDGER	C8060 - C8061
PRECAST WALL CONTINUOUS FOOTING	C8080 - C8081
PRECAST WALL SPREAD FOOTING	C8082
PRECAST WALL BEAM SUPPORT AT WALL	C8100 - C8102
PRECAST WALL JOINT CONNECTION	C8120 - C8121
PRECAST WALL CORNER CONNECTION	C8140 - C8141
PRECAST WALL SHELF SUPPORT CONNECTION	C8160 - C8161
CONCRETE WALL WOOD STUD WALL INTERSECTION	C8180
8" CONCRETE END WALL & CONCRETE FLOOR SLAB	C8200
8" CONCRETE WALL BOND BEAM CORNER	C8201
10" CONCRETE END WALL & CONCRETE FLOOR SLAB	C8202
8" CONCRETE INTERIOR WALL & CONCRETE FLOOR SLAB	C8220
10" CONCRETE INTERIOR WALL & CONCRETE FLOOR SLAB	C8221
8" CONCRETE END WALL & STEEL ANGLES & PLATES TO CONC. SLAB	C8240
8" CONCRETE INTERIOR WALL & STEEL ANGLES & PLATES TO CONC. SLAB	C8241
CONCRETE BLOCK BOND BEAM INTERSECTION	C8260
CONCRETE WALL INTERSECTION CONNECTION	C8261

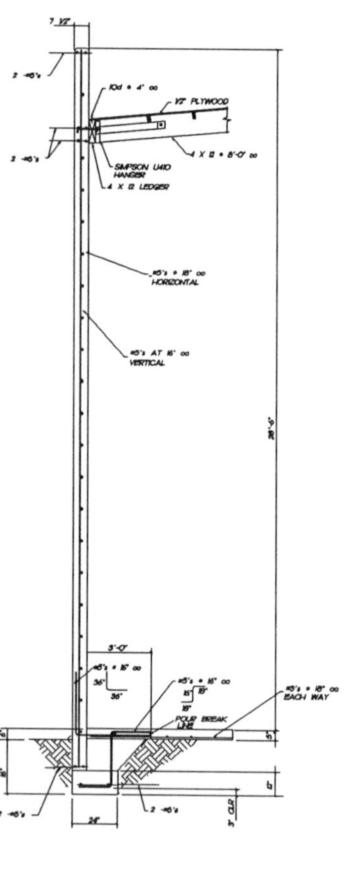

C8020

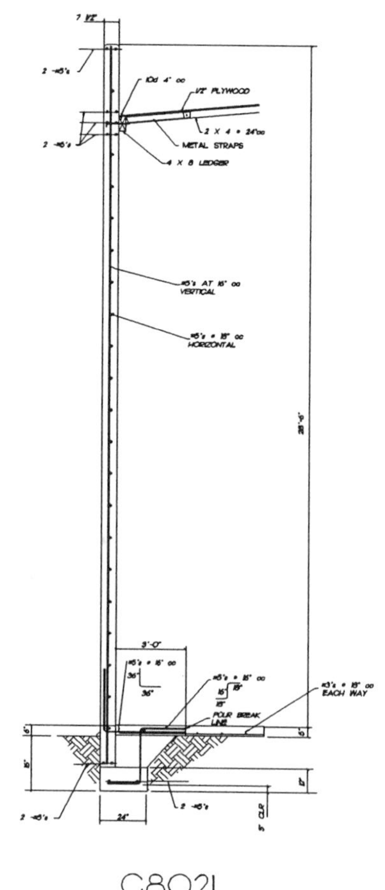

C8021

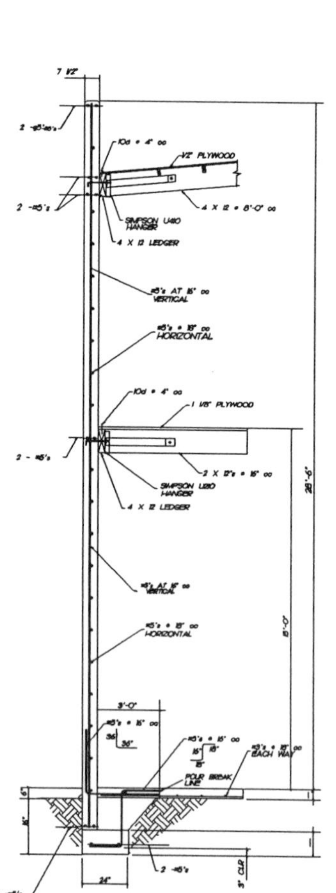

C8022

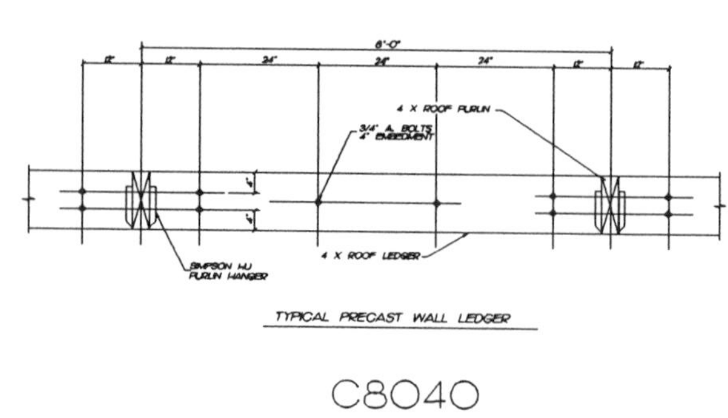

C8040

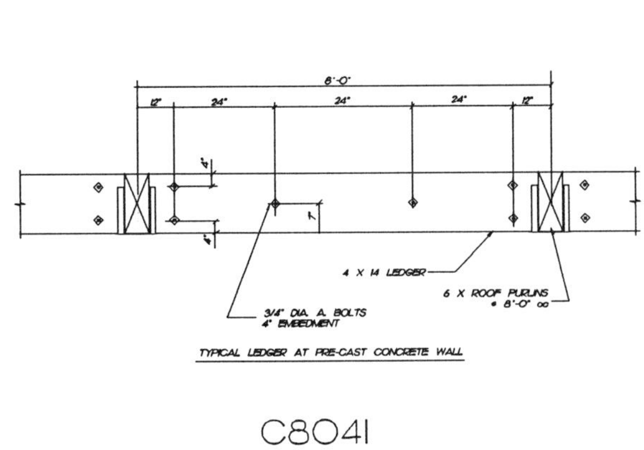

C8041

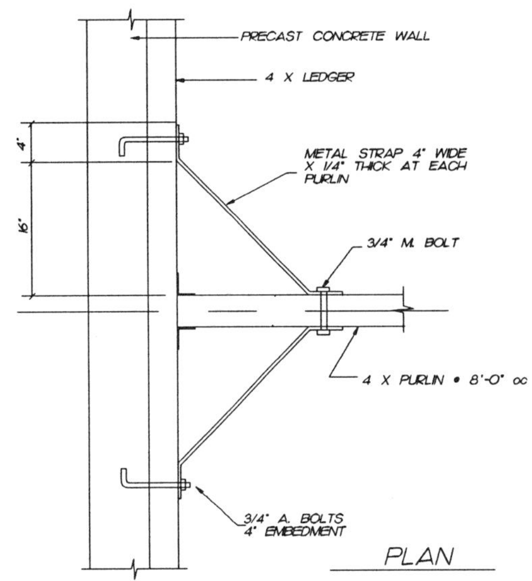

C8042

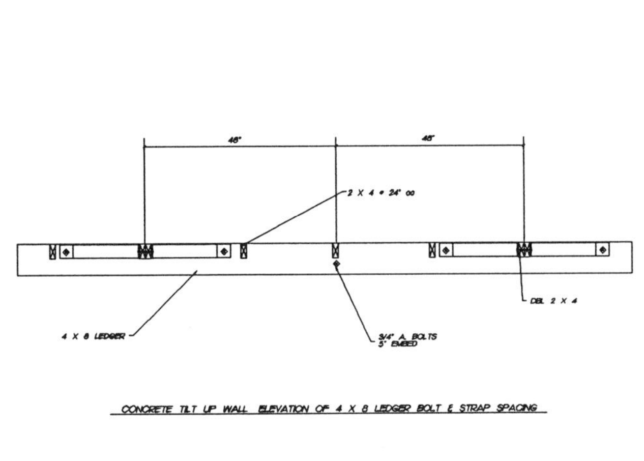

C8060

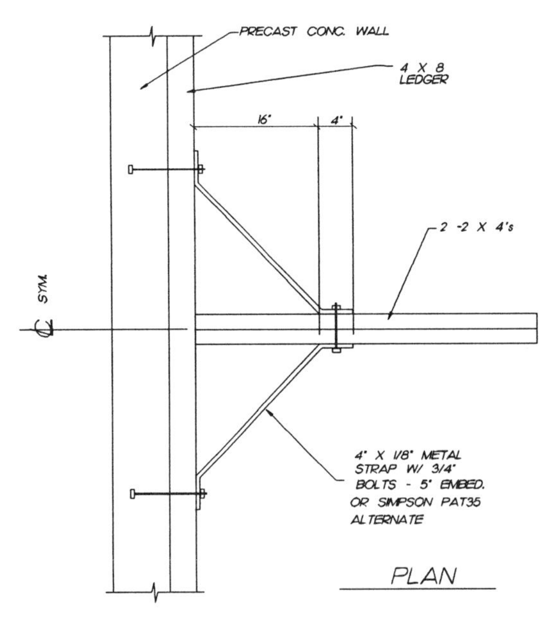

C8061

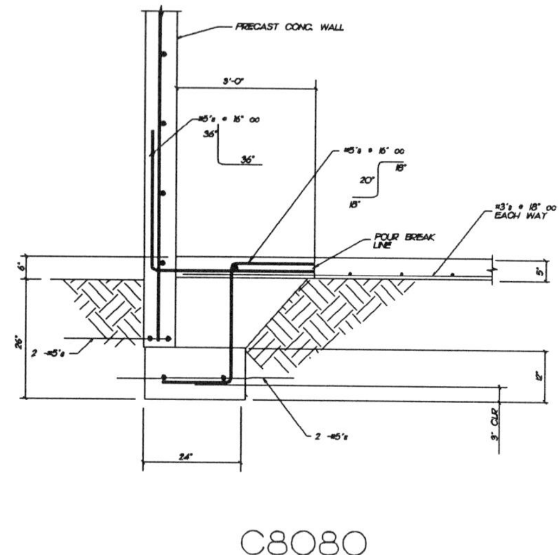

C8080

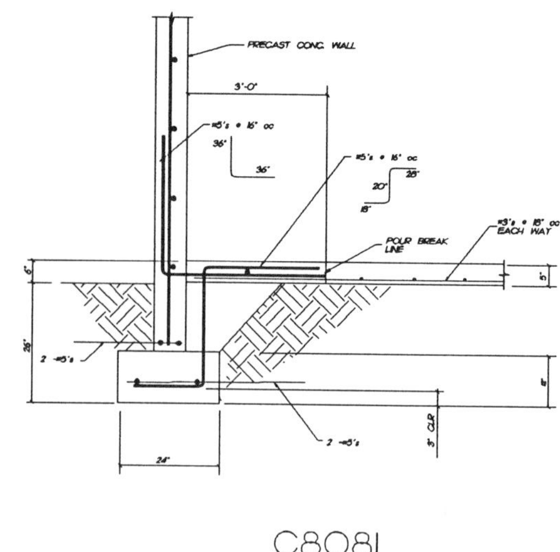

C8081

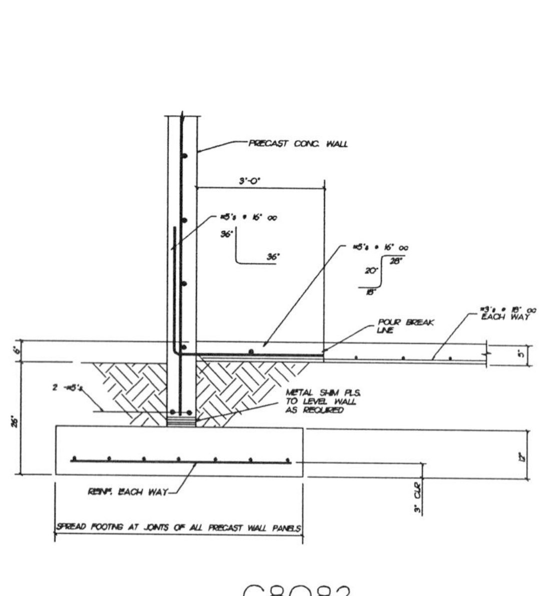

C8082

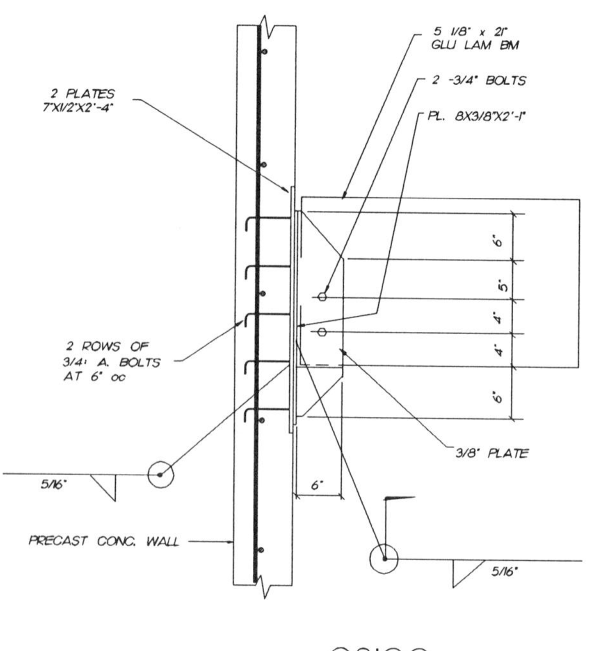

C8100

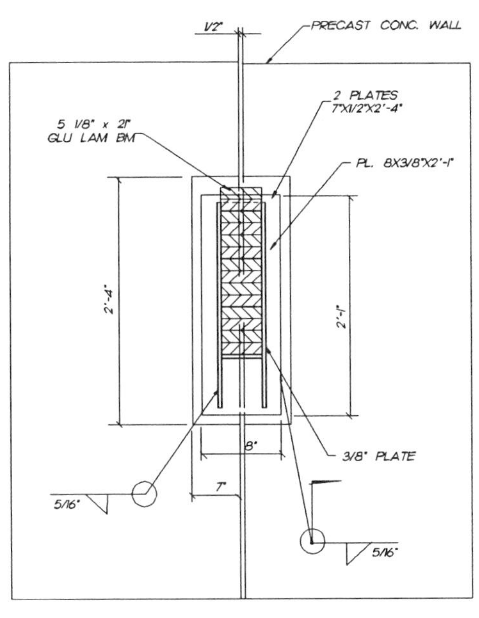

C8101

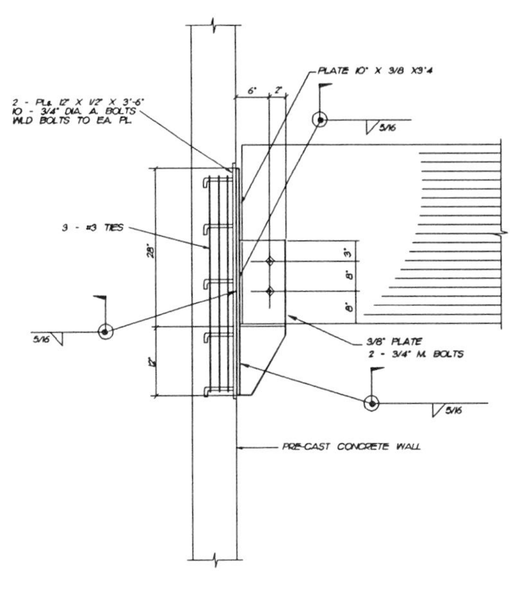

C8102

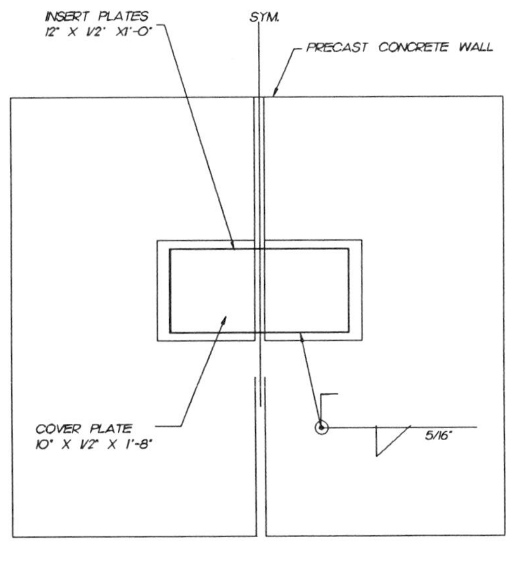

C8120

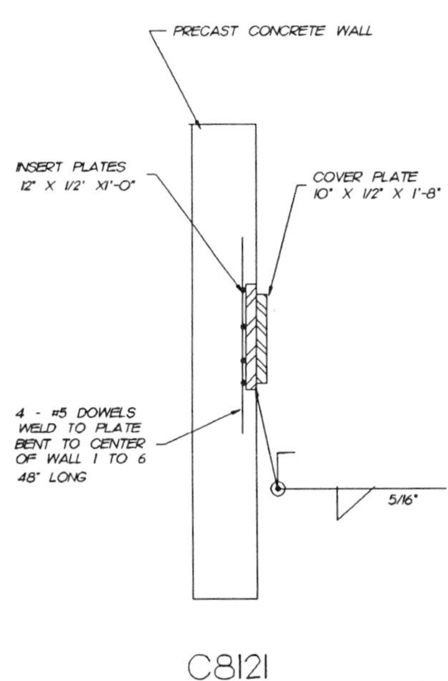

C8121

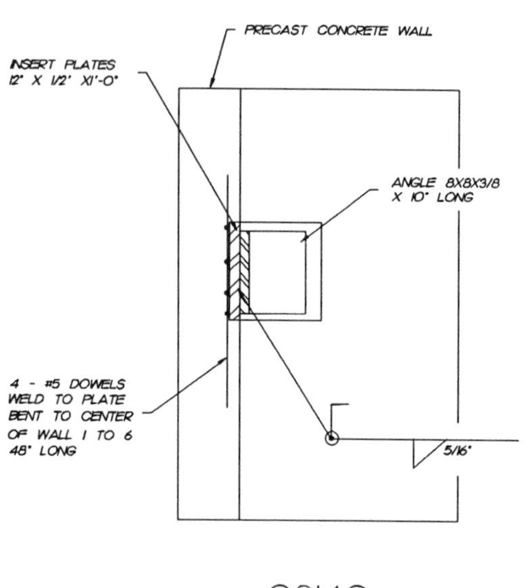

C8140

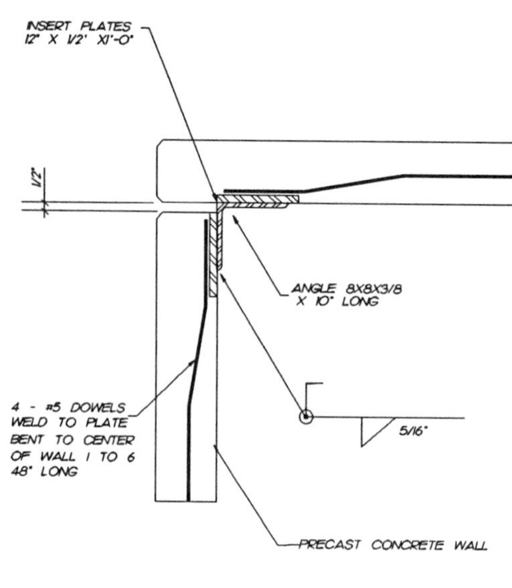

C8141

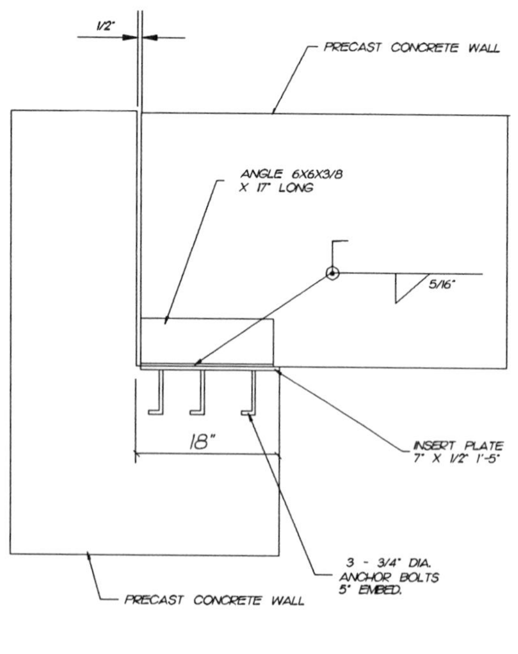

C8160

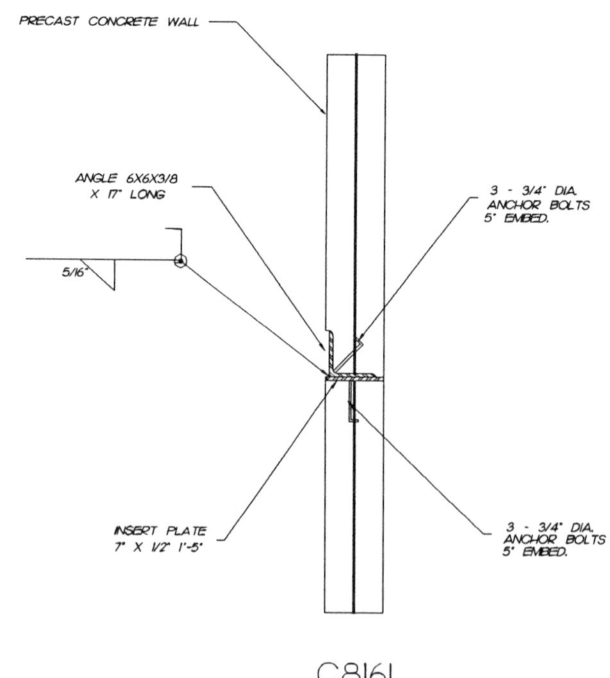

C8161

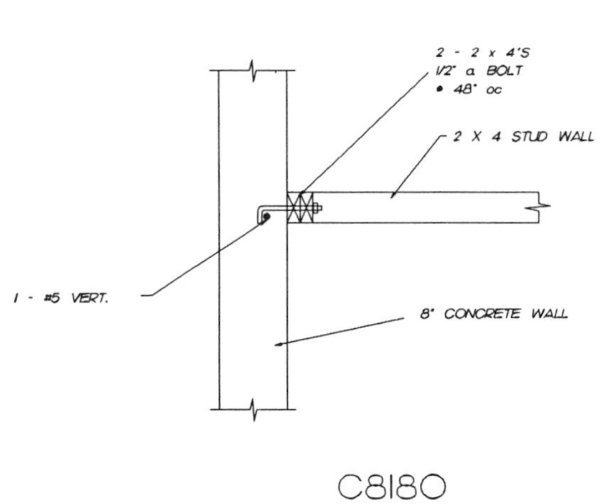

C8180

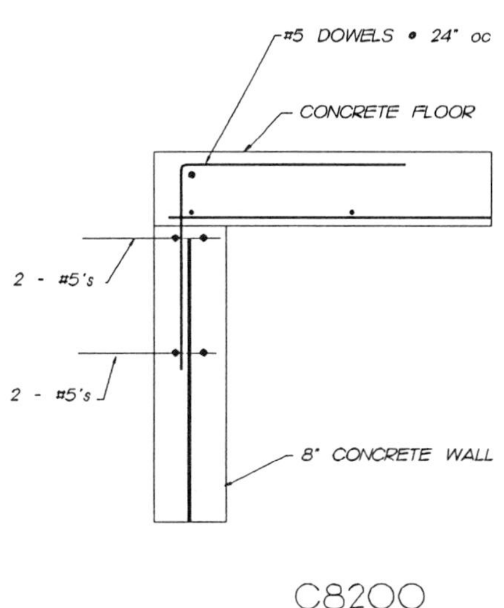

C8200

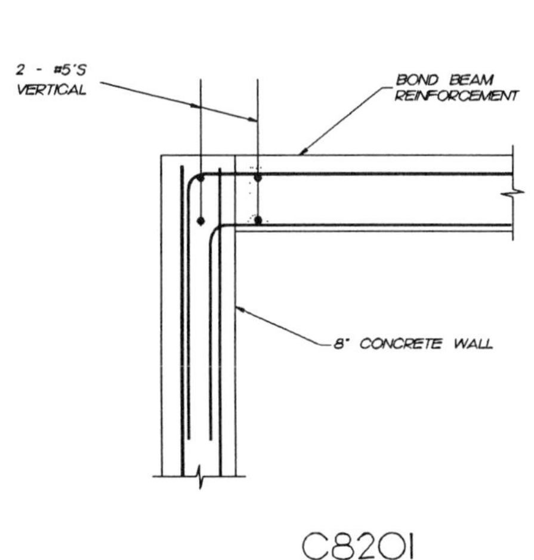

C8201

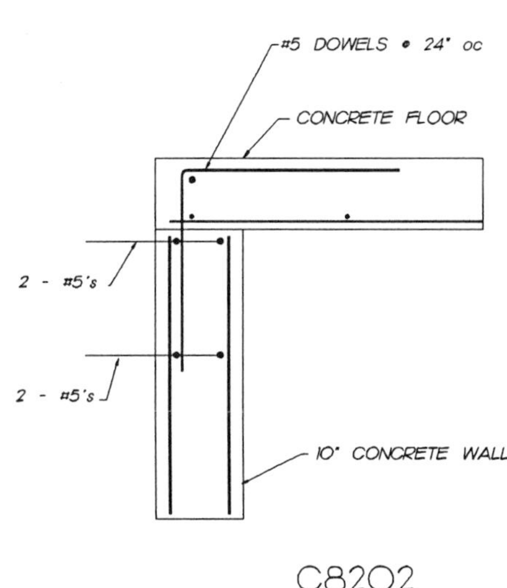

C8202

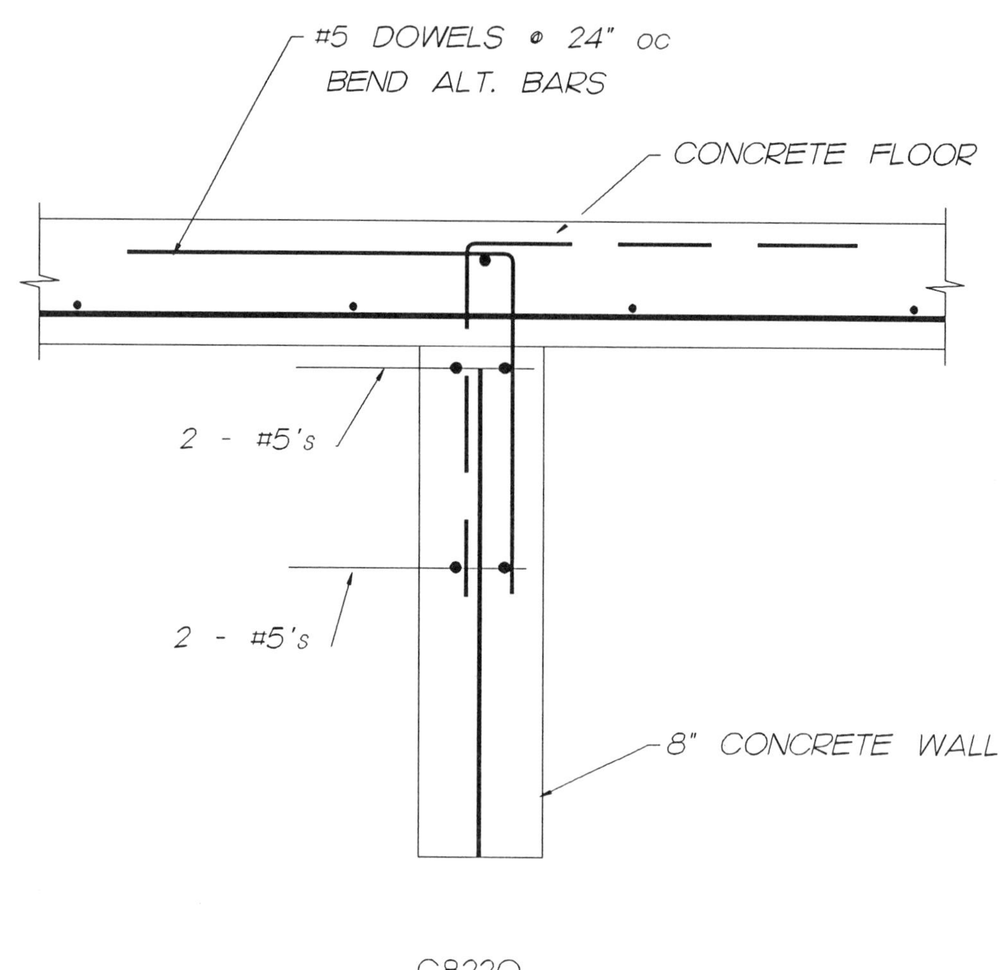

C8220

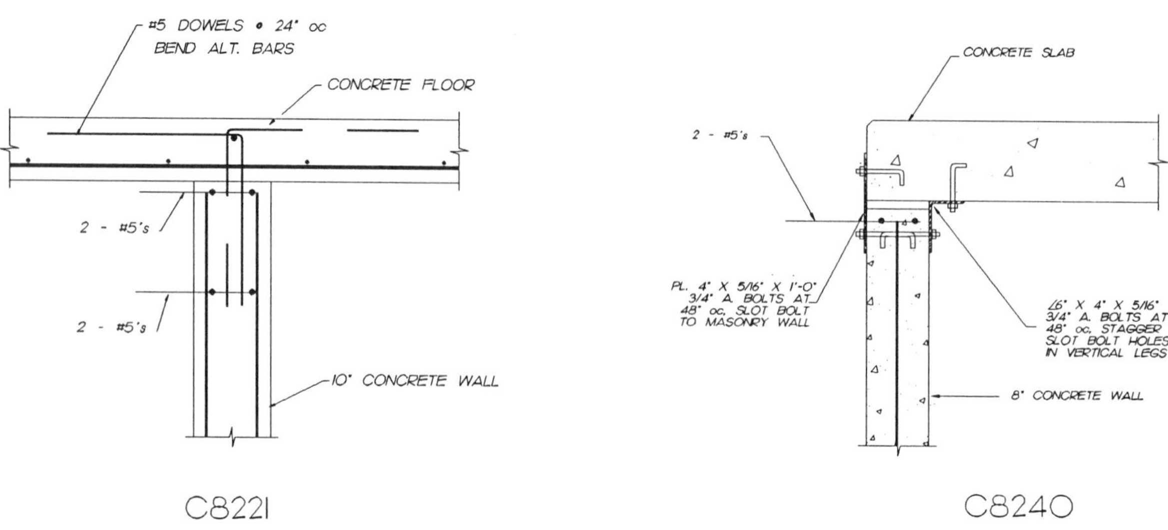

C8221

C8240

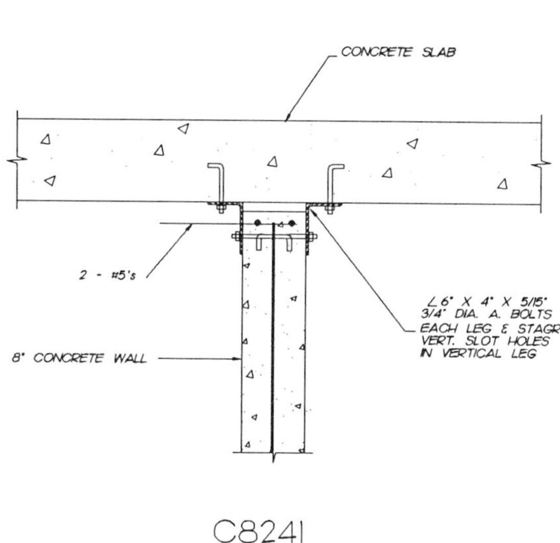

C8241

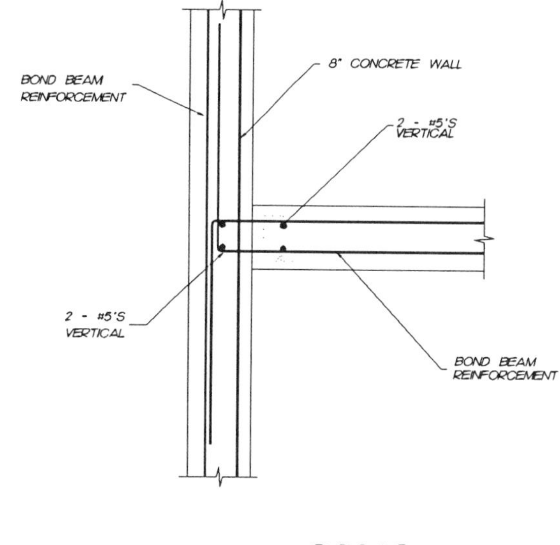

C8260

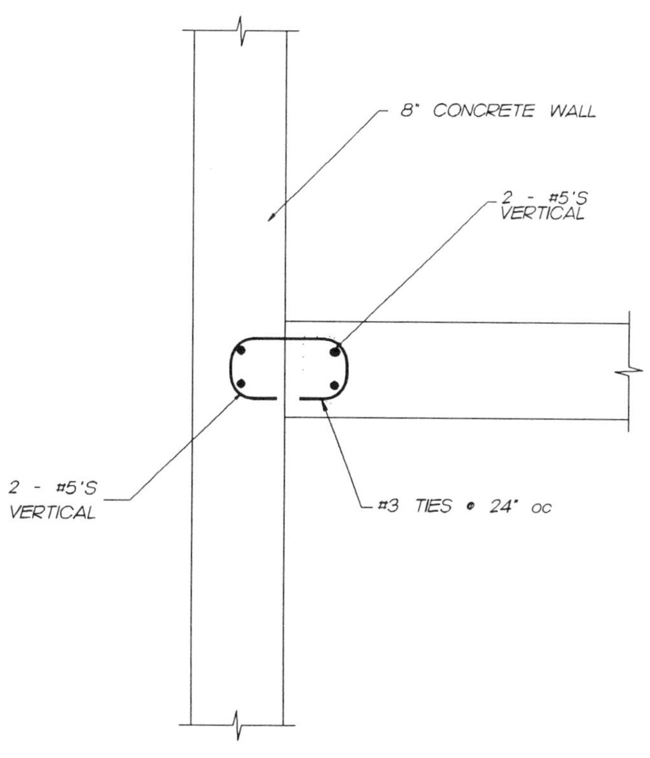

C8261

CHAPTER 3 CONCRETE SECTION 9	MISCELLANEOUS
REINFORCEMENT - TYPICAL BAR BENDS & HOOKS	C9020
CONCRETE BEAM - TYPICAL CROSS SECTION	C9040
CONCRETE BEAM INTERSECTION	C9041
CONCRETE BEAM - PIPE PASSING THROUGH HORIZONTALLY	C9060
CONCRETE BEAM - PIPE PASSING THROUGH VERTICALLY	C9061
CONCRETE SLAB REINFORCEMENT PLACEMENT	C9080
CONCRETE SLAB CONSTRUCTION JOINT	C9100
CONCRETE SLAB DEPRESSION	C9120
FOUNDATION - PIPE PASSING THROUGH	C9140
EDGE OF CONCRETE SLAB ON GRADE	C9160
DEPRESSON OF CONCRETE SLAB ON GRADE	C9161
WALL DOOR OPENING	C9180
WALL WINDOW OPENING	C9181
CONCRETE WALL JAMB & SILL SECTION	C9182
CONCRETE WALL LINTEL SECTION	C9183
WALL DOOR OPENING WITH STEEL CHANNEL JAMB & LINTEL	C9200
CONCRETE WALL STEEL CHANNEL JAMB & LINTEL	C9201
STEEL BEAMS IN CONCRETE WALLS	C9220
CONCRETE BEAMS IN CONCRETE WALLS	C9240 - C9241
VERTICAL REINFORCEMENT AT STEEL BEAM SUPPORT	C9260
VENEER SUPPORT ON CONCRETE WALL	C9280
CONCRETE WALL CONSTRUCTION JOINT	C9300 - 9301
CONCRETE SLAB SUPPORTING A MASONRY PLANTER WALL	C9320
VERTICAL REINFORCEMENT AT CONCRETE BEAM SUPPORT	C9340
CONCRETE WALLS & CONCRETE STAIRS	C9360
6'-0" HIGH FREE STANDING WALL	C9380
5'-0" HIGH FREE STANDING WALL	C9381

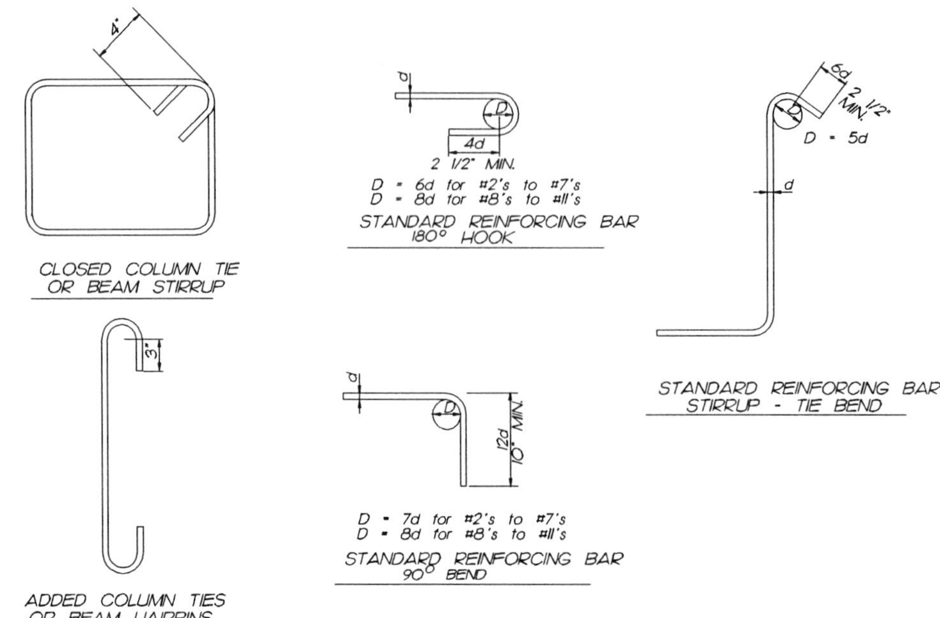

C9020

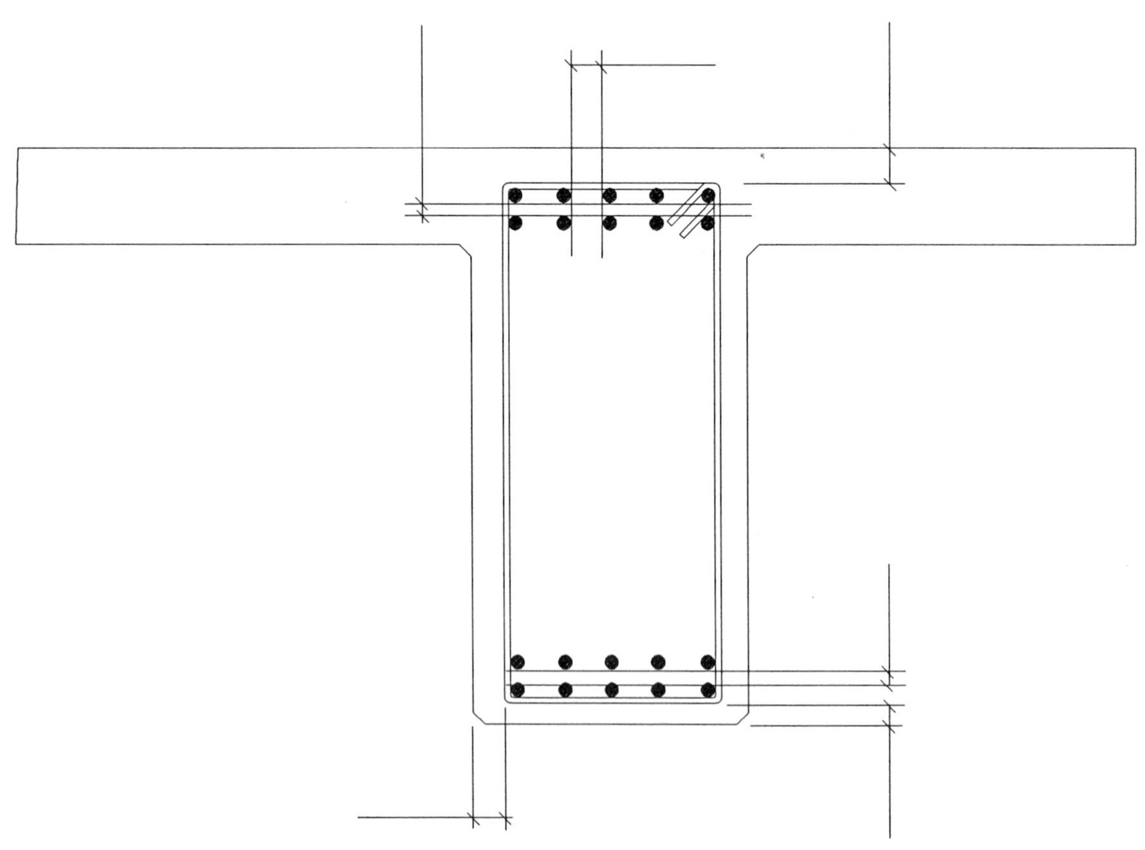

TYPICAL REINFORCED CONCRETE BEAM SECTION

C9040

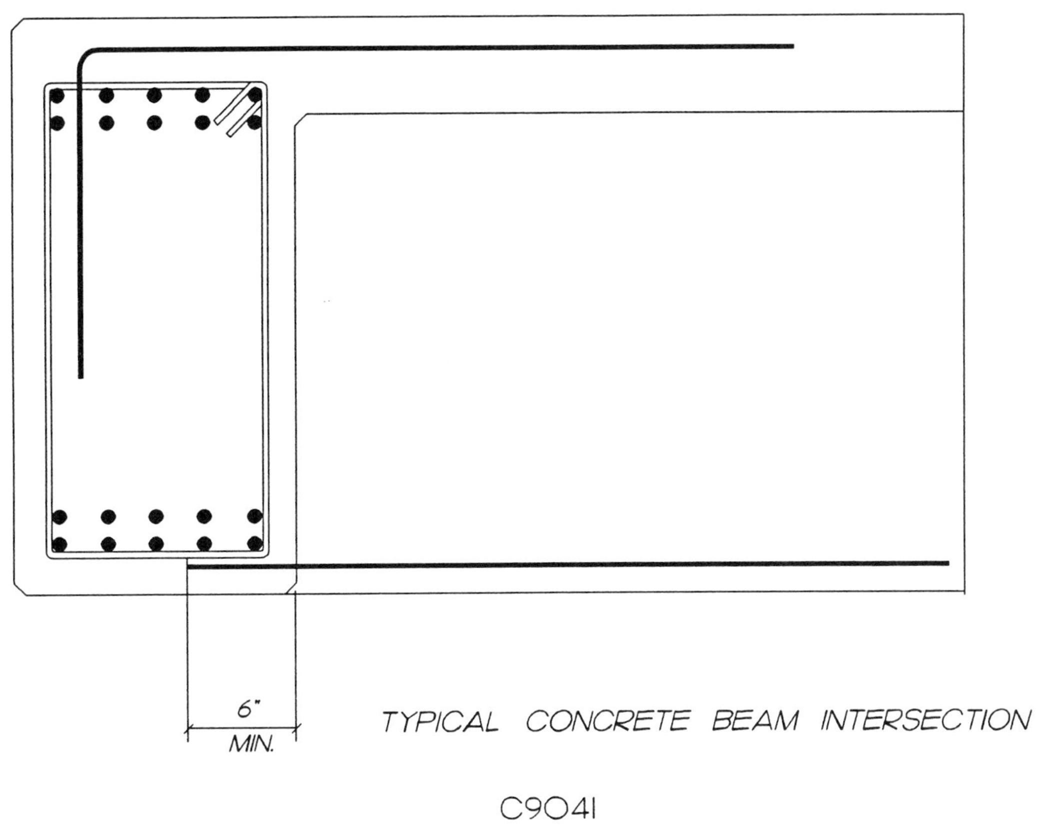
6"
MIN.
TYPICAL CONCRETE BEAM INTERSECTION

C9041

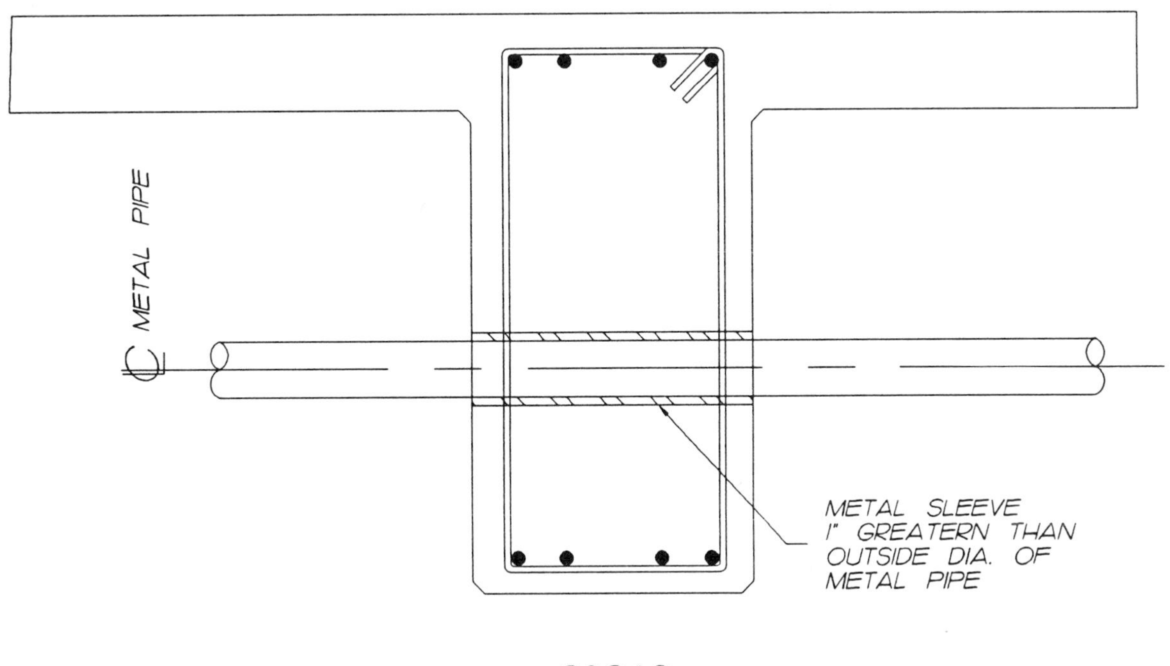
METAL PIPE
METAL SLEEVE
1" GREATERN THAN
OUTSIDE DIA. OF
METAL PIPE

C9060

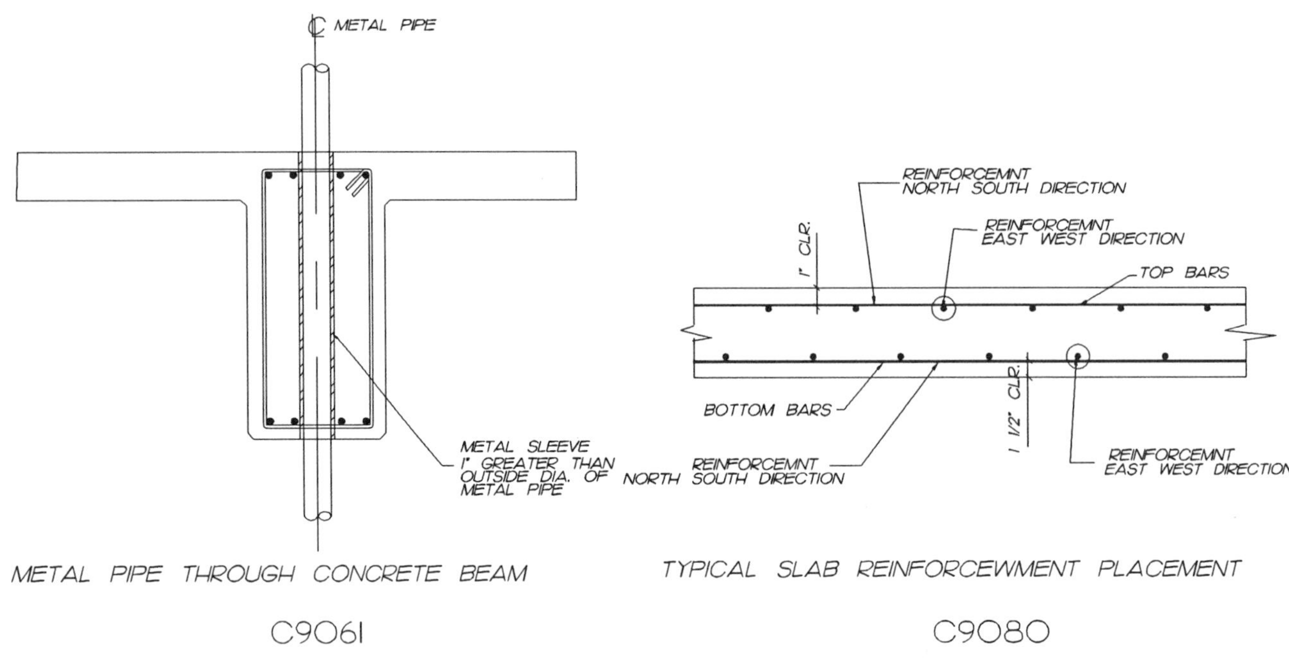

METAL PIPE THROUGH CONCRETE BEAM

C9061

TYPICAL SLAB REINFORCEWMENT PLACEMENT

C9080

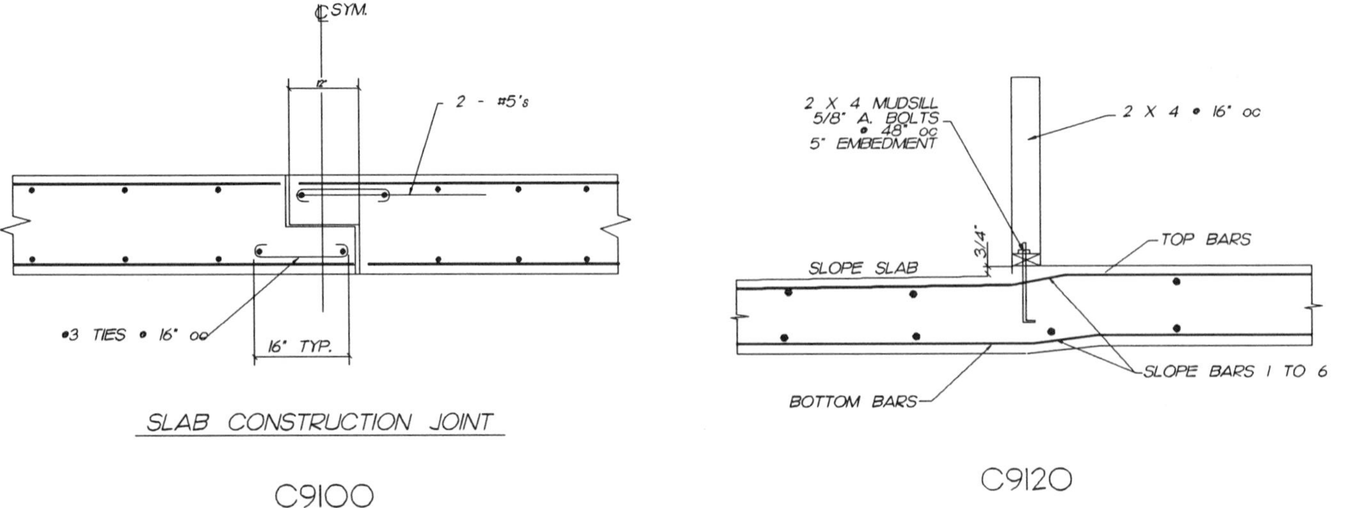

SLAB CONSTRUCTION JOINT

C9100

C9120

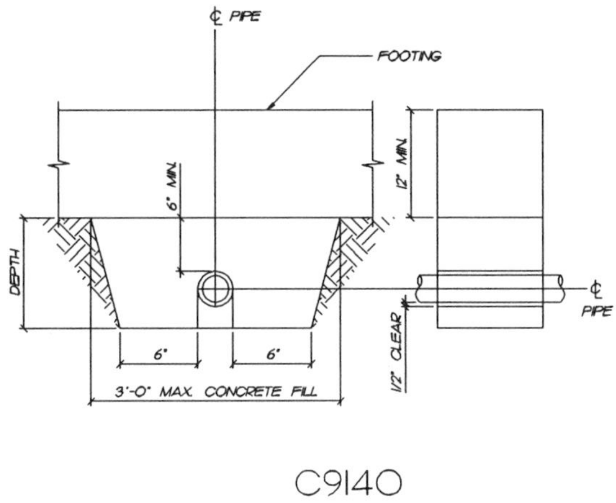

C9140

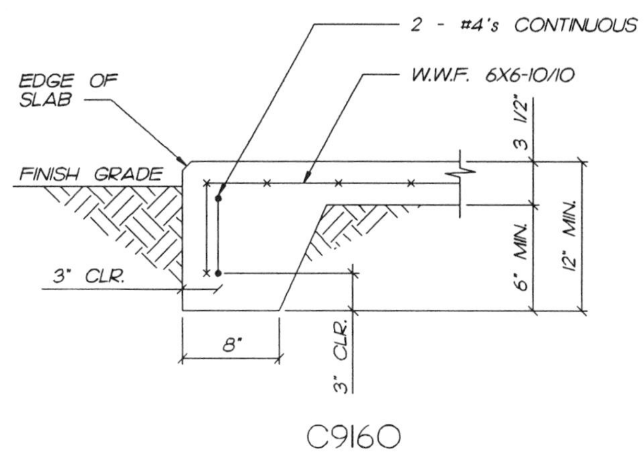

C9160

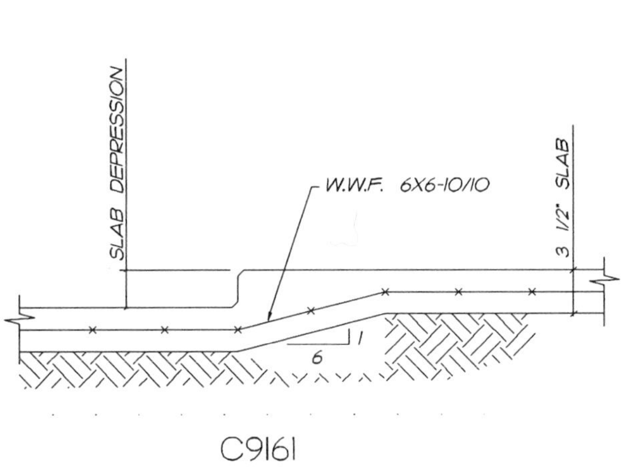

C9161

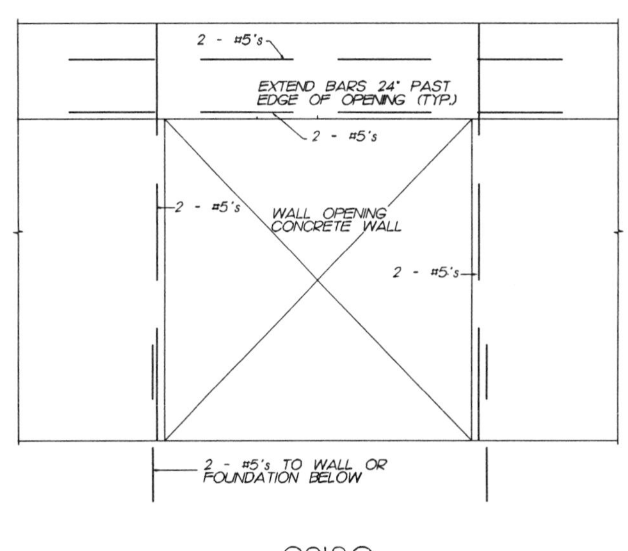

C9180

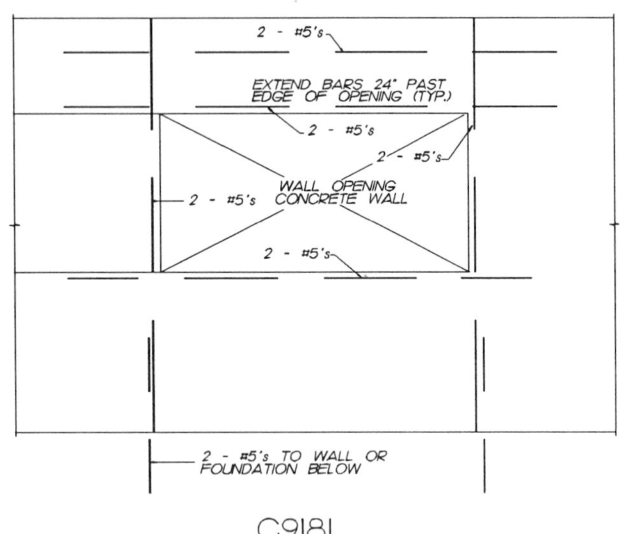

C9181

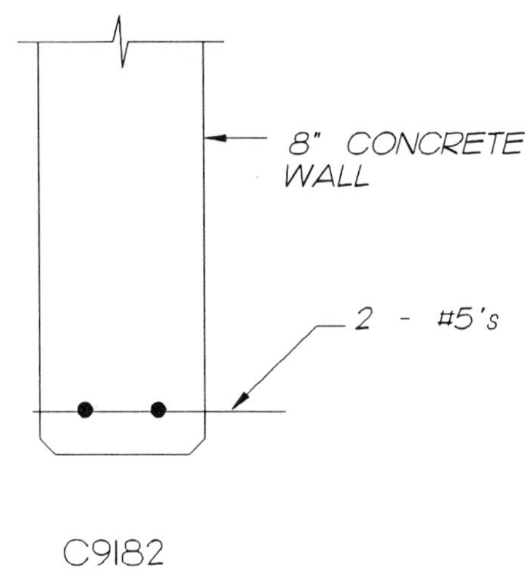

C9182

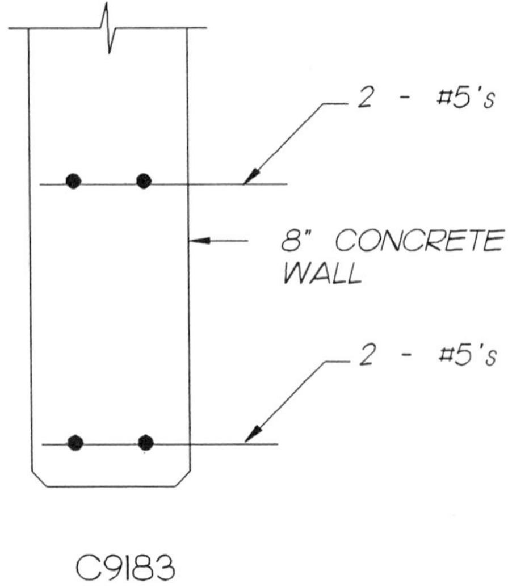

C9183

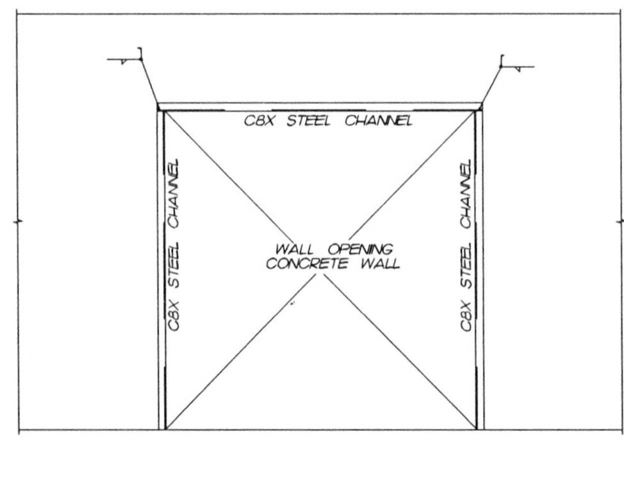

C9200

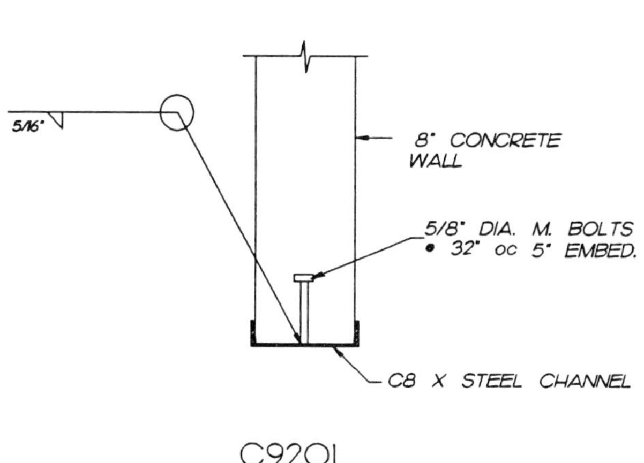

C9201

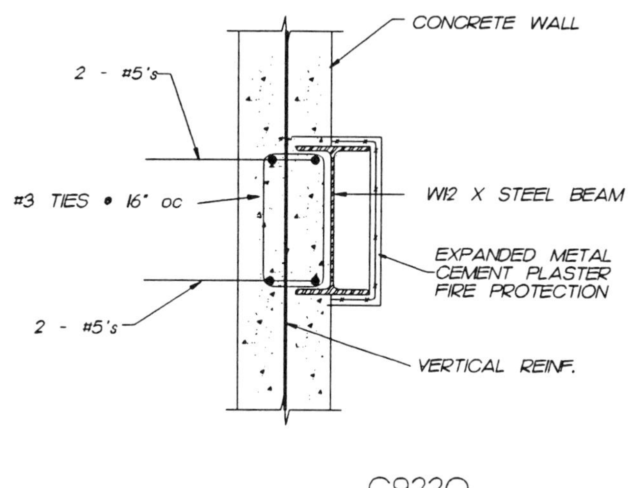

C9220

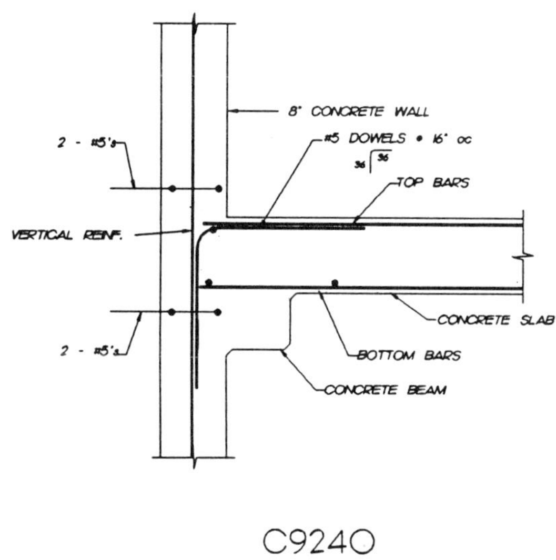

C9240

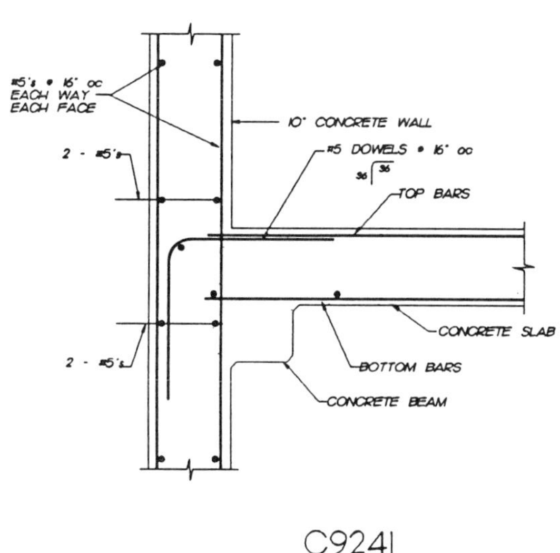

C9241

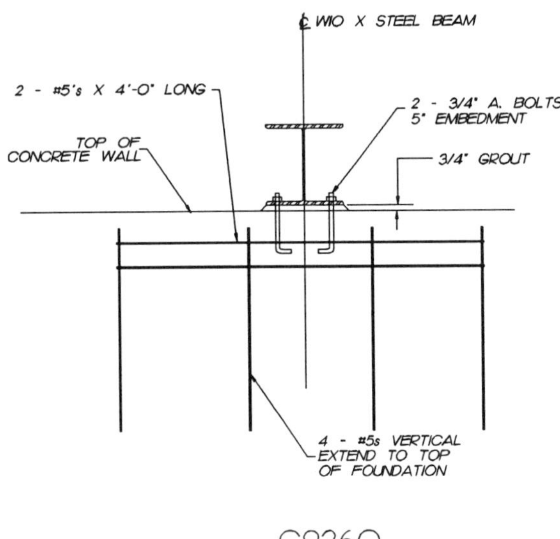

C9260

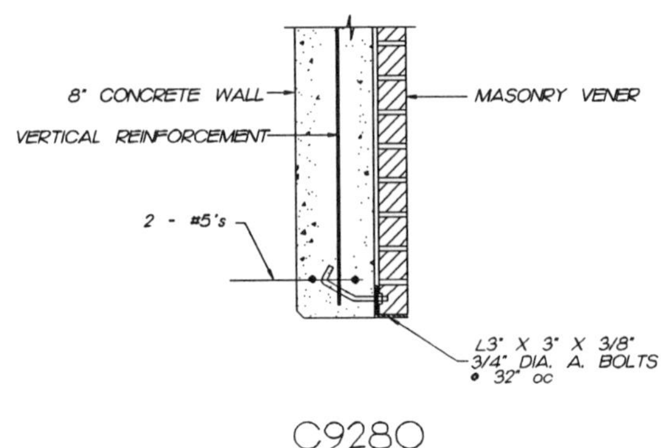

C9280

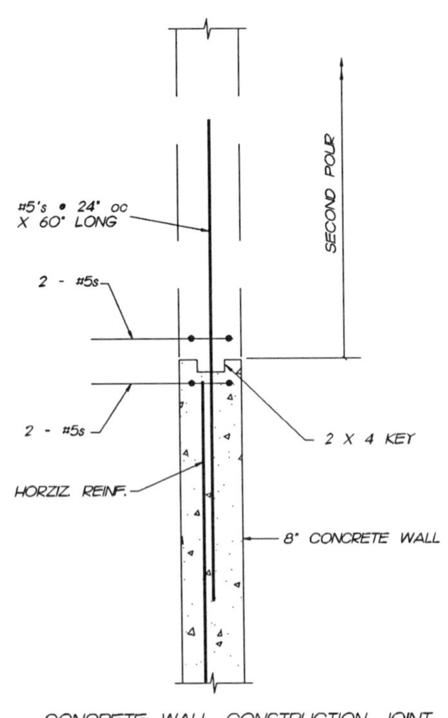

CONCRETE WALL CONSTRUCTION JOINT

C9300

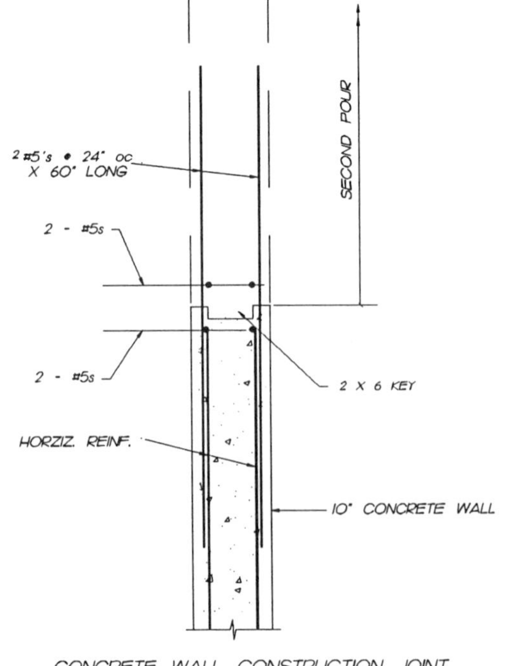

CONCRETE WALL CONSTRUCTION JOINT

C9301

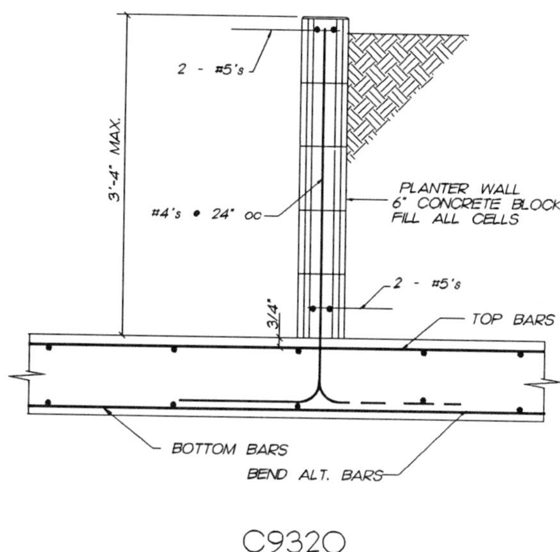

C9320

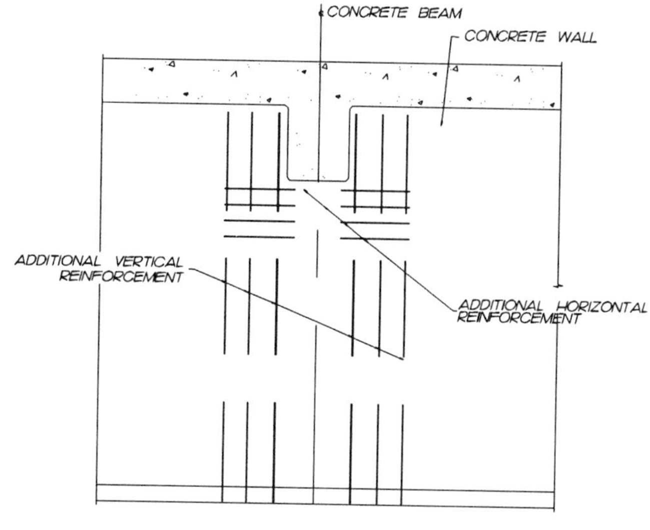

C9340

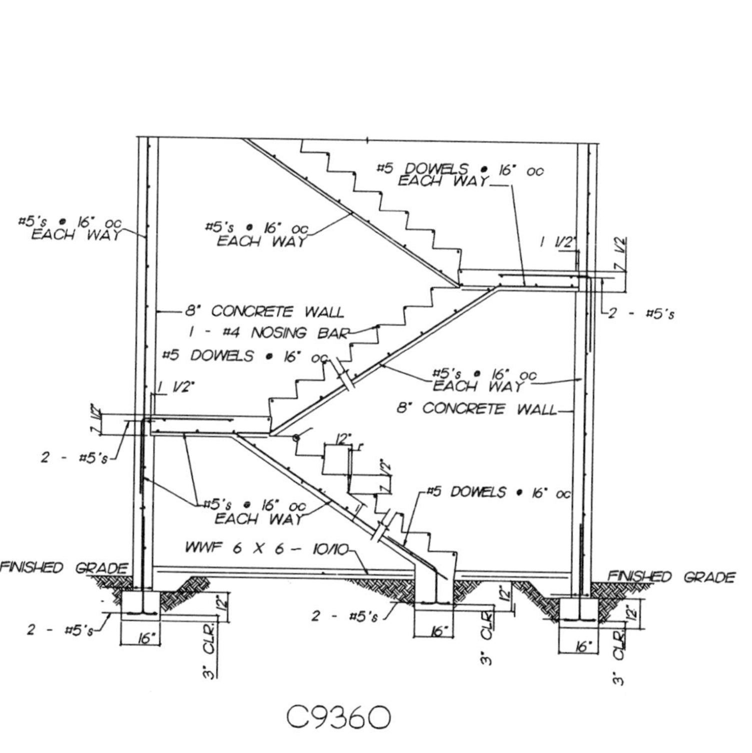

C9360

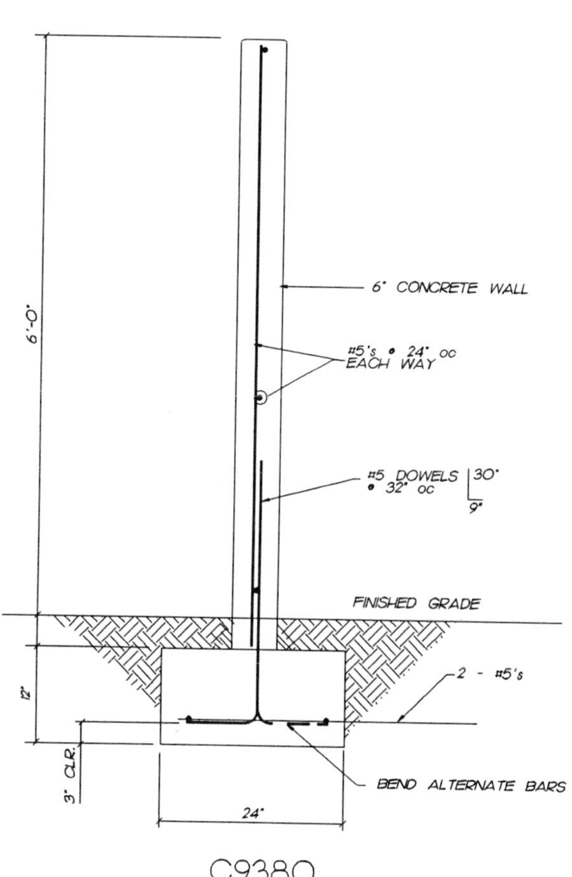

C9380

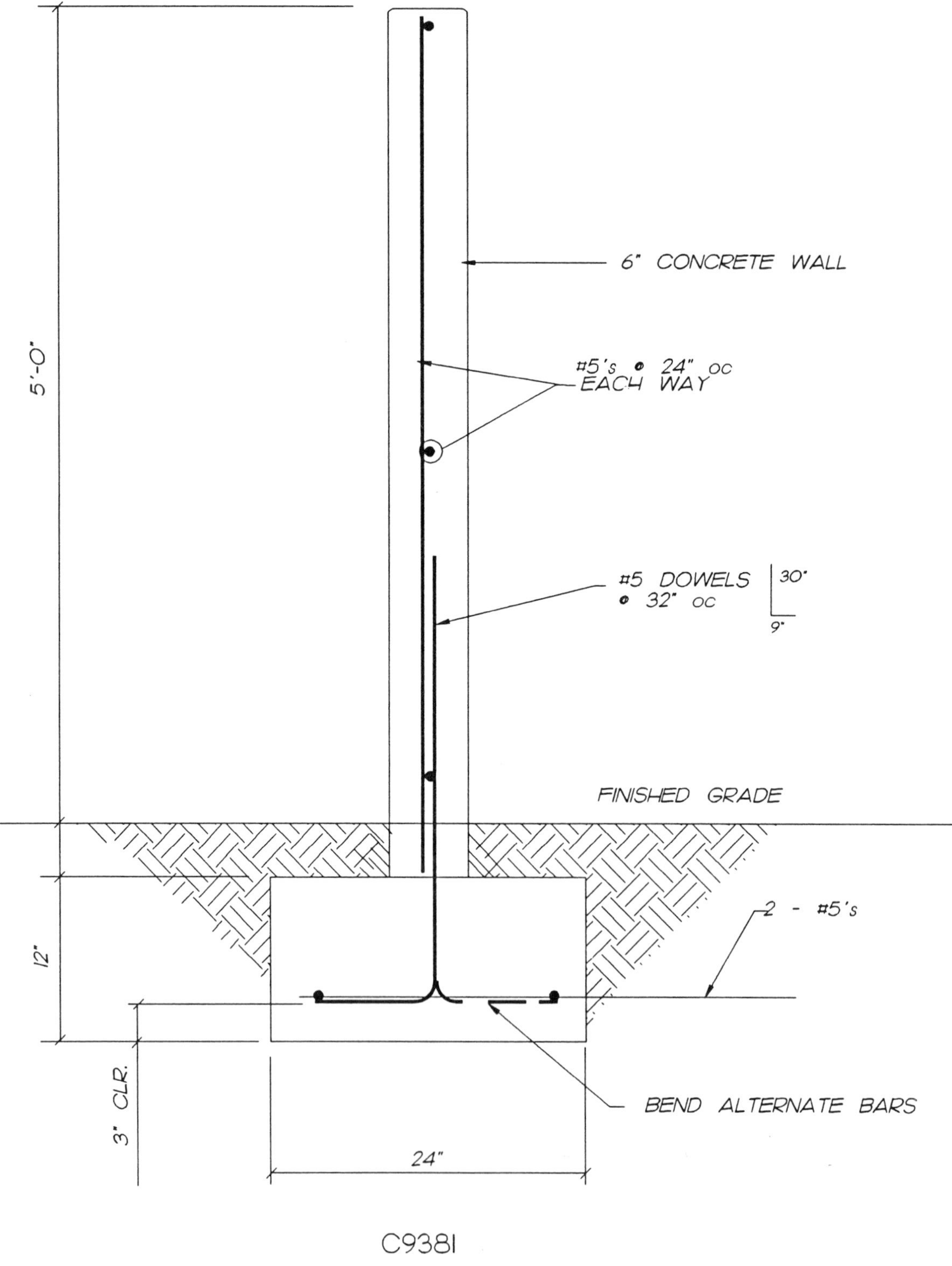
5'-0"
6" CONCRETE WALL
#5's @ 24" OC
EACH WAY
#5 DOWELS
@ 32" OC
30"
9"
FINISHED GRADE
2 - #5's
12"
3" CLR.
BEND ALTERNATE BARS
24"

C9381

Steel Structural Details

he structural steel details and assemblies presented in this chapter are xclusively concerned with steel construction. There are other details in nis book that require structural steel parts or cross sections; however, teel is not the primary element of the detail, and therefore these details re not included in this chapter. Most of the steel details and assemblies rovide specific data, such as the number and size of bolts, the sizes of teel members, and the size and type of welds. This information is included o make the detail a more helpful example of the use of steel as a con-truction material. Standard structural details should be used with an in-ormed concern for the strength of the design. This prerequisite is etermined by the structural engineer.

The use of structural steel requires a high degree of dimensional ac-uracy in fabrication and erection. Structural parts are fabricated in the hop in such a way that they can be interconnected at the job site with a ninimum of adjustment. In this respect structural steel differs from other onstruction materials, which can be more easily altered at the job site to chieve a specific configuration. Reworking steel members in the field is xpensive and time-consuming. The constraint of dimensional precision n the fabrication and erection of structural steel members and parts has ecessitated the adoption of many standard industry methods and prac-ces. The American Institute of Steel Construction specifications provide omprehensive criteria for the design, fabrication, and erection of steel onstruction. This code is the generic basis for steel requirements in most uilding codes.

This chapter presents drawings of the various alternative accepted met ods of connecting the component members of a structural steel fram Each type of connection, and its combinations of different sizes of ste members, is detailed and shown individually. The accessibility of the i formation in the chapter is facilitated by arranging the material in a logic relevant sequence so that the reader may readily locate a particular deta The basic concept of this arrangement places the drawings in an assigne hierarchy which starts with the most general condition, then progress to a discrete number of associated explicit conditions.

Each drawing is identified by a coded designator starting with a materi identification character and followed by a four-digit number. Since th chapter is concerned with steel details, the first character of the det designator is the letter S. The detail numbers which follow are coded represent a construction function section and the number in that particul section. The last digit of the detail number represents the number of th variation of the original detail. This arrangement allows for some degre of parallelism between the different chapters. For example, the numbe for the steel columns are similar to those for the concrete or mason columns. However, an exact comparison between chapters will reveal th the drawings are not always identical.

The details are organized in a sequence of sections, each of whic pertains to a particular function in the construction process. The first dig of the detail number is used to describe the particular construction pu pose on function of the detail. This chapter consists of seven functic sections, which are defined as follows:

Foundations	S1020 to S1xxx
Columns	S2020 to S2xxx
Walls and wood floors	S3020 to S3xxx
Walls and steel floors	S6020 to S6xxx
Stairs	S7020 to S7xxx
Beam connections	S8020 to S8xxx
Miscellaneous	S9020 to S9xxx

The following is an example of the number coding system:

Given No. S2164

The number 2 indicates that the detail is in the Columns section; th number 16 indicates that it is the sixteenth drawing in this section; an the number 4 indicates that it is the fourth variation of the original co figuration. A look at Detail S2164 shows a steel wide-flange column su ported by a steel beam. The catalog charts prior to the drawings a presented to assist the reader in locating details. The construction functio section numbers at the top of the charts are arranged in numerical orde as outlined above. The detail descriptions and their respective particula are stated in the left-hand columns, and the designation numbers are give in the right-hand column of the chart.

The primary considerations in specifying structural steel members fc construction are strength, deflection, cost, and availability. The capacity a steel member to resist stress and deflection is a function of its size, i shape, and the yield point stress F_y of the steel. Since steel is general purchased by weight, it is important to use the lightest sections that wi e capable of sustaining the imposed loads and forces within the limits f the allowable stress and deflection criteria. The type and strength of ructural steel used in the drawings shown on the following pages are ot specifically designated, since they are not primary factors in the con- guration of the details.

Manufacturers and fabricators of rolled steel sections are required to onform to the American Institute of Steel Construction specifications for ie fabrication, erection, and design of structural steel. The AISC code lentifies each type and strength of structural steel by an American Society f Testing Material Laboratory (ASTM) number which indicates the material ield point F_y. The allowable stress criteria for tension, compression, and hear for the various uses of the steel members, such as columns, beams, truts, and plates, are calculated as determined percentages of the value f F_y. At the present time there are two basic methodologies used in ructural steel design. Until recently (1978), most building codes required ie allowable stress design (ASD) method. This procedure is analogous o the working stress design (WSD) method used for reinforced concrete. stipulates that the stresses of structural members shall not exceed the alculated allowable stress criteria determined by a safety factored value f F_y. The ASD method uses the actual calculated dead loads, live loads, nd applied lateral forces. A structure designed by this method will have certain amount of reserve strength.

The load resistance factor design (LRFD) method is gaining acceptance. his procedure is analogous to the ultimate strength design (USD) method sed for reinforced concrete. The concept of LRFD is based upon designing ie members to an adjusted limit of the material yield point F_y by using ertain load factors for dead loads, live loads, and lateral forces. The load ctors are based upon probable load occurrence; however, they are not ecessarily equal. The safety factor in the LRFD design method is derived om the increased values of the applied loads. The LRFD design method roduces designs that are more economical and uniformly reliable. With ither method, the vertical and lateral deflections of structural members hould be considered in decisions regarding the design and strength of iaterials.

The equations used to calculate working stress values for structural steel re given in the AISC *Manual of Steel Construction* and the *Specifications or the Design, Fabrication, and Erection of Structural Steel for Buildings.* STM No. A36 steel is currently the steel most widely used with the ASD iethod. The allowable flexural stress criterion for this design method is alculated as $0.66 \times F_y$ ($0.66 \times 36{,}000 = 24{,}000$ psi). The AISC code also ecreases working stress values based on the beam's flange-to-thickness atio and the unbraced length of the span. The allowable compression tress criterion for columns is calculated as a function of the slenderness atio of the unbraced length in either the X or Y direction and by a factor epending on the column's end bending restraints.

Structural steel designated as ASTM No. A529 has $F_y = 42{,}000$ psi. The exural stress criterion for ASD is calculated as $0.66\,F_y = 27{,}720$ psi. This pe of structural steel can be used in ASD design; however, in view of its igher value of F_y and modulus of elasticity E, it would be better used ith the LRFD method. High-strength low-alloy steels are usually used in RFD design of structural steel members. These types of steel are desig- ated as ASTM No. A441 and ASTM No. A572, with respective minimum ield point stresses F_y of 40,000 psi and 42,000 psi. ASTM No. A242 steel highly corrosion-resistant; ASTM No. A441 steel is a relatively corrosive material and is used primarily in the construction of bolted or riveted structures and for welded structures that may be subjected to high impact loads.

The allowable working stresses that are derived as a percentage of the yield point F_y of high-strength steel are also governed by the thickness of the material. That is, a thickness not greater than ¾ in has a yield point stress of 50,900 psi; thicknesses from ¾ to 1½ in have a yield point stress of 46,000 psi; and thicknesses from 1½ to 4 in have a yield point stress of 42,000 psi. These are only general statements concerning the types and strengths of the structural steels that are presently available for use in construction. Complete data regarding the strength and use of structural steel can be found in the *Manual of Steel Construction*, published by the AISC. Mill test reports may be supplied to the fabricator to verify material quality control. These reports must be requested prior to the time the steel members are fabricated at the rolling mill.

Structural steel members are manufactured by extruding steel billets between a set of rolling dies in a series of forward and reverse passes until a uniform, standard-size shape is produced. The nomenclature for steel shapes describes their cross-sectional shape, for example, wide-flange beams, I beams, sections, angles, and channels. Structural steel sections are fabricated to standard sizes and weights; however, each manufacturer's product may vary from another's by a small degree. The AISC *Manual of Steel Construction* lists the major manufacturers of standard structural steel shapes in the United States. The manual also lists the shapes that are most frequently used, special series shapes, the manufacturers of each shape, and availability. Certain structural steel sections are manufactured exclusively by only a few steel-producing companies, such as Jones & Laughlin Steel Corporation, Inland Steel Company, and Northwestern Steel and Wire. For use in the United States, imported structural steel must meet the standards of the ASTM. It is important that the product also have the approval of the controlling authority in the location in which it is to be used. Obtaining approval of a new product from a local department of building and safety can be a long and tedious process. The engineer who specifies structural steel that is produced in foreign countries or is not within the AISC specifications should take into consideration that these members do not have the same physical properties as domestic structural steel.

Section 1: Foundations. This section shows a series of continuous concrete footings supporting a variety of steel stud walls. It was decided to place these drawings in this chapter and section, since their most basic material is steel. The details could have been placed in Chapter 1, "Wood"; however, they would then be placed with wood stud walls, which could cause some confusion in retrieving the drawings.

Section 2: Columns. This section contains a set of drawings that depict various types of column base plates and beam-to-column connections. The function of a base plate is to transfer the column load uniformly to a concrete foundation; therefore, the base plate should have sufficient surface area and cross-sectional thickness to spread the superimposed loads. The contact surface between the top of the plate and the steel column should be flat to create a completely even bearing surface. Bearing ends of steel columns and base plates over 2 in thick are required to be milled to ensure a smooth contact surface; however, the AISC Code does allow plates between 2 and 4 in thick to be straightened by a pressing process. The bearing pressure on the bottom of the base plate is assumed to be equally distributed over its entire area. This uniform pressure will cause the plate surface to bend upward (dish) by cantilever action. The thickness of the plate is used to calculate the resistance to this bending.

The column is attached to the base plate by either shop or field welding; in either case, the weld must be strong enough to resist the column's combination of vertical and lateral loads. Anchor bolts connect the concrete base plates to the concrete foundation. The capacity of the anchor bolt to act as a resisting element depends on the size of the bolt, the thickness of the plate, and the depth of embedment of the bolt into concrete. The anchor bolts shown in details S2043 and S2044 are also used to transfer the column bending moment to a foundation by acting as resisting elements of a force couple. Anchor bolts are required to develop sufficient bond between the concrete and the bolt surface to prevent their slipping or being pulled out from tension.

The placement of base plates and anchor bolts is the initial step in the field work to coordinate the foundation with the steel superstructure; therefore, it is necessary that the plates and anchor bolts be accurately set so that no dimensional adjustments will be required to align the building. Quite often the steel erector will set plywood or light metal template plates to maintain the accuracy. Even small dimensional variations in plate and anchor bolt location may result in time-consuming field modification. The steel columns are vertically plumbed to their final positions by setting the base plates on a relatively thin layer of nonshrinkable grout to level the plates. The structural steel erector should verify the elevations of the tops of base plates to be sure that the steel columns are cut to the exact length.

Details S2100 to S2143 are examples of the various methods of making steel column splices. These details are good examples for the need for dimensional accuracy and also for the milling of the ends of the column bearing surfaces. In general, column splices are used in the construction of multistory buildings to coordinate the column sizes with the progressively increasing loads. The splices should be made several feet above the floor so that rigid connections of beams or girders to the column will not be affected and to allow enough clearance to erect the upper column. The splice connection must be capable of transferring the upper column moment, shear, and axial load to the lower column. The axial load is transferred to the lower column by bearing. The bearing surfaces of the members and connecting parts should be milled to ensure that they will have complete bearing on the cross sections. The shear on the upper column is usually transferred to the lower column through a bolted or riveted connection between the webs of the two members; however, the webs can be welded, depending on the conditions of erection and the shear to be transferred. Column moments are transferred through splice plates attached to the upper and lower column flanges. These plates are bolted, riveted, or welded, depending on the moment load and the method of erection. The designer should endeavor to arrange all column splices in a multistory structure so that they occur at the same stories and to make the columns standard lengths. The length of steel columns is governed by fabrication, erection, and convenience of delivery to the job site.

Details in Section 2 also show many types of simple beam-and-column connections. These types of connections are classified generally as nonrigid, or simple, connections, since it is assumed that they have no restraint against rotation or bending moment. Nonflexible connections are classified as rigid frame connections. This type of connection is capable of resisting bending and shear; each connected member is restrained from rotating independently. With rigid connection joints, by definition, the angle between the connected members will remain the same after the membe are stressed from flexure or shear. Details S2460 to S2543 show rigi column-and-beam connections. It can be seen in these details that th connections are fabricated to resist both shear resulting from vertical loac and bending by restraining the end of the beam. In both rigid connectior and simple connections, the vertical shear reaction of the beam is su ported with either angles or plates; however, in rigid connections th rotational bending in the joint is restrained. Resistance to joint deformatio is achieved by using welds, rivets, or bolted fasteners that will not perm slippage between the connected parts.

Simple beam-and-column connections are designed to permit bendin at the joint to occur. The connection is not restrained against rotation deformation; therefore, some horizontal and vertical slippage in the joi is allowed. Simple joints are usually constructed with a combination c welds and machine bolts. Joints classified as simple connections are usuall made with unfinished, or machine, bolts of low-carbon steel, designate as ASTM Specification A307. Machine bolts can be easily identified by the square head and nut and the designation "A307" on the bolt head. I standard simple connections, the bolt shank reacts in bearing and shea to transfer the load to the connected parts. The nuts are tightened in plac with a spud wrench with sufficient force to prevent the bolt from becomin loose, but not enough to overstress the washer in bearing. This type c connection can also be made by using hot-driven rivets or by using bol that are manufactured with ribbed or serrated shanks. The *Manual of Ste Construction* lists and diagrams the standard connections for the variou size structural beams and the connection capacity of the different types c mechanical fasteners. It is important to use the minimum specified siz angles in this type of connection. Thicker angles will reduce the flexibilit of the joint and restrain the end of the beam as it bends.

High-strength carbon steel bolts are used in rigid frame joints to resi slippage caused by shear between connecting parts. These bolts are know as friction-type connectors and are designated as ASTM Specification A32 or ASTM Specification A490. Friction resistance in the connection i achieved by mechanically tightening the nuts to develop a specified tensio in the bolt shank that will clamp the contact surfaces of the connectin parts together. The friction resistance of the high-strength bolt depend on the tension in the bolt and the condition of the contact surfaces, whic must be free of paint or any coating that may permit slippage. A325 an A490 bolts can be identified by their hexagonal-shaped heads and nut and by their respective ASTM number and manufacturer's symbol on th bolt head. The AISC specification for installing high-strength bolts require that they be tightened by using a calibrated torque wrench or by the "tur of nut method." The calibrated torque wrench may be power-operate and set to cut off at a torque resistance that will correspond to the require tension in the bolt shank. The turn of nut method is performed manuall with a wrench; the bolts are successively tightened by rotating the nuts certain amount after they are brought to a snug, tight condition. High strength bolt assemblies are required to have two hardened steel washer however, A325 bolts do not require a hardened steel washer when the are installed by the turn of nut method. Where the connecting part surfaces slope more than 1 in 20, the washers should be beveled to ac commodate a snug fit of the bolts. The AISC recommends that bolts o rivets be installed through holes that are $\frac{1}{16}$ in larger in diameter tha the bolt.

The details in this book specifically identify the type of bolt used for a connection; that is, A307 bolts are called out as M. bolts (machine bolts), and A325 or A490 bolts are called out as HS (high-strength) bolts.

Bolt or rivet holes may be reamed, punched, or drilled. The dimension between centers of bolts or rivets should be not less than $2\frac{2}{3}$ times the nominal shank diameter of the bolt. The minimum recommended distance between bolt or rivet holes is 3 in o.c. The minimum distance between the edge of a structural steel part and the center of a bolt hole depends on the bolt diameter and on whether the edge is sheared or the edge of a rolled section or plate. AISC specifications give the edge distance requirements for bolt diameters for sheared and rolled edges of plates and structural sections.

Many of the details in this chapter require that the members or parts of joints be connected by welding, particularly those members connected by rigid joints. Welding technology is a vast and complex subject and will not be completely covered in this text. The specific information that is ordinarily required in the structural design and drawing of steel buildings can be found in Sec. 1.17 of the AISC specifications and in the *Structural Welding Code—Steel* of the American Welding Society. Basically, structural welding is the process of uniting two metal surfaces by fusion through the heat of an electric arc. The electric-arc welding process consists of applying intense heat generated by a low-voltage, high-amperage electric arc to the steel parts to be joined. The arc is maintained across the steel by an electrode which deposits a small amount of weld metal on the fused surfaces. Welding electrodes consist of a coated steel rod composed of a metal with the same chemical and mechanical properties as the steel to be joined. The strength of the weld also depends on the chemical composition of the rod coating.

When the shielded arc welding process is used, the arc deposits the weld metal on the surface, and the rod coating burns to create a gaseous shield immediately around the weld and at the same time introduces a flux material into the molten metal. The gaseous shield prevents the hot weld material from combining with the oxygen and nitrogen in the air. Oxidation will make the weld porous and therefore weak, while nitrogen in the weld material will make it less ductile and therefore brittle. The flux in the rod coating is used as a purifying element to raise any impurities to the surface of the cooled weld. These impurities will take the form of a slag coating on the finished weld, which should be chipped off before the welds are field painted. The shielded arc welding process is usually performed manually. Another welding process that is extensively used is the submerged arc welding process. Submerged arc welding refers to the fact that the arc is submerged in a layer of powdered flux material that is deposited as the work progresses. As in shielded arc welding, the flux raises the impurities to the surface of the weld as a slag. The slag that is formed by the submerged arc welding process is a loose scale that can be removed without chipping. In detailing welded connections, the designer should consider the direction and accessibility of the electrode relative to the work.

The strength and labor costs of welding can be affected by the position from which it is performed. There are four positions from which a weld can be made: flat, or down hand; horizontal; vertical; and overhead. The flat position is the most convenient position from which to weld and the least expensive. The overhead position is slow, expensive, and inconvenient; this position will permit the molten weld metal to flow from the surfaces that are to be joined. Welds are designated on the drawings b standard symbols which approximate their cross-sectional configuration

Section 3: Walls and Wood Floors. This section is a collection c details depicting the methods of supporting wood floors by either stee studs or steel beams. There are many instances in which these two materia are used in combination, and since steel is the predominant material, was decided to place these details in this chapter. Connecting wood t steel requires that the materials be joined in such a way that loads an stresses will be transferred and the materials will not separate. Detail S3020 to S3049 show that lateral diaphragm forces are resisted by stee stud walls through the metal clips to the wood double plates. In S3060 t S3089, diaphragm forces are resisted by ledger angles welded to the stee studs. The balance of the details in this section depict steel beams sup porting wood floors. The floor and the steel beam are joined by a stee clip angle connection.

Section 6: Steel Floors. This shows the various methods of connectin metal decking to steel beams and bar joists. Details S6160 to S6184 sho metal decking supported by a steel beam and connected with steel shea studs. This method of attaching the two different materials allows them t react as a single structural element. This type of construction is known a composite beam construction. AISC Specification Sec. 1.11.1 states: "Com posite construction shall consist of steel beams or girders supporting reinforced concrete slab, so interconnected that the beam and slab ac together to resist bending." Part 2 of the *Manual of Steel Constructio* gives many design examples of composite construction and tabulates th properties of sections of 4-in, 4½-in, and 5-in concrete slabs that ar constructed integrally with the various steel beam standard sections. Sec tion 1.11 of the AISC specifications outlines the design and constructio criteria for composite beams. This method of construction was first de veloped to be used for the deck slabs of highway bridges required t sustain heavy traffic loads. As the composite method of construction wa developed for bridge design, it was also found to be applicable to th design of steel buildings. The use of composite beams in building struc tures will result in lighter weight and smaller size steel beams, less floo deflection from live loads, and floors that are capable of supporting greate loads through a more efficient use of the materials.

Section 7: Steel Stairs. This section shows a series of details of pre fabricated steel stairs and connections. Most architects have their ow special configuration of stairs which originates from a sense of design These details are presented to display the various possibilities.

Section 8: Connections. This section displays a series of beam-to-bean connections. Notice that the connections are made with either clip angle or a shear tab plate. In either case the types of bolts used for the connectio are also specified as either machine bolts (M. bolt) or high-strength bolt (HS bolts). A variety of combinations of sizes of beams are shown t demonstrate the method of connection.

Section 9: Miscellaneous. This is very much as the section title in dicates. This section contains details that are often needed but do not fa into the other section categories of this chapter. The reader should perus this section to become familiar with what is available.

Structural steel members are fabricated to conform to shop drawings which are prepared from information obtained from the engineer's work ing drawings. Working drawings that are not clear or complete will cause delays in preparing the shop drawings and increase the cost of fabrication of the members. The engineer should check the shop drawings prior to shop fabrication to see that they agree with the design. Any discrepancies or variations in member sizes or dimensions should be brought to the attention of the fabricator and corrected before the shop work begins. Structural steel is manufactured and fabricated to meet rigid standards for quality and workmanship. Dimensional accuracy is crucial for both of these requirements. It is of little value for the manufacturer and fabricator to take great care in these considerations, only to find that the working drawings of either the architect or the engineer were not correctly dimensioned. This can be the source of much extra time and expense. An incorrect dimension is the least forgivable error at the job site.

Structural steel members are delivered to the job site with one coat of prime paint. Also, all steel members and connection parts will be marked in the shop with numbers and letters to correspond with a field erection diagram. Before the members are shop-painted, they are cleaned with a wire brush to remove rust, mill scale, and dirt. Areas of steel members that will be encased in concrete after the structure is completed should not be painted. This will permit the surrounding concrete to bond to the surface of the member. The shop coat of paint is only a temporary protection of the steel from weather exposure during erection. Some field painting may also be necessary for this purpose; in particular, field welds should be chipped free of surface slag and painted. The steel members should be shipped from the fabricator's shop in a sequence that will be convenient to expedite their erection.

CHAPTER 4	STEEL SECTION 1		FOUNDATIONS
STEEL STUD WALL EXTERIOR WALL CONTINUOUS FOOTING SLAB ON GRADE	RECTANGULAR SHAPE	3 5/8" STUDS	S1020 - S1023
		5 5/8" STUDS	S1024 - S1027
STEEL STUD WALL EXTERIOR WALL CONTINUOUS FOOTING SLAB ON GRADE	RECTANGULAR SHAPE WITH A CURB	3 5/8" STUDS	S1040 - S1043
		5 5/8" STUDS	S1044 - S1047
STEEL STUD WALL EXTERIOR WALL CONTINUOUS FOOTING SLAB ON GRADE	RECTANGULAR SHAPE WITH VENEER	3 5/8" STUDS	S1060 - S1063
		5 5/8" STUDS	S1064 - S1067
STEEL STUD WALL EXTERIOR WALL CONTINUOUS FOOTING SLAB ON GRADE	RECTANGULAR SHAPE WITH A CURB WITH VENEER	3 5/8" STUDS	S1080 - S1083
		5 5/8" STUDS	S1084 - S1087
STEEL STUD WALL EXTERIOR WALL CONTINUOUS FOOTING SLAB ON GRADE	L SHAPE	3 5/8" STUDS	S1100 - S1103
		5 5/8" STUDS	S1104 - S1107
STEEL STUD WALL EXTERIOR WALL CONTINUOUS FOOTING SLAB ON GRADE	T SHAPE	3 5/8" STUDS	S1120 - S1123
		5 5/8" STUDS	S1124 - S1127
STEEL STUD WALL INTERIOR WALL CONTINUOUS FOOTING SLAB ON GRADE	RECTANGULAR SHAPE	3 5/8" STUDS	S1140 - S1143
		5 5/8" STUDS	S1144 - S1147
STEEL STUD WALL INTERIOR WALL CONTINUOUS FOOTING SLAB ON GRADE	T SHAPE	3 5/8" STUDS	S1160 - S1163
		5 5/8" STUDS	S1164 - S1167

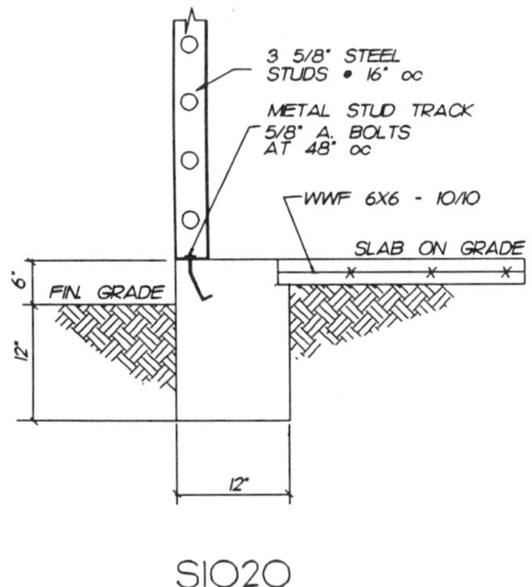

S1020

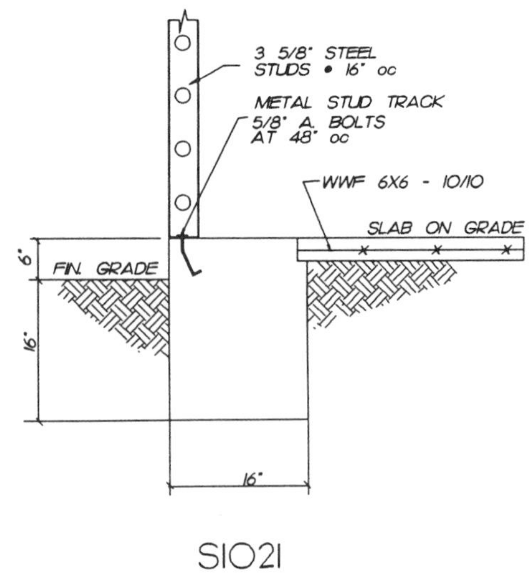

S1021

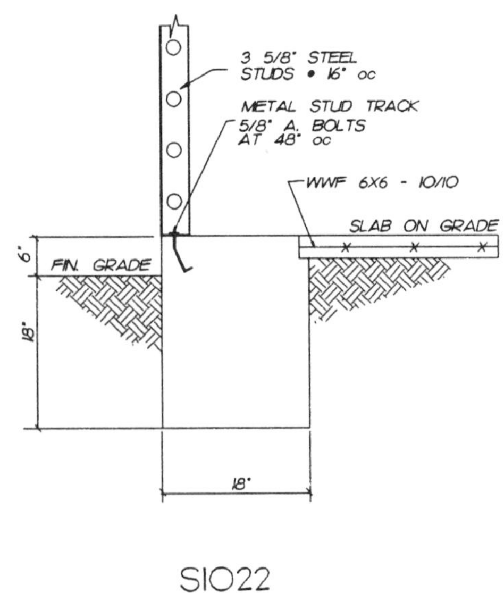

S1022

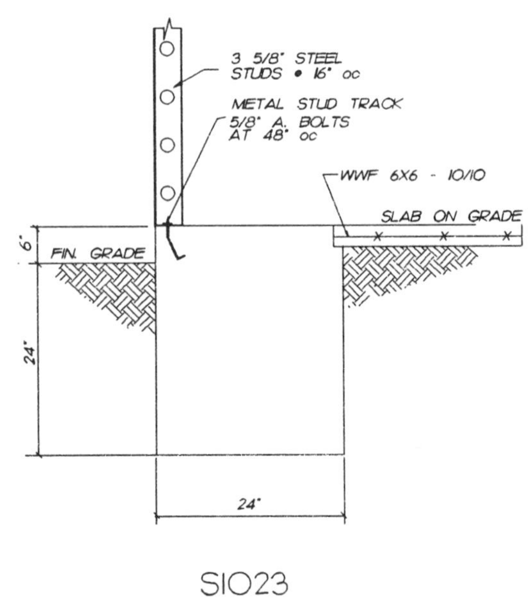

S1023

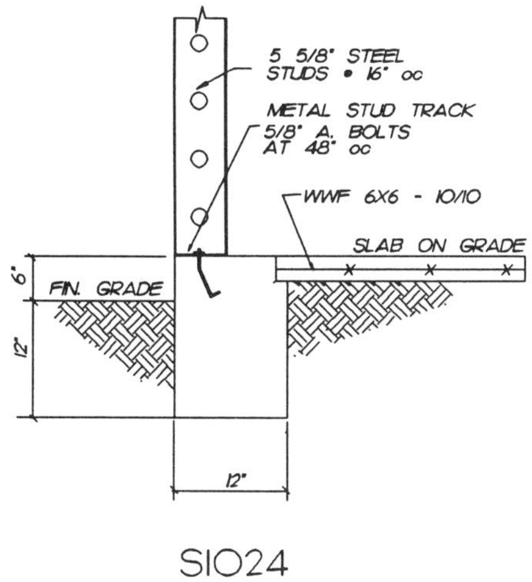

SIO24

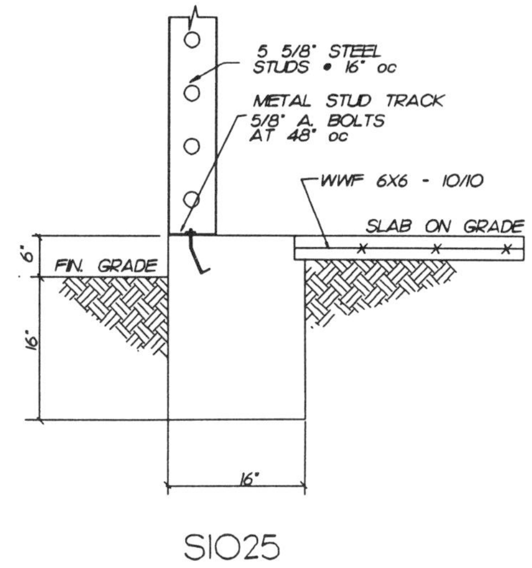

SIO25

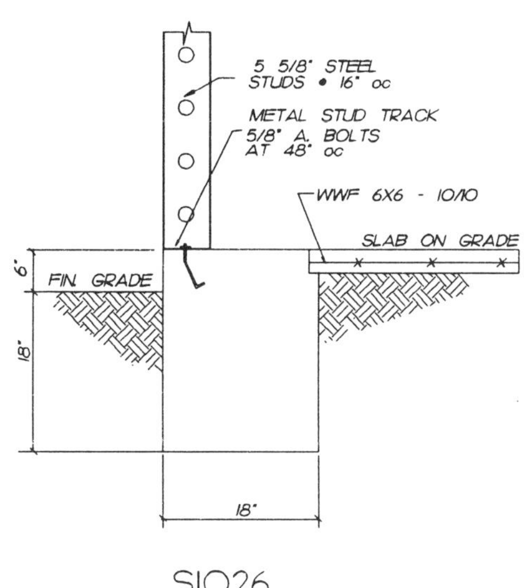

SIO26

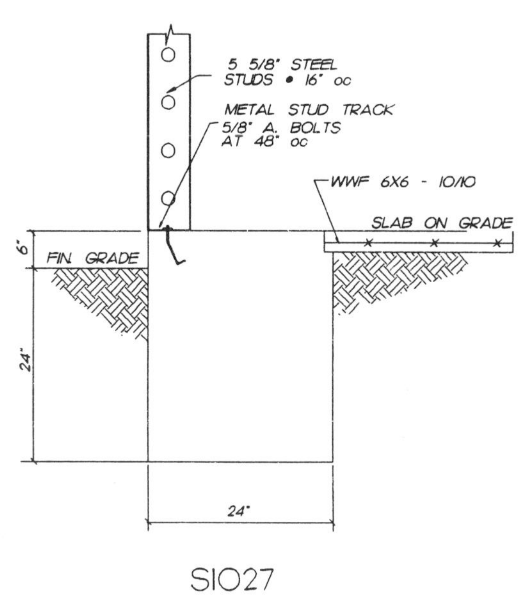

SIO27

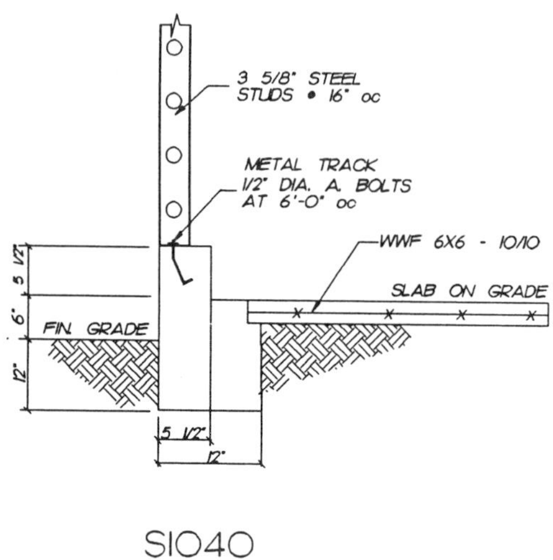

S1040

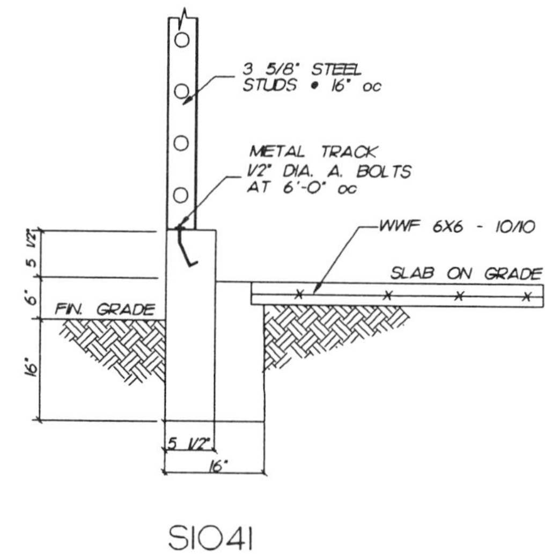

S1041

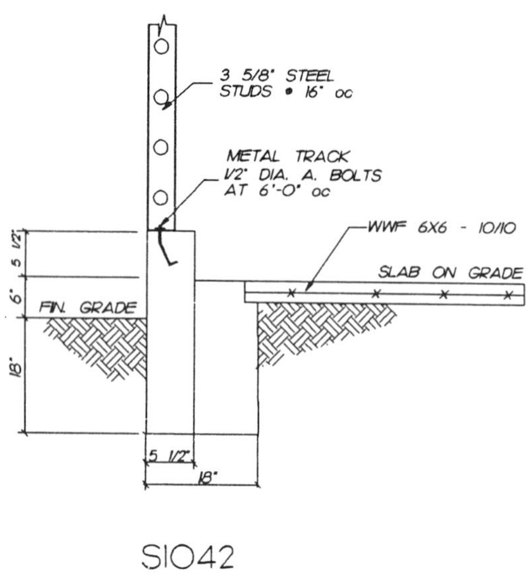

S1042

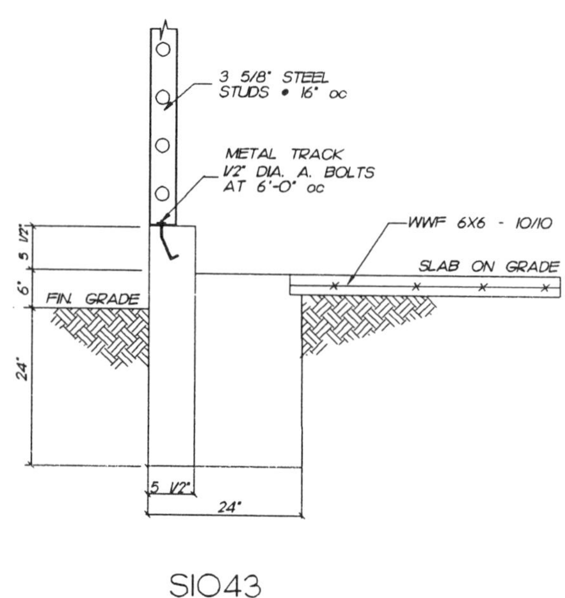

S1043

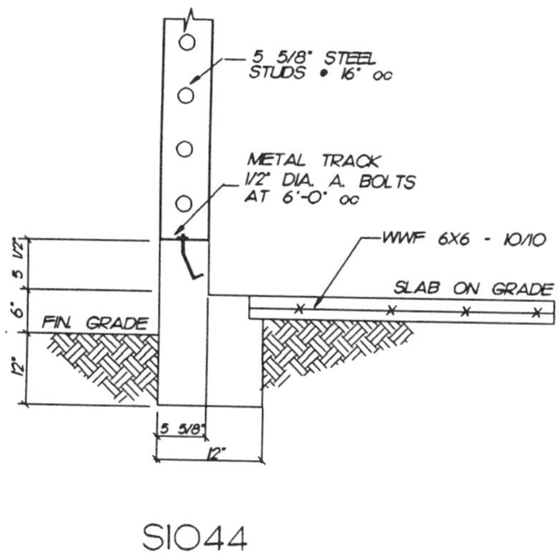

SIO44

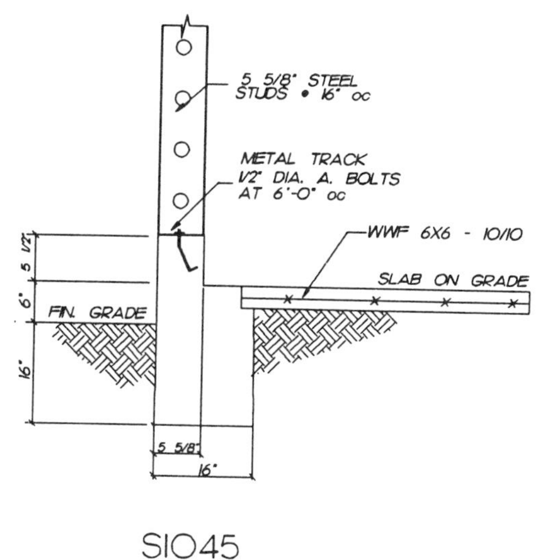

SIO45

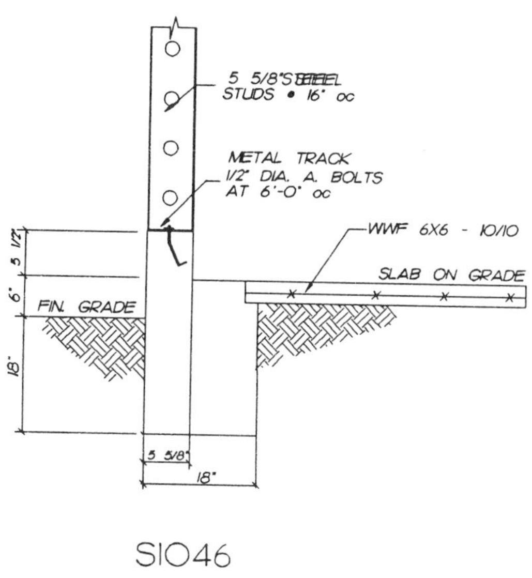

SIO46

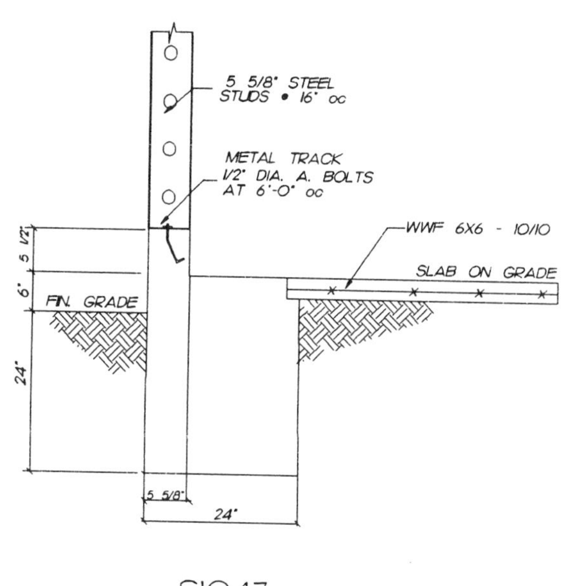

SIO47

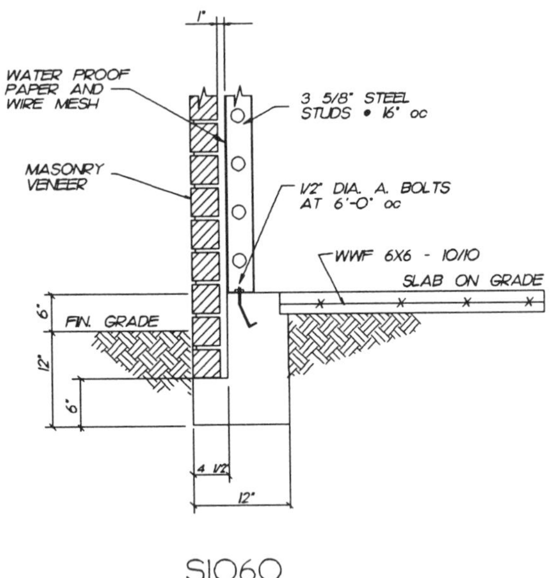

S1060

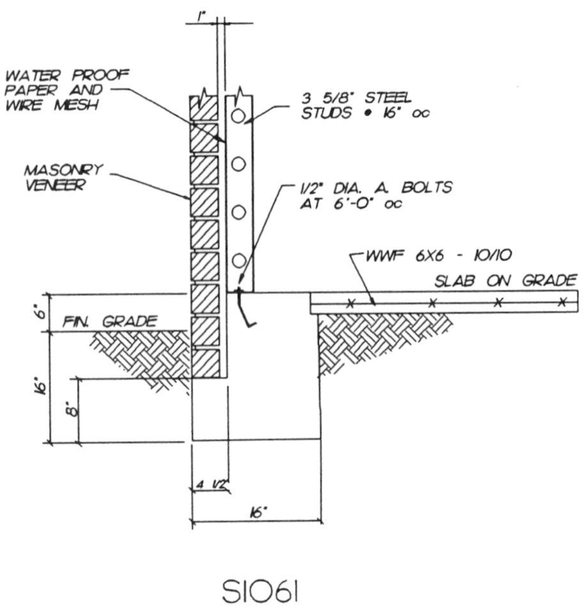

S1061

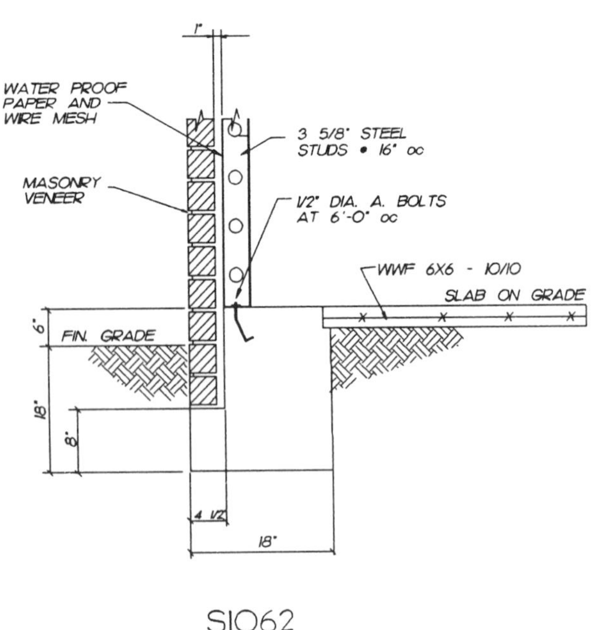

S1062

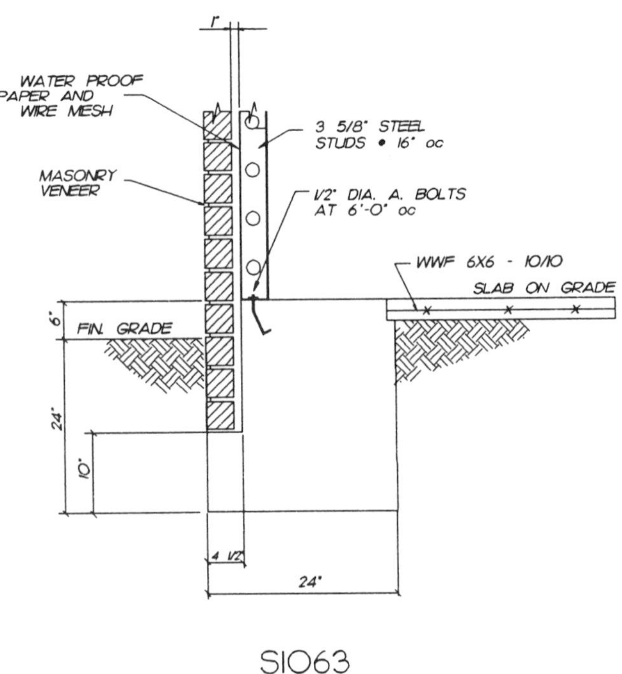

S1063

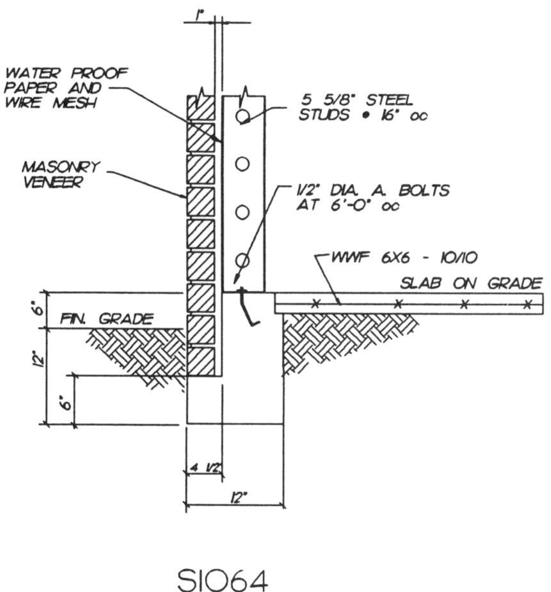

S1064

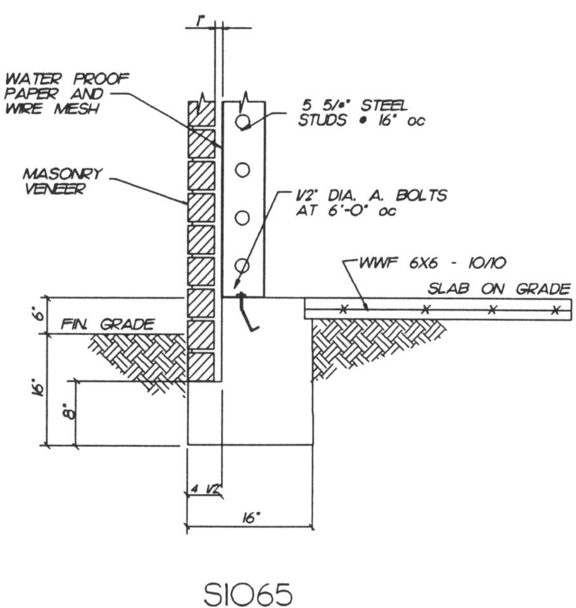

S1065

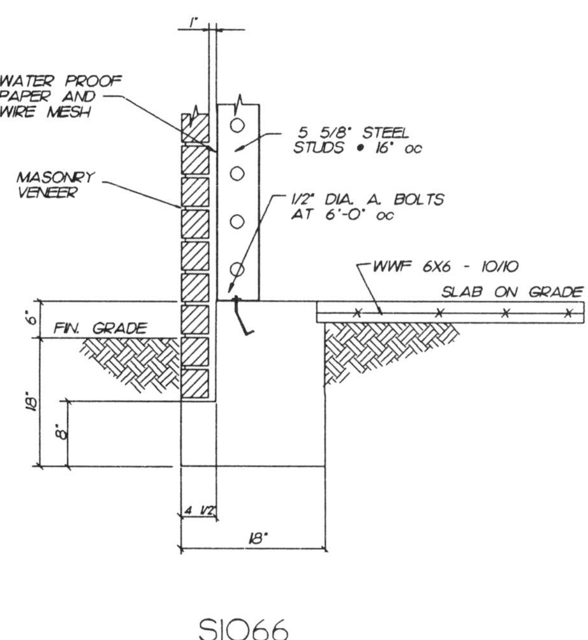

S1066

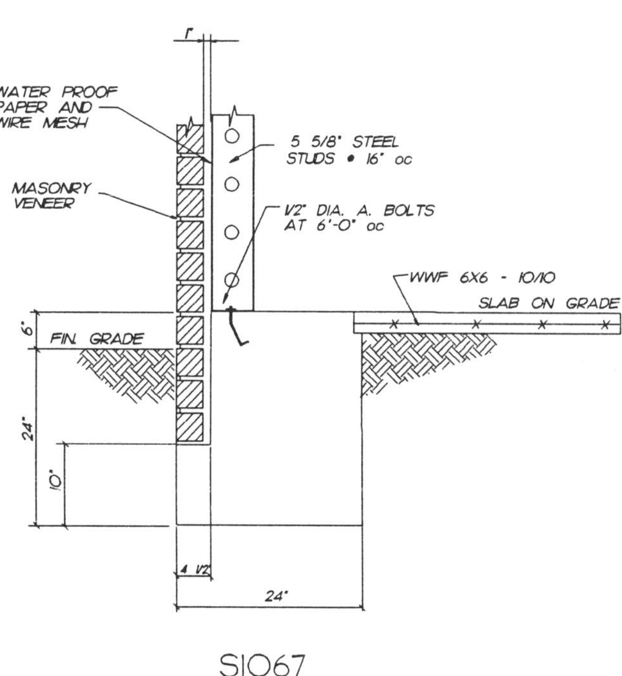

S1067

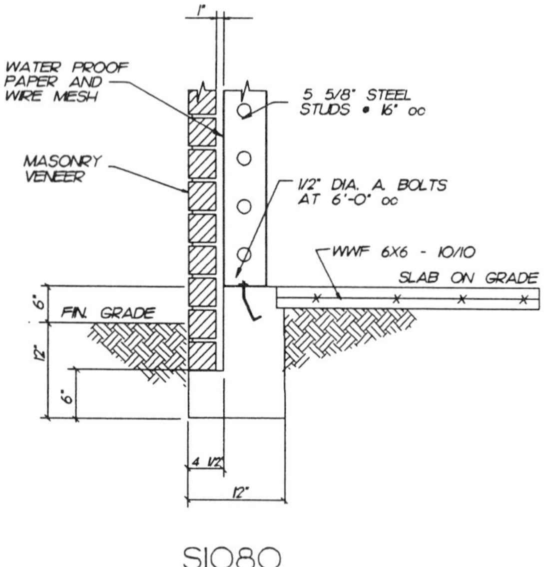

S1080

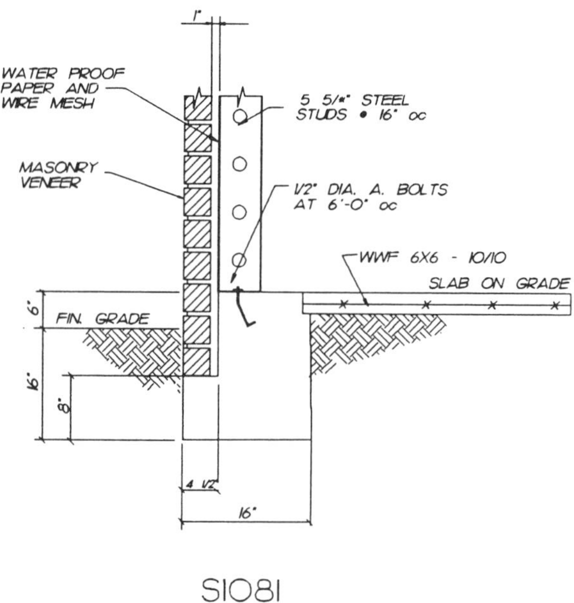

S1081

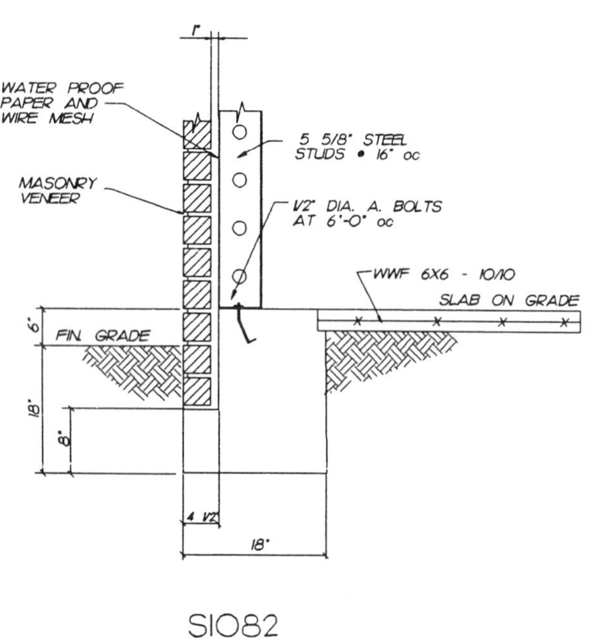

S1082

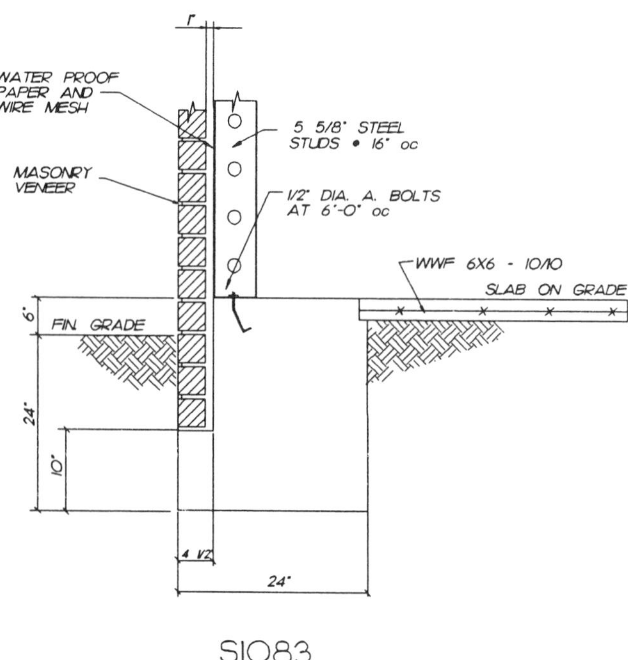

S1083

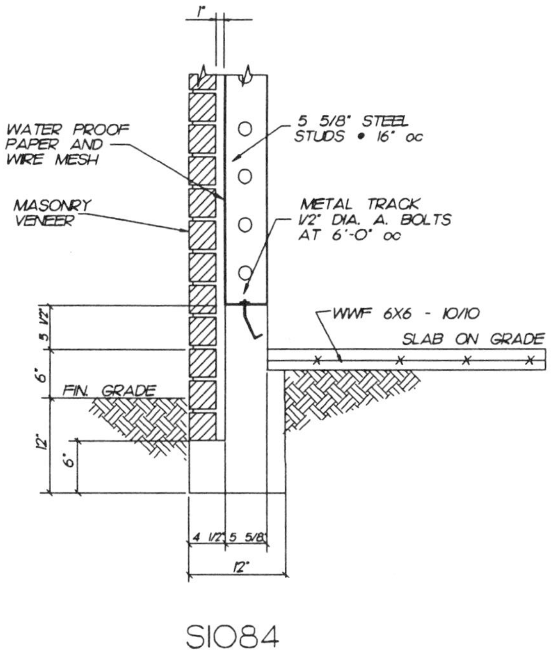

S1084

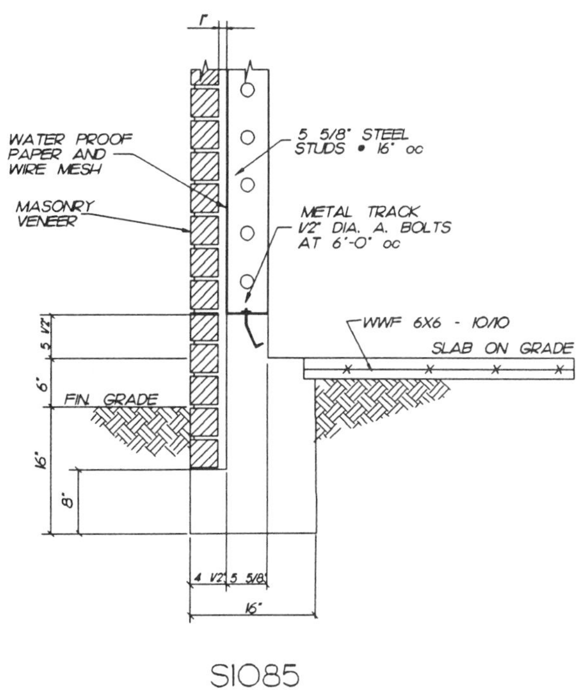

S1085

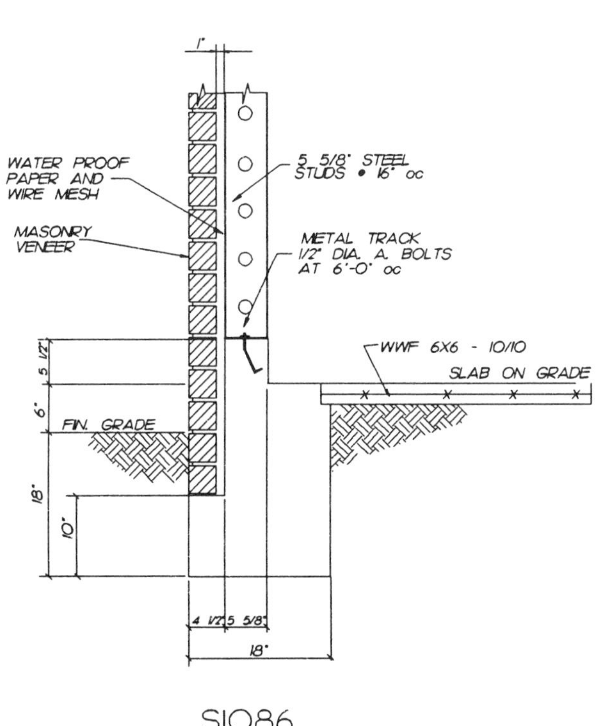

S1086

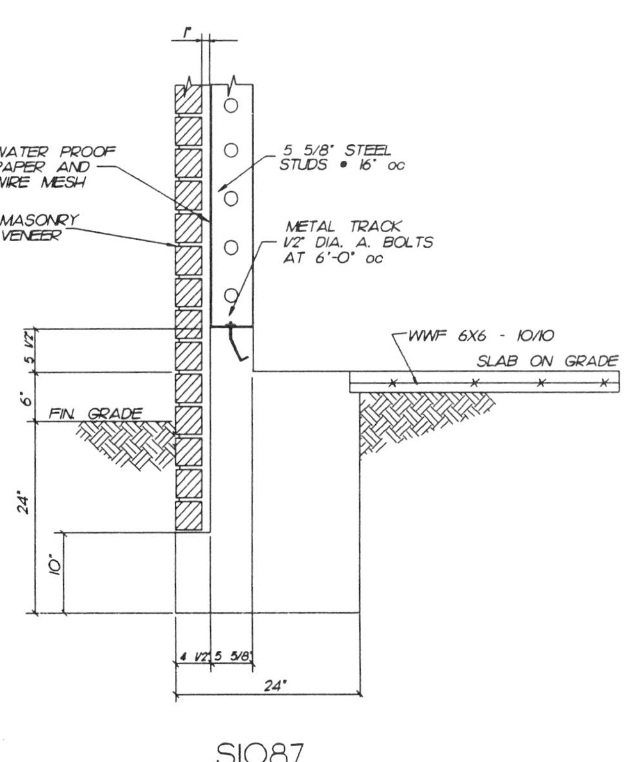

S1087

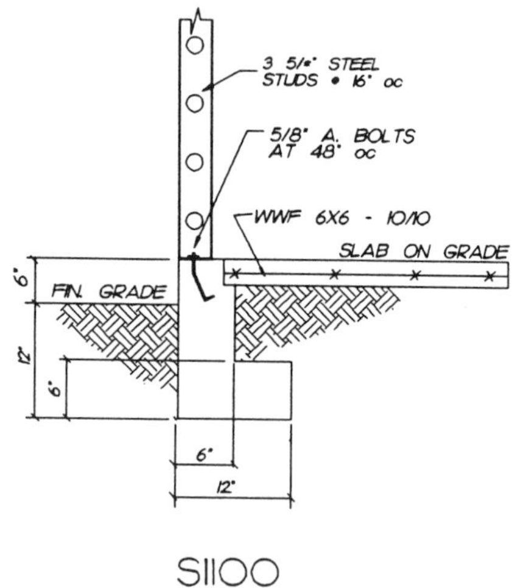

S1100

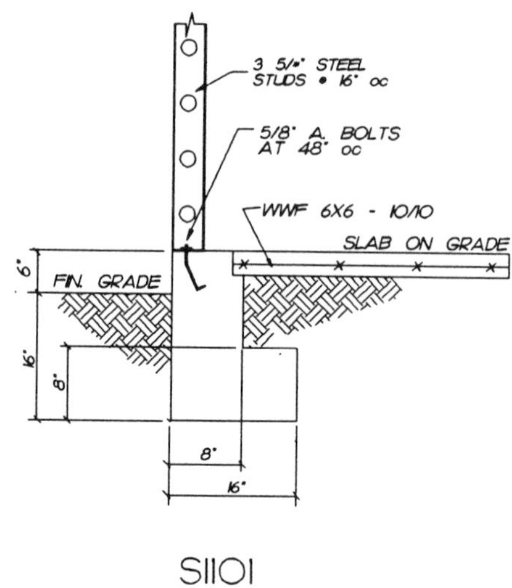

S1101

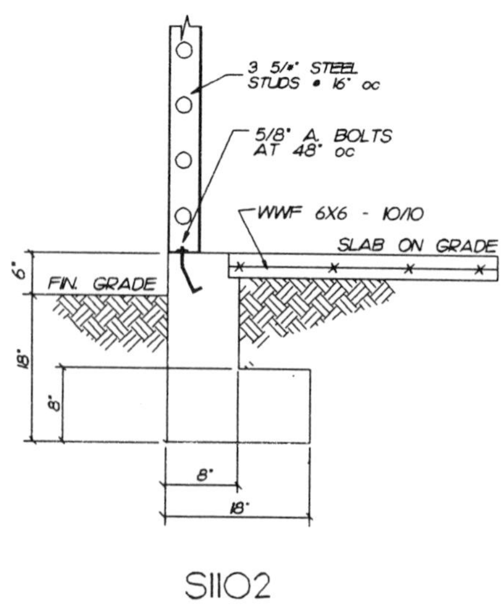

S1102

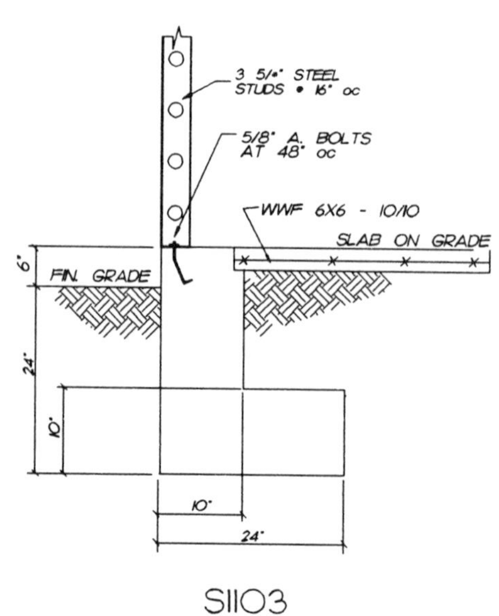

S1103

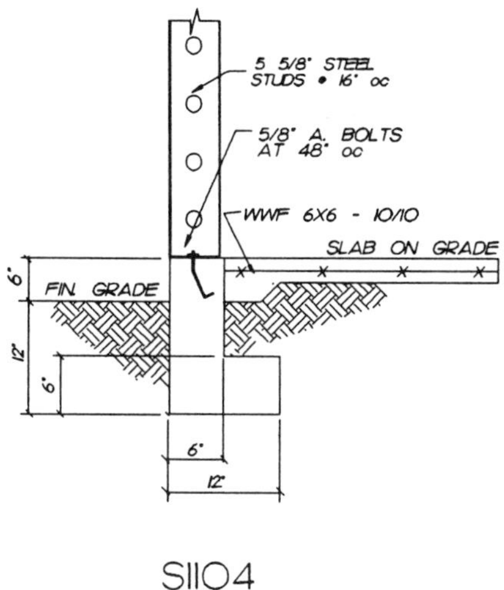

SI104

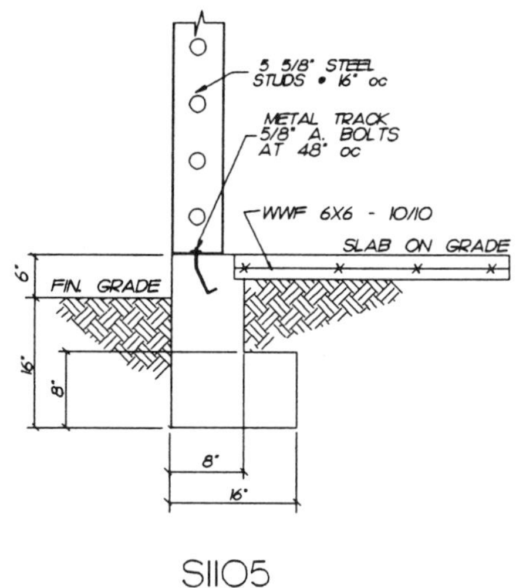

SI105

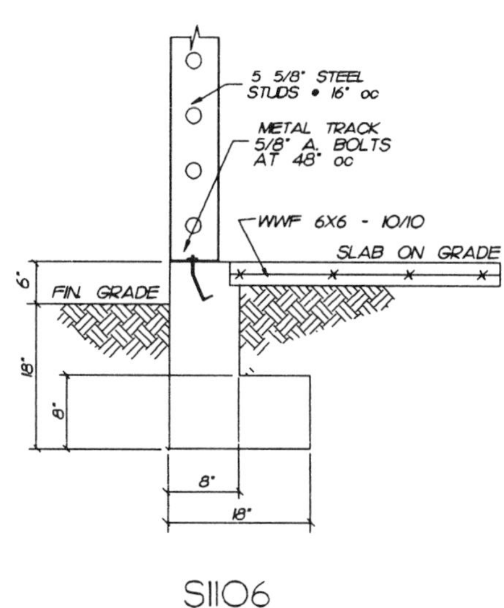

SI106

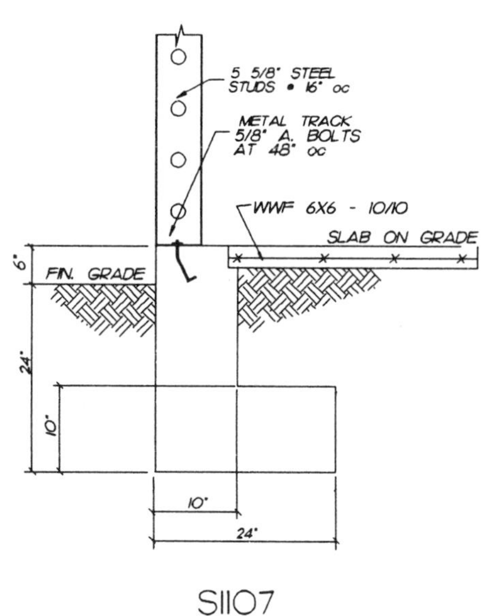

SI107

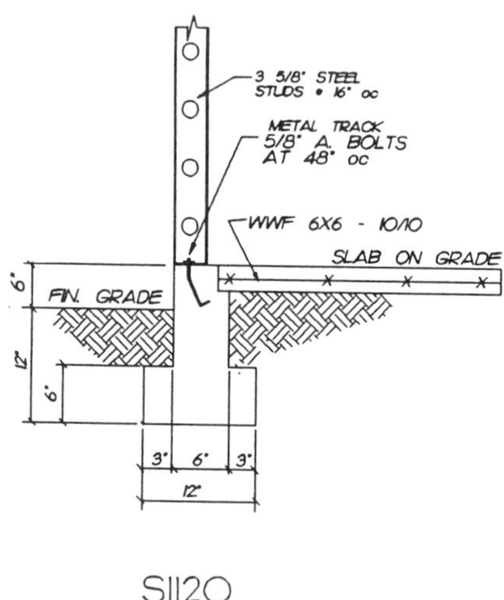

S1120

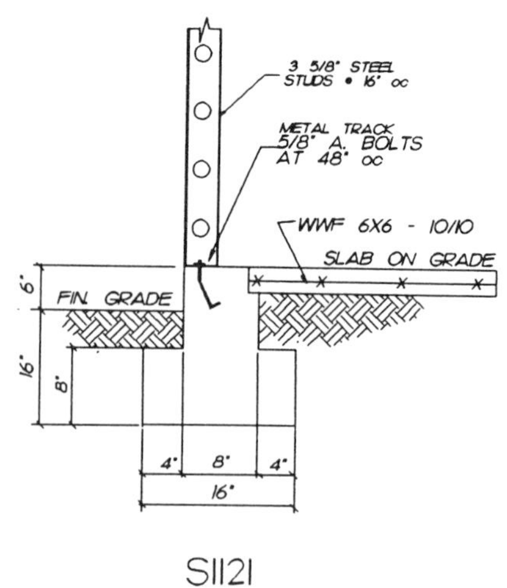

S1121

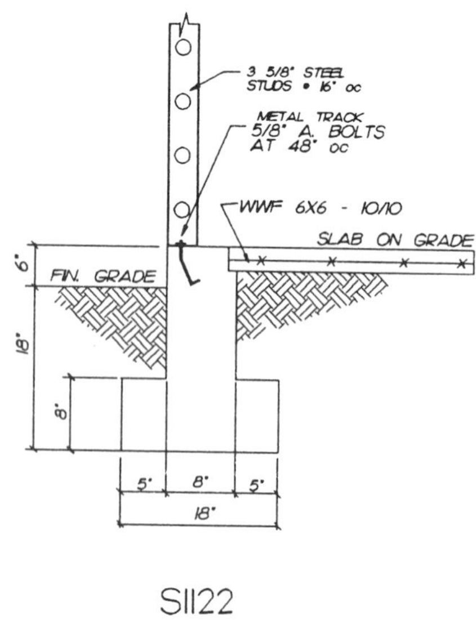

S1122

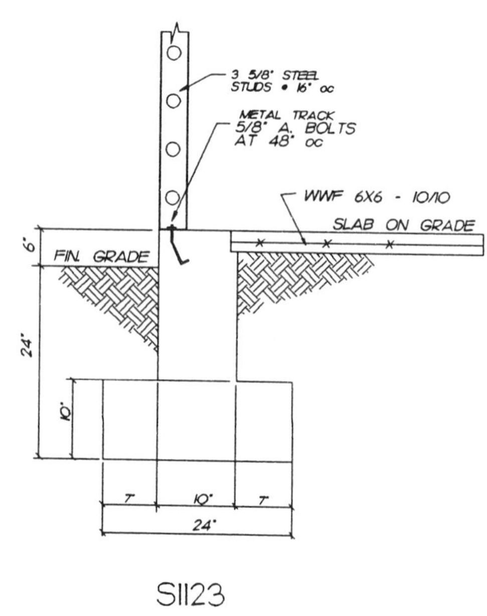

S1123

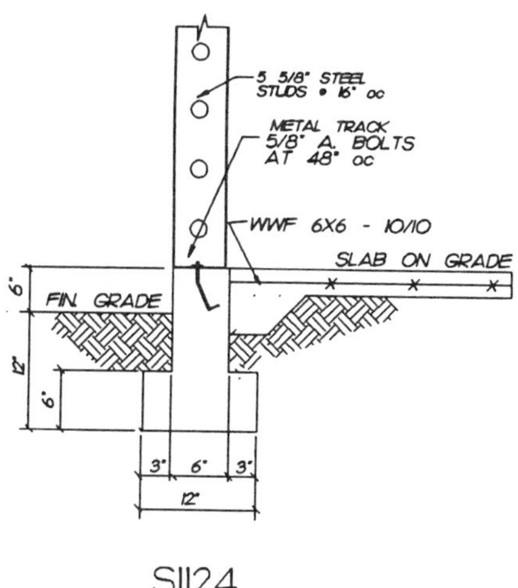

S1124

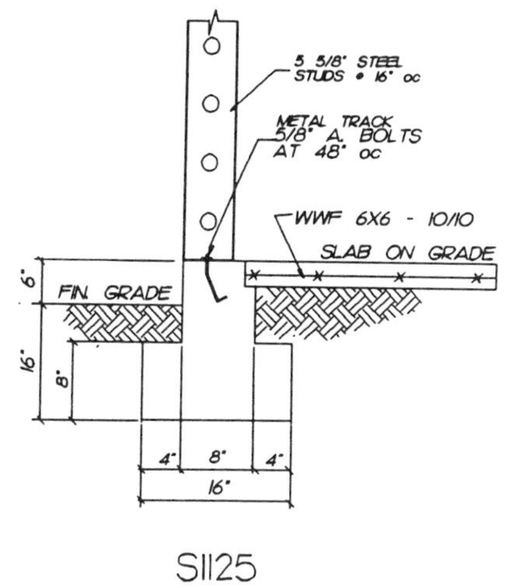

S1125

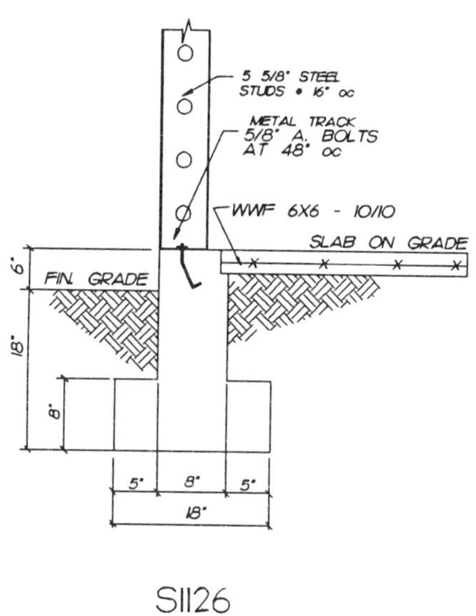

S1126

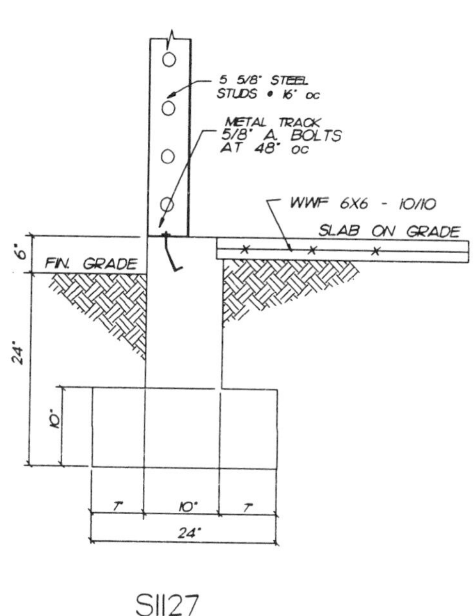

S1127

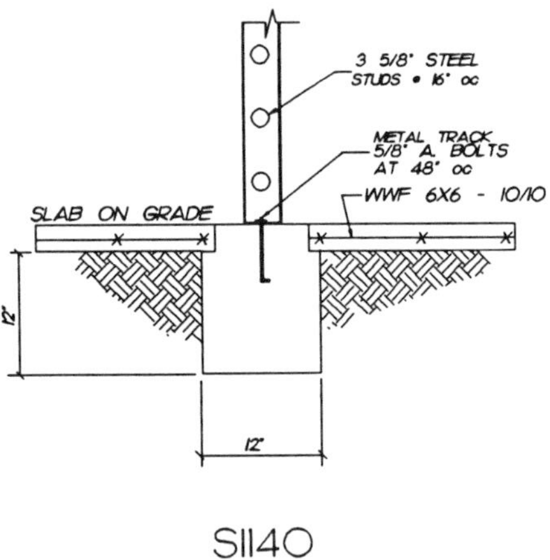

S1140

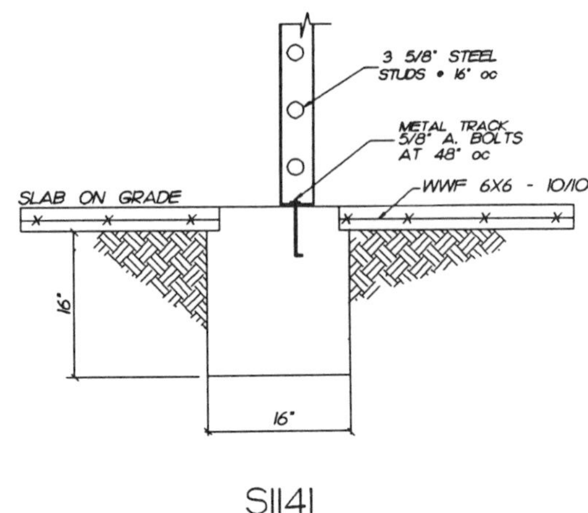

S1141

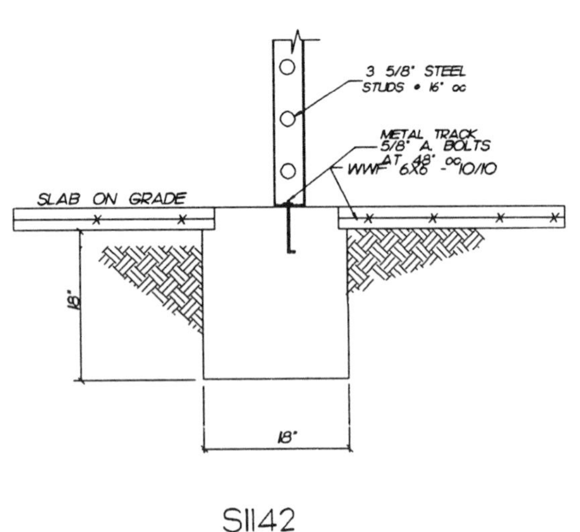

S1142

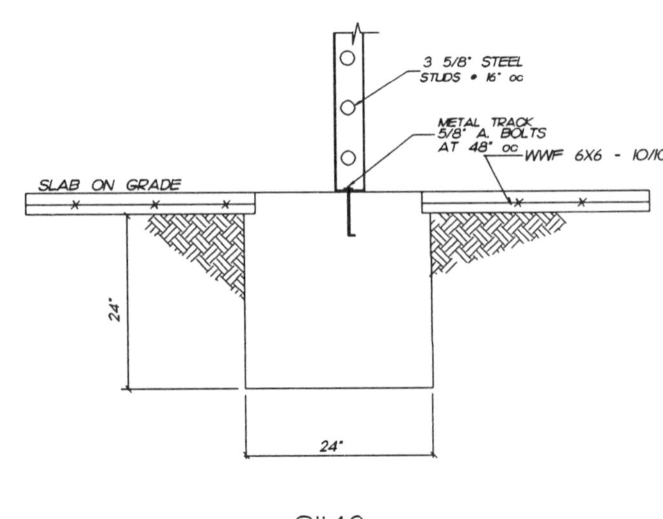

S1143

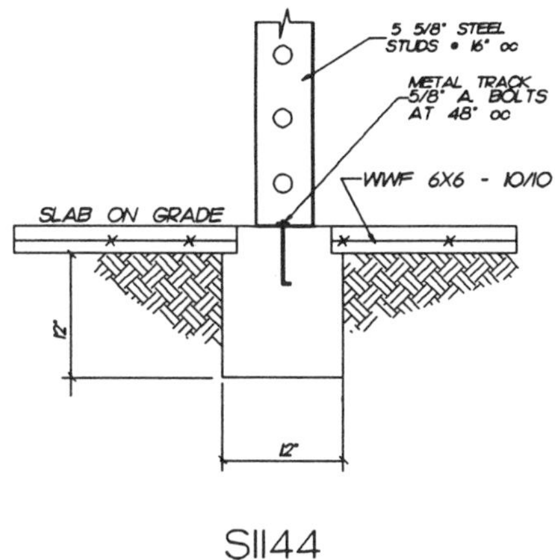

S1144

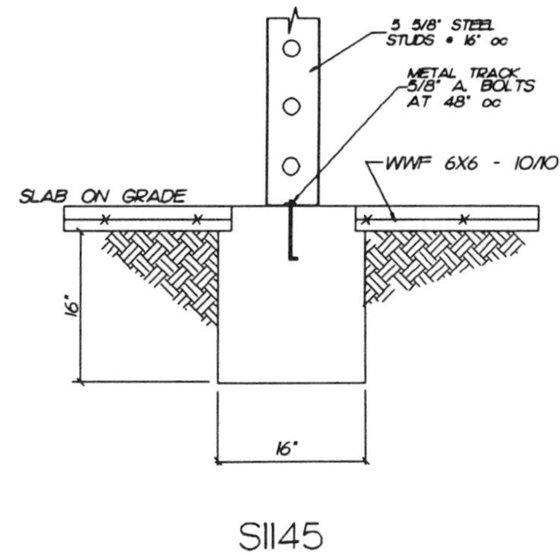

S1145

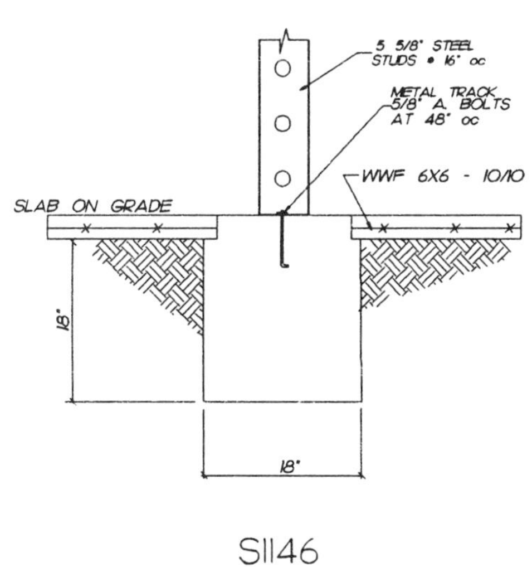

S1146

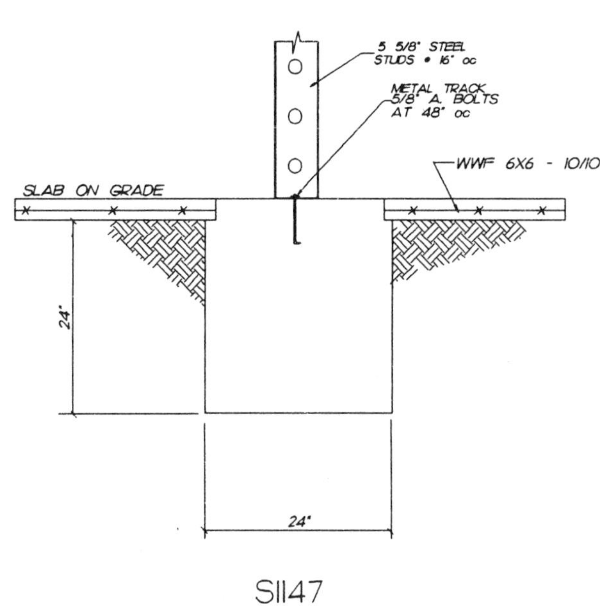

S1147

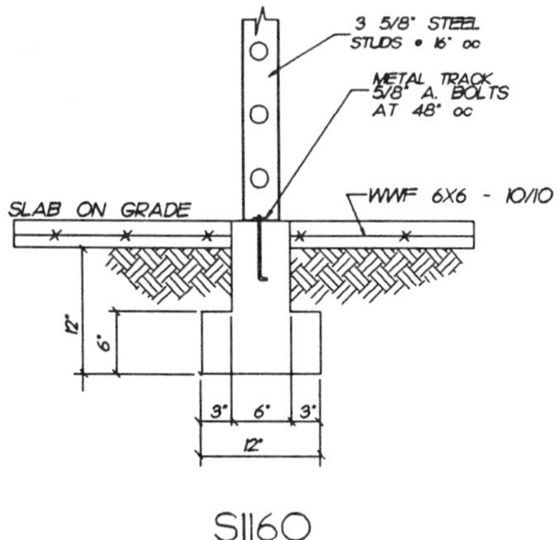

S1160

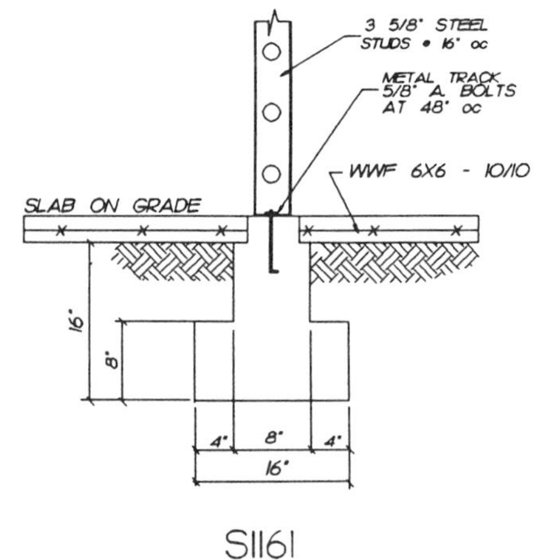

S1161

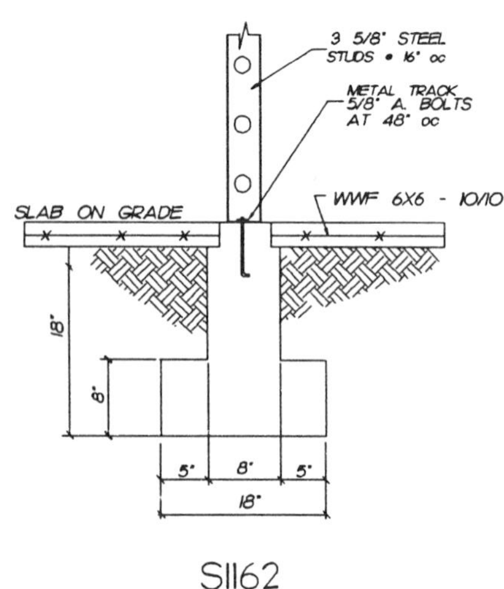

S1162

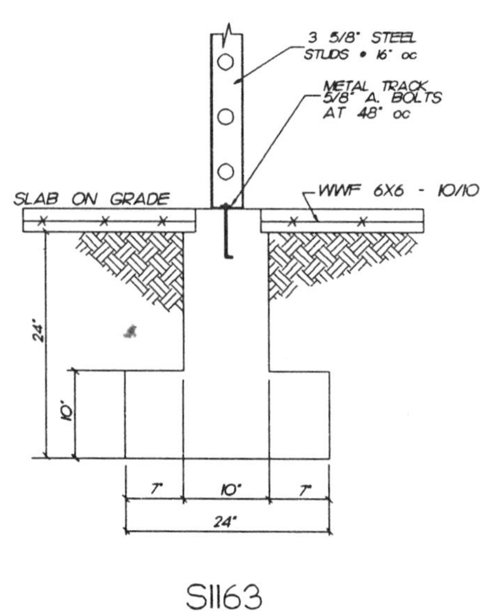

S1163

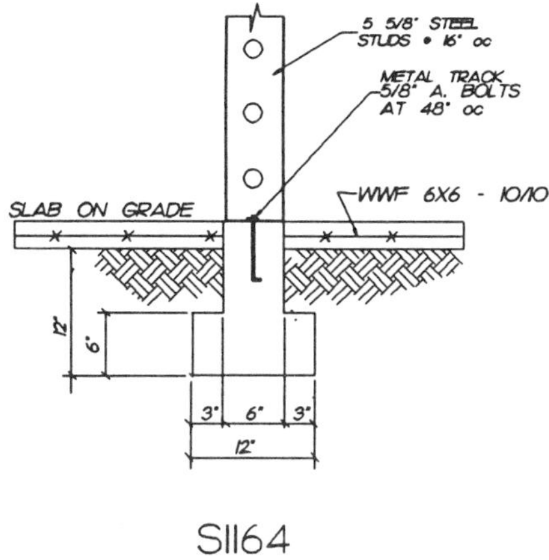

S1164

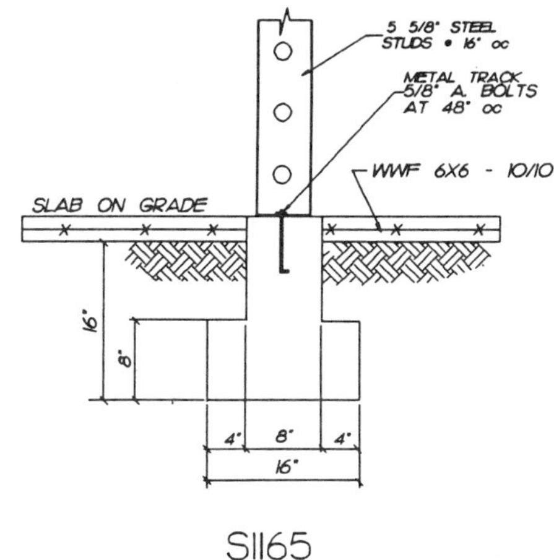

S1165

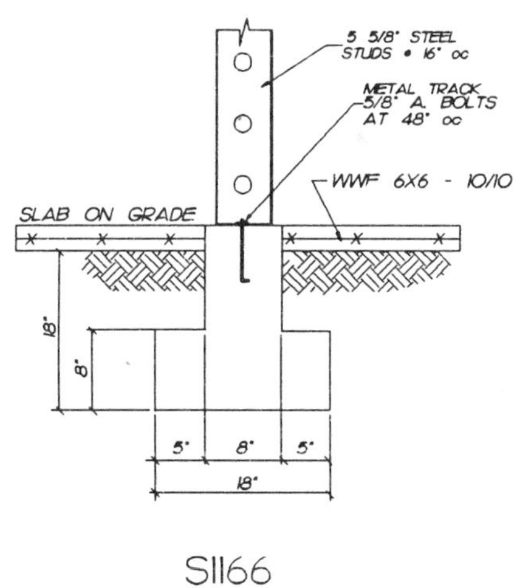

S1166

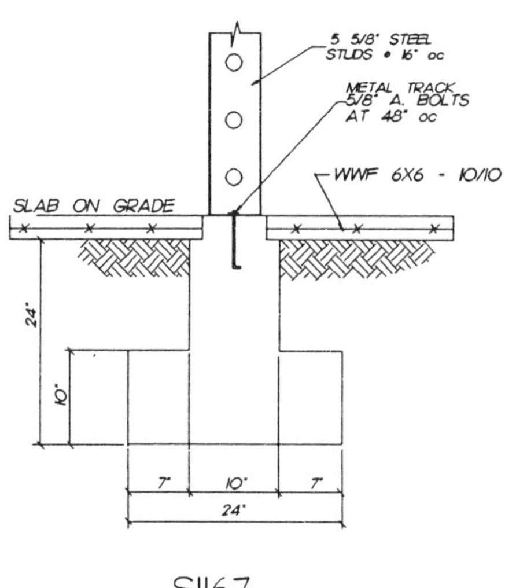

S1167

CHAPTER 4	STEEL SECTION 2	COLUMNS
BASE PLATE - SQUARE	PIPE COLUMNS & SQ. TUBES	S2020 - S2022
	WIDE FLANGE COLUMNS	S2023 - S2027
BASE PLATE - RECTANGULAR	WIDE FLANGE COLUMNS	S2040 - S2044
	PIPE COLUMNS & SQ. TUBES	S2060 - S2083
COLUMN SPLICE BEARING JOINT WELDED	PIPE COLUMNS & SQ. TUBES	S2100 - S2102
	WIDE FLANGE COLUMNS	S2120 - S2122
COLUMN SPLICE BEARING JOINT SPLICE PLATES HIGH STRENGTH BOLTS	WIDE FLANGE COLUMNS	S2140 - S2143
BEAM SUPPORT OF A COLUMN	WOOD POST	S2160
	PIPE COLUMNS	S2161 - S2163
	WIDE FLANGE COLUMNS	S2164 - S2167
ONE LEVEL END COLUMN END OF WOOD BEAM	PIPE COLUMNS	S2180
	WIDE FLANGE COLUMNS	S2181 - S2182
CONTINUOUS END COLUMN END OF WOOD BEAM	PIPE COLUMNS & SQ. TUBES	S2200
	WIDE FLANGE COLUMNS	S2201 - S2203
ONE LEVEL INTERIOR COLUMN WOOD BEAM	PIPE COLUMNS & WIDE FLANGES	S2220 - S2226
CONTINUOUS COLUMN INTERIOR WOOD BEAM	PIPE COLUMNS	S2240
ONE LEVEL END COLUMN STEEL BEAM	PIPE COLUMNS	S2260
	WIDE FLANGE COLUMNS	S2261
CONTINUOUS COLUMN STEEL BEAM BOLTED CONNECTION	PIPE COLUMNS	S2280
	STEEL TUBE	S2281
CONTINUOUS COLUMN END STEEL BEAM BOLTED CONNECTION SEAT ANGLE	WIDE FLANGE COLUMNS MACHINE BOLTS	S2300 - S2308
	WIDE FLANGE COLUMNS HIGH STREGNTH BOLTS	S2320 - S2327
END OF BEAM SUPPORT STEEL BEAM PIPE COLUMN ABOVE & BELOW	————	S2340 - S2344
CANTILEVER BEAM SUPPORT STEEL BEAM	PIPE COLUMN	S2360
	WIDE FLANGE COLUMN	S2361
CONTINUOUS BEAM SUPPORT STEEL BEAM	PIPE COLUMN	S2380 - S2384
TWO BEAMS SUPPORT STEEL BEAMS	PIPE COLUMN	S2400
TWO BEAMS SUPPORT STEEL BEAMS - UNEQUAL SIZE	PIPE COLUMN	S2420 - S2443
		NEXT PAGE

CHAPTER 4	STEEL SECTION 2	COLUMNS
END COLUMN RIGID CONNECTION STEEL BEAM - ONE SIDE	HIGH STRENGTH BOLTS	S2460 - S2463
	WELDED CONNECTION	S2480 - S2483
END CONTINUOUS COLUMN RIGID CONNECTION STEEL BEAM - ONE SIDE	WELDED CONNECTION	S2500 - S2505
	————————	
INTERIOR COLUMN RIGID CONNECTION STEEL BEAM EACH SIDE	HIGH STRENGTH BOLTS	S2520 - S2523
	WELDED CONNECTION	S2540 - S2543
COLUMN FIRE PROTECTION	————————	S2560 - S2561

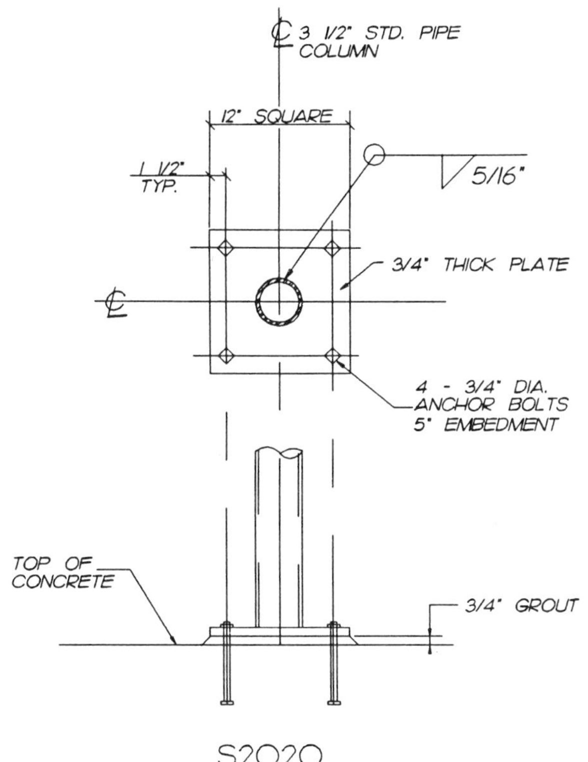

S2020

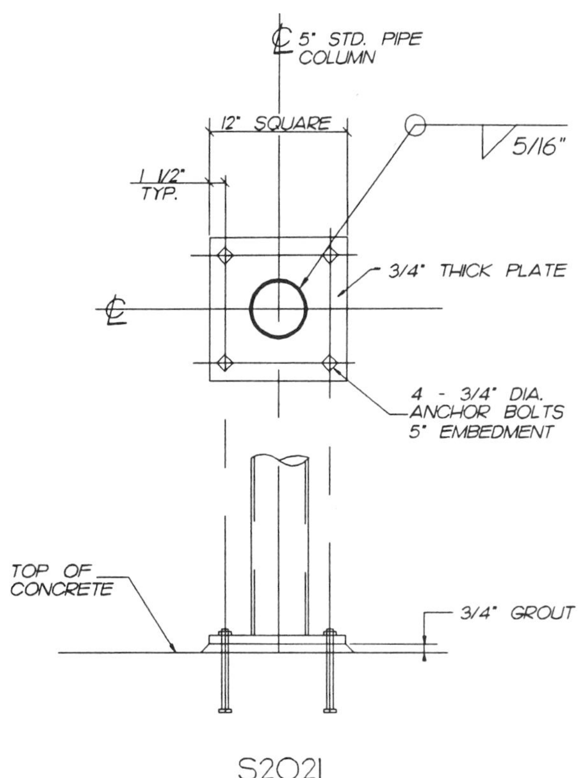

S2021

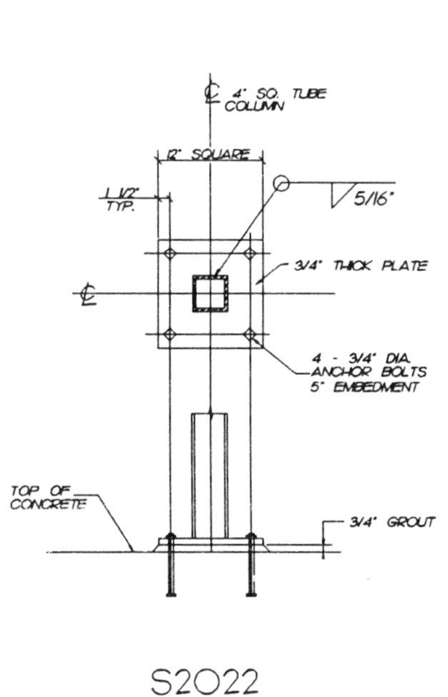

S2022

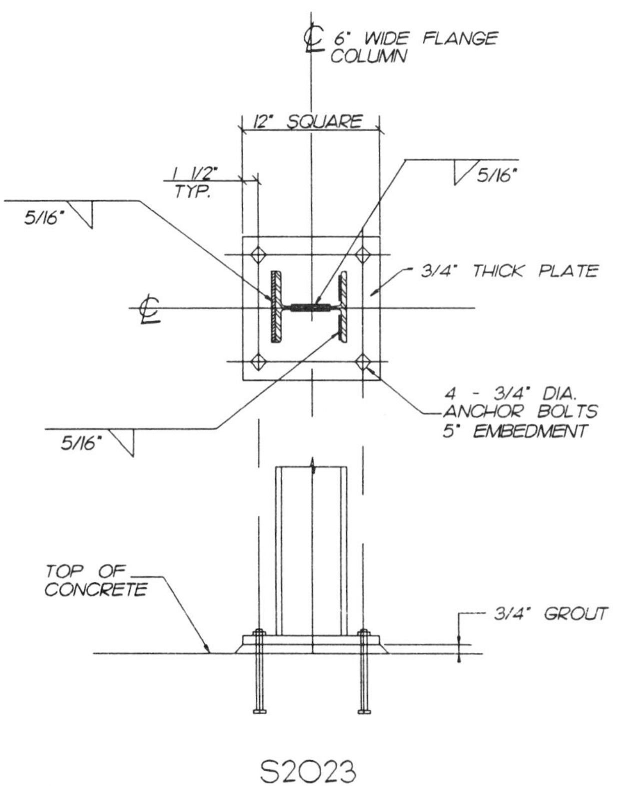

S2023

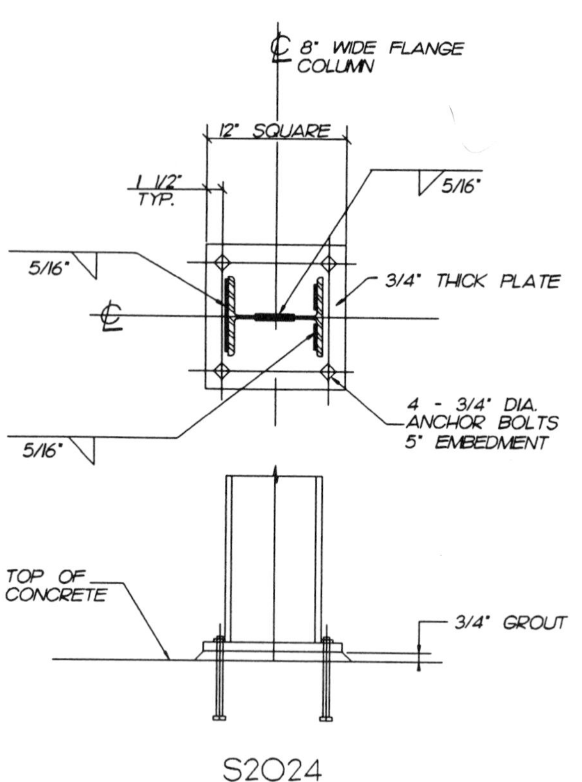

S2O24

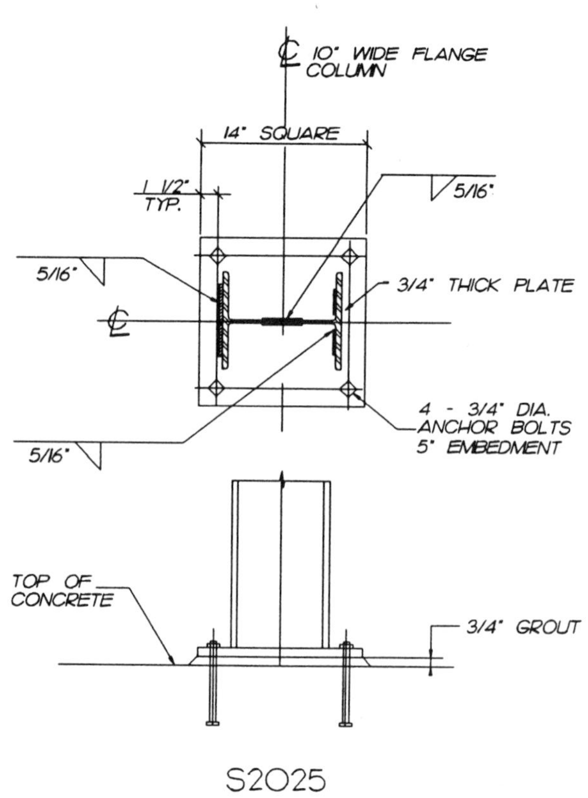

S2O25

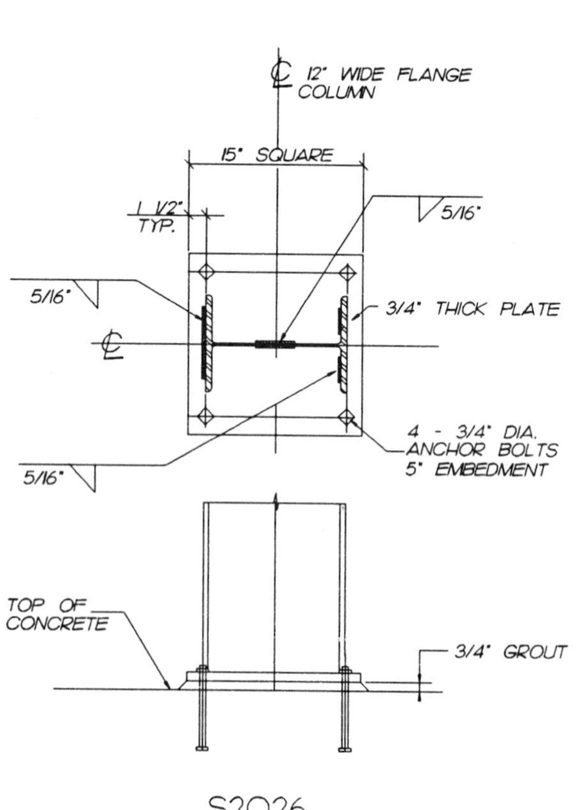

S2O26

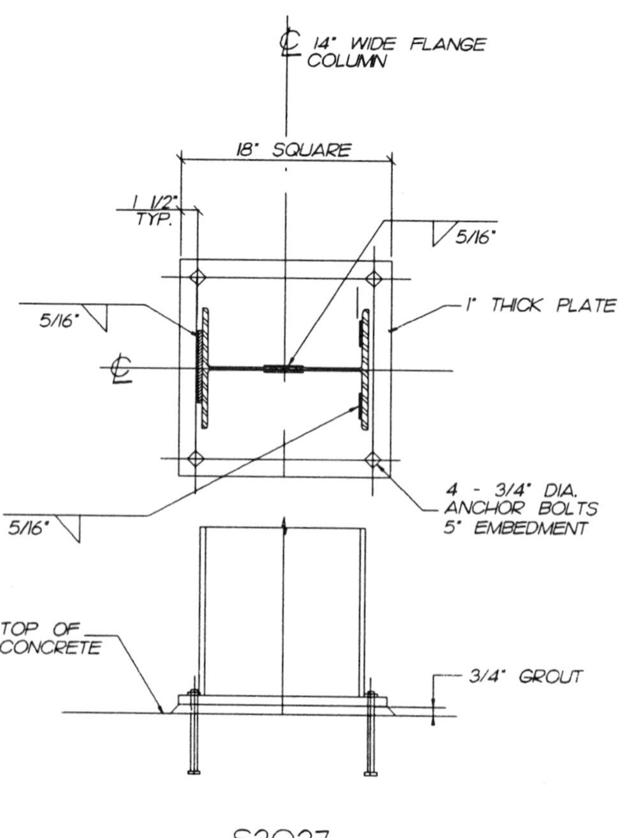

S2O27

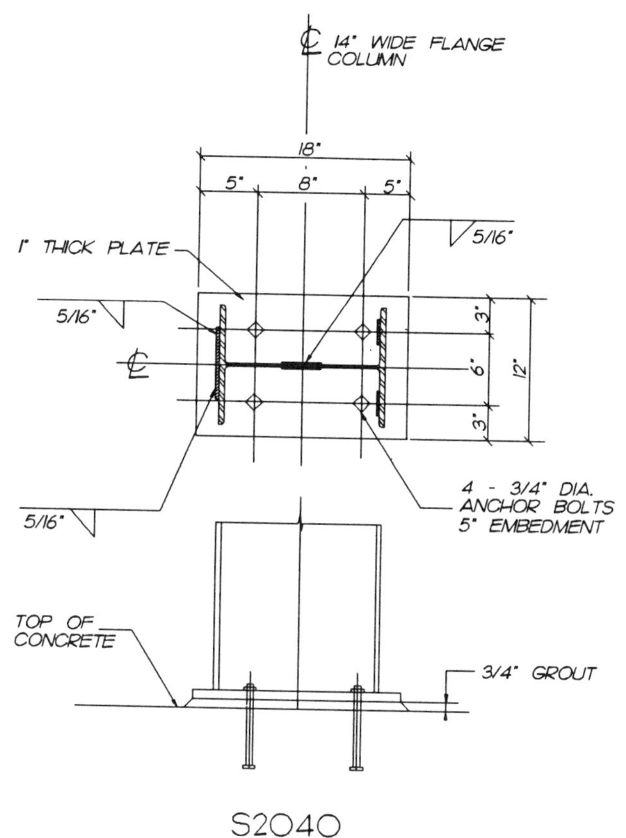

S2040

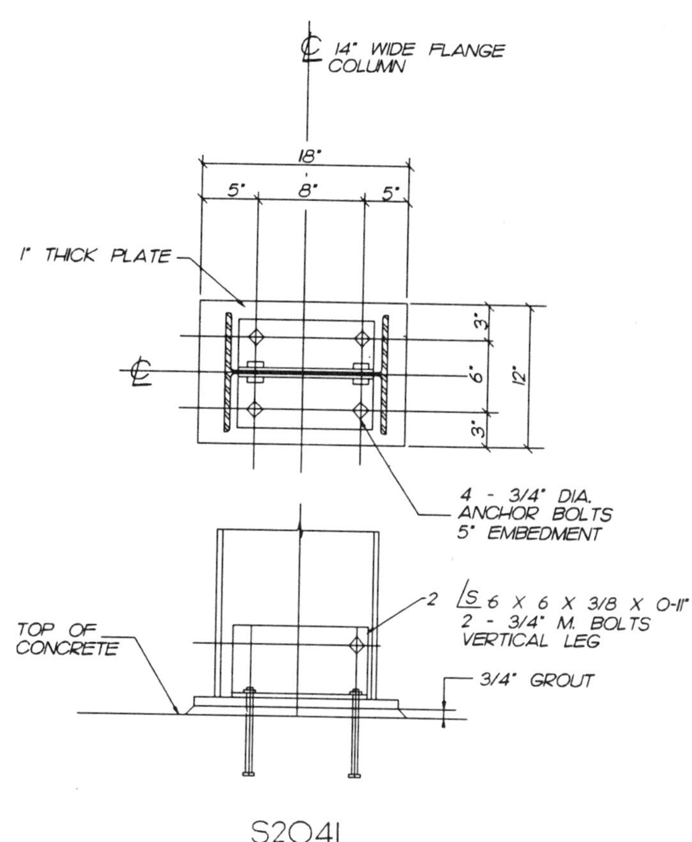

S2041

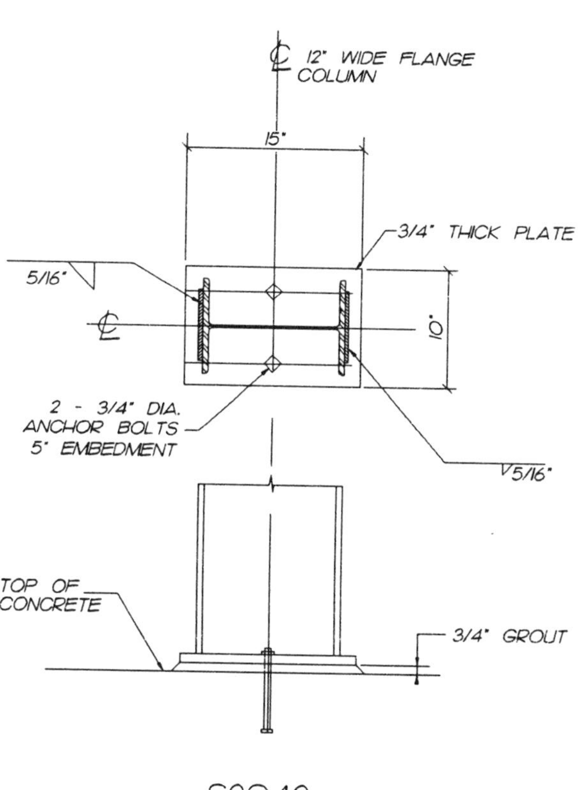

S2042

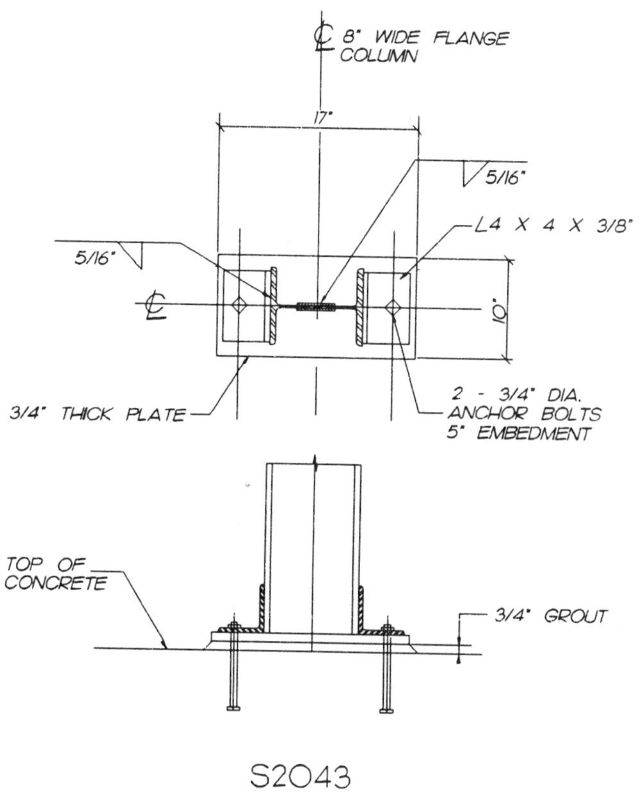

S2043

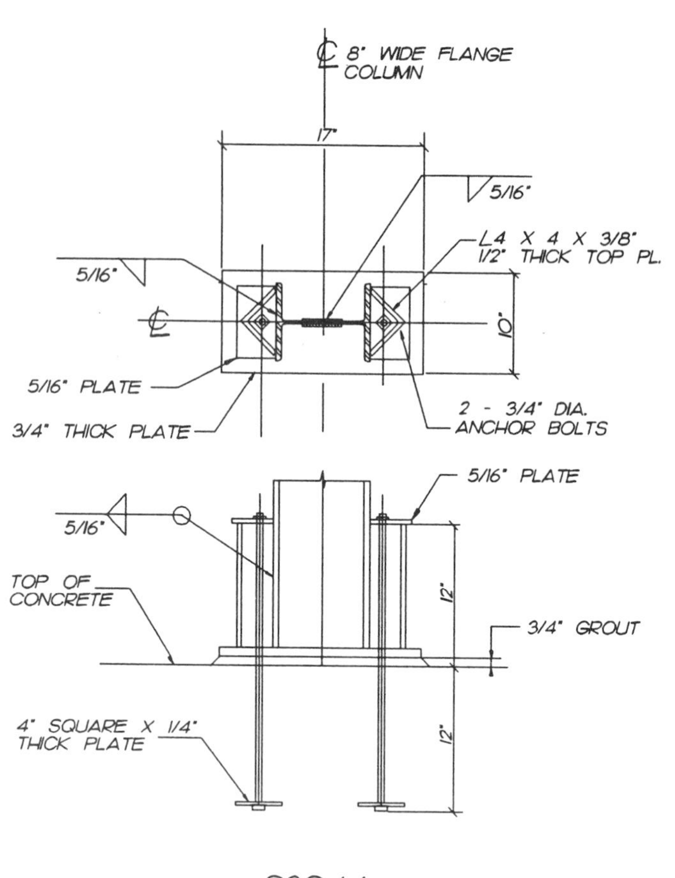

8" WIDE FLANGE COLUMN
17"
5/16"
L4 X 4 X 3/8"
1/2" THICK TOP PL.
5/16"
10"
5/16" PLATE
3/4" THICK PLATE
2 - 3/4" DIA.
ANCHOR BOLTS
5/16" PLATE
5/16"
TOP OF CONCRETE
12"
3/4" GROUT
4" SQUARE X 1/4" THICK PLATE
12"

S2044

3" SQ. TUBE COLUMN
12"
1 1/2" TYP.
5/16"
3/4" THICK PLATE
WALL
3 1/2"
2 - 3/4" DIA.
ANCHOR BOLTS
5" EMBEDMENT
SYM.
TOP OF CONCRETE
3/4" GROUT

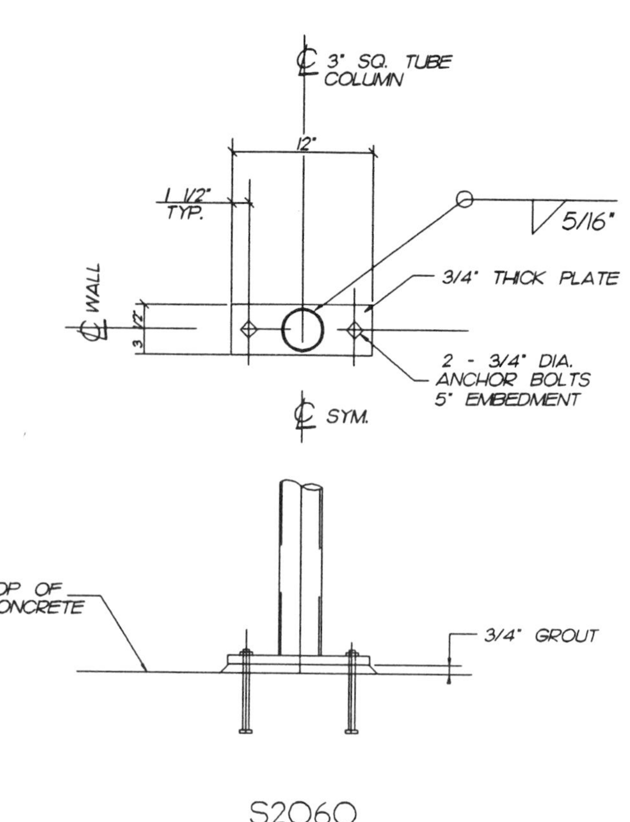

S2060

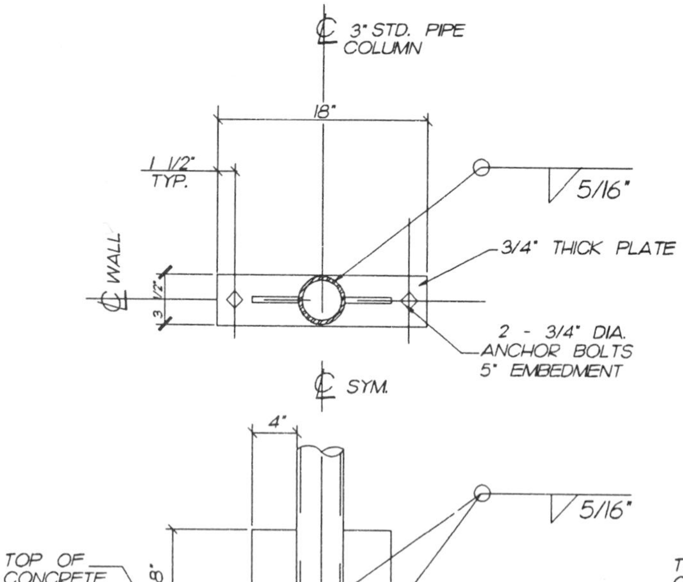

3" STD. PIPE COLUMN
18"
1 1/2" TYP.
5/16"
3/4" THICK PLATE
WALL
3 1/2"
2 - 3/4" DIA.
ANCHOR BOLTS
5" EMBEDMENT
SYM.
4"
5/16"
TOP OF CONCRETE
8"
3/4" GROUT

S2061

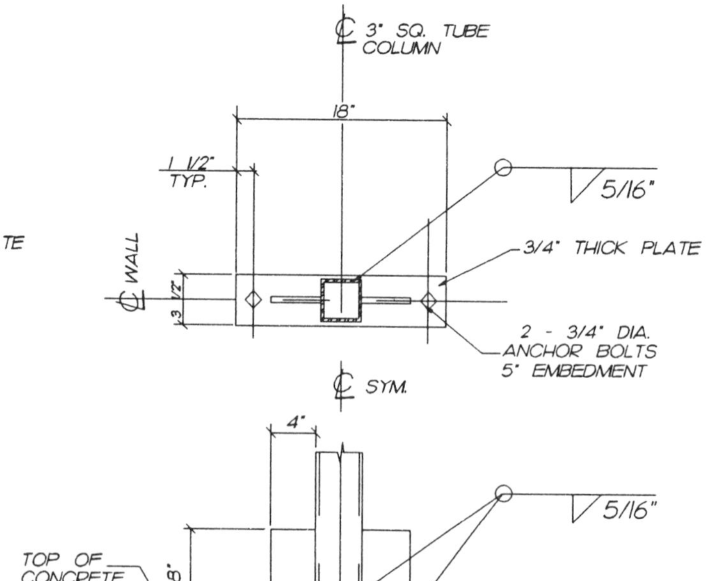

3" SQ. TUBE COLUMN
18"
1 1/2" TYP.
5/16"
3/4" THICK PLATE
WALL
3 1/2"
2 - 3/4" DIA.
ANCHOR BOLTS
5" EMBEDMENT
SYM.
4"
5/16"
TOP OF CONCRETE
8"
3/4" GROUT

S2062

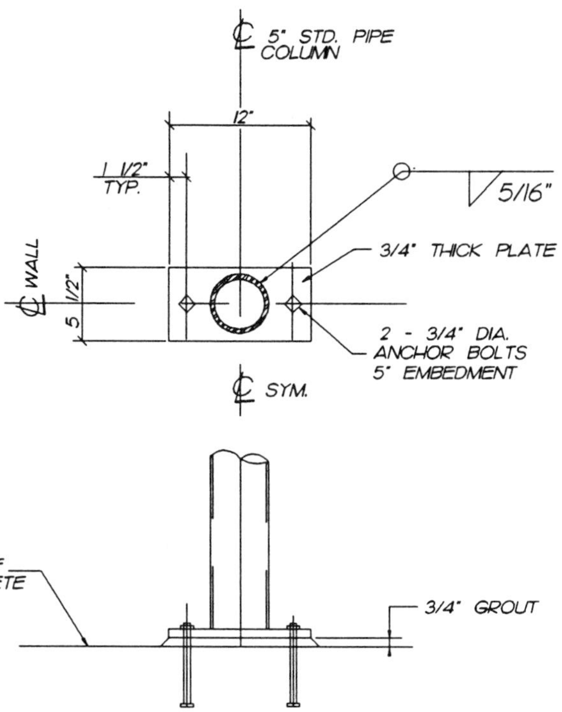

℄ 5" STD. PIPE COLUMN
12"
1 1/2" TYP.
5/16"
3/4" THICK PLATE
℄ WALL
5 1/2"
2 - 3/4" DIA. ANCHOR BOLTS 5" EMBEDMENT
℄ SYM.
TOP OF CONCRETE
3/4" GROUT

S2080

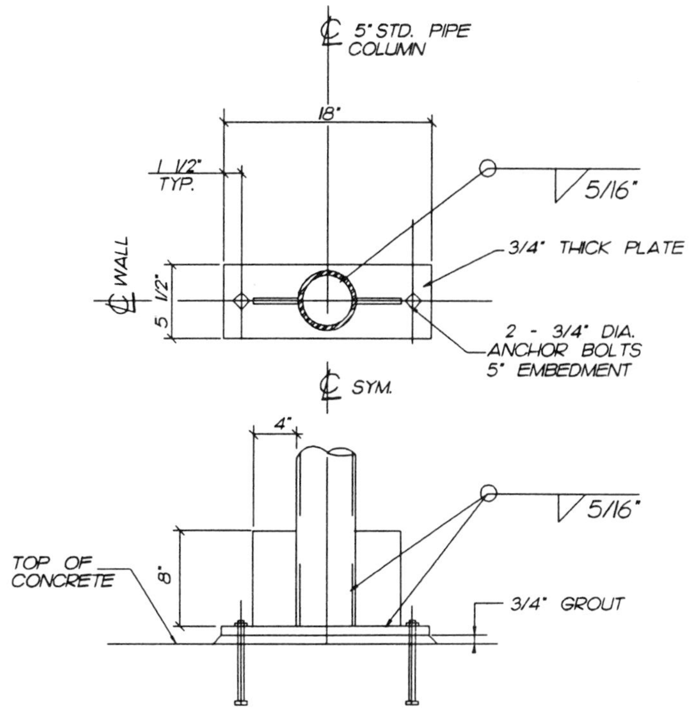

℄ 5" STD. PIPE COLUMN
18"
1 1/2" TYP.
5/16"
3/4" THICK PLATE
℄ WALL
5 1/2"
2 - 3/4" DIA. ANCHOR BOLTS 5" EMBEDMENT
℄ SYM.
4"
5/16"
8"
TOP OF CONCRETE
3/4" GROUT

S2081

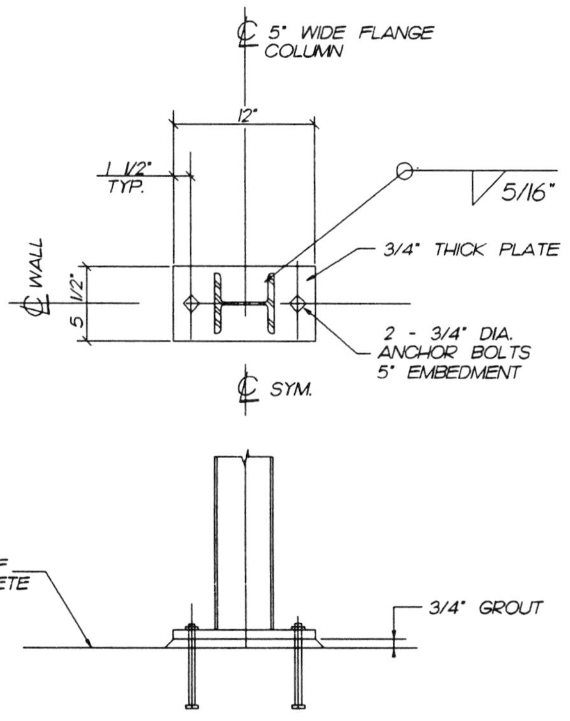

℄ 5" WIDE FLANGE COLUMN
12"
1 1/2" TYP.
5/16"
3/4" THICK PLATE
℄ WALL
5 1/2"
2 - 3/4" DIA. ANCHOR BOLTS 5" EMBEDMENT
℄ SYM.
TOP OF CONCRETE
3/4" GROUT

S2082

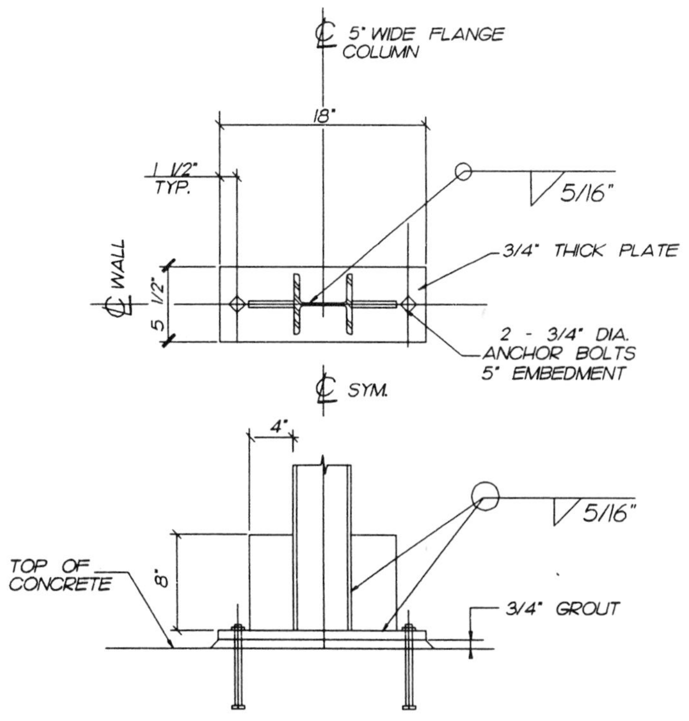

℄ 5" WIDE FLANGE COLUMN
18"
1 1/2" TYP.
5/16"
3/4" THICK PLATE
℄ WALL
5 1/2"
2 - 3/4" DIA. ANCHOR BOLTS 5" EMBEDMENT
℄ SYM.
4"
5/16"
8"
TOP OF CONCRETE
3/4" GROUT

S2083

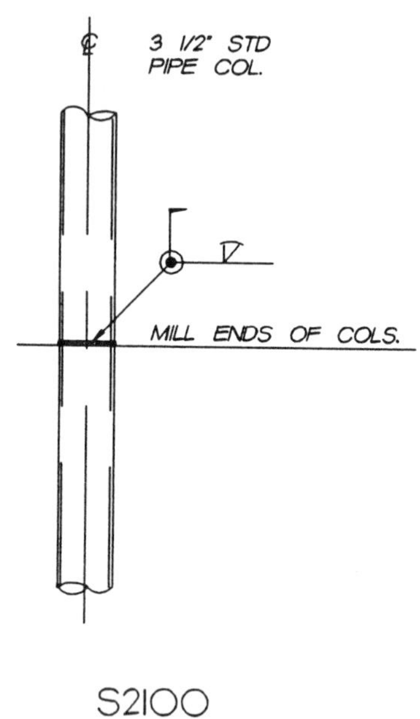

S2100

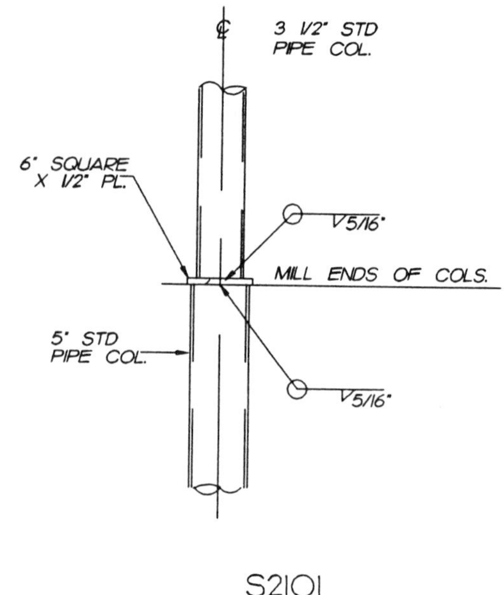

S2101

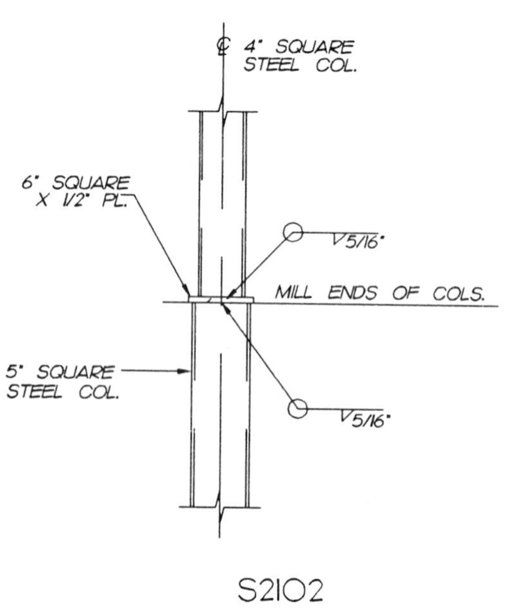

S2102

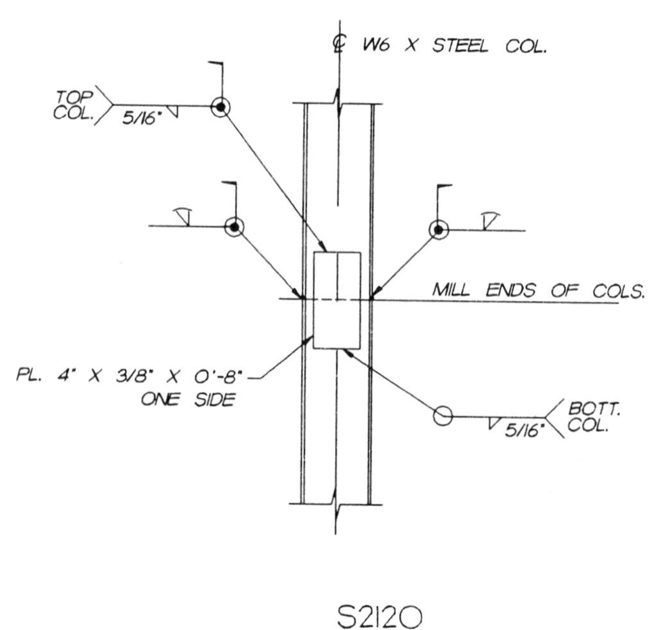

S2120

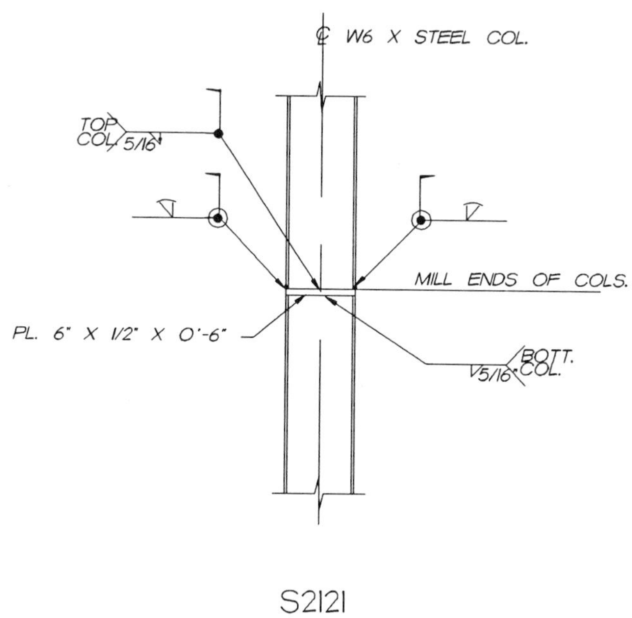

S2121

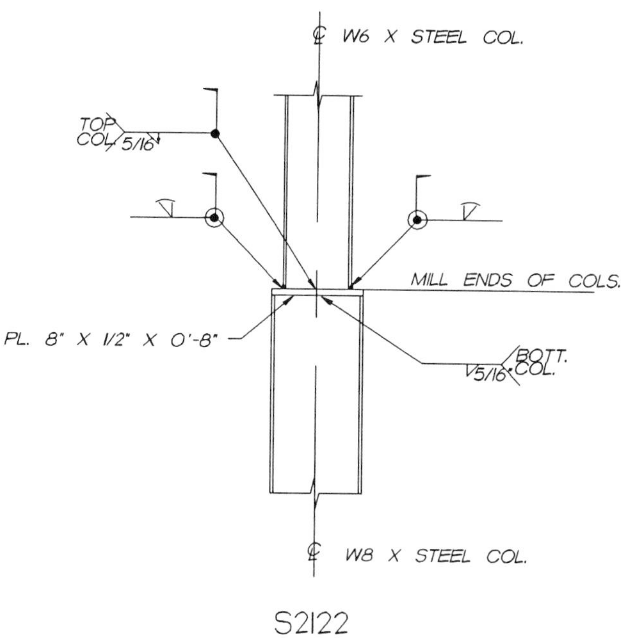

S2122

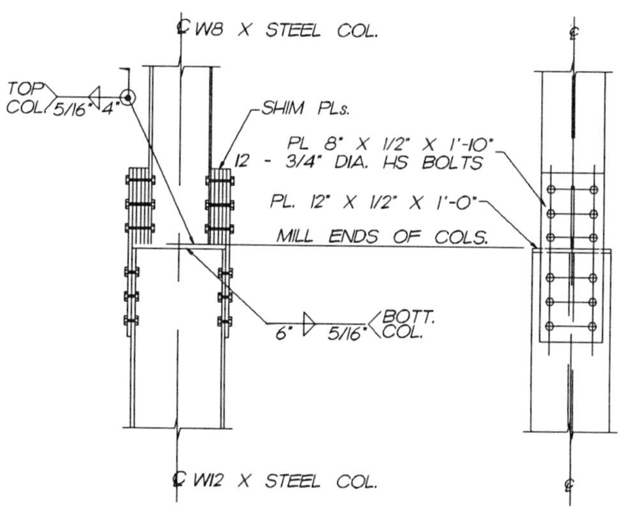

S2140

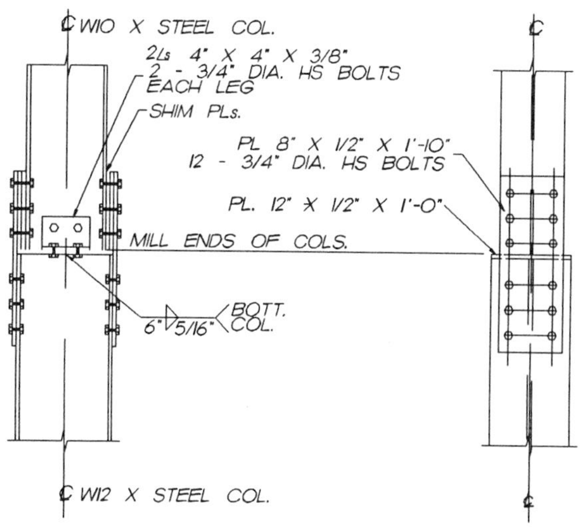

S2141

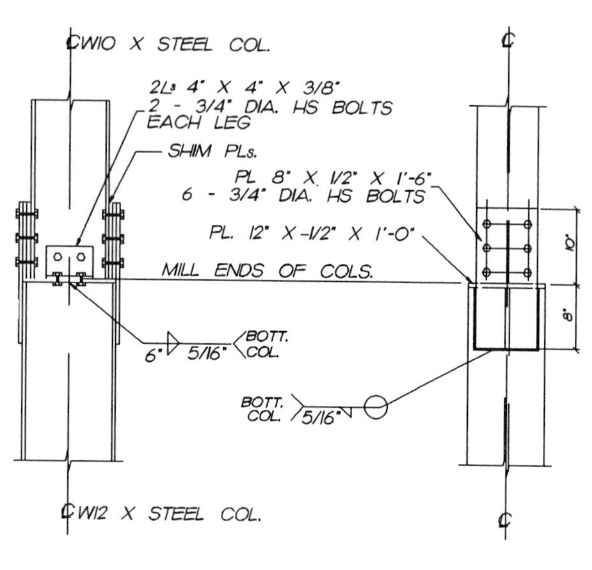

S2142

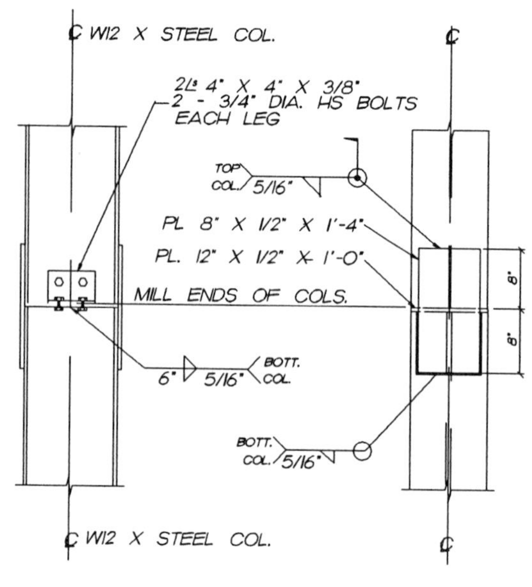

S2143

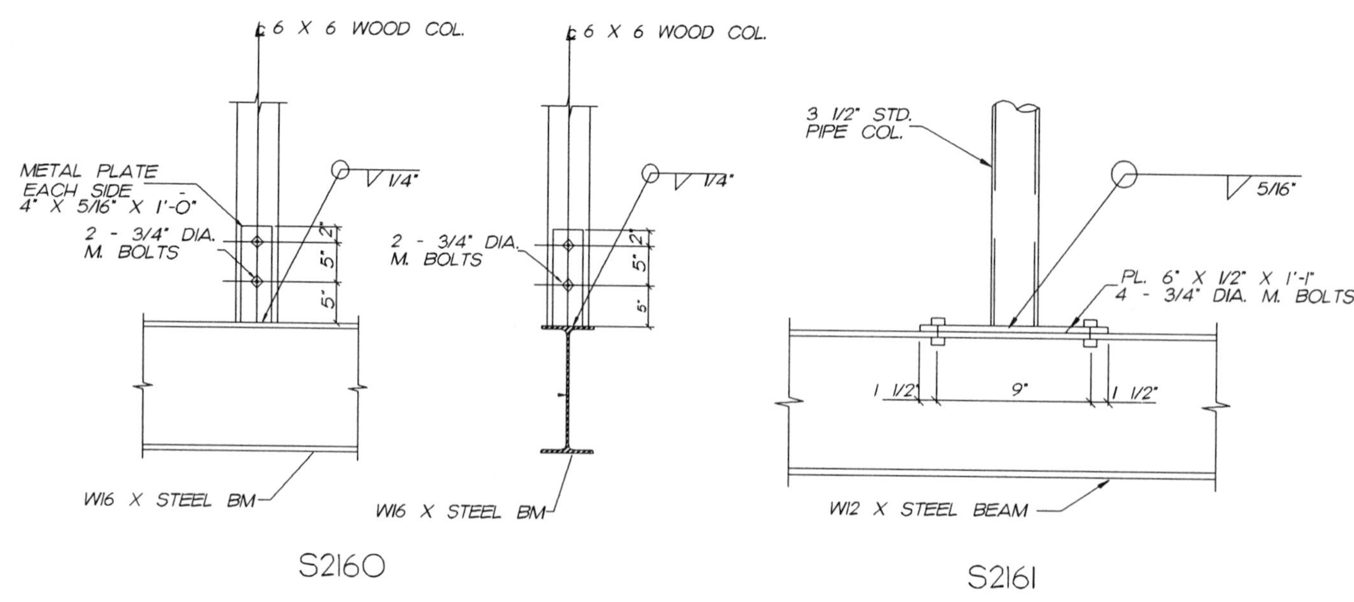

S2160

S2161

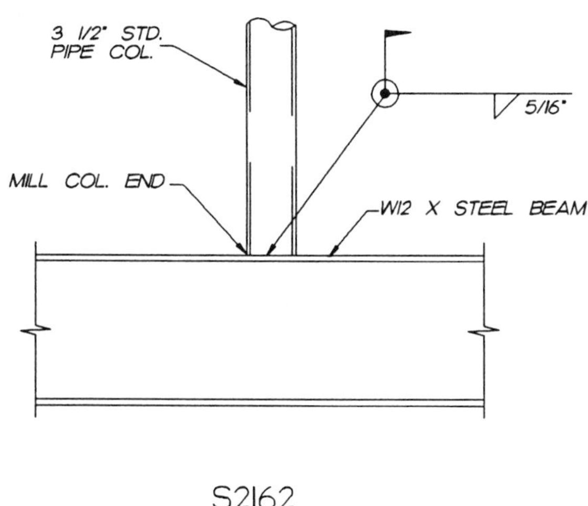

S2162

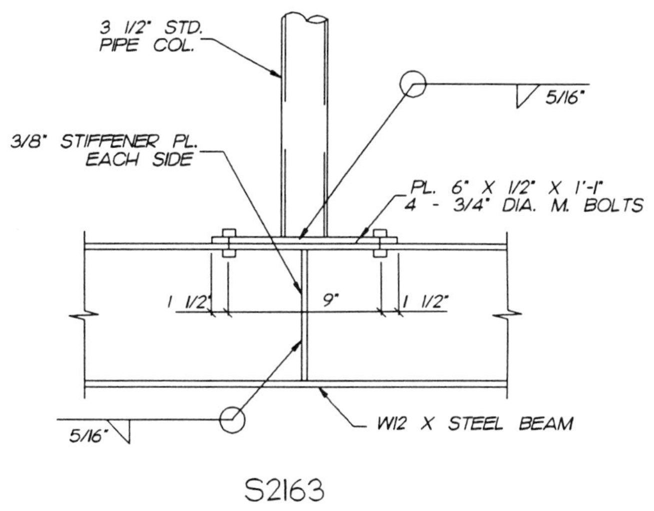

S2163

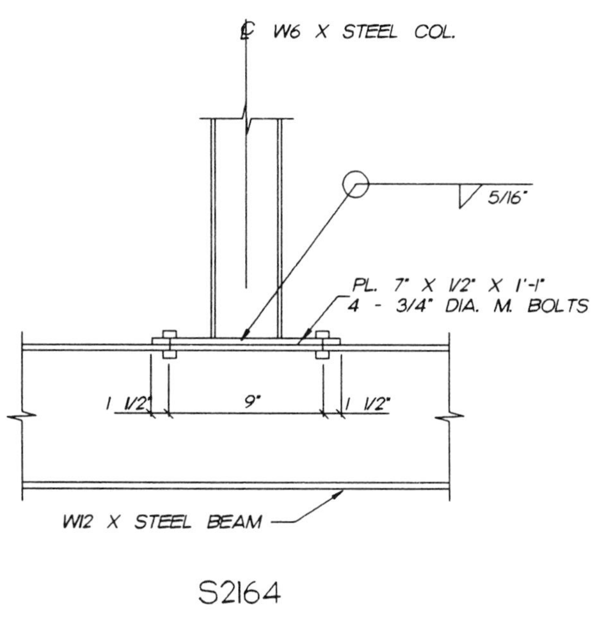

S2164

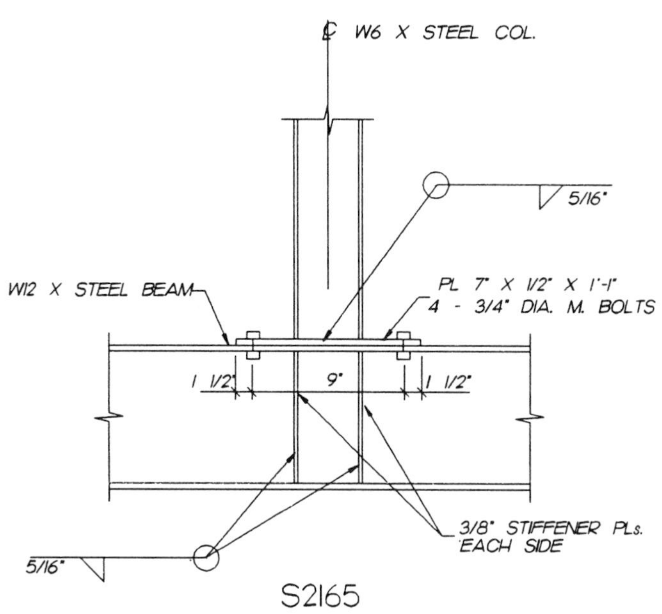

S2165

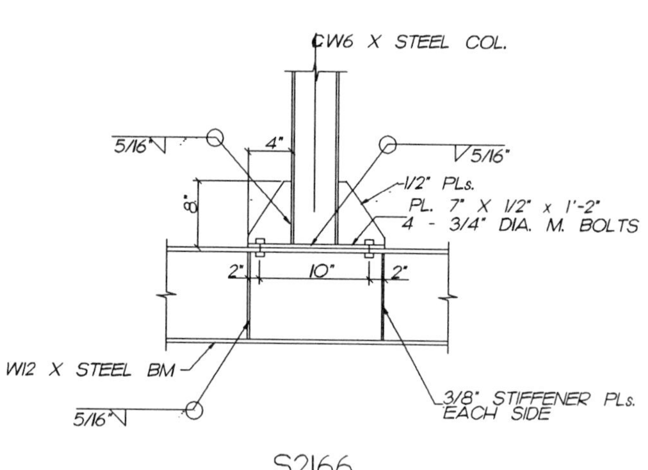

S2166

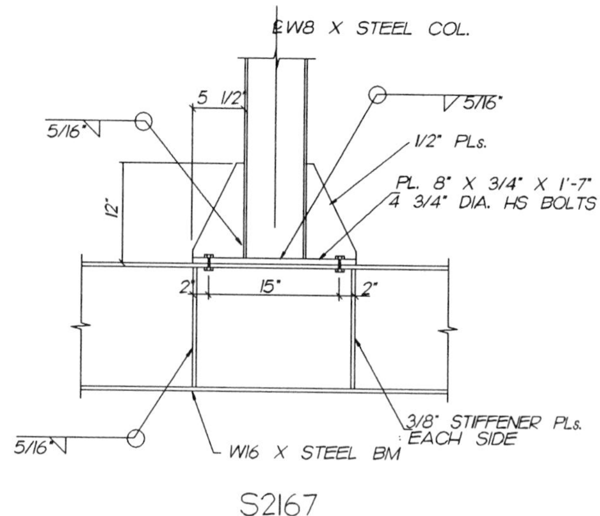

S2167

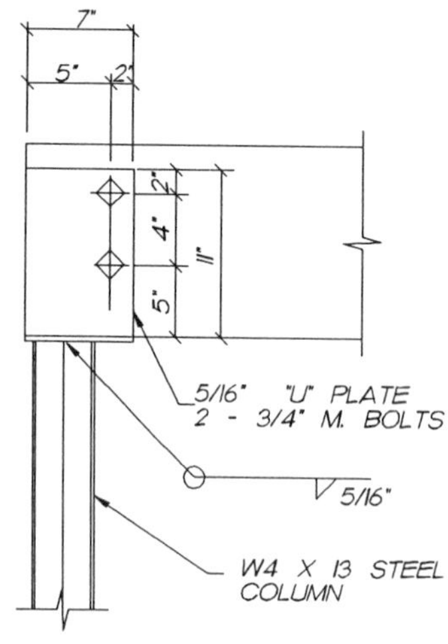

S2180

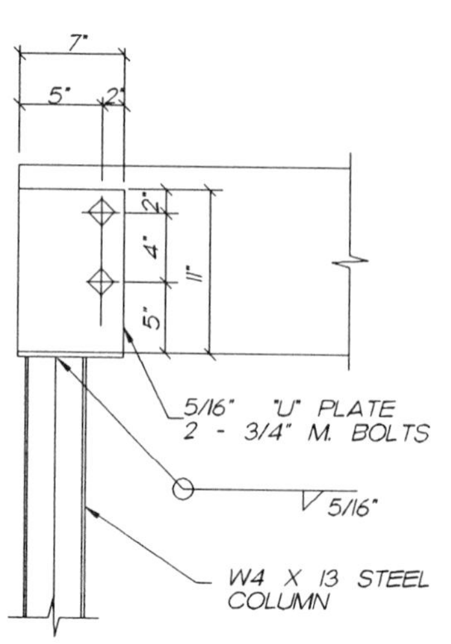

S2181

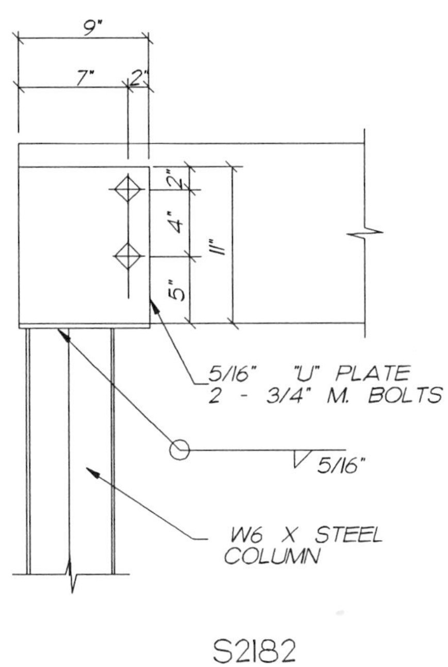

S2182

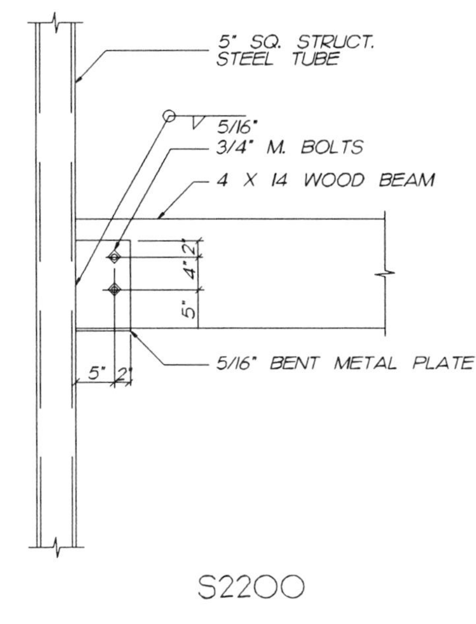

S2200

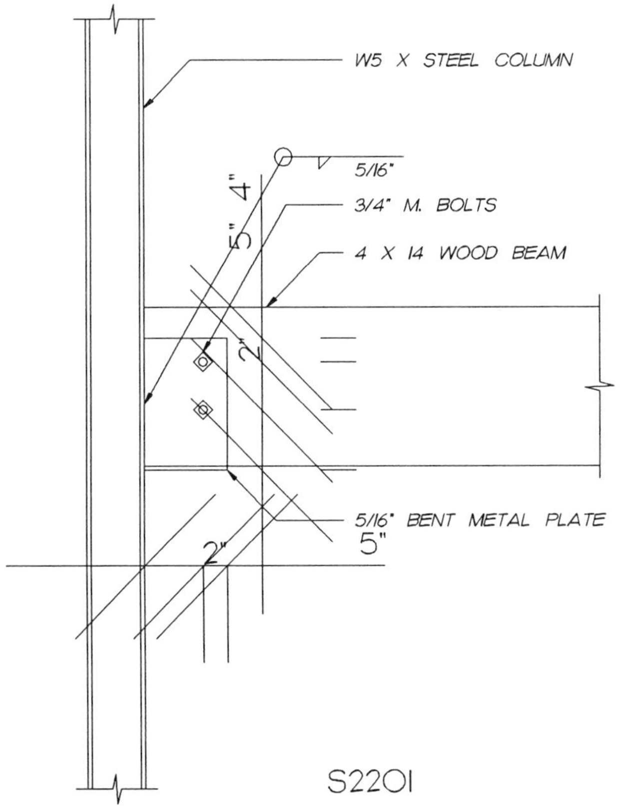

S2201

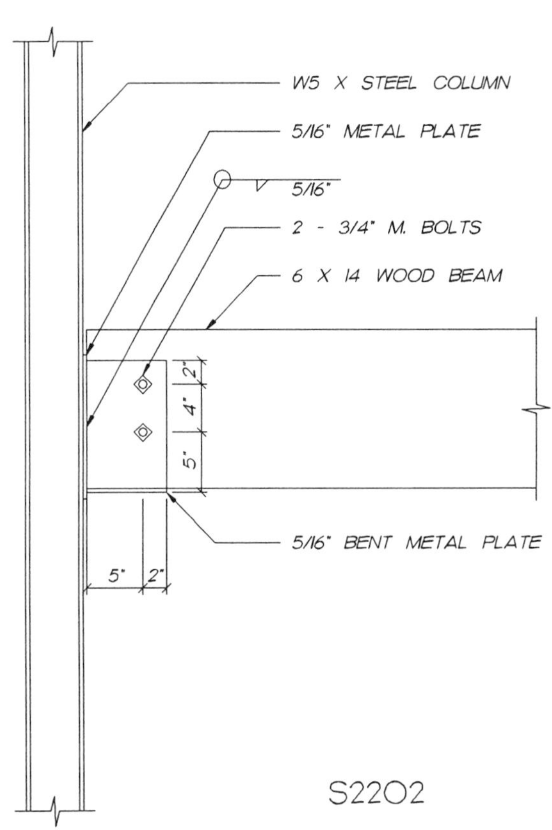

S2202

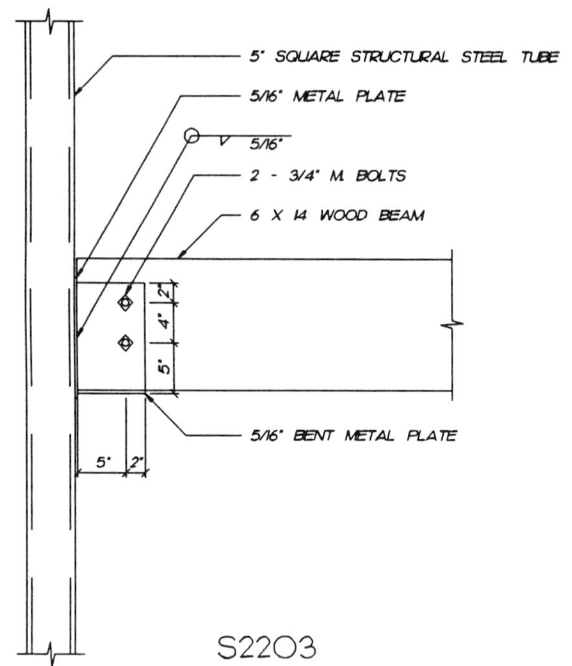

S2203

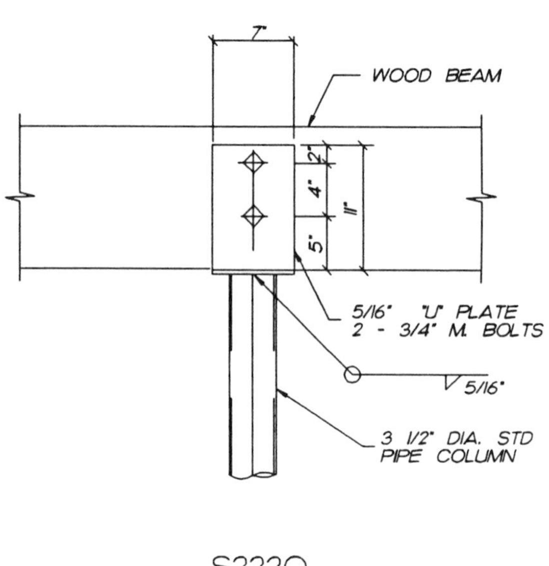

S2220

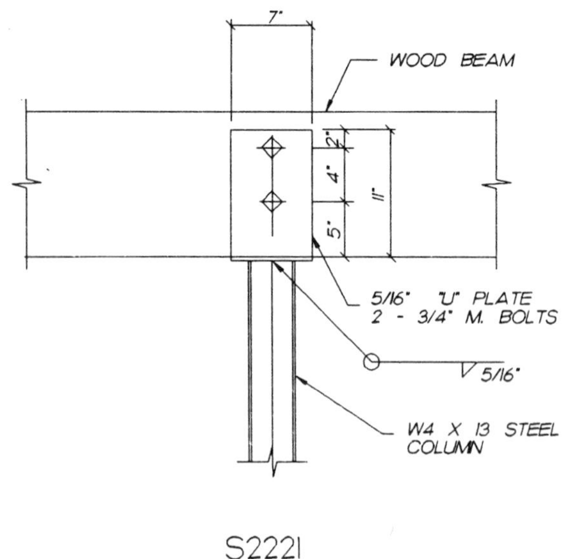

S2221

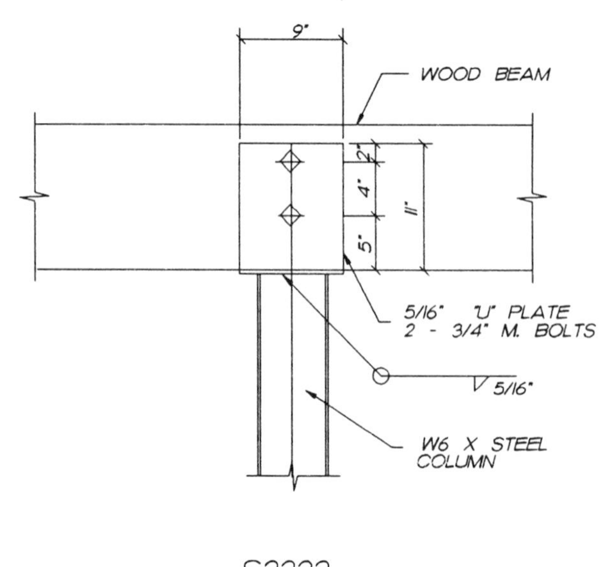

S2222

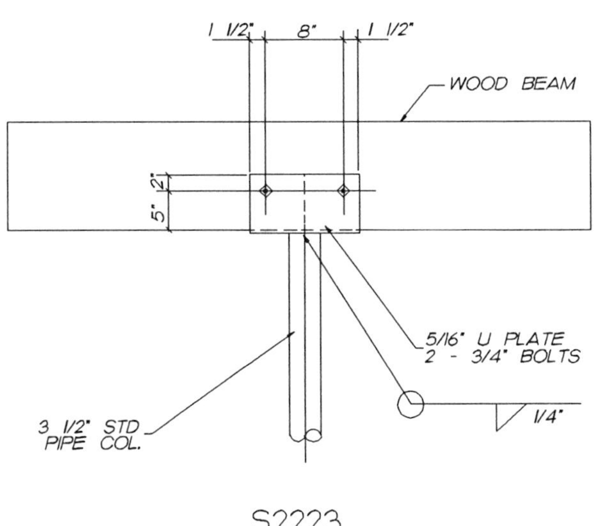

S2223

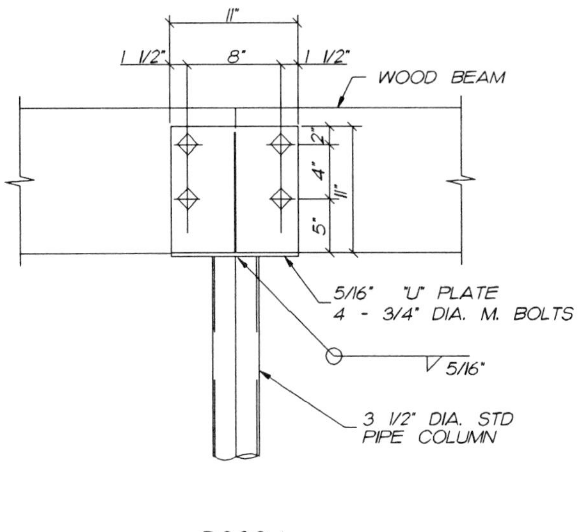

S2224

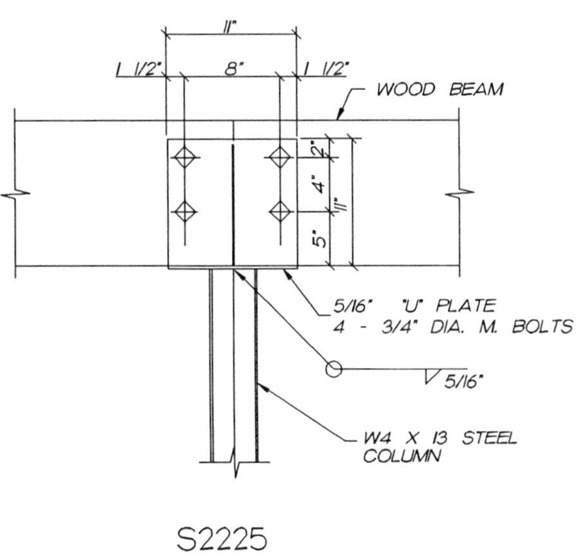

S2225

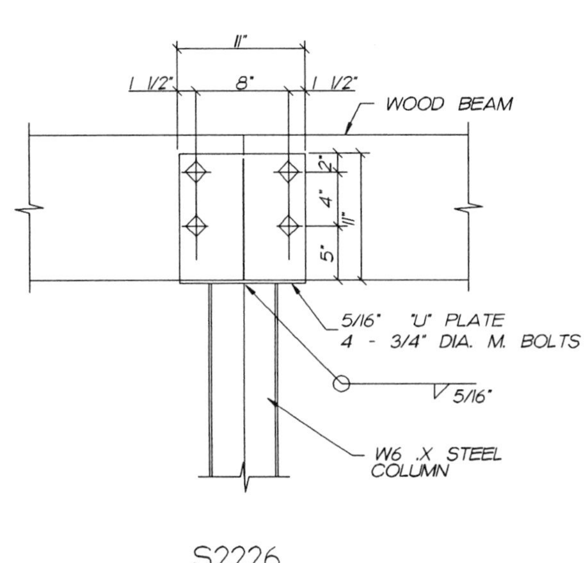

S2226

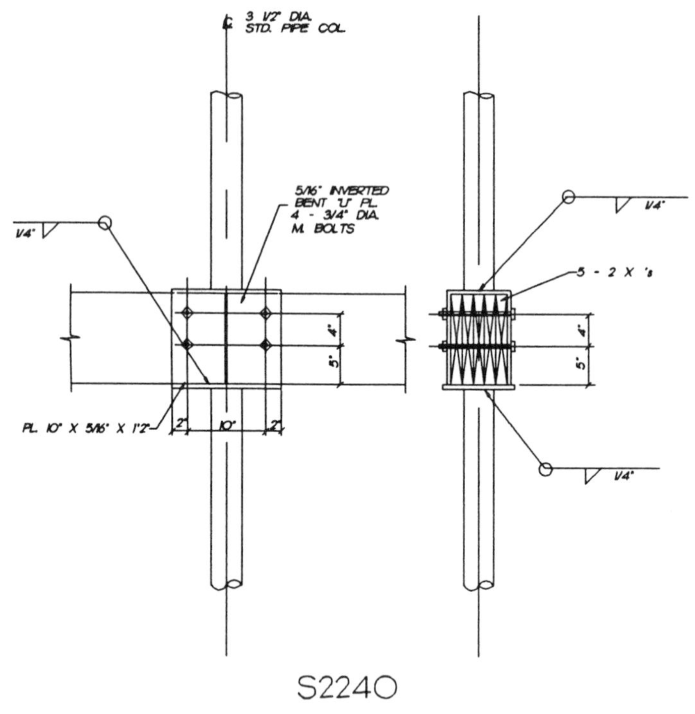

S2240

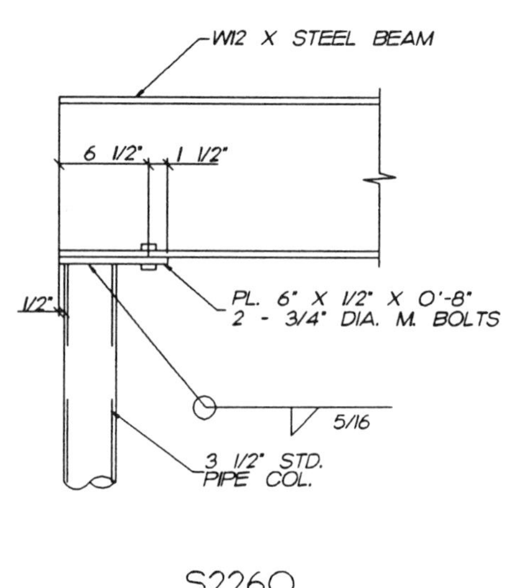

S2260

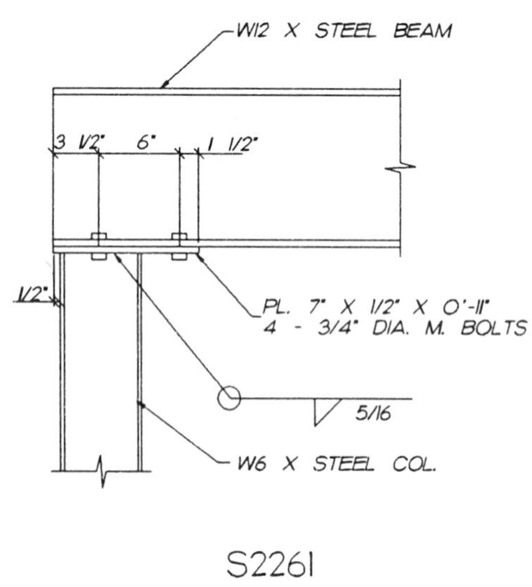

S2261

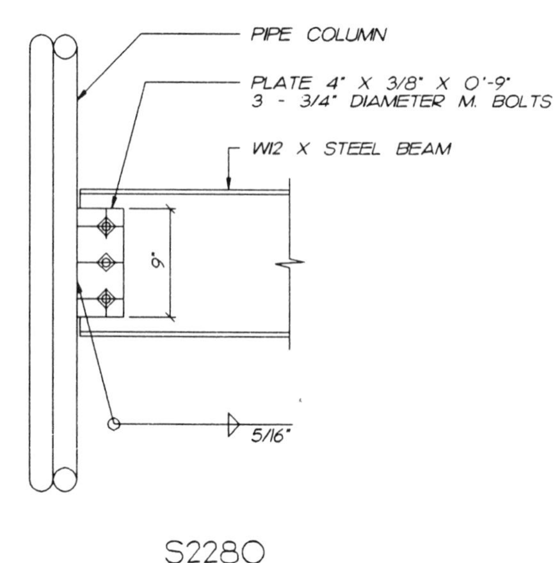

S2280

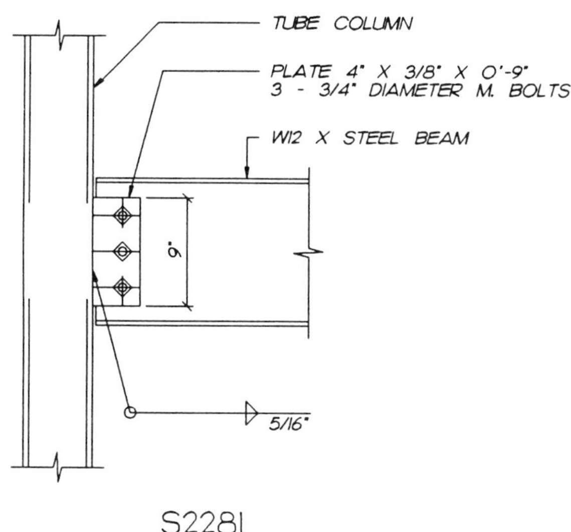

S2281

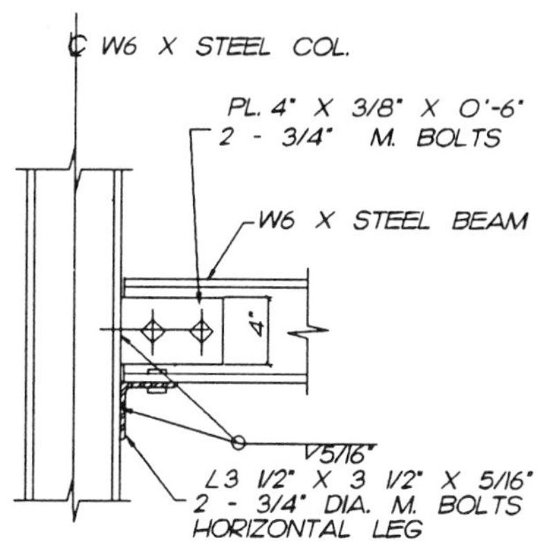

S2300

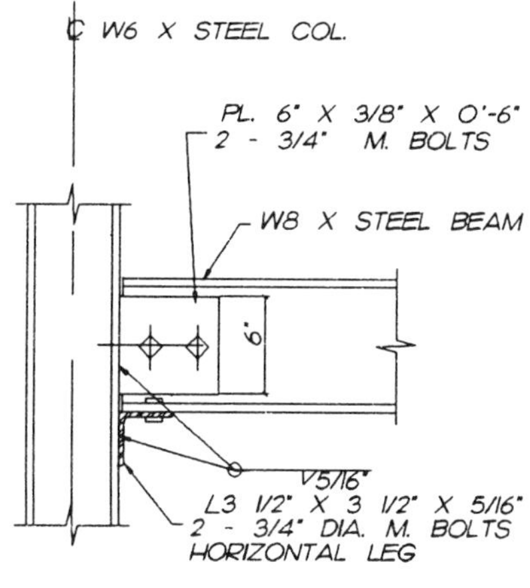

S2301

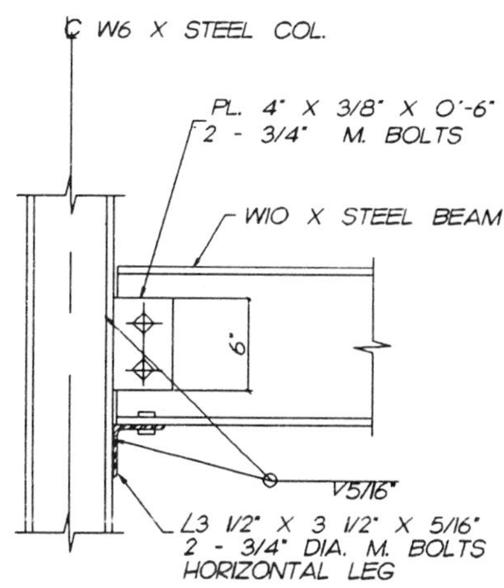

S2302

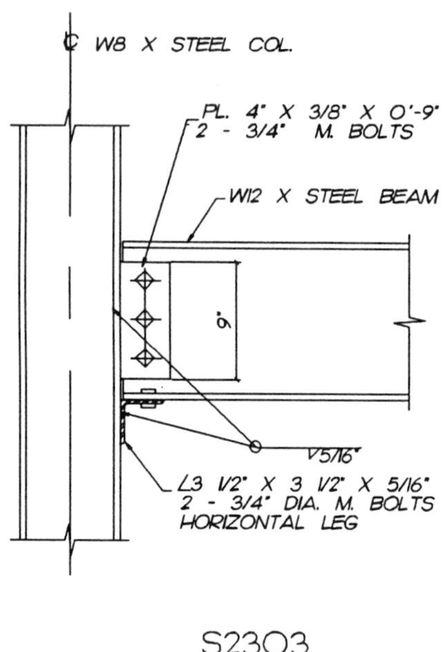

S2303

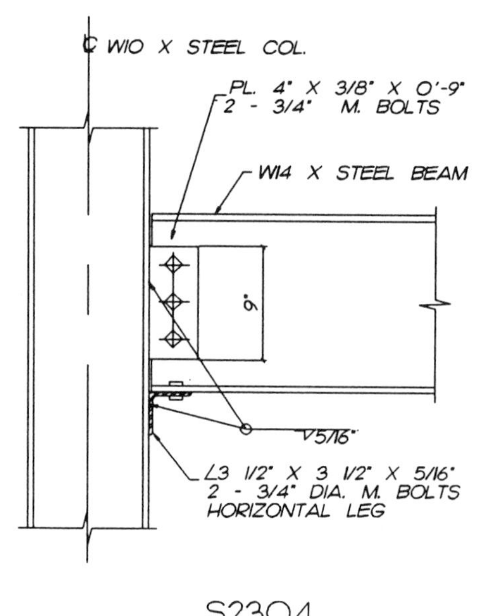

S2304

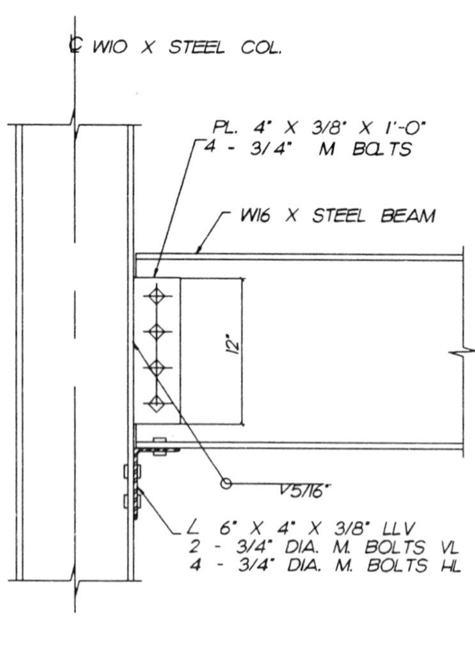

S2305

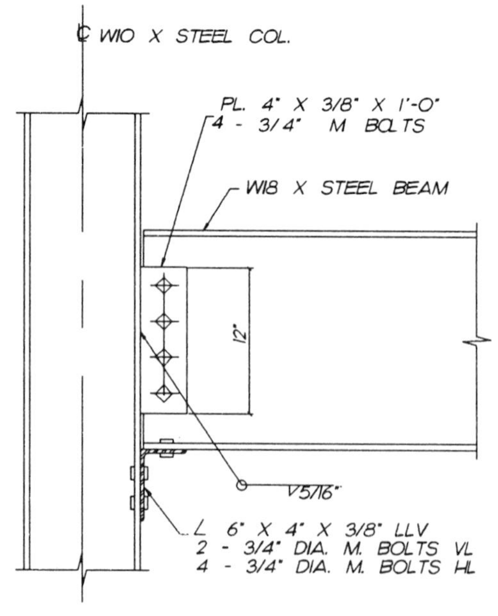

S2306

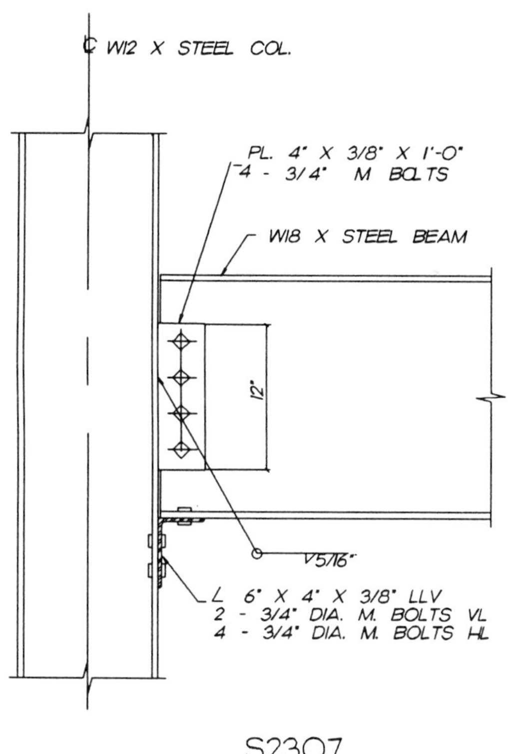

S2307

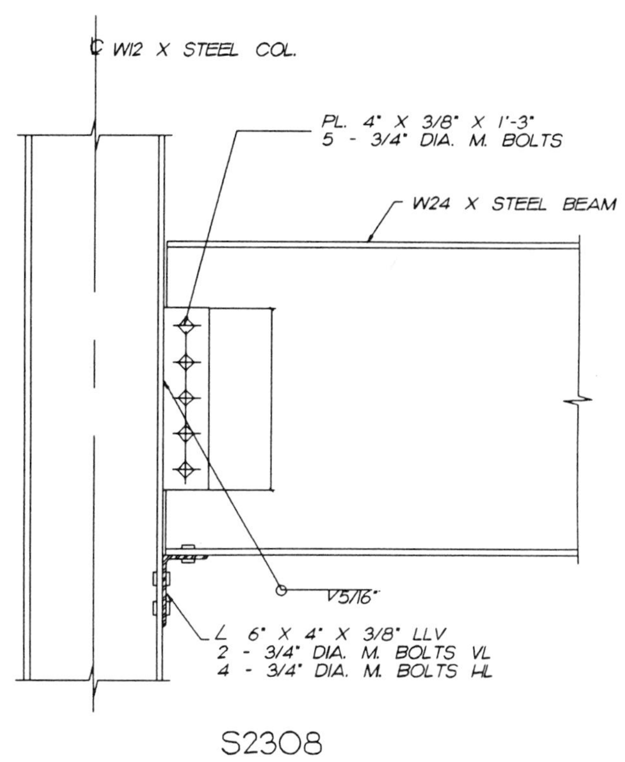

S2308

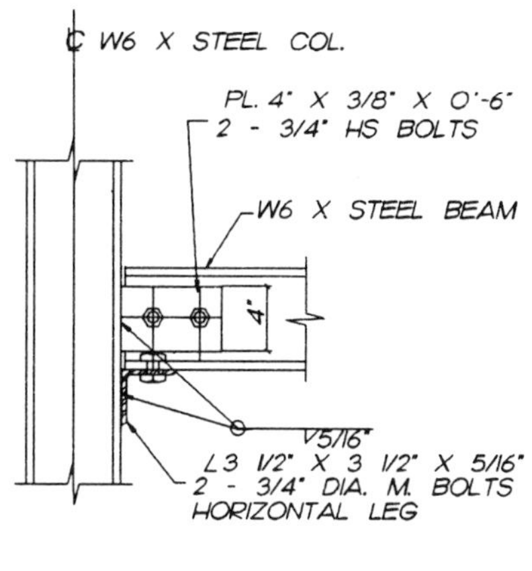

S2320

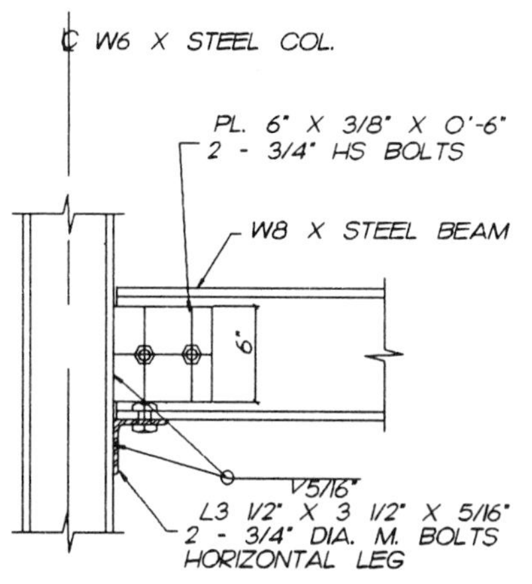

S2321

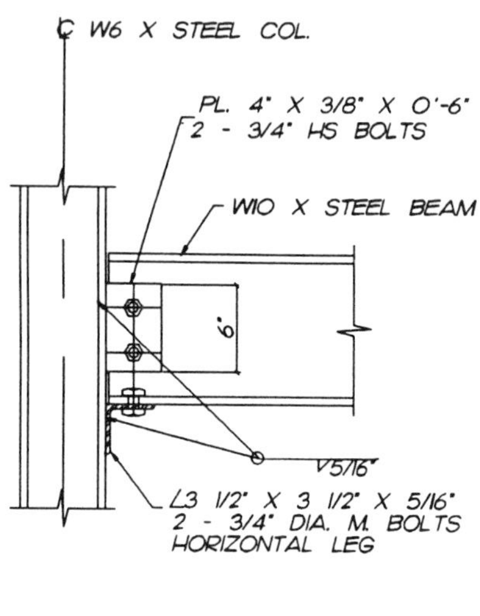

S2322

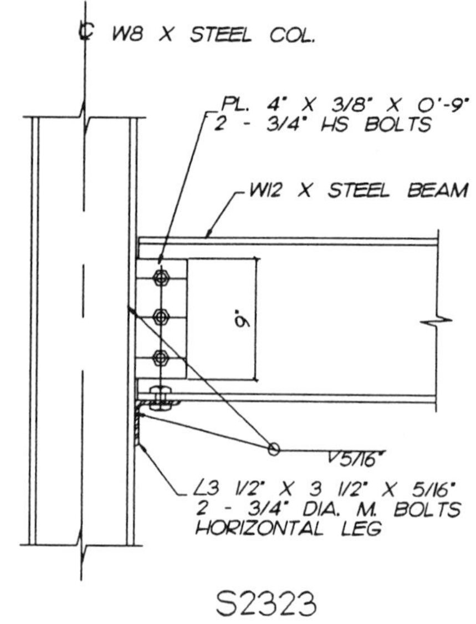

S2323

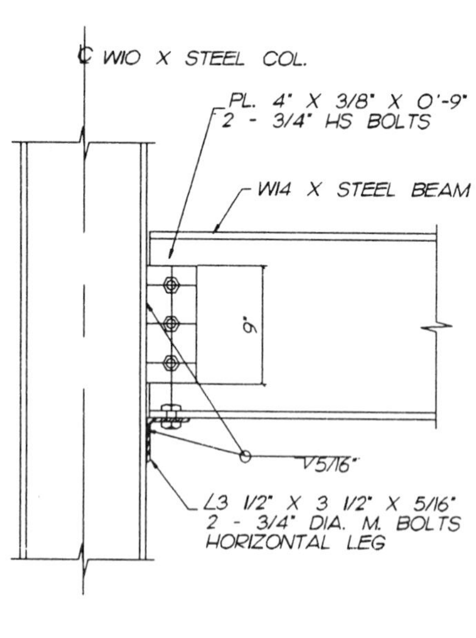

S2324

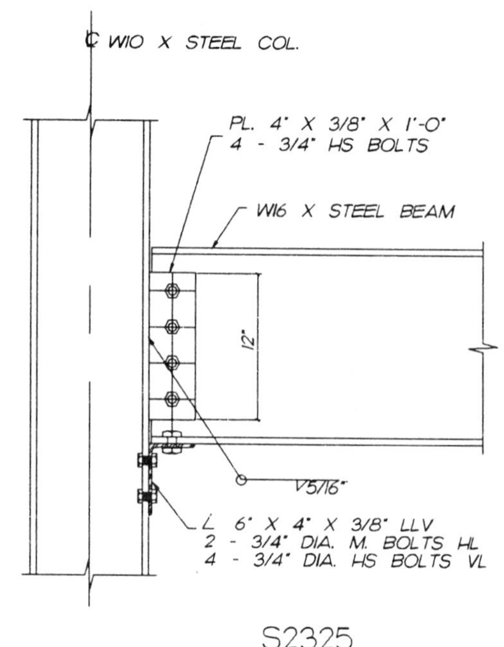

S2325

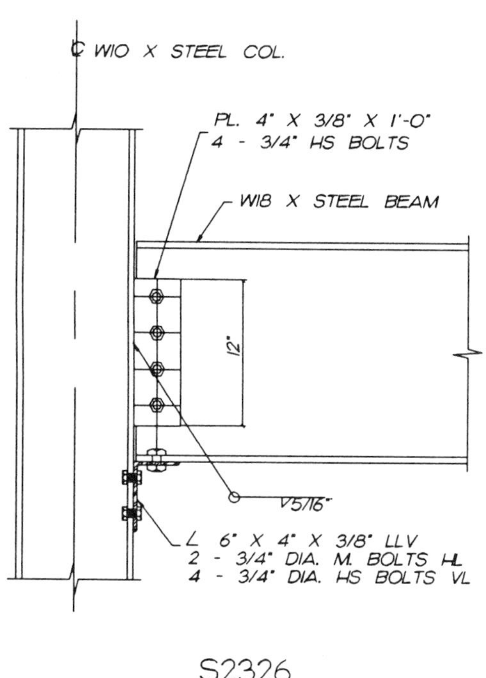

S2326

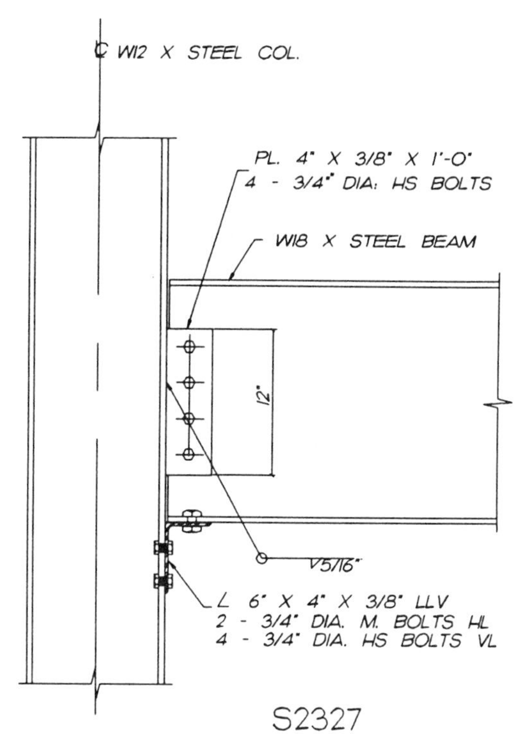

S2327

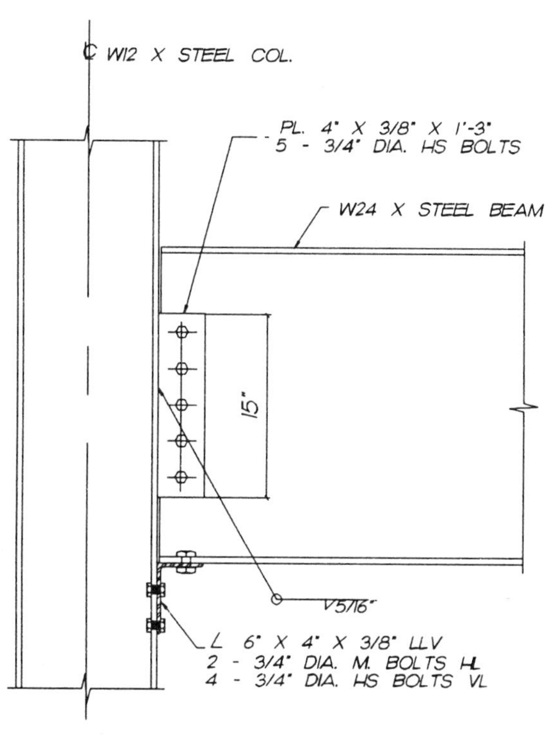

S2328

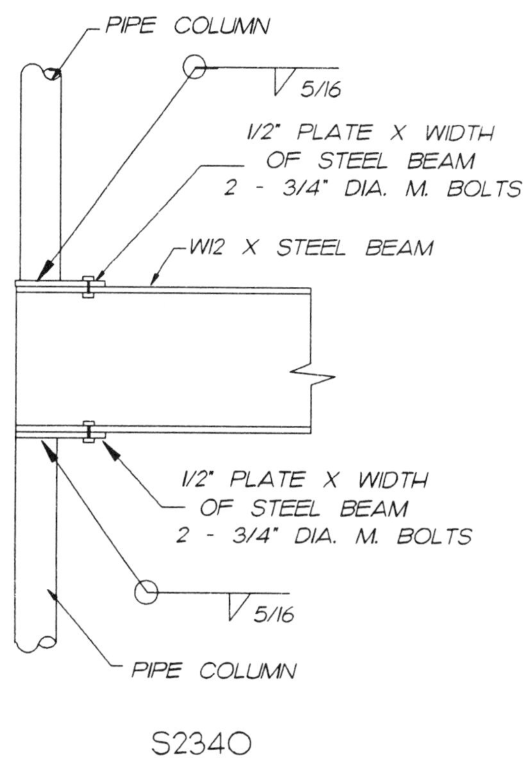

S2340

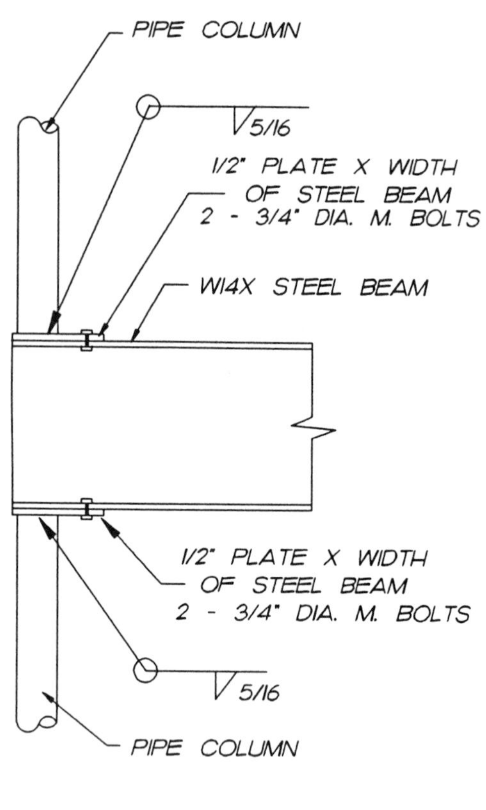

S2341

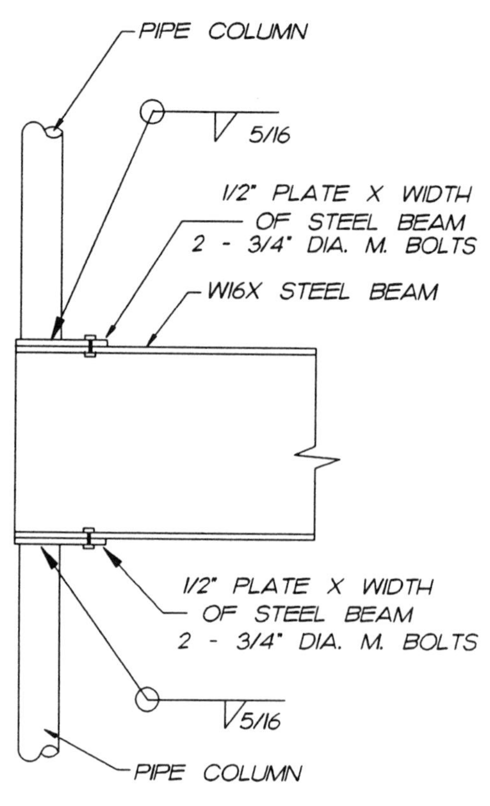

S2342

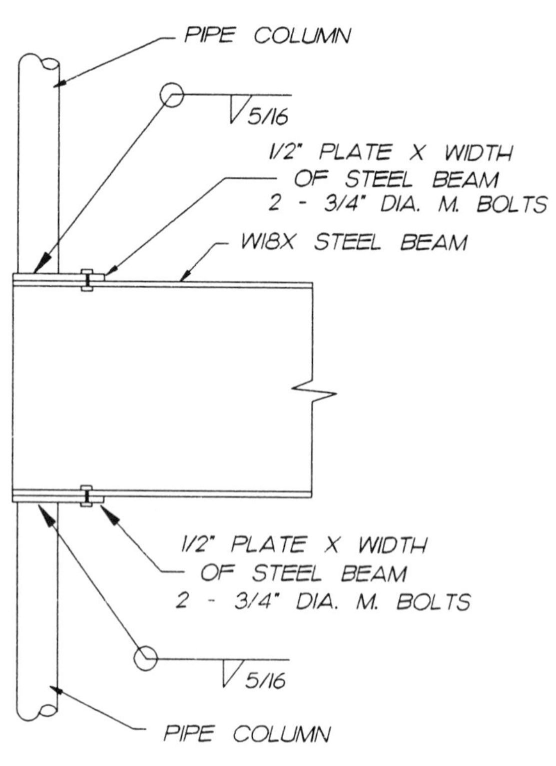

S2343

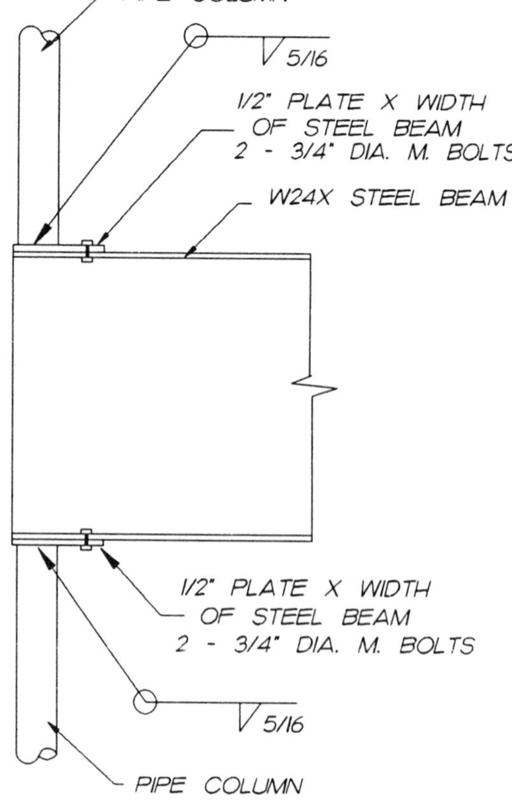

S2344

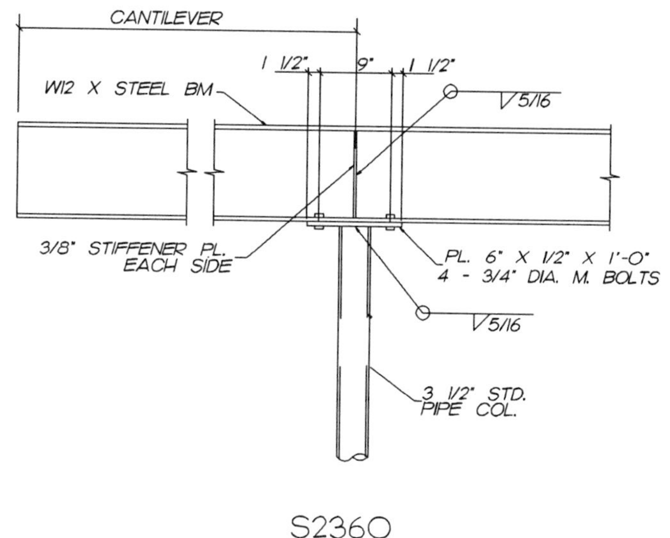

S2360

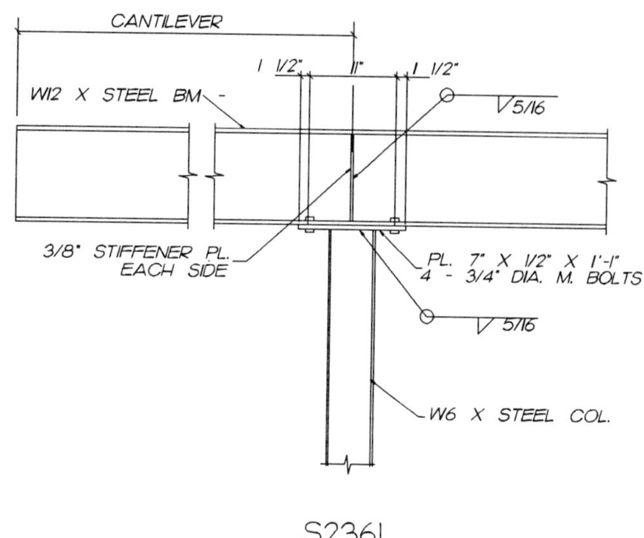

S2361

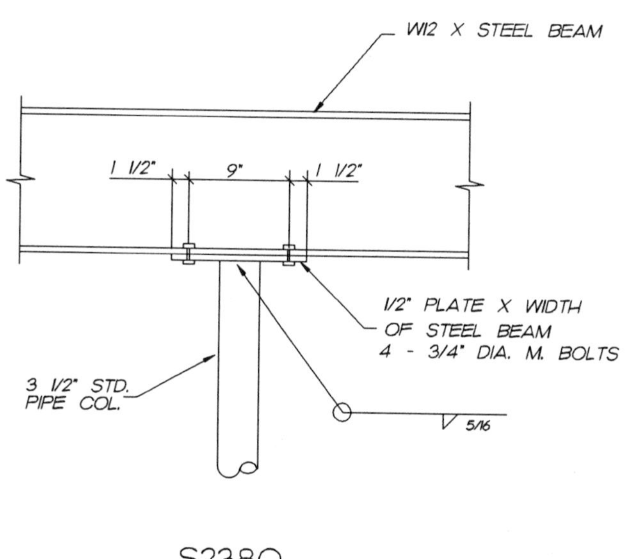

S2380

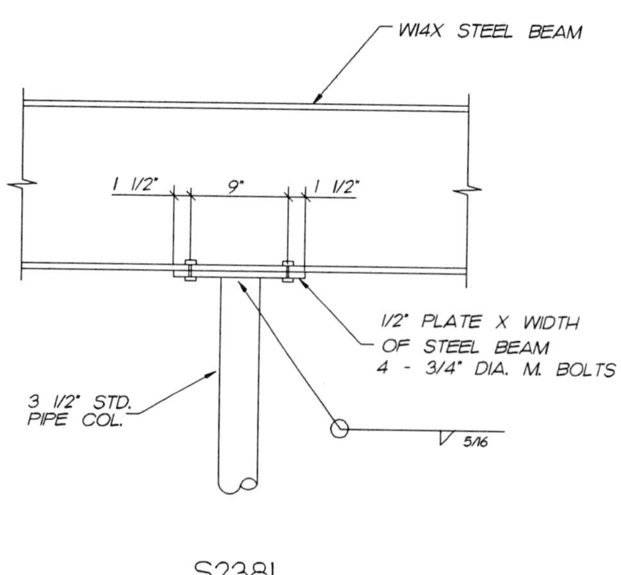

S2381

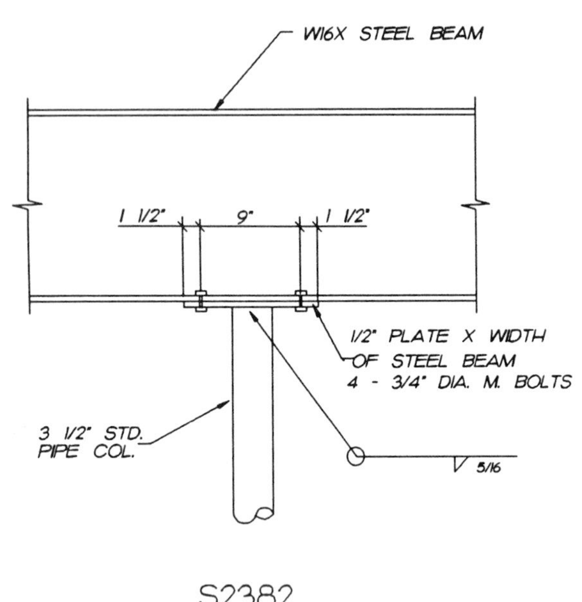

S2382

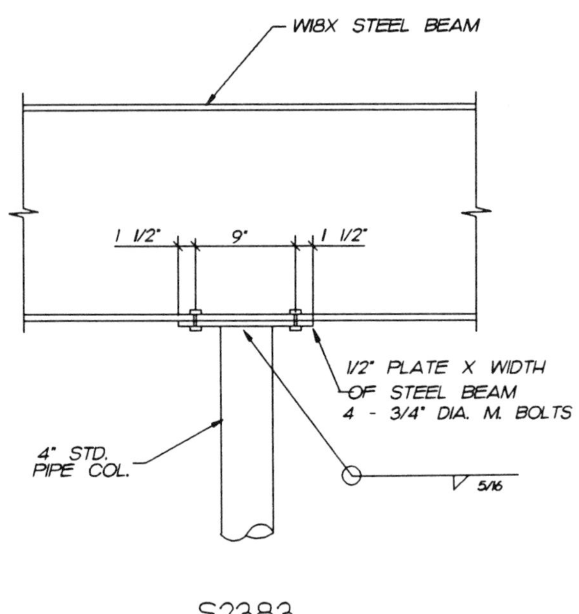

S2383

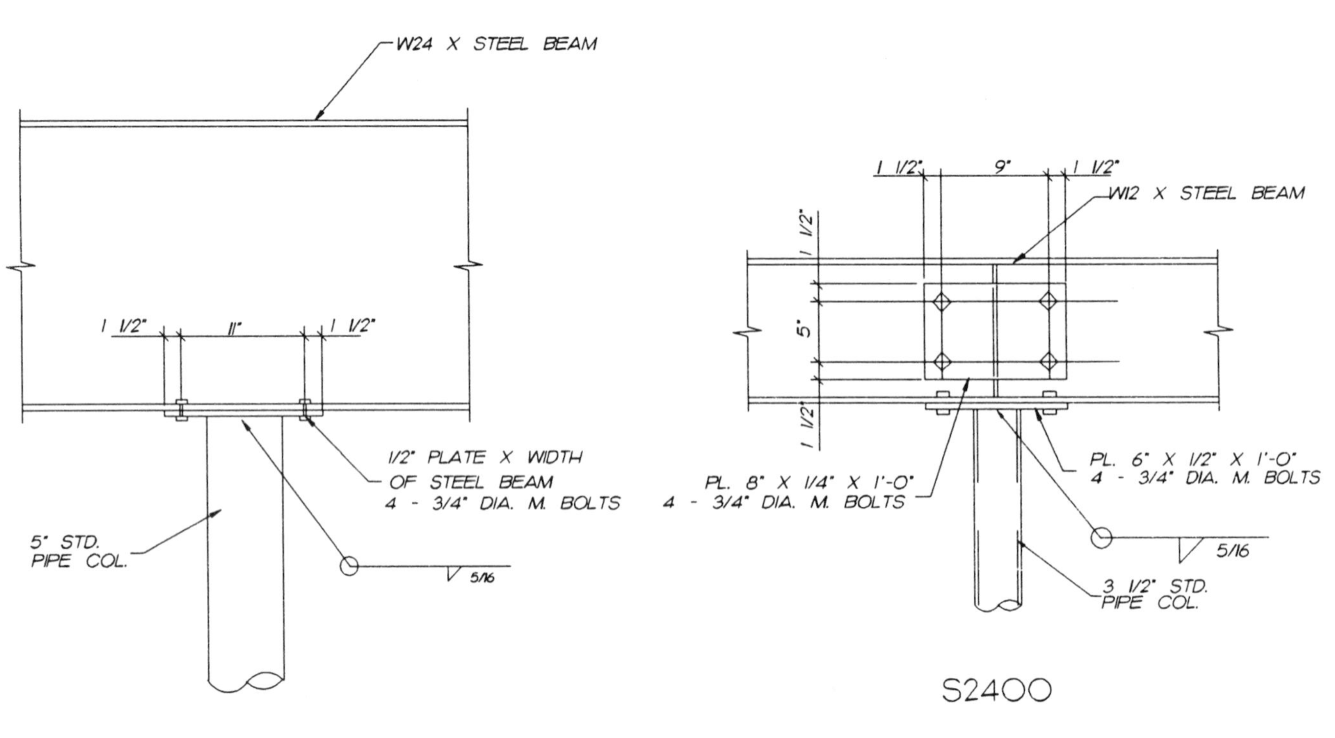

S2384

S2400

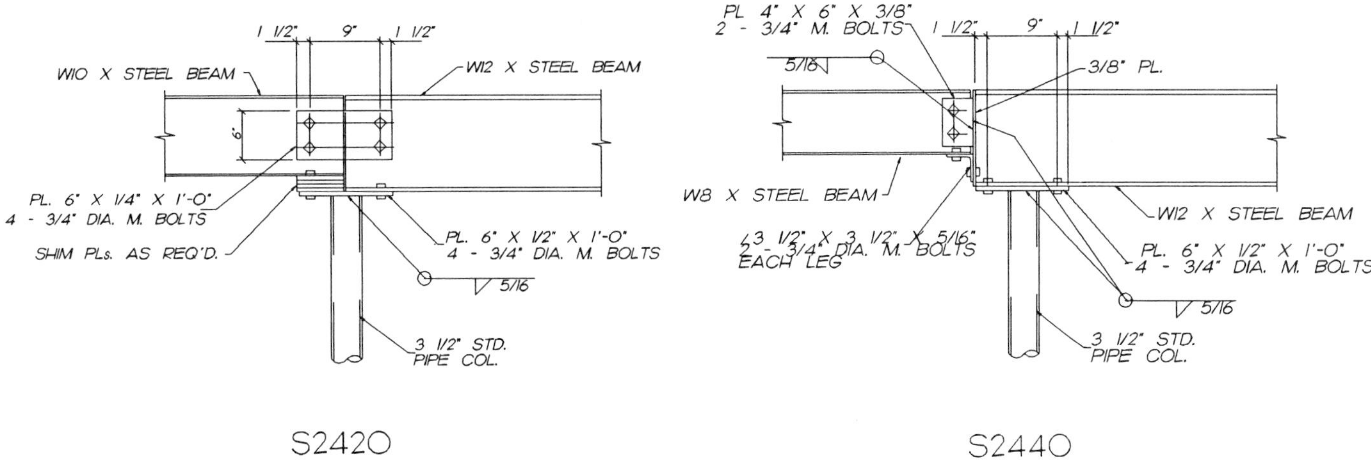

1 1/2"
9"
1 1/2"
W10 X STEEL BEAM
W12 X STEEL BEAM
6"
PL. 6" X 1/4" X 1'-0"
4 - 3/4" DIA. M. BOLTS
SHIM PLs. AS REQ'D.
PL. 6" X 1/2" X 1'-0"
4 - 3/4" DIA. M. BOLTS
5/16
3 1/2" STD.
PIPE COL.
S2420
PL 4" X 6" X 3/8"
2 - 3/4" M. BOLTS
1 1/2"
9"
1 1/2"
5/16
3/8" PL.
W8 X STEEL BEAM
W12 X STEEL BEAM
∠3 1/2" X 3 1/2" X 5/16"
2 - 3/4" DIA. M. BOLTS
EACH LEG
PL. 6" X 1/2" X 1'-0"
4 - 3/4" DIA. M. BOLTS
5/16
3 1/2" STD.
PIPE COL.
S2440

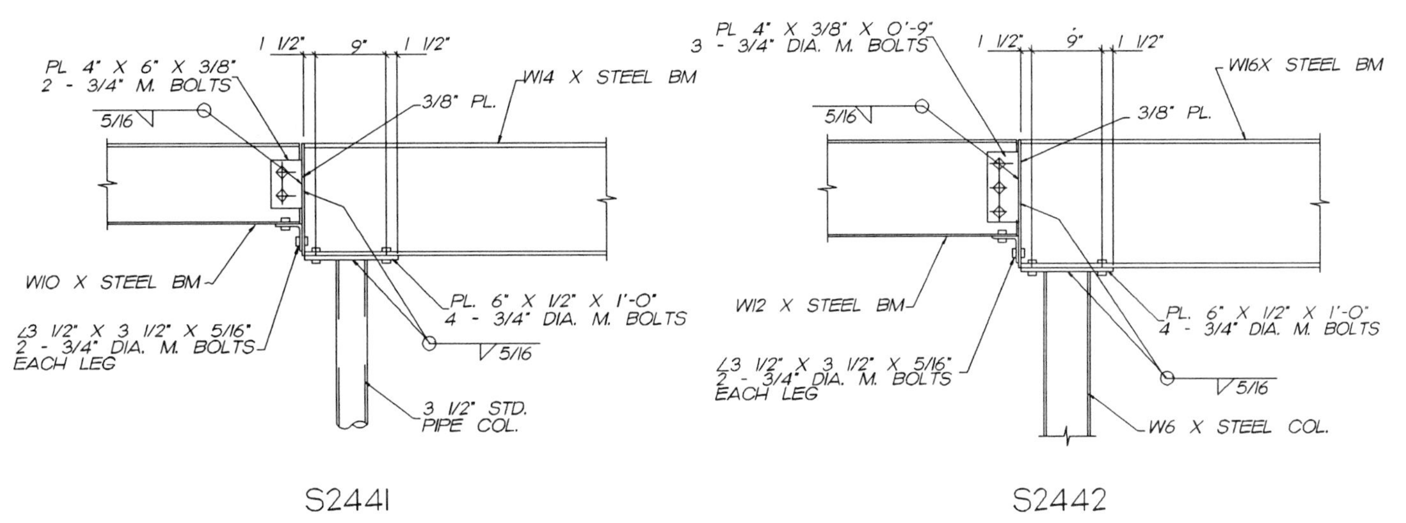

1 1/2"
9"
1 1/2"
PL 4" X 6" X 3/8"
2 - 3/4" M. BOLTS
W14 X STEEL BM
5/16
3/8" PL.
W10 X STEEL BM
PL. 6" X 1/2" X 1'-0"
4 - 3/4" DIA. M. BOLTS
∠3 1/2" X 3 1/2" X 5/16"
2 - 3/4" DIA. M. BOLTS
EACH LEG
5/16
3 1/2" STD.
PIPE COL.
S2441
PL 4" X 3/8" X 0'-9"
3 - 3/4" DIA. M. BOLTS
1 1/2"
9"
1 1/2"
W16X STEEL BM
5/16
3/8" PL.
W12 X STEEL BM
PL. 6" X 1/2" X 1'-0"
4 - 3/4" DIA. M. BOLTS
∠3 1/2" X 3 1/2" X 5/16"
2 - 3/4" DIA. M. BOLTS
EACH LEG
5/16
W6 X STEEL COL.
S2442

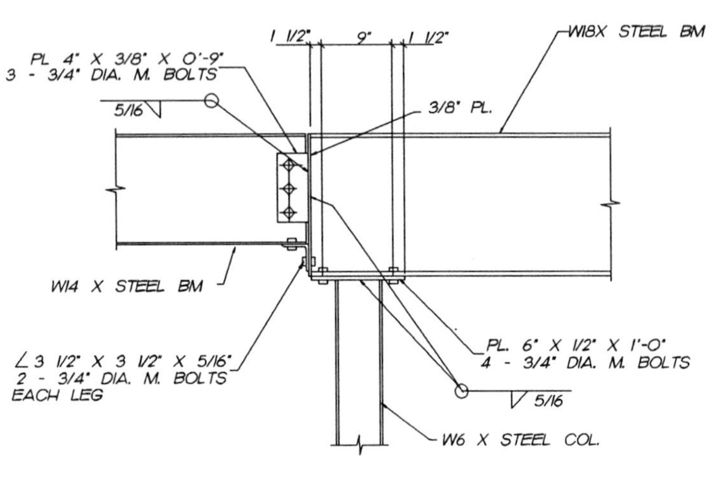

S2443

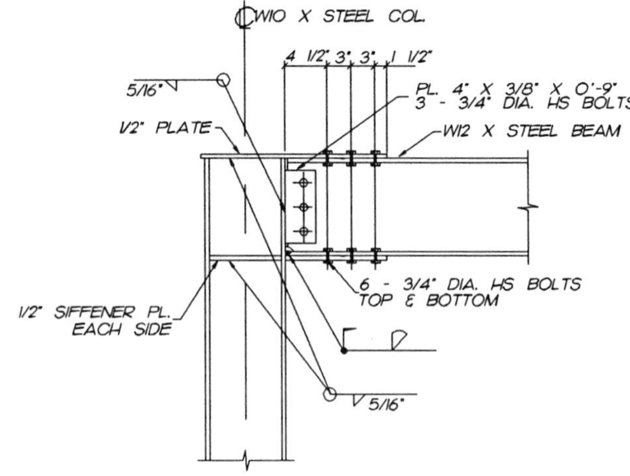

S2460

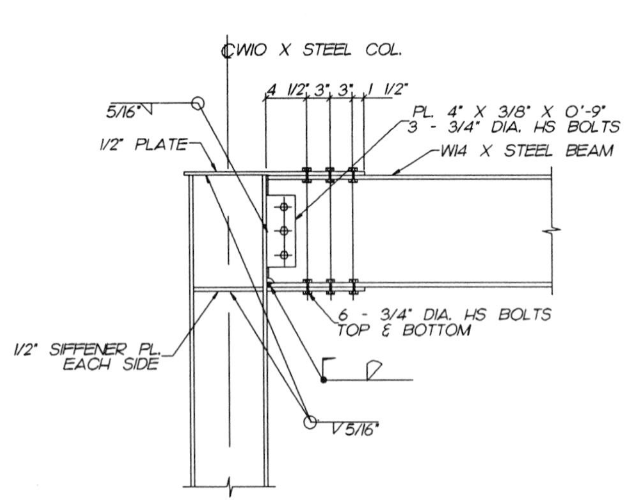

S2461

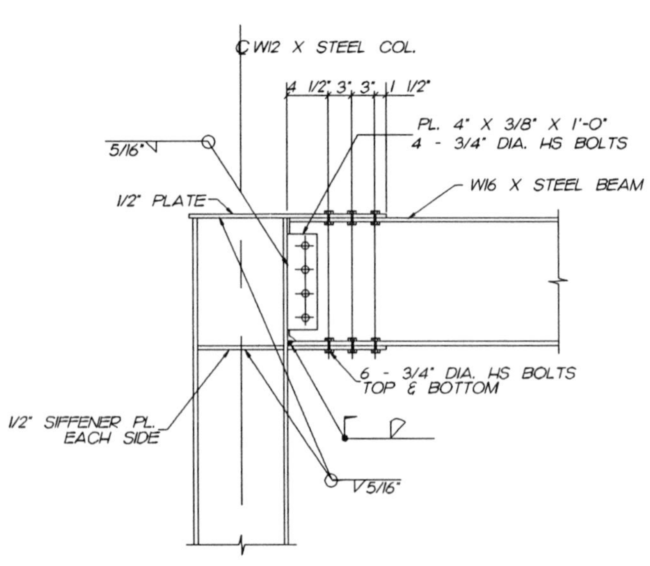

S2462

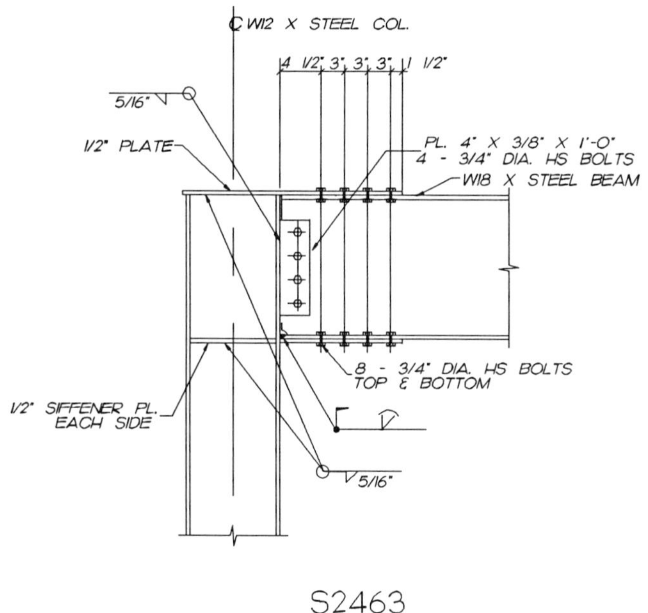

S2463

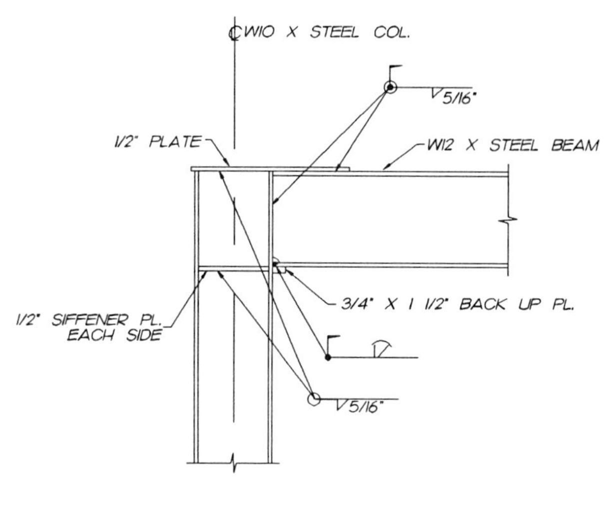

S2480

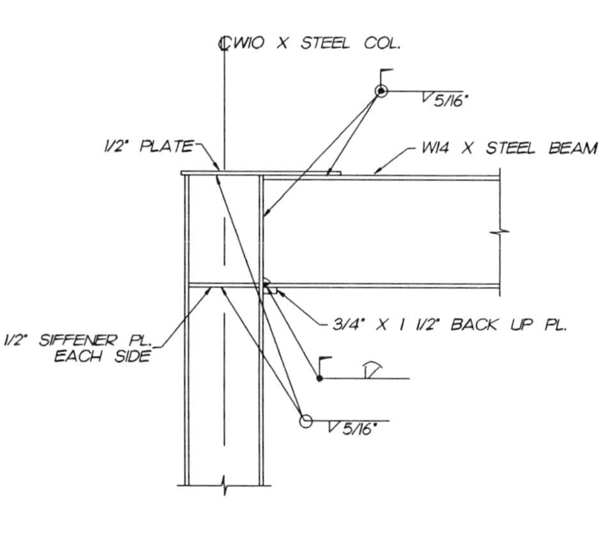

S2481

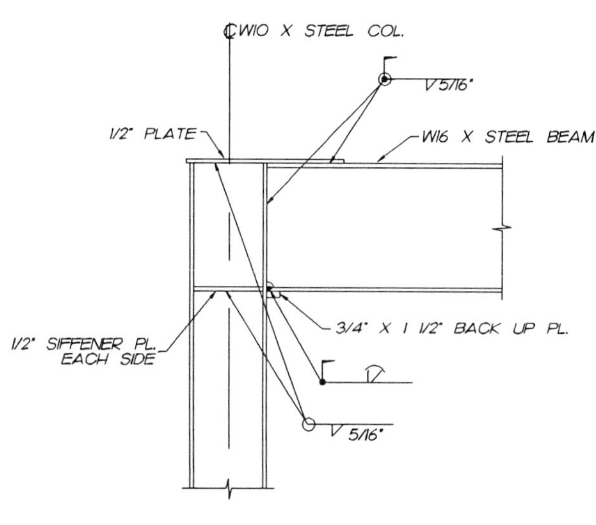

S2482

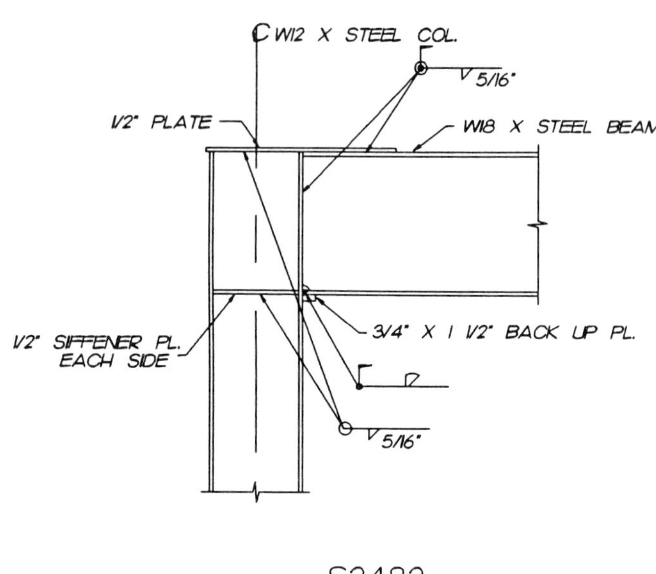

S2483

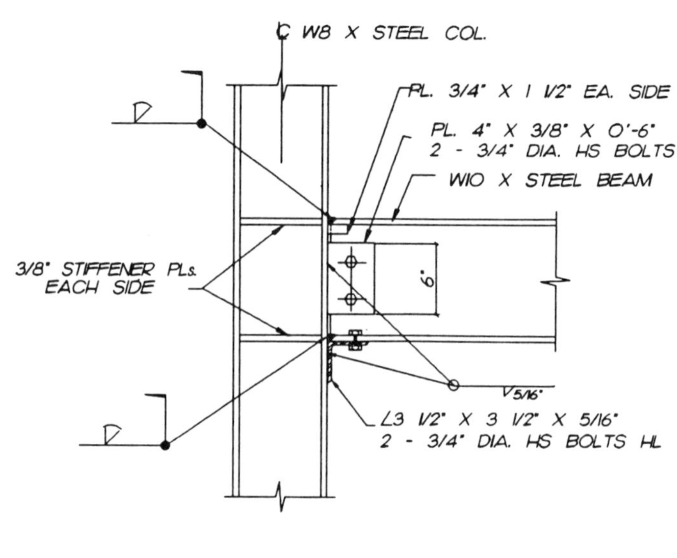

S2500

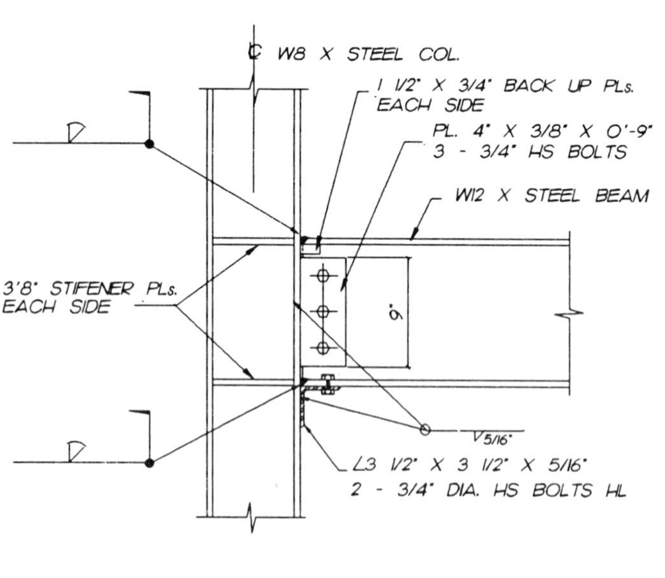

S2501

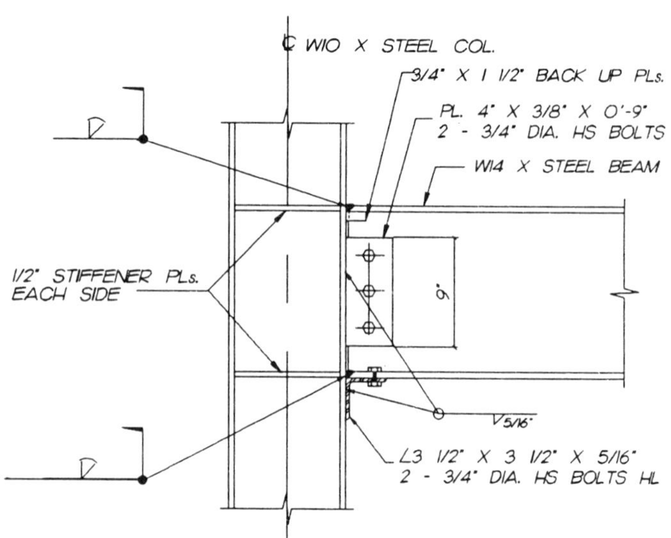

S2502

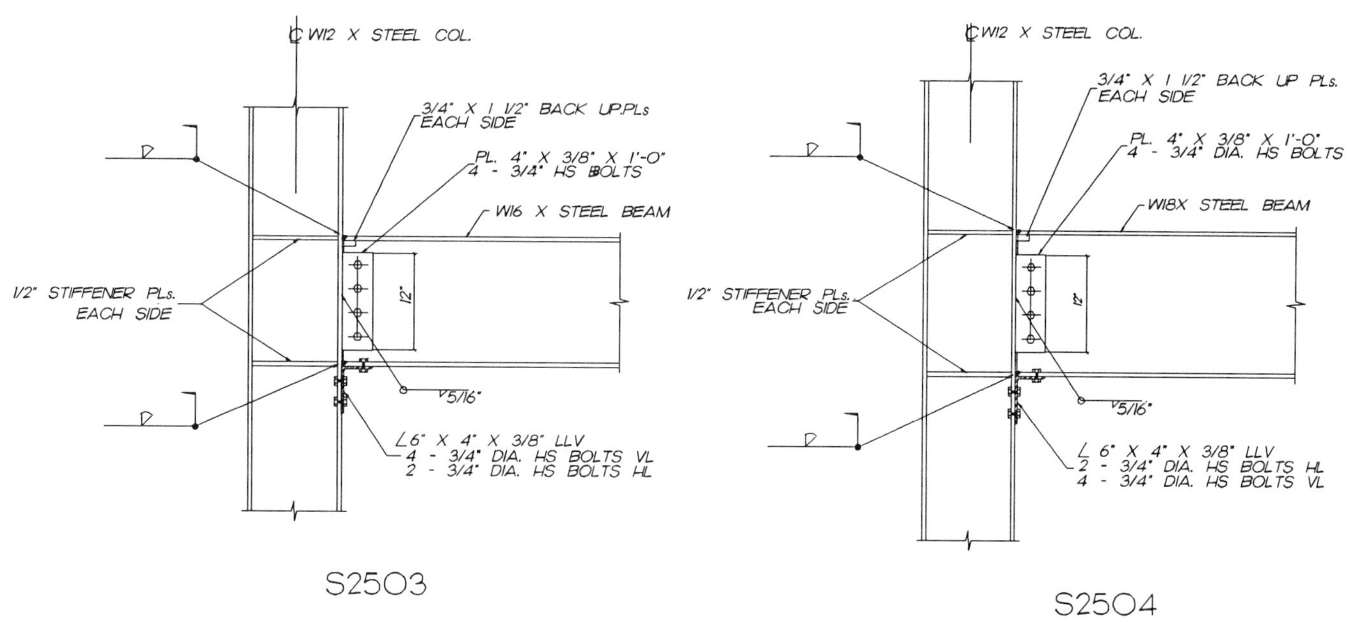

℄ W12 X STEEL COL.
3/4" X 1 1/2" BACK UP PLs EACH SIDE
PL. 4" X 3/8" X 1'-0"
4 - 3/4" HS BOLTS
W16 X STEEL BEAM
1/2" STIFFENER PLs. EACH SIDE
12"
5/16"
∠6" X 4" X 3/8" LLV
4 - 3/4" DIA. HS BOLTS VL
2 - 3/4" DIA. HS BOLTS HL
S2503
℄ W12 X STEEL COL.
3/4" X 1 1/2" BACK UP PLs. EACH SIDE
PL. 4" X 3/8" X 1'-0"
4 - 3/4" DIA. HS BOLTS
W18X STEEL BEAM
1/2" STIFFENER PLs. EACH SIDE
12"
5/16"
∠ 6" X 4" X 3/8" LLV
2 - 3/4" DIA. HS BOLTS HL
4 - 3/4" DIA. HS BOLTS VL
S2504

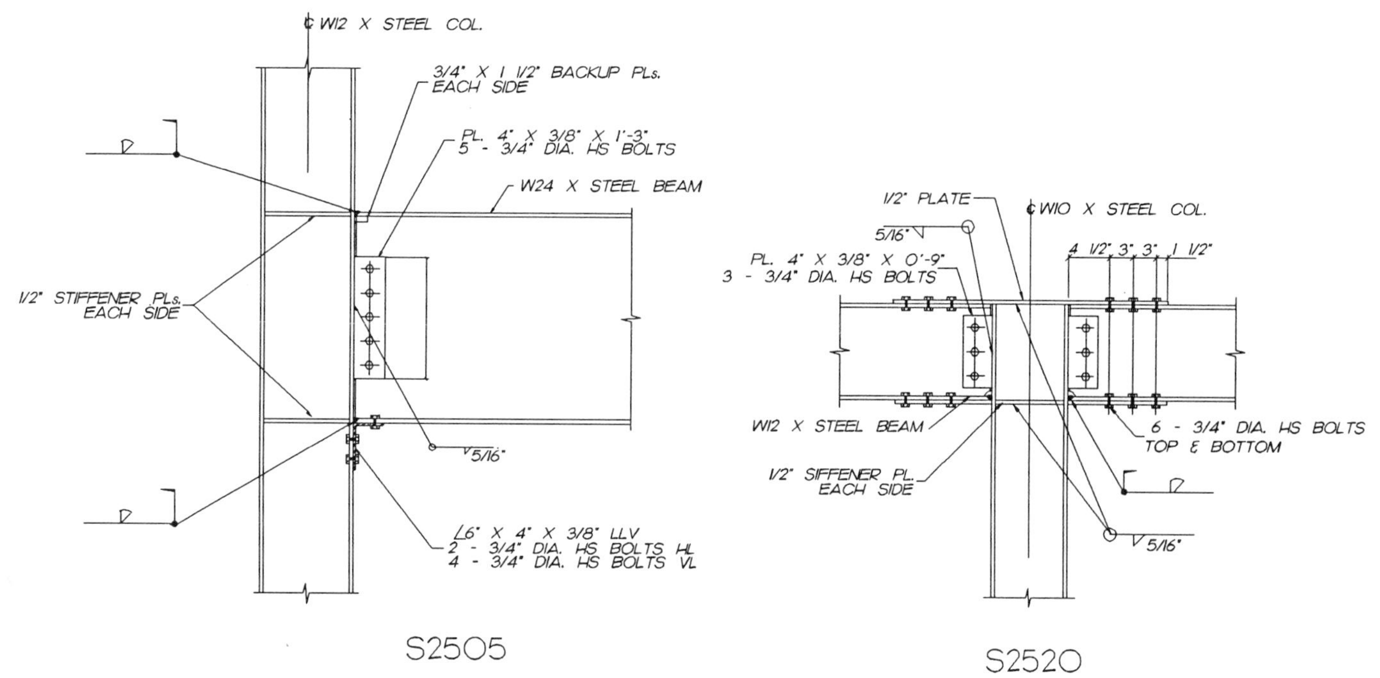

℄ W12 X STEEL COL.
3/4" X 1 1/2" BACKUP PLs. EACH SIDE
PL. 4" X 3/8" X 1'-3"
5 - 3/4" DIA. HS BOLTS
W24 X STEEL BEAM
1/2" STIFFENER PLs. EACH SIDE
5/16"
∠6" X 4" X 3/8" LLV
2 - 3/4" DIA. HS BOLTS HL
4 - 3/4" DIA. HS BOLTS VL
S2505
1/2" PLATE
℄ W10 X STEEL COL.
5/16"
PL. 4" X 3/8" X 0'-9"
3 - 3/4" DIA. HS BOLTS
4 1/2" 3" 3" 1 1/2"
W12 X STEEL BEAM
6 - 3/4" DIA. HS BOLTS TOP & BOTTOM
1/2" SIFFENER PL. EACH SIDE
5/16"
S2520

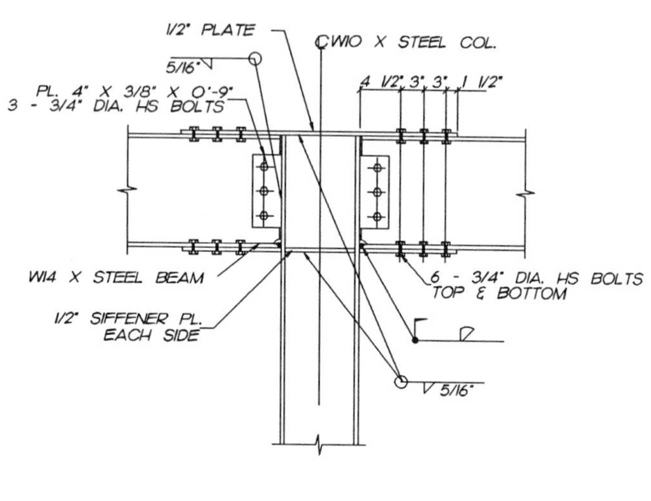

S2521

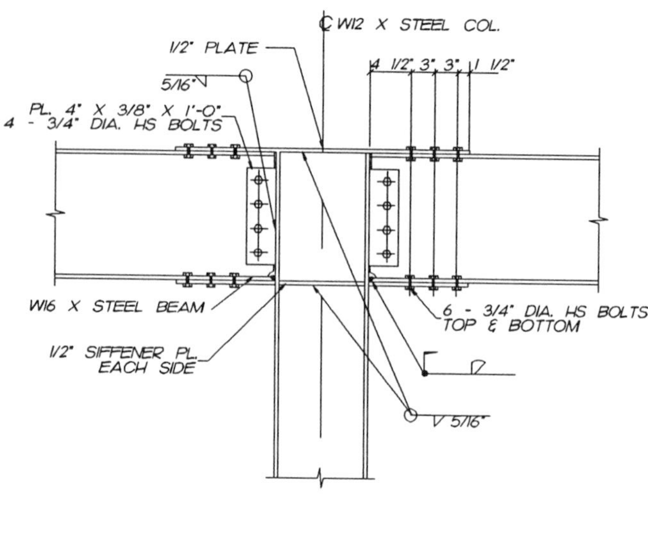

S2522

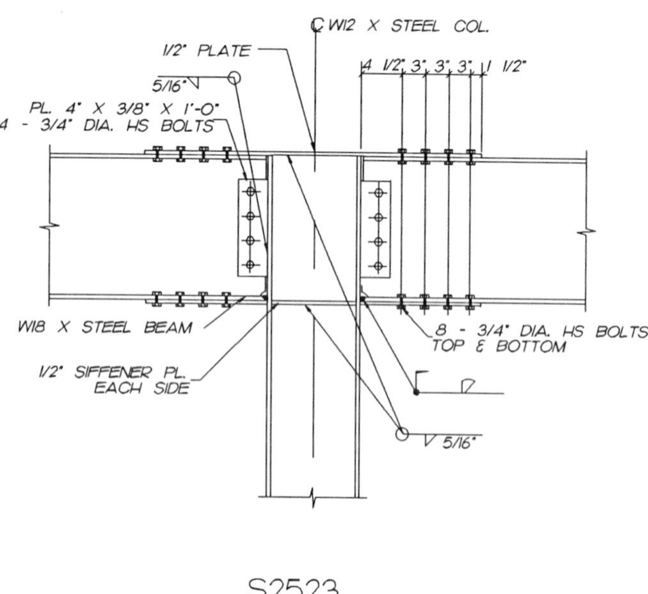

S2523

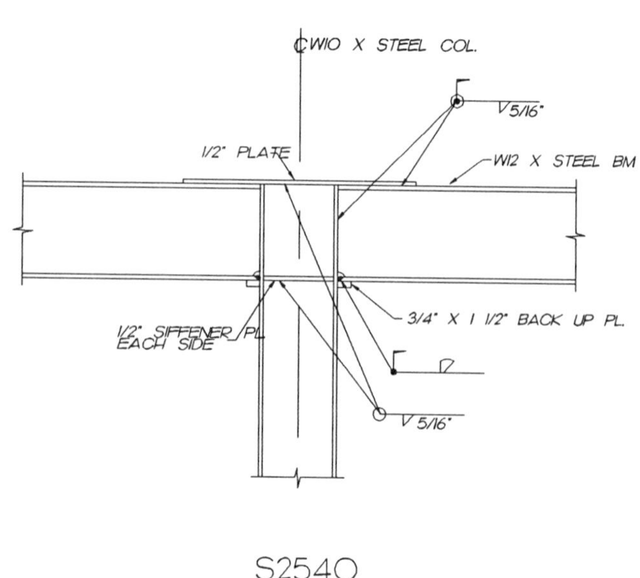

S2540

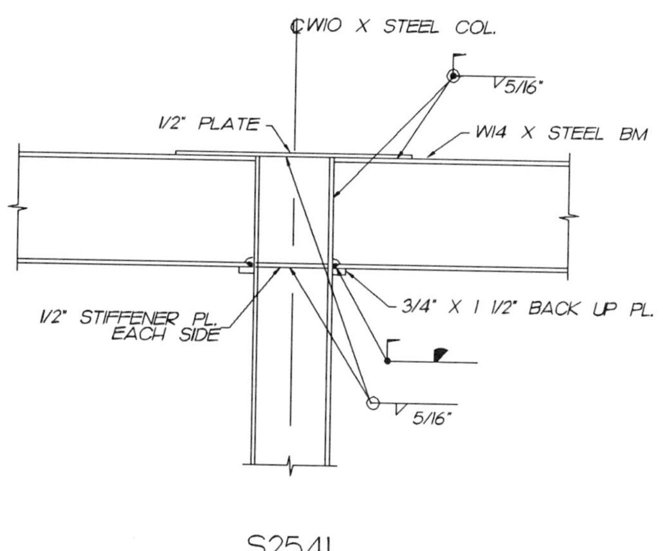

S2541

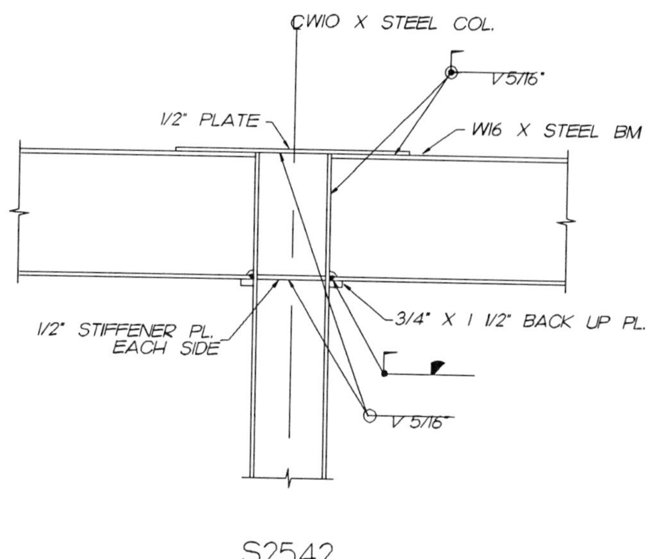

S2542

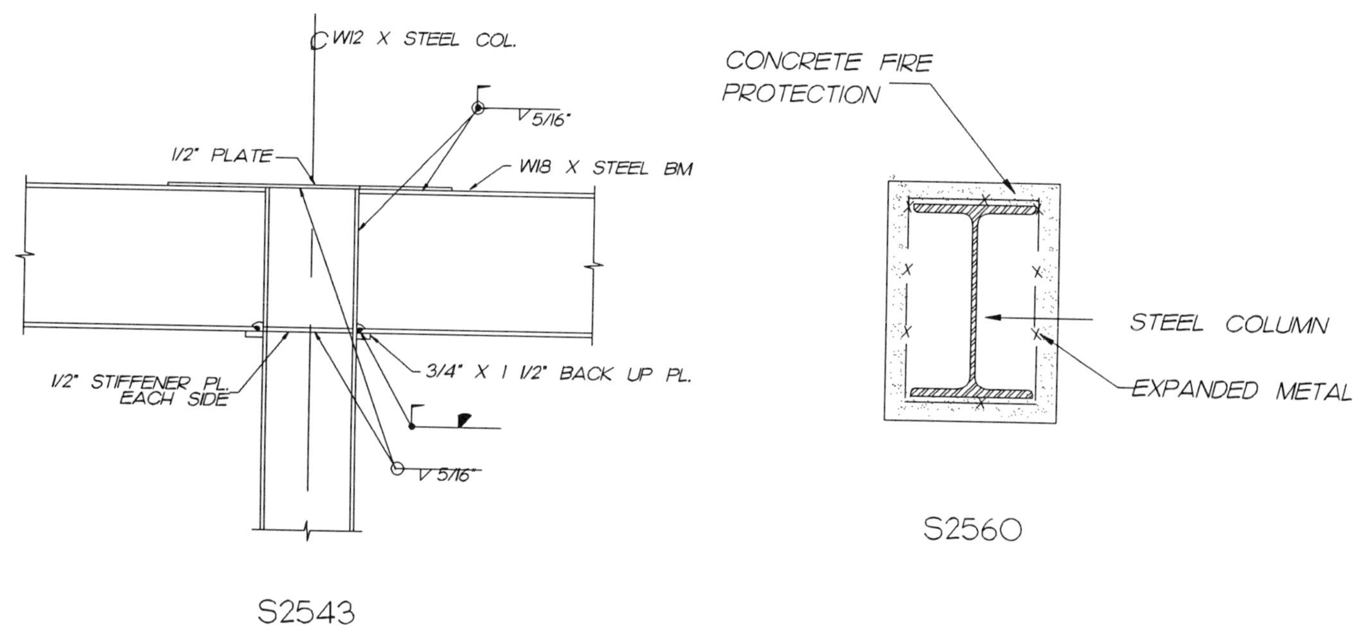

S2543

S2560

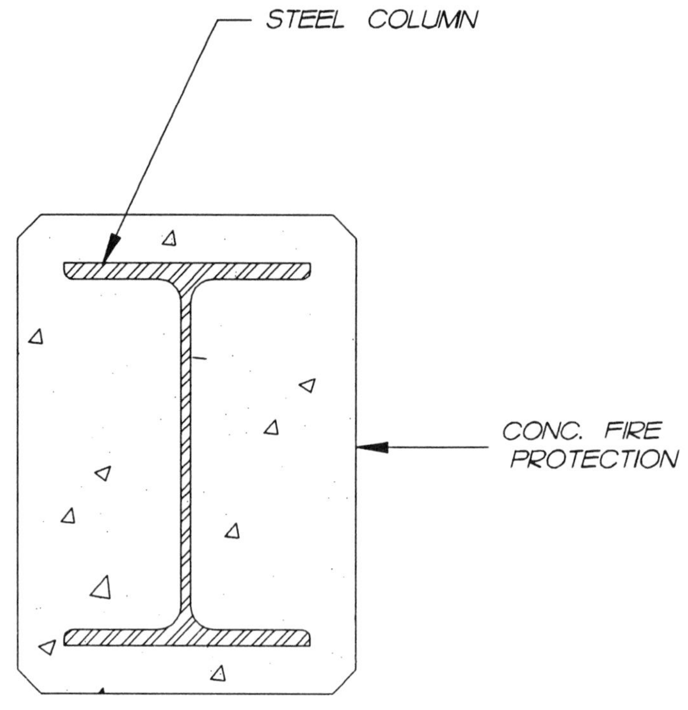

S256I

CHAPTER 4 STEEL SECTION 3 WALL - WOOD FLOOR		
EXTERIOR WALL - STEEL STUDS JOISTS PERP. TO WALL LT. WT. CONCRETE METAL SHEAR RESIST. CLIPS	3 5/8" STUDS ABOVE & BELOW	S3020 - S3024
	3 5/8" STUDS ABOVE - 5 5/8" BELOW	S3025 - S3029
EXTERIOR WALL - STEEL STUDS JOISTS PARALELL TO WALL LT. WT. CONCRETE METAL SHEAR RESIST. CLIPS	3 5/8" STUDS ABOVE & BELOW	S3040 - S3043
	3 5/8" STUDS ABOVE - 5 5/8" BELOW	S3044 - S3049
EXTERIOR WALL - STEEL STUDS JOISTS PERP. TO WALL NO LT. WT. CONCRETE ANGLE & 4 X LEDGER	3 5/8" STUDS	S3060 - S3064
	5 5/8" STUDS	S3065 - S3069
EXTERIOR WALL - STEEL STUDS JOISTS PARALELL TO WALL NO LT. WT. CONCRETE ANGLE & 4 X LEDGER	3 5/8" STUDS	S3080 - S3084
	5 5/8" STUDS	S3085 - S3089
STEEL BEAM EXT. WOOD WALL JOISTS PERP. TO BEAM JOISTS ON THE TOP FLANGE METAL SHEAR RESIST. CLIPS	————	S3100 - S3104
STEEL BEAM EXT. WOOD WALL JOISTS PERP. TO BEAM JOISTS ON THE TOP FLANGE METAL SHEAR RESIST. CLIPS	————	S3120 - S3124
STEEL BEAM EXT. WOOD WALL JOISTS PERP. TO BEAM JOISTS ON THE TOP FLANGE BOLTED ANGLE SHEAR RESIST.	————	S3140 - S3144
STEEL BEAM EXT. WOOD WALL JOISTS PARALELL TO WALL JOISTS ON THE TOP FLANGE BOLTED ANGLE SHEAR RESIST.	————	S3160 - S3164
STEEL BEAM EXT. WOOD WALL JOISTS PERP. TO BEAM JOISTS FLUSH WITH BOTTOM FL. METAL STRAP JOISTS TO BEAM	————	S3180 - S3182
STEEL BEAM EXT. WOOD WALL JOISTS PARALELL TO WALL JOISTS FLUSH WITH BOTTOM FL. METAL STRAP JOISTS TO BEAM	————	S3200 - S3202
STEEL BEAM - INT. WOOD WALL JOISTS PERP. TO BEAM JOISTS ON THE TOP FLANGE BOLTED ANGLE SHEAR RESIST.	————	S3220 - S3223
STEEL BEAM - INT. WOOD WALL JOISTS PARALELL TO WALL JOISTS ON THE TOP FLANGE BOLTED ANGLE SHEAR RESIST.	————	S3240 - S3243
STEEL BEAM - INT. WOOD WALL JOISTS PERP & PARALELL ALT SIDES JOISTS ON THE TOP FLANGE BOLTED ANGLE SHEAR RESIST.	————	S3260 - S3263
STEEL BEAM - INT. WOOD WALL JOISTS PERP. TO BEAM JOISTS FLUSH WITH BOTTOM FL. METAL STRAP JOISTS TO BEAM	————	S3280 - S3283
STEEL BEAM - INT. WOOD WALL JOISTS PARALELL TO WALL JOISTS FLUSH WITH BOTTOM FL. METAL STRAP JOISTS TO BEAM	————	S3300 - S3303
STEEL BEAM - INT. WOOD WALL JOISTS PERP & PARALELL ALT SIDES JOISTS FLUSH WITH BOTTOM FL. METAL STRAP JOISTS TO BEAM	————	S3320 - S3323

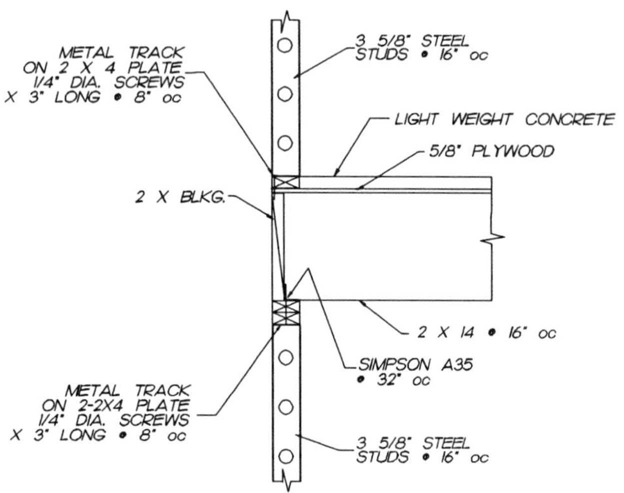

S3020

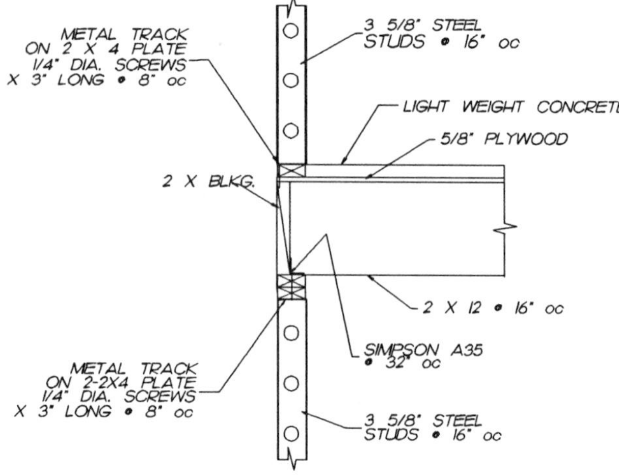

S3021

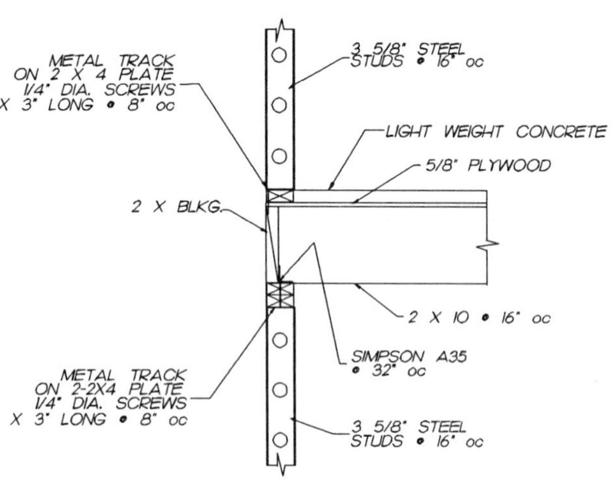

S3022

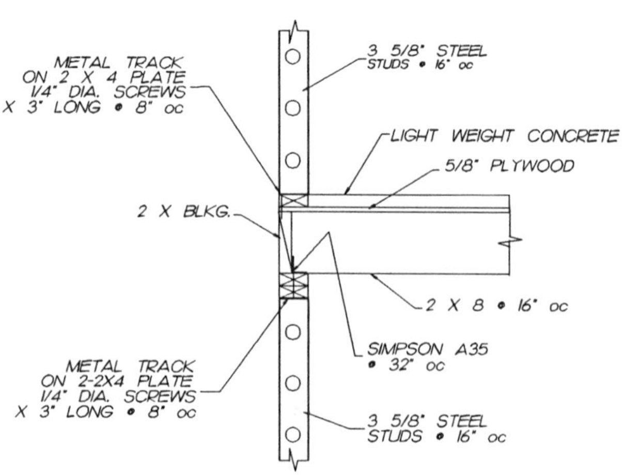

S3023

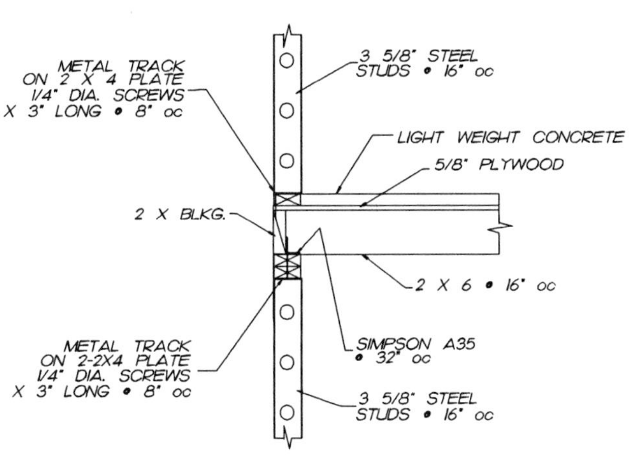

S3024

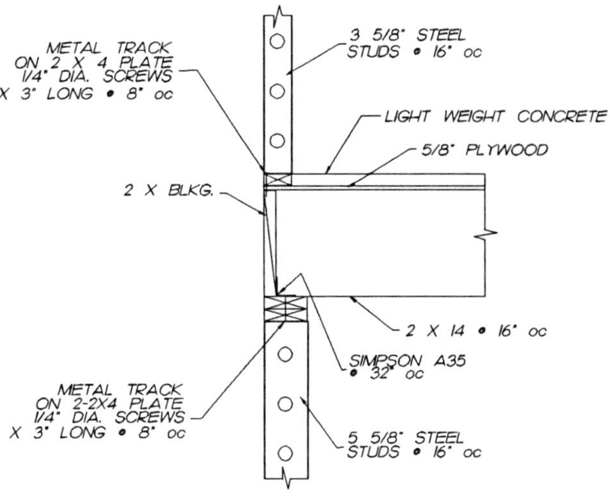

S3025

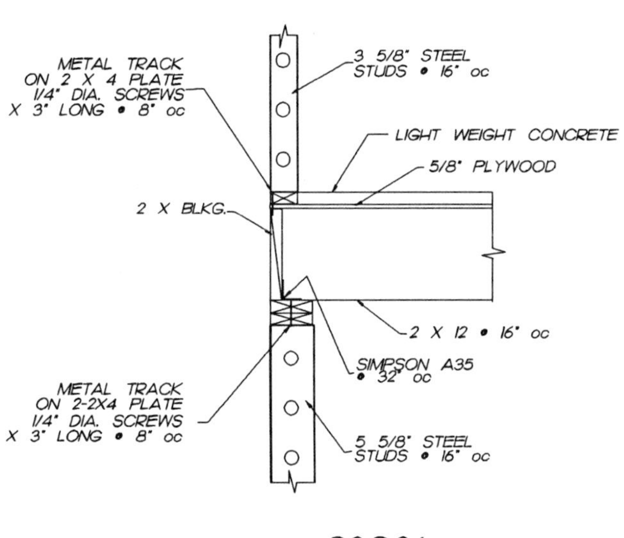

S3026

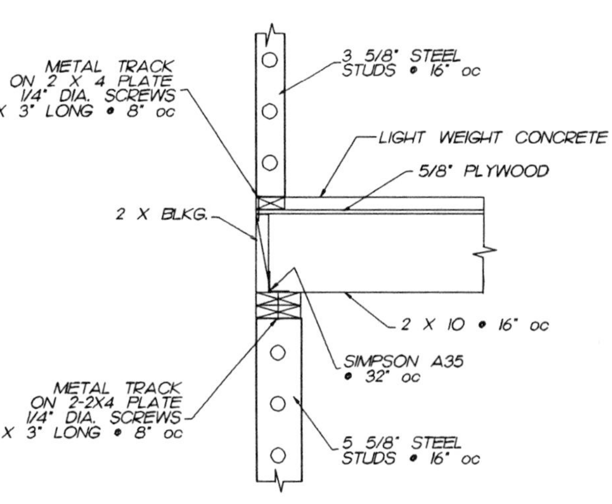

S3027

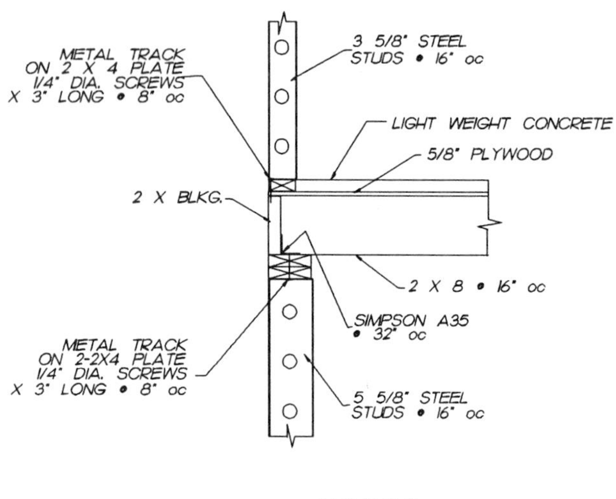

S3028

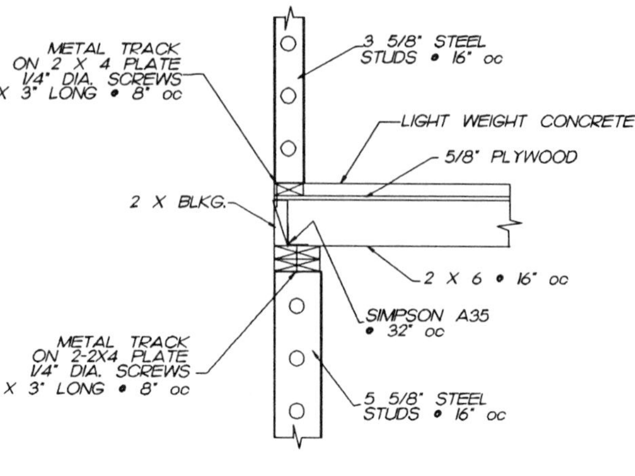

S3029

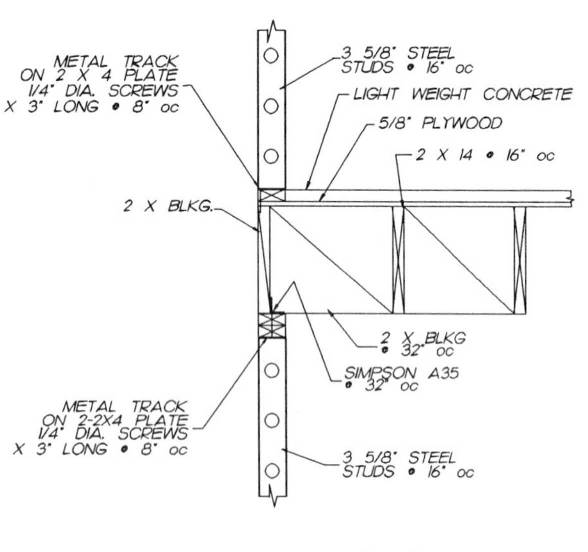

S3040

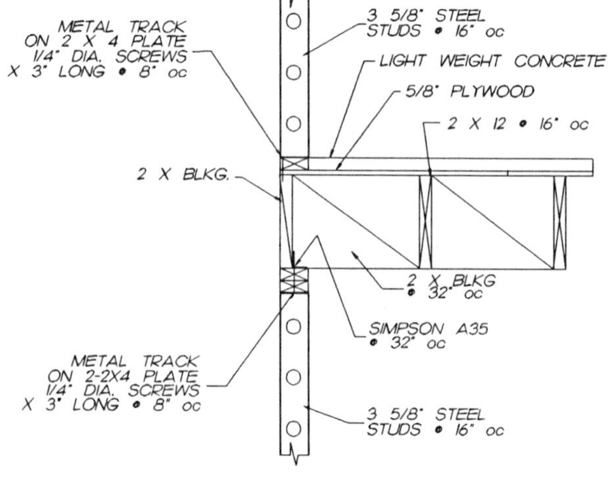

S3041

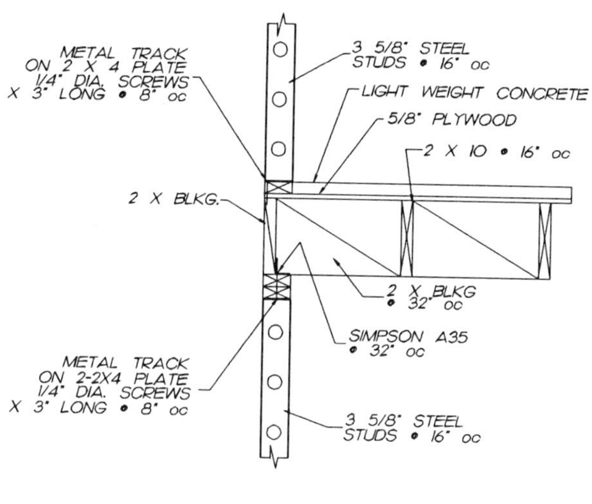

S3042

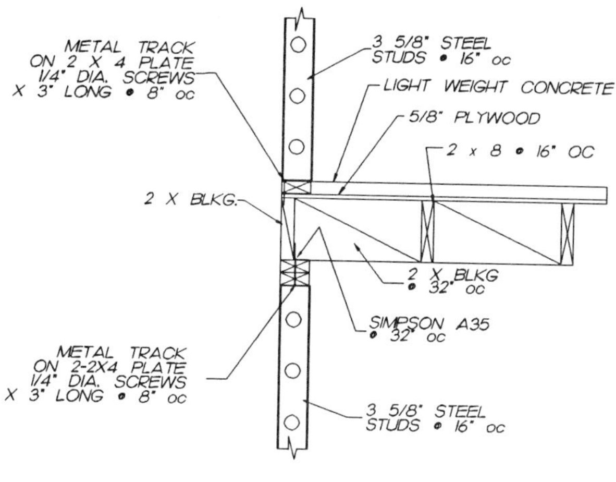

S3043

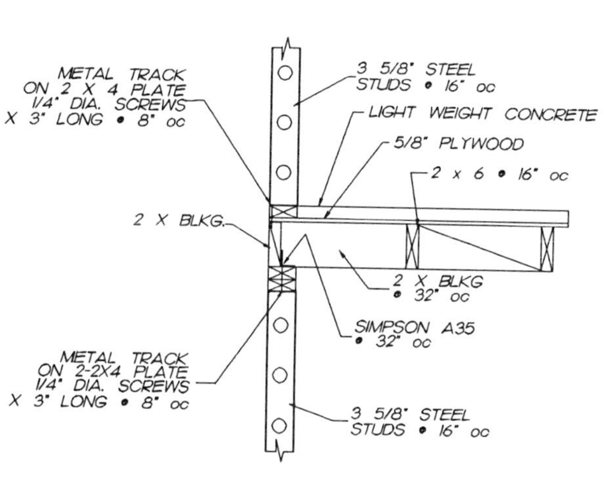

S3044

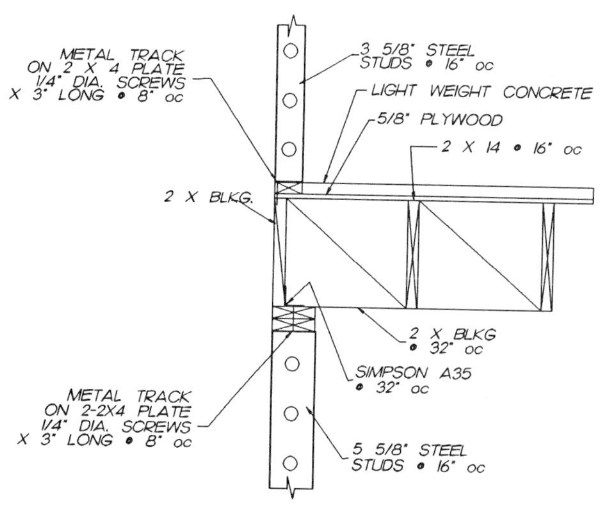

S3045

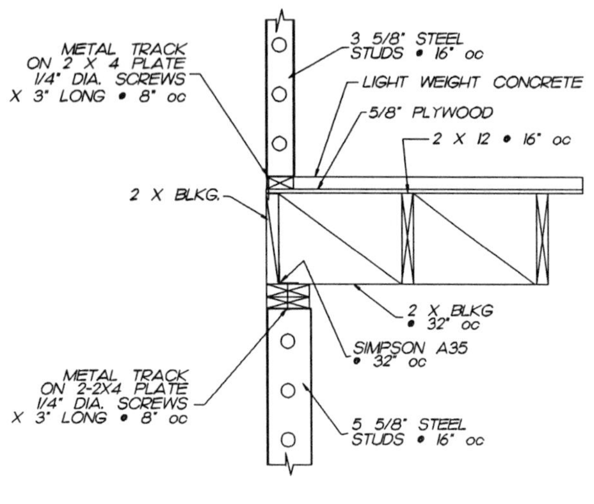

S3046

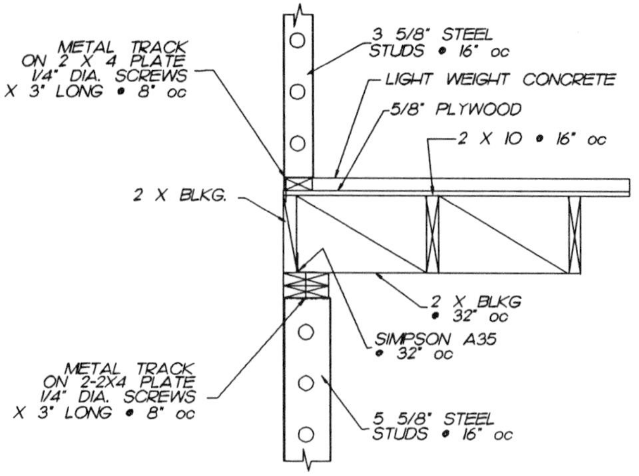

S3047

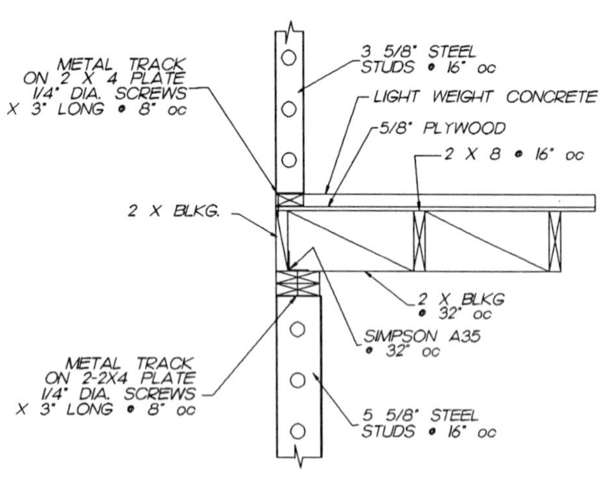

S3048

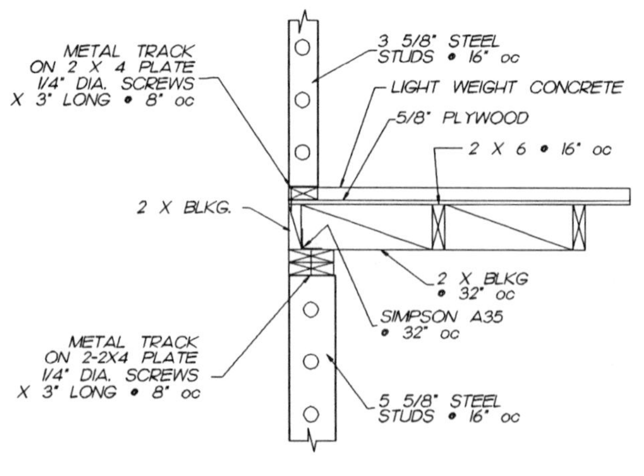

S3049

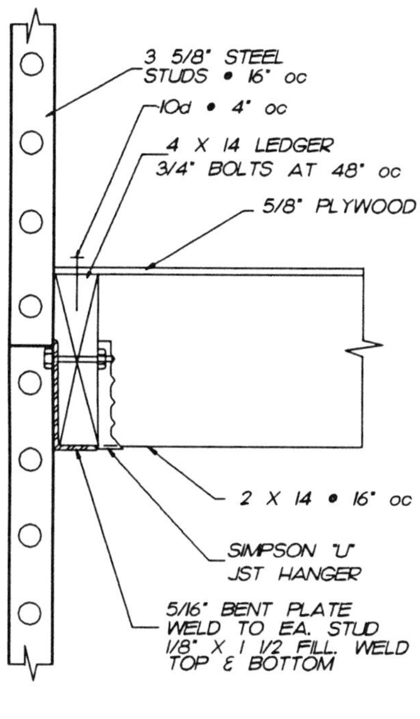

S3060

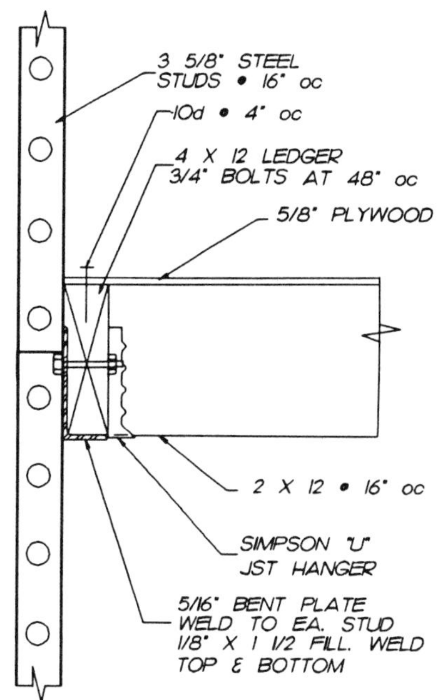

S3061

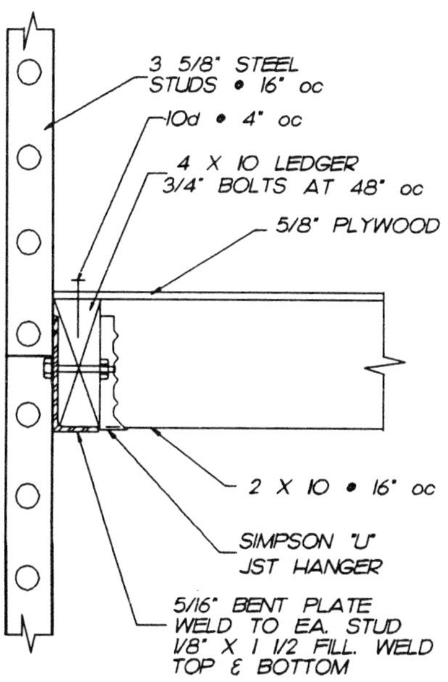

S3062

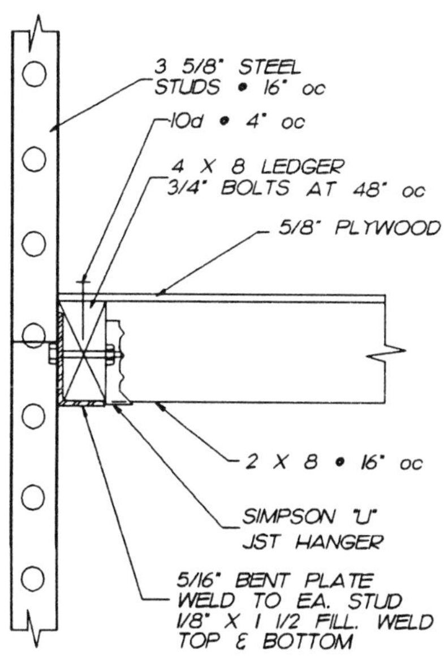

S3063

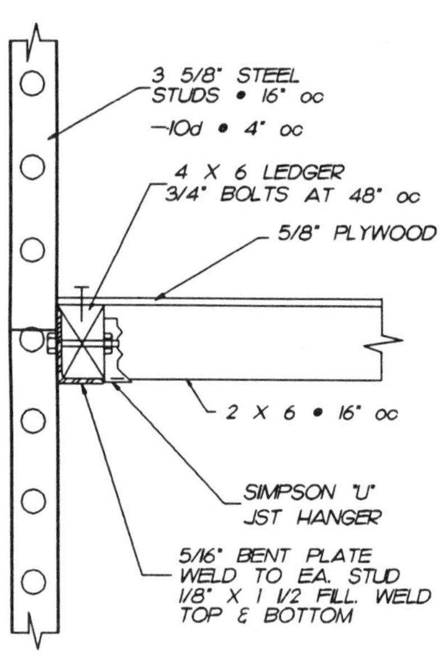

S3064

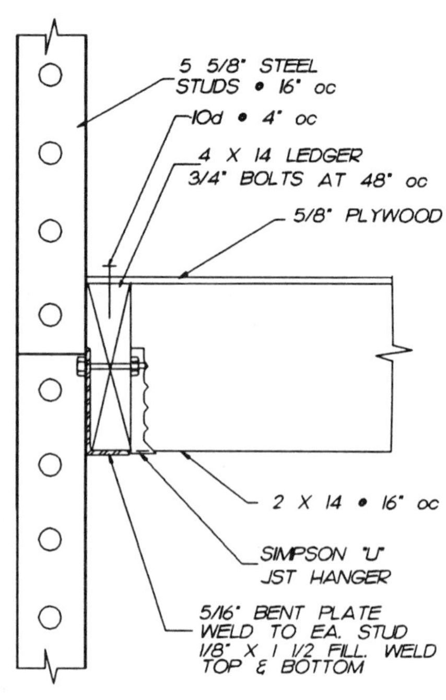

S3065

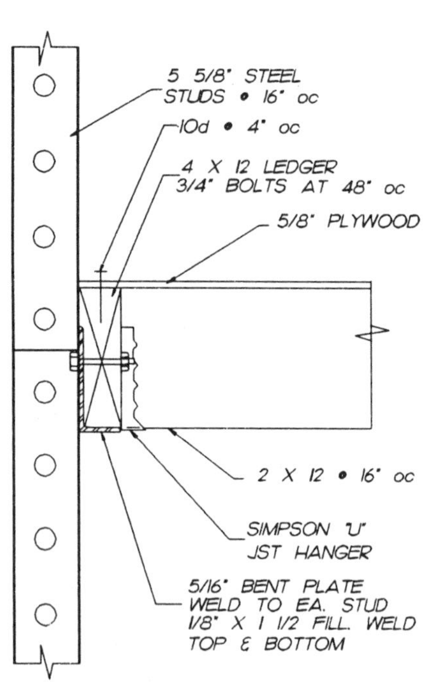

S3066

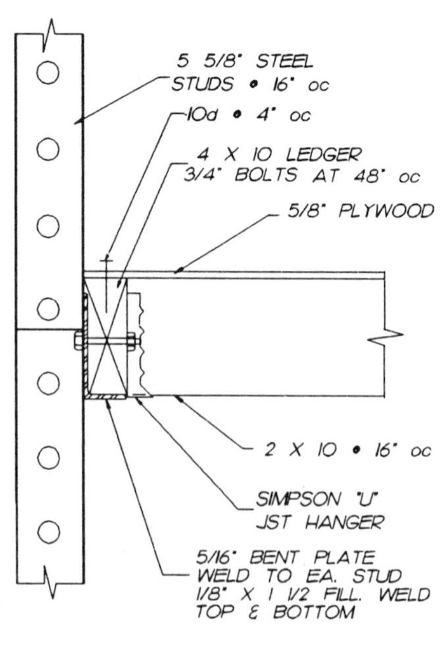

S3067

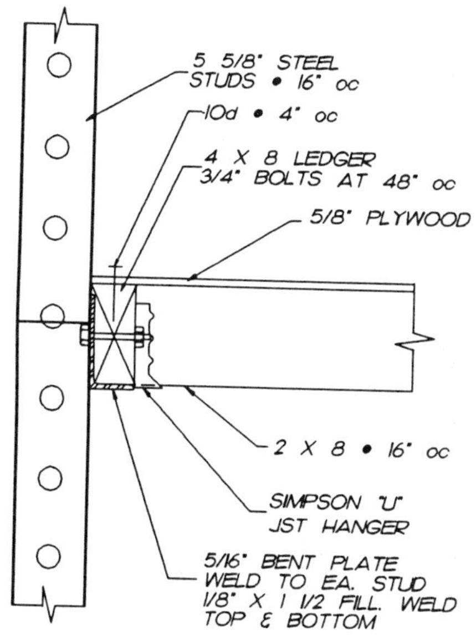

S3068

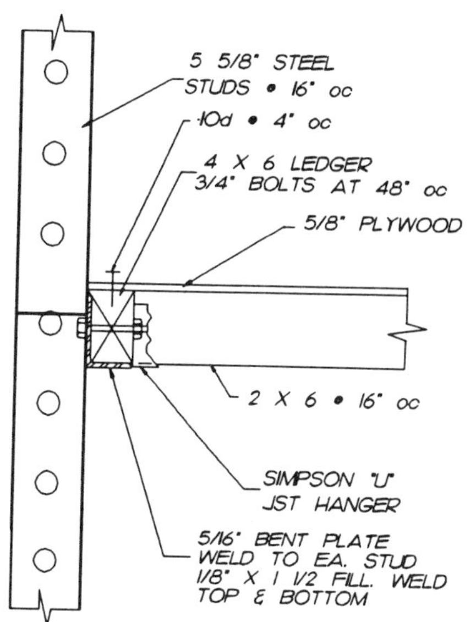

S3069

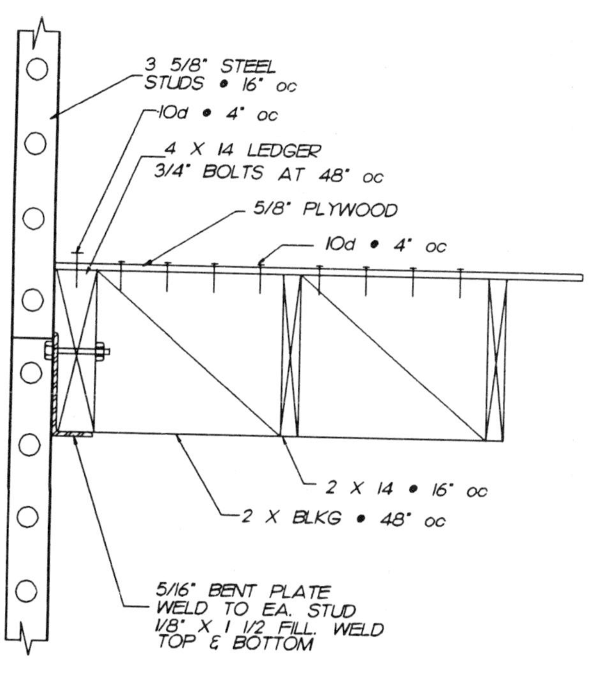

S3080

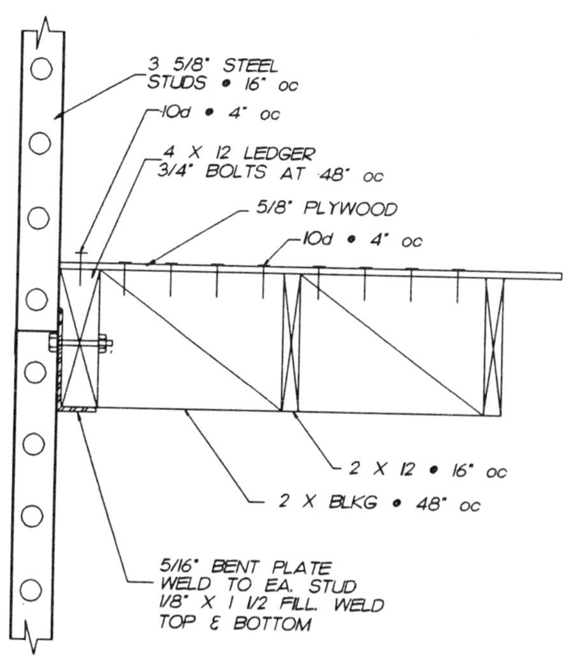

S3081

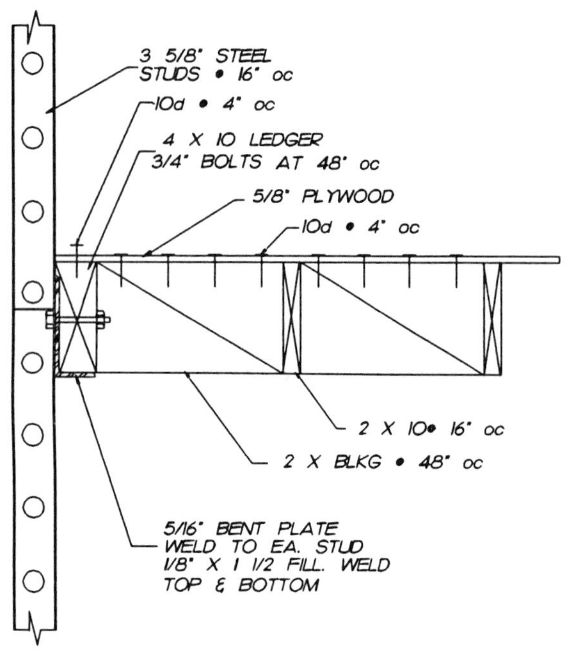

S3082

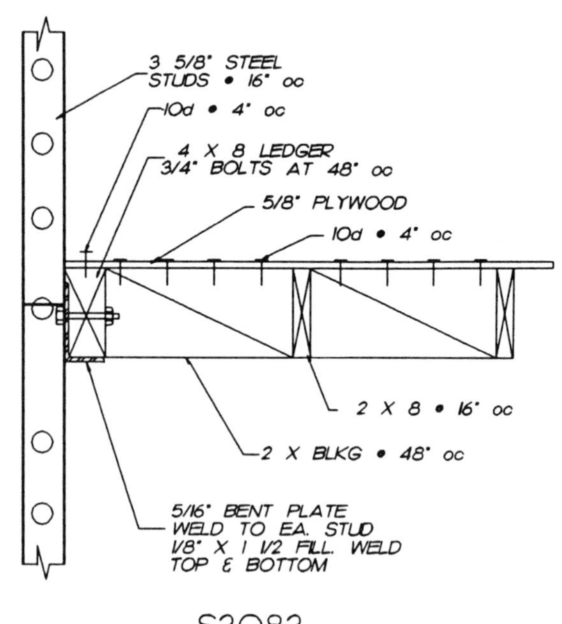

S3083

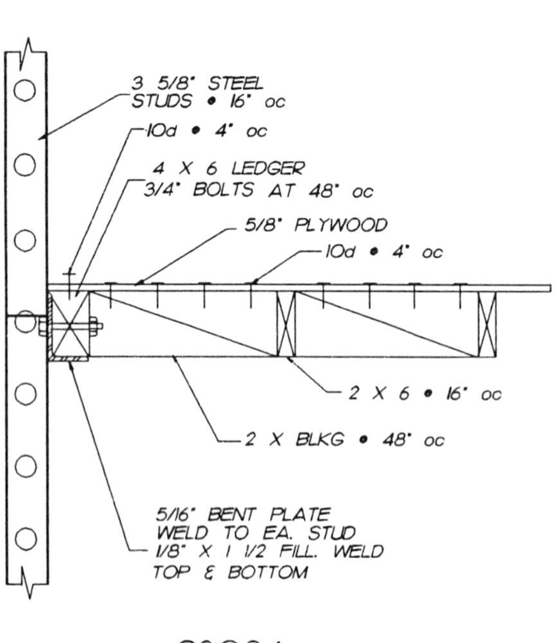

S3084

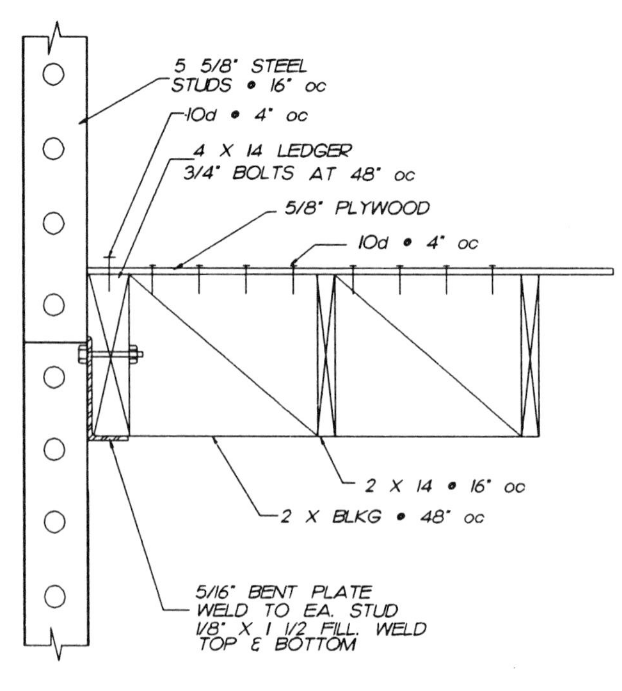

S3085

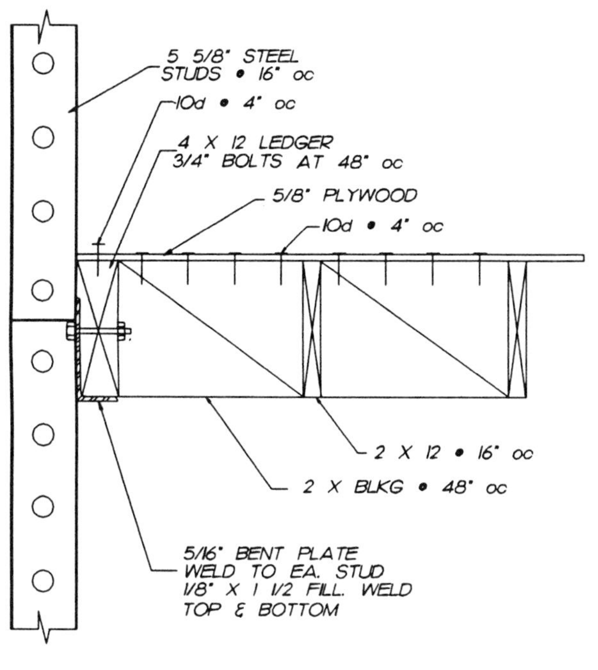

S3086

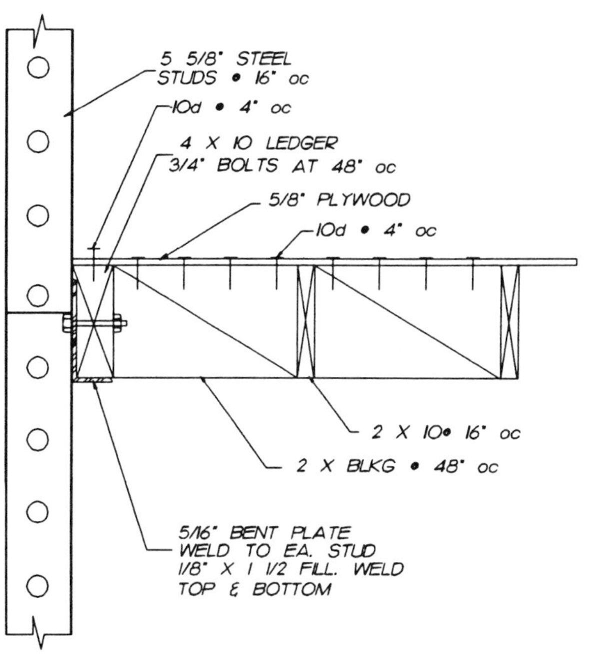

S3087

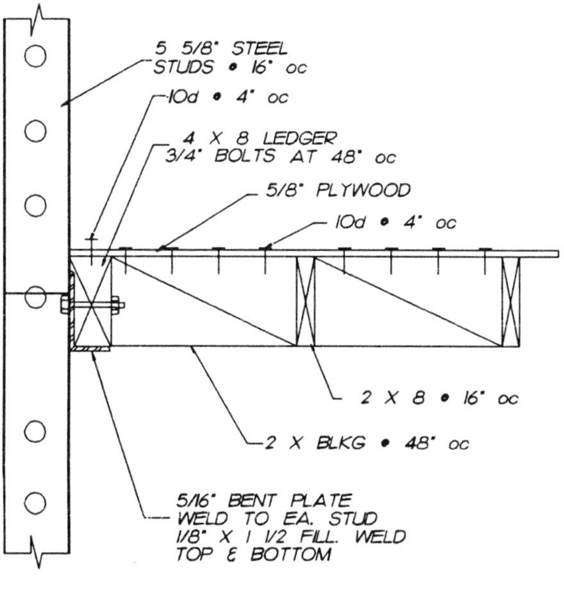

S3088

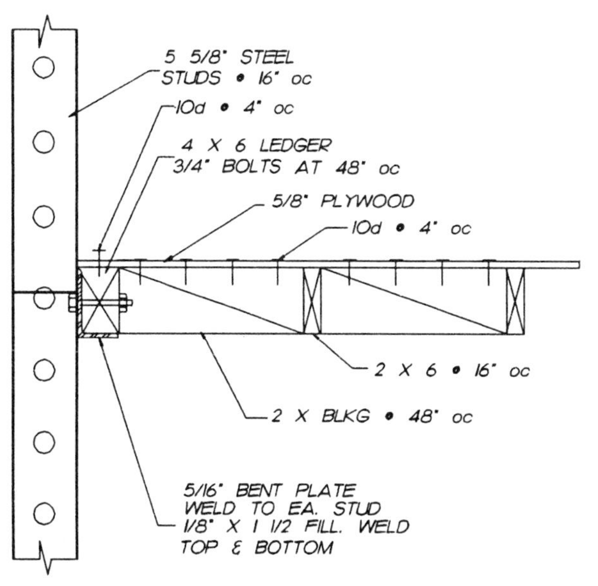

S3089

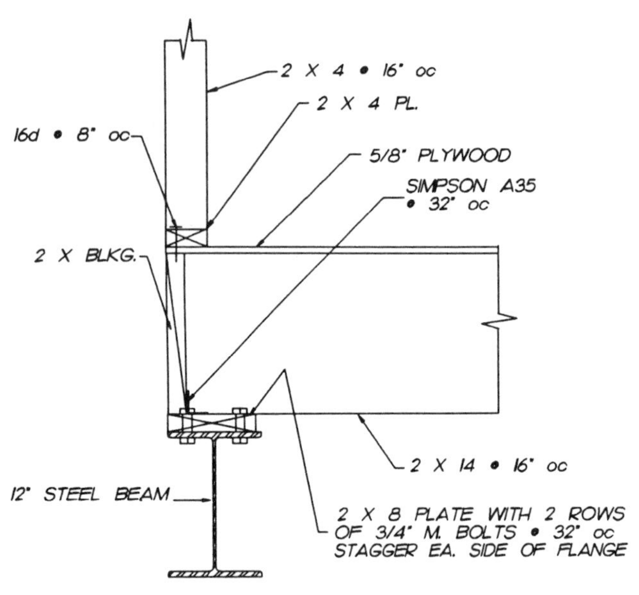

S3100

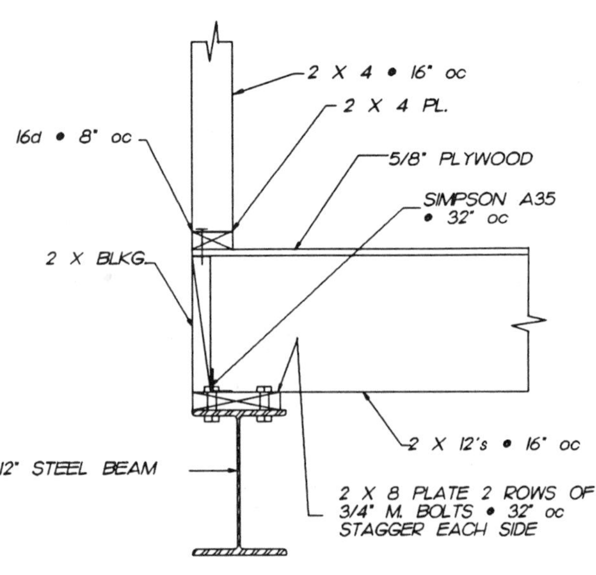

S3101

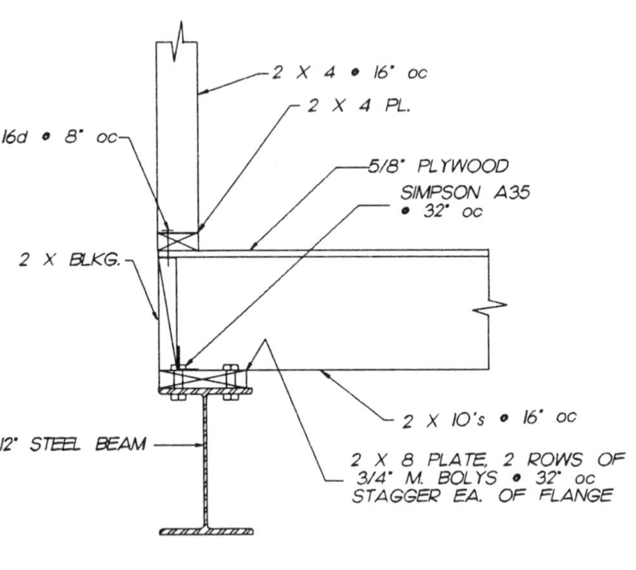

S3102

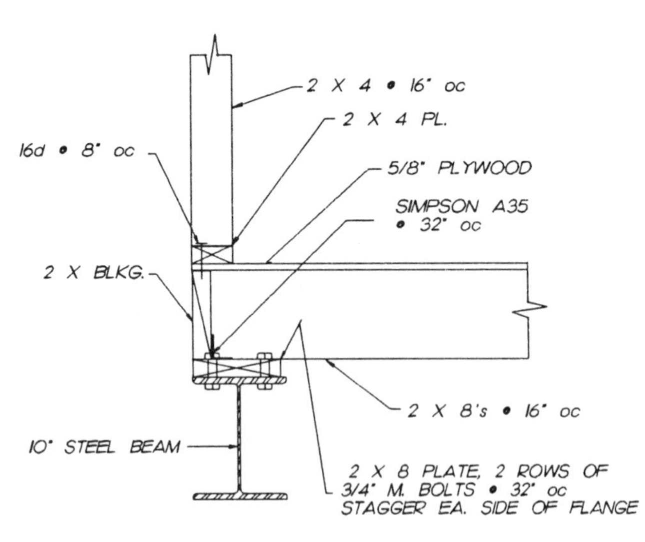

S3103

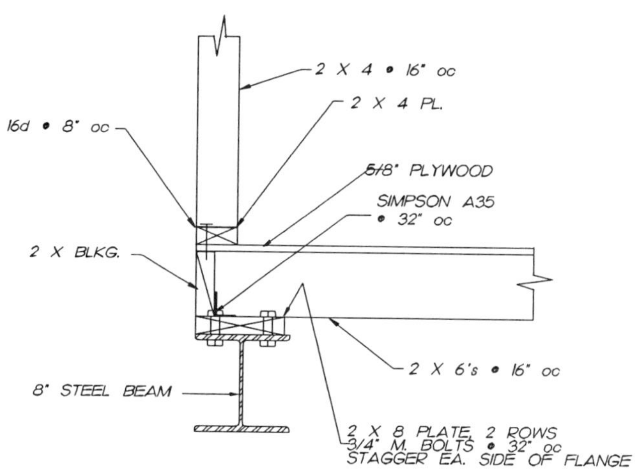

S3104

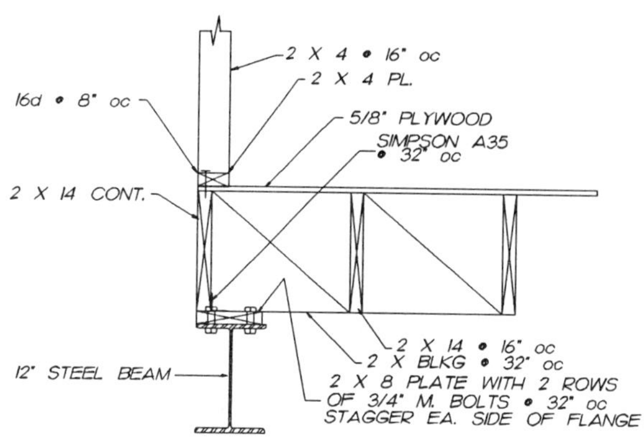

S3120

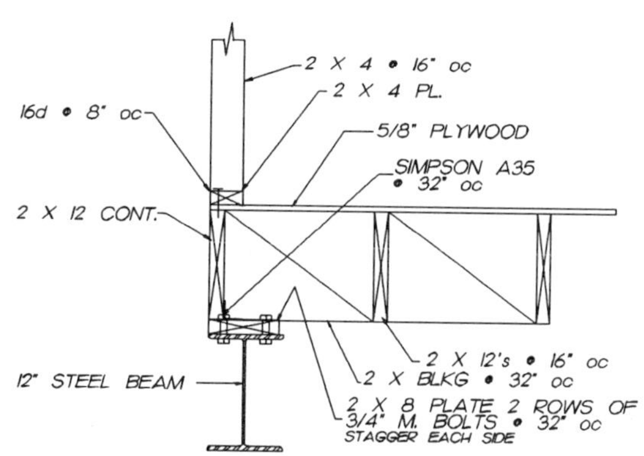

S3121

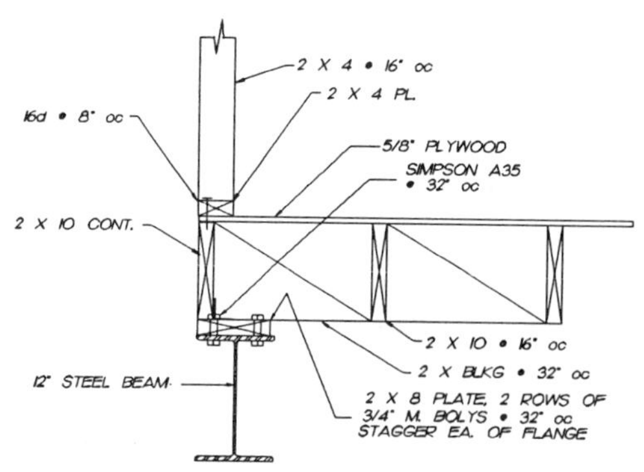

S3122

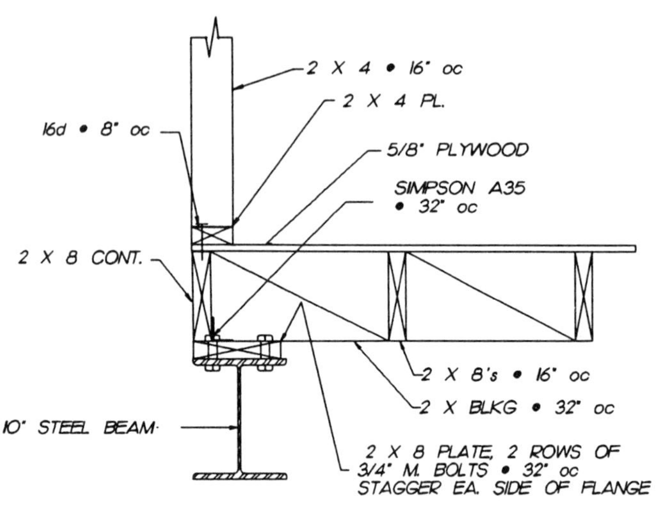

S3123

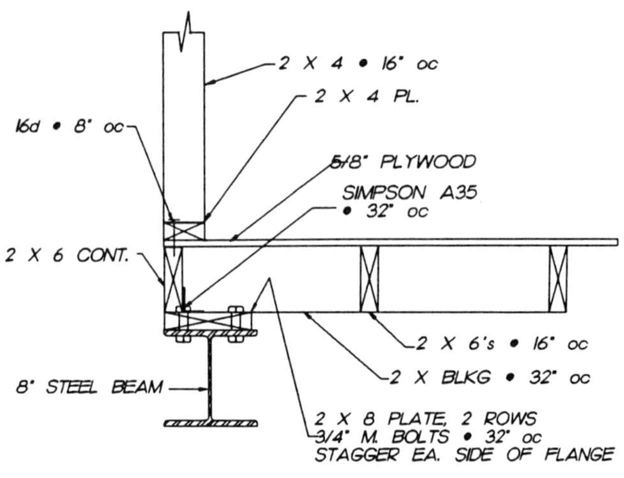

S3124

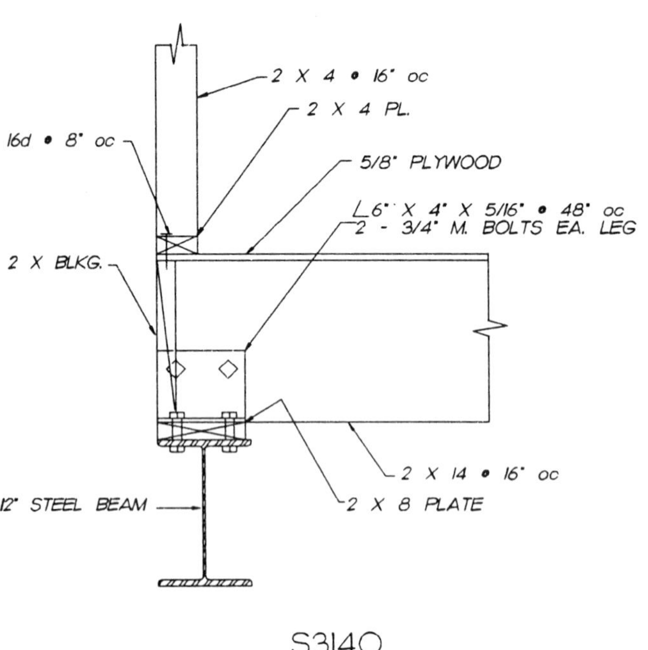

S3140

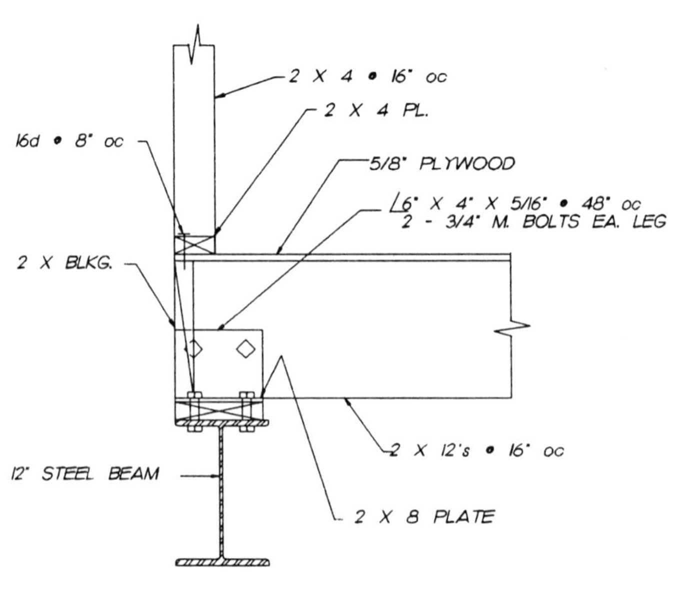

S3141

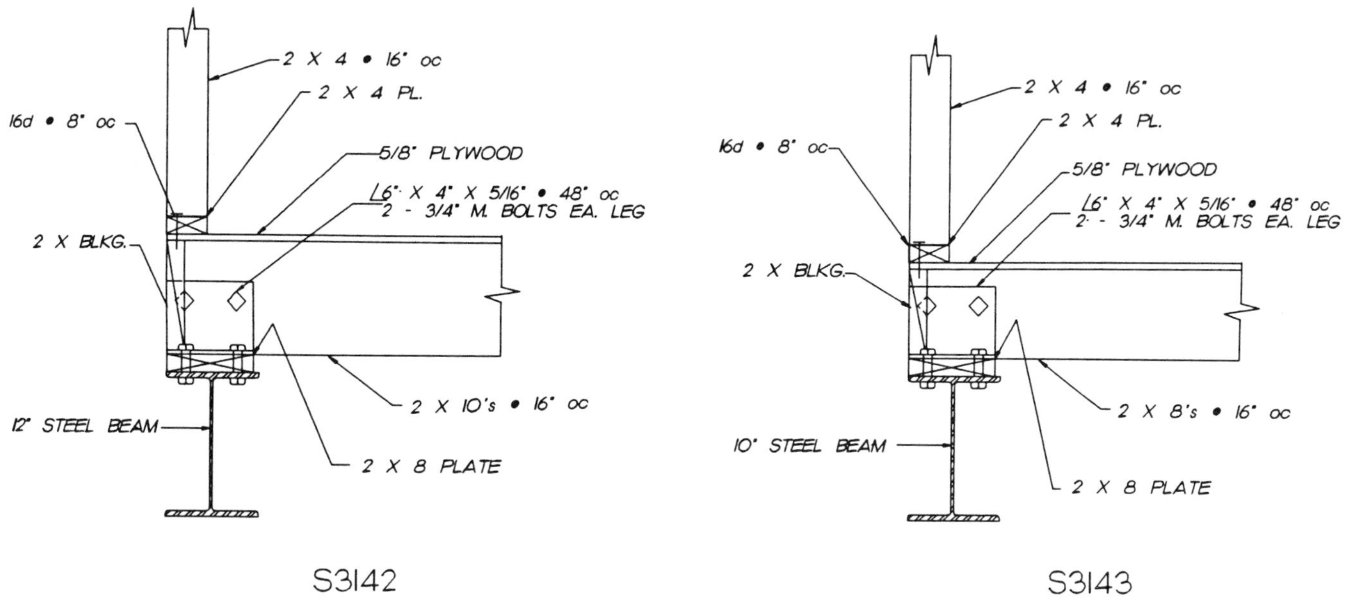
2 X 4 • 16" oc
2 X 4 PL.
16d • 8" oc
5/8" PLYWOOD
L6" X 4" X 5/16" • 48" oc
2 - 3/4" M. BOLTS EA. LEG
2 X BLKG.
2 X 10's • 16" oc
12" STEEL BEAM
2 X 8 PLATE
S3142
2 X 4 • 16" oc
2 X 4 PL.
16d • 8" oc
5/8" PLYWOOD
L6" X 4" X 5/16" • 48" oc
2 - 3/4" M. BOLTS EA. LEG
2 X BLKG.
2 X 8's • 16" oc
10" STEEL BEAM
2 X 8 PLATE
S3143

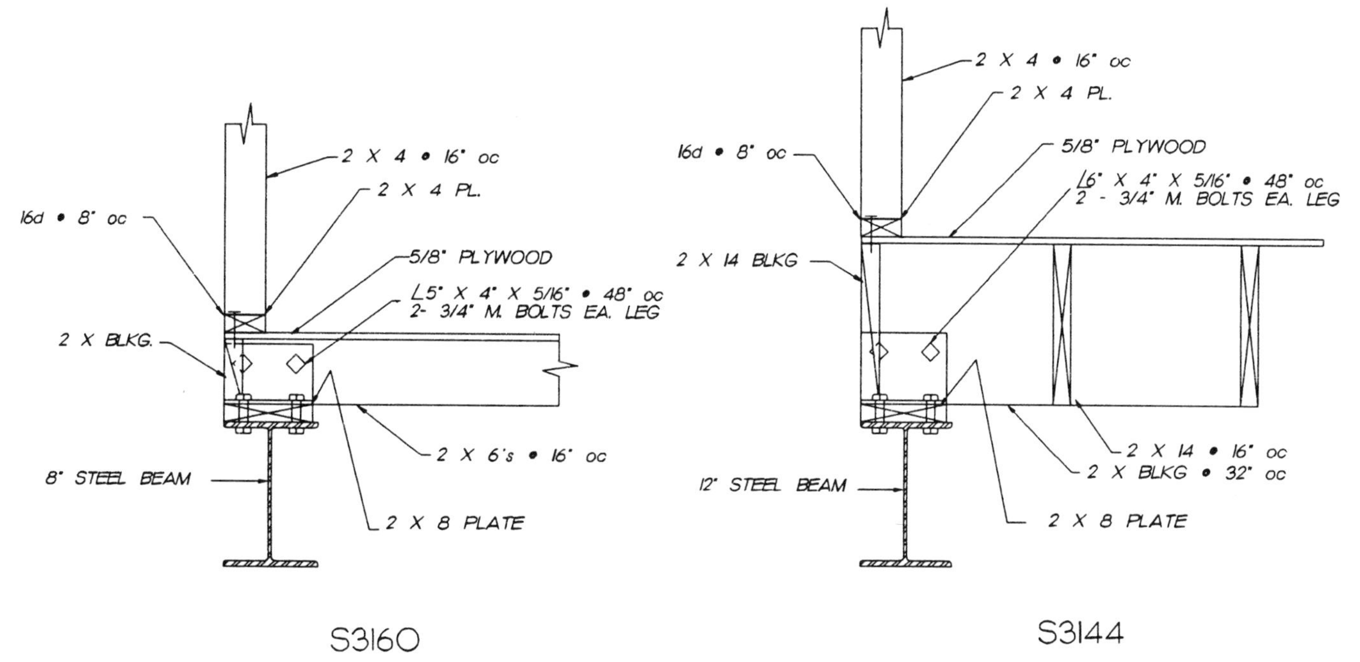
2 X 4 • 16" oc
2 X 4 PL.
16d • 8" oc
5/8" PLYWOOD
L5" X 4" X 5/16" • 48" oc
2- 3/4" M. BOLTS EA. LEG
2 X BLKG.
2 X 6's • 16" oc
8" STEEL BEAM
2 X 8 PLATE
S3160
2 X 4 • 16" oc
2 X 4 PL.
16d • 8" oc
5/8" PLYWOOD
L6" X 4" X 5/16" • 48" oc
2 - 3/4" M. BOLTS EA. LEG
2 X 14 BLKG
2 X 14 • 16" oc
2 X BLKG • 32" oc
12" STEEL BEAM
2 X 8 PLATE
S3144

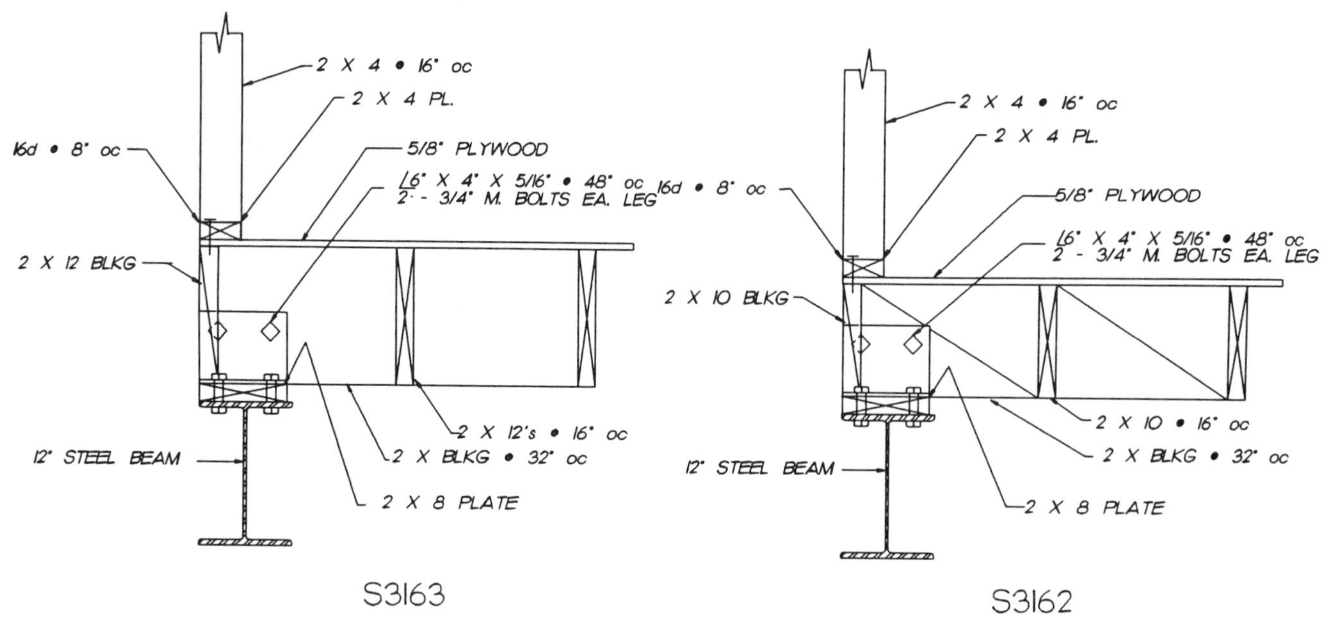
2 X 4 • 16" oc
2 X 4 PL.
16d • 8" oc
5/8" PLYWOOD
L6" X 4" X 5/16" • 48" oc
2 - 3/4" M. BOLTS EA. LEG
2 X 12 BLKG
2 X 12's • 16" oc
12" STEEL BEAM
2 X BLKG • 32" oc
2 X 8 PLATE
S3163
2 X 4 • 16" oc
2 X 4 PL.
16d • 8" oc
5/8" PLYWOOD
L6" X 4" X 5/16" • 48" oc
2 - 3/4" M. BOLTS EA. LEG
2 X 10 BLKG
2 X 10 • 16" oc
12" STEEL BEAM
2 X BLKG • 32" oc
2 X 8 PLATE
S3162

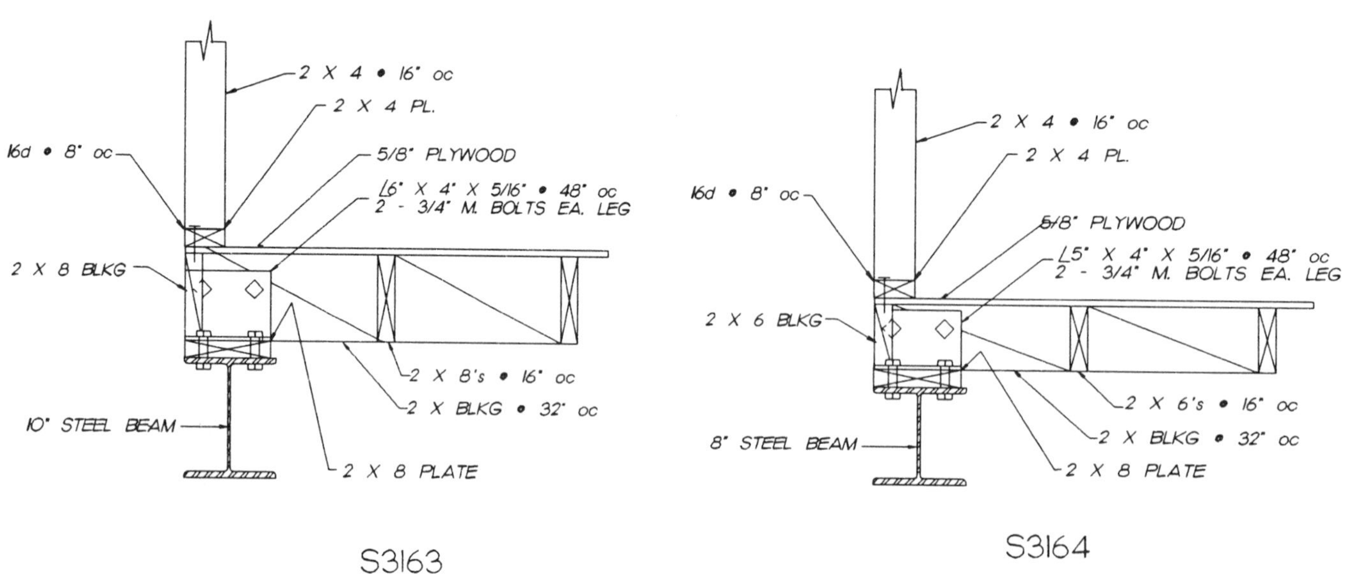
2 X 4 • 16" oc
2 X 4 PL.
16d • 8" oc
5/8" PLYWOOD
L6" X 4" X 5/16" • 48" oc
2 - 3/4" M. BOLTS EA. LEG
2 X 8 BLKG
2 X 8's • 16" oc
2 X BLKG • 32" oc
10" STEEL BEAM
2 X 8 PLATE
S3163
2 X 4 • 16" oc
2 X 4 PL.
16d • 8" oc
5/8" PLYWOOD
L5" X 4" X 5/16" • 48" oc
2 - 3/4" M. BOLTS EA. LEG
2 X 6 BLKG
2 X 6's • 16" oc
2 X BLKG • 32" oc
8" STEEL BEAM
2 X 8 PLATE
S3164

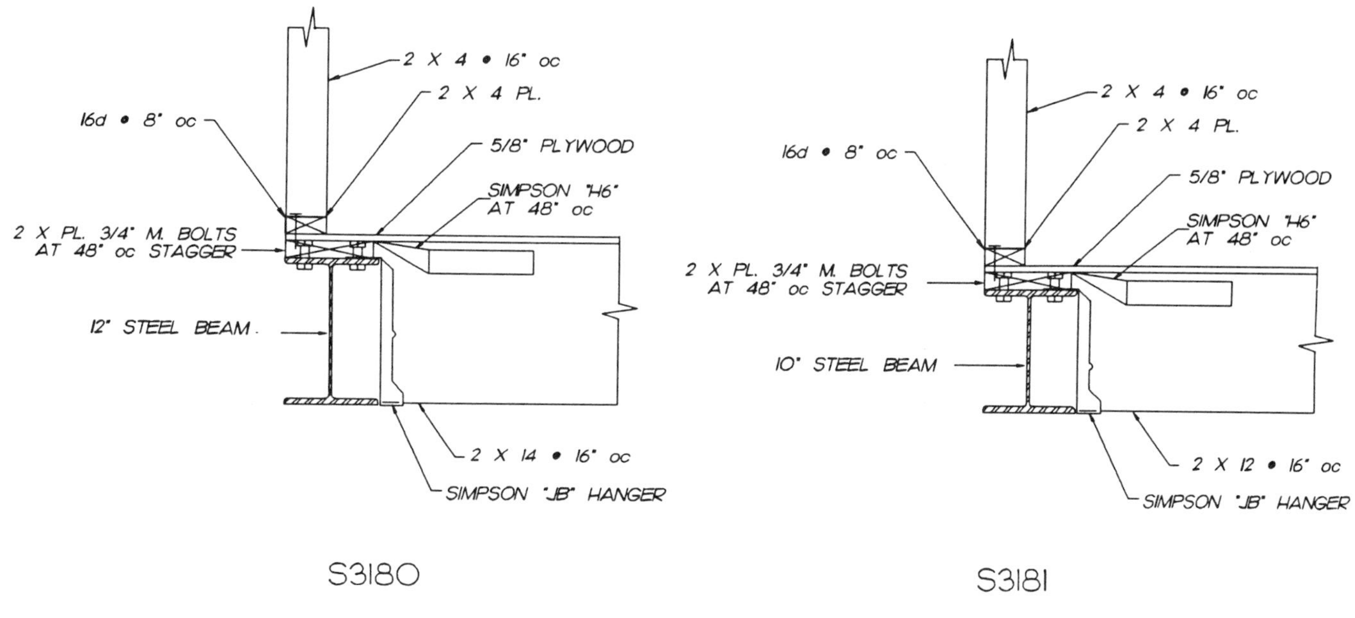
2 X 4 • 16" oc
2 X 4 PL.
16d • 8" oc
5/8" PLYWOOD
SIMPSON "H6"
AT 48" oc
2 X PL. 3/4" M. BOLTS
AT 48" oc STAGGER
12" STEEL BEAM.
2 X 14 • 16" oc
SIMPSON "JB" HANGER
S3180
2 X 4 • 16" oc
2 X 4 PL.
16d • 8" oc
5/8" PLYWOOD
SIMPSON "H6"
AT 48" oc
2 X PL. 3/4" M. BOLTS
AT 48" oc STAGGER
10" STEEL BEAM
2 X 12 • 16" oc
SIMPSON "JB" HANGER
S3181

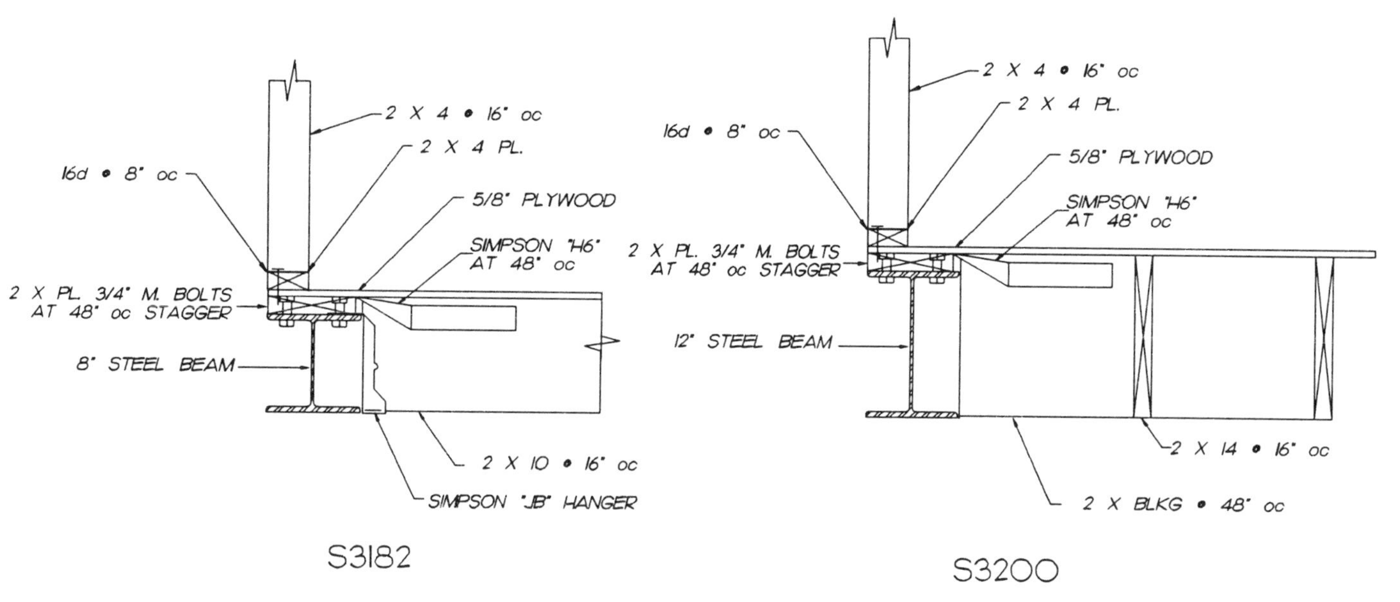
2 X 4 • 16" oc
2 X 4 PL.
16d • 8" oc
5/8" PLYWOOD
SIMPSON "H6"
AT 48" oc
2 X PL. 3/4" M. BOLTS
AT 48" oc STAGGER
8" STEEL BEAM
2 X 10 • 16" oc
SIMPSON "JB" HANGER
S3182
2 X 4 • 16" oc
2 X 4 PL.
16d • 8" oc
5/8" PLYWOOD
SIMPSON "H6"
AT 48" oc
2 X PL. 3/4" M. BOLTS
AT 48" oc STAGGER
12" STEEL BEAM
2 X 14 • 16" oc
2 X BLKG • 48" oc
S3200

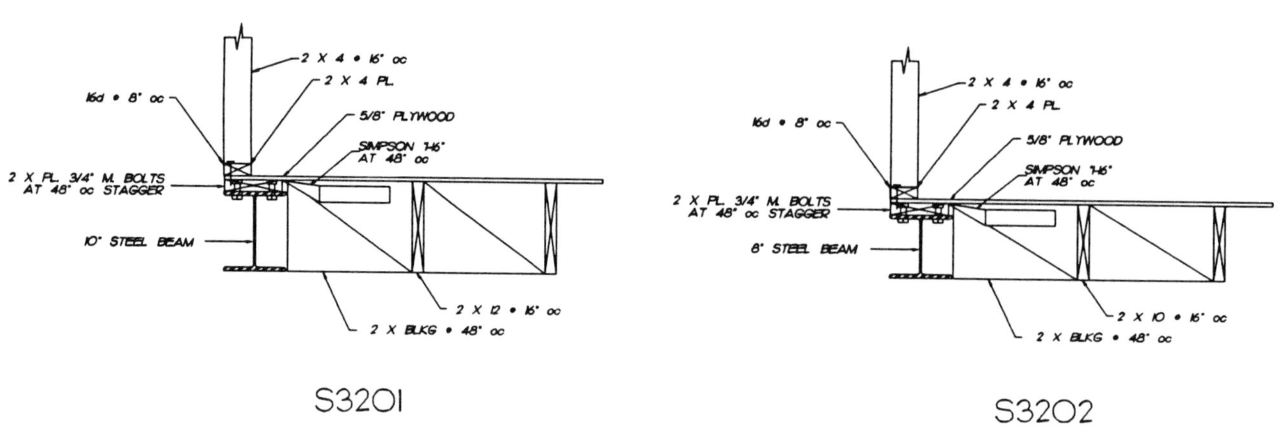
2 X 4 • 16" oc
2 X 4 PL
16d • 8" oc
5/8" PLYWOOD
SIMPSON T-16"
AT 48" oc
2 X PL 3/4" M. BOLTS
AT 48" oc STAGGER
10" STEEL BEAM
2 X 12 • 16" oc
2 X BLKG • 48" oc
S3201
2 X 4 • 16" oc
2 X 4 PL
16d • 8" oc
5/8" PLYWOOD
SIMPSON T-16"
AT 48" oc
2 X PL 3/4" M. BOLTS
AT 48" oc STAGGER
8" STEEL BEAM
2 X 10 • 16" oc
2 X BLKG • 48" oc
S3202

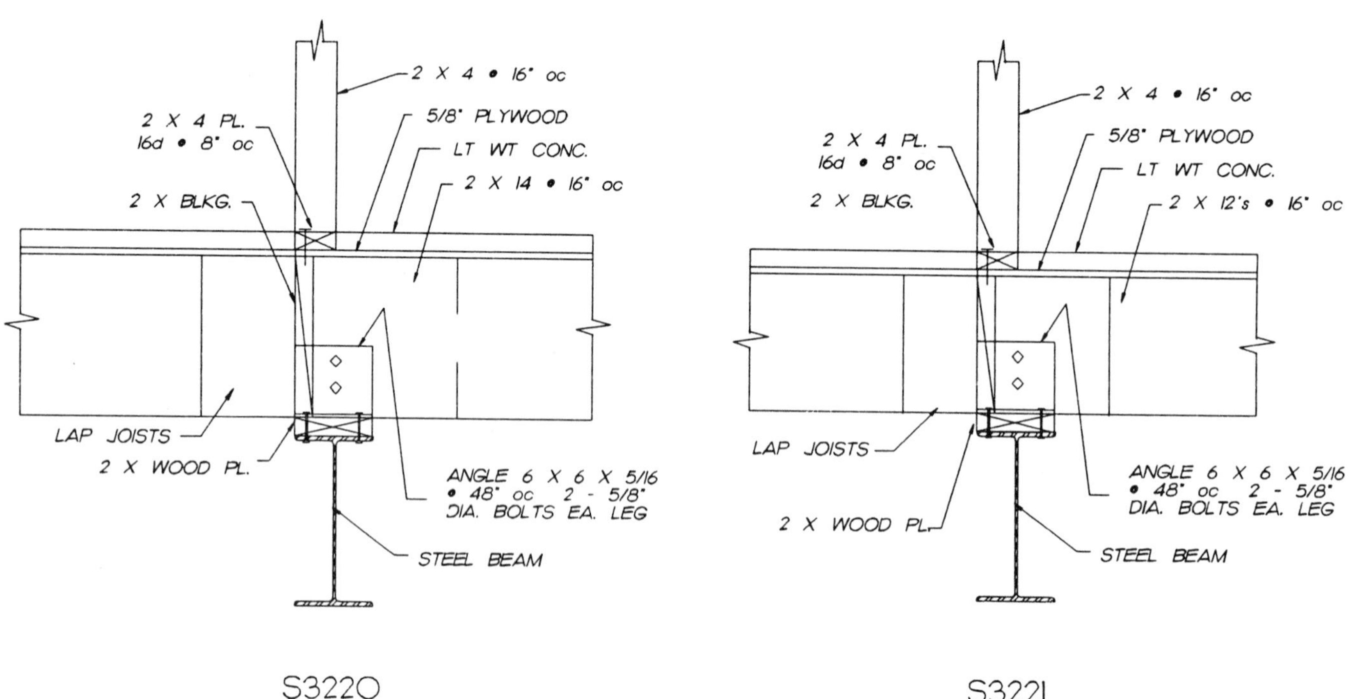
2 X 4 • 16" oc
2 X 4 PL.
16d • 8" oc
5/8" PLYWOOD
LT WT CONC.
2 X 14 • 16" oc
2 X BLKG.
LAP JOISTS
2 X WOOD PL.
ANGLE 6 X 6 X 5/16
• 48" oc 2 - 5/8"
DIA. BOLTS EA. LEG
STEEL BEAM
S3220
2 X 4 • 16" oc
2 X 4 PL.
16d • 8" oc
5/8" PLYWOOD
LT WT CONC.
2 X 12's • 16" oc
2 X BLKG.
LAP JOISTS
ANGLE 6 X 6 X 5/16
• 48" oc 2 - 5/8"
DIA. BOLTS EA. LEG
2 X WOOD PL.
STEEL BEAM
S3221

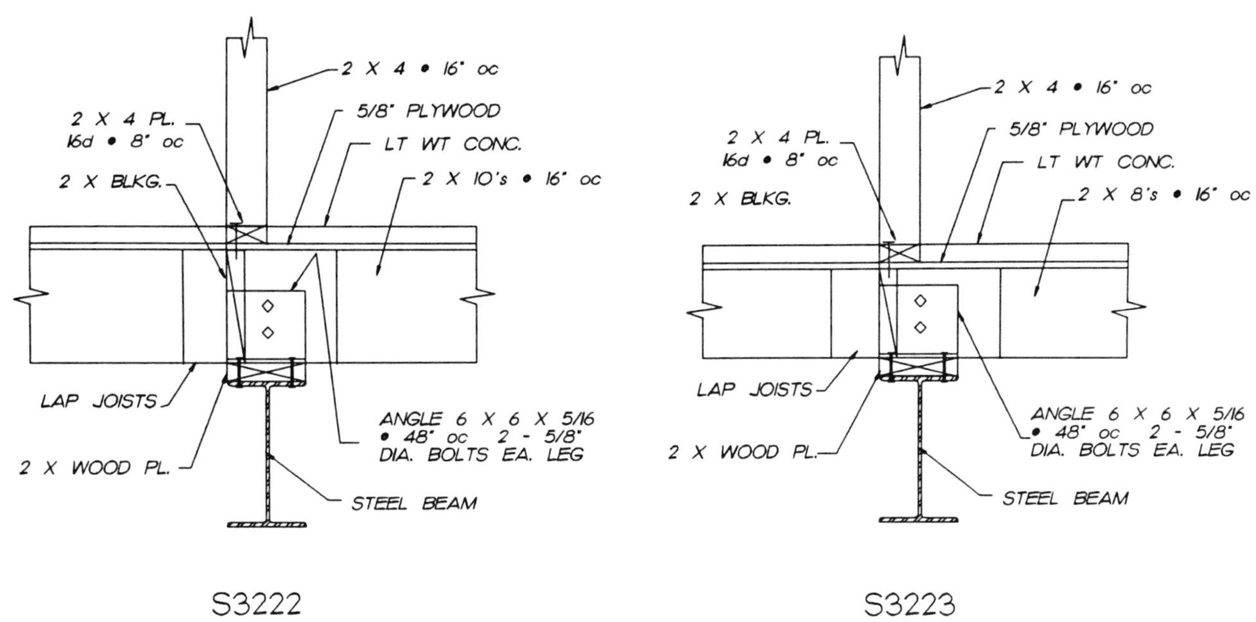
2 X 4 • 16" oc
2 X 4 PL.
16d • 8" oc
5/8" PLYWOOD
LT WT CONC.
2 X BLKG.
2 X 10's • 16" oc
LAP JOISTS
ANGLE 6 X 6 X 5/16
• 48" oc 2 - 5/8"
DIA. BOLTS EA. LEG
2 X WOOD PL.
STEEL BEAM
S3222
2 X 4 • 16" oc
5/8" PLYWOOD
2 X 4 PL.
16d • 8" oc
LT WT CONC.
2 X BLKG.
2 X 8's • 16" oc
LAP JOISTS
ANGLE 6 X 6 X 5/16
• 48" oc 2 - 5/8"
DIA. BOLTS EA. LEG
2 X WOOD PL.
STEEL BEAM
S3223

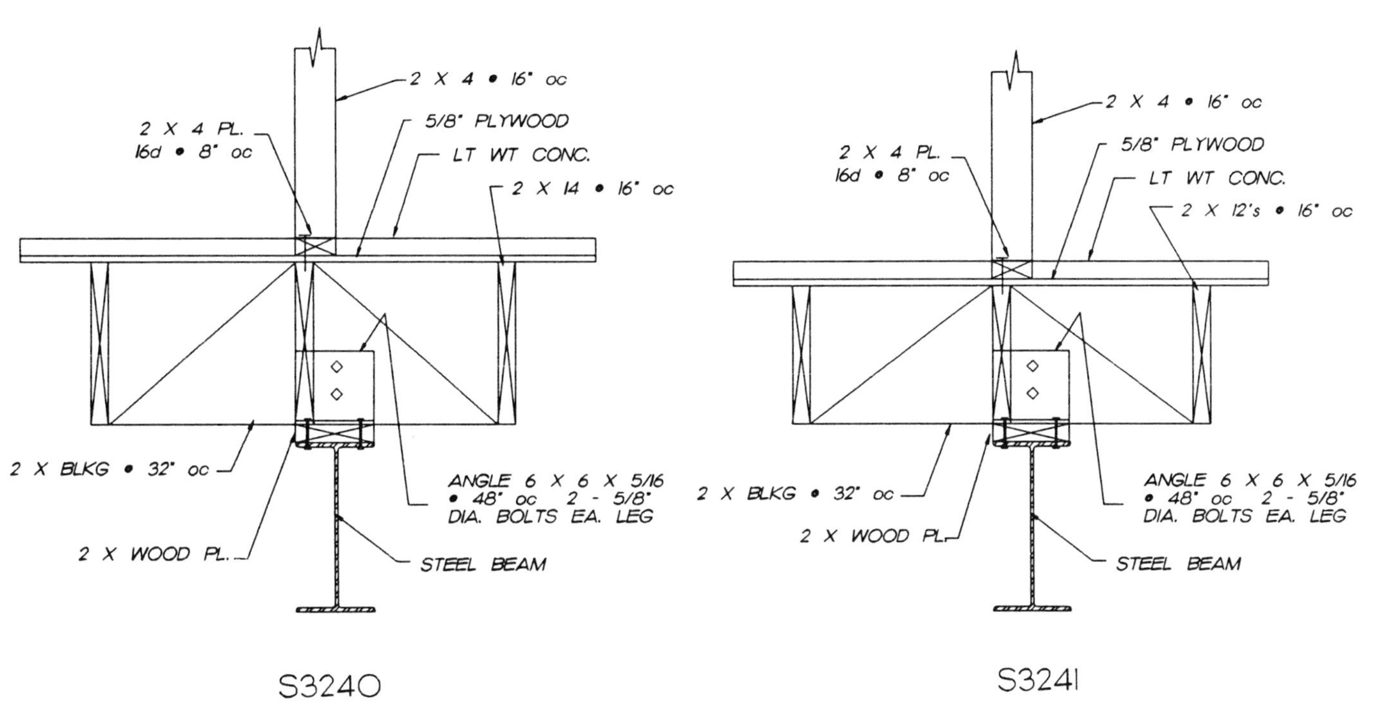
2 X 4 • 16" oc
5/8" PLYWOOD
2 X 4 PL.
16d • 8" oc
LT WT CONC.
2 X 14 • 16" oc
2 X BLKG • 32" oc
ANGLE 6 X 6 X 5/16
• 48" oc 2 - 5/8"
DIA. BOLTS EA. LEG
2 X WOOD PL.
STEEL BEAM
S3240
2 X 4 • 16" oc
5/8" PLYWOOD
2 X 4 PL.
16d • 8" oc
LT WT CONC.
2 X 12's • 16" oc
2 X BLKG • 32" oc
ANGLE 6 X 6 X 5/16
• 48" oc 2 - 5/8"
DIA. BOLTS EA. LEG
2 X WOOD PL.
STEEL BEAM
S3241

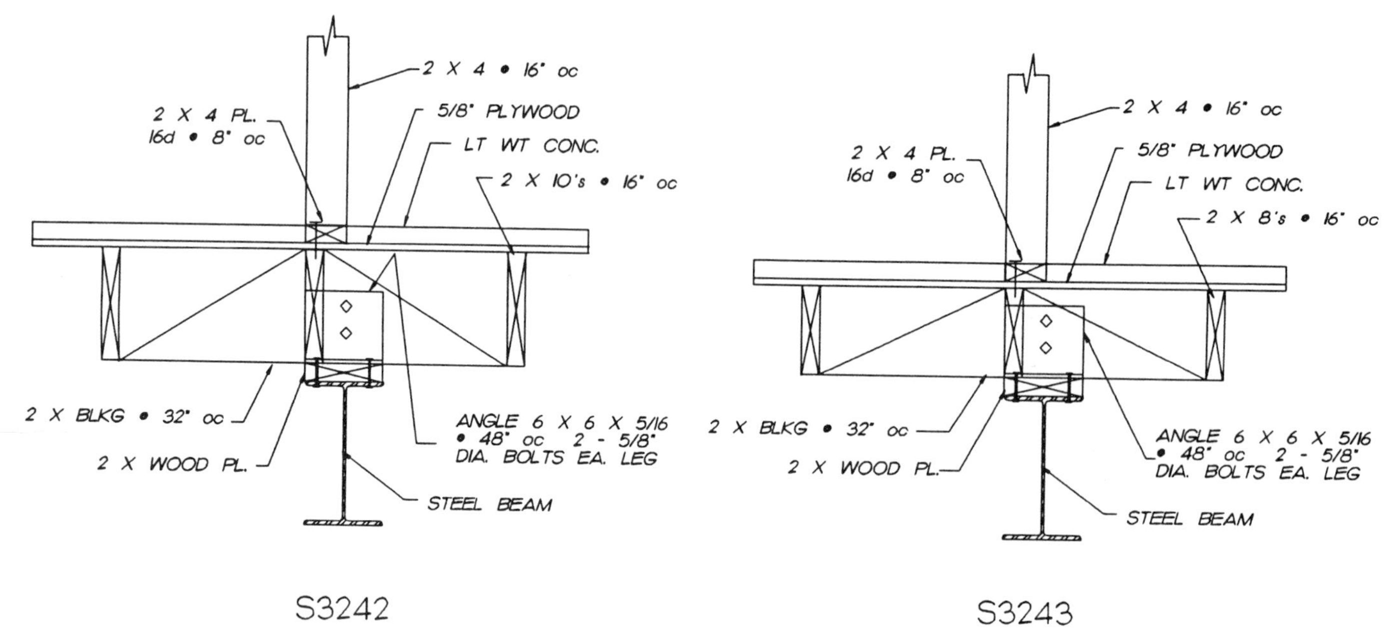
2 X 4 • 16" oc
2 X 4 PL.
16d • 8" oc
5/8" PLYWOOD
LT WT CONC.
2 X 10's • 16" oc
2 X BLKG • 32" oc
2 X WOOD PL.
ANGLE 6 X 6 X 5/16
• 48" oc 2 - 5/8"
DIA. BOLTS EA. LEG
STEEL BEAM
S3242
2 X 4 • 16" oc
2 X 4 PL.
16d • 8" oc
5/8" PLYWOOD
LT WT CONC.
2 X 8's • 16" oc
2 X BLKG • 32" oc
2 X WOOD PL.
ANGLE 6 X 6 X 5/16
• 48" oc 2 - 5/8"
DIA. BOLTS EA. LEG
STEEL BEAM
S3243

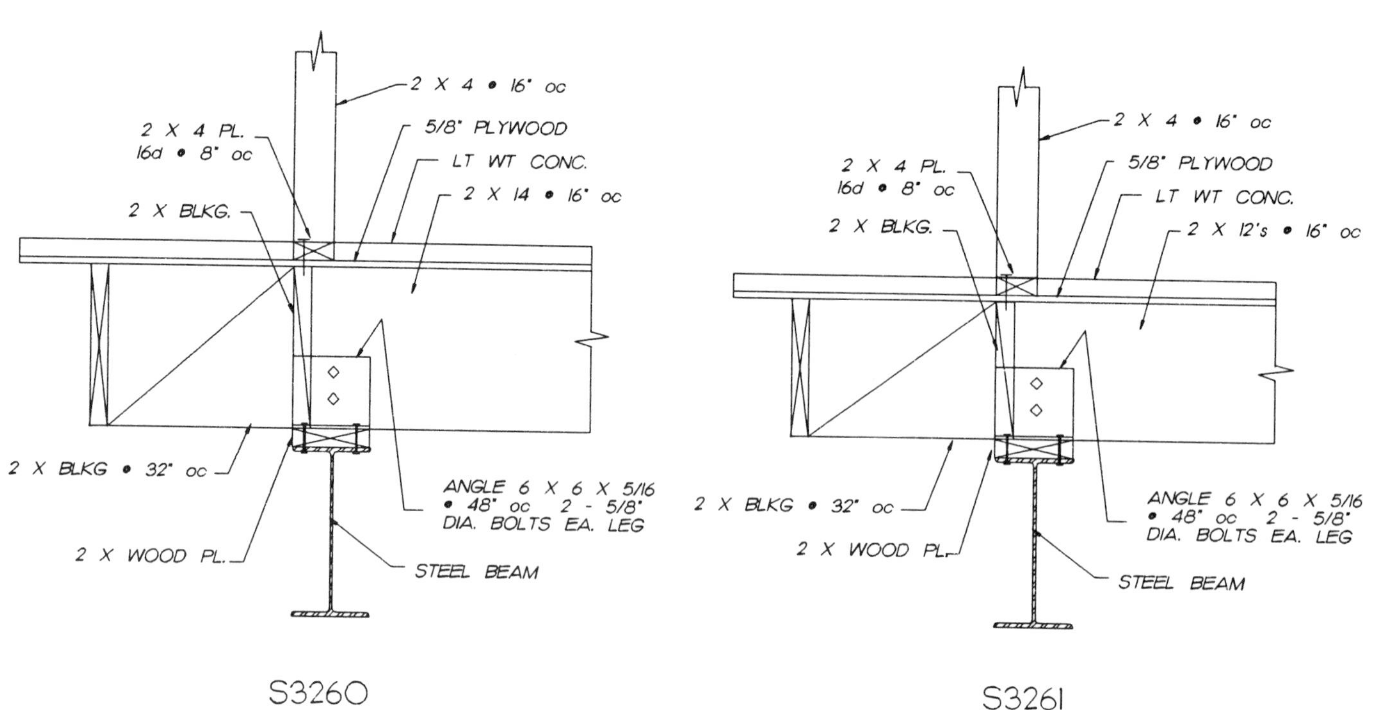
2 X 4 • 16" oc
2 X 4 PL.
16d • 8" oc
5/8" PLYWOOD
LT WT CONC.
2 X 14 • 16" oc
2 X BLKG.
2 X BLKG • 32" oc
2 X WOOD PL.
ANGLE 6 X 6 X 5/16
• 48" oc 2 - 5/8"
DIA. BOLTS EA. LEG
STEEL BEAM
S3260
2 X 4 • 16" oc
2 X 4 PL.
16d • 8" oc
5/8" PLYWOOD
LT WT CONC.
2 X 12's • 16" oc
2 X BLKG.
2 X BLKG • 32" oc
2 X WOOD PL.
ANGLE 6 X 6 X 5/16
• 48" oc 2 - 5/8"
DIA. BOLTS EA. LEG
STEEL BEAM
S3261

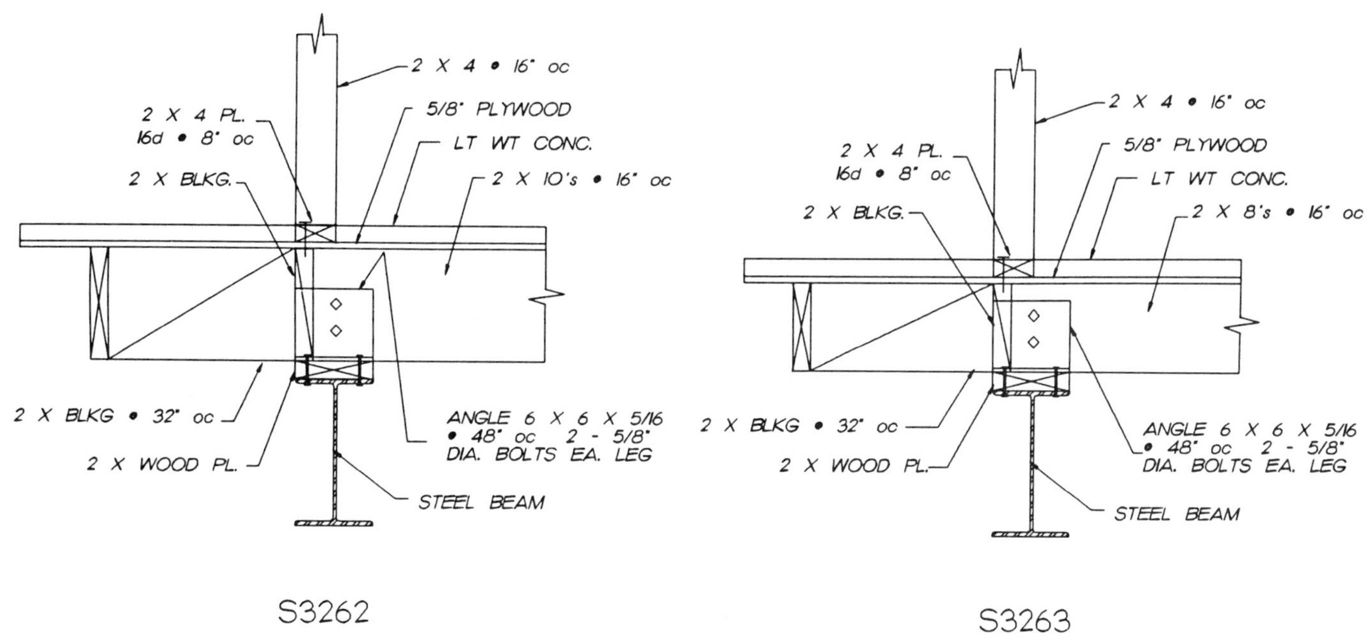
2 X 4 • 16" oc
2 X 4 PL.
16d • 8" oc
5/8" PLYWOOD
LT WT CONC.
2 X BLKG.
2 X 10's • 16" oc
2 X BLKG • 32" oc
2 X WOOD PL.
ANGLE 6 X 6 X 5/16
• 48" oc 2 - 5/8"
DIA. BOLTS EA. LEG
STEEL BEAM
S3262
2 X 4 • 16" oc
2 X 4 PL.
16d • 8" oc
5/8" PLYWOOD
LT WT CONC.
2 X BLKG.
2 X 8's • 16" oc
2 X BLKG • 32" oc
2 X WOOD PL.
ANGLE 6 X 6 X 5/16
• 48" oc 2 - 5/8"
DIA. BOLTS EA. LEG
STEEL BEAM
S3263

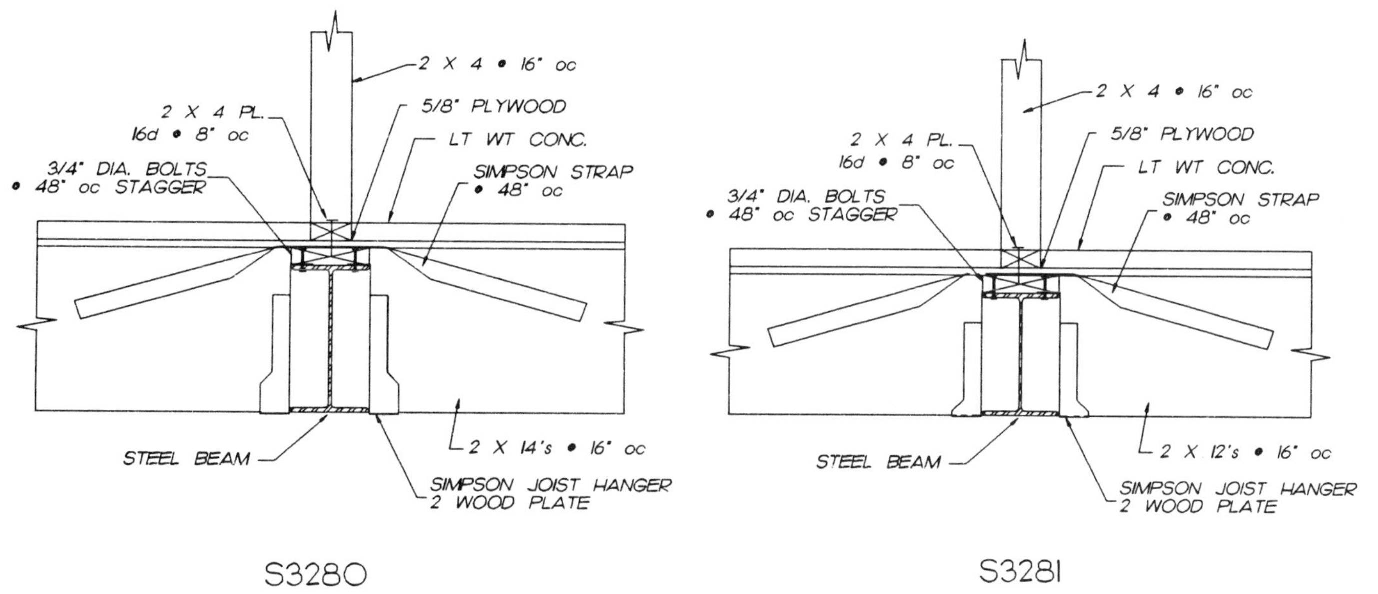
2 X 4 • 16" oc
2 X 4 PL.
16d • 8" oc
5/8" PLYWOOD
LT WT CONC.
3/4" DIA. BOLTS
• 48" oc STAGGER
SIMPSON STRAP
• 48" oc
STEEL BEAM
2 X 14's • 16" oc
SIMPSON JOIST HANGER
2 WOOD PLATE
S3280
2 X 4 • 16" oc
2 X 4 PL.
16d • 8" oc
5/8" PLYWOOD
LT WT CONC.
3/4" DIA. BOLTS
• 48" oc STAGGER
SIMPSON STRAP
• 48" oc
STEEL BEAM
2 X 12's • 16" oc
SIMPSON JOIST HANGER
2 WOOD PLATE
S3281

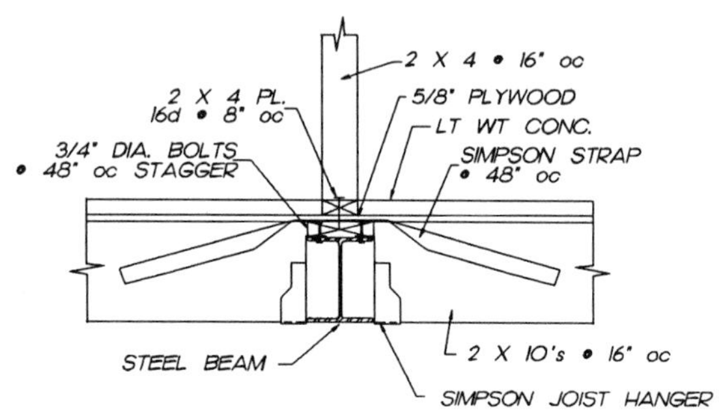

S3282

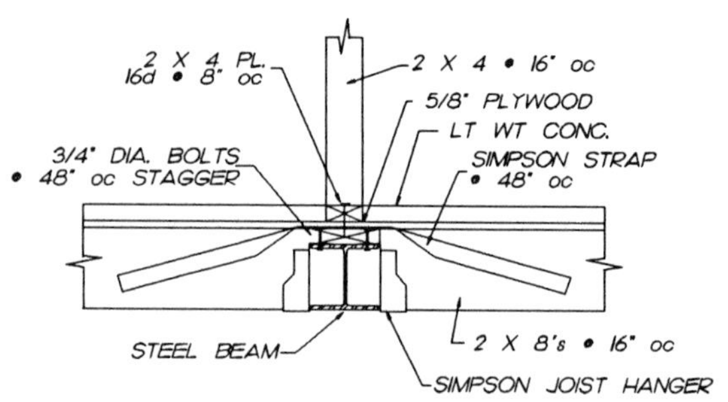

S3283

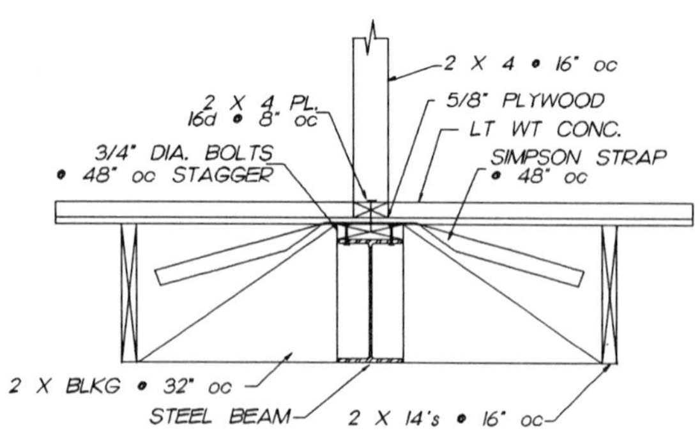

S3300

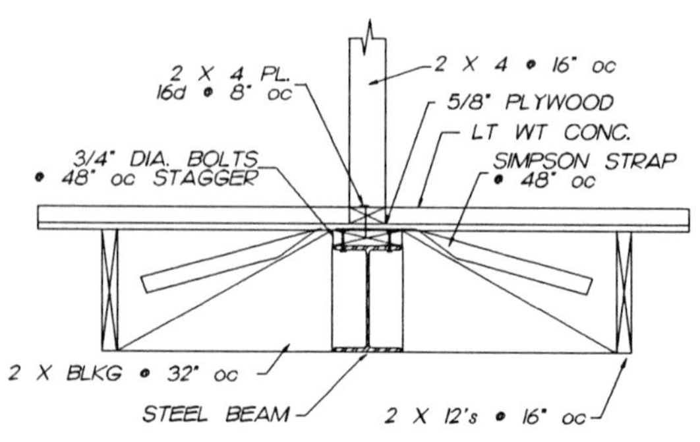

S3301

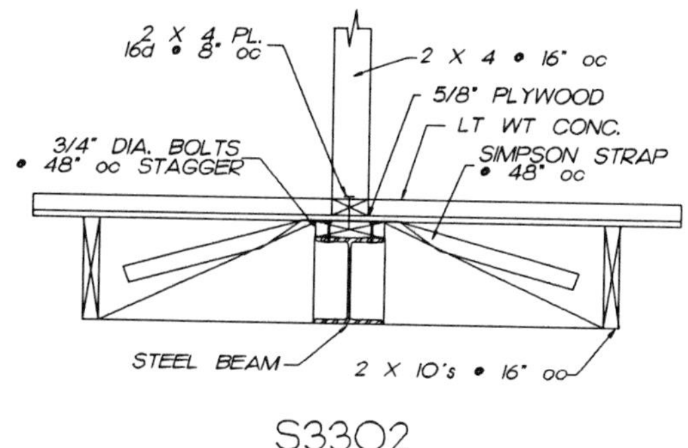

S3302

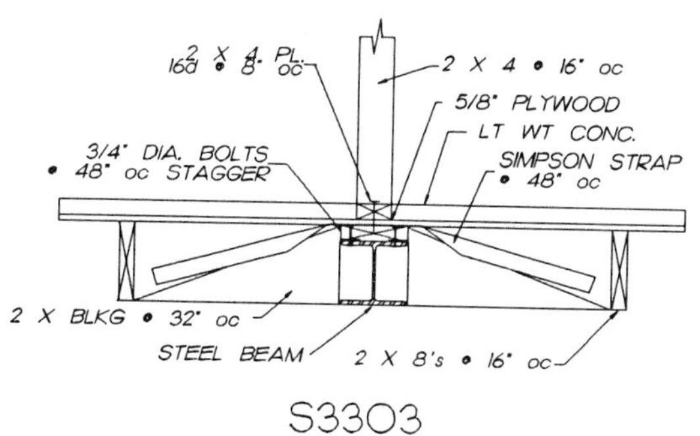

S3303

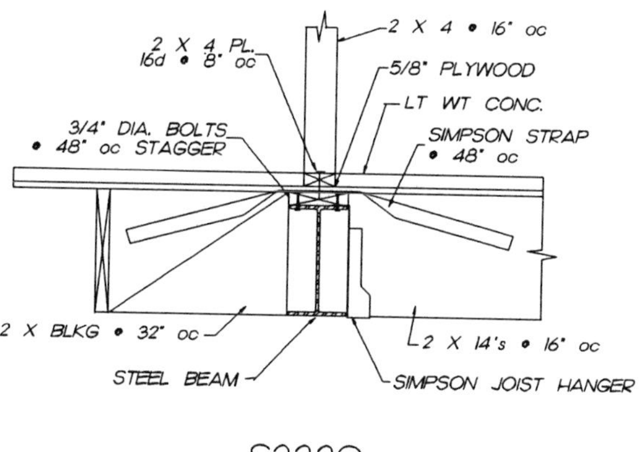

S3320

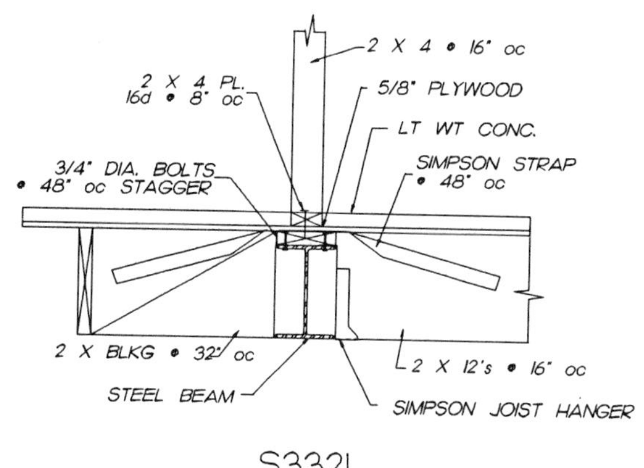

S3321

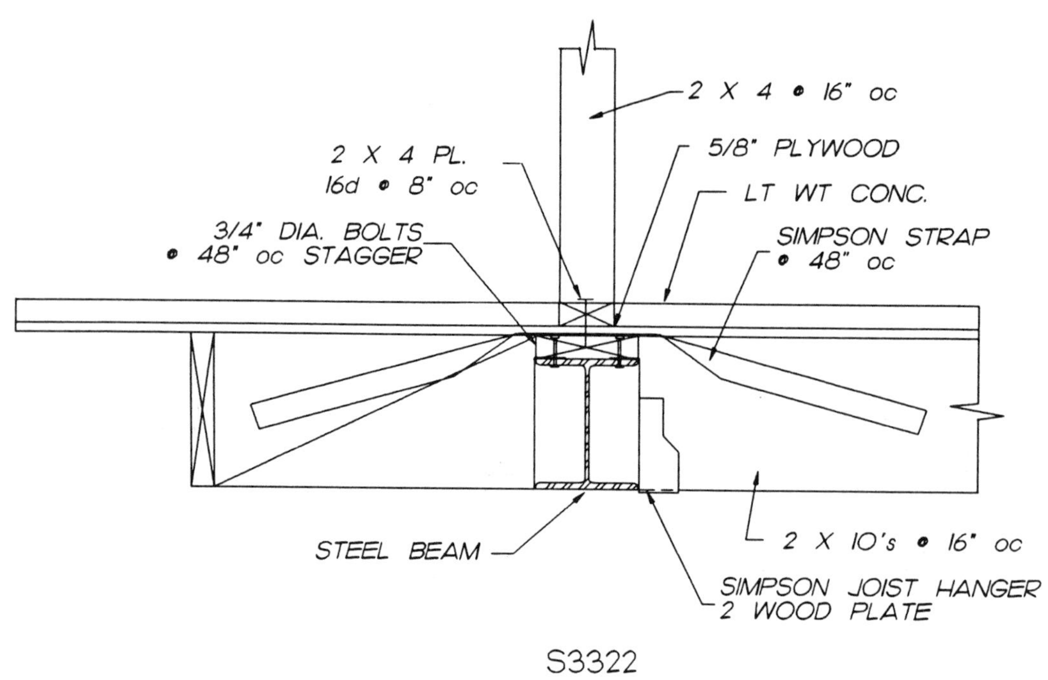

2 X 4 @ 16" oc
2 X 4 PL.
16d @ 8" oc
5/8" PLYWOOD
LT WT CONC.
3/4" DIA. BOLTS
@ 48" oc STAGGER
SIMPSON STRAP
@ 48" oc
STEEL BEAM
2 X 10's @ 16" oc
SIMPSON JOIST HANGER
2 WOOD PLATE

S3322

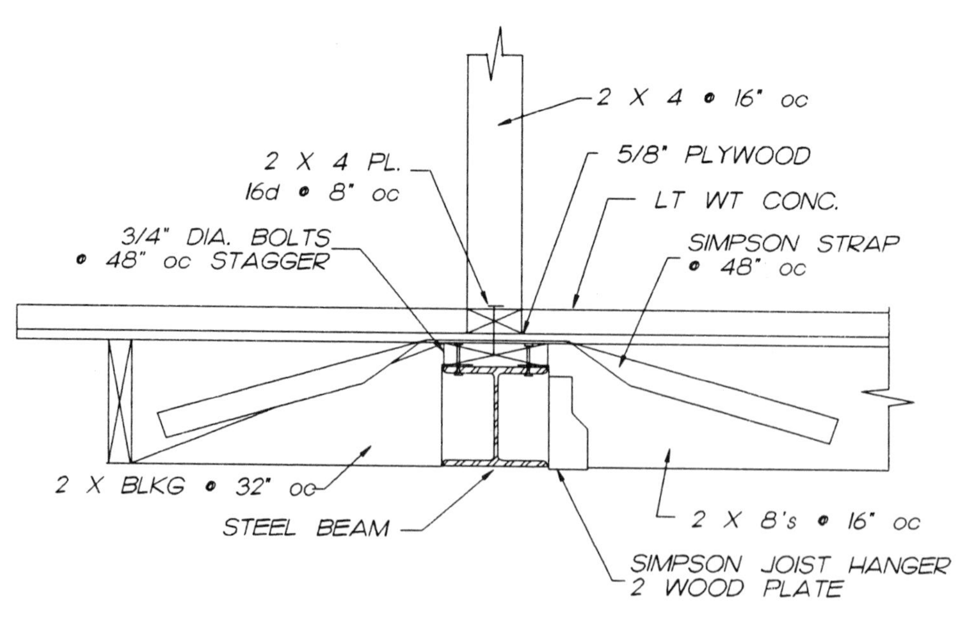

2 X 4 @ 16" oc
2 X 4 PL.
16d @ 8" oc
5/8" PLYWOOD
LT WT CONC.
3/4" DIA. BOLTS
@ 48" oc STAGGER
SIMPSON STRAP
@ 48" oc
2 X BLKG @ 32" oc
STEEL BEAM
2 X 8's @ 16" oc
SIMPSON JOIST HANGER
2 WOOD PLATE

S3323

CHAPTER 4 STEEL	SECTION 6	WALL & STEEL FLOOR
EXTERIOR WALL - STEEL STUDS STEEL BEAM METAL DECKING	DECKING PARALELL TO BEAM	S6020
	DECKING PERP. TO BEAM	S6021
EXTERIOR CURTAIN WALL STEEL BEAM METAL DECKING STUD BOLTS ON BEAM	DECKING PARALELL TO BEAM	S6040 - S6044
	DECKING PERP. TO BEAM	S6060 - S6064
STEEL BAR JOISTS METAL DECKING & SLAB INTERIOR STEEL STUD WALL	DECKING PERP. TO BAR JOISTS	S6080
	DECKING PARALELL TO BAR JOISTS	S6100
STEEL BAR JOISTS PERP. EACH SIDE METAL DECKING & SLAB SUPPORTED BY A STEEL BEAM	DECKING PERP. TO BAR JOISTS	S6120
	DECKING PARALELL TO BAR JOISTS	S6140
COMPOSITE STEEL BEAM 2 ROW OF STEEL STUDS METAL DECKING & SLAB	DECKING PARALELL TO BEAM	S6160 - S6164
	DECKING PERP. TO BEAM	S6180 - S6184

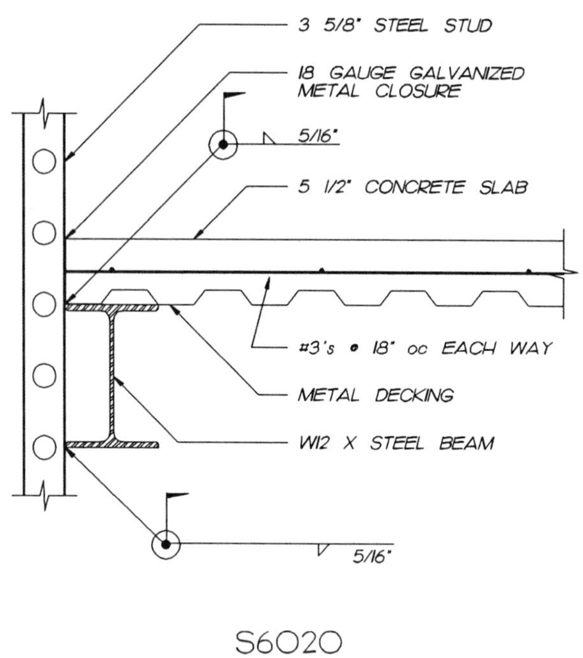

S6020

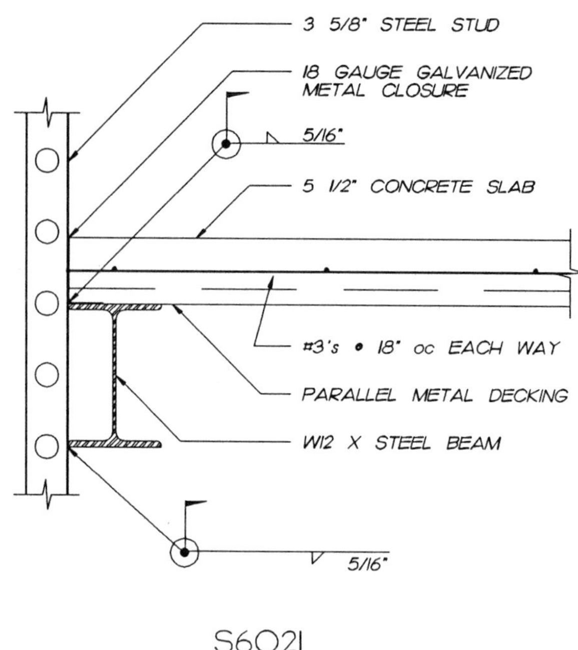

S6021

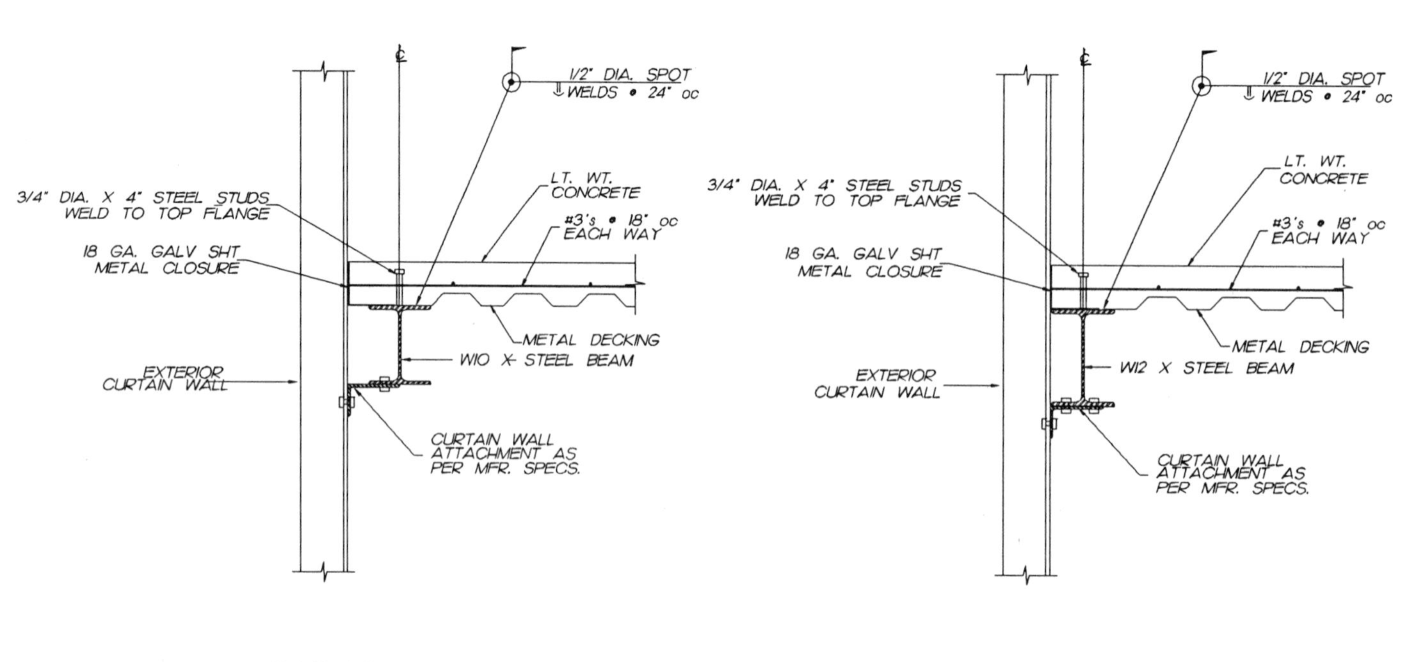

S6040

S6041

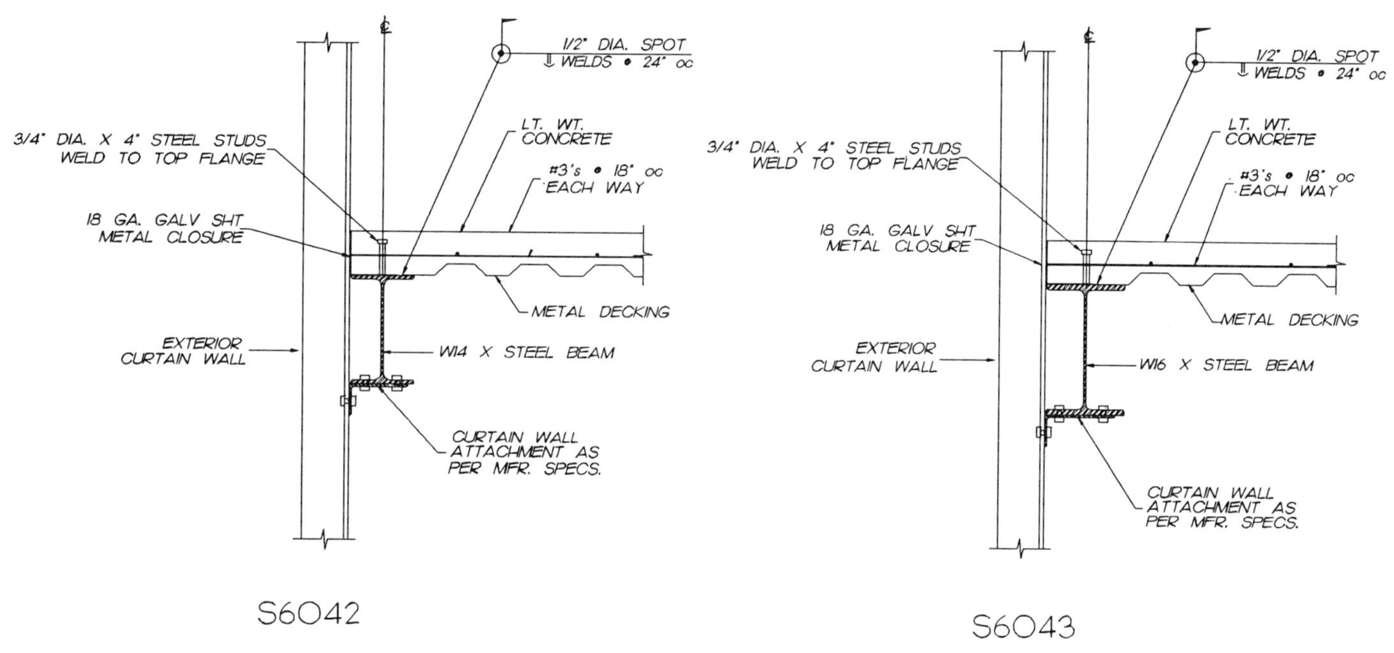

1/2" DIA. SPOT
WELDS @ 24" oc
LT. WT. CONCRETE
3/4" DIA. X 4" STEEL STUDS WELD TO TOP FLANGE
#3's @ 18" oc EACH WAY
18 GA. GALV SHT METAL CLOSURE
METAL DECKING
EXTERIOR CURTAIN WALL
W14 X STEEL BEAM
CURTAIN WALL ATTACHMENT AS PER MFR. SPECS.
S6042
1/2" DIA. SPOT
WELDS @ 24" oc
LT. WT. CONCRETE
3/4" DIA. X 4" STEEL STUDS WELD TO TOP FLANGE
#3's @ 18" oc EACH WAY
18 GA. GALV SHT METAL CLOSURE
METAL DECKING
EXTERIOR CURTAIN WALL
W16 X STEEL BEAM
CURTAIN WALL ATTACHMENT AS PER MFR. SPECS.
S6043

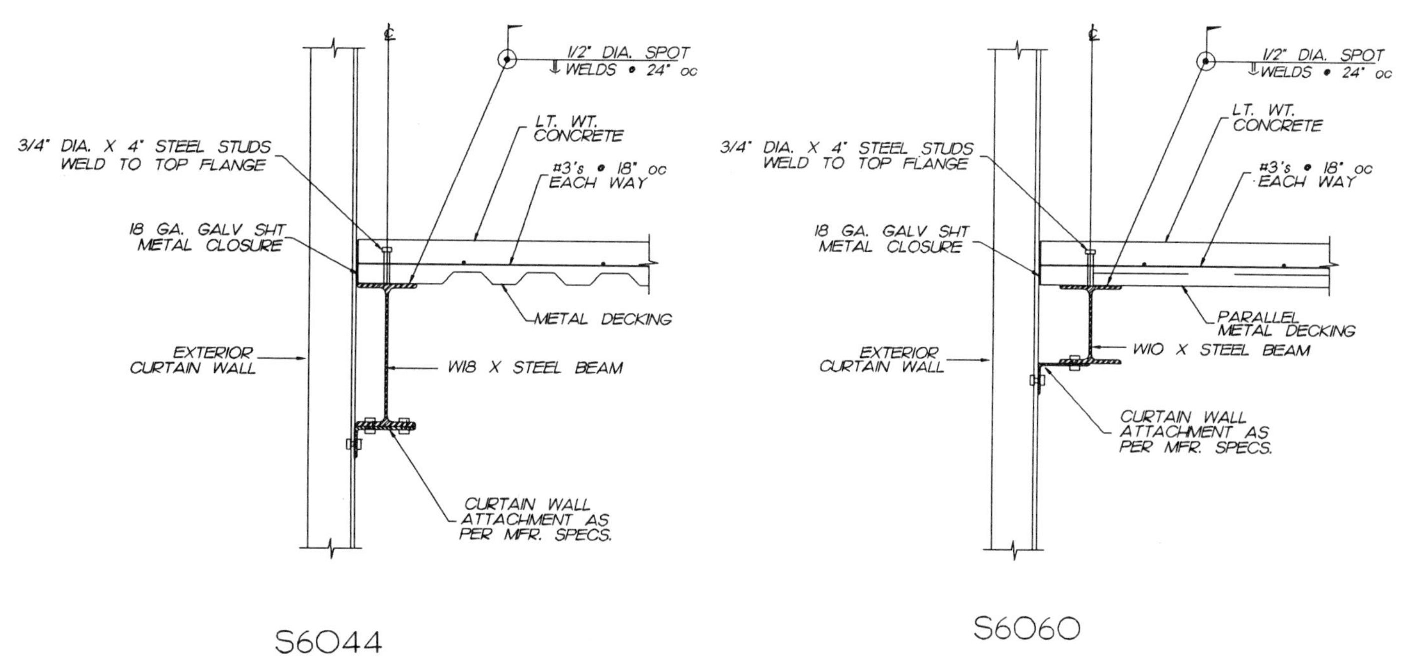

1/2" DIA. SPOT
WELDS @ 24" oc
LT. WT. CONCRETE
3/4" DIA. X 4" STEEL STUDS WELD TO TOP FLANGE
#3's @ 18" oc EACH WAY
18 GA. GALV SHT METAL CLOSURE
METAL DECKING
EXTERIOR CURTAIN WALL
W18 X STEEL BEAM
CURTAIN WALL ATTACHMENT AS PER MFR. SPECS.
S6044
1/2" DIA. SPOT
WELDS @ 24" oc
LT. WT. CONCRETE
3/4" DIA. X 4" STEEL STUDS WELD TO TOP FLANGE
#3's @ 18" oc EACH WAY
18 GA. GALV SHT METAL CLOSURE
PARALLEL METAL DECKING
EXTERIOR CURTAIN WALL
W10 X STEEL BEAM
CURTAIN WALL ATTACHMENT AS PER MFR. SPECS.
S6060

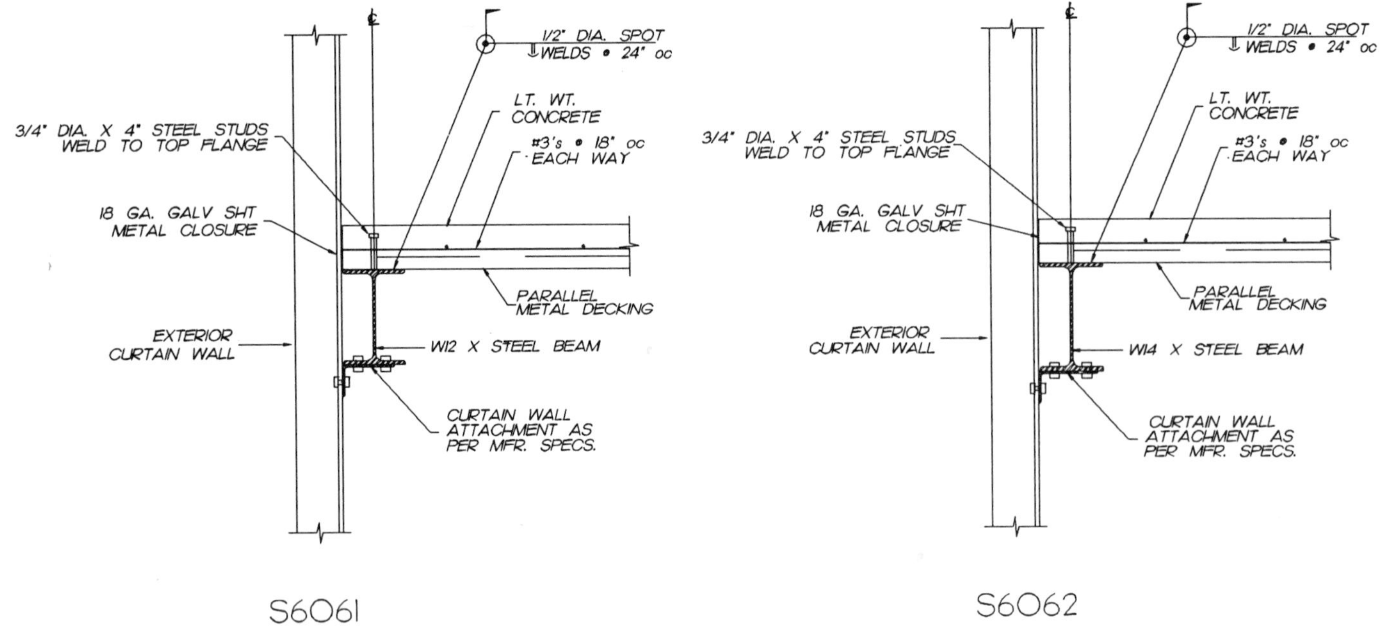

1/2" DIA. SPOT
WELDS • 24" oc
LT. WT.
CONCRETE
3/4" DIA. X 4" STEEL STUDS
WELD TO TOP FLANGE
#3's • 18" oc
EACH WAY
18 GA. GALV SHT
METAL CLOSURE
PARALLEL
METAL DECKING
EXTERIOR
CURTAIN WALL
W12 X STEEL BEAM
CURTAIN WALL
ATTACHMENT AS
PER MFR. SPECS.
S6061
1/2" DIA. SPOT
WELDS • 24" oc
LT. WT.
CONCRETE
3/4" DIA. X 4" STEEL STUDS
WELD TO TOP FLANGE
#3's • 18" oc
EACH WAY
18 GA. GALV SHT
METAL CLOSURE
PARALLEL
METAL DECKING
EXTERIOR
CURTAIN WALL
W14 X STEEL BEAM
CURTAIN WALL
ATTACHMENT AS
PER MFR. SPECS.
S6062

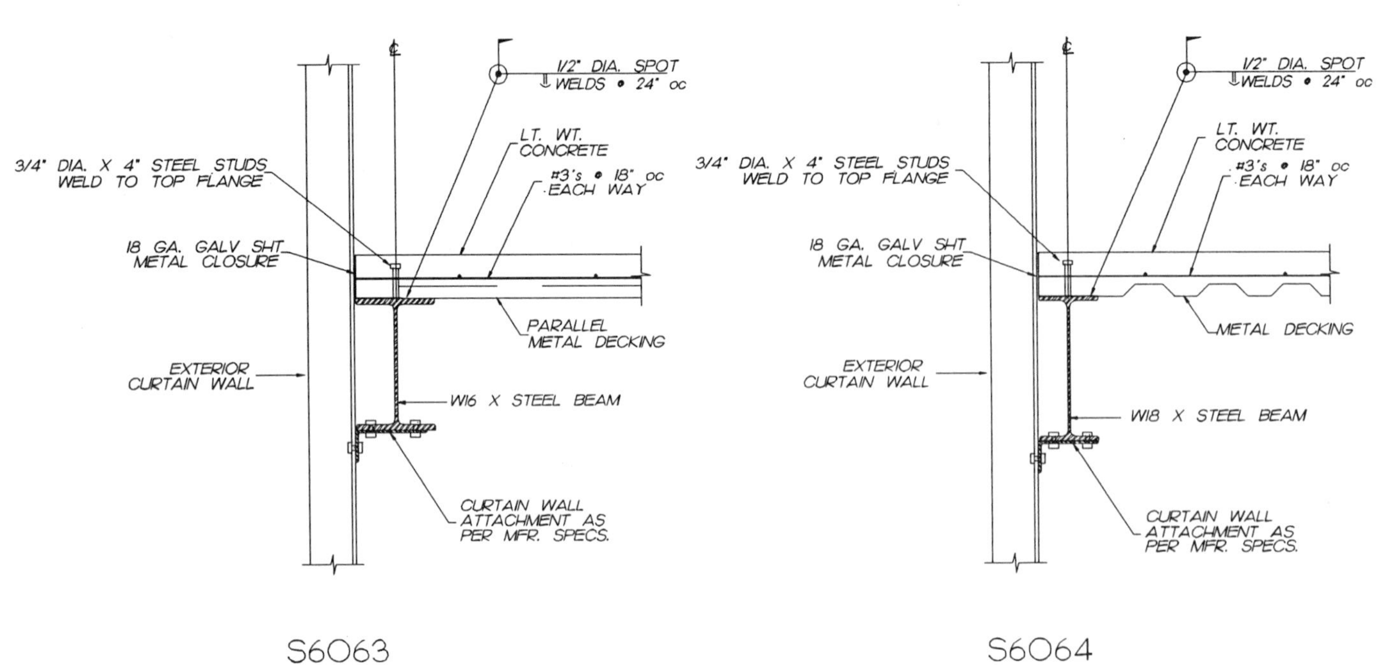

1/2" DIA. SPOT
WELDS • 24" oc
LT. WT.
CONCRETE
3/4" DIA. X 4" STEEL STUDS
WELD TO TOP FLANGE
#3's • 18" oc
EACH WAY
18 GA. GALV SHT
METAL CLOSURE
PARALLEL
METAL DECKING
EXTERIOR
CURTAIN WALL
W16 X STEEL BEAM
CURTAIN WALL
ATTACHMENT AS
PER MFR. SPECS.
S6063
1/2" DIA. SPOT
WELDS • 24" oc
LT. WT.
CONCRETE
3/4" DIA. X 4" STEEL STUDS
WELD TO TOP FLANGE
#3's • 18" oc
EACH WAY
18 GA. GALV SHT
METAL CLOSURE
METAL DECKING
EXTERIOR
CURTAIN WALL
W18 X STEEL BEAM
CURTAIN WALL
ATTACHMENT AS
PER MFR. SPECS.
S6064

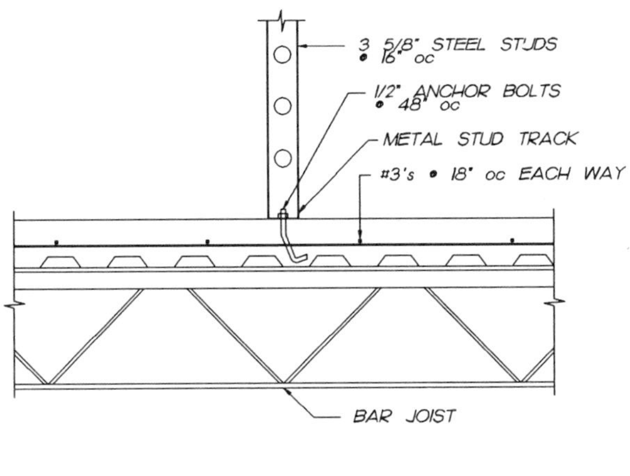

S6080

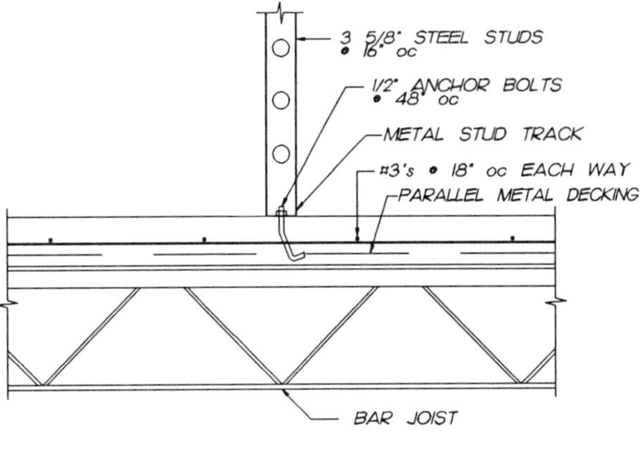

S6100

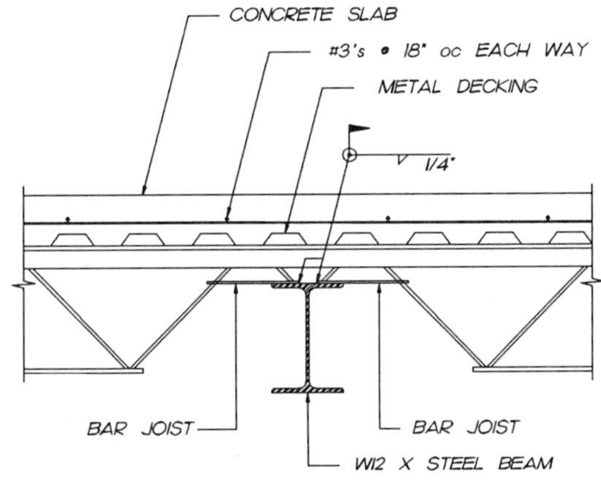

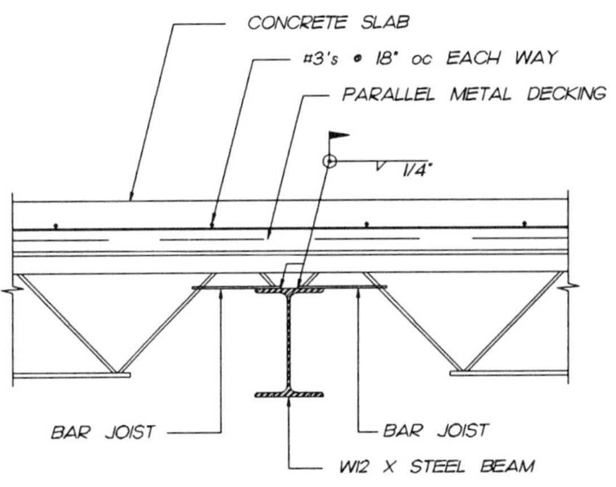

S6140

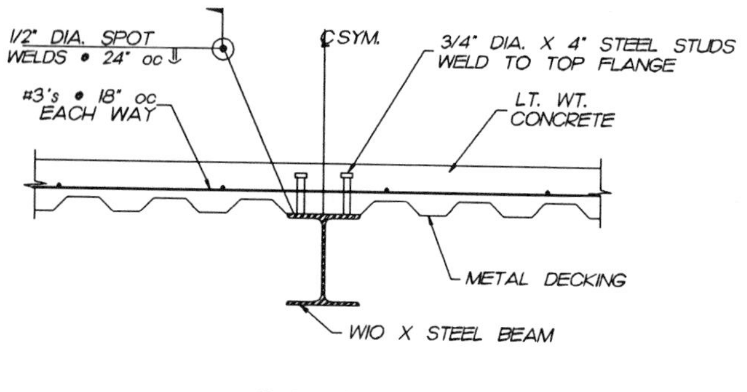

S6160

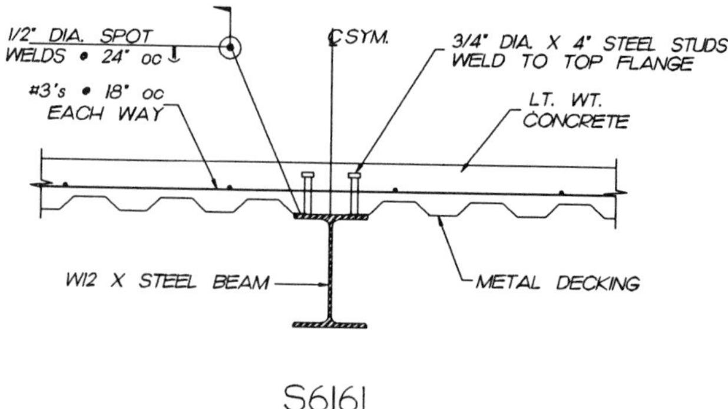

S6161

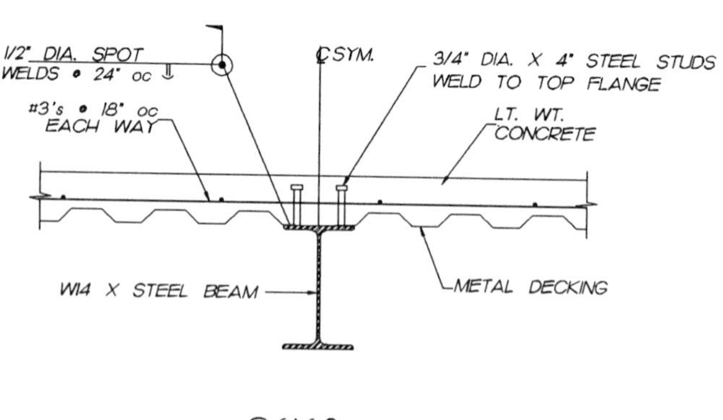

S6162

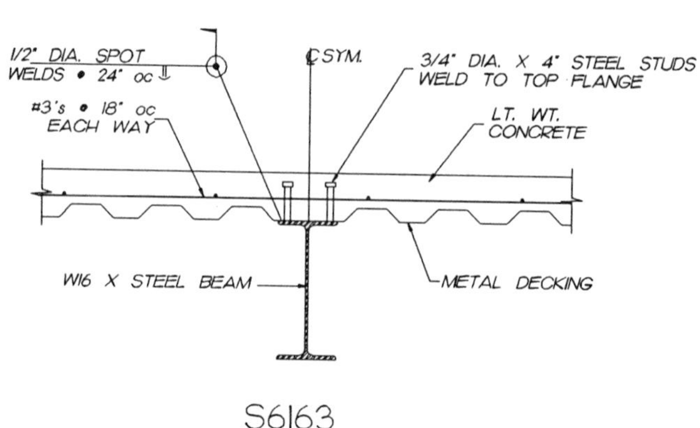

S6163

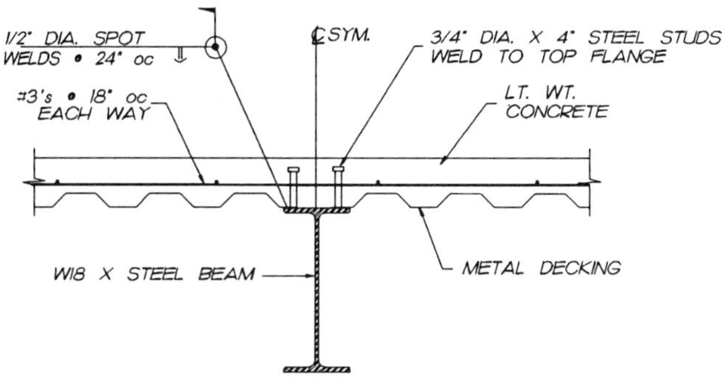

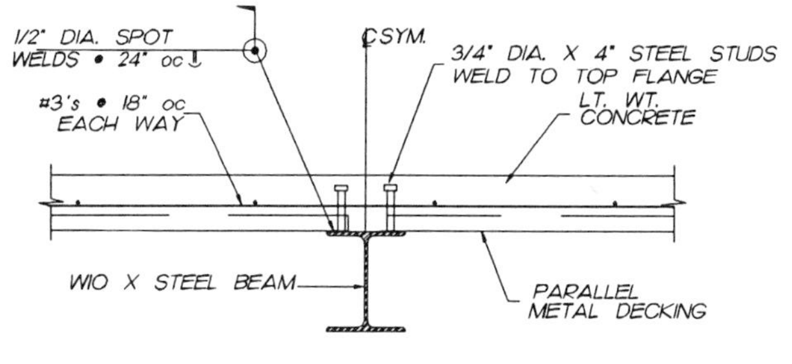

S6180

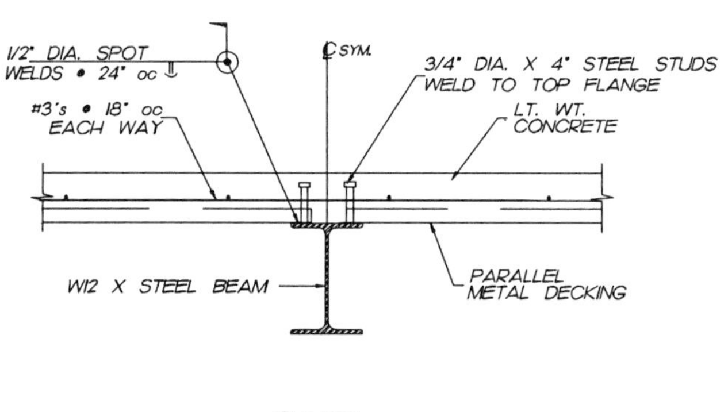

S6181

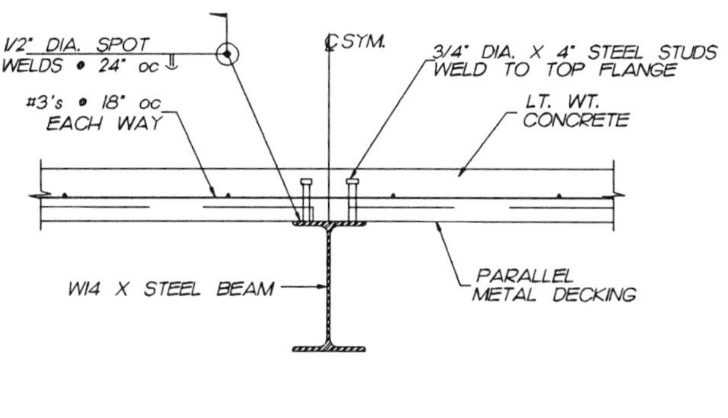

S6182

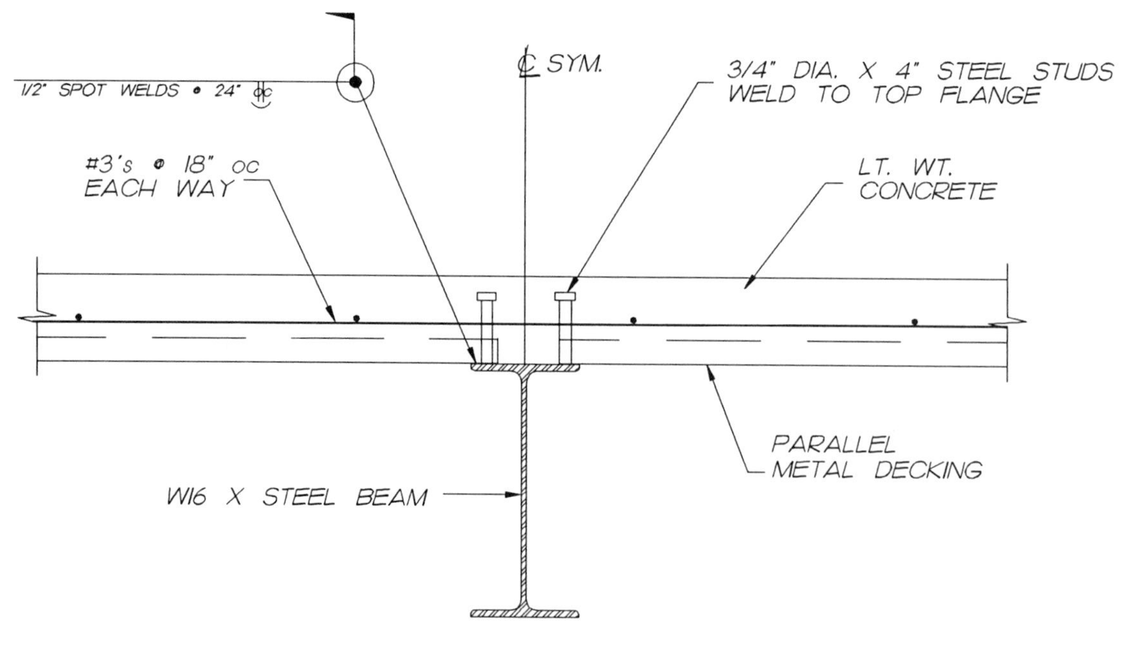
1/2" SPOT WELDS @ 24" oc
℄ SYM.
3/4" DIA. X 4" STEEL STUDS
WELD TO TOP FLANGE
#3's @ 18" oc
EACH WAY
LT. WT.
CONCRETE
PARALLEL
METAL DECKING
W16 X STEEL BEAM

S6183

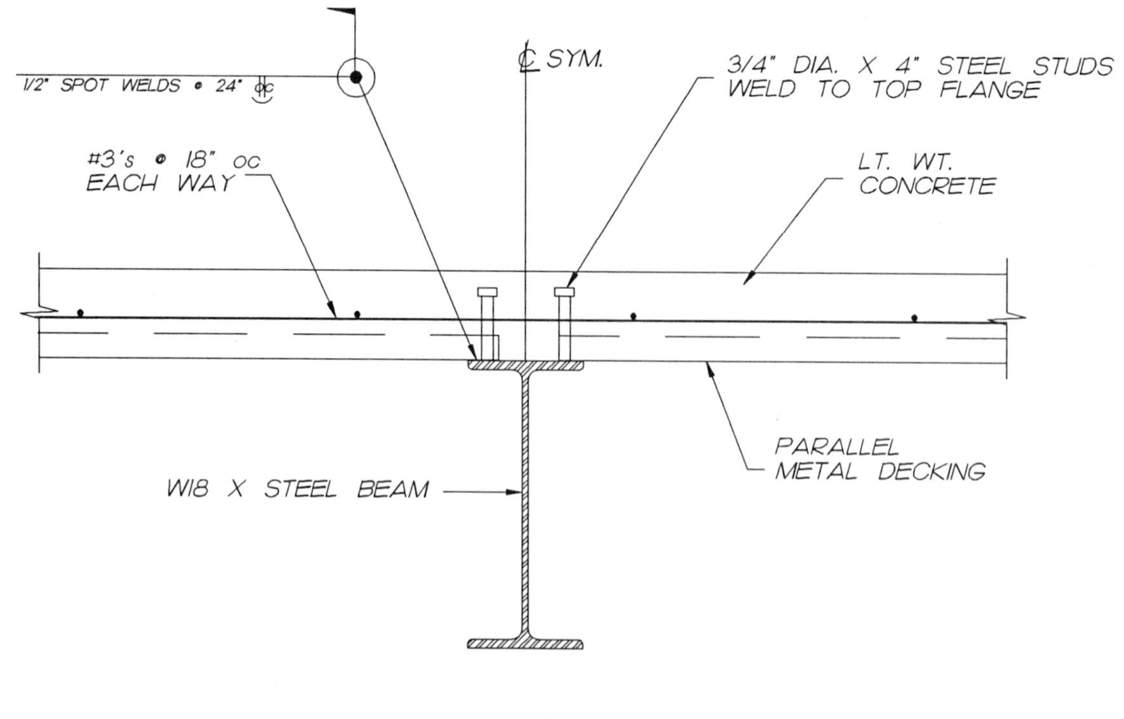
1/2" SPOT WELDS @ 24" oc
℄ SYM.
3/4" DIA. X 4" STEEL STUDS
WELD TO TOP FLANGE
#3's @ 18" oc
EACH WAY
LT. WT.
CONCRETE
PARALLEL
METAL DECKING
W18 X STEEL BEAM

S6184

CHAPTER 4 STEEL	SECTION 7	STEEL STAIRS
STRINGER ELEVATION FOUNDATION TO LANDING	————	S7020
STRINGER ELEVATION MIDDLE LANDING	PIPE COLUMN SUPPORT	S7040
	MASONRY WALL SUPPORT	S7060
	STEEL BEAM SUPPORT	S7080
STRINGER BASE PLATE	————	S7100
STRINGER ANGLE CONNECTION TO SLAB	————	S7120
TREAD & STRINGER CONNECTION	————	S7140, S7160, S7180, S7200 & S7220
STRINGER CONNECTION AT CONCRETE LANDING	————	S7240
STRINGER CONNECTION STEEL LANDING	————	S7260
STRINGER CONNECTION WOOD LANDING	————	S7280
CHANNEL TREAD & STRINGER	————	S7300
PLAN OF CIRCULAR STAIR	————	S7320
CIRCULAR STAIR TREAD	————	S7340

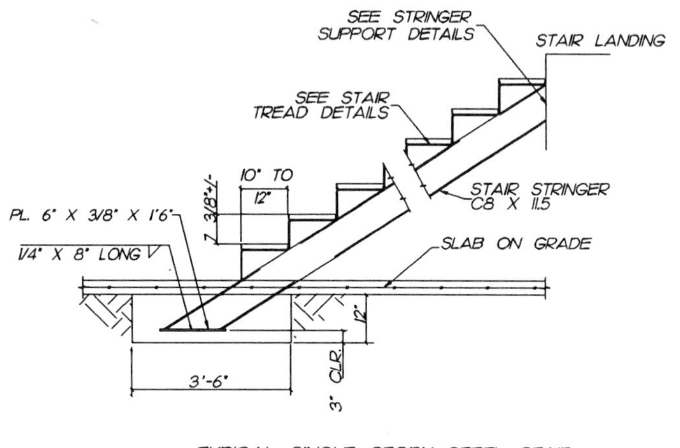

TYPICAL SINGLE STORY STEEL STAIR

S7020

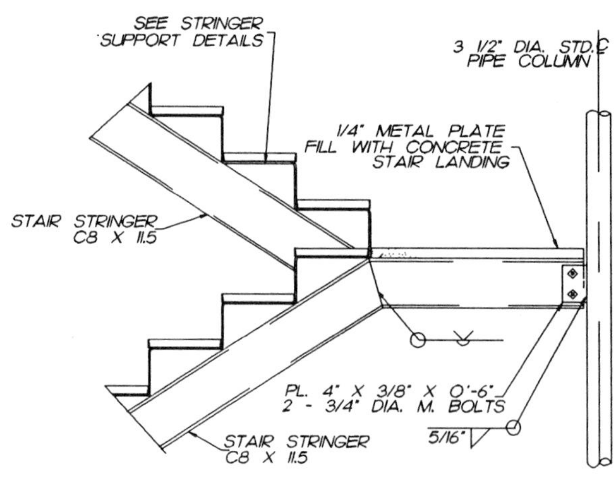

TYPICAL STEEL STAIR AT LANDING

S7040

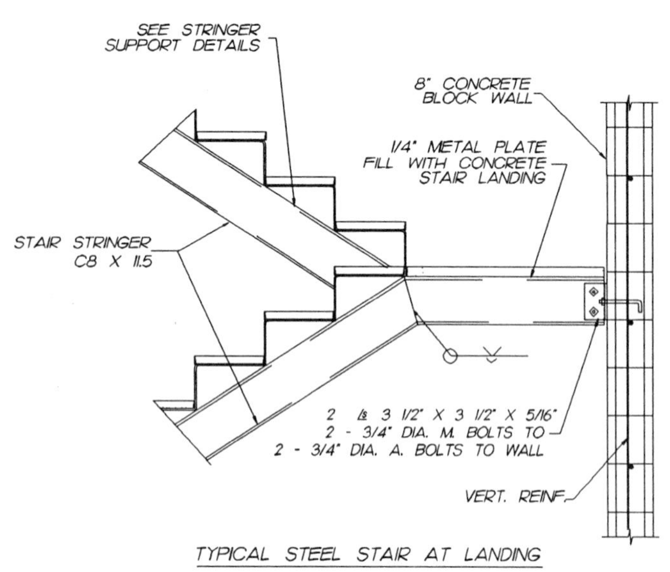

TYPICAL STEEL STAIR AT LANDING

S7060

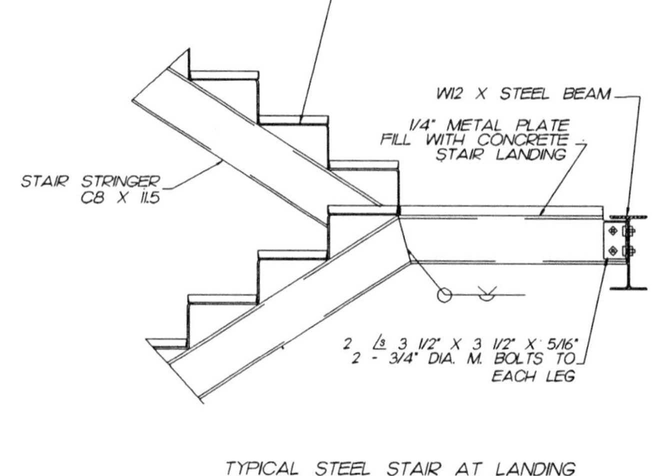

TYPICAL STEEL STAIR AT LANDING

S7080

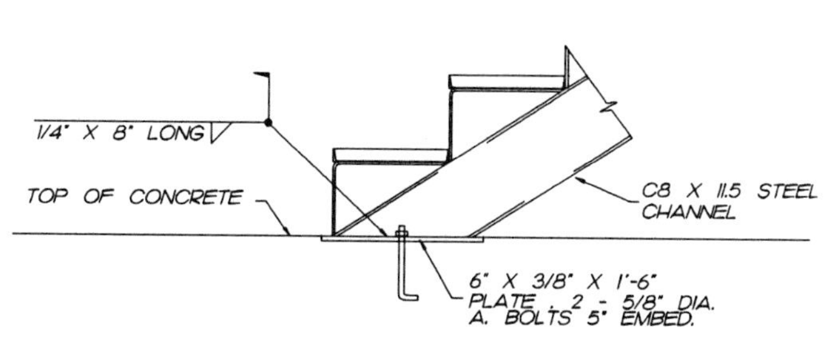

S7100

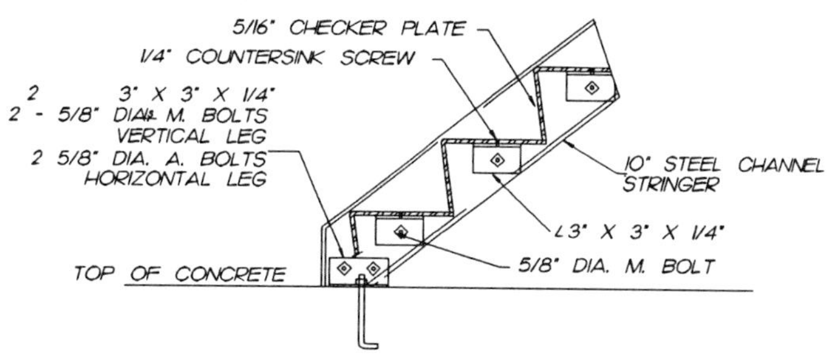

S7120

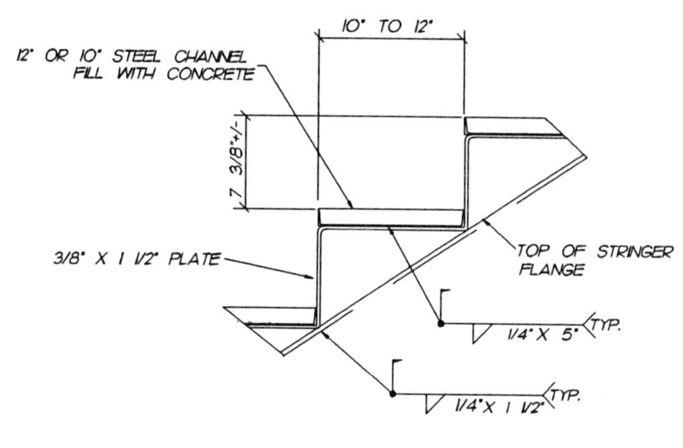

S7140

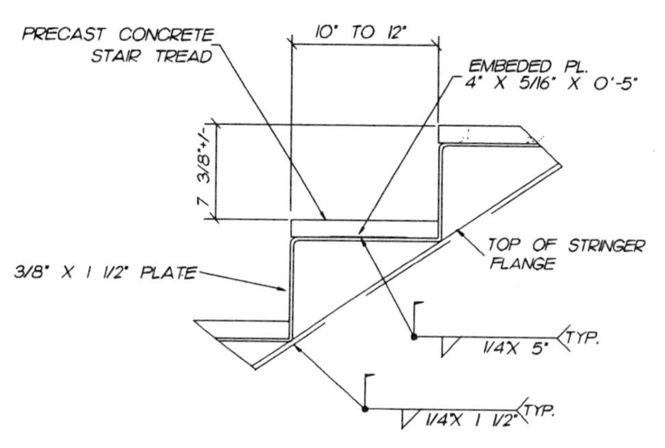

S7160

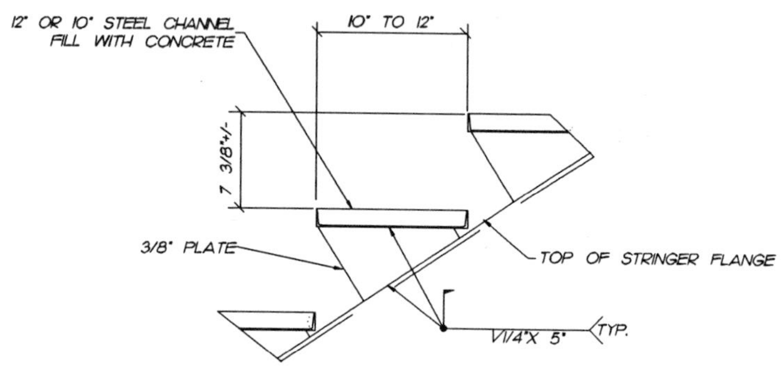
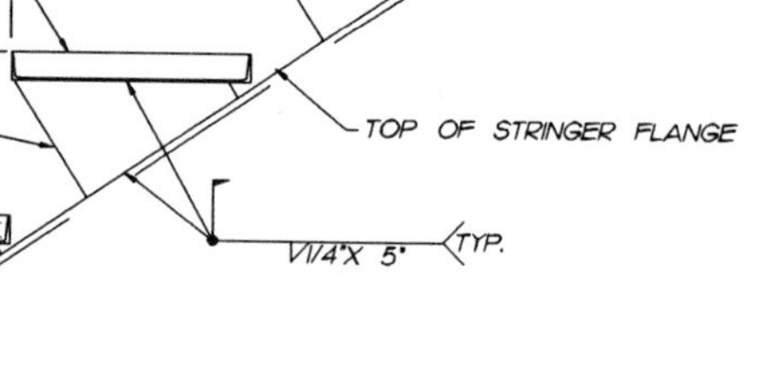

S7180

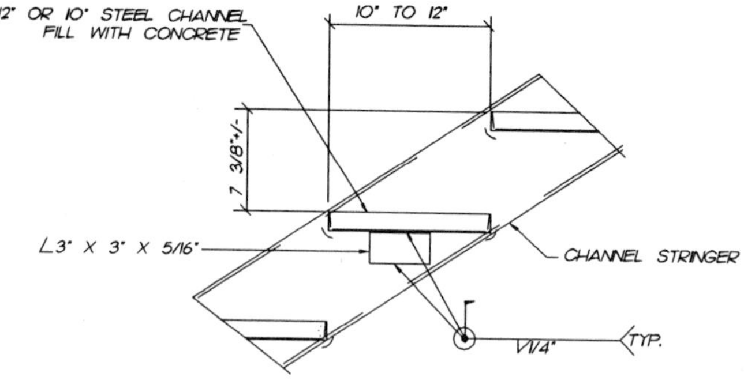

S7200

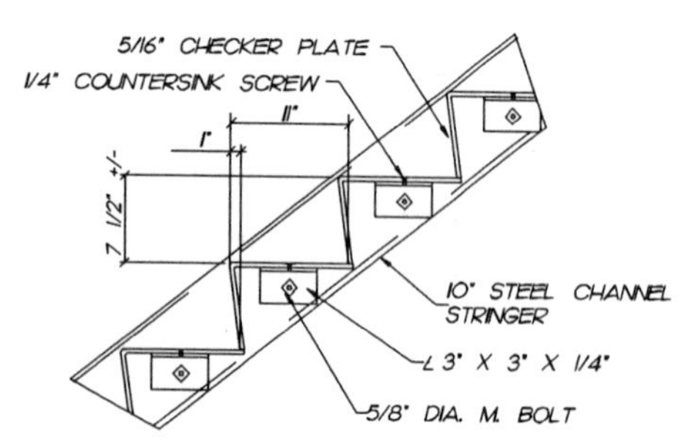

S7220

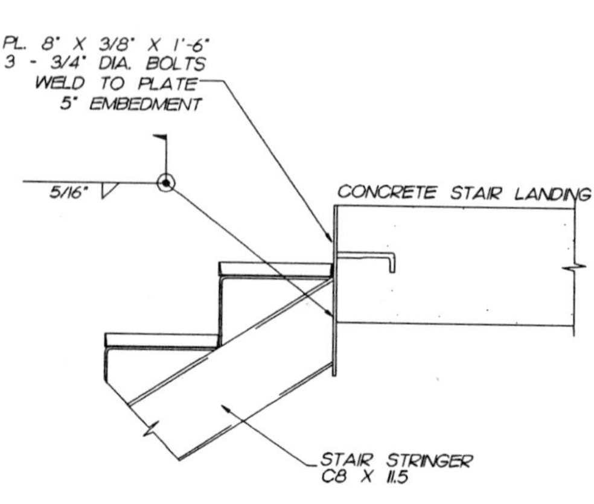

S7240

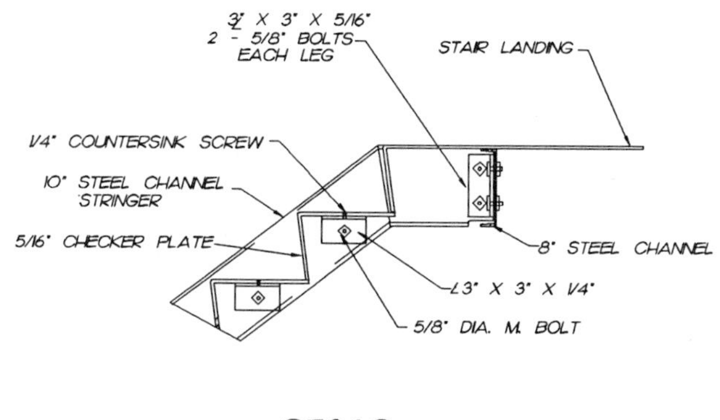

S7260

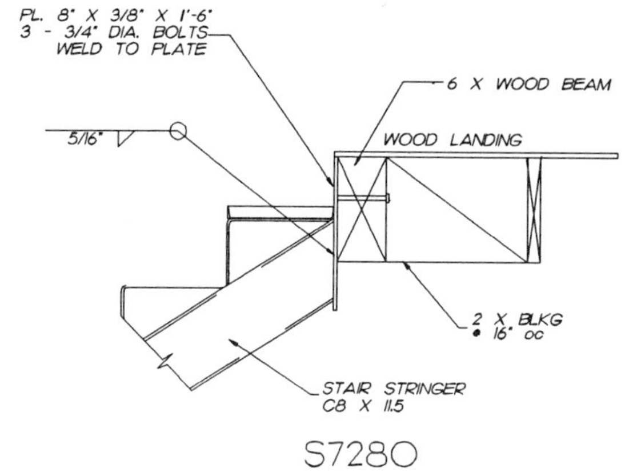

S7280

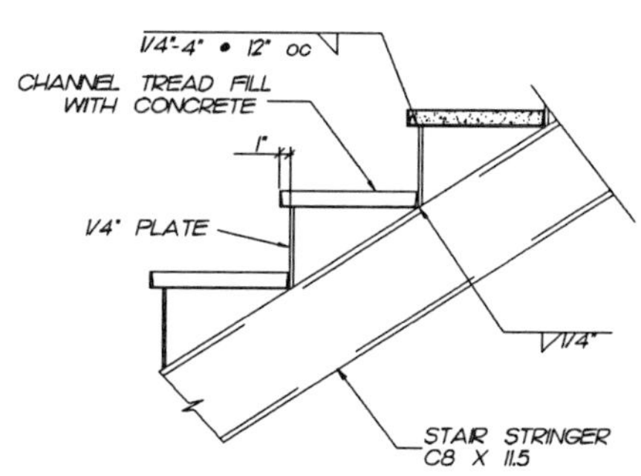

S7300

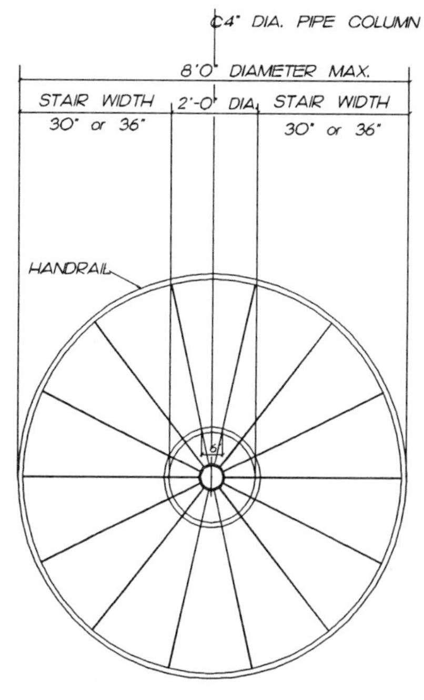

S7320

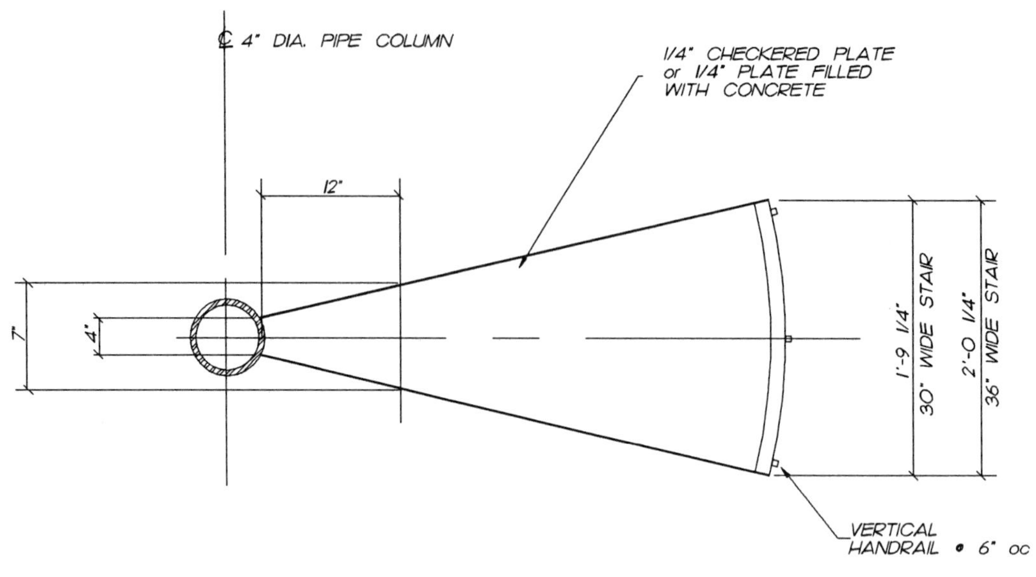

PLAN

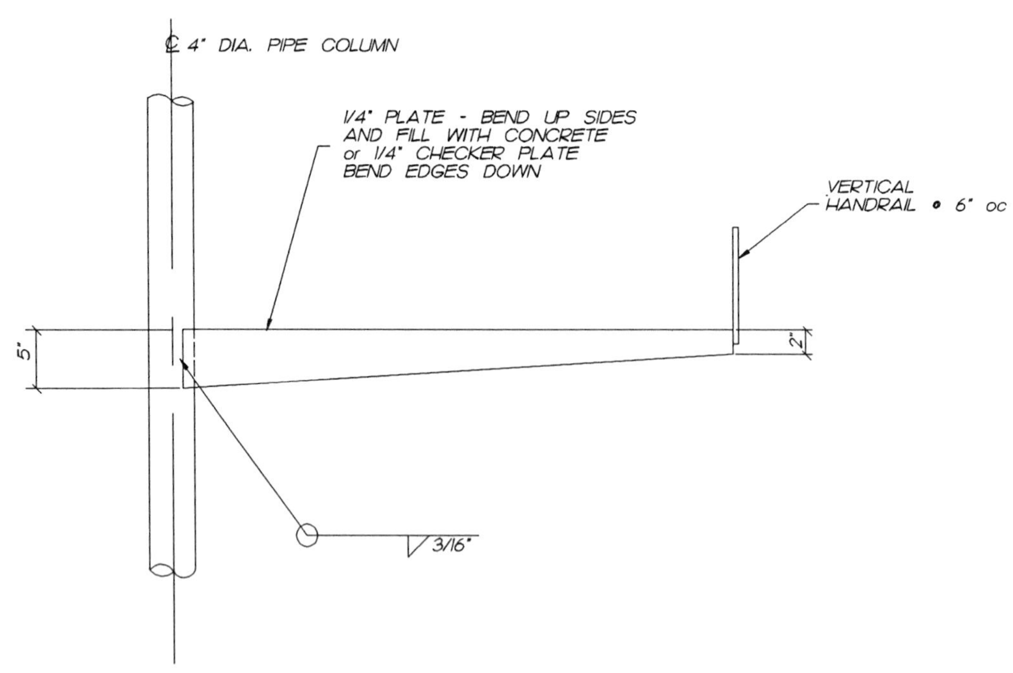

CIRCULAR STEEL STAIR TREAD

S7340

CHAPTER 4	STEEL SECTION 8	CONNECTIONS
W6 X STEEL BEAM SUPPORTING A STEEL BEAM ONE 1 SIDE CLIP ANGLE CONNECTION	MACHINE BOLTS	S8020 - S8021
	HIGH STRENGTH BOLTS	S8200 - S8201
W8 X STEEL BEAM SUPPORTING A STEEL BEAM ONE 1 SIDE CLIP ANGLE CONNECTION	MACHINE BOLTS	S8040 - S8042
	HIGH STRENGTH BOLTS	S8220 - S8222
W10 X STEEL BEAM SUPPORTING A STEEL BEAM ONE 1 SIDE CLIP ANGLE CONNECTION	MACHINE BOLTS	S8060 - S8063
	HIGH STRENGTH BOLTS	S8240 - S8243
W12 X STEEL BEAM SUPPORTING A STEEL BEAM ONE 1 SIDE CLIP ANGLE CONNECTION	MACHINE BOLTS	S8080 - S8084
	HIGH STRENGTH BOLTS	S8260 - S8264
W14 X STEEL BEAM SUPPORTING A STEEL BEAM ONE 1 SIDE CLIP ANGLE CONNECTION	MACHINE BOLTS	S8100 - S8105
	HIGH STRENGTH BOLTS	S8280 - S8285
W16 X STEEL BEAM SUPPORTING A STEEL BEAM ONE 1 SIDE CLIP ANGLE CONNECTION	MACHINE BOLTS	S8120 - S8126
	HIGH STRENGTH BOLTS	S8300 - S8306
W18 X STEEL BEAM SUPPORTING A STEEL BEAM ONE 1 SIDE CLIP ANGLE CONNECTION	MACHINE BOLTS	S8140 - S8148
	HIGH STRENGTH BOLTS	S8320 - S8328
W21 X STEEL BEAM SUPPORTING A STEEL BEAM ONE 1 SIDE CLIP ANGLE CONNECTION	MACHINE BOLTS	S8160 - S8168
	HIGH STRENGTH BOLTS	S8340 - S8348
W24 X STEEL BEAM SUPPORTING A STEEL BEAM ONE 1 SIDE CLIP ANGLE CONNECTION MACHINE BOLTS	MACHINE BOLTS	S8180 - S8188
	HIGH STRENGTH BOLTS	S8360 - S8368
W6 X STEEL BEAM SUPPORTING A STEEL BEAM ONE 1 SIDE TAB PLATE CONNECTION	MACHINE BOLTS	S8380 - S8381
W8 X STEEL BEAM SUPPORTING A STEEL BEAM ONE 1 SIDE TAB PLATE CONNECTION	MACHINE BOLTS	S8400 - S8402
W10 X STEEL BEAM SUPPORTING A STEEL BEAM ONE 1 SIDE TAB PLATE CONNECTION	MACHINE BOLTS	S8420 - S8423
W12 X STEEL BEAM SUPPORTING A STEEL BEAM ONE 1 SIDE TAB PLATE CONNECTION	MACHINE BOLTS	S8440 - S8444
W14 X STEEL BEAM SUPPORTING A STEEL BEAM ONE 1 SIDE TAB PLATE CONNECTION	MACHINE BOLTS	S8460 - S8465
W16 X STEEL BEAM SUPPORTING A STEEL BEAM ONE 1 SIDE TAB PLATE CONNECTION	MACHINE BOLTS	S8480 - S8486
W18 X STEEL BEAM SUPPORTING A STEEL BEAM ONE 1 SIDE TAB PLATE CONNECTION	MACHINE BOLTS	S8500 - S8508
W21 X STEEL BEAM SUPPORTING A STEEL BEAM ONE 1 SIDE TAB PLATE CONNECTION	MACHINE BOLTS	S8520 - S8528

W24 X STEEL BEAM SUPPORTING A STEEL BEAM ONE 1 SIDE TAB PLATE CONNECTION	MACHINE BOLTS	S8540 - S8548
W10 X STEEL BEAM SUPPORTING BEAMS EACH SIDE CLIP ANGLE CONNECTION	MACHINE BOLTS	S8560 - S8561
	HIGH STRENGTH BOLTS	S8660 - S8661
W12 X STEEL BEAM SUPPORTING BEAMS EACH SIDE CLIP ANGLE CONNECTION	MACHINE BOLTS	S8580 - S8581
	HIGH STRENGTH BOLTS	S8680 - S8681
W14 X STEEL BEAM SUPPORTING BEAMS EACH SIDE CLIP ANGLE CONNECTION	MACHINE BOLTS	S8600 - S8601
	HIGH STRENGTH BOLTS	S8700 - S8701
W16 X STEEL BEAM SUPPORTING BEAMS EACH SIDE CLIP ANGLE CONNECTION	MACHINE BOLTS	S8620 - S8621
	HIGH STRENGTH BOLTS	S8720 - S8721
W18 X STEEL BEAM SUPPORTING BEAMS EACH SIDE CLIP ANGLE CONNECTION	MACHINE BOLTS	S8640 - S8641
	HIGH STRENGTH BOLTS	S8740 - S8741
W10 X STEEL BEAM SUPPORTING BEAMS EACH SIDE TAB PLATE CONNECTION	MACHINE BOLTS	S8760 - S8761
W12 X STEEL BEAM SUPPORTING BEAMS EACH SIDE CLIP ANGLE CONNECTION	MACHINE BOLTS	S8780 - S8781
W14 X STEEL BEAM SUPPORTING BEAMS EACH SIDE CLIP ANGLE CONNECTION	MACHINE BOLTS	S8800 - S8801
W16 X STEEL BEAM SUPPORTING BEAMS EACH SIDE CLIP ANGLE CONNECTION	MACHINE BOLTS	S8820 - S8821
W18 X STEEL BEAM SUPPORTING BEAMS EACH SIDE CLIP ANGLE CONNECTION	MACHINE BOLTS	S8480 - S8481

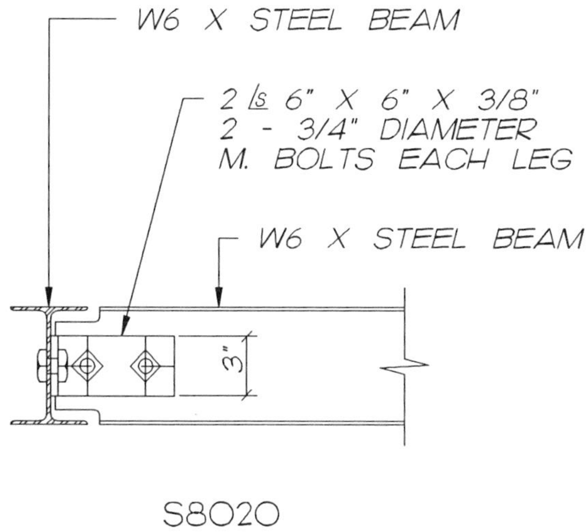

S8020

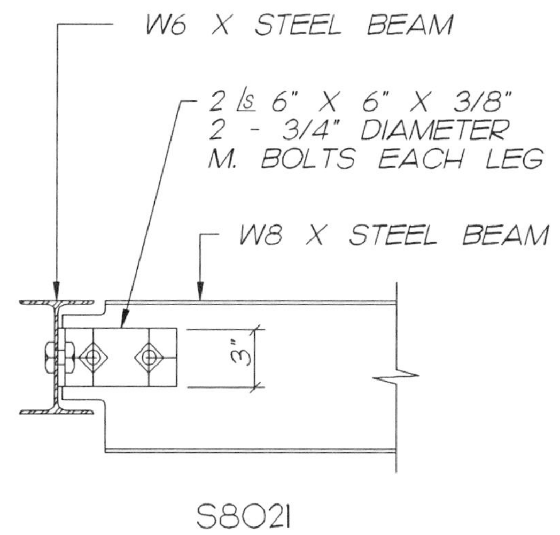

S8021

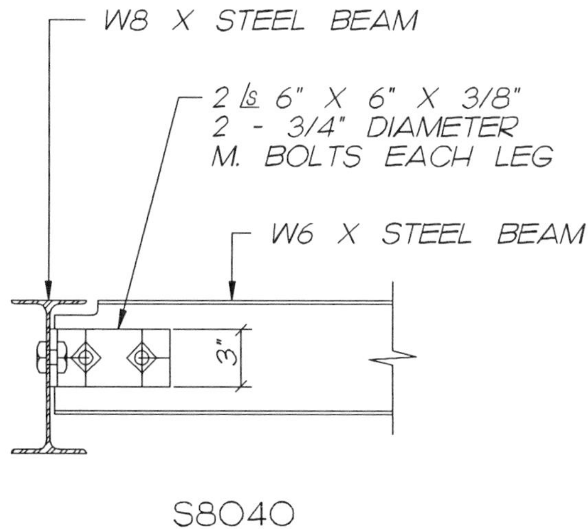

S8040

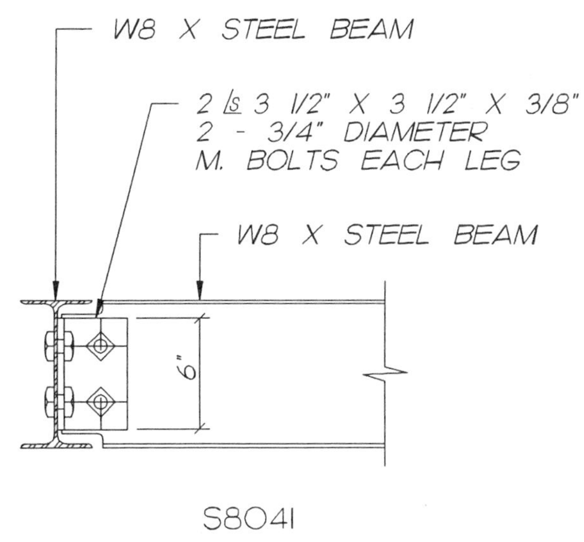

S8041

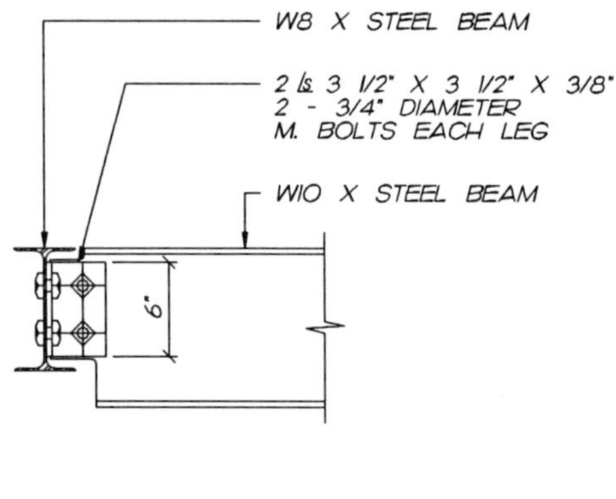

S8042

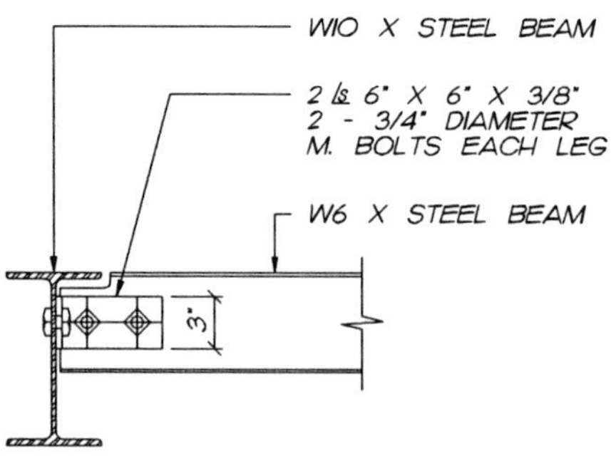

S8060

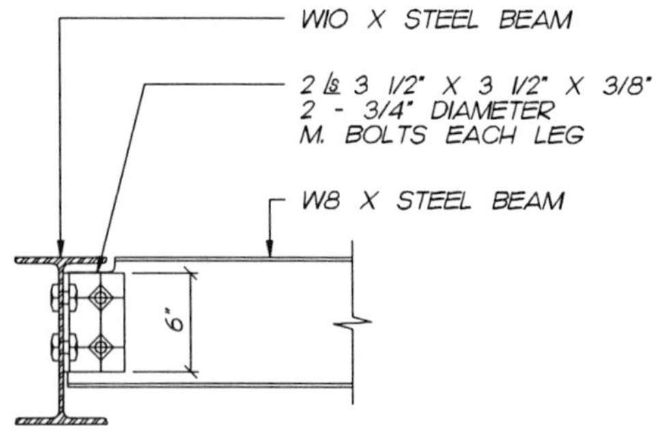

S8061

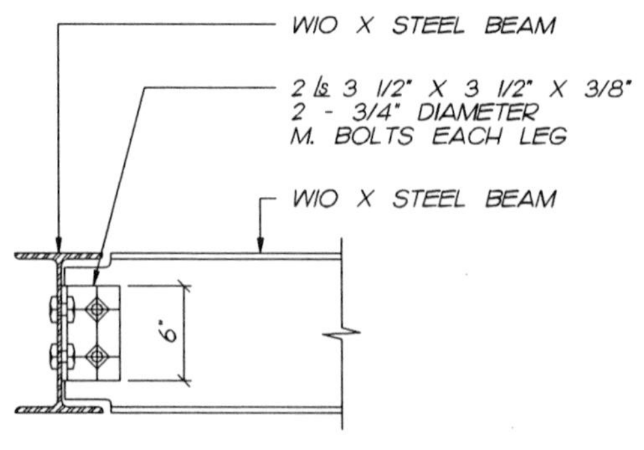

S8062

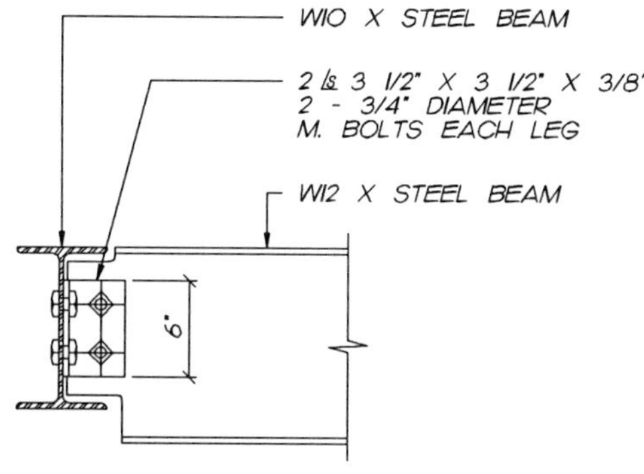

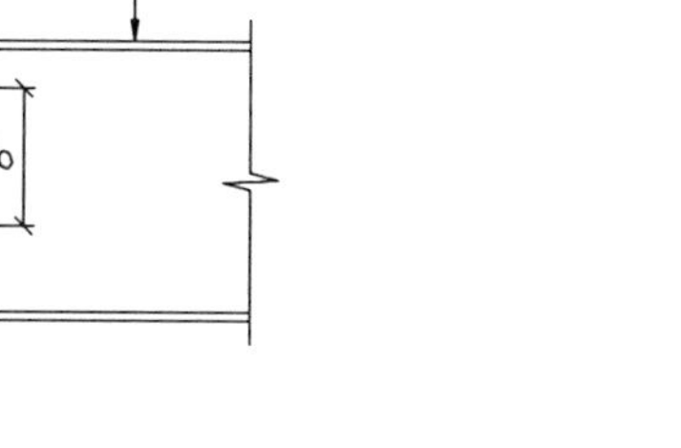

S8063

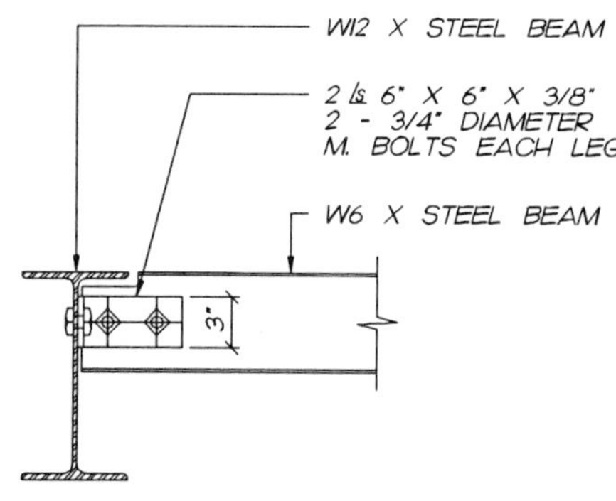

S8080

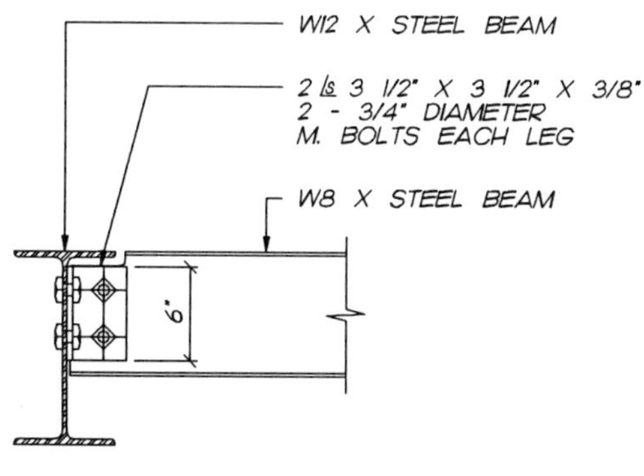

S8081

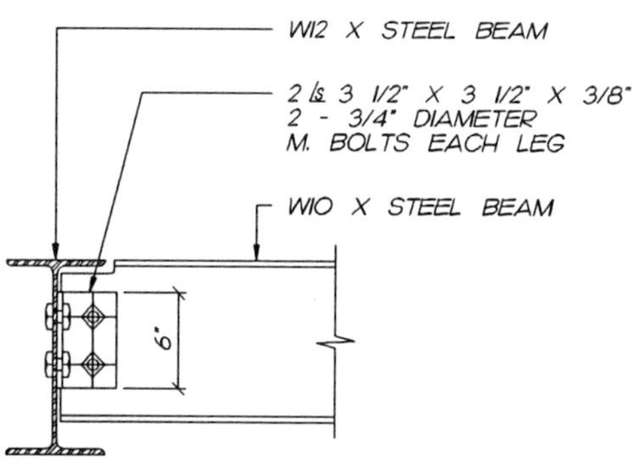

S8081

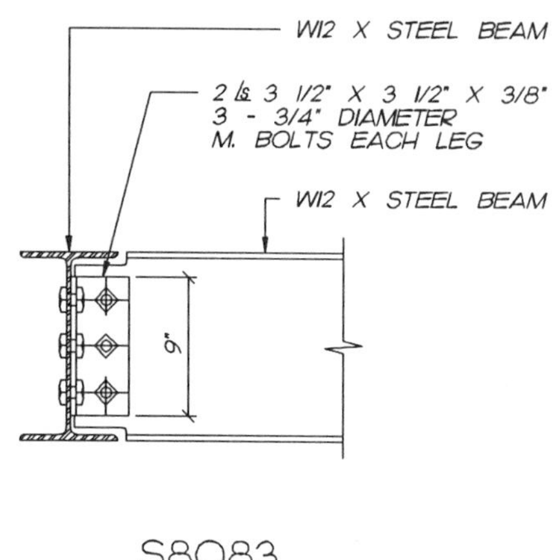
W12 X STEEL BEAM
2 Ls 3 1/2" X 3 1/2" X 3/8"
3 - 3/4" DIAMETER
M. BOLTS EACH LEG
W12 X STEEL BEAM
9"
S8083

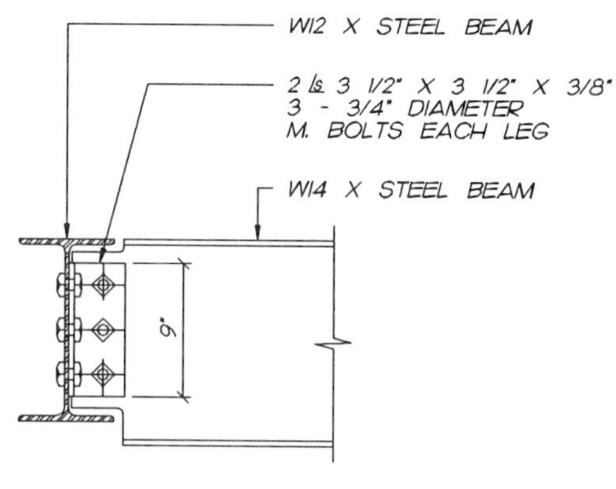
W12 X STEEL BEAM
2 Ls 3 1/2" X 3 1/2" X 3/8"
3 - 3/4" DIAMETER
M. BOLTS EACH LEG
W14 X STEEL BEAM
9"

S8084

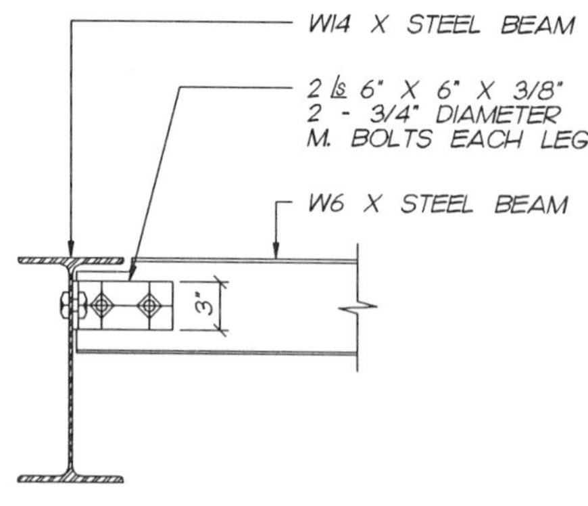
W14 X STEEL BEAM
2 Ls 6" X 6" X 3/8"
2 - 3/4" DIAMETER
M. BOLTS EACH LEG
W6 X STEEL BEAM
3"

S8100

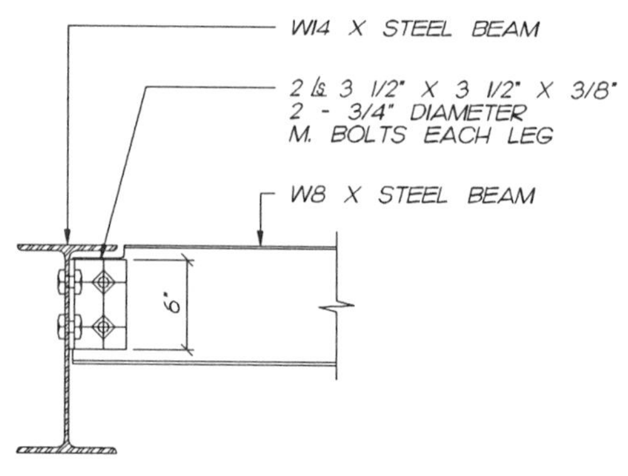
W14 X STEEL BEAM
2 Ls 3 1/2" X 3 1/2" X 3/8"
2 - 3/4" DIAMETER
M. BOLTS EACH LEG
W8 X STEEL BEAM
6"

S8101

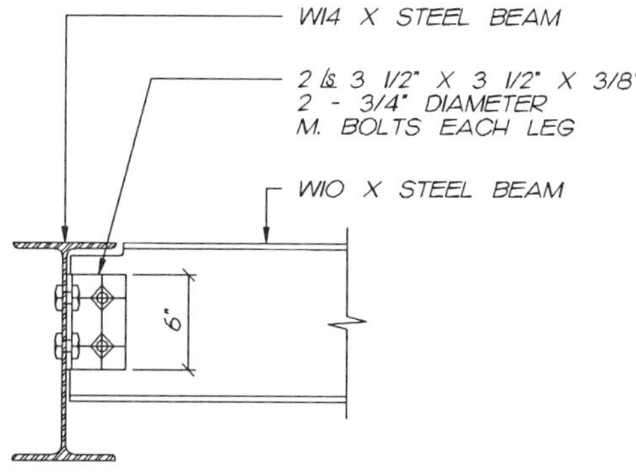

S8102

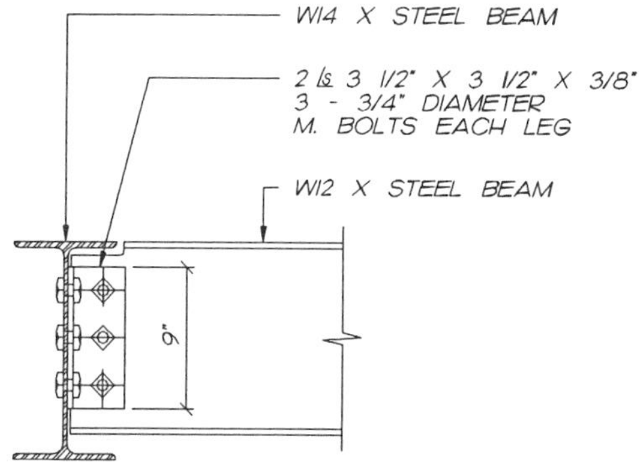

S8103

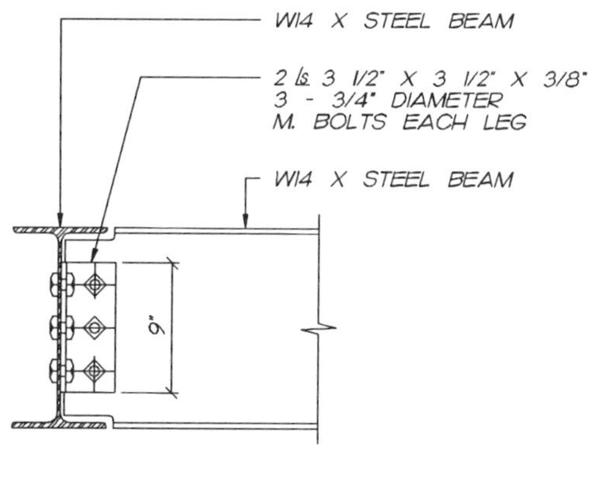

S8104

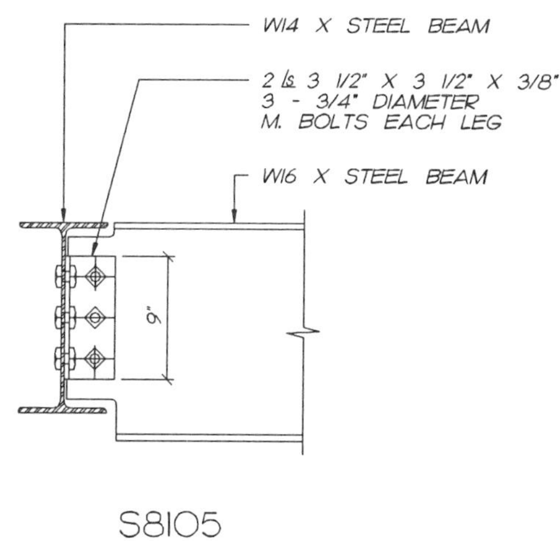

S8105

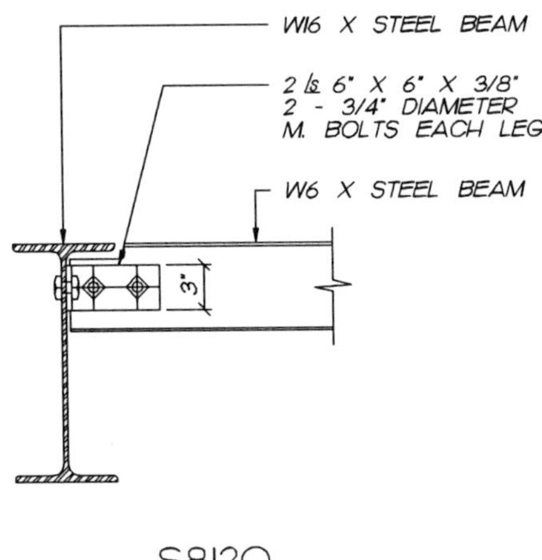

S8120

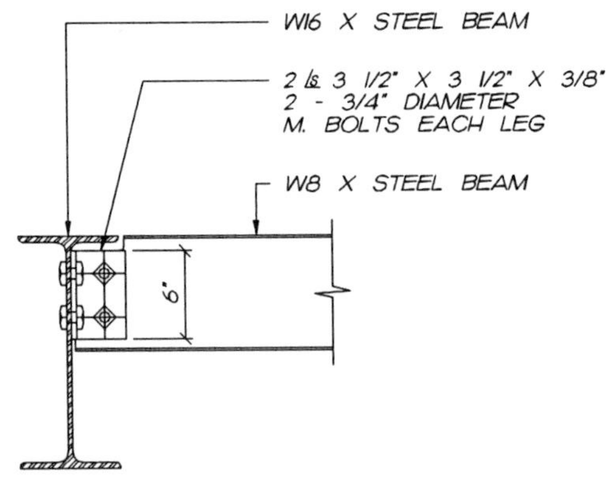

S8121

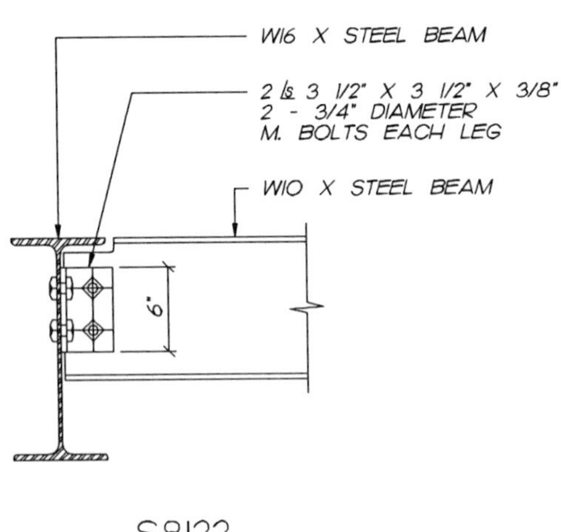

S8122

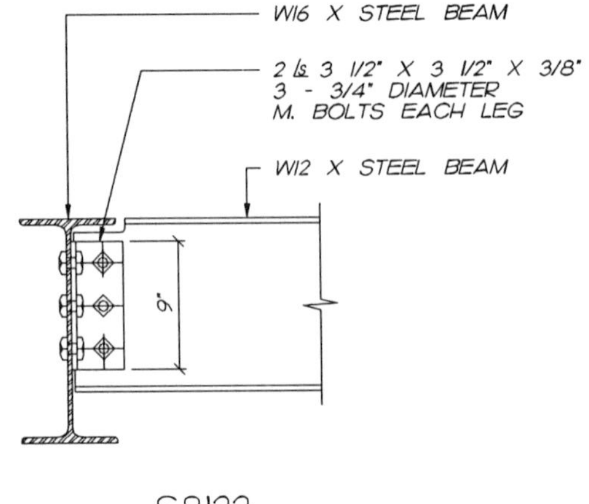

S8123

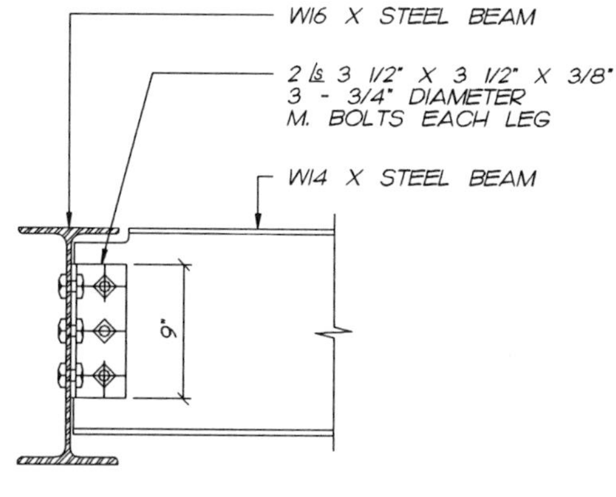

S8124

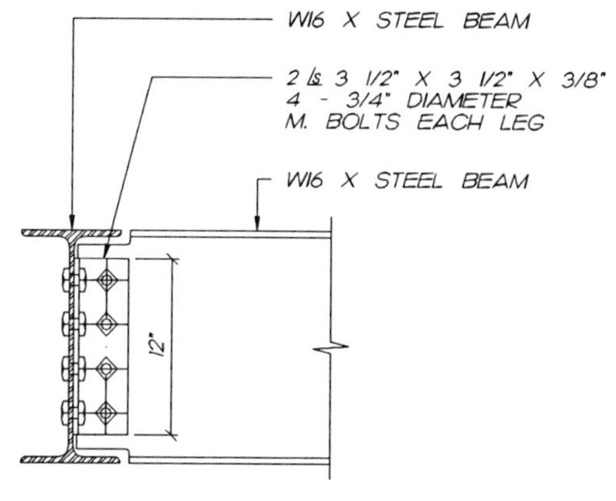

S8125

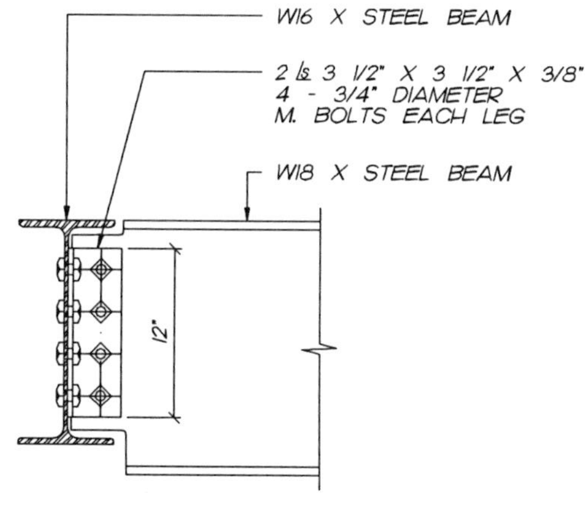

S8126

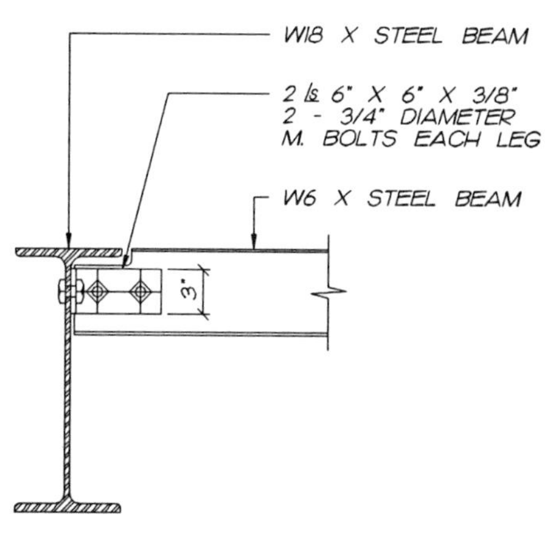

S8140

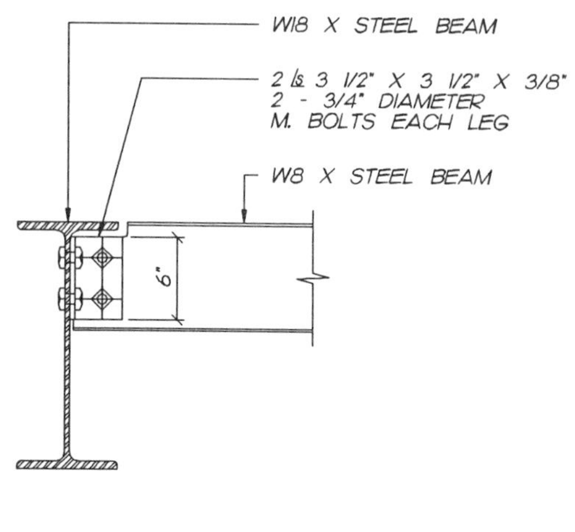

S8141

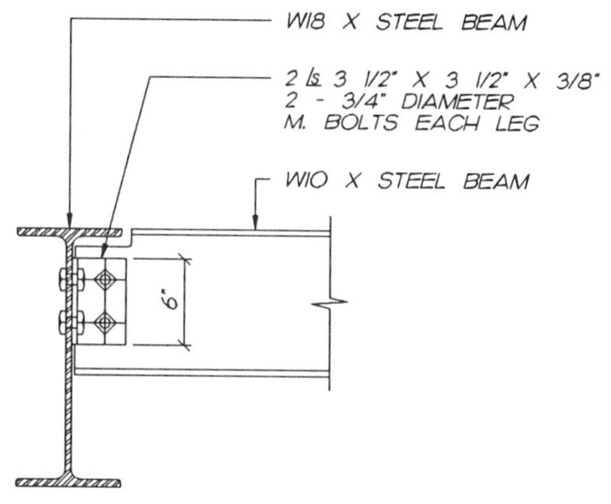

S8142

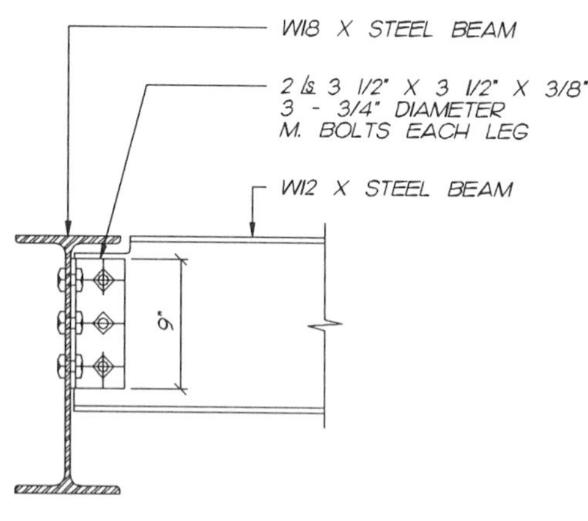

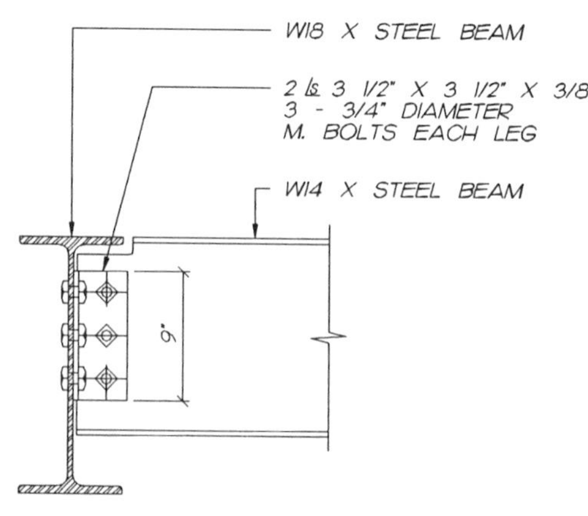

S8144

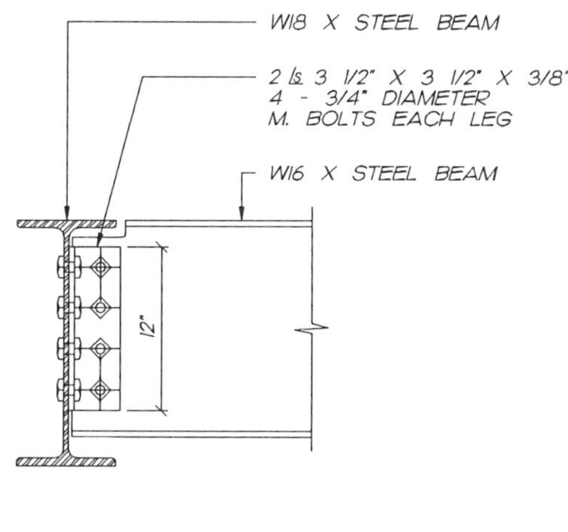

S8145

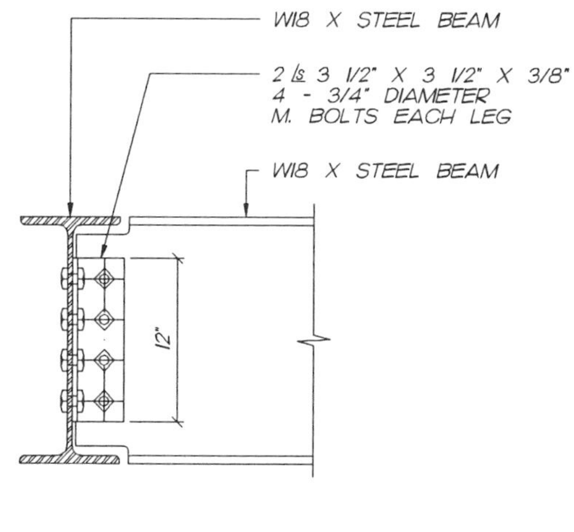

S8146

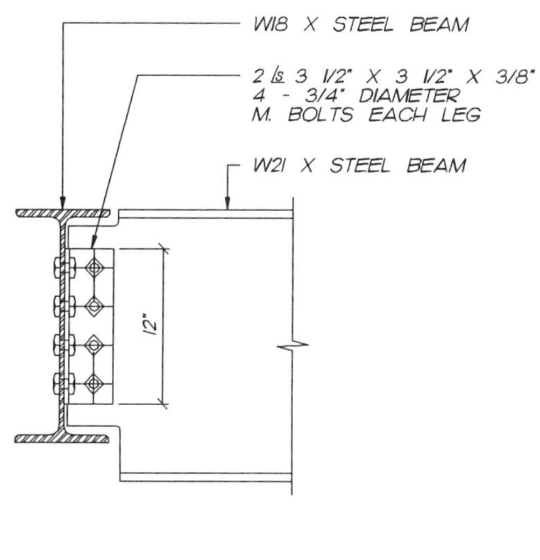

S8147

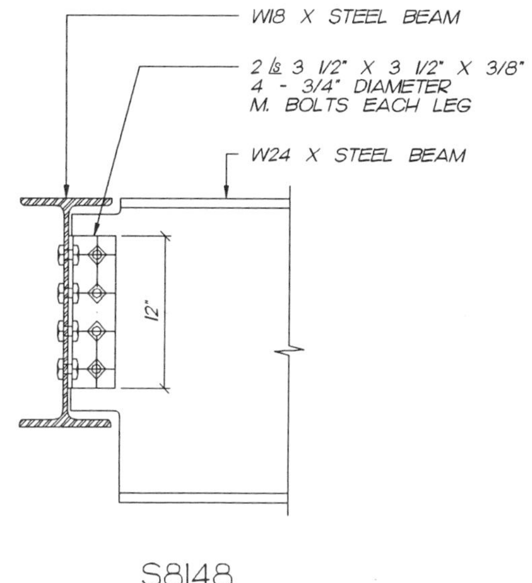

S8148

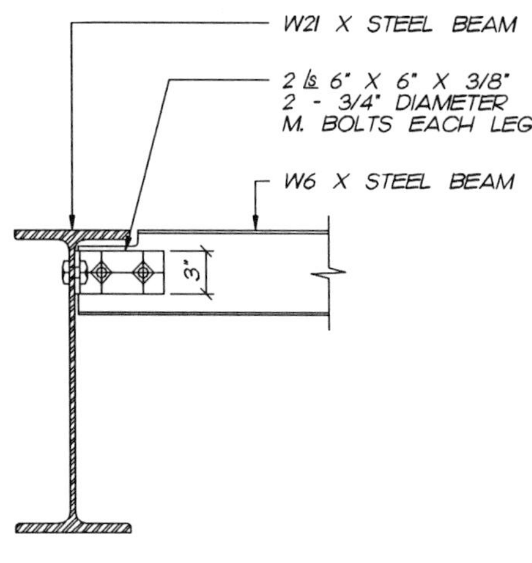

S8160

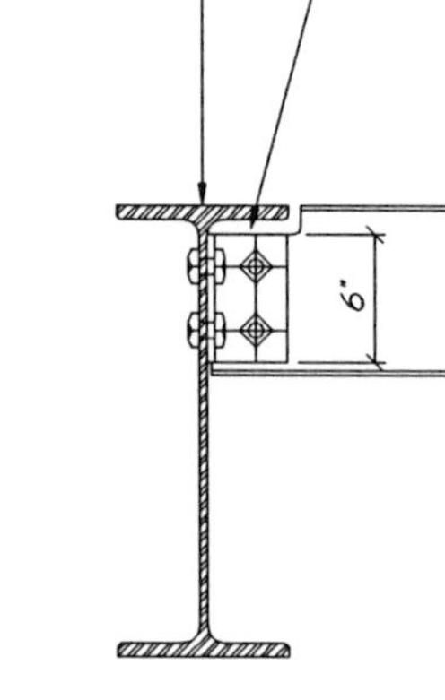

S8161

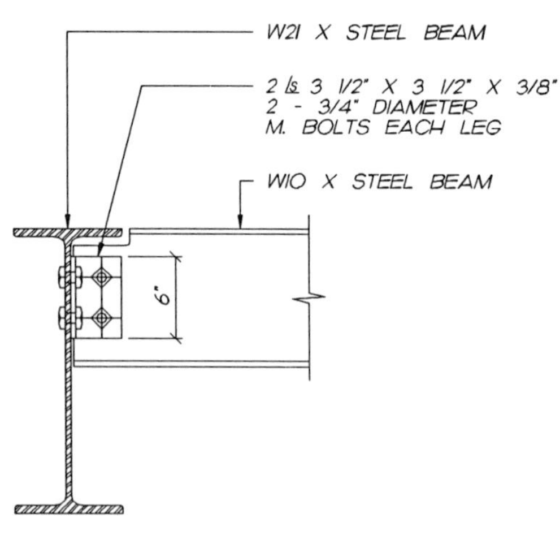

S8162

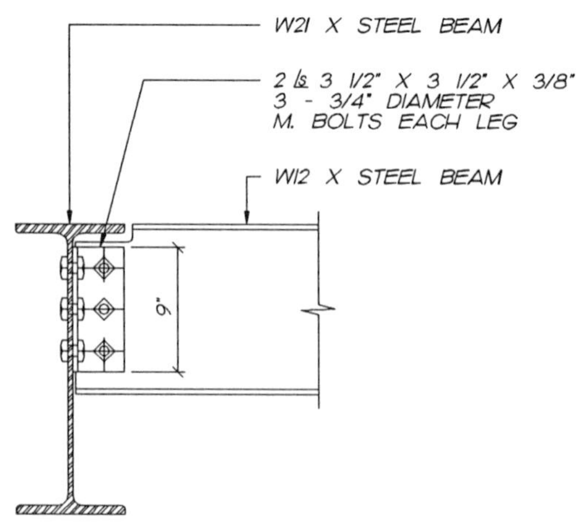

S8163

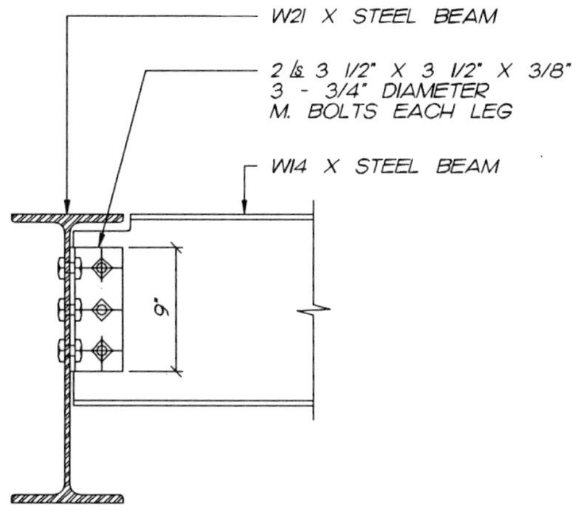

S8164

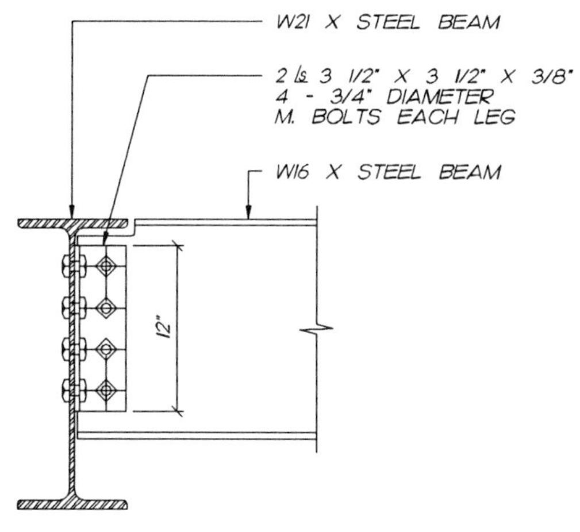

S8165

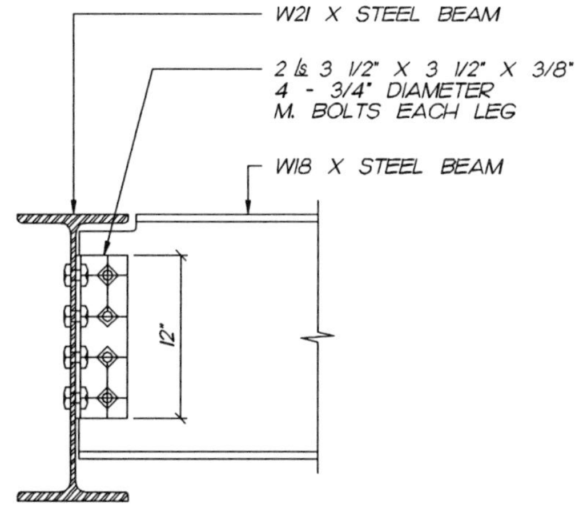

S8166

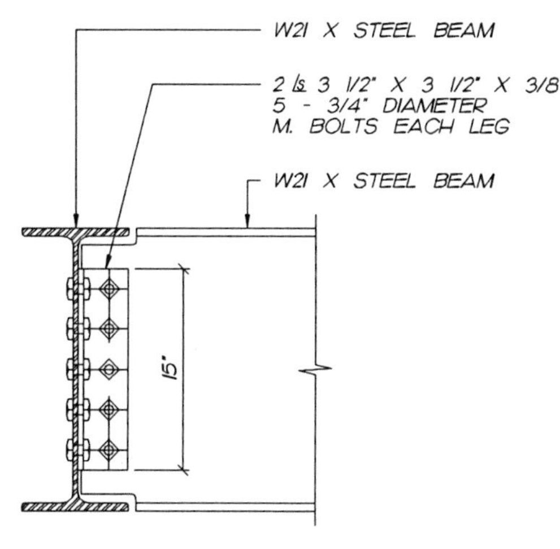

S8167

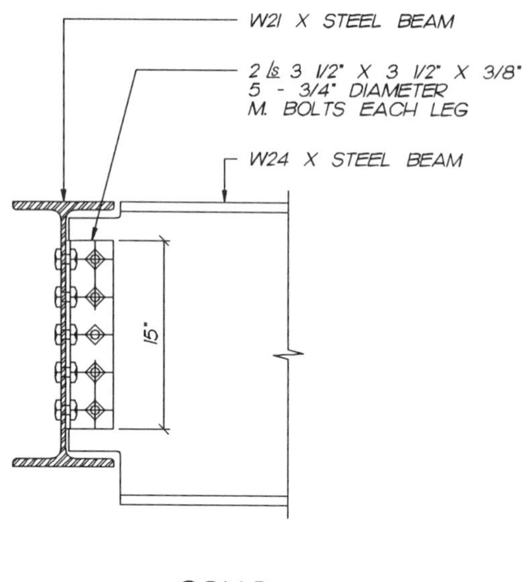

S8168

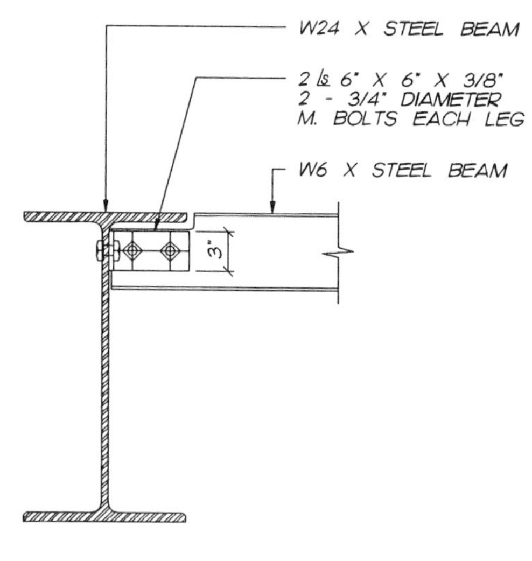

S8180

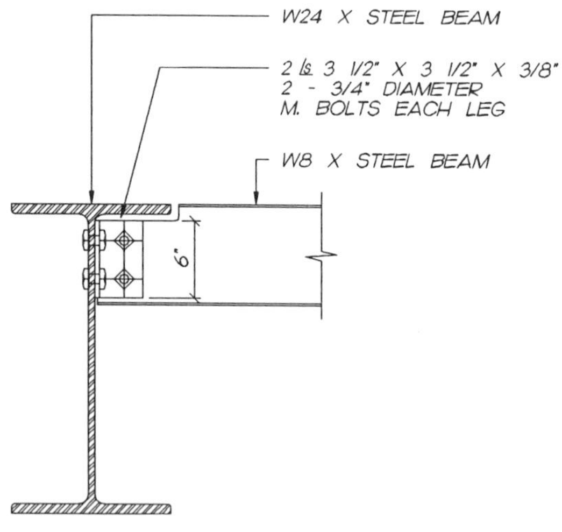

S8181

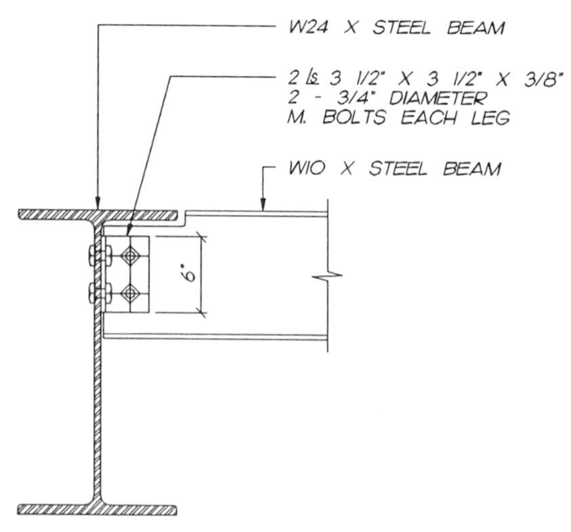

S8182

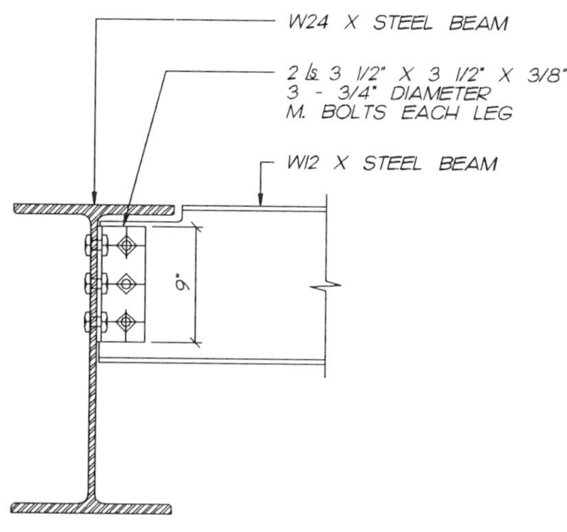

S8183

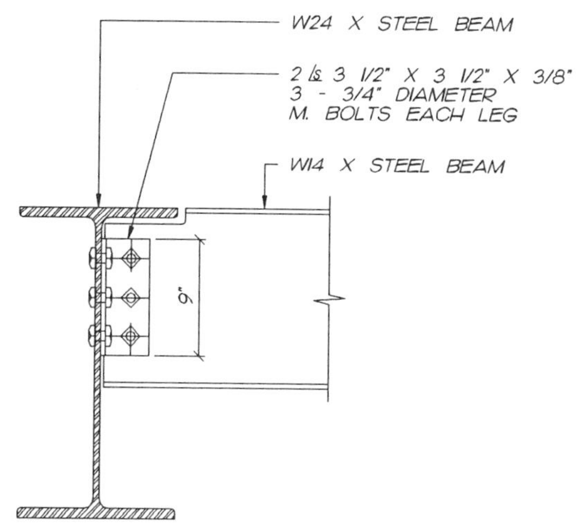

S8184

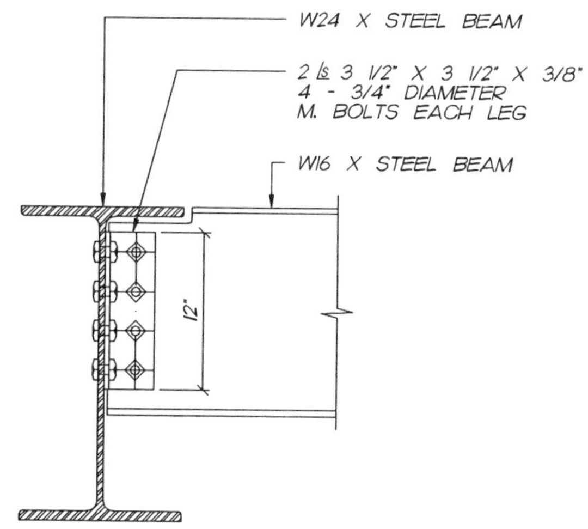

S8185

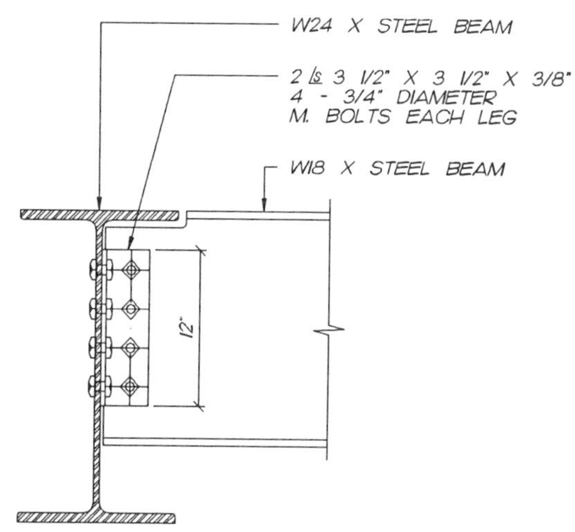

S8186

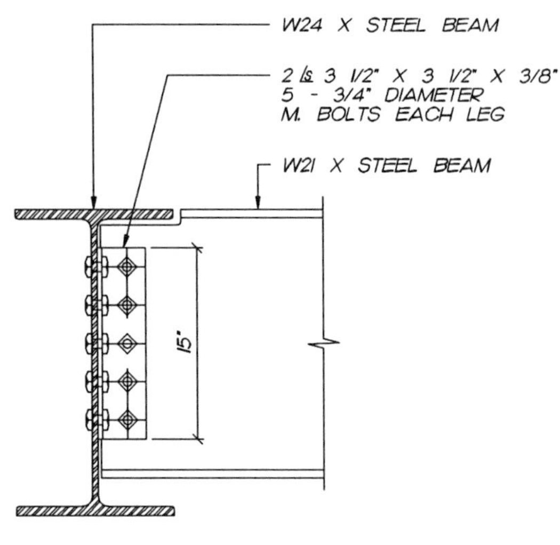

S8187

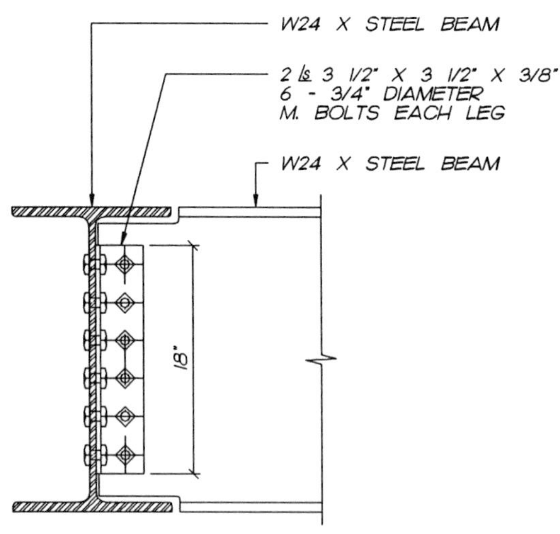

S8188

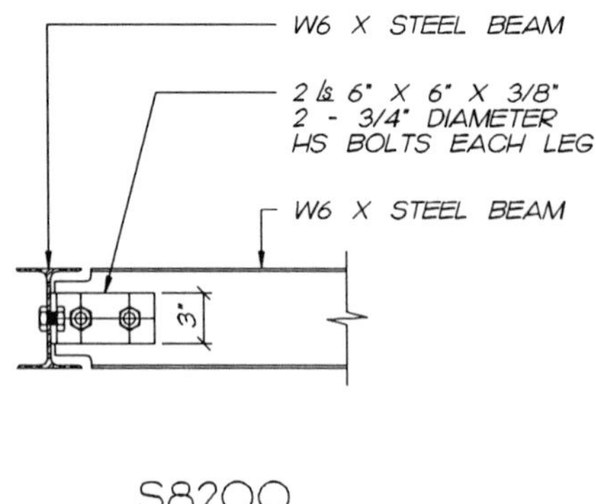

S8200

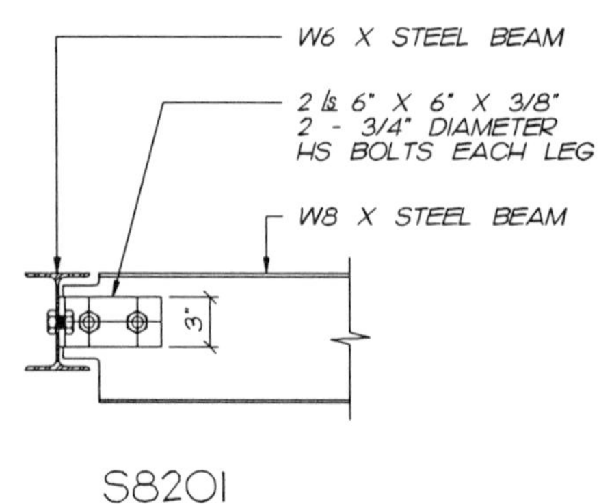

S8201

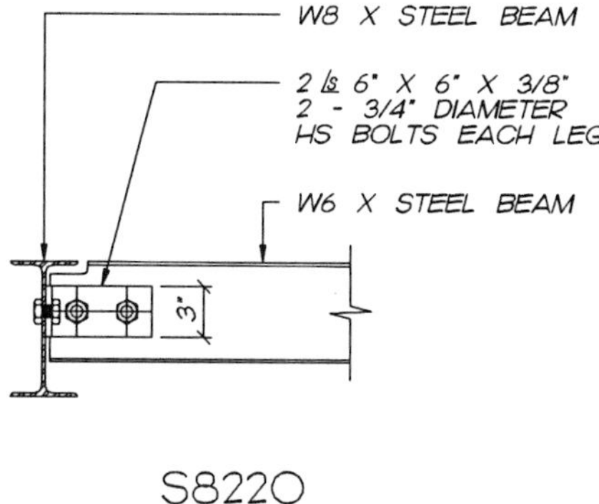

S8220

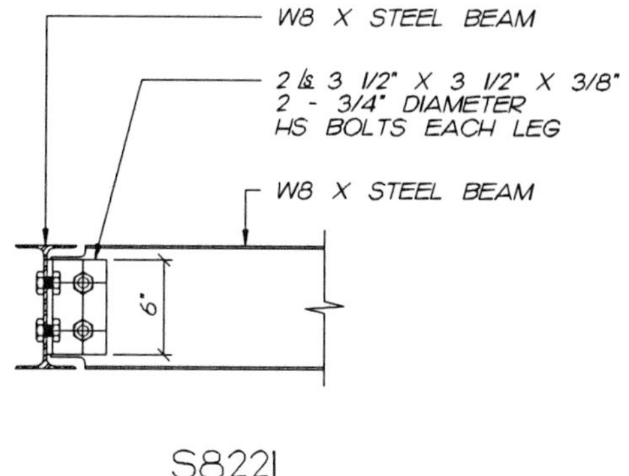

S8221

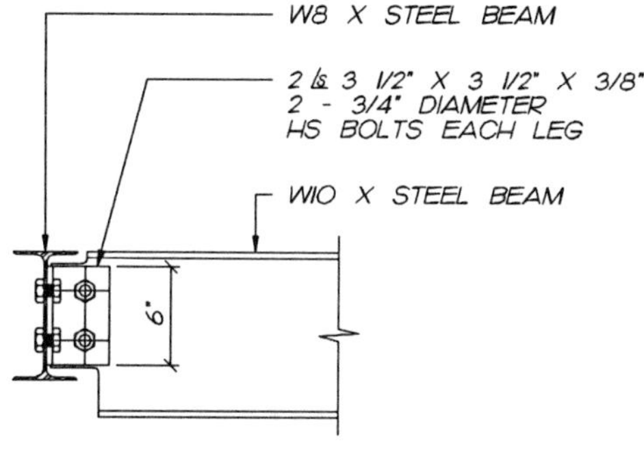

S8222

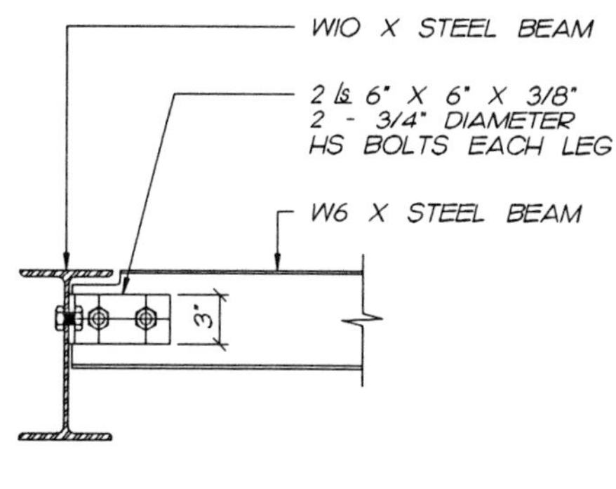

S8240

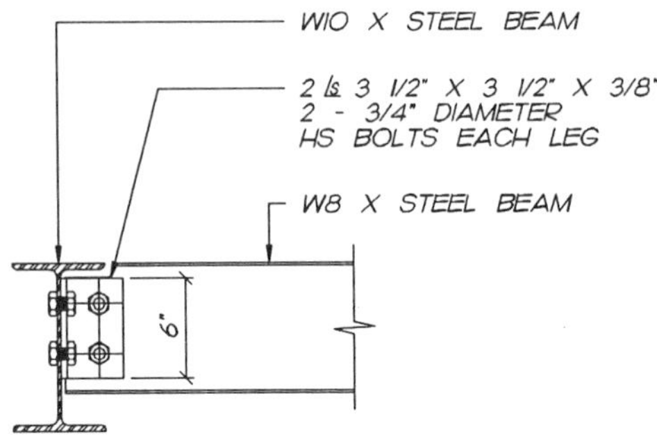

S8241

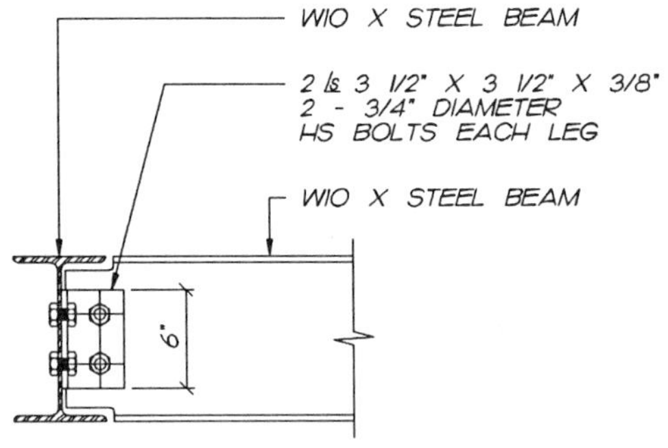

S8242

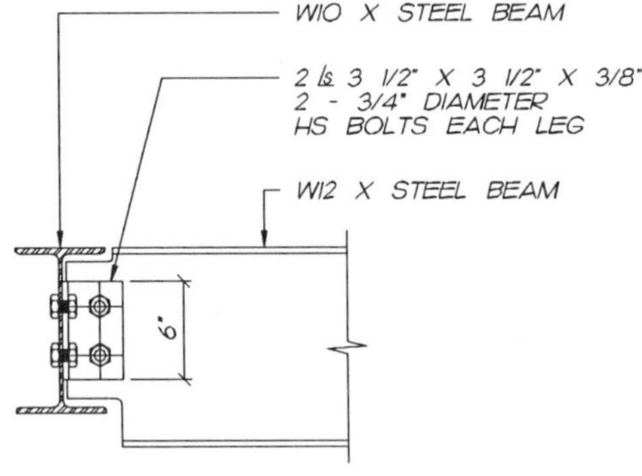

S8243

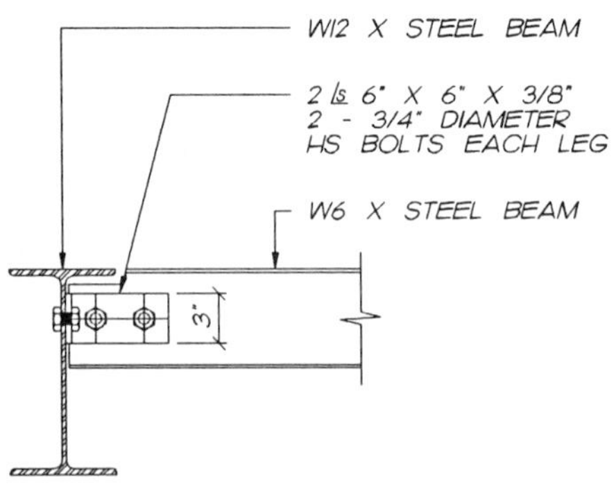

S8260

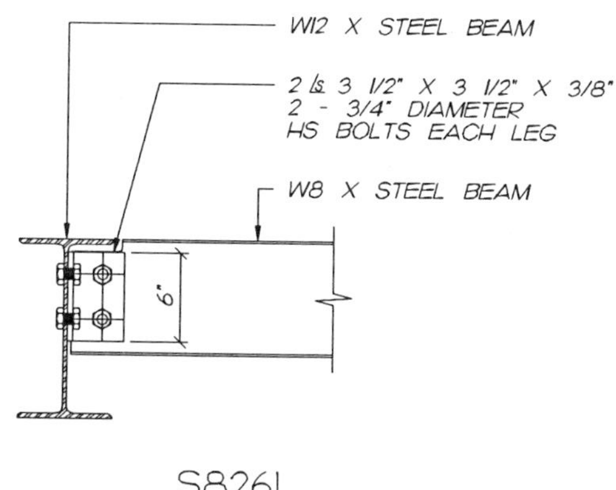

S8261

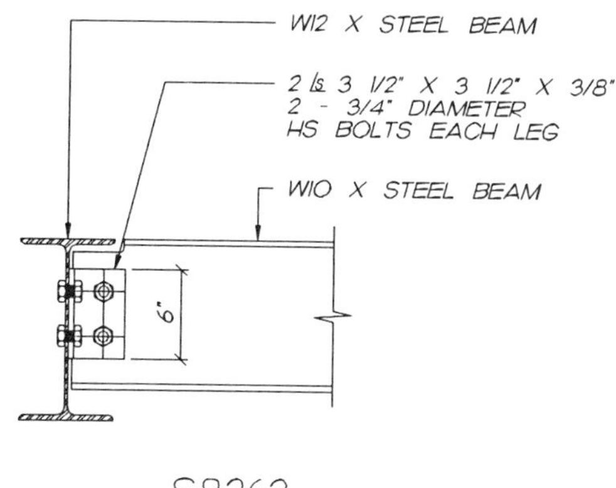

S8262

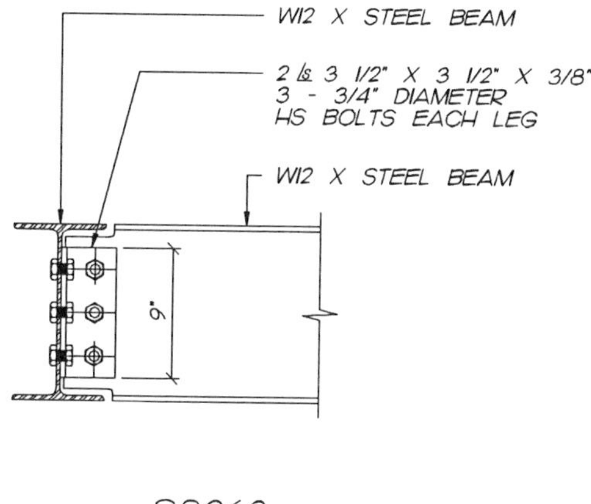

S8263

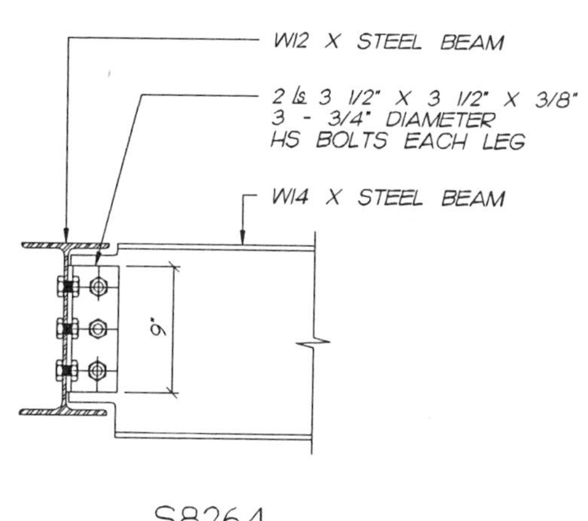

S8264

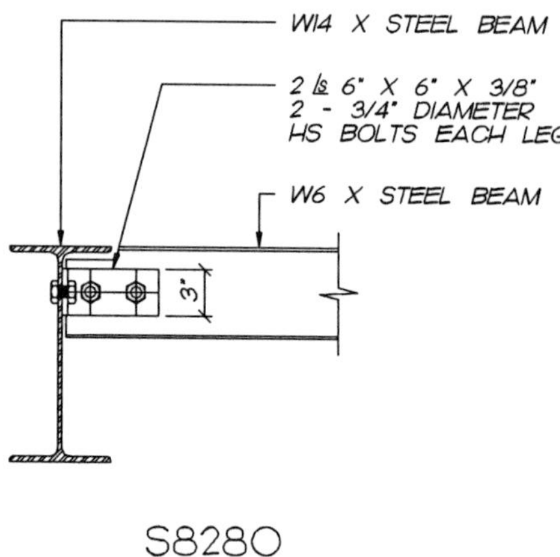

S8280

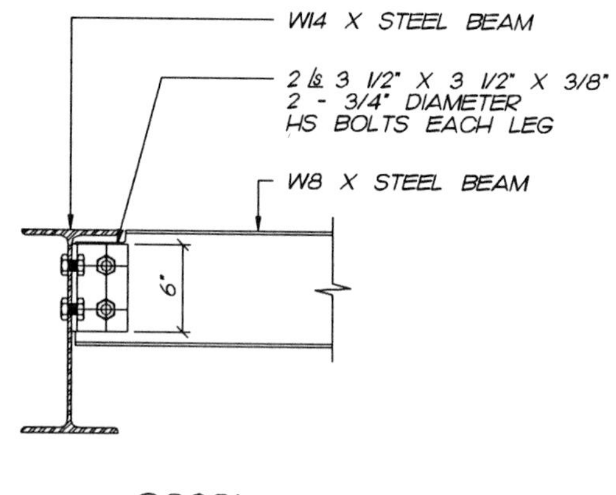

S8281

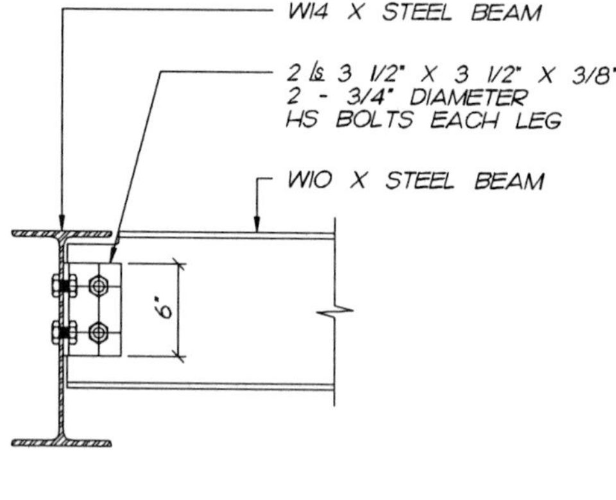

S8282

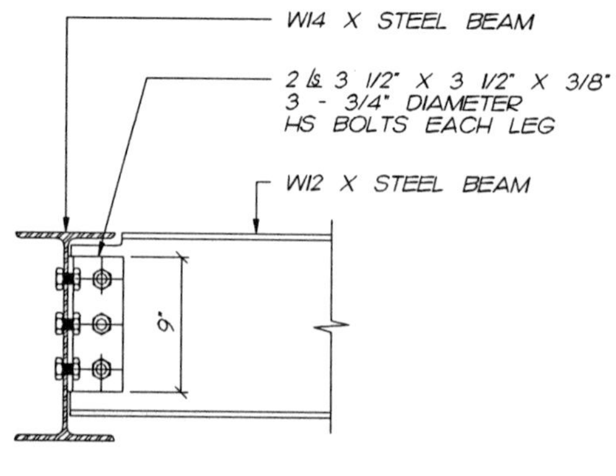

S8283

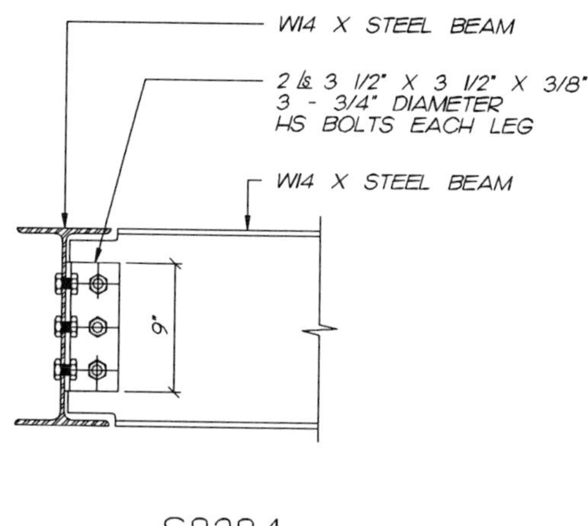

S8284

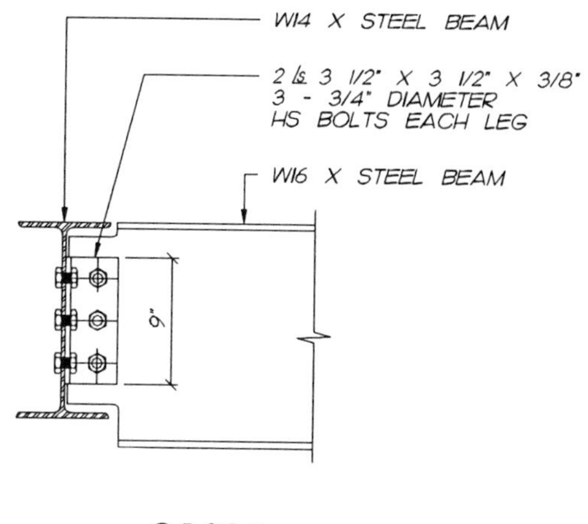

S8285

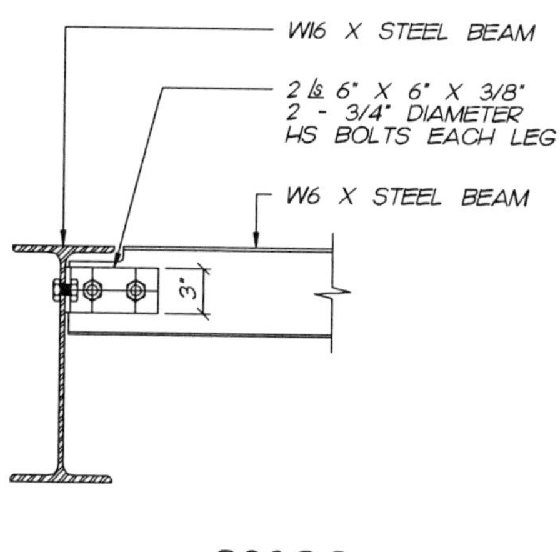

S8300

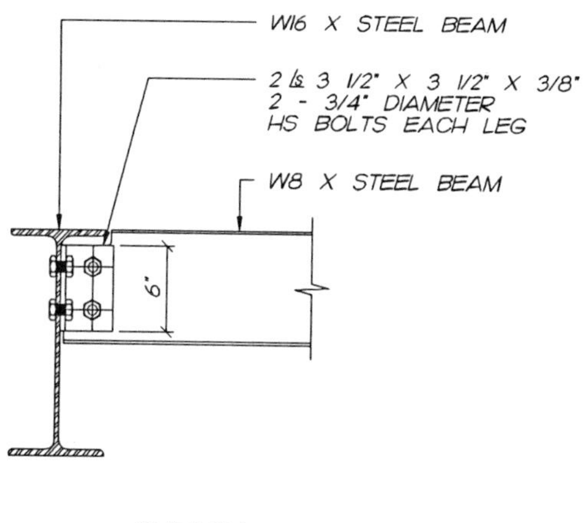

S8301

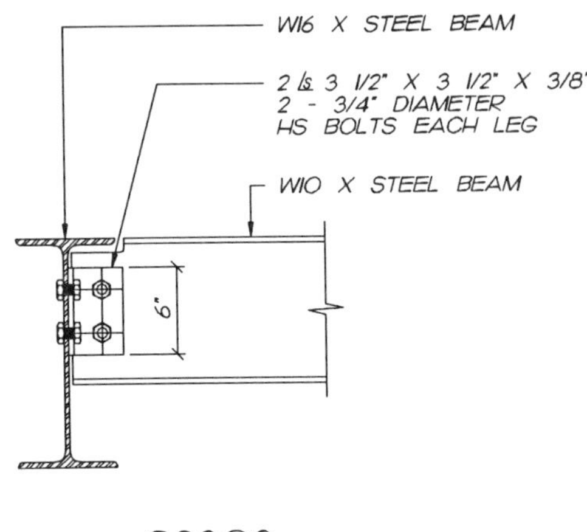

S8302

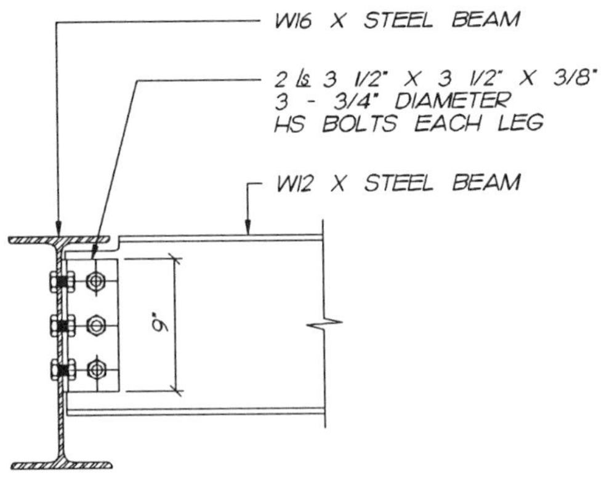

S8303

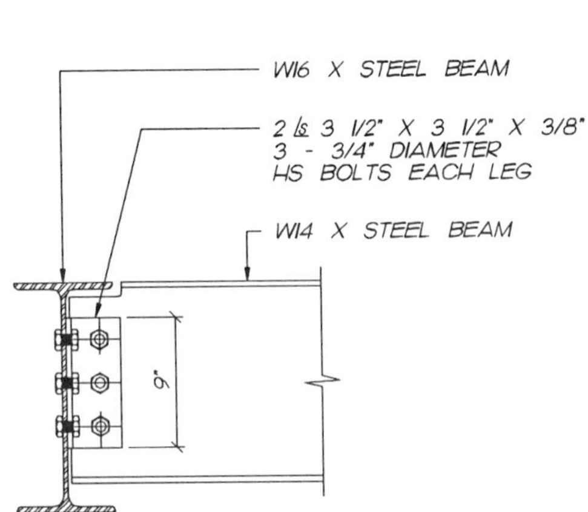

S8304

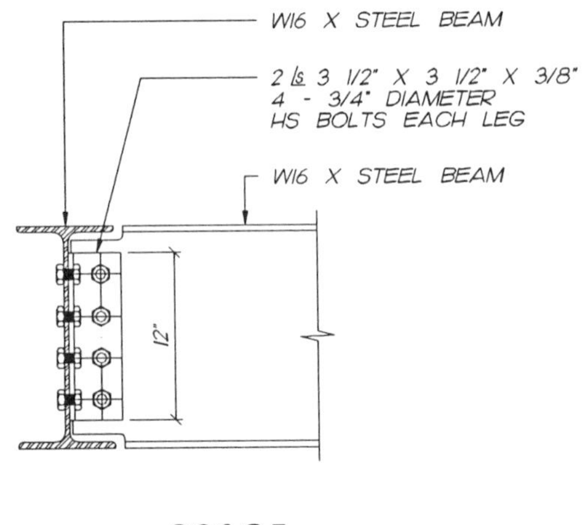

S8305

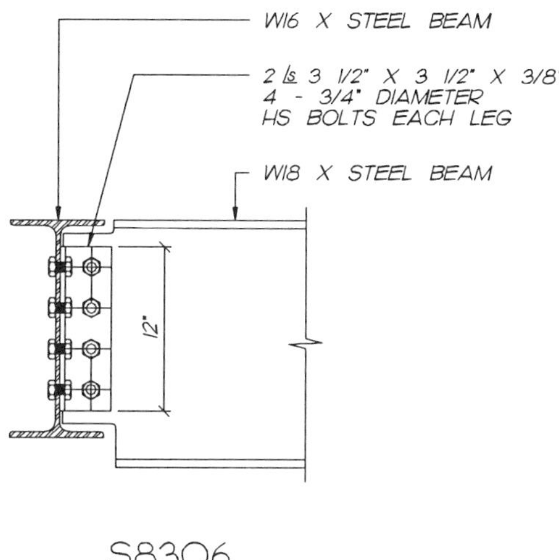

S8306

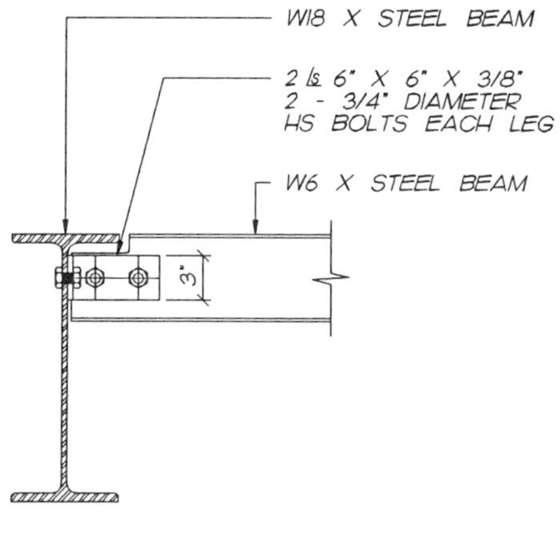

S8320

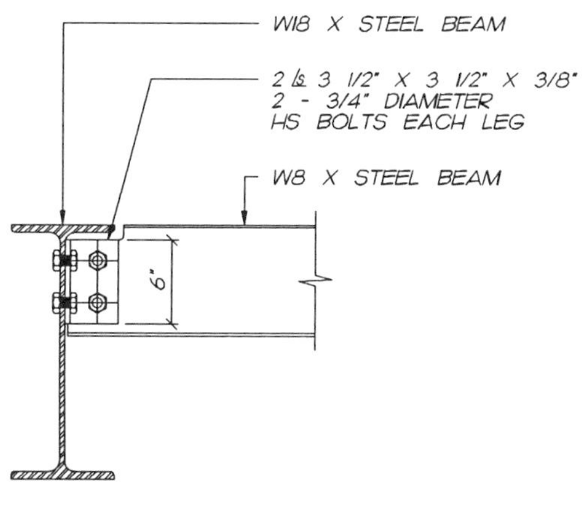

S8321

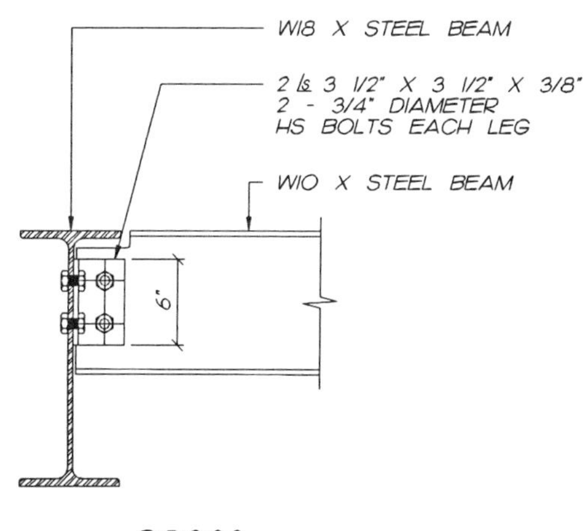

S8322

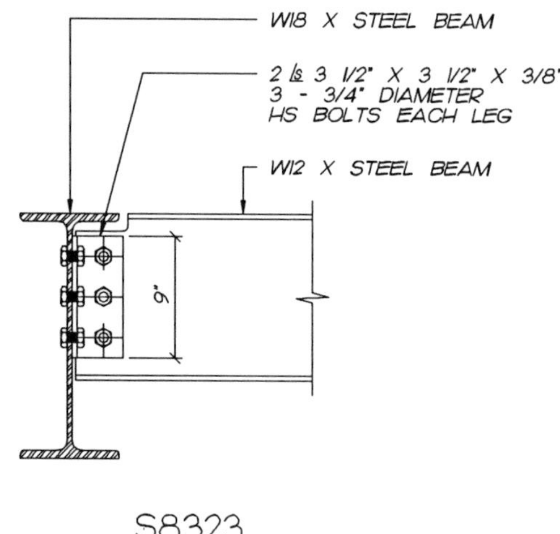

S8323

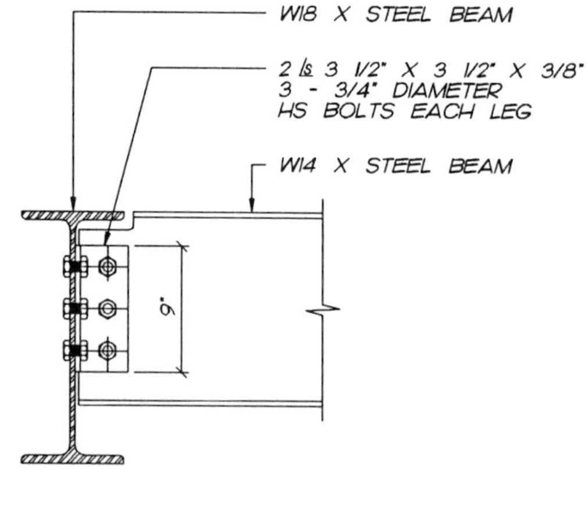

S8324

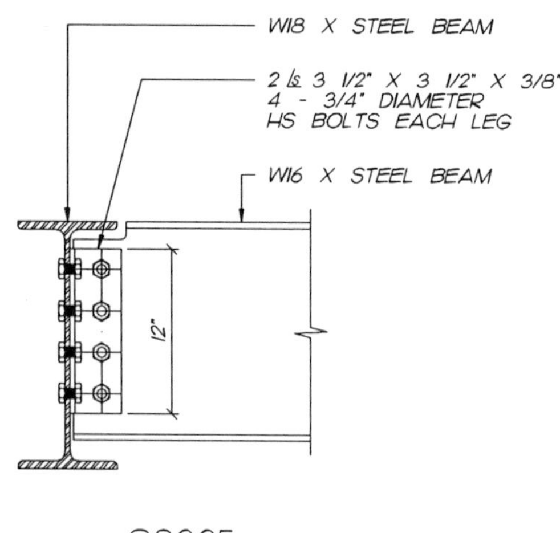

S8325

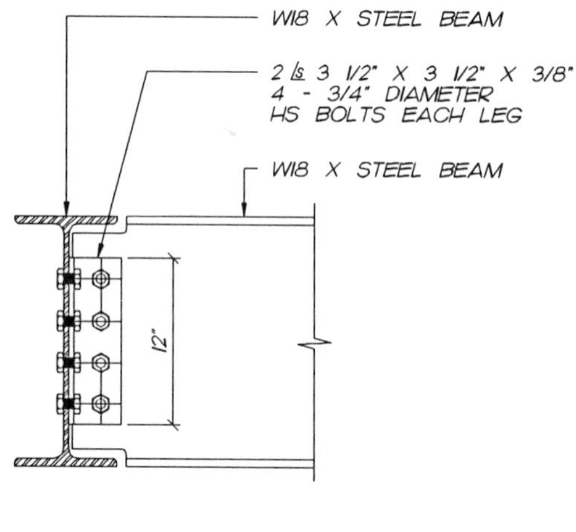

S8326

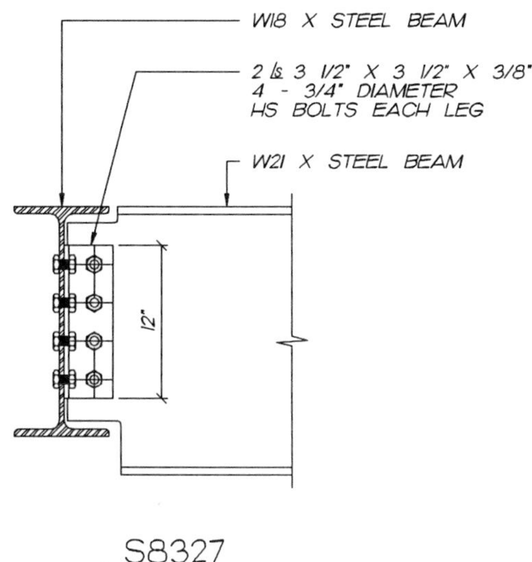

S8327

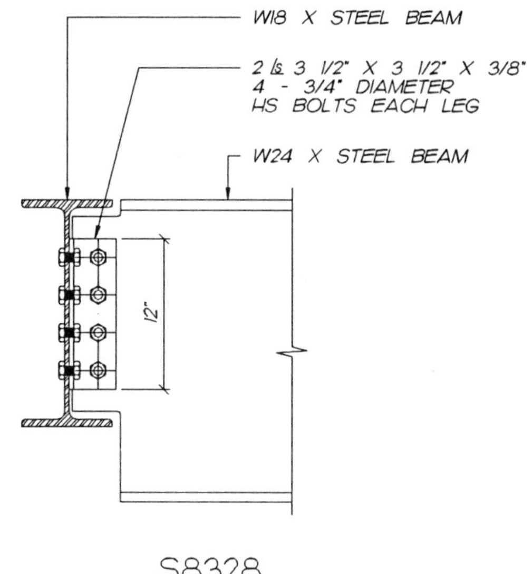

S8328

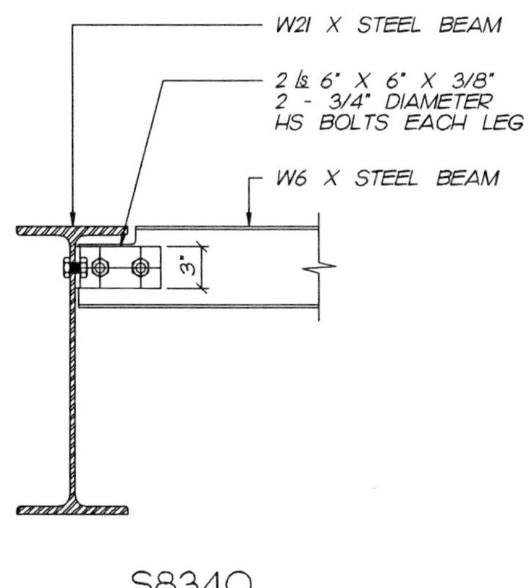

S8340

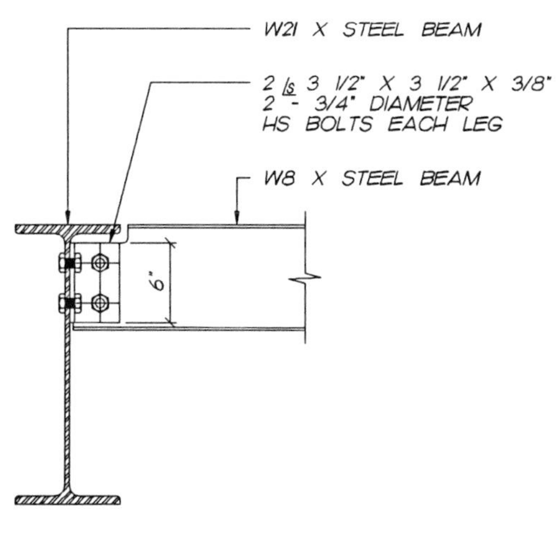

S8341

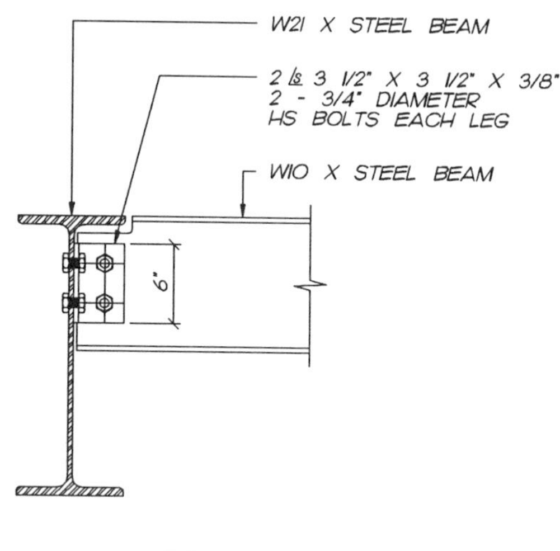

S8342

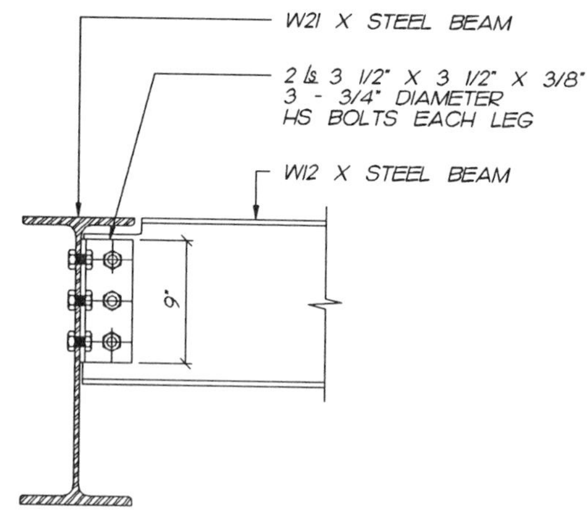

S8343

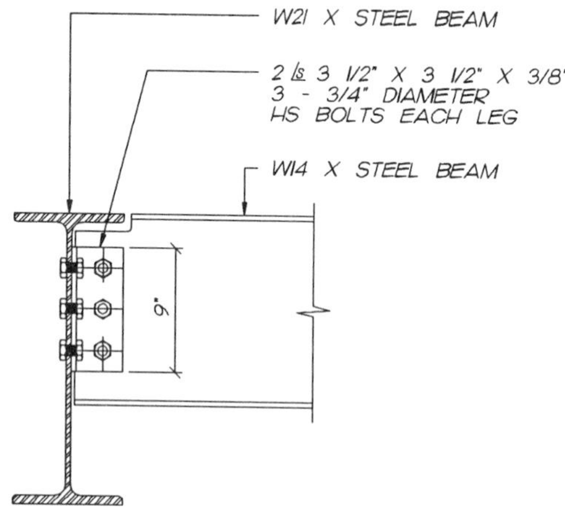

S8344

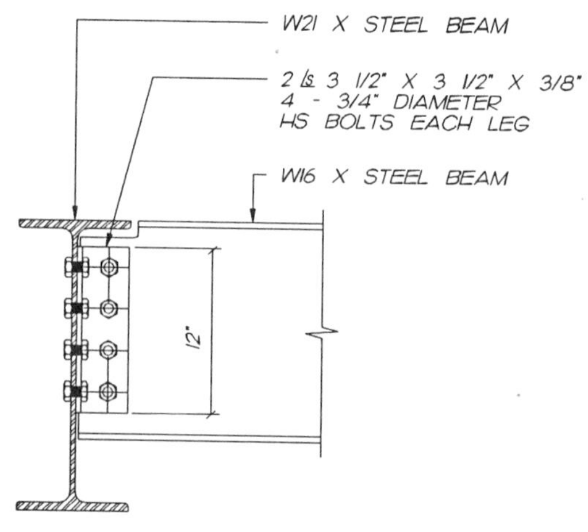

S8345

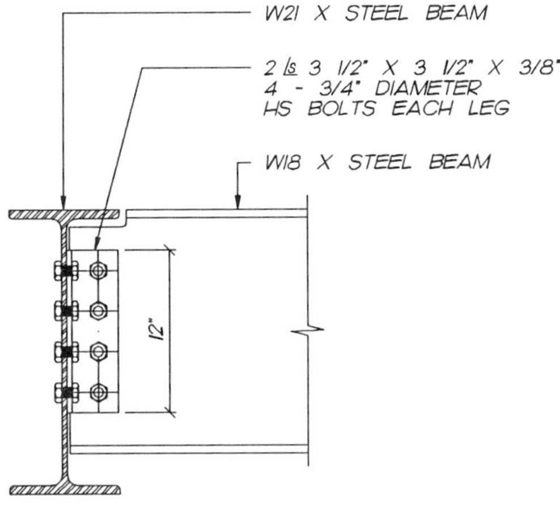
W21 X STEEL BEAM
2 ∟s 3 1/2" X 3 1/2" X 3/8"
4 - 3/4" DIAMETER
HS BOLTS EACH LEG
W18 X STEEL BEAM
12"

S8346

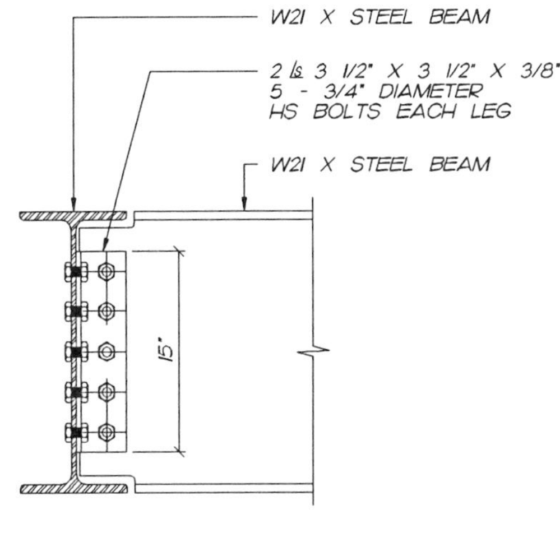
W21 X STEEL BEAM
2 ∟s 3 1/2" X 3 1/2" X 3/8"
5 - 3/4" DIAMETER
HS BOLTS EACH LEG
W21 X STEEL BEAM
15"

S8347

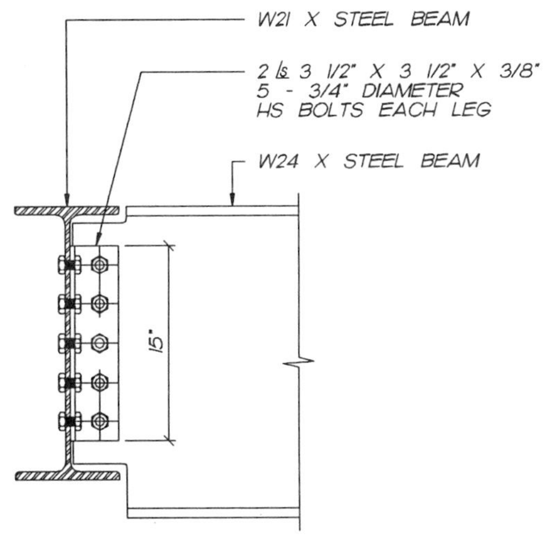
W21 X STEEL BEAM
2 ∟s 3 1/2" X 3 1/2" X 3/8"
5 - 3/4" DIAMETER
HS BOLTS EACH LEG
W24 X STEEL BEAM
15"

S8348

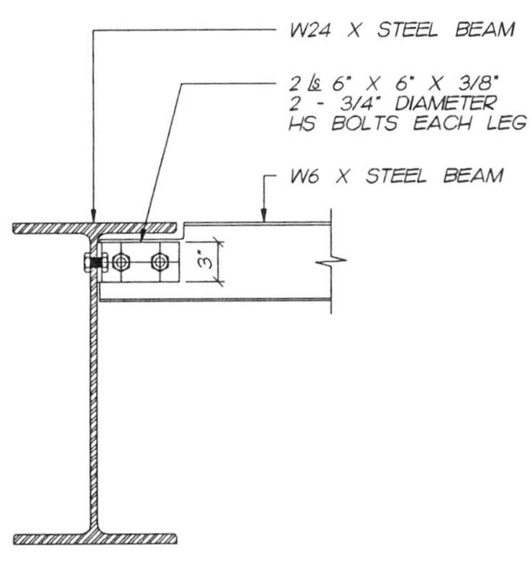
W24 X STEEL BEAM
2 ∟s 6" X 6" X 3/8"
2 - 3/4" DIAMETER
HS BOLTS EACH LEG
W6 X STEEL BEAM
3"

S8360

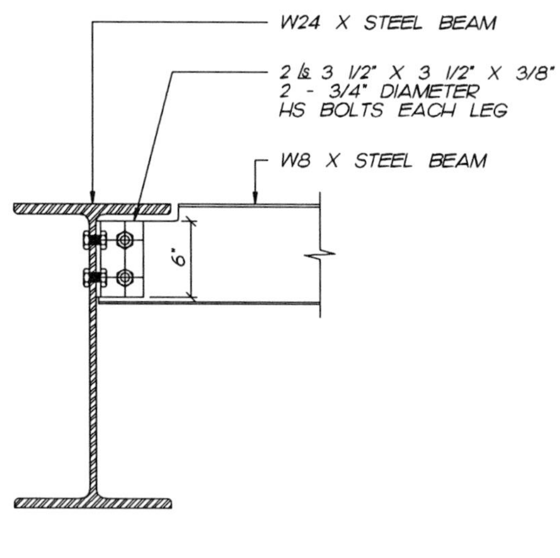

S8361

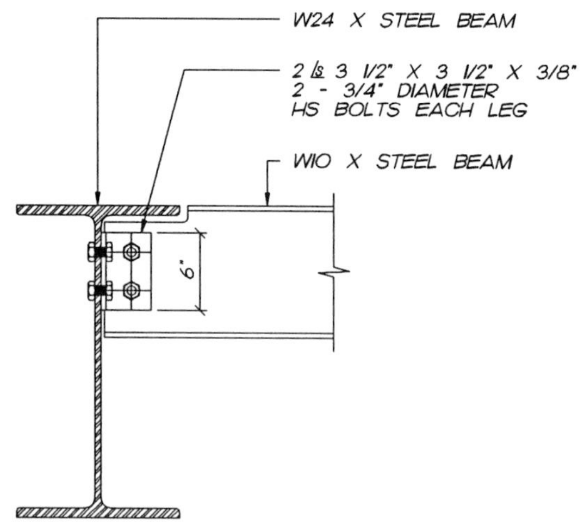

S8362

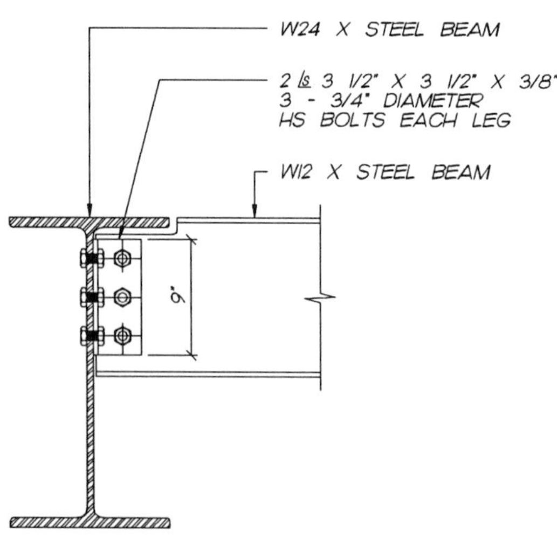

S8363

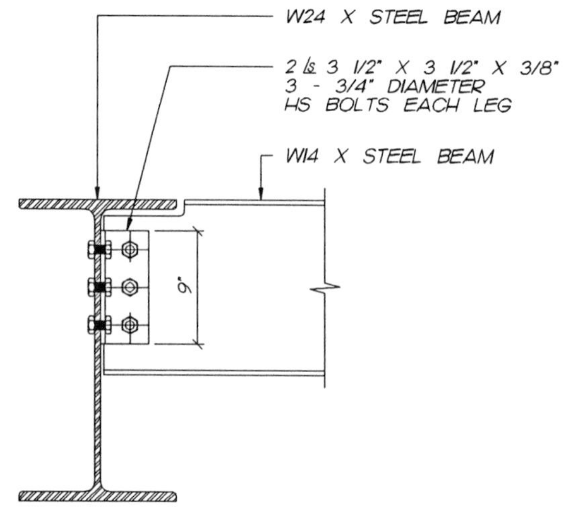

S8364

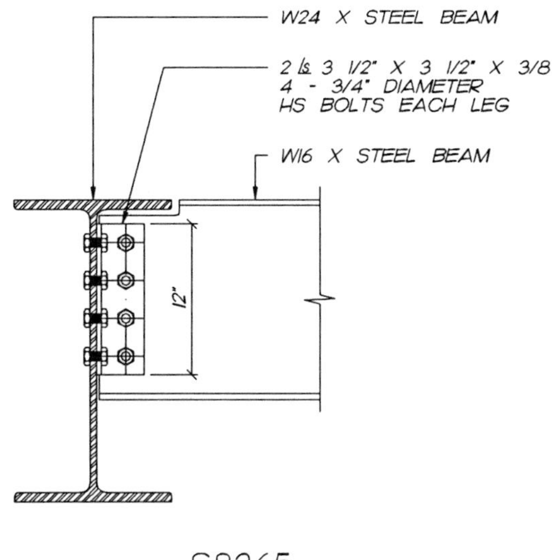

S8365

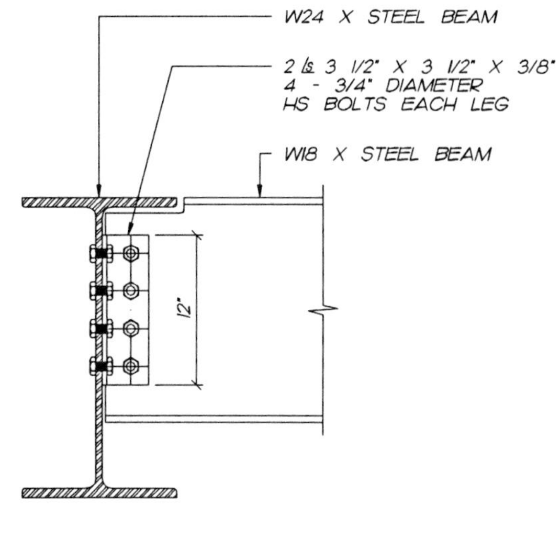

S8366

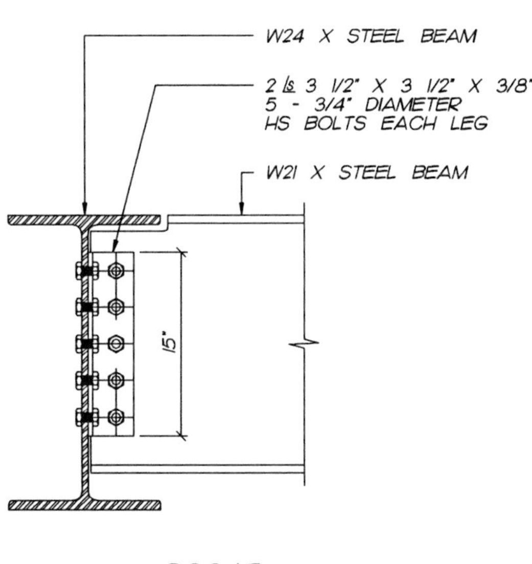

S8367

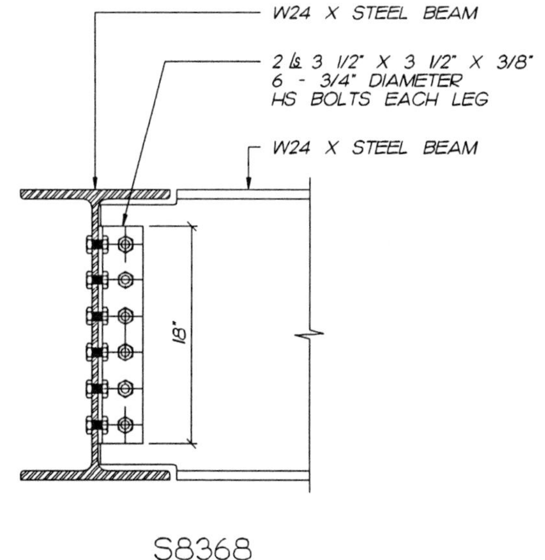

S8368

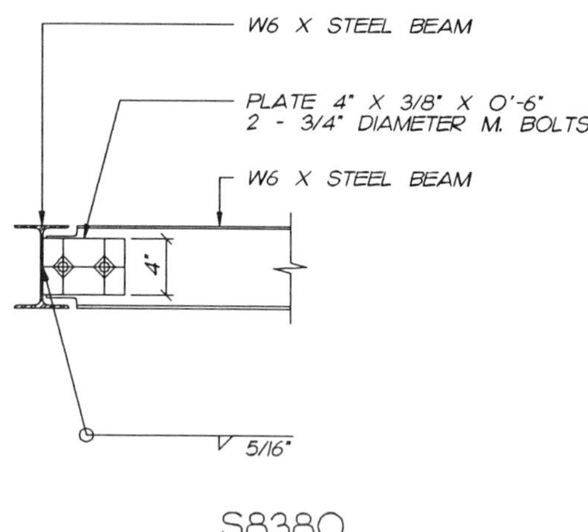

S8380

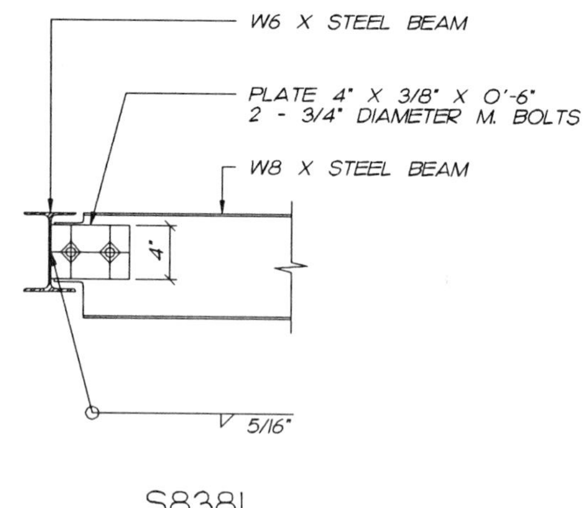

S8381

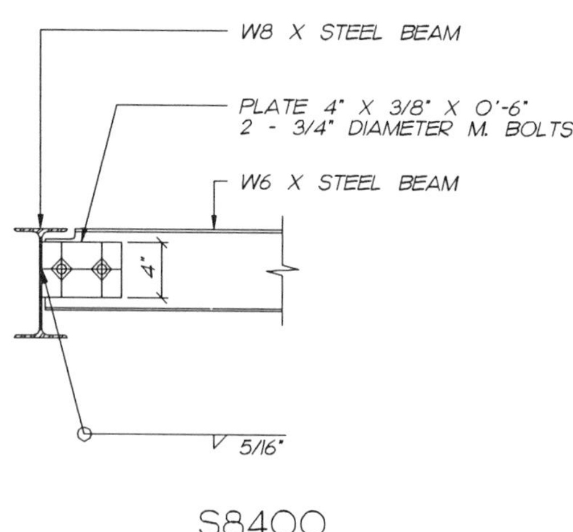

S8400

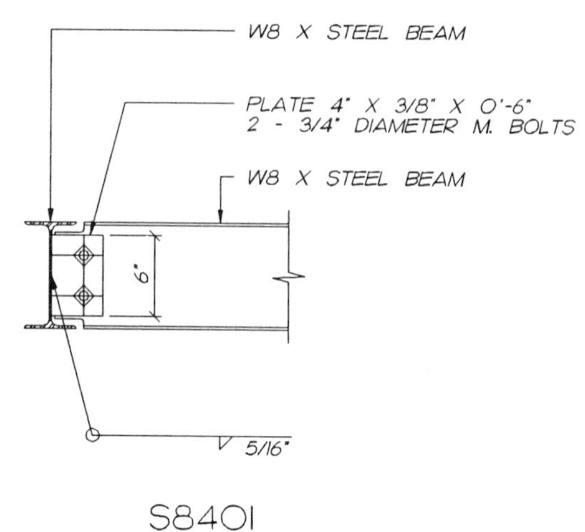

S8401

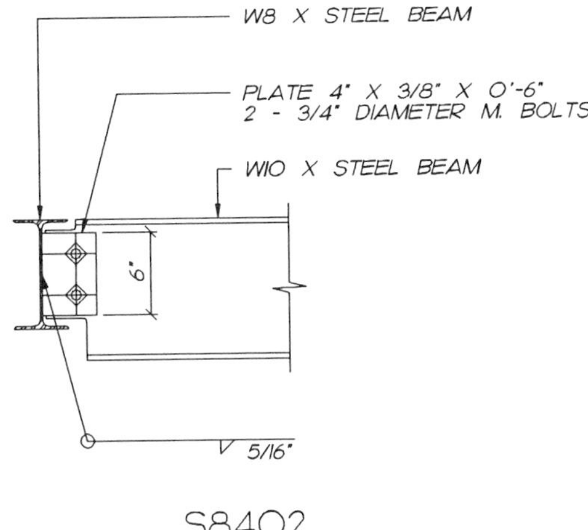

S8402

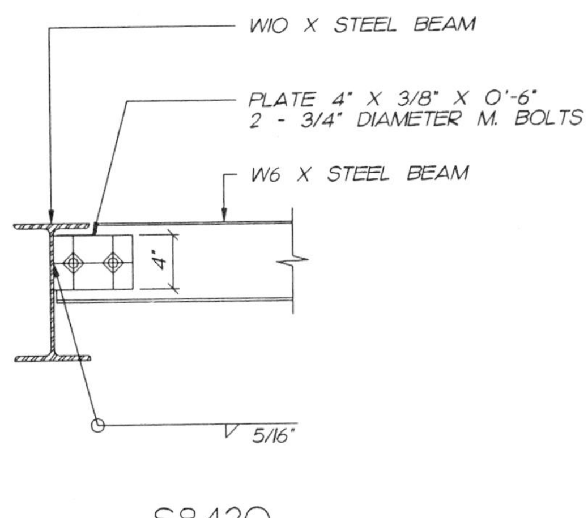

S8420

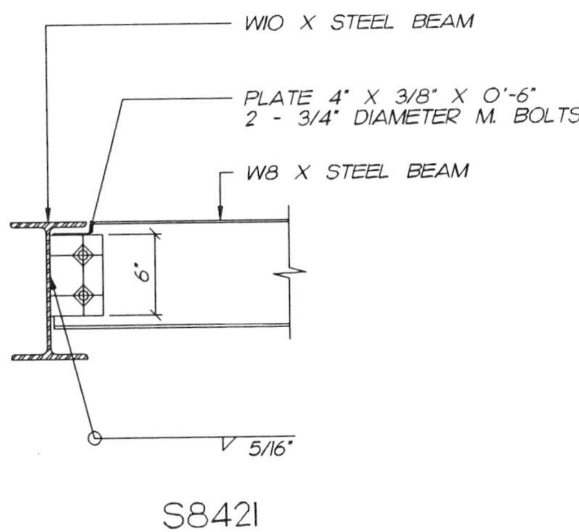

S8421

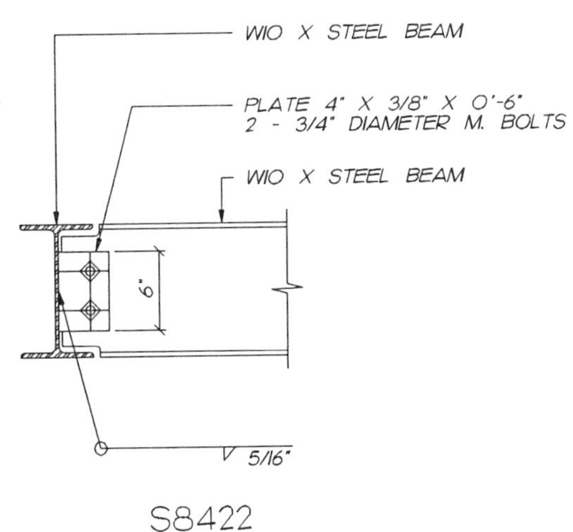

S8422

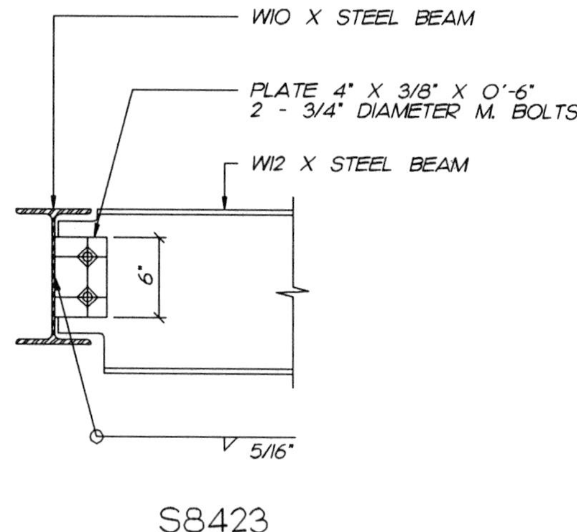

S8423

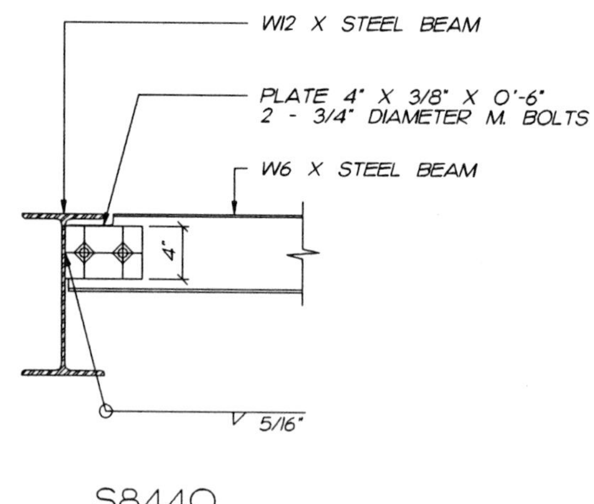

S8440

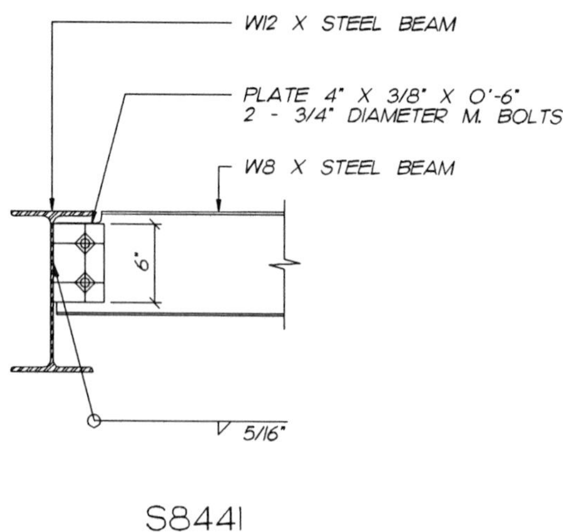

S8441

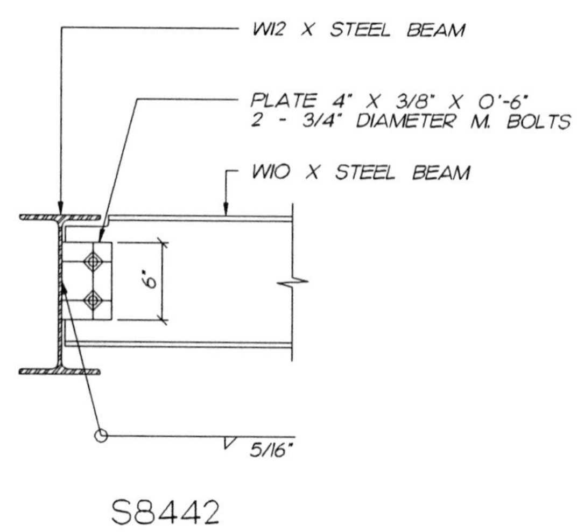

S8442

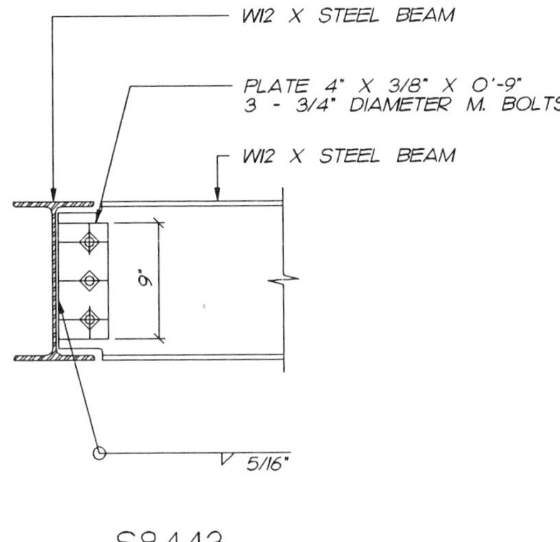

S8443

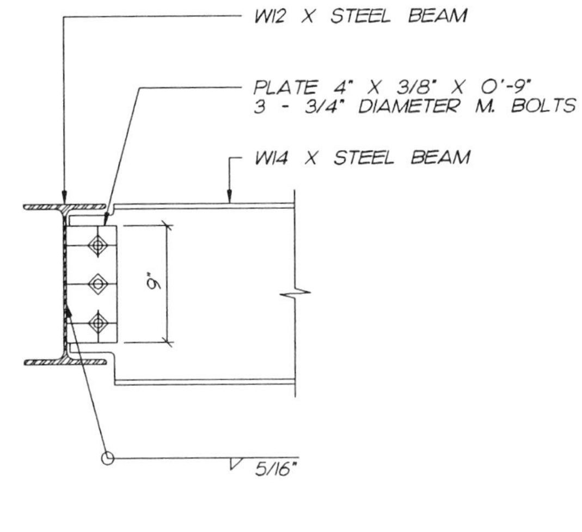

S8444

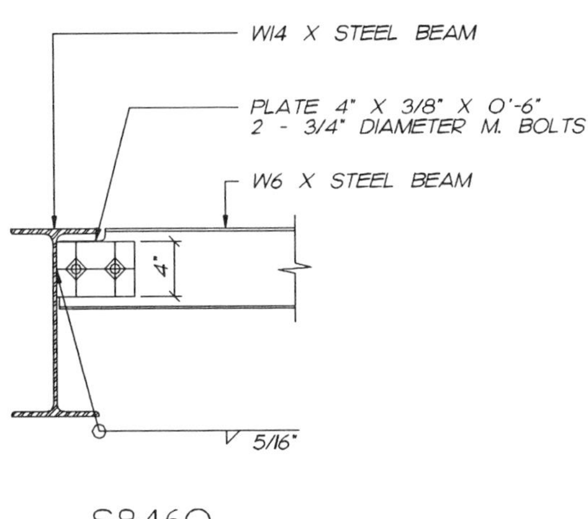

S8460

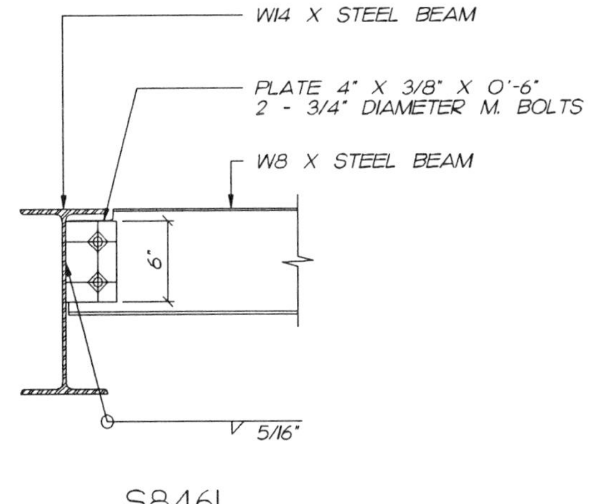

S8461

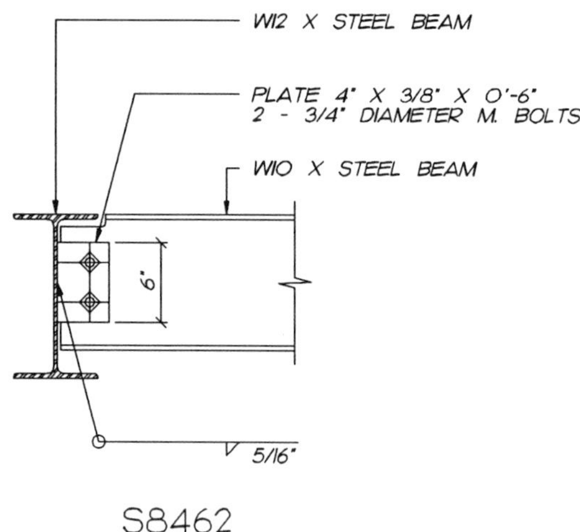

S8462

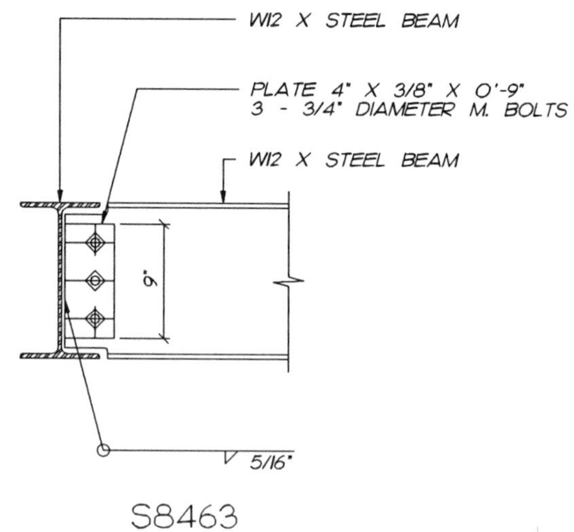

S8463

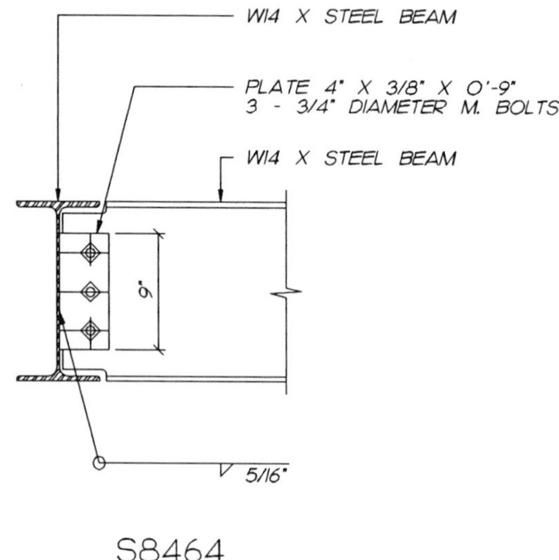

S8464

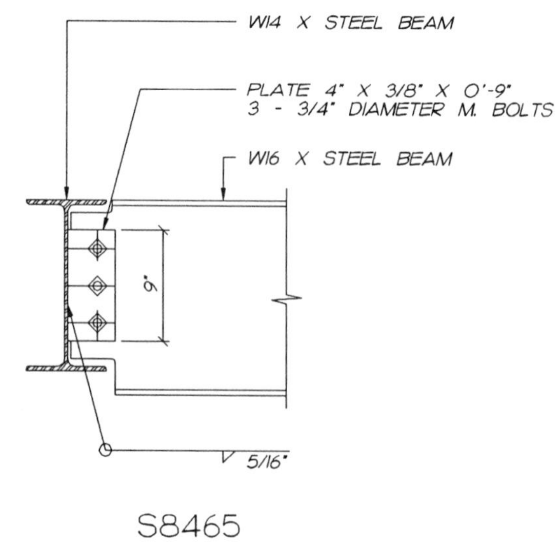

S8465

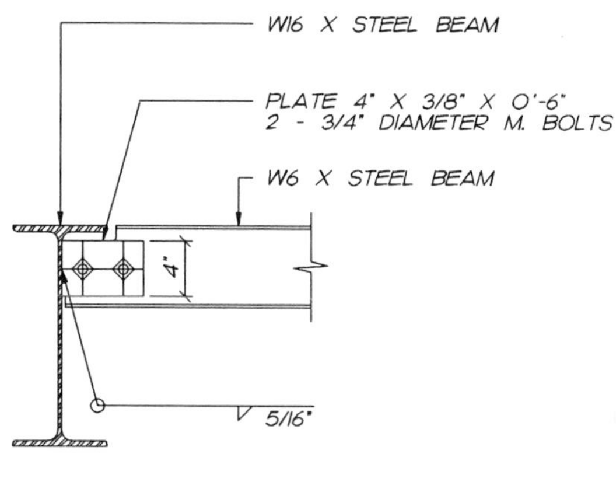

S8480

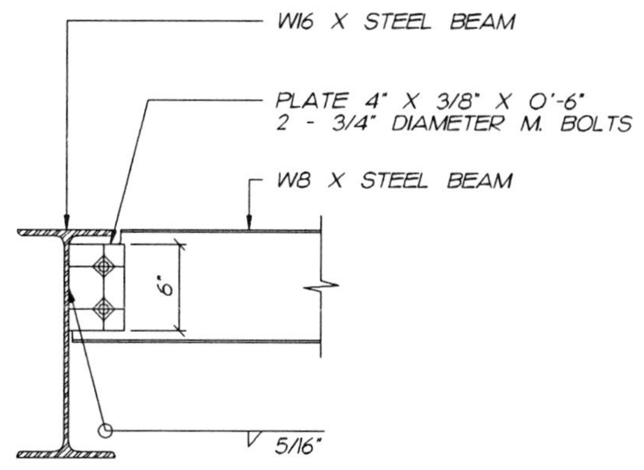

S8481

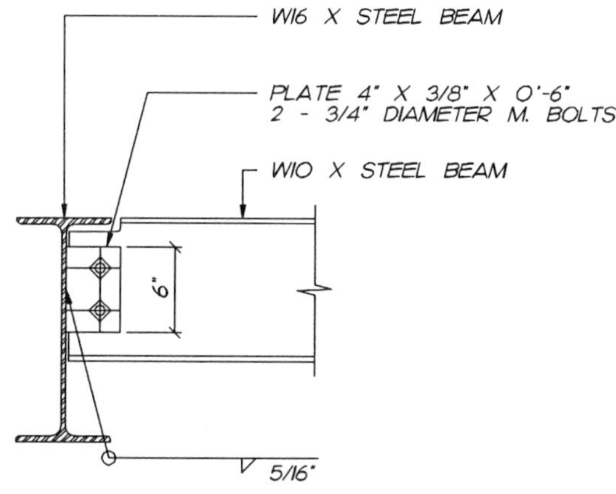

S8482

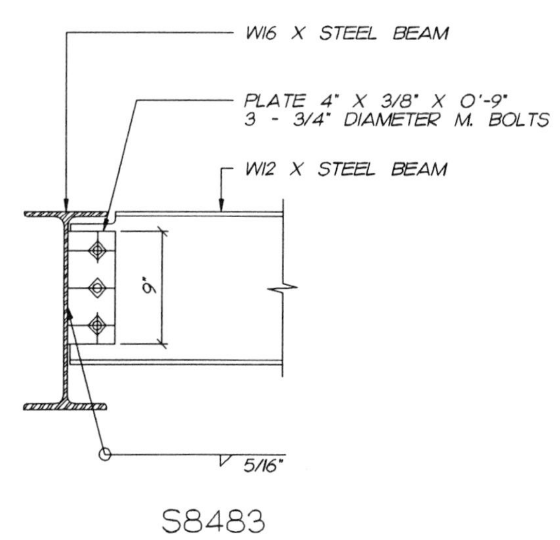

S8483

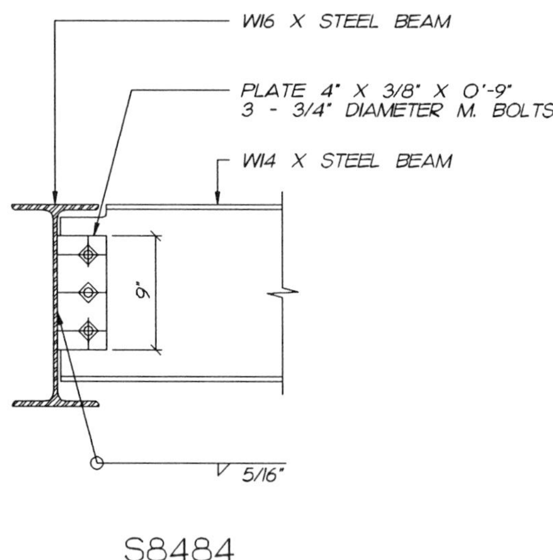

S8484

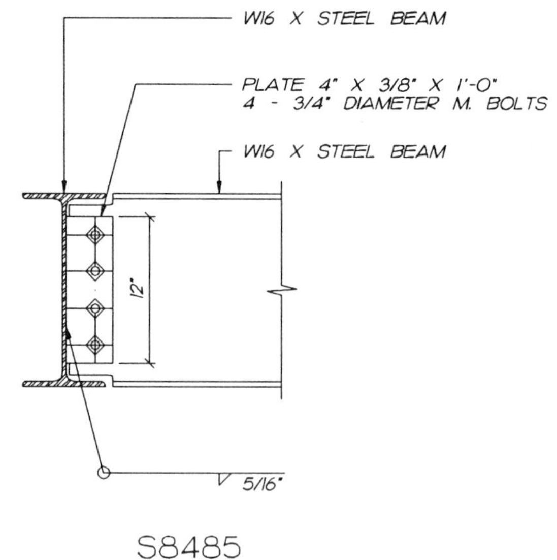

S8485

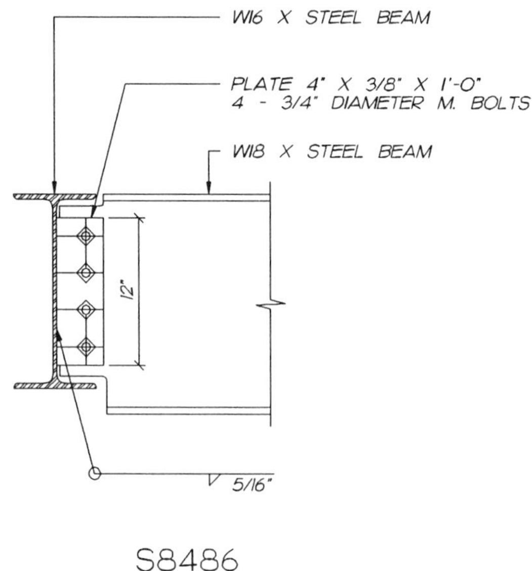

S8486

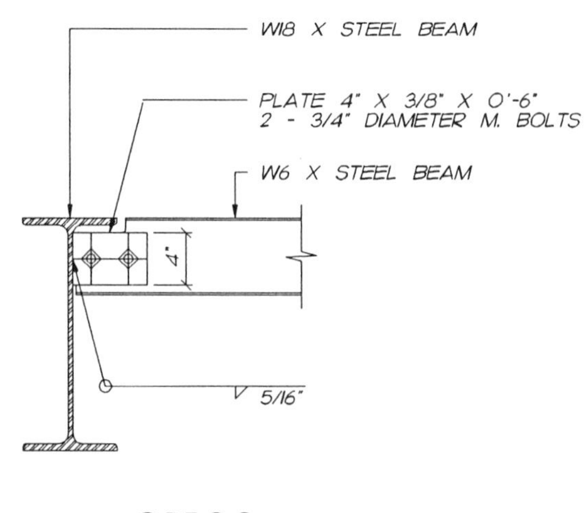

S8500

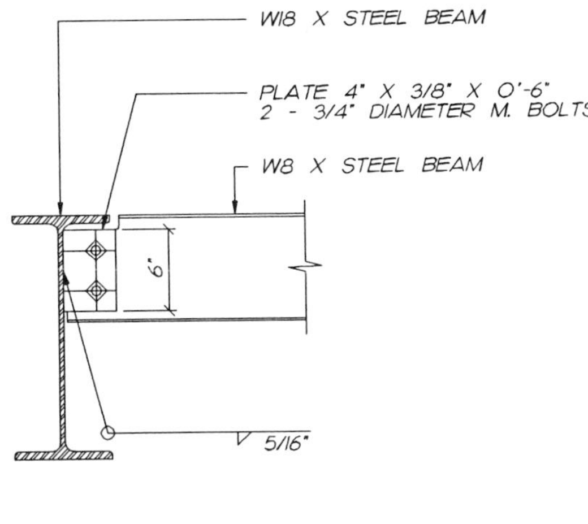

S8501

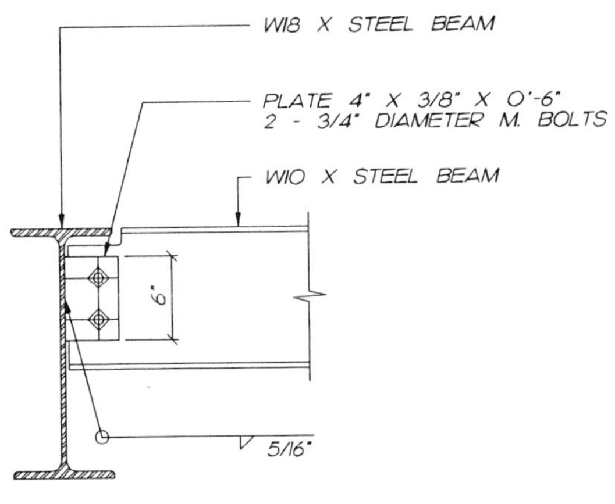

S8502

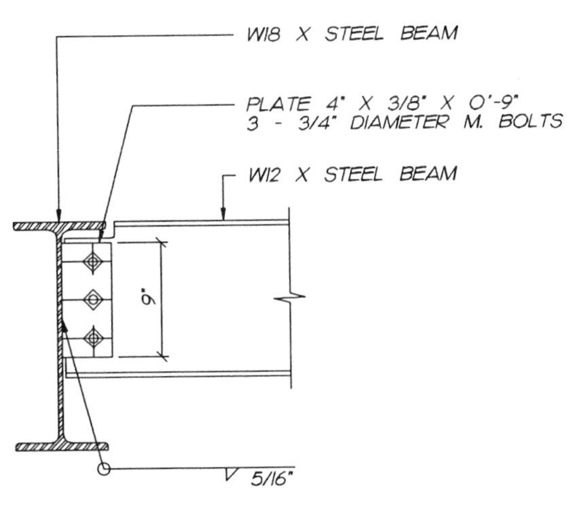

S8503

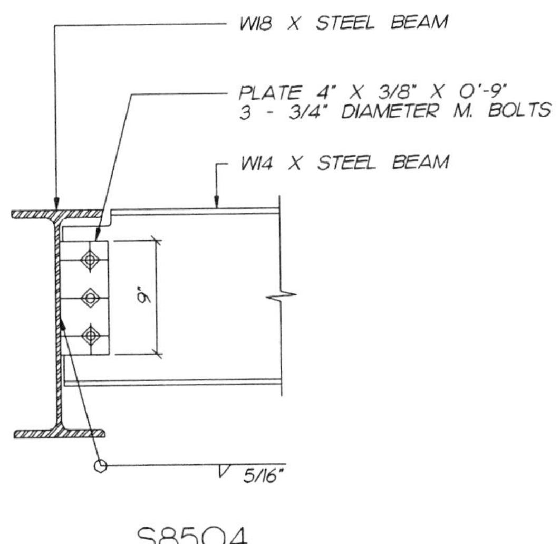

S8504

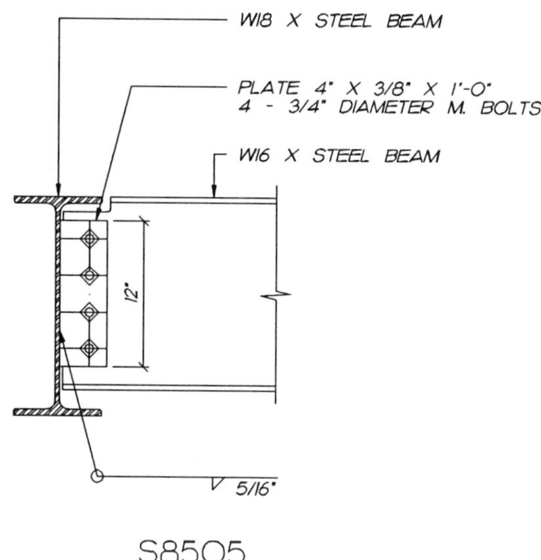

S8505

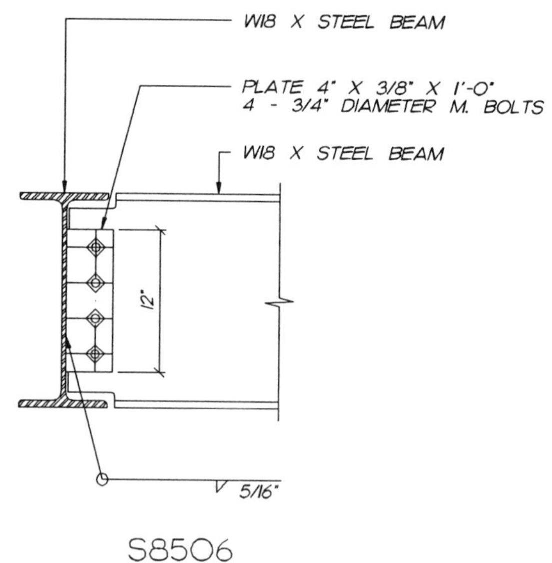

S8506

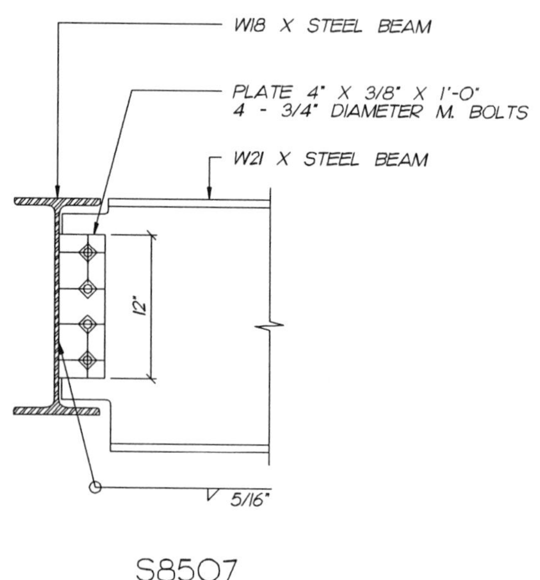

S8507

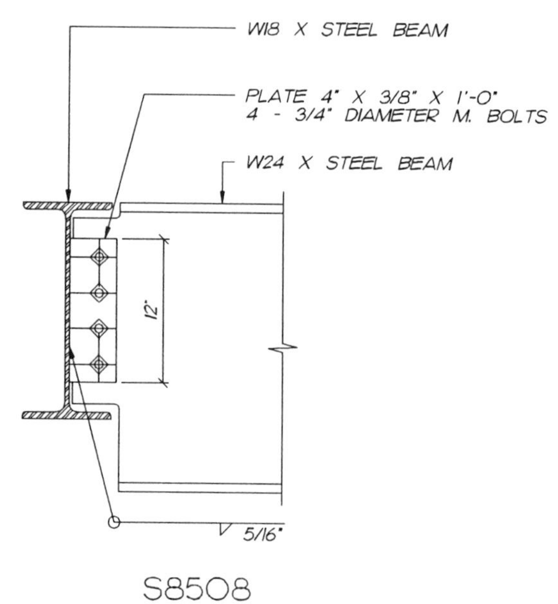

S8508

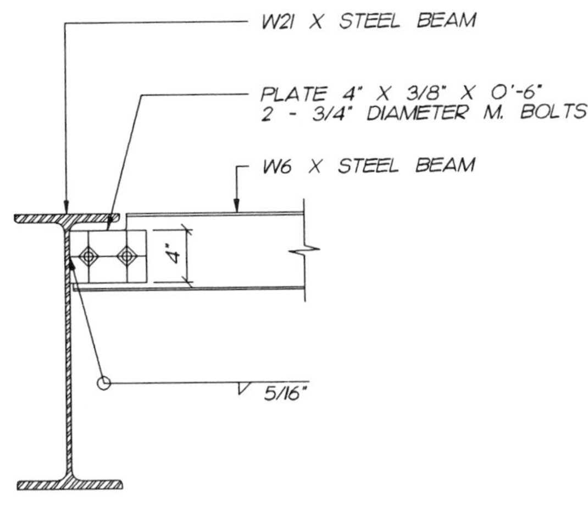

S8520

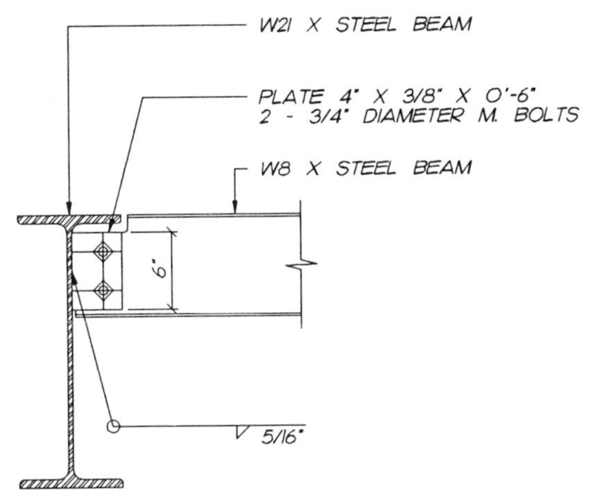

S8521

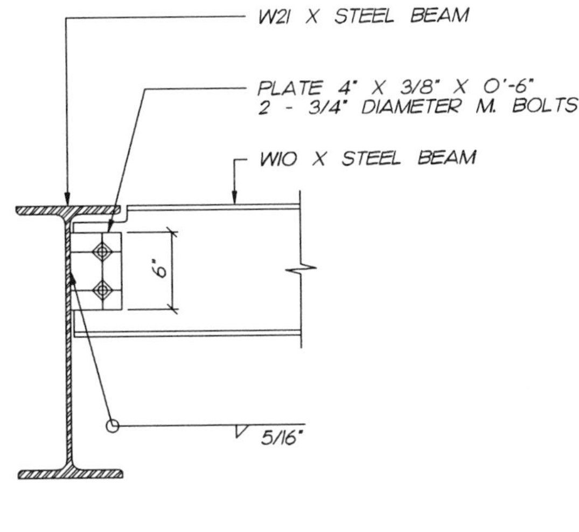

S8522

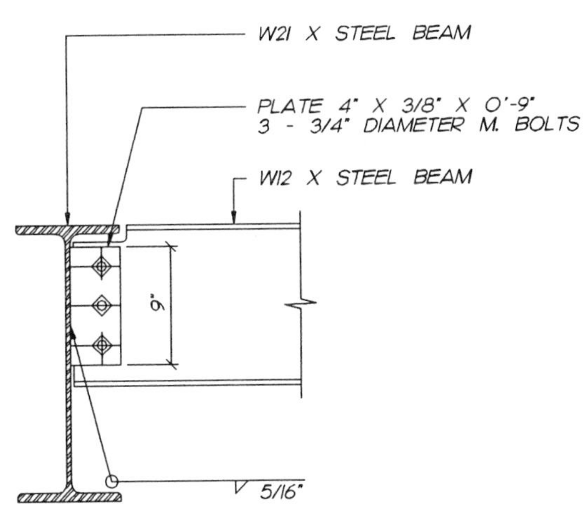

S8523

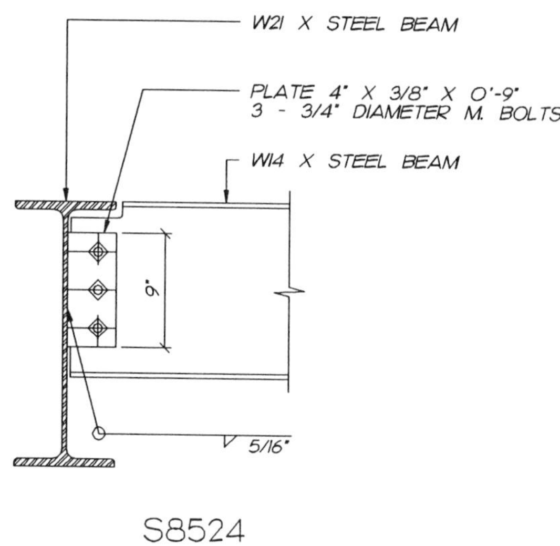

S8524

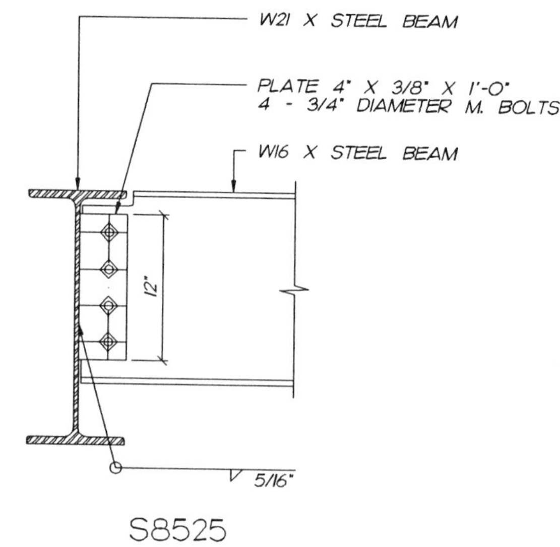

S8525

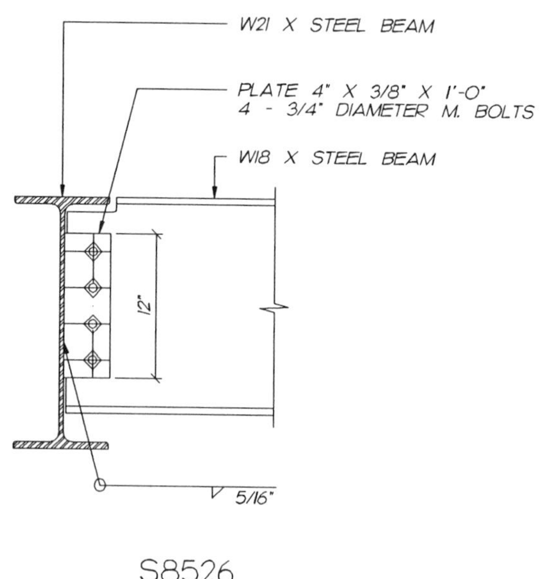

S8526

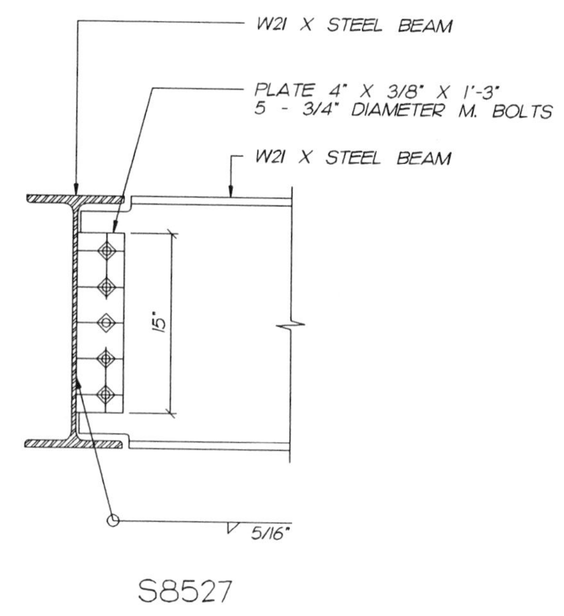

S8527

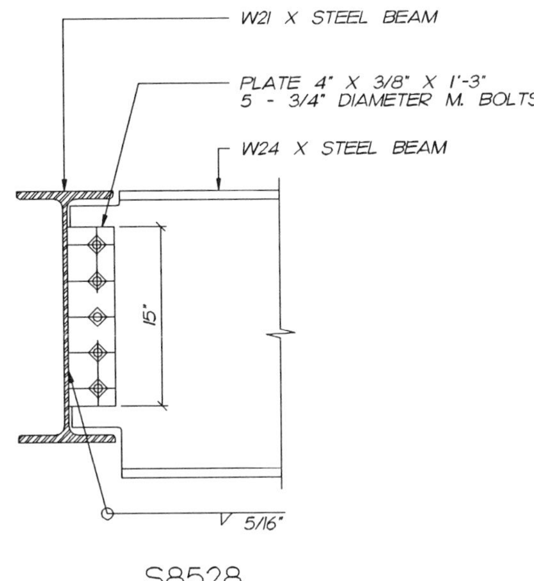

S8528

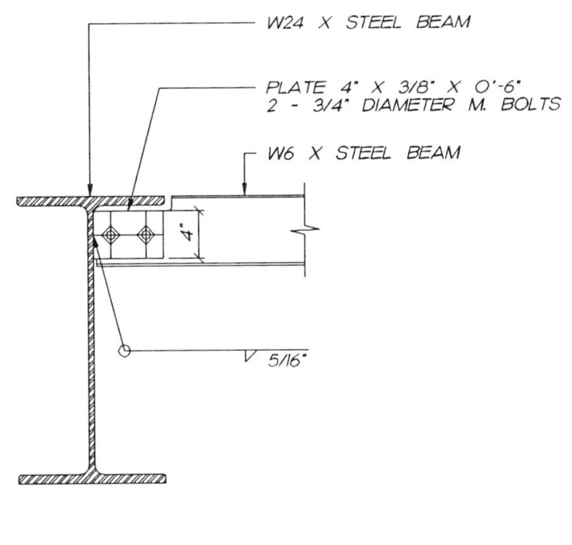

S8540

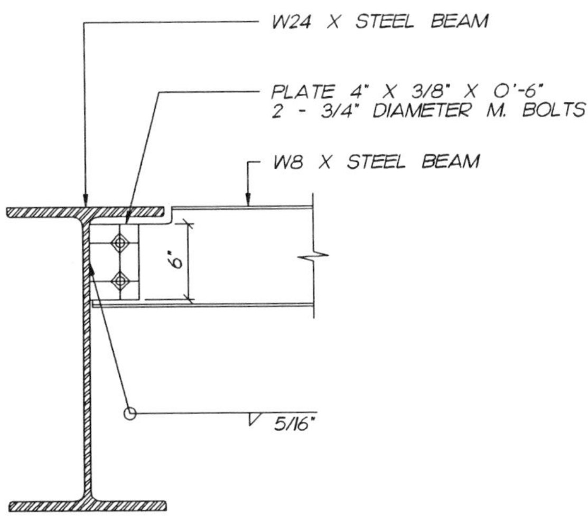

S8541

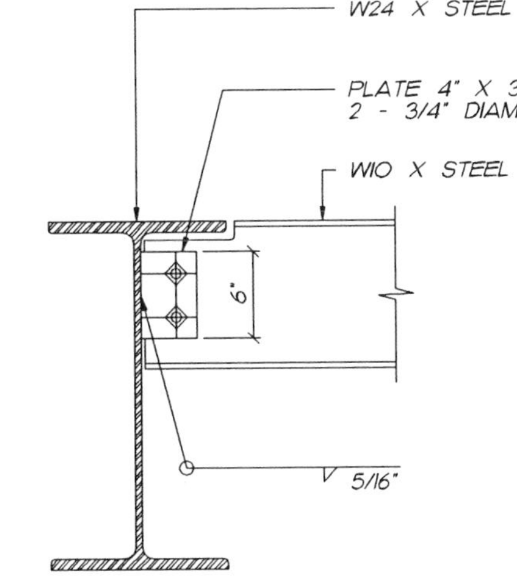

S8542

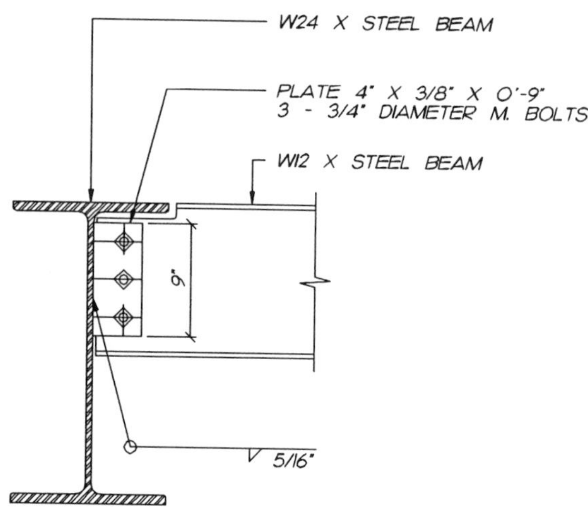

S8543

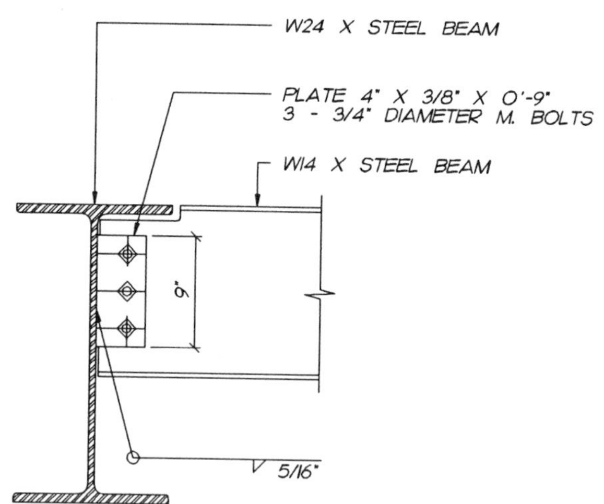

S8544

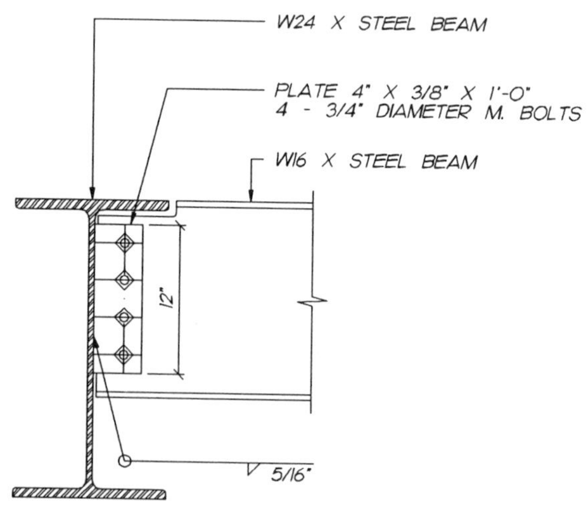

S8545

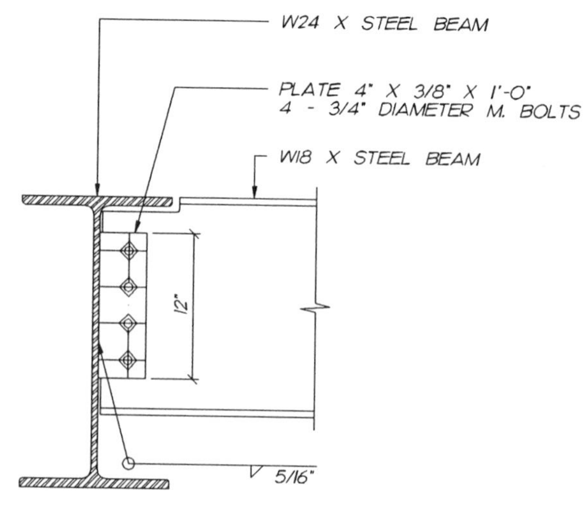

S8546

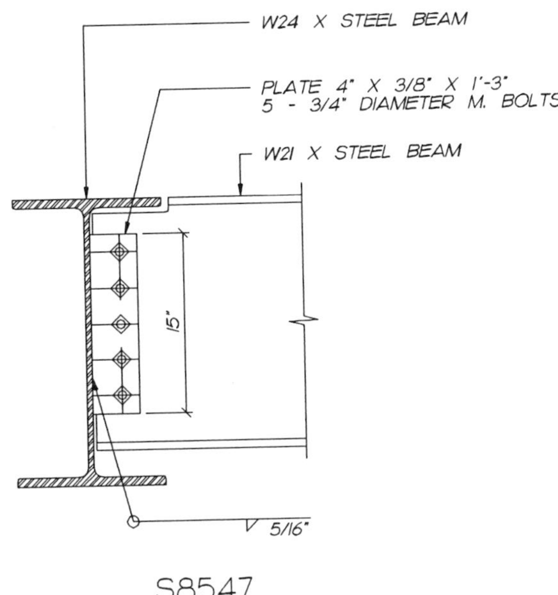

S8547

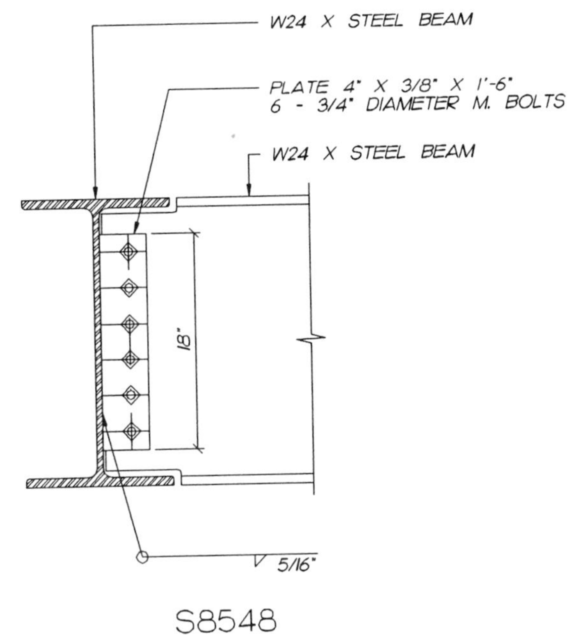

S8548

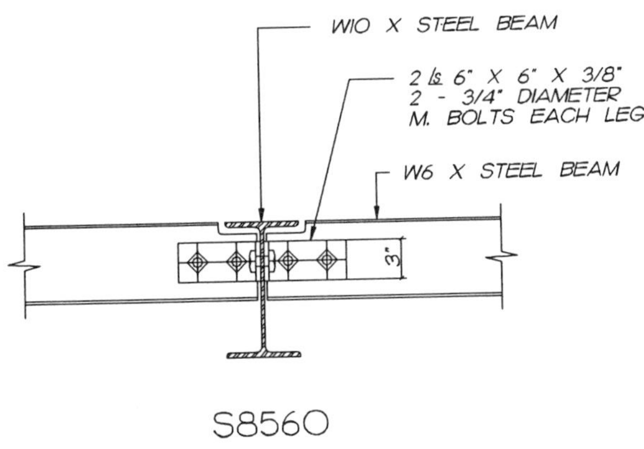

S8560

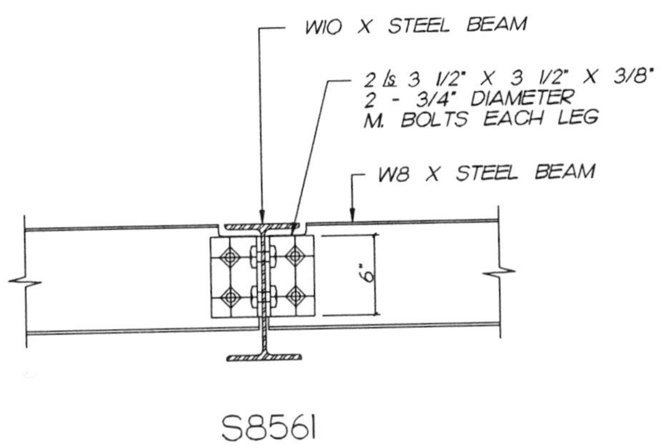

S8561

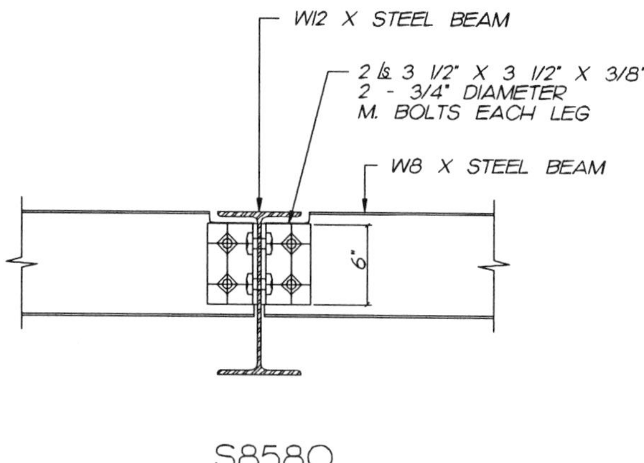

S8580

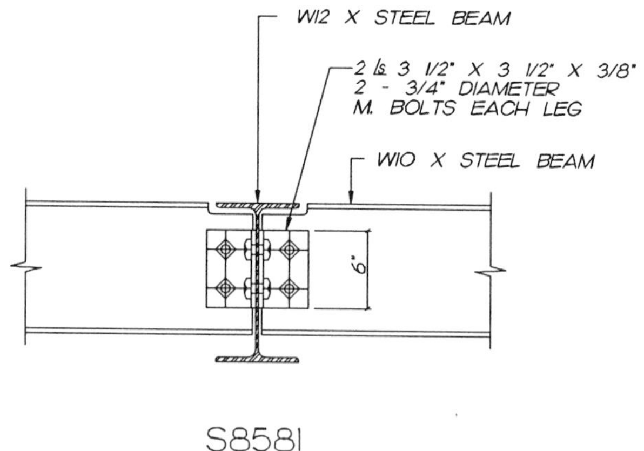

S8581

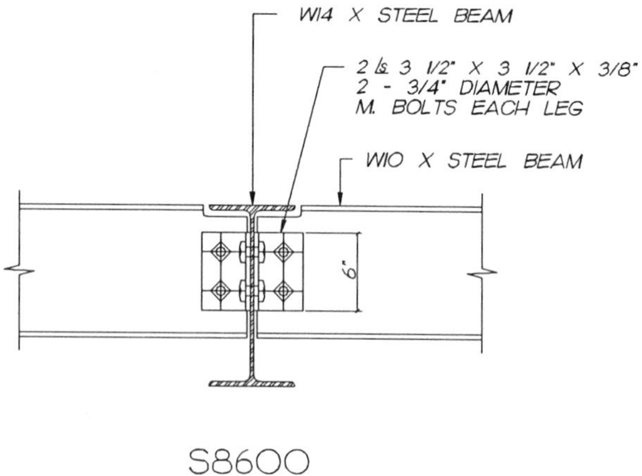

S8600

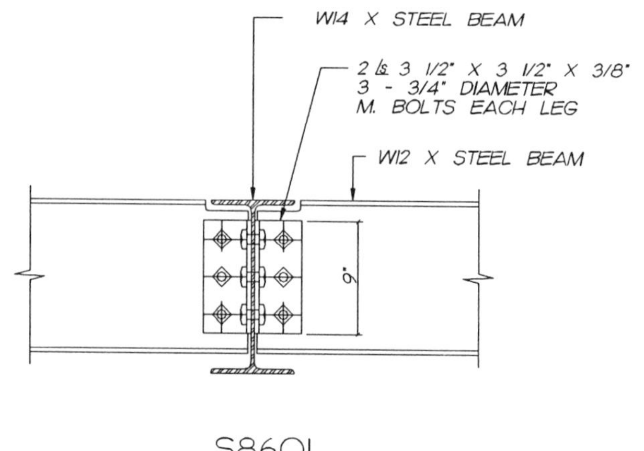

S8601

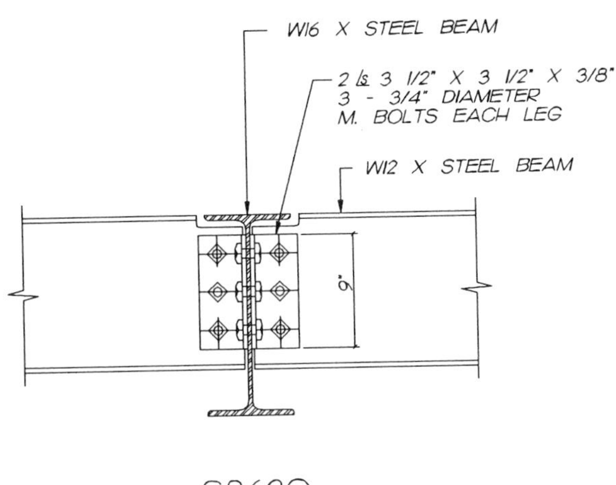

S8620

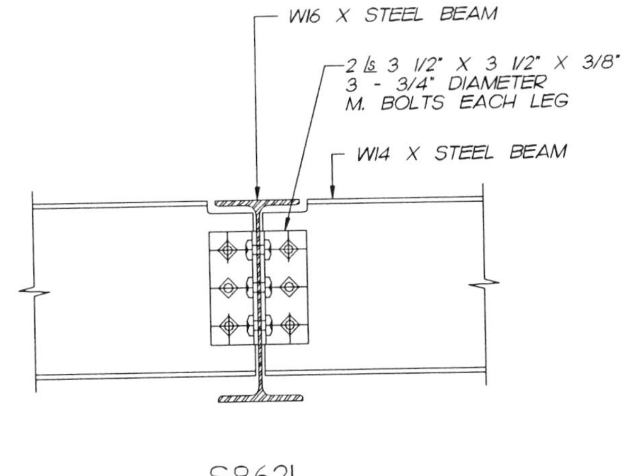

S8621

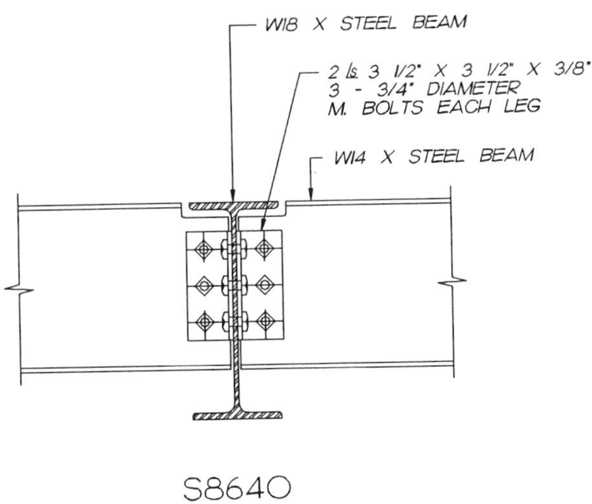

S8640

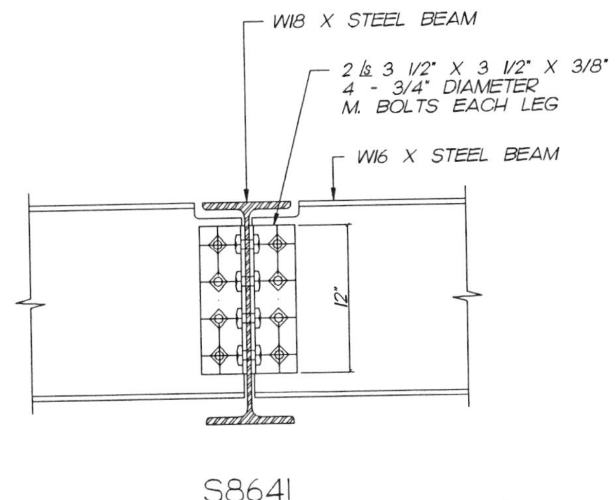

S8641

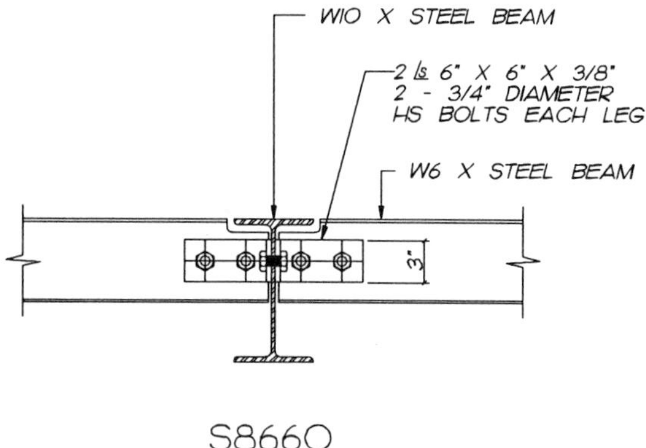

S8660

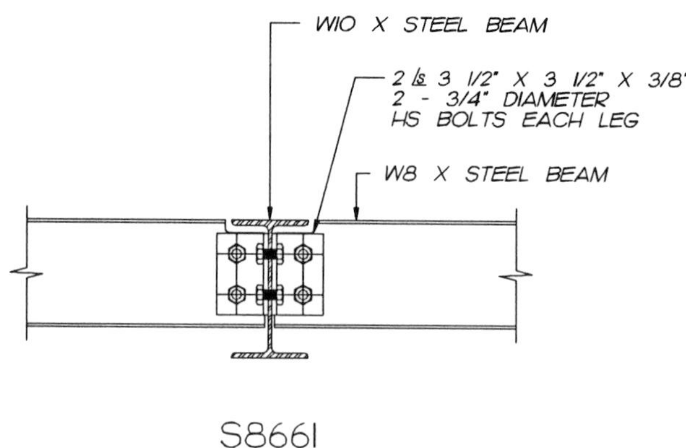

S8661

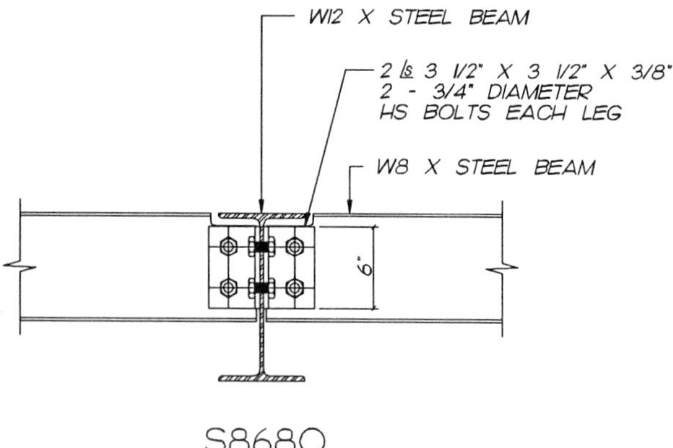

S8680

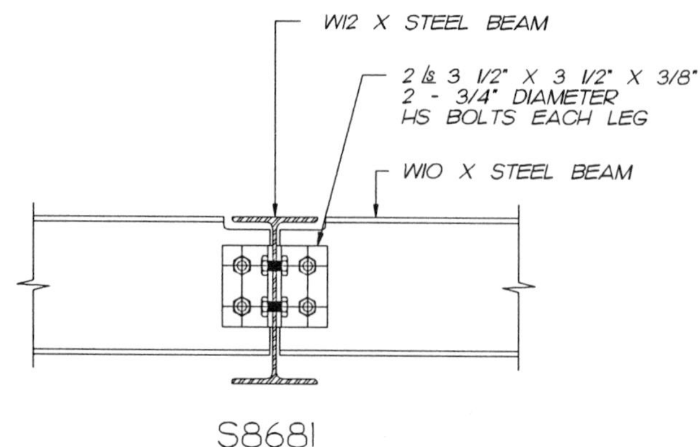

S8681

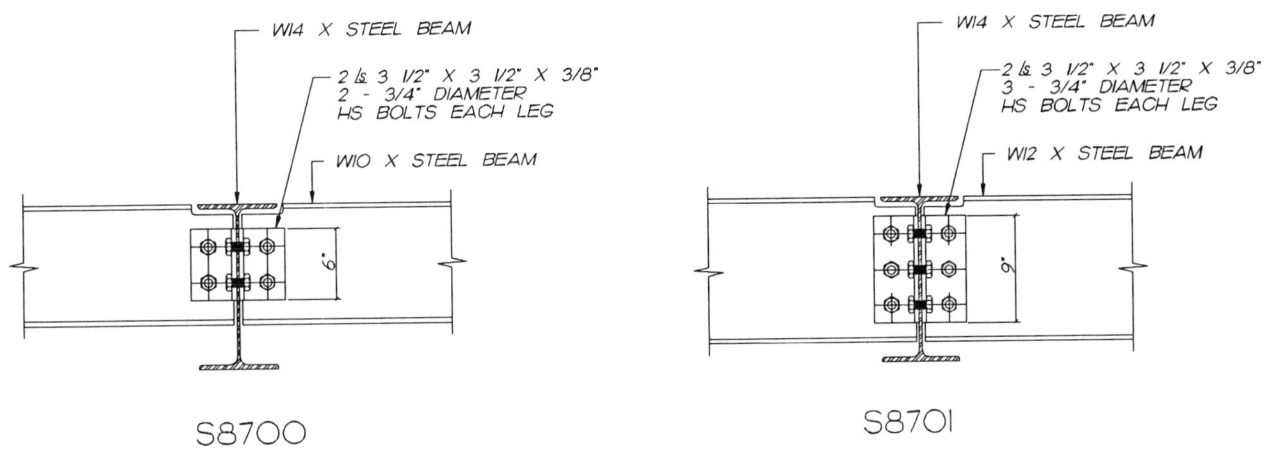
W14 X STEEL BEAM
2 ∟s 3 1/2" X 3 1/2" X 3/8"
2 - 3/4" DIAMETER
HS BOLTS EACH LEG
W10 X STEEL BEAM
6"
S8700
W14 X STEEL BEAM
2 ∟s 3 1/2" X 3 1/2" X 3/8"
3 - 3/4" DIAMETER
HS BOLTS EACH LEG
W12 X STEEL BEAM
9"
S8701

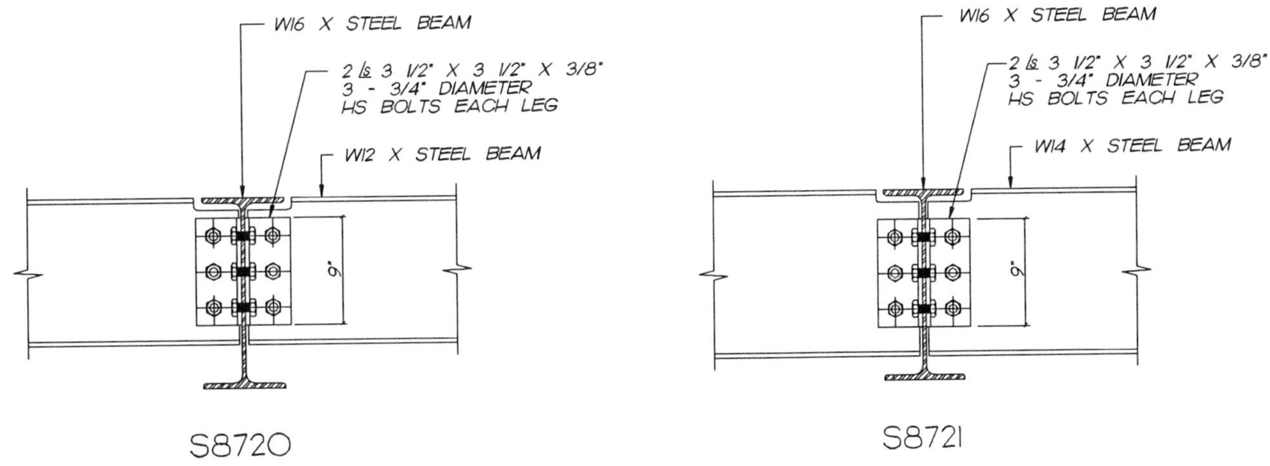
W16 X STEEL BEAM
2 ∟s 3 1/2" X 3 1/2" X 3/8"
3 - 3/4" DIAMETER
HS BOLTS EACH LEG
W12 X STEEL BEAM
9"
S8720
W16 X STEEL BEAM
2 ∟s 3 1/2" X 3 1/2" X 3/8"
3 - 3/4" DIAMETER
HS BOLTS EACH LEG
W14 X STEEL BEAM
9"
S8721

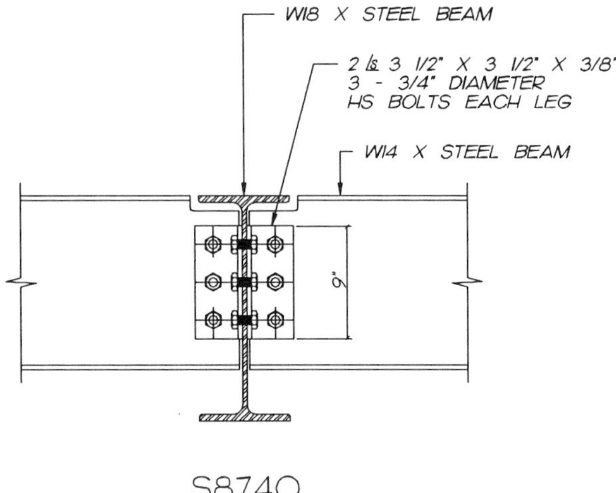

S8740

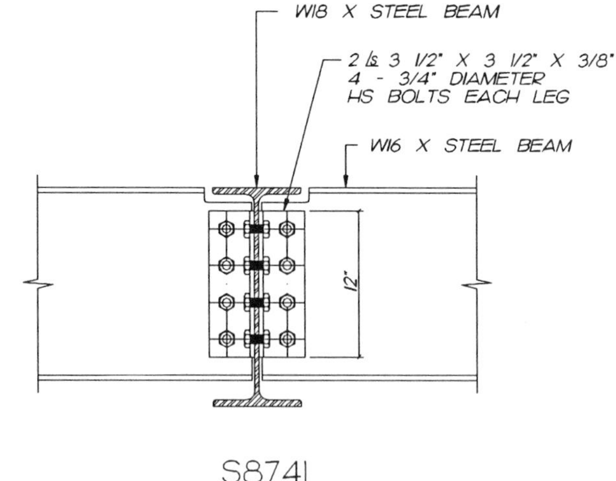

S8741

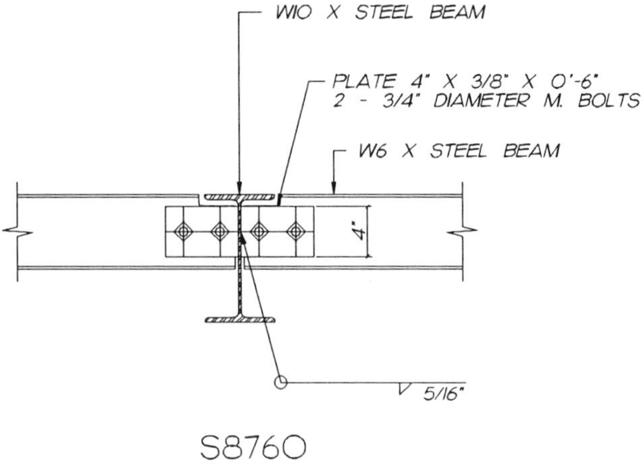

S8760

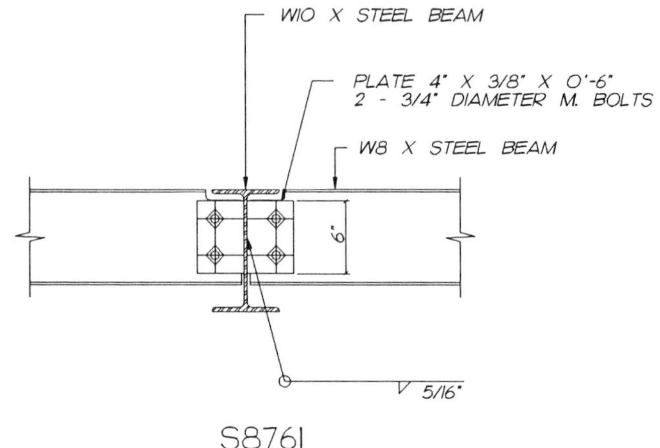

S8761

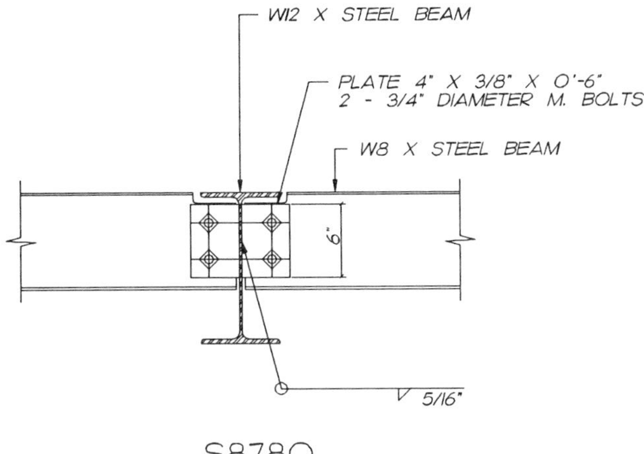

S8780

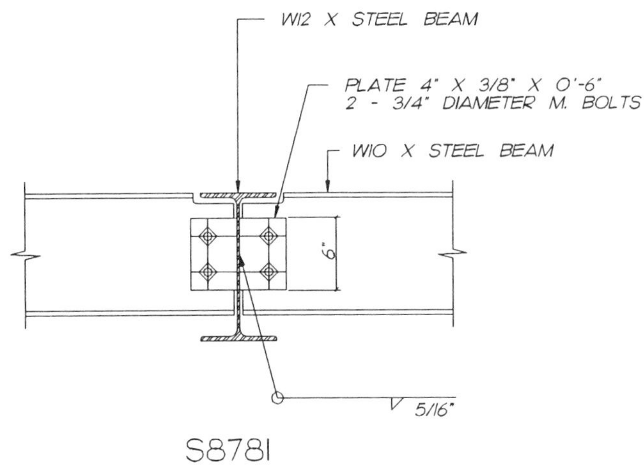

S8781

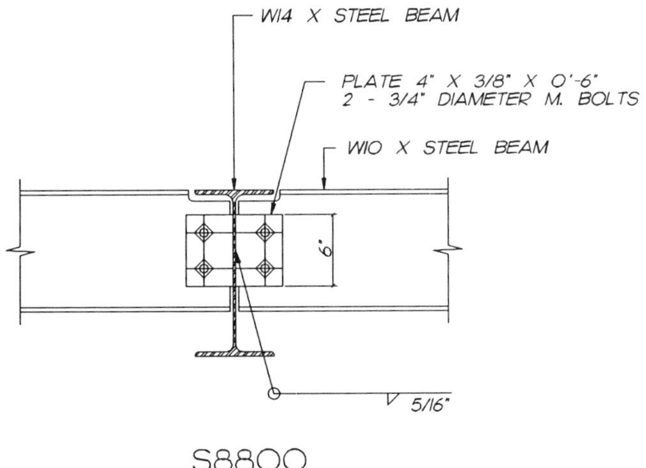

S8800

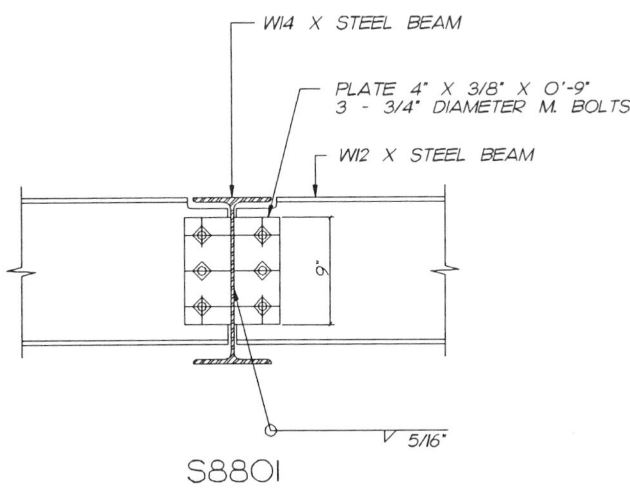

S8801

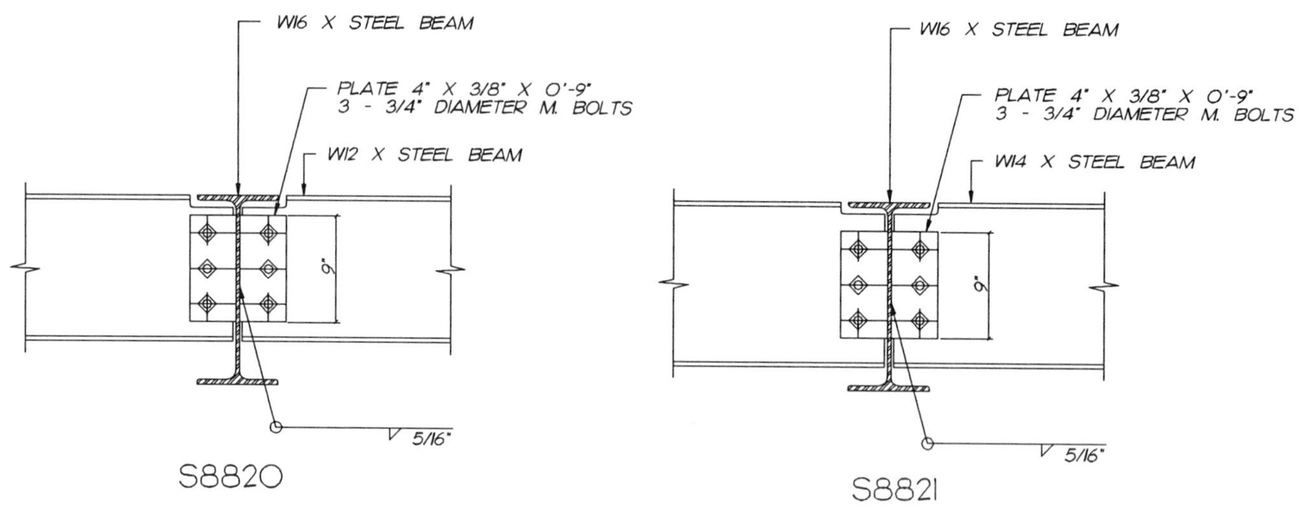
WI6 X STEEL BEAM
PLATE 4" X 3/8" X O'-9"
3 - 3/4" DIAMETER M. BOLTS
WI2 X STEEL BEAM
9"
5/16"
S8820
WI6 X STEEL BEAM
PLATE 4" X 3/8" X O'-9"
3 - 3/4" DIAMETER M. BOLTS
WI4 X STEEL BEAM
9"
5/16"
S8821

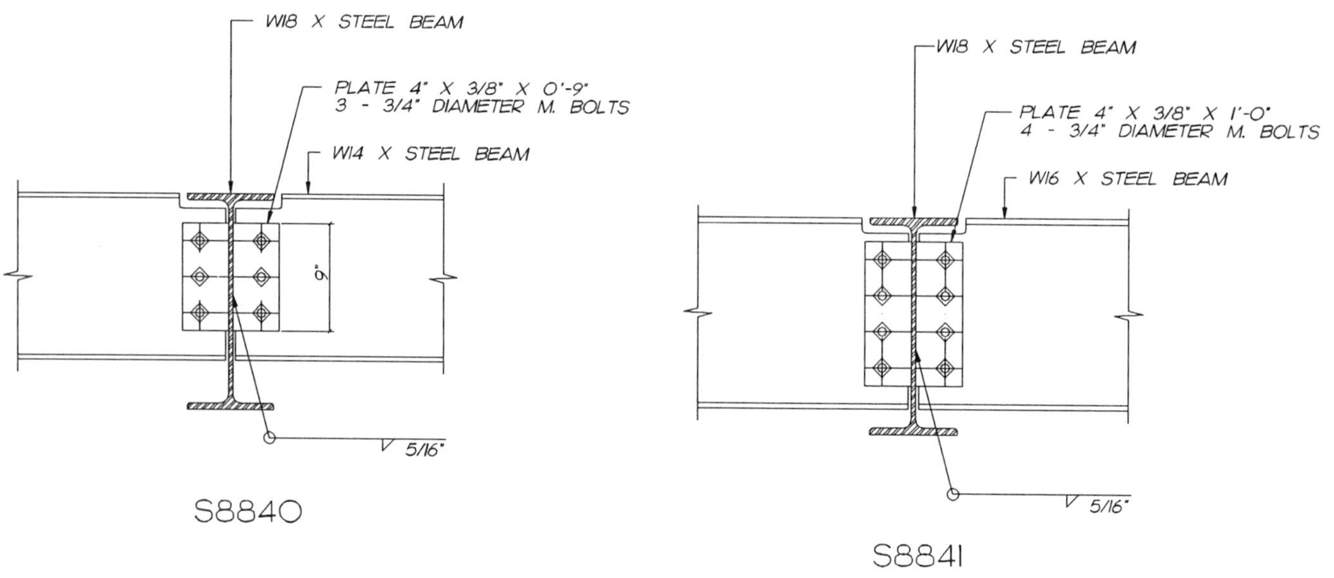
WI8 X STEEL BEAM
PLATE 4" X 3/8" X O'-9"
3 - 3/4" DIAMETER M. BOLTS
WI4 X STEEL BEAM
9"
5/16"
S8840
WI8 X STEEL BEAM
PLATE 4" X 3/8" X I'-O"
4 - 3/4" DIAMETER M. BOLTS
WI6 X STEEL BEAM
5/16"
S8841

CHAPTER 4	STEEL SECTION 9	MISCELLANEOUS
STEEL BEAM FIRE PROTECTION	————————	S9020 - S9026
STEEL FRAMED CANOPY	————————	S9040
CLEVIS & TURNBUCKLES	————————	S9060, S9080, S9100

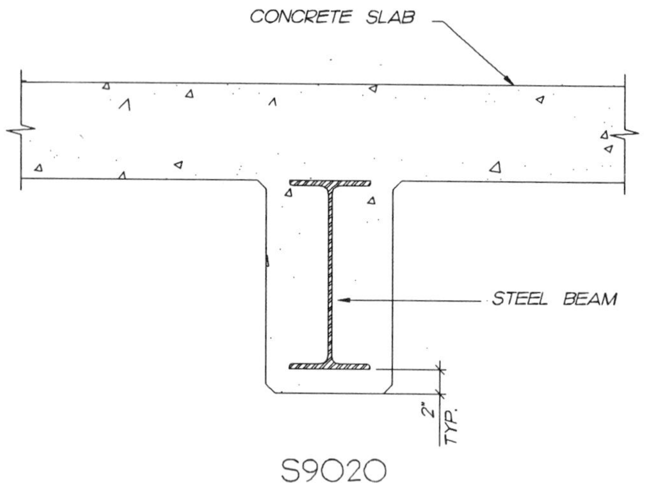

S9020

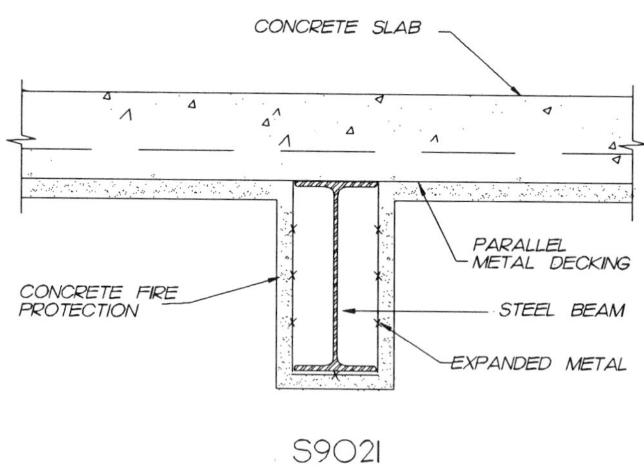

S9021

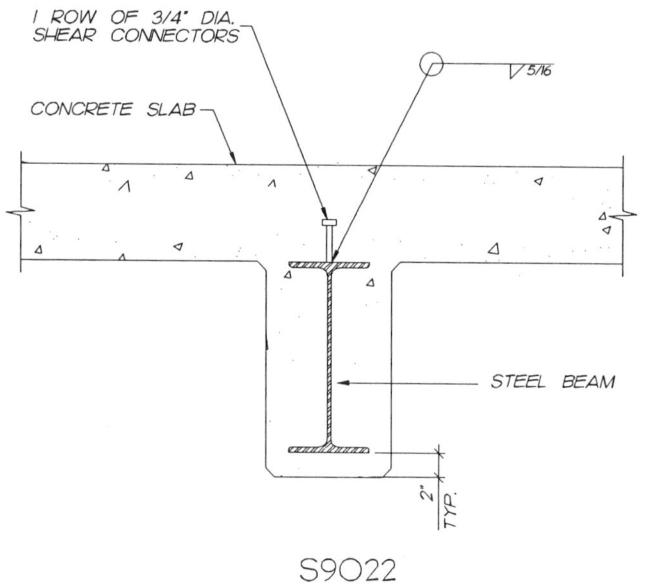

S9022

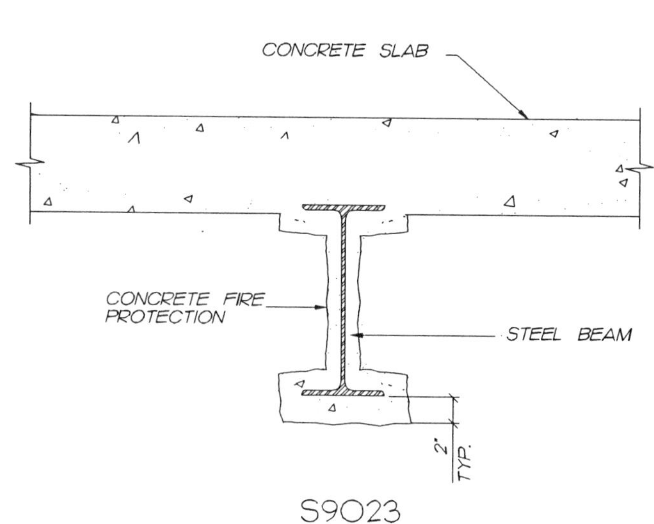

S9023

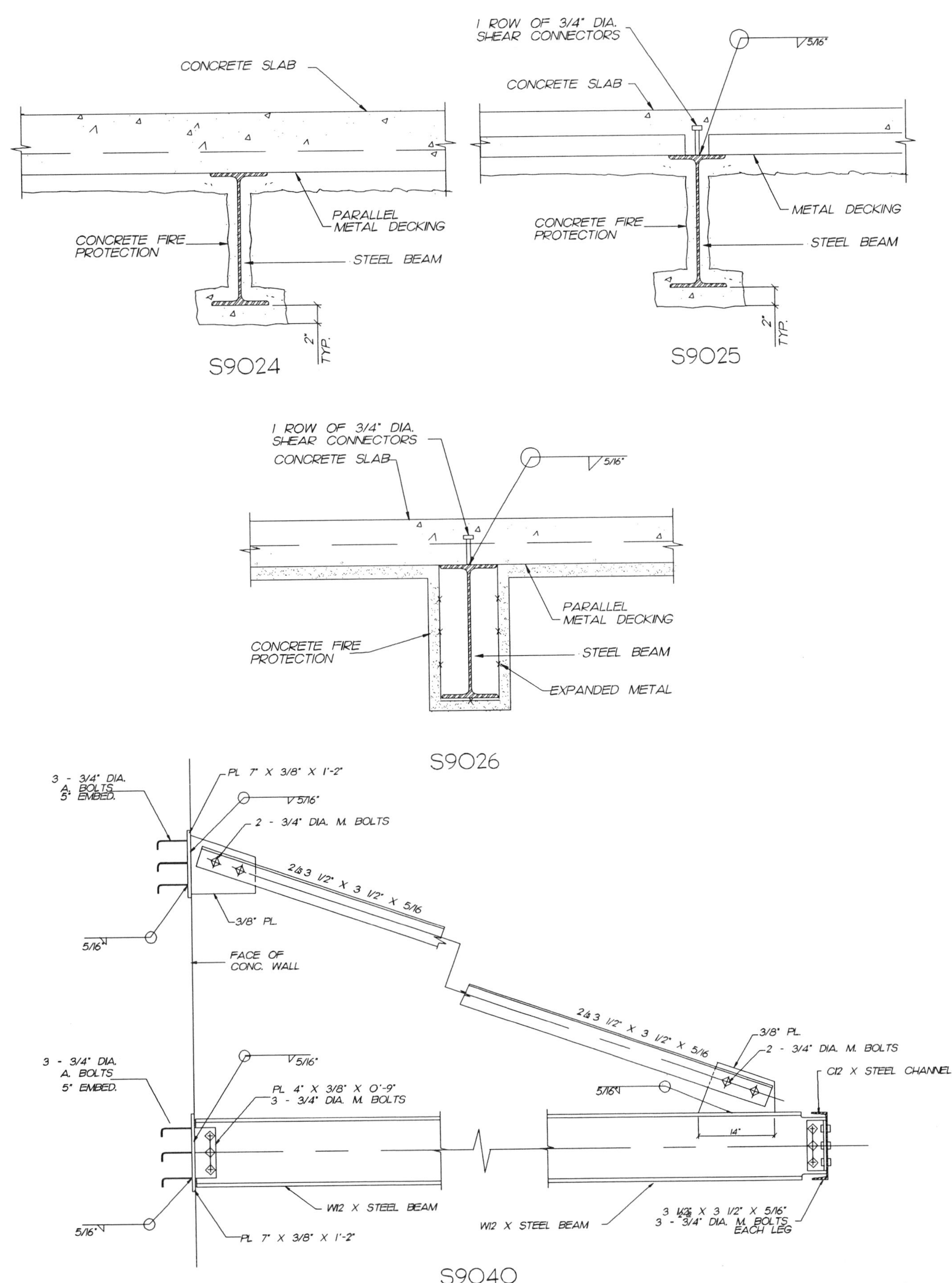
CONCRETE SLAB
PARALLEL METAL DECKING
CONCRETE FIRE PROTECTION
STEEL BEAM
2" TYP.
S9024
1 ROW OF 3/4" DIA. SHEAR CONNECTORS
5/16"
CONCRETE SLAB
METAL DECKING
CONCRETE FIRE PROTECTION
STEEL BEAM
2" TYP.
S9025
1 ROW OF 3/4" DIA. SHEAR CONNECTORS
CONCRETE SLAB
5/16"
PARALLEL METAL DECKING
CONCRETE FIRE PROTECTION
STEEL BEAM
EXPANDED METAL
S9026
3 - 3/4" DIA. A. BOLTS 5" EMBED.
PL 7" X 3/8" X 1'-2"
5/16"
2 - 3/4" DIA. M. BOLTS
2Ls 3 1/2" X 3 1/2" X 5/16
3/8" PL.
5/16"
FACE OF CONC. WALL
2Ls 3 1/2" X 3 1/2" X 5/16
3/8" PL.
2 - 3/4" DIA. M. BOLTS
C12 X STEEL CHANNEL
3 - 3/4" DIA. A. BOLTS 5" EMBED.
5/16"
PL 4" X 3/8" X 0'-9"
3 - 3/4" DIA. M. BOLTS
5/16"
14"
W12 X STEEL BEAM
5/16"
PL 7" X 3/8" X 1'-2"
W12 X STEEL BEAM
3 Ls X 3 1/2" X 5/16"
3 - 3/4" DIA. M. BOLTS EACH LEG
S9040

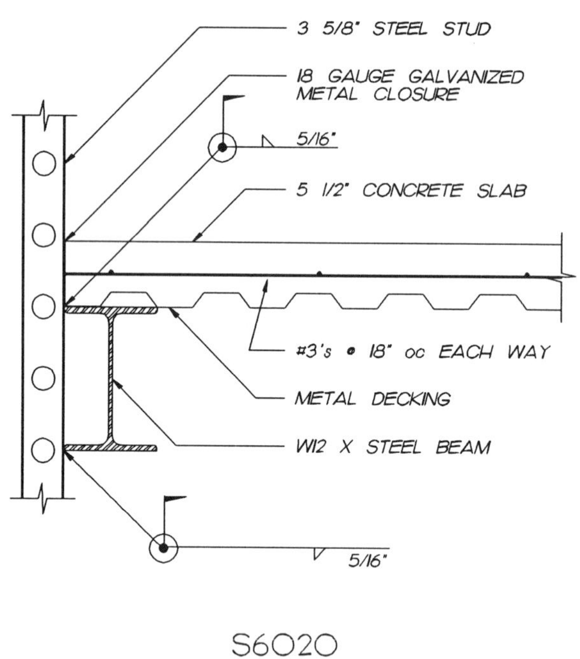

S6020

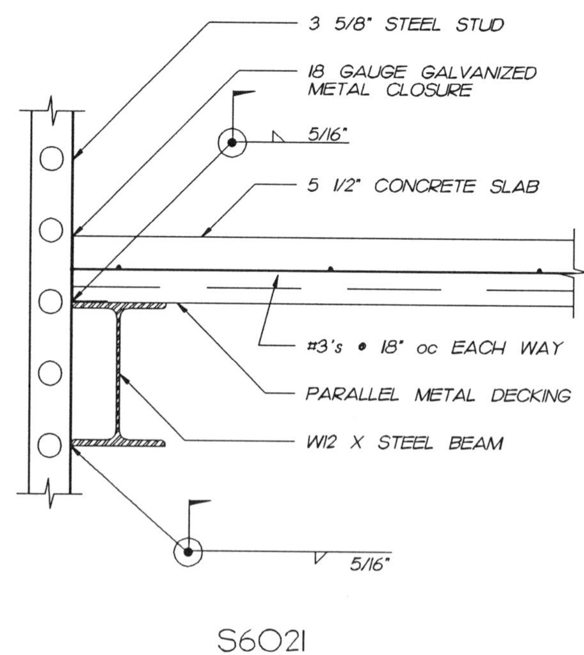

S6021

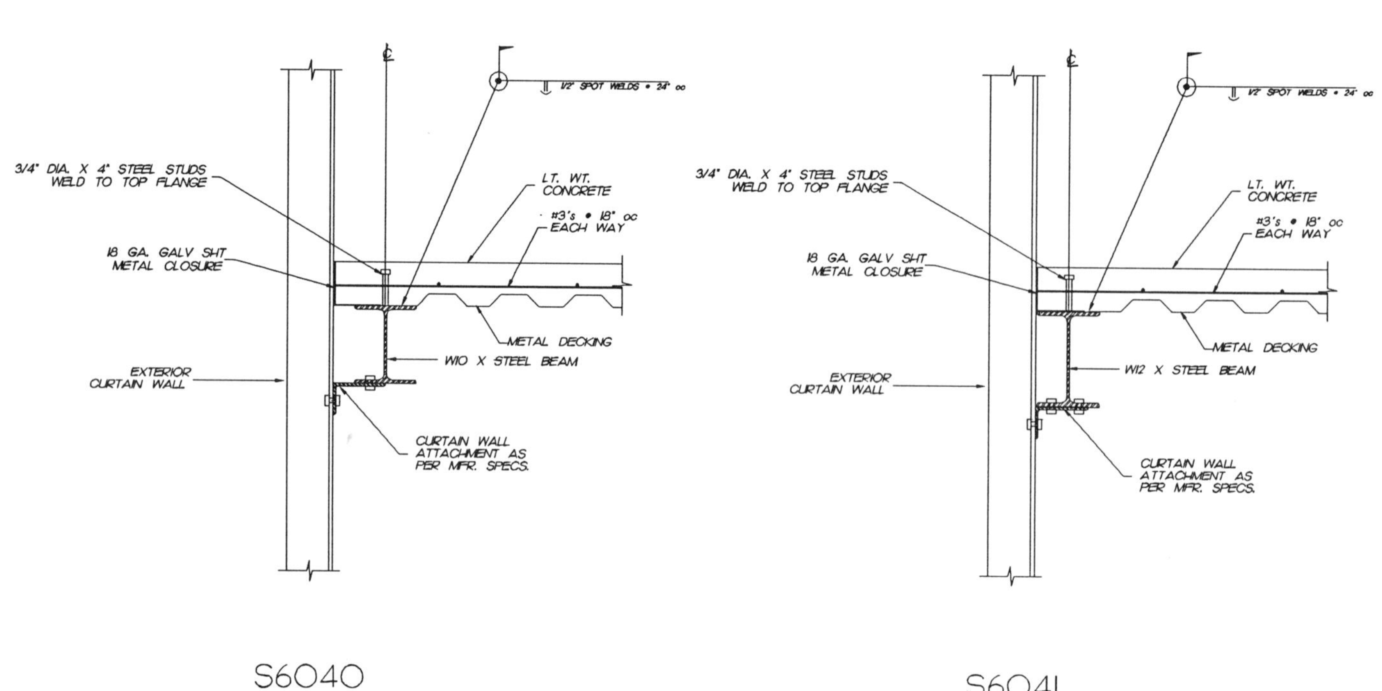

S6040

S6041

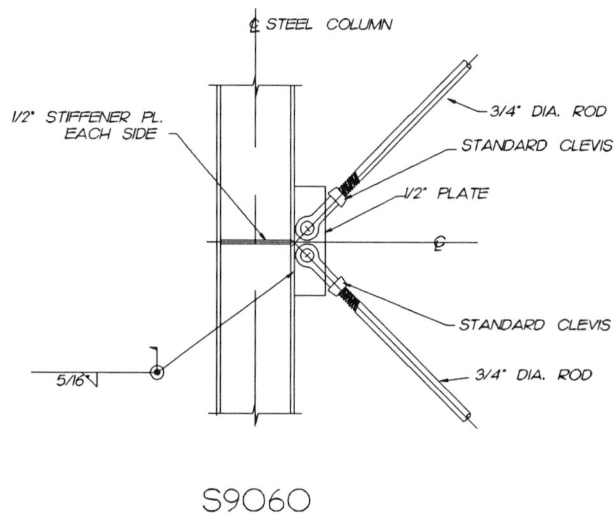

℄ STEEL COLUMN
1/2" STIFFENER PL.
EACH SIDE
3/4" DIA. ROD
STANDARD CLEVIS
1/2" PLATE
℄
STANDARD CLEVIS
3/4" DIA. ROD
5/16"

S9060

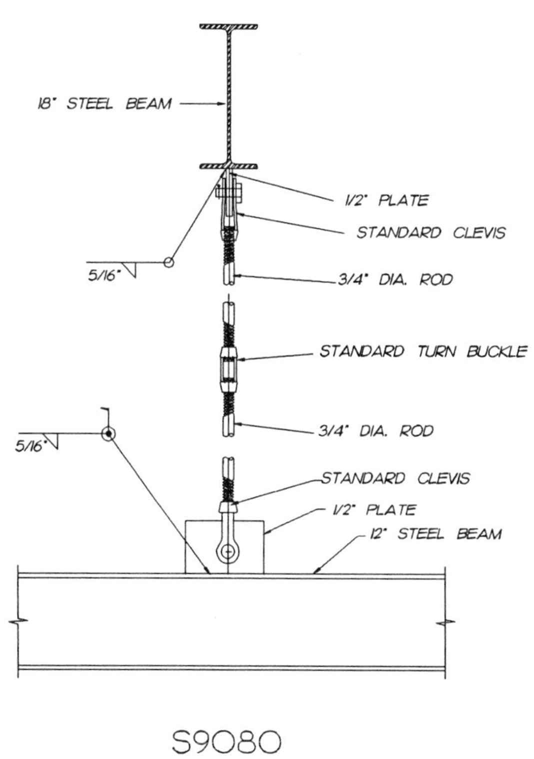

18" STEEL BEAM
1/2" PLATE
STANDARD CLEVIS
5/16"
3/4" DIA. ROD
STANDARD TURN BUCKLE
3/4" DIA. ROD
5/16"
STANDARD CLEVIS
1/2" PLATE
12" STEEL BEAM

S9080

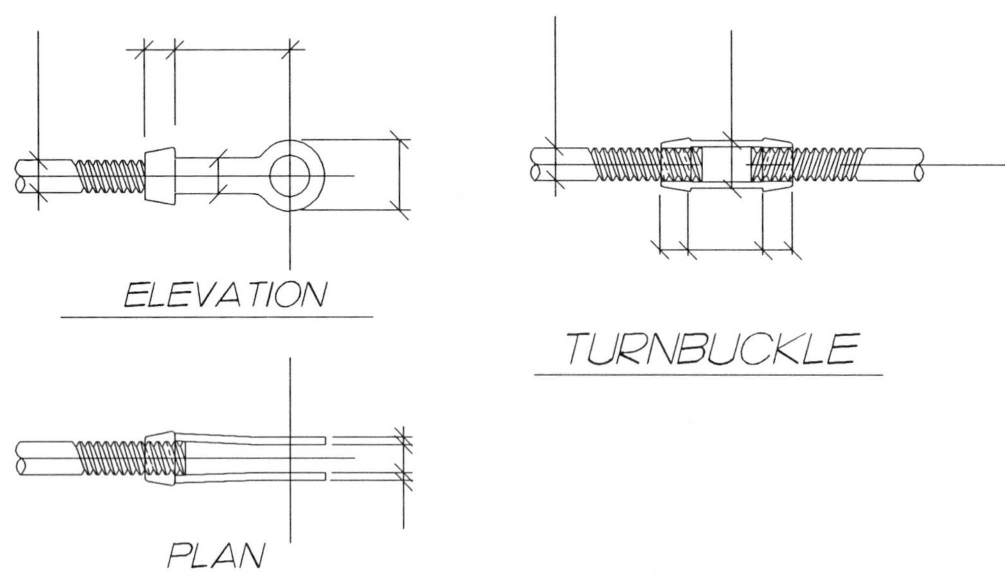

CLEVIS CONNECTOR

S9100

SUBJECT INDEX

ABOUT THE AUTHOR

Morton Newman founded The Newman Engineering Co. of Los Angeles, California, in 1958. His firm has designed many types of facilities—from industrial parks and apartment buildings to marinas and motion picture studio sound stages—using combinations of all types of construction materials. Mr. Newman has also worked at U.S. Steel; Welton Becket and Associates, Architects; and Perriera and Luckman, Architects. He holds a degree in civil engineering from the University of Southern California.